Pam Vandiver
1 Feb 1993

Materials Issues in Art and Archaeology III

MATERIALS RESEARCH SOCIETY SYMPOSIUM PROCEEDINGS VOLUME 267

Materials Issues in Art and Archaeology III

Symposium held April 27-May 1, 1992, San Francisco, California, U.S.A.

EDITORS:

Pamela B. Vandiver
Smithsonian Institution
Washington, D.C., U.S.A.

James R. Druzik
The Getty Conservation Institute
Marina Del Rey, California, U.S.A.

George Segan Wheeler
Metropolitan Museum of Art
New York, New York, U.S.A.

Ian C. Freestone
The British Museum
London, United Kindom

MRS

MATERIALS RESEARCH SOCIETY
Pittsburgh, Pennsylvania

Single article reprints from this publication are available through University Microfilms Inc., 300 North Zeeb Road, Ann Arbor, Michigan 48106

CODEN: MRSPDH

Published by:

Materials Research Society
9800 McKnight Road
Pittsburgh, Pennsylvania 15237
Telephone (412) 367-3003
Fax (412) 367-4373

Library of Congress Cataloging in Publication Data

Materials issues in art and archaeology III: symposium held April 27-May 1, 1992, San Francisco, California, U.S.A./editors, Pamela B. Vandiver, James R. Druzik, George Segan Wheeler, and Ian C. Freestone.
p. cm. -- (Materials Research Society symposium proceedings, ISSN 0272-9172; v. 267)
Includes bibliographical references and index.
ISBN 1-55899-162-X
1. Antiquities--Collection and preservation--Congresses. 2. Art objects--Collection and preservation--Congresses. 3. Cultural property, Protection of--Congresses.
I. Vandiver, Pamela B. II. Druzik, James R. III. Wheeler, George Segan
IV. Freestone, Ian C. V. Series: Materials Research Society symposia proceedings; v. 267.
CC79.3.M36 1992 92-33338
700-dc20 CIP

Manufactured in the United States of America

Plate 1

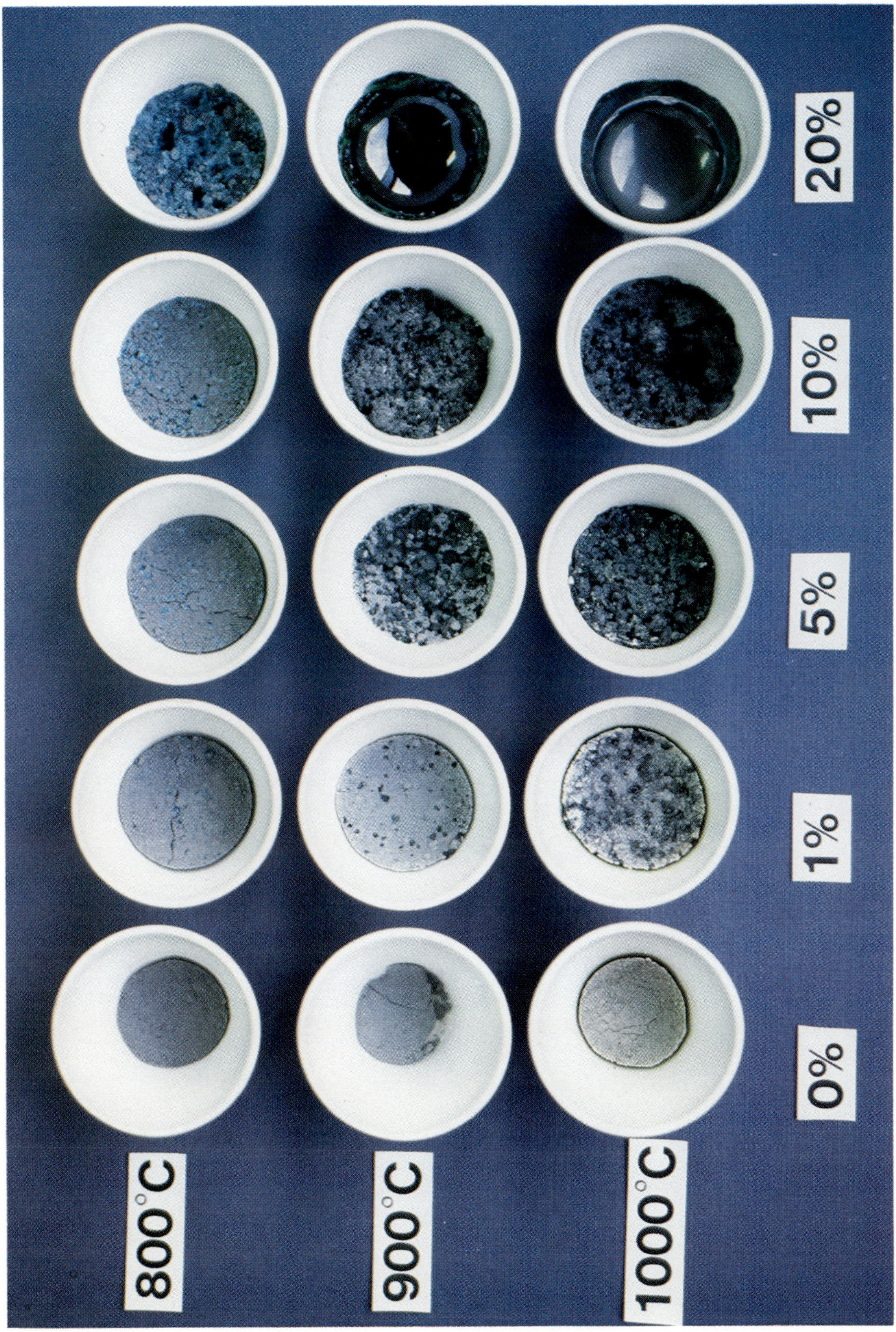

See Figure 2 on page 522.

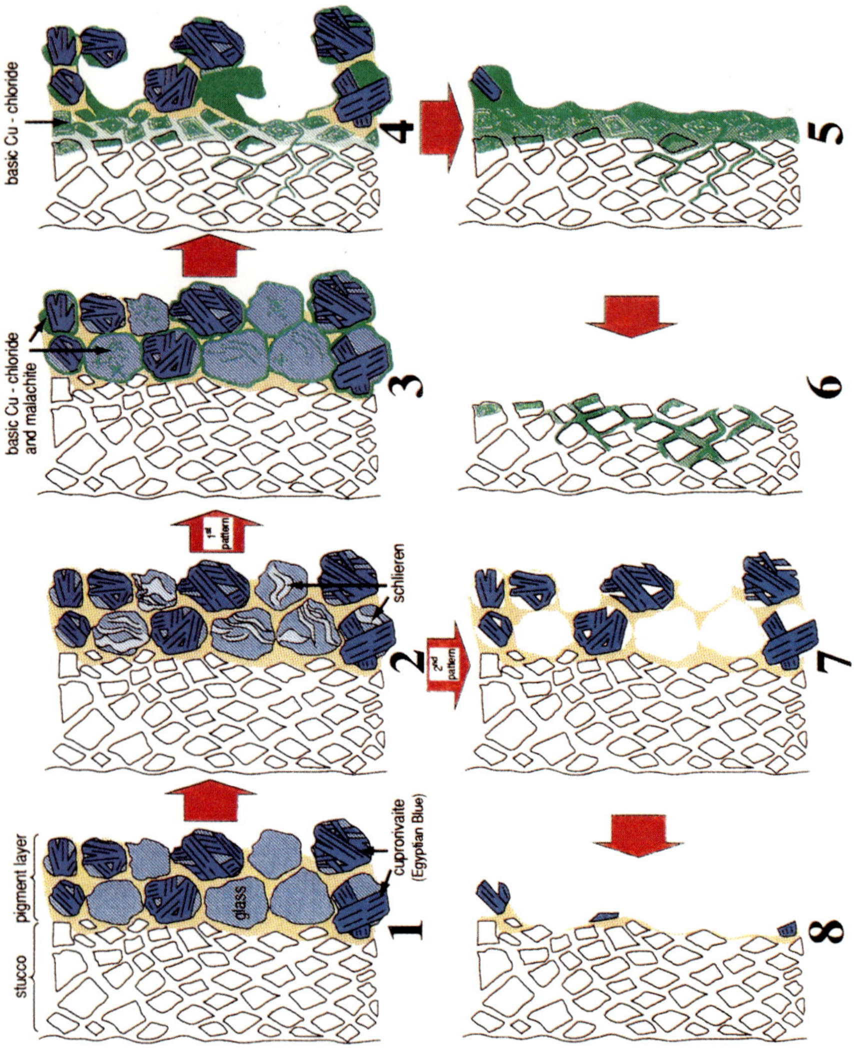

See Figure 8 on page 841.

Plate IA

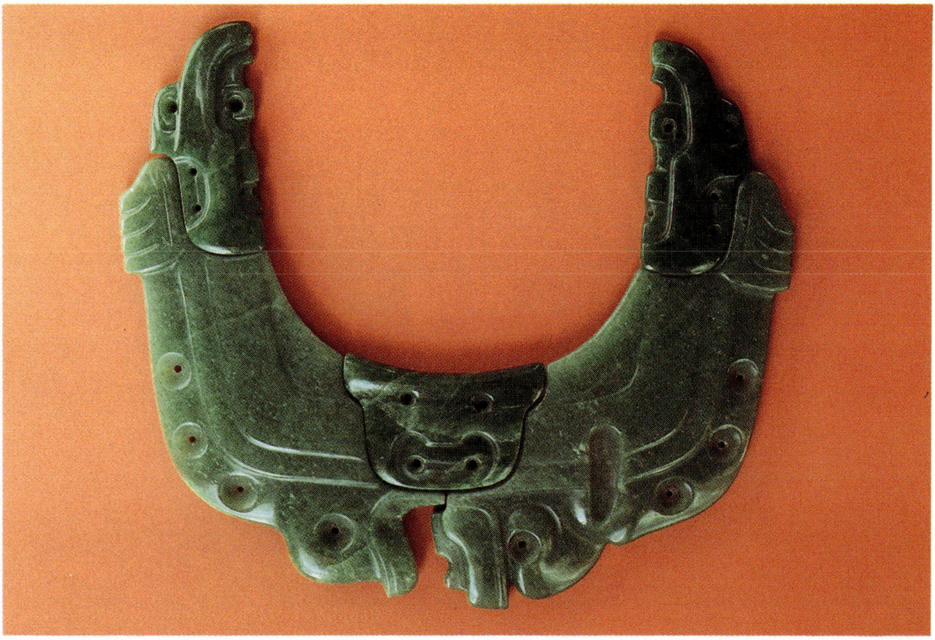

Plate IB

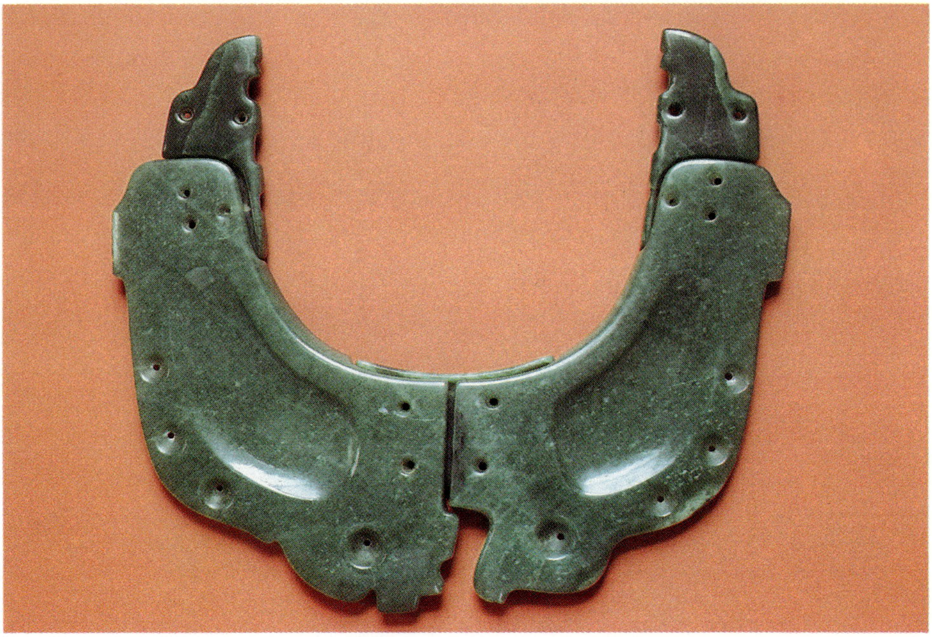

See page 892.

Plate IIA

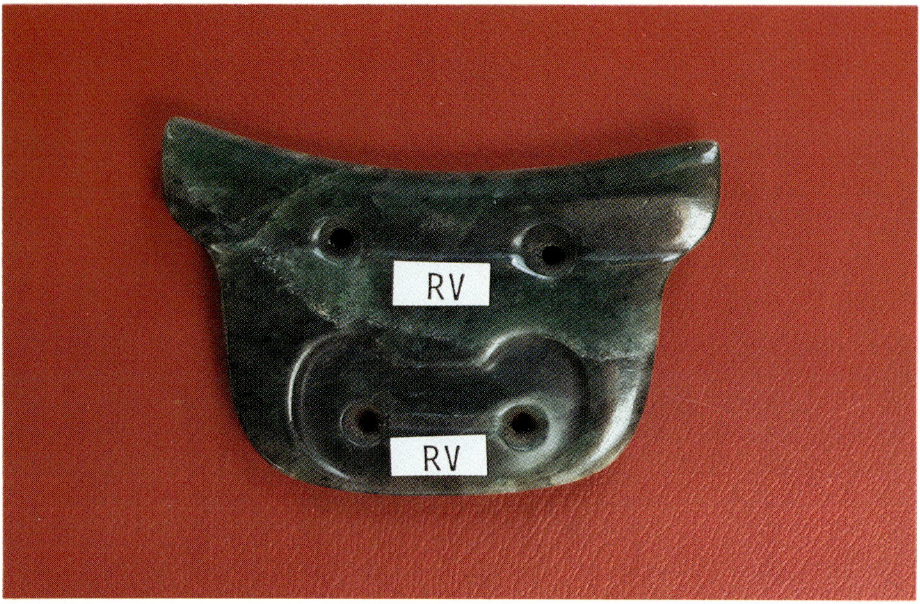

Plate IIB

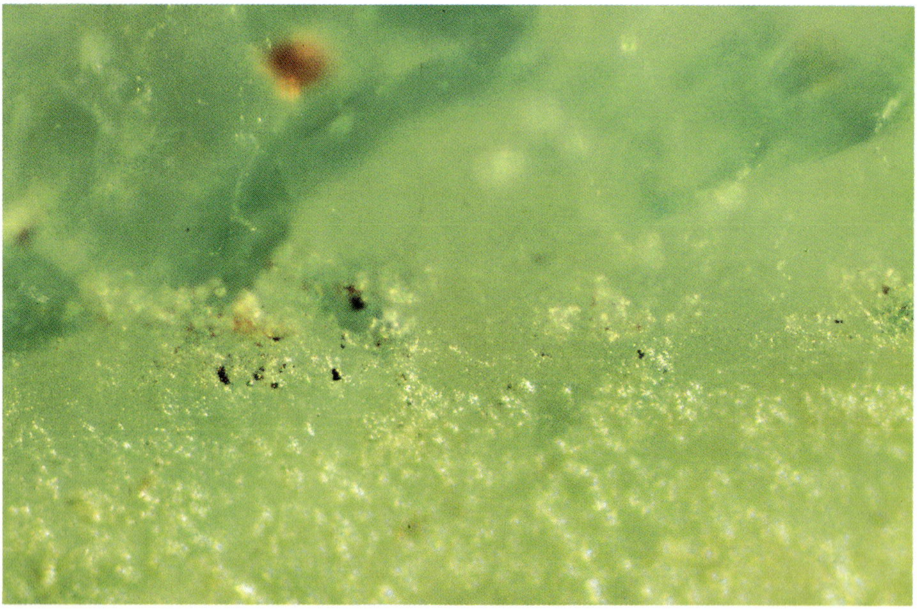

See page 893.

Contents

*Invited Paper

*Invited Paper

C. PROPERTIES OF MATERIALS USED BY THE ARTIST, ARTISAN AND CONSERVATOR

*Invited Paper

PART III: ANCIENT TECHNOLOGY, CRAFT AND THE ROLE OF TECHNOLOGY IN MATERIAL CULTURE

A. CRAFT RECONSTRUCTION FROM EVIDENCE OF WORKSHOPS, INDUSTRIAL DEBRIS AND ARTIFACTS

1. Production Events, Tools and Techniques

2. Production Sites and Process Reconstruction

*Invited Paper

B. PATTERN OF PRODUCTION, RAW MATERIAL CONSTRAINTS AND STYLE OF TECHNOLOGY

C. TECHNICAL ANALYSIS, ETHNOGRAPHY AND THE WRITTEN RECORD

D: THE USE OF ARTIFACTS AND THEIR REUSE

PART IV: CONSERVATION SCIENCE AND PRESERVATION

A. IDENTIFICATION AND PHYSICAL CHEMISTRY OF POST-DEPOSITIONAL PROCESSES

*Invited Paper

*Invited Paper

Preface

This symposium, Materials Issues in Art and Archaeology, is the third in a series in which the goal is to present recent work on (1) ancient materials, (2) the technologies of selection, production and usage by which they are transformed into the objects and artifacts we find today, (3) the science underlying their deterioration, preservation and conservation, and hopefully, although this goal is achieved infrequently, (4) socio-cultural interpretation based on an empirical methodology of observation and measurement.

Our call for papers read: "This symposium will provide a multidisciplinary forum for reporting and interpreting new developments in technical studies of material culture and in the conservation science required to preserve that heritage. In order to promote a dialogue among empiricists who examine the materials science and cognitive and cultural behavior underlying art, archaeology and technology, those who develop the materials science and technology underlying modern artifacts, and those who attempt to elucidate the underlying mechanisms of deterioration and means of stabilization, we solicit contributions in the following areas:

THE ROLE OF TECHNOLOGY IN MATERIAL CULTURE:

Processing evidence from workshops and industrial debris.
Bone, ivory and stone technologies.
Lead-based processes, arsenical copper, wootz.
Innovation and technology transfer vs. conservativism and craft tradition in ceramics.
Analysis of properties to interpret function.
Cultural, historical and technological reconstruction through artifact analysis.

CONSERVATION SCIENCE:

Mechanisms of deterioration.
Assessment and monitoring of condition.
Evaluation of methods of cleaning and consolidation.
Model tests of accelerated aging procedures.

CHARACTERIZATION THROUGH COMPOSITIONAL AND STRUCTURAL ANALYSIS:

New methods and applications including case studies using surface and/or bulk analysis, isotope analysis, nondestructive evaluation or techniques applicable in archaeological fieldwork.
Comparison of empirical data with textual or ethnographic accounts.

MECHANICAL AND PHYSICAL PROPERTIES OF MATERIALS AND/OR ARTIFACTS:

Optical properties and visual appearance.
Strength, wear and fracture.
Interfacial phenomena in composites, especially treated objects.
Replication studies.

Papers should report completed case studies and new applications, rather than potentially useful methods or work in progress. Our intent is to promote research in ancient technology, archaeology, art and conservation which shares the knowledge, methods and tools of materials science and engineering."

Each of the papers was reviewed by three peers, some by more, according to the following criteria.

Significance: The research should be of strong interest to those interested in ancient materials, reconstruction of their technologies, and conservation of artifacts made from them.

Originality: Results reported should stem from a new line of research or a new approach or present original results and should not be merely an extension of previously published material. What original thought does this paper contribute, or how does this paper extend current knowledge in the fields of conservation science, ancient materials analysis or interpretation of material culture?

Technical Validity: Data and analytical methods should clearly support the conclusions drawn. Does the paper "over-interpret" the results? Is there a clear separation between strict interpretation of experimental data and results and the more conjectural, less substantive interpretation? Are there flaws in the experimental part of the paper?

Conciseness: Invited papers are supposed to be 14 pages; contributed papers are to be about seven. If the authors have expanded their papers in order to explain results, rather than just announce them, is the length justified? If the paper exceeds the limits, what in your opinion might be shortened or cut?

We wish to thank the conveners of the sessions for their contributions, especially E.V. Sayre, Charles Selwitz and J. Peter Northover. The Materials Research Society staff and the organizers of this conference are also thanked for making the symposium and proceedings volume possible. We gratefully acknowledge the assistance of the many unnamed reviewers who gave their time and invested this manuscript with their energy and expertise. Our special thanks to members of each of our several labs who served as reviewers; we are sure they are glad this proceedings is finally complete. Lastly, since volume II was published Cyril Smith has passed away; we here include further information on his accomplishments.

We welcome the participation of each of you who reads this book or part of it in planning for and attending the next conference, which will be held in San Francisco in April of 1994. Most especially we welcome people who would be interested in organizing a session of this symposium which would focus on a particular topic.

Pamela B. Vandiver
James R. Druzik
George S. Wheeler
Ian C. Freestone

September 1992

INTRODUCTION

by Pamela Vandiver

That the purpose of this symposium is more than anything an interdisciplinary dialogue has been stated, but the work of organizing this symposium has been rewarded amply by recent reviews of the previous proceedings. Four reviews were published in three separate fields: conservation, archaeology and history of science—they are proof that we are earning an interdisciplinary label for our intentions and activities. They each point out the shortcomings of the proceedings, but also the potential of using the methodology and instrumentation of modern materials science to investigate questions of materials, technology, preservation and conservation by observing, measuring and analyzing artifacts and their contexts. The paradigmatic example of results obtained by investigation of structure, processing, properties, performance and their interrelationships is shared by most of the eighty-one papers in this proceedings which represent the efforts of 171 authors. The four reviews of the previous proceedings are reprinted below with the permission of the journals in which they appeared and in the spirit of hope that we may learn and grow from such criticism.

A CONSERVATION PERSPECTIVE

A review by Dr. Hisao Mabuchi of the Tokyo National Institute for Research on Cultural Properties, Tokyo, Japan, of Materials Issues in Art and Archaeology, edited by E.V. Sayre, P.B. Vandiver, J. Druzik and C. Stevenson, Materials Research Society Proceedings Vol. 123 (1988), xii + 321 pp., illus., price $45.00, ISBN 0-931837-93-6, as published in Studies in Conservation, Vol. 36 (1991) 121-126.

A new trend of looking at ancient objects in the context of materials science has developed among American scientists, probably having its roots more than 10 years ago but becoming visible over the past five years. The publication of the journal Archeomaterials in 1986 may be taken as a symbolic event in this development. Such a trend will, I expect, give a new perspective on interdisciplinary study between natural sciences and art history/archaeology, just as Archaeometry, published from the Research Laboratory for Archaeology and this History of Art, Oxford University, gave rise to that new field some 30 years ago.

Materials Issues in Art and Archaeology is the 123rd volume of the Materials Research Society Symposium Proceedings. The symposium, which well reflects the above-mentioned new trend, was held on 6-8 April 1988 in Reno, Nevada, USA. Forty-one articles were presented, of which 39 were published in this volume. The editors set up three main sections each divided into two subgroups:

Part I. Structural and Compositional Analyses
- A. Studies of Structure
- B. Compositional Analyses

Part II. Ancient Materials Technology
- A. Technology of Silicates
- B. Technology of Ancient Metals

Part III. Processes of Deterioration and Conservation
- A. Glasses: Natural, Ancient and Modern Replicas
- B. Evaluation of Processes in Art and Architecture

In the paper entitled "Tapping the Memory in Archaeological Materials" (Part IIB), Michael R. Notis el al. mention that "In the areas of archaeology, archaeometry, museum management and conservation science conflict often exists between the desire to preserve, and the desire to understand an artifact." From the content of this volume, it can be said that the organizers (editors) of the symposium intended to reconcile such a conflict on the ground of materials science. I feel personally that the study to understand an artifact is as important for conservators as the study of deterioration, since both approaches are closely related and information from academic curiosity might be fed back to the studies for conservation and enrich the scope of conservation science. Notis et al. are quite right to mention that "conservation can best be done when the materials and techniques that were used for making the object originally are understood."

In this sense, most of the papers included in this volume are useful for conservators and conservation scientists. Particularly instructive are those review papers which are generally put at the head of each section.

Pieter Meyers surveys the various methods which may make clear, mainly through images, the macro- and microstructures of works of art and historic artifacts, while Edward V. Sayre discusses compositional analyses, both elemental and isotopic, as applied to glass, pottery, bronze, human bones, cloth, pigments and manuscripts. On reading these papers, one understands how structural and compositional analyses can be carried out in more or less nondestructive ways as well as why the two kinds of analyses have to go hand in hand.

Pamela B. Vandiver's paper "Reconstructing and Interpreting the Technologies of Ancient Ceramics" is a unique presentation of the author's philosophy and strategy to make a breakthrough into archaeology from the side of materials science, with examples ranging from the neolithic potteries up to modern high-tech ceramics. W. David Kingery's paper, "A Role for Ceramic Materials Science in Art, History and Archaeology," treats the subject in nearly the same spirit as Vandiver but in a more pedagogic way, stressing plaster technology in the pre-pottery neolithic Near East.

R. Maddin's paper "Technical Studies—Early Use of Metals" reviews many of the methods which provided insight into early metallurgical technology. However, the uninitiated may regret the lack of a table of technological development and a map showing the important sites, like those which appear in Vandiver's and Kingery's papers, in order to understand the historical background of this interesting paper.

Apart from the review papers and other contributions dealing with methodology, there are several papers in Part I and Part II which report experimental studies of various artifacts from regional cultures: Greek marble sculpture by Stanley V. Margolis et al., Roman ceramics in Britain by Ian C. Freestone and Val Rigby, Greek roof tiles by Vassilis Kilikoglou et al., Maya blue pigment by Luis M. Torres, terracotta warriors of the Qin dynasty by Hans G. Wiedemann et al., early Christian Irish glass and second millennium B.C. glass beads from Britain by Julian Henderson, and medieval window-glass by Helen I. Alten. It is interesting to read these papers as topics, but for the conservation scientist the greatest interest would lie in knowing whether or not the apparent conflict about the firing temperature of the Qin terracotta between the present authors (who concluded that they had not been fired) and Chinese scientists (who said that they had been fired at 800-1000°C) came from improper sampling of a degraded part in the case of the former.

Six papers out of eight grouped in Part IIIA concern obsidian hydration dating. This already classic method, first proposed by Friedman and Smith in 1960, aims at dating implements from the stone age. Since the hydration rate is dependent on environmental variables, the method requires various kinds of theoretical and experimental investigations, as explained in two review papers by Jonathon E. Ericson and William B. White. Here conservation scientists will find a field in common with archaeological scientists in studying the causes and processes of deterioration.

Part IIIB is, notwithstanding its attractive heading, rather miscellaneous, with such articles as "The Fading of Some Traditional Pigments," "Recent Salinization of Ancient

Egyptian Temples" and "Physicochemistry of the Tomb of Nefertari." Conservators will find more of interest in the last two papers, by Mary W. Colby et al., on "Protective Coating for Stained Glass" and J.P. Jackson et al. on "A New Tool for Cellulose Degradation Studies".

In future, not only conservators but restorers will become more and more involved in materials science, because they are in a field where "archaeomaterials" and new materials overlap. This book is an attempt to introduce a new area of materials science.

Unfortunately there are technical drawbacks, due to the photo-offset printing of the authors' manuscripts, that make it difficult to read and understand very interesting academic papers. First, several misspelled words and incorrect figures were found in some papers, mistakes which could have been prevented by a careful rereading. Second, use of color photographs would have allowed a better understanding of such papers as that by Wiedemann et al., since the authors specifically discuss differences in color. Third, the proceedings as a whole would have been much easier to read if some of the papers had been typographically less crowded.

AN ARCHAEOLOGICAL PERSPECTIVE

A review by David Killick, newly appointed Professor of Anthropology and Materials Science at the University of Arizona, Tucson, Arizona, who is known for his insightful studies of African ironworking, of Materials Issues in Art and Archaeology II, edited by P.B. Vandiver, J. Druzik and G.S. Wheeler, Materials Research Society Symposium Proceedings Vol. 185 (1991), xxviii + 844, illus., price $52.00, ISBN 1-55899-074-7. This review was published in American Antiquity, Vol. 57, No. 3 (1992) 560-561.

This leviathan among symposium proceedings (70 papers, 853 pages) is the product of a meeting held in April 1990 on the application of materials science to the interpretation and conservation of prehistoric and historic artifacts. The editors and publisher of this volume are to be congratulated for bringing the product to market so promptly from camera-ready copy, although at some cost of editorial control. The type in some chapters is tiny, others are in barely legible dot-matrix, and several are much too long (37 pages in one case) or have errors such as a missing page, figures without scales, or unformatted tables.

The papers in this volume range over a very wide field indeed. Twenty-six of these report research on the conservation of masonry, adobe, glass, oil paintings, paper and metals. Since most readers of this review will be archaeologists, I will not consider these papers further except to note that the scientific standards on display are consistently high. An excellent section on the conservation of stone and adobe should, however, be read by any archaeologist who is charged with the stabilization of standing structures.

The remaining papers fall into several loose groups. There is a short section on new techniques, two of which promise to be exceptionally useful. These are the industrial CAT scanner, which can provide digital radiographs and computed tomographic sections of metal, wooden and ceramic artifacts. This technique will become indispensable for the study of complex objects that cannot be disassembled or physically sectioned. There are also two papers on the environmental scanning electron microscope (ESEM). This modification of the conventional SEM dispenses with high vacuum in the sample chamber; samples need not be dried or coated. It should therefore prove particularly useful for examination of botanical and zoological specimens.

Most of the papers in this volume are case studies of metals, ceramics or glasses. There are also three papers on textiles and one wildly speculative essay on the environmental effects of lime plaster production in the Neolithic of Jordan. Interspersed among these are five papers on provenience analysis (one each on bricks, copper alloys, obsidian, marble and pottery), and two papers on the detection of fake ceramics and jades by analysis of the surface patina. Finally, all of this heavy but nourishing technical matter is leavened by several non-technical chapters. These include a short but incisive review of the history of archaeometallurgy, a summary of the early development of systems of weights and measures, and a delightful essay on ceramics as metaphor in the literatures of the ancient Near East.

The case studies in this volume range from the pedestrian to the truly remarkable. The best provide a clear demonstration of the ability of materials science, when closely integrated with history and archaeology, to reveal aspects of past social knowledge and behavior. The study of single objects brings to light the technical knowledge and the level of skill of its maker—the artisan behind the artifact. When studies of many artifacts and their manufacturing debris are combined with field studies of workshops, kilns, quarries, mines and other such loci, a great deal may be inferred about the organization of production. When many such studies have been done, it becomes possible to detect regional technological traditions and their mutual interaction. This volume contains some excellent papers at each of the hierarchical levels of analysis, but there is unfortunately not space to list them here.

The range of institutions represented at this symposium indicates that there is now widespread appreciation among archaeologists of the value of materials analysis. What is equally clear from this volume is that only a fortunate few have access to appropriate technical expertise and advanced analytical equipment. The dominant players in this field are the well-funded public and private research laboratories, such as the Conservation Analytical Laboratory of the Smithsonian Institution, the British Museum Research Laboratory, and the Getty Conservation Institute. Archaeologists who can arrange collaborative projects with these centers reap the benefits of a wide range of specialist expertise and equipment, often at highly subsidized rates. Other archaeologists are blessed with sympathetic colleagues in science or engineering, who are willing not only to undertake the analyses but also (to judge from the acknowledgments) to absorb the cost as well. These saintly individuals are still rare, though their numbers are increasing.

These formal and informal arrangements fall a long way short of meeting the rising demand for materials analysis of archaeological materials. Many other archaeologists lack access to appropriate expertise and equipment and do not have adequate funding to contract out the work. Under these adverse circumstances they do the best they can with whatever they have at hand. There are several such studies in this volume. Rather than criticizing these papers for their relative lack of sophistication, it is more appropriate to view them as symptomatic of a major problem in American archaeology today. Enormous advances have been made in archaeometry in the last fifteen years while research funding in archaeology has remained essentially flat. Thus while most archaeologists are aware of the potential benefits of new archaeometric techniques, few can afford to use them in their own research. Unless there is marked improvement in the funding of archaeology in the U.S.A., many of the valuable but expensive techniques exhibited in this volume will never be widely used in archaeological interpretation.

A PERSPECTIVE FROM THE HISTORY OF TECHNOLOGY

A review by Robert B. Gordon, professor of geophysics and applied mechanics at Yale University, who is well-known for his study of colonial American iron technology, of Materials Issues in Art and Archaeology II, which appeared in Technology and Culture, Vol. 33, No. 3 (July 1992), pp. 602-603.

Archaeologists and museum conservators are enthusiastically using new instruments for microscopy and chemical analysis that have been developed by applied physicists for materials scientists to probe and measure artifacts. For several decades now, archaeometric methods, such as radiometric dating, dendrochronology, and microprobe analyses, have given archaeologists information to use in revising older interpretations of, for example, the peopling of the American continents, the development of copper metallurgy in southeastern Europe, or the technological capabilities of ancient South Americans. Art historians have also found archaeometric methods useful in discovering aspects of the techniques used by individual artists, in detecting fakes, and establishing dates. Many practitioners of archaeometry attended a symposium at a meeting of the Materials Research Society in 1990 to present the results of their instrumental examinations of stones, ceramics, alloys, paints, and fibers in prehistoric, ancient, and a few historical artifacts. The editors of the symposium proceedings managed the remarkable feat of publishing sixty-nine of the articles within a year of the meeting. They are a guide to current research in archaeometric materials analysis, and they demonstrate the range of techniques that can be brought to bear in the study of material evidence from the past.

The articles in this volume (with only a few exceptions) are four pages long, stand alone, and report work in progress. Many of the authors have emphasized the sophisticated instrumentation and methods they used; the introductory comments for the different sections of the proceedings bristle with adjectives like stunning, advanced, cutting edge, high-tech, tour de force modern technology, or high-tech understanding. The stated objective of the conference was assessment of premodern behavior and cognition based on technical studies. In their summary of the proceedings, the editors describe some of the problems that arise in interpreting archaeometric data this way. In the brief space allotted them, however, most authors were unable to go beyond the presentation of their observations. There are some exceptions. Karen Foster's extended essay on ceramic imagery in ancient Near Eastern literature makes no use of archaeometry. Jack Ogden's article questioning the commonly held view that ancient gold jewelry was cast rather than hand-wrought shows that important research can still be done with the aid of a hand lens alone. Nevertheless, a nonspecialist reader of this volume may gain the impression that the capacity of modern analytical technologists to produce data is outrunning the capacity of scholars to use and digest the results. In part this is because of the increasingly common preference of funding agency and museum administrators in the United States for novelty over reflection. It is a paradox, then, that this volume is dedicated to Cyril S. Smith, the scholar who has done so much to interpret the wider context of materials in art and technology.

Most historians of technology have used material evidence sparingly, while curators at museums of technology often display items in their collections to illustrate conclusions drawn from documents rather than as sources of information. In the culture of archaeometry, artifacts identified as "ancient" or "fine art" are accorded higher status than recent or vernacular ones; something that is "earliest" or "first" commands the highest respect. Only one of the sixty-nine articles in this proceedings volume deals with a North American industrial material. There are opportunities for a convergence of interest between historians of technology and practitioners of archaeometry that could be mutually beneficial.

A SECOND CONSERVATION PERSPECTIVE

A review by Andreas Burmester of Materials Issues in Art and Archaeology II which appeared in Studies in Conservation, vol. 37 (1992) pp. 211-213.

A book review may be associated with the homely atmosphere of long winter evenings, a fireplace, a pot of tea, a warm sweater and a nicely laid out and illustrated book that grips the reviewer from the first page to the last. The reviewing of Materials Issues in Art and Archaeology II was in sharp contrast with this scenario, because its perusal began in a copy shop. Here the reviewer tried to enlarge individual articles whose deciphered titles showed promise. Without this, the tiny print of most parts of the book would have made reading a real torture. First point: I do not think that a copy shop is the right introduction to a book.

Browsing through the thick volume, all sorts of fonts, font sizes, formatting and layouts can be found, as well as handwritten alterations and even subsequent editorial changes in fonts other than that of the original paper. This may reflect the colourful variety of our world or the simple fact that many different printers are on the market and the editors have given no guidelines to establish uniformity. Second point: Nothing has been done to make Materials Issues in Art and Archaeology II easy to read. This is not a new observation: these technical drawbacks were reported by Hisao Mabuchi in respect of the first volume of Materials Issues in Art and Archaeology [Studies in Conservation 36 (1991) 122-124].

An introductory note by Pamela Vandiver and George Wheeler presents the goals of the symposium, held on 17-21 April 1990 in San Francisco. The symposium asked for contributions related to ancient technology, materials degradation and conservation science, characterization through compositional and structural analysis, and physical properties of art materials. The note mentions "communication" problems between people around the world and among the disciplines involved in the investigation, preservation and conservation of works of art. The object is seen to be surrounded by archaeologists, art historians, materials and conservation scientists and conservators. They all speak different languages, their mother-tongues as well as the languages of their disciplines, and unfortunately only a few of them have received an interdisciplinary education. Therefore they all have different ways of approaching the object. This is not a new observation, and has led in the past to the staking of various claims such as "archaeometry" and "conservation science". Both areas have their own journals, symposia, heroes, gurus and fan clubs, as well as their characteristic topics with, unfortunately, little overlap. Third point: It is a good thing to overcome this division between our curiosity about ancient materials on the one hand and our task of preserving or conserving objects of cultural importance on the other. Symposia and publications such as Materials Issues in Art and Archaeology II aim to bring these two artificially separated areas together again. Unfortunately a number of authors did not follow this wise philosophy. They have produced papers with (sometimes) good scientific results which have lost any relationship they may once have had to the objects under consideration. Fourth point: It is arguable whether papers obviously written by scientists for scientists in the field should be excluded, or should be left as possible seeds for future fruitful feedback.

To review symposia proceedings of this kind is unpleasant because of the wide range of topics, including those with which the reviewer is unfamiliar. Thus this review cannot hope to cover all the more than 800 pages. In practice, the enlarged and deciphered contributions were skimmed through and some were finally studied in detail. In other words, I decided to skip many, many pages and I how have to live with the knowledge that most authors who contributed to this volume may say that I skipped the wrong pages. Sorry for this! The decision will be more comprehensible if one recognizes that Materials Issues in Art and Archaeology II covers New Methods and New Applications of Technical Analyses, Painting, A Problem in Composite Materials, Substrates, Binding Media and Pigments; New Approaches from the Mechanics of Materials; New Methods of Non-Destructive Analysis Compared with Analyses of Microsamples; Stone, Adobe and Architectural Glass: Monitoring and Stabilization; Characterization of Stone Sculpture and Stone Tools; The Role of Technical Studies in Establishing Cultural Functions of Ceramics; The Cultural Uses of Ceramics; Ceramics as Cultural Indicators which Reveal Exchange Patterns of Goods and Technologies; Reconstruction of Ceramic Technology; Alteration during Use, Burial and Testing of Treatment in the Field; Refractories and Glasses: Materials Choices, Workshops and Industrial Debris; Metals: The Unnatural Indicators of Cultural Differences; Innovation,

Cultural Differentiation and Properties; Theoretical and Practical Approaches to the Study of Metals; Textiles, Paper and Polymers; Characterization, Technology, Conservation and Maintenance of Fragile Materials, a summary, an author index and (very useful) a subject index. Phew! Each of the parts is introduced by a short and well-written overview of the papers within the section. This is particularly useful because it guides one through the jungle of these thousands of papers and it allows the editors (mainly Pamela Vandiver) to explain their philosophy; to link areas so far separated, such as conservation science, ancient materials science and engineering. Luckily, the unfortunate term "archaeometry" is exorcised!

In Part I on new methods and new applications of technical analysis F.H. Seguin gives an excellent introduction to the disadvantages and benefits of classic and digital X-ray imaging techniques. The paper is followed by a short section on applications. One would have enjoyed seeing more relevant applications to works of art, which would have forced a critical evaluation of image processing techniques applied in the field of X-ray techniques.

Two of the most fascinating papers deal with the use of environmental electron microscopy, a new technique presented by D. Stulik et al. and applied to actual conservation problems such as dissolution and crystallization processes in stone and other porous materials; this technique is also applied by E. Doehne et al. to humidity cycling in adobe and dynamic lead corrosion. The outstanding advantage over conventional SEM is that wet or outgassing samples can be observed directly under very high magnification, without the limiting sample preparation procedure necessary for SEM where such samples cannot be investigated at all. However, the method is limited to liquids with low volatility because there is still a partial vacuum. The new technique opens the large "dark" area of liquid-solid or gaseous-solid interactions for future "close" observation by conservation scientists. Congratulations!

It is not easy to see why a paper on cation-selective reagents is included in part I. These molecules are indeed a fascinating topic and it can certainly be expected that some of these compounds will one day serve in conservation campaigns. Thus far one fully agrees with the author, who gives a review of the former use of chelating agents (such as EDTA) as well as an introduction (or rather, a compendium of facts which can be found in any university textbook), the paper lacks any practical case histories of cleaning or solid experimental results (what about residues left on the surface?). The paper simply seeps away. This type of preliminary contribution can have disastrous results when used by fascinated but inexperienced readers. Fifth point: Today, why do we read more preliminary than final papers?

Another maddening example: in a paper on "Ion beam analysis of pigments" conducted with proton-induced X-ray emission (PIXE), one might expect results based on the investigation of at least two pigments. One might even expect a few critical comments about the benefits or disadvantages of the method, based on applications to real works of art. Something like "there are not too many sites where you may find PIXE but, in any case, you have to bring your object to the equipment." No, the paper reports the investigation of a single pigment, cadmium red, "mixed with lithopone ($BaSO_4$)". Do the authors mean baryte ($BaSO_4$) or lithopone ($BaSO_4$ + ZnS)? Sixth point: this type of limited paper based on a dummy sample or samples considerably inflates the number of pages in the volume and suggests the timid question whether everything written is worth printing.

In Part II the valuable input from materials science to all the dynamic processes to be observed on paintings is obvious. This is true of the work of Gustav A. Berger, well known to readers of this journal, as well as the contributions of Marion F. Mecklenburg. Mecklenburg's computer-aided application of engineering principles to paintings yields surprising results which contribute definitive new aspects to the problem of art transit. One of the conclusions is that "vibration may not be a serious problem for stretched canvas paintings" and vibration is therefore not seen to be a major cause of cracking. This solid paper, with its language of materials science and its cool objectivity, does not, however, banish my personal concern that scientific approaches like this will one day be abused to increase the number of paintings and objects that go whoring around the world from one exhibition to another.

The luminescence of painting materials, most of them used as media, has again attracted the scientist's attention. With the kind of laser spectroscopy presented by Londa J. Larson et al., very small samples can be investigated. The authors describe the method as being "nondestructive for the sample", which is not what we mean by non-destructive testing in our field. It is, however, an advantage to have a method that is non-destructive of the sample since the sample may then be reused for further investigations. The emission spectra shown do not fully convince me that laser spectroscopy may serve one day to identify, or even differentiate, organic mixtures of varying composition which, in addition, are altered by aging. Any of the very sensitive separation techniques available nowadays will finally yield more reliable results. As long as the study includes only a few samples, as in this case, the conclusion that the "emission peak maxima and shape can be used to differentiate between samples" may be correct. But as soon as more and aged samples of this complexity are investigated, or simply those (mis)treated by several cleaning cycles, reliable differentiation will become extremely difficult. A second paper by A. Larson and I. Zink, has obviously been submitted to the wrong publication. This contribution does not show the slightest effort on the part of the authors to make a connection between their work on the "Luminescence of alizarin and its metal complexes" and Materials Issues in Art and Archaeology II. One can only hope that this paper will one day serve as one of the seeds mentioned above (my fourth point).

The paper by D. Scott on two silver plates in the collection of the J. Paul Getty Museum impressed by the way in which the analytical work is integrated into the information about other Byzantine silver objects so far investigated (and by the choice of a readable font size). The discussion shows how valuable scientific results can be, but also that they can never provide absolute proof of the authenticity of an object.

H.G. Wiedemann et al. journey into the world of forgotten weights and money, and make the reader feel like one small part in an endless series of human generations. The authors point out that they are using thermoanalysis as an important experimental technique. However, it is not made all that clear what this technique has been used for or what the experimental results have been. The only application described is the identification of $PbCO_3$ as a corrosion product on a Roman lead ounce. X-ray diffraction has also been used, and this, in fact, is the only method of choice.

Prudent closing remarks by Pamela Vandiver and George Wheeler summarize Materials Issues in Art and Archaeology II. To learn from the past and to prevent a further loss of cultural heritage and technological knowledge is a key task for scientists. It is unimportant whether they are materials or conservation scientists. In their neglect of traditional boundaries, books like Materials Issues in Art and Archaeology II, which are based on very recent results and on the philosophy expressed by the editors throughout this thick volume, are valuable tools for the fulfillment of our daily obligations. Bearing this in mind, the very heterogeneous quality of the papers and the maddening make-up of the book are of minor relevance.

A VIEWPOINT OF THIS PROCEEDINGS

The editors thank these reviewers for their incisive comments and their thoroughness. Our response is to endeavor to do a better job with this volume and the next one, and we have made several changes. In an effort to focus on larger issues of cultural significance which pose both physical and philosophical problems, Druzik assembled data on management of buildings and collections during times of risk due to armed conflict. From presentations in the session, Cultural Heritage in Conflict, we learned of the necessity of detailed planning and preparedness. With the hope that more planning will take place, we reproduce a model publication from the British Museum, Air Raid Precautions in Museums, Picture Galleries and Libraries. The overwhelming conclusion is that behavior based on strong emotional dislike will cause the most damage to cultural heritage because that heritage is symbolic of the control of the "other side." Such destructive behavior can come from raiding by the "authoritarian" side in a conflict or looting by the "weaker" side. Protection afforded by the Hague Convention seems to be minimal.

Some of the problem may reside in the five-sided figure on a white background which is supposed to be a protective insignia. The green color is difficult to read from a distance. The figure which consists of a square motif with a point on the lower is perhaps not as effective a shape as might have been designed. The lower side seems to be pointing downward, almost as if to point to the structure like an arrow to a target. In a test which asked 50 high school juniors to associate seven symbols (cross, star with 5 points, hexagon, octagon, circle, pointed square, arrow) with protection or direction to a target and to order them from one extreme to the other, 88% of the students placed the Hague pentagon next to the arrow as indicating direction to a target.

We would like to present more interpretive and reflective articles on the wider context of materials issues in art and archaeology. While this proceedings is not as successful as we would wish, there are several other recent attempts which should be cited to provide a wider context for this proceedings. The January issue of the MRS Bulletin concentrated on the theme of "Art and Technology" with articles spanning modern technologies for art production by Otto Piene of M.I.T. to Northern and Southern European styles of painting during the Renaissance, Old World and New World styles of majolica production, the connection between the metallurgy of harpsichord wire and Columb's pendulum, new approaches to silk preservation, to production of Neolithic period plaster sculpture from Jericho, and to refinements in the interpretation of paintings which accelerator radiocarbon dating of very small samples will make possible. Steven Lubar and W. David Kingery are publishing a proceedings of a conference held at the Smithsonian Institution, entitled Learning from History (Smithsonian Institution Press, Washington, D.C., expected 1993) in which interpretation of the social context of technology is stressed. Another such conference is to be held next spring. David Scott of the Getty Conservation Institute is editing the symposium on New World archaeometry as presented at the 1992 Archaeometry Conference in Los Angeles. The American Ceramics Society will continue its biennial symposia on the history of ceramics next April and the publications in the series entitled Ceramics and Civilization (American Ceramics Society, Westerville, OH) which now numbers six. Archeomaterials, under the new management of Robert M. Ehrenreich and J. Emlen Myers, will continue publication of articles of both a technical and interpretive nature which are too long for many other archeometric or conservation science journals. The same excellent quality of illustration will continue. The new editors welcome submissions (P.O. Box 4190, Rockville, MD 20849, Tel. (301) 294-3430).

In the second part of this proceedings, "Examination, Technical Analyses and Discovery," we have grouped papers into three sections: (1) the geological and geochemical context in which materials are found, manipulated and degrade, (2) the characterization of composition and structure of art objects and archaeological artifacts, and (3) properties of materials used by the artist, artisan and conservator. We hope that the section on geological context will grow in future volumes, as we become more aware of the environmental constraints on production and use. A recent book, Life as a Geological Force: Dynamics of the Earth by Peter Westbroek (Norton, New York, 1991) sets forth the perspective of biogeochemistry in a popular style.

With time, fewer studies of artifacts can be divided purely into either a compositional or a structural category, as many questions depend on combining analytical techniques of both types, so we have discontinued this division, perhaps ill-advisedly. However, one special book on compositional analysis will undoubtedly be acknowledged in future contributions to this symposium as promoting both accuracy and precision of measurement as well as interpretation, Chemical Characterization of Ceramic Pastes in Archaeology, edited by Hector Neff (Prehistory Press, Madison, WI, 1992). Another volume has been published which may further similar studies of stone, Ancient Stones: Quarrying, Trade and Provenance, edited by M. Waelkens, N. Herz and L. Moens (Leuven University Press, Leuven, Belgium, 1992). Another book on structural analysis should have a similar effect in its particular area of specialty, Phytolith Systematics: Emerging Issues, edited by George Rapp, Jr. and Susan C. Mulholland (Plenum, New York, 1992). This effort is the first publication in a new series promoted by the Society of Archaeological Sciences.

A relatively new area in museum studies which is yielding significant results is the measurement of mechanical properties. This area has grown considerably with each symposium, as participants have investigated more varied materials. The presentations in this proceedings are just a part of the increase in recent studies of the mechanical properties of art objects and archaeological artifacts. Another conference proceedings has added to the growing literature, Art in Transit, edited by Marion Mecklenburg (National Gallery of Washington, 1991).

In the third part, "Ancient Technology and Craft: The Role of Technology in Material Culture," we group studies according to craft reconstruction from evidence provided by workshops, industrial debris or the artifacts themselves. Production events, tools and techniques, production processes and sites, or styles of production and use are three areas where most papers are individual case studies. This is the first time that we have not had to group papers in the style of objects studied in the last century, that is by material. Instead there are themes of inquiry clearly independent of material. Most of these papers depend on "deconstruction" of particular artifacts into minute observations and data, and then arguments are made on the basis of "structuralism" to compare and contrast particular artifacts or groups of artifacts within an all-important historical or archaeological context.

Another group of papers expresses a range of perspectives on the interaction among technical analyses, ethnographic evidence and the written record. This is an old query and theme in the study of artifacts, but one which is being revitalized by books such as Ceramic Ethnoarchaeology, edited by William A. Longacre (University of Arizona Press, Tucson, 1992), The Ceramic Legacy of Anna O. Shepard, edited by Ronald L. Bishop and Frederick W. Lange (University Press of Colorado, Niwot, CO, 1991), and The Historic City of Nara: An Archaeological Approach by Tsuboi Kiyotari and Tanaka Migaku and translated into English by David Hughes and Gina Barnes (The Center for East Asian Cultural Studies, Tokyo, and U.N.E.S.C.O., Paris, 1991). Another group of papers question limits to technological change provided by constraints of raw materials and past ways of working with materials. One symposium dedicated to Donald Harden (Roman Glass: Two Centuries of Art and Invention), edited by M. Newby and K. Painter, Society on Antiquaries of London, 1991) has this same theme implicit in many of the papers. Other papers in this volume investigate the use of artifacts as visual objects of display or as tools, and in so doing discuss cross-cultural interactions and cultural diversity which is reflected in the making and use of artifacts. One study by Lechevallier, Inizan and Plumet postulates cultural contact and communication based on similarities and differences in technology and technological style, or patterns of activities recognizable by study of residual material culture.

The fourth part of the proceedings consists of "Conservation Science and Preservation," and is divided into three parts. The identification and physical chemistry of post-depositional processes is the foundation upon which conservation treatment relies. Identification of weathered and degraded products is also necessary before any studies of technology or materials identification can take place. Treatment, deterioration and structure in architectural and building materials is a concern which has been a focus of many participants, particularly during the second meeting. Corrosion of metals and ceramics and their protection is also featured as a section. One of the participants, David E. Clark, with his colleague, Bruce K. Zoitos, has edited a significant contribution to the conservation science literature, Corrosion of Glass, Ceramics and Ceramic Superconductors: Principles, Testing, Characterization and Applications (Noyes Publications, Park Ridge, NJ, 1992). In the interest of promoting further understanding among scientists of the field of conservation, a book which presents case studies and is edited by Andrew Oddy, keeper of conservation at the British Museum, is recommended (The Art of the Conservator, Smithsonian Institution Press, Washington, D.C., 1992).

The last section, "Experiments in Education and Artifact Study," is new this year. Several researchers see the study of artifacts as a way to promote interdisciplinary education or as a means of intriguing more students with materials science and other science and engineering disciplines as careers or simply for their knowledge and enjoyment. In addition, replications of artifacts are a way of understanding the process of making objects and gaining firsthand knowledge of craft processes as well as making study copies widely available.

The quality of this book still is somewhat lacking, but again it is being published in a timely fashion being available within months of the conference. The articles may not be as deep, or as wide, or as long, or as tall as the possible dimensions of our field which many can envision, but this is the best the editors and participants can do. As you, the readers, criticize this proceedings, as you try to find something of use in it, please consider the possibility of contributing to the next proceedings for which the meeting will be held in April-May of 1994.

IN MEMORIAM

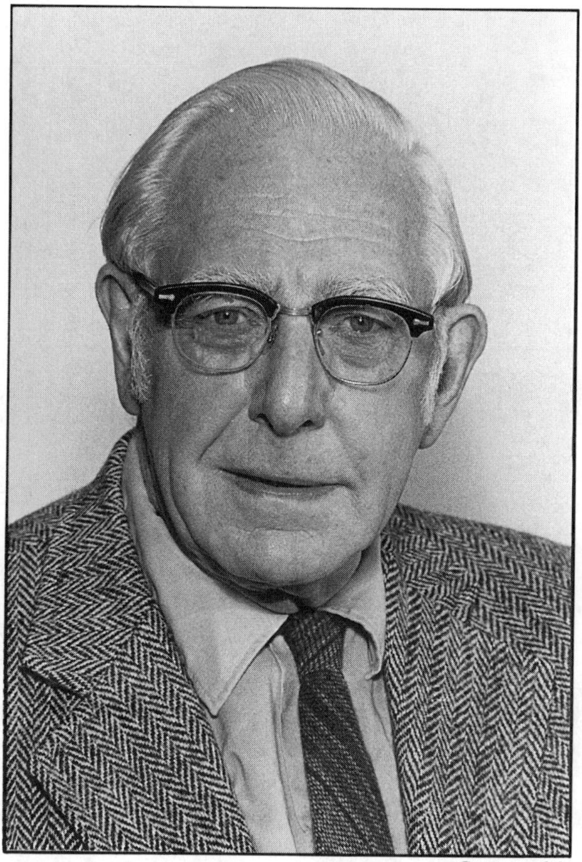

Courtesy of the M.I.T. Museum

Cyril Stanley Smith
1903-1992

In Memoriam

CYRIL STANLEY SMITH

1903-1992

Cyril Stanley Smith died peacefully at his home in Cambridge, Massachusetts, on August 25, 1992, after a long illness. He was born on October 4, 1903, in Birmingham, England. He had an early interest in experimental science. He was allowed by his parents to have a home laboratory, and worked for a year as lab assistant at King Edward's Grammar School in Camp Hill before entering the University of Birmingham in 1921 with advanced standing. There he briefly considered studying philosophy before receiving the BSc degree in 1924.

An article in a French magazine extolling the Bell Telephone Laboratory inspired him to look for a job in the United States, but immigration restrictions led instead to graduate study at the Massachusetts Institute of Technology, where in 1926 he was awarded the Doctor of Science degree. He remained at MIT as a research associate in the x-ray laboratory, working under John Norton of the Physics Department, before obtaining a summer job in 1927 with the American Brass Company in Waterbury, Connecticut. This became a permanent position, where his main task was alloy development. He became head of the Copper Alloys Research Laboratory and received some 20 patents. He married Alice Kimball in 1931, and they had two children. He returned to England to spend a year at Cambridge (1933-34) and another in London (1955-56). The occasion of his last visit to England was the annual conference of the Historical Metallurgy Society in 1984, which met a few miles from his birthplace in Birmingham.

When the war came he went to Washington to work with the War Metallurgy Committee, and in 1943 was invited to join the Manhattan Project at Los Alamos, New Mexico, to direct the preparation of fissionable metal for the atomic bomb. The decision to go to Los Alamos was not easy for him. He told me that he went that evening to the tranquil and inspiring surroundings of the Lincoln Memorial to think it through. The years at Los Alamos were recorded by Alice Kimball Smith, who had received her doctorate in history from Yale.[1] Professor Smith was awarded a Presidential Medal of Merit by President Truman for his work on the atomic bomb.

In January 1946 he founded and became first director of the Institute for the Study of Metals at the University of Chicago. This was the first academic interdisciplinary research organization in America dealing with materials. President Truman appointed him to the original nine-member General Advisory Committee to the Atomic Energy Commission that was chaired by Robert Oppenheimer. He resigned from this committee in 1952.

He served on the President's Science Advisory Committee under Dr. James Killian and Professor George Kistiakowsky, on the Advisory Committee for the Oak Ridge National Laboratory, and was a member of the Committee on Science and Public Policy of the National Academy of Sciences. The ten years from 1966 to 1976, he served on the Council of the Smithsonian Institution.

He was the first chairman of the board of governors of Acta Metallurgica, a member of the founding board of editors of Metallography, the journal of the International Metallographic Society (now Materials Characterization), and a member of the editorial board of the Bulletin of Atomic Scientists. He was a founder and an early president of the Society for the History of Technology.

In 1961 he left Chicago to return to MIT, where he became Institute Professor with appointments in the departments of humanities and metallurgy, in order to "encourage the understanding of human history and human activity through the scientific investigation of the material record of the past." At MIT he established the Laboratory for Research on Archaeological Materials in 1967. Its success led to the founding with Professor Heather N. Lechtman ten years later of the Center for Materials Research in Archaeology and Ethnology (CMRAE), a consortium of eight Boston-area universities and museums devoted to research and graduate education in archaeometry.

Meanwhile his wife had a distinguished academic career, and became Dean of the Institute of Independent Study, later called the Bunting Institute, of Radcliffe College at Harvard University.

Professor Smith was a member of the National Academy of Sciences, the American Philosophical Society, the American Academy of Arts and Sciences and the Académie Internationale d'Histoire des Sciences as well as holding honorary membership in the Institute of Metals, the Akademie der Wissenschaften, the Indian Institute of Metals and the Institute of Metals of Japan. He received two John Simon Guggenheim Memorial Foundation fellowships, one in a 1955 historical study of the inter-relation between pure science, applied science, and technology using the development of metallurgy as a particular example. The subject of the second, in 1978, was "A structural metaphor for matter and history."

He received many honors such as the Francis J. Clamer Medal of the Franklin Institute in 1952, the Pfizer Medal of the History of Science Society and the Gold Medal of the American Society of Metals in 1961, The Douglas Medal of the American Institute of Mining, Metallurgical, and Petroleum Engineers in 1963, the Leonardo da Vinci Medal of the Society for the History of Technology in 1966, the Platinum Medal of the Institute of Metals in 1970, the Sorby Medal of the International Metallographic Society in 1977, and in 1991 the Gemant Award from the American Institute of Physics for "pioneering the use of solid state physics in the study of ancient art and artifacts to reconstruct their cultural, historical, and technological significance."

Some of his papers on this subject were collected in a volume whose title, A Search for Structure, summarized the thrust of his work.[2] He wrote a number of books and monographs[3] and edited an unprecedented series of technological treatises given reliable translation into English.[4] All these are essential tools in the practice of archaeometallurgy.

He was a prolific and thought-provoking correspondent. Those fortunate enough to receive letters from Cyril have been known to frame rather than file them. He had a gift for inspiring people. He encouraged Theodore Wertime in the writing of The Coming of the Age of Iron;[5] later Professor Smith joined the technological expeditions led by Wertime in 1962 and again in 1966 in search of traditional technology in Iran. Wertime, with James Muhley, organized a festschrift for Cyril's 80th birthday, The Coming of the Age of Iron,[6] for which I was technical editor. A student of his at Chicago, Radomir Pleiner, was inspired to found the Comité pour la Sidérurgie Ancienne under the UNESCO-sponsored International Union of Prehistoric and Protohistoric Sciences.

His lectures at MIT emphasized a sensual and aesthetic understanding of materials. He often brought an object to his class to be experienced and understood on many levels. He shared his own experiences of materials, as when he would tell the story of his going out into the desert outside Los Alamos, finding ore, and smelting copper on a bet with Enrico Fermi. He became Institute Professor emeritus in 1969, but continued to lecture on the history of materials and to inspire others.

His thought has had a penetrating influence on popular culture. The television series and book, The Ascent of Man by Jacob Bronowski, was influenced by his view of the history of metallurgy, as was the volume in the Time-Life series on The Emergence of Man, The Metalsmiths,[7] for which he was the consultant.

Professor Smith is survived by his wife, Alice Kimball Smith; a daughter, Anne Smith Denman, chairman of the Department of Anthropology at Central Washington University, Ellensburg, Washington; a son, Stuart Marchant Smith of Del Mar, California, a marine geologist at the Scripps Institution of Oceanography in La Jolla; a sister, Mary Smith, of Sutton Coldfield, England; and four grandchildren.

The Cyril Stanley Smith Memorial Fund has been established at MIT to foster studies relating ancient and contemporary art and science in the materials field.

Martha Goodway
Conservation Analytical Laboratory
Smithsonian Institution
Washington DC 20560

NOTES

1. Alice Kimball Smith, A Peril and a Hope, Chicago 1965.
Alice Kimball Smith and Charles Weiner, Robert Oppenheimer, Letters and Recollections, Cambridge, USA 1980.

2. A Search for Structure: Selected Essays on Science, Art, and History, Cambridge, USA and London (MIT Press) 1981. A bibliography of his publications to 1980 is given on pages 391-399. See also his "Retrospective Notes on a Changing Profession" in Archeomaterials 1 (1) Fall 1986, 3-11.

3. Professor Smith's books and monographs include:
Of Typecasting in the Sixteenth Century, New Haven (Carl Rollins for the Columbiad Club of Connecticut) 1941.
Robert Gomer and Cyril Stanley Smith, editors. Structure and Properties of Solid Surfaces, Chicago (University of Chicago Press) 1953.
A History of Metallography: The Development of Ideas on the Structure of Metals before 1890, Chicago (University of Chicago Press) 1960. Reprinted with additions, Cambridge, USA and London (MIT Press) 1988.
Four Outstanding Researches in Metallurgical History, Philadelphia (American Society for Testing Materials) 1963.
(Editor) The Sorby Centennial Symposium on the History of Metallurgy, New York (Gordon and Breach) 1965.
From Art to Science: Seventy-two Objects Illustrating the Nature of Discovery, Cambridge, USA (MIT Press) 1980.

4. Translations edited by Professor Smith include:
(With Martha Teach Gnudi) The Pirotechnia of Vannoccio Biringuccio, New York (American Institute of Mining and Metallurgical Engineers and Yale University Press) 1942. Reprinted Cambridge, USA (MIT Press) 1966.
(With A.G. Sisco) Bergwerck- und Probierbüchlein, New York (American Institute of Mining and Metallurgical Engineers) 1949.
(With A.G. Sisco) Lazarus Ercker's Treatise on Ores and Assaying, translated from the German edition of 1580, Chicago (University of Chicago Press) 1951.
(With A.G. Sisco) Réaumur's Memoirs on Iron and Steel, Chicago (University of Chicago Press) 1956.
(With John G. Hawthorne) On Divers Arts: The Treatise of Theophilus, Chicago (University of Chicago Press) 1963. Reprinted New York (Dover) 1979.
(Editor) Sources for the History of the Science of Steel, 1532-1786, Cambridge, USA (Society for the History of Technology and The MIT Press) 1968.

(With John Hawthorne) *Mappae Clavicula, A Little Key to the World of Medieval Techniques*, Philadelphia (American Philosophical Society) 1974.

(Editor, with W. Rozanski) *Officina Ferraria, A Polish Poem of 1612 Describing the Noble Craft of Ironwork* by Walenty Rozdzienski, Cambridge, USA (Society for the History of Technology and MIT Press) 1976.

(Editor) *Kodo Zuroku, Illustrated Book on the Smelting of Copper* by Masuda Tsuna, Norwalk, Connecticut (Burndy Library) 1983.

5. Theodore A. Wertime, *The Coming of the Age of Steel*, Chicago (University of Chicago) 1962.

6. Theodore A. Wertime and James D. Muhley, *The Coming of the Age of Iron*, New Haven and London (Yale University Press) 1980.

7. Jacob Bronowski, *The Ascent of Man*, London (British Broadcasting Corporation) 1973; Boston (Little Brown) 1974.

8. Percy Knauth, *The Metalsmiths*, New York (Time-Life Books) 1974.

DEDICATION

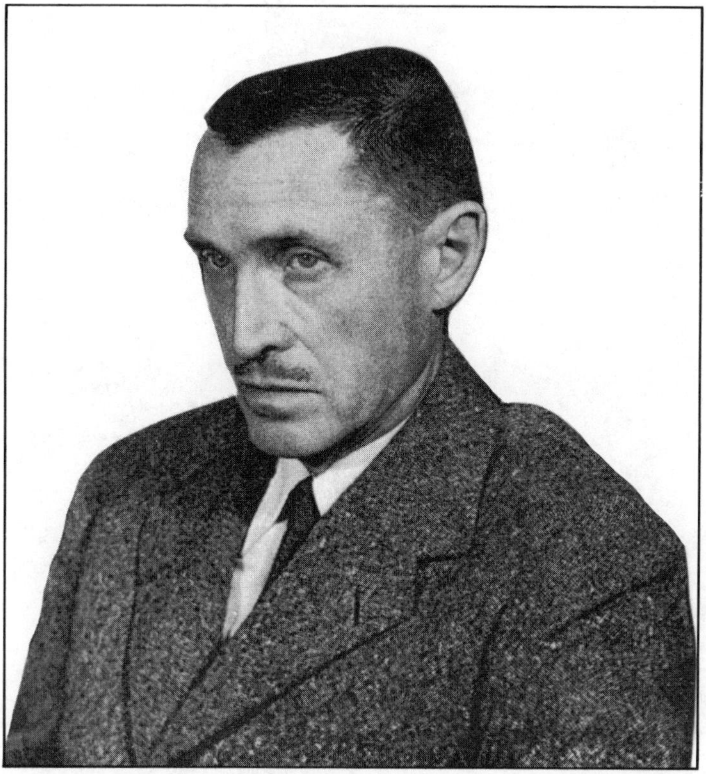

George L. Stout

DEDICATION

GEORGE L. STOUT

George L. Stout has been eulogized many times before. In the late 1920's he played a seminal role in the technical studies of art-works at the Fogg Art Museum at Harvard along with Edward Forbes, Alan Burroughs and Rutherford John Getten. From 1932 to 1942, "Technical Studies in the Field of Fine Arts" was issued quarterly by the Fogg, and George Stout was both its founder and a name that appeared frequently as author.

Stout was a strong proponent of the "three-legged stool" approach to conservation—science, history, and practice. Marjorie Cohn, commenting on George Stout's legacy, quotes from a 1931 writing, "What is needed," said Stout, "...is not only technical competence, but something for want of a better name I shall call aesthetic judgment, and beyond that a genuinely scientific point of view."

Stout never over-emphasized the value or maturity of the corpus of conservation knowledge and what we fashionably call today "conservation science." Even as late as 1977 he described it as "...little more than a wobbly skeleton," at another time a "...mongrel pup that had clawed through the academic fence." In the 1920's, the number of people engaged in technical studies could be counted on one's fingers. The difficulties in finding a scientist for the Institute of Fine Arts, New York University, in the late 1950's illustrates how microscopic the technical support side of conservation was. Robert L. Feller was securely fixed at the Mellon Institute; Edward Sayre was loath to leave Brookhaven; and John Gettens would only give a single seminar on microscopy the first year.

George Stout would pass away just as a new cutting edge was emerging. He would just miss the advances collectively made by the Canadian Conservation Institute, The Conservation Analytical Laboratory of the Smithsonian Institution, and The Getty Conservation Institute, not to mention the work carried out by a host of other major institutions, national and international: museums, libraries and archives, universities and other conservation programs including the training academies.

All who speak the name of George Leslie Stout do so in reverence for his vision of a futuristic conservation at a time when most craftsmen preferred to call themselves restorers. However, being a great pioneer in this field is not the reason for this dedication.

Just after the Japanese bombed Pearl Harbor on December 7, 1941, there was a meeting at the Metropolitan Museum of Art. We know that it was attended by Francis Henry Taylor, the Director of the MET, Paul J. Sachs, Professor of Fine Arts at Harvard and George Stout, among others. It was at this meeting that Stout's concern for the protection of cultural patrimony was clearly heard. In January of 1942, a paper entitled, "The Preservation of Paintings in Wartime," by Stout appeared in *Technical Studies*. In November 1942, Taylor, according to Craig Hugh Smyth, went to Washington to take up their concerns with the Director of the National Gallery of Art, David Finley. Finley made an appointment to see Supreme Court Justice Harlan F. Stone, who happened to be an *ex officio* Chairman of the Board of Trustees of the Gallery. Stone turned around and pleaded the case to a receptive President Roosevelt, who on August 23,1943 established the American Commission for the Protection and Salvage of Artistic and Historic Monuments in War Areas. This commission, which was an unprecedented government agency, created the instruments of implementation—The Monuments, Fine Arts and Archives Service (MFA&A) in the American and British armed forces.

The best description of the activities of the MFA&A Services is given by Craig Hugh Smyth in his wonderful little book *Repatriation of Art from the Collecting Point in Munich after World War II*. The MFA&A had as its mandate to oversee, as well as it could, the protection of cultural patrimony during conflict and subsequent occupation. This was no small task in the heart of so vast a war—and there were human losses. The MFA&A was chronically under-manned and officers were not given the rank they required to fulfill their duties easily. One enlisted man, later to become an officer, was Sheldon Keck. Still, Eisenhower was a supporter of the MFA&A. In his message to his commanders just before D-Day, Eisenhower said,

> "Shortly we will be fighting our way across the Continent of Europe in battle designed to preserve our civilization. Inevitably, in the path of our advance will be found historical monuments and cultural centers which symbolize to the world all that we are fighting to preserve. It is the responsibility of every commander to protect and respect these symbols wherever possible."

There was never any doubt in Sachs' mind whose idea this all was. Sachs' letter to Stout acknowledged his "brain child" and his clear thinking during the Metropolitan meeting after Pearl Harbor. He said, "It is my deliberate opinion that the appointment of this Commission is due to your initiative, imagination and energy." Stout tried to downplay Sachs' rich complements by indicating that the idea was one thing but to carry it out required far greater powers. In an almost surreal comparison, Stout remarked in a letter to Sachs dated September 13, 1943 that, "A long time ago I figured out that it would be nifty to have a machine that would wash dishes, but that didn't help anyone." The same could have been said by Leonardo di Vinci about many of his inventions.

The demise of *Technical Studies* was caused by Stout's entry into the Navy where, ironically, he was subsequently recruited to assist the MFA&A in Europe. Stout wrote the MFA&A's field manual *First-aid Protection of Arts and Monuments*, and was with the army from Normandy. After hostilities ended, Collecting Points were established to assist in the repatriation of artworks retrieved from the Germans. Because of his expertise, Stout was responsible for packing and removing of artworks from the various locations then being discovered, where they had been sequestered away. He shipped artworks from salt mines in Austria to the Collecting Point in Munich, the largest of these facilities. By October 1945, he was finished in Europe and off to Japan.

George L. Stout is not commemorated here for his contributions toward leading conservation out of the Dark Ages and into a bright new world of unified aesthetics, craftsmanship, and science; rather, he is commemorated for his services to mankind—past, present and future, and in helping to verbalize the objectives of the Allied Powers during the Second World War in recording and limiting damages done by combat operations to the cultural heritage of man.

This is, arguably, the single most important contribution any conservator or scientist has ever made to the heritage of Mankind.

JAMES R. DRUZIK

NOTES FOR A HISTORY OF MUSEUM SCIENCE

IN THE UNITED STATES

Martha Goodway
Conservation Analytical Laboratory
Smithsonian Institution
Washington DC 20560

The smashing of the Portland Vase by a deranged person in the British Museum seems hardly an auspicious beginning for museum conservation in the United States but, as in so many other things, truth proves stranger than fiction. The Portland Vase is the most notable surviving example of Roman cameo glass, and had been on loan to the British Museum from the fourth Duke of Portland since 1810. When it was vandalized in 1845 Scotland Yard was called in to investigate. Among the investigators was a forensic chemist who became grandfather to William J. Young, founding director of the Research Laboratory of the Museum of Fine Arts in Boston.

The association with the British Museum that began with the case of the Portland Vase eventually led the chemist's son, Young's father, to join the staff of the British Museum where for three or four years he trained in conservation. In 1900 he went to Oxford University to be the conservator at the Ashmolean Museum. There the association with the Portland Vase continued. The restoration of glass is among the most difficult tasks in conservation and, because of the visible yellowing of adhesives, it must be repeated from time to time. So the Portland Vase has been reglued more than once. The second time this was done by J.H.W. Axtell, who had been trained at the Ashmolean by Young's father.

During his career, Young's father restored a number of Greek vases. He signed his restorations "Neos," Greek for "recently born" or young. Years later a scholar who had been studying the "Neos painter" of Greek vases in collections in England and Europe came to the Museum of Fine Arts, where Young had great difficulty convincing him that the Neos painter was not an ancient Greek but none other than his father.

William Young was educated at Oxford, where Balliol College was particularly noted for its well-equipped laboratories. He was trained in conservation by his father, who, in the British tradition, expected his son to succeed him at the Ashmolean. This is still the custom in some English museums, such as the Armoury at the Tower of London. So Young's father built a laboratory for his son at the Ashmolean.

Hogarth was then the director. It was the heyday of imperial archaeology in the Near East and many well known objects came to the Ashmolean to be treated, such as the large reliefs from Assyria. There was no lack of work to do. Thousands of cuneiform tablets were being unearthed and sent to the Ashmolean for preservation, where the unfired clay tablets were "baked" (fired) and the soluble salts removed. Stephen Herbert Langdon of the Field Museum in Chicago was one of the archaeologists who sent the tablets they excavated to the Ashmolean for baking and desalinization.

Along with the tablets came all the famous archaeologists of the era. Wooley and Evans came, and Garstang and Gertrude Bell, even Lawrence of Arabia. When Young's father remarried, Maspero's niece, Ethel, became his stepmother. There were also visitors from America, including Caskey from the Museum of Fine Arts in Boston, and Dows Dunham, the curator of their Egyptian collection.

With his experience in the latest methods of conservation, Young received a number of offers. He excavated with Alexander Keiller, of the marmalade fortune, at Avebury. Keiller had a laboratory in London and wanted him to start an institute there. Young also worked with E.P. Warren in Lewes in Sussex. Warren was a classical archaeologist, an American and a trustee of the Museum of Fine Arts. Langdon invited Young to Iraq. Wooley wanted him at Ur.

These offers became even more attractive when Hogarth was no longer director at the Ashmolean. So plans had been made for Young to go to Baghdad in 1928 to join Langdon at Kish but before he could leave Langdon died of malaria, as did a friend in the oil business in Iraq. As it happened, Young had also received an invitation from Warren to come to Boston for six months to establish a laboratory at the Museum of Fine Arts. This seemed a safer thing to do, and so he set sail for America.

Meanwhile at the Fogg Art Museum of Harvard University, under the directorship of Edward Waldo Forbes, the scientific examination of paintings was being practiced. Burroughs' book on the methods used at the Fogg, *Art Criticism from a Laboratory*, gives the details (1). The methods were limited to radiography, ultra-violet and infra-red photography and microchemistry. The text was elementary, with little exposition of the physical principles upon which the methods were based, but reported the most advanced use of scientific methods in museums in the United States at that time.

When Young arrived in Boston all he had to begin with was "a bench and three jam jars." Adequate instrumentation was an obvious priority. He proposed buying a spark-gap spectrograph, then the state-of-the-art instrument for elemental and trace analysis, but the museum authorities were frightened and thought such a device too dangerous for a museum. Permission was granted for its acquisition only after being assured of its safety by several letters to the trustees from professors at the Massachusetts Institute of Technology. Thus began a fruitful association with the universities of the Boston-Cambridge area that was later formalized in the Scientific Advisory Committee for the Research Laboratory. Many prominent scientists served on this committee and often offered assistance as well as advice.

Equipment that could not be bought was built or borrowed. For X-ray diffraction he went to the geology department at Harvard where he received generous help from Berman and the Frondels. He built a thermoluminescence unit, later replaced by one built at MIT. Neither worked as well as he had hoped, and eventually he was able to buy a unit from Oxford. The tradition of practicing science in the museum was firmly established at Oxford. In this favorable climate E.T. Hall and Martin J. Aitken had founded the Research Laboratory for Archaeometry there after World War II where the methods of thermoluminescence dating had been developed.

When the laser was invented Young immediately saw the possibility of substituting it in the spectrograph for the spark that had so frightened the trustees. He developed a method of using the laser to take samples in situ, for example ink from a print, at the same time vaporizing the sample for analysis spectrographically.

Another important innovation was the series of international seminars on the Application of Science in Examination of Works of Art. These were organized in 1958, 1965, 1970, and 1974. Each of these seminars resulted in a publication that is still referred to (2). The success of these seminars was due in no small part to Young's gift for raising funds, and to substantial help from the Scientific Advisory Committee. Invited speakers were offered the then-princely sum of six hundred dollars, enough to allow many European authorities to come to the United States for the first time. These seminars were the first such international meetings in the United States. Young also organized a group of curators, collectors and patrons interested in Asian ceramics who called themselves the Far Eastern Ceramics Group. The purpose of the group was the technical examination and analysis of oriental ceramics. Many came to Boston from England and Europe to attend the Group's conferences, which were held once or twice a year. The publications that resulted from these meetings (3) are among the earliest technical studies of ceramics. As a result, to this day most Asiatic art departments distinguish themselves from other museum departments by their interest in and fostering of technical studies.

In collaboration with the Fogg Museum at Harvard, Young established a course for art historians and conservators, giving them practical experience of technical analysis and understanding of the strategies, functions and limitations of various methods of analysis as well as their procedures. The course is still being taught by staff at the Museum of Fine Arts, presently by Mimi Leveque and Richard Newman.

The founding of the Research Laboratory at the Museum of Fine Arts and the examination methods being used at the Fogg Museum mark the beginning of the transition in United States museums from restoration workshops of skilled craftsmen to conservation laboratories backed by the necessary scientific support. Other museums in the United States soon followed suit.

The history of this development is still to be written, but it is not too soon to gather the materials for such a history (4). Prompt publication while these materials can still be amplified and corrected by the participants is to be encouraged, and additions and corrections to this paper will be most welcome. Best of all would be accounts by the participants themselves.

ACKNOWLEDGMENT

The author would like to thank William J. Young and Florence Whitmore Young for their help and encouragement, in this and many other projects.

NOTES

1. Alan Burroughs: **Art Criticism from a Laboratory**, Boston (Little, Brown) 1938.

2. Museum of Fine Arts, Boston: **Application of Science in Examination of Works of Art**, Boston 1959.
Museum of Fine Arts, Boston: **Application of Science in Examination of Works of Art**, Boston 1967.
William J. Young, editor: **Application of Science in Examination of Works of Art**, Boston (Museum of Fine Arts) 1973.
William J. Young, editor: **Application of Science to the Dating of Works of Art**, Boston (Museum of Fine Arts) 1978.

3. William J. Young and F.E. Whitmore, editors: **FECG, Far Eastern Ceramic Group**, Boston (Museum of Fine Arts), annual volumes 1948-1956.

4. Willa K. Baum: **Oral History for the Local Historical Society**, Third Edition, Revised (American Association for State and Local History) 1987, is a brief (68 pp.) and practical guide to interviewing and comes highly recommended. It is available from the AASLH (172 Second Avenue North, Suite 102, Nashville TN 37201), the Oral History Association (P.O. Box 926, University Station, Lexington KY 40506-0025), or through libraries.

EXAMINATION AND ANALYSIS OF OBJECTS IN THE MUSEUM LABORATORY

EDWARD V. SAYRE
Conservation Analytical Laboratory, Smithsonian Institution,
Washington, DC 20560

Objects are examined and analyzed in a museum laboratory for two primary purposes, 1) to determine their conditions and 2) to seek evidence to establish or confirm provenances for the objects. The questions concerning the condition of an object might relate to its preservation, e.g., to what extent is the object in a deteriorated condition and what should be done to stabilize the deteriorated portions, or to what degree is the object undergoing active deterioration and what should be done to halt the process of deterioration. Other questions concerning the condition of an object might relate to its intactness, e.g., have portions of it been added or replaced in past restorations or has the surface of the object been repainted or repatinated. I remember a Tang dynasty Chinese ceramic horse, which was brought to the Research Laboratory of the Museum of Fine Arts in Boston for examination. The horse had been published as having been authenticated by thermoluminescence dating, and indeed we found that the barrel of the horse, from which the sample for dating had been taken, was indeed of the Tang dynasty. However, examination also revealed that all of the appendages of the horse, the head, mane, tail and legs, were modern additions. Therefore, almost all portions that defined the style and appearance of the horse had been recently added to it. Under these circumstances it becomes very moot whether the horse, in its present condition, could be exhibited as a authentic art object, and whether, stripped of its modern additions, it would be worth exhibiting. The museum is much concerned about the condition of objects when considering 1) their acquisition, 2) whether unauthentic additions to them should be removed and 3) what should be done best to preserve them.

A museum is constantly concerned about the accuracy of the provenance attributed to objects. The origins of some objects, i.e., their provenance, may be well established by historical record. An exact record of the creation of a work of art by a particular artist might exist, or an object might have been excavated within a well documented archaeological context. Many objects reach a museum, however, without a well established history, and, in such cases, their origins are inferred by comparing them to those objects whose origins are well known. The curator usually makes this comparison upon the basis of visual characteristics of the objects, e.g., conformation, style, details of decoration or treatment, etc. In the laboratory the comparison is usually made upon the basis of structure and composition, characteristics that may well be invisibly hidden within the object. Both sets of characteristics can supply significant evidence of provenance, and their interpretation is basically similar. One is well advised to include them both in arriving at conclusions. Indeed, most major museums require that important objects being considered for acquisition receive both a curatorial and a laboratory examination. Accordingly, a museum laboratory is constantly examining and analyzing objects of known provenance to understand and record their characteristics, and objects of unknown provenance in the endeavor to infer their origins from their characteristics.

Examination in a museum laboratory starts with a careful overall visual observation of the object in question combined with observation of details with a low power microscope. The object is then usually examined under ultraviolet light. The surface fluorescence, under ultraviolet radiation, of regions where the surface has been altered by cleaning, repair or surface treatment, such as overpainting or repatination, is usually discernibly different from the fluorescence of an unaltered original surface. For some materials, such as marble or bronze, a natural patination acquired over a long period of time will tend to fluoresce in a characteristic way, that is often distinguishable from the fluorescence of artificially applied patination. Also observation under infrared illumination, using an infrared scanner or photography to make the image visible, is fairly often resorted to, particularly if the surface is painted. Infrared scanning of a painting or painted surface will sometimes penetrate the surface layer, permitting one to see details of drawing or painting underlying it.

The next step is to examine the inner structure of the object, should this be thought necessary. The two techniques commonly employed for this are radiography and microscopic examination of cross sections of the object, or its surface. X-ray radiography is, by far, the most popular radiographic technique, although neutron radiography or radiography with isotope emitted or accelerator generated radiation are sometimes used. The problem in radiography, of course, is to match the penetrating power of the radiation to the density and amount of material it must penetrate, e.g., high energy X-rays or gamma rays are needed to radiograph massive bronze objects and very low energy photons or beta rays are required to radiograph thin, low density materials like paper. Radiography can often reveal how an object was made, e.g., what casting or working technique was used to form a metal or a ceramic object, and how it has been altered subsequently. Cross sections are usually examined microscopically: 1) to determine how the body of a object was formed, e.g., whether a metal object has been cast and whether it has been otherwise heat treated or worked, or 2) to determine the materials used in or the conditions under which an object was made, e.g., what temper was used in a ceramic object or at what temperature was it fired, or 3) to determine how the surface of the object has been treated or altered with time. The application of gilding to a surface and the preparation of a surface for painting or the application of paint layers to it are examples of surface treatments that might be examined in cross sections. It is also sometimes possible to distinguish between prolonged natural corrosion and false patination or rapid artificial corrosion of a surface through cross sectional examination. For example, natural corrosion on bronze usually has a very characteristic microstructure, a thin bright red layer immediately on top of the uncorroded metal that is covered with a thicker green layer. The red layer is a monovalent copper oxide, cuprite, and the green layer is usually a divalent copper basic carbonate, malachite, although basic copper chlorides or sulfates may also be present. That is, the layers progress from unoxidized metal to partially oxidized copper oxide to fully oxidized copper salts. Most false patinations do not show this characteristic structure. A prolonged natural corrosion of bronze usually also results in a characteristic penetration of the corrosion into the metal along the boundaries between individual crystal grains in the metal structure, so called "intergranular" corrosion. Rapid corrosion of a bronze surface seldom achieves such penetration.

Some types of structural examination are, in effect, analytical procedures in that they lead to the identification of materials present within objects. The museum laboratory is often called upon to identify woods, hairs and a variety of other fibers. Such identification is usually based upon microscopic examination of the morphological structures of these materials. Microscopic examination of petrographic thin sections of rocks and ceramic bodies is often used to identify the individual minerals present within them.

Chemical analysis may then be used to confirm or amplify what has been inferred from structural examination. The analytical methods are of two types, 1) those that reveal the presence of chemical elements and determine their concentrations and 2) those that identify the presence of specific chemical compounds. Museum objects are usually unique and irreplaceable and, hence, only analytical methods that leave the object essentially unaltered may, as a rule, be employed. If samples must be removed from the object, they must be very small ones, taken from inconspicuous positions. This requirement can often lead to significant sampling errors when the region being analyzed is nonuniform in composition. However, it is seldom necessary that the analysis of a museum object be highly quantitative. All that is usually required of analysis is that the materials present be correctly identified, and, if they contain impurities or are an intimate mixture of elements, such as a metal alloy, that the order of magnitude of the concentrations of the individual components be correctly estimated. For example, one is usually satisfied to be able to determine whether a metal object, predominately composed of copper, should be considered to be copper metal, brass or a tin, arsenic or antimony bronze and whether the concentrations of impurities within it are high, as they frequently are in ancient objects, or low, as they usually are in modern objects. X-ray fluorescence, either X-ray, electron or proton beam induced, emission spectroscopy, either arc, spark, laser microbeam, or plasma induced, and atomic absorption are the instrumental methods of elemental analysis most often employed. Neutron activation may be used to determine characteristic patterns of impurities but it is seldom available as an in house technique. Chemical spot tests, done on plates or under a microscope, are still sometimes used for qualitative analysis, but have largely been supplanted by more convenient and more quantitative instrumental methods. All of these methods require, at most, the removal of very small samples. X-ray fluorescence and neutron activation can be carried out without removing any sample. They all also are sensitive methods for determining many elements. None of them, however, can be used for all elements. An X-ray fluorescence unit that functions with a thin collimated external X-ray beam, which can analyze small areas on large objects, without the removal of any material, is the most popular instrument for elemental analysis in museum laboratories. For analyzing microscopic details, a scanning electron microscope equipped for X-ray fluorescence detection or an electron microprobe is often used.

Elemental chemical analysis is used: 1) for identification of materials present, which need not require highly quantitative accuracy, and 2) for the establishment of characteristic patterns of impurities that might relate to the origins of the materials, which requires fairly high quantitative accuracy. For example, in the late first millennium A.D. glass making in northern europe changed from a "Roman" formulation, in which sodium was the predominate alkali, to a "medieval" formulation in which potassium was the major alkali. The difference resulted from the use of a relatively pure sodium carbonate, natron from the Wadi Natrun in Egypt, as an alkali in Roman glass and the use of a potassium rich alkali derived from wood ashes in medieval glass. Some feel that both kinds of glass are to be found in some early medieval church windows. A simple semiquantitative analysis of the alkaline metals present in these glassed will be adequate to distinguish between them. However the wood ash alkali is relatively impure and carries with it a pattern of minor and trace chemical elements, characteristic of the particular wood source that is used for its production, which is transferred to the glass formed from it. The compositional differences among these source patterns may be fairly small yet definitely recognizable. Therefore, a careful quantitative determination of a number of elements in medieval glasses may be required to relate the glasses to the output of individual workshops.

The methods most frequently used for the identification of specific chemical elements or minerals are microscopy, particularly with a polarizing optical microscope, X-ray diffraction and infrared, visible and ultraviolet spectrophotometry. For inorganic materials X-ray diffraction or thin section microscopy are the usual methods employed. Organic materials are more usually identified with one or more forms of spectrophotometry. Chromatography and even mass spectrometric spectrophotometry are used very effectively for organic analysis but are not often available for in house use.

A museum laboratory may well be equipped in other ways. It may be equipped to do absolute dating, such as carbon 14 or thermoluminescence dating. It may also be equipped to carry out a wide range of research projects dealing with preservation or provenance. In an overall sense, what most characterizes a museum laboratory is diversity. Museum objects are composed of all types of materials, both organic and inorganic. The laboratory must be prepared to deal with all of this great variety of diverse materials. For example, it should be as fully prepared to cope with all of the materials found in a painting mounted on textile or wood as with those found in a gilded or enameled metal statue inlaid with glasses and various minerals.

UNIVERSITY LABORATORIES AND ANCIENT MATERIALS SCIENCE

PAMELA VANDIVER
Conservation Analytical Laboratory, Smithsonian Institution,
Washington, D.C. 20560 USA

In a university laboratory, the product is educated students rather than the maintenance and display of collections. Many mottos from various colleges summarize this task; one that I like is "The paramount obligation of a college is to develop in its students the ability to think clearly and independently and the ability to live confidently, courageously and hopefully" (1). A university lab is thus constrained to produce reasonable research while at the same time educating and using the capacities of individual students.

When comparing a university to a museum laboratory, the methods are just as varied as described above by Edward Sayre, but the process of formulating questions is different. Laboratories for ancient materials science are tied to various departments, such as art history and conservation, anthropology or archaeology, ancient studies, history, materials science and engineering, geology, biology, biogeochemistry, applied physics, chemistry, etc. Many disciplines are interested in the history of their discipline within the wider framework of humanistic studies, especially intellectual history. Other disciplines are interested in artifacts brought from fieldwork or found in museum or university collections or in the application of recently developed techniques used to investigate modern materials but which can also be applied to the study of artifacts. Some labs are established for a special or limited purpose; some have very little equipment other than a low-power microscope, while others are multi-purpose, high-tech facilities. Certainly what can be accomplished depends on access to equipment, as has been noted by David Killick in the introduction to this volume, but it primarily depends on the intent of the faculty and students.

At the undergraduate level students may be asked for literature surveys or a written evaluation of particular characteristics of an object or group of artifacts. Often no more than a low-power microscope or a hand-lens is used, and these are not to be discounted, as stated by Robert Gordon in the introduction. A recent trend in the Humanities has been to ask students to examine an idea or event in the light of artifactual evidence, but far more common are projects requiring written records which are collected and evaluated much as in the method of French historians in the "Annales" school. Most of these projects do not penetrate the problems in sufficient depth for publication.

Science students, on the other hand, are regularly asked to examine and test samples, usually replicating results which are already well-known. These students are usually given the problem and use already-calibrated and established equipment. For any of these students to be creative and insightful is quite difficult, because they are limited by constraints of time since most projects last no more than part of one semester, by access to and knowledge of current literature or equipment, and by the experience of the student and the ability and confidence to attack a problem. The student also is limited by the organization of learning at American universities into departmental disciplines. The problem of having too little time to do in-depth research is a similar problem to that in most museum laboratories, where the pace of work often is exhausting due to the demands of the necessary "service" work for the many curatorial departments, so that researchers often have difficulty in pursuing their own research goals.

The advantages of graduate education are the possibility of long-term and intensive study of a problem of the student's choosing and/or formulation, under the supervision of professors chosen by the student. The period of study is similar to an apprenticeship, in that the student often works on several projects initiated by others, masters the methods of study, the process of data evaluation and the paths of interpretation. At a more advanced level of graduate education, students are expected to find and pose problems, rather than choose them. Instead of showing the capacity for original research within an established discipline, as is common at the master's level, the student is expected to conduct original research. At the level of a doctoral thesis, the possibilities for research into materials, artifacts, collections and sites are unlimited; the thesis arguments are data-driven and empirically-derived. Most studies emphasize new applications or new interpretations.

One source of information which summarizes the work of many researchers in several labs involving many different disciplines related to archaeological collections has been the Society of Archaeological Sciences. In the *SAS Bulletin* vignettes describing several laboratories have been featured over a number of years (2). Another source of information are the research directories, published by most universities, that summarize the projects of each faculty member or laboratory.

In addition, many professors are interested in the interdisciplinary aspects of humanities, arts, science and technology. Some courses are taught, often by several faculty members, which emphasize problem-solving with actual samples to derive information from artifacts and to interpret material culture. Many centers for science, technology and society only concern twentieth century technologies and mainly study documents rather than artifacts, but several groups focused on history, in fact, make extensive use artifactual evidence, so that one cannot judge a program by its description alone. A partial list of several professors interested in such interdisciplinary research is included below (3). We would be interested in hearing talks and publishing work of students involved in such courses or projects at the next MRS meeting.

In these university labs and courses, because students are the product, their learning to solve problems is what is of paramount importance. Many problems in ancient materials science are easier to comprehend and require less sophisticated equipment to demonstrate or investigate. The particular techniques of analysis and research questions that students attempt to answer tend to be as ephemeral as styles of decoration on pottery. The structural building of pottery, however, has been found to be extremely conservative, and likewise the structural fabric of thought processes aimed to solve problems is also a conservative and learned craft which requires lots of practice.

To learn and practice this skill, students can use some problems in ancient technology, conservation science or materials identification. A few of these problems are simpler than problems in modern technology, but an understanding of the solutions to many of these problems gives students respect for the complexity and sophistication of some ancient technology and a firsthand knowledge of the capable, skilled and resourceful people who made prestige objects for elites as well as objects of common usage by people of low status, and even the specialty goods for periods and places with a middle class. Many students gain a new perspective on how time and skill are valued and used in preindustrial societies. Most of these problems have a well-articulated cultural context, so their relevance is easy for a student to understand. For instance, Rustum Roy has stated that it is simpler for students to address problems of agriculture, environment, energy and communication, than it is to study with any passion for the first time such subjects as biology, chemistry, physics and math. Students are attracted and enrapt by problems they recognize as immediate within their own life or the lives of their families. Thus, teaching by identifiable, some say "relevant", example is a way of compelling student interest. Likewise, examples from the lives and activities of previously living people is a similar way of holding student's attention; "living history" or replicative archaeology are related strategies for learning.

Some other problems have a human dimension, and even a political or emotional one, that tends to overshadow our comprehension of those aspects of a problem best isolated and attacked in terms of science and engineering. The destruction of cultural heritage or the theft and illicit trade in antiquities are such problems. Schemes for the prevention of such destruction during conflict as well as the marking and recording of antiquities are engineering problems which are solvable with available technology and manpower in industrialized and traditional cultures. The field identification of a large statistical sampling of materials on-site in even remote field excavations is likewise fully within the range of well-known technologies, many of which are nineteenth century practices, yet only rarely are these practiced sufficiently to reconstruct activities or technologies from more than a small percentage of recovered remains. These problems, and many others as well, have entrenched ways of being handled which makes them very difficult to tackle.

Some problems of complex or sophisticated ancient technology which have been used to give students an appreciation of history, culture, science and engineering are:

The manufacture and heat treatment of Japanese swords
Neolithic Southwest Asian gypsum and limestone plasters
Peruvian weaving and the chemistry of dyes and mordants
Indonesian and Indian textile decoration
European Renaissance flute and lute manufacture
Chinese ceramic manufacture, especially of the Song dynasty
Greek red-on-black or black-on-red pottery
Flint tool knapping, retouch, refit and use-wear
Ground stone tool and bead replication
Mesopotamian composite ceramic tool manufacture
Fresco, egg tempera, and oil media painting styles
Identification of traditional paint pigments and media, and synthesis of modern ones
Egyptian quartz-bodied faience manufacture by efflorescence, cementation and application
Glass vessel fabrication on a ceramic core
Roman glass blowing, molding and empontilling
Glass finger rings according to Theophilus
Champleve and cloissonne enamels and thermal expansion
Blacksmithing
Raising and repousse of bowls or vessels in copper, brass and silver
Prince Rupert drops and stored-strain energy
Low-thermal expansion Bottger stoneware and thermal shock
Tiffany phase-separated glass compositions and coatings demonstrating superplasticity
Ancient copying and photographic methods
Papyrus, vellum and paper making
Maya blue pigment and other intercalation compounds

There is a direct experience with the behavior of materials which compels belief in a predictable physical world and learning of a skill to manipulate materials, but also identifies the raw material and processing variations and vagaries, while at the same time trying to understand and replicate some of the most wonderful examples of craft with those materials. This experience is humbling before the objects and the ancient artisans who made them; it challenges one and at the same time develops a judgmental skill which allows one to recognize defects in ancient objects or when and why a technological performance or visual appearance is a little "off" or is really an exceptional tour-de-force. For instance, we may ask, "What are the criteria of structure, composition, processing and properties which separate a masterpiece worth millions from an example from the same workshop or another worth very little?" Or similarly, we may ask through artifact analysis, "What is the physical basis of visual appearance," or "What are the geological, cultural and historical constraints on appearance?"

Cyril Smith has emphasized that such direct, immediate experiences are interdisciplinary in nature and require a meeting of art, science and history in the hands of the doer or student experimenter. By playing and interacting with materials, one develops a set of expectations about their behaviors as one sees and feels changes in color, texture, density, shape, reactivity, temperature, viscosity, surface tension and begins to realize that equations which describe sintering, nucleation and growth, stress relaxation during annealing and other changes of state have meaning.

What is common to student education is the methodology which requires empiricism, the practice or experimentation of performance, along with thinking and rethinking, or "reflecting". Having the luxury of time and a properly conducive atmosphere to promote the self-discipline, involvement and commitment which such study requires is rare, even in universities long renown for such conditions. Many believe that what happens in this process is a kind of magic at which some students are more successful than others. Some scientists say they know how they solve problems and cannot be bothered with an analysis of what they do; some say it might disrupt the success of their own process. However, David Kingery teaches a course which many of his students value more than any other course in their education in which he tries to show students how to plan, propose, initiate and evaluate research. Some of his teaching is based on the book, *How To Solve It* by George Polya(5), a highly recommended treatise regularly extolled in the pages of *Scientific American*. This text analyzes the process into several steps:

Understanding the problem in terms of identifying the unknowns, data and sufficient conditional relationships.
Making the problem visual, kinesthetic and notational.

Devising a plan using what you know, similar problems or parts of the problem to analyze and synthesize a solution.

Carrying out the plan, checking each step and evaluating possible negative evidence; and

Re-evaluating the method, data and results and any other similar evidence or research so that the most direct path to the solution is found rather than the one taken in the solution. The conclusions are well integrated into the fabric of knowledge, sometimes by renaming or reconceptualizing some aspect of what is known in light of the newly devised and solved problem.

The last step can take as long as the previous ones taken together. It is always a temptation to publish first and reevaluate, integrate, reflect and interpret later, but we cannot afford not to take the time, make the energy and do a better job. The preservation of cultural heritage, reconstruction of ancient materials science, discovery of ancient technological knowledge and education of at least some students all depend on it.

REFERENCES

(1) Ellen Browning Scripps, Scripps College, Claremont, CA, 1922.

(2) The Society of Archaeological Sciences reports:
"The Center for Materials Research in Archaeology and Ethnology" (Massachusetts Institute of Technology), SAS Newsletter 1, no. 4 (1978) 2-4.
"Center for Archaeological Sciences, University of Georgia", SAS Newsletter 10, no. 1 (1987) 4, 7.
"Archaeological Metallurgy at Yale", SAS Newsletter 10, no. 2 (1987) 2-3.
"Environmental Archaeology Unit, University of York", SAS Newsletter 12, no. 3 (1987) 6-7.

"The Archaeometry Laboratory, University of Minnesota", SAS Newsletter 10, no. 4 (1987) 1-3.

"The Amber Research Laboratory at Vassar College", SAS Newsletter, 11, no. 2 (1988) 2-4.

"The Archeochemisches Laboratorium, Institut fur Urgeschite, University of Tubingen", SAS Newsletter 11, no. 3 (1988) 3.

"Archaeometric Laboratory, Department of Physics and Astronomy, University of Maryland", SAS Newsletter 11, no. 4 (1988) 2-3.

"Accelerator Mass Spectrometry Laboratory, Institute of Nuclear Sciences of the New Zealand Department of Scientific and Industrial Research", SAS Bulletin 12, no. 2 (1989) 2-3.

"Conservation Analytical Laboratory, Smithsonian Institution", SAS Bulletin 12, no. 3 (1989) 8-10.

"Archaeometric Research Laboratory, Research Reactor Facility, University of Missouri", SAS Bulletin 12, no. 4 (1989) 2-3.

"The British Academy Group Research Project: An Aspect of Science and Archaeology in Great Britain", SAS Bulletin 13, no. 1 (1990) 3-5.

"Laboratory for Archaeological Chemistry, Department of Anthropology, University of Wisconsin-Madison", SAS Bulletin 13, no. 1 (1990) 6-9.

"Paleoethnobotanical Laboratory, Texas A and M University", SAS Bulletin 14, no. 1 (1991) 2-3, 5.

For further information about the Society for Archaeological Sciences, contact Prof. R.E. Taylor, Radiocarbon Laboratory, University of California at Riverside, Riverside, CA 92521 USA.

(3) David Kingery and David Killick, University of Arizona; Mark Kenoyer, University of Wisconsin-Madison, Norman Muller, Princeton University; Richard Newman and Mimi Levique, Boston Museum of Fine Arts and Harvard Fine Arts Department; David Hounshell, University of Delaware; Jules Prown, Yale University; Arthur Steinberg, Concourse Forum, M.I.T.; Stuart Fleming, Patrick McGovern and Vincent Pigott, MASCA, University Museum, University of Pennsylvania; Donald Avery, Brown University; Nicholas van der Merwe, Peabody Museum, Harvard University; Eugene Farrell, Fogg Art Museum, Harvard University; Stanley Taft, Cornell University; Barbara Purdy and David Clark, University of Florida; various faculty, New York University Institute of Fine Arts; and Institute of Archaeology, London, among many others.

(4) R. Roy, "Materials Education: The Second Time Around," *MRS Bulletin*, Vol. 17, No. 9 (1992) 22-26.

(5) G. Polya, *How to Solve It*, Princeton University Press, 1945, 1957, 1971, 1973.

PART I

Cultural Heritage in Conflict

Amidst a symposium dedicated to materials used to produce cultural artifacts and their conservation, I thought it appropriate to step back and consider for a moment human processes frequently dedicated to the destruction of those materials—armed conflict.

In the course of the next few hours, I have assembled about as broad a "collage" of papers as I could reasonably put together. They are all united only in this one central theme—loss and preservation; death and birth; material culture as defenseless innocence in war.

The papers that follow discuss the principle international accord for the protection of cultural monuments and objects—The Hague Convention of 1954; precautions taken by both the British Museum and the National Gallery of Art during World Wars I and II; the fate and contemporary efforts to rectify the internment of the Aleuts also during World War II; a discussion of the effects of the Gulf War on the cultural heritage of Iraq; two discussions on the recent war in Croatia; lessons to be learned on natural disaster planning; and of particular value is the discussion on entropy and catastrophe.

Critics of the Hague Convention note that signatories avoid implementing Chapter II during peacetime. This section calls for the compilation of national lists of protected monuments that shall be deposited with the Secretary-General of UNESCO. Evidently, few nations have attempted to comply because the urgency of such measures seem distant in peacetime. The poignant reality contained in "entropy and catastrophe" is that this complacency by national inexorable fact is that the deterioration and destruction of international cultural and natural heritage can be described by a mathematical function that may not discriminate between peace and war in establishing long-term statistical losses. Peace and war are human agencies: natural disasters are God's. All march to tempos that are independent of political conveniences. Therefore, the message is to fear inactivity, because monuments are not as eternal as we imagine them to be, and the clock ticks away.

J. DRUZIK

PART I A

Historical Perspectives

THE HAGUE CONVENTION OF 1954: HISTORY, SIGNIFICANCE AND COMPLIANCES

MIGUEL ANGEL CORZO
The Getty Conservation Institute, 4503 Glencoe Avenue, Marina del Rey, California

The UNESCO Convention for the Protection of Cultural Property in the Event of Armed Conflict, signed at the Hague in 1954, is a document that reflects 20th century thinking on the means to safeguard the world's cultural heritage. It is our task to transform it into one that anticipates the challenges of the 21st century. First, then, we should pay homage to those individuals who had the spirit and the resolve to formulate the Convention and its Protocol. Second, we should admit that the Convention's effectiveness has been minimized in the past, largely due to a Euclidean conceptualization of the problem when in fact during war the axioms become spontaneously non-Euclidean, non-linear and highly chaotic. Clearly there is a need to reevaluate its premises in fresh ways, and to strengthen it in the context of the New Age that shall define the future.

The Hague Convention evolved out of a number of precedents, a brief review of which will follow shortly. But the direct impetus behind it was the massive and unprecedented destruction of the European cultural heritage in WW II. The technological advances in arms production that had made such devastation possible went on, as we all know, to achieve a more global dimension even as East-West tensions intensified over the decades that followed. Thus, for over forty years not only western Europe, but the entire world lived in mortal fear of a nuclear holocaust—the ultimate technological "progress" in the manufacture of the machinery of destruction.

That a period of our contemporary history should have been defined in terms of war—albeit a 'cold' war—should in itself alarm every reflecting soul on the fallacy of our notions of progress and the misuse of a term that since the Age of Reason had become synonymous with industrialization and 'the triumph of civilization over savagery'. Nor is the irony of the misnomer or the sophistry of the Eurocentrism that it suggests lost on anyone, for during this period many nations outside of the western orbit have, like Ceasar's spirit (but not always with like cause) cried "Havoc!" and 'unleashed the dogs of war' across many blazing killing fields: Vietnam, Cambodia, Zimbabwe, Lebanon, Ethiopia, Israel/the Arab world, Cyprus, India/Pakistan, Afghanistan, Iran/Iraq,

Nicaragua . . . and since the 'post-cold war' period, the Persian Gulf, Azerbaijan, Romania, Georgia . . . and Yugoslavia. The tragedy of countless human lives lost in these wars has been compounded by the parallel devastation of our cultural heritage in the process. Needless to say, the devastation has happened in spite of the signing of the Hague Convention at the end of the Second World War.

It is not our business at this meeting to comment on the means by which the ultimate war—the war against war itself—may be won. But it is valid to ask why we should spend time and resources to safeguard art and culture in periods of armed conflict when precious human lives are at risk. Let me state that my personal and primary concern as a human being is the survival of human life. Our task as conservation professionals, on the other hand, is the protection of the human spirit as it manifests itself in the tangible, artistic products of culture. To the Devil's advocate who would mischievously repeat with Oscar Wilde that art is useless, I would reply that useless it is indeed—if we were to judge the matter from a purely utilitarian perspective. Art is in fact as useless to society as the spirit is to the body. The human species as a biological organism *can* survive without art and culture. But the definition of what it is to be human cannot. Culture is to human civilization as trees are to the environment. That is the justification for our preoccupation with its survival.

The roots of the Hague Convention go back to the mid-18th century when a reaction was first voiced to the old assumption that the spoils of war belong to the victor, with no restrictions imposed in kind or quantity. In 1758, the Swiss jurist Emmerich de Vattel promoted a concept in his *Law of Nations* that the transfer of objects of value should be limited to what was deemed fair compensation for war expenses. He further introduced the notion that works of art were the common heritage of mankind. Not that such idealism reflected on the reality of the Napoleonic period which was characterized by wholesale plunder. Napoleon's sense of legitimacy in hoarding other people's art was in fact such that he once signed an armistice with the Pope with the proviso to carry off large quantities of art works, and then sealed the deal by sending the Vatican a bill for shipping costs as well!

The concepts of Vattel and his followers took over a hundred years to be codified. In 1863, regulations were drawn up by an international lawyer, Dr. Francis Lieber, for the U.S. armed forces. The instructions listed churches, hospitals, educational facilities, museums and scientific establishments as privileged and exempt from possession by occupying powers. These notions were internationalized in Brussels in 1874, and further elaborated by the Hague Conventions of 1899 and 1907. The latter two stated clearly that damage to cultural property could not be justified by war. The ideological premises that they shared with the Brussels document became the cornerstone of all subsequent humanitarian laws regarding cultural property. The premises were as follows:

- culture belongs to all mankind;
- artistic treasures are irreplaceable; and
- nations are merely stewards of this heritage.

Then came World War I, which dramatized the inadequacy of these noble, intellectual constructs. In fact, systematic violations of the Conventions became as commonplace as in the Napoleonic period.

A few attempts were made between the two World Wars to improve the earlier documents. In 1919, the preliminary conference to the Treaty of Versailles explored the question of punishments and sanctions for Hague violations; the 1922 Washington Conference formulated rules to be followed by air forces in wartime; and the 1935 Roerich Pact created a cultural protection policy among the states of the Americas; and at the outbreak of World War II, the League of Nations was working on a similar document.

World War II brought incalculable devastation to the European cultural heritage. When it was over, something substantial was needed to avert the horrors of Dresden, Coventry, Warsaw, and Montecassino. What might that be?

The United Nations Educational, Scientific, and Cultural Organization (UNESCO) was founded in 1945. As early as the fourth session, the agenda included seeking ways to strengthen earlier laws protecting world cultural heritage. The result was the 1954 Hague Convention for the Protection of Cultural Property in the Event of Armed Conflict.

The main document contains forty articles that define the operating environment of the Convention, singling out its obligations, limitations and jurisdictions. Separate from this, and equally important are the Regulations for the Execution, the Protocol, and the Rules Concerning the International Register of Cultural Property under Special Protection. Collectively, they provide for preventive measures during peacetime (such as identification, registration, and education. . .) and operational directives during periods of armed conflict.

The Director-General of UNESCO is, by definition, the highest executive authority in all matters relating to the execution of the Convention. He is followed by members of the international community from among whom Commissioners-General are selected. The latter are UNESCO's field officers and are assisted by Inspectors and Experts. They coordinate work with specialized personnel within the military and civilian sectors in the geographic regions of conflict.

In order for the system to work, the Convention calls for the preparation, distrubution, and updating of inventories of nationally significant monuments and objects.

Under Article 8, the Convention provides for the establishment of a select number of Refuges to be granted special protection for storage of moveable cultural property. It also grants special protective status to centers that contain immovable monuments of significance. In order to be eligible for special protection, Refuges and centers have to be situated at a safe distance from industrial and military complexes. Furthermore, they have to be entered in the "International Register of Cultural Property under Special Protection," which is kept and distributed by the Director-General of the UNESCO. In a related provision, every country is expected to have at least one central refuge, as Germany does for its vast historical microfilming project.

Under Article 17, a distinctive emblem in the form of a pointed shield has been devised to identify cultural property. A triple emblem indicates monuments that are under special protection as well as objects being transported to safe depots. A single emblem identifies cultural property not under special protection, and is also worn by personnel engaged in the protection of cultural property. The emblem, in theory, grants immunity from seizure, capture and prize.

The Convention realizes that these precautions are meaningless without an educated military. Therefore, it is mandated that the text of the convention be a course of instruction to the military forces, officers and high commanders of signatory nations. The Convention also provides for the continuous education of the general public. There have been some heroic examples of massive and innovative educational programs in the past, notably that of Austria.

One of the ennobling qualities of the Convention is its ability to identify and close loopholes. Chapter 6 covers the scope of application, embracing total or partial occupation, and situations where one or more of the belligerent forces refuses to recognize a formal state of war. It binds signatory parties to the safeguard not only of their own national cultural property, but also that of the opposing party or parties—*whether or not they are signatories.* This chapter also addresses civil wars. The cultural patrimony of every distinctive entity or "breakaway republic"—whether it is a formal administrative unit or a culturally homogeneous group within a signatory country—is protected.

A reporting mechanism has been devised to gauge the effectiveness of the Convention over time. Reports are to be submitted by signatories to the Director-General of UNESCO every four years. Those provided to date reveal many different standards and little or no consistency. Extensive efforts have been undertaken, for example, by Austria, the Netherlands, Germany and France in this respect. Other reports are more rhetorical on broad cultural policies and restoration efforts and short on specifics. One finds mention of monuments that have been added to the "World Heritage List," for instance, but there is near total silence on the Convention's Register or application for Refuges. Many nations have not reported even once. Some reports, on the other hand, contradict later behavior. For instance, in 1989, Yugoslavia reported that their military staff was informed about the Convention in writing, and that a textbook on related responsibilities formed part of the curriculum in all military schools. It went on to claim that,
"These Instructions prescribed by the Supreme Commander of the Armed Forces of Yugoslavia are still in force."

Still others contain what many have been better left unsaid. For example, a 1989 report proclaims, "It pleases us to state that no armed conflict has occurred in Kuwait since Kuwait ratified the Convention." Barely a year after that contented declaration, the treasures of the Kuwait National Museum were trucked off by the Iraqi army, and the building itself was set on fire. Many of the objects that have been returned since are damaged; others are missing. A notorious example of the violation of standards was the Iraqis' use of 16th century royal Mamluke linen shirts as "packing material".[1]

Other blatant cases of the failure to comply with the Hague Convention include the destruction of monuments in Cyprus in the mid-1970s, and the shelling of the 8th-to-14th century Cambodian temples of Angkor Wat in the late 1970s. The greatest damage to Cambodia occurred, however, as an indirect result of the war, with the looting of

---

[1]*Art News.* Feb. 1992: 36.

statues—conveniently chopped up for easy transport—by art thieves working out of Thailand when the war had ended.[2]

The assault on the Lebanese cultural heritage is yet another case in point. First, belligerent forces encroached upon Tyre in 1982-83. UNESCO intervened, but failed to put a stop to the Israeli army's carrying out earthworks along the site. Then came a worse onslaught by thieves, treasure hunters and unscrupulous dealers. Priceless Phoenician, Græco-Roman and Byzantine art treasures were stolen by Christian militiamen from the Lebanese Department of Antiquities and smuggled out of the country. And amateur treasure hunters dug up a staggering number of artifacts from archæological sites—which made their way to the art markets of Europe, the US and Japan via local dealers and international middlemen. A report describes the state of Roman cemeteries east of Tyre bulldozed for gold jewelry and pottery as "a series of massive bomb-craters gouged into the earth".[3]

Other examples could be cited as well: there is the Iran-Iraq war, for instance, with damage to medieval monuments in Iran, and to 4500-year old archæological sites in Iraq; museum collections in Romania and Georgia; more damage to ancient Mesopotamian ruins during the Persian Gulf war in 1991; and others still.

This is by no means an exhaustive list of war damaged cultural property in the post-Hague Convention period. Targeting cultural property with the aim to obliterate a people's identity is an historical commonplace, particularly in ideological wars. Such intentions drove Babylonians to destroy the First Temple; the Ptolemies to efface Egyptian religious imagery; Romans to raze Carthage; Charlemagne to destroy Saxon monuments; Thomas Cromwell to deface Catholic icons; Spaniards to burn Maya codices; French Revolution zealots to behead statues representing the *acien régime,* and so on. The Serbian attack on Croatia is the most conspicuous example in recent history. The Ministry of Education & Culture, Institute for Protection of Cultural Monuments of the Republic of Croatia reported that by November 5, 1991, "358 individual monuments registered or filed as cultural treasure, including 26 museums, 7 archives and 13 library buildings" had been damaged or totally destroyed. Twenty days later, the numbers had grown by 114. Not least disturbing were news out of Croatia that Yugoslav armed units were intentionally targeting monuments bearing the emblem of the Hague Convention.

---

[2]*New Scientist.* 14 Oct. 1989: 56.

[3]*The Independent.* London. 30 July 1991: 10.

One obvious lesson is that the official positions of nations in time of peace need to be reviewed in light of their behavior in times of war. The latter is proof of the credibility of the former. Seen from another perspective, it is proof of the incompatibility of the provisions of the Hague Convention and its appended documents with reality. Whereas the Convention

- defines what is to specifically be protected;
- calls for an infrastructure with limited immunity to provide that protection;
- severely restricts the interpretation of when that protection does not apply;
- calls for the education of the military and civilians alike; and for
- reports on its successes or failures,

it has not succeeded in effectively preventing hostile actions against cultural property.

It is nearly forty years since the Convention was first formulated. It is clearly due for a review. I would like to share some thoughts with you on a number of short- and long-term solutions for strengthening and improving this document.

(1) Under the auspices of UNESCO, an international community of policy makers, thinkers, educators, executives, and managers could be reconvened to review the Convention and its Regulations and Protocol. The lessons of the past need to be creatively applied to the complexity of the problem, and to the difficulties of implementation.

(2) 1994 could be established as "Cultural Amnesty Year" with the objective to seek universal ratification of the Hague Convention. To date, only seventy-six states—less than half the member states of the United Nations—have deposited instruments of ratification or accession to the Convention. As an essential first step, the United States, Canada, Great Britain, Japan and other member states of the United Nations could be invited to join the rest of the High Contracting Parties in ratifying the Convention. Radical changes to the political landscape of the 1950s, and the reduced threat of an East-West nuclear confrontation—issues which had impeded some nations from ratification in the past—warrant a favorable revision of their positions. We have reason to believe that the international community is currently more responsive to ratification, in

that since 1972, one-hundred and fifteen nations have become parties to the World Heritage Convention.

(3) Standards for preparing four-year reports need to be set, and a control system will have to be devised if the reports are to be meaningful and functional. Incentives must be found for accurate and regular reporting, perhaps by means of greater access to international monetary funds and fact-finding mechanisms for cultural preservation projects.

(4) National inventories, and the International Register of Cultural Property under Special Protection need to be energetically developed. Past failure can be generally ascribed to poor management strategies, inefficient data collection capabilities, and lack of funding. It is conceivable that private institutions could be induced into providing matching grants and electronic data processing systems to ministries of culture. The benefit of such an undertaking both to the home country and to our collective knowledge of the human cultural heritage is self-evident.

(5) Every means must be sought to encourage the establishment and expansion of Refuges for the safeguard of moveable cultural property. To date, only four countries have applied, and received recognition, for the following nine Refuges:

Vatican City;
The Steinbergstollen Refuge at Alte-Aussee,
Upper Austria;
Six Refuges in the Netherlands;
The Oberieder Stollen Central Refuge in Germany.

There is a need to investigate the reasons for lack of action in this direction by the seventy-two remaining countries that are signatories to the Convention.

(6) Adequate "early warning systems" need to be established to detect danger, and to intervene before armed conflict breaks out. As is, the Director-General of UNESCO is notified and the Convention called into action only after violations have begun. Past events have shown that such a point of departure is too late for the work of UNESCO Inspectors and Experts to be effective.

To deal with this predicament, a private non-profit organization, for instance the World Monuments Fund (WMF) and its national affiliate organizations, could provide the Director-General of UNESCO and the public with frequent reports on the safety of world cultural patrimony. The WMF sees itself as "a catalyst, which focuses attention on major works in peril..." Perhaps it would see fit to install a uniquely funded program in its structure that would become the official "watchdog" and information outlet for the Convention.

(7) A meeting of legal experts could be called to establish international laws and regulations to which violators of the Convention would be held accountable. Aside from representation by diplomats and legal experts, this meeting needs the active participation of experienced members of the military if the Convention is to be strengthened in realistic terms. Their ability to forecast probable predicaments would be a vital input in the process.

To cite an authority in support of the expediency of this notion, "From a doctrinal point of view, the newly emerging concept in international law of the "common heritage of mankind" provides a good foundation for the panoply of measures to be taken by states individually or collectively in order to protect the cultural heritage of mankind."[4] That statement was made 15 years ago. It is time to explore its promises.

(8) A permanent Cultural Rescue Team—a 'cultural Red Cross'—could be established to provide the means for more aggressive stewardship of cultural property. The Team could be funded by dues collectively paid by signatory countries. It would be composed of conservators, civil and mechanical engineers, historians, archaeologists, and documentation specialists. It would have a small permanent staff and would recruit extra personnel as needed. The permanent staff would work closely with the "watchdog" group described above to help it keep abreast of facts, and develop appropriate plans. In the field, the Team would work closely with the local support structure. The Cultural Rescue Team would be entitled to the same rights and protection as the International Red Cross. It would remain in the field for as long as the signatory country could justify the need.

There remain two other important long-term goals.

(9) One is that education regarding humanitarian cultural property laws needs to be stepped up—at all levels, and in every part of the world. Educators should be induced to introduce the concept of conservation in

---

[4]Sharon A. Williams. *The International and National Protection of Movable Cultural Property: a comparative study*. New York 1978: 201.

courses dealing with the arts and sciences. New contributions to textbooks could combine discussions of the protection and respect for the natural environment and the world cultural heritage.

Conservation professionals and institutions can also play a significant role in educating the people. An educated public is our best ally in the war against the destruction of cultural property. All too often, the expert and the professional underestimate the role of the non-specialist in matters deemed scientific and technical. But since the mid-1970s, the trend in scientific inquiry has shown a dramatic reversal away from increasing specialization and reductionism, and toward holistic approaches to the conceptualization and analysis of systems. We must search within these progressive notions for new ideas to deal with the unpredictable, chaotic complexities that inform war. We must strengthen institutionalized forms of defence, but we must learn from the lessons of the past not to rely on state-controlled institutions to always react in predictable ways at times of crisis. We must therefore expend human and financial resources toward building a collective media campaign to reach people at the grass roots level. We must capitalize on the capabilities offered by computer technology to communicate not only information, but particularly knowledge, with the rest of the world. We may be unable to win the great war against the destructive forces of time, but we can win the battle against the destruction that is brought upon by human ignorance, neglect, lack of interest, and greed. The battle will be won the day every man and woman on earth will become a conservator; when every man and woman feels that he or she is a custodian of our common cultural heritage.

Conservation is a science and a skill. But above all, conservation is a concept. It is a concept that has to be promoted as a way of life, a conviction, and a universal mission. It has the virtue of shaping the definition of what it is to be human. It has the potential of defining the New Age that should, by right and necessity, come to be known as the Age of Preservation. Never before has world consciousness been so ready to embrace such a proposition. We must reach out to the world, to the individual human being, and look especially to the young generation as equal partners in our mission.

(10) Another thought—and this is my final one—is based on the realization that ideas can better materialize into actuality when given public recognition, direction, and support. In view of the global ramifications of the issue at hand, we believe that the promotion of conservation as a universal cause will be greatly enhanced if it receives impetus from an institution that is held in respect the world over.

Perhaps, if all the conditions are fully met, there will come a time when such an organization will understand the importance that conservation holds for the world's cultural heritage and will acknowledge the profound influence that an internationally recognized award in conservation would have for the good of humanity. We suggest that such an award be given to individuals, groups, or institutions, and defined inclusively in terms of the cultural and the natural environment.

In conclusion, I would like to emphasize that

- We need to apply the lessons of history to a critical review of the Convention;

- We need to redefine our notion of progress, and our goals, within a new, 'preservationist' worldview;

- We need to underscore the need for putting technology at the service of mankind, not against it, and help provide appropriate technologies where it is absent;

- We need to elicit creative responses to the socio-economic problems of population growth and mobility that affect cultural property; and

- If we are to survive as wholesome human beings, if we are to attain the goals that we have set ourselves as upholders of humanity's values, we need to build upon past strengths, and look to the future with optimisim. We need to encourage ways and means to raise public awareness and a sense of responsibility for the preservation of the cultural and natural masterpieces that are the very source of our vitality, identity, and survival. We need to ensure that in the future, our children and our children's children will be able to experience the awe, the respect, the satisfaction, the pleasure, and the joy that a painting, a sculpture or a monument has brought to our lives, because without them their lives would be barren, desolate and lonely.

CATASTROPHES AND ENTROPY OF CULTURAL PROPERTIES

MICHEL BENARIE* and JAMES R. DRUZIK**
* 20, Blvd. Jean Pain, 38000 Grenoble, France
** The Getty Conservation Institute, 4503 Glencoe Avenue, Marina del Rey, California 90292, USA

0. Introduction

"The wear of iron teeth of time..."
"nothing is eternal"
"everything ends up in ashes"

Such poetic locutions have been with us since Biblical and Homeric times. Lately it has become fashionable to clad this commonplace insight in more scientific garb such as the Second Law of Thermodynamics --

"The natural tendency of all (closed) systems is to maximize randomness and to arrive at a state of energetic minimum."

But so long as such assertions remain purely verbal they are of little scientific value. In this paper we will attempt to provide some definitions, and to apply an exponential decay curve together with the concept of entropy to conservation science, thus proceeding towards a unifying quantitative assessment of the lifetime of artifacts. This approach should show that unless more effort is applied to isolate precious cultural heritage from manmade disaster, they will disappear in the same inexorable march as all that has proceeded them into extinction.

1. The Definition of Identity

To evaluate change, whether you call it alteration, decay or an increase in randomness, we must first define the starting point, a theoretical point of "changelessness" called Identity. The problem of identity was formulated and answered by Thomas Hobbes (1588-1679) in his "De Corpore" (1655) through "Theseus' ship", which not only attempts to define identity but also anticipates a basic view of restoration. Theseus' ship (we will call it T_1) is being continuously repaired by immediately replacing worn out planks with new ones, until each piece of wood has been exchanged thus yielding a "new" ship, T_2. (This example is similar to all thousand year old wooden Asiatic temples and not substantially different than the human bodies replacement of cells.) The removed old planks, however, are not discarded but transported elsewhere and assembled to form another ship, T_3. Hobbes posited the question: Which ship, T_2 or T_3 shall be considered identical to T_1? Hobbes' answer, shared by us, was that as T_2 preserves the continuity, the spatial unity (form) and the function, T_2 must be the same ship as T_1, while T_3 is merely the preservation of a numerically identical collection of planks.

Instead of planks, we may look in a more modern way, at atoms and molecules. Consider an object like a painting, weighting overall 10 kg and made up of many chemical substances. We shall assume the painting has an average molecular weight of 100, a fair approximation for the pigments but low for the canvas support or binder. Thus this painting will contain between 10^{24} and 10^{26} molecules -- simplifying to 10^{25}. Since we have not defined the type of painting, how it was painted, or even if it was painted with current materials, from the past or a 1000 years in the future; there are perhaps 10^5 common molecules from which to choose for each site. Thus, at least in principle, we have 10^{30} options.

Neglecting Heisenberg's uncertainty principle, let's further assume we can pinpoint the location of each molecule fairly precisely at 10^{-8} to 10^{-9} cm, within the approximately 10,000 cm^3 volume of the painting. Thus one could place at approximately 10^{30} point any of 10^{30} molecules, yielding 10^{60} possibilities. Spatial and chemical correlation's may greatly reduce the possibilities, so we can define a redundancy factor of 10^{10}, leaving us still with 10^{50} combinations. To fully describe the position and the kind of all these (groups) of molecules, one needs roughly 10^{39} bytes or 10^{27} terabytes of information.

Obviously this must remain a "thought experiment" since 10^{27} terabytes is more than all the information stored on all digital optical disks now in existence in the world. Using the equivalence between information and energy proposed by Dennis Gabor, that the minimum "cost" per bit is $4x10^{-14}$ erg, something like 10^{23} megawatt-hours would be required to generate such an "ultimate copy". At 8 cents per kilowatt-hour, the cost would be near $\$10^{10}$.

The ultimate copy or ultimate fake remains a fiction, even if not considered as mere object, but an incredible amount of digitized information. To paraphrase Marshal McLuhan, the artifact is the message. As such it may be taken as reference for all subsequent deterioration.

2. Definition of Decay

Any non-purposeful, non-anthropogenic departure from the initial state (which is expressed digitally by the ultimate copy) towards any other less ordered state is labeled as decay -- or deterioration, disintegration, crumbling, collapse, decomposition, destruction, corrosion or another synonym. In other words, all change is decay by definition.

Decay may be anthropogeneous -- purposeful destruction, attack by air or water pollutants, vandalism, act of war or the like. It may also be pre-meditated, with an appreciation of the cause and effect relationship, or inadvertent when the causal relationship was not visualized beforehand.

Non-anthropogeneous decay may proceed either because the substance(s) proceed intrinsically towards an energetically more stable state (radioactive decomposition, degradation of cellulose by acidity) or linked to external causes (corrosive environments, photochemical processes or mechanical forces from wind erosion to earthquakes.) Let us combine all entropy-inducing factors into a single spectrum.

The causes or triggering mechanisms may cover a wide energy scale from those involving a few electron-volts within an atom to the multimegaton power of a major earthquake or trillions upon trillions of times larger in nova. Following their magnitude, they may be called shocks or catastrophes, characterized by randomness in time. Starting from the random succession of shocks (nuclear, molecular, mechanical), Benarie [1] proposed the description of decay by "first-order" reaction law, which may be expressed by an exponential formula with one constant, which can be the "lifetime". The true mechanism of decay reactions is certainly more complex, but the rate-determining step can usually be approximated by a first order reaction. Thus even the diffusion-limited corrosion of metals fits an exponential law.

3. Expected Lifetime

Shocks may be characterized either by the energy involved, or by the number (order magnitude) of atoms which take part in the event. The number of atoms obviously corresponds to the mass. The number of atoms (or mass) scale is so wide, that a classification on logarithmic basis is useful. As a further simplification we introduce "degree" which is factored by 1000.

Table 1
"Shock" Scale as a Function of the Number of Atoms

Degree	Number of Atoms	Mass
0	1,000	$<10^{-24}$ g
1	10^{6}	10^{-21} g
2	10^{9}	10^{-18} g
3	10^{12}	10^{-15} g (femtogram)
4	10^{15}	10^{-12} g (picogram)
5	10^{18}	10^{-9} g (nanogram)
6	10^{21}	microgram
7	10^{24}	grams
8	10^{27}	kilograms
9	10^{30}	ton
10	10^{33}	>1,000 tons

Degree 0 affects less than one thousand atoms. Potassium-argon, fission track and amino acid racemization dating techniques are based upon counting such shocks. The range around degree 6 (where the nanogram to microgram cm^{-2} sec^{-1} shocks are located, and which involves 10^{18} to 10^{21} atoms/molecules sec^{-1}), is the domain of surface corrosion phenomena usually progressing at the rate of grams m^{-2} yr^{-1}. Degree 9 (kilograms to tons) includes much of cultural heritage, while degree 10 covers outdoor monuments and whole buildings. Destruction of the latter, in human terms, is more easily described as catastrophe or cataclysmic destruction than shocks, but from the perspective of the planet Earth (Degree 16), mere shocks are still adequate descriptors. (If the Sun were to become a nova it would be a Degree 18 shock involving 10^{27} metric tons. Fortunately stars of the Sun's type have nice long lives.)

Knowledge about the distribution function and/or the randomness of any category or degree of shock allows one to formulate the mathematical model for the decay by that process. Having the means to define the average lifetime for an object or a structure, related exclusively to the shocks under consideration (including civil or international warfare), we have a rational basis for risk assessment.

Thus, if the order of magnitude of radioactive disintigration within a potassium containing matrix is a few events per second, and each ionization path breaks less than one thousand molecules (degree 0), it will take 10^{19} to 10^{21} seconds (10^{13} years) until the object significantly crumbles. Thus from this point of view, the object is fairly secure. But if degree 6 shocks occur (the rate of acidic deterioration in paper, surface corrosion on stone or metal), it will take only 10^8 seconds or only a relatively few years until significant damage begins to show. At the upper end of the scale, in seismic regions perhaps three major (>7.5 Richter) earthquakes occur century^{-1} region^{-1} (Degree 10) so the lifetime of non-earthquake proof structures may be assessed at 30 years. The reverse conclusion can be made that where there are high towers or structures like gothic spires or minarets made from unreinforced masonry, still standing after several centuries, the frequency of earthquakes must be correspondingly low. Seemingly exceptional cases are Byzantine churches in the Yugoslavian Republic of Macedonia which is highly active seismic area. Research has shown however that these structures are engineered to withstand these levels of shocks. [2]

The above considerations however, were based on an assumption of homogeneity. That is the thousands of tons of stone or other structural material constituting a monument are equally stable (or unstable), and that the destruction of the object implies the dislocation of the major part of the mass. While this may be common in massive earthen movement, for most objects, this is very far from the real situation. A few percent of the whole structure of a monument -- some mortar, a few vault stones -- represent the "critical bonds". The destruction of these bonds brings on the ruin of the whole structure. Based on the idea of the critical bond, Benarie [3] deduced correct values for the natural weathering rate of stones, later experimentally confirmed by the United States' National Acid Precipitation Assessment Program findings [4].

The "critical bond" concept applies to almost any object. Even though 90% of the paint and canvas of a painting may exist, nevertheless the work of art could be irretrievably lost. When the degree of polymerization of cellulose in a book degrades to the level of individual crystallites, all functionality of the book is lost although its corporeal form remains. Thus instead of scrutinizing the thermodynamic instability of the whole, one must specify the location and extent of critical bonds, and concentrate on them. Selwitz has shown this for materials made of cellulose nitrate. [5] The critical bond concept even holds up for materials displaying very complex behavior. Selwitz found that under accelerated aging the critical bond for cellulose nitrate in one situation was thermal ring decomposition, shifting the conditions, it became faster nitrate group dissociation, and under surges of humidity -- hydrolytic, acid-catalyzed ester cleavages.

4. Application to War Damages

From strickly a thermodynamic point of view, there is no difference whether a shock bringing down a struture was due to an earth tremor or to man-made explosives. The question is only the energy of the shock -- how much material was brough from a metastable state (high position in a wall) to a more stable condition (heap of stones.)

L. F. Richardson [6] established a record of wars which occurred in the world during the interval 1825 to 1945 (see also [7]) and classified them according to the number of war dead. Moreover, he found a relationship representing all conflicts from world wars to those which caused a single death. However he did not link loss of life to a related risk to cultural heritage.

Richardson's enquiry and results concern the whole world, thus making difficult their application to narrowly circumscribed areas containing a specific number of monuments and/or other heritage. However, he paid special attention (Chapter VI) to civil wars, insurrections, revolts, rebellions and mutinies. These lamentable events have the property to afflict repetitively the same regions, thus useful for the purposes of our analysis. In his definition, insurrection, revolt, rebellion and mutiny form together a subclass of civil war in which one belligerent had previously exercised authority over the other. "The time which elapsed before a common government became an effective bond of peace, may be gathered.. from the median value...half the revolts and civil wars broke out after less than 23.5 years of common government. The suggestion (Richardson's) is that such a time is too short for a new type of loyalty to become habitual."

A main war objective in civil wars is the destruction and appropriation of the opponent's symbols [8]. **Thus, in those unlucky territories where civil strife is woefully endemic, the mean lifetime (or repair time) of monuments is less than one quarter of a century.**

5. Digital Conversion of Cultural Heritage

Where other means fail, copying by any method, from manual reproduction to the most sophisticated casting, holographic, photographic, digital or other recording and publishing techniques, insures the survival of our cultural heritage. Conservation has as an integral concept the dictum "...to preserve in perpetuity." This idea is purely a well meaning intention when applied to cultural objects. The information about objects, on the other hand, because of its independence of a single physical form, has an opportunity for survival as long as the species that spawned it survives. Thus culture can be said to have a "soul," infinitely less perfect than the Christian concept and not in a religious sense, but a soul nevertheless. So long as humanity can maintain its copies of "Theseus' ships" much about history endures. It has been shown previously that the perfect record, the ultimate copy, is not feasible. Neither is it required for fair recording and safeguarding purposes. The relaxation of too stringent criteria, such as that formulated in section 1 can be approached from two sides.

5.1 Redundancy in the "Ultimate Copy"

The molecule-by-molecule description is highly redundant and data compression can be effectively applied. First, there is chemical redundancy. The same chemical compounds makes up specks, grains and crystals from the microgram size up to brushstrokes and bricks. Thus from 10^{21} to 10^{27} bits of information may be compressed into only a few bits. Second, almost always there is a strong spatial correlation among extended elements. As there must be cross-correlation of the chemical and spatial aspects, it is more difficult to estimate the latter. This means that probably 10^4 or less "molecules" will be really discernible. A resolution on a molecular scale is obviously an ideal luxury; even micro-scale resolution will be, in most cases, too fine. Thus one would have to record something like 10^{13} sites for 10^4 molecules. This 10^{17} possibilities, which can be described by 10^9 bytes could be accommodated on a single optical disk.

5.2 Channel Capacity of the Receptor

If the receptor is the human eye with a channel capacity of about 10^4 per image [9], there is no point in transmitting any picture of better resolution. Many tens of thousands such images are presently accommodated on a single disk. Correspondingly, resolution of each image could be multiplied by ten or one hundred and practical recording is still possible. David Saunders at The National Gallery in the United Kingdom, using the VASARI system, fully describes the full spectrophotometric color range of a painting in 700 megabytes [10]. This can be taken as the upper boundary for the color assessment of easel paintings.

In summation it should be said that we are recording only the technical part of the conservation record. But from the silex sliver to Angkor Vat, the soul of each object making up our cultural heritage contains three kinds of information, of which only one is purely technical. The other two are the cultural facet which combines all the information about the people who made the artifact and the purpose to which it was used, and the artistic/aesthetic sets of values. These latter two facets are beyond the range of this paper.

6. Entropy and Management

A critical question can now be asked: What does this paper contribute? The answer is really very simple. Critics of the Hague Convention note that signatories avoid implementing Chapter II during peacetime. This section calls for the compilation of national lists of protected monuments that shall be deposited with the Secretary-General of UNESCO and establishing refuges for movable cultural property. Evidently, few nations have attempted to comply with Chapter II because the urgency of such measures seem distant in peacetime. The poignant reality contained in 'entropy and catastrophe', is that this complacency by national leaders and agencies is a dangerous and foolish tactic. The inexorable fact is that the deterioration and destruction of international cultural and natural heritage can be described by a mathematical function that does not discriminate between natural disasters in peacetime and human destruction in war when calculating long-term statistical losses. Peace and war are human agencies: natural disasters are God's. All march to tempos that are truly independent or semi-independent of human, political conveniences. Men of wisdom may possess no greater effect on the actions of Man than they do on natural events. Therefore, the message is to fear inactivity in Hague Convention compliance, since monuments are not the eternal messages of past civilization we imagine them to be, rather, they are fragile bubbles, awaiting the moment their existence is shown to be, yet another time -- metastable .

The clock ticks away as each flickering light blinks out.

References

1. Benarie, M. , American Chemical Society Miami Beach Meeting (1989), Paper C-108; accepted for publication in Mathematical Modeling.

2. Gavrilovic, P., and L. Sumanova, "Aseismic Strengthening, Conservation and Restoration of Churches Dating from the Byzantine Period (9- Century) in Macedonian," Final Report to The Getty Conservation Institute, Six Volumes (1991).

3. Beanie, M., European Cultural Heritage Newsletter on Research, Vol. 1, No. 1, page 5 (1987)

4. European Cultural Heritage Newsletter on Research, Vol. 5, No. 1, page 18, footnote (1991)

5. Selwitz, C., Cellulose Nitrate in Conservation, Research in Conservation Series No. 2, The Getty Conservation Institute, Marina del Rey, CA (1988)

6. Richardson, L.F., "Statistics of Deadly Quarrels", The Boxwood Press, Pittsburgh, 327 pages (1960)

7. Richardson, L.F., Journal of the American Statistical Association, **43**, pp. 523-546 (1949)

8. Benarie, M., and J.R. Druzik, European Cultural Heritage Newsletter on Research, Vol. 5/5-6/1, page 3 (1991/2)

9. Tolraldo di Francia, G., Optica Acta, 2, page 5 (1955)

10 Saunders, D., private communication, February, 1992.

THE PROTECTION OF NATIONAL TREASURES AT THE BRITISH MUSEUM DURING THE FIRST AND SECOND WORLD WARS

MARJORIE L. CAYGILL
The British Museum, Great Russell Street, London W.C.1, England.

ABSTRACT

With the emergence in the 20th century of long-range methods of destruction, museums and galleries have had to develop appropriate measures to protect their collections. The British Museum first took such measures on site in 1914 and followed this by evacuating some of its most valuable objects in 1918. Similar procedures were followed in the Second World War and, despite major damage to the building, no significant items (apart from books) were lost.

Introduction

The destruction of national treasures, and their preservation, accidental or deliberate, goes back far into antiquity. The British Museum has, for example, ivories thrown into a well at the time of the sack of Nimrud in 612 BC, silver hoards buried in the 4th and 5th centuries AD in the unsettled period before and after the Roman Army left Britain. There is part of the immense war chest of a Viking Army hidden in the 9th century, innumerable coin hoards from the 17th-century English Civil War and, most recently, a rich collection of Celtic gold torcs from Snettisham, Norfolk. The 20th century has added a new element - the air raid, sometimes discriminating but more often no respecter of non-combatants both human and man-made. The first mechanised flight was in 1903, explosive devices were dropped before the First World War and bombing raids took place during it.

That the British Museum was so successful in protecting its collections can be attributed both to luck and judgement. Part of the luck stems from Britain being an island which was not invaded; the judgement from careful planning and, at least in the Second World War, the high value placed from the outset on scientific advice as to the best conditions in which to store objects. It should, however, be borne in mind that because of the varied nature of its collections a museum's problems as regards packing, transport and environment are more complex than those experienced by a picture gallery.

The title of this paper is something of a misnomer since the British Museum in both world wars decided that the best means of protection for many of its greatest treasures was to remove them from the Museum premises. This was done for the first time in 1918 and the then Director, Sir Frederic Kenyon, wrote afterwards:

> If a Museum official had been asked in cold blood before the war whether he could dismantle the whole Museum without considerable injury to its contents, he would probably have hesitated to undertake it.

However, the Museum never regretted the decisions to move the collections out. Although during the First World War the only damage was to the spines of two unimportant German books (and this from a fragment of anti-aircraft shell which fell through a glass roof of the main book-store), bombs did fall within 150 yards of the building. By contrast, it has been calculated that during the Second World War in the area of London in which the British

Museum is situated there was an average of two high explosive bombs to every three acres, not to mention incendiaries, mines and flying bombs. The Museum was hit by seven high explosive bombs - slightly under its fair share of eight (fortunately three were ineffective and two fell in the forecourt).

The first bomb to hit the British Museum building struck on 18 September 1940. A 1,000 kilo high explosive bomb fell through the roof of the King Edward VII building and passed through the Print Room to the Oriental Basement. The Museum was fortunate - the bomb's tail was torn off by a girder which it struck, pulling out the lead-cased detonator. On 23 September, another unexploded bomb was found in the same area by the dawn patrol. The Director of the British Museum was rather proud of these two bombs which came through the same hole and wrote that they:

> deserve special notice as a rare or perhaps unique phenomenon, appropriately occurring in a Museum. But it is, ironically a phenomenon of which no material evidence can exist.

On 16 October 1940 the dome of the great Reading Room where Marx wrote Capital, Lenin pondered the land question, Hardy, Kipling, Shaw and other great authors researched their novels and plays, Sun Yat Sen studied to transform China and Gandhi India, was burning and flickering 'like a Christmas pudding'. Again the Museum's luck held: an incendiary fell through to the floor below but left most of its burning oil behind. The resulting fire was extinguished within seventeen minutes by the Museum's staff. On 14 October a large bomb fell immediately in front of the Research Laboratory, putting it out of action. The Newspaper Library at Colindale in North London was very badly damaged on 20 October 1940, with the loss of some 30,000 volumes of English provincial newspapers - about a third of the collection.

On the night of 10/11 May 1941 700 tons of bombs were dropped on London, starting 2,200 fires. The Museum's luck ran out when a stick of incendiaries hit the south-west corner of the building. Around 150,000 books were destroyed: some by fire as the old cast iron floors buckled in the intense heat, other volumes succumbed to water used to extinguish the burning bookstacks. It was a terrible night - the London Fire Brigade was called to assist the Museum's staff (for the only time during the war) but water was in short supply and much of London was burning.

Many of the galleries on the south-west side lost their roofs as incendiaries lodged in the attic space between the outer copper shell and the ornate plaster ceilings. Fires broke out ahead of the fire fighters and for years afterwards it was necessary to carry an umbrella when walking through what had been the upper Greek vase rooms. Of the antiquities, however, the Museum lost only a few duplicates - flints, pots, coins - featured in an exhibition officially dealing with recent archaeological research and aspects of British and European prehistory, but more aptly known to the staff as 'the suicide exhibition', which had been put together as a hostage to fortune in August 1940.

The Museum did not lightly embark on the evacuation of its collections. In its archives there is a comment, written in 1941 in a letter from one of the wartime repositories which refers to the gulf between those who try to preserve the past and others who destroy it:

> If they [the collections] perish or are ruined after all we can only say that mankind just has not wanted them so much as his own follies and has justly lost them. ... We can but do our best...

This paper indicates something of how this 'best' was achieved.

The First World War (1914-18)

During the First World War there was less urgency than in the Second when lethal waves of bombers were expected within hours of the declaration of war. However, at a time when objects were never loaned outside the Museum building there was limited expertise in packing and transport and certainly no expertise in wartime removal. As one of the Keepers put it:

> the whole training of the Museum staff has always aimed at setting security before speed, and taking no risks of any kind that can be avoided...

Also, scientific conservation was in its infancy. There was practical knowledge of the effects of environment on museum objects but no theoretical base. The establishment of the British Museum's Research Laboratory (the first such institution in the world in continuous existence) did not come until 1920.

The British Museum has the advantage of inhabiting a very solid Victorian building, largely built between 1823 and 1847, with the massive round Reading Room of 1857, and a large northern wing - the King Edward VII Building - opened (and immediately closed) in 1914. There are extensive basements, sub-basements and sub-sub-basements rejoicing in names such as 'the spider', 'the Sepulchral basement', 'the old arms basement', 'the cage' and 'the Mausoleum basement'. Although much of the roof is of glass and therefore vulnerable, the walls and floors are thick and solidly constructed of Portland and York stone and the structure is of cast iron and steel.

At the beginning of the First World War, therefore, protection was initially 'at' the Museum since it was assumed that with existing technology any bombs which happened to fall on the building would almost certainly explode before they penetrated the solid floors of the galleries. Further, as the Director wrote:

> There was never any real likelihood that the Museum would form a deliberate object of attack...because, if they [the Germans] could deliberately choose targets in London, there were many others which would serve their purpose better.

Initially, objects of special value were taken from the upper galleries to the ground floor and basement, portions of which were converted into strong rooms. As the war lengthened and destructive technology advanced it was decided to reduce risks by protecting some objects to a considerable depth of sandbags. Some idea of the scale of this operation is given by the Department of Egyptian Antiquities' requirement: 15,000 sandbags, 200 tons of dry sand, 150 wooden cases.

The first air raids on London took place in 1915. In 1917 the British government began a determined attempt to find office space for the Air Board in the British Museum. Since this would make the building a legitimate target the Museum resisted strongly and successfully. Increasing pressure to protect national treasures by removal came from government departments, accompanied by warnings of bigger and more effective bombs to be expected in Spring 1918. This culminated in January 1918 in a demand from the Prime Minister and War Cabinet that the most valuable objects be evacuated. In any case, the Museum had already decided at the end of 1917 that the time had come to move some of the collection to greater safety outside the building.

The most convenient repository, within range of 'pantechnicon vans' and hand trollies was the Holborn Post Office tunnel 40 to 70 feet below ground, consisting of two parallel tubes, about 170 feet in length with a diameter of 25 feet and smaller tubes projecting from each end of each of the larger tubes. The total length was about 225 feet with a diameter of 14 feet. Flooring was placed across each of the larger tubes at its greatest width and ballast was laid in the smaller tubes so as to give a fairly level surface for the packing cases. The ends of the smaller tubes were blocked by concrete walls. Ventilating apparatus and electric radiators were installed, with hygrometers and thermometers. There was a permanent patrol of two Museum clerks.

It was recognized that, because of the humidity, conditions were unsuitable for books, drawings, prints, manuscripts and other delicate objects. There were, however, robust items which would not suffer and into this tunnel went 333 cases of Egyptian and Assyrian Antiquities, 422 cases of Greek and Roman antiquities, the whole collection of coins and medals in their cabinets (occupying an area of 160 square feet, 6 feet high) and the more portable objects from the British and Medieval Department. The more delicate items went by special train to a new building at the National Library of Wales at Aberystwyth, far out of the range of aircraft from the Continent: 21 cases of printed books and maps, 47 cases of manuscripts, 22 of oriental printed books and manuscripts, 19 of office records (registers and the like). This was followed by two similar consignments. Only a small selection of the rarest and most valuable books from the 'finest working library in the world' could be moved since there were, even then, 50 linear miles of shelved volumes. A smaller consignment went to the strong room of a country house at Davenham, Malvern. The main work of removal and protection was completed by the end of May.

The Museum remained unscathed in the raids of 1918. Armistice was declared on 11 November 1918 and orders were given immediately to commence the removal of sandbags. The return of the collections began on 3 December and was completed in March 1919.

It is a sad reflection on human affairs that twenty years afterwards the collection was on the move again.

Between the Wars: A Time of Preparation

In the interim, in 1920, the Museum established its Research Laboratory - which although commonplace today was a significant departure - modern science was for the first time introduced to antiquities within the Museum and the old 'trade secrets' of generations of proto-conservators were doomed. The task of the new Department was to assess the effects on objects of the recent evacuation and also to consider generally optimum conditions for the care of antiquities. As its first head, Dr Alexander Scott, FRS wrote:

> Those at the head of the various Departments are primarily and naturally more directly interested in the artistic, ethnographic, linguistic and historical problems on which light is thrown by the objects in their keeping...although they have undoubtedly required considerable knowledge and insight into the methods used by those whom we may call the artificers employed to clean, restore and mount for exhibition the specimens, they cannot be expected to devise new and more scientific methods...the processes employed...are kept more

> or less as trade secrets by these artificers...I do not believe much in the secret processes and have no doubt whatever that a little careful work and research will soon discover and elaborate methods more efficient and safer than those at present in use.

Dr Scott was as good as his word and when the time came to look for safe havens for the collections in the Second World War scientific advice was taken from the outset.

What is interesting about the second move is the early date at which planning began. In January 1933 Adolf Hitler became Chancellor of Germany. In July of that year the British Ministry of Works convened a conference of directors of national museums and galleries 'to consider the safety of British national art treasures in time of war'.

Sir John Forsdyke, Classical scholar, Director and Principal Librarian of the British Museum 1936-50, was one of the main activists in the Second World War evacuation. In the 1930s there prevailed in some quarters a certain fatalistic view that 'the bomber will always get through' and indeed the conference of directors recorded on 15 November 1933 that:

> the crisis, should it arise, would be decided so quickly that all arrangements for such distribution [i.e. evacuation] would be futile, and that nothing less than the provision of bomb-proof repositories in immediate contiguity with the museums would be of any service.

Forsdyke would have none of this. One of his most influential contributions was to edit an extremely practical sixty-page booklet entitled Air Raid Precautions in Museums, Picture Galleries and Libraries* published by the British Museum in 1939. He wrote after the War, with a fully justified lack of modesty:

> In reading these pages now, one finds nothing (except the protection against poison gas) that might not have been written in the light of experience after the war.

With hindsight, the booklet appears at times to have a tendency to excessive detail and on occasions to state the obvious. However, given the limited experience of wartime evacuation it must have been in many ways reassuring and much of it is sensible advice. The preliminary chapters provide general information from government sources but in the chapter on 'Material: Methods of Protection' Forsdyke deals with problems peculiar to museums and galleries. First there is protection in place. There are detailed comments, for example, on sand bags - the sand should be thoroughly dried in order to avoid damage through damp and moulds, foamed slag is lighter and drier but not so effective in stopping missiles. Books on shelves need to be properly treated: they should be packed closely or the spaces between filled with bricks or other dry and fireproof substances to reduce air between and therefore lessen damage by fire. Forsdyke comments on sealed cases and sculpture, pictures and showcases, bookcases and books and manuscripts.

As a second alternative, since there was no point in leaving objects in galleries if they could be moved, Forsdyke advocated protection on the premises provided careful consideration was given, in particular, to relative humidity. Where high explosives were concerned, objects and humans had similar requirements, although objects did not need a second exit from a basement shelter and humans could cope with damp.

The third possibility was evacuation. Forsdyke perhaps states the obvious when he writes:

*Note: See complete text of this publication in appendix to this paper.

> Evacuation will only be advisable if the Repository chosen is definitely less exposed to air attack than the Museum, Picture Gallery, or Library itself, and if adequate provision is made for the security of the evacuated material from damage by fire and damp, and from loss by accident in transit or by theft. These considerations are particularly important in view of the fact that the objects chosen for evacuation will normally be the most valuable in the collections.

The Museum began looking for suitable repositories as early as 1933. Dr H J Plenderleith, Head of the Research Laboratory set out the common-sense approach:

> storage space which affords safe protection from the effects of bombing can always be utilized; if the temperature and humidity can be controlled at reasonable levels its value will be infinitely greater; and if constant atmospheric conditions are obtainable in the right degree it will satisfy the exacting requirements of a permanent repository for the most delicate material.

Forsdyke later (June 1941) made a more exasperated and unofficial comment:

> it is more difficult than you think to find suitable repositories. I get letters from both sides: from people asking where they can send their precious belongings and from others offering unsuitable accommodation. One lady who wrote recently said that she had a large cellar which was dry as cellars go. This is a sound statement: no cellars...are naturally dry enough for perishable objects and not many can be made so. A result of indiscriminate cellar storage would be destruction by mildew, slower but no less certain than that by bombs...

In 1933, having dismissed a suggestion that all the nation's treasures be gathered together at Hampton Court Palace - which would, if attacked, have made a spectacular bonfire, a number of country houses were assigned by the Ministry of Works for wartime use by museums and galleries. These were near enough to London for quick transport, but away from areas likely to be bombed. By March 1938 the British Museum also secured storage space in the Aldwych Underground Railway tunnel (the 'Tube') quite near the Museum, on a little-used spur line. Its two country houses were Boughton near Kettering, a seat of the Duke of Buccleuch, and Drayton House, also in Northamptonshire, much of which dated from the 14th and 15th centuries and which was owned by the Stopford Sackvilles.

An arrangement had also been made to occupy again some of the rooms at the National Library of Wales in Aberystwyth. In addition the Museum and Library had agreed in 1938 to share the costs of a bomb-proof tunnel constructed underneath a humpbacked hill (known as Grogyhan 'The Hangman's Hill') a few hundred yards from the main building. Owing to the distorted formation of the slate seams and other obstacles the cost was greater than estimated; all tunnelling had to be done by blasting. (The tunnel was completed in October 1939 after 14 months' work but had to await the installation of air conditioning and months of tests to ensure that it was suitable for the Museum's collections.)

Having selected these repositories the problem was how to get the objects there. With hindsight we know that there was a considerable breathing space following the formal declaration of war by Britain on 3 September 1939 and major air attacks on London which did not begin until September 1940. At the time, however, it was feared that great waves of bombers would appear immediately and therefore speed was essential.

The solution to many of the British Museum's problems was the 'no-nail box'. Advertised as being suitable for, inter alia, baby foods, chilled fish, whisky, soda syphons and sports goods, it was also admirable for antiquities. Made of plywood with flexible metal edging, the cases were designed to be stored flat and later reassembled. They could therefore be piled in thousands in very little space; 500 could be stored in the same area as 40 conventional boxes. By 1939 the Museum was to have 3,300 of them, crammed floor to ceiling in the Duveen Basement.

For small or non-fragile objects ordinary millboard cases in several uniform sizes could be used, if necessary secured with riveted steel bands - 600 of these cases were ordered for the coin cabinets. Large amounts of other packing materials were acquired - 1,450 yards of steel tape for binding coins and medals cases, 30,000 sheets of tissue paper for Printed Books.

To move vast quantities of these boxes the Museum used an early version of the container principle which has revolutionised todays's ocean transport - the 'A' type container measured 7 x 6 1/2 x 6 1/2 feet with a capacity of 2 1/2 tons, the larger 'B' type was 13 feet 10 inches x 6 1/2 x 6 1/2 fet and could take 4 tons. They could be supplied in advance, loaded with no-nail or millboard boxes, sealed by British Museum staff, loaded on to trucks and transported to the railway station, placed on a carriage and off-loaded at the other end - an ideal method for transporting priceless material which would both minimise the possibility of theft and ensure that the Museum's experts controlled the packing of fragile objects.

The Munich crisis of September 1938 provided a rehearsal - in the British Museum's case largely on paper - which highlighted some deficiences in planning which could be resolved before the real thing.

The Second World War (1939-45)

On Wednesday 23 August 1939 the news broke of the formal signing in Moscow of the Nazi-Soviet pact. In Berlin the British Ambassador, Sir Nevile Henderson warned Hitler that the guarantee to Poland stood. Although war was still a week and a half away and the comings and goings between Berlin and London continued, this seems to have been the catalyst which set many of the preparations for war in train, among them, in the late evening, the notification from the Home Office to the Director of the British Museum, Sir John Forsdyke, that war was imminent and that it was time to move the nation's treasures.

Some preliminary steps had already been taken. The railway companies had been asked to deposit 30 'A' and 30 'B' containers in readiness and early on the 23rd the Librarian at Lambeth Palace (the residence of the Archbishop of Canterbury) had delivered to the British Museum seven cases of priceless books and manuscripts - the 'cream' of the Archbishop's collection. Packing began at 7 o'clock on the morning of the 24th.

Forsdyke's master plan was put into operation. The choice of objects for the initial evacuation had been made long before, schedules of 'the most precious objects in the Museum with a view to rapid removal in case of war' having been requested as early as November 1933. Each Department had a list of portable priorities based on that prepared in case of fire - in Greek and Roman Antiquities, for example, the list (which was a lengthy one) was headed by the Portland Vase, the Elgin jewellery and the Chatsworth head of Apollo. Prominent on the Medieval Antiquities lists were the Lewis chessmen, the treasure from the Sutton Hoo ship burial (only just excavated), the royal gold cup of the Sovereigns of France and England, the bust of the dashing Prince Rupert.

There were six loading points on the Museum Colonnade, with various loading ramps allocated according to destination and type of material. Routes to them were specified so as to avoid a tangle of objects and packers.

The Museum's galleries were closed. Packing proceeded rapidly with empty assembled boxes being delivered to departmental packing points by a team of labourers. The boxes were already stencilled with their destination, e.g. T for tube, departmental initials and an individual number to enable their contents to be located at the repository from master lists. Antiquities took longer to disperse but in due course the Parthenon, Phigaleian, Nereid and Mausoleum friezes went into the Aldwych tube, as did the Assyrian reliefs.

At the same time as objects were leaving the Museum, items too heavy to move immediately were being protected either by sandbags (advice having been taken from staff who remembered the First World War precautions) or by removal to the basements, some sections of which had been reinforced. Objects too fragile to travel were also placed in the basements.

Considering the great value of the material and its complexity, the amounts despatched are impressive. On the first day 10 tons of library material was sent to Aberystwyth, 12 tons of perishable antiquities to Boughton and Drayton and a quantity of imperishable antiquities to the Aldwych underground tunnel. On the second day 12 1/2 tons of coins, 12 tons of perishable antiquities, and 15 tons of library material went out of London and more objects went into the underground tunnel. By noon on the third day the whole national collection of coins and medals, then valued at at least £3 million, was in a Butler's Pantry in the English Midlands and the most valuable material of every Department was in safety. In all, in the weeks before the declaration of war 100 tons of library material (around 12,000 books, a similar number of volumes of manuscripts, together with three-quarters of the collection of prints and drawings), had been moved, all without loss or damage. Among the treasures evacuated were the Codex Sinaiticus and the Codex Alexandrinus, Gutenberg bibles, drawings by Leonardo, Michelangelo and Raphael and Shakespeare first folios. The trains travelled with minimal escort - one unarmed curator and one railway company representative. Soon, only heavy sculpture and antiquities of minor importance were left brooding in the Museum's basements.

At Boughton neatly numbered cases of portable antiquities from the Egyptian, British & Medieval, Greek & Roman, Coins & Medals Departments now covered an area the size of a tennis court in the Great Hall and flowed into the Drawing Room, Loggia, Little Hall and Servants Hall. At Drayton were Oriental Antiquities including Ethnography. Books, manuscripts, prints and drawings were in rather more conventional surroundings at the National Library of Wales, Aberystwyth. Where there was access to the collections work continued on cataloguing and indexing as it would do throughout the war.

On 10 May 1940 the German offensive in the west began. There seem to have been no plans to move any of the Museum's collections out of the country as the threat of invasion increased. Had invasion actually occurred a few staff would have stayed behind with the objects to alert invading troops to their importance.

We now come to what until recently was the British Museum's one major secret - a west country quarry whose existence was known but whose name was erased from all records made available to the public.

Boughton and Drayton which had seemed so safe in 1939 were less so in view of the violence of the air raids on Britain and particularly the intensive use of incendiaries. Their old timber roofs and floors were now an unacceptable fire risk. While the Aldwych tunnel which was shared with the Victoria & Albert Museum, Queen Mary's china from Buckingham Palace and Westminster Abbey, was considered reasonably safe, the environment was not suitable for some objects. At Aberystwyth the new tunnel which was completed in 1940 provided excellent, if limited, sanctuary but the National Library of Wales building was on a route used by enemy aircraft and there were fears that it would lie in the path of an invasion from Ireland.

At the end of October 1940 the Museum learned with horror of a project to build a bomber airfield at Grafton Underwood, at its nearest point just over a mile from Boughton House. Even if, as the Ministry of Works said, the Luftwaffe were unlikely to miss the airfield by that range, the risk was unacceptable. It was therefore decided in December 1940 that the collections must move again and that the best solution would be properly air conditioned underground storage. It had been suggested to the Museum that some worked-out stone quarries in the west country might be suitable if one of them could be converted from its normal use for growing mushrooms. A quarry at Westwood, near Corsham, Wilts, was earmarked at the beginning of 1941 for the British Museum and the Victoria & Albert Museum. Ninety feet below ground, with a ceiling height of 12 feet, its horizontal shaft was accessible by lorry and 25,000 square feet of storage space was available. It took six months and a considerable sum of money to convert it: clearing and levelling the floors, smoothing the walls and roofs, adding some supporting piers, sealing the whole stone surface with a waterproofing compound before the air conditioning machinery and fittings could be installed. The Museum insisted on a stable environment within the range 60-75 degrees Fahrenheit, 60-65% R.H. and a reserve air conditioning unit in case of breakdown - in other words 'perfection or nothing'.

In the interim the Museum accepted the use of three more country houses: Compton Wynyates in Warwickshire, Northwick Park, Gloucestershire and Haigh Hall, Lancashire. The Banqueting Hall and Toddy Room of Skipton Castle in Yorkshire were offered for material from Aberystwyth. Because of its massive stone walls Skipton was damp even in summer and special heating equipment had to be installed.

The quarry was ready for occupation at the end of 1941 but as objects would have to travel by road at 20 mph it was decided to delay the move until more daylight was available and the journey could be made in a single day. Interestingly, although the war was far from over, Forsdyke was already, in March 1941, warning the repositories to make plans for the eventual return of objects to London. The move began on 24 February 1942 - rail consignments from Aberystwyth, Boughton and Drayton and trucks from Compton Wynyates. On arrival the plywood boxes of antiquities were stacked on duckboards to ensure free ventilation between them and the floors. Books, manuscripts and prints and drawings, mostly packed in uniform millboard cases, were placed in wooden racks. Ethnographical textiles were hung in a separate room, where they could be sprayed with insecticides.

By the end of the war the quarry was to contain a fabulous collection - one of the greatest collections of treasures ever brought together in one place and, it is hoped, never to be repeated.

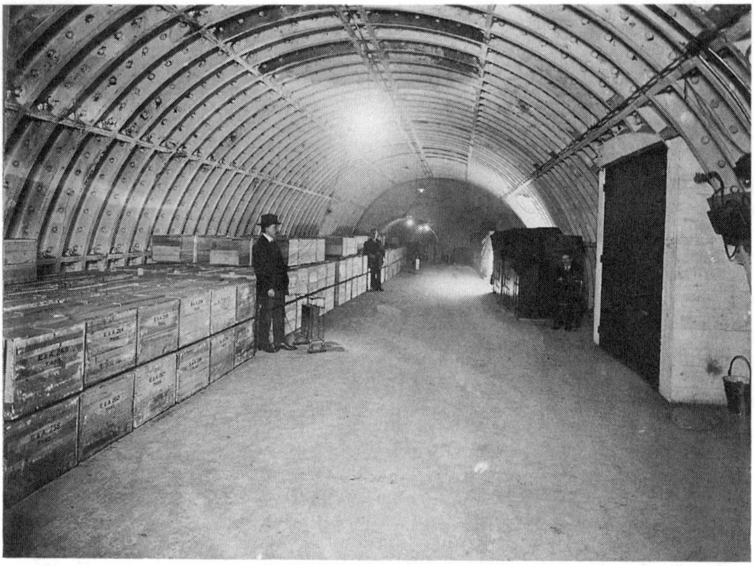

Fig. 1 British Museum objects in storage in the Holborn Post Office Tunnel 1918.

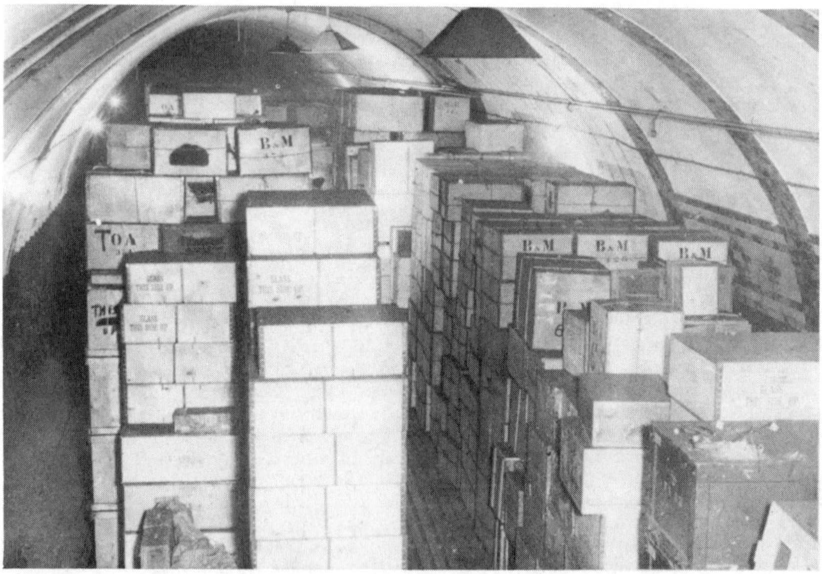

Fig. 2 Objects return underground during the Second World War. The unused Aldwych 'Tube' Tunnel.

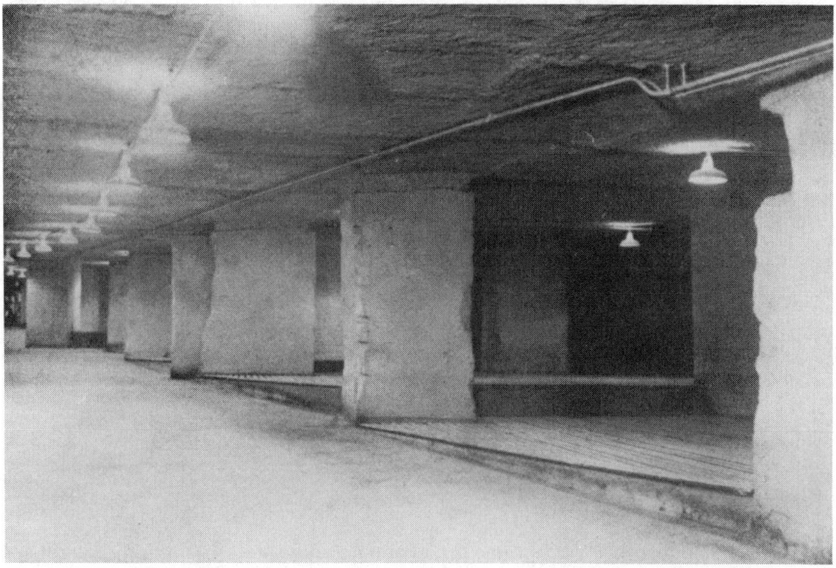

Fig. 3 The quarry in the West of England ready for the storage of the nation's treasures at the beginning of 1942

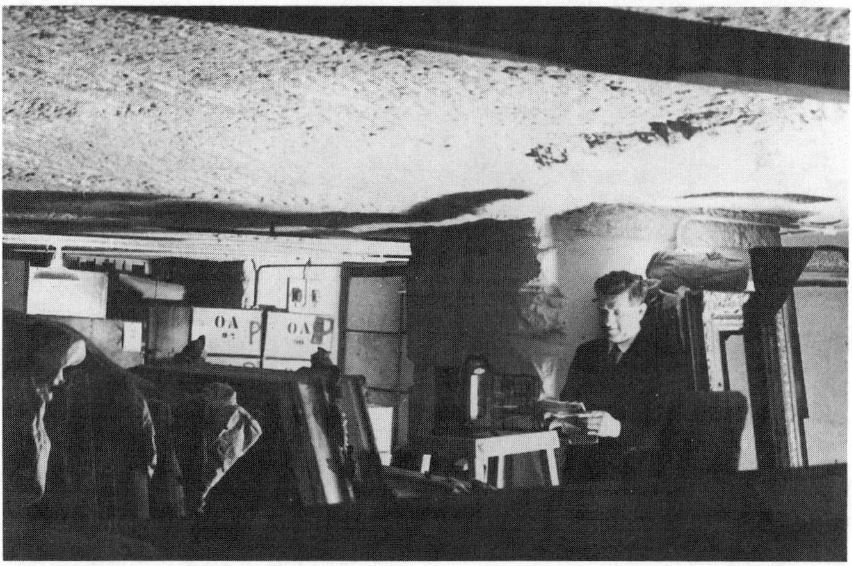

Fig. 4 A curator at work in the quarry during the Second World War

Among the objects were not only the cream of the British Museum, the Victoria & Albert Museum and the National Portrait Gallery collections, but also items from private collections and from some 30 other institutions. They included the Canterbury Gospels, the Chatsworth Head of Apollo, the Benedictional of St Aethelwold, the Winchester Bible, the Shakespeare Deed from the Guildhall, St Chad's Gospels from Lichfield Cathedral, the portrait of Richard II and King Henry VII's altarpiece from Westminster Abbey, the Aberdeen Bestiary, the Exeter Book, stained glass and the Magna Carta from Salisbury Cathedral, Hereford's Mappa Mundi, the Savernake Horn, the Lloyd coin collection, the Stonyhurst Gospels and the Burns manuscripts from Edinburgh.

An inscription, which neatly sums up the exodus, was left on the walls of the quarry by the Assyriologist, Cyril Gadd. It reads:

> In the year of our lord 1942, the sixth year of George, king of all lands, in that year everything precious, the works of all the craftsmen, which from palaces and temples, were sent out, in order that by fire, or attack by an evil enemy they might not be lost, into this cave under the earth, a place of security, an abode of peace, we brought (them) down and set (them).

Bibliography

Air Raid Precautions in Museums, Picture Galleries and Libraries, (Trustees of the British Museum, London, 1939). (See appendix to this paper)

Jamie Ambrose, 'Heritage Under Fire', British Heritage, August/September 1989, pp. 47-53.

Marjorie Caygill, '1939: Evacuating the BM's treasures', British Museum Bulletin, 62, Winter 1989, pp. 17-21.

Marjorie Caygill, 'The British Museum at War', BM Magazine, Summer 1990, pp. 35-39.

Sir John Forsdyke, 'The Museum in War-time', British Museum Quarterly, XV 1941-50, pp. 1-9.

Sir Frederic Kenyon, The British Museum in War-time, David Murray Foundation Lecture, University of Glasgow, (Jackson, Wylie & Co, 1934).

Jacob Leveen, 'The British Museum Collections in Aberystwyth', Transactions of the Honourable Society of Cymmrodorion, 1946.

H. J. Plenderleith, 'The Preservation of Museum Objects in War-Time', Proceedings of the Royal Institution of Great Britain, XXXII, III, 149, Meeting 21 May, 1943, pub. 1945, pp. 538-548.

Standing Commission on Museums and Galleries, Third Report, (HMSO, London, 1948).

AIR RAID PRECAUTIONS
IN MUSEUMS
PICTURE GALLERIES AND LIBRARIES

PRINTED BY ORDER OF THE TRUSTEES
THE BRITISH MUSEUM
1939

PREFACE

THIS Handbook embodies the Report of a Committee representing the National Museums, Picture Galleries and Libraries, set up at the instance of the Standing Commission for Museums and Galleries.

All, or nearly all, the instructions and recommendations prescribed in Chapters I and II (*Protection of Staffs and Buildings*) are taken from the series of Air Raid Precautions Handbooks and Memoranda issued by the Home Office, to which these two chapters will form an introduction. Chapter III (*Protection of Material*) has been compiled mainly from the experience of the Directors of the National Museums, Picture Galleries and Libraries, and the researches of their Laboratory staffs.

CONTENTS

LIST OF ILLUSTRATIONS

INTRODUCTION

The protective measures described in the following chapters are designed to serve as a guide to Air Raid Precautions for Museum, Picture Gallery and Library authorities, who would in time of emergency be responsible for safeguarding their Collections, their Premises, and their Staffs while on duty.

It is assumed that Exhibition Galleries and Reading Rooms will be closed to the general public at the outbreak of hostilities, in order that the staffs may give their whole attention to protecting the material in their charge. The question of shelter accommodation for members of the public caught on the premises by an air raid need not therefore be considered separately. It may, however, be said here that visitors whom special duties or occupations may bring to such institutions in war time ought not to be admitted unless they bring gas-masks with them. Members of the staff must also bring the gas-masks which they will have obtained from their Local Authorities.

The public services organised by the Local Authorities for defence will be generally available, but these arrangements will not be sufficient to meet every requirement. In establishments employing large staffs, and particularly in Museums, Picture Galleries and Libraries, where special risks to material are involved, it is necessary to design self-contained Air Raid Precautions schemes, embracing as many as may be practicable of the various services which will be required. The organised local services may be expected to assist the Museums if they are able to do so in case of need, and the Museum services may be a similar reinforcement to the Local Authorities. It is important, therefore, that all such institutions should work in close co-operation with the Local Authorities in whose areas they are situated.

In so far as any of the premises concerned come within the scope of the provisions of the Civil Defence Bill now before Parliament which apply to 'commercial buildings,' the guidance given in this Handbook is subject to any specific directions regarding the measures to be taken in such buildings which may subsequently be issued by the Home Office, Air Raid Precautions Department.

It is recognised that the full range of practical precautions prescribed here is beyond the resources of any but the largest establishments. But the intention has been to define the greatest needs, so far as they can be foreseen, and to devise the fullest means of meeting them. Smaller establishments will be able to adapt these measures to their own needs and capacities.

CHAPTER I

STAFF

Personal Protection

The efficacy of any scheme of Air Raid Precautions will depend largely on the extent to which panic is avoided through the intelligent co-operation of staffs. All persons employed on the premises should be acquainted with the effects of the three kinds of bomb which may be used, High-Explosive, Incendiary and Gas, and of the methods of protection adopted by their own institutions.* The latter will include systems of air raid warnings and communications, shelter accommodation, and the provision and use of protective equipment. It is particularly important that staffs should be aware both of the dangers involved, and of the ways in which to avoid becoming a casualty.

No practicable measures of protection against direct hits from heavy high-explosive bombs are possible in an ordinary building, but the direct hit is a disaster which need not be expected. Adequate protection can be prepared against the commoner and lesser dangers of blast, splinters, fire and poison gas.

Staff Shelters

In suitable buildings rooms should be assigned for the accommodation of the staff during air raids. The rooms chosen should preferably be in a basement or sub-ground floor, provided that in addition to access there is separate means of escape. Their outside walls should have a minimum thickness of either 13½ inches of brickwork or 12 inches of reinforced concrete, and if the floor above is not strong enough to bear the weight of debris in case of the collapse of the superstructure, it must be made so. There should be the least possible window space, and the rooms should be free from risk of flooding through broken water-mains or sewers. If no such accommodation can be found on the premises, trenches or other external shelters must be made.†

---

* *A.R.P. Handbook No.* 6, *1st edition* (Air Raid Precautions in Factories and Business Premises), gives particulars of possible forms of air attack including the three kinds of bomb mentioned above, gas-spray and machine-gun fire. A revised edition of this Handbook is in preparation.

† Selection of shelters and their reinforcement are discussed in Chapter II (BUILDINGS), pp. 29, 30. For gasproofing see Chapter II (BUILDINGS), p. 31, and *A.R.P. Handbook No.* 6, *1st edition*, Appendix E.

Accommodation

To minimise casualties from a direct hit of a high-explosive bomb, no shelter party should consist of more than 50 persons together. Shelter accommodation should be capable of being made gasproof without undue difficulty should the need arise. It is essential that every member of the staff should have his respirator in his possession at all times.

The amount of space to be allowed in the shelters for each person should follow the requirements laid down in the Shelter ' Code ' (Clause 10 of the Civil Defence Bill) to be issued by the Minister.

Equipment

Members of the staff should be familiarised as early as possible with their own shelters and with the most expeditious routes to them. Nominal rolls of persons allocated to shelters should be posted in each, shelter-wardens appointed and made responsible for discipline, communications and equipment. Provision must be made for independent lighting, sanitation, water and food, and it is desirable that means of quiet recreation be supplied. Great care must be taken in an unventilated shelter to economise the supply of air; for this reason physical exertion should be avoided by the occupants, and needless to say, there must be no smoking. The staff should also be acquainted with emergency measures for dealing with minor damage which may be done to shelters by bombs, and each shelter should be provided with picks and shovels to enable the occupants to dig themselves out if necessary.*

General Training

Although it is possible to obtain a knowledge of the means of general protection against air attack from the A.R.P. Handbooks,† it is recommended that all staffs should receive a short course of instruction in the methods to be adopted. This course may be given either by instructors who have qualified at one of the Home Office A.R.P. Schools at Falfield or Easingwold, or by instructors possessing certificates awarded by local authorities, or by Grade I instructors (A.R.P.) of the St. John Ambulance Brigade, the St. Andrew's Ambulance Association or the British Red Cross Society.

---

* A list of equipment recommended for staff shelters is printed in APPENDIX A.

† Especially *A.R.P. Handbook No.* 1 (Personal Protection against Gas); *A.R.P. Handbook No.* 6, *1st edition* (Air Raid Precautions in Factories and Business Premises); and *A.R.P. Handbook No.* 9 (Incendiary Bombs and Fire Precautions).

The cost of the training and of the provision of any necessary equipment will fall on the institution concerned.

The general training should include a knowledge of the various types of war gas: tear gases, nasal irritants, lung irritants, and blister gases, and their effects, and of the general distinction between persistent and non-persistent gases. Staffs should be acquainted with the methods of detecting the presence of gas, either by sensory perception or by chemical means, and their training should include practical experience in the use of respirators.* All members of the staff will be in possession of a 'Civilian' respirator issued by their local authorities, but members of A.R.P. services, who may have to work in gas, should be provided with 'Civilian Duty' respirators.

A.R.P. SERVICES

Selected parties and individuals will be assigned to special duties in the application of protective measures, and most of these persons will need special training in addition to the general staff instruction described above.

In large establishments an *Air Raid Precautions Control Officer* should be appointed to take charge of the protective organisation. A certain number of *Key Men* and *Messengers* will be required to remain on duty at important points for observation and communications during an air raid. The emergencies for which specially trained parties will be needed are: Incendiary bombs and fires (*Fire-fighting*); Damage to buildings or collections and clearance of combustible material or obstructions (*Salvage*); Poison Gas in liquid and vapour form (*Decontamination*); and Personal Injuries (*Air Raid Casualties*). In large institutions, these special duties should be assigned separately. In smaller institutions, however, division of labour on these lines will not be possible, and the functions of single squads or persons will have to be multiplied.

Respirators

As indicated above, members of A.R.P. services who may have to work in gas should be provided with the 'Civilian Duty' respirator. This respirator, which provides a larger

---

* The subject of anti-gas training is dealt with at length in *A.R.P. Memorandum No. 5, 2nd edition* (Anti-gas Training), a revised edition of which, including syllabuses of instruction in incendiary bomb control and elementary methods of protection against high-explosive bombs is to be published shortly. The latest approved syllabuses for anti-gas training are contained in *A.R.P. Dept. Circular No.* 26/1939, of the 21st March, 1939 (which may be purchased through H.M. Stationery Office, price 1d.). The 'Short Course' given on pages 4-5 of that Circular would appear to be appropriate for the general training of staffs.

measure of protection than the 'Civilian' respirator, has been specially designed for the use of the members of certain of the Civil Air Raid Precautions Services. Licenses have been granted to certain manufacturers to affix a Government certification mark to such respirators made to the approved pattern.*

Protective Clothing

Steel helmets should be worn by all members of the staff who may have to be on duty out of doors or in remedial work in damaged parts of buildings during or after an air raid, and protective clothing should be available for those who may have to enter gas-contaminated areas.†

A.R.P. Control Officer

The Air Raid Precautions Control Officer will supervise the practical application of the scheme adopted by his institution. He should be fully conversant with the organisation and equipment of the staff shelters, and with the details and training of the A.R.P. services. In an air raid he would take charge of operations and co-ordinate the necessary activities of the staff.

Control Centre

The Control Centre from which the A.R.P. Officer would work, should be centrally situated, should provide a maximum degree of protection against blast, splinters and fire, and should be capable of being made gasproof without undue difficulty. The room chosen should be large enough to provide sleeping accommodation for the officer and the assistants whom he may need. Provision will be needed for food and water, lavatory accommodation should be accessible, and telephone communication with key-points within the premises and with fire and police stations and hospitals outside must be available. In large buildings a system of independent or auxiliary telephones should be installed, providing two-way communication between the control centre, service posts and key-points, and one-way communication with staff shelters. It would probably be found undesirable in the stress of an air raid for staff shelters to have direct access by telephone to the control centre.

---

* The 'Civilian Duty' respirator is described in detail in *A.R.P. Handbook No.* 1, Chapter V.

† A suggested scale of personal protective equipment for trained staff is given in *A.R.P. for Government Contractors*, Appendix 8.

Air Raid Warnings

Warnings of the approach of hostile aircraft will be given by means of sirens or hooters, which will be sounded in some places by a 'warbling' note changing every few seconds, and in other places by a succession of short blasts. The warning signal will be timed so as to give from 5 to 7 minutes' warning to the district which is threatened. Arrangements should be made for a member of the staff to be stationed where he can hear the warning signal, and for the warning to be conveyed to the staff throughout the premises. The signal that the raiders have passed will be given by the sirens or hooters sounding continuously for two minutes on the same note. The presence of gas will be indicated by means of hand rattles; and hand-bells will be rung when the danger of gas is past.*

Key-Men

Important points throughout the building, where damage can be quickly observed and communications or mechanical apparatus controlled, should be scheduled as Key-points, at each of which men responsible for maintaining contact with the Control Centre will be posted. The Key-men will be on duty during the progress of an air raid. They should be posted in pairs, or, if possible, in threes, or a Key-man may be incorporated with the nearest Fire Watcher Party. Splinter-proof protection must be provided for them, preferably in the corners of buildings. This can be done with sandbags (2 feet 6 inches thickness) on the outside of a ¼-inch steel plate or a portion of a 5-foot steel shelter, or built with the necessary specifications of brick or concrete.†

Messengers

A certain number of the staff should be detailed as Messengers. Those chosen should have intimate knowledge both of the premises and of the district in which they are situated. Where possible, they should also attend the courses of instruction for A.R.P. Messengers provided by local authorities. Messengers will be needed at the Control Centre, Shelters and Service Posts (First Aid, Decontamination, Fire and Salvage) and will act as runners in case of the absence or breakdown of telephones, conveying messages within the premises and between the control centre and fire stations, local hospitals, local authorities, etc., outside. It

* For full details of the Air Raid Warning System see *A.R.P. Memorandum No.* 8.

† Designs for small protected posts are given in *A.R.P. for Government Contractors*, Appendix 6.

will be desirable to have experienced cyclists for external communications.

Fire-Fighting

It is probable that the difficulties of movement caused by damaged streets would greatly impede the activities of the public fire-brigades, and in view of the many calls likely to be made on them when incendiary or high-explosive bombs are used, institutions should as far as possible rely upon their own fire-fighting organisations.*

Although the public brigade might be unable to afford full assistance in an emergency, the brigade should be called directly a fire is discovered. All concerned should be familiarised with the emergency fire brigade organisation of the local authority.

Organisation and Training

Selected members of the staff should be detailed and trained as firemen, and regularly drilled with the appliances which they require. Provision should be made for the attendance of some of the trained staff throughout the twenty-four hours in war time, and the numbers trained should be sufficient to provide for continuous duty in three eight-hour shifts. Provision must be made for the accommodation of the night staff on the premises.

It cannot be too strongly emphasised that the fire caused by an incendiary bomb is more dangerous than the bomb itself, and that knowledge of fire-fighting and general behaviour in the presence of fire is essential for the trained staff.† The following distribution of Firemen for a large building during an air raid is recommended :—

(1) *Fire Watcher Parties*, inside the building. Appliance : Stirrup hand-pump. One party for every 8,000 square feet of roof area is a rough guide.

(2) *Local Fire Parties*, inside or outside the building, in support of Fire Watcher Parties. Appliance : 2-man manual. About one for every three to eight Watcher Parties according to the nature of the fire risk.

(3) *Group Fire Parties*, outside the building, in support of Local Fire Parties. Appliance : Trailer-pump.

---

* In addition to the guidance given in *A.R.P. Handbook No.* 9, it will be useful to refer to *Safety Pamphlet No.* 13 (Fire Protection in Factories), issued by the Factory Department of the Home Office, which gives information regarding precautionary methods and apparatus which could well be used in Museums, Galleries and Libraries.

† For fire-fighting apparatus see Chapter II (Buildings), p. 32, and for detailed information on this subject *A.R.P. Handbook No.* 9. The list of equipment for fire parties of various sizes given there is reprinted below in Appendix B.

The Fire Watcher Parties are needed to deal promptly with incendiary bombs. Their protected accommodation will be best sited in the corners of buildings and may be improvised with sandbags, as in the case of the Key-posts. The limiting factor is the range of vision, and this should be borne in mind in placing them.

As regards the training of the personnel of Fire Parties, where hand appliances only are provided it should be sufficient for the personnel to attend the courses in incendiary bomb control organised by the local authorities for their A.R.P. Services. In the case of the larger establishments where the equipment includes a trailer-pump, it is desirable that arrangements for training should be made with the local fire brigade.

Incendiary Bombs

Incendiary bombs are of various kinds, and range in weight from 1 to 25 kilos. The lightest of these (about 2¼ lbs.) is the one which is most likely to be used upon civilian objectives. It is generally called the light magnesium (electron) bomb, and consists of a thick-walled tube 9 inches long and 2 inches in diameter made of magnesium with a small admixture of aluminium. The tube is filled with a priming composition of the thermite type and has an igniter which functions on impact. The bomb does not explode, and the thermite serves only to ignite the magnesium. The priming composition burns for 40 or 50 seconds at a temperature of about 2,500° C., after which the molten magnesium burns for 10 to 20 minutes at a temperature of about 1,300° C., that is to say, at a heat which will set fire to anything inflammable within a few feet. Jets of flame are emitted while the priming composition is burning, and molten particles may be thrown off to a distance of 50 feet. The burning magnesium gives off thick white smoke, which, though acrid and unpleasant, is not dangerous.

One large bomber could carry between 1,000 and 2,000 of these bombs. They would be released from containers holding 10 or 20, and the contents of several containers can be released simultaneously. The bombs spread out as they fall, and a group dropped simultaneously from 5,000 feet would cover an area of about 100 yards square.

The kilo bomb has poor powers of penetration, being designed to pass through an ordinary roof of slates, tiles, etc., but to remain on the floor immediately below, thus starting fires in places which are difficult of access. It is of the utmost importance that the situation should be dealt with as soon as possible after the bomb has functioned in order to prevent an incipient fire from becoming a general conflagration. Staffs should be organised, trained and properly equipped for this special service.

Methods of Control

The thermite composition contains its own oxygen and cannot be extinguished by smothering, but the magnesium must get its oxygen from the air or from surrounding materials in order to burn. The effect of water is not to extinguish the burning magnesium, but to accelerate its combustion. Alternative methods of control are described in Chapter II of *A.R.P. Handbook, No.* 9. The equipment recommended by the Home Office consists of a stirrup hand-pump, 30 feet of hose with dual-purpose (jet or spray) nozzle, and at least three large buckets or other containers, for control with water; and a 'Redhill' container and a long-handled scoop and hoe for control with sand.

Use of Sand

If the bomb comes to rest on a floor which will not burn readily it should be easy, when the priming composition has finished burning, to approach close enough to smother the bomb with sand from the scoop, and either to let it burn out in place or to carry it out of the building in the container, into which it will have been raked with the hoe. Earth, ashes or foamed slag will do instead of sand; foamed slag has the advantage that it is considerably lighter in weight. The smothering substance does not extinguish the magnesium, but makes it burn less fiercely by cutting off the supply of free air. It must be remembered, however, that the metal is still burning underneath the sand, and if left, may burn through a boarded floor in a few minutes. Even if the floor-covering is fireproof, it may have combustible elements beneath it, and the heat which is driven downwards may ignite these.

Use of Water

If the bomb has already started a fire in the floor or in adjacent objects, it may be necessary to get this under control before dealing with the bomb. The jet from the stirrup-pump will effectively deal with the fire if used within the first few minutes. But if the jet is turned on the bomb, its effect will be to scatter the burning magnesium to a considerable distance. This is avoided by the use of the spray, and its application will cause the magnesium to be consumed in 1½ minutes, meanwhile keeping the surroundings wet and preventing the fire from spreading. About 5 or 6 gallons of water are generally required to control a bomb and extinguish the fire which it has produced in an ordinary room.

Outside the building, or even inside if no greater damage will be done by flooding, a jet from a hydrant or trailer pump may be directed at the bomb. The result will be that the magnesium will flare up and particles will be thrown off violently, but the bomb will be consumed in about 15 seconds.

Equipment

The conditions in the proximity of a bomb burning indoors are such that it would be unwise to employ any appliance requiring the operator to stand up or undertake any physical exertion. It is for this reason that 30 feet of hose is prescribed for use with the stirrup hand-pump, so that the person working the pump may stand well clear of the heat and smoke. When approaching a burning bomb in the open, dark glasses may be worn, but it is thought that indoors the glare will be effectively masked by the smoke. Fireproof clothing is not needed, but a shield of any light board (asbestos or plywood) may be useful to protect the face from flying incandescent particles. An electric hand-lamp or hurricane oil lamp should be available for night operations, and each Fire Watcher Party should have a fireman's axe, preferably with an insulated handle for protection against electric current.

Salvage and Rescue

Provision is required for the rescue of persons and the removal of valuable material from damaged buildings, for the clearance of obstructions and for such urgent repairs as can be carried out in an emergency. These duties can normally be performed by the reserves of the Fire and First Aid Parties. In large institutions, however, separate Salvage Squads, each consisting of six or eight men under a foreman, should be organised.

No special training in clearance or demolition need be given, if intelligent men can be selected. But the foreman should have practical knowledge of building mechanics or structural engineering, and the others should understand the manipulation of tools and tackle. All the members of the party should be able to appreciate the risks of movement in damaged buildings, so as to avoid incautious action which might endanger them as well as other persons.

In regard to rescue work which they might be called upon to perform, it would be helpful if some of the party were trained in First Aid ; and since they might have to operate in gas concentrations or on gas-contaminated debris, they should be provided with protective clothing and have a knowledge of anti-gas precautions.

Provision may be necessary for the salvage of material contained in the collections, or for emergency measures of protecting them in place. Persons detailed for this service should be used to handling the objects in the collections, and equipped with tools for clearance and materials for the protection of objects and cases from fire, water and falling wreckage.*

The water used in extinguishing fire and particularly the liquids ejected by chemical extinguishers, may do as much damage to Library and Museum material as the fire itself. Members of the Salvage Squads who will work in conjunction with the Fire Parties should, therefore, carry waterproof and fire-resisting covers. Materials which would serve the latter purpose are asbestos cloth or mineral boards, corrugated iron, or plywood impregnated or painted with non-inflammable compositions.

Decontamination

Buildings and localities into which liquid gas has entered will be dealt with by Decontamination Squads.† Decontamination is done either by chemical destruction or physical removal of the gas, or by sealing off the vapour from areas affected. A single squad normally comprises five men and a foreman, all of whom require special training, clothing and equipment; the number trained should be sufficient to provide adequate reserves. The course of special training must be given by an instructor who has qualified at one of the Home Office A.R.P. Schools.‡ The work of decontamination is exhausting, involving strenuous effort and the wearing of heavy protective clothing. Members of staffs chosen for Decontamination Squads should therefore be as young as possible, of good physique, and medically certified as fit.

Cleansing Centre

Where Decontamination Squads are formed in large establishments it may be desirable to provide a Cleansing Centre.

---

* Suggested lists of equipment for Rescue and Salvage Parties are printed below in Appendix C.

† The care of persons injured by liquid gas will be carried out by the Air Raid Casualties service (see below). The treatment of contaminated Library, Picture Gallery or Museum material, if any treatment is possible, will be laboratory work.

‡ The syllabus for the course required is given on p. 6 of *A.R.P. Dept. Circular No.* 26/1939. A list of equipment necessary for Decontamination Squads is printed in Appendix D of this Handbook. For methods of decontamination, reference should be made to *A.R.P. Handbook No.* 4 (Decontamination of Materials).

The decontaminators would assemble there, put on their working clothes in readiness for action, and receive cleansing treatment and change back into their ordinary clothes after action. The facilities required for cleansing consist of an open shed or covered space for the removal of contaminated outer clothing, and three adjacent rooms (or one or two large rooms divided up) to provide for the three separate operations of undressing, washing and dressing. Undressers (themselves wearing respirators and protective clothing) should be available to remove the contaminated clothing of the men as they come off duty, and one attendant trained in giving bleach and eye treatment and minor first-aid should be on duty. It is essential that these rooms should not become contaminated; all outer clothing must therefore be left outside, and an air-lock must be provided between the external shed and the undressing room.

Equipment

In the External Shed metal or wooden seats should be provided and should have covers of oil-cloth or other impervious fabric. Metal bins with close-fitting lids will also be needed to receive contaminated clothing. At the entrance there should be a bleach tray, in which each person entering must decontaminate his boots. The Undressing Room should be large enough to accommodate the whole Decontamination Squad, capacity being assessed on a basis of about 15 square feet per person. This room must also be provided with seats and bins and should contain a latrine bucket, in order to avoid contamination of water-closets. The Washing Room should be fitted with two shower-baths separated by sufficient space to enable a man to wet his body under one shower, soap himself in the intervening space and to wash the soap off under the other shower. Facilities should be installed for washing out the eyes in warm water, and for further eye-treatment or bleach treatment, if required in particular cases. There should also be sufficient space for the men to dry themselves. The Dressing Room, in which the men would change into their working clothes before going on duty, as well as dressing after their baths, should contain cupboards or lockers in which they can leave their ordinary clothes, and in which clean protective clothing can be kept until required. The whole station (apart from the external shed or internal space used instead) must be capable of being made gasproof as far as is practicable, and must be protected against splinters and blast.

It is thought to be unnecessary, however, for any but the largest establishments, if the premises are in a town, to provide themselves with Cleansing Centres. Use would be made in case of need of the nearest Cleansing Centre as of the First Aid Post of the Local Authority.

Air Raid Casualties

In view of the difficulties which may be expected in obtaining professional medical attendance or hospital transport during an air raid, every effort should be made to give first aid for minor or even major injuries, and to deal with cases of gas contamination within the premises. To this end all Museums, Galleries and Libraries should organise First Aid parties, the members of which need instruction both in First Aid and Anti-gas, followed by practical training in the handling of air raid casualties.

Organisation and Training

First Aid parties should consist of four persons, who will also act as stretcher-bearers. An establishment of say 250 persons should have at least two such parties, and as a rough guide, it is advised that an additional party should be provided for every 250 staff or fraction thereof. To cover all emergencies it is generally desirable that a few extra members of the staff, in addition to the First Aid Parties, should be trained in First Aid. For the preliminary First Aid instruction the Government has made arrangements with the St. John Ambulance Brigade and the British Red Cross Society and the St. Andrew's Ambulance Association, whereby the services of lecturers and instructors attached to these societies are available to give courses of lectures at moderate charges. It is desirable that all members of First Aid parties should possess the certificate of qualification issued by these societies. The instruction in the First Aid treatment of gas casualties may be given by any instructor qualified to give general anti-gas training (see page 11). In the case of establishments in private ownership the advice of the local authority should be sought in relation to arrangements for the collective training of First Aid personnel.

Equipment

Each First Aid party will need to be equipped with one First Aid Haversack per party and one Pouch per person, and with stretchers.*

* For details of the equipment see APPENDIX E.

First Aid Rooms and Posts

A First Aid Room will be necessary in any establishment, and large institutions may find it desirable to provide a specially equipped First Aid Post to serve as a centre for the reception of casualties other than those for which treatment on the spot by First Aid parties would be sufficient. It will be the task of First Aid parties to convey such casualties to the First Aid Post. Means of transporting serious cases to hospital would be an advantage, but the Local Authority will normally be responsible for the evacuation of casualties to hospital.

The First Aid Room should be capable of being made gas-proof without undue difficulty, and should be, as far as practicable, protected against blast, splinters and fire. The equipment should include facilities for storage of dressings and medicaments, and means of boiling water for sterilising instruments. A supply of running water (preferably hot and cold) and a sink or lavatory basin should be accessible, and separate lavatory accommodation, including water-closets, should be available for those awaiting treatment or removal by ambulance. A small section of the room should be curtained or screened off for use by patients requiring rest.

A First Aid Post must be similarly protected and equipped, but should provide fuller accommodation, including facilities for cleansing cases of personal contamination by liquid gas. This may be done by combining it with the Cleansing Centre. It is essential that contaminated casualties should have their clothing removed before entering the First Aid Room for further treatment. The Cleansing Centre will contain an undressing room and a washing room equipped with a shower or other bathing facilities, and it must be separated from the First Aid Post by air-locks.* Lavatories provided with hot water facilities may often be utilised as Cleansing Centres.

In a small combined First Aid Post and Cleansing Centre, one trained person would be required to take charge, two trained persons to give First Aid, one semi-trained person to act as undresser for contaminated cases, and two semi-trained or untrained persons for general duties.†

---

* See above, p. 19, (DECONTAMINATION).

† Suggestions for the lay-out of First Aid Posts will be included in the forthcoming revised edition of *A.R.P. Handbook No.* 6. See also *Home Office Circular, First Aid Posts* (26th April, 1938).

CHAPTER II

BUILDINGS

Survey of Premises

No ordinary building is proof against the direct effects of attack from the air, and Museums and Picture Galleries, with roof lights and large windows, are particularly vulnerable. It will be necessary to survey buildings in relation to the dangers to which they are exposed, and having noted their weak points, to consider the means of reinforcement which may be afforded. This survey would be a necessary preliminary to the choice and preparation of shelter accommodation for the protection of the staff during air raids and for the storage of such material from the collections as can be suitably placed there. Other measures which will involve some adaptation of existing buildings are the darkening of windows, the installation of fire-fighting apparatus, independent systems of lighting and telephones, and, in large institutions, the provision and equipment of First Aid Posts, Decontaminating Stations and other A.R.P. Service Posts.*

High Explosive Bombs

High explosive bombs vary in weight from 20 lbs. to 2,000 lbs. or more. They may be fitted with instantaneous fuses which cause them to explode and do much damage on the surface, or they may have delayed-action fuses, which enable them to sink into the target before exploding. These latter are designed for use against military objectives.

Direct Hits

Bombs arrive at an angle which may vary from 15 degrees to 45 degrees to the vertical and with a velocity which may be from 400 to 1,000 feet per second. These vary according to the height and velocity of the plane at the time when they are dropped.

Protection against a direct hit from a heavy bomb designed for penetration can only be obtained by going 60 to 80 feet below ground or, alternatively, by interposing a very thick layer of reinforced concrete. Protection against a bomb exploding on the surface can be obtained by means of a

* For Service Posts, see Chapter I (STAFF), p. 12 *sq.*

specially reinforced concrete slab, varying in thickness according to the size of the bomb. Such slabs must, however, be strongly supported in a manner not usually practicable in an existing building.*

Blast and Splinters

It is evident that it is not possible to protect ordinary buildings against direct hits by high-explosive bombs. Protection can be obtained, however, against blast and splinters from bombs bursting outside.

Protection of Walls

If blast pressures were prolonged and consistent, few walls in existence could resist them. But they are instantaneous and fluctuating, and investigations tend to show that the walls of buildings which will resist penetration by splinters will not be affected by the blast of bombs exploding not less than 50 feet away. The following thicknesses of materials will afford protection against splinters from bombs exploding at this distance or more :—Brickwork, 13½ inches ; reinforced concrete, 12 inches ; sand or earth, 2 feet 6 inches ; shingle, 2 feet ; mild steel, 1½ inches. Of such materials in combination, 9 inches of brick and 10 inches of earth together will resist penetration.† The most convenient and the most effective external reinforcement is earth in sandbags, the earth being dug from the nearest vacant ground. If sandbags are not available, earth or shingle held up by wood or metal shuttering is effective.

Sandbagging

The standard Army pattern sandbag measures 33 inches by 14 inches. It should be filled only three-quarters full and beaten with a shovel to a flat rectangular shape. The sandbag when filled measures, roughly, 20 inches by 10 inches by 5 inches, and contains a little over half a cubic foot of earth. If harder materials, such as hard ground, gravel, bricks, etc., are used, these must be broken up into small fragments, in order that the sandbag may afterwards be shaken into a compact, pliant mass. Sandbags filled with such materials should not be beaten. The bond used in sandbagging consists of alternate courses of 'headers' and 'stretchers' (English bond). The first course should be 'headers,' which should be laid with 'chokes' (tied ends) inside. If the

* For the roof-construction required to withstand the direct hit of a 500 lb. delayed-action bomb, see *Air Raid Shelters* (Report of the Lord Privy Seal's Conference, 1939), p. 12.

† For further information regarding material likely to provide protection against blast and splinters, reference should be made to the shelter 'Code' to be issued by the Minister (see above, p. 10).

bag has only one seam, this should also be turned towards the inside when 'stretchers' are laid. Care should be taken to break vertical joints, and the outer surface of the sandbag wall should be battered to a slope of 1 in 4. To resist penetration by bomb splinters, the sandbag wall should be of an average thickness of 3 feet (FIGURE 1.)

If sandbags are not readily available, grain sacks or bags may be used. These usually contain 2½ cubic feet of grain, and should therefore not be more than half filled, as they are otherwise too heavy to handle.

In estimating the number of sandbags required, it may be assumed that a sandbag wall 8 feet high, constructed of

Fig. 1.—SPLINTER-PROOF EXTERNAL SANDBAG WALLING

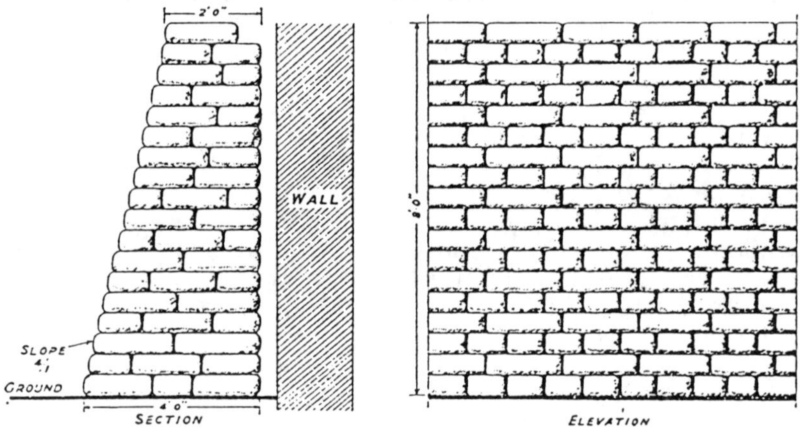

Note.—The average thickness of sand or earth must not be less than 3 feet, and an air space of 1 or 2 inches should be left between the sandbags and the wall of the building.

Army pattern sandbags in the manner described above, would require 40 to 50 sandbags per foot run. Since the fabric of sandbags will deteriorate if exposed to the weather, the outside surface of sandbag walling should be covered with tarpaulin or brushed over with cement wash.

The personnel required for sandbagging operations consists of a filling party of three men (two holding and tying and one shovelling), a building party of two men, and a carrying party to take the bags from the filling party to the building party. The filling and building parties should fill and lay 60 bags in one hour, and the carrying party should be sufficiently large to carry this number in each hour from the former to the latter.

Storage of Sandbags

Sandbags can be stored for periods up to twenty years provided that the following conditions are observed :—

(1) The store must be absolutely dry.
(2) The bales should be piled on grids a few inches above the floor, and there must be a free passage of air around each pile.

Protection of Windows

Windows are the most vulnerable points. Complete protection can only be provided for them by a sandbag wall built on the ground outside, of a thickness not less than 2½ feet (see above), or by steel shutters not less than 1½ inches thick. But protection against flying debris and splinters from more distant bursts may be afforded by less formidable reinforcements, such as sheets of steel ½ inch thick, slipped into slots or on to bolts in the wall, or stout wooden shutters 2 inches thick. These may also give some protection against the effects of blast, but they will not prevent the window-glass from being shattered. Even curtains of heavy material or mattresses inside the windows may be of use in keeping out debris, but will naturally afford no protection against splinters and blast. They would, however, lessen the danger of broken glass being thrown by the force of an explosion into the room. If casement windows are left open, the breakage of the glass may be lessened.

Fragments of glass may be kept from scattering by sheets or strips of transparent foil (' cellophane ') or paper attached to the panes, or better, by wire-netting fastened to the sashes. Where it is desirable, as in staff shelters, to make the windows gas-proof, the paper or foil should cover both sides of the glass. The foil can be made to adhere to glass with a mixture of gum arabic and glycerine ; it can also be bought ready coated with special adhesive.

But the blast of a bomb bursting at close quarters will drive the glass of a window out of its frames, or the frames out of the walls. The effects of blast are incalculable. The glass may fall outwards, or it may be thrown inwards with such violence as to stick into the wall on the opposite side of the room, and the only effect of ' cellophane ' or paper reinforcement in such conditions would be that the glass would fly in larger pieces.

Protection of Corridors

The blast of a high-explosive bomb will be considerably lessened inside the building if its course is checked by closed doors. Where doors do not exist, baffle-walls of brickwork or sandbags projecting from the entrance-walls of galleries or corridors and overlapping in the middle of the gangways will give protection against blast.

Incendiary Bombs

Incendiary bombs may weigh anything from 1 kilo (2 lbs. 3 oz.) up to 25 kilos.* They do not explode, and their penetration depends upon their weight and design and the height from which they are dropped. The kilo bomb is likely to be the type most commonly used. When dropped from a normal height, this bomb will penetrate any ordinary roof of tiles or slates, but it will be arrested by a reinforced concrete roof 4 inches thick, one layer of sandbags, or ¼ inch of mild steel. It does not appear to be practicable to provide protection against penetration of heavier bombs.

Protection of Roofs

Where roof supports are strong enough to bear the weight of such reinforcements, protection may be afforded by sandbagging or steel sheeting, but nothing less than the prescribed thicknesses will stop the penetration of the smallest bomb, and it will therefore be necessary to take protective measures inside the building.

Protection of Skylights and Windows

A special weakness of top-lit galleries will be the skylights. These can be covered, if their frames are strong enough, by sheets of ¼-inch mild steel laid on planking. An incendiary bomb falling obliquely may penetrate into a building through a side window, and exposed windows should therefore be protected with sandbags, heavy wooden shutters, or steel sheets.

Protection of Woodwork

It will be difficult or impossible to secure all top-lights and non-concrete roofs against the entry of incendiary bombs, and provision must be made for protection against bombs which have penetrated into the building. The light incendiary bomb would penetrate a lath-and-plaster ceiling after passing through an ordinary roof, but would be arrested by the first boarded or solid floor encountered.

Joists and boarding between roofs and ceilings should be treated with some fireproof material such as asphalt, or

* For details of incendiary bombs, their effects and methods of dealing with them, see Chapter I (FIRE-FIGHTING), p. 15, and *A.R.P. Handbook No.* 9.

with several coats of flame-resisting paint or plaster. The fireproofing of existing wooden structures cannot be accomplished with the same efficiency as if the wood were impregnated with chemical fireproofing agents prior to building. Where the wood surface is porous and can absorb material applied to it, impregnating solutions or flame-resisting paints may be applied, but their efficacy in actual contact with an incendiary bomb is dubious.

Impregnating Solutions

The following materials have been used for fireproofing wood: ammonium salts and in particular ammonium phosphate, zinc salts, boric acid and solutions of silicates such as sodium silicate (waterglass). There are various proprietary mixtures on the market, the most effective of which appear to be those containing silicates or ammonium phosphate. Several coats are required, and the efficacy of the treatment is directly proportionate to the degree of penetration.

Flame-resisting Paints

Paints are only effective when in direct contact with natural timber. No means can be recommended of protecting painted woodwork short of removing the paint and applying a flame-resisting coating in its place. It is questionable whether the results obtainable would justify the expense involved. In practice, therefore, the question of protecting timbers with paints or solutions refers only to the problem as it arises under a roof or in other places where raw timber would be exposed to the full effect of heat from fire.

Protection of Floors

Attics or roof spaces should be made easily accessible, so that bombs may be dealt with there by the methods described above (p. 16). Precautionary measures should be taken in such places, and on the floors of top-lit galleries if the skylights are not protected. Two inches of foamed slag or dry sand or earth or 2½ inches of household ash, will afford protection against burning, provided that the material is not scattered by the fall of the bomb. This risk may be lessened by laying small-mesh wire netting over the protective material. Pumice slabs 2 inches thick would be more effective than the loose material, and are light in weight. Complete protection would be given by corrugated iron (20 gauge) laid on sand, earth or foamed slag. Iron of this thickness will not be burned through by a magnesium bomb, though it will get red-hot, and the insulating material will prevent the floor from catching fire.

Certain types of asphalt laid about ¾ inch thick will also stop the heat of an incendiary bomb from igniting a wooden floor.

Care must be taken, however, not to impose a greater weight of protective material than the floor will bear. Sand is the heaviest of those mentioned here (1¾ in. thick, 13½ lbs. per sq. ft.), and foamed slag the lightest (2 in. thick, 5 lbs. per sq. ft.)*

Gas Bombs

The explosion of a gas bomb will be of sufficient force only to break the casing and scatter the contents in the immediate neighbourhood. Its impact is therefore the only force to be considered and its penetration will depend upon the resisting power of the surface on which it falls. It may reasonably be anticipated that the bombs used will be of heavy types (i.e., from 100 lbs. upwards). No ordinary roofs could resist such weights, nor could any kind of reinforcement be improvised. Protective measures would consist, in case of a direct hit, of neutralising the liquid gas by the usual methods of decontamination†.

Protection of Windows

The only risks of liquid gas entering a building otherwise than by a direct hit, are by penetration through the windows from a burst outside, or by spray from an aeroplane. Wooden or steel shutters are the most effective protection against these risks, but sheets of galvanised iron will serve the purpose, or even curtains of impervious material, such as tarpaulin, oiled fabrics or rubberised cloth.‡ The presence inside a Museum, Picture Gallery or Library of gas in vapour form is not likely to have any harmful effect on the structure or its contents.

Shelters

The conditions in which the basement of a building would be proof against the impact and detonation of a high-explosive bomb have been already indicated (p. 23). Shelter accommodation of this kind is expensive and cannot be improvised. In default of it authorities should provide selected rooms in which such structural precautions have been taken as will

---

* A table of comparative weights and efficiencies of various materials is given in *A.R.P. Handbook No.* 9, p. 52.

† See *A.R.P. Handbook No.* 4 (Decontamination of Materials).

‡ For the resistance of various fabrics to liquid gas, see Chapter III (Material), p. 40.

provide protection against the effects of blast, splinters, and the collapse of the floors above, and will ensure that the rooms are capable of being made gasproof without undue difficulty should the need arise.

If the existing buildings are unsuitable, and adjacent land is available, protection may be provided by building separate shelters or constructing trenches. Trench systems, if properly designed, afford the same standard of protection as built shelters, but require skilled setting out.*

Choice of Rooms

Basement accommodation is preferable for all shelters, because it affords the greatest degree of lateral protection. Where a building has two or more basement floors, the sub-ground (i.e., the floor immediately below the ground floor) should generally be chosen, in order to reduce the risk of flooding from burst water-mains or sewers. Where such risk is unavoidable, the room should be fitted with entrances above flood level. In districts liable to flooding from damaged river banks or river walls consideration should be given to preventing the access of water to the building at all points. This may require that external basement courts should have solid parapets, and that external doors or other openings which extend below the level to which floods may rise should be provided with slots in the structure of the openings for the rapid insertion of tide boards in an emergency.

Service pipes conveying hot water, steam, gas or compressed air would be a source of danger unless they could be emptied rapidly. If this cannot be done, and the only available shelters are exposed to them, such services must be diverted.

Where the walls are of sufficient thickness to afford lateral protection (13½ inches of brickwork, 12 inches of reinforced concrete, or the equivalent of these in other materials), the window space of moderate extent, and effective gas-proofing is possible, the ground floor of a building may be considered as a suitable site for a staff shelter, but the door and window openings should be protected by sandbag walling or earth between shuttering to prevent the entry of splinters (see above, p. 24).

Basement and ground-floor shelters should have at least two exits; these should be as far apart as possible, to avoid the risk that both may be blocked by debris at the same time. Suitably placed windows may be used for the purpose as well as doorways. The doors of doorway exits should preferably

* Drawings and descriptions of the various types of shelter, including the Home Office 'Block System' of trench shelters, are given in the Shelter 'Code' (see above, p. 10). For basement shelters see *A.R.P. Memorandum No.* 10 (Provision of Air Raid Shelters in Basements).

open inwards. Exits should have as much lateral protection as possible ; if necessary, a sandbag traverse should be erected about 3 feet away from the exits either inside or outside the refuge. Where necessary, a short ladder should be provided to give access from basement refuges to exits at ground level. Where no alternative exits exist, they can sometimes be arranged by means of large concrete pipes or steel tubes (about 2 feet 9 inches diameter), communicating with a vertical shaft with a light manhole cover.*

Shelter accommodation on upper floors should generally be not higher than the second floor, and should have a minimum cover of two floors and the roof of the building. The room chosen should have the least number of windows ; the sills should be as high as possible and the individual windows as small as possible. The floors of shelters, where not on solid ground, should be of reinforced concrete not less than 8 inches to 10 inches thick, to resist splinters striking upwards through the windows of the floor below, and should have reinforcement running in two directions at right-angles.

Strength of Shelter Roofs

The effect of blast from a large bomb bursting just outside a building may be to force the lower walls outwards, so that the floors and upper walls may fall inside. It is therefore essential that the floor forming the roof of shelter accommodation shall be strong enough to carry the weight of debris which may fall on it, and all such floors should be surveyed and, if necessary, reinforced. The following figures are given as a guide to the various strengths of shelter roofs required for buildings constructed in solid brickwork :

200 lbs. per ft. super. for buildings up to two complete storeys above the shelter.
300 lbs. per ft. super. for buildings up to four complete storeys above the shelter.
400 lbs. per ft. super. for buildings with more than four complete storeys above the shelter.

These loads are in addition to the live load which the immediate roof of the shelter is calculated to stand for normal purposes, and for which additional allowance must be made. For buildings of steel frame construction or reinforced concrete buildings, the debris load may be taken as 200 lbs. per ft. super., irrespective of the number of floors above it.

Where the superstructure above the shelter is inconsiderable in respect of overhead protection as well as debris loading,

* For diagrams and specifications of emergency exits, see *A.R.P. Memorandum No.* 10, Appendix IV.

as, for example, a top-lit exhibition gallery, the roof of the basement shelter should provide protection of the following standards :—

Concrete not less than 6 inches thick (or of structural quality not less than 4 inches) ; brickwork or masonry arches where the arch ring is not less than 8½ inches deep ; earth, or sand in bags, ballast or broken stone not less than 1 foot 6 inches thick, or combinations of these materials in suitable thicknesses.

Shelter accommodation should not be situated immediately below large water-storage tanks, safes, or heavy machinery which might be dislodged by a bomb and crush the roof. The walls and roof should not be pierced by pipes or conduits, but if this is unavoidable, the holes should be efficiently caulked by means of some elastic material such as bitumen. Facilities for shutting off gas and water and electricity services from the shelter should also be provided.

The shelter roof should also provide sufficient resistance to fire in the superstructure to allow the occupants time to escape before conditions become intolerable.*

Gasproofing of Shelters

Window glass should be pasted on both sides with strong paper or foil, or the sashes covered with wire-netting, to prevent the fall of broken pieces (see p. 25). The inside of the window-frame should be covered with anti-gas cloth or with improvised substitutes such as oilskins, linoleum, carpet or blanket. Linen or cotton sheets, backed by pasting strong paper on one side, will also prevent the penetration of gas.

Doors should be protected by a blanket or a covering of gasproof material fastened to the outside of the frame so as to leave a flap which can be turned up to allow entrance. If there is a crevice under the door, a thin strip of wood should be nailed to the floor to form a step against which the door will fit tightly, and a strip of rubber tubing should be nailed to the foot of the door. An air-lock will be necessary if any number of people require to enter or leave the shelter while the gas is present outside. If reliance is placed on a single door or curtain, every time it is opened some gas will enter.

Chimneys and other openings, however small, through which air can enter must be closed by wood, cardboard or similar material. For stopping up cracks and holes, a mash of newspaper in water will be found useful.†

---

* Recommendations for strengthening floors are made in *A.R.P. Memorandum No.* 10, p. 5, and for fire-protective treatment, *ibid.*, p. 8.

† The methods of gas-proofing rooms are explained at length in *A.R.P. Handbook No.* 6, *1st edition*, Appendix E.

Fire-fighting Apparatus

Water Supplies

In view of the danger that public mains may be broken, or pressure seriously reduced through water being drawn off to fight fires elsewhere, a reserve of water should be available independent of the public supplies. Storage tanks of water, preferably underground tanks, should be provided, or it may be possible to store supplies by flooding a basement or a cellar; but a civil engineer should be consulted before any such plan is decided upon.

Independent supplies for the smaller appliances should also be provided. The stirrup hand-pump requires about 2 gallons a minute, and while three buckets (say 6 gallons) should be posted with each pump, there should be a reserve of at least 10 gallons per pump in buckets, tanks or other receptacles.

Where internal hydrants are installed and a reserve water supply is available for use with power fire pumps, valves should be fitted so that, in the event of failure, the inlet from the public supplies can be closed, and water under pressure delivered directly into the pipes supplying the internal hydrants. Where suitable hydraulic services are available, these should be adapted for use in the event of failure of public mains.

An effective internal hydrant system may be provided by 'dry risers.' These consist of vertical piping not connected to the water-main, but with an external inlet at ground level to which Fire Brigade pumps (or trailer-pumps) can be coupled. Hydrants provided with hose and branch pipes should be connected to the riser at the various floor levels inside the building.

In small buildings not equipped with hydrants, but which have an internal supply of water under pressure of not less than, say, 25 lbs. per square inch, sufficient rubber hose (1 inch or ¾ inch) to cover the area to be protected, and fitted with ¼-inch or $\frac{5}{16}$-inch nozzle, may be of service in dealing with fires in their early stages. Rubber hose attachable to ordinary water taps may be of use for filling buckets and hand-pumps; but it would be unwise to rely on supplies from the service mains in a war-time emergency.

Hand-pumps and Extincteurs

The stirrup hand-pump and its use have been described above (p. 16). More powerful mobile hand-pumps, such as the '2-man manual,' can be obtained, and should be part of the normal equipment of all institutions in which valuable material has to be protected from fire. Such institutions ought also to possess adequate numbers of chemical extinguishers, but it must be understood that the use of these may have disastrous effects on perishable materials. It is, however, possible to obtain a type of extincteur which throws clean water.

Trailer Pumps

Trailer pumps are small fire-engines which can be drawn by man-power to convenient points outside a building. The pump is driven by a petrol engine and can be used with water supplies from hydrants or storage tanks. A small pump can deliver 150 gallons or more per minute, and more powerful types up to 500 gallons. The standard size of the hose for trailer pumps is 2¾ inches. Canvas hose should be used, and sufficient hose should be provided to reach to any part of the building from the various tanks, hydrants and points of entry.

Fire-engines, pumps and hoses are of vital importance, and should be housed in splinter-proof and incendiary bomb-proof shelters.*

Lighting, Heating and Power Supplies

Electricity

The supply of electricity may be needed during an air-raid, particularly for ventilation in staff shelters and for lighting there and in the various Service Posts. Whether possessing their own generating plant or dependent on outside supplies, institutions should be prepared to meet the danger of a failure of electric power, and be prepared with an emergency lighting system. Where facilities for charging accumulators exist, emergency lighting will be done with accumulator lamps. Otherwise, stores of dry-battery lamps should be provided, and accumulator lamps should in any case be supplemented by dry-battery lamps, hurricane oil lamps and candles. Electrically-driven plant used to ventilate staff shelters should be adapted for manual operation in the event of failure of electric power. Electric lifts should not be used during an air-raid, in view of the risk of the power failing and passengers being trapped between floors.

Telephones

The system of independent or auxiliary telephones mentioned above (p. 12) may take the form either of an inter-communi-

---

* For fire-fighting apparatus in general, see *A.R.P. Handbook No. 9*, and for training and organisation of Firemen, see above, p. 14 (Staff).

cations system, in which the installation is fixed and permanent, or of a field system, in which the wires lie on the floor or are loosely hung on walls or ceilings. The field system is more adaptable and cheaper, but if it is used, wires should lie where they cause least obstruction, and are thereby exposed to least risk of damage. Whatever system is used, a small red indicator lamp should be installed close to each instrument.

Gas

While the danger of explosion and fire resulting from burst gas-pipes would be reduced if supplies of gas throughout the premises were cut off at the controlling point for the duration of a raid, this procedure would involve the risk of gas escaping from individual fires and burners when the supply was again turned on. Gas should therefore not be turned off at the controlling point unless it has been ascertained that all the consuming points are already closed; or if it is necessary to cut off supplies in emergency without this precaution, the gas must not be turned on again until all individual fires and burners have been inspected.

Heating

Fires and furnaces should be drawn and the steam valves of boilers opened immediately after warning of an air-raid is received. Arrangements should be made to cut off the supply of hot water flowing through radiators or pipes in Staff Shelters by means of control valves, and to empty the pipes.

Obscuration of Lights

Windows and roof-lights from which any kind of illumination used within a building might be visible to aircraft must be darkened, by fitting blinds or shutters or by painting the glass black.

Light-proof blinds may be improvised by painting ordinary blinds black, and by pasting a strip of black-painted brown paper at least 6 inches wide round the edge of the glass to cover the joint between blind and window-frame.

In working rooms, where daylight is necessary, brown paper or light-proof cloth may be stretched on frames of battens which are made to fit the windows. Or the whole of the window-glass may be covered at night with brown paper pasted round the edges only; the unpasted area of paper may be cut out on the following morning and replaced by pasting the edges on the following evening. This process may be repeated as long as the paper is usable.

Where glass is permanently obscured the medium should be applied externally, thus showing a matt surface and avoiding the reflection of moonlight and sunlight. Internal panels should be made of light weather-proofed material such as plywood or millboards, which can be used for stopping the window-openings if the glass is broken by blast. They should be fixed lightly, so that they will come away whole under blast.

All internal lights should so far as is practicable be shaded so as to prevent direct light being cast above the horizontal or in any direction in which lighting is not required for working purposes. Particular care should be taken to shade lamps so that no direct light is cast on windows. This will be important if brown paper only is used on windows.

An additional safeguard which it is desirable to take is to arrange for all lights to be connected to a single control point, so that they can be instantly extinguished in the event of damage to the premises causing exposure of lighting.*

---

* For further guidance on this subject see *A.R.P. Pamphlet*, '*War-time Lighting Restrictions for Industrial and Commercial Premises.*'

CHAPTER III

MATERIAL

Methods of Protection

In planning the protection of the collections in their charge, Museum, Picture Gallery and Library authorities should consider three methods. Material may be protected in place, transferred to safer storage on the premises, or removed to repositories elsewhere. In the general description of these methods which follows, separate consideration is given where necessary to Museum collections, Library material (books, manuscripts and prints) and Pictures.

The policy of particular institutions will depend upon the nature and value of their collections, the design and structure of their buildings, the risks to which their localities are exposed and the alternative accommodation and means of transport which are available. In most cases it will be found expedient to divide the risks by adopting protective measures of all three kinds. Since evacuation will only be adopted if a perfectly safe repository is available, and since the packing and transport of even a small amount of material may be particularly difficult at the outbreak of hostilities, it is evident that the evacuated material will be the most valuable part of the collections. Movable objects which come next in value will be stored in the places of greatest security on the premises. But when these precautions have been taken there will probably remain a large bulk of the collections, particularly in Libraries, which cannot be removed, and which must receive such protection as is possible in place.

PROTECTION IN PLACE

Rooms containing objects which are too numerous, too large or too fragile to be removed must be given such protection as is practicable, and the material should be moved as far as possible to the safest parts of the rooms. That is to say, direct exposure to blast or missiles through windows or roof-lights should be avoided. Bookcases, pictures and other objects may best be protected by sandbagging. The construction of sandbag walling has been described above (p. 24), but special precautions are necessary when sand is used to protect objects indoors.

Sandbagging

The sand should be thoroughly dried in order to avoid damage through damp and moulds, and ample space must be left for ventilation between the objects and the protective covering. Light steel bearers, lengths of iron tubing or wooden

Fig. 2.—Section of Splinter-proof Sandbag Protection for Wall-cases or Pictures.

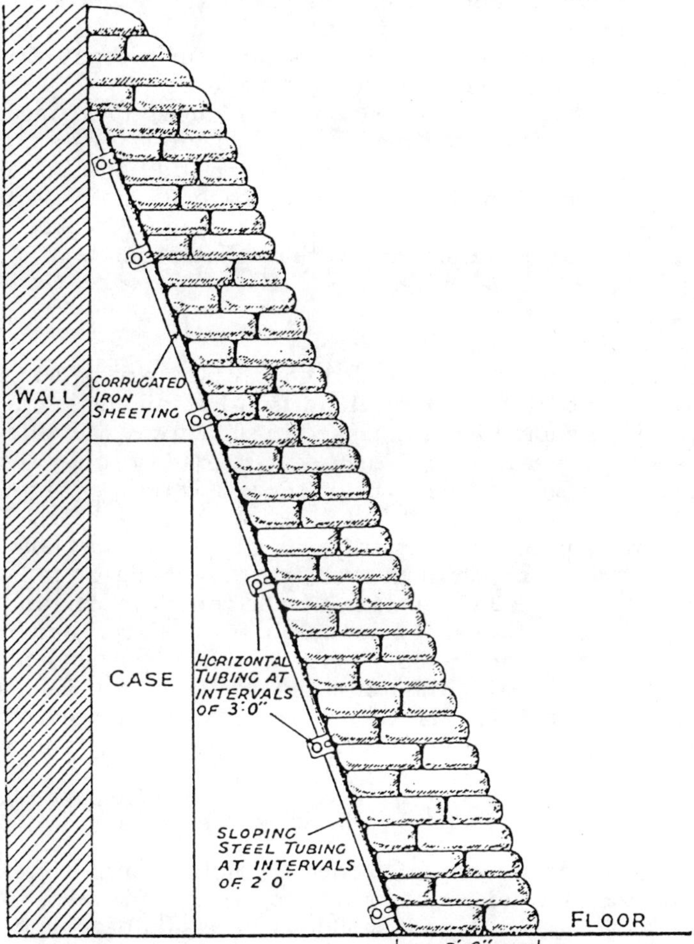

Note.—The supporting framework is made of corrugated iron sheeting laid on tubular steel scaffolding. With the measurements given here, the iron sheets (2 feet 3 inches wide) will overlap on the upright supports; but if the slope is less steep than this, it may be necessary to place the uprights at lesser intervals.

struts should be inclined against the walls over the tops of the cases, and sheets of corrugated iron laid against these to form a support for the sandbags. The protection should be carried well above the tops of cases (see diagram, FIGURE 2). A special danger to which Museum, Picture Gallery and Library material is exposed is damage from collapsed structural elements. The sandbagging should therefore be solid enough to bear the shock of falling roofs or girders, even in places to which bomb splinters could not penetrate. But care must be taken not to introduce a heavier weight of protective material than the floors can bear without strutting or other reinforcements in the rooms below.

The weight will be considerably reduced if foamed slag is used as the filling for the bags. This substance, which is a by-product from the blast furnace, has an additional advantage for this purpose in that it will not harbour moisture to the same extent as sand or earth. But its resistance to penetration by missiles is substantially less than that of sand.

Sealed Cases

The atmosphere inside sealed glass cases may be controlled to some extent by chemical units; but all unsealed cases and objects should be sandbagged in such a way that adequate and natural ventilation is secured. Humidity in an enclosed space can be controlled by the presence of certain salt hydrates, such as hexa- and hepta-hydrates of zinc sulphate. These salts are exposed in zinc trays as dry substances and they remain dry in appearance even when emitting or taking up water vapour, which they do to effect a condition of equilibrium representing (in the case of zinc sulphates) a relative humidity of 55 per cent. at ordinary room temperature. The scheme can only hope to be effective when an adequate supply of the salts is present, the actual quantity depending on the air-tightness of the case. Zinc sulphate is cheap and chemically stable.

Sculpture

Large objects such as statues or architectural pieces may be dismounted, laid in the angles of floors and walls, covered with asbestos wool and shielded with sandbags laid on a supporting framework. But it is undesirable to use wood in the supporting frames, in case of fire. In placing objects for protection against walls it should be borne in mind that outside walls are structurally the strongest and give the best shelter from blast and fragments entering the windows, to which interior walls are generally exposed.

Pictures

Large paintings may be similarly screened with sandbag walling, but the picture to be sandbagged should be protected with boards, after removal of the glass, and enough space should be left between the supports of the sandbags and the picture to allow access for inspection from time to time. A limited amount of protection can be given to pictures by hanging mattresses in front of them. A mattress covered with sheets of corrugated iron, or of steel and asbestos composition such as 'dura steel,' would be a good protection against minor shocks, incendiary bombs and liquid gas. Asbestos boards or curtains would protect for a short period against fire and liquid gas.

Experiments have so far shown that pictures suffer no damage from vapour gas in any of the concentrations likely to be encountered. It is practically impossible, however, to decontaminate pictures splashed with liquid gas, and where no protection against high explosives or fire, which would also protect against liquid gas, is provided, pictures should be placed in bags of oilskin (linen or silk), or 'cellophane,' and the bags sealed with 'aeroplane dope' (a cellulose adhesive). Bags of this type will also provide protection against damage by water.

Showcases and Bookcases

Glass-fronted cases will protect their contents against splashes of liquid gas, but the glass will not withstand the heat of an incendiary bomb nor the shock of an explosion. Paper pasted on both sides of the glass may prevent breakage to some extent, but it is advisable to shield all such objects, which are not protected by sandbags, with fireproof and waterproof covers. These are, for example, sheets of corrugated iron, asbestos and other mineral boards, plywood painted or impregnated with flame-resisting solutions, asbestos blankets and various protective cloths.*

If the angle which such case-coverings form with the floor is filled with sandbags, these will be an additional protection against the spread of fire or water.

Books and Manuscripts

Damage by fire may be lessened if bookcases are kept as free as possible from air-spaces. Books which cannot be removed to safe storage should therefore be packed closely on their shelves, and it may be possible to fill up the shelves in sheltered positions from those which are more exposed.

* For fireproofing paints and solutions, see above, p. 27, and for protective cloths, below, p. 40.

If empty spaces cannot be filled with books, asbestos or pumice-blocks, bricks, bags of foamed slag, or other dry and fireproof substances should be used.

The projecting galleries which are a common feature in Libraries are generally not strong enough to bear the weight of sandbag constructions, and it may be easier to transfer the books to cases on the floor of the room than to prop the galleries.

Books and manuscripts will quickly absorb splashes of liquid gas, and their decontamination, if at all practicable, could only be effected by expert laboratory treatment. Wrappings of transparent foil ('cellophane') or waxed paper would protect single books or packages against this danger, and also against accidental damage by water used in fire-fighting.

There is no reason to suppose that gas in vapour form will damage books or other Library material. Even coloured inks and watercolour pigments will not be likely to suffer if the paper on which they are applied is reasonably dry.

Protective Fabrics

Different types of protective fabrics show considerable variation in the resistance which they offer to penetration by liquid blister gases. Transparent foils of viscose type ('cellophane') are more efficient in this respect than oilskins, and the latter are more resistant than rubberised fabrics. But the susceptibility of the materials to damage by creasing or tearing must also be taken into account. Oilskin may be improvised by painting sailcloth or linen sheets with linseed or hempseed oil.

Waterproof and Fireproof Papers

The principal types of waterproof papers used for packing purposes are bitumen papers, wax papers, and Willesden papers. The strongest wrapping papers are waterproofed by coating with bitumen, a layer of scrim cloth being attached at the same time to afford added strength and to protect material from the possibility of being stained by the bitumen. Another kind of paper takes the form of two sheets rolled into one very strong sheet with a bitumen adhesive. In this case there is no possibility of staining. Such papers are quite waterproof and afford a certain protection against contamination of the contents of packages by liquid gases.

Paper can be rendered waterproof by impregnating with beeswax or paraffin wax dissolved in turpentine or by immersion in liquid wax. A convenient form of commercial paper is the waxed sulphite paper used for wrapping bread. While such papers are entirely waterproof, they are not so effective in excluding liquid gas because of the tendency of the wax film to fracture when the paper is folded.

Willesden paper is quite a strong paper rendered waterproof by treatment with cuprammonium hydroxide. Thick sheets can be obtained by sticking together thinner ones during manufacture and rolling. In an emergency any paper can be waterproofed by coating with oils, preferably of a drying nature like linseed oil, but these are not very permanent because the oxidation of the oil causes the paper in time to become brittle. Oiled papers are not suitable for packing, but might be useful as emergency screens.

Ordinary brown paper can be made fireproof by immersing in a solution composed as follows :—

Ammonium sulphate	8
Boric acid	3
Borax	2
Water	100 parts by weight.

STORAGE ON THE PREMISES

The most important movable material which cannot be evacuated will be assigned to the places of greatest security on the premises. These will generally be rooms which have a sheltered situation and are strongly built or capable of effective reinforcement.

Selection of Storage Rooms

The principles of security to be considered in the choice and reinforcement of shelters have been indicated above (pp. 29–31). The standards of resistance to blast and splinters of high-explosive bombs and the impact of incendiary bombs should be the same for the accommodation of staff and the storage of material. But storage shelters should have better protection against fire in the superstructure and adjacent rooms, and from danger of flooding or other damage from broken service pipes. Good ventilation and freedom from damp are necessary for the storage of perishable materials. On the other hand, alternative access is not necessary, and there is no need of proofing against gas in vapour form.

Perishable Materials

Perishable materials include paper, vellum and leather; paintings, textiles, wood and unfired clay; Ethnographical specimens other than those in stone, metal, or fired clay; Natural History specimens with the exception of most fossils and some minerals ; bone and ivory ; corroded metals, and terracotta or stones impregnated with hygroscopic salts.

Humidity, Temperature and Ventilation

In selecting the storage-rooms, therefore, particularly if a basement must be chosen, careful consideration should be given to the relative humidity of the atmosphere (i.e., the ratio of the amount of moisture actually present in any volume of air, to the amount needed for saturation at the same temperature, air being saturated when it is unable to take up any more water in the form of vapour).

Certain moulds absorb moisture particularly readily, and have been found to grow in a relative humidity as low as 68 per cent. ; in general, the relative humidity of the atmosphere should not be allowed to rise above 70 per cent., and authorities should aim at a relative humidity of 60 per cent. combined with a temperature of 60°F. Humidity may be reduced by raising temperature and increasing ventilation. Extra heating alone may be sufficient, and apart from possible damage to bindings, books will stand comparatively high temperatures. But vellum cannot be exposed to heating above 80°F. without danger of becoming brittle and discoloured. Extra heating without adequate ventilation, however, is seldom effective, and every effort should be made to ensure the free circulation of a current of air. Doors should be perforated or fitted with ventilators, and partition walls should, if necessary, be pierced.

Floors and Fittings

It is most undesirable to place perishable material on basement floors, and storage rooms should be fitted with shelves, or furnished with bookcases, preferably of metal. Space should be left between shelves and the walls behind them to ensure that air circulates freely. If shelves are not available, boxes or packages must be raised off the floor on grids or battens, so that air can circulate underneath them.

A general protective treatment against attacks by insects, bacteria or moulds can be given to woodwork used in storage with a 10 per cent. solution of zinc naphthenate or copper naphthenate in paraffin. The zinc solution is colourless, the copper is more efficacious but is green.

Library Material

Boxes of books or manuscripts, prints or drawings (other than solander boxes), must not be left closed. If the lids, or better, lids and bottoms, of the boxes are removed, and the boxes stacked on their sides, they will be effective substitutes for bookcases. This use should be borne in mind if boxes are specially made for storage (see below p. 48).

Pictures

The method of storage will vary according to the number of pictures to be stored and the space available. Where space is of no consideration, pictures can be left in their frames and hung on screens. Where space is limited, they should be removed from the frames in their backboards and placed side by side in wooden racks raised off the ground. The canvases must not be allowed to touch one another, but should be separated by sheets of 'cellophane' or pads (see below, pp. 51–52). Miniatures and water-colours should be stored away from light.

Inspections of all pictures in storage should be carried out at regular intervals. Each should have its name or other means of identification clearly marked on the end of its stretcher, so that it may easily be identified and removed from the rack when required. Photographic records will be found useful for checking any deterioration.

Furniture

Wood, however old or well seasoned, is liable to warp in changes of temperature and humidity. Veneered and marquetry furniture (especially 'Boulle' work) needs particular care in this respect. In storing such material special attention should therefore be given to the control of atmospheric conditions in the storage rooms. The presence of textiles in the same room will prevent to some extent acute fluctuations of humidity.

Spirit Collections

The protection of Natural History specimens preserved in spirit presents a special problem on account of the inflammability of the spirit and the fragility of the glass containers which are normally used. If the building in which such collections are contained is considered to be reasonably proof against incendiary bombs, and the collections themselves are in comparatively small self-contained sections (such as small rooms with no direct inter-communication, or better still, closed cupboards), the likelihood of the spirit giving rise to a serious fire is not very great. In many cases, however, spirit collections will be kept in rooms to which there is an ample supply of air, and where also there may be other objects such as perhaps stuffed animals which, with the spirit, would be very dangerously inflammable.

The specimens themselves represent but a small proportion of the weight and bulk of a spirit collection, about 90 per cent. of the space occupied being taken up by bottles and spirit.

Elimination of bottles in an emergency might be possible by either of the following methods :—

(*a*) By individual tagging of the specimens, each to have a tag bearing an embossed register-number, details being entered in a register. With delicate specimens some protective wrapping would be needed. A press-operated numbering machine would be necessary to emboss the tags. A fibre waterproof paper can be obtained suitable for the tags.* A number of wide-mouthed metal containers would be required in suitable sizes, but averaging probably about 2 ft. cube. The ideal arrangement would be the provision of one container of suitable size for the contents of each cupboard or room. When the emergency arose the contents of all the bottles in each cupboard could be simply poured into the container, the surplus alcohol filtered off, and the tank sealed for removal or perhaps burial. If removed they could be stored in a very small space, and the only attention they would require for the duration of hostilities would be a periodical inspection of the containers and the spirit-level in them. By the use of non-corrosive metal tanks the choice of suitable storage places is enlarged ; they could be housed in any cellar or cave or, as suggested above, buried. Registers could be stored in the same way as Library material.

(*b*) Where individual tagging is not possible, or is particularly difficult, an alternative is the provision of paper labels bearing the necessary data, and a supply of muslin bags of sizes corresponding with the bottles. Other equipment would be as outlined above. The contents of each bottle, including the label, would be poured into a bag of suitable size and the resultant package stowed in the metal container.

Of these two schemes (*b*) is probably more generally applicable, would be cheaper and would require less advance preparation. But (*a*) has one considerable advantage. The individual tagging of the specimens would be a valuable piece of work in itself as an improvement on the more usual external labelling of bottles.

Methods of Removal

Since the work of removal to storage will have to be done in conditions of urgency and probably with depleted staffs, it will be advisable to devise beforehand means of handling material in bulk with the greatest possible facility and speed.

Where there is direct access from the exhibition galleries

* 'Stock No. 25 K 14938,' made by the Dennison Company.

to the storage rooms, barrows or trolleys of suitable size, that is to say, not too large for use in the lifts or passages, should be provided. If the access is by staircases on which ramps cannot be improvised, hand-carriage will be necessary, and trays or baskets should be available for this purpose. These ought not to be too large for one person to manage easily. A tray measuring 2 feet by 1½ feet packed 8 or 9 inches deep with books weighs about 50 pounds.

It will facilitate handling and avoid confusion in replacement if books and manuscripts are tied in packages with webbing bands. Packages or separate books can be conveniently carried in square sailcloth bags, or, with careful handling, in ordinary sacks. Prints and drawings should be placed in their mounts in solander boxes, and left so in the storage room (see below, p. 53).

EVACUATION

Evacuation will only be advisable if the Repository chosen is definitely less exposed to air attack than the Museum, Picture Gallery or Library itself, and if adequate provision is made for the security of the evacuated material from damage by fire and damp, and from loss by accident in transit or by theft. These considerations are particularly important in view of the fact that the objects chosen for evacuation will normally be the most valuable in the collections.

Since even slight vibration will be harmful to pictures of certain classes, and may cause serious damage to any painted surface in delicate condition, pictures will often be exposed to greater risk by being moved to safer storage than by being left in place and there provided with whatever protection may be possible. For certain classes of material, such as wall-paintings or large altar pieces, such protection may be the only possible safeguard (see above, p. 39).

Selection and Preparation of Repositories

The Repository should be at a safe distance (not less than two miles) from any aerodrome, factory, railway junction or other place likely to constitute a target for air attack. Notwithstanding the safe overhead cover offered by deep excavations, caves or disused railway tunnels, these places ought not to be used for books, pictures, or works of art in any way alterable by damp, though they may serve to shelter certain classes of ceramics, sculpture and the like. They will, however, usually involve grave difficulties of invigilation.

The most desirable Repositories are country houses or

public buildings situated in localities remote from danger of air attack. But any arrangements made must be considered in relation to the Government scheme for the evacuation of the civilian population ; and in every case the specific permission of the Ministry of Health, or of the Department of Health, or of the Department of Health of Scotland, as the case may be, must be obtained before any particular building is selected.

If the premises chosen have not previously been used for these purposes, special consideration must be given to the following points :—

(1) Control of access, security of doors and windows, and invigilation. The latter service will probably necessitate the transfer of certain members of the staff to residential duty at the Repository.

(2) Proper ventilation and freedom from damp of all rooms which will be used for storage of perishable material. There should be periodical tests for damp and variations of temperature.* Old timber floors should be examined for dry-rot.

(3) Fire-fighting arrangements. The existing facilities must be reviewed and, where necessary, supplemented to meet the new risks involved.

Transport and Packing

It is probable that transport would be severely restricted immediately before and on the outbreak of hostilities. Museum, Picture Gallery and Library authorities should therefore estimate their requirements in advance and make provisional arrangements with railway companies or haulage contractors for the necessary railway accommodation or road vehicles. In the case of road vehicles the arrangements must be agreed by the appropriate Chairman of Traffic Commissioners. Local conditions and advice must also determine the choice of railway or road transport. It should be understood that in any case transport facilities cannot be guaranteed in the early days of hostilities, and that as soon as a crisis arises movement would be difficult and hazardous owing to the demands on transport for the evacuation of children and women from the larger towns, for mobilisation and other military purposes, etc. Wherever practicable, therefore, it is advisable that particularly valuable Museum, Gallery or Library material should be removed to places of safety before conditions of emergency exist.

For short distances (up to, say, 25 miles) motor vehicles

* For the principles of safe storage, see above, p. 42.

will probably be found to be the most suitable form of transport, but in view of the special danger to pictures from engine vibration, it is advisable to use the railways for these when greater distances have to be covered.

Railway Containers

Where considerable quantities of material have to be moved, the door-to-door service provided by the railways may be advantageous. Van-bodies ('containers') are sent on lorries by the railway to the Museum, Picture Gallery or Library, and packed there with the material; they are carried by rail on special trucks, and delivered by lorry at the Repository. Two types of covered railway container are particularly suitable for the kinds of material which would be evacuated.

The 'A' type Covered Container is a wood or iron box measuring about 7 feet long by 6 feet 6 inches wide by 6 feet 6 inches high, with a weight capacity of 2½ or 3 tons. A full load of books or manuscripts in boxes would weigh about 2½ tons. Two of these can be carried on a railway truck, and if they are placed with the opening ends together, it is impossible to get into them during transit.

The 'B' type Covered Container measures about 14 feet long by 6 feet 6 inches wide by 6 feet 6 inches high, and has a weight capacity of 4 tons. In some patterns the sides are fitted with strong vertical battens and horizontal slats, to which objects can be tied. These are more suitable for the carriage of Pictures and Museum material not separately packed. Covered containers are designed to be weatherproof, but cracks may occur in course of usage, and it is advisable to have the containers sheeted when valuable material is carried in them.

Road Vehicles

Road vehicles must be equally weatherproof, and must always be accompanied by members of the staff as guards. Both guards and drivers should be provided with rations and other facilities for spending days or nights on the road in case of breakdown or delay. Whatever kinds of vehicles may be preferred, it is important to examine them beforehand, and to make sure that they will be available when required.

Packing-cases

Rapid removal is an essential part of the evacuation process, and since it may be impossible to obtain enough packing-cases at short notice or in the stress of war, it is advisable to hold a stock in readiness. Ordinary packing-cases are bulky and inflammable, and involve difficulties of storage for both reasons.* Collapsible packing-cases can be obtained, made either of plywood with iron bindings or of cardboard,

Fig. 3.—Packing-case with Details of Construction

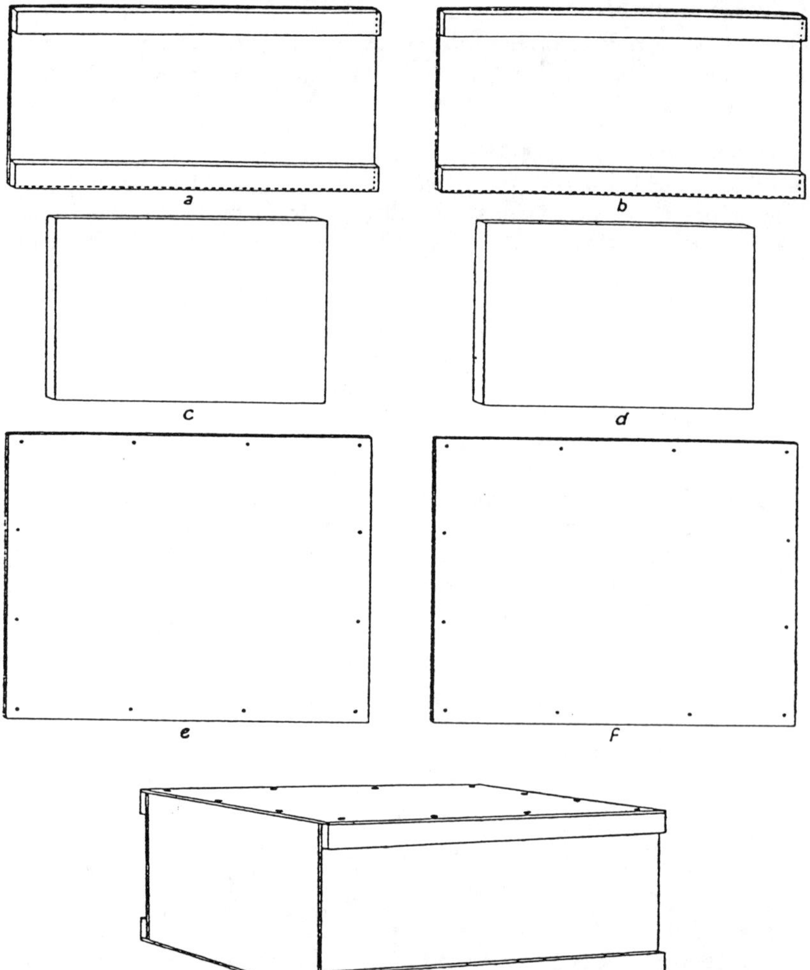

Details: *a*, *b*, Plywood sides with battens at top and bottom. *c*, *d*, Solid ends (or plywood edged with battens). *e*, *f*, Plywood lid and bottom, to be attached with screws if the box is intended for storing Library material or for repeated use.

and these are easily stored. Cardboard cases may be found safe enough to hold even valuable material if they are expertly

* For preservation of wooden boxes, see above, p. 42, and for fire-proofing, p. 27.

packed and unpacked by members of the staff in the door-to-door service. But ready-made wooden cases can seldom be bought in small quantities or in large or odd sizes, and it may be desirable to have these made locally. A design for a light plywood case with solid ends, which can be stored flat and put together when needed for use, is shown in the accompanying diagram (FIGURE 3.)

Cases which are not designed to fit particular objects should be made in standard sizes, and should not be too heavy when filled for two men to carry easily, or for one man to handle in an emergency. A convenient size for Library material measures 25 inches long by 19 inches wide by 12 inches high inside. This will take prints and drawings in solander boxes of royal size (24 inches by 18 inches) with the necessary waterproofing, and when filled with books or manuscripts weighs about 80 lbs. The weight of books varies considerably with the quality of the paper or vellum and the nature of the binding.

Packing of Museum and Library Material

Museum material is so various in kind that no particular rules for packing it can be given. Delicate objects must be first wrapped in cotton-wool, or if there is danger of the cotton-wool adhering to the object, in tissue-paper inside cotton-wool. The separate packages are placed in coarse packing material in the packing-cases. Resilient substances such as straw, kapok, or crumpled paper should be used for the coarser packing. Series of small objects are best packed together in cardboard boxes, of which several will go into one packing-case.

If the objects to be carried are susceptible to damp, the cases should be lined with waterproof paper, so as to withstand accidental wetting by rain or otherwise in transit. Cases containing books, manuscripts, prints or drawings should always be thus lined. Two strips of waterproof paper cut slightly wider than the length and width of the box, so as to overlap in the corners, and long enough to fold back under the lid, will serve this purpose most efficiently. The books themselves, which will always be valuable examples and may have fine bindings, should have a thick inner wrapping of tissue-paper and an outer wrapping of stouter paper or corrugated card. The boxes should be packed solid to prevent movement, spaces being filled with any packing material which is free from damp and grit.

Lists of the articles contained in each box should be made and duplicated, one copy being retained in the records of the Museum or Library, the other forwarded to the Repository for reference in unpacking there. Numbers and other identification marks should be stencilled on the boxes, and the boxes may be bound with riveted steel tape or sealed for despatch.

All Library material, and any Museum material which may be damaged by damp or moulds, must be unpacked at the Repository and stored on shelves. If these are not available, the tops and bottoms of the boxes may be removed and the boxes piled on their sides to serve as open shelves.

Pictures

Evacuation of pictures should be avoided in cases when they are likely to suffer damage in transit. Paintings on panel, for example, are both heavier and much more liable to damage from flaking, cracking or warping than paintings on canvas. Pastels should not be moved about more than is absolutely necessary, because of the risk of flaking. Water-colours are particularly liable to deterioration from damp or exposure.

Removal from Frames

When a large number of pictures has to be evacuated speed becomes of primary importance. Frames may be adapted, by the method shown in the accompanying diagram (FIGURE 4), so that the pictures may be quickly removed from the front, leaving the frames on the wall. These can be taken down at leisure after the pictures have been despatched to a place of safety. The adaptation need not apply to small pictures, which can be moved quite easily in their frames, with paper pasted over the glass to minimise dangers of breakage.

To secure blisters and other weaknesses in the paint of a picture, thin paper can be pasted on the surface with flour-and-water paste, which can later be removed by warm water and cotton-wool.

Canvases should not be rolled, unless every other means of removal has been found impossible. The diameter of the drum should then be at least 3 feet, and the picture should be rolled with the painted surface outwards.

Packing of Pictures

Large numbers of pictures are best packed in railway containers, so that they travel from door to door without disturbance. The floor of the container should be covered with sponge rubber matting $\frac{1}{2}$ inch in thickness, or with mattresses. The pictures are then loaded vertically, in the direction of travel. In order to avoid abrasion, they must be kept apart. This may be done by means of pads of sponge rubber of dimensions 3 inches by 3 inches by ½ inch, joined by

Fig. 4.—Picture-frame Fitted for Rapid Removal of Canvas.

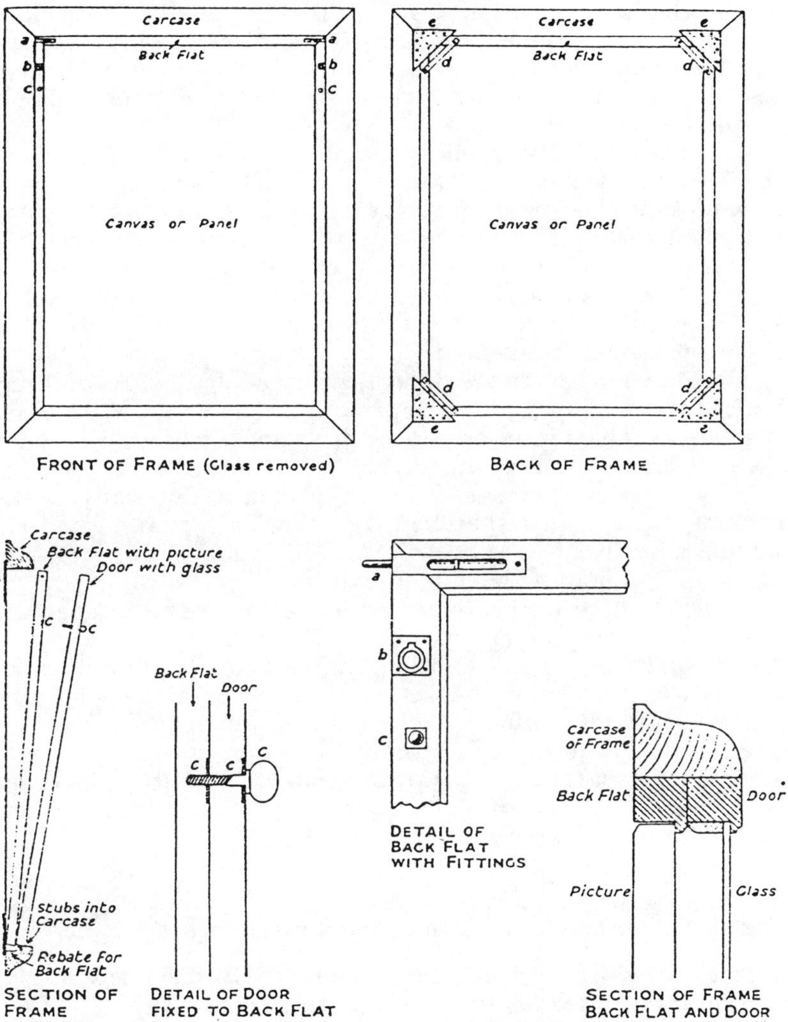

Details of Fittings : *a*, Flush bolt fastening back flat to carcase of frame. *b*, Lifting ring for removal of back flat. *c*, Thumbscrew with guide-plate and socket, fastening door to back flat. *d*, Metal strips screwed to back flat to hold picture. *e*, Metal plate screwed to carcase to hold back flat.

adhesive tape and hung loosely over the corners of pictures. Rubber contains a certain amount of sulphur and should therefore not come into direct contact with the painted surface. If pictures are packed in their back flats, this will not occur.

Between every three or four pictures a light wooden framework partition should be placed and attached by rails at each end to the next partition. This will assist speed of unpacking and prevent pressure. Between the ends of the pictures and the back and front of the container sacks stuffed with some resilient material, such as straw or crumpled paper, should be placed to absorb shocks.

Another method of packing paintings on canvas is to attach corks of about ½ inch in thickness to each corner of the front of a picture with panel pins (i.e., pins pointed at both ends), and then to place it face to face with another. Bundles can be made of, say, six canvases protected with two sheets of plywood, and covered in waterproof paper. This method should only be used where there is a number of pictures of the same size to be packed.

The above methods are dependent upon intelligent handling, both in the packing and the unpacking of pictures. Where this is not available, it is advisable to pack pictures in wooden cases or crates suitably lined.

Very large pictures should be placed in specially made packing cases and despatched by open lorry or railway well-wagon. Twelve feet is the maximum height permissible, and cases up to about 16 feet long can be taken by rail.

At least three check lists should be made of the contents of each container as it is packed. One should be retained at the despatching end, another forwarded to the place of storage, and a third packed into the container with the pictures. Where sufficient clerical assistance is available, it is advisable that an index should be compiled, each picture having its separate card, upon which its condition, both at time of packing and on subsequent inspection, and any other information may be entered.

Tempera, Water-colours and Miniatures

Tempera paintings on plaster should be packed separately, each in its own frame and in a separate crate.

Water-colours should, if possible, be kept permanently in standard-size mounts and frames. They can then be easily unframed and transferred to solander boxes for transport. To protect the surface from scratches, a sheet of transparent foil ('cellophane') should be placed over the water-colour, and held in position by the outer part of the hinged mount. To minimise the danger of moulds, a sheet of thymolised paper* should be packed in each solander box. A contents list should be placed in each box, and two others kept for reference (see above). If the boxes cannot be kept in locked cupboards or a safe, each should at least be fastened with a seal.

Very large water-colours are safest if left in their frames, the glass being protected by pasting paper over the surface. Where there is a number of them, those of approximately the same size may be removed from their frames and packed together as above or in bundles wrapped in waterproof paper.

Miniatures should be packed in wooden or metal boxes fitted with flat baize or felt-lined drawers. If these are not available, they may be packed in trays in strong boxes, with thick sheets of cotton-wool between each layer. The boxes should be fitted with locks, and lists of the contents should be kept for reference.

---

* Thymolised paper is prepared by crushing some thymol crystals and ironing them into sheets of tissue, blotting or other paper, which when placed in a solander box from which there will not be particularly rapid evaporation, should retain their disinfectant property for about six months.

APPENDICES

APPENDIX A

(See page 10.)

Equipment for Staff Shelters

1. A list of those who are to use the shelter, so that a roll-call can be made.
2. Tables and chairs or benches.
3. Blankets or other warm coverings.
4. Water in covered receptacles for drinking, washing, etc.
5. Food in air-tight tins or jars.
6. Plates, cups, knives, forks, etc.
7. Means of occupying the time without exertion (books, playing-cards, etc.).
8. Emergency lighting arrangements (independent electric lighting or candles and matches or electric torches or hurricane lamps and matches).
9. Basins for washing, soap and towels.
10. Chemical closets or latrine-pails, toilet paper, and screens.
11. First aid outfit.
12. Bucket or box of sand with a shovel or a fire extinguisher.
13. Means of rendering shelter gasproof, if gasproofing should be needed.
14. Pickaxe and shovel for clearing debris.

APPENDIX B

(See page 14)

Equipment for Fire Parties

Fire Watcher Parties

Composed of 3 persons

Equipment required :

1 stirrup hand-pump.
2 (or more) 2½ to 3 gallon buckets.
1 axe (preferably with insulated handle).
1 hand lamp, in case of failure of electric mains.

Local Fire Parties

Composed of 7 persons (Full Party).
5 persons (Reduced Party).

Equipment required :

2-Man manual.
2 60 ft. lengths of 1¾ in. unlined canvas hose with 2½ in. instantaneous couplings.
2 2½ in. branches with ¼ in. nozzles.
1 2½ in. dividing breeching.
1 canvas dam (not required with container type of pump).
20 canvas buckets in packets of 10.
1 hand or felling axe (insulated handle preferred).
1 hand-lamp.
80 ft. lowering line.

Group Fire Parties

Composed of 6 to 8 men.

Equipment required :

1 light trailer pump.
4 100 ft. lengths of 2¾ in. hose.
2 75 ft. lengths of 1¾ in. hose.
2 8 or 10 ft. lengths of suction hose.
1 dividing breeching.
2 branches with ⅝ in. nozzle.

Spare nozzles, one ¾ in. and one ½ in.
1 canvas dam.
20 canvas buckets.
1 preventer.
1 felling axe.
1 copper suction strainer.
1 basket suction strainer.
1 two-way suction collecting head.
Lines and small gear.

NOTE.—(1) All couplings both for 1¾ in. and for 2¾ in. hose should be 2½ in. instantaneous.

(2) If the trailer is towed behind a vehicle, all gear may be carried therein, but if not towed a suitable hand-cart is necessary.

APPENDIX C

(See page 18)

Equipment for Salvage Squads

The following lists of rescue equipment are put forward for guidance. The 'light' set has been recommended to Local Authorities, and many large institutions will already be in possession of the equipment here indicated. The list entitled 'extra light' has been devised for smaller establishments.

	'Light' set (for a party of 6 men)	'Extra light' set (for a party of 6 men)
Iron shod levers, 10 ft. long	2	2
Heavy block for fulcrums	2	1
Crowbars	3	2
Picks	3	2
Shovels	3	2
½ round files	2	2
Sledge hammer	1	1
Heavy axes	2	1
Light (timbermen's axes)	2	1
Two-handled cross-cut saws	1	—
Hand saws	2	1
Wheelbarrows (iron)	1	—
40 ft. lengths 1½ in. manilla lashing lines	3	2
Length (up to 100 ft.) 3 in. manilla rope	1	1
Chain tackle (or set of rope blocks, 3 sheave and 2 sheave)	1	1
Single sheave snatch block	1	1
Hydraulic jacks with 10 or 15 ton lift	2	—
35 ft. ladder (extending)	1*	1*
9 in. by 3 in. deals (about 12 ft. long)	3	—
Oil flood-light projectors or acetylene flares	1	1
Hand electric lamps	4	2
Hurricane lamps	4	2
Tarpaulins or stout canvas sheets (to protect trapped persons from falling debris until released)	2	—
'Fire devil' (or fire basket) for warming trapped persons in winter	1	—
Box of miscellaneous tools, spikes, timber dogs, etc.	1	1

* A smaller ladder may suffice, or sufficient scaling ladders in 6 ft. sections, which can also be used as bridges.

APPENDIX D

(See page 18.)

Equipment for Decontamination Squads

Each member of the squad will require two ' Civilian Duty ' respirators (one in use and one in reserve) and two suits of full protective clothing, consisting of oilskin jacket, oilskin trousers, oilskin hood, oilskin gloves, and rubber gumboots. As the gloves soon wear out, four or six additional pairs should be kept in reserve, and in addition canvas mittens should be provided to protect the gloves when worn. Each man will also require two suits of underclothing in addition to those he is already wearing, to be worn under the protective clothing.

The apparatus required in a small institution is as follows :—

2 buckets for water.
2 buckets for mixing bleach paste.
2 whitewash brushes for applying bleach paste.
2 long-handled scrubbers or hard brushes.
1 shallow tray (about 30 in. by 18 in. by 3 in. deep) for bleach for the treatment of boots or tools contaminated during work.
2 bins (*e.g.*, galvanised iron sanitary bins) for contaminated clothing.
1 bin for contaminated tools.
Supply of bleaching powder and of paraffin oil for swabbing metal surfaces.
1 vehicle to carry equipment.
Suitable lengths of hose (as pressure supplies of water are not necessary for decontamination work, the hose used need not be in perfect condition. Old canvas hose, no longer fit for use in fire fighting, would be quite suitable).
1 standpipe and adaptor (if necessary).
2 spades.
2 picks.
Rope and pickets for enclosing contaminated areas.

APPENDIX E

(See page 20)

Equipment for First Aid Parties

First Aid Haversack: Suggested Contents

(One haversack per party)

Bandages, triangular		36
Canes for tightening improvised tourniquets (6 in. long)		8
Clasp knife, large		1
Cotton wool, 1 oz. pkts.		6
Dressings, first aid, large		18
,, ,, ,, medium		12
Lint, unmedicated, squares about 8 in. by 12 in.		6
Ointment, bleach (and clean, washed rags), 2 oz. tins		4
Safety pins, large, cards of	cards	8
Tourniquets, St. John type		2
,, Samway		2
Splints, sectional, wooden	sets	2
Straps, webbing, with metal fasteners for securing splints, length about 18 in., width about 2 in.		18
Tannic acid jelly, ½ or ¾ oz. tubes		4
Torch or lamp, with stand and shield		1

First Aid Pouch: Suggested Contents

(One pouch per man)

Bandages, triangular		9
Dressings, first aid, large		6
,, ,, ,, medium		6
Canes for tightening improvised tourniquets (6 in. long)		3
Labels, casualty identity. Book of 15 (with indelible pencil)		1
Ointment, bleach (and clean, washed rags), 2 oz. tin		1
Safety pins, large, cards of	cards	3
Scissors, 7 in., with lanyard (at least one blade pointed)		1
Tannic acid jelly, ¾ oz. tubes		2
Torch, electric, hand		1
Tourniquets, St. John type		1

(Each member of a First Aid Party should carry a water-bottle.)

[The above lists replace those given in *A.R.P. Handbook No.* 6, Appendix J.]

THE NATIONAL GALLERY AT WAR

DAVID SAUNDERS

The National Gallery, Trafalgar Square, London, WC2N 5DN
United Kingdom

With the outbreak of the First World War in 1914, Britain found herself involved in a European war for the first time in nearly a century. The National Gallery had thus been immune to the possibility of attack by hostile forces, although it is interesting to note that when war was declared, the Gallery had only recently reopened to the public following an attack on a Bellini by a suffragette.

There were two concerns for the Gallery, aerial attack, from Zeppelin airships, and a German invasion. Advice was taken from the Office of Works on methods of protecting the Collection from bombing while a search for safe houses and museums outside London appears to have been motivated by a fear of invasion, at least at the beginning of the war. It is clear from the discussions which took place throughout this period, that no plans had been made prior to the outbreak of war; each offer of alternative accommodations for the paintings was considered separately as was each recommendation for strengthening the building against attack.

In September 1914 letters were received by the Director from the Chief Inspector of Explosives at the Home Office and from the Office of Works. Both recommended storing the most valuable paintings in the cellars and placing between two and four feet of sand upon the floors above the cellar. This advice was later modified when it was discovered that the weight of four feet of sand would cause the ceiling to collapse. By the end of September, 50 of the most important paintings were housed on racks in cellars with a sand-covering. The racks were installed to prevent damage in the event of a flood.

In order to catch falling bombs before they could damage the Gallery, the installation of torpedo nets was considered. In the event, the cost seems to have been too great and they were not erected.

The fire arrangements in the Gallery were reviewed and additional buckets of sand provided. Other types of extinguishers were not considered suitable. The staff were given additional fire training. Sand bags were placed before the windows. By the end of October, 217 paintings were in store in the cellars and the new fire arrangements had been made. In addition, a list of other paintings it was considered should be removed from exhibition had been drawn up.

Meanwhile, in early October 1914, the Trustees began to consider removing paintings from the building to suitable places of safety away from London. The Director visited a number of provincial galleries during the latter part of that year. In June 1915, despite protects from the government, a number of the important paintings were sent "on loan" to provincial museums, or to be stored in the cellars of country houses, while the remainder stayed in the secure cellar rooms at Trafalgar Square.

Throughout 1915 and 1916, the fabric of the Gallery was strengthened as the size of the bombs increased. The Gallery was an obvious target from the air and the roof was hit by shrapnel twice during 1916. On the second occasion glass was broken and some masonry damaged. By the end of the year, however, the size of the bombs being used was such that it was reported that no measure would be sufficient to prevent damage to the Collection.

Fortunately, by this time the Gallery has another, deeper, storage area in central London. In July 1916, an unfinished underground railway station (now Aldwych) was put at the disposal of The National Gallery by the London Electric Railway Company. The space was rapidly converted to a painting store for small and medium sized paintings. Heating and security were provided to allow the temperature to be maintained at 60 to 65°F (15 to 18°C), while additional fans and heaters could be used to reduce the problem of damp. Fortunately, the tunnel was above the level of the River Thames. Many of the most important items in the collection were moved to Aldwych Station in January 1917.

In all, 675 paintings were removed from the Gallery during the war, to Aldwych, provincial galleries, and private homes.

In the years between the first and second world wars, the Gallery laid plans for the evacuation of the paintings to safety in the event of future conflict. It was clear that more efficient aircraft and bombs necessitated the removal of the Collection from London. In consultation with other national museums and galleries, a list of country houses to which paintings could be sent was drawn up. The Gallery had been advised to send the Collection to the northwest and accordingly sent Martin Davies, later to become the Director of the Gallery, to North Wales to survey likely repositories. His notes made during this tour suggest that the selection was as much influenced by the character of the owner or custodian of each potential store as the nature of the space available.

Back in London, a series of practices were conducted to familiarize staff with the removal of the paintings. In one such exercise, Van Dyck's large equestrian portrait of Charles I was taken from display, packed and loaded onto a vehicle. After a journey of a few hundred yards toward Leicester Square, the process was reversed. In another practice, a number of small paintings were sent by rail to North Wales and back.

By the time war was declared on the 3rd of September 1939, the entire collection had left the Gallery. Over the previous three weeks the well-laid plan, rehearsed since the Munich crisis, had been put into operation. The co-ordination of this procedure owed a great deal to Ian Rawlins, who in addition to being the Gallery's first scientific advisor, was a keen railway enthusiast.

By means of the London Midland Scottish and Great Western railways, the Collection was evacuated via Bangor and Aberystwyth to a number of locations.

From Bangor, paintings were taken for storage in the Pritchard Jones Hall at the University College of North Wales (Figure 1). The size of the building made the maintenance of a stable climate difficult, but the relative humidity was kept above 40%, even with winter heating. The other stores in Bangor were the dining room and garages at Penrhyn Castle (Figures 2). Here, there were initially great problems in preventing large fluctuations in relative humidity and biological activity in the stagnant air. These effects were overcome by improvising better seals of doors and windows and by installing some office-type fans to increase air circulation. At Aberystwyth, a large number of paintings were stored at the National Library of Wales in rather good conditions.

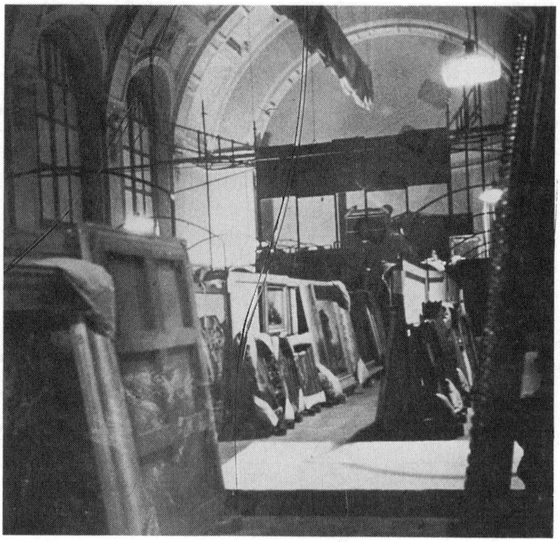

Figure 1: Pritchard Jones Hall at the University College of North Wales

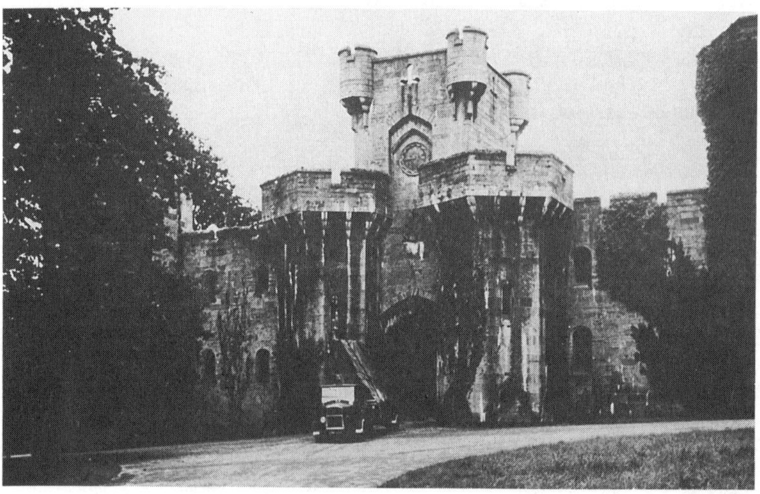

Figure 2: Penrhyn Castle

In mid-1940, the fear of invasion was such that it was decided further to disperse the Collection to more remote locations, including Caernarvon Castle (Figure 3). The conditions at Caernarvon and at another of the stores, Plas-yr-Bryn, were satisfactory. Unfortunately, the antiquated heating systems in the library at Crosswood near Aberystwyth produced a very low relative humidity in winter, less than 30%. The solution was to soak old blankets and felts in a nearby stream and hang them in the library to increase the moisture content of the air.

The decision to evacuate the paintings proved well founded. During 1940, the Gallery suffered severe bomb damage. The room in which the Raphaels had been displayed was completely destroyed, while the roof glazing of many of the other rooms was shattered.

By the summer of 1940, the number of enemy bombers passing over North Wales on their way to and from the industrial centres of Liverpool and Manchester was a cause of concern. The principal threat was that these planes might discard their remaining bombs to reduce weight for the return flight.

In response to this, Martin Davies and Ian Rawlins began a search for suitable underground storage. After a tour of a number of caves, tunnels and slate quarries in north and mid Wales they discovered Manod Quarry near Ffestiniog. Manod offered several advantages over the other sites visited. First, a reasonable road ran to the quarry entrance. Second, there were electricity and water supplies already available to the quarry. Most important, the natural temperature of the interior was a fairly even 47°F (8°C) throughout the year. It would thus be possible to maintain a steady relative humidity be heating alone.

Over the next eleven months the Ministry of Works and the quarry's own labor force made the necessary modifications to convert the slate workings into a picture store.

The first stage was to enlarge the quarry entrance from six by six feet to thirteen by ten feet. Since this entrance tunnel was some 200 feet long, a considerable volume of rock had to be removed, totaling about 5,000 tons. Other galleries were widened and rubble cleared from the floor. To accommodate the paintings, six brick storage sheds were then constructed within the underground chambers (Figure 4). The storage areas were linked to each other

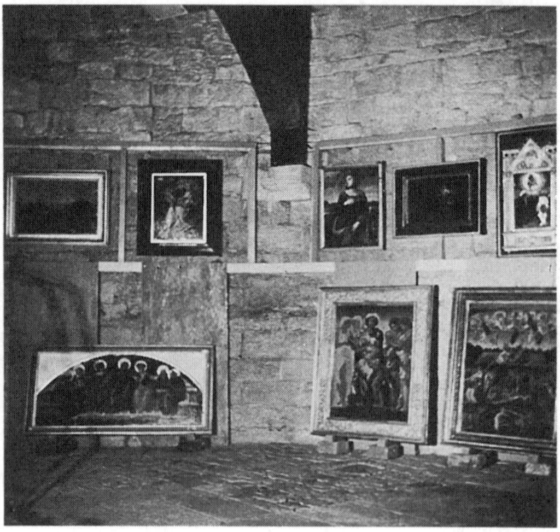

Figure 3: Caernaryon Castle

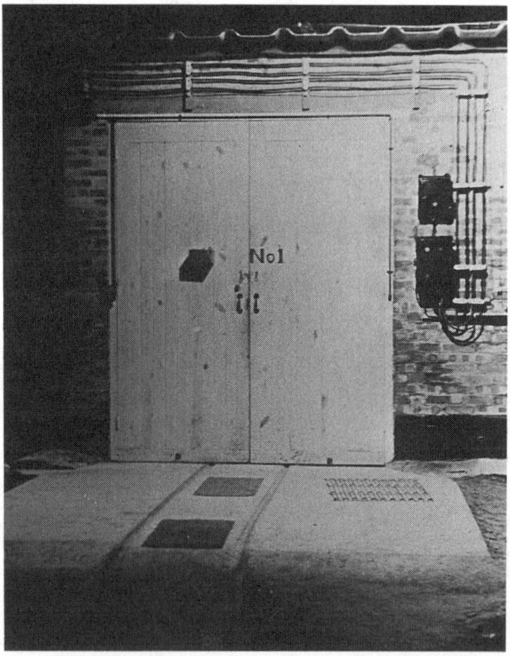

Figure 4: Storage sheds deep within the Manod Quarry

and to the entrance tunnel by a narrow-gauge railway system (Figure 5).

Each of the six buildings had an independent heating and ventilation system to ensure that the relative humidity was controlled and that the air was circulated to prevent local microclimates from developing. During the four years that the paintings spent in Manod the relative humidity was maintained at around 58% by controlled heating of the air to a temperature of 63°F (17°C). A contemporary set of publicity photographs show attendants checking the thermostat and the recording thermohygro-Graph (Figure 6). An emergency 140 horsepower diesel-electric generator was installed to ensure that conditions could be maintained in the event of a disruption of the electricity supply from the North Wales Power Company.

Once the modifications were complete, in the summer of 1941, the paintings were gathered from their scattered storage locations and reassembled at Manod. The transportation was again undertaken using vehicles supplied by the railway companies. The Great Western Railway conveyed paintings from Aberystwyth area while the London Midland Scottish transported the paintings from the University College of North Wales and Penrhyn Castle to Manod. At Manod, the cases were transferred to the railway system to be transported to the storage racks in the sheds (Figure 7). Once in store, the paintings had an uneventful sojourn for the rest of the war. A considerable number of Gallery staff worked at Manod. A daylit conservation studio was set up at the entrance to the quarry. The Gallery library had also been transferred from Trafalgar Square allowing Martin Davies to continue his research. Davies, Rawlins and two of the administrative staff lived nearby in Pengwern Old Hall.

In London, The National Gallery found new uses. Throughout the war lunch-time concerts were held in the rooms, inspired by the pianist Dame Myra Hess (Figure 8). A canteen, staffed by volunteers, was established to provide lunch for the audiences. The Gallery ran a series of exhibitions of contemporary artists. As the war progressed and the threat of bombing in London diminished, individual paintings were brought from Manod to London to feature as the "Picture of the Month". The first of these exhibitions was held in March 1942 when Titian's "Noli me tangere" was displayed. In 1945, when the end of the war seemed inevitable, a number of paintings returned to the Gallery in advance of the rest of the

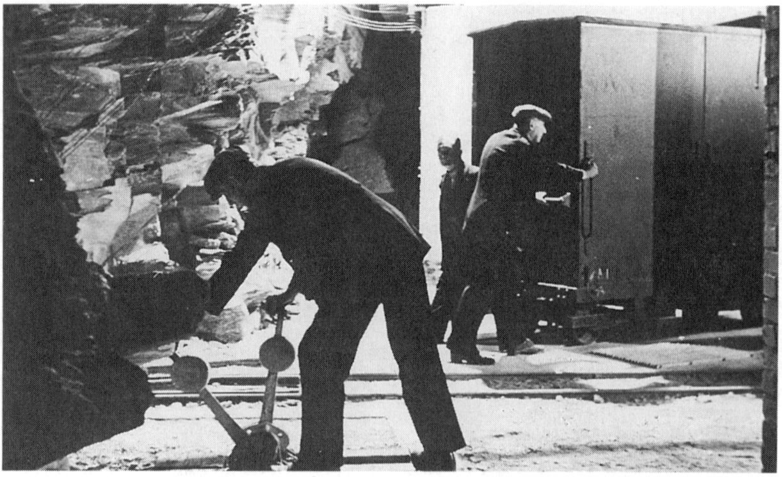

Figure 5: Section of 1,320 yards of Narrow-gauge Railway

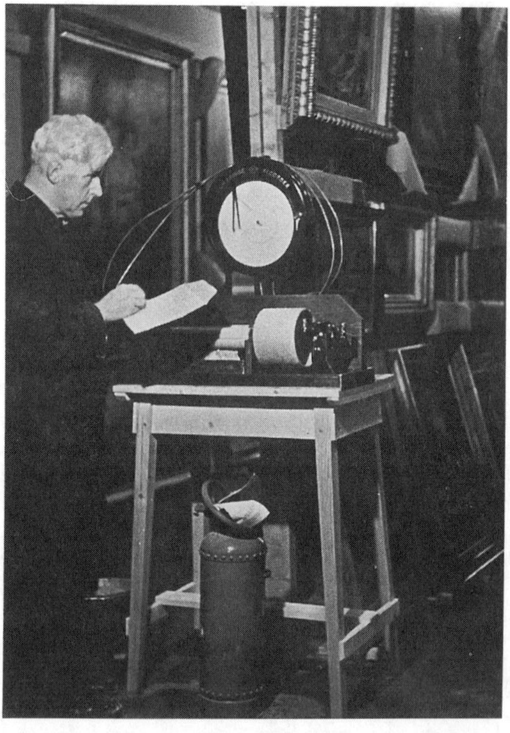

Figure 6: Checking Environmental Conditions

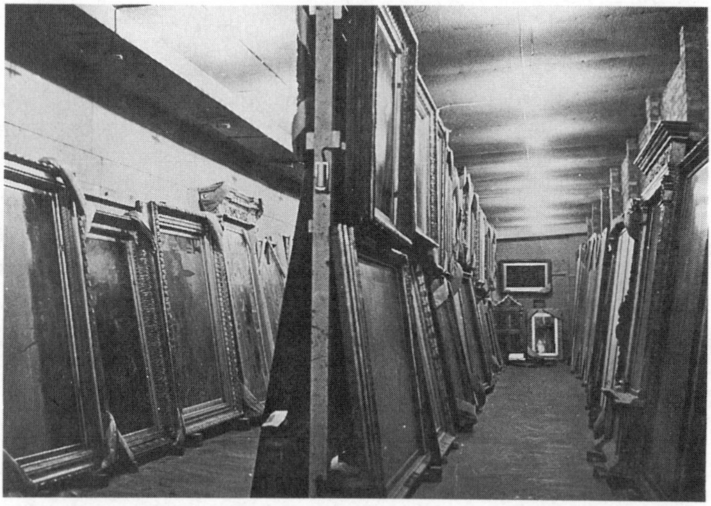

Figure 7: Storage Racks in Manod

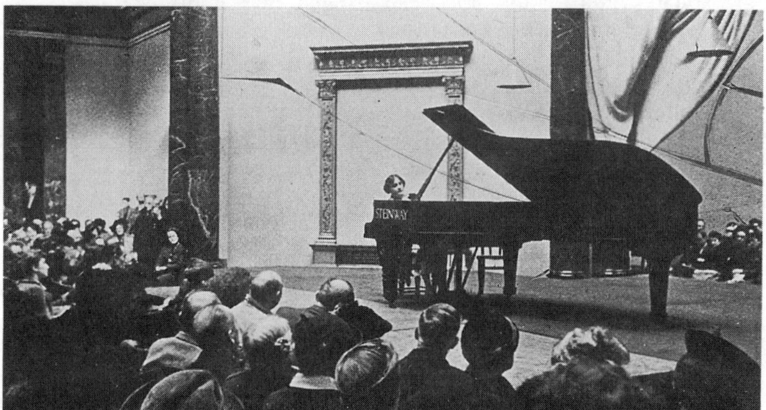

Figure 8: Luncheon Concert by Dame Myra Hess during the War

collection to be shown in an exhibition entitled "Forty-Three Selected Masterpieces."

By the end of 1945, the entire collection has returned to Trafalgar Square and much of the superficial damage to the rooms had been remedied. The work of reconstructing the rooms devastated by bombing was to take longer, since the experience of the stable climatic conditions at Manod combined with the experiments on air pollution made by Rawlins before the war led to a decision to install air conditioning in these rooms. In the words of Rawlins himself, January 1943:

> "A sizable body of data relating to temperature, relative humidity, condition and reaction of materials is being assembled, and may well take its place in contributing towards the post-war design of museums and galleries, and to the choice of environments considered best for works of art. It is possible that full air-conditioning of such institutions in large cities and certain other places might be found to be financially desirable, when sums spent annually in restoration and repair are critically re-viewed. Careful inspection will always be needed, but the experience so far of housing a collection of pictures below ground under controlled conditions, scientifically planned, is decidedly encouraging. It would be a pity if some of this could not be translated into terms appropriate to times of peace. Many great pictures are probably now going through severe hardships and many vicissitudes. Those for which the Trustees of the National Gallery are responsible, however, are presently enjoying a climate of such salubrity that the greatest problem for the future is to foresee how they will react when they leave it."

As a postscript, the Gallery maintained the lease on the storage spaces at Manod until the early 1980's.

PART I B

Contemporary Perspectives

STATUS REPORT: ASSESSMENT OF CULTURAL LOSSES SUFFERED BY AMERICAN ALEUTS DURING WORLD WAR II

BARBARA SWEETLAND SMITH
Aleutian/Pribilof Islands Association, 401 E. Fireweed Lane, Suite 201, Anchorage, AK 99503.

ABSTRACT

In 1988, the U. S. Congress provided $1.4 million to six Aleut communities in Alaska as restitution for damage done to their churches by U. S. forces stationed in the Aleutians during World War II. The residents of these communities had been forcibly evacuated and interned for up to three years in camps in southeast Alaska where many died. At the request of the Aleuts, restitution to the Orthodox churches was given a special place in the U. S. decision to make reparations. The Aleutian/Pribilof Islands Association, a non-profit organization, has been charged by Congress with assessing the damage and estimating the cost of restoration. The assessment project, begun in November 1991, has utilized the expertise of an architect and art conservator to examine structures and church furnishings of these churches and has reached some preliminary conclusions: 1) The stress of travel for those icons and liturgical vessels which accompanied the evacuees was significant and damage related to that travel is evident today; 2) The churches themselves were often targets of vandalism and looting by the American troops; 3) the churches were left without regular maintenance in the harsh climate of the Aleutians for three years; 4) the Aleuts were destitute on their return and could not restore the buildings to their former condition; 5) the results of makeshift repairs in the immediate post-war era are evident today in a general state of deterioriation. The final report to Congress on this project is due in the Spring 1993.

INTRODUCTION

In 1942, the United States was invaded by a foreign power for the first time since the War of 1812. During the so-called "Forgotten War" of World War II, Japan bombed the Aleutian Islands in Alaska and occupied two of them. In a panic reaction to the assault, the U. S. government decided to evacuate the 881 Aleut residents from the islands and to send them to makeshift camps in southeastern Alaska, 1500 miles from their homes. There the Aleuts were virtually abandoned, without sanitation or health facilities. Over the next three years, one-tenth of their population died--mostly the elders and the young. In the 1980s, the decisions leading to this evacuation and lengthy internment were reviewed by the President and

Congress and found to have been in error. A Presidential apology and money from Congress for reparations followed.

An important feature of the reparations--or restitution --settlement is a $1.4 million trust fund established at the request of the Aleuts for the restoration of churches and church property in the six affected communities. The Aleutian/Pribilof Islands Association, for which I work, is a non-profit organization which serves the education, health, and cultural needs of the Aleutian people. It has been charged by Congress with conducting an assessment of the damages to the churches and the cost of restoration. This assessment process began six months ago in November 1991.

It will be immediately apparent that this is a formidable task--for the events in question occurred almost 50 years ago to this day. Despite the obstacle of time, it is, nonetheless, possible to reach some conclusions about damage and loss, for churches are avid record-keepers. Inventories and yearly journals provide clues to the pre-war condition. And surviving elders contribute photographs and information about what they found when they returned and repairs which were necessary. It is, nonetheless, a daunting project.

HISTORIC BACKGROUND

Who are these Aleuts? And why have their churches figured so prominently in the reparations settlement?

The numerous islands of the Aleutian Archipelago extend in a graceful arc across the North Pacific, dividing it from the Bering Sea. Covering some 1100 miles the islands have been home for 8000 years to the Aleut people. In 1741, their islands were first exposed to European culture, as the exploratory vessels of Vitus Bering and Alexei Chirikov, serving the Russian Tsar, made call in search of water and sea mammal skins. Over the next 126 years, the Aleutian Islands became part of what was known as Russian America.

Being consummate maritime hunters, the Aleuts became the major labor force for the Russians. About forty years after Russia first encountered the rich waters of the Aleutians, they discovered two volcanic islands in the Bering Sea. These islands, called the Pribilofs, were the northern breeding grounds for the Pacific fur seal which congregated by the hundreds of thousands each year. The Aleuts had deemed the islands uninhabitable. But the Russians thought otherwise and removed able-bodied men and then their families from several islands to become permanent residents on the Pribilofs--on St. George and St. Paul islands.

Although Russian rule was harsh, there is one facet of the colonial era which has endured: Orthodox Christianity. Under enlightened leadership, the church became a Native institution, with Native clergy and toleration of many indigenous customs. The church also encouraged literacy and education and, most significant for the survival of culture, the use of Native language in its worship services. The advent of American rule in 1867 did not change the fundamental loyalty of the Aleuts to the Orthodox church. Thus one finds today the 3000 members of the Aleut community are almost entirely Orthodox. Within the

16 Aleut villages, which varied in population from a score to several hundred, the economy was oriented around the riches of the sea--fishing, whaling, and the fur seal harvests--and cultural life around the Orthodox church. Some communities enjoyed modest wealth, which was reflected in the furnishing of their churches, which were resplendent with icons and liturgical vessels from Russia. Life was governed by the sea and the church holidays--until World War II.

The Impact of War

During June and July 1942, war unleashed traumatic change for the Aleuts. On June 3, Japan bombed a U.S. Naval installation at Dutch Harbor, next door to Unalaska. Five days later the Japanese had occupied two of the western-most islands of the Aleutian chain--Kiska, which was uninhabited, and Attu, home to a small Aleut community. Forty-two Aleuts were taken prisoner by the Japanese and removed to Japan.

Until early June 1942, Atka Island had a community of 81 Aleuts which enjoyed a self-sufficient lifestyle from the sea. Their Orthodox church was the center of social and cultural life, along with the school. On June 12, an order was given by the U. S. military command to evacuate Atka as well as both of the Pribilof Islands. U. S. naval personnel razed the town of Atka, burning the homes and possessions of the Atkans and their church with its legacy of Orthodox liturgical art--and all of its records. The Atkans were forcibly taken to a camp in southeast Alaska with only the clothes on their backs. Three years were to elapse before they were allowed to return to a desolate landscape.

The people of the two Pribilof Island villages were given only a few hours notice to gather one suitcase of belongings before being shipped out to an unknown destination on June 15 and 16. On June 29, an order went out from the U. S. military command to evacuate all of the Aleuts west of Unimak Island. In all, 881 Aleuts were forcibly taken from six villages. They spent the rest of the war in unsanitary, decrepit, crowded bunkhouses in abandoned canneries and a gold mine in southeast Alaska.

While the Aleuts were away, their villages were occupied by U. S. military personnel. Servicemen lived in the private homes of the Aleuts in four of the five villages--Atka, of course, having been destroyed. The presence of the large military base at Dutch Harbor near Unalaska spared the private dwellings from occupation, but as with all the homes, they became prey to the whims of the occupying forces. When the Aleuts finally returned, they found their homes in shambles, their photo albums ripped apart, their clothes torn and scattered, their personal treasures gone. Most particularly, they reported that their churches had borne the brunt of looting and off-duty sport. Windows had been broken on the inside, even where boards protected the outside, bullet holes riddled the walls and cupolas, and those religious objects which they had been forced to leave behind had been desecrated or stolen.

For almost 40 years the story of the Aleut evacuation and the losses these people suffered at the hands of their fellow countrymen was untold and virtually unknown. A news blackout on the evacuation had hidden it even from other Alaskans. In 1980, Congress created a Commission on Wartime Relocation and Internment of Civilians to consider the claims of Japanese-Americans and also Aleut Americans. In 1981 hearings were held in Alaska to take testimony from survivors and for the first time, Alaskans learned of the fate of the Aleuts during World War II. During the testimony by the Aleuts, the number one priority (even ahead of personal property) was restitution to their beloved churches. In 1988, Congress made reparations to the Aleuts for the ill-advised policies of evacuation and internment, as well as for the damage done by the troops. Restitution for the churches was an integral part of the settlement.

THE PROJECT

To assess the damage suffered by the churches, the Aleutian/Pribilof Islands Association has taken the following approach:

1) An architect will provide a current condition assessment and cost estimates for repair of the church;

2) An inventory of present objects in the church will be made and compared with the last pre-war inventory;

3) A search is on for historic photographs both of interior and exterior views of the churches before and immediately after the war;

4) An art conservator, Vera B. Espinola, will visit each church and assess the current condition of the objects and attempt to identify damage related to the war-time conditions; she will also provide an estimate for conservation and/or restoration.

5) Elders who experienced the evacuation are being interviewed for their recollections.

As of April 1992, we have interviewed most of the elders. We have also completed all of the architectural surveys and the conservator's assessment of two churches. The objects in these two churches, however, comprise about half of the total under consideration. The remaining four churches will be inventoried and then surveyed by Ms. Espinola in summer 1992. The final report is planned for March 1993.

Turning to the two churches which we have fully surveyed, it should be noted that both have been designated National Historic Landmarks:

Unalaska: Church of the Holy Ascension

Unalaska is a predominately Aleut community associated with its neighbor Dutch Harbor, which is today the second busiest fishing port in the United States. The Orthodox Church of the Holy Ascension in Unalaska dates from 1825. The present structure, built in 1896, is the third on the site. A large nave is flanked by two side chapels. The basic structure of the building is sound, but its roof leaks badly and the foundation is rotten. The building is not insulated. In the Aleutian Islands, the climate is maritime, with dense fog, rain, snow, high winds and average temperatures from 10 to 70 degrees Fahrenheit with humidity ranges from 70 to 100%.

At the time of the evacuation, the residents had enough warning to be able to remove all the canvas icons from their frames and to pack them with the wooden icons, candelabra, altar furnishings, and vestments to take to the internment site at the southern tip of Alaska. The climate there was temperate, with as much as 400 inches of rain a year--drastically different from the Aleutians. The icons were stored in an unheated warehouse and began to mold. To protect them, the Unalaskans brushed them periodically with "3 in 1" oil. The conservator, Vera Espinola, has concluded that the transit of these 18th and 19th century works of art was extremely damaging. The loving care given by the Aleuts and their desire to conserve these objects was an ameliorating factor but could not compensate for the conditions of travel, the ship off-loading at least 8 times, and the poor storage conditions. Many objects clearly were damaged by the transit.

When the Unalaskans returned home, they discovered that the paint was flaking from many of the icons and to stabilize them, they covered the canvas icons with shellac. Over time the shellac and oil have reacted and darkened to such an extent that the images are almost invisible. A test cleaning in a small area of one icon shows that the shellac is soluble in ethanol, and therefore the canvas icons probably can be successfully cleaned.

The extreme conditions in the church, occasioned by periods of intense cold and brief bursts of heat, have combined with the shellac to make the canvas icons very brittle and subject to cracking and tears.

During the conservator's review, an important discovery was made. Ms. Espinola was able to identify a number of canvas icons as probably of local origin. These are of such fine workmanship as to be very like those from Russia, but bear certain tell-tale signs of provincial creation. Among these are the coarseness of the underlying canvas (resembling sailcloth), the thin layer of paint, the absence of conventional Cyrillic superscripts on the face of the icon identifying the subject, and distinctive local idiosyncracies. Among the latter are background mountains much like those in the Aleutians, a sheep which looks like a white dog with a long tail, and a similarity of style in the rendering of faces and hands. In one case, the artist's touch is clearly visible in an icon of St. Nicholas who has four feet--the artist's correction of his design not completely disguised.

All in all, the Church of the Holy Ascension at Unalaska is a remarkable edifice, treasured by the 150 Aleuts who continue to live and worship there. In overwhelming numbers those who survived the evacuation have asked that the first priority of restitution be given to restoring their church. This building has been designated as number 1 among the most endangered buildings in the United States by the National Trust for Historic Preservation and the National Park Service. An architectural plan for restoration places the cost at $1.4

million. Conservation for 200 icons and frames, liturgical vessels, paper icons and some books is estimated at $147,000.

St. Paul Island: Church of Saints Peter and Paul

The other church which we have examined is on the island of St. Paul. This is the larger of the two Pribilof communities, with a population of 650 Aleuts.

The Pribilof Islands are in the middle of the Bering Sea and are shrouded for much of the year in fog and are whipped in winter by fierce storms. Temperatures range from 0 to the mid-50s Fahrenheit, but are often much colder due to the wind chill factor. St. Paul is today a busy fishing port, with a large fish processing plant. Until 1972, however, the island economy was exclusively the harvesting of seals, under the administration of the U. S. Fish and Wildlife Service. Every aspect of the Pribilof Aleut's life was under the scrutiny of the agent of the U. S. Fish and Wildlife Service; the Aleut residents were not free to leave the island without permission. While deplorable from a humanitarian standpoint, this government overlordship produced volumes of records concerning community life and thus has permitted us to know more about many facets of the Pribilof Island communities than other Aleut villages.

The present Church of Saints Peter and Paul was built in 1906, being the third Orthodox house of worship on the site. The church was well constructed to withstand the climate of the Bering Sea, but its exterior shell has changed over the years as new siding and a new roof have been added, the old cornices removed, and the elegant cupola replaced by a wire frame. The furnishings reflect the devotion of the St. Paul Aleuts to the church, which, during American rule, was the one place on the island where their native language could be spoken.

The people of St. Paul were given only 24 hours to gather their possessions for the evacuation. While not yet confirmed, it seems clear that at least the portable icons were taken with them to their internment in southeast Alaska, as were the service utensils for the sanctuary. They suffered the same trauma from travel and relocation in an alien climate as those treasures from Unalaska. Theft and vandalism accompanied the military occupation of the island. A few years ago, the church received in the mail a small cross with cloisonne and a portion of a quadriptych with a note expressing apologies to the community for their theft during the war. At present, we are conducting a comparison of the items in the last inventory of the church in 1915 with our current inventory to assess the extent of loss.

A particular concern for the St. Paul church is its lack of insulation. During the three days of the conservator's survey, the temperature inside the church did not rise above 52 degrees Fahrenheit while a gale outside caused a windchill of -30 degrees. While most of the year is more moderate, the wind seeps into the building creating drafts and uneven heating. A relatively new oil forced-air furnace heats the church modestly for services but is turned off for most of the week. The resulting extremes of temperature have taken a toll on the icons. An icon of St. Innocent of Irkutsk dates from the previous church structure, built in 1877. Its gilding reflects the lavish attention given to the church by the Aleuts during the heyday of the seal harvests. The heaving of the canvas is caused by cracking of the underlying gesso due to fluctuations in the temperature and humidity of the building.

A special conservation concern for this church arises from a fire which occurred behind the iconostas in 1932, searing the icons on the icon screen and destroying most of those within the altar area. Behind a 20th-century canvas icon of the Mother of God on the iconostas is the original oil and gilt painting from 1877, so severely damaged by the fire that it had been covered. Fortunately, it very likely can be successfully restored. Its companion, the icon of Christ, already has been restored by the Lucas Conservation Laboratory of Vancouver, Washington.

At St. Paul, 64 icons and liturgical vessels were examined for conservation attention; the cost for repair and stabilization is estimated to be $59,000. The architect estimates the cost of structural restoration at $325,000.

SUMMARY

To summarize the project to date, six churches are under study, to assess their current condition and attempt to determine what happened to them as a result of U.S. policies during World War II. These are the churches at Nikolski on Umnak Island, Akutan, Unalaska, St. Paul and St. George. The post-war church at Atka, built from scrap, green lumber, is a special concern. The residents there wish to have this church replaced by a new structure which would replicate the one burned in 1942.

From the start, it has been apparent that determining specific damage and loss after 50 years under the harsh climate of the North Pacific is next to impossible. My assistant, whose mother was in the evacuation from Atka, and I are combing the testimony which led to the reparations settlement and we have interviewed more than 50 survivors, who are old enough to have some knowledge of the church 50 years ago. Needless to say, their memories are imperfect of events so long ago. But while they have little recollection of specifics, they are adamant that the church was in sad condition on their return and it should be restored because it is their principal cultural institution today.

Although somewhat daunted by proving the near impossible, we are convinced that there is sufficient evidence in the church records, the photographs, and personal recollections to document loss and damage. At the core of my understanding of this project is the knowledge that the Aleut evacuees were destitute when they returned to their homes and no money was provided to them for losses associated with their church. They struggled to restore the buildings and the church life to its pre-war vigor, with varying degrees of success. But repairs could not be made sufficient to overcome the damage of up to three years of absence. The wind and rain and seeping fog play havoc with all structures in the Aleutians and buildings require constant maintenance. Three years absence would, without question, cause serious erosion. The deteriorating buildings we see today are a legacy from those days.

Decisions about the church and its property will be made by a Committee of representatives from each community, appointed to administer the trust created by Congress. One project the Trustees may wish to consider is an idea proposed by Ms. Espinola, who has suggested that a conservation laboratory be established at Unalaska. This lab could undertake the cleaning and some repair of damaged pieces, while those which required restoration could be sent away. Such a project might attract conservators who would be interested in short-term stints working in a true "museum without walls". This might also be a way for Unalaska to resume its historic role as a cultural center for the whole Aleutian region and even beyond, since there are no such conservation facilities in the Russian Far East--only a hop, skip, and jump from Unalaska, which is a sister city with Petropavlovsk on the Kamchatka Peninsula.

Although daunting in its aim, we have hope that the Church Restitution Project of the Aleutian/Pribilof Islands Association soon will bring tangible benefit to the six communities which fifty years later still bear the scars of the Forgotten War in the Aleutians.

THE GULF WAR AND ITS AFTERMATH ON THE CULTURAL HERITAGE OF IRAQ.

SELMA M. S. AL-RADI
29 Claremont Avenue, New York, N.Y.10027.

ABSTRACT

The damage caused during and in the aftermath of the Gulf War to the cultural monuments and heritage of Iraq resulted directly from the bombing. Some of the effects were immediate, the rest were due to delayed reactions to the vibrations. However, the most serious loss to the country took place during the insurgencies immediately after the war was over when nine provincial museums were looted and robbed of their objects. In all some 4000 objects are missing. Although many will have been destroyed during the upheavals, others have begun to appear on the antiquities markets in the West.

In all, the Gulf War lasted six weeks (January 16 - February 18, 1991) but its effects will no doubt remain and be felt in Iraq for a long time to come. Iraq, Ancient Mesopotamia - the so-called "Cradle of Civilization", is a land rich in cultural monuments, archaeological sites dating from the pre-historic to medieval Islamic times, historic cities and quarters, as well as numerous museums. This land that lies between the two rivers, the Tigris and the Euphrates, has been inhabited for so long that the remains of past cultures have been literally piled up one on top of the other to form giant mounds (tells). The mound of the Assyrian city of Nineveh became engulfed by the medieval and modern city of Mosul, and the 11th century Abbassid town of Samarra with its famous mosque and twisted minaret is now part of the modern industrial town of the same name. Cultural remains have become inextricably interwoven, the ancient into the new, the archaeological into the rural, and the historic into the growing urban fabric of the modern country of Iraq. It is virtually impossible to bomb or do battle anywhere in Iraq without hitting or damaging some cultural artifact.

During the massive Allied bombing raids over Iraqi military and not so military targets, areas adjacent to these targets were also affected. Hitting the bridges in Baghdad caused damage to the Mustansiriya School (13th century), while bits of flying shrapnel blew the top off the Kaplanya minaret nearby (18th century). Two mosques were destroyed by rockets fired from Allied aircraft in Basra, including the 17th century Al Kawwaz Mosque; both are in residential areas. In Mosul a historic quarter was bombed, as were three churches, that of Saint Thomas' - one of the oldest standing churches in Iraq today - was badly damaged. It must be presumed that these were mistakes and not intentional hits.

Another kind of damage was inflicted by the Allied land forces when they entered and occupied parts of Southern and Northern Iraq. Tell el Laham, a Third Millennium BC site in the south, suffered from US army trenching. These trenches ressemble illicit diggings. In fact, it is difficult to assess the full amount of destruction done to the archaeological mounds in Iraq. That would require a physical survey of hundreds of sites - not a top priority in a country with more pressing problems.

The most difficult damage to assess is that produced by the delayed effects of the bombing, that is by the impact and vibrations made by tons of bombs hitting the ground. How much shaking can old or even new buildings withstand before they begin to suffer structurally? In the second of two trips that I undertook to Baghdad last year, I noticed cracks that were not there during my first trip eight months previously. Some settling is therefore already beginning to happen. In fact, almost all the modern houses that I visited are developing cracks. Vibrations also caused a number of the larger objects that could not be moved in the Iraq Museum to fall off their pedestals and break. What adverse effects objects packed away in storage suffered is yet to be estimated.

Would the immediate damage to the built environment have been worse if the terrain had been rocky rather than alluvial? Did Iraq's alluvial soil diffuse or absorb the impact, and blunt the pounding and the shock tremors of the bombs? During the 1983 earthquake in Yemen, which measured 6.3 on the Richter scale, the houses built on rocky outcrops came down like nine pins, those built on the alluvial plains were at first virtually unscathed but later developed cracks. Are the effects of earthquakes similar to those produced by intensive bombing ? One needs the opinion of experts on this issue.

The Arch of Ctesiphon, the world's largest and oldest brick span dating to the 3rd century AD, lies close to a major military base that was bombed daily. New cracks are already apparent in the arch and some bricks have fallen down. The fragile state of the monument is very obvious. Cracks are also beginning to appear in some old buildings in Baghdad; these were not apparent immediately after the war. A missile arsenal near the ancient city of Hatra was bombed and exploded, raining shells that missed the city walls by feet. Only minor chipping of the sculpture resulted. Luck also saved the zigurat at Ur when four shells exploded in its immediate vicinity leaving huge craters in the ground. However, the ferocity of the battles waged nearby are very visible on the walls of the ziggurat which are pock-marked with over 400 hits from small arms fire.

It was, however, in the aftermath of the war that the most damage was done to Iraq's cultural heritage. Historic sites and cities were damaged during the insurgencies, and at the time when government troops retook control of the towns that had revolted against them. The holy cities of Kerbela and Nejef in the south were badly shelled, the facades and domes of the mosques lost most of their glazed tile decorations. The damage to these shrines has now been repaired by the central government.

During the chaotic times of the insurgencies nine museums were looted, and trashed. In the north, the museums at Kirkuk, Dohuk and Suleimaniyah were literally stripped bare. At about the same time in the south the Kufa, Diwaniyah, Amara and two Basra museums were pillaged and ransacked. These museums lost not only the objects on exhibition, some 4000 all told, but also lost their libraries, catalogue registers and official papers - the last three were wantonly destroyed and burnt. The objects, which included such diversities as terra-cotta figurines, marble statues, cylinder and stamp seals, prehistoric figurines and tools, jewelry from the Royal Cemetery at Ur, pottery vessels, bronzes and coins, carved Nimrud ivories and hundreds of inscribed clay tablets, as well as gold and silver coins and Islamic medieval objects were stolen or broken during these violent upheavals. The damage in most cases was quite willful. When objects were too large to remove they were broken in situ, and many were simply trampled underfoot. Small objects made from fragile materials such as copper tools, bone awls, and unbaked clay tablets would simply not have survived the rampage. Those that were stolen intact, for example, the stone sculptures, seals, baked clay tablets and carved ivory plaques, objects that could be put into one's pockets, were either smuggled across the border into Iran (some may perhaps show up in Turkey and Syria), or they were apparently sold for $10 and $20 bills to the Allied occupying troops stationed in the country.

These regional museums displayed a comprehensive range of objects that were intended to teach modern-day Iraqis about their history from pre-historic times to the present day. All the objects were excavated from archaeological contexts, all had provenance, and all were historically or art historically important. Individually these pieces may not be considered as great art works, but their importance comes from the fact that unlike many Mesopotamian objects in western museums, their excavation history is well-documented. In order to reconstruct an artistic and cultural tradition, it is imperative that the interrelation of different objects be studied from their stratigraphic building context. Objects from documented excavations therefore provide chronological dating information that cannot be obtained by stylistic or epigraphic analysis alone. Luckily, a large proportion of these lost objects have been published by their excavators.

A descriptive catalogue of some 2600 of these objects was compiled by the Department of Antiquities in Baghdad and sent to Unesco in Paris. Photographs were only available for 10% of the objects because supplies of chemicals and photographic paper were insufficient to print more; the embargo forbids all imports into Iraq except those related to food and health. Unesco, slow at the best of times, sent the list to Interpol, Sotheby's and other agencies that deal with stolen objects some six months after they had received it. Meanwhile, the objects were beginning to appear on the international antiquities market, offered by Iranian dealers to private collectors and museums in London and in New York. The Iraq Museum numbers have been scraped off the objects, but they can still be identified from the list.

The problems associated with this particular group of missing objects are threefold. The first is that there are simply too many of them. The second is that the market for these objects lies in the countries that waged the war against Iraq. There is therefore no official incentive to aid a "pariah" state, which is Iraq's position today. The third problem lies with the looters. The Hague Convention specifies that booty taken during wartime must be returned. But what happens when it is the local people who loot their own cultures and sell it to middle-men? Who is responsible, and how can these objects be returned to their country of origin when and if they are found and identified ?

How can one save the world's heritage in times of war? Is it enough that nations sign the Hague Convention or ratify the additional Protocols I and II proposed by Unesco to protect cultural property and monuments? These protocols state that cultural properties worthy of protection must be identified, inventoried and protected from attack prior to the start of hostilities, and that pillaging is illegal; however, no measures were established to enforce these good intentions. As of December 1991, 105 and 95 nations respectively have signed these two protocols, but neither the United States nor the United Kingdom have ratified the Convention or the Protocols, which means they are not legally binding in those countries.

During the wars and civil upheavals that have plagued the world in recent times, in Vietnam and Cambodia, in Afghanistan, Lebanon, Iran/Iraq, Panama (which also had its National Museum looted after the American invasion), and most recently in the countries that were affected by the Gulf War, rules and protocols do not seem to apply, and good deeds are not foremost in the minds of the warriors. The world's heritage and its cultural monuments must be protected from man's destructive instincts.

Postscript: A similar situation occured in Lebanon during the civil war. The Beirut Museum was looted on numerous occasions, and the antiquities therefrom sold on the European market. Many archaeological sites were pillaged, for instance, the ancient city of Tyre by the Israeli occupation army and by locals alike. In the mid-1980's Kammet el-Loz, an important Bronze Age site in the Bekaa Valley, was virtually razed to the ground. As recently as 1990 a Phoenecian cemetery near Tyre, with very rare children's burials, was pillaged. The local authorities have little power and no incentive to act, while the international agencies, UNESCO and Interpol, were and are equally careless. The looting continues...

MONUMENTS AS TARGETS

ZELIMIR KOSCEVIC

Museum of Contemporary Art, Zagreb, Croatia

My paper is a video presentation, but before I show it, allow me just a few words. My contribution is not a scientific paper. In this video we shall be faced with the facts. Facts are always that starting point in the process of thinking, and thought is the basis of human activity. These facts are given concisely in 25 minutes, which is -- frankly -- nothing compared to the hours and hours I spent watching the raw documentary material, selecting it, and making the final version of the video that you are about to see.

The Yugoslav crisis, a term which is an understatement, and therefore a completely inaccurate one, for the disintegration of Yugoslavia and the cruel war in Croatia, resulted in 538 destroyed or severely damaged cultural sites on the Croatian territory. Besides the strategic targets, such as industry, communication lines and information relays, Yugoslav Army and joined Serbian para-military forces turned their wepons against cultural heritage. When you see how many cultural sites have been destroyed and when we see how it was done, it becomes obvious that the cultural sites were the strategic targets as well.

This military novelty is frightening, and few precedents in history, and this is something which we should bear in mind in the future. The message of Dubrovnik, Vukovar, Zadar, Osijek, Karlovac, St. Martin Chapel near Nasice is one and only one; The Hague Convention of 1954 is not efficient enough to protect and save the cultural sites against military outrage. After all that happened in Croatia, one can well imagine a Patagonian air raid on Machu Pichu, or a Mongolian attack on the Great Wall. UNESCO must have much more radical means of defending cultural heritage than petitions, declarations, admonisions and the Hague Convention flag.

The strategic aim of the enemy was to destroy the cultural identify of Croatia. And even if we can say that some assaults on the cultural sites were of military nature and in fact the general consequence of the conflict, most of them without doubt involved a deliberate destruction of cultural heritage which not only belongs to Croatia, but to the world as well.

Once this strategy has become clear, we were confronted with the dilemma whether to remove the distinctive emblem of the Hague Convention from cultural sites, or to leave it there. Of course, we did not remove it, and now our cultural patrimony is demilished, and we still have legal and moral right to accuse the enemy of the crimes of war. I hope this will be done sone -- but this is the duty of other civic services.

I remember very well the dead silence of my colleagues from The Getty Consrvation Institute after seeing the video about the damages of cultural sites in Croatia. The crucial question in the air then -- and now -- is WHY? I am not ready to accept benevolent interpretation of all these crimes on cultural monuments as the irrational aspect of the conflict. There is an undeniable evidence that shows that the background of this barbaric attack on cultural heritage was,and still is, very rational, and that action was planned. I know that this conclusion sounds incredible. But the list of 538 destroyed or damaged cultural sites, all properly marked with the sign of the Hague Convention, forces us to accept this unbelieavable fact, and to see what are the consequences when cultural monuments become the target.

And as this is being written -- the tragedy continues in Bosnia-Hersegovina...where will it every end?

Croatian Monuments as Targets, 1991/92

Joško Belamarić,
The Regional Institute for the Protection of Monuments
Split - Croatia

The purpose of this talk is to give a brief account of the current situation in Croatia—a small country located between the Adriatic sea and the Danube, along the south-eastern spur of the Alps. I would like to begin with some general remarks about the Croatian cultural heritage and the damage that has been inflicted on it by nine months of exceptionally brutal warfare. I will give some rather dry statistics, but it is important to keep in mind that each of these statistics represent an act of appalling vandalism of a kind that has not been seen in Europe for a generation. I will then focus on the specific cases of Dubrovnik and Zadar. Since both of these cities are under the jurisdiction of my Institute I can report about them in some detail from personal experience. Finally I have some concrete proposals for what could be done to mitigate the effects of the damage already caused to cultural monuments by the war in Croatia. Croatia desperately needs immediate international help to preserve its cultural heritage. I will report on some initiatives already under way.

Part of the tragedy of the current situation in Croatia is that our richly varied Croatian cultural heritage is being literally destroyed before our very eyes. Much of our architecture, painting, and sculpture fully deserves the kind of careful, scrupulous documentation and study given to masterpieces of world art. Every time one of our cities is shelled or bombed another bit of this heritage is irretrievably lost.

Despite its position in the very center of Europe, for many Westerners Croatia is just another peripheral region, an "exotic" country. Gibbon once astutely remarked that while the division between North and South in Europe was real and palpable, any demarcation between East and West, especially in the Mediterranean region, has always been more or less arbitrary. In the late 3rd century the Emperor Diocletian, himself an Illyrian from Dalmatia by birth, formally divided the Roman Empire into a Western and an Eastern part. The line he drew between areas directly subject to Rome and those under Constantinople eventually came to mark the boundary between the West and the Byzantine Empire, and finally that between Europe and the Ottoman world. Croatia lies precisely on this line. This location on the "geological fault" between West and East contributes to the richness and variety of Croatia's cultural life and to its specific cultural profile, but it also represents a constant threat to its continued existence. Even the odd concave shape of Croatia's borders bears witness to the intensity of the external forces that have been impinging on it for the past ten centuries.

The present shape of Croatia does not approximate that of the original Croatian state, which reached its apogee in the period between the 9th and 12th centuries. During that period Croatia's dukes and kings were able to maintain a fragile independence against their powerful neighbors: Franks and Hungarians, Venice and Byzantium.

Croatia, then, shows a surprisingly dense layering of cultures. Although the "golden moments" of its history have been brief and infrequent, it has made its own unique contribution to the common European cultural heritage. The ceramic dove found near Vukovar is acclaimed as one of the finest prehistoric artefacts from the prehistoric period, and there are few lands in the Mediterranean which can boast such abundance of well-preserved Hellenistic and Roman sites as Croatia. The Palace in Split, built by the Emperor Diocletian in the late 3rd century, and to which he retired when he abdicated in 305, is one of the most significant architectural remains of the late Roman Empire. It is registered on the World Heritage List. Without the early Christian monuments in Dalmatia and Istria it would be virtually impossible to reconstruct the development of Roman art during the late Empire. The maturity of style and iconographic originality displayed by the sculptures produced in the period of the last Croatian kings in the 11th century mark one of the most significant stages in the genesis of Romanesque art in Europe. Works of Croatian artists, from Andrija Buvina, Master Radovan, Juraj Matejev, Ivan Duknović (Iohannes Dalmata), the brothers Lucijan and Franjo Vranjanin (Laurana) and many others down to the 20th century sculptor Ivan Meštrović form a distinct Croatian strand in the history of European art.

One of the most salient features of Croatian culture is its strongly urban character. Many towns and cities have prehistoric or at least classical roots and are sited in cultivated landscapes from which they arise organically. At Poreć, Pula, Osor, Rab, Zadar, Trogir, Split, Dubrovnik, Kotor and elsewhere the world of antiquity did not die. The artistic language of the ancient world was gradually transformed into a new national vernacular. The classical tradition is a continuing presence and remains the basis for Croatian national culture.

What is being destroyed in Croatia is the heritage of a highly and self-consciously urban civilization. Croatia's towns and cities with their extraordinarily thick layering of history and culture are the expression of the collective memories and aspirations of generations and their destruction is an attack on the very identity of the people who have inherited them.

Dubrovnik can serve as a metaphor for the plight of Croatia's historical and cultural heritage as a whole. It may lack the almost metaphysical grandeur of cities like Vukovar or Osijek, spreading along the mighty Pannonian rivers, or the diversity and inner mystery of Split, or the intimacy and refined beauty of detail to be found in Trogir, or the dramatic local topography of Sibenik, or the monumentality of Zadar's Late Antique and Medieval episcopal center, but Dubrovnik's unique greatness lies in its ability to preserve and express a fragile liberty in a harsh and unyielding environment. Surrounded by larger and more powerful neighbors, each ruthlessly pursuing its own advantage, Dubrovnik succeeded in maintaining its independence for ten centuries while balancing individual freedom and social control, creativity and system, justice and coercion.

Very early in its history Dubrovnik defined the limits of its growth, but it never stopped expanding its mercantile and intellectual contacts with the rest of the world. Dubrovnik has no single outstanding structure that could stand serious comparison with the Romanesque belfry of the Cathedral in Split, for instance, or the Cathedral of Šibenik, or St. Donatus at Zadar. What is uniquely impressive about Dubrovnik is the city itself as a unitary urban configuration. For centuries Dubrovnik was a piece of Croatian utopia on the edge of the Balkans, the archetypical image of "the town" we dream of even after we have seen dozens of others. It is Dubrovnik's shell, its "armour", not its internal urban design that first fascinates us. The walls of Dubrovnik however, are the expression of a paranoid fear. The confidence they are intended to inspire surpasses their actual protective power. They are yet another symbolic line, like the one the Emperor Diokletian drew between East and West, a line dividing the inner urban space, the sphere of civilization and order from what was thought to be a realm of amorphous and chaotic evil outside.

The fate of Dubrovnik illustrates an aspect of this war that is of central importance in understanding it. It is utterly essential to realize that the destruction of monuments in Croatia is not an accidental by-product of the war. Cultural and historical monuments are being consciously and systematically targeted as part of an attempt to eradicate Croatia's identity. Romanesque houses, Gothic churches, and Baroque palaces are not being demolished because they happen to stand in the line of fire, but because they are recognized as the most enduring and authentic documents of our national identity. A few minutes after occupation of Drniš, a small town in the Dalmatian hinterland, the Federal Army burnt the local Court House and the Land Registry Office to the ground. Why? The aim of this action was to destroy all records of ownership of the land and thus make it impossible for Croats to prove that they had legal title to any property in the region around the town.

After the shelling of Dubrovnik on December 6, 1991 a team of UNESCO experts made a careful map of the damage. If one looks at this map it becomes clear that the shelling followed a systematic and carefully conceived plan. Contrary to the official statements issued by the Federal Army that some anonymous soldiers irresponsibly violated their orders and fired some random rounds into the city, the UNESCO map proves that the destruction was carefully prepared and planned and quite systematic. The cannons were moved methodically over the city, sector by sector. If this could happen in a few hours of one December morning, in cold blood and in front of television cameras from all over the world, it is horrifying even to imagine what has taken place in the third of Croatia that has been under enemy occupation for months—in the Dalmatian Hinterland, in Lika, and in the Croatian parts of the Pannonian Lowlands. The tragedy for members of my profession is that there are now hundreds of monuments which we will be able to study only as archaeologists, not as art historians.

What I have said up to now has been rather abstract, but the statistics of war damage in Croatia are gruesomely concrete. I must begin by saying that the information available to the Institute for the Protection of Monuments of the Republic of Croatia is incomplete because a sizeable part of the territory of the Republic is outside the control of the legally constituted authorities and inaccessible. However, according to the information available, 245 settlements of historic importance have been subjected to artillery and air attack (47 historic city centres, 187 historic rural entities, 10 archaeological sites). The old city centres of Vukovar, Vinkovci, Lipik, Kostajnica and Petrinja have been more or less completely destroyed, and those of Osijek, Pakrac, Karlovac, Gospić and many others have been heavily damaged.

Since October 1, 1991, the area of Dubrovnik and the immediate surroundings of its old city centre, and since October 24, the old city centre itself—a cultural property entered in the World Heritage List—have been subjected to intense attacks. Several Renaissance Summerhouses of Dubrovnik patricians, the Franciscan Monastery and many other sacred and secular buildings in Rijeka Dubrovačka and Rožat, and in the entire surrounding area were ravaged and some were completely destroyed prior to October 1. The central area of Čilipi was destroyed and burnt down, including the Regional Museum there and the parish church. After November 9 a major escalation took place and the entire historic centre within the ancient walls of Dubrovnik came under direct artillery fire, with targets including churches, monasteries, palaces, fortifications, among others the second oldest synagogue in Europe. Almost all the hotels in Dubrovnik and the surroundings (virtually the only source of revenue for the town) have been devastated or burnt down. Perhaps we should add that there is no Serbian minority, or any garrisons, or military installations of any kind in the area of Dubrovnik. On December 6, 1991, a particularly heavy shelling of the historic center was initiated. Missiles fired from Yugoslav Navy vessels struck the city's main street, the Cathedral's cupola, the famous Gothic Fountain. Thirty percent of the historic building stock of Dubrovnik's old centre was heavily damaged, of which 10% was completely destroyed in fires caused by incendiary missiles. For months Dubrovnik was kept under the naval blockade, without water or electricity.

On November 15 attacks were launched against Split, including old city core, and the Diocletian Palace, a cultural treasure entered in the World Heritage List. In Zadar the historical core has been subjected to cannon fire from Yugoslav Navy vessels and shelling by land artillery. Even parts of the city spared in the heavy bombardments of World War Two have been ravaged. Thus a series of monuments in the highest category of importance there were seriously damaged: the Romanesque church of St. Krševan, the ancient church of St. Simon, the Renaissance Land Gate, sections of the 13th

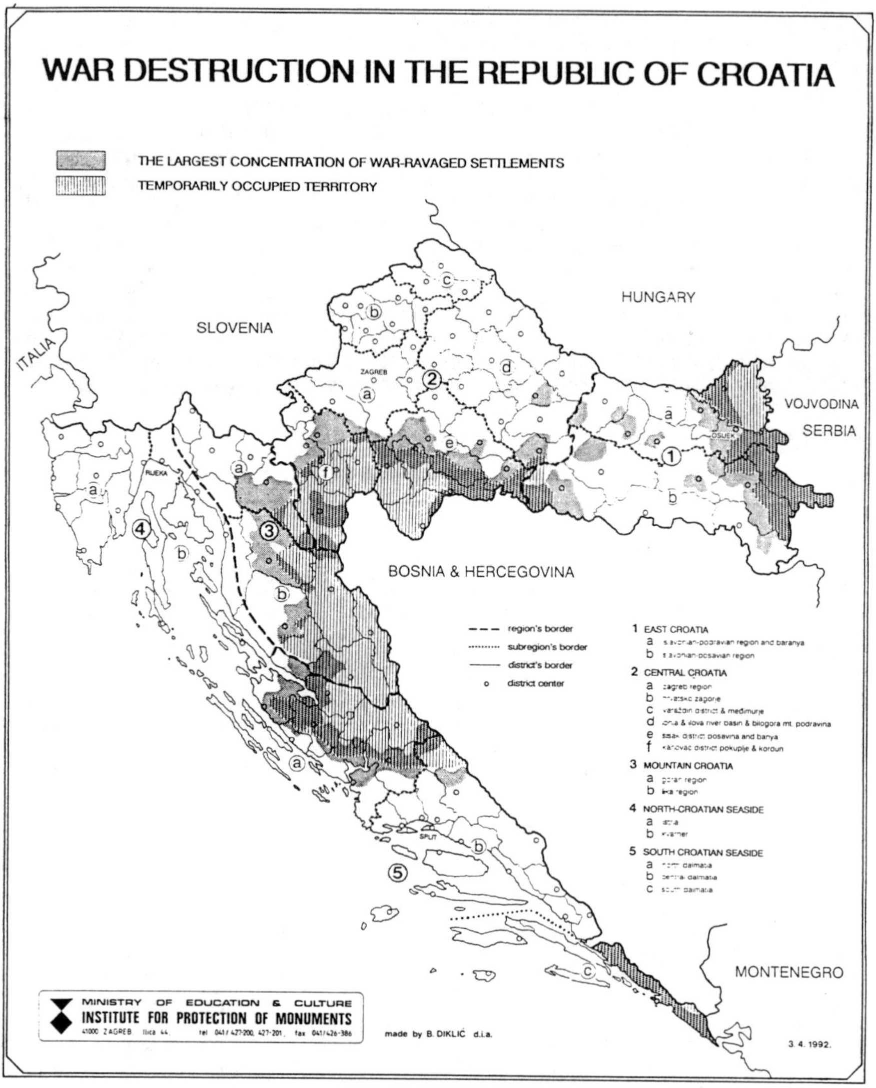
WAR DESTRUCTION IN THE REPUBLIC OF CROATIA
THE LARGEST CONCENTRATION OF WAR-RAVAGED SETTLEMENTS
TEMPORARILY OCCUPIED TERRITORY
HUNGARY
SLOVENIA
ITALIA
ZAGREB
VOJVODINA
SERBIA
OSIJEK
RIJEKA
BOSNIA & HERCEGOVINA
SPLIT
MONTENEGRO
region's border
subregion's border
district's border
district center
1 EAST CROATIA
a region and baranya
b region
2 CENTRAL CROATIA
a region
b zagorje
c district & međimurje
d & ilova river basin & bilogora mt. podravina
e district posavina and banya
f district pokuplje & korđun
3 MOUNTAIN CROATIA
a region
b region
4 NORTH-CROATIAN SEASIDE
a
b
5 SOUTH CROATIAN SEASIDE
a dalmatia
b dalmatia
c dalmatia
MINISTRY OF EDUCATION & CULTURE
INSTITUTE FOR PROTECTION OF MONUMENTS
41000 ZAGREB Ilica 44. tel 041/427-200, 427-201, fax 041/426-386
made by B. DIKLIĆ d.i.a.
3. 4. 1992.

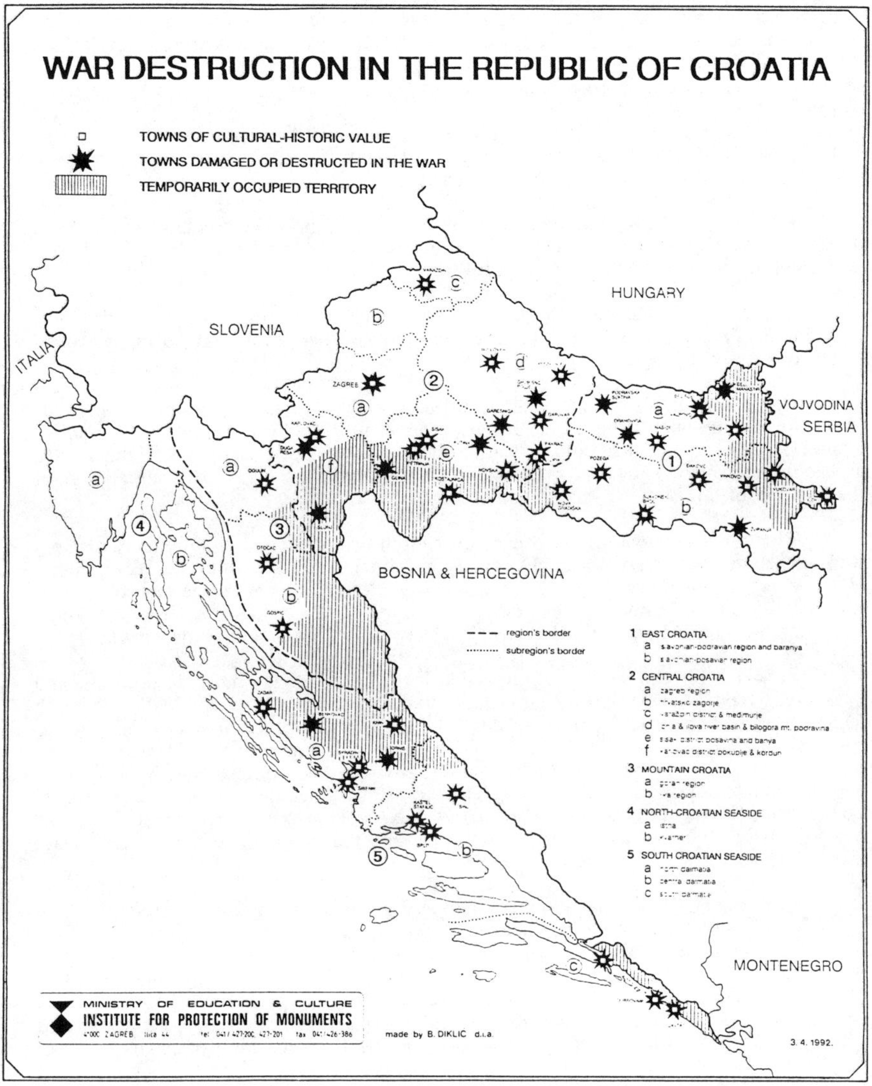
WAR DESTRUCTION IN THE REPUBLIC OF CROATIA
TOWNS OF CULTURAL-HISTORIC VALUE
TOWNS DAMAGED OR DESTRUCTED IN THE WAR
TEMPORARILY OCCUPIED TERRITORY
HUNGARY
SLOVENIA
ITALIA
VOJVODINA
SERBIA
BOSNIA & HERCEGOVINA
MONTENEGRO
region's border
subregion's border
1 EAST CROATIA
2 CENTRAL CROATIA
3 MOUNTAIN CROATIA
4 NORTH-CROATIAN SEASIDE
5 SOUTH CROATIAN SEASIDE
MINISTRY OF EDUCATION & CULTURE
INSTITUTE FOR PROTECTION OF MONUMENTS
made by B. DIKLIC d.i.a.
3. 4. 1992.

century fortifications, the Governor's Palace, the University Library, the History Archive and many other monuments. In one of its repeated air raids on the city (on November 18) the Army bombed the magnificent and just recently restored Late Antique—Romanesque Cathedral with episcopal complex. Rockets destroyed a part of the original Zadar Forum pavement.

The Zagreb Upper Town was rocketed and bombed in an air raid launched on October 10, 1991. Nearly all the buildings in the area were damaged; the Vice-Roy's Courts, in particular, were heavily damaged.

Including museum, archive and library buildings 111 structures have been destroyed and 186 heavily damaged. The extent and nature of damage to 137 buildings is not yet known due to their inaccessibility.

The churches and monasteries have been especially badly hit and are the most frequent targets. Out of 359 affected churches and monasteries 88 have been destroyed and 106 heavily damaged.

Not to mention here more than 200,000 modern houses, apartments and flats (don't forget that at this moment there are more than 700,000 refugees from occupied territories), the damaged buildings include those of all historical periods. Neither have archaeological sites, especially in Slavonia, been spared by shells, fortification work or the passage of tanks.

The movable monuments, especially church inventories, although many of them were evacuated are also being destroyed or looted. In many churches (21) highly valuable organs have been heavily damaged. Cemeteries, too, have been the objects of wanton destruction; some of these have exceptional historic value, such as St. Ann's Cemetery in Šibenik and the old Jewish cemeteries in Cernik and Dubrovnik.

Shocked by the extent of the destruction in Croatia many institutions and individuals from Europe and overseas have lodged protests, expressed solidarity with the victims and provided concrete help. Getty Conservation Institute has already sponsored a seminar with Prof. Radovan Ivančević, Chairman of the Croatian Society of Art Historians, who brought a report on the two-day international meeting dedicated to the reconstruction of the cultural heritage of Croatia sponsored by the ARCH Foundation. I can inform you also about UNESCO's preliminary report on the destruction in Dubrovnik and further activities and plans for its reconstruction. I am using here some of their materials.

Our principal aim is to mobilize as many international agencies as possible and to obtain foreign support for systematic elaboration of a comprehensive plan for reconstruction including the assessment of the nature and extent of damage. The problems of reconstruction are discussed in different categories: architecture and town planning, villages and environment, movable cultural property and institutions (museums, galleries, archives, libraries). I will report here some of the conclusions and proposals. First for monuments, architecture and the reconstruction of towns:

1. An international information center should be established for the collection of facts related to the devastation of the cultural heritage of Croatia.

2. An adequate supply of materials for the preventive protection of architectural monuments (plastic sheeting, timber, roof, tiles, glass, bricks, iron for reinforced concrete, cement, lime, scaffolding, etc.) will be needed.

3. Considering the current state (in terms of the number of personnel, available equipment, etc.) of the services and other institutions involved in the protection of

Preliminarni izvještaj o bombardiranju Dubrovnika 6. 12. 1991. (UNESCO) / Releve preliminaire des destructions bombardement du 6 decembre 1991. (UNESCO)

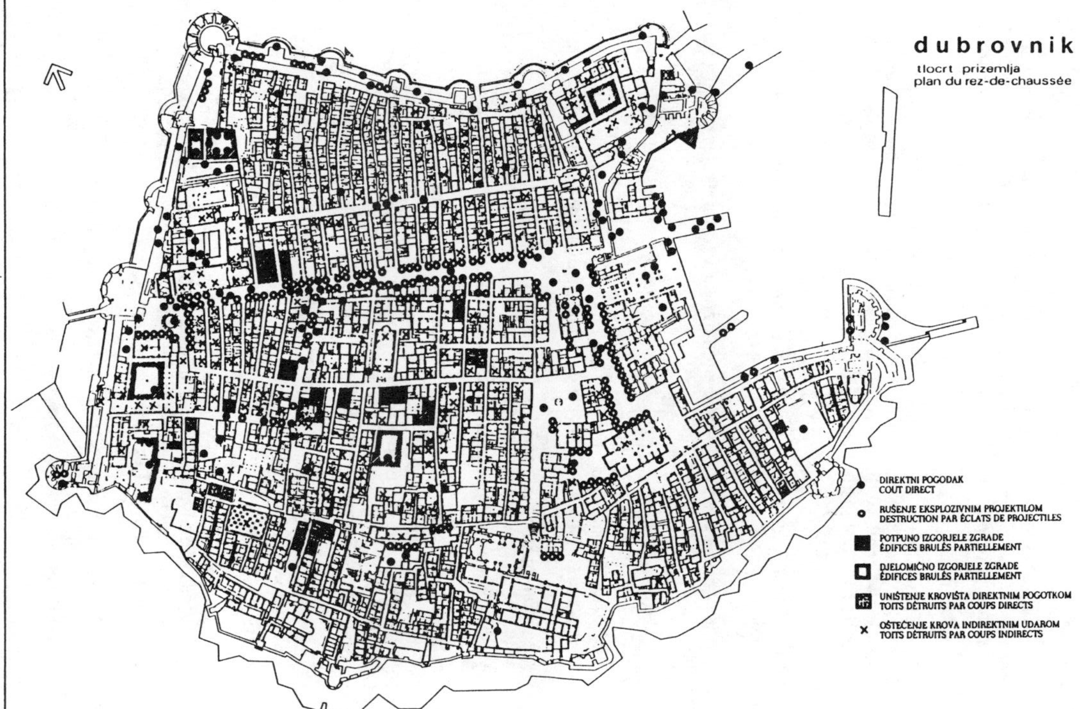

cultural heritage (museum, archives, libraries) in Croatia, special attention should be paid to the promotion of these service and institutions. For that reason:

3.1. Proper conditions should be created and arrangements made for the training of personnel specializing in restoration and the conservation of monuments through organization of courses in Croatia and provision of scholarships for young people adequately qualified to attend courses in specialized institutions abroad.

3.2. Help should be given for the procurement of equipment essential for the work of the institutions charged with the protection of the cultural heritage of Croatia. This applies to equipment designed for wartime conditions as well as for regular use after the war.

3.3. Procurement of essential construction and other materials should be facilitated for preventive protection of damaged monuments. The Institute for Protection of Monuments of Croatia will see to it that lists of the required equipment are prepared and submitted to all the institutions concerned.

The following assistance by the international community will be needed on a long-term and on a continuous basis:

1. Formation of teams for the preparation of architectural and other documentation.

2. Special training, through:

- visits to Croatia by foreign experts who will help with the training of restoration specialists and with the preparation of documentation,

- specialist training abroad for Croatian personnel,

- establishment of specialist schools and workshops in Croatia,

- establishment of a permanent workshop for traditional crafts

3. Assistance in the course of restoration and reconstruction work:

- formation of joint professional teams for the preparation of conservation proposals, guidelines and projects,

- professional and material assistance in the course of reconstruction and restoration work, including technical expertise, equipment, specialist craftsmen, and materials.

The process of reconstruction and restoration of our cultural heritage cannot, and should not, be confined to representative buildings and urban wholes, but should encompass the entire cultural heritage of Croatia.

I — Small settlements, villages and archaeological sites

A - small settlements and villages

1) It is necessary to record and assess the damage caused to the monuments of traditional architecture. This will be done by teams of experts, and the results will be used to complete the existing records and reclassify some of the monuments.

2) For monuments of traditional architecture belonging to one of the higher categories of importance, careful studies will have to precede reconstruction, to make sure that the monument is restored to its original state, i.e. the one before it was damaged.

3) In restoring monuments belonging to a lower category and in erecting new buildings in rural areas and small towns, all reconstruction and building work should ensure the continuity of traditional architecture in order to preserve the original regional identity and ecosystems. This means that while respecting the principles of traditional architecture, we can insure that every house will be a genuine product for the needs of the community.

Aid for architectural and archaeological work is expected in the following:

- A precondition for all activities is carefully prepared documentation. This will include the preparation of an atlas of traditional architecture encompassing characteristic examples organized by regions for Croatia, separately for villages and for small towns. The purpose of the atlas is to serve as an model for architectural design along traditional lines.

- Financial aid will be needed for research, evaluation and publication of the atlas and for organizing a competition for the best designs of regional houses.

B - Archaeological sities

The activities to be undertaken should include the following:

1) Making a field survey of all damaged areas containing archaeological sites, assessing the damage to archaeological monuments and preparing documentation about archaeological sites and damage;

2) Working out a program for the protection of archaeological sites;

3) Organizing the protection of moveable archaeological finds and recording them;

4) Applying archaeological methodology to historic urban wholes.

II. Restoration, conservation and protection of movable cultural property

In trying to save the cultural heritage in Croatia restorers are meeting with three main groups of problems:

a) the lack of trained experts

b) rather poor technical equipment

c) the lack of working space

In addition to these the war has caused a lack of adequate storage facilities.

Up to now over 3,000 pieces of art have been dismantled, evacuated and stored in the inadequate shelters. This will definitely make their condition even worse, so after their return to their original place they will require a restoration treatment.

Therefore it is necessary:

a) To organize specialist courses for restorers and Art Academy students with the aim of achieving secure conservation and restoration skills. These courses should become the basis of regular education in the following years.

b) To establish the close cooperation of international experts within organized courses in activities directed at saving war-damaged items in the Croatian cultural heritage.

c) To obtain the necessary financial means for carrying out conservation and restoration treatments, for packing and transporting of art pieces and for necessary technical equipment in the workshops and shelters.

d) To establish a specialized restoration center for wooden art pieces due to the enormous number of evacuated wooden art pieces. This center should provide for adequate storage, conservation research and restoration of wooden art pieces.

e) To form the mobile restoration studios in the threatened regions.

It is necessary to collect further information about:

- Damaged collections
- Collections "evacuated" by force
- Collections in areas where hostile military activity is still taking place
- Collections in occupied regions
- General threat of total damage in the future.

Finally, I have to say that the Institute for the Protection of Monuments has elaborated detailed recommendations about the protection of each specific category of monuments. It has recently completed photogrametric documentation of most important architectural monuments.

The Ministry of Culture has done everything it could to organize procedures for the protection of monuments following the guidelines of the Hague Convention about the special protection, transport and signalization of the protected monuments. Unfortunately, the Yugoslav Army has not been respecting the principles of the Convention. In this context the Croatian Ministry of Culture has a whole list of remarks which might be very useful in future situations. For instance, the fact that Republic of Croatia had to wait for months for formal political recognition meant that any direct contact with UNESCO was impossible. Contacts about the shelling of Dubrovnik and Split were exceptional. All normal contact had to be made through the Yugoslav Federal Government in Belgrade, and they of course were the opposite side in the war; while planning further attacks, they consistently denied that destruction was taking place.

WAR DAMAGE ON THE CULTURAL HERITAGE IN CROATIA
(Data collected by April 1, 1992)

INDIVIDUAL CULTURAL MONUMENTS

REGION/SUB-REGION	DEGREE OF DAMAGE				TOTAL
	A[1]	B	C	E	
EAST CROATIA					
Slavonian-Podravian Region and Baranya	14	14	3	16	47
Slavonian-Posavian Region	16	22	20	30	88
SUBTOTAL	30	36	23	46	135
CENTRAL CROATIA					
Zagreba Region	59	16	1	1	77
Hrvatsko Zagorje	0	0	0	0	0
Varaždin District Podravina and Međimurje	5	1	0	0	6
Lonja and Ilova River Basin and Bilogora Mt. Podravina	2	4	6	6	18
Sisak District Posavina and Banya	3	22	22	18	65
Karlovac District Pokuplje and Kordun	6	17	11	3	37
SUBTOTAL	75	60	40	28	203
MONTANE CROATIA					
Goran Region	1	0	2	1	4
Lika Region	5	6	14	2	27
SUBTOTAL	6	6	16	3	31
NORTH-CROATIAN SEASIDE					
Istra	0	0	0	0	0
Kvarner	0	0	0	1	1
SUBTOTAL	0	0	0	1	1
SOUTH-CROATIAN SEASIDE					
North Dalmatia	27	28	22	19	96
Central Dalmatia	7	4	1	3	15
South Dalmatia	40	30	5	27	102
SUBTOTAL	74	62	28	49	213
TOTAL	185	164	107	127	583

---

1. DEGREE OF DAMAGE:
A - minor damage
B - major damage
C - destruction
E - information not available

WAR DAMAGE ON THE CULTURAL HERITAGE OF CROATIA

THE DEGREE OF DAMAGE CAUSED

(Data collected by April 1, 1992)

1.
INDIVIDUAL CULTURAL MONUMENTS

DEGREE OF DAMAGE	Category					Type of potection			TOTAL
	O[1]	A	B	C	N	R[2]	P	E	
Minor damage	0	59	25	67	34	80	1	104	185
Major damage	0	43	33	50	38	77	8	79	164
Destruction	0	8	18	31	50	26	6	75	107
Information not available	0	9	15	23	80	25	5	97	127
TOTAL	0	119	91	171	202	208	20	355	583

2.
OLD CITIES AND PLACES

Destroyed	0	0	1	0	26	1	3	23	27
Burnt down	0	0	1	0	31	0	1	31	32
Partially destroyed	1	5	12	1	65	17	2	66	84
Damaged	1	5	8	5	74	17	6	70	93
Information not available	0	3	3	0	3	4	0	5	9
TOTAL	2	13	25	9	199	39	12	195	245

---

1. CATEGORY: O - entered in the World Heritage List; A - national and world importance; B - regional importance; C - local importance; N - non-categorized items

2. KIND OF PROTECTION: R - registered; P - preventively protected; C - filed

WAR DAMAGE ON THE CULTURAL HERITAGE OF CROATIA

MUSEUMS, ARCHIVES, LIBRARIES

(Data collected by April 1, 1991)

1.
TYPE OF INSTITUTION

TYPE OF MONUMENT	Category					TOTAL
	O[1]	A	B	C	N	
Museums, Galleries, Collections	0	9	16	11	1	37
Archives	0	0	0	0	9	9
Libraries	0	0	0	0	22	22
TOTAL	0	9	16	11	32	68

2.
DEGREE OF DAMAGE

Minor damage	0	5	6	4	17	32
Major damage	0	4	5	4	9	22
Destroyed	0	0	0	1	3	4
Data not available	0	0	5	2	3	10
TOTAL	0	9	16	11	32	68

---

1. CATEGORIES: O - entered in the World Heritage List; A - national and world importance; B - regional importance; C - local imporatance; N - non-categorized items

WAR DAMAGE ON THE CULTURAL HERITAGE OF CROATIA

THE TYPES OF MONUMENTS AFFECTED

(Data collected by April 1, 1992)

1.
INDIVIDUAL CULTURAL MONUMENTS

TYPE OF MONUMENTS	Category					Type of potection			TOTAL
	O[1]	A	B	C	N	R[2]	P	E	
Palaces and mansions	0	42	25	81	27	63	3	109	175
Fortresses	0	29	6	1	1	34	0	3	37
Sacred buildings (I) (churches)	0	30	46	84	171	82	15	234	331
Sacred buildings (II) (monasteries)	0	15	9	2	2	22	1	5	28
Monuments in public places	0	1	1	0	0	0	0	2	2
Mausoleums, tombs and cemeterys	0	1	3	3	1	5	1	2	8
Commercial buildings	0	1	1	0	0	2	0	0	2
TOTAL	0	119	91	171	202	208	20	355	583

2.
OLD CITIES AND PLACES

TYPE OF MONUMENTS	O	A	B	C	N	R	P	E	TOTAL
Historic and memorial sites	0	0	1	0	0	1	0	0	1
Old cities	2	9	16	6	14	32	6	9	47
Historic villages	0	0	5	0	182	0	6	181	187
Archaeological sites and monuments	0	4	3	0	3	5	0	5	10
TOTAL	2	13	25	6	199	38	12	195	245

---

1. CATEGORY: O - entered in the World Heritage List; A - national and world importance; B - regional importance; C - local importance; N - non-categorized items

2. KIND OF PROTECTION: R - registered; P - preventively protected; C - filed

Saving Grace: Can Cambodia's Ancient Site of Angkor Be Preserved?

Donald Garfield, Senior Editor, Museum News, 1225 Eye St. N.W., Suite 200, Washington, D.C. 20005.

For 1,200 years Cambodia's historic royal complex of Angkor, one of the world's architectural wonders, has survived all threats from man and nature. But the greatest pressures on Angkor have come during the late 20th century, especially the devastation caused by more than two decades of war and civil strife. Now, with increased conservation and restoration efforts initiated by the World Monuments Fund and other international cultural institutions and coordinated by the United Nations Educational, Scientific and Cultural Organization (Unesco), three new threats have emerged. One is an increase in the looting of the sculptures and bas-reliefs decorating the sides of temples, galleries, and terraces. The second is the lack of a comprehensive, coordinated plan for conservation. The third, with potentially the most destructive results, is the uncontrolled development of the sacred site as a major tourist attraction.

Angkor's history is a rich and complex one. For more than 600 years Angkor served as the hub of the Khmer Empire, which at its height controlled an area stretching from Thailand to Vietnam. While other world capitals sprang up aside important water sources, the rise of Angkor followed the creation of reservoirs begun by the late-9th-century king Yasovarman I. A series of canals and moats irrigated the land for extensive agriculture and encircled the capital's architectural monuments, while also serving as the means whereby the buildings' sandstone blocks weighing up to six tons were delivered to the work sites.

Angkor consists of seven distinct capitals occupying 200 square kilometers, one-third the size of New York City. Today the monuments that remain are temples, monasteries, and city walls. Secular architecture reflecting the daily life of Angkor was built primarily of timber, so no longer remains. Angkor Wat, built in the 12th century, is the most famous monument at the site. It exemplifies the so-called mountain-temple with its multistory core and towers. Both Hindu and Buddhist images decorate the temples, gates, and terraces providing a precious record of the Khmer culture. By the early 13th century the era of ambitious building activity drew to a close and gave way to a period of decline that led ultimately to the termination of the Khmer hold on Angkor in 1432, when the neighboring Thais stormed the site.

Angkor, unlike the Mayan ruins of Central America, never totally disappeared into the jungle and lost contact with the outside world. Records show that it was visited in later centuries by the Japanese, who drew the first plans of Angkor Wat, and by Europeans, first the Spanish and Portuguese, then the Dutch, Germans, and French. In the course of constructing its Indochinese empire, the French turned the country into a protectorate in 1863, which lasted until the granting of independence in 1953. For the century that the French dominated the cultural life of Cambodia and undertook maintenance and architectural conservation, the Ecole Française d'Extrême Orient compiled documentation of the site and researched Khmer culture and art. All this was abandoned in the early 1970s when they left the country due to the outbreak of war.

Cambodia became virtually inaccessible after the outbreak of war set in motion by the U.S. bombing of Cambodia in 1969. A series of events followed that caused the Cambodian people immense suffering. The government of Prince Norodom Sihanouk collapsed when Lon Nol took power. That military regime fell to the Khmer Rouge in 1975. In one of the most tragic chapters in the region's long history, the Khmer Rouge force-marched the citizens of the capital Phnom Penh to resettlement in the countryside, causing the deaths of an estimated million and a half of the country's eight million inhabitants. In 1979 the Vietnamese invaded and installed the regime of Hun Sen. The country endured another 13 years of civil war and embargoes imposed by the major non-Communist countries of the world. The four warring factions vying for control of the country's future signed a peace treaty in October 1991, and United Nations peace-keeping forces arrived. In early April this year a trickle of Cambodian refugees housed in camps inside Thailand began to return home to a countryside strewn with land mines, while military action between government forces and the Khmer Rouge continued. Although elections are scheduled for mid-1993, no one is making any predictions about the immediate return of peace.

For the past two decades no one knew exactly what toll the nation's warfare had taken on the monuments of Angkor. According to conservation architect John Sanday, news would come out usually from reporters who were covering the war and had made side trips to Angkor. Their reports suffered from a lack of knowledge of conservation principles. They exaggerated the "damage" being done by a team of Indian restorers, the only outside group allowed into the country to work on Angkor Wat, the result of a six-year contract negotiated by the late Indian premier Indira Gandhi. A minor controversy resulted among western conservationists about the Indian efforts. Sanday says they have been unfairly criticized but admits that their work was largely superficial and the materials and techniques employed of questionable value. What arose from a political gesture lacked continuity and coordination and the vital component of involving the Cambodians. The Indians did not speak Khmer or French and a different team went in each year.

"The Indians deserve praise," says Sanday, "for undertaking an enormously complicated task. So far, they are the only group that has gone in and actually done something. Sadly, there is mismanagement and money spent on cosmetic work is wasted."

Aside from the Indian work on Angkor Wat, the world had to await the end of the last decade to appreciate the true state of affairs at Angkor. In the fall of 1989 a hurricane swept through Angkor. This natural disaster led to the government repeating its call to outside groups like the World Monuments Fund. A message from the Ministry of Culture, titled "An Appeal for Angkor," reported the uprooting of 750 large trees and consequent damage to monuments such as the Temple of Bayon and the Terrace of the Elephants. The government of Hun Sen appealed to "countries and specialized institutions such as Unesco to take urgent steps to help to repair, to restore, and to preserve one of the world's wonders.... [We] welcome all kinds of assistance, cooperation, and suggestions aimed at preserving this common cultural heritage for future generations."

That December, after two years of planning and an invitation by the Cambodian government, the World Monuments Fund sent an investigative team to Angkor. Bonnie Burnham, the fund's executive director, writing in the Winter 1990 issue of *Art International*, summarized what they found: "Although there is isolated evidence of shelling, of attempted looting and defacement of sculpture, and of damage caused by small bands of soldiers who used the Angkor structures as shelter, the greatest detriment brought on by the war was the loss of continuity in the conservation process that is so necessary to the survival of these magnificent temples. Their imposing appearance conceals an extreme fragility that will, unless action is taken, turn rapidly to irreversible decay."

In the years since this initial visit, WMF drafted a report outlining the steps necessary to ensure the protection of Angkor. Unesco has convened two round tables of experts to consider the recommendations, the first in June 1990 in Bangkok, Thailand, and the second in September 1991 in Paris. A major boost came in October 1991 with the signing of a peace treaty by the four warring parties of Cambodia and the arrival of United Nations peace-keeping forces.

Burnham, Sanday, and other visitors to Angkor continue to worry about the three major threats to the site: tourism, looting, and the lack of a comprehensive conservation plan.

Tourism only became a potential problem during the '60s with the explosion of international travel made possible by inexpensive air fares, the proximity of hundreds of thousands of Western soldiers, and the decade's adventurous *zeitgeist*. According to the WMF's draft Angkor report, "a sizeable tourist infrastructure had begun to develop around Angkor to receive the 45,000 tourists anticipated in 1969." The outbreak of war brought development, which had not progressed very far, to a halt. The two hotels on the site were bombed. The WMF report continues, "There are devastating examples in Asia of where the tourist industry dominates the historic site. Cambodia should not be provoked into using Angkor as a foreign exchange earner without giving careful thought to the controlled development of a sympathetic tourist infrastructure." A delicate balance currently exists between the architectural monuments and its natural and human environment. Farmers and small groups of Buddhist worshippers have returned to Angkor, while the surrounding jungles retard wind erosion and give precious shade to the monuments. Clearing the forest either for commercial logging or tourist development poses a major threat as does the danger of environmental pollution.

But uncontrolled development may be difficult to forestall, given the current financial conditions in Cambodia, which reportedly has been unable to pay its government workers or army. The golden goose of the tourist trade may be a temptation impossible to resist. So speculates Eleanor Mannikka, professor in the Department of the History of Art at the University of Michigan in Ann Arbor and noted authority on Khmer art and culture. She visited Angkor with a group of graduate students last December and January. With the removal of the U.S. embargo, real estate speculators and money have streamed into Phnom Penh.

"It's like the Wild West" says Mannikka. As for the touristic exploitation of Angkor, it has already begun. "Unlike other heritage sites in Thailand that have not been inundated with visitors, at Angkor one is no longer alone. When we were there last December the hotel at Siem Reap didn't have enough room for the tourists, so they were farming them out to individual homes. They had up to five planes landing a day with 50 people on average aboard. In other words, it was total chaos. Tourists came and left whenever they wanted and were not to be controlled by the Cambodian guides whom they treated like servants."

Evoking memories of 18th-century aristocratic visitors to the Roman ruins lounging atop exedrae overgrown with vegetation, Mannikka described a group of Japanese tourists parking themselves and a gigantic lunch cooler on the floor of one of Angkor's sacred temples. This profanity sent a French scholar who happened to be nearby into fits. He asked whether they would do the same in a Buddhist temple at home. "Soda pop sellers have set up shop. Pepsi cans are being jettisoned and Angkor is becoming littered." Another threat reported by Mannikka, to both monuments and visitors, is the free access tourists have to the site. "There is no way to predict when a section of a monument is going to collapse." And then, finally, there is the rumor she heard that plans are afoot to build a hotel right at the heart of Angkor.

Unknown at this stage is whether these depredations will continue and grow. The WMF has recommended controlled development targeted at limited, high-end (as opposed to mass) tourism with proceeds reinvested in conservation and training.

Removal of cultural property probably began in the 15th century with the fall of the Angkor to the Thais. European visitors, explorers and tourists, carried off works of art beginning in the late 19th century. André Malraux, France's first minister of culture and an influential museum theorist, was convicted in Phnom Penh in the 1920s for trafficking in antiques from one of Angkor's temples. In survey books on

Cambodian art, much of the illustrated material is credited to museums in France and the U.S. This is part of museum collecting history, a fact of life.

The climate changed in 1972 with ratification by the U.S. Senate of Unesco's "Convention on the Means of Prohibiting and Preventing the Illicit Import, Export, and Transfer of Ownership of Cultural Property," which, however, took another decade to put into effect until passage of the 1983 Convention on Cultural Property Implementation Act. Because of the unsettled state of internal affairs at that time, Cambodia was not among the signatories nor is Angkor covered by the World Heritage Convention, designed to protect the world's recognized heritage sites.

In interviews with curators of Southeast Asian art at the Walters Art Gallery and the Asian Art Museum of San Francisco, the Avery Brundage Collection, a clear distinction exists between material legally acquired prior to 1972 and works of art on the market since then, which could not have been brought into the country anyway because of the U.S. embargo on Cambodia. Still there is a potential conflict should the Asian country at a later date seek repatriation of artworks that left the country, legally or not.

All who speak of Angkor and Cambodia tell a sad tale of looting, either of sculptures from the Conservation d'Angkor, where statues were brought for protection, or from the site itself—sawn off heads and limbs trucked out to Thailand for sale. An apocryphal episode narrated in Unesco's *Source* of January 1992 makes the relevant point: "In one case, so the story goes, villagers in the north of the country were paid $100,000 for a statue they had uncovered at Banteay Chmar. This sum represents several centuries of revenue for someone who earns probably less than five dollars a month, which is about what the few guards around Angkor are paid." In addition to decapitated monumental statues and missing movable pieces, figures and sections of the ubiquitous sculptural relief have disappeared into the maw of a voracious art market, visible to anyone wandering through the antique shops of Bangkok, says John Sanday. Muddying the situation is the fact that unrecorded artwork can have an ambiguous provenance and come from a private collection in Thailand as well as from an illicit smuggling ring in Cambodia. Nancy Hock, curator of Southeast Asian art at the Asian Art Museum of San Francisco, offers a suggestion that Cambodia may wish to pursue once the political climate is more settled: "When Cambodia is in a position to do so, it should look to Japan for legislative models to inventory their national artistic patrimony and set up categories of significance so that the most important artwork is protected without banning altogether the sale of art."

Photo documentation of the approximately 150 items missing from storage and others lost from the Angkor site is being forwarded by the Ecole Française d'Extrême Orient for a brochure to be published and distributed by the International Council of Museums in Paris. According to Hiram Woodward, curator of Asian art at the Walters Art Gallery, such publications can be used to identify stolen material and lead to their repatriation. But the ultimate solution likely hinges on the success of the comprehensive conservation management plans taking shape in New York, Paris, and Phnom Penh. Part of this plan is a thorough inventory and photographic documentation of the site, its sculptures, and bas-reliefs. WMF and Tokyo's Sophia University team are customizing a database management system for site inventory field records. Its format can accommodate other forms of documents and holds the potential for image libraries, storage data, and other applications.

The third threat to Angkor involves the protection and stabilization of the architectural monuments. Two factors conspire to wreak havoc on Angkor's buildings. One is inherent design flaws in the construction of the temples. Particularly unstable are the so-called mountain temples like Angkor Wat built on a raised stepped pyramid with one central and four corner towers. Deep undercutting of the walls faced with sandstone blocks to create plinths and cornices places the structure in jeopardy. Structural instability also results from inadequate drainage and the precipitous angle of the temple's heavenward reach. Finally, the columns and foundations carry an enormous burden of galleries and heavy corbelled arches that can at any moment send the whole edifice crashing to earth.

The second factor is nature itself. Rampant vegetation has sent roots and trunks through, over, and around walls and gates. The wet climate and improper drainage attack the buildings' stability and encourage the formation of salts on surfaces, marring the sculpted reliefs and weakening the facing's bond to the building's structural support.

The report prepared by the WMF and submitted to the Unesco consortium of experts and the Cambodian government sets out an overall blueprint for action that resembles a cultural heritage Marshall Plan. It is grounded in realism and sympathetic to the national pride of Cambodia, whose cultural identity depends on the glories of the historical Khmer empire and its flowering at Angkor. At the top of the organizational chart stands the government of Cambodia and its several ministries and offices through which the planning must pass. When the WMF sent its first team to the country in 1989 they had a long talk with Premier Hun Sen, and Prince Sihanouk is serving as liaison between the Supreme National Council and Unesco.

The key to success, according to Sanday, rests on the transfer of appropriate technology and the training of Cambodians in the basics of conservation technology, methodology, and philosophy. Indigenous expertise either did not survive the war or moved away from Angkor. At the outset, therefore, the WMF team took 25 university students to the site to begin the long process of creating from scratch the education system and the trained individuals needed to realize the ambitious plans delineated in the Angkor project report.

Under Unesco's umbrella all aspects of the plan are being coordinated to avoid an ad hoc and fragmented approach. The WMF Angkor report urges all countries and organizations participating in the Angkor project, including the Archaeological Survey of India, the Ecole Française d'Extrême Orient, Japan's Sophia University, a Polish group, and the WMF, to follow the guidelines of architectural conservation spelled out in the 1964 Venice Charter—i.e., minimal intervention, reversibility, historical faithfulness, and documentation of procedures. Each participant has selected one of the structures to concentrate on, but the documentation and inventory process is being systematized with field data entered into a computer.

What role can museums play in the salvation of Angkor? Initially they can reach out to the National Museum in Phnom Penh, spared military-related damage and pilferage owing to the pan-Cambodian importance of its collections, and participate in the transfer of expertise required by the Cambodians and interwoven into the proposed WMF plan. If one assumes that the cultural fabric of the country is seamless, support for Cambodia's museums contributes to the overall heritage superstructure, all parts of which rely on a vigorous educational system, government support, and management coordination of limited resources.

According to Mannikka, the museum in Phnom Penh is a shambles, with paint flaking from the walls and bats entering the roof and fouling sculpture. Statues placed in the basement for protection during the war have stood in as much as 12 inches of water because of inadequate drainage. At last report workers were putting them on concrete blocks to raise them above the water level during the rainy season.

Asked what U.S. museums might do, Burnham replied that the museums of Cambodia are so destitute that any assistance would be very meaningful. "Individuals and institutions," she said, "can help out in a sisterly fashion to rebuild archives, libraries, and conservation facilities and consult in areas like paper conservation and the display of materials." She recalls that after the devastating earthquake that struck

Mexico City a number of years ago, two U.S. museums that happened to be presenting an exhibition of the work of Mexican muralist Diego Rivera channeled help through the WMF.

Burnham also cites the need for museums to be aware of the looting. They could assist the Cambodians in understanding the workings of the art market and in setting up mechanisms to protect the country's artistic patrimony. Those institutions with Southeast Asian collections are in a position to contribute to the fund raising and advocacy efforts, informing the public of the desperate situation at Angkor and in Cambodia. Through educational programs and exhibits, museums can draw attention to the wonders of Khmer culture and convey the fact that Angkor belongs not only to the Cambodians, but to the entire world.

As one example of what part museums can play, New York's Asia Society, in collaboration with the WMF, will hold a symposium June 5-7 to review the current status of Cambodia's architectural treasures and to form a dialogue among specialists in the areas of art, history, conservation, and anthropology as well as among representatives of the major players in the Angkor project.

Ultimately, the salvation of Cambodia's museum and historic sites like Angkor depend on a resolution of the political situation and financial rebuilding of the nearly bankrupt country. While some observers to the peace process are pessimistic, the director of the national museum in Phnom Penh sounds a note of optimism despite personal tragedy. Pich Keo, profiled in Unesco's *Source* of January 1992, survived four years of forced labor in the country's rice paddies, cut off from his family and profession as an archaeologist. He and his wife currently draw a combined salary of about $25 a month.

Keo preserves his passion for Khmer culture and its preservation in his museum and at Angkor. His assessment is realistic, but at the same time hopeful: "We are faced with the enormous task of saving our national and international heritage and we are forced to face the world empty-handed. Some of us don't even have hands to hold out for help and many of us are thin and hungry. But we must save Angkor." Some help has come from Unesco, which gave $35,000 to clean and repair the museum's basement and its contents. Convinced that he and his colleagues are working not only for their own country but for the citizens of the world, he expresses a sanguine view of the fragile peace process. "I am sure peace will come," he writes. "I am very optimistic—things are moving in the right direction, and as long as the international community does not abandon us Angkor will not perish."

Addenda:

The World Monuments Fund

The World Monuments Fund was founded in 1965 by James Gray, a retired Army colonel who once had a scheme to help the Italians prop up Pisa's leaning tower. Today the fund describes itself as "the only private nonprofit organization that sponsors worldwide preservation of cultural heritage." Its work began in earnest with the 1966 Venice flood, and scores of endangered sites have benefited since. The WMF has succeeded in joining conservation expertise with advocacy, including fund raising and training, while nurturing local commitment by governments and individuals.

The organization is now under the leadership of Bonnie Burnham, who came to the fund in 1985 after seven years at the International Foundation for Art Research, a stolen-art documentation center in New York. The WMF prides itself on responding quickly when help is urgently needed. Recently, for example, it helped underwrite the curation of a photo exhibition on the Yugoslav military bombardment of historic sites and structures in Croatia. The Samuel H. Kress Foundation supported this and much of the fund's work in Europe.

As part of its mission to keep the public informed, the fund publishes international preservation reports. In addition to one on the Angkor temple complex, these include *The Razing of Romania's Past*, describing the depredations of Nicolae Ceausescu on his country's historic vernacular architecture, and *The Future of Jewish Monuments,* which documents the current condition, use, and ownership of Jewish sites worldwide.

To help accomplish its global mission, the fund has opened affiliate organizations in France, Italy, Spain, and Mexico. As a complement to its conservation and renovation of the Church of St. Ann and the Holy Trinity in Brooklyn, N.Y., the fund in cooperation with the St. Ann Center for Restoration and the Arts founded the American Heritage Center for Conservation and Training to impart technical expertise.

Burnham speaks eloquently about the need to join historic preservation to the global environmental movement: "Cultural heritage and its protection are part of the environment, and the issues that confront the conservation of the earth—pollution, population crowding, uncontrolled tourism, and global warming—also threaten the built environment. WMF's goal is to bring a concern for the built environment to the global conservation agenda."

Reprinted from *Museum News*, Vol. 71, No. 3 (May/June 1992) pp. 42-47.

A STRATEGY FOR PLANNING AND PROTECTING IMMOVABLE HERITAGE: LESSONS FROM NATURAL DISASTER PLANNING

Charles C. Thiel Jr.
Consultant to the Getty Conservation Institute, Marina del Rey, California

ABSTRACT

Immovable cultural properties, whether buildings, archeological sites, museums or libraries, are at substantial risk in the event of armed conflict. In many ways, natural disasters pose many of the same threats to such structures as does war. Earthquakes impose horizontal loads on structures, their weak direction, tornadoes and hurricanes impose large substantial over pressures on walls and roves, just as ground shock and air blast from explosions do. Fire spreads when there is both fuel and ignition, regardless of whether the ignition is purposeful or accidental. A clear message from the natural hazards literature is that you cannot do anything during the emergency period that you have not carefully thought through before the emergency. This paper reviews approaches to protection of cultural properties and the existing Hague Convention. It is proposed that a concerted effort be mounted by the professional conservation community to: 1) prepare guides on how to identify and document cultural properties important to the community; 2) descriptive manuals on how to protect cultural properties using easily available materials and techniques; 3) distributed them to the local communities likely to be impacted; and 4) provide limited technical resource teams to stabilize or repair after hostilities.

THE PROBLEM

The images from the recent disintegration of Yugoslavia are vivid. You don't have to be a preservationist to feel the agony as a medieval enclave appears to be systematically destroyed, its monuments reduced to rubble, or when serious damage is done to the Diocletion summer palace at Split. The collective loss is clear and apparent to the most casual observer. Indeed it is so vivid and so wrenching that the print and electronic images often invoke support for those who suffered the loss, outside the context of the conflict.

How can such images have such power? Could it be that there are some physical structures that are so compellingly part of a community's collective consciousness that they are part of the fabric of identity for the community? That their destruction stiffens the resolve to resist for those suffering the loss, rather than causing the capitulation? There is more than some evidence to support this thesis. How else can we explain the fact that statuary on the Winter Palace was removed and buried while Leningrad was under its historic siege, or that Kyoto was not saturation bombed during World War II. In the latter case when such an attack was planned, the observation was made that it was of little military importance and that if we hoped to win the peace, we should not strike a lethal blow to the identity of the Japanese people. When a monument is used as a center for military purposes the tables are turned. We showed little sympathy for the combatants who used Angkor Watt and forced their advisories to attack and seriously damage its irreplaceable structures.

The power of images of the destruction of significant historical heritage structures is so strong that it can be used as part of the war waging process to curry favor from non combatants, swaying opinion and military support.

The occurrence of a war or armed conflict has a wide range of impacts on a community. It is natural to focus on the immediate political, social and economic interests that are the root cause of the conflict; but, there are others. Often one of the longest lasting impacts suffered by a community is the lost of structures that are at the core of the community's self-identification. The purpose of this paper is to begin the process of identifying approaches to protecting immovable heritage—buildings, archaeological sites, museums, libraries, statuary—that are within a war zone and for which movement out of harms way is not feasible or practical. The focus is on learning from experience in mitigating and responding to natural disasters. In many ways, the occurrence of a natural disaster is similar to war. Earthquakes impose horizontal loads on structures, their weakest direction, tornadoes and hurricanes cause substantial over pressures on walls and roofs, just as ground shock and air blast from explosions do. When a key structural element is destroyed, progressive collapse can proceed from gravity loads alone; it matters not whether the member was destroyed by a shell or a natural event. Fire spreads when there is both fuel and ignition, regardless of whether the ignition is purposeful or accidental.

There is considerable world-wide experience in the issues of protecting structures from the ravages of natural disasters—whether earthquake, fire, wind or flood—and in responding to their occurrence to reduce long term consequences, Drabeck 1986, Petak and Attkisson 1982, Anderson and Thiel 1979. From this rich literature there is one overwhelmingly clear message:

> *You cannot do anything during the emergency period that you have not carefully thought through before the emergency, .*

The truth of this is repeated over and over and, in the experience of the author, is the most fundamental of precept in planning for natural or man-made disasters. The occurrence is always characterized by surprises. It is this very property that makes planning the key issue in moderating or mitigating consequences. Many people trust their wits to improvise under pressure; yet examination of almost every positive improvisation in a natural disaster uncovers the simple fact that there was a prior intellectual examination of a comparable situation followed by formulation and evaluation of alternative approaches to mitigating the unacceptable consequence.

In a sense, the spreading of a conflict to a particular community differs from the occurrence of a natural disaster in one important aspect. While there is usually little warning of an earthquake, the likelihood of conflict reaching a particular community is usually signaled over a period of time, sometimes months.

Most of our physical images of natural disasters come from photographs of the damage done. Earthquake images tend to leave the viewer with the impression that everything was destroyed. It is not surprising that these press images convey a sense of complete destruction—the news worthiness of an undamaged building is small. Yet when an objective censuses of damage is completed, usual that only a small proportion of structures are destroyed and only a modest proportion are severely damaged. Interviews with individuals who have been in war zones suggest that the same observation is true for most combat areas. A prime recent example is the 1989 Loma Prieta Earthquake. If you trusted the impressions of television reporting, you may have concluded that the entire Marina District was destroyed. Yet the fires were confined to a few buildings and as many as 90% of the buildings had little or no structural damage, Thiel 1990.

Table 1 illustrates typical proportions of damaged buildings for unreinforced masonry buildings (URM) in regions of strong earthquake ground motions. URM are a particularly earthquake vulnerable building type. Undamaged buildings are excluded, which would make the numbers of damaged buildings somewhat lower. Recently available data for the Croatian civil war, Table 1 abstracted from Belamaric 1992, suggest comparable rates of damage to those for the earthquakes listed. This supports the thesis that damage is not always total as might be concluded from news reports.

War damage is often concentrated where military assets (e.g., personnel, communications, supply) are concentrated. When mass destruction weapons (including saturation bombing) are not used or available the likelihood is that many structures can and will be spared. Thus, it is reasonable to presume that there is a substantial likelihood that a cultural property, or any non military property, can survive a conflict without being severely damaged. Clearly, the question before us is how to enhance this likelihood either by preventing the targeting of the property, see the section on the Hague Convention, or how to decrease the structures vulnerability.

Table 1. *Comparison of the numbers of buildings with minor and major damage and destroyed in different circumstances. The earthquake data are for unreinforced masonry buildings and were taken from data collected by Thiel and Zsutty (1987); the Croatian civil war data through April 1,, 1992 are for cultural monuments reported by Belamaric, 1992.*

	Minor	Major	Destroyed
1933 Long Beach Earthquake, pre-1930 unreinforced masonry buildings	496 (35%)	556 (39%)	359 (25%)
1906 San Francisco Earthquake, downtown unreinforced masonry buildings	37 (34%)	55 (50%)	18 (16%)
Croatian cultural monuments, 1991-92 civil war	185 (41%)	164 (36%)	107 (23%)

Table 2. *Actions taken to respond to the treat of a natural disaster compared to those that could be taken to the possibility of war, Thiel 1981.*

Disaster Phase	**Time Period**	**Typical actions during disaster phase**	
		Natural Disasters	**Armed conflict**
Hazard identification	Ongoing	Identification of the threats, where they are greatest, structures that are most vulnerable	Identify and document the most important cultural heritage structures
Mitigation	Ongoing and before event warning	Physical changes to structures to decrease either their risk or the consequences of their damage	Remove elements as practical and undertake protective steps to reduce the structures vulnerability; notify enemy of structures description, location and importance
Prepared-ness	Before and after warning	Preparation of response plans, training, testing of plans, pre-positioning of supplies.	Protective markings for important structures; instruct defenders to avoid their military use
Emergency response	During and immediately following event	Provision of emergency life saving services, evacuation of hazardous damaged structures, actions to prevent further destruction; rubble removal, restoration of critical services	Fight fires; remove debris loading; inspect to determine if damaged and shoring required to maintain current condition; protect damaged structures from rain and snow
Recon-struction	After event	Restoration of the community, including steps to make the next natural disaster less severe in its impacts	Identical

Planning for natural disasters typically focuses on the five distinct phases or types of actions as given in Table 2. The actions taken before and during a conflict could be quiet similar. Some are suggested in the Table.

A key observation from successfully managing the consequences of say an earthquake when compared to war is that the consequences of the earthquake's occurrence are determined by the actions that are taken when there is only a threat, not when it has become clear that the earthquake is about to occur, as might be inferred by a fore shock. For most natural disasters, hazard identification, mitigation and preparedness before the event—not when it is eminent—are the only practical ways to protect structures, Thiel 1988. For war situations the time frame after warning may be significantly longer, say weeks, than for a natural disaster. War seldom occurs without warning, and the conflict rarely engulfs an area immediately without substantial warning. This provides a major opportunity to protect important structures. It is during the build-up phase before the conflict directly threatens an area when the most productive steps can be taken to protect significant structures.

NATURE OF THE THREAT

A structure can be subject to damage from a variety of damage agents during the course of a conflict:

- explosions on or within the structure
- ground shock from nearby explosions
- air blast from nearby explosions
- direct hits from projectiles, shrapnel, or small arms fire
- debris from other failing structures
- fire

Damage can either be purposeful, that is targeted, or accidental. Purposeful attacks can be prompted either by the presumed use of the facility, say as a command post or shelter for combatants, or as a psychological blow against the community, that is, an attempt to break the will of resistance by destroying a valued symbol of the community.

When a conflict is raging, the focus is on the immediate life safety for your side and on wining. It would be surprising if combatants were to consider preserving a particular structure if they had cause to believe that their action could endanger their own lives. When fire comes from a structure, the immediate response it to throw every thing you have at it.

Thus, it is of paramount importance that every effort be made to make structures we desire to protect *unattractive* for use during the conflict by the combatants. One historic strategy has been to identify hospitals by big bold makings and to depend on the humanitarian instincts of the combatants to make them off limits to hostilities—either as targets or for offensive and defensive operations. But as technology has allowed war to be fought at impersonal distances, the control and targeting of weapons may make it much more difficult to select specific targets.

While it is certainly widely believed that with the advent of "smart weapons" the avoidance of damage to important cultural heritage is practical, it is also true that the number of combatants having these resources are small. Undaunted, it still suggests the same issue, prior identification of targets of cultural heritage importance (description, latitude and longitude) would allow a combatant to avoid them when using such weapons. When there is a disparity of technological capability between the combatants, the superior force may want to avoid damage to such structures so as to not make the post-conflict environment any more difficult than necessary.

Some caution in considering vulnerability is warranted. One of the discouraging aspects of protecting cultural property is that there is a natural tendency to think that the an inability to provide maximum protection makes protection not worth while. Such is not the case; some protection is almost always better than non at all. The explosive charge could be set of far enough away that the minimal protection is just adequate. Half way measures may be sufficient to same critical damage.

WHAT MAKES A STRUCTURE PART OF OUR HERITAGE?[1]

Speaking to the broad subject of seismic retrofitting of cultural monuments, the architectural conservation program coordinator of ICCROM in 1988 said: "The principle is to define the limits

---

[1] This section is based in part on the paper by Edna Kimbro "Conservation Principles and Seismic Retrofitting Technology for Culturally Significant Adobe Buildings" presented at the 1991 Seismic Retrofit of Historic Buildings Conference, Kimbro 1991. This is an element of the Getty Seismic Adobe Project (GSAP)

Protection of culturally significant structures in time of war poses difficult problems for conservators and designers as well as an urgency. Architectural conservators would prefer that a magic wand be waved and an invisible veil of protection appear without any alteration or intervention whatsoever. Designers are understandably frustrated by such unreasonable expectations fueled by what seem to be, if not esoteric, at least subjective notions. An explanation of the relevant principles of architectural conservation may be useful and improve communication between representatives of all disciplines involved in preservation planning. Further complicating this dialog is the conflict among too many structures to protect, resources, and the immediacy of the conflict, all spiced with a deep despair.

It is useful to consider how one might approach the issue of structural interventions in a more leisurely environment. Three basic conservation principles should guide the design of interventions contemplated for historically or culturally significant structures:

1. knowledge of the "object" or building;
2. minimal intervention; and,
3. reversibility.

The first of these fundamentals instructs the conservator to know the object; to study its materials, systems, condition, cultural and physical context, history and alterations, etc. before contemplating any intervention. The object or building is examined as a whole in context, and each of its parts individually assessed. It is evaluated for aesthetic merit, unity or disunity as a formal work of art and historical values as a document of the passage of time or the "flow of history." Any perceived conflicts surfacing between these fundamental values are resolved at this juncture through the evaluation of all available information, and the contextual weighing and balancing of evidence to reach reasoned, justified and non-arbitrary decisions. As a part of this information gathering and analysis process, the need for alterations to the structure is assessed, including anything from replacing broken window panes to seismic retrofitting.

Once the need for alteration has been identified, the conservationists precept of "minimum intervention" comes into play, influencing the design and extent of changes to the structure. The goal of minimizing alterations is to preserve as much of the culturally significant fabric of the building as possible, thereby safeguarding its authenticity or genuineness, while accomplishing its protection from war damage.

The third guiding principle of "reversibility" holds that the alterations made to the object or building should be able to be undone in the future. This allows for the use of improved technologies as they are developed. The precept of reversibility encourages alterations of an additive nature and discourages the removal of material or features. Also, the documentation and storing of any removed material or feature is important to provide the opportunity for future replacement. Since all alterations become part of the history of the building, failure to achieve reversibility results in permanent alteration.

The scholarly approach to identification of significant structures and development of strategies to protect them that are mindful of their characteristics requires thoughtful analysis. Analysis that is better suited to the more leisurely pursuits of peace time. It is no doubt best pursued then, rather than waiting until the need is apparent, when it is beyond the pall of accomplishment except by luck. Reference has been made to the importance of identifying the values inherent in culturally significant structures. (The term culturally significant is preferred to "historic" because it is broader and inclusive of architectural significance). Much literature deals with this subject and values have been categorized in various ways. For purposes of discussion here, values are divided into tangible (possessing obvious visible manifestations) and intangible. Architectural, aesthetic or artistic values of structures fall in the former category, while spiritual, symbolic, and historical or documentary values fall in the latter. Archaeological values and research potential have feet in both camps depending upon the nature of the archaeological resources.

<u>THE HAGUE CONVENTION</u>

The Convention on Protection of Cultural Property in the Event of Armed Conflict signed in the Hague in 1954 provides the world a statement of obligations of combatants to respect and preserve cultural properties. It is interesting to note that the preamble to this Convention includes the observation that "being of the opinion that such protection cannot be effective unless both national and international measures have been taken to organize it in time of peace." This is consistent with the guiding observation for mitigation of natural hazards.

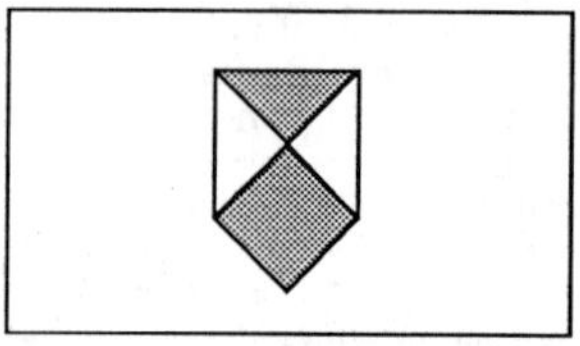

Figure 1. *The Hague Convention marking for cultural properties. The shaded areas are royal blue, while the field is white. It is repeated three times in a triangular formation for properties under special protection.*

The Convention provides, Nobelcourt, 1958:

1. General provisions for the protection of cultural property and obligations of states, including a definition of "cultural property."
2. Provisions for "Special Protection" in a limited number of refuges intended to shelter movable cultural property and immovable cultural property of very great importance.
3. Provisions for the transport of cultural property and its protection in transit.
4. An emblem to designate cultural property under the protection of the Convention, see Figure 1.
5. That its provisions are to be applied in all armed conflicts, whether declared or undeclared, and to cases where an area is occupied even if there is no armed resistance.
6. Administrative execution of the Convention.

The Convention has been adopted by most countries; they are termed contracting parties. UNESCO has been assigned certain implementation responsibilities, Coroz 1992.

Noblecourt has written an excellent text, *Protection of Cultural Property in the Event of Armed Conflict,* that provides details on the Convention, its implementation and substantial discussion of technical issues for protecting properties. This is a monumental, scholarly text recommended to anyone interested in the problem. Its five parts review all the relevant issues. Considering when it was written, early 1950's, it is not surprising that its focus is on nuclear weapons, and other large impact weapons. The technical provisions discussed focus on traditional civil engineering analysis and construction, not on quickly implementable strategies for protection.

A limited review of the literature suggests that there is little available technical information that might be used by a community at risk to guide its actions. Much of the emphasis seems to be on establishing administrative procedures and obligations of signatories and on large scale protective efforts. Following the signing of the Hague Convention there was a spat of activity, including writing and distribution of technical pamphlets, Nobelcourt 1958, but the impetus seems to have waned and little appears to be currently available. While it is not the purpose of this paper to comment on the Hague Convention, it should be noted that its implementation appears to have all the trappings of world legal commitments, and appears to focus on elitist definitions of what is important. Its implementation seems to have maintained this focus, rather than on the practical issues of communities actually protecting their cultural resources.

As is suggested by recent events, when a country signs the Convention, as most have, there is serious question when civil war or insurrections occur whether they feel bound by the Convection, or whether the combatants even know about it.

PHYSICAL PROTECTION

Four specific approaches to protecting important structures are discussed below and illustrated in adjoining diagrams.

1. Avoid making the structure a target.

If the structure can be marked as a non-combatant structure, like hospitals and clinics today, then they are afforded some added protection. The Hague Convention provides for such a mechanism, see Figure 1.

What is required is to identify the few culturally significant structures in the community and provide documentation to the appropriate international agencies. The Hague Convention does not appear to be widely known. It would be prudent to provide combatants with information to aid them in making decisions to protect their cultural heritage. The structures which are most important to a community are usually well known to the local populous and should require very little effort to determine. The problem, of course, is that the defenders would want to identify everything, making the entire area a non-target. This is unreasonable unless both sides are prepared not to dispute control of the subject area.

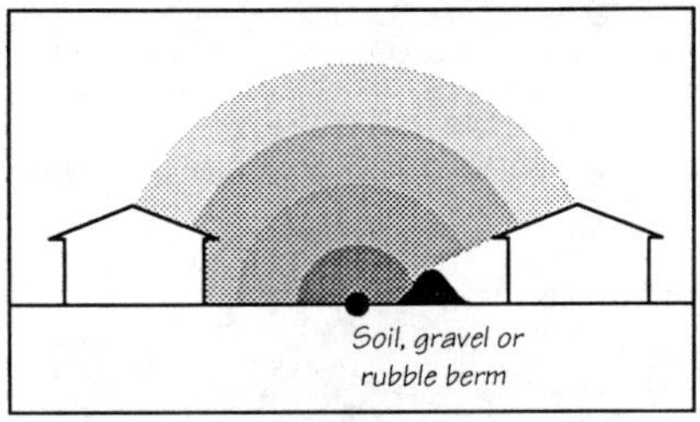

Figure 2. *An effective way to shield a structure from blast effects (air pressure, shell fragments, and rubble) is to place a soil (or gravel, or other non-flammable debris) berm around the perimeter of the building. It need not be in contact with the building or as high as the walls to provide protections.*

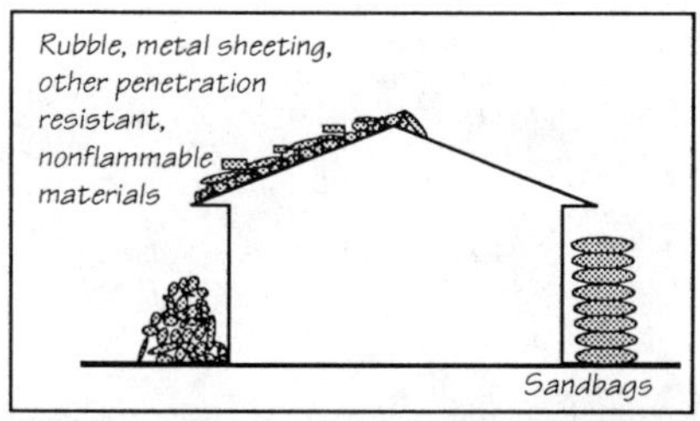

Figure 3. *If the roof is of sufficient strength, then it can be protected by placing a layer of rubble materials over it. When a projectile explosion impacts the roof, then its energy is dissipated by the rubble before it penetrates the roof. Similarly, sandbags, the ban of every foot soldier, are a well tested protective breastworks and should be used when ever and where ever practical.*

The conservation ethic for architectural and cultural monuments is sharply at variance with the preservation ethic. Preservation is inclusive, while conservation tends to be very selective, focusing on those objects that have inherent value beyond merely being old. While it is clear that an argument for preservation of everything in a community, say because it is old, will have little influence on battle plans and actions; it is clear, however, that identification and marking of a few structures or limited areas as non-targets is practical.

2. Protect the structure and/or important elements.

Experience in natural hazards mitigation, particularly earthquakes, winds and accidents, provides considerable understanding of the performance of structures under extreme loading. Even more valuable are the lessons gained in the military protection of bunkers and facilities. The prevalent approach in the latter case is to cover the structure with dirt, and the use of sandbags.

Figures 2 through 9 provide a few ideas on how to physically protect structures. They are self explanatory. The focus is on protecting with earthen materials (burying or building up a berm to deflect air blast and shrapnel) and provision of bracing. These techniques require little technical sophistication. Almost every area with an agricultural and/or artisan capability is well equipped to implement these. Furthermore, they are the types of actions that would be taken to protect a military command post, weapons dump or communications center. There is little doubt that the expertise is available to implement simple protective strategies in every community at risk *if* it can be marshaled.

The damageability of a structure can be decreased by removal of contents that pose a hazard to the structure (say flammable materials) and/or clearing adjacent areas of hazards. The later could include removing flammable landscaping or removing parapets of adjacent buildings that could fall and damage the structure. Any utilities that service the structure (e.g., natural gas, electricity) or hazardous materials (e.g., stored cooking gases) should be removed to a safe distance.

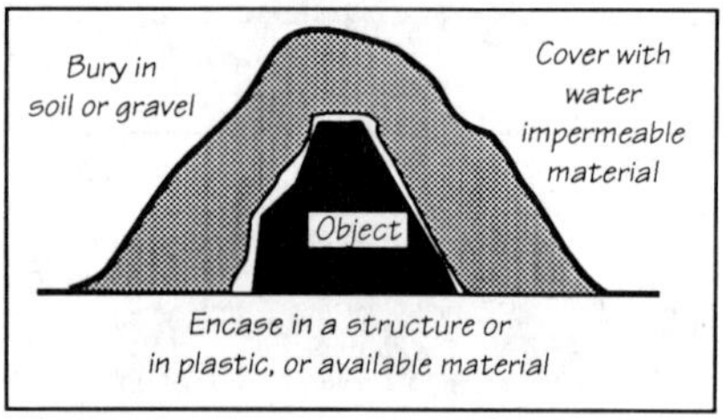

Figure 4. *In many cases the object to be protected is a statue, archeological ruin, or removed architectural detail. Often the most effective method to protect such objects is to bury them in soil or gravel. The principal thereat to the object then is water infiltration. Therefore it is advisable to cover them with a waterproof membrane, including both the object and the surface of the berm.*

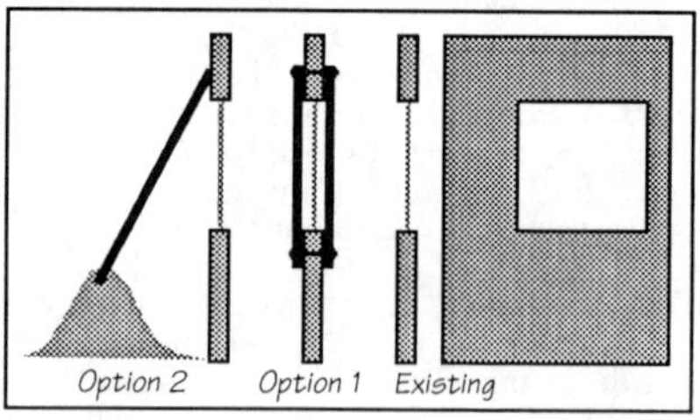

Figure 5. *Openings (e.g., windows, doorways) should be shielded so that projectiles do not enter the structure and damage the interior. In option 1 the opening is physically covered with sheathing, which should extend over the full opening so that physical loads are carried by the wall. In the second option, a "lean to" covers the opening to deflect low energy projectiles. Heavy sheathing is required to protect against high velocity projectiles.*

3. If the structure is damaged, protect the structure from further damage while awaiting repairs.

When considering damageability of structures, water must always be considered. When the exterior of a structure is cracked or damaged, then it takes only a few rain storms to substantially increase the damage. Often cracked and damaged elements will progressively fail under gravity loads unless the damaged members are shored.

4. Repair the damaged structure.

The first problem posed is to identify damaged structures and arrest their further damage from natural causes, Thiel 1984. The key information that is needed to repair the structure is an accurate characterization of the structure before it was damaged. This often means measured drawings and historical investigations, but in the age of video recorders, a simple strategy is to spend a few hours taping in great detail the structure from every point of view. The tape could then be archived out side the combat area. It may be a good strategy for every community to systematically identify and record their important structures so that if and when, for what ever reason, they are damaged they can be repaired. Sometimes there is extensive documentation that should be copied and deposited in the same safe location.

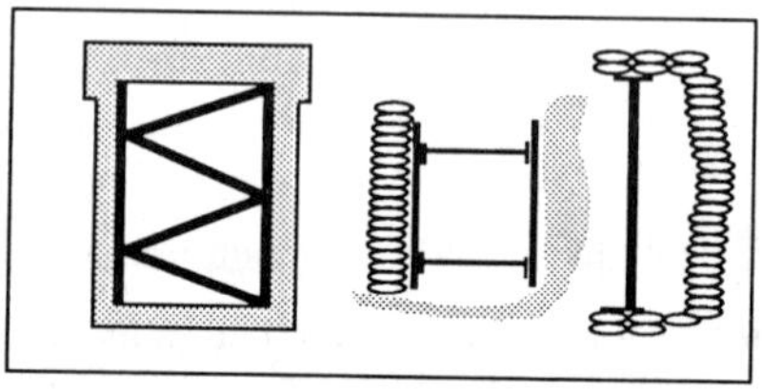

Figure 6. *Shoring of either undamaged or damaged structural elements and openings is an effective way to increase their strength and reliance.*

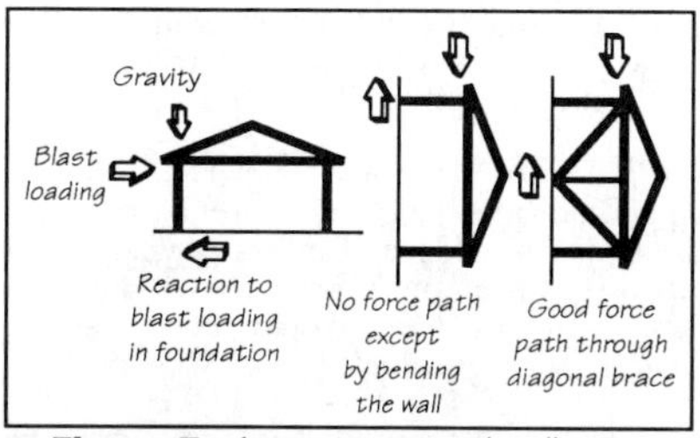

Figure 7. *An easy way to visualize whether a structure needs lateral support is to visually rotate it 90°, holding it by its base. A structure needing bracing to support gravity loads will be apparent, see the middle diagram, while those with bracing, which could be solid walls, are immediately apparent and need no further support. See also Figure 9.*

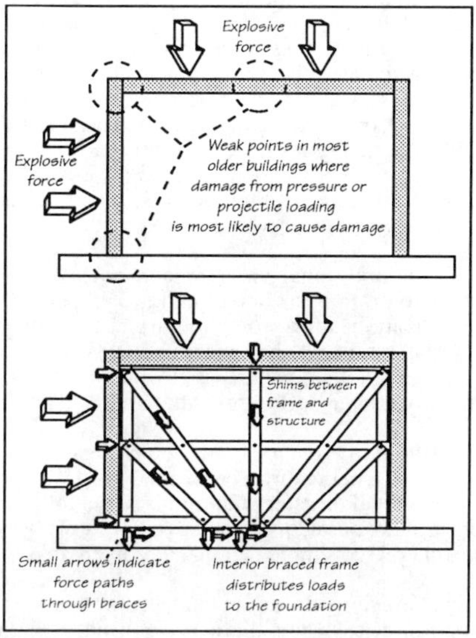

Figure 8. *Most structures that are culturally significant within a community are older structures. They are often very vulnerable to explosive overloads, particularly lateral loads, in which case the walls and ceilings tend to buckle and fracture where shown. The addition of an interior frame provides a means of taking explosive loads from the wall or roof element into the foundations, without fracturing the overloaded elements.*

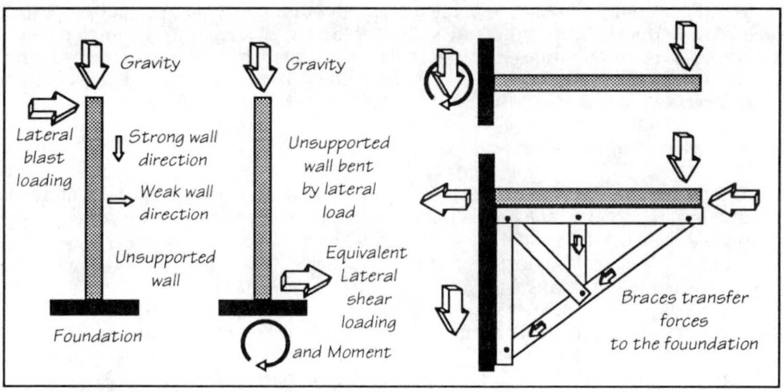

Figure 9. *Most structures are more vulnerable to lateral, that is horizontal, loads, than they are to vertical loads, that is gravity. These diagrams emphasize the need for diagonal structural members to carry the lateral loads to the foundation. The space between the brace and the wall should be filled with shims to provide positive contact.*

A STRATEGY FOR PROTECTING IMMOVABLE HERITAGE

There is no question that

> *The destruction of cultural heritage properties during military conflicts is recognized as unacceptable.*

Most countries of the world have signed the Hague convention. But it is not clear to the author that it is effectively communicated or implemented. It provides that base from which any additional initiatives should and could be developed. For civil wars, insurrections, or terrorism, being a signatory is of little practical import.

The Convention provides for the identification and documentation of important structures within signatory countries. Clearly much of this already has been done. But it is unclear that the process has been inclusive enough; its apparent focus has been on "monumental" structures. One of the issues is that the perspective needs to be more focused on the community than on a world consensus of what is important.

This preliminary assessment indicates that experience with natural hazards mitigation has much to offer protection of cultural property in time of war. The most important of which is that the techniques for protection are not technically difficult, but do require focused thought and effort at a time when thought is very sharply focused on other political and military needs. From a conservation point of view it would be nice if heritage issues could some how mold combatants military strategy. But this is both unrealistic and not consistent with the chaos of a battle field.

Today, there appears to be no clear effort to enhance the likelihood of significant heritage properties being protected during conflicts. And nothing is likely to be done unless the fundamental disaster planning observation is acknowledged: you cannot do anything during the emergency period that you have not carefully thought through before the emergency. To this we might add the simple fact that a problem that is not perceived cannot be addressed.

There is little doubt that protective steps can be taken for important structures by the communities when there is a military threat. What is in doubt is whether the technical information has been presented to local communities at risk on how to identify what should be protected and how to do it. This is not to suggested that a central authority be established to do the protecting. Rather procedures, practices and limited technical assistance should be available that provide to community members guidance on how to identify and protect properties, both before and after conflicts. The communities themselves are envisaged as the implementors. The current Hague Convention strategy seems to focus actions from the top of the governmental hierarchy, with a focus on organizational responsibilities rather than from where something will actually be done.

What is needed are four resources:

1. A clearly written, local language, guide on how to identify and document cultural properties important to the community. Preferably this would be done in peace time, but realistically it could be performed when the threat of conflict is imminent. The resulting listings and documentation (photographs, etc.) should be deposited out of harms way.
2. Simple, descriptive manuals should be written on how to protect cultural properties using easily available materials and techniques implementable by non-experts. The focus should be on actions that can protect structures to the degree practical, not necessarily definitively. These are not likely to cover massive, monumental structures, such as cathedrals, which are likely to be beyond the capacity and capability of local, individual initiative efforts to protect.
3. When conflict is imminent, manuals should be distributed to the local communities likely to be impacted so that the technical information is available to those who may be able to utilize them.
4. Limited technical resource teams should be organized and made available by the international community to enter a ravaged area after hostilities cease to identify what damage has been done to cultural properties, advise on how to stabilize or repair them, and identify what actions could be taken to protect them if the area is again within the conflict zone.

These are not very profound proposals. The machinery exists to implement such actions. The only thing missing is a recognition that the problems of protecting cultural properties during

4. Limited technical resource teams should be organized and made available by the international community to enter a ravaged area after hostilities cease to identify what damage has been done to cultural properties, advise on how to stabilize or repair them, and identify what actions could be taken to protect them if the area is again within the conflict zone.

These are not very profound proposals. The machinery exists to implement such actions. The only thing missing is a recognition that the problems of protecting cultural properties during armed conflict is a matter of most importance to the local communities where they are. The reliance, as appears to be the case now, on centralized government to act as the principal agents for their protection seems by recent example, ill advised as the principal line for their defense.

The United Nations, particularly UNESCO, is a logical institution focus for such an undertaking. They could serve as the logical repository for archival of information that may be collected in this initiative. However, for them to assume overall programmatic responsibility would require institutional redefinition of their goals, possibly a process of some difficulty.

The more likely agent for change is the conservation community itself. The professional community could well be the focus for development of the resources and provide the impetus for implementation of the recommended program. The professional community has the knowledge and capability to develop simple, implementable protection strategies and descriptive documents and is unfettered by all the exogenous political and social issues that an international body is likely to bring to the table. After development is completed then traditional international institutions could assume the principal role of implementation during in time of crisis.

The nature of the problem is clear as are the actions necessary to resolve it. The technical information and resources are available to protect cultural properties. The only thing lacking is the resolve to protect our cultural heritage. It is time for the professional community to act.

ACKNOWLEDGMENT

This paper owes its origins to provocative questions of Mr. James Druzik of the Getty Conservation Institute. He posed the question of whether there are lessons to be learned from mitigation of natural disaster impacts that can be applied during time of war. The development of the paper benefited from conversations in recent years with Edna Kimbro, upon whose work the section on conservation principles is based.

REFERENCES

Anderson, W.A., C.C. Thiel, 1979. "The Response of Social Institutions to Earthquake Prediction," Invited theme lecture, *Proceedings*, UNESCO Conference on Earthquake Prediction, Paris, France.

Belamaric, J. 1992. "Croatian Monuments as Targets, 1991-92," *Proceedings*, Materials Research Society Meeting, Cultural Heritage in Conflict: Session 15, San Francisco, April 1992.

Corozo, M.A., 1992. "The 1954 Hague Convention: Protection of Cultural Properties in the Event of Armed Conflict," *Proceedings*, Materials Research Society Meeting, Cultural Heritage in Conflict: Session 15, San Francisco, April 1992.

Drabeck, T.E. 1986. *Human System Responses to Disaster; An Inventory of Sociological Findings,* Springer-Verlag, New York.

Jokilehto, Jukka 1988. "Modern Principles of Architectural Conservation," *Proceedings*, First International Seminar on Modern Principles in Conservation and Restoration of Urban and Rural Cultural Heritage in Seismic-Prone Regions, Skopje, Yugoslavia.

Kimbro, E.E 1991. "Conservation Principles and Seismic Retrofitting Technology for Culturally Significant Adobe Buildings" *Proceedings*, Seismic Retrofit of Historic Buildings, San Francisco, California, November, 1991.

Continuing Actions to Reduce Potential Losses from the Future Earthquakes in Arkansas and Nearby States, USGS Open File Report 83-846.

Thiel, C.C. 1981. "An Approach to Seismic Safety for the Central United States," in *Earthquakes and Earthquake Engineering–Eastern United States*, J.E. Beavers, ed., Ann Arbor Science, Ann Arbor, Michigan.

Thiel, C.C., editor, 1990. *Competing Against Time,* Report to the Governor from The Board of Inquiry on the 1989 Loma Prieta Earthquake, George W. Housner, Chairman, State of California, Office of Planning and Research, Sacramento, California, May.

Thiel, C.C. and T.C. Zsutty, 1987. "Earthquake Characteristics and Damage Statistics," *Earthquake Spectra*, Volume 3, No. 4, November 1987.

Thiel, C.C., E.L. Tolles, E.E. Kimbro, F.A. Webster, and W.S. Ginell, 1991. *GSAP: Getty Conservation Institute Guidelines for Seismic Strengthening of Adobe Project Report of First Year Activities*, Getty Conservation Institute, Marina Del Rey, California, October 1, 1991, 280 pp.

PART II

Examination, Technical Analyses and Discovery

PART II A

The Geological, Geochemical and Biogeochemical Context

ROCK VARNISH CATION-RATIOS MAY NOT BE A RELIABLE METHOD FOR DATING LITHIC ARTIFACTS

PAUL BIERMAN* AND KAREN HARRY**
*Department of Geology, University of Washington, Seattle, WA 98195
**Statistical Research, Inc., 2500 N. Pantano Road, Tucson, AZ 85751

ABSTRACT

Previous research suggested that the chemical composition of rock varnish, a micrometer thick Si/Al/Mn/Fe coating, changed predictably through time and so could be used as a chronometer. In particular, the cation ratio, (Ca+K)/Ti, was posited to decrease as the varnish aged. However, many earlier varnish studies were not well documented and were based on chemical analyses now shown to be inaccurate.

This paper describes a test and the first direct comparison between two different analytic methods used previously to "cation-ratio date" archeologic and geologic materials. It provides methodology sufficiently detailed for replication of analytical and sampling techniques.

Our analyses of rock varnish from a prehistoric chert quarry in California show that the two principal methods used to "cation-ratio date" rock varnish generate consistently different elemental abundances and are therefore not comparable. Neither analytic method produced a significant time-dependent trend in varnish chemistry, suggesting rock varnish cation ratios are not a reproducible and thus probably not a reliable way to date lithic artifacts.

INTRODUCTION

Rock varnish is found most commonly on exposed rock surfaces in arid regions. It is a thin (<200 μm) coating made up predominately of Si, Al, Mn, and Fe. There is general agreement that the bulk of the varnish is derived not from the underlying substrate but from dust falling on the rock surface. There is evidence that microbes are an important factor in the concentration of Fe and perhaps Mn at abundances far greater than those in the ambient dust flux [1].

Rock Varnish Cation Ratio Dating

Rock varnish could be an important dating tool for archeologists and geologists because it is frequently found in circumstances where other dating methodologies cannot be applied. For instance, varnished artifacts are often found on the land surface with no stratigraphic context to constrain their age and many geologic deposits are not amenable to ^{14}C or K/Ar dating because of their age or composition.

It has been suggested that the chemistry of rock varnish changes predictably through time [2-4]. Specifically, the varnish cation ratio, (K+Ca)/Ti, appears to decrease as the age of the varnished rock surface increases. This trend has been calibrated and rock-varnish cation ratios have been used to assign numerical ages to landforms, artifacts, and rock art (see review in [3]).

There is no agreement regarding the cause of previously observed cation-ratio changes. Some [3-5] attribute the apparent time-dependent decrease in cation ratios to leaching of more "mobile" Ca and K. Others [6] find no evidence of such cation leaching and suggest that the observed decrease in cation ratios is due to inadvertent contamination of varnish analyses by the underlying rock substrate. To test these hypotheses, we specifically analyzed varnish on a chert (Si-rich, Ti-/Ca-/K-poor) substrate.

Recent publications have questioned not only the cause of reported cation-ratio changes, but their veracity. For example, many published rock-varnish chemical analyses and resultant cation ratios have been shown to be inaccurate [7, 8]. The assumption that micron-thick varnish and the underlying substrate remain stable for hundreds of thousands to millions of years is commonly not valid [9]. There are significant flaws in prior varnish sampling strategies [10] and simple statistical tests show that many archeologic conclusions previously drawn from varnish data may not be robust [11,12].

Methods Used Previously for Varnish Analysis

Most rock varnish analyses used for cation ratio dating have been conducted by one of two methods: in situ analysis or analysis of varnish scraped from the underlying substrate. In situ analysis of varnished surfaces has been done primarily by SEM/EDS [2] and PIXE/EDS [13]. Analysis of scraped samples has been done primarily by PIXE and ICP [3].

Although proponents of each method claim it minimizes the incorporation of substrate in varnish analyses [2, 3], it has not been demonstrated that either method eliminates substrate incorporation. The in situ SEM method is inherently surface biased, analyzing at most the upper few microns of the varnish coating [14]. The scraping method homogenizes the entire varnish coating and analyzes a larger volume of varnish. There is no indication, in the refereed literature, that these two disparate analytic methods have ever been applied to the same varnish sample.

This study applies both the scraping and in situ methods to similar, and in three cases, the same samples. Because the varnish we analyzed coats a chert substrate containing very low levels of Ti, K, and Ca, substrate incorporation probably does not significantly affect the cation ratios we present in this paper.

CA-KER-140 Prehistoric Quarry Site

CA-KER-140 is a Native American quarry site located in the Mojave Desert, south-central California at an elevation between 760 and 820 m (2500 and 2700 feet, Fig. 1, 2, 3). Present vegetation is sparse desert scrub and average yearly precipitation is < 25 cm (10 inches). The site consists of three buttes of predominately dark colored chert (Si-rich) mantled by colluvial debris. There are extensive alluvial fan and colluvial surfaces at the base of the buttes. Chemical analyses show that most of the chert contains <0.15% Ti, Ca, and K, the important components in varnish cation ratio dating (Table 1). Samples for this study were collected from the top of eastern butte because there was a higher density of varnished artifacts on this part of the CA-KER-140 site and because this part of the site was least disturbed by human activity.

The chert bedrock is frequently well-varnished where it crops out, especially on the eastern butte (Fig. 2). Detrital chert fragments on alluvial and colluvial surfaces are less well varnished than the bedrock, but are better varnished than artifacts. Most artifacts have only thin and patchy varnish on the worked or cultural faces and many,

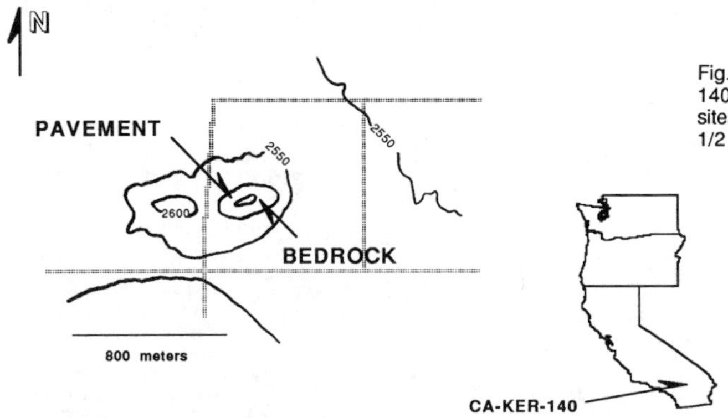

Fig. 1 Location of CA-KER-140 and specific sampling sites (USGS Galileo Hills 7 1/2 minute quadrangle).

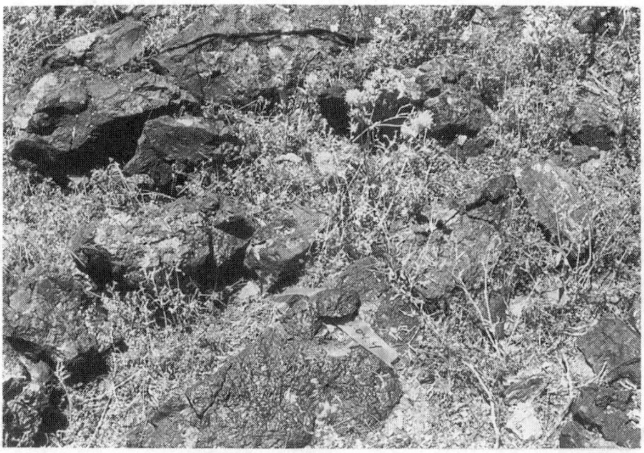

Fig. 2 Heavily varnished bedrock outcrop from which sample 1.04.4 was collected. Labeled tape in center of photograph is about 3 cm wide.

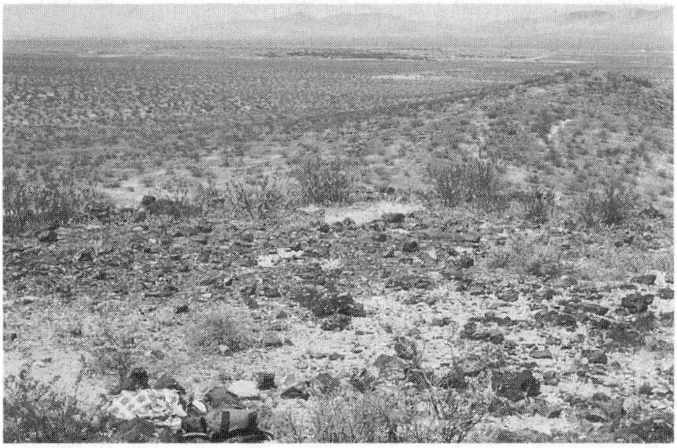

Fig. 3 Desert pavement from which artifacts were collected.

TABLE 1. Summary of Rock Varnish Analyses

Sample	N	Type	(K+Ca)/Ti	SiO2	TiO2	Al2O3	Fe2O3	MnO
In Situ Substrate								
KER-140	36	Mean		94.98 +/- 2.94	0.11 +/- 0.12	2.08 +/- 1.68	1.40 +/- 0.93	0.18 +/- 0.33
Substrate		Median		94.98	0.09	1.82	1.33	0.06
In Situ Varnish Analyses								
1.04.10	10	ROCK	9.51 +/- 1.37	27.27 +/- 2.41	0.37 +/- 0.05	19.38 +/-1.02	12.45 +/-0.84	30.39 +/- 3.81
1.04.08	8	ROCK	8.44 +/- 1.65	30.94 +/- 1.23	0.38 +/- 0.09	20.56 +/-0.99	13.64 +/-1.17	25.00 +/- 1.40
1.04.04	10	ROCK	9.71 +/- 1.81	30.50 +/- 1.80	0.42 +/- 0.09	17.76 +/-0.97	11.77 +/-1.37	28.83 +/- 2.13
1.03.02C	9	NCUL	6.29 +/- 1.30	29.13 +/- 3.64	0.58 +/- 0.13	15.09 +/-1.84	12.40 +/-1.81	32.16 +/- 5.69
1.03.02B	9	NCUL	8.21 +/- 1.21	36.79 +/- 3.81	0.55 +/- 0.11	17.54 +/-1.97	10.98 +/-2.06	23.47 +/- 6.24
1.03.02A	8	NCUL	7.85 +/- 1.44	32.66 +/- 2.90	0.62 +/- 0.17	16.74 +/-2.93	13.58 +/-2.59	26.59 +/- 3.24
1.03.03D,O	9	OMIX	8.20 +/- 1.20	33.43 +/- 4.05	0.53 +/- 0.08	18.07 +/-1.30	14.49 +/-1.85	23.71 +/- 2.75
1.03.03IY1,O	8	OMIX	16.30 +/- 5.62	34.78 +/- 3.69	0.39 +/- 0.08	9.82 +/-1.64	4.36 +/-0.44	34.30 +/- 3.34
1.03.03S,O	8	OMIX	9.95 +/- 1.89	28.59 +/- 3.46	0.44 +/- 0.08	14.87 +/-1.42	9.03 +/-0.92	34.43 +/- 3.97
1.03.03I,O	8	OMIX	10.02 +/- 1.72	30.53 +/- 2.76	0.57 +/- 0.10	14.38 +/-0.96	11.17 +/-0.78	31.70 +/- 3.07
1.03.03I,Y1	13	YMIX	16.03 +/- 8.85	25.60 +/- 5.22	0.29 +/- 0.09	7.92 +/-0.62	3.53 +/-0.41	46.58 +/- 4.32
1.03.03S,Y	10	YMIX	9.39 +/- 2.09	55.92 +/- 9.10	0.35 +/- 0.09	9.79 +/-1.99	5.15 +/-1.36	21.39 +/- 4.78
1.03.03I,Y2	10	YMIX	9.61 +/- 1.88	36.71 +/- 6.49	0.31 +/- 0.05	7.06 +/-1.20	2.99 +/-0.69	39.41 +/- 4.00
1.03.03D,Y	10	YMIX	15.35 +/- 5.95	43.48 +/- 14.62	0.24 +/- 0.05	3.98 +/-0.93	2.66 +/-0.49	42.66 +/- 13.19
1.03.01F	10	CULT	5.39 +/- 0.70	38.30 +/- 8.66	0.42 +/- 0.05	12.73 +/-1.33	7.77 +/-0.93	30.08 +/- 7.32
1.03.01E2	10	CULT	13.28 +/- 4.75	47.45 +/- 6.53	0.35 +/- 0.08	14.48 +/-1.29	6.61 +/-0.62	20.63 +/- 4.92
1.03.01E1	10	CULT	5.35 +/- 1.63	33.84 +/- 4.54	0.59 +/- 0.18	11.36 +/-2.41	12.92 +/-3.71	29.23 +/- 5.30
1.03.01B	13	CULT	8.78 +/- 2.30	39.24 +/- 5.93	0.53 +/- 0.11	15.99 +/-2.05	8.78 +/-1.40	24.27 +/- 6.87
1.03.01A2	10	CULT	20.65 +/- 5.33	43.48 +/- 11.56	0.25 +/- 0.05	7.53 +/-2.00	3.68 +/-1.01	36.80 +/- 11.03
1.03.01A1	10	CULT	9.81 +/- 1.69	36.44 +/- 13.95	0.41 +/- 0.10	12.66 +/-1.08	5.40 +/-0.51	32.09 +/- 10.08
Scraped Varnish Analyses								
1.04.10	9	ROCK	5.48 +/- 0.29	28.90 +/- 1.80	0.79 +/- 0.05	18.09 +/-1.42	13.26 +/-1.00	28.58 +/- 0.73
1.04.08	9	ROCK	6.06 +/- 0.55	34.28 +/- 3.11	0.70 +/- 0.05	19.72 +/-1.68	11.22 +/-0.73	24.61 +/- 1.06
1.04.04	5	ROCK	5.75 +/- 1.22	29.59 +/- 2.34	0.67 +/- 0.17	15.61 +/-2.15	9.97 +/-0.72	34.89 +/- 1.18
1.03.3O,O	11	OMIX	5.59 +/- 0.50	29.97 +/- 2.57	0.88 +/- 0.04	16.79 +/-2.76	11.84 +/-0.45	28.85 +/- 0.60
1.03.3B,O	11	OMIX	3.79 +/- 0.80	32.85 +/- 1.03	1.08 +/- 0.24	19.03 +/-0.91	13.05 +/-0.49	24.54 +/- 0.31
1.03.3F,O	6	OMIX	4.36 +/- 0.45	56.16 +/- 1.09	0.59 +/- 0.05	11.72 +/-0.51	7.66 +/-0.23	17.31 +/- 0.59
1.03.02D	11	NCUL	3.98 +/- 0.51	47.81 +/- 4.51	0.73 +/- 0.08	17.11 +/-1.58	11.60 +/-1.06	16.41 +/- 1.70
1.03.3F,Y1	7	YMIX	7.34 +/- 0.51	36.69 +/- 1.43	0.65 +/- 0.03	17.15 +/-0.49	10.74 +/-0.59	24.63 +/- 0.72
1.03.3B,Y	9	YMIX	4.91 +/- 0.35	43.79 +/- 3.14	0.79 +/- 0.07	15.87 +/-1.47	8.88 +/-0.59	22.66 +/- 0.87
1.03.3F,Y2	6	YMIX	8.70 +/- 0.20	33.27 +/- 1.00	0.61 +/- 0.02	16.46 +/-0.61	8.93 +/-0.48	29.38 +/- 0.39
1.03.3M,Y	8	YMIX	5.97 +/- 0.44	32.78 +/- 1.71	0.67 +/- 0.02	20.24 +/-1.52	11.48 +/-0.55	25.51 +/- 0.59
1.03.03P	13	CULT	6.05 +/- 0.26	38.39 +/- 0.90	0.57 +/- 0.01	19.95 +/-0.65	9.73 +/-0.29	23.17 +/- 0.32
Other Varnish								
Potter and Rossman, 1979		on Andesite		21.89	1.07	21.13	23.67	16.63
Potter and Rossman, 1979		on Quartz		36.83	0.93	21.56	18.87	13.70
Potter and Rossman, 1979		on Rhyolite		35.02	0.80	20.40	12.40	23.07
Reneau et al., 1991		on Basalt		28.92	0.89	22.63	24.20	16.96
Reneau et al., 1991		on Granite		37.17	0.71	24.15	12.60	18.70
Reneau et al., 1991		on Rhyolite		37.98	1.01	24.18	17.49	11.17

Note: Type of Sample: ROCK = Bedrock, NCUL = Noncultural, CULT = Cultural
YMIX = Mixed, young (cultural) surface, OMIX = Mixed, old (noncultural) surface.
Uncertainties in table are one sample standard deviation about the mean and represent variability in sample or glass.
Analytic uncertainties (counting statistics) are stated in text

MgO	CaO	Na2O	K2O	BaO	SO3	P2O5
0.34 +/- 0.29	0.08 +/- 0.24	0.15 +/- 0.07	0.05 +/- 0.20	0.09 +/- 0.12	0.06 +/- 0.07	0.43 +/- 0.20
0.27	0.01	0.13	0.01	0.07	0.03	0.35
1.80 +/- 0.22	1.49 +/- 0.18	0.36 +/- 0.04	1.20 +/- 0.10	1.94 +/- 0.42	1.03 +/- 0.20	2.38 +/- 0.30
1.91 +/- 0.12	1.39 +/- 0.28	0.31 +/- 0.06	1.02 +/- 0.05	1.41 +/- 0.20	1.16 +/- 0.15	2.27 +/- 0.23
1.95 +/- 0.13	1.78 +/- 0.22	0.40 +/- 0.05	1.32 +/- 0.10	1.93 +/- 0.46	1.36 +/- 0.25	1.98 +/- 0.19
1.44 +/- 0.18	1.24 +/- 0.10	0.28 +/- 0.04	1.45 +/- 0.12	4.51 +/- 1.11	0.51 +/- 0.17	1.19 +/- 0.22
2.31 +/- 0.35	1.61 +/- 0.21	0.41 +/- 0.07	1.81 +/- 0.27	3.48 +/- 1.59	0.30 +/- 0.06	0.73 +/- 0.05
2.19 +/- 0.23	2.03 +/- 0.47	0.45 +/- 0.10	1.62 +/- 0.21	1.45 +/- 0.21	0.77 +/- 0.12	1.27 +/- 0.25
1.84 +/- 0.10	2.04 +/- 0.33	0.25 +/- 0.05	1.32 +/- 0.12	1.40 +/- 0.19	0.58 +/- 0.09	2.30 +/- 0.23
2.12 +/- 0.38	3.86 +/- 1.25	0.34 +/- 0.05	1.04 +/- 0.23	7.66 +/- 0.79	0.23 +/- 0.08	0.94 +/- 0.66
1.82 +/- 0.24	1.89 +/- 0.14	0.42 +/- 0.06	1.44 +/- 0.13	5.15 +/- 1.46	0.66 +/- 0.23	1.25 +/- 0.38
1.91 +/- 0.14	2.74 +/- 0.38	0.39 +/- 0.06	1.67 +/- 0.16	1.82 +/- 0.24	1.33 +/- 0.22	1.77 +/- 0.23
1.55 +/- 0.22	2.50 +/- 0.40	0.29 +/- 0.07	0.73 +/- 0.08	10.38 +/- 1.06	0.13 +/- 0.03	0.28 +/- 0.11
1.79 +/- 0.29	1.08 +/- 0.28	0.40 +/- 0.10	1.35 +/- 0.28	1.94 +/- 1.25	0.15 +/- 0.05	0.44 +/- 0.07
1.34 +/- 0.26	1.68 +/- 0.76	0.36 +/- 0.04	0.73 +/- 0.16	8.68 +/- 0.85	0.21 +/- 0.07	0.35 +/- 0.10
0.91 +/- 0.23	1.28 +/- 0.54	0.49 +/- 0.29	1.44 +/- 0.67	0.88 +/- 0.32	1.20 +/- 0.42	0.64 +/- 0.11
1.47 +/- 0.14	0.76 +/- 0.12	0.36 +/- 0.03	0.98 +/- 0.15	5.64 +/- 1.41	0.36 +/- 0.08	0.93 +/- 0.17
2.40 +/- 0.22	1.69 +/- 0.32	0.36 +/- 0.09	1.62 +/- 0.20	3.61 +/- 1.18	0.14 +/- 0.08	0.48 +/- 0.15
1.25 +/- 0.31	1.08 +/- 0.10	0.29 +/- 0.10	1.18 +/- 0.18	5.71 +/- 1.15	0.66 +/- 0.12	1.70 +/- 0.42
2.19 +/- 0.43	1.59 +/- 0.37	0.30 +/- 0.06	1.91 +/- 0.39	4.40 +/- 1.79	0.12 +/- 0.03	0.56 +/- 0.12
1.34 +/- 0.22	2.33 +/- 0.81	0.50 +/- 0.19	1.66 +/- 0.38	1.23 +/- 1.23	0.64 +/- 0.43	0.52 +/- 0.15
1.84 +/- 0.17	1.43 +/- 0.27	0.32 +/- 0.06	1.58 +/- 0.21	7.03 +/- 3.37	0.19 +/- 0.03	0.41 +/- 0.10
1.48 +/- 0.12	1.88 +/- 0.09	0.09 +/- 0.01	1.51 +/- 0.10	2.97 +/- 0.17	0.11 +/- 0.20	2.33 +/- 0.16
1.78 +/- 0.13	1.77 +/- 0.09	0.13 +/- 0.01	1.52 +/- 0.13	2.36 +/- 0.20	0.13 +/- 0.10	1.78 +/- 0.14
1.11 +/- 0.18	1.54 +/- 0.12	0.14 +/- 0.02	1.34 +/- 0.04	2.42 +/- 0.14	0.37 +/- 0.32	2.34 +/- 0.17
1.46 +/- 0.17	2.40 +/- 0.15	0.12 +/- 0.02	1.50 +/- 0.16	4.14 +/- 0.29	0.02 +/- 0.02	2.02 +/- 0.14
1.73 +/- 0.11	1.50 +/- 0.05	0.16 +/- 0.02	1.53 +/- 0.10	2.62 +/- 0.13	0.12 +/- 0.06	1.79 +/- 0.09
1.22 +/- 0.09	1.38 +/- 0.05	0.09 +/- 0.01	0.65 +/- 0.05	2.42 +/- 0.18	0.04 +/- 0.03	0.79 +/- 0.06
1.57 +/- 0.17	1.26 +/- 0.15	0.16 +/- 0.07	0.99 +/- 0.12	1.03 +/- 0.11	0.06 +/- 0.07	1.26 +/- 0.11
2.01 +/- 0.07	1.79 +/- 0.09	0.11 +/- 0.01	1.94 +/- 0.25	2.96 +/- 0.25	0.08 +/- 0.11	1.24 +/- 0.05
1.81 +/- 0.12	1.36 +/- 0.04	0.22 +/- 0.03	1.62 +/- 0.11	2.20 +/- 0.17	0.05 +/- 0.08	0.74 +/- 0.03
1.66 +/- 0.10	2.72 +/- 0.12	0.08 +/- 0.01	1.50 +/- 0.09	4.26 +/- 0.14	0.13 +/- 0.06	1.00 +/- 0.05
1.82 +/- 0.15	1.65 +/- 0.11	0.10 +/- 0.02	1.48 +/- 0.15	2.14 +/- 0.15	0.14 +/- 0.11	1.99 +/- 0.09
2.22 +/- 0.09	1.04 +/- 0.03	0.08 +/- 0.01	1.60 +/- 0.08	2.56 +/- 0.14	0.06 +/- 0.06	0.62 +/- 0.02
2.21	0.72	0.12	1.36	0.76	NR	2.45
2.90	0.74	0.12	1.84	0.61	NR	1.88
2.48	1.30	0.23	1.93	1.24	NR	1.12
2.10	0.72	0.06	1.55	0.40	0.12	1.45
2.88	0.58	0.08	1.67	1.15	BDL	0.30
3.42	0.80	0.06	2.30	0.32	0.12	1.15

BDL = Below detection limit
NR = Not reported
In situ Na analyses include a machine background of 0.15 to 0.30 % as described in Bierman and Kuehner, 1992.
Analyses have been normalized to 100%.

especially those on the topographically lower sections of the buttes, are smooth and glossy but unvarnished. None of the worked faces had varnish as continuous or well developed as that on non-cultural or bedrock surfaces; this suggests that the varnish on cultural surfaces is the youngest varnish we sampled.

There are desert pavements present both on the summits of the buttes and on the fan surfaces (Fig. 3). These pavements contain a surface layer of stones and artifacts supported by an underlying matrix of silt and clay. The pavements are flat and support only scant vegetation. Recent findings suggest that stones mantling these silt-supported desert pavements float on the surface, increasing the likelihood of continuous subaerial exposure [15, 16].

SAMPLE COLLECTION AND TEST DESIGN

This study was designed to compare the two principal methods of cation ratio dating [2, 3] and to determine, using both methods, whether cation ratios of relatively younger varnish differed systematically from cation ratios of relatively older varnish. Samples were collected from the CA-KER-140 site on June 24 and 25, 1991 by the authors. Samples were assigned identification numbers in the field and were placed in plastic bags for transport. Statistical Research survey teams mapped the locations from which the samples were collected.

We collected subaerially-exposed varnished samples from both bedrock outcrops (Fig. 2) and desert pavements (Fig. 3). All of the samples collected from outcrops were dark and well-varnished and showed no evidence of human-induced flaking (Fig. 2). Bedrock samples were collected from uniformly varnished, gently sloping or flat-lying rock faces. There was no indication that these surfaces had recently eroded or that the varnish on them had not formed subaerially.

From the desert pavement, we collected samples with two types of varnished surfaces, cultural and non-cultural (Fig. 4). Cultural surfaces showed evidence of percussion and had varnish on the fresh or worked face. Non-cultural surfaces had varnished portions which showed no evidence of percussion or working. Some samples we collected had both cultural and non-cultural surfaces, both of which were varnished. We termed these "mixed" samples.

If varnish forms continuously, as suggested by [3, 4], superposition mandates that varnish on worked (cultural) surfaces is younger than varnish on unworked (noncultural) surfaces. Similarly, varnish on exposed bedrock surfaces should be older than varnish on the worked (cultural) material. On "mixed" objects, varnish on the worked (cultural) portion is no older than varnish on the unworked (noncultural) portion. Even if our decision to label a sample "cultural" is in error, a broken face should have an exposure time no longer than a natural face whether or not that break is anthropogenic. The thickness and extent of varnish coverage (bedrock>noncultural>>cultural) support the relative chronology we present and suggest that varnish on bedrock is much older than varnish on artifacts.

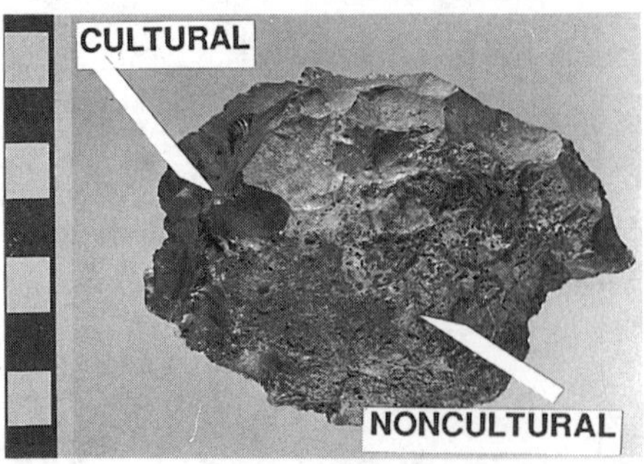

Fig. 4 Mixed sample 1.03.03 I on which cultural and noncultural varnish was analyzed by SEM/EDS. Sampling sites are labeled. Cm intervals on scale bar.

There is an uncertainty intrinsic to dating varnished artifacts which applies both to published studies [17] and to this test. Because most artifacts (and in this study, varnished stones) are relatively small (5-20 cm), they are prone to disturbance on the ground surface. For instance, animals and humans visiting this or any other quarry, kill, or habitation site may have overturned varnished materials exposing previously unexposed, unvarnished sections and burying varnished portions of an artifact.

There is also the possibility that artifacts now exposed on the surface were at some time in the past buried and then exposed by erosion. By selecting samples from desert pavements, where the material is believed to be continuously floated at or near the ground surface [15, 16], we minimized the likelihood that samples were buried and then exposed to the varnishing process long after manufacture.

SAMPLE PREPARATION AND CHEMICAL ANALYSIS METHODS

Samples were returned to the University of Washington, lightly scrubbed with a toothbrush under running water, rinsed in deionized water, and air-dried. Each sample was photographed, measured and described. A subset of samples was selected for chemical analysis based on the certainty with which the cultural and non-cultural surfaces could be identified.

All analyses were performed on a JEOL 733 superprobe equipped with four wavelength-dispersive spectrometers (WDS) and one energy-dispersive spectrometer (EDS). This instrument also functions as a scanning electron microscope (SEM) allowing us to image the sample using both secondary and backscattered electron detectors. All analyses were made using beam stabilization to minimize changes in beam current. Collection of EDS spectra and all data reduction was performed on a Tracor Northern 5500 computer.

In Situ SEM/EDS Analysis

Samples used for in situ SEM/EDS varnish analyses were, if necessary, cut to appropriate size (<4 cm) with a rock saw. These samples were again rinsed in deionized water, air dried, and then carbon coated. In situ analyses of the varnish surface were made during June and July 1991 using SEM/EDS and followed the calibration and analysis protocol detailed in [18]. Spectra were collected for 800-1000 s using a beam current of 15 nA. Uncertainties (1σ, counting statistics only) for each analysis are typically less than < 2 % for Ca, Ba, and K and < 5% for Ti.

Our SEM/EDS analyses were made by rastering the electron beam using magnifications between 100 x and 3000 x, resulting in areas of varnish being analyzed between 1 and 10^{-3} mm^2. We chose these high magnifications in order to minimize the amount of substrate incorporated in our analyses. At lower magnifications, we were unable to select areas of varnish that did not include patches of unvarnished substrate. The specific magnification for any one sample was chosen by first scanning the entire sample and then selecting the least magnification at which we could make ten analyses free of visible substrate.

Areas for analysis were selected using both secondary and backscattered electron detectors in order to minimize the amount of substrate included in varnish analyses (Fig. 5). Areas selected for varnish analysis were specifically chosen to avoid exposed rock and microcolonial fungi visible on the varnish surface. Eight to twelve varnish analyses were made on each sample; on most samples, two substrate analyses were also made. For each varnish analysis, we recorded such

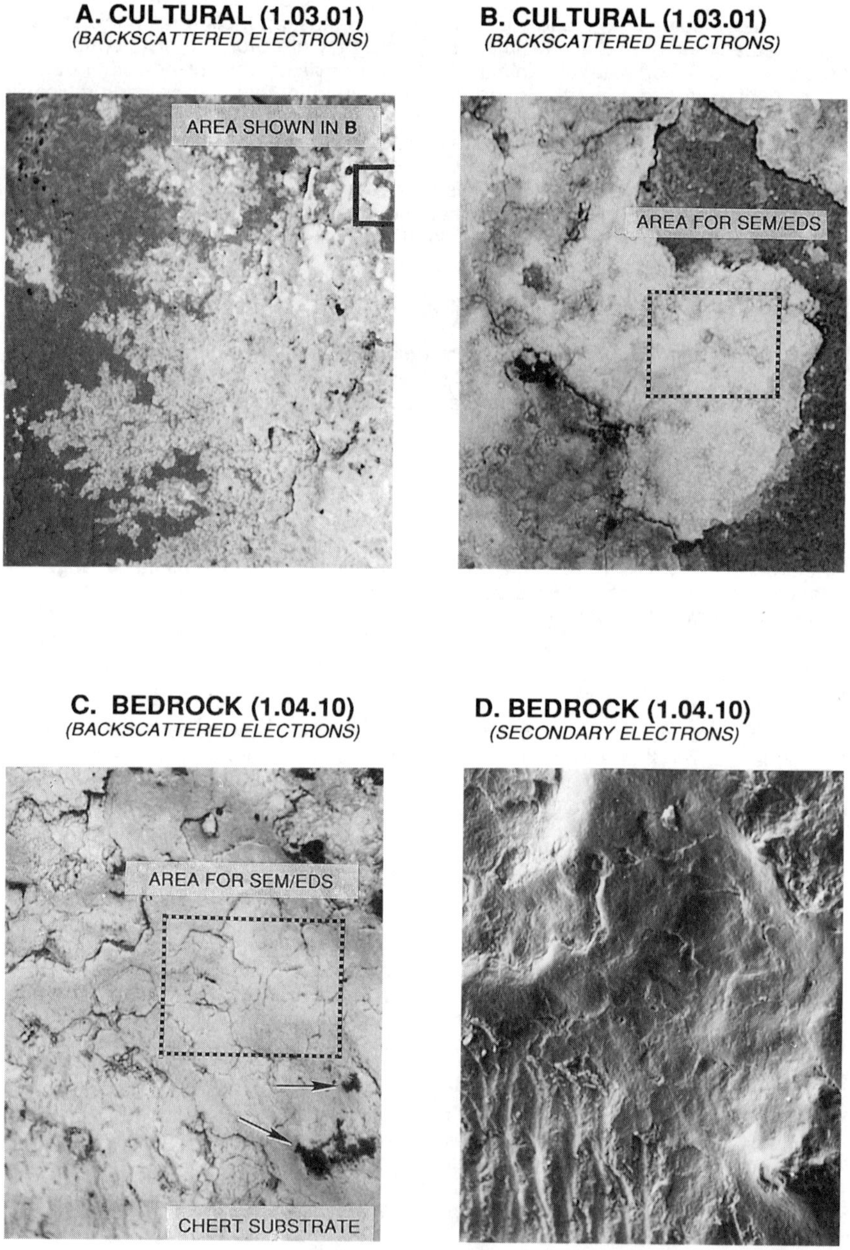
A. CULTURAL (1.03.01)
(BACKSCATTERED ELECTRONS)
AREA SHOWN IN B
B. CULTURAL (1.03.01)
(BACKSCATTERED ELECTRONS)
AREA FOR SEM/EDS
C. BEDROCK (1.04.10)
(BACKSCATTERED ELECTRONS)
AREA FOR SEM/EDS
CHERT SUBSTRATE
D. BEDROCK (1.04.10)
(SECONDARY ELECTRONS)

Fig. 5

Fig. 5 Scanning electron micrographs of bedrock (1.04.10) and cultural (1.03.01) varnish. Secondary electron images show topography. Backscattered electron images show both topography and average atomic number of material; Fe-, Mn-rich varnish is light and Si-rich chert is dark. A, C, and D show approximately 2 x 2.2 mm of varnish; B shows 0.2 x 0.22 mm. A) Backscattered electron image shows thin, patchy, light colored varnish extending radially over dark gray chert cultural surface. B) Same cultural sample as A magnified 10x shows texture and relative thickness of varnish; thicker varnish is white. SEM/EDS analysis would be conducted in boxed area to minimize substrate contamination. C) Backscattered electron image of varnished bedrock surface shows almost continuous varnish cover compared to A. Outcrops of dark gray chert substrate are indicated by arrows. SEM/EDS analysis would be conducted in the boxed area. D) Secondary electron image of the same area as C shows smooth varnish in center. Furrowed area in lower left-hand corner was scraped for analyses.

characteristics as surface texture, relative brightness of backscattered electron image, and surface topography.

Our SEM/EDS technique differs from that used previously [2, 19] because these workers used lower magnification to analyze larger areas of varnish. Thus, our analyses may include a lesser contribution from substrate than previous in situ measurements, although differences in varnish age and thickness make such a comparison uncertain. Unlike [2] and similar to [19] we made our analyses using a single accelerating voltage, 15 keV.

Scraping and SEM/WDS Analysis

Samples from which the varnish was scraped received the same initial preparation. The varnish was scraped by hand from the chert substrate using a carbide drill bit provided by R.I. Dorn. Scraping was done at various magnifications under a binocular microscope avoiding the few microcolonial fungi that were present. In most cases, the varnish could be removed cleanly from the hard underlying chert. In cases where the underlying chert was weathered, separating the varnish from the substrate was more difficult and resulted in the incorporation of substantial amounts of substrate in varnish scrapings. We used a tweezers to remove visible pieces of chert substrate from the scrapings. Between 10 and 100 mm^2 of varnish was scraped.

Between 0.5 and 2 mg of varnish scrapings were collected on glassine paper, stored in silver crucibles, and transferred to dimpled 0.013 mm thick molybdenum strips for fusion. Samples were fused, under an argon atmosphere, by passing a current through the strip for 20 s. No flux was used and each strip was used only once. Fusion temperature was probably between 1200 and 1400 °C. The resulting glass was detached from the strip, mounted in epoxy, polished, and carbon-coated.

The fused varnish scrapings were analyzed using SEM/WDS with an accelerating voltage of 15 keV, a beam current of 20 nA, and a beam diameter of 20 µm; Bence-Albee matrix correction was used. Standards and operating conditions are reported in Table 2. Counting time for each element was set at 130 s or 2 % precision, whichever came first. Uncertainties (1σ, counting statistics only) are typically < 2 % for Ca, Ba, Ti, and K for each analysis. Five to ten analyses were made of each glass in order to reduce the dominant uncertainty in the WDS analyses, the inhomogeneity of the fused glasses. Because some of the glasses, especially those high in SiO2, were quite vesicular, most totals ranged between 85 and 95%.

Accuracy of Analyses

To ensure the accuracy of our analyses, we utilized synthetic varnish standards. Before each analytical session, three varnish standards [8] were each analyzed repeatedly as unknowns. All varnish analyses in this report have been corrected by a factor representing the ratio of the observed and nominal compositions of these standards [8]. All corrections were < 8% (relative) of the reported values; most corrections, including Ti, Ca, and K were < 2% (relative). We believe our SEM/ EDS and WDS analyses are directly comparable because they were corrected to the same three standards, and because surface roughness appears not to bias cation ratios determined with SEM/EDS [18].

To determine whether fusion could affect the composition of varnish samples, the synthetic varnish standards were powdered and fused in a similar fashion to the CA-KER-140 varnish samples. Fusion did not alter significantly the composition of the varnish standards, nor the calculated cation ratios, so fusion presumably did not alter the composition of CA-KER-140 varnish samples.

Raw analytic data were minimally edited. Analyses with low totals (<70%), extremely inhomogeneous glasses, and samples containing extremely large amounts (>60%) of Si (presumably substrate contamination) were omitted before varnish summary statistics were calculated. No analyses were omitted because of Ti, Ca, Ba, or K "anomalies".

DATA AND DISCUSSION

CA-KER-140 varnish is chemically similar to rock varnishes from the American southwest (Table 1), suggesting that our results are probably widely applicable. CA-KER -140 varnish is dominantly Si (30-40%), Mn (20-40%), Al (10-20%), and Fe (5-12%). Ba, K, Mg, P, and Ca are usually present at concentrations between 1 and 3 wt%. Ti, S, and Na are present at concentrations < 1%. The chert substrate is almost exclusively Si (95%) containing only low levels of Fe and Al (1-3%) and Ca, Ti, and K (<0.15%). We report all analyses as oxide weights normalized to 100%.

SEM images support the initial field observation that varnish on cultural (younger) surfaces is both thinner and less extensive than varnish on either bedrock or noncultural surfaces (Fig. 5). Varnish on cultural surfaces appears to be spreading radially away from sites where it first began to form. Because the varnish is thinner and patchier on cultural surfaces, we used higher SEM magnifications to analyze cultural varnishes than we used to analyze noncultural varnishes (Fig. 5). The use of high magnifications explains, at least in part, the higher variance of in situ-measured cation ratios on cultural surfaces. Because analytic uncertainties are only a few percent, the variance in cation ratios reflects intrinsic variability in varnish chemistry.

SEM images show that scraping removed most but not all of the varnish from the scraped areas (Fig. 5). On cultural surfaces, we scraped the entire varnished surface so the mean value reported is the population mean. On noncultural surfaces, we scraped only part of the surface and so the reported mean is that of the subsample we gathered. Because we scraped all the varnish on cultural surfaces, the high variability in scraped cultural varnishes (Fig. 6B) reflects artifact to artifact variability; it is probably not the result of our analytic protocol. Means of scraped samples are less variable than those measured in situ; nevertheless, analysis of scraped samples demonstartes that varnish is chemicaly heterogeneous on the artifact to artifact scale and that cation ratios and attendent dates are, at best, noisy signals.

Lack of Time Dependence in Cation Ratios

Our data suggest that the in-situ SEM/EDS method is not useful for cation-ratio dating rock varnish on chert substrates at CA-KER-140 . In-situ analyses cannot be used to date individual varnished samples because varnish on cultural (young) surfaces had both the highest (20.65 +/- 5.33, 1σ) and lowest mean cation ratios (5.35 +/- 1.63, 1σ) (Fig. 6A). Nor can in-situ analyses be used to date populations of CA-KER-140 varnished chert clasts because grand means of all analyses in an age class did not change significantly as a function of relative varnish age (Fig. 7). These findings are supported by the lack of a consistent relationship between cation ratios of relatively older and younger varnish on mixed samples (Table 1).

Similarly, analyses of scraped varnish cannot be used to date individual CA-KER-140 artifacts, because varnish on some cultural (younger) surfaces had cation ratios lower than or equal to that on older noncultural or bedrock surfaces (Fig. 6B).

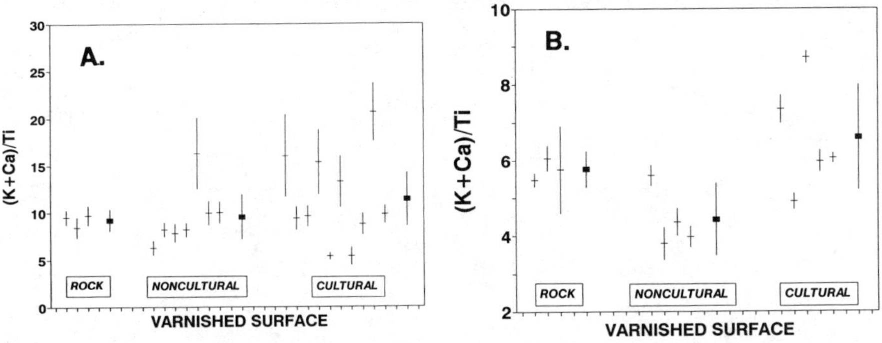

Figure 6. Mean and grand mean cation ratios for individual bedrock, noncultural, and cultural varnish samples analyzed in situ and after scraping. Thin line symbols are means for individual samples. Error bars represent 90% confidence intervals. *Rock* refers to bedrock samples. *Noncultural* refers to both varnish on pieces of unworked rock and to varnish on the unworked cortical portions of mixed samples. *Cultural* refers to both varnish on pieces of worked rock and to varnish on the worked faces of mixed samples. Grand mean for all bedrock, noncultural, and cultural analyses is shown by square with 90% confidence interval about the grand mean shown by error bars. A. In situ varnish. B. Scraped varnish fused into glass.

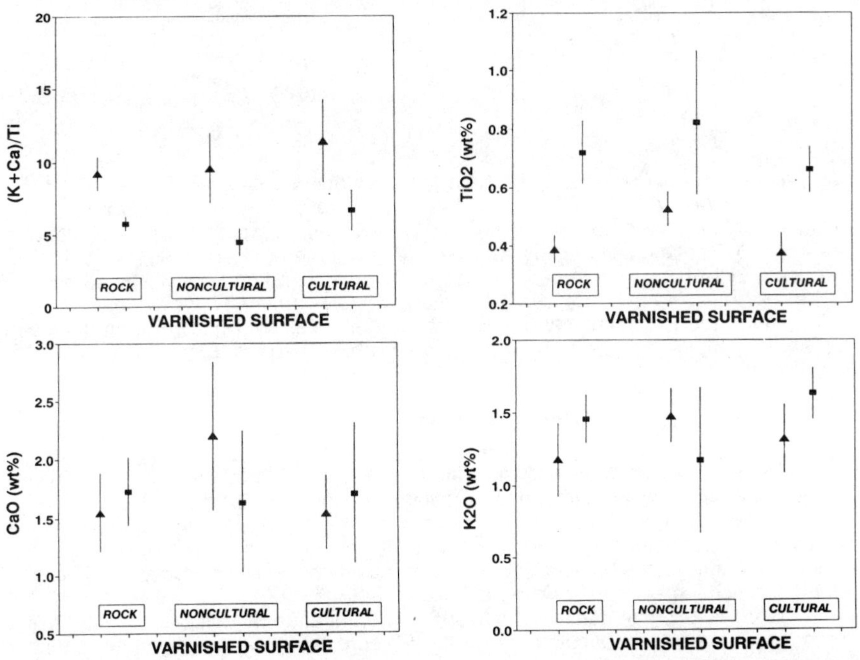

Figure 7. Grand means of Ti, Ca, K, and (K+Ca)/Ti for bedrock, noncultural, and cultural samples show difference between in situ analysis and analysis of varnish scrapings. Uncertainties are 90% confidence intervals. Triangles represent in situ analyses. Squares represent scraped varnish.

Table 2. Protocol for SEM/WDS Analysis

OXIDE	CRYSTAL USED FOR ANALYSIS	BACKGROUND OFFSET (sin θ)	PEAK	STANDARD	CONCENTRATION OF ELEMENT IN STANDARD (WEIGHT %)
SiO_2	TAP	+/- 0.03929	k_α	Basalt Glass	23.28
Na_2O	TAP	+/- 0.01786	k_α	Albite, Tibron	8.75
Al_2O_3	TAP	+/- 0.03571	k_α	Basalt Glass	6.88
MgO	TAP	+0.03571 - 0.03929	k_α	Basalt Glass	4.89
P_2O_5	TAP	+0.01429 - 0.01071	k_α	Apatite, Wilberforce	17.77
K_2O	PET	+0.00571 - 0.00857	k_α	Orthoclase, OR-1	12.39
CaO	PET	+0.01429 - 0.02143	k_α	Basalt Glass	8.08
TiO_2	PET	+/- 0.01071	k_α	TiO_2	59.95
BaO	PET	+/- 0.00714	L_α	Barite	58.84
Fe_2O_3	LIF	+/- 0.02143	k_α	Basalt Glass	8.53
MnO	LIF	+/- 0.02857	k_α	Garnet, Nuevo	16.31
SO_3	TAP	+/- 0.02143	k_α, n = 2	Barite	13.73

Nor can analyses of scraped varnish be used to date populations of CA-KER-140 artifacts, reliably. Although the grand mean cation ratio of five cultural (younger) surfaces is marginally higher (p=0.045) than the grand mean of four noncultural (older) surfaces, it is not significantly higher than the grand mean of three older bedrock surfaces (Fig. 6B). On both of the mixed samples, cultural (younger) varnish had higher cation ratios than adjacent noncultural (older) varnish. However, one of the cultural (younger) surfaces had a cation ratio (4.91 +/- 0.35, 1σ) lower than the mean for adjacent older bedrock (5.76 +/- 0.29, 1σ) (Table 1 and Fig. 6B).

Comparison of In Situ and Scraping Results

Unless the varnish is extremely thin (<1 μm), in-situ and scraping methods sample different levels of the varnish coating. The in-situ SEM method is surface-biased, analyzing only the upper few microns of varnish [14]; the scraping method samples and homogenizes the entire thickness of the varnish. Therefore, measured cation-ratio differences between the two methods most likely reflect varnish thickness and differences in elemental distribution with depth.

Interpreting differences in absolute elemental abundances between the two methods is less certain because substrate (Si) contamination can affect apparent elemental abundance. If the abundance of major elements such as Si and Mn in cultural and non-cultural varnish is similar, then a Si excess or a Mn deficiency would indicate contamination by the chert substrate. For instance, cultural varnish analyzed in situ appears to contain more Si than noncultural varnish. However, average Mn concentration is similar for cultural and non-cultural varnish. If Mn indicates the degree of contamination, then there is little incorporation of chert substrate and the apparent abundance of other elements in cultural varnish has not been lowered. If Si indicates the amount of substrate contamination, then the abundance we report of other elements in cultural varnish is uniformly 10-30% too low compared with varnish on bedrock surfaces. In either case, substrate incorporation will not affect cation ratios because there is little Ti, Ca, and K in the chert.

Cation ratios measured on scraped varnishes were consistently and significantly lower than those measured in situ (Fig. 7), suggesting that deeper layers of CA-KER-140 varnish have lower cation ratios. Since the concentrations of K and Ca measured by both methods did not differ significantly, the difference in cation ratios appears to be caused by the presence of >50% more Ti at depth than in the upper microns of varnish (Fig. 7). The inference that Ti is not uniformly distributed in varnish could imply variable rates of Ti input or incorporation or it could imply a previously unrecognized diagenetic process. This finding has not been reported before although other studies have found Ti-rich bands of detritus within varnish [20, 21]. Our findings could be explained by the presence of Ti-rich layers below the surface of CA-KER-140 varnish; layers which are too deep to be detected by in situ SEM/EDS measurements.

Cause of Previously Reported Cation-Ratio Variations

Our data suggest that substrate contamination could be a more important factor than cation leaching in previously reported cation-ratio variations because: 1) Cation ratios for varnish on CA-KER-140 chert, which contains little Ti, Ca, or K, show no consistent change corresponding to relative age. 2) Neither in-situ nor scraped varnish analyses shows a consistent and significant decrease in K or Ca with increasing relative age. 3) Comparison of in-situ and scraped varnish suggests that Ti is more abundant but K and Ca are not depleted in deeper, presumably older CA-KER-140 varnish.

CONCLUSIONS

This study used two previously proposed techniques [2, 3] to analyze the chemical composition of rock varnish at the CA-KER-140 site in the Mojave Desert, CA. The study was specifically designed both to compare and test the utility, for dating lithic artifacts, of these two cation ratio methods. Our study showed the following:

1. The composition of rock varnish at the CA-KER-140 site is similar to that of other arid-region rock varnishes. CA-KER-140 varnish is predominately Si, Al, Fe, and Mn with lesser amounts of P, Ba, Ca, K, and Mg and trace levels of Ti, Na, and S. Because the chert substrate is almost entirely Si, its incorporation should not significantly affect the varnish cation ratio, (Ca+K)/Ti.

2. Cation ratios determined from analysis of scraped varnish [3] and from analysis of in situ varnish [2] are significantly different indicating that these two previously proposed analytic and dating techniques are not sampling the same populations and are not directly comparable at CA-KER-140. Cation ratios measured in situ were consistently higher than those measured in scraped varnish.

3. Varnish on artifacts showed great variance in composition and cation ratios suggesting such scatter is an intrinsic characteristic of varnished artifacts from the KER 140 site and precluding simple or meaningful chronologic interpretation of cation ratio measurments.

4. Neither scraped nor in-situ analyses of varnished rock from the CA-KER-140 site show cation ratios changing predictably and consistently with time, as would be expected if rock-varnish cation ratios were a consistently reliable dating tool.

Considered along with other recent findings [6-12] these data suggest that rock-varnish cation ratios provide neither a well understood nor a consistently reproducible basis for dating subaerially exposed lithic artifacts.

ACKNOWLEDGMENTS

Research supported by Statistical Research, Inc. under contract to the U.S. Bureau of Land Management and by NSF EAR-9004252 and USGS 14-08-001-G1783 grants to Bierman and A.R. Gillespie (UW). Harry (Statistical Research) directed sample collection and characterization. Bierman performed sample preparation, chemical analyses, and data interpretation. S. Kuehner (UW) assisted with chemical analyses. A draft of this or a similar paper was reviewed by T. Oberlander, B. Atwater, A.R. Gillespie and L. McFadden. S. Reneau made extensive suggestions to several drafts of this manuscript.

REFERENCES CITED

1. J. T. Staley, J. B. Adams and F. E. Palmer, Soil Biochemistry, 7, 173 (1992).
2. C. D. Harrington and J. W. Whitney, Geology, 15, 967 (1987).
3. R. I. Dorn, Physical Geography, 13, 559 (1989).
4. R. I. Dorn, Quaternary Research, 20, 49 (1983).
5. R. I. Dorn and D. H. Krinsley, Geology, 19, 1077 (1991).
6. S. L. Reneau and R. Raymond, Geology, 19, 937 (1991).
7 C. D. Harrington, D. J. Krier, R. Raymond and S. L. Reneau, Scanning Microscopy, 5, 55 (1991).
8. P. Bierman and A. Gillespie, Geology, 19, 196 (1991).
9. P. Bierman and A. Gillespie, Geology, 19, 641 (1991).
10. S. Wells and L. McFadden, Geology, 15, 1178 (1987).
11. M. Lanteigne, Rock Art Research, 6, 145 (1989).
12. M. Lanteigne, Rock Art Research, 8, 127 (1991).
13. C. A. Pineda, M. Peisach and L. Jacobson, Nuclear Instruments and Methods in Physics Research, B35, 463 (1988).
14. S. L. Reneau, R. C. Hagan, C. P. Harrington and R. Raymond, Scanning Microscopy, 5, 47 (1991).
15. L. D. McFadden, S. G. Wells and M. J. Jercinovich, Geology, 15, 504 (1987).
16. S. G. Wells, L. D. McFadden and C. T. Olinger, Geological Society of America Abstracts with Programs, 23, A207 (1991).
17. R. I. Dorn and 11 others, Science, 231, 830 (1986).
18. P. Bierman and S. Kuehner, Chemical Geology, 95, 283 (1992).
19. D. P. Dethier, C. D. Harrington and M. J. Aldrich, Geological Society of America Bulletin, 100, 928 (1988).
20. C. D. Harrington, S. L. Reneau, R. Raymond, and D. J. Krier, EOS, 71, 1341 (1990).
21. R. J. Raymond, S. L. Reneau and C. D. Harrington, Scanning Microscopy, 5, 37 (1991).
22. R. M. Potter and G. R. Rossman, Chemical Geology, 25, 79 (1979).

A HOLISTIC APPROACH TO THE ANALYSIS OF ARCHAEOLOGICAL DEPOSITS, ILLUSTRATED USING A LATE ROMAN URBAN SEQUENCE FROM NORTHWEST EUROPE

ORPAH S. FARRINGTON* AND **RICHARD M. BATEMAN****

*Department of Geology, Birkbeck College (London University), Malet Street, London, WC1E 7HX, U.K.

**Departments of Earth and Plant Sciences, Oxford University, Parks Road, Oxford, OX1 3PR, U.K.

ABSTRACT

Dark Earth deposits immediately overlie late Roman urban stratigraphy across northwest Europe, representing the crucial but poorly documented collapse of provincial Roman urbanism. Deficient in artefacts, they have proved recalcitrant to traditional methods of archaeological interpretation. Here, they are used as a vehicle to promote a more integrated, holistic approach to scientific archaeology. This recognises the great potential value of applying a wide range of geoanalytical techniques to the finer-grained matrices that enclose (or, more precisely, grade into) artefacts in archaeological sequences. As this multifaceted approach is time-consuming, deposits chosen for analysis should contain potential answers to profound historical questions. Comparative studies are necessary, in which samples of known age, provenance, environment and mode of deposition outnumber those of a more equivocal nature. Considerable knowledge is required to select the optimal range of complementary techniques for application to a particular suite of materials; this case study outlines the relative merits of analyses for fabric orientation, particle size, micromorphology, bulk geochemistry (ICP), particle geochemistry (microprobe), heavy mineralogy, plant remains (pollen, phytoliths, wood), animal remains, macroscopic artefacts, and radiometric dating. The resulting large bodies of data are best summarised by multivariate analyses (notably ordination algorithms), together with semivariograms for spatial data. Interpretations should take full account of the range of anthropogenic processes and products inherent in archaeological deposits.

INTRODUCTION

In this paper, we use a specific urban case-study as a vehicle to promote a broad-based, integrated, **holistic** approach to archaeological investigation that employs concepts and techniques from history, the earth sciences, and the life sciences. We compare and contrast in outline a range of geoanalytical techniques, discuss their inter-relationships, and attempt to show that a potentially accurate picture of the past can emerge only when several appropriate techniques are brought to bear. We argue that interpretation of former landscapes requires geoanalysis of fine-grained matrices as well as the coarser artefacts that they enclose[1-4]. In this case-study, the techniques were applied primarily by a single researcher over an extended time-span; they could equally be applied by a team of researchers, pooling their collective expertise and resources, over a much shorter period.

Roman territorial expansion across northwest Europe was completed by ca. 100 AD. Subsequent pragmatic 'romanisation' of the indigenous late Iron Age peoples involved the establishment of an administrative and economic infrastructure, a communications network centred on roadways. The key to romanisation was the imposition of urbanism on previously non-urban societies during the following two centuries, which generated the classic Roman urban sequences; thick, complex, and

Mat. Res. Soc. Symp. Proc. Vol. 267. ©1992 Materials Research Society generally rich in artefacts[5-7]. The age and function of most horizons in the sequences are readily interpreted, and comparable patterns of urban development have been traced across Europe. This process of stratigraphic data acquisition has been accelerated by rescue excavations prompted by recent urban redevelopment.

These have revealed a dramatic and ubiquitous change in urban stratigraphy that approximates the contraction of the Roman Empire. In urban centres across the region, the latest Roman buildings and roadways are blanketed with a thick (<3 m), homogeneous, dark grey deposit termed Dark Earth; only the outer defensive walls generally stood proud of the deposit. Dark Earth apparently lacks internal stratification and is poor in contemporary artefacts; coarse particles in the dark matrix appear to be reworked from the underlying stratigraphy. Evidence of contemporaneous occupation is similarly lacking; successive occupation layers generally date from at least the late Early Anglo-Saxon period (>650 AD). Dark Earth is by far the dominant urban stratigraphic evidence of a 200-300 year period of European history; a period also much poorer in unequivocal historical evidence than the preceding Roman period. Thus, the enigmatic Dark Earths play a pivotal and controversial role in late Roman/'Dark Age' archaeology, and have been the subject of much speculation[6-9].

The present study has taken advantage of recent excavations to sample at narrow stratigraphic intervals Dark Earth and all underlying deposits at 24 sections from seven urban centres: in England at London, Colchester, Canterbury, Dorchester, Staines, and Baldock, and in France at the Gallo-Roman capital of Paris. Given the recalcitrant nature of Dark Earth, a battery of techniques was applied with varying expectations of success, as much to test the applicability of the techniques and the value of matrix-based studies as to solve the many problems posed by Dark Earth. Key questions include (1) provenance, (2) mode and environment of deposition, (3) period of deposition and burial, and (4) post-depositional modification[10]. Together, answers to these questions allow a fifth and more important question to be addressed: what is the historical significance of Dark Earth?

The body of data acquired cannot possibly be covered within the scope of this paper. Consequently, we have elected to focus on a single section. The chosen site, Culver Street in Colchester, offered a well sampled but relatively simple and readily interpreted stratigraphy.

ROMAN/POST-ROMAN STRATIGRAPHY AT COLCHESTER

The documented Roman/post-Roman history of Colchester is well reflected in the stratigraphic column from a 15 m-long, north-facing section at the 1980-82 Culver Street excavation[11]. The Roman settlement was established on the floodplain of a minor river close to its discharge into the southern North Sea in what is now Essex, 80 km northeast of London along the northern margin of the London Basin.

The simplified lithological log of the sampled section (Figure 1) shows seven deposits of four basic types: natural substratum, in situ building, building debris, and Dark Earth. The anthropogenic succession rests on a natural substratum of Quaternary 'brickearths' that were largely reworked from the Palaeogene London Clay Fm, a shallow marine deposit. Above this are in situ building materials representing the construction of a Colonium: a colony of Roman veterans imposed on the tribal capital of the indigenous Belgic Trinovantes in 49 AD, only three years after the Roman invasion of Britain. This was razed with heavy loss of life during the rebellion of Queen Boudicca (Boadicea) in 61 AD, leaving a horizon of building debris known as the Boudiccan destruction layer[11-13]. A more elaborate city was rebuilt on the ruins in Flavian-Antonine times (ca. 100-200 AD); in the study section, this Flavian reconstruction layer is capped by a characteristic Opus Signinum floor of a substantial building. The destruction of this second phase of building is much more difficult to date; it is most likely to have occurred in the last few decades before the final Roman withdrawl from Britain (traditionally accepted as ca. 410 AD). Again, this resulted in a distinct

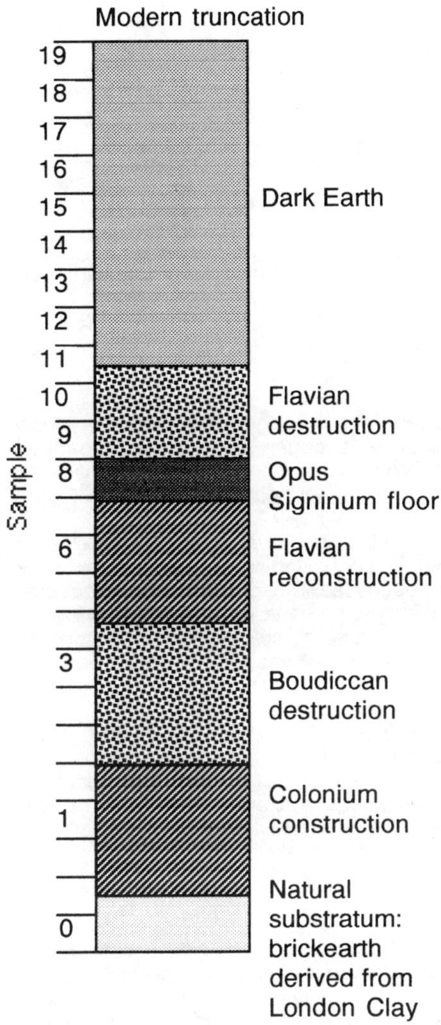

Figure 1. Simplified Roman stratigraphy at Culver Street, Colchester, showing two cycles of construction-destruction underlying the enigmatic Dark Earth deposit. Left-hand ticks indicating sampling intervals are 10 cm apart.

layer of building debris. This in turn is overlain by the thickest and most enigmatic deposit in the section: a truncated 85 cm of Dark Earth, penetrated at one point by a mid-7th century Anglo-Saxon hut. The section was sampled at 10 cm intervals, and many of these samples (at least one from each horizon: Figure 1) were subjected to a range of geoanalyses.

CLAST ORIENTATION FABRICS

Clast orientations were the only geoanalytical data collected in the field. Three-dimensional orientations of samples of 50 lithic clasts with long axes exceeding 1 cm were summarised in two parameters: the dip and the compass orientation of the long axis (both in degrees of arc). The resulting data are used to summarise the directionality and strength of the stress fields operating during deposition of the material, thereby inferring mode of deposition[14].

14 fabrics were measured at Culver Street. Here, discussion focuses on a lateral set of six fabrics, spaced 1-1.5 m apart, taken from the Dark Earth, and a comparable set of three fabrics from the Boudiccan destruction deposit (the western-most Dark Earth fabric and easternmost Boudiccan fabric are shown in Figures 2A and 2B respectively). These two sets of fabrics allow interpretation of variation within each deposit, as well as comparison of the two deposits. Whereas the mode of deposition of Dark Earth is speculative, that of the Bouddican destruction layer is more certain; active redistribution of building rubble by man.

Both of the deposits have fabrics with moderate mean dips (16-33°) and high dispersion; consequently, most are polymodal with low vector magnitudes (only two Dark Earth fabrics and two Boudiccan fabrics attained the 25% threshold necessary for statistical significance by the Rayleigh test). Thus, resultant vectors for most of the fabrics (interpreted as the direction of greatest stress) are void, and the fabrics are regarded as statistically random.

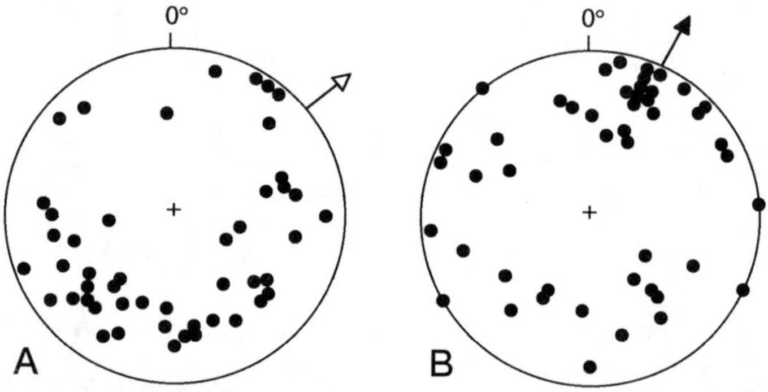

Figure 2. Polar diagrams for clast orientations at Colchester. (A) Typical Dark Earth fabric: resultant vector 52°, vector magnitude 20% (random by Rayleigh test), mean dip 27°. (B) Typical Boudiccan destruction layer fabric: resultant vector 18°, vector magnitude 30% (non-random by Rayleigh test), mean dip 22°.

Nonetheless, similar modal vectors in adjacent fabrics suggest slight and very local preferred orientation in the Dark Earth, generally favouring the southwest quadrant (e.g. Figure 2A). Although weak, Boudiccan fabrics are stronger and more consistent, with both resultant vectors and modes favouring the northeast quadrant (e.g. Figure 2B).

We conclude that both the Dark Earth and the Boudiccan destruction layer were deposited by man, and that such anthropogenic deposition is characterised by weak directionality relative to most natural strata deposited in hydraulic media (aqueous, glacial, or aeolian). Here, the main stress acting on the particles is likely to have been gravity, as materials were shovelled, dumped, or passively collapsed onto the contemporary topographic surface. Whereas the long axes of naturally stressed particles typically dip toward the source of the dominant stress, those subject only to gravity are likely to dip downslope or lie transverse to the slope, showing little dip. The greater (though still weak) directionality of the Boudiccan deposit could indicate a subtly different mode of deposition from the Dark Earth, but we believe that it merely reflects the greater proportion of coarse clasts in the former. The matrix of the Dark Earth deposits would have reduced the opportunity for gravity-induced re-orientation. If so, the closest natural analogues for Dark Earth fabrics should be screes s.l. and diamictons.

MICROMORPHOLOGY

Deposits viewed in section have a definable macromorphology, allowing the physicochemical identification of the larger particles in the deposits and discussion of their spatial and boundary relationships. Micromorphological analysis of petrographic thin-sections allows similar interpretations at a smaller scale and in much greater detail[15,16]. The flexibility of thin-section analysis often aids the integration of otherwise disparate information from a range of geoanalytical techniques. For this study, exceptionally large (8 x 6 cm) vertical thin-sections were prepared from crystic resin-impregnated Kubiena samples and cover-slipped for light microscopy[17]. A few duplicate sections were left naked for microprobe analysis.

As developed to its full potential by pedologists and Quaternary geomorphologists, micromorphology has two main applications. The first considers composition and provenance via semi-quantitative (or, occasionally, fully quantitative) descriptions of porosity and of the relative proportions of different types of particle, classified according to size and petrographic composition. The second provides potentially complex accounts of post-depositional change, whether chemical (e.g. geochemical weathering, authigenesis), physical (e.g. cracking, particle translocation), or biotic (e.g. zooturbation, root penetration, microbial activity). For near-surficial materials such as archaeological deposits, these processes effectively constitute pedogenesis.

The Colchester profile provided little evidence of pedogenesis. Minor authigenic mineral growth in some voids and local low-key mottling indicate slight post-depositional modification, probably related to some groundwater movement under mildly reducing conditions; this was insufficient to cause significant particle translocation. Root material is rare, and at least some of the observed invertebrate activity followed exposure of the section during excavation. The overall impression is that the succession accumulated rapidly, affording little opportunity for pedogenesis.

Micromorphology proved more informative about the composition of the Colchester deposits. It is most effective at summarising a range of characteristics in a single view, and can be used to quantify both particle size and composition (though these criteria are more accurately assessed using specialised techniques discussed below. Coarser materials found in varying proportions throughout the section can be categorised as non-anthropogenic lithic (e.g. flint, quartz, rock fragments), anthropogenic lithic (e.g. brick, tile, plaster, mortar, pottery, metals, glass, waste products of the manufacture/processing of the aforementioned), and biotic (e.g. bone, teeth, shell, wood, seed/grain). This simple classification conceals a more complex set of underlying concepts. In particular, it defines particles on their primary mode of origin rather than their most recent provenance, cannot take into account agent of deposition as well as origin, and can be undermined by composite particles. Consider for example a fragment of a fired brick within the Dark Earth. This could be an aggregate of particles exclusively from a single natural source rock, but brought together by man in an anthropogenic object. This in turn has been deposited by man in the Flavian buildings, and reworked by man into the Dark Earth.

COARSE FRACTIONS

Of all the geoanalyses described in this paper, sieving out of coarse (>2 mm) particles is both the most simple and the most commonly applied to archaeological deposits. The greatest amount of information can be obtained by using phi-interval sieves (see below) to assign each particle to a narrow size category, and then measure the composition by weight of each lithology[18]. It should be remembered that weight percentages do not correspond precisely to relative volumes of different components in the original deposit, which can vary greatly in density (e.g. compare wood/seeds with lead artefacts). Other problems are the difficulty of identifying small particles and the lack of a statistically valid sample for large particles; a large bulk sample is needed even for medium sized clasts, thereby complicating attempts to sample at narrow stratigraphic intervals. One benefit of careful sieving is the recovery of unusually informative artefacts such as small datable coins.

Thus, only one bulk sample was collected from the Colchester profile for coarse fraction analysis, spanning the entire thickness of the Dark Earth. The 8 kg sample was 15% coarse clasts; these are dominantly natural stones (especially flint) with subordinate fired clasts (ceramics s.l.). Glass and metallic fragments do not exceed 5%. Biotic fragments (<10%) are concentrated in the finer size fractions; bone/teeth predominate over plant fragments (mostly charcoalified wood) and invertebrate shells.

BIOTIC PARTICLES

Thin-section studies in particular revealed a wide range of biotic fragments in the Colchester profile. Little work has been done on the animal remains. Vertebrate bone/teeth (e.g. pig, cow) and invertebrate shells (e.g. oysters, mussels) are dominant, and indicate the presence of domestic refuse. Insufficient attention has been paid to other groups such as arthropods. More work has been done on the plant remains although these too have not yet been quantified, some of the provisional results are significant. Three main categories are present. Charcoalified wood and seeds occur in both the coarse and fine fractions, and were studied in both the sieved coarse fractions and thin-sections. Woods are dominantly oak and chestnut, hardwoods that are typically used for structural components, though other fragments (e.g. rosiids) could represent fuel. The larger wood fragments, together with the better preserved invertebrate shells, offer the possibility of high-precision radiocarbon dating.

Opaline phytoliths are confined to the fine fraction and were observed primarily in the thin-sections, often associated with concentrations of fine-grained amorphous ash. Morphologies were diverse but phytolith identification is still somewhat handicapped by insufficient study in situ in plants and by the likelihood of convergent morphologies in different taxonomic groups[19]. At Colchester, phytoliths appear dominantly derived from trees such as oak, though herbs such as grasses, nettles, and composites are also present.

Spores/pollen are also confined to the fine fraction, and were extracted using standard acid-based treatments[20]. Quantities are low and preservation mediocre, presumably due to the alkaline pH of all the deposits at Colchester (see below). Such preservation weakens the likelihood of obtaining a reliable and representative pollen spectrum. Spores/pollen recovered thus far include typical components of disturbed wasteland or pastureland floras, such as *Urtica*, *Taraxacum*, and *Chenopodium*[21].

PARTICLE SIZE DISTRIBUTION

Particle size analysis by aqueous sedimentation in laboratory columns provides a purely physical quantitative description of poorly consolidated deposits[22]. Data are usually expressed as percentages of the <2 mm fraction, in phi (ø) categories based on $-\log_2$ of particle diameter. Two alternative levels of precision are recommended: the more simple involves three categories (sand: -1-4 ø; silt: 4-9 ø; clay >9 ø), whereas the more complex involves the ten whole-phi intervals between -1 ø and 9 ø plus the clay.

Sand-silt-clay analyses for the Colchester profile reveal little of interest. Clay is subordinate to sand and silt throughout; 5-10% in the lower part, rising to 15-20% in the Flavian reconstruction layer and above. Sand and silt occur in approximately equal proportions (40-50%) up to the top of the Flavian destruction layer; thereafter, there is a progressive increase in sand at the expense of silt through the Dark Earth.

More information can be gleaned from the more sophisticated phi-interval data. Most naturally deposited samples are unimodal and therefore can be compared by applying moments statistics to clay-free phi-interval data[23,24], but most of the archaeological deposits analysed in the present study are polymodal (even the natural substratum was bimodal, with peaks at 1-2 ø and 5-6 ø). Consequently, the data were summarised using a Genstat[25] multivariate software package that included principal coordinates analysis, a robust form of ordination representing each sample in multivariate hyperspace[26]. Gower similarity coefficients were also calculated for all pairwise combinations of samples, and used to construct minimum spanning trees representing maximum overall similarity[27,28] (the same analytical procedure was used for other complex data matrices discussed in subsequent sections).

Figure 3 shows the ordination and superimposed minimum spanning tree for phi-interval particle size data from Colchester. The Dark Earth samples form a tight cluster relative to the underlying deposits. The first coordinate (PC1) accounts for a large proportion (48%) of the total variance, separating the Dark Earth samples on the basis of their greater content of coarse and fine sand but relative deficiency in coarse and medium silt. Within the Dark Earths, two stratigraphically distinct groups are

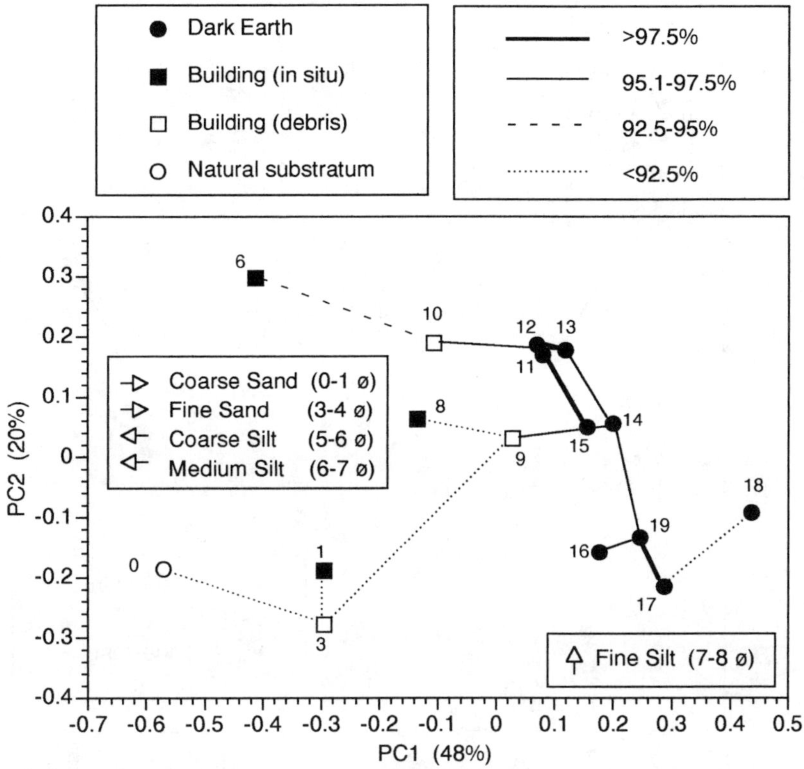

Figure 3. Principal coordinates plot and superimposed minimum spanning tree for particle size data from Colchester. Sample points are numbered as in Figure 1 and the four main deposit types are differentiated by symbols (see key). Samples are projected onto the plane of maximum variance. Major variables contributing to the first two coordinates are listed, together with the direction of increase of the variable; the proportion of the total variance accounted for by each coordinate is also given. The minimum spanning tree links samples by maximum Gower similarity; the percentage strengths of links are denoted by solid, dashed and dotted lines (see key). The keys and data-analytical methodology also apply to Figures 4 and 6.

revealed by the weaker second coordinate (PC2: 20%); the lower samples (11-15) possess less fine silt than the upper (16-19). The non-Dark Earth samples also show more evidence of clustering according to stratigraphic position than deposit type; the substratum and early Colonium materials (samples 0-3) contain less fine silt than the overlying Flavian materials (6-10), with the building debris much closer in composition to the Dark Earth than are the in situ buildings.

Overall, the data reveal that particle size composition increasingly diverged from that of the substratum as the anthropogenic deposits accumulated at Colchester. This could simply indicate the anthropogenic addition of non-local natural materials to the Flavian buildings that were reworked into the overlying destruction layer and Dark Earth, but the continued increase in sand through the Dark Earth suggests that this explanation alone is inadequate. The additional sand particles may have been anthropogenically formed as well as anthropogenically introduced.

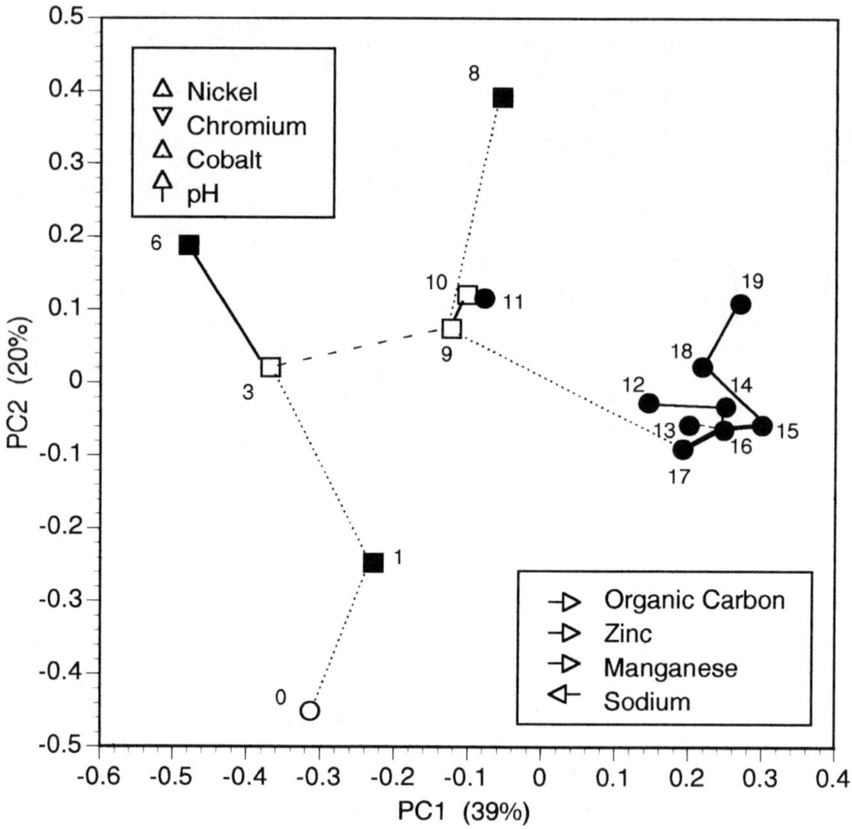

Figure 4. Principal coordinates plot of geochemical data from Colchester (see Figure 3 for keys and explanation).

At Rangoon Street in London, Dark Earth was sufficiently well exposed to allow three-dimensional sampling at 1 m lateral intervals and 10 cm vertical intervals of a body 10 m long, 8 m wide and up to 1 m deep (yielding 253 samples). Sand-silt-clay data for these samples were used to address a different question, namely whether textural variation within the Dark Earth showed stronger trends at particular spatial scales. Data were used to generate semivariograms[29,30] for each particle size category at each stratigraphic level, testing assumptions of both isotropy (adirectional variation) and anisotropy (directional variation). However, trends were few and weak. Semivariance showed little change over the analysed spatial range of 1-10 m; any spatial trend should have been revealed as a initial progressive increase in semi-variance followed by levelling off at the spatial scale of greatest variance. Three conclusions are possible regarding Dark Earth at Rangoon Street: (1) textural variation occurs at a scale greater than 10 m (highly improbable), (2) it varies at a scale less than the 1 m sampled grid (possible), or (3) there is no overall pattern to its admittedly subtle textural variation (probable; for example, sand content at 49.2±1.1% gives a coefficient of variation of only 2.3%).

GEOCHEMISTRY

Geochemical data are a bulk measure of material composition. In this study, inductively-coupled plasma-arc spectrometry (ICP[31]) was applied to the <2 mm matrix to obtain data on the ten major elements (P, K, Ca, Mg, Mn, Na, Fe, Al, Ti, Si) plus eight minor elements (Zn, Cu, Ni, Cd, Cr, Pb, Mo, Cb). These data were supplemented with four partially interrelated measures influenced by the amount and nature of carbon present: pH, loss-on-ignition, organic carbon, and carbonate content[22]. Semi-quantitative colour data collected in the field[32] are also most appropriately included under this heading, though they were excluded from the principal coordinates analysis that forms the basis of this discussion (Figure 4).

The pattern of samples in Figure 4 resembles that for particle size (Figure 3). Most of the Dark Earth samples form a tight cluster toward the positive end of the strong first coordinate (PC1: 39%), distinguished primarily by their greater contents of organic carbon, together with zinc and manganese (both readily bound to organic matter). The non-Dark Earths are again separated into two stratigraphic groups on PC1, this time between the Flavian reconstruction layer (sample 6) and its capping Opus Signinum floor (sample 8: Figure 1). Interestingly, the lowest Dark Earth sample (11) clusters with the underlying Flavian building materials, suggesting substantial reworking of the latter into the base of the Dark Earth.

The much weaker second coordinate (PC2: 20%) slightly separates the uppermost Dark Earth sample from the remainder, largely on its high molybdenum content

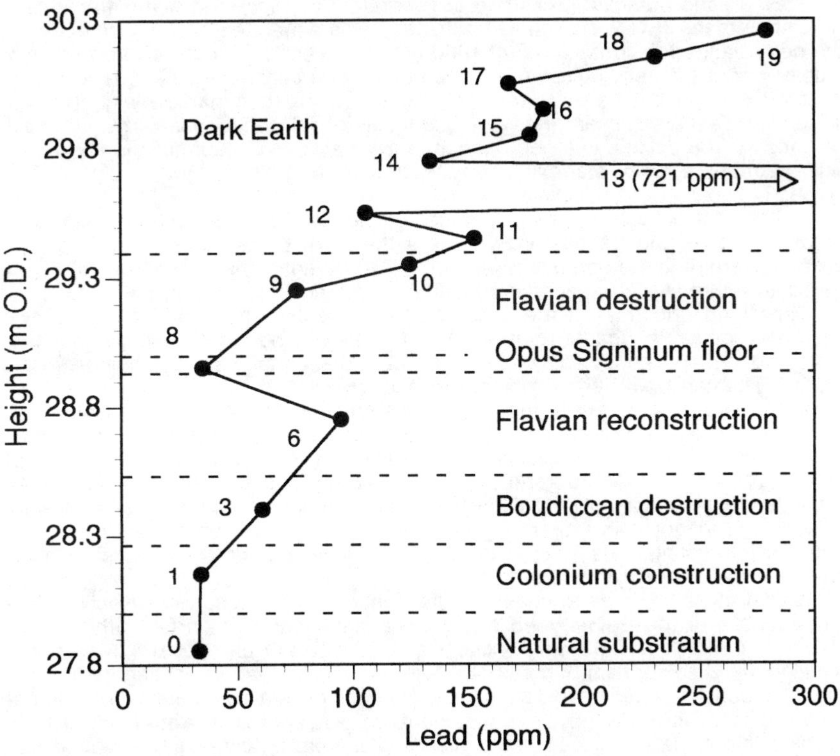

Figure 5. Variation in ICP-assessed lead content through the Culver Street profile, Colchester.

(78 ppm, probably a modern contaminant of this uppermost horizon of the section). The main effect of this coordinate is to contrast the relatively low pH, carbon and minor element contents of the natural substratum and overlying Colonium construction layer with high values for these parameters in the Flavian reconstruction layer and capping Opus Signinum floor, probably indicating greater proportions of mortar/cement.

Additional interpretations require consideration of individual elements. To illustrate this point, Figure 5 plots lead levels against sample depth. The background level of 35 ppm, typical of natural sediments in southern England, is reflected in the Colonium construction and Opus Signinum layers. Above each of these baselines, values rise progressively to maxima of 100 ppm in the Flavian reconstruction layer and 130 ppm in the Flavian destruction layer, values typical of mildly contaminated urban soils in present-day Europe and here ascribed to passive buildup of lead in the urban environment[33]. However, this lead level is itself the baseline for the Dark Earth, which exhibits even higher values of 280 ppm at the top of the profile and a startling, highly toxic 720 ppm in sample 13. We suggest that the Dark Earth contains substantial proportions of industrial waste, as other particulate analyses reveal a wide range of materials potentially representing local, small-scale manufacture (or, more probably, secondary processing) of glass, enamels and/or metals.

FINE SAND MINERALOGY

Fine sand particles (2-4 ø) were separated into light and heavy fractions in tribromomethane (SG 2.9). The heavy fraction was separated optically into opaque and non-opaque fractions, and 500-1000 grains of each of the three fractions were counted under the petrological microscope to yield percentage data[22,24]. The results revealed a recurrent basic suite: light fractions were dominated by quartz with subordinate feldspars, chert and varying amounts of calcite/aragonite, opaque heavy fractions by iron oxides, and non-opaque heavy fractions by tourmalines and zircon with subordinate rutile, staurolite, kyanite, apatite, hornblende, augite and fayalite olivine.

The principal coordinates plot for sand minerals (Figure 6) contrasts with those for particle size and geochemistry, where the Dark Earth samples formed tight clusters. Here, it is the non-Dark Earths that form a tight cluster, whereas the Dark Earths form three widely spaced groups that reflect their stratigraphic positions. They are separated primarily on the second coordinate according to contents of two calcareous minerals: apatite (mostly bone fragments) and calcite (mostly shell and chalk fragments). The lowest group is richest in these materials, the highest group poorest. The non-Dark Earths compare most closely with the middle and upper Dark Earth clusters; within this group, the brickearth substratum and the Boudiccan destruction layer are least calcareous. The first coordinate separates the Dark Earths from the non-Dark Earths according to the greater augite, fayalite, and synthetic glass content of the former. Microprobe data from selected grains of these minerals (see Farrington, contribution J8.13 to this volume) support optical evidence that they were formed anthropogenically rather than naturally, as a result of contemporaneous light industry. Interestingly, as for geochemistry, the lowest Dark Earth sample closely resembles the underlying Flavian destruction layer.

Overall, a basic sand mineral suite dominates throughout the section. It is characteristic of the Palaeogene deposits of the London Basin[24,34], which were reworked to form the underlying brickearths at Colchester and apparently were the dominant sand component of the bricks, tiles and mortar that constituted the Roman buildings. Superimposed on this suite are calcareous mineral components from mortar and especially foodstuff-derived household refuse, most prevalent in the lower Dark Earths, and materials produced during glass and/or metal processing, which particularly characterise the Dark Earth (basalmost sample excepted).

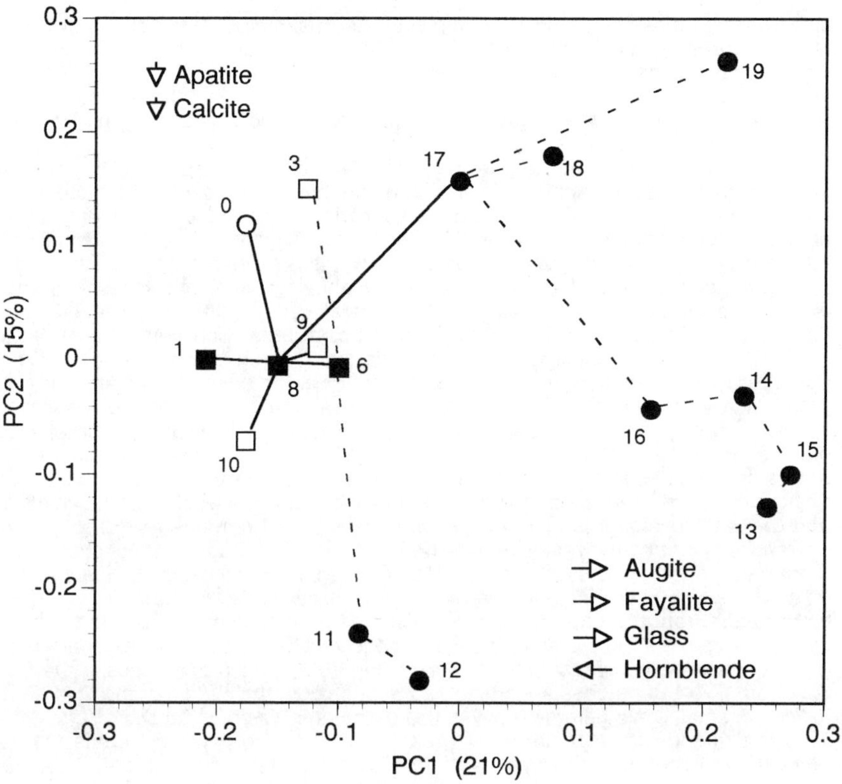

Figure 6. Principal coordinates plot for fine sand (2-4 ø) mineral grain counts from Colchester (see Figure 3 for keys and explanation).

CONCLUSIONS

We will now return to the original questions of provenance, mode and environment of deposition, period of deposition, and post-depositional modification of the horizons in the Colchester profile, with particular reference to Dark Earth. At the same time, we will emphasise the advantages of a **holistic** approach that integrates information from a range of geoanalytical techniques (dominantly matrix-based rather than artefact-based) to address specific archaeological questions.

All of the techniques applied have provided information on provenance. The dominant lithology, most clearly shown by the sand mineralogy, is the local solid geology of shallow marine/estuarine Palaeogene sediments. This has been naturally reworked into the Quaternary 'brickearths' that immediately underlie the Roman stratigraphy in Culver Street, and was clearly also the main component of the bricks, mortar and tiles used to construct the Roman urban buildings. Mineralogically, the brickearth and overlying Roman horizons rich in building materials show relatively little variation (Figure 6); in contrast, their variation in the bulk properties of particle size (Figure 3) and geochemistry (Figure 4) is considerable. However, the variation reflects the stratigraphic position of the samples more than whether they represent construction or destruction layers, showing a gradual upward transition (with greatest change perhaps within the Flavian construction horizon) from compositions similar to the substratum to those more similar to the Dark Earth samples. This trend could simply reflect increased input from more distant sources into Flavian relative to the cruder Colonium buildings (certainly, more calcareous building materials are evident in the Flavian reconstruction/Opus Signinum horizon), but we suspect that it also indicates the gradual buildup in the environment of anthropogenic waste resulting from prolonged urban occupation.

Previous discussions of Dark Earths have emphasised apparent homogeneity and lack of clear stratigraphic subdivisions within these relatively thick archaeological deposits[11,35]. At Culver Street, putative subdivisions based on apparent laterally extensive concentrations of coarse clasts occur approximately between samples 11-12, 13-14, and 16-17. However, these inferred hiatuses receive at most limited support from the geoanalytical data, which primarily show gradual progressive changes through the Dark Earth that result in far less overall variation than is observed among the non-Dark Earths. Geochemical data show a clear break between samples 11 and 12, implying extensive incorporation of material from the underlying Flavian horizons. Particle size data are similarly Flavian in character at the base of the Dark Earth, but gradually diverge upward as sand increases at the expense of silt; the greatest change occurs between samples 15 and 16. In contrast, mineralogical data show much greater variation among the Dark Earths than the non-Dark Earths, but again the Dark Earths become increasingly dissimilar from the non-Dark Earths when traced upward. The changes are interpreted as greater amounts of industrial waste above sample 12, and decreasing amounts of mortar and/or animal food remains (with the greatest decreases occurring between samples 12-13 and 16-17). Thus, clear-cut stratigraphic breaks supported by several types of data are absent. Nonetheless, the Dark Earth is demonstrably not wholly homogeneous at a large scale, gradually diverging in composition from the underlying stratigraphy as it accumulated. Admittedly, not all individual variables show such clear progressive change; a good example is the generally upwardly increasing lead content (Figure 5), which decreases in the Opus Signinum floor but reaches a startling 720 ppm peak in the middle of the Dark Earth, indicating small-scale heterogeneity in this parameter. Overall, the Dark Earth differs from the underlying deposits in its greater content of organic materials (notably household refuse) and industrial waste (particularly metallurgical); as the Dark Earth accumulated, the latter increased at the expense of the former.

Several types of analysis provided information on the much-discussed cause of the very dark grey colour that characterises Dark Earth. As a by-product of particle size analysis, sample 12 was subdivided into five size fractions, and organic carbon content was determined for each. This revealed concentrations of amineralised organic matter in the sand fraction and in the fine silt-clay fraction. Micromorphological study suggests that the coarser peak represents recognisable plant fragments (often charcoalified), whereas the finer material appears amorphous and often loosely coats mineral grains, a conclusion supported by microprobe study of thin-sections. However, even these organic carbon peaks do not exceed 2.2%, normally insufficient to so deeply colour a deposit of this texture. Organic complexing with metallic ions, perhaps largely derived from the industrial waste, is suspected.

Provenance interpretations can be seriously modified by several forms of post-depositional alteration associated with pedogenesis[3,15,16,36]. However, none of the geoanalyses used suggests substantial soil formation. Mineral weathering is minimal, and neither geochemistry nor micromorphology suggest significant translocation of materials in suspension or solution. Some of the larger voids are lightly and loosely coated with thin discontinuous translocated silt, and there is very local growth of authigenic phosphatic minerals, often nucleated on bone and charcoal fragments. These could be explained by groundwater movement rather than prolonged subaerial exposure; indeed, rapid accumulation is implied. The most problematic criterion to assess is the degree of bioturbation experienced by the deposit, which could have been a major contributor to the homogeneity observed in the Dark Earth (for example, earthworm activity could in theory have reoriented clasts and thereby generated the random fabric shown in Figure 2A). This is compounded by the difficulty of distinguishing contemporary from recent, post-exhumation bioturbation by soil fauna. Current evidence suggests that neither was sufficiently intense to have obscured former stratification in the Dark Earth.

The relative homogeneity of the Dark Earth in turn bears on its mode of deposition. It is readily distinguished from the relatively well documented anthropogenic formation of the repeated Roman construction and destruction layers beneath, and is too thick and homogeneous to represent an upward continuation of the Flavian destruction layer. Its large and diverse content of anthropogenic materials belies previous suggestions of natural accumulation. The low abundance but fairly high diversity of plant remains, lack of evidence of root penetration, and the toxically high levels of elements such as lead bode ill for hypotheses of cultivation, though both non-cereal cultivation and animal husbandry are notoriously difficult to detect in the urban archaeological record[7]. Rather, the Dark Earth appears to be a poorly sorted admixture of all the potential waste products of Roman urban life: building debris, food waste, hearth ash, and industrial by-products. However, it cannot be ascribed simply to dumping in areas of urban decay within a still densely populated settlement, as it is ubiquitous, coating without prejudice the remains of all Roman constructions except the city walls. The only logical conclusion is that the Dark Earth was deliberately deposited by man, presumably after the urban centres had become depopulated. As in the previous destruction layers, building rubble and household refuse on site were the preferred infilling material, but as supplies dwindled, greater amounts of local industrial waste were incorporated into the Dark Earth. Deposition in small, manually transported 'aliquots' is consistent with the fabric and particle size evidence[9].

The question of **when** Dark Earth was deposited is linked to the even more perplexing question of **why** this unintuitive event occurred in Roman urban centres throughout northwestern Europe. In the current absence of radiocarbon dates and contemporary documentary sources, the best dating comes from the otherwise uninformative traditional source of coarse-fraction artefacts. Although many of those found in Dark Earth are poorly preserved and reworked, the most reliable coin and pottery evidence together date from 350-400 AD, the period of decline of Roman influence in urban Britain that preceded their final withdrawal in ca. 410 AD[9]. The evidence from Dark Earth suggests that this decline was especially marked in the formerly prosperous and populous urban centres. However, understanding the historical significance of Dark Earth and its depositional context require comparison of Roman urban centres across Europe. Although they are the keystone of this project, such comparisons are beyond the scope of this short paper.

We conclude with the following general recommendations, which together constitute holistic geoarchaeology: (1) Apply several geoanalytical techniques chosen with the aim of complementing each other to address specific questions. Narrow approaches have a high probability of generating seriously misleading conclusions. (2) Sample profiles in detail to detect subtle trends, also ensuring that thoroughly understood deposits are well represented for comparison with more ambiguous materials. In the absence of such 'standards', interpretations are performed in an undesirable vacuum. (3) Analyse the matrices of archaeological deposits as well as the macro-artefacts that they enclose. In this study, the former proved much more informative than the latter. (4) Summarise the resulting large bodies of data using computerised multivariate algorithms. These are adept at detecting subtle trends in data that are not readily detected in the raw data by the human eye. (5) Qualify use of the ambiguous term 'anthropogenic.' It can apply to an object made by man of synthetic materials, an object made by man of natural materials, or a wholly natural object merely transported by man. (6) Confine a holistic approach to archaeological profiles capable of answering key historical questions. It is too expensive, both financially and (especially) in research time, to apply to less profound topics.

ACKNOWLEDGMENTS

John Catt, Jim Rose and Brin Roberts offered much helpful advice during the long history of this project. Specific technical problems were overcome through the generosity and expertise of Peter Bullock, Peter Loveland, James Rayner, Lynn Parry, Bob Norrish, Paul Grant, Dick Giddins, and Fran Hueber. Philip Crummy, Director of the Colchester Archaeological Trust, kindly permitted field sampling at Culver Street. Most of the analytical work was performed in the laboratories of the Soils Division and Soil Survey of England and Wales, then at Rothamsted Experimental Station.

REFERENCES

1. P. Wilson and R.M. Bateman, Boreas 15, 137-153 (1986).
2. P. Wilson and O.S. Farrington, Proc. R. Irish Acad. B 89, 1-23 (1989).
3. J.A. Catt, Paleopedology Manual (Quat. Int. 6, Pergamon, 1990).
4. J.K. Stein and P.A. Teltser, Geoarchaeology 4, 1-30 (1989).
5. M. Millett, The Romanization of Britain (Cambridge Univ. Press, Cambridge, 1990).
6. A.G. Vince, Saxon London: An Archaeological Investigation (Seaby, London, 1990).
7. R.F.J. Jones (editor), Roman Britain: Recent Trends (Collis, Sheffield, 1991).
8. M. Jones, England Before Domesday (Batsford, London, 1986).
9. O.S. Farrington, 28th Int. Geol. Cong. Abstr. 1, 473 (1989).
10. R.M. Bateman, 28th Int. Geol. Cong. Abstr. 1, 97-98 (1989).
11. P. Crummy, Current Arch. 103, 238-245 (1987).
12. Tacitus, Agricola (AD98).
13. R. Dunnett, The Trinovantes (Duckworth, London, 1975).
14. D.J. Briggs, Sources and Methods in Geography: Sediments (Butterworth, London, 1977).
15. P. Bullock, N. Federoff, A. Jongerius, G. Stoops, and T. Tursina, Handbook for Soil Thin Section Description (Wayne, Wolverhampton, 1985).
16. R.A. Kemp, Soil Micromorphology and the Quaternary (Quat. Res. Ass. Tech. Guide 2, Cambridge, 1985).
17. C.P. Murphy, Thin Section Preparation of Soils and Sediments (Pudoc, Wagenigen, 1985).
18. D.R. Bridgland (editor), Clast Lithological Analysis (Quat. Res. Ass. Tech. Guide 3, Cambridge, 1986).
19. D.R. Piperno. Phytolith Analysis (Academic Press, New York, 1988).
20. P.D. Moore, J.A. Webb, and M.E Collinson, Pollen Analysis (Blackwell, Oxford, 1991).
21. R.G. Scaife, Pollen Analysis of some Dark Earth Samples (unpubl. Ancient Monuments Laboratory Rep. 3001, 1980).
22. B.W. Avery and C.L. Bascomb, Soil Survey Laboratory Methods (Soil. Surv. Tech. Monogr. 6, Harpenden, 1974).
23. R.L. Folk and W.C. Ward, J. Sed. Petr. 27, 3-26 (1957).
24. R.M. Bateman and A.J. Moffat, Tertiary Res. 8, 75-103 (1987).
25. A.W. Payne, et al., Genstat 5 Reference Manual (Oxford Univ. Press, Oxford, 1987).
26. J.C. Gower, Biometrika 52, 325-338 (1966).
27. J.C. Gower, Biometrics 27, 857-872 (1971).
28. J.C. Gower and G.J.S. Ross, J. R. Stat. Soc. C 18, 54-64 (1969).
29. E. Zubrow and N. Harbaugh, in Simulation Studies in Archaeology, edited by I. Hodder (Cambridge Univ. Press, Cambridge, 1978) pp. 109-122.
30. S.R. Vieira, J.L. Hatfield, D.R. Nielsen, and J.W. Biggar, Hilgardia 51, 1-75.
31. R.L. Dahlquist and J.W. Knoll, Appl. Spectrom. 32, 1-30 (1978).
32. J.M. Hodgson, Soil Survey Field Handbook (Soil Surv. Tech. Monogr. 5, Harpenden, 1976).
33. I. Thornton and J.S. Webb, in Applied Soil Trace Elements, edited by B.E. Davies (Wiley, New York, 1980) pp. 381-439.
34. A.C. Morton, Proc. Geol. Ass. 93, 263-274 (1982).
35. D. Perring and S. Roskams, Early Development of Roman London West of the Walbrook (Council Br. Arch. Res. Rep. 70, London, 1991).
36. R.M. Bateman and J.A. Catt, Catena 12, 1-21 (1985).

PART II B

Characterization of Composition and Structure of Art Objects and Archaeological Artifacts

UNDERSTANDING ART THROUGH TECHNICAL ANALYSIS INFERENCES FROM PIGMENT IDENTIFICATION

BARBARA H. BERRIE
Scientific Research Department, National Gallery of Art,
Washington, DC 20565.

ABSTRACT

Pigment identification and technical analysis of works of art provide the historian and conservator with information that allows a new view of an object. The methods the analyst chooses depends on the nature of the sample and the materials and method expected for the age of the object being examined. Several analytical methods are used frequently in the museum field with great success; these include x-ray fluorescence spectroscopy, x-ray powder diffraction methods and optical microscopy. Results from the analysis of pigments in paintings by Derain, Beckmann, El Greco, G. David, and Gentileschi are given. The results include discovery of an unpublished formulation of cobalt violet, example of the use of particle size to change the appearance of color, and the use of x-ray fluorescence to determine artists' palettes without removing samples from the works of art. The study of the El Greco and the Gentileschi show how technical investigations can complement art historical research.

INTRODUCTION

Identification of the pigments has been one of the most common types of analysis undertaken on works of art. Merely knowing the pigments used by an artist is no longer novel; however, it does provide important information to the art historian and conservator. To the layman, the most obvious application is authentication. However, other applications that are far more frequent in museums are understanding why the work of art appears as it does today, assisting in choices for the conservation of a work of art, identifying later additions, distinguishing authorship, revealing artists' working practices, and helping identify how the transfer of new materials and techniques proceeded through the art community. Two aspects of the analysis of pigments used in a work of art are not always appreciated. One is the technical difficulty in identification of some pigments. The number and diversity of compounds on the market as artists' colors by the end of the nineteenth century can be a source of difficulty in that the first step of analysis may be asking the question "Animal, Vegetable, or Mineral?" The dates of introduction and occurrence of various new colors has not been greatly studied and their characterization is very difficult. A second aspect of pigment identification is varied applications that the information has to the understanding of a work of art. Some of these applications will be illustrated by the following descriptions.

INSTRUMENTATION AND THE NATURE OF THE RESULTS

There are many methods used to identify pigments. The choice of the method is, of course, critical in obtaining pertinent information. A thorough knowledge of the historical context of a certain painting or art object can help in the judicious choice of analytical method. Previous studies of artists' materials and methods can provide a critical starting point for analysis. Thoroughness in the analytical method is essential to obtain new information on pigments. The list of analytical methods that has been used in the pursuit of pigment identification is voluminous. It includes particle induced x-ray emission spectroscopy, x-ray fluorescence spectroscopy (XRF), x-ray powder diffraction methods (XRD), neutron autoradiography, infrared spectroscopy, thermal gravimetric analysis, laser microprobe analysis, and the optical microscope. The latter has been used most frequently and is perhaps the most commonly found instrument in museum laboratories. Most information can be obtained by using more than type of analysis and indeed multiple techniques are often required for a comprehensive analysis. As in any field, artists' techniques have changed considerably over time. For example, at different times pigments fell from general use. It has been through careful analysis of the pigments in paintings that artists' erstwhile materials have been identified. The rediscovery of lead-tin yellow is an example of this.

There are many pitfalls for the analyst along the way. This paper will illustrate some of the difficulties as well as some of the achievements of the endeavors of the analyst. One great difficulty is that pigments to be analyzed from works of art are rarely pure. They have been bound in an organic medium, they have been mixed with extenders to give a workable paint, and they have been mixed with other pigments to provide the desired paint color. All these factors militate against easy and categorical identification of particles.

In some cases one technique can be used to provide some information that is valuable though not complete. The best example of this is the air-path x-ray fluorescence spectrophotometer. This important instrument is widely used in museum science because no samples need to be removed from a work of art to obtain information on the elemental content of an area being studied. There are several drawbacks to XRF analysis used this way. X-rays penetrate more than the surface paint film and the information we have on lower paint films, specifically their number and colors, is limited. The elements, but not the compounds, are being analyzed. Thus, lead in a red paint could indicate the presence of red lead, white lead mixed with red pigments, massicot mixed with red pigments, or it could indicate a lead white underpaint where the lead was unrelated to the color of the top layer of paint. In such a case, optical microscopy is a useful way to pursue complete elucidation of the results. Only a very small sample is required and red lead can be easily identified from its distinctive anomalous interference colors. Pigments such as organic red lakes and ultramarine, which do not contain elements with $Z > 18$, cannot be detected using air-path instrumentation. Some pigments contain a very small amount of a detectable element, such as Prussian blue (Fe) and smalt (Co). These pigments are difficult to detect using XRF and other instrumental methods are often required to complement the information obtained.

Paintings that have single layers of paint are most amenable to pigment identification using XRF. The Fauve artist Derain, painted using streaks of color laid directly on the grounded canvas. Mountains at Collioure, painted in 1905, depicts the scene in short bold strokes of intense color that frequently do not overlap one another (see figure 1). Because there is only one layer of paint, it is possible to infer the pigments used from the elements present. The elements found are listed in the table 1. Although the paint layers do not overlap, one may infer several possibilities for the pigments Derain chose. The presence of copper together with arsenic suggests that

Derain was using emerald green, $Cu(CH_3CO_2)_2.3Cu(AsO_2)_2$. However, although it is unlikely, Scheele's green, $Cu(HAsO_3)$, cannot be ruled out from the data. The presence of chromium on the spectra of yellow and green areas of the painting suggest the use of chrome yellow. Chrome yellow is mixed with Prussian blue to give the commercial pigment chrome green. In the yellow sky there is also iron, implying the admixture of a yellow iron earth pigment. Mercury in the salmon-pink paint shows the vermilion was used. From the absence of certain elements, it is possible to note some pigments Derain did not use when he painted Mountains at Collioure. He used neither cadmium yellow nor orange nor Naples yellow ($Pb_3(SbO_4)_2$). Unfortunately, we cannot determine from XRF which pigments Derain used to make his violet paint.

Table 1. Results of x-ray fluorescence spectroscopy on Derain's Mountains at Collioure. Predominant elements asterisked, minor in parentheses, trace in double parentheses. The Kevex 0750A was used equipped with a $BaCl_2$ secondary target. The anode voltage and current were 60 kV and 0.7 mA.

Location	Elements	Possible Pigments
A Lilac between tree trunks	Zn*, (Ca, Pb, Sr) ((Fe))	zinc white, Prussian blue or ultramarine and red lake
B White area between trees	Pb*, Zn*, (Sr), ((Fe))	zinc white, lead white
C Salmon-pink tree trunk	Zn*, Hg, Pb, (Sr), ((Fe, Ca))	zinc white, vermilion, lead white
D Light green between tree branches	Zn*, Pb*, Cr, (As, Sr, Cu), ((Fe))	zinc white, lead white, emerald green, chrome green or chrome yellow and chromium oxide green or viridian with Prussian blue
E Yellow in grass	Zn*, Pb, (Cu, Cr, As, Sr), ((Fe, Co,))	zinc white, chrome yellow, emerald green
F Dark blue outline on mountain	Zn*, Pb, (Fe, Sr), ((Ca, Cu))	zinc white, lead white, Prussian blue
G Dark green leaf	Cr*, Zn*, Pb*, Cu, As ((Ca))	chrome green or chromium oxide green or viridian, emerald green, zinc white, lead white
H Yellow sky	Zn*, (Pb, Cr, Fe, Sr)	zinc white, chrome yellow, iron oxides
I Orange-yellow in sky	Zn*, Cr, Pb, (Sr)	zinc white, chrome yellow
J Orange in mountain	Pb*, Zn, Cr (Fe, Sr)	lead white, zinc white, chrome yellow (possibly red lake)

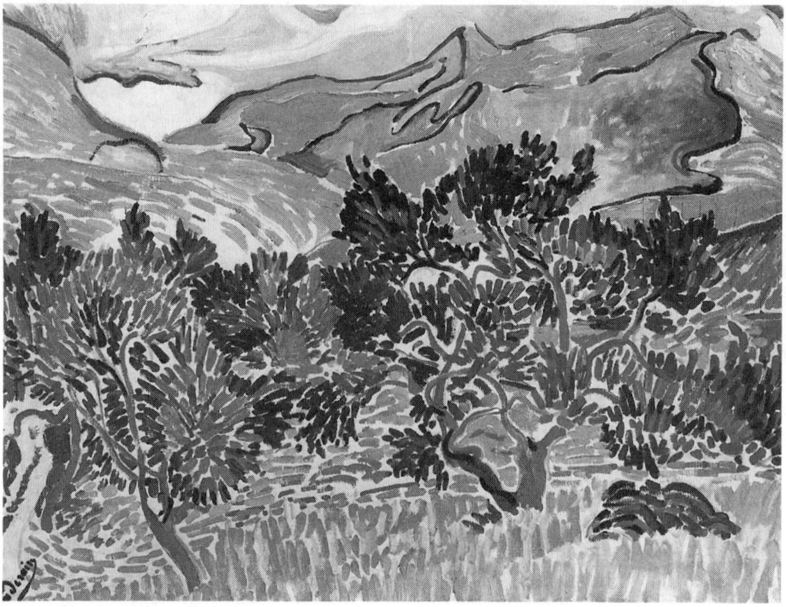

Figure 1. André Derain, Mountains at Collioure, 1905. National Gallery of Art, Washington, DC. John Hay Whitney Collection (1982.76.4; 0.813 x 1.003 m).

It is crucial for the analyst to have the opportunity to carefully study the work of art under investigation and to be well-acquainted with our best understanding of artists' practices and procedures. An example of how various analytical methods and knowledge of artists' materials converge to give new information on a work of art is identification of the pigments in the painting Falling Man by Max Beckmann. XRF showed that cobalt was present. However, this result did not allow us to distinguish whether Beckmann had used cobalt blue mixed with a red lake or cobalt violet. A small sample was removed and examined using XRD. The powder diffraction pattern (obtained by Michael Palmer, [1]) is given in table 2. The pattern indicates that $Co(NH_4)PO_4.H_2O$ and ZnO are present. The results are given in the table below. This compound is undoubtedly the colorant although it does not have the formula usually quoted for cobalt violet, which is given as $Co_3(PO_4)_2$ or $Co_3(AsO_4)_2$ [2]. This result demonstrates that the analyst should be aware of the limitations of the currently accepted history of pigments. The alteration in formulas of cobalt violet and dates of use of variants have yet to be thoroughly described. In such circumstances the analyst can be the catalyst for art historical investigation.

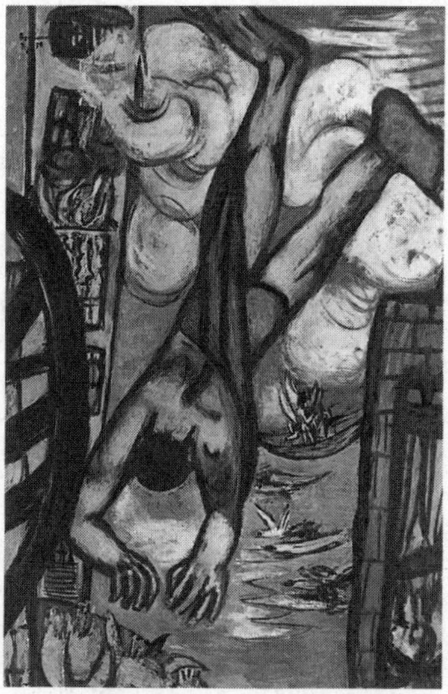

Figure 2. Max Beckmann, Falling Man, 1950. National Gallery of Art, Washington, DC. Gift of Mrs. Max Beckmann.

Table 2. Observed d-spacings and intensities of lines in the powder diffraction pattern of the violet paint in Falling Man by Max Beckmann and the values for zinc oxide and ammonium cobalt phosphate hydrate as listed in the JCPDS files [3].

dark violet		ZnO JCPDS 5-664		$Co(NH_4)PO_4.H_2O$ JCPDS 21-793	
d,Å	I	d,Å	I	d,Å	I
8.7	100			8.77	100
7.5	vw				
4.7	w			4.73	2
4.2	w				
4.05	vw				
3.8	vw				
3.7	vw				
3.4	10				
3.35	8				
2.80	80	2.82	71	2.80	14
2.59	40	2.60	56		
2.46	60	2.48	100		
2.43	vw				
2.40	vw				
2.33	vw				
				2.28	2
2.12	vw				
2.04	vw				
1.90	w	1.90	29		
1.82	10				
1.68	vw				
1.62	20	1.63	40		
1.475	15	1.48	35		
		1.46	28		
1.375	10				
1.35	10				

The scanning electron microscope coupled with energy dispersive x-ray fluorescence spectroscopy (SEM-EDX) is becoming more widely used by conservation scientists. The advantages of SEM over optical microscopy at first seem irresistible but, as the following example shows, SEM should not be relied on as a sole method of analysis in all situations. An example of the difficulties is given by the following which again emphasizes the care required when dealing with known--let alone unknown samples. Dry chrome green (Kremer Pigments, New York) was mounted on carbon paint on a carbon stub and carbon-coated. The image was featureless. The sample appeared homogeneous. The EDX spectrum contains lead, chromium, titanium and silicon. Iron was not detected under the condition used. Chrome green is a mixture of two pigments, chrome yellow and Prussian blue. The information obtained by SEM-EDX did not provide incontrovertible evidence that the sample is chrome green because iron from Prussian blue was not detected. This example shows that it is not possible to standardize an analytical procedure for general pigment analysis and the protocols must be varied according to what might be expected in a work of art. Thus, to find a predictive element in a highly tinting pigment such as Prussian blue, the analyst must be thorough. When the chrome green is viewed through the microscope, yellow rods of chrome yellow and blue clumps of Prussian blue are easily visible. The two pigments in the mixture can be readily identified from their optical properties and the sample identified as chrome green. This is one example of the power of the optical microscope. Aside from being practical, an important characteristic of a pigment--its color--can be discerned. The SEM does, however, find indispensable uses, especially when coupled with EDX.

Technical analysis of paintings, which includes but is not restricted to pigment identification, can provide the answers to questions that may not be approachable in other way. For instance, the question of which of two versions of the painting of Saint Martin and the Beggar by El Greco came first is hard to determine from a stylistic approach since the two versions are so similar. The smaller version (1937.1.84) could have been a study for the larger one (1942.9.25) or it could have been a copy made after the large version was completed. The use of cross-sections showed that the larger version has many changes in its layout prior to completion. The simplicity of the layer structure of the smaller version shows that its composition was already established when it was painted. Although the paintings now look similar, there were many refinements made to the composition during the painting of the larger version of Saint Martin and the Beggar and it is possible to infer that it was painted first. This has been discussed by Griswold [4]. Cross sections from the green landscape behind the horse are shown in figure 3.

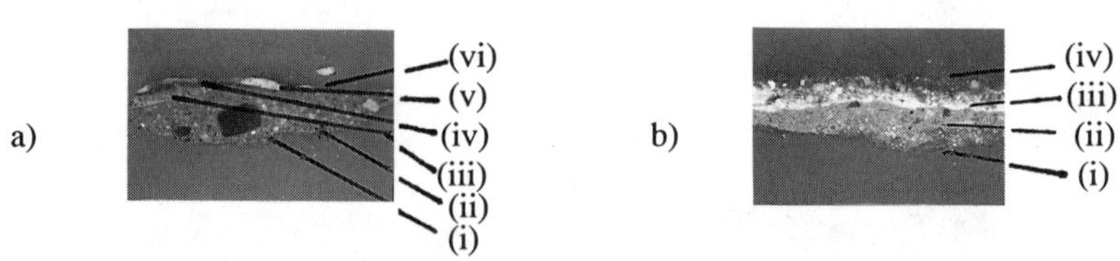

Figure 3. (a) Cross section from 1942.9.25 showing (i) the gesso ground, (ii) red ground, (iii) grey, (iv) black, (v) white and (vi) green paint. (b) the section from 1937.1.84 showing the (i) gesso, (ii) red ground, (iii) white and (iv) two layers of green paint. Both sections x132.

Aside from chemical nature of pigments, their physical state is important for determining the appearance of paintings. One factor that has a profound effect on color is particle size. Undoubtedly particle size was a major discriminant for Medieval and Renaissance artists. We can often see the use of the same pigment mixture giving differing hues where the particle size is different. An example of the use of this by an artist is seen in Gerard David's St. Anne Altarpiece. This triptych was recently cleaned by Catherine Metzger and an art historical and technical investigation undertaken. The results of this will be published in full [5]. In the three panels there are many different tones of blue. The hue ranges from the petrel blue of Saint Nicholas's glove to the light smokey blues of the floor tiles and the rich blue of the Virgin's robe. The pigments in these different places were identified using optical microscopy. In all the areas the desired color was achieved using a mixture of the same three pigments. These are azurite, white lead, and red lake. The difference is not limited to the proportion of the pigment but also to the size of the particles of azurite. The deeper blue paints contain large-particled azurite. Sometimes the particles are as large as 50 μm while the particle size in the lighter or greyer areas averages around 15 μm. While the use of various grades of the expensive pigment ultramarine is generally acknowledged, the motivation is often considered to be economic and not aesthetic. The Medieval and Renaissance artists knew well their materials and their properties and the observations in the panels by David attest to skilful use of particle size to expand the palette. This practice was perhaps the norm and accounts for the availability of several grades of azurite to the artist [6].

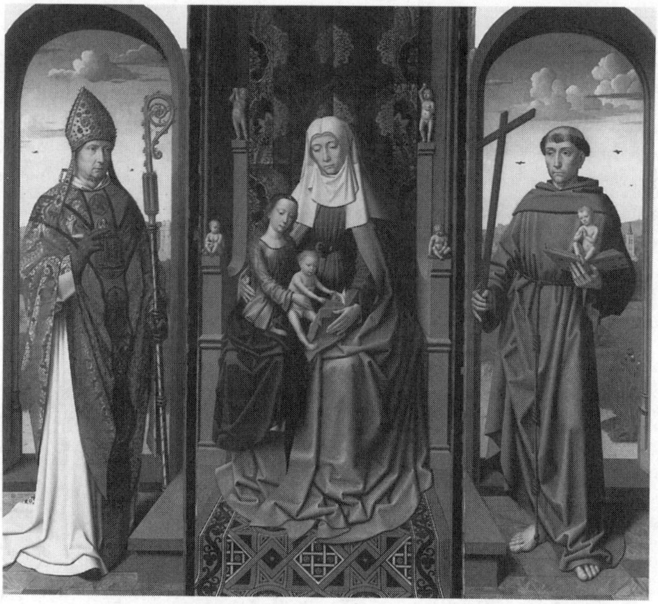

Figure 4. Gerard David, The Saint Anne Altarpiece, c. 1505. National Gallery of Art, Washington, DC. Widener Collection.

The interaction between technical investigation and art historical research on a work of art is well demonstrated by the work that has been accomplished on a painting called St. Cecilia and Angel. This painting, formerly ascribed to Orazio Gentileschi, has been recently reattributed to him and Giovanni Lanfranco on stylistic grounds [7]. The painting and its x-radiograph are shown in figure 5. From the x-radiograph, it can be seen that the picture was painted on four pieces of canvas stitched together. While the part of the painting that is attributable to Gentileschi occupies the center piece of canvas, technical evidence shows that all four pieces of canvas were prepared using the same mixture of lead white, calcium sulfate, calcium carbonate, and carbon black. It can be deduced therefore that the painting was always the larger size and not enlarged by Lanfranco. Optical microscopy shows that St. Cecilia's bodice was painted using a red iron earth pigment but her skirt was painted using vermilion. The use of different pigments rather than the same pigment in mixtures shows the effort put into obtaining the desired color effects. X-ray fluorescence spectroscopy of the painting shows that the parts that are attributed to Gentileschi contain lead and tin together which suggests the use of lead-tin yellow. In contrast the parts of the painting that have been attributed to Lanfranco contain antimony in addition to lead and tin. It is not possible from XRF to determine if this is due to lead-tin yellow mixed with Naples yellow or if the spectrum is due to a single compound where tin and antimony substitute for one another. X-ray powder diffraction methods will be used to investigate this. We hope to be able to determine the distribution in the painting of the variations in the formulation of the yellow pigment to provide information that is complementary to the stylistic evaluation of the work of art and provide corroborative evidence for the attribution based on connoisseurship.

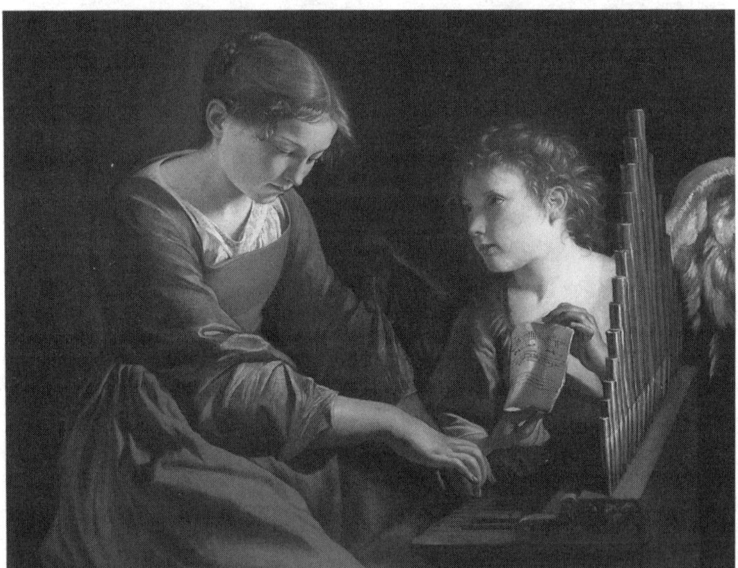

Figure 5a. O. Gentileschi and G. Lanfranco, St. Cecilia and an Angel, c. 1510. National Gallery of Art, Washington, DC. Widener Collection.

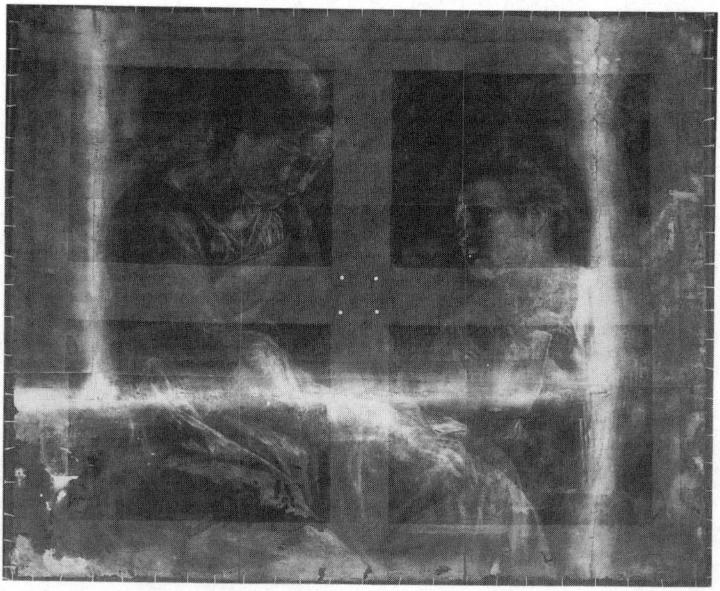

Figure 5b. X-radiograph of St. Cecilia and an Angel.

SUMMARY

New information on pigments and their use in art is frequently being brought to light by analytical methods. The results change how we see a painting. Our knowledge of the materials allows us to appreciate the early Renaissance paintings when they are cleaned and we can enjoy the brilliance of the treated paintings rather than be offended by it because we know the materials chosen were pure bright colors. The careful analysis of artists' materials and methods helps open a new window for our appreciation of paintings and can help return to us the enjoyment intended by the artist. The results also help us understand artists' working methods more clearly and can be a valuable adjunct to more traditional art historical research on works of art.

ACKNOWLEDGEMENTS

I thank Michael Palmer for sharing his results found in Beckmann's Falling Man and Melanie Feather of the Smithsonian Institution for her willing assistance using the scanning electron microscope. The Division of Visual Services provided excellent photographs for which I am grateful.

REFERENCES

1. Michael Palmer, Analytical Report, National Gallery of Art, November 19, 1986.

2. R. J. Gettens and G. L. Stout, Painting Materials A Short Encyclopedia (Dover Publications, Inc, New York, 1966), p.109.

3. International Centre for Diffraction Data, Swarthmore, PA (1991).

4. S. P. Griswold, "Two Paintings by El Greco: Saint Martin and the Beggar Analysis and Comparison," in Studies in the History of Art, vol. 41, Monograph Series II Conservation Research (National Gallery of Art, Washington, DC, 1992).

5. John Hand, Catherine Metzger, and Barbara Berrie, an art historical and technical investigation into Gerard David's St. Anne Altarpiece, in preparation.

6. D. V. Thompson, The Materials and Techniques of Medieval Painting (Dover Publications, Inc, New York, 1956).

7. D. De Grazia, entry in a forthcoming catalogue, National Gallery of Art, Washington, DC.

CHINESE BUDDHIST BRONZES IN THE FREER GALLERY OF ART: PHYSICAL FEATURES AND ELEMENTAL COMPOSITION

PAUL JETT AND JANET G. DOUGLAS
Freer Gallery of Art/Arthur M. Sackler Gallery
Department of Conservation and Scientific Research
Smithsonian Institution, Washington, D.C. 20560

ABSTRACT

Technical studies were performed on one hundred and seven Chinese copper-alloy sculptures dating from the fifth to the nineteenth centuries A. D. in the collection of the Freer Gallery of Art. X-radiography, stereomicroscopy and metallography were used to determine various physical features; atomic absorption spectroscopy was used to characterize the metal composition. All sculptures were determined to be cast by the lost wax method, and details were tooled by engraving and chasing. Metallographic examination and microprobe analysis showed that the gilding technique used was mercury gilding. Bronzes from the Northern Wei to the Song dynasty are similar in their physical characteristics and alloy type, leaded tin bronze. Beginning in the Ming dynasty, an increase in the amount of zinc is found, both as an impurity in tin bronzes and as a brass alloy constituent.

INTRODUCTION

Buddhist bronze sculpture forms a category of Chinese art unto itself. The beginning of this form of sculpture coincides with the introduction of Buddhism into China from India. Along with the tenets of Buddhism came sculptural images of the Buddha. The earliest account of a gilt bronze Buddhist image from China is a passage from the history of the Three Kingdoms which states: "He erected a buddha shrine, making a human figure of bronze whose body he coated with gold and clad in brocades." The sculpture which the passage describes dates to about 190 A.D. and was probably made in Jiangsu Province [1]. While initially owing much to Indian Buddhist sculpture, the Chinese works soon became indigenous in style.

The earliest Chinese Buddhist bronzes extant date to the Six Dynasties period, the period between the end of the Eastern Han dynasty in 220 A. D. and the beginning of the Sui dynasty in 589 A. D.. During this time, China was divided with numerous short-lived dynasties. The large number and often short duration of these dynasties complicate the study of Buddhist bronzes from this period. Inscriptions and style permit confident attribution of some sculpture to the Northern Wei (386 - 534 A. D.) and Northern Qi (550 - 577 A. D.) dynasties; other works may only be generally attributed to the Six Dynasties period.

By the Sui (581 - 618 A. D.) and Tang (618 - 907 A. D.) dynasties, Buddhist sculpture was thought to reach its apogee in China [2]. Tibetan Buddhism and Taoism influenced Chinese Buddhist sculpture during later dynasties, particularly the

Ming dynasty (1368 - 1644 A. D.) when, at times, Tibetan and Chinese works are almost indistinguishable. Thus, from the introduction of Buddhism into China until the late dynasties, gilded bronze Buddhist images were widespread.

Prior to just a few years ago, there had been little technical study of Chinese Buddhist bronzes in particular and "later" Chinese bronzes in general. Only isolated analyses of Chinese Buddhist bronze figures were to be found [3,4,5,6], and the few previous analyses of sculpture do not give a clear picture of the copper alloys used during these periods. More in-depth studies of the copper alloys during the later Chinese periods, particularly of coinage, have since been undertaken [7,8,9,10].

The lack of technical or compositional studies of Buddhist bronze figures is understandable as such studies are fraught with problems from the outset. Without the benefit of a collection with an archaeological provenance, judgments must be made about the authenticity of works based on their stylistic and general physical attributes. Even after such a selection is made, the results of compositional analyses for particular objects may seem so anomalous that one feels the authenticity of those objects should be rejected in spite of the lack of a large body of comparative analyses. Other analyses may fall into some middle ground, not seeming either typical nor greatly unusual; in these cases, it may, at that time at least, be impossible to find a correlation between the composition of the metal and the chronological and regional attribution of the object. Further complications may arise with this class of objects because it is known that a number of periods of Buddhist persecutions occurred during which bronze figures were destroyed or melted down [11]. When these persecutions ended, new bronzes were cast, and it is possible that newly mined metal, old scrap, or a combination of the two, were used.

The role of zinc as an alloying ingredient is of particular interest in the study of late Chinese copper alloys. Recent studies indicate that brass was not used in the production of coins until the early 16th century [7,8]. However, brass wire and shavings have been found in the decoration of a Chinese lacquer scroll tray of a style attributed to the mid to late fourteenth century [12]. This isolated example may not indicate the general use of brass in China prior to the 16th century.

To better understand the nature of the copper alloys and gilding techniques used during the later periods in China, a technical study was undertaken of the later Chinese bronzes in the collection of the Freer Gallery of Art. One hundred and seven works of sculpture dating from the 5th through the 19th century were studied. About eighty percent of these date to the Tang dynasty or earlier. Also, almost eighty percent of the objects are gilded. The technical features and the compositional analyses of some of these bronzes are described below.

PHYSICAL FEATURES

Visual and X-radiographic examination of all the sculpture indicate they are structurally similar in that they were cast, presumably by the lost-wax method. The smallest figures were typically cast solid, with the bases and mandorlas cast integrally (Figure 1). Larger works (figures greater than 15 centimeters in height) are either solid or hollow; generally the mandorlas and bases were cast separately. As the complexity of the sculptures increase, the number of separately cast elements also increases (Figure 2). Separate elements, in virtually all cases, are attached by physical means. The most common attachment is the use of phalanges which pass through openings in the attached parts and are pinned to secure the separate pieces.

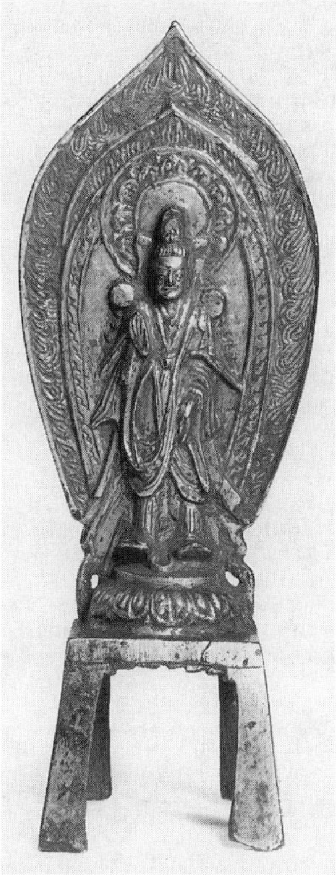

Figure 1. Gilt bronze, inscribed 518 A.D., no. 11.132.

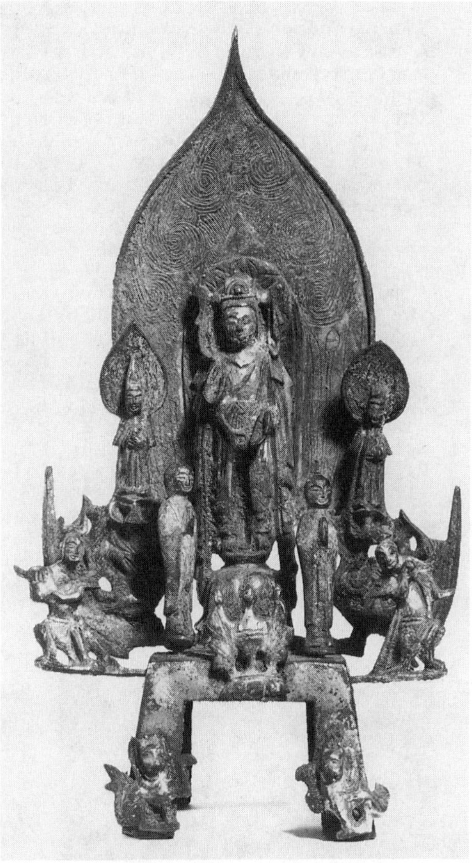

Figure 2. Gilt bronze, inscribed 596 A.D., no. 13.43.

Details of the sculptures often were tooled to sharpen cast features and to smooth undecorated surfaces. Engraving was the most common method employed for details, but chasing was also used. Inscriptions were engraved, and it often appears that the inscriptions were not considered in the original design as they frequently do not conform well to the available space (Figure 3). The tooling usually appears to have been applied prior to gilding.

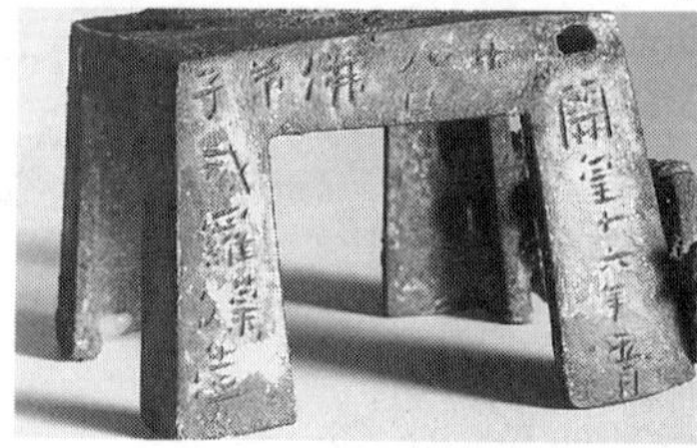

Figure 3. Inscription on base of bronze, no. 13.43.

Visual examination of the gilding suggests application by fire gilding, also known as mercury gilding or amalgam gilding. A pasty, granular layer of gold, indicative of unburnished amalgam gilding, was often found in the recesses of engraved lines. Gilding was often found on the underside of the feet of the bases, a feature consistent with amalgam gilding due to this material's tendency to flow during the heating which drives off the mercury. Visual evidence of leaf gilding, such as overlapping borders of leaves, or loose, ragged edges of leaves, was not found.

To confirm the method of gilding used, metallographic samples were removed from seven of the sculptures and gilded bronze coupons prepared in the laboratory. The sculptures sampled were two attributed to the Northern Wei dynasty, two to the Northern Qi dynasty, and three to the Tang dynasty. The samples were examined using light microscopy, scanning electron microscopy, and electron microprobe analysis. A full account of this portion of the study has been previously presented [13], and consequently only a summary will be given here.

Exemplary of the gilding features found is the metallographic sample taken from the sculpture illustrated in Figure 1. The gilding layer of the sample exhibits the grainy structure that often results from the amalgam gilding process (Figure 4). A typical cast leaded bronze structure was observed beneath the gilding layer, a feature that was true of all the samples. An electron microprobe scan across the sample moving from the substrate towards the surface shows the concentration of mercury within the gilding, present in roughly consistent proportions to the gold concentration (Figure 5).

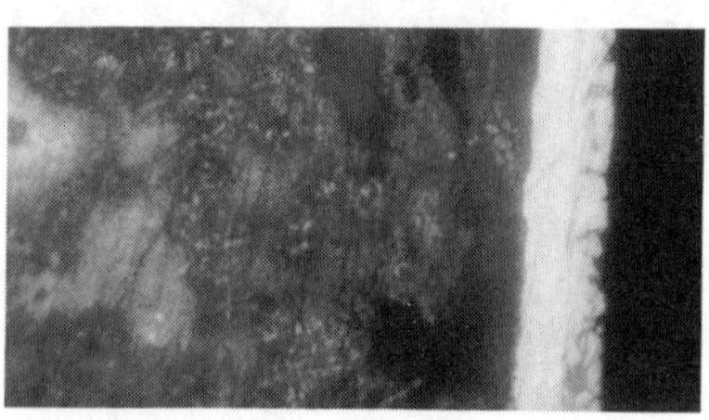

Figure 4. Metallographic sample from gilt bronze, no. 11.132, showing variations in thickness and granular nature of gilding layer, 1000 X. Bronze substrate is to left and gilding to right in Figures 4 &5.

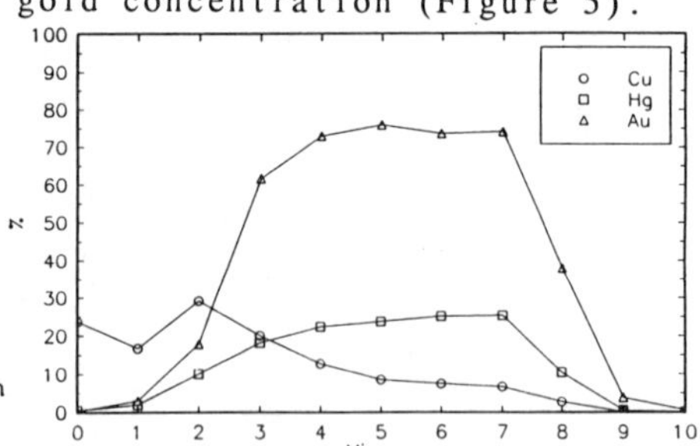

Figure 5. Electron microprobe scan of metallographic sample shown in figure 4.

All of the samples from the Buddhist figures exhibited similar features indicative of mercury gilding and showed the same physical structure and electron microprobe scan as samples from the prepared coupons of fire gilded bronze. From these results, it seems apparent that mercury gilding was the technique of choice for Chinese gilt bronze Buddhist images dating from the sixth to the tenth century A.D.

In addition to gilding, painted surface decorations are found on both gilded and ungilded work. One example of a painted, gilded surface is seen on the Buddha figure (no. 11.121) whose inscription dates it to 451 A.D. and attributes it to the Southern Song dynasty (Figure 6). Some of the late Tang pieces, such as three works which depict a Buddha and two attendants (nos. 12.92 - 12.94), display painted, ungilded surfaces, a trait common among late Tang bronze figures [14]. Little remains of the painted decorations however, so the extent to which this decorative technique was employed is not known.

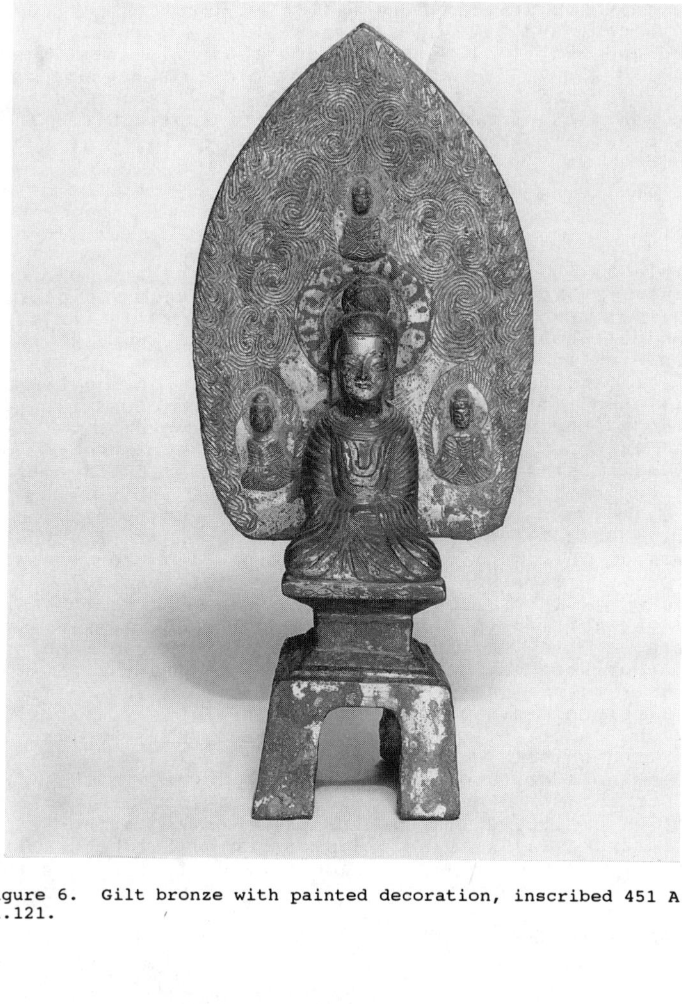

Figure 6. Gilt bronze with painted decoration, inscribed 451 A.D. no. 11.121.

ELEMENTAL COMPOSITION

Experimental

Compositional analysis was performed using atomic absorption spectroscopy. Samples weighing approximately 30 milligrams were removed by drilling. Individual samples were taken from separate parts on the sculpture where it was practical; in all cases where a sculpture was comprised of eight or fewer pieces, samples were taken from each piece. About half of each sample was used for the analysis. Each run typically consisted of three standards and 12 to 17 samples, one of which had been previously analyzed. The British Museum Research Laboratory procedure was followed [15]. A nitrous oxide/acetylene flame was used for the measurement of tin. Twelve elements were measured: copper, tin, lead, zinc, iron, arsenic, antimony, nickel, bismuth, silver, cobalt, and gold.

The results for the objects analyzed by atomic absorption spectroscopy are presented in Table I. Results are given in weight percent and have an accuracy of $\pm$ 2% for copper, + 5% for lead and zinc, + 10% for tin and those elements in amounts between 1.0 and 0. 1%, and $\pm$ 50% for elements present in amounts less than 0. 1%. Objects are listed by period and include analyses of multiple parts where applicable. Those objects considered recent or suspect are included at the end of the table.

Discussion

Table II is a summary of the analyses by time period. This table is based on a single result for each sculpture. Only the most prominent part (designated as a) is included for those objects consisting of multiple parts. Because the number of sculptures in each category is often too low for meaningful statistical analysis, rigid characterization of the elemental composition of objects by time periods is not possible at this time. Some general trends are apparent, however.

The earliest 14 pieces attributed to the Northern Wei dynasty are most similar in composition. The average major element composition is Cu = 87%, Sn = 5.6%, and Pb = 5.1% (Table 2). Two possible reasons for this consistency in composition are 1) the pieces were made by a small number of craftsmen or workshops, or 2) the recipes for bronze casting were rigidly prescribed in the Northern Wei dynasty.

The works attributed to the Northern Qi dynasty show a greater variability in elemental composition. A majority of the Northern Qi objects have a higher tin or lead concentration than the Northern Wei bronzes. The objects in the "Six Dynasties -- other" category also do not show any particular elemental groupings. This is not surprising given the length of time and number of dynasties covered by these works. However, one object in this category (no. 16.709) is so similar to the Northern Wei bronzes that it may belong to that group.

After the Six Dynasties period, much variability in the composition of leaded tin bronzes is observed, although a general trend towards higher lead contents that began with the Northern Qi works continues. The variety of alloys employed could be due to a number of reasons. As Buddhism spread, it is possible that the number of craftsmen and workshops producing sculpture increased, with a corresponding increase in the number of alloy recipes. It is possible that the melting of earlier pieces to cast new ones became more common, resulting in a greater variety of alloy compositions. (This possibility however seems less feasible given the results of lead isotope ratio analyses. See following paper, Sayer et al.)

The works attributed to the Song dynasty are also found to vary greatly in their copper, tin and lead contents. Interestingly, two sculptures depicting Guanyin (nos. 13.115 and 17.17, Figure 7) are virtually the same in composition. Though the two sculptures appear unalike initially because of their different surface conditions and bases, close visual inspection reveals that the figures have the same shape and size, and bear identical mold marks [14]. No previous use of replicate casting methods for figures has been found among Chinese sculpture prior to the modern period. This may indicate a far later date of manufacture for the figures than the Song dynasty. The Song dynasty Guanyin figure (no. 46.10) from the Da Li kingdom, now Yunnan Province, is unique because it is cast from an arsenical copper alloy (Cu=94%, As=2.10%). This composition is consistent with similar figures bearing the same attribution now found in numerous collections [16]. These works are probably representative of a regional style and technical approach removed from the mainstream of Chinese culture at that time.

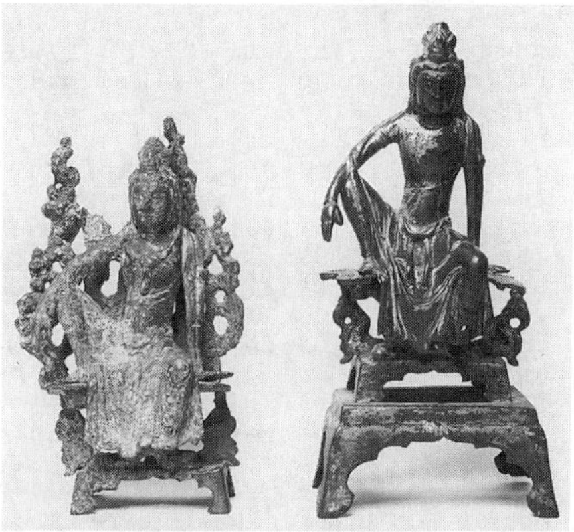

Figure 7. Two bronzes which appear to be made using the same mold. Nos. 13.115 & 17.17.

The first significant amount of zinc as an alloying element is found in the Head of a deity (no. 16.218) from the Ming dynasty. The Qing dynasty animal figure (no. 17.262) is also composed of brass, though leaded tin bronze is the most common alloy among the objects analyzed in this period.

Another trend which seems apparent is that the amounts of gold and silver impurities within the bronzes decreased over time. In the case of gold, which is virtually never present in more than trace amounts, the alloys from the latter periods typically contained amounts below the limits of our detection capabilities. With only one exception, a work included in the questionable category but attributed to the Wei dynasty (no. 15.60), gold was never found to measure more than 0.03% after the Tang dynasty.

It is clear from the compositional analyses that the individual parts of certain sculptures, and some groups of sculptures, were cast at the same time. Among these are a Tang dynasty group of three figures consisting of a Buddha and two monks (nos. 12.92 - 12.94) whose lead contents are well above 20 percent; two lion figure (nos. 47.13 and 47.14); and most interestingly, two Northern Wei Buddha figures (nos. 11.127 and

11.128) which have virtually identical compositions but whose inscriptions bear dates that differ by six years.

In other cases, when the analyses of parts from one work are compared, it appears that the amounts of a single major element vary considerably, thus effecting the percentages of the other major constituents (for example no. 11.139). When greater differences in the compositions of separate parts sometimes occur, it suggests that separate batches of metal were used or that the parts are not original to one another (nos. 13.43 a-h or 14.21 a-h). The physical aspects of the junctures between the parts can be important in making a determination. In some cases, the joins, and thus the originality, of separate parts are unquestionable (no. 49.4). Where joins appear rough or ill fitted, three explanations are possible. First, the join could be original but poorly executed. Second, the join may be unoriginal, perhaps a repair, but the parts are original to one another. Finally, the parts may not be original to each other. It is certainly possible that parts from different authentic works were assembled together, either in antiquity or in recent restorations. It may also be inherent in the nature of the sculpture that parts were made to be interchangeable or that the various parts were manufactured independently and later assembled. Finally, there are some cases where parts are clearly a later addition because of their anomalous composition, such as no. 15.106h, where a brass part is used while the other parts are leaded tin bronze. As a general rule, however, multiple parts from the same sculpture are fairly similar in composition.

Works Considered Recent or Suspect

The last category in Table I, entitled "Works Considered Recent or Suspect", includes objects whose previous attributions became extremely doubtful during the course of this study. These judgments were made primarily on the basis of compositional analysis, but typically stylistic attributes, or other anomalous features, reinforced the doubts raised by the composition. The brass sculptures attributed to dynasties earlier than the Ming were felt to be especially suspect. Among these were four complete works. Two of these figures (nos. 52.28 and 55.16) are Northern Wei in style; one (no. 55.16) is most similar in appearance to a work in the collection of the Yurinkan, Kyoto, Japan [17]. A small figure attributed to the Tang dynasty (no. 11.591) also is composed of brass. In addition, this work has an illegible inscription and is one of a group similar in appearance whose inscriptions are problematic [18]. The Song dynasty style lion (no. 16.394) also is suspect because of the composition. It contains 4.93% zinc, which is uncharacteristic of objects from the Song dynasty; the relatively high percentage of antimony is an inexplicable anomaly.

Parts of three objects dated prior to the Ming dynasty are also brass. Two of these parts (nos. 15.106h and 13.45b) are listed with the sculptures to which they belong under their attributed dynasties although the brass parts are certainly modern. The third part (no. 13.41b) is listed with the suspect works as is the figure to which it belongs. The figure (no.

13.41a) appears to be modern, is composed of a very pure 85/15 copper-tin alloy and is believed to be Japanese in style and origin [14]. The last object included in this group is a Northern Wei style figure (no. 15.60); the analysis is not unusual but concerns about style and the inscription have led to its inclusion among the suspect works.

A few pieces are not included in this category, but seem suspicious because of their unusually low lead content. For instance, one Guanyin figure (Six Dynasties -- Northern Qi, no. 13.42a) and three animal figures (Six Dynasties -- Other, nos. 16.251 and 17.318 and Song dynasty, no. 16.445) all have a lead content of less than three percent. However, the authenticity of these pieces cannot be discounted on this criteria alone. The Guanyin from Yunnan province dating to the 12th century (no. 46.10), for instance, has a low lead content consistent with other works of a similar provenance.

Multivariate Statistical Analysis

The elemental compositions of the whole body of Chinese Buddhist bronzes were examined using multivariate statistical analysis [19]. Five alloy groups were found based on the concentrations of copper, tin and lead. The relative positions of the groups are shown in a lead-tin scatter plot (Figure 8). Multivariate probability calculations based upon each group showed significant probabilities of inclusion of group specimens within the group and exclusion of other specimens from the group. The ellipses surrounding the groups in Figure 8 represent a 90 percent probability of containment for the groups. It has been suggested that the mean elemental concentration of the N. Wei group is very similar to a basic bronze formulation of 90/5/5. This N. Wei group, which contains 10 of the 14 bronzes from this time period are surprisingly consistent in elemental composition. The standard deviation for copper is about 2%, and about 20 to 25% for the

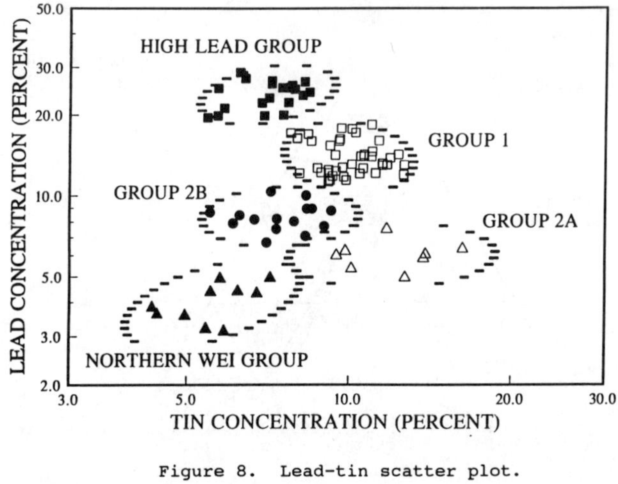

Figure 8. Lead-tin scatter plot.

minor elements. Alloy group 1 corresponds to a 75/10/15 bronze formulation. Groups 2A and B correspond to 80/12.5/7.5 and 82.5/7.5/10 bronze formulations. Group 2A contains only eight samples and thus may be too small to represent a truly separate group as is indicated by multivariate analysis. The physical or metallurgical differences between bronzes with these two compositions would be minuscule and in spite of the distinction made by multivariate analysis between the two groups, it is tempting to speculate that 2A and 2B derive from the use of an 80/10/10 bronze formulation. The high lead group corresponds to a 67.5/7.5/25 bronze formulation. This formulation suggests the use of two parts copper to one part of a mixture of tin and lead. Interestingly, ten of the fifteen objects represented in this group are ungilded, in contrast to the majority of bronzes which are gilded.

Multivariate consideration of the minor elements arsenic, antimony, silver, bismuth and nickel show some interesting trends in their concentrations. The concentrations of these elements were quite consistently matching in more than seventy percent of the specimens in four of the alloy groups, 1, 2A, 2B, and N. Wei. The average percent concentrations from the combined matching specimens from all four groups are: As = 0.36 $\pm$0.12, Sb = 0.29 $\pm$0.06, Ag = 0.14 $\pm$0.05, Bi = 0.041 $\pm$0.017, & Ni = 0.078 $\pm$0.021. In more than eighty percent of the high lead group, the concentrations of these elements were consistent within the group and significantly different from the other four groups, with a higher level of bismuth (Bi = 0.086 $\pm$0.025) and lower levels of the remaining elements (As = 0.23 $\pm$0.07, Sb = 0.13 $\pm$0.05, Ag = 0.070 $\pm$0.027, & Ni = 0.039 $\pm$0.025). The higher bismuth content may be due to its presence as an impurity in the lead. The lower arsenic and antimony contents imply that these elements were present as impurities in the copper. The differences between the high lead group and the other groups are so great in this respect, however, that it suggests that metals from different sources were used or that the metals used were more highly refined.

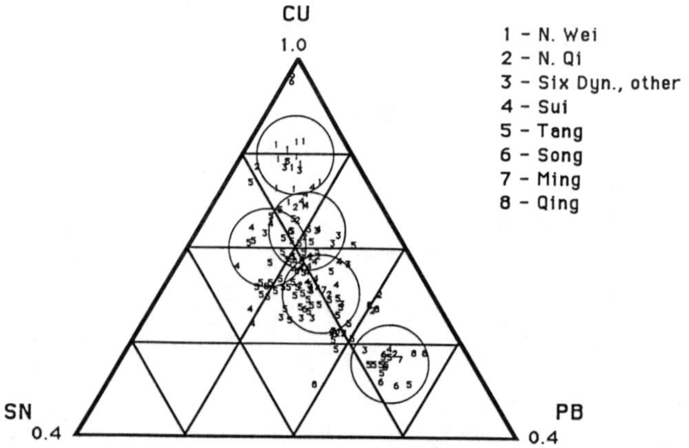

Figure 9. Ternary diagram for copper, lead, and tin contents showing five proposed basic alloying formulations $\pm$ 3 percent.

The results of multivariate analysis prompted us to plot the elemental composition data for all the bronzes on a ternary diagram showing copper, tin and lead, with these three elements normalized to 100 percent (Figure 9). In Figure 9, circles are drawn for the five bronze formulations ± 3 percent. The majority of samples fit within these five bronze formulations, and as mentioned earlier, the general trend is for the later period bronzes to contain more lead.

SUMMARY

Chinese Buddhist bronze figures dating from the Northern Wei to the Song dynasty are similar in their physical construction and alloy type; the Northern Wei objects are particularly consistent in these respects. The use of cast, leaded tin bronze with mercury gilding was found to be the rule. After the Song dynasty, the majority of the sculptures analyzed also consisted of cast, leaded tin bronze, but with the Ming dynasty an increasing amount of zinc as an impurity was found in the bronzes. The use of brass alloys, and a decrease in the amounts of gold and silver are seen. A few fundamental ratios of the major alloying elements copper (tin and lead) appear to have been used over the course of time. Brass alloys, presumed to be of recent origin, are quite distinctive because of their relative purity.

A succeeding paper will describe the lead isotope ratio analyses of samples taken from approximately sixty of the works of sculpture.

ACKNOWLEDGEMENTS

The authors wish to thank Eugene Jarosewich and Joseph Nelen, Department of Mineral Sciences, National Museum of Natural History, Smithsonian Institution, in whose laboratory the atomic absorption analyses was performed, and Edward V. Sayre, Emeritus Research Fellow, Conservation Analytical Laboratory, Smithsonian Institution, for his statistical treatment and analysis of the compositional data.

TABLE I

Accession #	DESCRIPTION	CU	SN	PB	ZN	FE	AS	SB	NI	BI	AG	CO	AU	TOTAL
SIX DYNASTIES -- NORTHERN WEI														
09.264	Guanyin. inscribed: 520 AD	87	6.2	4.4	0.06	0.50	0.36	0.38	0.07	0.04	0.17	0.03	0.04	99.58
09.265	Buddha.	85	7.1	4.9	0.12	0.68	0.45	0.29	0.09	0.06	0.12	0.01	0.02	99.19
09.266	Guanyin. inscribed: 453 AD	88	5.8	4.9	0.03	0.21	0.40	0.33	0.08	0.03	0.14	0.02	0.04	99.57
09.268	Buddha (?). 520-530 AD	89	5.9	3.1	0.03	0.09	0.29	0.25	0.07	0.03	0.14	0.02	0.01	98.46
09.269	Guanyin. 520-530 AD	85	7.0	6.7	0.09	0.41	0.35	0.32	0.07	0.04	0.16	0.02	0.01	99.98
11.125	Sakyamuni and Prabhutaratna. inscribed: 503 AD	87	5.0	3.6	0.19	0.18	0.34	0.23	0.05	0.03	0.15	0.02	0.80	97.31
11.126	Sakyamuni and Prabhutaratna. inscribed: 473 AD	88	4.7	5.8	0.01	0.16	0.44	0.35	0.09	0.04	0.13	0.03	0.01	99.49
11.127	Sakyamuni and Prabhutaratna. inscribed: 500 AD	89	4.4	3.6	0.01	0.05	0.45	0.27	0.05	0.03	0.13	0.02	0.02	98.49
11.128	Sakyamuni and Prabhutaratna. inscribed: 506 AD	91	4.3	3.8	0.05	0.14	0.38	0.16	0.05	0.03	0.09	0.02	0.00	100.27
11.129	Guanyin. inscribed: 541 AD	84	7.3	7.5	0.04	0.11	0.41	0.33	0.09	0.04	0.13	0.02	0.01	99.93
11.130	Sakyamuni and Prabhutaratna. inscribed: 516-520 AD	90	5.4	3.2	0.04	0.15	0.26	0.22	0.07	0.04	0.16	0.02	0.51	99.93
11.131	Guanyin. inscribed: 482 AD	81	6.4	7.3	0.01	2.40	1.04	0.44	0.05	0.05	0.19	0.02	0.04	99.04
11.132	Guanyin. inscribed: 518 AD	85	4.0	8.3	0.01	0.02	0.39	0.26	0.08	0.05	0.17	0.01	0.00	98.39
11.137	Buddha.	85	5.5	4.4	0.20	0.15	0.43	0.35	0.07	0.05	0.14	0.03	0.00	96.70
SIX DYNASTIES -- NORTHERN QI														
09.262a	Buddhist Trinity. inscribed: 556 AD, 2 parts, a = figures.	72	5.8	21.1	0.02	0.02	0.28	0.11	0.04	0.09	0.05	0.01	0.00	99.51
09.262b	See 09.262a. b = base	74	3.9	21.1	0.01	0.02	0.27	0.11	0.05	0.09	0.05	0.01	0.00	99.19
11.122a	Guanyin or Maitreya. inscribed: 555 AD, 2 parts, a = figures	83	8.5	8.9	0.02	0.13	0.48	0.25	0.06	0.03	0.16	0.01	0.16	101.85
11.122b	See 11.122a. b = base	76	11.5	13.0	0.02	0.09	0.52	0.26	0.06	0.03	0.11	0.02	0.03	101.89
11.133	Buddha.	83	8.3	7.1	0.06	0.45	0.34	0.30	0.07	0.03	0.14	0.03	0.08	99.53
11.139a	Sakyamuni. inscribed (possibly spurious), 2 parts, a = figure	74	9.2	15.4	0.02	0.07	0.67	0.23	0.07	0.00	0.50	0.01	0.09	99.79
11.139b	See 11.139a. b = base	68	5.7	25.0	0.41	0.01	0.19	0.16	0.03	0.09	0.05	0.00	0.00	99.40
13.42a	Guanyin. 4 parts, a = figure	87	8.3	2.2	0.02	1.67	0.24	0.08	0.09	0.02	0.03	0.05	0.01	99.77
13.42b	See 13.42a. b = lower base	74	12.4	10.2	0.02	0.51	0.12	0.16	0.08	0.05	0.12	0.03	0.03	97.57
13.42c	See 13.42a. c = mandorla	78	8.8	12.2	0.03	1.05	0.46	0.10	0.08	0.05	0.13	0.05	0.04	100.57
13.42d	See 13.42a. d = upper base	78	8.1	12.1	0.03	0.98	0.00	0.14	0.08	0.04	0.14	0.04	0.04	99.75
19.81a	Guanyin or Maitreya. inscribed: 561 AD, 3 parts, a = figure	69	9.9	20.1	0.00	0.03	0.32	0.28	0.05	0.04	0.12	0.02	0.01	101.02
19.81b	See 19.81a. b = base	71	10.7	18.9	0.01	0.19	0.28	0.27	0.06	0.04	0.13	0.02	0.00	101.09
19.81c	See 19.81a. c = mandorla	71	10.7	19.9	0.01	0.05	0.38	0.32	0.05	0.05	0.13	0.02	0.00	102.42

Accession #	DESCRIPTION	CU	SN	PB	ZN	FE	AS	SB	NI	BI	AG	CO	AU	TOTAL
SIX DYNASTIES -- OTHER														
09.267	Bodhisattva. Eastern Wei Dynasty, inscribed: 550 AD	78	5.7	17.0	0.07	0.13	0.40	0.29	0.09	0.04	0.14	0.03	0.02	101.55
11.121a	Buddha. Southern Song Dynasty, inscribed: 451 AD, 2 parts, a = figure	80	9.8	6.2	0.03	0.13	0.32	0.34	0.06	0.06	0.15	0.02	0.02	96.97
11.121b	See 11.121a. b = mandorla	78	10.7	4.8	0.02	0.57	0.23	0.32	0.08	0.03	0.14	0.03	0.34	95.58
13.96	Animal figure.	77	8.3	10.8	0.02	0.03	0.49	0.16	0.05	0.03	0.18	0.01	0.18	97.00
15.38	Animal figure.	72	7.8	17.1	0.00	0.00	0.48	0.38	0.09	0.03	0.16	0.01	0.00	98.21
15.213	Animal figure.	71	12.3	14.2	0.02	0.88	0.62	0.40	0.11	0.07	0.16	0.03	0.00	100.29
15.214	Animal Figure.	71	11.3	16.0	0.00	0.44	0.35	0.29	0.07	0.06	0.16	0.04	0.00	99.74
15.215	Animal Figure.	71	14.5	13.4	0.00	0.61	0.17	0.20	0.07	0.02	0.14	0.01	0.03	100.07
16.251	Animal Figure.	87	5.6	2.4	0.00	0.05	0.37	0.42	0.06	0.07	0.14	0.01	0.00	95.86
16.444	Animal Figure.	78	11.2	10.4	0.00	0.17	0.20	0.52	0.49	0.02	0.21	0.39	0.01	101.87
16.446	Animal Figure.	66	7.7	22.2	0.04	0.13	0.97	1.24	0.14	0.07	0.29	0.08	0.01	99.09
16.709	Animal Figure.	86	6.7	4.3	0.01	0.54	0.43	0.15	0.07	0.05	0.39	0.07	0.01	98.54
16.710	Animal Figure.	74	10.1	13.1	0.03	0.71	0.45	0.16	0.09	0.06	0.30	0.05	0.00	99.39
17.318	Animal Figure.	88	5.9	2.6	0.01	0.04	0.42	0.27	0.08	0.03	0.12	0.01	0.01	96.37
19.77	Animal Figure.	79	6.8	13.3	0.02	0.21	0.30	0.35	0.10	0.03	0.15	0.07	0.03	100.62
19.78	Animal Figure.	80	4.3	13.5	0.02	0.04	0.20	0.36	0.10	0.03	0.17	0.05	0.02	98.63
SUI DYNASTY														
09.263	Figure. inscribed: 587 AD	65	6.8	22.1	0.01	0.13	0.41	0.21	0.05	0.04	0.12	0.01	0.05	95.33
11.140	Guanyin. inscribed: 583 AD	78	6.7	14.7	0.02	1.02	0.30	0.17	0.05	0.02	0.07	0.03	0.03	101.12
11.141	Guanyin. inscribed: 599 AD	76	8.0	16.3	0.06	0.18	0.30	0.24	0.07	0.03	0.11	0.03	0.04	101.61
13.43a	Buddhist Altar. inscribed: 596 AD, 8 parts, a = main figure	77	9.3	11.8	0.02	0.14	0.34	0.27	0.07	0.02	0.13	0.02	0.04	99.13
13.43b	See 13.43a. b = mandorla of main figure	78	9.1	12.4	0.03	0.15	0.28	0.30	0.06	0.03	0.14	0.02	0.04	101.06
13.43c	See 13.43a. c = Proper Right bodhisattva	71	17.1	8.6	0.03	0.20	0.77	0.35	0.07	0.01	0.13	0.02	0.04	98.31
13.43d	See 13.43a. d = Proper Right monk	78	9.1	11.3	0.04	0.04	0.39	0.32	0.08	0.00	0.15	0.02	0.05	99.07
13.43e	See 13.43a. e = Proper Right apsara	77	9.6	11.4	0.01	0.14	0.37	0.30	0.08	0.00	0.13	0.02	0.06	98.85
13.43f	See 13.43a. f = Proper Left bodhisattva	72	16.5	8.9	0.01	0.20	0.70	0.37	0.06	0.04	0.17	0.02	0.03	99.18
13.43g	See 13.43a. g = Proper Left monk	79	9.1	12.1	0.03	0.06	0.42	0.31	0.08	0.02	0.14	0.01	0.04	100.97
13.43h	See 13.43a. h = Proper Left monk	77	9.4	12.3	0.00	0.15	0.42	0.29	0.08	0.01	0.12	0.02	0.04	100.22
14.21a	Sakyamuni and Bodhisattvas. inscribed: 597 AD, 8 parts, a = main figure	84	6.1	7.9	0.00	0.94	0.28	0.22	0.10	0.00	0.11	0.12	0.00	99.69
14.21b	See 14.21a b = lower section of base	74	10.6	14.0	0.00	0.41	0.25	0.20	0.07	0.00	0.10	0.03	0.02	99.85

Accession #	DESCRIPTION	CU	SN	PB	ZN	FE	AS	SB	NI	BI	AG	CO	AU	TOTAL
14.21c	See 14.21a. c = mandorla of main figure	76	10.5	14.0	0.00	0.41	0.02	0.21	0.06	0.04	0.11	0.01	0.03	101.19
14.21d	See 14.21a. d = upper section of base	84	6.2	8.5	0.01	3.24	0.17	0.24	0.09	0.01	0.11	0.22	0.00	102.68
14.21e	See 14.21a. e = Proper Right figure	80	7.1	10.3	0.00	1.14	0.25	0.31	0.08	0.03	0.10	0.11	0.00	99.34
14.21f	See 14.21a. f = Mandorla of Proper Right figure	75	10.6	14.2	0.02	0.40	0.04	0.22	0.06	0.04	0.10	0.01	0.04	101.29
14.21g	See 14.21a. g = Proper Left figure	85	7.3	8.2	0.00	0.51	0.19	0.29	0.08	0.00	0.12	0.05	0.00	101.74
14.21h	See 14.21a. h = Mandorla of Proper Left figure	86	5.1	6.8	0.01	2.24	0.00	0.23	0.09	0.04	0.10	0.18	0.02	100.73
16.328	Guanyin.	82	12.7	5.0	0.01	0.25	0.37	0.24	0.07	0.05	0.04	0.01	0.00	100.57
45.30a	Sakyamuni and Prabhataratna. inscribed: 609 AD, 3 parts, a = figure	77	16.2	6.3	0.01	0.17	0.22	0.25	0.06	0.02	0.10	0.02	0.00	100.72
45.30b	See 45.30a. b = base	77	8.7	12.6	0.00	0.23	0.11	0.23	0.05	0.03	0.11	0.01	0.04	99.26
45.30c	See 45.30a. c = mandorla	81	10.1	5.4	0.02	0.16	0.29	0.25	0.06	0.02	0.11	0.02	0.00	97.21
79.51a	Guanyin. 2 parts, a = figure	76	10.1	17.5	0.01	0.22	0.11	0.17	0.05	0.01	0.10	0.02	0.03	104.24
79.51b	See 79.51a. b = base	79	11.6	12.1	0.04	0.30	0.42	0.22	0.11	0.06	0.17	0.06	0.05	103.95
TANG DYNASTY														
07.521	Guanyin.	71	10.8	12.6	0.00	0.12	0.44	0.34	0.07	0.07	0.61	0.04	0.00	96.36
09.270	Statue of a Monk.	73	8.0	16.4	0.00	0.02	0.28	0.20	0.10	0.04	0.12	0.02	0.01	98.34
11.123	Guanyin.	79	10.6	9.6	0.00	0.19	0.68	0.29	0.25	0.05	0.21	0.07	0.00	100.77
11.124	Padmapani.	83	8.3	8.9	0.00	0.02	0.00	0.04	0.12	0.04	0.12	0.00	0.08	100.22
11.136	Guanyin.	70	6.9	19.8	0.00	0.05	0.17	0.13	0.06	0.04	0.10	0.00	0.02	97.71
11.150	Guanyin.	74	8.4	16.9	0.00	0.00	0.91	0.60	0.13	0.07	0.16	0.00	0.04	101.63
11.499a	Seven Buddhas. 2 parts, a = figures	77	7.4	15.5	0.02	0.77	0.57	0.39	0.12	0.07	0.34	0.12	0.00	102.46
11.499b	See 11.499a. b = base	80	9.8	11.4	0.06	0.26	0.43	0.28	0.13	0.06	0.24	0.06	0.04	102.62
12.90	Buddha.	73	9.6	16.2	0.01	0.10	0.51	0.34	0.09	0.07	0.18	0.02	0.03	100.18
12.91	Guanyin.	80	7.9	8.0	0.02	0.05	2.01	1.10	0.11	0.10	0.21	0.01	0.05	99.38
12.92a	Sakyamuni. 3 parts, a = figure. Forms a group with 12.93 & 12.94.	67	8.4	24.2	0.01	0.05	0.15	0.08	0.02	0.10	0.05	0.00	0.00	100.28
12.92b	See 12.92a. b = base	66	7.7	25.3	0.01	0.11	0.29	0.06	0.02	0.11	0.05	0.00	0.00	99.78
12.92c	See 12.92a. c = mandorla	66	7.1	26.7	0.00	0.13	0.00	0.07	0.01	0.10	0.04	0.00	0.01	100.02
12.93a	Figure of a monk. 2 parts, a = figure. Forms a group with 12.92 & 12.94.	67	8.2	23.6	0.03	0.03	0.21	0.13	0.03	0.08	0.07	0.01	0.05	99.19
12.93b	See 12.93a. b = base	65	7.9	24.9	0.01	0.03	0.21	0.14	0.04	0.11	0.09	0.01	0.00	98.43
12.94a	Figure of a Monk. 2 parts, a = figure. Forms a group with 12.92 & 12.93.	66	7.8	25.7	0.03	0.06	0.17	0.15	0.02	0.09	0.06	0.01	0.00	99.62

Accession #	DESCRIPTION	CU	SN	PB	ZN	FE	AS	SB	NI	BI	AG	CO	AU	TOTAL
12.94b	See 12.94a. b = base	65	7.1	26.1	0.04	0.09	0.32	0.16	0.04	0.10	0.08	0.01	0.00	98.65
13.44a	Guanyin. 3 parts, a = figure	73	14.6	12.8	0.02	0.54	0.63	0.30	0.13	0.06	0.17	0.05	0.00	101.84
13.44b	See 13.44a, b = base	81	8.3	10.0	0.00	0.53	0.28	0.12	0.11	0.06	0.37	0.04	0.01	101.14
13.44c	See 13.44a. c = mandorla	72	11.0	14.6	0.01	0.24	0.53	0.34	0.12	0.09	0.11	0.03	0.01	99.56
13.45a	Guanyin. 3 parts, a = figure	82	9.5	6.0	0.01	0.54	0.47	0.13	0.05	0.04	0.08	0.02	0.02	98.38
13.45b	See 13.45a. (Recent addition ?) b = base	71	7.2	16.4	2.24	0.12	0.50	0.29	0.07	0.08	0.09	0.01	0.01	98.17
13.45c	See 13.45a. c = mandorla	87	10.8	2.4	0.00	0.11	0.19	0.07	0.05	0.01	0.03	0.01	0.02	101.03
13.95	Animal figure.	79	7.1	12.5	0.13	0.21	0.59	0.30	0.07	0.03	0.11	0.02	0.05	100.56
15.37a	Guanyin. 2 parts, a = figure	75	12.7	7.8	0.02	0.14	0.51	0.17	0.12	0.04	0.18	0.06	0.09	97.16
15.37b	See 15.37a. b = base	73	9.7	12.7	0.05	0.16	0.84	0.42	0.08	0.06	0.16	0.02	0.06	97.55
15.89	Guardian figure.	68	10.2	17.2	0.02	0.53	1.12	0.61	0.08	0.06	0.27	0.03	0.01	98.15
15.106a	Sarcophagus. many parts, eight sampled, a = Proper Right front figure	75	14.3	8.0	0.00	0.29	0.54	0.23	0.12	0.05	0.23	0.07	0.02	99.42
15.106b	See 15.106a. b = lower base	73	14.2	8.8	0.01	0.63	0.55	0.24	0.10	0.05	0.21	0.07	0.02	98.24
15.106c	See 15.106a. c = upper base	74	14.9	9.0	0.01	0.40	0.61	0.24	0.11	0.06	0.24	0.07	0.03	100.03
15.106d	See 15.106a. d = outer sarcophagus	74	14.1	9.9	0.03	0.66	0.59	0.26	0.10	0.06	0.22	0.06	0.03	99.59
15.106e	See 15.106a. e = lid of outer sarcophagus	71	14.4	9.2	0.01	0.38	0.76	0.27	0.12	0.05	0.25	0.06	0.02	96.65
15.106f	See 15.106a. f = inner sarcophagus	79	13.7	5.8	0.03	0.12	0.45	0.21	0.11	0.04	0.28	0.04	0.03	99.42
15.106g	See 15.106a. g = feline decoration on outer sarcophagus	79	13.8	6.0	0.05	0.11	0.36	0.22	0.11	0.03	0.28	0.04	0.03	99.71
15.106h	See 15.106a. (Recent repair) h = balustrade	88	0.0	0.8	9.40	0.39	0.00	0.01	0.15	0.00	0.05	0.00	0.00	98.43
15.111	Animal figure.	74	11.9	13.8	0.04	0.15	0.31	0.21	0.05	0.03	0.11	0.02	0.04	100.87
15.203	Bodhisattva on a lion.	75	12.9	10.7	0.01	0.28	0.51	0.20	0.09	0.05	0.27	0.05	0.03	100.10
15.212	Animal figure.	74	12.6	12.0	0.03	0.35	0.25	0.27	0.10	0.04	0.20	0.08	0.03	99.88
15.216	Animal figure.	67	11.0	18.4	0.01	0.35	0.17	0.42	0.06	0.08	0.13	0.04	0.04	98.07
16.15	Padmapani.	74	13.2	11.0	0.02	0.34	0.17	0.17	0.06	0.04	0.23	0.02	0.16	99.82
16.214	Statue of a monk.	69	9.9	19.8	0.02	0.29	0.27	0.22	0.07	0.08	0.24	0.05	0.03	99.53
16.249	Guardian figure.	75	11.2	12.1	0.00	0.09	0.34	0.20	0.05	0.05	0.03	0.00	0.01	98.95
16.250	Guanyin.	72	8.5	16.0	0.09	0.12	1.88	1.15	0.08	0.08	0.15	0.02	0.09	99.86
16.321	Figure of a Monk.	84	6.7	8.2	0.00	0.04	0.19	0.22	0.04	0.05	0.13	0.00	0.03	99.19
16.326	Attendant figure (?).	70	9.6	20.4	0.02	0.09	0.36	0.38	0.10	0.06	0.17	0.10	0.04	101.32
16.327	Buddhist figure.	81	11.7	7.5	0.01	0.75	0.14	0.15	0.06	0.08	0.51	0.07	0.10	101.81
16.329a	Guardian figure. 2 parts, a = figure	73	10.6	12.6	0.03	0.64	0.30	0.20	0.05	0.05	0.47	0.04	0.07	98.24
16.329b	See 16.329a. b = base	80	4.1	16.2	0.04	0.03	0.26	0.32	0.08	0.00	0.16	0.00	0.00	101.09
16.338	Animal figure (lion).	66	6.2	28.7	0.01	0.02	0.00	0.06	0.03	0.05	0.05	0.00	0.00	101.50

Accession #	DESCRIPTION	CU	SN	PB	ZN	FE	AS	SB	NI	BI	AG	CO	AU	TOTAL
16.344	Animal figure (hare).	81	7.2	13.2	0.19	0.03	0.17	0.19	0.11	0.04	0.10	0.00	0.00	102.73
16.353	Animal figure.	74	12.6	13.0	0.01	0.09	0.47	0.43	0.08	0.03	0.26	0.02	0.17	101.33
16.465	Animal figure.	73	9.6	16.0	0.01	0.07	0.33	0.23	0.07	0.03	0.05	0.00	0.03	99.07
16.467a	Figure of a Arhat. 2 parts, a = figure	76	10.4	12.1	0.07	0.16	0.24	0.22	0.05	0.03	0.10	0.01	0.04	99.27
16.467b	See 16.467a. b = base	75	11.5	13.2	0.00	0.21	0.59	0.43	0.05	0.10	0.15	0.02	0.01	101.68
17.159	Padmapani.	72	11.0	14.0	0.02	0.15	0.26	0.22	0.06	0.05	0.07	0.00	0.02	97.93
17.407	Animal figure.	79	9.8	11.7	0.00	0.02	1.31	0.05	0.03	0.46	0.29	0.00	0.00	102.27
18.51a	Buddha. 2 parts, a = figure	76	12.1	10.8	0.00	0.27	0.49	0.38	0.11	0.04	0.12	0.04	0.04	99.90
18.51b	See 18.51a. b = base	74	16.4	10.3	0.00	0.24	0.18	0.20	0.09	0.03	0.12	0.05	0.05	101.73
45.31	Bhaisjyaguru.	79	9.3	8.8	0.06	0.76	0.98	0.59	0.09	0.04	0.20	0.03	0.03	100.31
47.13	Animal figure (lion). one of pair with 47.14	80	10.5	11.7	0.00	0.07	0.00	0.00	0.09	0.00	0.18	0.01	0.00	102.60
47.14	Animal figure (lion). one of pair with 47.13	80	11.2	9.9	0.00	0.04	0.00	0.00	0.09	0.01	0.16	0.01	0.04	101.05
49.4a	Plate with a figure. 3 parts, a = figure	72	14.2	13.2	0.01	0.01	0.04	0.00	0.01	0.00	0.06	0.01	0.02	99.70
49.4b	See 49.4a. b = base	64	19.9	10.8	0.03	0.10	0.07	0.16	0.29	0.00	0.07	0.03	0.00	95.86
49.4c	See 49.4a. c = plate	72	13.1	13.4	0.02	0.01	0.09	0.00	0.02	0.01	0.08	0.00	0.00	98.74
SONG DYNASTY														
13.115a	Guanyin. 2 parts, a = figure	65	8.2	26.5	0.03	0.10	0.00	0.11	0.04	0.10	0.00	0.01	0.00	100.42
13.115b	See 13.115a. b = base	67	7.1	23.1	0.01	0.00	0.00	0.14	0.08	0.12	0.00	0.01	0.00	97.55
15.335	Animal figure (elephant).	81	9.0	7.7	0.01	0.37	0.14	0.12	0.17	0.04	0.19	0.11	0.03	99.22
16.445	Animal figure (water buffalo).	98	1.0	0.6	0.00	0.02	0.05	0.02	0.16	0.00	0.26	0.01	0.03	100.39
17.17a	Guanyin. 2 parts, a = figure	63	7.5	25.2	0.04	0.08	0.25	0.09	0.02	0.07	0.06	0.00	0.00	95.84
17.17b	See 17.17a. b = base	66	7.6	25.2	0.04	0.03	0.00	0.12	0.03	0.08	0.09	0.00	0.00	99.30
17.261	Animal figure (ox).	72	11.7	13.0	0.01	0.24	0.91	0.29	0.12	0.05	0.12	0.03	0.00	98.35
46.10	Kuanyin. Yunnan Province, mid 12th century	94	0.1	0.0	0.00	2.90	2.10	0.00	0.12	0.00	0.02	0.00	0.00	99.58
MING DYNASTY														
15.118	Head of a deity.	73	9.4	14.2	1.69	0.11	0.44	0.30	0.08	0.03	0.09	0.02	0.01	99.24
16.217	Head of a deity.	69	10.1	17.7	0.26	0.74	0.42	0.24	0.12	0.04	0.09	0.03	0.01	98.63
16.218	Head of a deity.	80	1.4	10.4	6.20	1.16	0.12	0.09	0.10	0.00	0.01	0.01	0.00	99.79
16.219	Head of a deity.	66	6.4	27.3	0.07	0.13	0.16	0.18	0.10	0.12	0.12	0.01	0.00	100.82
QING DYNASTY														
07.31a	Animal figure (dog). 2 parts, a = body	72	5.4	19.5	0.12	0.01	0.35	0.21	0.05	0.06	0.04	0.01	0.00	98.11

Accession #	DESCRIPTION	CU	SN	PB	ZN	FE	AS	SB	NI	BI	AG	CO	AU	TOTAL
07.31b	See 07.31a. b = head	67	3.8	27.1	0.04	0.04	0.37	0.21	0.05	0.09	0.07	0.00	0.01	99.22
13.106	Animal figure (bull).	69	9.6	17.8	0.02	0.07	0.24	0.14	0.03	0.09	0.03	0.02	0.00	97.39
14.9	Animal figure (dog).	64	14.0	18.7	0.02	0.25	0.80	0.18	0.06	0.03	0.07	0.02	0.00	98.08
17.259	Animal sculpture.	72	5.7	19.8	0.23	0.00	0.45	0.21	0.09	0.06	0.11	0.01	0.00	98.65
17.262	Animal figure.	69	0.3	2.6	26.74	1.01	0.09	0.07	0.35	0.00	0.02	0.01	0.00	100.40
19.74	Animal figure.	71	7.5	19.9	1.02	0.05	0.16	0.19	0.05	0.01	0.02	0.01	0.00	100.06
WORKS CONSIDERED RECENT OR SUSPECT														
11.591	Maitreya. Tang Dynasty style	91	0.2	1.6	8.31	0.21	0.11	0.00	0.04	0.03	0.14	0.00	0.00	101.57
13.41a	Figure. attributed to Sui Dynasty, 2 parts, a = figure	84	15.0	0.1	0.01	0.07	0.04	0.02	0.00	0.01	0.11	0.01	0.00	99.66
13.41b	See 13.41a. b = base	90	0.2	0.9	7.43	0.21	0.03	0.01	0.04	0.01	0.04	0.00	0.00	99.28
15.60a	Maitreya. Northern Wei Dynasty style, inscribed: 539 AD, 4 parts, a = figure	85	5.5	8.7	0.13	0.19	0.42	0.29	0.07	0.02	0.13	0.03	0.01	100.16
15.60b	See 15.60a. b = base	78	5.4	17.5	0.01	0.29	0.87	0.45	0.08	0.03	0.13	0.08	0.00	102.44
15.60c	See 15.60a. c = mandorla	80	5.2	14.2	0.02	0.25	0.90	0.47	0.09	0.03	0.16	0.08	0.00	101.11
15.60d	See 15.60a. d = apsara	81	6.0	11.9	0.03	0.08	0.47	0.27	0.08	0.02	0.18	0.02	0.02	100.34
16.394	Animal figure (lion). Song Dynasty style	89	0.6	1.7	4.93	0.28	0.49	2.86	0.17	0.19	0.10	0.07	0.00	100.45
52.28a	Buddha. Northern Wei Dynasty style, 2 parts (figure and base are one), a = figure	94	0.0	1.1	5.97	0.26	0.18	0.22	0.07	0.03	0.06	0.01	0.00	102.18
52.28b	See 52.28a. b = base	92	0.0	1.0	5.88	0.26	0.19	0.19	0.07	0.03	0.06	0.01	0.00	99.89
52.28c	See 52.28a. c = mandorla	92	0.4	1.2	5.54	0.46	0.09	0.21	0.07	0.02	0.06	0.01	0.01	100.44
55.16a	Buddha. Northern Wei Dynasty style, 3 parts, a = figure	91	0.3	0.8	7.80	0.16	0.00	0.00	0.06	0.00	0.05	0.00	0.00	100.29
55.16b	See 55.16a, b = base	91	0.4	0.6	7.70	0.17	0.00	0.00	0.06	0.00	0.04	0.01	0.00	99.46
55.16c	See 55.16a, c = mandorla	90	0.3	0.7	7.15	0.15	0.06	0.00	0.09	0.00	0.07	0.00	0.00	98.94

TABLE II

Period	Cu	Sn	Pb	Zn	Fe	As	Sb	Ni	Bi	Ag	Co	Au
Six Dynasties, N. Wei Dynasty, 386 - 534 A. D. (14 objects)												
average, wt.%	87	5.6	5.1	0.06	0.38	0.43	0.30	0.07	0.04	0.14	0.02	0.11
range, wt.%	81-91	4.0-7.3	3.1-8.3	0.01-0.20	0.02-2.40	0.26-1.04	0.16-0.44	0.05-0.09	0.03-0.06	0.09-0.19	0.01-0.03	0.00-0.80
st. dev.	3	1.0	1.7	0.06	0.59	0.18	0.07	0.01	0.01	0.02	0.01	0.23
Six Dynasties, N. Qi Dynasty, 550 - 577 A. D. (6 objects)												
average, wt.%	78	8.2	12.5	0.02	0.40	0.39	0.21	0.06	0.04	0.17	0.02	0.06
range, wt.%	69-87	5.8-9.3	2.2-21.1	0.00-0.06	0.02-1.67	0.24-0.67	0.08-0.30	0.04-0.09	0.00-0.09	0.03-0.50	0.01-0.05	0.00-0.16
st. dev.	7	1.2	6.9	0.02	0.59	0.15	0.08	0.02	0.03	0.16	0.01	0.06
Six Dynasties, other, 220 - 589 A. D. (15 objects)												
average, wt.%	77	8.5	11.8	0.02	0.27	0.41	0.37	0.11	0.04	0.19	0.06	0.02
range, wt.%	66-88	4.3-14.5	2.4-22.2	0.00-0.07	0.00-0.88	0.17-0.97	0.15-1.24	0.05-0.49	0.02-0.07	0.12-0.39	0.01-0.39	0.00-0.18
st. dev.	6	2.8	5.6	0.02	0.28	0.19	0.25	0.10	0.02	0.07	0.09	0.04
Sui Dynasty, 581 - 618 A. D. (8 objects)												
average, wt.%	77	9.5	12.7	0.02	0.38	0.29	0.22	0.07	0.02	0.10	0.03	0.05
range, wt.%	65-84	6.1-16.2	5.0-22.2	0.00-0.06	0.13-1.02	0.11-0.41	0.17-0.27	0.05-0.10	0.00-0.05	0.04-0.13	0.01-0.12	0.00-0.20
st. dev.	5	3.2	5.6	0.02	0.35	0.09	0.03	0.02	0.02	0.03	0.03	0.06
Tang Dynasty, 618 - 907 A. D. (42 objects)												
average, wt.%	75	10.1	14.0	0.02	0.21	0.46	0.28	0.08	0.06	0.18	0.03	0.06
range, wt.%	66-84	6.2-14.6	6.0- 28.7	0.00-0.19	0.00-0.77	0.00-2.01	0.00-1.15	0.01-0.25	0.00-0.46	0.03-0.61	0.00-0.12	0.00-1.21
st. dev.	5	2.2	5.2	0.04	0.22	0.44	0.24	0.04	0.07	0.12	0.03	0.18
Song Dynasty, 960 - 1279 A. D. (6 objects)												
average, wt.%	78	6.2	12.2	0.02	0.62	0.58	0.11	0.11	0.04	0.11	0.03	0.01
range, wt.%	63-98	1.0-11.7	0.0-26.5	0.00-0.04	0.02-2.90	0.00-2.10	0.00-0.29	0.02-0.17	0.00-0.10	0.00-0.26	0.00-0.11	0.00-0.03
st. dev.	14	4.2	10.6	0.02	1.03	0.75	0.09	0.06	0.04	0.09	0.04	0.01
Ming Dynasty, 1368 -1644 A. D. (4 objects)												
average, wt.%	72	6.8	17.4	2.06	0.54	0.29	0.20	0.10	0.05	0.08	0.02	0.01
range, wt.%	66-80	1.5-10.1	10.4-27.3	0.07-6.20	0.11-1.16	0.12-0.44	0.09-0.30	0.08-0.12	0.00-0.12	0.01-0.12	0.01-0.03	0.00-0.01
st. dev.	5	3.4	6.3	2.47	0.44	0.15	0.08	0.01	0.04	0.04	0.01	0.01
Qing Dynasty, 1644 - 1911 A. D.(6 objects)												
average, wt.%	70	7.1	16.4	4.69	0.23	0.35	0.17	0.11	0.04	0.05	0.01	0.00
range, wt.%	64-72	0.3-14.0	2.6- 20.0	0.02-26.74	0.00-1.01	0.09-0.80	0.07-0.21	0.03-0.35	0.00-0.09	0.02-0.11	0.01-0.02	-
st. dev.	3	4.2	6.2	9.87	0.36	0.23	0.05	0.11	0.03	0.03	0.00	0.00

REFERENCES

1. Alexander Soper, Literary Evidence for Early Buddhist Art in China, (Ascona, Switzerland, 1959).

2. Mizuno Seiichi, Chinese Sculpture, (Nihon keizai shimbun sha, Tokyo, 1960), p.7.

3. G. Beguin and J. Liszak-Hours, Annales du Laboratoire de Recherche des Musees de France, 28-82 (1982).

4. P. T. Craddock, in Application of Science in Examination of Works of Art, edited by Pamela England and Lambertus van Zelst, Museum of Fine Arts, Boston, 1985) pp. 59-67.

5. A. F. Dubrovin, Kinzoku Hakubutsukan Kiyo, (Bulletin of the Metals Museum) 13, 22-30 (1988).

6. H. Uhlig, Das bild des Buddha, (Safari Verlag, Berlin, 1979).

7. S. G. E. Bowman, M. R. Cowell, and J. Cribb, Historical Metallurgy 23, (1), 25-30 (1987).

8. M. R. Cowell, J. Cribb, S. G. E. Bowman, and Y. Shashoua, in Proceedings: Techniques of Coin Production, (Royal Numismatic Society, London, 1988).

9. M. Fabrizi, Report on Later Chinese Copper Alloys, Ms. on file, Victoria and Albert Museum, London, 1988 (unpublished).

10. Yu Dwun-Ping, in Recent Advances in Conservation, edited by J. Black, (Summer School Press, London, 1987) pp. 119-124.

11. Hugo Munsterberg, Chinese Buddhist Bronzes, (Charles E. Tuttle, Rutland, Vermont, 1967).

12. Paul Jett, Archives of Asian Art, XLIII, 59-60 (1990).

13. Paul Jett, in The History of Metal Plating and Patination, edited by S. La Niece and P.T. Craddock, (Butterworth, London, 1993, forthcoming).

14. Kazuharu Asai (personal communication).

15. M. J. Hughes, M. R. Cowell, and P. T. Craddock, Archaeometry, 18, 19-26 (1976).

16. Paul Jett, in Der Goldschatz der Drei Pagoden, edited by Albert Lutz, (Gebunden, Zurich, 1991) pp. 68-74.

17. Mizuno Seiichi, op. cit. pl. 117.

18. Alexander Soper, Chinese, Korean, and Japanese Bronzes, (Istituto Italiano per il Medio ed Estremo Oriente, Rome, 1966), Figs. 27 a-d.

19. Edward V. Sayre (personal communication).

A TECHNICAL EXAMINATION OF THE CHINESE BUDDHIST BRONZES IN THE FREER GALLERY OF ART, PART B: STABLE LEAD ISOTOPE ANALYSIS.

EDWARD V. SAYRE*, PAUL JETT** AND EMILE C. JOEL*
*Conservation Analytical Laboratory, Smithsonian Institution, Washington, DC 20560
**Freer Gallery of Art, Smithsonian Institution, Washington, DC 20560

ABSTRACT

Stable lead isotope ratios were determined on sixty-four samples taken from sixty of the Buddhist bronze images which were examined and chemically analyzed by Jett and Douglas [preceding paper]. The data were grouped using multivariate techniques and were compared with an accumulated data base of lead isotope measurements on more than 700 ancient Chinese artifacts and more than 200 Chinese and Korean ores. One group of ten images which were all datable to a period of less than one century, Eastern Wei through Sui dynasties, form a nearly unique isotopic group which coincides only with an unusual group of Lengshuichong bronze drums. A second large group of thirty-two isotopically matching images contains almost all, fifteen out of sixteen, of the images attributed to the Northern Wei dynasty. The remaining members of this group are of later date, extending into the Tang dynasty, or are animal figures attributed more generally to the Six Dynasties. This group partially overlaps somewhat the isotope fields of some Western and Eastern Han mirrors and some of the Sackler ritual bronze vessels of the Shang and Zhou dynasties, but appears to have been derived from a different ore source from them. The remaining specimens break up into some very tentatively defined small groups of two to four specimens each and a small residue of four unique, unmatched specimens.

INTRODUCTION

This is a continuation of the technical study of the Chinese Buddhist bronzes in the collection of the Freer Gallery of Art, whose structures and elemental chemical compositions were determined and reported on by Jett and Douglas (the preceding paper). In this paper the stable lead isotope ratios of specimens from the same group of sculpture are considered as they relate to themselves, to other, previously reported, isotopic analyses of Chinese ores and lead containing bronzes and to the elemental chemical analyses of Jett and Douglas.

The complete group of Freer late bronzes that were examined range in date from 451 A.D. to the late Qing dynasty, early twentieth century A.D. The great majority of these bronzes depict Buddhist figures or are part of Buddhist altar sets. Some works, such as some of the early animal figures, are also presumed to be from altar sets. A few pieces, such as the Tang figure atop a plate, Accession No. 49.4, or some of the Qing animal figures, e.g. Accession No. 07.31, do not appear to relate to Buddhist iconography.

Because practical considerations prohibited the comprehensive isotopic analysis of the entire collection of late bronze sculptures the analysis was limited to objects that had been stylistically attributed, with the reservation that a few of them may be more modern copies of earlier styles, either to the period from the middle of the Northern Wei dynasty to the middle of the Tang dynasty or, more generally, to the Six Dynasties, the time between the end of the Eastern Han dynasty, 220 A,D,, to the beginning of the Sui dynasty, 589 A.D. From the group of one hundred and six works of sculpture, sixty-four samples from sixty objects were used for lead isotope ratio analysis. Single samples were taken for analysis from forty-two sculptures consisting of a single piece of metal and from fifteen works of sculpture that consisted of multiple parts. Samples were taken from two parts of two Sui dynasty sculptures, 14.21 and 79.51, both of whose separate parts were found to differ significantly in isotope ratios, and samples were taken from three parts of a Wei Maitreya figure, 15.60, of which only two of the parts had matching isotope ratios.

ISOTOPIC GROUPING OF THE SPECIMENS

The specimens can be divided upon the basis of their stable lead isotope ratios into two reasonably well defined groups, containing ten and thirty-three specimens respectively, two relatively poorly defined tentative groups, containing four specimens each, three pairs of objects that are both isotopically reasonably matched and related to each other in other ways, and a small residue of three isotopically unique, nonmatching specimens. Three gilt brass images from the collection also have been analyzed and found to tentatively form an isotopically similar group. The distributions of isotopic values for most of these objects are shown in Figures 1 and 2. One of the matching pairs of specimens, 47.13 and 47.14, and some of the isotopically unique specimens are extremely divergent from the remaining specimens. These two groups have not been plotted because their inclusion would result in a undue contraction of the plotting space of the main body of specimens.

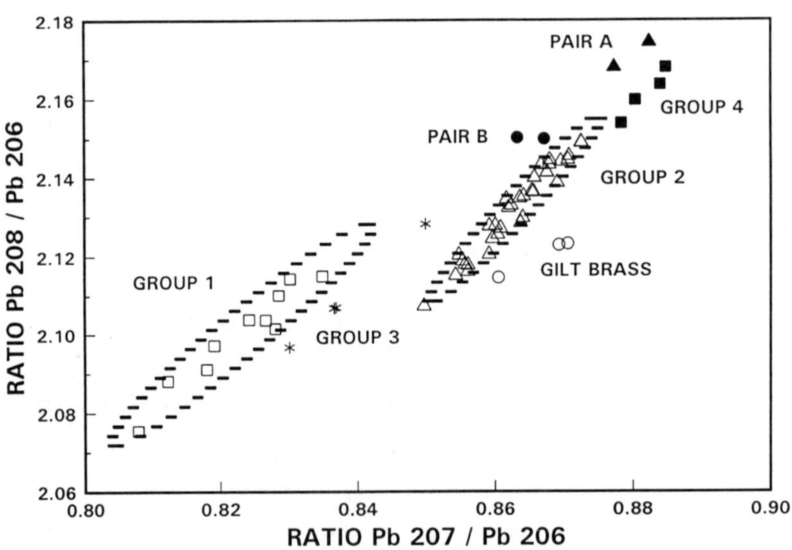

Figure 1 Standard Isotope Ratio Plot of the Freer Buddhist Bronze images.

Figure 1 is the Pb 208/Pb 206 ratio versus the Pb 207/Pb 206 ratio plot of these data, which is the customary way of presenting them. Such a plot is not, by itself, adequate to fully describe the stable lead isotope data that have been obtained because it disregards the data relative to Pb 204, which may show differences between specimens not apparent in the Pb 208/Pb 206 and Pb 207/Pb 206 data. We are dealing with a three dimensional set of independent measurements and although any group of specimens that truly lie close to each other in the full three dimensional space defined by these measurements will, of course, be close to each other in any two dimensional projection of the data, such as Figure 1, specimens that are actually well separated in the three dimensional space might accidentally overlap each other in such a two dimensional projection [1 and 2]. The data can be presented as ratios of isotopic concentrations or as the concentrations of the isotopes within the lead present themselves. The percent concentrations within the lead of the four isotopes concerned, Pb 208, Pb 207, Pb 206 and Pb 204 are not independent of each other, since they must always add up to 100 percent. However scatter plots of pairs of these concentrations are every bit as informative as similar plots of isotope ratios and sometimes provide greater insight into the distribution of data points. Many such plots of pairs of isotope ratios or pairs of isotopic concentrations can be drawn for a data set, and among these we have chosen a plot of Pb 204 concentration versus Pb 208 concentration, Figure 2, to supplement Figure 1 because it brings Pb 204 into consideration and because, of all the plots, it most clearly shows the relationship between the two main groups of specimens and the deviant specimens. In both figures the ellipses around the two main groups represent 90 percent probability of group enclosure calculated assuming Normal rather than Hotelling's distributions.

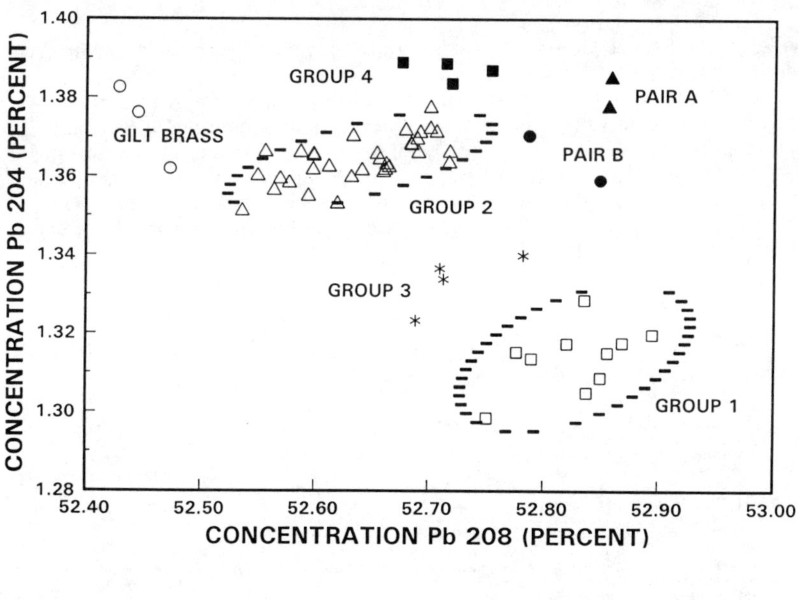

Figure 2 Isotope Concentration Plot of the Freer Buddhist Bronze Images

One of the sizable groups of isotopically similar gilt bronze specimens, Gilt Bronze No. 1, has a nearly unique isotopic signature. In the entire database of lead isotope ratio measurements on Chinese objects and ores only one group of specimens overlap them, a group of Lengshuichong bronze drums which have been notable for the degree to which they differed isotopically from other contemporary drums. The Gilt Bronze No. 1 group is composed of nine Buddhist images all datable to the relatively brief period following the end of the Northern Wei dynasty into the beginning of the Tang dynasty, and one of the animal figures attributed to Six Dynasties. This group is, therefore, also unique among the Chinese bronze groups that have been characterized through isotopic analysis in the brevity of the time period to which its specimens are attributed. The individual specimens are described in Table I.

Table I. Gilt Bronze Group No. 1 Specimens

Accession	Attribution	Description	208/206	207/206	204/206
09.267	Wei (Eastern)*	Bodhisattva, inscribed 550 A.D.	2.10368	0.82657	0.052467
11.133	Northern Qi	Buddha	2.11503	0.83485	0.053186
11.141	Sui	Guanyin , inscribed 599 A.D.	2.11435	0.83017	0.052753
13.42a	Northern Qi	Guanyin (a = figure)	2.08808	0.81227	0.051568
14.21c	Sui	Sakyamuni (c = mandora) and Bodhisattvas, 597 A.D.	2.10149	0.82799	0.052368
15.215	Six Dynasties	Animal figure	2.09107	0.81804	0.052030
16.467a	Tang	Arhat (a = figure)	2.09717	0.81910	0.051930
19.81a	Northern Qi	Guanyin or Maitreye (a = figure), insc. 561 A.D.	2.07547	0.80791	0.051086
45.30b	Sui	Sakyamuni and Prabhataratna (b = base), 609 A.D.	2.11006	0.82850	0.052587
79.51a	Sui	Guanyin (a = figure)	2.10381	0.82416	0.052342

*Note. A number of the objects are attributed in the Accession Register of the Freer Gallery to the Wei Dynasty without differentiation to a specific Wei dynasty. We have added in parentheses our estimate as to which best applies, usually derived from the inscribed dates. We realize, of course, that an object may be inscribed with a date that does not precisely define its date of fabrication.

The large Gilt Bronze Group No. 2 is strongly associated with the Northern Wei Dynasty in that it includes all but one, i.e., fifteen out of sixteen, of the Buddhist images in the Freer Gallery Collection attributed to this period for which stable lead isotope measurements have been made. The group also contains a Northern Qi, a Western Wei, a Sui and eight Tang Buddhist images and seven animal figures attributed to the Six Dynasties and Tang periods. Hence the metal source to which the group relates was used, at least, from early Northern Wei into the Tang Dynasty. The individual objects within the group are described in Table II.

Table II. Gilt Bronze Group No. 2 Specimens

Accession	Attribution	Description	208/206	207/206	204/206
		Buddhist Images Attributed to the Period of the Northern Wei Dynasty			
09.264	Wei (Northern)	Guanyin, inscribed 520 A.D.	2.13656	0.86537	0.055245
09.265	Wei (Northern)	Buddha	2.14904	0.87258	0.056166
09.266	Wei (Northern)	Guanyin, inscribed 453 A.D.	2.13294	0.86236	0.055129
09.268	Wei (Northern)	Buddha (?) ca. 520-530 A.D.	2.14558	0.87079	0.055764
09.269	Wei (Northern)	Guanyin ca. 520-530 A.D.	2.14417	0.86961	0.055587
11.121a	Six Dynasties	Buddha (a = figure), inscribed 451 A.D.	2.12779	0.86003	0.054970
11.125	Wei (Northern)	Sakyamuni and Prabhutaratna, inscribed 503 A.D.	2.13533	0.86418	0.055269
11.126	Wei (Northern)	Sakyamuni and Prabhutaratna, inscribed 473 A.D.	2.13639	0.86562	0.055413
11.127	Wei (Northern)	Sakyamuni and Prabhutaratna, inscribed 500 A.D.	2.11800	0.85562	0.054714
11.128	Wei (Northern)	Sakyamuni and Prabhutaratna, inscribed 506 A.D.	2.13488	0.86363	0.055226
11.129	Wei (Northern)	Guanyin, inscribed 541 A.D.	2.14340	0.86790	0.055804
11.130	Wei (Northern)	Sakyamuni & Prabhutaratna, inscribed 516-520 A.D.	2.14331	0.86673	0.055540
11.131	Wei (Northern)	Guanyin, inscribed 482 A.D.	2.14462	0.87064	0.055843
11.137	Wei (Northern)	Buddha	2.14140	0.86758	0.055603
45.31	Wei (Northern)	Bhaisjyaguru ca. 500 A.D.	2.12448	0.85959	0.054999

Table II. Gilt Bronze Group No. 2 Specimens (Continued)

Accession	Attribution	Description	208/206	207/206	204/206
	Buddhist Images Attributed to Dynastic Periods following Northern Wei				
09.262a	Northern Qi	Buddhist Trinity (a = figures), inscribed 556 A.D.	2.11887	0.85526	0.054585
15.60a	Wei (Western)	Maitreya (a = figure), inscribed 539 A.D.	2.14919	0.87256	0.055912
79.51b	Sui	Guanyin (b = base)	2.13235	0.86195	0.055249
11.124	Tang	Padmapani	2.14455	0.86797	0.055463
11.150	Tang	Guanyin	2.12055	0.85911	0.055124
12.91	Tang	Guanyin	2.12035	0.85472	0.054519
12.92a	Tang	Sakyamuni (a = figure), in a group with 12.93	2.11521	0.85424	0.054581
12.93a	Tang	Monk (a = figure), in a group with 12.92	2.11784	0.85604	0.054760
15.203	Tang	Bodhisattva on a lion	2.13457	0.86166	0.055436
16.250	Tang	Guanyin	2.11620	0.85604	0.054769
49.4a	Tang	Plate with a figure (a = figure)	2.12980	0.86406	0.055294
	Animal Figures Attributed to the Six Dynasties and Tang Periods				
13.96	Six Dynasties	Animal	2.12846	0.86373	0.055291
15.213	Six Dynasties	Animal	2.12709	0.86080	0.055083
15.214	Six Dynasties	Animal	2.12766	0.85914	0.055030
15.38	Six Dynasties	Animal	2.13879	0.86912	0.055678
16.251	Six Dynasties	Animal	2.10730	0.84962	0.054192
17.318	Six Dynasties	Animal	2.12558	0.86037	0.055158
15.212	Tang	Animal	2.14010	0.86574	0.055687

One hesitates to conclude that an assemblage of only three or four specimens, no matter how related in isotopic signature they may be, will provide a characterization of a specimen group likely to be derived from a particular ore source. However, since each of the following tentative groups of objects and have isotope ratios that lie reasonably close to each other and along regression lines that are reasonably parallel to those of well characterized groups, the possibilities that they might constitute the nuclei of isotopic groups, yet to be well defined, should be considered. The specimens belonging to these small tentative groups, to pairs of probably related objects and to the residue of unmatched specimens are listed in Table III.

Table III. Tentative Small Groups and Residue Specimens

Accession	Attribution	Description	208/206	207/206	204/206
	Tentative Gilt Bronze Group No. 3 Specimens				
11.122a	Northern Qi	Gaunyin or Maitreya (a = figure), ins. 555 A.D.	2.09673	0.83009	0.052660
12.90	Tang	Buddha	2.10692	0.83684	0.053427
13.43c	Sui	Buddhist Altar (c = P.R. bodhisattva), 596 A.D.	2.12823	0.84984	0.054027
14.21a	Sui	Sakyamuni (a = figure) and Bodhisattvas, 597 A.D.	2.10675	0.83656	0.053311
	Tentative Gilt Bronze Group No. 4 Specimens				
11.132	Wei (Northern)	Guanyin, inscribed 518 A.D.	2.16376	0.88388	0.056998
15.60b	Wei (Western)	Maitreya (b = base), inscribed 539 A.D.	2.15985	0.88031	0.056682
15.60d	Wei (Western)	Maitreya (d = apsara), inscribed 539 A.D.	2.16807	0.88470	0.056999
16.710	Six Dynasties	Animal	2.15394	0.87832	0.056793
	Probably Related Pair A				
16.444	Six Dynasties	Animal figure	2.17443	0.88231	0.056972
16.446	Six Dynasties	Animal figure	2.16807	0.87728	0.056506
	Probably Related Pair B				
19.77	Six Dynasties	Animal figure	2.15004	0.86719	0.055819
19.78	Six Dynasties	Animal figure	2.15036	0.86325	0.055303
	Matched Pair of Lions				
47.13	Tang	Animal (lion, one of pair with 47.14)	2.20034	0.90712	0.059431
47.14	Tang	Animal (lion, one of pair with 47.13)	2.19996	0.90683	0.059428
	Tentative Gilt Brass Group Specimens				
11.591	Tang (Recent?)	Maitreya	2.12330	0.87062	0.055992
52.28a	Wei (Recent?)	Buddha (a = figure) ca. 520-540 A.D.	2.11471	0.86048	0.054890
55.16a	Wei (Recent?)	Buddha (a = figure) ca. 520-540 A.D.	2.12298	0.86933	0.055706
	Residue of Exceptional Specimens				
09.263	Sui	Figure, inscribed 587 A.D.	2.11500	0.81099	0.051396
13.41a	Sui (Recent?)	Figure (a = figure)	2.08629	0.83815	0.052877
15.106a	Tang	Sarcophagus (a = P.R. front figure)	2.14103	0.87053	0.056119
16.709	Six Dynasties	Animal figure	2.15363	0.87025	0.055645

The first of these tentative small groups is composed of two Sui dynasty, one Northern Qi and one Tang Buddhist images, which have isotope ratios that fairly closely parallel those of the Gilt Bronze No. 1 group. Figure 2 shows the separation of these specimens from that group more clearly than does Figure 1. We call these four objects the Tentative Gilt Bronze No. 3 group. One notes that in Figures 1 and 2 this group lies between the Gilt Bronze No.1 and the Gilt Bronze No. 2 groups. Other such plots confirm that this tentative group is truly situated intermediately between these groups in both in the three dimensional isotope ratio or the four dimensional isotope concentration space, and that, therefore, it may be composed of specimens in which the metal of the Gilt Bronze No.1 group may have been mixed with Gilt Bronze No. 2 metal. The mixing of two materials can only produce specimens whose compositions are intermediate between the compositions of the materials being mixed, and therefore lie in the compositional space that connects the original materials. The mixing of metals from different ore sources, that were being used at the same time in the same region, would not be unlikely, and it is a possibility well worth considering in this instance.

The second small group, Tentative Gilt Bronze No. 4. is composed of a Northern Wei and a Western Wei Buddhist statuette group, two separate parts of the latter having been analyzed, and a Six Dynasties animal figure. In Figure 1 it lies parallel to and a little above and to the side of, Gilt Bronze No.2. The last small group, the Tentative Late Brass group, is composed of three brass, rather than bronze, objects. These brass objects are just sufficiently similar isotopically that they may have been derived from the same source, but there are various reasons for suspecting that they may have been made much more recently than their styles would suggest.

A stylistically matched pair of Tang lions, 47.13 and 47.14, have nearly identical isotope ratios and also closely match each other in chemical composition. These coincidences clearly confirm that they must have come from the same workshop. Their isotope ratios are notably different from those of the other gilt bronze objects measured. Another pair of animal figures, attributed to the Six Dynasties, 19.77 and 19.78, are similar to each other both isotopically and chemically. Although the analytical matching for this pair is not nearly as close as it was for the Tang lions, they also seem likely to have come from a common source. A second pair of Six Dynasties animal figures, 16.444 and 16.446, have isotope ratios that fall within the field of a group of Western Han mirrors, which will be described later. The possibility that the metal sources used during the Western Han dynasty might have continued to be used for metal castings during the Six Dynasties seems sufficiently great to merit pointing out that this pair of figures might be so related to the Western Han mirrors.

The remaining objects, with individually exceptional isotope ratios, include two Sui Dynasty figures, a Tang Dynasty Sarcophagus and an additional Six Dynasties animal figure.

ISOTOPIC COINCIDENCE WITH OTHER CHINESE OBJECTS AND ORES

At present we have accumulated in our data base more than 700 stable lead isotope analyses of ancient Chinese metal artifacts and more than 200 such analyses of metallic ores from Chinese and Korean mine sites. For adequately large isotopically matching groups of artifacts, such as, the Gilt Bronze Groups Nos. 1 and 2, we determine which of the specimens in our data base might relate to the groups by considering the groups to be samples of statistical populations and calculating the three dimensional probabilities that the individual data base specimens might belong to these populations. The core group, i.e. population sample, must contain at least four members for three dimensional probabilities to be calculated relative to it. The probabilities calculated for such small groups, of course, have very uncertain quantitative significance. However, the calculations can be relied upon to reveal those specimens values overlap or come close to the isotopic field covered by the small core group, and we use them to identify such potential matchers. For individual or pairs or triplets of nonmatching specimens we rely upon methods that more simply estimate the degree of similarity between specimens, such as, the calculation of similarity parameters between specimens and clustering dendrograms. The probability calculations are made both with and without the Hotelling's T correction for the additional uncertainty that arises from the use of a small population sample. Without using Hotelling's T distribution one obtains an estimate of the Normal distribution probabilities that would apply if the group were a large one, providing that the population sample at hand is a reasonably random sample of it. Experience with the accumulation of expanding data bases has convinced us that quite often this condition is met and that the greater probabilities calculated with Hotelling's T often do tend to converge to the smaller probabilities calculated without Hotelling's T as the size of a group is increased. Accordingly we have found it useful to base our inference upon consideration of both probabilities and will report both of them in the text, using the notation (PH - PN) in which PH is the probability based upon Hotelling's T distribution and PN is the probability based upon a Normal distribution.

Probability calculations using Gilt Bronze Group No.1 as a core population sample showed that out of the entire data base of artifacts only six Lengshuichong bronze drums showed significant probabilities of matching the group. The probabilities of all of the drums matching Group 1 specimens were high and in all scatter plots of the isotopic data, such as Figure 3, the fields of the two groups of specimens closely overlap. These drums were part of a large group of bronze drums of the Beiliu, Lingshan and Lengshuichong types, mainly excavated in Guangxi Provence, that have been analyzed for lead isotope ratios by Peng Zicheng and his collaborators [3 and personal communications to W.T. Chase]. Overall the drums are considered to date from Western Han (1st cent. B.C.) to Tang (7th cent. A.D.) and have been found over a wide area extending to Szechwan in the north, Guangdong in the east, Yunnan in the west and Hainan Island in the south. Of the drums, the Lengshuichong are reported to have had the widest geographic and chronological distributions. This perhaps explains their isotopic diversity. All of the Beiliu and Lingshan drums, and more than two-thirds of the Lengshuichong drums, that have been analyzed group together in a moderately tight isotope ratio field that is well offset from the Gilt Bronze 1 field, see Figure 3. This field is strongly overlapped by ores from mining sites within Guangxi Province, and Peng and his coworkers have concluded that the metal for these drums was derived from these ores. In our own analysis of these data, to be reported in detail later, we agree that most of the drums do seem to be derived from these Guangxi sources but feel that some of them relate more closely to some Hunan Province mines with a similar, but slightly different, isotopic signature.

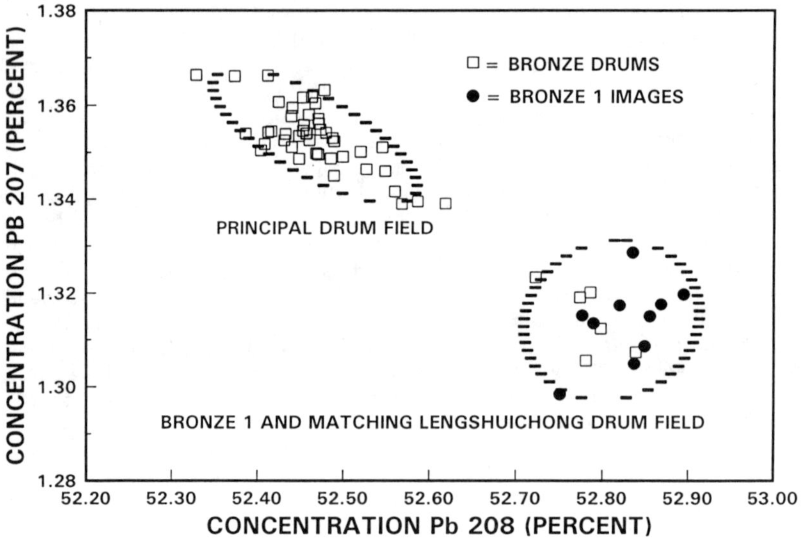

Figure 3 Comparison of the Principal Field of the Chinese Bronze Drums with the Combined Field of the Gilt Bronze 1 Images and Lengshuichong Drums

Using the combined Gilt Bronze 1 statuette and matching drums as a core group one finds two individual ore specimens with significant probabilities of matching it, which were analyzed by Chen Yuwei, Mao Cunxiao and Zhu Bingquan [4]. They are a galena ore from Chayu in Tibet with high probabilities of matching, (68.9 - 62.6) and a galena ore from Renhuai in Kweichow Province with moderate probabilities of matching (14.2 - 5.3). The Tibetian source would seem to be too remote geographically to be considered seriously, despite the good match. The Kweichow source is very central to the distribution of the drums and should certainly be considered as a potential source for these isotopically exceptional objects. However, one would want to have this mine characterized much more fully, by additional and better calibrated isotopic measurements, before evaluating the probability that it, indeed, was the source of their metal.

The large Gilt Bronze Group No. 2 strongly overlaps a group of Eastern Han mirrors and slightly overlaps a group of Western Han mirrors analyzed by Mabuchi and Hirao [5]. Figure 4 is a representative scatter plot showing the degree of these overlaps. One of the Western Han mirrors whose isotopic position lies in the overlap region actually has a higher probability of belonging to the Gilt Bronze 2 group than to the Western Mirror group and ten Eastern Han mirrors have higher probabilities of belonging to Gilt Bronze 2 than their own group. While it is possible that some of these mirrors may have been made from Gilt Bronze 2 metal and, conversely, that some of the Gilt Bronze statuettes could have been made from one or the other of the Han mirror metals, the three groups of objects do appear, more likely, to have been derived from separate sources that overlap, in part, in isotope ratios. We have experimented with regrouping some of the objects strictly upon the basis of isotope ratio values but have not found alternate divisions that appear to make more sense than the present ones, which respect the stylistic and chronological differences between the objects.

Mabuchi and Hirao also measured the isotope ratios of some Kai-Yuan Tong-Bao coins, which were first issued in the year A.D. 621 during the Tang Dynasty. Two of these coins had isotope ratios fully consistent with the Eastern Han mirrors and the Gilt Bronze 2 images.

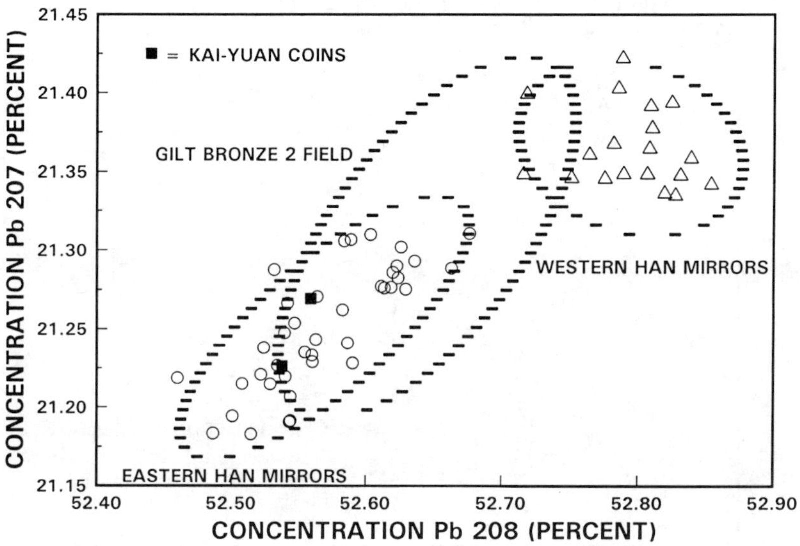

Figure 4 Comparison between Gilt Bronze Group 2 and the Han Mirror Groups

A sizable number of Shang and Zhou dynasty bronze ritual vessels in the Arthur M. Sackler collections, which were extensively analyzed for lead isotope ratios [6], showed significant probabilities of matching the Gilt Bronze No. 2 isotope field. However, plots of these matching specimens relative to the Gilt Bronze 2 field, Figure 5, show that they all tend to lie to one side of the field and, therefore, may better relate to other source fields that overlap the Gilt Bronze 2 field. Indeed one observes that eight of these matching specimens are well centered in and relatively uniformly distributed over the Eastern Han mirrors field. They are the ones in Figure 5 that are surrounded with the 90 percent probability ellipse for the Eastern Han field. It would seem more likely that the metal of these was derived from the same ore source as the Eastern Han mirrors objects than that of the Gilt Bronze 2 images. A ninth vessel consistently lies just outside of the Eastern Han ellipse and also may well be a less well fitting member of that group. These nine Sackler vessels are about equally attributed to the Shang, Western Zhou and Eastern Zhou dynasties, which, if their correlation with Eastern Han mirrors is correct, would imply that their common ore source was one that continued to be used for many centuries. It should be mentioned that a majority of the eight vessels best matching the Eastern Han were originally assigned to two small tentative groups, Sackler Groups 9 and 11 [6], which were similar to the small, tentative Gilt Bronze Groups 3 and 4 proposed here. At the time of the previous assignment there were too few data points in this isotopic region to perceive them as being part of the larger group to which they now seem to belong.

Mabuchi and Hirao also analyzed two more Kai-Yuan Tong-Bao coins [5] which have significant probabilities of belonging to Gilt Bronze 2. Their positions in Figure 5, show that they fairly well match the Extended Sackler Group 10 field but are quite peripheral to the Gilt Bronze 2 field.

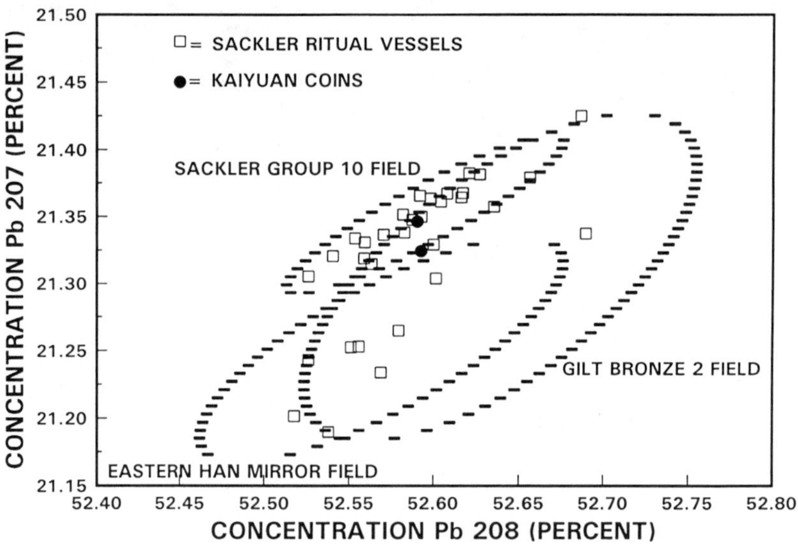

Figure 5 Comparison between Gilt Bronze Group 2 and some Sackler Ritual Vessels

The remaining Sackler specimens that overlap the Gilt Bronze 2 field all lie at the extreme edge of the field and accordingly have relatively low probabilities of belonging to it. They contain most of the original small and tentative Sackler Group 10, which we now feel can be expanded to form a larger Group 10, which includes a tailing end portion of the very large Sackler Group 3, which we now believe was a little over extended. The members of this expanded Group 10 are plotted in Figure 5, where it can be seen that they are all quite peripheral to or outside of the Gilt bronze 2 field.

Thus it is apparent that the Gilt Bronze 2 field lies in a region of overlapping source fields, and hence it often may not be possible to assign specimens to it on the basis of lead isotope ratios alone. However, the overall placement of the Gilt Bronze 2 field is sufficiently offset from those other fields that do overlap it to permit the conclusion that it was derived from a separate ore source that was not used in earlier metal fabrication.

The observation that the Gilt Bronze 2 field overlaps the isotopic fields of other artifact groups is reinforced by the observation that ores from several mining regions throughout China are found to have significant probabilities of matching it. There are a pair of lead ore specimens from a mine at Tao-ling in Hunan province, a lead carbonate ore, FG1080, and a lead sulfide ore, FG1081, which were collected by Chase and analyzed at NIST, that show very high average probabilities of matching Gilt Bronze 2, (72.5 - 69.7). The same specimens have average probabilities of matching the Eastern Han mirrors field of only (24.6 - 20.7). One may consider the ore of this Hunan mine to be a possible source of Gilt Bronze 2 metal, but the mine must be much more fully characterized before one can realistically judge the probability of this. Another Hunan mine, at Shaodong, reported by Chen, has probabilities of matching Gilt Bronze 2 of (25.5 - 20.3) but probabilities of matching the Eastern Han mirrors of (80.0 - 78.8), and four additional Hunan ore specimens have had significant probabilities of matching Eastern Han but insignificantly low probabilities of matching Gilt Bronze 2. The possibility of separate Hunan sources for the two artifact groups should be considered. A group of four ore specimens from the province of Zhejiang with about equal average probabilities of matching Gilt Bronze 2 and Eastern Han, (22.3 - 17.1) and (24.8 - 21.0), were analyzed by Mabuchi and Hirao. Two of these specimens were from the site Wubu and two were from Huangyan Wubu. Finally there are two ores from sites in Guangdong, one analyzed by Chen and one by Mabuchi and Hirao, and an ore specimen from Guangxi, analyzed by Peng, that have reasonable probabilities of matching Gilt Bronze 2, but these sites are probably too geographically distant to be considered a potential sources.

Neither of the two tentative groups of four specimens each, Tentative Gilt Bronze Groups 3 and 4, overlapped extensively with other specimens within the data bank. Three dimensional probability calculations made without Hotelling's T, i.e., calculations of the probabilities of belonging to hypothetical large groups with the same means and spreads as the small ones, found no specimens in the entire data bank with two percent or greater probabilities of matching Group 3. Group 4 lies in a somewhat sparsely filled isotope region between the Gilt Bronze 2 and the Sackler Group 3 fields. A first probability calculation selects two ritual vessels, both in the Sackler collection, a Western Zhou jue, V-251, and an Eastern Zhou ding, V-352, and a Han dynasty glass cicada in the Royal Ontario Museum, 930.170.9, with significant probabilities of matching Group 4. A second calculation with a core group formed by combining these three specimens with the original Group 4, adds an additional Western Zhou li ding, Sackler V 156, and a "recent" bronze cylindrical jar, Freer Gallery of Art 04.356. It is not obvious these specimens are indeed significantly related, although the match between the Han dynasty glass and Tentative Bronze Group 4 gilt images may well be. Such calculations for small core groups should pick up the specimens in the data bank that are closely similar to the groups, even though the quantitative values of probabilities calculated may be quite unreliable. There is no great point in making three dimensional probability calculations using Hotelling's for groups with only four specimens, because the spread of uncertainty generated by the small sample size is so great it includes nearly all the specimens in the data bank. Such a calculation for Group 4 produced probabilities of match of greater than five percent for nearly 90 percent of the specimens in the data bank.

Our experience has been that, when sizably large groups of specimens have been found, the standard deviations of the individual ratios for each of the groups are approximately the same. Expressed as a percentage of the values themselves, the standard deviations of the Pb 208 / Pb 206 ratios for average groups are about 0.5 percent, those of the Pb 207 / Pb 206 ratios are about 0.7 percent and those of the Pb 204 / Pb 206 ratios are about 0.9 percent. These average expectancies allow us to define a region around the determined ratios of a single specimen, or a matched pair or triplet of specimens, within which one might expect to find other specimens that might also match it. By using two standard deviations of average group spread around data points to define such a region, or by noting when individual, or a few matching specimens, fall within the fields of established groups we have attempted to determine how the remaining specimens relate to the overall data base.

As has already been noted, the Pair A specimens are moderately compatible with the Western Han mirrors with average probabilities of belonging of (15.4 - 10.5). It is quite reasonable that a metal source used during the Western Han dynasty might have continued to be used during the Northern Wei. Two specimens from a mine at Fengcheng, Jingzhengzi in Liaoning province, that were analyzed by Chen, also have good probabilities of matching the Western Han mirrors and, therefore, the metal for the Pair A specimens may have come from there.

Five of the remaining specimens stand out as having relatively unusual isotope ratios. The matched pair of bronze lions have such an unusual set of isotope ratios. The only object that is similar to them in isotope ratios is a Western Zhou zun, Sackler Collection V 387. Another gilt bronze object with unusual isotope ratios is the Sui dynasty figure, 09.263, listed among the exceptional specimens which is matched only by a glass Pi disk, Corning Museum of Glass 51.6.548, attributed to 4th to 3rd Centuries B.C. Yet another object among the exceptional specimens, the Tang dynasty sarcophagus 15.106a, has a unique set of isotope ratios that are not closely matched by anything in the data base.

The Pair B of Six Dynasties animal figures and the similar figure 16.709, which has somewhat like isotope ratios, lie close to the Western Han mirror field, but do not have significant probabilities of belonging to it. The three gilt brass figures have isotope ratios quite similar to the Sackler Group 2 specimens but are not fully compatible with that group.

CORRELATION BETWEEN ELEMENTAL COMPOSITIONS AND ISOTOPIC RATIOS

It was noted by Jett and Douglas in the preceding paper that all fourteen of the pieces dating to the Northern Wei Dynasty were all very similar in composition. We further note that ten of these objects have the same isotopic signature, i.e., they are all members of the Gilt Bronze No. 2 isotope group. The close agreement in both elemental and isotopic concentrations among these specimens strongly suggests that they were formulated with raw materials from the same sources used in the same proportions. Also one finds only four other objects in the compositional database that have concentrations of the basic alloying metals, copper, tin and lead, at all closely similar to them. Their formulation, therefore, is almost unique for this period. The conclusion expressed by Jett and Douglas that these pieces may have been made by "the same or a small number of craftsmen or workshops" appears to be quite probable.

However, specimens conforming to the large Gilt Bronze No. 2 isotopic group also occur in all of the five alloy groups described by Jett and Douglas. Therefore, the lead source to which it relates must have been widely used over a long period of time. The correlation between the other isotope groups and the alloy groups is, however, too occasional and sporadic to permit any conclusions to be drawn from it.

REFERENCES

1. K.A. Yener, E.V. Sayre, H. Özbal, E.C. Joel, I.L. Barnes and R.H. Brill, Journal of Archaeological Science 18, 541 (1991)

2. E.V. Sayre, K.A. Yener, E.C. Joel and I.L. Barnes, Archaeometry 34, 73 (1992)

3. Peng Zicheng, Wan Fubin and Yao Shun'an, Kexue Tongbao 33, 1027 (1988)

4. Chen Yuwei, Mao Cunxaio and Zhu Bingquan, Geochimica 3 No. 9, 215 (1980)

5. H. Mabuchi, Y. Hirao and M.Nishida, Archaeometry 27, 131 (1985)

6. L.I. Barnes, W.T. Chase, L.L.Holmes, E.C. Joel, P. Meyers and E.V. Sayre, in The Beginning of the Use of Metals and Alloys, Edited by Robert Madden (The MIT Press, Cambridge, Mass., 1988) pp. 296-306.

ANALYSIS OF MATERIALS CONTAINED IN MID-4TH TO EARLY 7TH CENTURY A. D. PALESTINIAN KOHL TUBES.

William D. Blanchard*, E. Marianne Stern**, and Leon P. Stodulski*
*Conservation Services Laboratory, The Detroit Institute of Arts, 5200 Woodward Avenue, Detroit, MI 48202.
**The Toledo Museum of Art, Box 1013, Toledo, OH 43697.

ABSTRACT

The powdered contents of 17 Late Roman and Byzantine glass vessels used to hold cosmetic eye-paints (kohl) were analyzed. The principal analytical techniques employed were X-ray diffraction, Fourier-transform infrared spectrometry, X-ray fluorescence, and scanning electron microscopy. The materials detected are described, indicating those which may have been used as kohl. Also reported are analytical results from a solid chunk of material surrounding a fragment of a metal rod, also obtained from a kohl vessel.

INTRODUCTION

The following is a report on a study of the materials contained in Late Roman and Byzantine glass vessels used to hold cosmetic eye-paints (kohl). The objective of the study was to determine what materials were used for kohl during this period. A typical glass vessel for this purpose might have one, two, or four individual tubes, each about 4-5 inches long and about 3/4 inches in diameter (Figs. 1A and 1B). The initial phase of the study involved seven vessels from The Toledo Museum of Art (TMA), and the study was later expanded to include another ten vessels from The Detroit Institute of Arts (DIA).

Studies of kohl previously reported in the literature have dealt primarily with Egyptian material and are summarized by Lucas[1]. The materials most widely used in Egypt for this purpose were galena (lead sulfide) and the basic copper carbonate, malachite. Both saw use in the Predynastic period, but the use of malachite had ceased by the end of the New Kingdom. The use of galena continued into the Coptic period. Reports that stibnite (antimony sulfide) was also commonly used for kohl in Egypt have proved to be erroneous. Unlike malachite and galena, a very common mineral, stibnite is not native to Egypt, but deposits are known elsewhere in the Near East. Reports of the use of stibnite for kohl in Mesopotamia remain unsubstantiated.

The glass vessels whose contents have been analyzed for this study are all of more recent date - mid-4th to early 7th centuries A.D.. All were

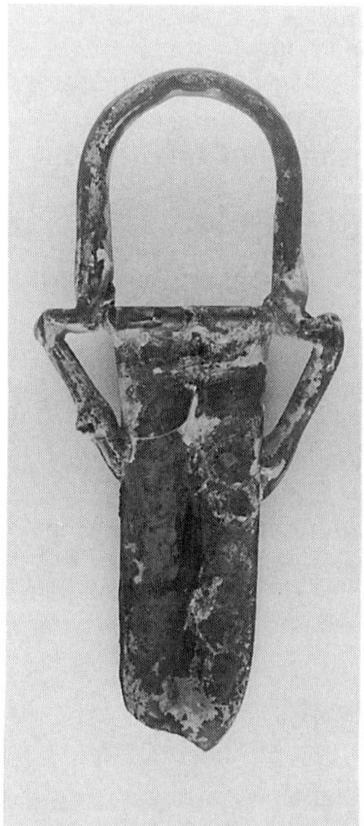

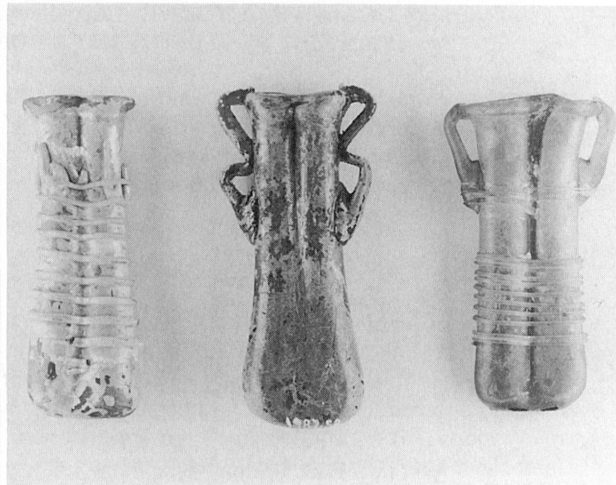

Figs. 1A and 1B. Fourth century A.D. kohl tubes from the collection of The Detroit Institute of Arts.

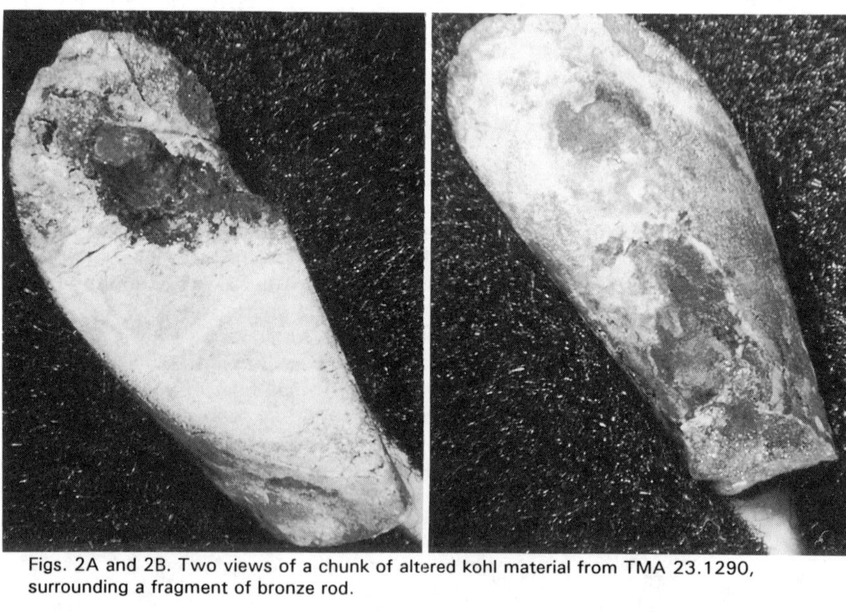

Figs. 2A and 2B. Two views of a chunk of altered kohl material from TMA 23.1290, surrounding a fragment of bronze rod.

purchased from dealers and are not well provenienced in an archaeological sense, though they are reasonably well dated on typological grounds. The vessels from the TMA are all from Palestine. Two vessels with single tubes and three with double tubes date from the mid-4th to late 5th centuries A.D., while the other two are more elaborate with four tubes each and date from the 6th to early 7th centuries A.D.. Five of the DIA vessels are from the vicinity of Tyre; one each from Syria, Egypt, and the "Eastern Mediterranean"; and two have no proveniences given. All are double tube vessels dating from the mid-4th to late 5th centuries A.D..

EXPERIMENTAL PROCEDURES

Detailed analyses were carried out on samples obtained from six glass vessels from the TMA and four glass vessels from the DIA. Preliminary analyses carried out on six other DIA glass vessels indicated the presence of no more than trace amounts of materials potentially associated with kohl, and further analyses of the contents of these vessels were not carried out. Studies on the contents of the seventh TMA vessel, one of the three with double tubes, is described below.

At least one sample of powdered material was removed from each individual tube of the glass vessels studied. In some cases, several samples were taken from an individual tube if visually distinct materials were observed in separate areas within the tube. After preliminary analysis of each sample using X-ray fluorescence, the samples were separated mechanically under a microscope into individual components. These components were chosen on the basis of visual characteristics (most often color), and have no necessary relationship to the actual number or identities of the materials present.

The two most important analytical techniques used were X-ray diffraction (XRD), using both a diffractometer and a Gandolfi camera, and Fourier-transform infrared analysis (FTIR). Polarized light microscopy was most useful in identifying galena which cannot be detected by normal FTIR and whose XRD powder pattern is often very weak and not detectable unless the galena is well crystallized and present in large quantities. X-ray fluorescence (XRF) and scanning electron microscopy with both energy dispersive spectrometry (SEM/EDS) and wavelength dispersive spectrometry (SEM/WDS) were used to identify the chemical elements present. Back scattered electron (BSE) images of properly prepared specimens were also obtained using SEM. Instruments used in this study included a Phillips X-ray diffraction spectrometer, a Nicolet 5DXB Fourier-transform Infrared Spectrometer, a Kevex X-ray fluorescence spectrometer, an International Scientific Instruments scanning electron microscope incorporating a Kevex energy-dispersive X-ray analyzer, a CamScan scanning electron microscope incorporating a Microspec wavelength-dispersive X-ray analyzer, and a Zeiss Universal polarizing microscope.

ANALYTICAL RESULTS

The results of the analyses of the kohl vessels are presented in Table I. The first column of the table identifies the vessels by museum and accession number, and gives the overall appearance of the contents and the number of components into which the contents had been separated for analysis. The second column identifies the materials detected by mineral name. The third column gives estimated relative amounts of the materials identified, based on microscopic observation and on the XRD and FTIR analyses. Trace usually means 1-2%. Minor could mean up to 20% depending on what else is present and in what quantities. If one component of a mixture made up about 80% of that mixture, and a second only about 20%, the first component would be considered major and the second minor. If a mixture involved five components, each at about 20%, all five would be considered major. The final column indicates the analytical techniques involved in the identification of each substance listed.

Although galena was observed in eight of the ten vessels given detailed analysis, its two most common alteration products, anglesite and cerussite[2], were more often present in major amounts. It is reasonable to assume that, with galena finely ground for use as kohl, 1400-1600 years would provide ample time for the conversion of most of the galena present into anglesite or cerussite. The basic lead carbonate, hydrocerussite, rather than the more common cerussite, was found in vessel DIA 90.1S14108.

The basic copper sulfate, brochantite, appears in TMA 23.1494 and, to a lesser extent, in TMA 23.1314. Although brochantite, like malachite, is a green material, its presence in only small quantities in tubes with major amounts of the dark grey galena or its alteration products makes it unlikely that it was a deliberate component of the kohl. It might have been inadvertently transferred over from another tube, but no tubes were observed with brochantite or any other green-colored species as a major component. It might also be attributed to corrosion products from a copper or copper alloy applicator rod no longer present in the tube, or possibly to the alteration of a copper containing accessory mineral for galena.

Sphalerite appears in TMA 23.1314. This is a very common accessory mineral for galena, and is frequently found with galena in mineral deposits[2]. It was probably unavoidably introduced into the tubes along with the galena.

Calcite, quartz, and gypsum are all common minerals and, the former two especially, frequent components of sand. The conditions of burial and circumstances of recovery of the glass vessels in this study are not known, but it would seem reasonable that these materials were introduced from the desert environment during burial rather than having been purposely added to

TABLE I - Analytical results from Late Roman and Byzantine kohl tubes from The Toledo Museum of Art and The Detroit Institute of Arts.

Description of Kohl-containing Vessels	Identities of Materials Detected (a)	Relative Amounts of Materials (b)	Analytical Techniques Employed (c)
TMA 23.1494			
single tube	Anglesite ($PbSO_4$)	M	a, b, d, e
dk. grey powder	Galena (PbS)	M-m	c, d
w. green particles	Cerussite ($PbCO_3$)	m-tr	a, b, d
3 components analyzed	Brochantite ($CuSO_4 \cdot 3Cu(OH)_2$)	m	a, c, e
	Calcite ($CaCO_3$)	m-tr	a, e
	Quartz (SiO_2)	m-tr	a, e
TMA 23.1255			
single tube	Anglesite	M	a, b, d
brown powder	Cerussite	m	a, b, d
5 components analyzed	Galena	m	c, d
	Quartz	tr	a, e
	Gypsum ($CaSO_4 \cdot 2H_2O$)	tr	a, e
TMA 23.1314			
tube 1	Anglesite	M	a, d
brown powder	Galena	m	c, d
6 components analyzed	Cerussite	m	a, d
	Sphalerite (ZnS)	m	a, d
	Calcite	m	a
	Quartz	tr	a
tube 2	Anglesite	M	a, b, d
brown powder	Galena	m	c, d
w. green particles	Cerussite	m	a, b, d
8 components analyzed	Sphalerite	m	a, d, e
	Brochantite	m-tr	a, c, e
	Gypsum	tr	a, e
tube 3 - empty			
tube 4	Anglesite	M	a, b, d
brown powder	Galena	m	c, d
8 components analyzed	Cerussite	m-tr	b, d
	Sphalerite	m	a, d
	Calcite	m	a
	Quartz	tr	a
	Apatite ($Ca_5(PO_4)_3(F,OH)$)	m-tr	a
	Brochantite	tr	a, c
	Gypsum	tr	a

TABLE I continued.

Description of Kohl-containing Vessels	Identities of Materials Detected (a)	Relative Amounts of Materials (b)	Analytical Techniques Employed (c)
DIA X1989.1435			
tube 1	Galena	M	a, c, d
grey-tan to dk. grey powder	Anglesite	M	a, d
5 components analyzed	Gypsum	M-m	a
	Pyromorphite ($Pb_5(PO_4)_3(F,OH,Cl)$)	m	a, d
tube 2 - empty			
TMA 23.1286			
tube 1	Cerussite	M	a, b, d
grey powder	Anglesite	M	a, b, d, e
2 components analyzed	Galena	M	a, c, d
	Quartz	tr	a, e
	Gypsum	tr	a, e
tube 2	Galena	m	a, c
white powder w. few black particles	Cerussite	tr	a
2 components analyzed	Anglesite	tr	a
	Calcite	M	a, e
	Pyromorphite	m	a
TMA 23.1293			
tube 1	Cerussite	M	a, b, d, e
grey powder	Galena	M-m	c, d, e
6 components analyzed	Anglesite	m	a, b, d, e
	Quartz	m	a, e
tube 2	Cerussite	M	a, b, d, e
grey powder	Anglesite	M	a, b, d, e
2 components analyzed	Galena	M	a, c, d
DIA 90.1S14108			
tube 1	Hydrocerussite ($2PbCO_3 \cdot Pb(OH)_2$)	M	a, b, d
grey-tan powder	Anglesite	M	a, b, d
3 components analyzed	Galena	m	c, d
	Calcite	m	a
	Quartz	tr	a
tube 2	Hydrocerussite	M	a, b, d
dk. grey-tan powder	Anglesite	M	a, b, d
not separated into components (XRD & FTIR of whole sample identical to tube 1)			

TABLE I continued.

Description of Kohl-containing Vessels	Identities of Materials Detected (a)	Relative Amounts of Materials (b)	Analytical Techniques Employed (c)
TMA 23.1313			
tube 1	Cerussite	M	a, b, d, e
tan powder	Anglesite	m	a, d
7 components analyzed	Galena	tr	c, d, e
	Apatite	m	a, d, e
	Pyromorphite	M-m	a, d, e
	Quartz	tr	a, e
tube 2	Apatite	M	a
tan powder	Pyromorphite	M	a
4 components analyzed			
tube 3	Anglesite	M	a
tan powder	Galena	m-tr	c
4 components analyzed	Calcite	m	a
	Apatite	m	a
	Pyromorphite	M	a
tube 4	Apatite	M	a, b, d, e
tan powder	Quartz	m-tr	a, e
7 components analyzed			
DIA 1987.58			
tube 1	Pyromorphite	M	a, b, d
brown powder	Apatite	m	a, b
3 components analyzed			
tube 2	Pyromorphite	M	a, b, d
brown powder	Apatite	m	a, b
5 components analyzed	Calcite	m	a
	Quartz	tr	a
DIA F80.161			
tube 1			
brown powder			
not submitted to detailed analysis (no evidence of kohl materials)			
tube 2	unidentified mineral (lead silicate?)	M	a, e
brown to grey-tan powder	Anglesite	m	a, d
9 components analyzed	Calcite	M-m	a, d
	Quartz	M-m	a

the tubes. A variety of more complex silicate minerals might be expected to have been introduced in the same way. Some of the so far unidentified materials which are present in some of the tubes might be silicates. Only one vessel, DIA F80.161, contained an unidentified material as a major component. Analysis by SEM/EDS gives lead and silicon as major elements in this material, indicating that it may be some kind of lead silicate. Other tubes containing much smaller amounts of unidentified materials are TMA 23.1314 - tube 4 and TMA 23.1286 - tube 1 (same species); TMA 23.1286 - tube 2; TMA 23.1293 - tube 1 (at least two unidentified species); TMA 23.1313 - tube 4 (two unidentified species); and TMA 23.1494. These unidentified materials are not indicated in Table I since they are present in only small quantities.

Apatite minerals were found in TMA 23.1314 - tube 4, in DIA 1987.58, and, in larger quantities, in TMA 23.1313. Apatite is a calcium fluorophosphate, although the fluoride can be partially or completely replaced by hydroxide or, less frequently, by chloride[3]. The X-ray diffraction patterns obtained for the apatite species in this study are consistent with the very similar patterns reported for fluorapatite and hydroxylapatite, but do not correlate well with the somewhat different pattern reported for chlorapatite. The apatite minerals are the most common phosphate minerals, and numerous deposits are known in the Near East where phosphates are mined commercially[4]. The presence of apatite in these vessels might then also be explained by introduction from the desert environment during burial. Hydroxylapatite can also occur as a decomposition product of bone or ivory, and might be introduced to a tube through decomposition of a bone or ivory applicator rod, but this would not explain apatite where present in a tube in large quantities. It remains a possibility that apatite was itself used as a kohl material.

Two of the glass vessels, TMA 23.1313 and DIA 1987.58, contain major amounts of pyromorphite minerals. Pyromorphite is similar to apatite except that lead replaces calcium, and chloropyromorphite is much more common that either fluoropyromorphite or hydroxypyromorphite[5]. Chloropyromorphite is a possible accessory mineral for galena. However, pyromorphite can also occur as an alteration product of galena in the presence of phosphoric acid[6], which could be formed from apatite under acidic conditions. The fact that the X-ray diffraction patterns for the pyromorphite species in the two vessels mentioned above are much closer to the patterns reported for the very rare fluoropyromorphite and hydroxypyromorphite than to the pattern of chloropyromorphite, and that apatite was also observed in these vessels, supports the hypothesis that the pyromorphite species were obtained through reaction of galena with apatite. Pyromorphite was also observed, in much smaller quantities, in TMA 23.1286 - tube 2 and in DIA X1989.1435 - tube 1. The diffraction patterns obtained for the pyromorphite species in these two tubes were less conclusive with regard to being able to distinguish between chloropyromorphite and its rare fluoride and hydroxide analogues.

In addition to the glass vessels mentioned above, one other from the TMA was also studied. This vessel, TMA 23.1290, had two tubes, one of which was empty. The other tube contained a solid chunk of material about 3/4 inch long, with the stub of a metal or alloy rod sticking out of the top (Figs. 2A and 2B). XRF analysis of several areas on the chunk located away from the metal rod indicated major amounts of lead, and trace amounts of iron, copper, and zinc. XRF analysis of an area which included the stub of the metal rod indicated a major amount of lead, minor iron and copper, minor to trace tin, and a trace of zinc. The large amount of lead observed is consistent with the materials in the chunk being primarily galena and its alteration products. The higher concentrations of iron, copper, and tin observed in the vicinity of the metal rod suggested iron or bronze as components of the rod. Several small samples were removed from various locations on the surface of the chunk and were analyzed by X-ray diffraction. Materials identified included galena, anglesite, cerussite, hydrocerussite, and gypsum, all materials that had been observed in the tubes previously studied. Also identified was cuprite (Cu_2O), a red colored copper corrosion product, obtained from an area adjacent to the metal rod.

The chunk was sectioned using a Buehler Isomet Plus saw with a 0.5 mm diamond coated brass blade. The cut was made about 1/8 inch below the top surface. Photomicrographs of the cut surfaces are shown for the top portion of the chunk (Figs. 3 and 4). The cut surface of this portion of the chunk was then studied using a scanning electron microscope. A BSE image of this surface showed two roughly circular areas near each other, both giving darker images than the surrounding material (Fig. 5). Elemental analysis showed the surrounding material to have a high concentration of lead. The lighter of the two circular areas, corresponding to the metal rod previously observed, had a core composed primarily of copper and tin, indicating a bronze alloy. The adjacent darker circular area showed a high concentration of iron. It appears that there may have originally been two metal rods in the tube, one of bronze and one of iron, the latter now completely corroded.

A BSE image of the area of the bronze rod shows a banded structure at and underneath the original surface of the rod (Fig. 6). This phenomenon of banding in corroded copper or bronze artifacts is reasonably common and has been discussed by Scott, who observed similar band formation in several small corroded bronze rod-like fragments[7]. A higher magnification BSE image (Fig. 7) and a series of elemental dot maps (Figs. 8A-E) are shown for an area extending from an uncorroded portion of the bronze rod, past the original surface, into the surrounding altered kohl material. The dot maps were obtained using an SEM/WDS instrument and show the distribution within this area of Pb, Cu, Sn, Fe, and Zn. Unlike the fragments studied by

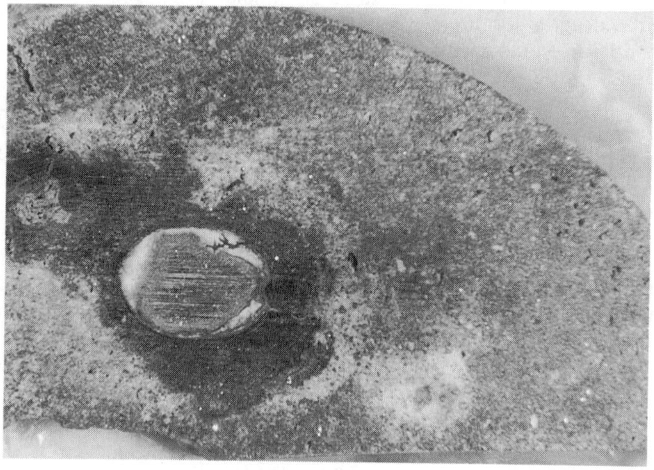

Fig. 3. Photomicrograph of the chunk from TMA 23.1290 after sectioning. Cut surface of the top portion. 10x.

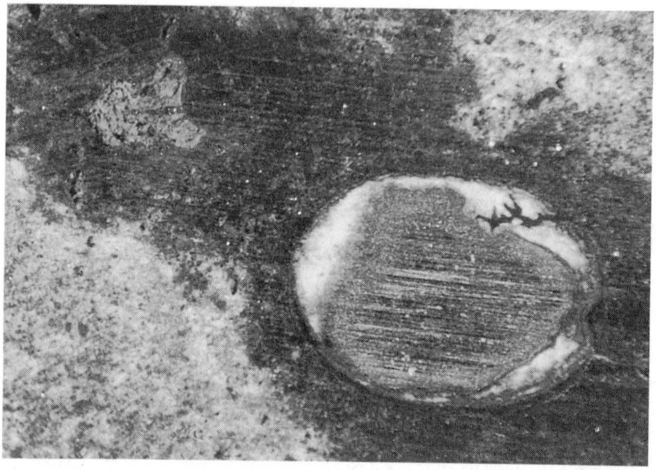

Fig. 4. Photomicrograph of the cut surface showing the area of the two rods. 20x.

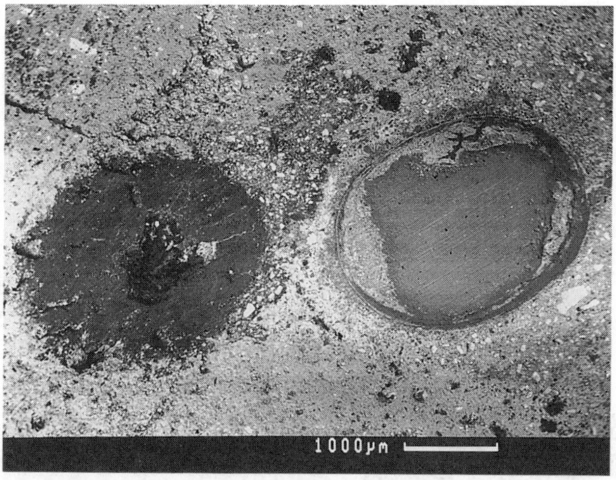

Fig. 5. BSE image of the area of the two rods. 17x.

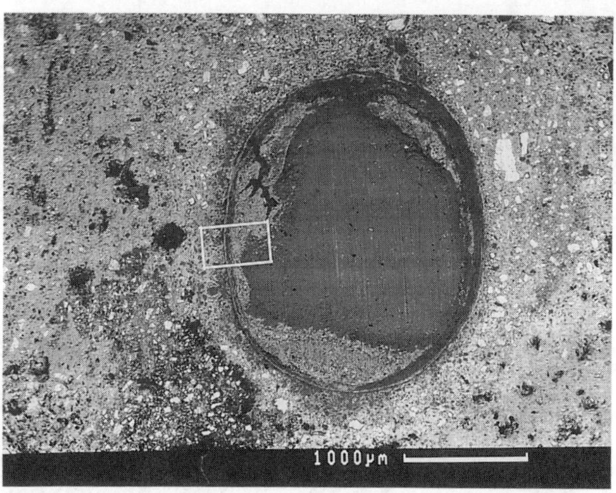

Fig. 6. BSE image of the area of the bronze rod. 22x. The area covered in Figs. 7 and 8 is indicated.

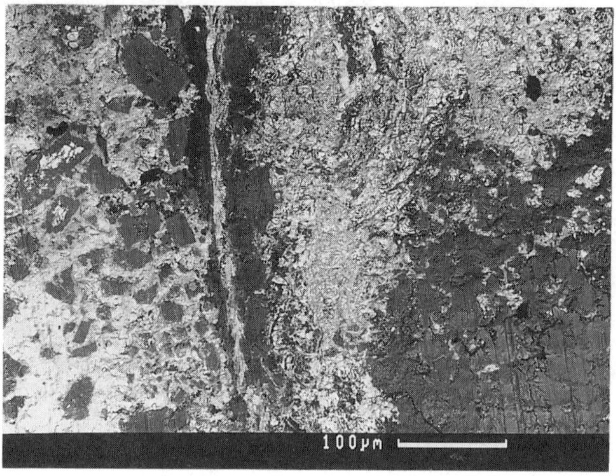

Fig. 7. BSE image of the area indicated in Fig. 6. 200x.

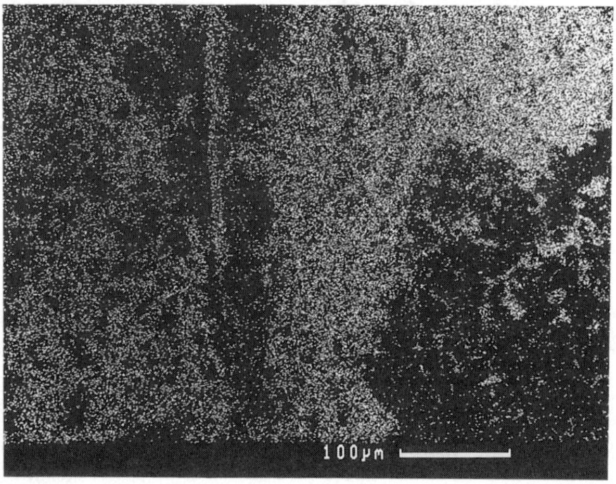

Fig. 8A. Elemental dot map for Pb. Area covered identical to that in Fig. 7. 200x.

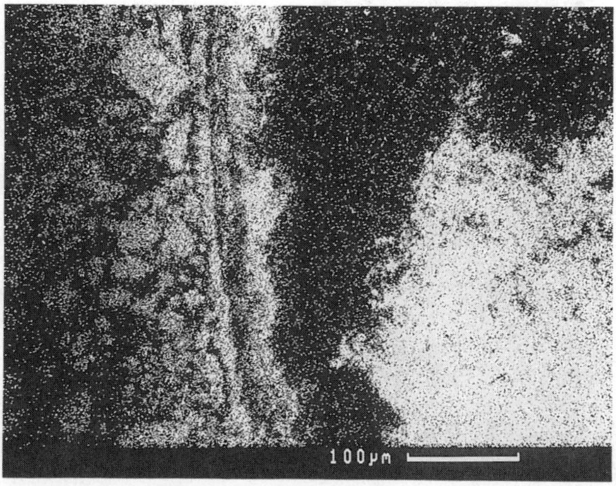

Fig. 8B. Elemental dot map for Cu. Area covered identical to that in Fig. 7. 200x.

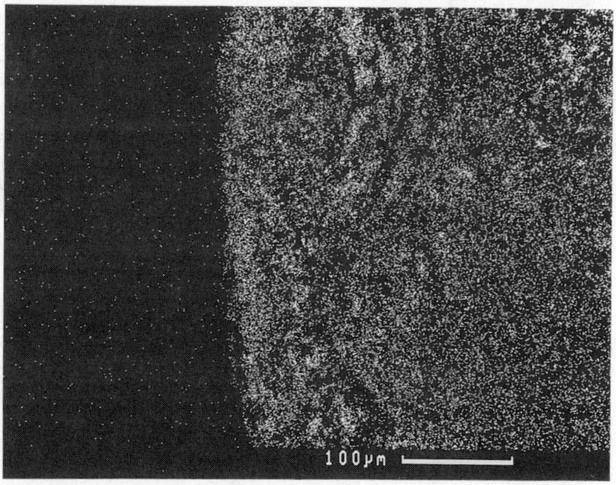

Fig. 8C. Elemental dot map for Sn. Area covered identical to that in Fig. 7. 200x.

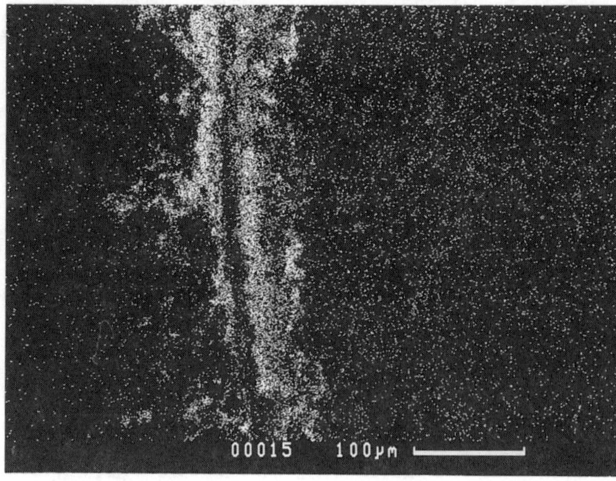

Fig. 8D. Elemental dot map for Fe. Area covered identical to that in Fig. 7. 200x.

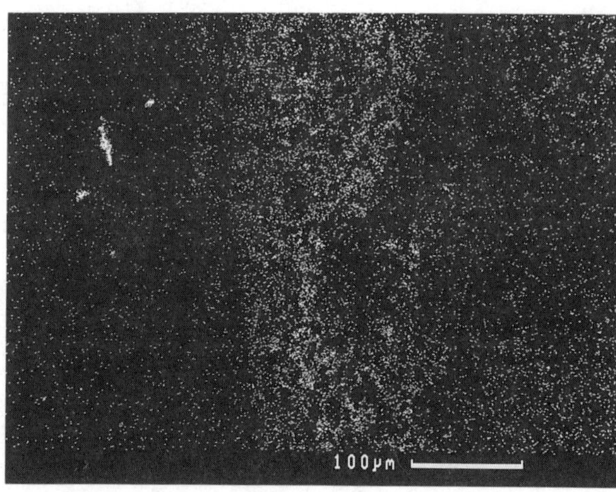

Fig. 8E. Elemental dot map for Zn. Area covered identical to that in Fig. 7. 200x.

Scott, the bronze rod described here is surrounded by the alteration products of the galena in which the rod had been placed, and the elemental dot maps show bands and areas of high Pb content underneath the original surface of the rod (Fig. 8A) as well as bands of high Cu content (Fig. 8B) and of high Sn content (Fig. 8C). Also of interest are two bands of high Fe content, the outer band being located approximately at the original surface of the bronze rod(Fig. 8D). These bands are well separated from the circular area of high iron concentration previously described, and may indicate that at least part of the bronze rod had been wrapped with either a thin iron foil or a very fine iron wire. The few small areas of high Zn concentration (Fig. 8E) are probably attributable to sphalerite.

CONCLUSIONS

The only previously reported kohl material for which evidence was found in this study was galena. Brochantite, the only copper mineral detected (other than cuprite which was associated with the corrosion of a bronze rod), was not found in sufficient quantity to indicate that it could have been an alteration product of a copper based kohl material, or that it could have been used as a kohl material itself. XRF analyses of the materials from the glass vessels showed a complete absence of the element antimony, thereby eliminating stibnite as a possible component in any of the materials analyzed. Apatite, found in substantial amounts in two tubes of one vessel, could have been used as kohl, but could also have been introduced from the desert environment during burial. Relatively few kohl vessels were available for this study. A considerably greater number would need to be included before any definitive statement could be made regarding the possible use of materials other than galena for kohl during this period.

Studies of the chunk of altered kohl material obtained from one tube of vessel TMA 23.1290 indicate that the tube may have originally contained two metal rods, one of bronze and one of iron. The elemental dot maps of the banded structure around the partially corroded bronze rod suggest the possibility that at least part of the rod had been wrapped with either a thin iron foil or a very fine iron wire.

ACKNOWLEDGEMENTS

The authors wish to thank the National Endowment for the Arts and the Andrew W. Mellon Foundation for their support. The authors would also like to express their appreciation to General Motors Corporation for assistance with the SEM analyses.

FOOTNOTES TO TABLE I

(a) Colors of materials detected are:
- Galena - dark grey to black
- Anglesite - white
- Cerussite - white
- Hydrocerussite - white
- Sphalerite - yellow to yellow-orange or black with lighter banding
- Brochantite - green
- Quartz - colorless
- Calcite - white
- Gypsum - white
- Apatite - light tan to orange-brown
- Pyromorphite - light tan to brown

(b) Relative amounts estimated microscopically and from XRD and FTIR line intensities.
- M = major amount
- m = minor amount
- tr = trace amount

(c) Analytical techniques used are:
- a. X-ray diffraction (XRD)
- b. Fourier-transform infrared spectrometry (FTIR)
- c. Polarized light microscopy
- d. X-ray fluorescence spectrometry (XRF)
- e. Scanning electron microscopy/energy dispersive spectrometry(SEM/EDS)

REFERENCES

(1) A. Lucas, Ancient Egyptian Materials and Industries, 4th ed., revised by J. R. Harris (Edward Arnold, London, 1962), pp. 80-84.

(2) E. S. Dana, A Textbook of Mineralogy, 4th ed., revised by W. E. Ford (John Wiley and Sons, New York, 1932), p. 417.

(3) L. G. Berry and B. Mason, Mineralogy: Concepts, Descriptions, Determinations (W. H. Freeman and Co., San Francisco, 1959), pp. 453, 454.

(4) Z. S. Altschuler (private communication).

(5) L. G. Berry and B. Mason, Mineralogy: Concepts, Descriptions, Determinations (W. H. Freeman and Co., San Francisco, 1959), p. 453.

(6) E. S. Dana, A Textbook of Mineralogy, 4th ed., revised by W. E. Ford (John Wiley and Sons, New York, 1932), p. 707.

(7) D. A. Scott, "Periodic Corrosion Phenomena in Bronze Antiquities", Studies in Conservation, 30, 49-57 (1985).

NON-DESTRUCTIVE X-RAY DIFFRACTION WITH 30-MICRON SPATIAL RESOLUTION: SOME EXAMPLES

M. Inaba, J. Miyata and R. Sugisita
Conservation Laboratory, Tokyo Geijutsu Daigaku
(Tokyo National University of Fine Arts and Music)
Ueno Koen, Taito-ku, Tokyo 110, Japan

ABSTRACT

We have been measuring x-ray diffraction spectra of small samples of layers in cross-sections using a Micro X-ray diffractometer (PSPC/MDG) system since 1988. In measuring with x-ray diffractometry, the position sensitive proportional counter (PSPC) covers 150 degrees of x-ray diffraction, continuously. The data are accumulated by the multichannel analyser (MCA) and monitored on a CRT during the measurement period. Mounted samples can rotate around three axes. The XRD spectra are very like those from a wide angle goniometer used in powder diffraction. Normally the spatial limits are around 30 microns in diameter, as we have found experimentallly in our samples. We feel this system is very feasible for art and archaeological studies. Some case studies are presented.

INTRODUCTION

X-ray diffractometry has been one of the most useful methods for identifying and determining the structure of crystalline compounds used by researchers in this century. However, the amounts of sample required for traditional x-ray diffraction techniques are large compared to what is usually available from important artistic or archaeological objects.

Often researchers are unsuccessful in obtaining an XRD spectra from powder samples using a wide-angle goniometer. To obtain XRD spectra from microamounts of samples, many researchers have been using the XRD camera. To compare a normal powder XRD pattern to a film pattern requires converting the spectrum on film into digitized values by using a densitometer. Lahanier classified the XRD equipment available in 1986 by the size of the samples to be analyzed(1). In his paper, a Gandolfi camera and a microdiffractometer equipped with a circular detector were classified as requiring the smallest samples. In this paper, we expand our recent paper(2) to describe the several years of experience we have now had with a new micro-XRD spectrometer used to analyze museum samples.

INSTRUMENTATION

Rigaku RAD-RC (12kw)
Rotating tube: Cu Filament: 10 x 1 mm (point configuration)
K filter: Ni
PSPC/MDG (position sensitive proportional counter /micro diffractometer)
Multichannel analyzer (MCA)
Personal computer (Casio, 16 bit CPU: V30)

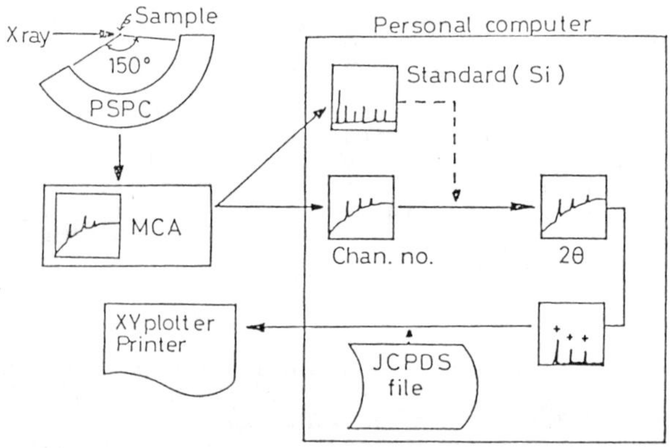

Fig. 1. Data flow in PSPC/MDG system.

Data flow in this system is shown in Fig. 1. Curved position sensitive proportional counter (PSPC) can cover 150 degrees of x-ray diffraction arc, continuously. PSPC can reduce measuring time to a great extent compared with scanning-type counters. These data are accumulated by multichannel analyzer (MCA) and monitored on a CRT during the measurement period. The advantage of this system is that we can know the favorableness of a particular placement of an object in a short time, and then we can decide easily the appropriate length of a measuring period. If a feature in a spectrum was confusing or had insufficient detail, the setting of the sample can be changed. This is a great advantage of micro diffractometry; in other systems, such as camera method, we find such a mistake in the setting of a sample only after finishing the time consuming measurement. Plate 1 shows an XRD spectrum of a silicone powder on MCA during a measurement.

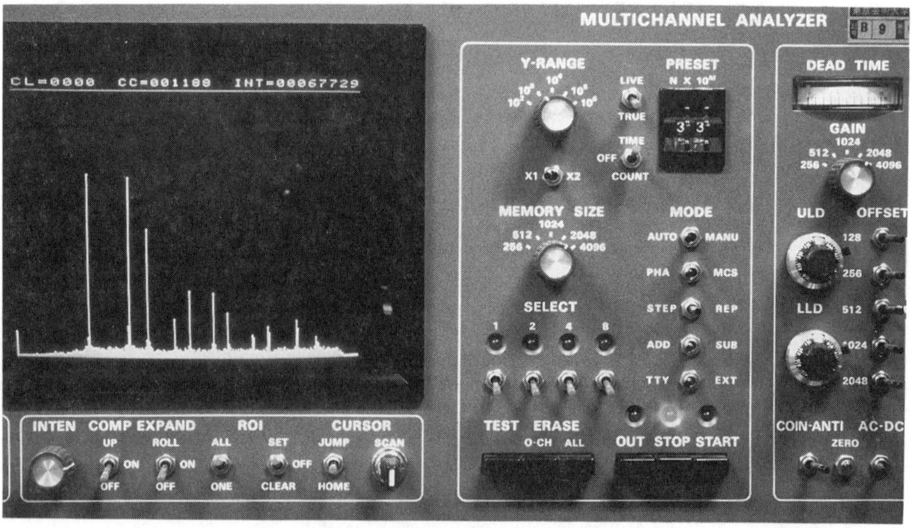

Plate 1. Multichannel analyzer

Accumulated data in the MCA are put into a personal computer. Channel number for a spectrum from the PSPC should be converted into 2 theta values comparing the known signals of a standard material. Powdered silicone is usually used as a standard material. The lowest 2-theta value of Si is 28 degrees (Cu K-alpha), so the values below it should be extrapolated. We had encountered a great deviation from correct angles in the lower angle region. It may be caused from a deformation of the wire in the PSPC. Then we use other standard materials, such as mica or talc, when we need to calibrate lower lines of x-ray diffraction.

After processing the spectra, that is, smoothing the shape, background subtraction and peak identification, d-values and intensities of peaks are calculated automatically and/or manually by the personal computer. The search-and-match software in our system is not particularly sophisticated, but the system can be upgraded. In comparing the pattern of XRD signals against the JCPDS (Joint Committee of Powder Diffraction Standards) on the CRT, the software is particularly helpful and user-friendly to operators. The result can be printed out by an X-Y plotter and a printer.

As regards measurements of smaller samples, the size and heterogeneity in the Debye-Scherrer ring of low angle x-ray scattering is increased by reducing the amount of powder particles in the sample. To solve this problem, the sample is rotated around two or three axes; chi (360°), phi (+-30°, +-40°), and/or omega (40°)(see Fig. 2). A Gandolfi attachment also rotates the sample around two axes. On the other hand, the old Rigaku micro XRD apparatus does not need to rotate the sample, since it has a circular detector to record the whole diffraction cone in a Debye-Scherrer pattern. However, this detector must scan the XRD spectrum on the sample. Rigaku has quit production of this type of apparatus.

Maximum size of the sample is 6mm x 6mm x 6mm. We currently measure small particles and cross-sections of paint layers, so we did not modify the micro XRD system for the study of works of art.

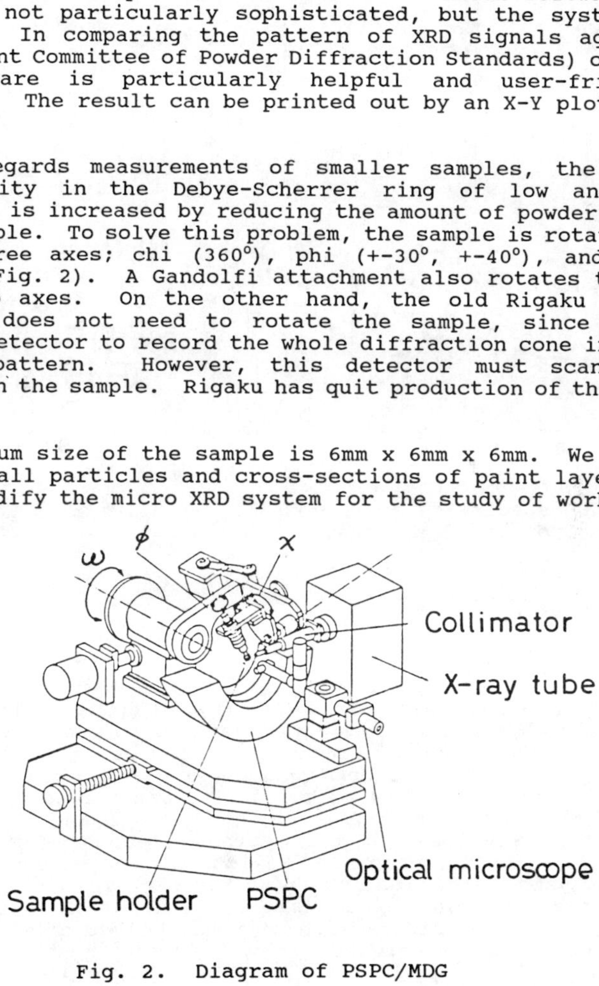

Fig. 2. Diagram of PSPC/MDG

INVESTIGATION OF ART OBJECTS

Measurement of a Small Particle

A small fragment of a pearl in a necklace from the Roman period (1st century B.C.-1st C. A.D., Plate 2) was analyzed(3) and the spectrum is shown as the middle curve in Fig. 3. In this case the time of accumulation of the spectrum was 2000 seconds. The spectrum was compared with standard crystals of calcium carbonate, calcite (upper spectrum) and aragonite (lower spectrum). The pearl was identified as containing aragonite.

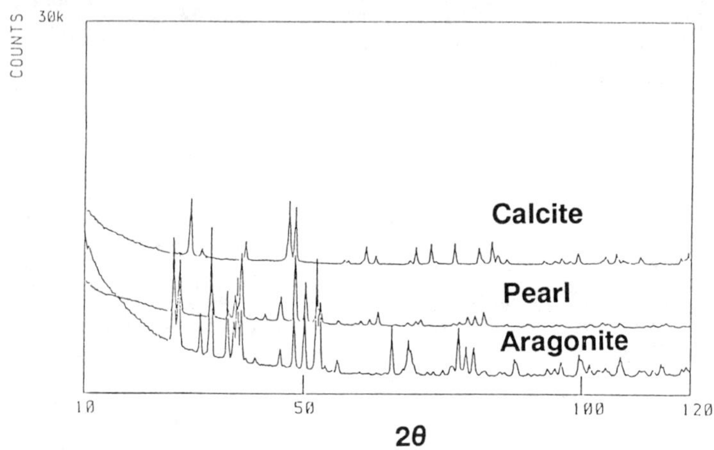

Fig. 3. XRD spectra of a fragment of a pearl from a Roman period necklace and of standard materials.

Setting a Sample

A sample is set under a microscope (magnification of 160x) using a cross wire (x, y) and focus (z). Focusing the sample is often difficult, especially in the case of a rough surface. We sometimes miss a proper setting of the sample.

A sample should be mounted on a small plastic rod with double-coated Scotch tape. In this technique, we can rotate the sample on the rod to get a flatter surface for the x-ray beam. In the middle of Plate 3 is a holder for this purpose. These are the normal sample holders for our machine. Maximum size of a sample is 6mm x 6 mm x 6mm, using the holder shown on the left side of Plate 3. We can measure a cross-section of a sample using this holder. Measurement of x-ray diffraction of a thin layer or a particle in a cross-section can be carried out by this system, but some limits are present.

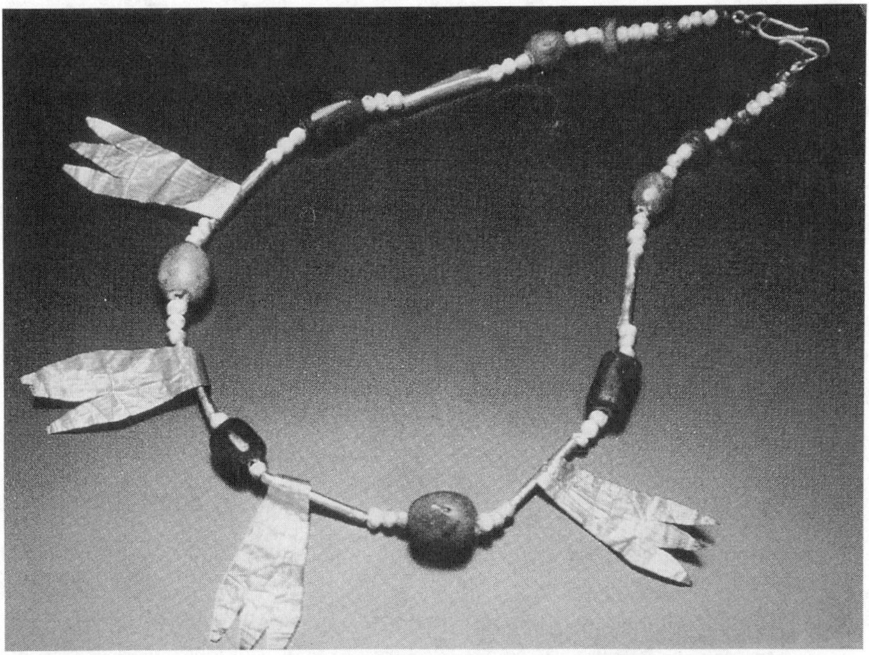

Plate 2. Necklace of the Roman period with pearls.

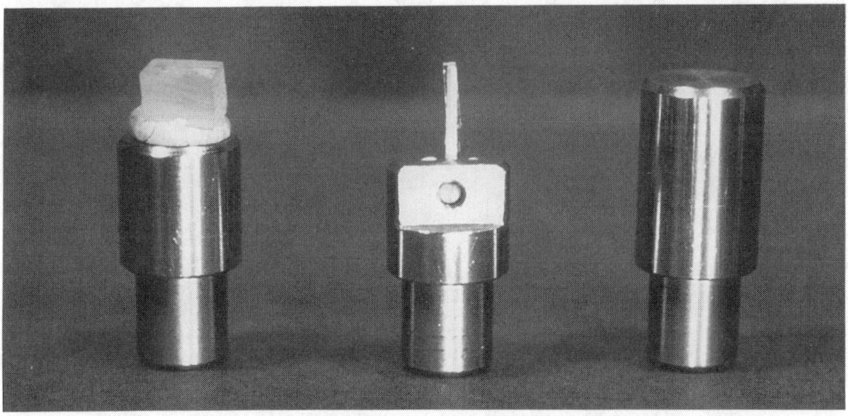

Plate 3. Sample holders.

Measurement of a Cross-Section

Omega axis of the sample holder can rotate 20 to 60 degrees at an angle to the incident x-ray beam (Fig. 4). So, an XRD measurement from the cross section can collect lines in the 2-theta more than 20 degrees. This means that we can collect up to 4.43 angstroms of d-spacing of the sample using a copper K-alpha beam, and cannot collect very specific peaks of some minerals. On the other hand, in the case of a small particle, we obtain lines in lower than 20 degrees of 2-theta region, as a transmittance spectrum. Plate 4 shows the measurement of a cross section of a sample.

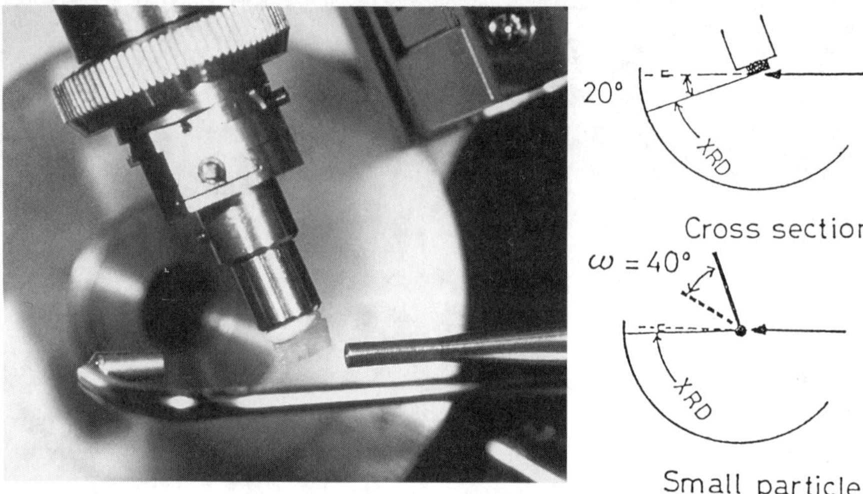

Fig. 4. Measurement angle of PSPC/MDS system.

Plate 5 shows a wall sample from Cave 55 (Dunhuang, China) built during the Song Dynasty (10th-11th C. A.D.). We made a cross-section from this sample(4). The green layer consists of atacamite and the white layer consists mostly of calcite (Plate 6, Fig. 5). Thickness of the green layer was about 30 micrometers.

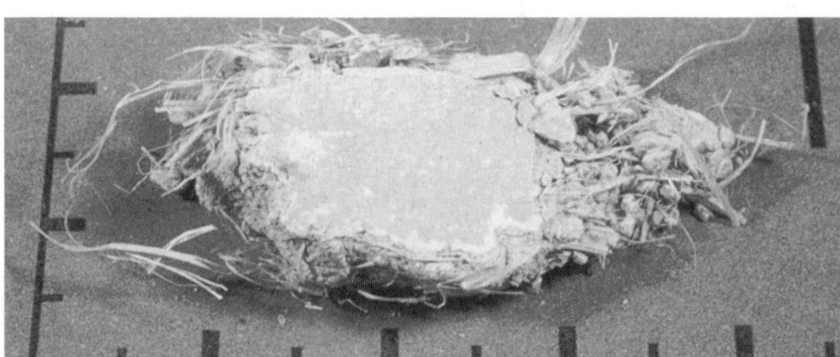

Plate 5. A fragment of wall painting from Dunhuang, Cave 55, 10th-11th C. A.D.

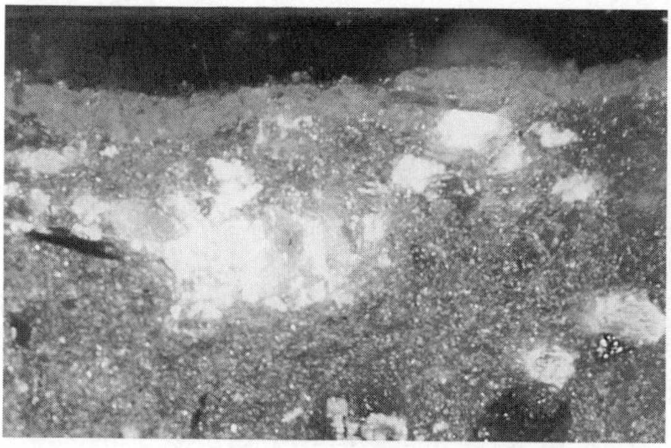

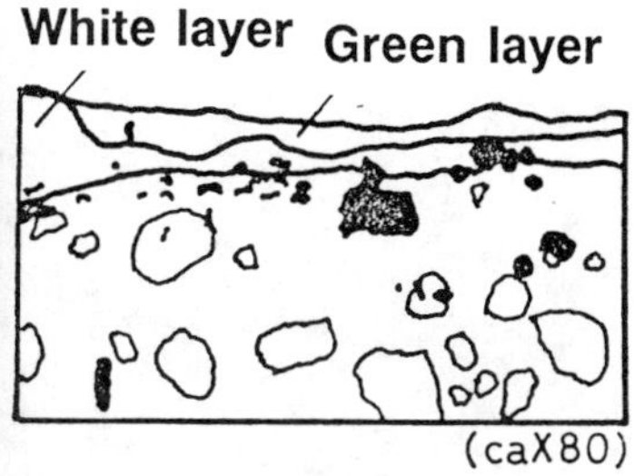

Plate 6. A cross-section showing green surface and white underlayer, and insert, Fig. 5. Diagram of the cross-section.

The lower spectrum is obtained from a green particle and the upper one is obtained from the green layer in the cross-section (Fig. 6). The spectrum of the particle has low angle lines less than 20 degrees but has less numbers of x-ray diffraction lines than the upper spectrum. The reason why the spectrum of the cross-section has so many lines is shown in Fig. 7. This figure shows the results of the second search-and-match iteration. This spectrum contains the lines of quartz and calcite, which is the underlayer of the green surface layer.

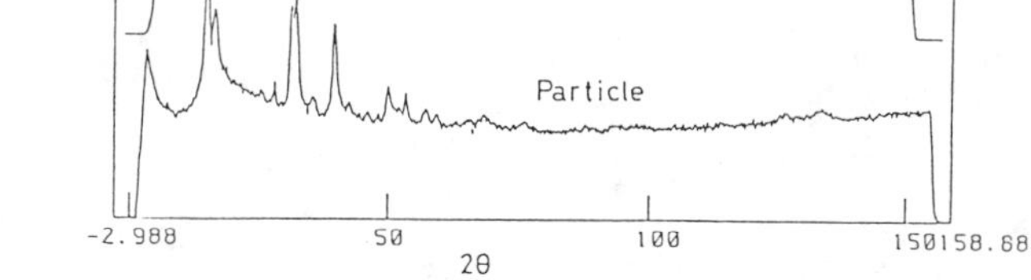

Fig. 6. XRD spectra from a cross-section and a particle of green pigment.

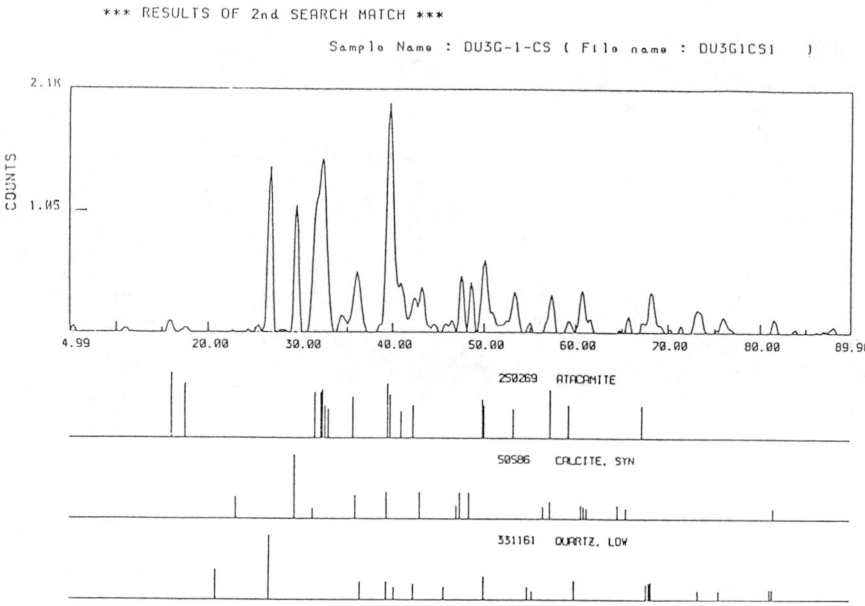

Fig. 7. Results of 2nd search-match of spectrum.

Fig. 8 shows a comparison of the thickness of layers to be analyzed. The pigment in each layer is different, but this figure shows a general tendency of our machine. In this case we used the same collimator of 30 micrometers. We cannot get any information from the lower spectrum, while the upper spectrum has a very large halo. We can, however, process the spectrum easily by background subtraction. This is one of the superior features of using a counter over a camera.

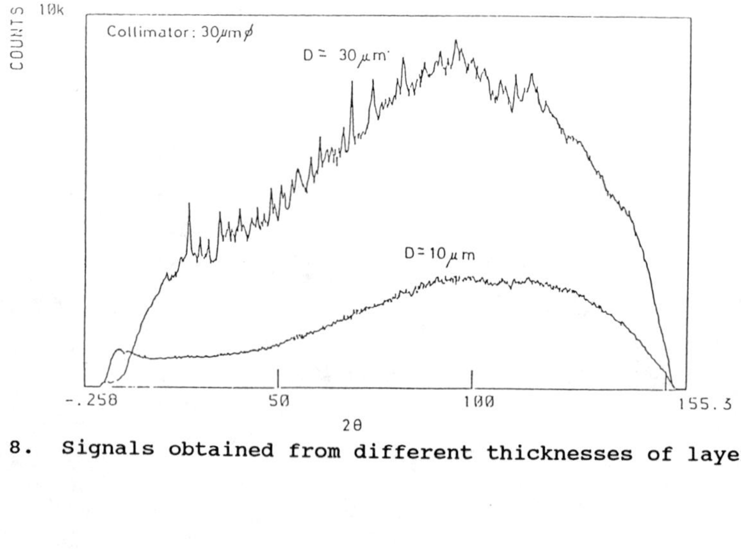

Fig. 8. Signals obtained from different thicknesses of layers.

In our experience in analyzing small particles in matrices or thin layered samples, 20-30 microns in particle diameter or in layer thickness is the smallest to get reasonable signals using a 30-micron collimator.

There are two reasons for this size limit. First, using a filament of normal size (1x10 mm) with point configuration at 12 Kw (60 kV, 200 mA), even the signals of silicon at 10 microns of collimator are very weak. Since the intensity of x-ray diffraction varies in proportion to the area of the x-ray beam, a fine filament is able to increase the intensity of the x-ray beam. However, we have not yet experimented with this parameter. Secondly, to set a sample at the right position at the center of rotation of the three axes within a deviation of less than a few microns is difficult. It also has a tendency to move from the center position mechanically due to temperature change in the chamber of the MDG system. A larger collimator thus is preferable to using a smaller one.

Using the Rigaku Micro diffractometer, accurate setting of the sample stage to less than a few micrometers of deviation from the right position is rather difficult, because each axis of the stage is adjusted by direct mechanical screwing. Future improvements must include a change to allow finer setting of the stage. JEOL's MDG system may be somewhat more like that of a micrometer.

Accumulation time is 1000-2000 sec (40 kV, 150 mA, 100 microns of collimator) for a normal sample and increases to 3000-10,000 sec (55 kV, 200 mA, 30 microns of collimator) for a small sample of 20 to 30 microns. Using a high intensity x-ray tube and a PSPC, acumulation time is shortened compared with a normal x-ray tube and Gandolfi-camera, or even a diffractometer.

Destruction or Non-destruction?

In Japan, sampling a cultural property is often limited, so some researchers have been using the Rigaku non-destructive XRD system(5). However, identification of small particles dispersed in large amounts of other materials is sometimes difficult by this system, since the XRD spectrum of the sample may be hidden by the strong XRD spectrum of the matrix. The XRD spectrum of the sample taken out of the matrix allows a more accurate identification (5,6). One other minor point is that direct irradiation of x-rays to some art objects may cause color changes in pigment, such as lead-containing pigments.

CONCLUSION

There are some small imperfections to the easy operation and productivity of this system, but we feel the PSPC/MDG system is feasible for analyses of very small particles and of particles in matrices.

REFERENCES

1. C. Lahanier, "The Powder Diffraction Pattern Applied to the Knowledge of Works of Art", Chemica Scripta, 26A, 47-55 (1986).

2. M. Inaba, J. Miyata and R. Sugisita, "Micro X-ray Diffractometer (PSPC/MDG) System for the Study of Works of Art", ICOM CC 9th Triennial Meeting Preprints, 1, 35-37 (Dresden, 1990).

3. Y. Matsuda and S. Matsuzuki, "Analysis of a Pearl Used in a Necklace in Roman Period", Preprint of Annual Meeting of Kobunkazai Kagaku Kenkyu-kai, 20 (Tokyo, 1988).

4. S. Duang, Y. Miyata, N. Kumagai and R. Sugisita, "Analysis of Pigments and Plasters from the Wall Painting of Buddhist Temples in North Western China", Kobunkazai no Kagaku, 32 (1987) 13-20.

5. M. Koyano and T. Kadokura, "A Preliminary Report on the Examination of Crystals Found on Oil Paintings in Japan Using Non-destructive X-ray Diffraction", ICOM CC 8th Triennial Meeting Preprints, 1, 59-65 (Sydney, 1987).

6. J. Miyata, "Analysis of the Crystalline-like Substances Observed on the Surface of Oil Paintings", Kobunkazai no Kagaku, 32 (1987) 21-30.

NON-DESTRUCTIVE SPECTROSCOPIC INVESTIGATIONS ON PAINTINGS USING OPTICAL FIBERS

M. Bacci, S. Baronti, A. Casini, F. Lotti, M. Picollo
Istituto Ricerca Onde Elettromagnetiche - CNR, Florence, Italy

and O. Casazza
Galleria degli Uffizi, Florence, Italy

ABSTRACT

The main objective of our research is the development of totally non-destructive methodologies for the investigation of art works. We propose the use of fiber optic reflectance spectroscopy (FORS) in the visible and near-infrared region as a tool for the identification of the pigments used in painting. We have considered several representative inorganic artists' pigments (mainly till the Renaissance period); powder X-ray diffractograms and diffuse reflectance spectra in the visible, near-, middle- and far-infrared range were recorded for pure pigments. Then we prepared suitable samples with the same pigments using fresco, tempera and oil techniques. The visible and near-infrared spectra of these samples were recorded using an optical fiber spectrum analyzer; color analysis (chromaticity, dominant wavelength and purity) was also performed. The obtained results were used in the interpretation of the spectra obtained from some paintings collected in the Uffizi Gallery, Florence (Giotto and Luca Signorelli). The pointwise information obtained by FORS analysis was also used to calibrate and tune an image spectroscopy system, based on sequences of band-pass filtered images in the near infrared range.

1. - INTRODUCTION

The main objective of our research is the development of totally non-destructive methodologies for the investigation of art works, in particular fiber optics reflectance spectroscopy (FORS) and imaging spectroscopy in the visible and near infrared region. Indeed these techniques are the most suitable ones when extended sampling and characterization of the whole work of art without any damaging are required.

Nowadays fiber optics technology is well grounded and low attenuation fibers, which allow the recording of spectra up to more than 2000 nm, are commonly commercialized. We used FORS as a powerful tool for: a) pigment identification; b) documentation of color changes induced in paintings by the environment (light, pollutants and so on) and c) help in calibration and specification of the most appropriate conditions for imaging spectroscopy. Here we deal mainly with aspect (a), giving only a hint of aspect (c), whereas studies concerning (b) have been reported elsewhere [1] and are still carried on in our laboratory.

It is obvious that pigment identification is of extreme importance in art historical studies, because it gives a contribution to the knowledge of painting materials and their diffusion and use by artists and also it can help in revealing possible retouches or even forgeries. Moreover the identification of the original pigment is essential in choosing materials to be used for restoration. As for this last aspect, particular care was devoted to the investigation on materials which are found at present on the market, with the aim both to collect a wide data base and to characterize products, which are traded under the same name, but actually have different composition. This fact must be taken in consideration, when a given material is utilized for restoration, and an appropriate knowledge of its optical properties is lacking. It is well known, for example, that the phenomenon of metamerism [2] may be responsible for color mismatching when pigments of different composition, but similar color, are observed under different light sources: an aesthetical damage could be produced by disregarding the chemical and optical properties of the material as well as the lighting conditions under which the restored painting has to be observed.

In our research we have first searched on the market for pigments used up to the Florentine Renaissance, followed by: i) mineralogical, chemical and spectrophotometric characterization using X-ray diffractometry, visible and infrared spectroscopy and, in some ambiguous cases, X-ray fluorescence techniques; ii) color characterization by means of chromaticity coordinates, dominant wavelength and purity; iii) application to real cases (paintings of Giotto and Luca Signorelli collected in the Uffizi Gallery, Florence) using FORS and imaging spectroscopy.

2. - MATERIALS AND METHODS

Inorganic pigments (see Table I) were purchased from two different firms. Since from our analysis a quality judgement could be drawn, which would bring credit or discredit to the firm, we prefer to label them as F1 and F2. Pigment powders were examined as purchased or were spread on wood panels, using a tempera technique: the panel was prepared with a thick layer of gypsum and animal glue and then the pigment, dispersed in egg white, was applied by means of a brush.

FORS measurements were performed on pure powders, laboratory panels and paintings, using a spectrum analyzer in the range 400-2200 nm (Guided Wave Mod. 260) equipped with 316SS optical fibers. The optical fiber bundle is constituted of 19 parallel fibers (9 sending and 10 receiving). A wand type probe was modified by screwing on the tip of the probe a suitably shaped cylinder made by blackened PVC, supplied with an O-ring. This end has the purpose of excluding external light, maintaining a safe and soft contact with the sample surface and keeping the probe about 3 mm from it. Indeed the beam from the sending fibers impinges perpendicularly upon the surface and, after back-scattering, is collected nearly on axis by the receiving fibers. The distance of 3 mm, due to the thickness of the fibers, allows collection of the maximum back-scattered light.

Reflectance spectra of the powders were also recorded by means of an internal integrating sphere connected to a Perkin-Elmer spectrophotometer (Mod. Lambda 19) in the visible and near infrared region (up to 2500 nm), and a Nicolet (Mod.740) FT-IR spectrophotometer in the medium and far infrared region (up to 50 cm^{-1}).

Chromaticity coordinates, dominant wavelength and color purity were evaluated for a standard illuminant D_{65} according to both the 1931 and 1964 CIE recommendations, making use of a home made program. X-ray diffractograms were recorded with a Philips PW1729

COLOR	N°	PIGMENT	FIRM	MINERALOGICAL COMPOSITION
RED	01	Hematite	F1	Hematite
	02	Morellone	F1	Hematite
	03	Caput Mortuum	F1	Hematite
	04	Vermilion Extrafine	F1	Barite, Cadmium/Mercury Sulfide
	05	Cinnabar	F2	Cinnabar
	06	Red Ochre	F1	Hematite, Goethite, Quartz, Kaolinite, Illite
	07	Indian Red	F2	Hematite
	08	Red Lead	F1	Minium
	09	Sinopia	F1	Hematite, Quartz, Kaolinite
	10	Mars Violet	F1	Hematite
YELLOW	11	Yellow Ochre	F1	Goethite, Kaolinite, Illite, Quartz
	12	Yellow Ochre (Light)	F1	Goethite, Gypsum, Anhydrite, Calcite
	13	Deep Ochre	F1	Goethite, Gypsum, Anhydrite, Calcite
	14	Raw Sienna	F1	Goethite
	15	Raw Umber	F2	Goethite, Quartz
	16	Yellow Ochre	F2	Goethite
GREEN	17	Malachite	F1	Malachite
BLUE	18	Azurite	F1	Azurite
	19	Ultramarine Azure (Deep)	F1	Lazurite plus minor amounts of not determined minerals
WHITE	20	White Lead	F1	Hydrocerussite
	21	White Lead	F2	Hydrocerussite
	22	Spain White	F2	Calcite, Quartz (traces)
	23	Vicenza White Earth	F1	Calcite

TABLE I

Table I - Mineralogical composition of the investigated pigments.

diffractometer and the identification was performed by comparison with the JCPDS powder diffraction file. The X-ray fluorescence analyses were made using a Philips PW1480 spectrometer.

Reflectance spectra can be collected on a limited number of significant points; however it could be useful to extract spectral information on the whole painting, both for complete documentation and research purposes. To this aim, complete *image spectroscopy* has also been started, using a Grundig videocamera (1" IR Vidicon sensor, 400-1800 nm), a set of 10 narrow band interference filters, with bandwidth = 10 nm in the range 750-1200 nm and 5 filters with bandwidth of 90 nm in the range 1300-1800 nm (by Oriel) and a computerized image grabbing and processing system. Image data acquisition is managed by software which controls lighting and A/D conversion parameters and performs offline geometric corrections and radiometric calibrations [3]. The obtained image sequence (one for each filter) represents a 3 dimensional set of data, where the first two dimensions reproduce the whole painting and the third one is the wavelength.

In order to make such kind of measurements reliable, it is necessary to solve a series of problems concerning sensor response, source and lighting conditions, focusing at different wavelengths, low signal to noise ratio due to limitations in either filter or the light source. The obtained spectrum profiles are in good agreement with corresponding data from FORS.

3. - RESULTS AND DISCUSSION

Table I reports the mineralogical composition of the pigments we have investigated. As can be seen, pure and mixed chemicals are sometimes sold under the same trade name: for example cinnabar F2 is pure cinnabar (HgS), while cinnabar F1 (Vermilion extrafine) is cadmium/mercury red [(Cd,Hg)S] mixed with barite. On the other hand, different pigments may mainly consist of the same mineral (like Caput Mortuum, Indian Red F2, Mars Violet, Morellone which contain mostly hematite). The ochres are the most complicated mixtures in this set and may contain variable amounts of quartz, kaolinite, illite, gypsum, anhydrite, calcite and goethite, the latter being the only colored component.

Depending on the electronic structure of the chemicals constituting the pigment, the electronic absorption (and the related reflectance) spectra, which are situated in the visible and/or near infrared region, are characterized by different shapes. In the case of ligand field transitions, that is electronic transitions within a metal ion center surrounded by coordinating ligands (for example Cu^{2+}, which is surrounded by 5 or 6 oxygen atoms in malachite and azurite), the bands are approximately bell-shaped. In insulators or semiconductors (like hematite, cinnabar, minium), instead, the bands are S-shaped, due to inter-band transitions [4]. In the latter case it can be hard to discriminate among different pigments by only looking at the reflectance spectrum. Therefore it is useful in such cases to report the first derivative of the absorption (reflectance) spectrum so that a minimum (maximum) peak of this curve corresponds to the inflection point. Figs. 1 and 2 report the absorption spectra and their first derivative, respectively, for cinnabar F2 and minium: the peaks at 585 nm and 562 nm of the first derivative spectra clearly discriminate the two compounds. It has to be noted that actually we measured reflected light, but in Fig. 1 (and the following ones displaying

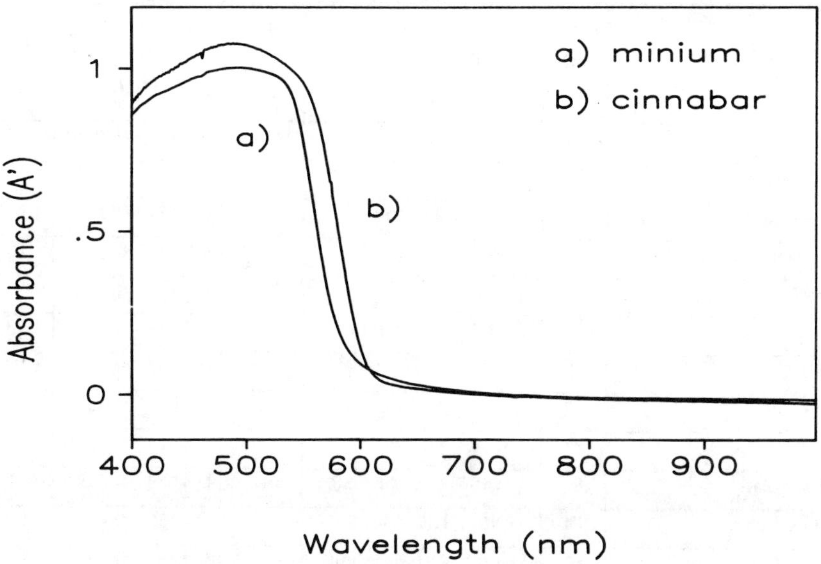

Fig. 1 - Absorption spectra of two commercial pigments: a) minium (F1); b) cinnabar (F2). A' = log(1/R), where R is the reflectance.

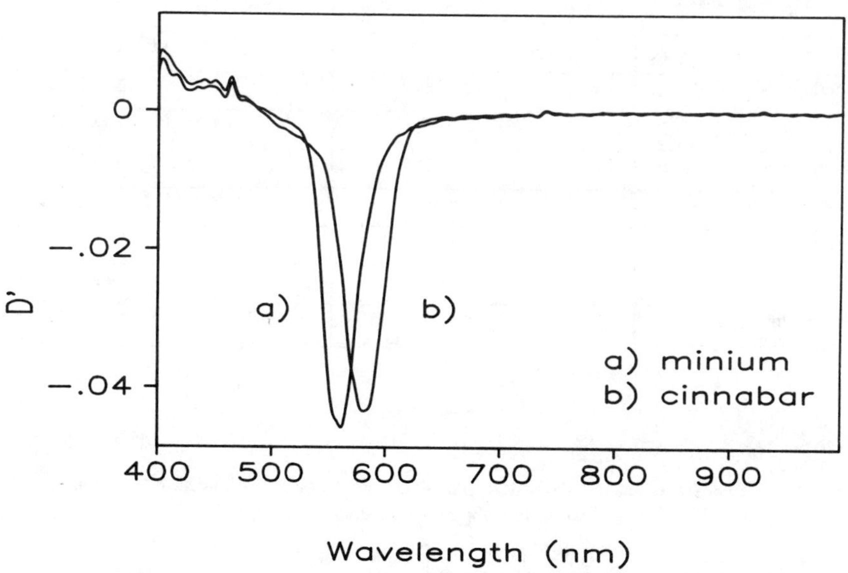

Fig. 2 - First derivative spectra of the same pigments as in Fig. 1.

PIGMENT	FIRM	ABSORPTION PEAK (nm)	1st DERIV. PEAK (nm)	CHROMATICITY COORDINATES (x,y)	λ_D (nm)
Hematite	F1	<532; 672(sh); 866		0.4671 - 0.3532	599
Caput Mortuum	F1	<552; 672(sh); 862		0.3811 - 0.3277	614
Morellone	F1	<554; 676(sh); 856		0.4144 - 0.3219	623
Vermilion extrafine	F1		586	0.5163 - 0.3276	613
Cinnabar	F2		585	0.5325 - 0.3375	607
Red Ochre	F1	<496; 686(sh); 884		0.4533 - 0.3690	593
Indian Red	F2	513; 672(sh); 860		0.4567 - 0.3260	614
Red Lead	F1		562	0.5167 - 0.3617	599
Sinopia	F1	<488; 668(sh); 870		0.4541 - 0.3624	595
Mars Violet	F1	<556; 674(sh); 858		0.4146 - 0.3190	632
Yellow Ochre	F1	<400; 466; 638; 938		0.4250 - 0.4069	582
Yellow Ochre (Light)	F1	<400; 470; 646; 906		0.4286 - 0.4184	580
Deep Ochre	F1	406; 468; 672(sh); 880		0.4288 - 0.3897	585
Raw Sienna	F1	<400; 466; 660(sh); 934		0.4442 - 0.4070	584
Raw Umber	F2	<404; 469; 652(sh); 911		0.4104 - 0.3930	582
Yellow Ochre	F2	413; 644; 647; 936		0.4423 - 0.4280	580
Malachite	F1	<400; 812		0.2931 - 0.3570	510
Azurite	F1	<400; 646; 830(sh)		0.2509 - 0.2764	481
Ultramarine Azure (Deep)	F1	<400; 602		0.2807 - 0.2997	480
White Lead	F1	none		0.3133 - 0.3294	582
White Lead	F2	none		0.3161 - 0.3330	575
Spain White	F2	<400		0.3184 - 0.3355	575
Vicenza White Earth	F1	<400		0.3431 - 0.3567	578

TABLE II

Table II - Absorption or first derivative peaks, chromaticity coordinates and dominant wavelength (λ_D) for the investigated pigments.

reflectance spectra) a fictitious absorbance A' is reported (A' = log(1/R), where R is the reflectance), for analogy of the absorption spectra commonly encountered in electronic spectroscopy.

Table II reports the absorption peaks or the first derivative peaks, the chromaticity coordinates and the dominant wavelength (λ_D) for the investigated pigments. It has to be stressed that the chromaticity coordinates depend on the size of the particles [5, 6], so that the values reported in Table II, which correspond to the powders as purchased without any grinding, have a mere documentation meaning.

The spectra of the pure powders and of the tempera samples do not differ appreciably. However in the near infrared region the tempera samples very often reveal the typical spectrum [7] of the preparatory gesso layer. In fact, infrared radiation can penetrate more deeply than visible radiation, so that information about underlying layers can be obtained (see also below).

Fig. 3 shows the near infrared spectra of the pigment azurite as powder and tempera: the typical three-band absorption of gypsum around 1450-1550 nm is quite evident for the tempera sample. Such an absorption is due to the overtones and combination bands of the OH stretching and bending vibrations [7]. Azurite also shows an absorption at 1500 nm, which is due to the hydroxyl groups present in the crystal lattice, and for this reason the intensity ratio of the three sub-bands observed in pure gypsum appear modified for the gesso (tempera) plus azurite system.

Various other pigments containing water or hydroxyl groups show absorption bands in the infrared region, but with different shapes and peak positions, which can be useful for identification [7-10]. The other bands reported in the same figure for gypsum around 1730-1750 nm and near 1900 nm [7] are also evident, but they can be encountered in other minerals as well.

The appearance of the gypsum spectrum in the near infrared has to be taken into consideration when FORS is performed on paintings with the aim of pigment identification. Poorly hiding pigments or thin layers can indeed reveal the preparatory stratum containing gypsum.

The paintings of the Uffizi Gallery that are under investigation are the "Madonna di Ognissanti" by Giotto (Maestà) and the "Holy Trinity predella" by Luca Signorelli. The latter painting has not been recently restored, while restoration on the Giotto Maestà was terminated during 1991. The studies reported here were performed just before and during the restoration.

In Fig. 5 the spectra of two different points of the Signorelli predella are displayed: the presence of gypsum is only evident in one of them, althrough in both cases visible reflectance spectra indicate the presence of the cinnabar. Accordingly, the pigment layer in one case should be thicker or be mixed with another highly hiding pigment, such as lamp or charcoal black.

As an example of pigment identification we report the spectrum of azurite (a main

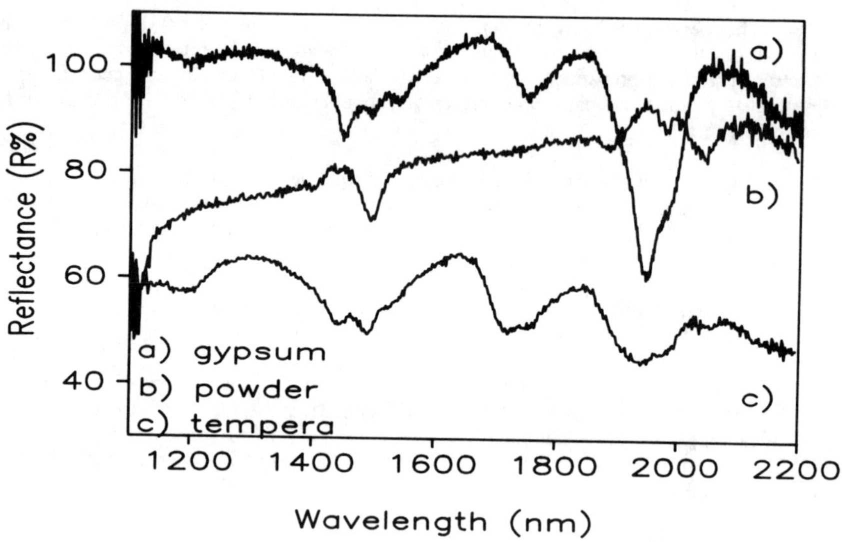

Fig. 3 - Near infrared spectra of the pigment azurite (F1): b) as powder; c) as tempera. The spectrum of gypsum (a) is also reported for comparative purpose.

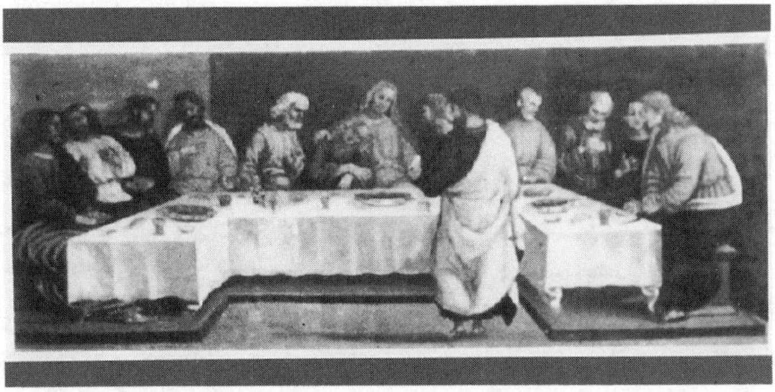

a)

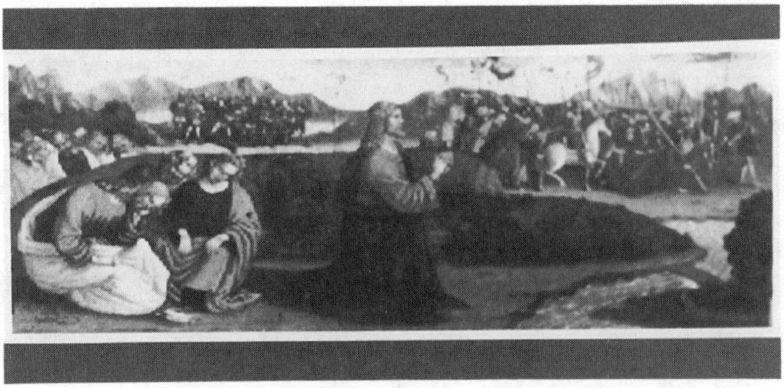

b)

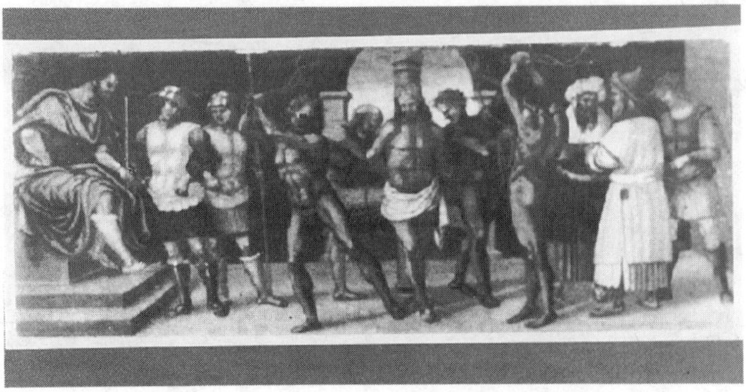

c)

Fig. 4 - The three scenes of the Holy Trinity predella by Luca Signorelli. (a) The Last Supper; (b) Jesus in the Garden; (c) Flagellation.

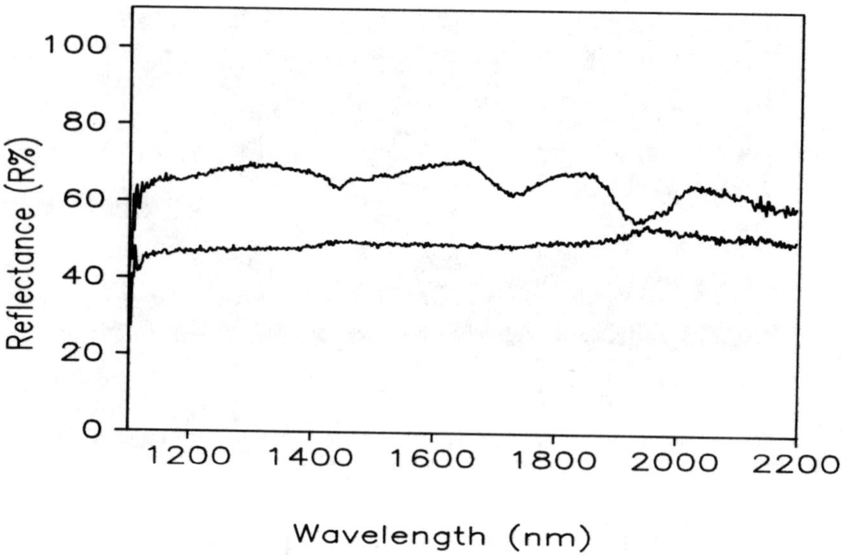

Fig. 5 - Near infrared spectra of two different red areas of the predella by Luca Signorelli containing the same pigment. In the upper spectrum the underlying gesso is evident.

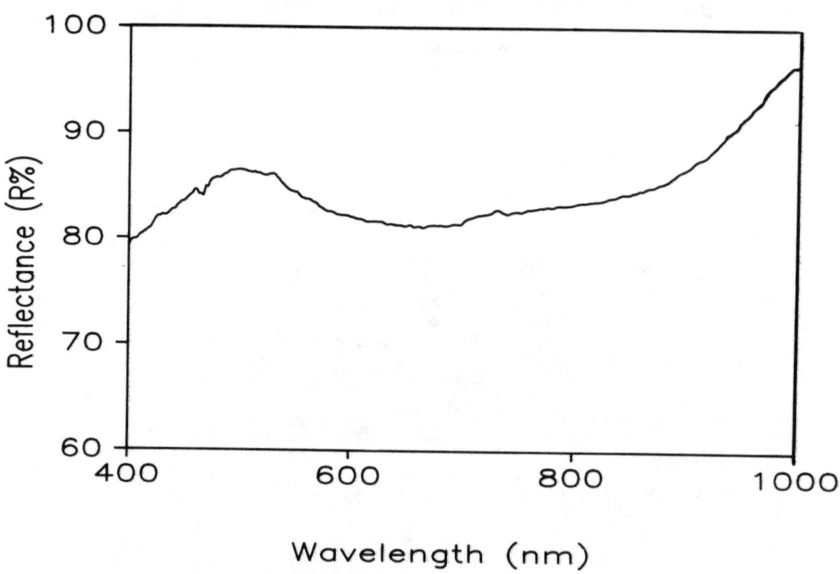

Fig. 6 - Reflectance spectrum of the light blue sky in the second scene of the predella.

absorption band with a shoulder in the red region) found in the light blue sky of the predella (Fig. 6). A quite similar spectrum is observed in Fig. 8 a) for the mantle of the Giotto Madonna (before restoration), even if the main reflectance peak is slightly red shifted with respect to the pure pigment. Such a shift could be due to some "dirt", which was present to a certain extent on the surface [1]. In any case FORS excludes with certainty the presence of lapis lazuli, the precious pigment often used for distinguished persons like the Virgin, which has a markedly different spectrum (Fig. 8 b). The presence of cinnabar is clearly evidenced by the first derivative spectrum in the red mantle of an angel in the Giotto Maestà (Fig. 9) as well as in some red garments of the Signorelli predella.

As an example of spectral information derived by image spectroscopy, Fig. 10 a) and b) show the plots of 15 reflectance values in the region 750-1700 nm, belonging respectively to two points of the first scene of the predella marked with a circle (red tunic of J. Christ) and an arrow (green mantle). There is qualitative agreement with the spectra measured with FORS, (see for example Fig. 10 a) and the upper curve of Fig. 5). A larger number of filters with narrower bandwidth in the lower frequency range would result in more precise spectrum profiles. Two images, among the acquired sequence of 15, are reported in Fig. 11 a) and b), taken at 800 and 1300 nm respectively. Differences in the intensities of the green parts are evident (see also Jude's tunic on the right), together with a remarkable modification of the second face from the right; in Fig 11 b) a dark area on the neck of the same person clearly reveals a successive retouch.

Finally, we would like to hint at the help that FORS can give, when reference data are fused with those coming from image spectroscopy, both to calibrate and make more objective the image data. Moreover, by looking at the position of the reflectance peaks and at the whole shape of the spectrum, it is possible to select appropriate narrow band images for a better evidence and a clearer definition of particular pigments or possible underdrawings.

In Fig. 12 a detail of the third scene (Flagellation) of the Signorelli predella is shown, as it appears in the visible (a) and in the whole near infrared region (b). Fig. 13 represents the same detail in the narrow band centered at 1000 nm. The distribution of lights and shadows is sharper, and the preparatory underdraw (see the perspective lines of the steps) is much more evident, as well as the retouches, in comparison with the broad band in Fig. 12b.

4. - CONCLUSIONS

Experimental results showed that fiber optic reflectance spectroscopy is a powerful and non invasive method to collect spectral information from paintings. When a complete reference data base is available, this technique can help in reliable pigment identification. Other remote sensing techniques, like near infrared image spectroscopy, are being developed and the related problems of reliability have been faced and partially solved. Preliminary results show that the two techniques can be integrated to give a help into the knowledge and documentation of works of art.

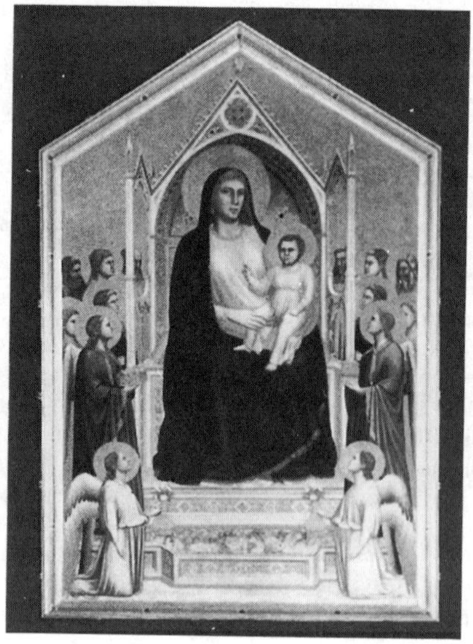

a)

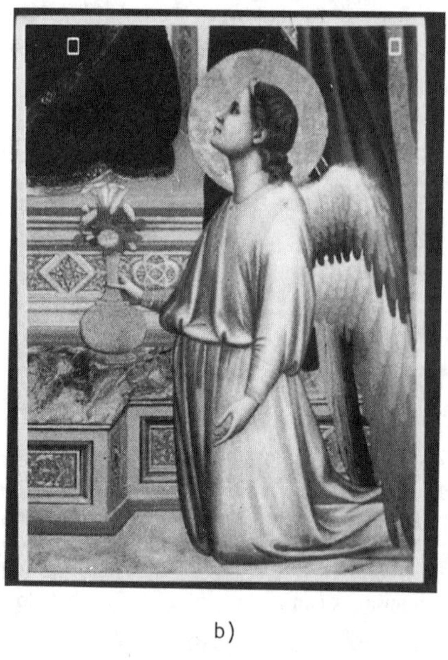

b)

Fig. 7 - (a) Madonna di Ognissanti (Maestà), by Giotto; (b) detail.

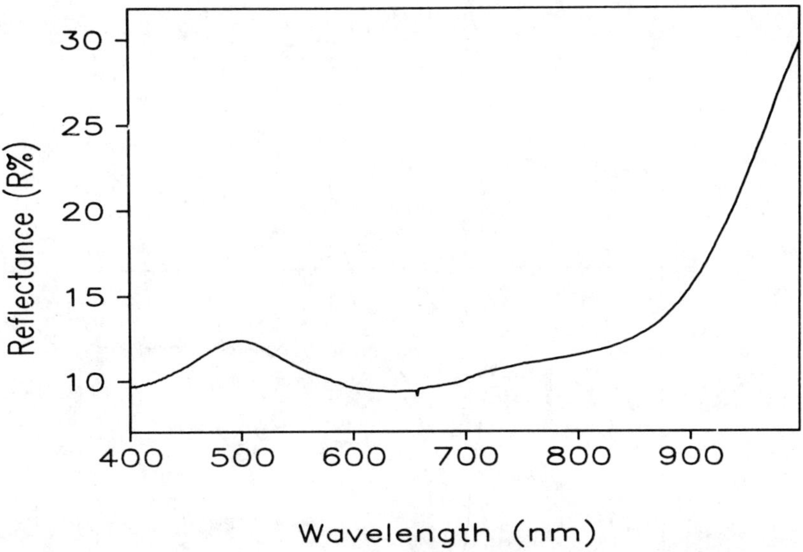

a)

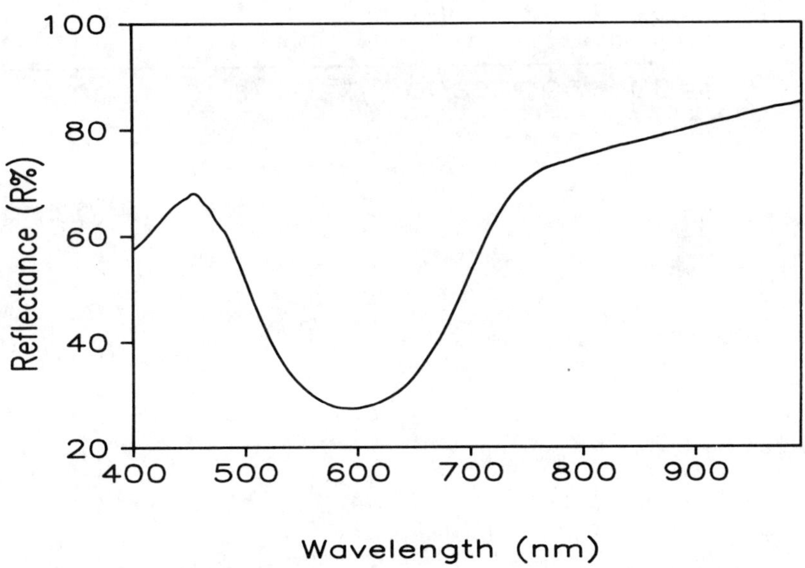

b)

Fig. 8 - (a) Reflectance spectrum of Virgin's mantle in the Giotto Maestà. See Fig. 7 b), top left square. (b) Reflectance spectrum of lapis lazuli (F1).

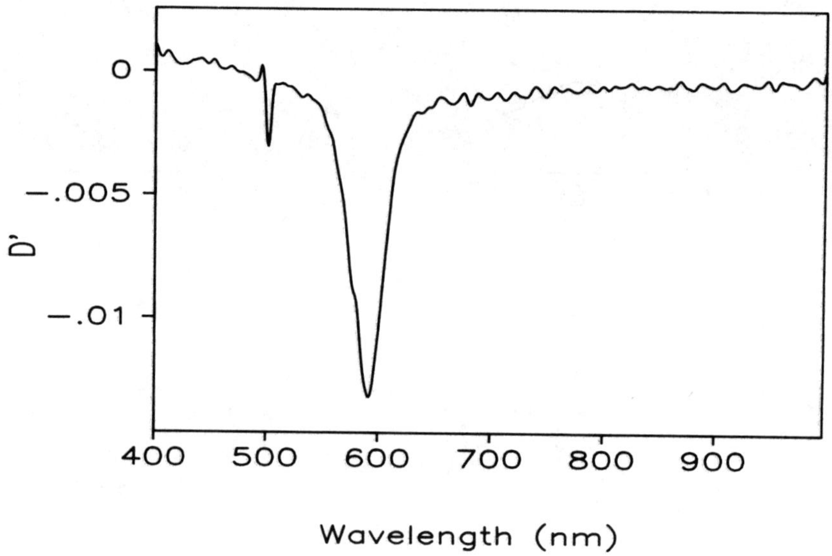

Fig. 9 - First derivative spectrum of the red mantle of an angel in the Giotto Maestà. See Fig. 7 b) top right square.

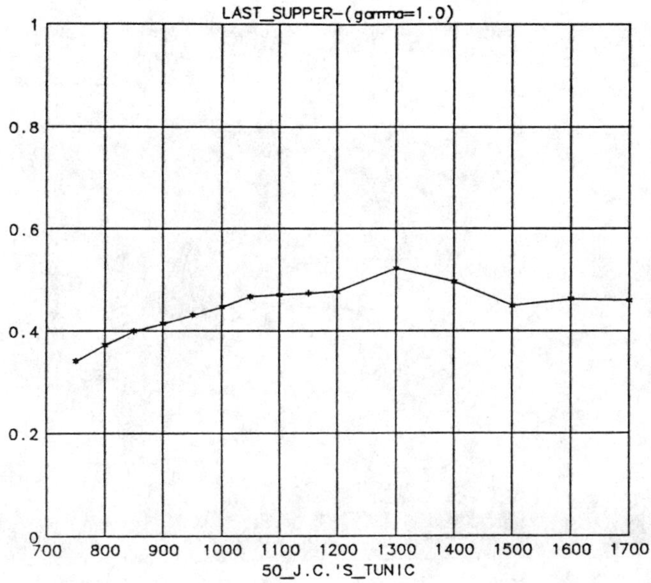

a)

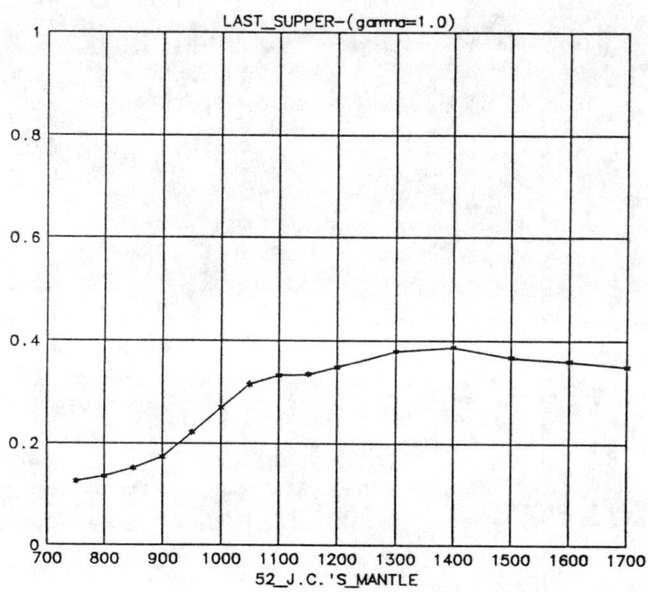

b)

Fig. 10 - Image spectroscopy. Spectrum profiles of two points in the first scene (Last Supper) of the Signorelli predella. (a) red tunic of J. Christ, (b) green mantle of J. C..

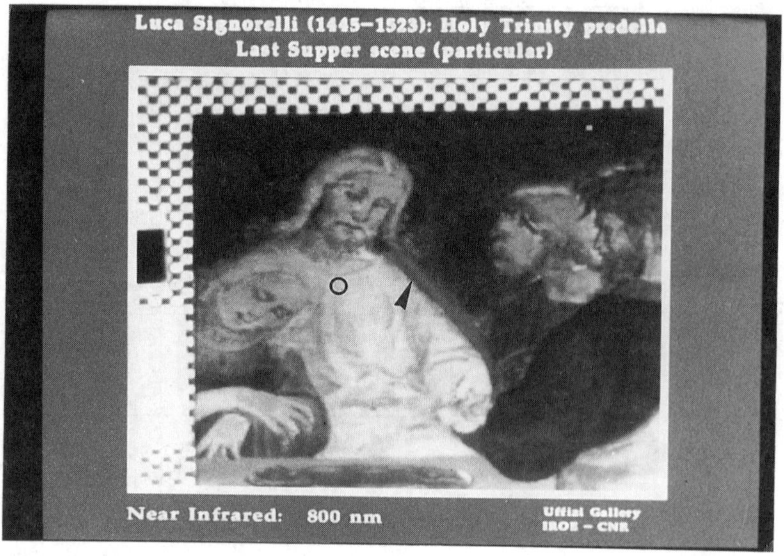

a)

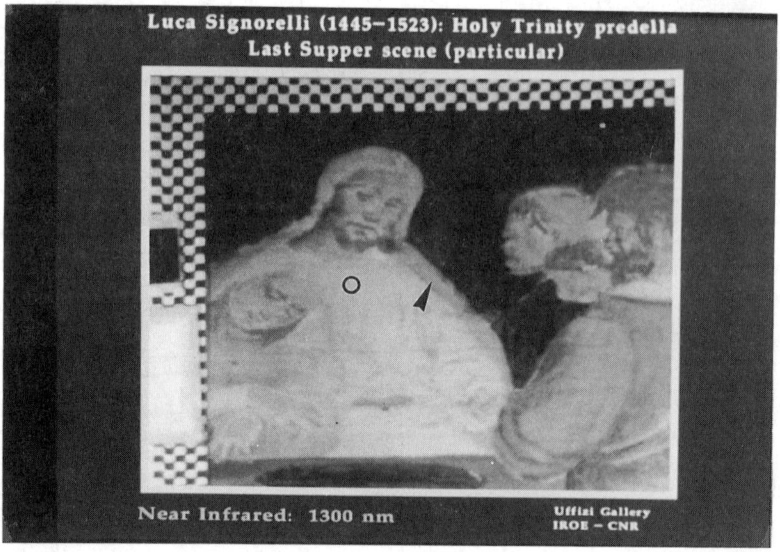

b)

Fig. 11 - Spectral images of a detail of the Last Supper scene in the Signorelli predella, at 800 (a) and 1300 nm (b).

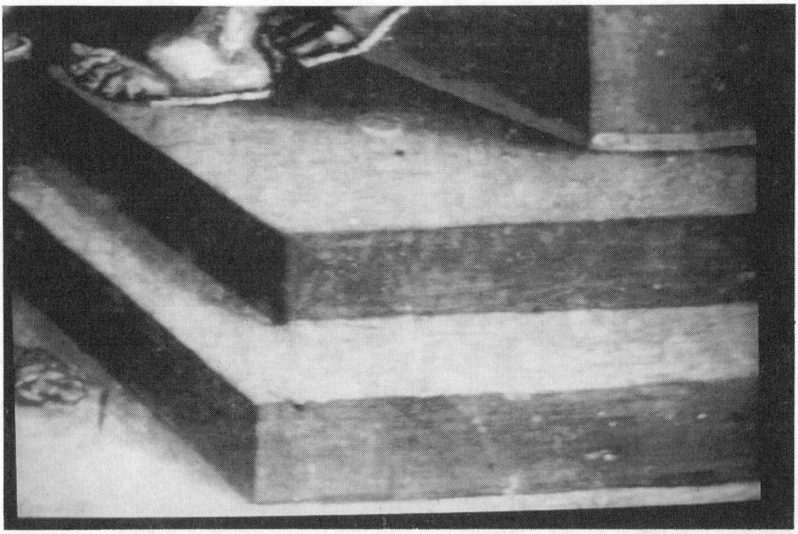

a)

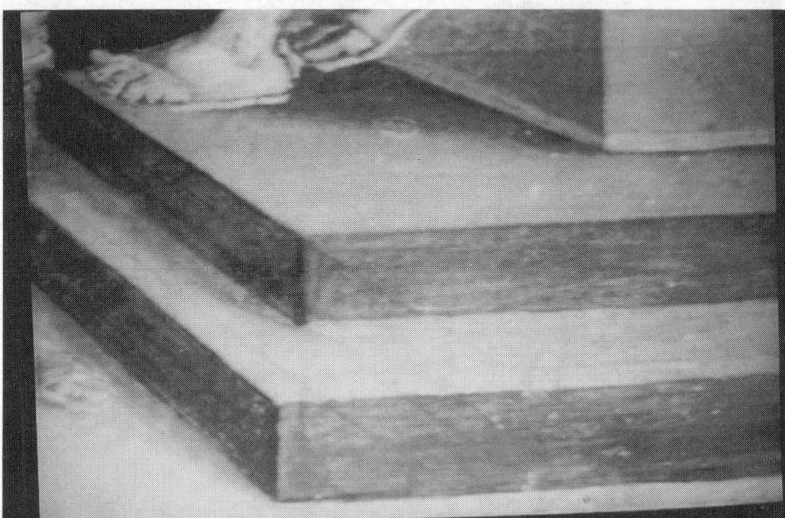

b)

Fig. 12 - Detail of the third scene (Flagellation) of the Signorelli predella, in the whole visible (a) and near infrared (b) region.

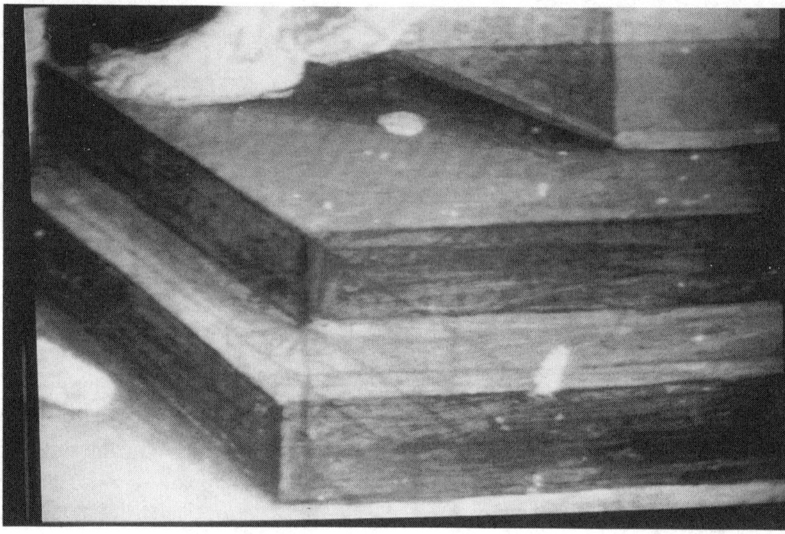

Fig. 13 - The same detail of Fig. 12 viewed in the band centered at 1000 nm.

ACKNOWLEDGEMENTS

The present investigation has been partially supported by the "Progetto Strategico Uffizi" of the Italian CNR. A particular acknowledgement to the Director of the Uffizi Gallery, Prof. A.M. Petrioli, who helped us in our work, with her authority and availability. Thanks are expressed also to the "Centro per la Conservazione delle Opere d'Arte", CNR, Florence and to the Department of "Scienze della Terra", University of Florence, for the help in recording X-ray diffractograms and in the X-ray fluorescence analysis.

REFERENCES

1. M. Bacci, F. Baldini, R. Carlà and R. Linari, Appl. Spectrosc., 45, 26 (1991).

2. S. Staniforth, Studies in Conservation, 30, 101 (1985).

3. M. Bacci, S. Baronti, A. Casini, F. Lotti and M. Picollo: Near Infrared Image Spectroscopy of Paintings, Proc. of the International Workshop on "Advanced Infrared Technology amd Applications", Florence, April 13-14, 1992 (in press).

4. K. Nassau, The Physics and Chemistry of Color (Wiley Interscience, New York,1983).

5. D. B. Judd and G. Wyszecki, Color in Business, Science and Industry (Wiley, New York, 1975).

6. G. Kortüm, Reflectance Spectroscopy (Springer-Verlag, Berlin, 1969).

7. G. R. Hunt, J. W. Salisbury and C. J. Lenhoff, Modern Geology, 3, 1 (1971).

8. G. R. Hunt, Geophysics, 44, 1974 (1979).

9. G. R. Hunt and J. W. Salisbury, Modern Geology, 1, 283 (1970); 2, 23 (1971).

10. J. K. Crowley and N. Vergo, Clays & Clay Minerals, 36, 310 (1988).

APPLICATION OF SURFACE AND BULK ANALYTICAL TECHNIQUES FOR THE STUDY OF IRON METALLURGY SLAGS AT TELL AFIS (N-W SYRIA)

G.M.INGO*@, L.SCOPPIO#, S.MAZZONIØ, G.MATTOGNO@ and A.SCANDURRA$

@) CNR-ITSE, C.P. 10, 00016 Monterotondo Stazione Roma (Italy)
#) Centro Sviluppo Materiali, CP 10747 Roma Eur Roma (Italy)
Ø) Università di Pisa, via Galvani 1, 56100 Pisa (Italy)
$) Consorzio Catania Ricerche, Dipartimento di Chimica della Università di Catania, via A. Doria 6, 95125 Catania (Italy)

ABSTRACT

By a combined use of small area X-ray photoelectron spectroscopy (SA-XPS),X-ray induced Auger electron spectroscopy (XAES), scanning electron microscopy (SEM) and electron probe for microanalysis (EPMA), the microchemistry of a group of slags from Tell Afis (North-Western Syria), dating back to Iron Age II (750 B.C.), have investigated. The results show and localize the presence of different iron oxides and hydroxides as well of various glassy and crystalline silicates, whose chemical structure and morphology have determined.

INTRODUCTION

A group iron-metallurgy slags, dating back to Iron Age II (Aramean period, ~ 750 B.C.), have been found at Tell Afis, N-W Syria). In order to investigate the microchemistry of the slags and to elucidate some aspects of the iron smelting process, surface and bulk characterizations have been carried out by means of the combined use of SA-XPS-XAES, SEM and EPMA. It is worth noting that SEM and EPMA are currently used in archaeometry for studying metallurgical slags. However, these latter bulk analytical techniques do not provide detailed information about the valence states and chemical bondings, that can obtained via XPS and XAES. Indeed, by XPS and XAES, chemical bonding fine structures are much easier to disentangle and in particular, by using small-area XPS-XAES,it is possible to extend the range of the analysis area downwards into the range 0.1-0.2 mm. Therefore, these surface techniques are particularly suitable to study the microchemistry of the different small phases of the slags.

EXPERIMENTAL

The SA-XPS study was carried out on an Escalab MkII spectrometer using Al Kα and Mg Kα radiations as excitation sources (hν= 1486.6 eV and hν= 1253.6 eV, respectively). The X-ray induced Auger transitions Si KLL and Al KLL were excited by the Bremstrahlung radiation. Due to the contamination of external surfaces of the slags, SA-XPS-XAES, SEM and EPMA analyses were carried out on the inner zones of the slags sections that were unaffected by soil contamination.

RESULTS AND DISCUSSION

The slags were found at Tell Afis during a stratigraphical excavation, on the floor of a building of Iron Age II (Aramean period, 750 B.C.). The slags were first crushed on-site: they show on the fracture surface an inhomogeneous phase distribution with zones characterized by a dark grey-brown metallic luster and zones with lighter colored phases. Chemical analyses, reported in Table I, suggest that these slags can be associated with an iron ore smelting process, the composition of the slags being similar to modern iron slag and to early iron slags from Palestine, Britain, Germany and Anatolia [1-5]. Furthermore, chemical analyses suggest to rule out a copper, tin or bronze smeltery origin, as copper and tin, are absent. It is worth noting that the two chemical analyses of Tab. I represent the range of variation of the FeO_x and SiO_2 amounts as measured in five different slag fragments.

SiO_2	Al_2O_3	FeO_x^{tot}	CaO	MgO	Na_2O	K_2O	TiO_2	P_2O_5	CuO	SnO_2	NiO_x
28.8	4.8	43.3	13	4.6	0.9	1	0.8	2.8	-	-	-
25.9	8.6	39.9	10	5.8	3.4	4.6	-	1.8	-	-	-

Table I Chemical analyses of slags from Tell Afis.

SEM+EDS and EPMA results help us to better investigate the microchemistry and phase morphology of the slags. In Figs 1-2, back-scattered electrons SEM images, EDS analyses of a section of the slag, are shown.

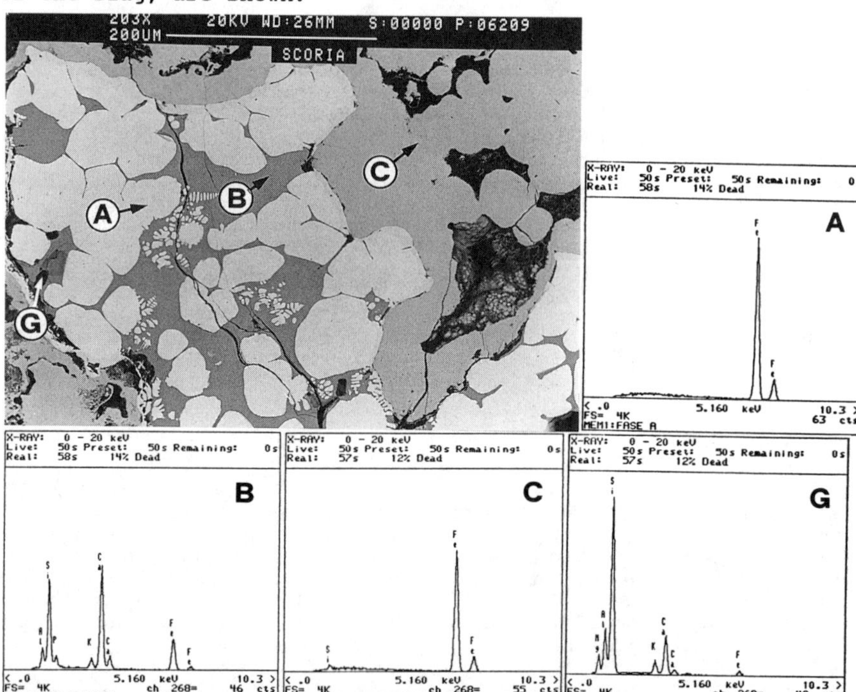

Figure 1. Back-scattered electron SEM image of a cross section of a slag showing : grains and dendrites of magnetite (A), an intergranular Ca-Fe-Si phase (B), a Ca-Al-Mg-Si phase (G) and lepidocrocite (γ-FeOOH,C). This latter is a variety of goethite

Table II reports the chemical analyses of the different phases obtained via EPMA. These latter analyses have been carried out on the same zones sampled by SEM+EDS; therefore, the slag phases have been labeled in the same way.

Phase	SiO_2	TiO_2	Al_2O_3	FeO_{tot}	MgO	CaO	K_2O	NiO_{tot}	Na_2O	CoO	S	P_2O_5
A	--	--	--	100	--	--	--	--	--	--	--	--
B	50.2	--	4.3	9.5	1.3	28.5	3.6	--	--	--	--	2.8
C	--	--	--	100	--	--	--	--	--	--	--	--
D	66.4	--	1.2	7.3	19.6	4.9	0.4	--	0.1	--	--	0.1
E	11.5	1.2	9.4	69.1	2.8	4.0	--	--	0.8	--	0.4	0.8
F	55.5	0.2	6.1	1.2	27.3	8.5	0.3	--	0.5	0.3	--	0.2
G	61.9	--	8.7	6.5	9.5	7.3	5.2	--	0.9	--	--	--

Table II. Quantitative EPMA results for different slag phases

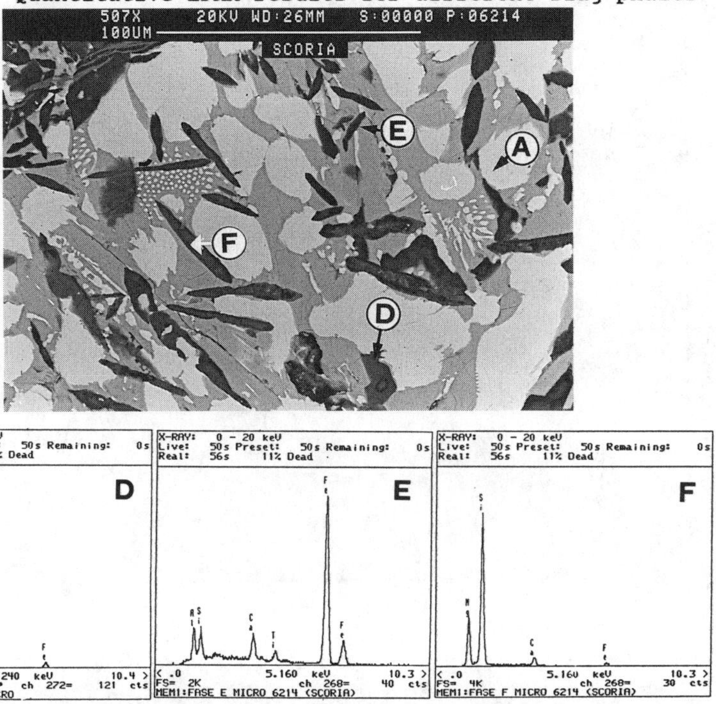

Figure 2. SEM micrograph of a section of the slag that shows the different chemical nature of the phases. EDS analyses reveal the presence of: magnetite (A); Al_2O_3-CaO-TiO_2 silicate (E) and CaO-MgO silicate (dark needle-shaped crystals, F and D).

SEM micrographs reveal the complex structure and the local chemistry of the slags. Indeed, Fig. 1 shows the coexistence of grains and dendritic particles of iron oxide (white zones) surrounded by a glassy intergranular calcium-silicon compound (dark zones, G). We propose the assignment of the iron oxide to magnetite, since this phase is inert when etched by a $SnCl_2$ solution. SEM and EPMA results show the presence also of lepidocrocite (γ-FeOOH, grey zones, C) and a Ca-Fe-Si phase (dark zones,B) that EPMA and EDS analyses indicate to be likely iron-akermanite, i.e. $CaFeSi_2O_7$. It is worth noting that in other zones of the slag, the presence of wustite (FeO_{1-x}), has been also disclosed by etching the slag surface with a $SnCl_2$ solution. Fig. 2 gives another, typical example of phase distribution, as shown by the SEM image and EDS analyses. These results clearly show that the chemical composition of the phases is very variable. Needle-shaped CaO-MgO-silicate crystals (F), CaO-MgO-Al_2O_3 silicate particles (D), CaO-TiO_2-Al_2O_3-FeO silicate (grey zones,E) and globular and dendritic magnetite (A), are present.

In order to gain a further insight in the microchemical composition of the slags, small-area XPS and XAES analyses were performed by selecting the same zones observed with SEM and EPMA. In Fig. 3, typical Fe 2p and O 1s XPS spectra, for different zones of the slag, are reported.

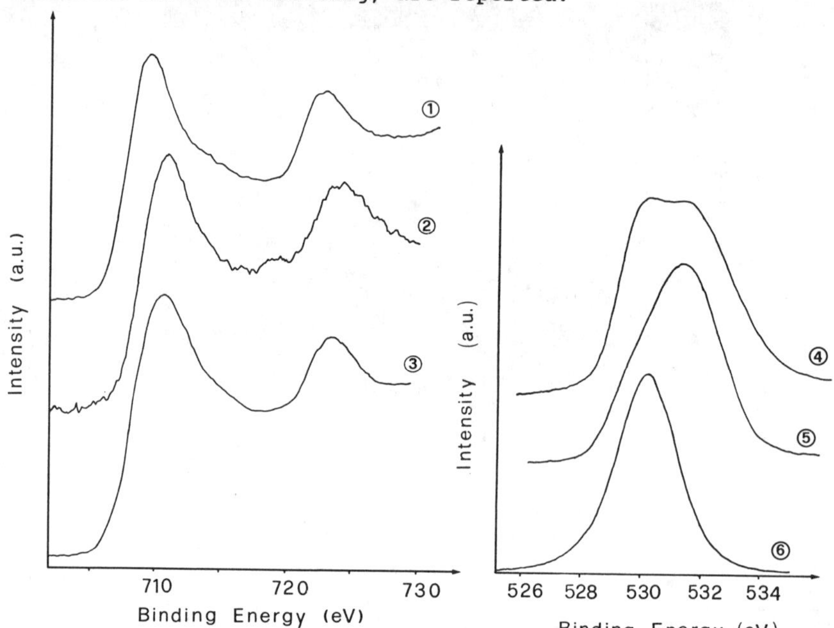

Figure 3. SA-XPS spectra in the Fe 2p and O 1s energy regions.

These spectra are assigned to Fe (II) oxide, as in wustite, i.e. FeO_{1-x}, (spectrum 1 of Fig. 3), to a hydrous Fe (III) oxide, as in goethite or lepidocrocite (spectrum 2 of Fig. 3 and phase C of Fig. 1) and to Fe_3O_4, i.e. magnetite (spectrum 3 of Fig. 3 and phase A in Fig. 1). Other interesting features are given by O 1s spectra that show dramatic differences as a function of the sampled area. In effect, the O 1s spectra 4, 5 and 6, are characterized by a twofold line shape, consisting of a component centered at 530.2 eV, typical of oxygen bonded to iron, along with a component at about 532.0 eV, typical of silicates. Also we notice that the relative intensity of the low-binding energy (BE) component varies as the sampled area, which suggests that the distribution of iron oxides and silicates particles is not uniform. Turning to the

Si 2p and Al 2p lines, not shown, we found good agreement with the literature data, given for silicates and silico-aluminates [6]. Furthermore, from other not shown XPS results, we can assign the calcium and magnesium chemical bondings, to a calcium and to a magnesium silicate. Indeed, the binding energy of Ca $2p_{3/2}$ and the α' Auger parameter are 347.0 eV and 638.5 eV, respectively, as in wollastonite [6] (i.e. calcium silicate) and BE of Mg 2p and α' value are 50.5 eV and 1230.8 eV, respectively, as in talc [6] or enstatite (i.e. magnesium silicate). It is worth noting that the α' parameter is obtained by means of the combined use of both photoelectron and X-ray induced or Bremsstralung Auger line according to:

α' = KE of Auger peak - KE of photoelectron peak + $h\nu$

So conceived α' has the advantages of being independent of photon energy, of always having positive values and is independent of the charge referencing operation. Therefore, α' is found with higher accuracy than the determination of either line energy alone. In order to confirm these assignments and to more precisely identify the different chemical structure of silicon and aluminum compounds in the slag, we have used, the two dimensional chemical state plots proposed by Wagner [6]. In these plots, which give very detailed chemical state information, are reported the kinetic energy of the sharpest Auger peak which for silicon and aluminum are the KLL peaks, the binding energy of the most intense photoelectron line and the modified Auger parameter α'. In Fig 4 are reported the two chemical-state plots for silicon and aluminum, with the data obtained from the different phases of the slag as well as from various compounds presumed similar to the latter.

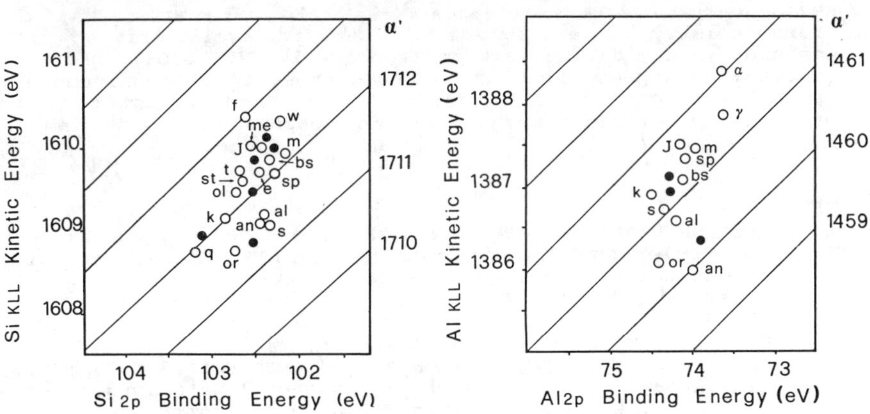

Figure 4. Chemical state plots for silicon and aluminum.
sp = spodumene $LiAl(SiO_3)_2$ (●) slag phases
s = stilbite $(Ca,Na_2)Al_2Si_6O_{16}\ 6H_2O$ or = ortoclasio $KAlSi_3O_8$
an = analcime $NaAl(SiO_3)_2$ k = kaolinite $H_4Al_2Si_2O_9$
w = wollastonite $CaSiO_3$ α= α-Al_2O_3 ol = olivine $(FeMg)_2SiO_4$
m = muscovite mica $H_2KAl_3(SiO_4)_3$ j = jadeite $NaAl(SiO_3)_2$
t = talc $H_2Mg_3Si_4O_{12}$ q = quartz al = albite $NaAlSi_3O_8$
e = enstatite $Mg_2(SiO_6)$ st = staurolite $(FeMg_2Al_9O_6[SiO_4](O)_2$
me= melilite $Na_2(Ca,Mg)_{11}(Al,Fe)_4(SiO_4)_9$ fa = fayalite $FeSiO_4$
γ= γ-Al_2O_3 bs=grinded modern blast furnace glassy slag

In these plots, the Auger parameter can be read from a series of diagonal lines drawn on the plot with a positive slope of 1. From the data of the silicon plot, we deduce that the silicon phases of the slag are silicon and silicon-aluminum simple compounds, such as quartz and wollastonite, as well as silicate phases with added metal ions whose α' parameters are slightly higher in kinetic energy. The presence in the slags of quartz is likely due to enclosing fragments of undissolved silica. Furthermore, we note that α' values are quite different from pure synthetic fayalite ($FeSiO_4$) and are similar to those of $MgSiO_4$ or $MgSiO_6$ compounds as well as to melilite (i.e. $Na_2(Ca,Mg)_{11}(Al,Fe)_4(SiO_4)_9$) and further, are similar to those of a blast furnace modern slag. Auger parameters for aluminum, indicate that the phases of the slags are alumino-silicates [6], where aluminum occupies either octahedral + tetrahedral sites (α' is ranging between 1461.3 eV and 1461.5 eV) or only tetrahedral sites (α' is 1460.3 eV).

On the base of these results as well as of the results of other authors [1-5], it appears likely that the minimum liquidus temperature of the worthless slag components is ranging between 1210 and 1250°C, that are the temperatures supposed to be attainable in an ancient iron furnace. Furthermore, we estimate that the viscosity at 1250°C, was ranging between 4 and 10 poise,therefore able to satisfy the metallurgical requirements of an early iron furnace.

CONCLUSIONS

By a combined use of surface and bulk analytical techniques such as small-area XPS, XAES, SEM and EPMA, the microchemistry of a group of slags from Tell Afis (North-Western Syria), dating back to Iron Age II (750 B.C.), has been studied. It appears that using these analytical techniques, it is possible to identify many of the chemical states and bondings of aluminum-oxygen-silicon compounds present in the slags.

REFERENCES

1. H.Liebowitz and R.Folk, J.Field Archaeology,11, 265 (1984)
2. T.S.Wheeler and R.Maddin, chap. 3 p. 99, "Metallurgy and Ancient Man", in "The Coming of the Age of Iron" ed. T.A. Wertime and J.Muhly,New Haven London Yale University 1980.
3. T.S. Wheeler, J.B. Muhly, K.R. Maxwell-Hyslop and R. Maddin, American Journal of Archaeology, 85, 245 (1981).
4. A.R.Williams and K.Maxwell-Hyslop,J.Arch.Sci.,3, 283 (1976).
5. Slag Atlas, by Committee for Fundamental Metallurgy, chap. 3 p. 103, Ed.Verlag Stahleisen M.B.H. Dusseldorf 1980.
6. C.D. Wagner, Appendix 4 p.477-509 "Auger and Photoelectron Energies and the Auger Parameters: a Data Set" in "Practical Surface Analysis by Auger and X-ray Photoelectron Spectroscopy" Ed. by D. Briggs and M.P. Seah, 1983, I.Wiley & Sons, Ltd.
7. D. Killick, N.J. van der Merwe, R.B. Gordon and D. Grebenart, J. Arch.Sci., 15, 367 (1988).
8. C.Forrieres, J-P.Petit and J. Schaub, Notes et Monographies Techniques n° 22. Ed. by C.N.R.S. Paris 1987, pages 1-125.

INVESTIGATION OF DECORATIVE COATING ON KONGO CERAMICS USING U. V. LIGHT AND FTIR ANALYSIS

MADELEINE R. HEXTER* AND WALTER R. HOPWOOD**
*Smithsonian Institution, National Museum of African Art, Washington, DC 20560
**Smithsonian Institution, Conservation Analytical Laboratory, Washington, DC 20560

ABSTRACT

The ethnographic and art historical literature reports the use of vegetable resins on Central African ceramics, specifically those made by the Kongo peoples. The purpose of this investigation was to develop a procedure for determining the nature of decorative coatings generally on African ceramics. The investigation focused on a group of Kongo ceramics. FTIR spectra of known resins were obtained and compared with several samples taken from Kongo ceramics. The ceramics were also examined under U.V. light. This paper discusses the results of the examinations and analysis.

Introduction

Although African ceramics have been studied for their form and function, their mode of manufacture, especially their decorative coatings, is less well documented. We do know that traditional African ceramics are handbuilt by women. These ceramics are fired at low temperatures outdoors, and they are not truly glazed. Many of the ceramics were burnished or decorated with a slip before firing.

References in the literature suggest that plant extracts such as tree bark, Congo copal, gum elemi and palm oil were applied to the ceramics as decorative coatings after firing. These decorative techniques present new issues in ceramic conservation. In the conservation field, ceramics are thought of as among the most stable objects in a museum. Yet if these decorative coatings are vegetable extracts, then they are vulnerable. Vegetable coatings are more sensitive to their environment than slip or other ceramic decorations.

This investigation began when a collection of eighty five Central African ceramics was being readied for an exhibition, "Purpose and Perfection: Pottery as a Woman's Art in Central Africa." These acquisitions, new to the National Museum of African Art (NMAfA), had been collected from Central Africa in the early 20th century. All of the ceramics had been in private European collections since their export from Africa. Some of them have European provenances dating as early as 1900. Of the more than twenty cultural groups represented in this collection, the Kongo peoples are the most heavily represented, with twenty one ceramics.

Literature Review

The 1907 Annales du Tervuren is possibly the most extensive single source of information for Central African ceramic technology [1]. The Museé Royal de l'Afrique Centrale, in Tervuren holds

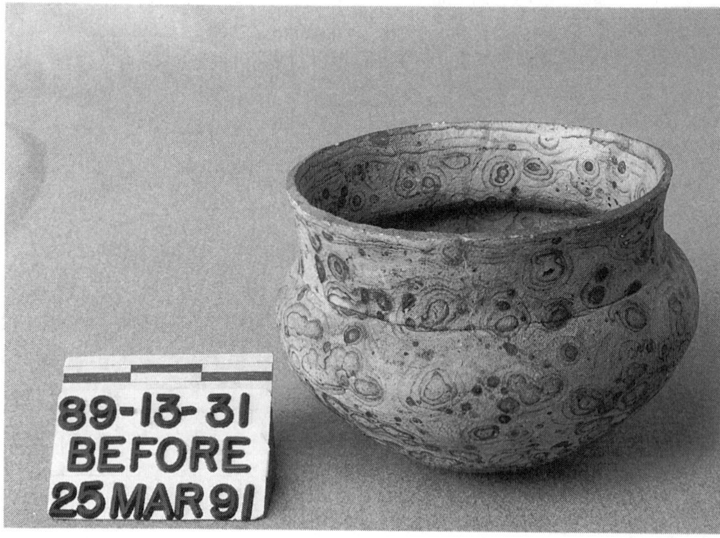

Fig. 1 Bowl. Kongo peoples, Congo, Zaire, and Angola. Collected before 1910. H. 4 7/16 (11.3 cm). NMAfA 89-13-31.

the largest collection of Central African ceramics outside Africa. The ceramics were collected in what was then the Belgian Congo, now mostly Zaire. Its authors Coart, and de Hauliville, present secondhand accounts of manufacturing techniques. The catalogue features drawings and poetic descriptions. It refers to research done at the Museum in Tervuren where evidence of the resins copal and elemi were found on ceramics as coatings, but it does not elaborate on the type of research done. The NMAfA collection parallels the Tervuren catalogue in several ways. Many of these ceramics were collected during the same time period and from the same peoples in Africa.

The three decorative techniques outlined in the Tervuren catalogue can also be used to classify the Kongo ceramics in the NMAfA collection. The first technique creates a spectacular marbleized effect (Fig. 1). The authors call this decorative technique the "vegetable matte slip." They describe a process in which palm oil and plant juices are applied to the ceramic while the ceramic is still hot from the firing [1]. Here they do not mean a true ceramic slip. Indeed, the marbleized design is matte and has concentric circles which look like tide lines left by an evaporating liquid. The second decorative process described in the Tervuren catalogue is called "vegetable slip." Here a two-step coating process is described as being applied to ceramics which are decorated with incised lines. After firing, a coloring agent such as wood powder or other vegetable matter is rubbed into the incised lines. Resins such as copal, elemi or bulungu are used to coat the ceramic [1]. In the third process, "metallic slip," mixtures of earth pigments or metallic iron are mixed with water or palm oil, applied and burnished before firing. Unlike the other two techniques the mixture could combine with the ceramic body to create a true slip. The result is described as being a contrast between the burnished areas and the matte areas [1]. After firing the authors say that the matte areas were covered with a vegetable resin. Ceramics in all three of these decorative categories are described as having been coated with vegetable matter or oil after firing.

A film included in the ceramic exhibition, The Hands of the Potter, shows the marbleized technique in action [2]. The hot ceramic is seen being immersed in a "solution" of tree bark. One sees wavy, concentric patterns form as this liquid evaporates from the surface of the ceramic. There are several written accounts of the marbleized effect having been achieved by splashing vegetable matter onto the hot ceramic after firing [3 & 4].

The wood powder described in the "vegetable slip" portion of the Tervuren catalogue was called tukula; camwood is also a term used for a wood powder. Tukula and camwood have long been referred to as red coloring agents for African objects. In her article "African Red Pigments" (1986), Jean Portell shows that indeed a great variety of coloring agents were used besides wood fibers. In the few objects she studied, she found non-bacterial iron and pigments such as red lead and wood fibers. The study indicated that assumptions based on unaided visual observation were often incorrect [5].

The museum world, in both published and unpublished work, supposes that organic coatings such as palm oil, copal, elemi, camwood and plant juices have been used as decorative coatings on African ceramics. We sought evidence of this vegetable matter on several Kongo ceramics.

Procedure

Samples of reference materials were collected so that their appearance under U.V. light and their infrared spectra could be compared with unknown coatings taken on Kongo ceramics. Over thirty five reference samples including copals, elemis, and palm oil were collected. Alcohol solutions of reference resins were evaporated onto four substrates: glass, wood, ceramic tile, and Mylar. Kongo ceramics and reference resin samples were viewed under U.V. light (365nm and 254nm wavelength). Their auto-fluorescence was noted. Color observations in U.V. light were noted and have been retained in the ceramic's Condition Reports stored in the NMAfA Conservation Department's records. Unknown specimens of approximately 2 mm^2 were taken from six Kongo ceramics where their coatings auto-fluoresced. Reference specimens of approximately 4 mm^2 were taken from the Mylar sheets. We subjected reference and unknown specimens to infrared spectroscopy by separately compressing tiny particles in a diamond anvil cell to flatten them. The instrument used was a Mattson Cygnus 100 Fourier-transform Infrared Spectrophotometer/Spectra Tech IR Plan Microscope (FTIR). Infrared spectra were obtained at 8 wavenumbers resolution in five-minute scanning times.

The first spectra obtained from ceramic coating samples had ill-defined peaks and showed mostly silica, so an extraction method was used to fractionate the samples. Each sample was placed in a small vial with ethylene dichloride. The closed vial was heated at 50°C for two to four hours. Much of the ethylene dichloride was then evaporated by blowing nitrogen into the open vial. Inorganic particles sank to the bottom of the vial with the concentrated organic solution above. The organic solution was transferred dropwise from the vial with a capillary tube onto the diamond cell. The last traces of ethylene dichloride were evaporated by heating the diamond cell. The sample was then subjected to FTIR analysis.

Results

U.V. light was not as helpful as had been hoped. When the study started the auto-fluorescence colors were noted thus: "bright yellow auto-fluorescence (like that of sandarac dust reference sample)." Initially it was hoped that through observation under U.V. light the trained, educated eye could identify and classify organic coatings on African ceramics. As the research progressed it was found that none of the ceramic coating's FTIR spectra matched any of the reference FTIR spectra. Though U.V. light observations continued to be noted, U.V. light observations were only felt to be useful in indicating the presence of an organic coating, and in this not wholly reliable. Not all the ceramics that appeared to have resin coatings auto-fluoresced in U.V. light.

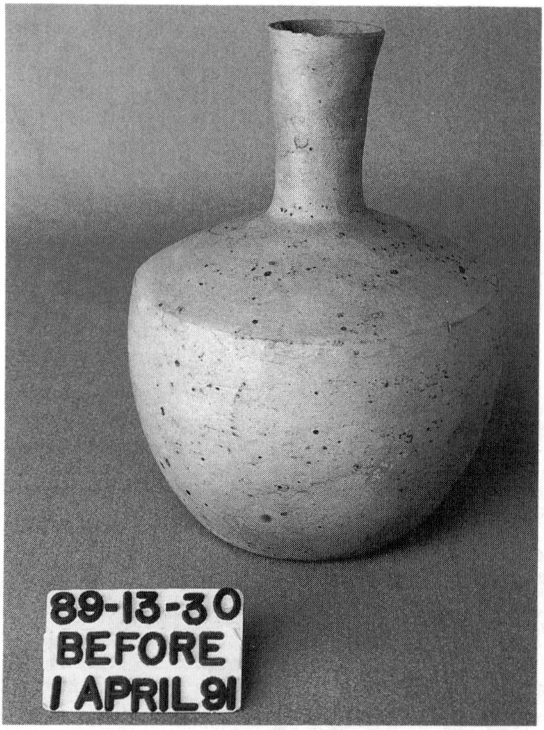

Fig. 2 Water bottle. Kongo peoples, Zaire. Collected before 1930. H. 9 in. (23 cm). NMAfA 89-13-30.

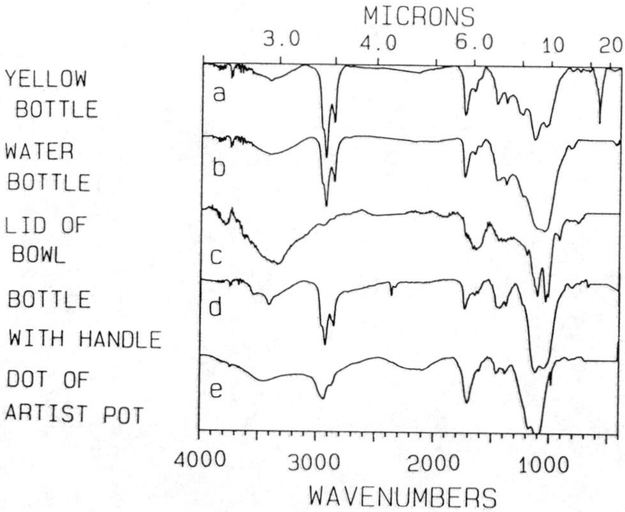

Fig. 3 Infrared spectra of samples from ceramics. Evaporates of ethylene dichloride, on diamond window. Note: individual spectra can be found on the following pages.

a. Sample of the coating on the water bottle, NMAfA 89-13-30, taken from an area that auto-fluoresced in U. V. light, and contained a dark marbleized spot. The ethylene dichloride extract gave a spectrum that suggested the presence of a mixture of organic material and inorganic phosphate.

b. Sample of the material in the incised lines on the water bottle, NMAfA 89-13-34, taken from an area in the incised lines that auto-fluoresced in U. V. light. The evaporate of this extract gave a spectrum consistent with the presence of a mixture of ester wax and inorganic phosphate.

c. Sample of the encrustation of the lidded vessel, NMAfA 89-13-23a, 23b, taken from an area on the lid that auto-fluoresced in U. V. light. The spectrum suggests the presence of a carbohydrate gum.

d. Sample of the coating on the water bottle with handle and copper alloy tacks 89-13-54, taken from an area on the register that auto-fluoresced in U.V. light. Upon extraction with ethylene dichloride, infrared spectroscopy revealed only silica, inorganic carbonate, and clay mineral.

e. Sample of green dot on Voania Muba's Stirrup handle spout vessel, NMAfA 89-13-47. The spectrum indicated the presence of inorganic sulfate and suggested the presence of an organic ester medium such as natural ester oil (drying oil).

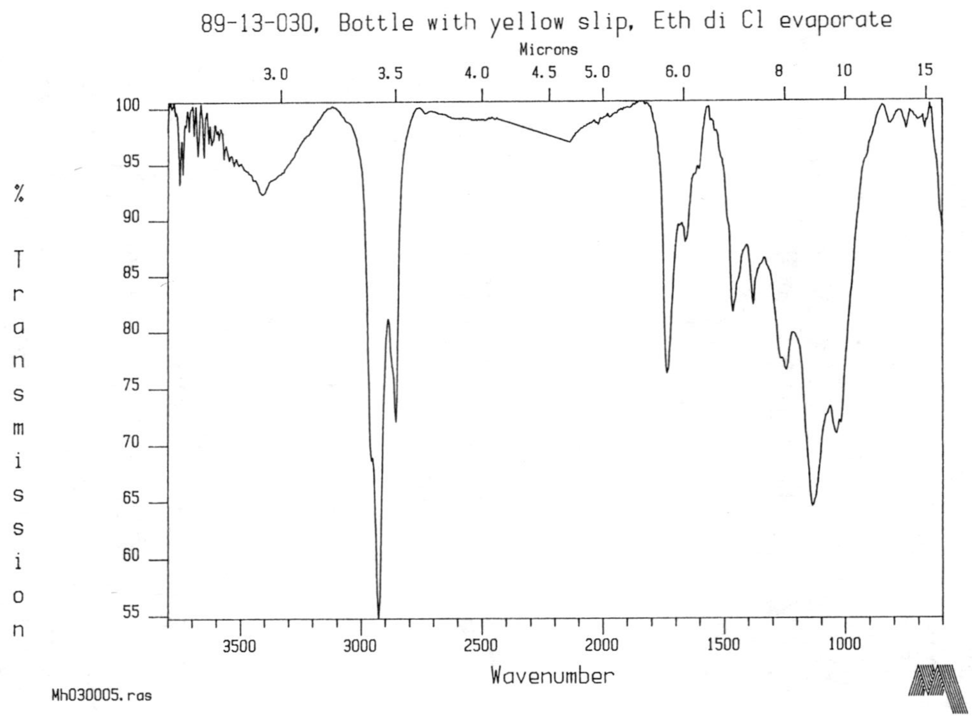

Infrared spectrum from the sample of the coating on the water bottle, NMAfA 89-13-30, taken from an area that auto-fluoresced in U. V. light, and contained a dark marbleized spot. The ethylene dichloride extract gave a spectrum that suggested the presence of a mixture of organic material and inorganic phosphate.

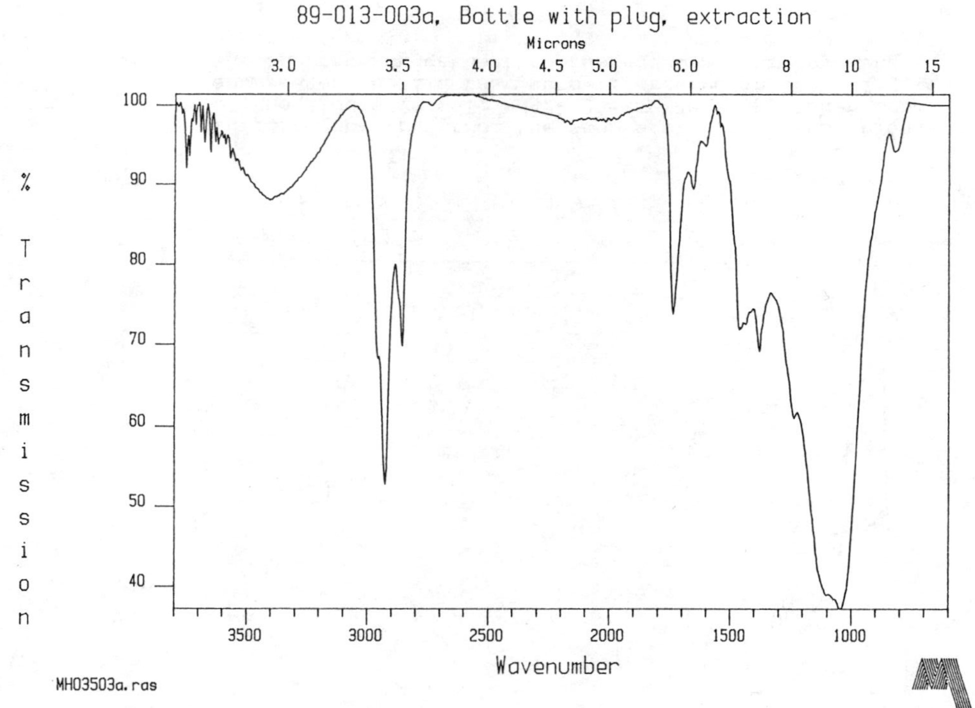

Infrared spectrum from the sample of the material in the incised lines on the water bottle, NMAfA 89-13-34, taken from an area in the incised lines that auto-fluoresced in U. V. light. The evaporate of this extract gave a spectrum consistent with the presence of a mixture of ester wax and inorganic phosphate.

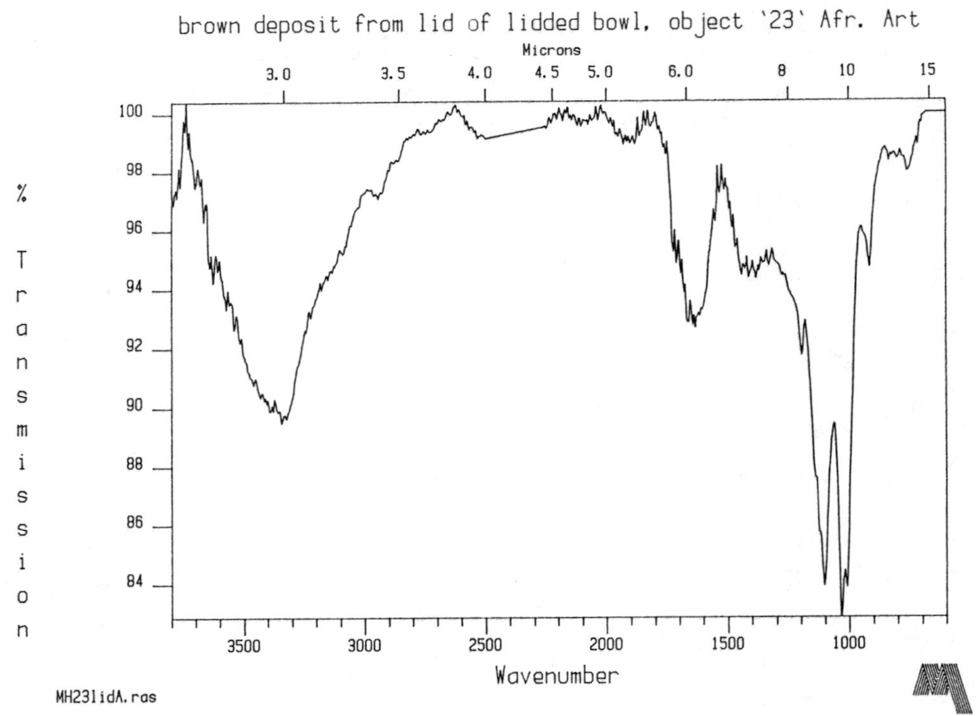

Infrared spectrum from the sample of the encrustation of the lidded vessel, NMAfA 89-13-23a, 23b, taken from an area on the lid that auto-fluoresced in U. V. light. The spectrum suggests the presence of a carbohydrate gum.

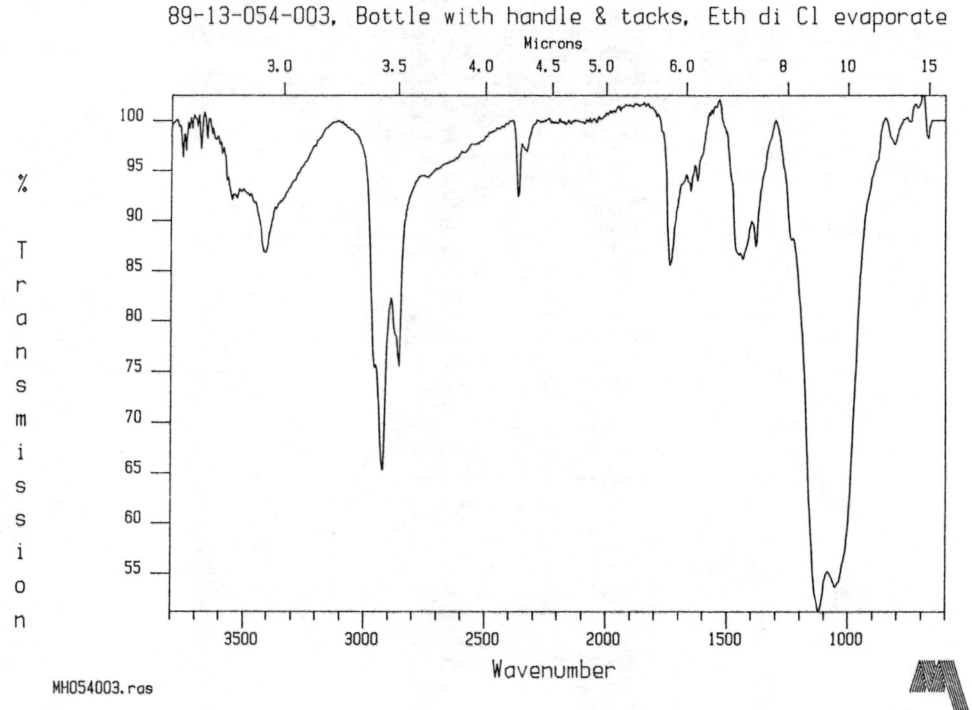

Infrared spectrum from the sample of the coating on the water bottle with handle and copper alloy tacks 89-13-54, taken from an area on the register that auto-fluoresced in U.V. light. Upon extraction with ethylene dichloride, infrared spectroscopy revealed only silica, inorganic carbonate, and clay mineral.

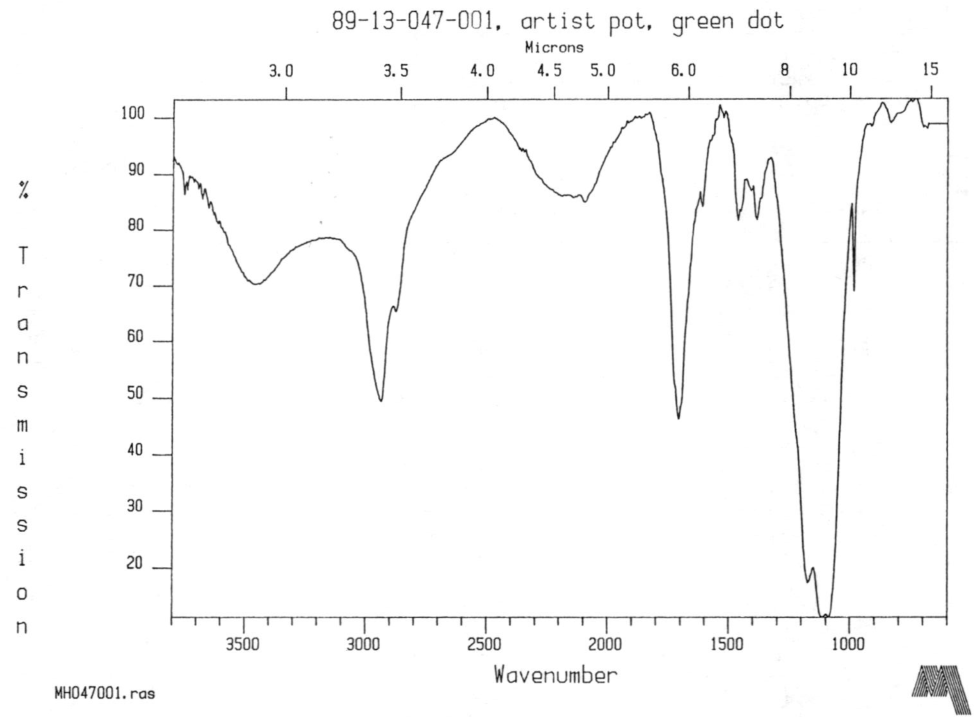

Infrared spectrum from the sample of green dot on Voania Muba's Stirrup handle spout vessel, NMAfA 89-13-47. The spectrum indicated the presence of inorganic sulfate and suggested the presence of an organic ester medium such as natural ester oil (drying oil).

TABLE 1

wavenumbers cm-1	chemical bond	substance
Spectrum a		
3405 w	OH	inorganic phosphate
2925 s	CH	organic material
1735 m	P=O?	inorganic phosphate
1655 w	OH	?
1460 w	?	inorganic phosphate
1380 w	?	inorganic phosphate
1240 w	?	?
1135 s	C-O?	?
1020 w	P-O	inorganic phosphate
Spectrum b		
3405 w	OH	inorganic phosphate
2925 m	CH	organic material
1735 m	P=O?	inorganic phosphate
1650 w	OH	?
1460 w	?	inorganic phosphate
1380 w	?	inorganic phosphate
1105 vw	?	?
1045 s	Si-O	silica or inorganic silicates
820 w	?	?
Spectrum c		
3320 m	OH	carbohydrate gum
2945 vw	CH	carbohydrate gum
1615 w	OH	carbohydrate gum
1380 vw	CH	carbohydrate gum
1195 w	C-O	carbohydrate gum
1100 s	C-O	carbohydrate gum
1030 s	C-O	carbohydrate gum
915 w	?	carbohydrate gum
Spectrum d		
3545 w	OH	clay
3405 w	OH	clay
2920 m	CH	organic material
2360 w	?	silica
1735 w	?	silica
1650 w	OH	?
1435 w	?	inorganic carbonate
1380 w	CH	inorganic carbonate
1120 s	?	?
1050 m	Si-O	silica
810 w	?	?
Spectrum e		
3435 w	OH	inorganic sulfate
2935 w	CH	organic ester medium
(2090 w)	(C=C)	(diamond anvil)
1705 m	C=O	organic ester medium
1460 w	CH	organic ester medium
1385 w	CH	organic ester medium
1170 s	S-O	inorganic sulfate
1110 s	S-O	inorganic sulfate
1085 s	S-O	inorganic sulfate
980 w	?	inorganic sulfate

This table summarizes the salient features of sample spectra [6 & 7]. The wavenumbers column indicates the wavelength, in cm-1, and strength (s=strong, m=medium, w=weak, and vw=very weak) of the various absorptions taken from the infrared spectrum itself. The chemical bond column lists those bonds which are most probably responsible for particular absorptions, while the substance column indicates in which compound the bonds are most likely to be found.

Two of the Kongo ceramics in the NMAfA collection have the marbleized appearance described in the Tervuren catalogue as "vegetable matte slip." We were able to sample one -- a water bottle (Fig. 2). This bottle appears to have a yellow slip applied overall, and over this the marbleized design appears. Under U.V. light, the bottle auto-fluoresced a bright yellow. The yellow slip could be seen to extend farther down into the neck of the bottle than did the auto-fluorescence. So the auto-fluorescence must represent another coating. The marbleized design was dark in U.V. light. The marbleized design on another ceramic (Fig. 1), was also dark under U.V. light. Unfortunately it was too heavily restored to sample.

A sample of the yellow bottle was taken from a cracked area that auto-fluoresced yellow under U.V. light and contained a dark marbleized spot. For FTIR analysis, the sample was subjected to the extraction technique described above. Because the literature describes the marbleized pattern as having been achieved by vegetable extract and/or palm oil drying on the surface of the ceramic, these materials were looked for by comparing the spectra of known resins and palm oil with the ceramic coating spectrum. The ceramic spectrum did not reveal the presence of vegetable oil or resin. The resulting spectrum indicated that an inorganic phosphate was present along with a fatty acid ester (Fig. 3a & Table 1a) The inorganic phosphate could be a calcium phosphate, one constituent of bone black. Although the FTIR spectrum suggests the presence of a phosphate, the phosphate may not be responsible for the marbleized design. Inorganic phosphate was found on light-colored incised decorated ceramics too. The bottle appears to have been decorated with a yellow slip, then the marbleized design applied. Possibly another coating was subsequently applied over all.

Three of the ceramics sampled in this study have incised decorations and appear to be of the "vegetables slip" variety described by the Tervuren catalogue. Two of these ceramics are of the water-bottle type (Figs. 4 & 5) Their FTIR results were very similar (Fig. 3b & Table 1b), The Tervuren catalogue states that the incised lines are filled with coloring agents and then vegetable resins are applied over the entire ceramic. Samples of the ceramic coatings were taken from the incised lines. Extracts from the samples gave spectra suggesting the presence of an inorganic phosphate and an ester wax. The ceramic coating spectra did not correspond to any of the known resin spectra.

The lidded vessel is a type adapted by the Kongo peoples [8]. The third incised decorated ceramic is this type (Fig. 6). Much of this ceramic is covered with a brown encrustation. Under U.V. light, the encrustation showed faint auto-fluorescence. Portions of the encrustation were sampled. FTIR spectra of encrustation from both the lid and body suggest the presence of a carbohydrate gum (Fig. 3c & Table 1c). Many trees and plants exude polysaccharide gums when accidentally or deliberately wounded [9]. Could this carbohydrate gum be evidence of the "resin" application applied to incised ceramics referred to in the Tervuren catalogue?

The third decorative category in the Tervuren catalogue is "metallic slips." The catalogue describes these ceramics as having a vegetable resin applied to them after firing. One of the ceramics sampled fits in this category (Fig. 7). It is a bottle with an applied handle. Its body appears to have been burnished, and it is decorated with metal tacks. It auto-fluoresced under

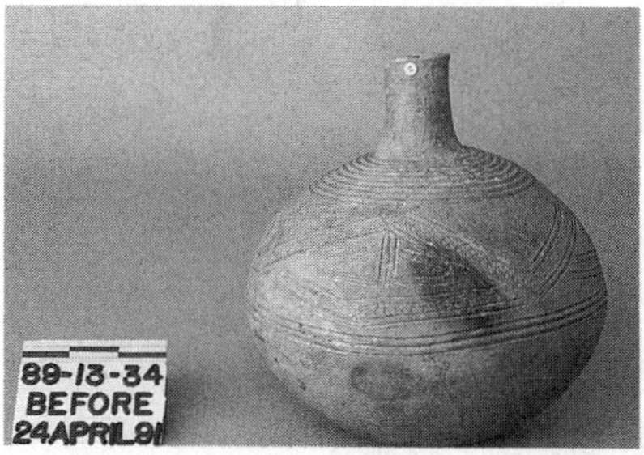

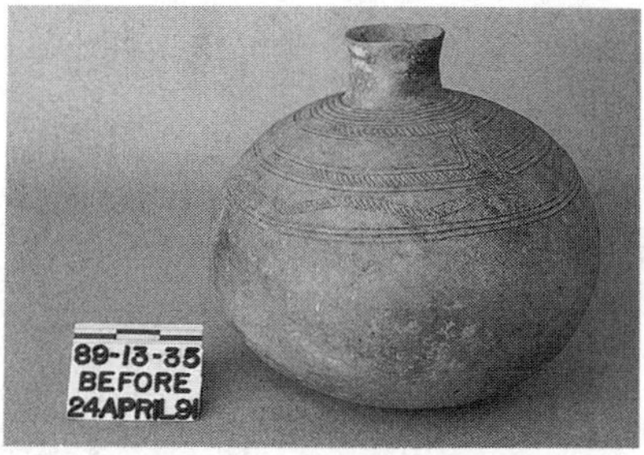

Fig. 4 Water bottle. Kongo peoples, Angola, Congo, and Zaire. Collected before 1930. H. 8 in. (20.3 cm). NMAfA 89-13-34.

Fig. 5 Water bottle. Kongo peoples, Angola, Congo and Zaire. Collected before 1930. H. 9 in. (23 cm). NMAfA 89-13-35.

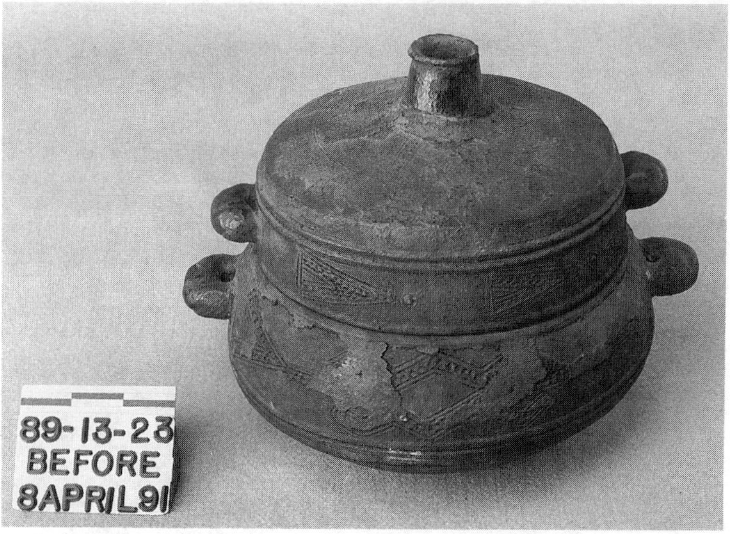

Fig. 6 Lidded vessel. Kongo peoples, Congo and Zaire. Collected before 1930. H. 8" (20.3 cm). NMAfA 89-13-23a, 23b.

U.V. light. The auto-fluorescence was assumed to indicate a coating. A sample was taken for FTIR analysis. Even after the sample was extracted, the spectra revealed little besides silica, inorganic carbonate, and clay (Fig. 3d & Table 1d).

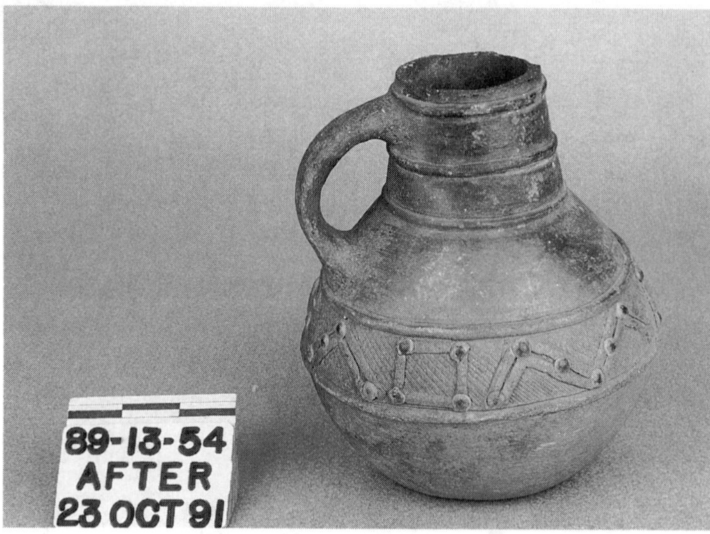

Fig. 7 Water bottle with handle and copper alloy tacks. Kongo peoples, Angola, and Zaire. Collected before 1930. H. 7¼ in. (18.5 cm). NMAfA 89-13-54.

The final ceramic to be considered in the investigation was one distinguished by its decorative resin dots (Fig. 8). This ceramic falls outside the Tervuren catalogue descriptions because it is not a traditional ceramic. Kongo ceramics were traditionally made by women for use in the home, but Voania Muba, a male potter, made this ceramic for the European market, in the 1920's. It is the only Kongo ceramic in the NMAfA collection signed by the maker. Under U.V. light, the decorative green dots auto-fluoresced. A sample was taken from a green dot. FTIR analysis showed the sample to be mainly composed of inorganic sulfates like barytes (barium white) in a ester oil medium (Fig. 3e & Table 1e). Barium white can be manufactured artificially or obtained from the barite mineral. It is found widely in Europe and the United States [10]. Maybe through contact with Europeans, Muba was able to obtain European materials for use on his ceramics.

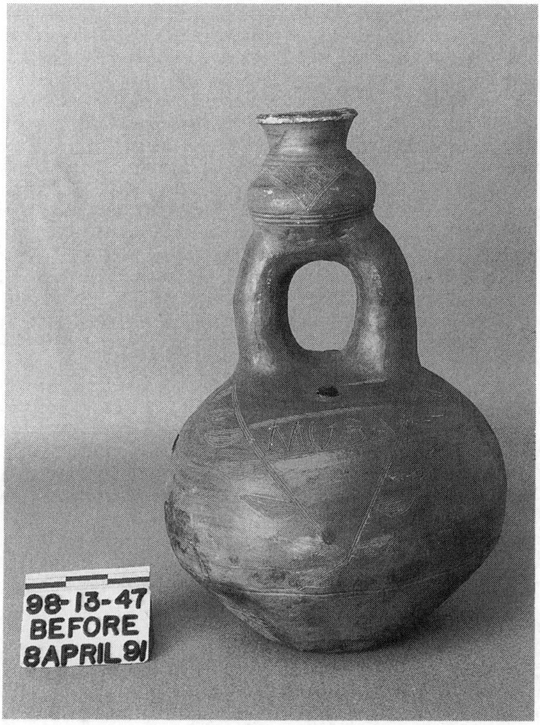

Fig. 8 Voania Muba's vessel with stirrup handled spout, Kongo peoples, Zaire. Collected before 1930. H. 12 ¼ in. (31 cm). NMAfA 89-13-47.

Conclusion

Finding inorganic phosphate in the FTIR spectra surprised the NMAfA curatorial department. Burned bone or ivory, two likely sources of inorganic phosphates, are not known as decorative additives on African ceramics. The inorganic phosphate may not indicate a decorative technique at all, since we found phosphates on both a darkly colored area on the marbleized water bottle and on the pale, incised ceramics (Fig. 2 & 4). The Tervuren catalogue does mention pulverized charcoal, which could include bone or ivory, as being added to the clay to increase its plastic qualities [1]. Could this be what the FTIR spectroscopy is detecting? Or, is it a more general phenomena of phosphates in the soil and therefore the clay?

U.V. light can not be relied upon to identify the presence of organic coatings much less to classify them. The auto-fluoresces colors of the reference resins seemed to be a function of their refinement and variables other than their biological source. This is unfortunate because U.V. light is an inexpensive readily available tool. FTIR was helpful in identifying the presence of organic coatings on the ceramics. However with this technique we were unable to classify the ceramic coatings by their biological source. It is probable that their original organic structure has altered considerably through oxidation and other degradation pathways. This probability is likely since the coatings have gone through nearly 100 years of wear, including cooking fires [9].

Many plants in Africa have waxes, oils, and gums associated with them. The Tervuren catalogue refers to "plant juices and oils" being applied during the vessels' use to lessen the porosity of the vessel and to create a "patina."[1] This might explain why we found evidence of fatty acid esters and ester waxes on both decorative types of water bottles (Figs. 2 & 4). The waxes could also have been applied while the ceramics were in Europe. Many other African objects now in the NMAfA collection were found to have been waxed while in European collections. Possibly the organic material indicated in the spectrum for the yellow water bottle (Fig. 2) is all that is left of an organic resin coating. Resins do degrade. The carbohydrate gum found on the lidded vessel could be a gum from a plant indigenous to Africa (Fig. 6). Ester oil was found along with pigments in the decorative green dots on Voania Muba's ceramic (Fig. 8). The oil could be indigenous to Africa. It also could be imported, since we believe Muba traded with Europeans. All the organic coatings we found could be indigenous to Africa.

The FTIR spectra for our ceramic coatings did not correspond with any of our reference resins spectra. The apparent absence of resin evidence in the ceramic FTIR spectra may suggest that the coatings have not survived, not that they were never present. The FTIR spectra indicates organic structures that may be remaining components of native resins and oils. Though we can not identify their biological source the organic substances on the ceramics are fragile. It is likely that their coatings have already started to degrade, and so are even more sensitive to environmental conditions such as heat, light, and relative humidity, as well as water and other solvents used in the practice of conservation.

REFERENCES

1. Coart and de Hauliville. Annales du Tervuren, Belgium Museé du Congo, (Museé Du Congo, Bruxelles, 1907);ibid., p.98-100; ibid., p. 100; ibid., p. 100-102; ibid., p. 35.; ibid., p. 52.
2. Dr. Raymond Lanfranchi and edited by Philip Ravenhill. The Hands of the Potter, filmed in the Peoples Republic of the Congo, Lower Zaire River region, 1983.
3. William Fagg and John Picton, The Potter's Art in Africa, (British Museum Publishing Ltd., London, 1970), p. 15.
4. Charlotte E. Speight, Hands in Clay, (Alfred Pub. Co., Sherman Oaks, 1979), p. 93.
5. Jean D. Portell in African Red Pigments, edited by R. Barclay, M. Gilberg, J.C. McCawley and T. Stone (Care and Preservation of Ethnological Materials, Symposium '86) pp. 119-129.

6. Federation of Societies for Paint Technology, Official Digest, Part 2 of March 1961 issue.
7. The Sadtler Commercial Spectra, Inorganics, 1965, Minerals 1973.
8. Andrea Nicolls (private communication, 1991).
9. John S. Mills and Raymond White, The Organic Chemistry of Museum Objects, (Butterworths, London, 1987), p. 66.; ibid., p. 89.
10. Rutherford J. Gettens and George L. Stout, Painting Materials, (Dover, New York, 1942), p. 96.

NOTES

The ceramics in this study were purchased by the National Museum of African Art with the support of the Smithsonian Collections Acquisition Program.

Madeleine Hexter was responsible for the photography. The FTIR spectra interpretation was done by Walter Hopwood.

We wish to acknowledge the support of the Smithsonian Institution's Conservation Analytical Laboratory, especially their Postgraduate Internship Fellowship under which this research was done. Donald Williams and Melvin Wachowiak, conservators at the Conservation Analytical Laboratory provided the resin samples. We are also grateful for the assistance provided by the National Museum of African Art, in particular Sylvia Williams, Director; Roy Sieber, Director of Research; Stephen Mellor, Chief Conservator; and Andrea Nicolls, Curator of the exhibition.

EXAMINATION OF METAL THREADS FROM SOME XV/XVI CENTURY ITALIAN TEXTILES BY SCANNING ELECTRON MICROSCOPY-ENERGY DISPERSIVE X-RAY SPECTROMETRY

D. Montegut*, C. Adelson, R.J. Koestler**, and N. Indictor***
*Conservation Center, Institute of Fine Arts, New York University, New York City, 10021
** Metropolitan Museum of Art, Objects Conservation Dept., New York City, 10028
*** Brooklyn College, CUNY and the Grad. Prog., Chemistry Dept., Brooklyn, NY, 11210

ABSTRACT

Scanning electron microscopy and energy dispersive x-ray spectrometry (SEM-EDS) has been employed along with the light microscope to describe some metal threads taken from some late Medieval and Renaissance textiles attributed to Italy in the collection of the Metropolitan Museum of Art, New York. Comparisons are made within the group studied and to those descriptions of metal threads found in recent literature. The results are discussed in terms of geographic and historic markers of the metal thread technology employed.

INTRODUCTION

Recent attempts to describe the use of metal in textiles[1-15] have suggested the following categories:

TABLE 1. The Five Categories for Inclusion of Metal in Textiles

I.	Metal applied (with adhesive) to already woven fabrics.
II.	Metal wire or flattened strips.
III.	Metal wire or strips wound around fiber core.
IV.	Metallic surface applied (with adhesive) to organic wrapping wound around fiber core.
	a. Organic = cellulosic.
	b. Organic = proteinaceous.
V.	Metallic surface applied (with adhesive) to organic strips.
	a. Organic = cellulosic.
	b. Organic = proteinaceous.

Previous studies[1-23] have reported composition and structure of metallic threads found in textiles from France, Italy, Spain, Germany, Switzerland, Egypt/Syria, Morocco, Turkey, Persia, India, Central Asia, Indonesia, Japan, and China.

Recently great strides have been made in the historical study of Italian textile production, especially that of the late Middle Ages and Renaissance. Many typologies of design and weaving techniques and structures can now be securely attributed to specific centers during a specific period. There are still numerous pieces in museum collections that are not identifiable as the product of a particular Italian workshop and there is uncertainty as to whether some 'Italian' textiles were woven in the Near East (especially Anatolia) or Spain. An effort to sort out questions of place and time of the manufacture of some of these textiles by examination of the metal threads used in their manufacture was the purpose of the present study. Difficulties in drawing decisive conclusions from this line of investigation will be underscored by the number of manufacturing techniques encountered, even among a small sampling of examples woven in a time period of about two hundred years.

Table 2 lists the textiles sampled, all of which are from the collection of The Metropolitan Museum of Art, New York, and the categories to which the specimens chosen for examination belong. The attribution of origin enclosed in square brackets[] are suggested possible alternatives to the museum's classification

Table 2. Textiles Sampled

Acc. No.	Catalog Description	Origin§	Category
44.54.2	Velvet Panel, voided cut brocaded, bouclé	Florence, XV	III
53.157	Velvet Panel, voided cut brocaded	Italy, early XVI [Anatolia]	III
*41.190.142(1)	Velvet Panel, voided cut brocaded, bouclé	Italy, Late XV [Venice]	III
*41.190.142(2)	--	--	III
88.3.9	Orphrey, embroidery, "or nué"	Florence, XV	III
46.109.26	Velvet Panel, voided cut brocaded, bouclé	Italy or Spain?,late XV[Venice]	III
46.156.72	Velvet Panel, voided, cut	Florence, XV (Spain? XIV/XV?)	IVb
46.156.98	Cope Hood, Assumption, brocatelle	Florence, XV/XVI	IVb
46.156.120	Velvet Panel, pile on pile, voided cut brocaded, bouclé	Venice, late XV	III
08.109.27	Orphrey Frag. Annunciation, tapestry	Florence?, Early XVI	III

*Samples from different places on same textile (not replicate).
§ According to Metropolitan Museum accession cards; possible alternative attributions are in square brackets [].

EXPERIMENTAL

Light microscopy and Scanning Electron Microscopy-Energy Dispersive X-Ray Spectrometry

Procedures for microscopy have been described earlier as well as energy dispersive x-ray spectrometry for the determination of elemental composition.[4-9,20,24-27]

EDS Analysis

Samples were scanned for 100-200sec at 500v-5kv at least twice at different spots at 20KeV, background corrected. Quantitative results are reported as averaged values normalized to exclude elements that appear in the analyses from corrosion products, dirt, soil, background wrapping, etc. Generally, agreement for replicate EDS analyses was within ca 10%. When disagreement was extreme, values are reported separately. On materials of extreme inhomogeneity, such as these samples, analysing a representative area becomes seriously problematic: At very low magnification, non-planar geometry and background elements may confound the analysis; at very high magnification, focusing on a small area may emphasize the presence of one element to the exclusion of others, again confounding the analysis of the whole.[27] Problems and limitations in the EDS analyses of gold-silver-copper mixtures of standards with respect to sample geometry, composition, and software analytical programs has been reported in another study.[27]

RESULTS AND DISCUSSION

Table 3 lists the results of light microscopic examination. Appearance of the surface is described, core fibers are identified, and the twists of core and wrapping are indicated. Dimensions of the threads and wrapping materials are also listed.

Table 3. Appearance (Light Microscope) and Dimensions of Metal Threads

Acc. No. Category	Core Fiber/Color Twist	Thread Diam. (μ)[a]	Outer Surface Appearance	Wrapping Width/ Thickness (μ)[b]	Substrate Adhesive Color	Wrapping Twist
44.54.2 III	Silk/Yellow S? or none	270	gold corros.	270/5-6 (200)	--	S
53.157 III	Silk/Yellow S-2ply	210	gold	320/8 (200)	--	S
41.190.142 III	Silk/Yellow S? or none	270	gold corros.	300/9 (240)	--	S
41.190.142 III	Silk/Undyed --	310	gold corros.	460/8 (390)	--	S
88.3.9 III	Silk/Undyed Z? or none	310	silver bl. corros.	230/12.5 (225)	--	S
46.109.26 III	Silk/Undyed S	440	gold	500/12 (530)	--	S over Z double
46.156.72 IVb	Silk/Or-Red Z	380	gold/black	500/400 (480)	leather? brown	Z
46.156.98 IVb	Linen/Undyed S-2ply	260	silver/gold	680/nd[c] (580)	parchment? membrane? bl brown, translucent	S
46.156.120 III	Silk/Lt Yellow S? or none	170	gold bl. corros.	270/10 (210)	--	S
08.109.27 III	Silk/Lt Yellow S? or none	500	gold bl. corros.	580/12 (520)	--	S

[a]Measured with cursor, light microscope.
[b]Width measured with cursor, light microscope./Thickness measured with cursor, SEM.
(Values in parentheses) = wrapping width measured with cursor, light microscope.
[c]nd = no measurement made.

The outer surface appearance using a light microscope indicated the presence of gold in each case consistent with the EDS analysis. Silver was not invariably observed but the observation of black corrosion (AgS) generally suggested correctly the presence of silver.

Thread diameters varied between ca 170-500μ as measured with a cursor using the light microscope. Wrapping width was also determined using the light microscope. It was possible to unwrap the lamellae and measure the width fairly easily. Comparable measurements done by cursor on the SEM were performed on samples that had been manipulated to expose both sides of the wrapping substance to the electron beam, followed by carbon coating. Widths were measured at several places and averaged. Values obtained for individual examples varied by as much as 20μ, clearly suggesting nonuniformity of the sample dimention - from manufacture, wear, handling, corrosion, etc. Differences between the light microscope measurements and SEM may be noted in Table 3. Differences of as much as 100μ were obtained (usually lower for the light microscope) possibly owing to the differences and difficulties of positioning the sample for measurement and partly owing to the nonuniformity of the samples themselves. Wrapping widths varied between just over 200μ to about 600μ. The two proteinaceous wrappings were among the widest, but two examples of metal wrappings were also about 500μ. The metal thicknesses measured by SEM cursor were ca 10±2μ. One sample, (44.54.2), was significantly thinner, ca 5-6μ. The proteinaceous wrappings were much thicker, ca 400μ. The wrapping twist for each of the metal wrappings was in the S direction. One example, (46.109.26), was double wrapped S over Z. One of the proteinaceous threads was wrapped Z with leather (46.156.72), one wrapped S with membrane (46.156.98). Some of the difficulties in identifying proteinaceous material microscopically have been described earlier.[4-7,28]

Table 4 shows elemental analyses obtained by EDS. Analyses were obtained on both sides of each wrapping material and for each core thread.

The interior and exterior of the metal wrappings gave different analyses for gold:silver:copper in five (44.54.2; 41.190.142(1); 41.190.142(2); 46.109.26; 08.109.27) of the eight category **III** examples tested, suggesting that in those examples gilding was limited to the outer surface, applied after the wire was drawn and flattened. Two examples, (53.157, 46.156.120), showed no substantial difference between outside and inside alloy content, suggesting that the alloy was fixed before the drawing and flattening of the lamellae (Cf descriptions of historic Indian gold wire drawing[8,29]). One example, 08.109.27, gave an inside and outside composition about the same (gold absent), but a spot on the surface clearly indicated the presence of gold, suggesting that gold was not (now) uniformly distributed on the exterior surface. How much this is attributable to wear, corrosion, manufacture, or accident, can only be conjectural. It should be stated that each of the samples showed evidence of corrosion on interior as well as exterior surfaces both in the microscopic imaging and in the elemental analyses. Chlorine and sulfur were usually detected on category **III** surfaces, and in the cases of the proteinaceous wrappings, many other elements were also detected on the outer surface. One unusual example that did not appear to be gilded at all (88.3.9), and whose core thread was very white (undyed or possibly bleached), as if the thread were intended to appear silvery rather than gilded[30], had a spot on the interior surface that analysed for gold, possibly an unintended workshop inclusion.

The interior surfaces of the proteinaceous threads gave a large array of elements, reflecting trace inclusions in the protein, adhesives, and corrosion products from the metal surfaces. The core threads also showed a wide array of elements, traces from the threads themselves, metal corrosion, and possibly mordant (Al,Fe) that may have been applied as part of the dying of the fibers. All the gilded threads showed Al,S,K as part of the analyses, sufficient to suggest that the threads may have been alum mordanted.[24-26]

Table 4. Elemental Analyses (EDS)[a] of Metal Threads

Acc. No.	Wrapping Exterior Surface (Ratios)[b]	Wrapping Interior Surface (Ratios)[b]	Core[c]
44.54.2	Au:Ag:Cu 48:43:9	Au:Ag:Cu 0:97:3	Ca,Cu,Al,Cl,Na,Si,K,S Mg,Fe,Ag,P
53.157	Au:Ag:Cu 11:89:0	Au:Ag:Cu 11:89:0	Ag,Ca,K,S,Si,Cl,Al
41.190.142 -1-	Au:Ag:Cu 36:53:11	Au:Ag:Cu 0:85:15	Ca,S,Cu,Ag,Al,K,Cl,Si,Fe
41.190.142 -2-	Au:Ag:Cu 17:72:11	Au:Ag:Cu 0:85:15	Ca,Al,S,Cl,Cu,Si,K
88.3.9	Au:Ag:Cu 0:98:2	Au:Ag:Cu 21:74:5 0:93:7	Cu,Ca,Ag,K,S,Cl,Si
46.109.26	Au:Ag:Cu 24:65:11	Au:Ag:Cu 0:83:17	Ag,S,Si,Ca Cl,Al,Cu,K
46.156.72	Au:Ag:Cu 100:0:0	*[c]Ca,Si,Au,S,K,Cl,Na Al,Mg,Fe	Ca,K,Si,Al,Cl,S
46.156.98	Au:Ag:Cu 30:70:0	*[c]Ca,Ag,Cl,K,Fe,S,Si	Ca,Cl,K,Si,S,Al
46.156.120	Au:Ag:Cu 11:83:6	Au:Ag:Cu 13:79:8	Ca,Al,Cl,S,Cu,Si,K,Ag
08.109.27	Au:Ag:Cu 0:85:15 16:71:13	Au:Ag:Cu 0:88:12	Ag,Cu,Al, S,Ca,K,Si,Cl

[a]Elements of atomic number >11; lists of elements are given In order of decreasing weight percent.
[b]Normalized values (excludes elements indicating corrosion,soil, etc, e.g. S,Cl,Ca,Si. etc
[c]Trace
*Proteinaceous substrates

CONCLUSIONS

It should be emphasized that the sample base for this study is small and that the conclusions are limited to the group of samples examined.

1. Category III [8 samples]

a. At least two different methods of production may be noted - one in which the wire is fashioned as a gold/silver alloy, then drawn and flattened [53.157;46.156.120]; another in which the gold is applied to only one surface (outer) so that analysis of the inner surface shows gold absent while the outer surface shows gold present [44.54.2; 41.190.142(1)&(2); 46.109.26].

b. No category III specimen analysed for gold with the exclusion of silver. The overall assay when both gold and silver were present showed silver>gold.

c. Silver was found in two examples [88.3.9; 08.109.27] without gold. In both cases gold was detected in some analyses.

d. Copper was a trace impurity detected in all but one sample [53.157].

e. All core threads were of silk.

f. The wrapping twist of the metal was in the S direction for all examples.

2. Category IVb [2 samples]

a. One sample had a pure gold leaf surface [46.156.72].

b. One sample had a silver/gold (silver>gold) leaf surface [46.156.98].

c. One sample had a substrate that appeared to be leather wrapped Z about a silk core [46.156.72].

d. One sample had a substrate that appeared to be membrane [or possibly parchment] wrapped S about a linen core [46.156.98].

3. A multiplicity of metal thread manufacturing methods is consistent with historic accounts which place Italy at a lively and intense explosion of textile technologies during the fifteenth and sixteenth centuries[28]. During this time metal threads made from proteinaceous substrates gilded with leaf and wrapped on silk or linen and said to have been imported from the East (Cypriot gold thread) via Cyprus was gradually being replaced by metal thread of local manufacture in which the drawn and beaten metal was wound around a silk core. Furthermore metalsmiths had learned to produce flattened strips gilded on one side only, a process unknown or at least unpracticed in earlier times. The ease with which it is possible to trade in metal threads and the likelihood that it was a much travelled commodity underlie the difficulty in ascribing a specific geographic locale to a textile exclusively on the basis of metal threads used, unless it is possible to establish a much broader data base for comparison or to find historic documentation that may give clues to the existence of workshops. No systematic study of metal thread technologies has been applied to Spanish or Anatolian (late Byzantine or early Ottoman) textiles of the fifteenth/sixteenth centuries. A cursory visual examination of Ottoman brocades, velvets, and embroideries indicates that most metal threads are of category III type (metal strip around fiber core) with occasional examples of category II type (metal without fiber core). It is not known whether these metal strips, when gilded, were manufactured by a process that permitted gilding on one side only or by drawing and flattening a silver/gold alloy (or by both processes). A systematic study of the metal threads used in Anatolian, as well as Spanish textiles of this period would obviously be useful.

REFERENCES AND FOOTNOTES

1. J. A. Darrah, Metal Threads and Filaments, in Recent Advances in the Conservation and Analysis of Artifacts, comp. by James Black, Summer School Pr., U. London, Inst. of Archaeol, London [1987].

2. J. A. Darrah, "The Microscopical and Analytical Examination of Three Types of Metal Thread," Conservation of Metals. Ed. M. Jaro, International Restorer Seminar (1-10 July 1989), Veszprém, Hungary, 1989, pp. 53-64.

3. E. Hoke, and I. Petrascheck-Heim, "Microprobe Analysis of Gilded Silver Threads from Medieval Textiles," Studies in Conservation, **22** (1977): 49-62.

4. N. Indictor and R.J. Koestler, "The Identification and Characterization of Metal Wrapping in Historic Textiles Using Microscopy and Energy Dispersive X-Ray Spectrometry - Problems Associated with Identification and Characterization," Scanning Electron Microscopy (1986): 491-497.

5. N. Indictor, R.J. Koestler, C. Blair, and A. Wardwell, "The Evaluation of Metal Wrappings from Medieval Textiles Using Scanning Electron Microscopy-Energy Dispersive X-Ray Spectrometry" Textile History, **19** (1) (1988): 3-22.

6. N. Indictor, R.J. Koestler, M. Wypysky, and A. Wardwell, "Metal Threads Made of Proteinacious Substrates Examined by Scanning Electron Microscopy-Energy Dispersive X-Ray Spectrometry," Studies in Conservation **34** (4) (1989): 171-182.

7. N. Indictor and M. Ballard, "The Effects of Aging on Textiles that Contain Metal: Implications for Analyses," Conservation of Metals, Ed. M. Jaro, International Restorer Seminar (1-10 July 1989), Veszprém, Hungary, 1989, pp. 67-76.

8. N. Indictor and C. Blair, "The Examination of Metal from Historic Indian Textiles Using Scanning Electron Microscope-Energy Dispersive X-Ray Spectrometry," Textile History **21** (2)(1990): 149-63.

9. N. Indictor "The Use of Scanning Electron Microscopy-Energy Dispersive X-Ray Spectrometry for the Examination of Metal Threads on a Karuppur Sari," Studies in Indian Textiles, a Socio-Cultural Survey, Ed. L. Varadarajan, Manohar, New Delhi, (in press).

10. M. Jaro, "The Investigation of the Metal Embroidery Threads of the Hungarian Coronation Mantle by Scanning Electron Microscope and Physical Methods of Analysis", ICOM 7th Triennial Meeting, Copenhagen, 10-14 Sept., Paris, l984, 84.1.22-84.1.24.

11. M. Jaro, "The technology and analytical examination of metal threads on old textiles," in Conservation-Restoration of Church Textiles and Painted Flags; Investigation of Museum Objects and Materials Used in Conservation-Restoration, Fourth International Restorer Seminar (2-10 July 1983), Budapest, 1984, pp. 253-264.

12. M. Jaro and E. Gondar, "Mediaeval Membrane Threads Used for Weaving and Embroidery," Archaeometrical Research in Hungary, Eds M. Jaro and L. Kolto, National Center of Museums, Budapest, 1988, pp. 255-266.

13. M. Jaro, E. Gondar and A. Toth, "Reconstruction of Gilding Techniques Used for Mediaeval Membrane Threads in Museum Textiles," Archaeometry '90, Birkhäuser Verlag Basel, (1990): 317-325.

14. M. Jaro, A. Toth and E. Gondar, "Determination of the Manufacturing Technique of a 10th Century Metal Thread," ICOM 9th Triennial Meeting, Dresden, 26-31 August., Los Angeles, l990, pp. 299-301.

15. M. Jaro,"Gold Embroidery and Fabrics in Europe: XI-XIV Centuries," Gold Bulletin **23** (2)(

l990): 40-57.

16. Genoa, Tessuti Preziosi del Giappone Kinran: Shokko-nishiki: Kinsha: secoli xvi/xix: Storia: Techniche: Analisi: Restauro. Museo d'Arte Orientale "Edoardo Chiossone"; Compagnia dei Librai, Genoa, 1985.

17. I.R. Harden, and F.J. Duffield, "Characterization of Metallic Yarns in Historic Persian Textiles by Microanalysis," CELL 44. Abstract of Papers. l88th ACS National Meeting, Phila., PA, Aug.26-31, l984. Historic Textiles and Paper Materials. Needles, H.L. & S.H. Zeronian, Eds. Adv. in Chem. Series **212**, Amer. Chem. Soc., Washington, D.C. 1986, pp. 231-252.

18. L. Lee-Whitman, and M. Skelton, "Where Did All the Silver Go?" Textile Museum Journal. **22** (l984): 33-52; M. Skelton and L. Lee-Whitman, "A Systematic Method for Differentiating Between 18th Century Painted-Printed Chinese and Western Silks," Historic Textiles and Paper Materials. Needles, H.L. & S.H. Zeronian, Eds., Adv. Chem. Series **212**, Amer. Chem. Soc., Wash., D.C. 1986, pp. 131-152.

19. L. Magagnato, et. al., Le stoffe di Cangrande. Alinari, Florence, 1983, pp. 190-193.

20. D. Montegut, N. Indictor, and A.&J. Summerfield, "Technical Examination of Metal Threads in some Indonesian Textiles of West Sumatra," (in prep).

21. B.A. Schreier and R.R. Bresee, History of Decorative Metallic Yarns. Book of Papers. 1979 National Technical Conf., AATCC Oct 3-5 Hyatt House, Cherry Hill, NJ, Host Delaware Valley Section.

22. L.P. Stodulski, H.F. Mailand, D. Nauman, and M. Kennedy, "Scanning Electron Microscope/Energy Dispersive X-Ray and Atomic Emission Spectrographic Studies of Precious Metal Threads from European, Middle Eastern, and Oriental Textiles," Abstr., Applications of Science in Examination of Works of Art. Mus.of Fine Arts, Boston, MA (1983); Eds. P.A. England & L. van Zelst, "Atomic Emission Spectrographic and Scanning Electron Microscope/Energy Dispersive X-Ray Studies of European, Middle Eastern, and Oriental Metallic Threads," Mus.of Fine Arts, Boston, MA, 1985, pp. 76-91. [Same as appendix of: P.S.Gilfoy, Fabrics in Celebration from the Collection. Indianapolis Mus. of Art, Indianapolis, 1983.]

23. Anne E. Wardwell, "Panni tartarici: Eastern Islamic Silk Woven with Gold and Silver (13th and 14th Centuries)." Islamic Art, **III**, (1988-89): 95-173.

24. N. Indictor R.J. Koestler, and R. Sheryll, "The Detection of Mordants by Energy Dispersive X-Ray Spectrometry Part I. Dyed Woolen Textile Fibers," Journal of the American Institute for Conservation **24** (1985):104-109.

25. R.J. Koestler, R. Sheryll, and N. Indictor, "Identification of Dyeing Mordants and Related Substances in Textile Fibers: A Preliminary Study Using Energy Dispersive X-Ray Spectrometry," Studies in Conservation **30** (1985): 58-62.

26. R.J. Koestler, N. Indictor, and R. Sheryll, "The Detection of Mordants by Energy Dispersive X-Ray Spectrometry Part II. Historical Silk Textiles," Journal of the American Institute for Conservation **24** (1985): 110-115.

27. S.M. White, R.J. Koestler, C. Blair, and N. Indictor, " Energy Dispersive X-Ray Spectrometry of Gold-Silver-Copper Standards for Comparison with Historic Gold Objects," Materials Research Society Symposium Procedings, San Francisco, 1990, Vol **185**, Materials Issues in Art and Archaeology II. Eds. P.B. Vandiver, J. Druzik, G.S. Wheeler, Pittsburgh, 1991, pp. 731-740.

28. Franz Bock, Goldstickereien und Webereien in alter und neuer Zeit und dazu verwandte

Goldgespinnst. Bayrischen Gewerbemusem, Nürnberg, 1884

29. J. Nissim, "A Monograph on Wire and Tinsel in the Bombay Presidency," in Art and Industry Through the Ages. Navrang, New Delhi, repr. (1895-97) 1976 pp. 287-302.

30. Cores of gilded threads often have yellow, yellowish or colored threads; cores of silver threads often have threads of uncolored or bleached fibers.

NEUTRON INDUCED AUTORADIOGRAPHY AND PIXE ANALYSIS

W. STANLEY TAFT*, JAMES W. MAYER*, HOWARD C. ADERHOLD*, MATT KELLER AND GIA RIZZO*
*Dept. of Art, Cornell University, Ithaca, NY 14853-7301
*Dept. of Materials Science, Cornell University, Ithaca, NY
*Dept. of Nuclear Science and Engineering, Cornell University, Ithaca, NY
*Undergraduates, Cornell University, Ithaca, NY

ABSTRACT

We have used the TRIGA reactor and the Ion Beam Facility at Cornell to illustrate neutron induced autoradiography and PIXE analysis of known pigments in demonstration paintings made for our undergraduate course, "Art, Isotopes and Analysis". A group of pigments were chosen containing elements with half-lives detectable by the neutron activation process. The pigments (suspended in a binding medium of linseed oil) were first painted on test swatches of acetate and their characteristic x-ray lines measured with a 5 Mev beam of protons directed through a thin Kapton film for external (in air) analysis (PIXE).

Three layers of images, each painted with a different pigment, overlay one another. The painting is then taken to the TRIGA reactor for neutron activation. After activation, the painting is placed in contact with sheets of Polaroid type AR positive transparency film at different times after exposure. The gamma spectra during the sequence are also measured to provide identification of the "active" isotopes during each film exposure.

Introduction

The techniques of painting and the analysis of painting materials and the construct of paintings are the subjects of our undergraduate course at Cornell University: "Art, Isotopes and Analysis". The presence of a source of neutrons on campus, the TRIGA reactor, allows us the opportunity to demonstrate neutron induced autoradiography on paintings made in our own studio with pigments whose composition has been analyzed by external-beam proton-induced x-ray emission (PIXE) [1]. Autoradiography allows us to discuss radioactivity, nuclear reactions, beta-ray emission and pigment analysis. This has led us to an investigation of activation and image formation by beta radiation of modern pigments [2].

We have made an eight minute video tape of "Neutron Induced Autoradiography at Cornell" in order to supplement our slide presentations. We show slides of autoradiographs taken on paintings at the Metropolitan Museum of Art in New York, as well as of our paintings and the autoradiographs generated by them. We have found the analysis of art to be a particularly effective method of presenting nuclear reactions and radioactivity to a general audience.

Sample Preparation

Sample paintings were produced for our research with pigments similar to those used in most oil paintings. Since this initial research is intended to establish a proceedure and to

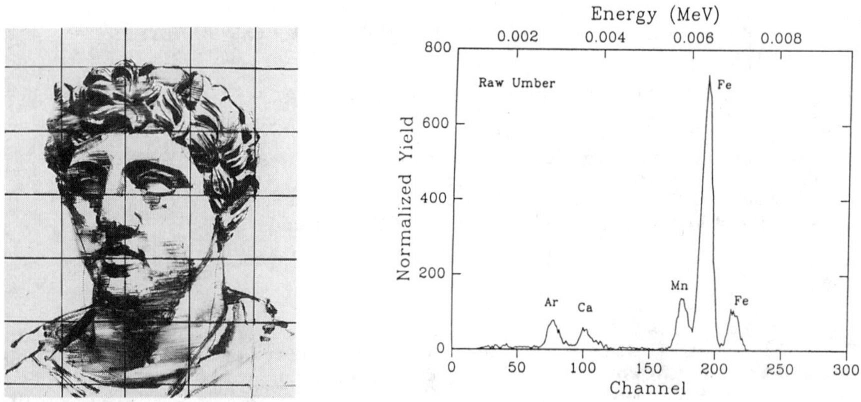

Fig. 1 Head painted in Raw Umber with PIXE spectrum.

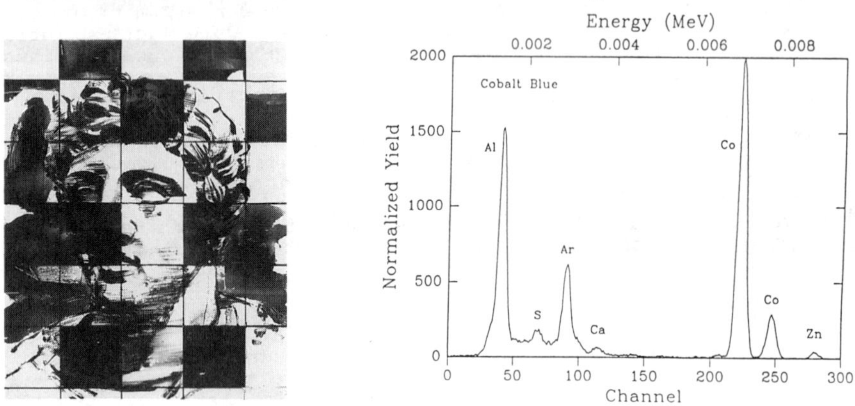

Fig. 2 Grid and squares painted in Cobalt blue with PIXE spectrum

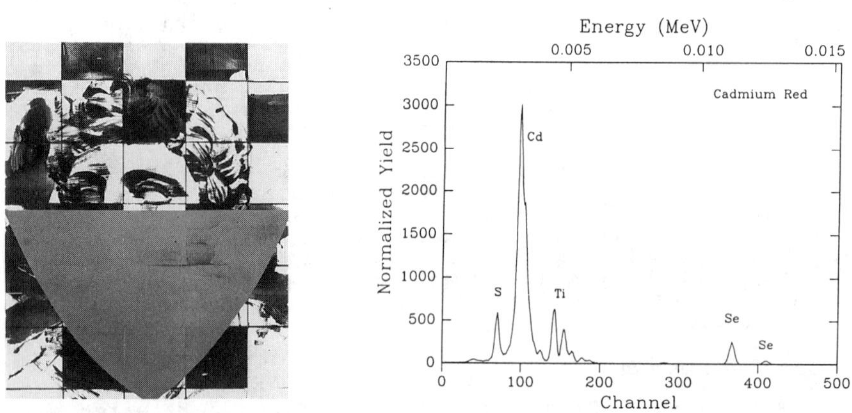

Fig. 3 Mask shape painted in Cadmium red with PIXE spectrum.

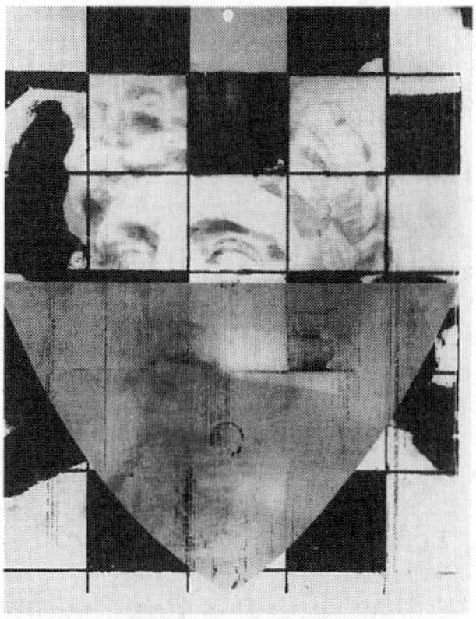

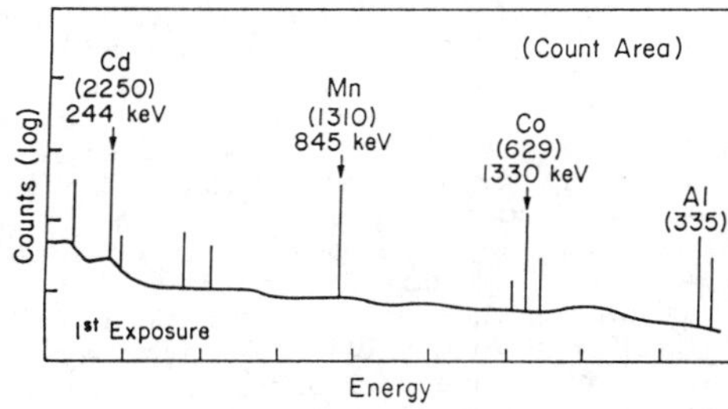

Fig. 4 First autoradiograph with Gamma spectrum of dominant isotopes.

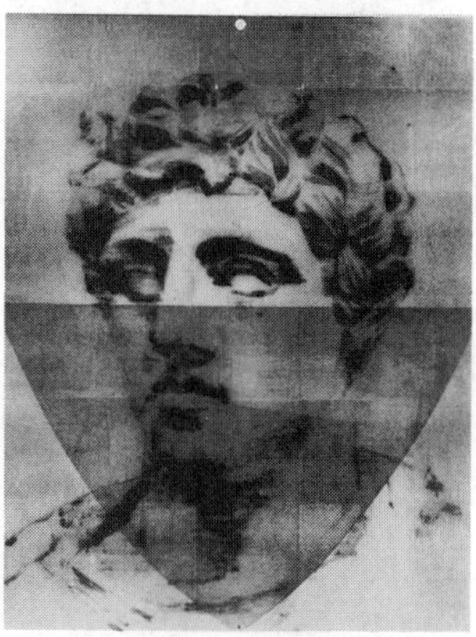

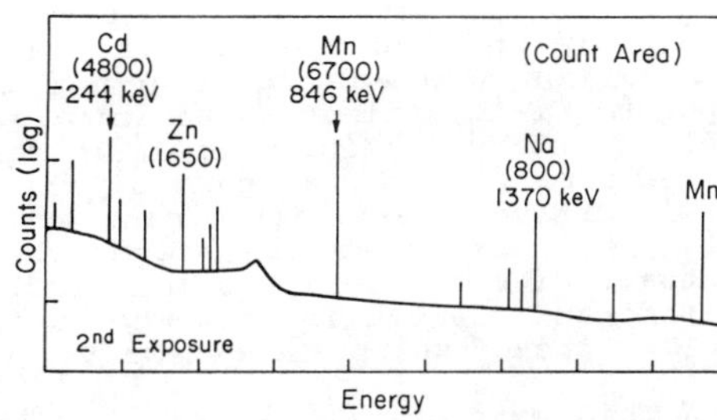

Fig. 5 Second autoradiograph with Gamma spectrum of dominant isotopes.

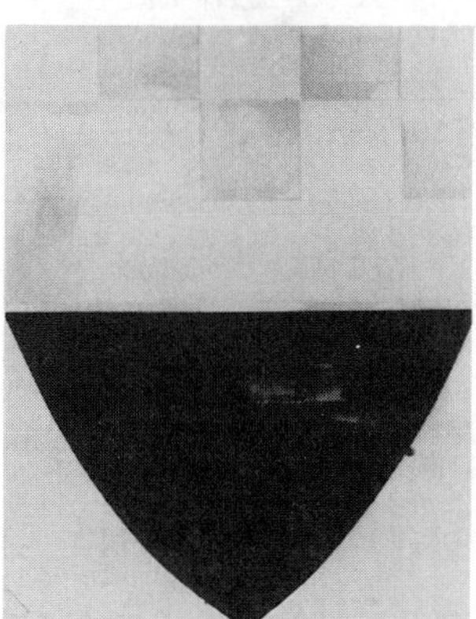

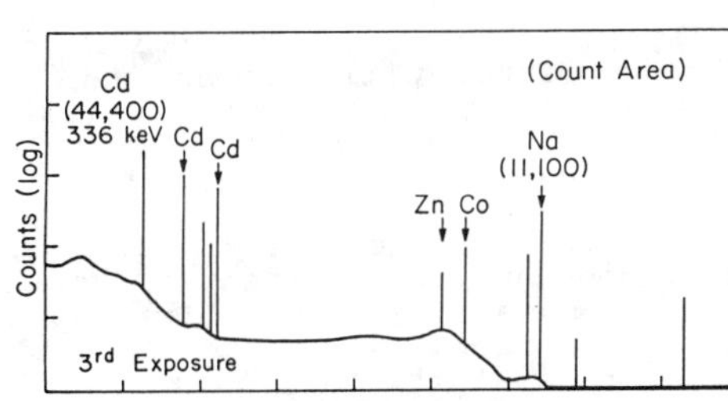

Fig. 6 Third autoradiograph with Gamma spectrum of dominant isotopes.

test the effectiveness of the photographic materials and pigments, painting methods applied to the samples do not follow historical models. Each discreet layer of paint contains an image distinct from that overlaying or underlaying it. This allows us to detect the sequence of layers in the autoradiographs. Although autoradiography locates the distribution of pigments accross the surface of a painting, it does not necessarily locate the position of the pigment in the sequence of layers making up the entire paint film. Because we know the sequence in the samples, we have been able to note the absorption of beta radiation by some overlaying pigments, producing vacancies in the image of otherwise active isotopes. The identification of the pigments in the autoradiographs is verified by cross-referencing with gamma spectra taken from the painting with each exposure in the sequence, gamma spectra taken from samples of each pigment separately, and by Proton Induced X-ray Emission (PIXE) spectra on the same samples.

The painting presented here was prepared as others have been for our research. In this case, the number of colors was limited to four. The support of 1/8" Lucite was coated with a layer of white lead oil paint. This gives a white ground for the images to follow and because of the inactivity of the Pb during the activation process, does not interfere with the images in exposure of the film. There is present in this layer however, a small quantity of Zinc Oxide. This produces a slight fogging of the entire surface for part of the exposure sequence, but does not interfere to any significant extent with the radiation of the other dominant elements. Over the white ground was painted a portrait of Pan. This was done exclusively with raw umber. The predominant element in raw umber which concerns us is manganese, the isotope of which (Manganese 56) initiates the exposure of the film.

This exposure has a position in the sequence distinct from those generated by isotopes of elements in the other pigments. The next layer of image is a grid pattern and checkerboard painted in cobalt blue. There are two cobalt isotopes which will produce images; Cobalt 60m and Cobalt 60. The short life Cobalt 60m appears clearly in exposure #1. The longer isotope will not appear until about 165 hours after activation. For this particular study we were interested in the short-life isotopes only. The third image, which overlays parts of both the Co and Mn images, is painted in cadmium red. This is the first sample we have tested with cadmium red pigment. Since most of the paintings analysed to date were painted before its manufacture, a "signature" for the pigment relative to autoradiography has not to our knowledge been established. This is an ubiquitous modern pigment, and its identification through autoradiography could be very useful in the analysis of modern works. Table 1 lists the three pigments, the elements identified in their composition and the isotopes responsible for generating the exposures. Both neutron induced gamma spectra and PIXE (proton induced X-ray emission) analysis were used to compile the table. Since there is some activity from all three pigments at any one time, the gamma spectra for any given exposure will show activity from a wide range of sources. The relative energies however help to identify those areas which most effectively expose the film.

A series of exposures were produced. We are fortunate to have been able to use a prototype film manufactured by Polaroid Corporation and developed for neutron activation techniques. The format used is 8"x10". This film produces images of very high resolution, and because of the instant processing, allows us to evaluate the results immediately. In addition, darkroom facilities are not necessary so processing can be done on site in the laboratory.

COLOR	Raw Umber	Cobalt Blue	Cadmium Red
ELEMENTS	Mn,Fe,Ar,Ca	Al,S,Ar,Ca,Co,Zn	Cd,Se,S,Ti
ISOTOPES	^{56}Mn	^{60}Co, ^{60n}Co, ^{65}Zn, ^{28}Al	^{75}Se, ^{115}Cd, ^{111m}Cd

Table 1

Summary

Neutron induced autoradiographs of paintings made with pigments selected for their nuclear properties is an effective way of demonstrating to undergraduate students, the structure of paintings, nuclear characteristics of painting materials and techniques of nuclear analysis. Our studies have led us to a continuing investigation of the nuclear characteristics of modern pigments.

Acknowledgements

We thank Professor David Clark of Ward Laboratory, Cornell University, for discussions of neutron activation; Peter Revesz and Nick Szabo for PIXE analysis and Eric Gasteiger for the videotape. In addition we thank Peter Roth and Philip Cravotta of Polaroid Corporation for their film, processing equipment and expertise.

References

1. B. Cho, J. Baum, P. Revesz, W.S. Taft, D. Mayer and J.W. Mayer, "Ion Beam Analysis of Pigments", Materials Issues in Art and Archaeology II, edited by P. B. Vandiver, J. Druzik and G. S. Wheeler (Materials Research Society, Pittsburgh, 1992), MRS Vol. 185, pp 125-132.

2. E.V. Sayre and H.N. Lechtman, "Neutron Activation Autoradiography of Oil Paintings", Studies in Conservation, Vol. 13, No. 4, 1968, pp 161-185.

3. Maryan Wynn Ainsworth et al, Art and Autoradiography: Insights into the Genesis of Paintings by Rembrandt, Van Dyck, and Vermeer, The Metropolitan Museum of Art, New York, 1982.

4. Susan Hobbs et al, Thomas Wilmer Dewing--A Look Beneath the Surface, Smithsonian Studies in American Art, Vol. 4, No. 3-4, Summer/Fall 1990.

5. Ralph Mayer, The Artist's Handbook of Materials and Techniques, Viking Press, New York, 1964.

FORGERY DETERMINATION BY FUNCTIONAL ANALYSIS

LEONARD GORELICK AND A.J. GWINNETT
SCHOOL OF DENTAL MEDICINE, SUNY AT STONY BROOK
STONY BROOK, NY 11794-8702

ABSTRACT

STONE STATUETTES FROM TURKEY IN SIMPLY SHAPED ANIMAL AND BIRD FORMS WERE ANALYZED USING SCANNING ELECTRON MICROSCOPY. OBSERVATIONS MADE IN DRILL HOLES IN SOME OF THE ARTIFACTS LED TO DEFINITE UNEQUIVOCAL EVIDENCE FOR FORGERY WHILE OTHERS WERE EQUIVOCAL.

INTRODUCTION

IT IS COMMON KNOWLEDGE THAT THERE ARE MANY AUTHENTICITY PROBLEMS IN MUSEUMS AND THE ART MARKET. SKILLED FORGERS SUCH AS WILHELM BECKER, GIOVANNI BASTIANINI AND ICILIO JONI HAVE LEFT A HERITAGE IN TERRACOTTA, COINAGE AND PAINTINGS RESPECTIVELY, THAT WE HAVE STILL TO FULLY PINPOINT AND EXTRACT FROM ANTIQUITY'S TRUE ARTIFACT RECORD [1,2]. IN RECENT YEARS, NEW SCIENTIFIC METHODS, PARTICULARLY DATATION BY THERMOLUMINESCENCE OF CERAMIC WARES, HAVE HAD A MAJOR IMPACT IN SORTING OUT SOME OF THESE PROBLEMS [3]. HOWEVER, MANY MATERIALS, INCLUDING GOLD, JADE AND, INDEED, MOST STONES USED IN SCULPTURE, LACK A RELIABLE AND/OR ROUTINE METHOD FOR THE DETERMINATION OF THEIR AUTHENTICITY.

THE PRINCIPAL PROBLEMS ARISE WHEN THE ARTIFACT UNDER SCRUTINY HAS LITTLE OR NO STYLISTIC PARALLEL, AS, FOR EXAMPLE, WAS THE CASE FOR THE SO-CALLED BAPHOMET STONE IDOLS [2] AND THE "STONE AGE" ENGRAVINGS AT AERTZEN [4]. UNLESS THE FORGER ACTUALLY ADMITS HIS OR HER INVOLVEMENT, THE AUTHENTICITY QUESTION MOVES INTO THE REALM OF RESEARCH.

SUCH A QUESTION WAS POSED BY A GROUP OF FIFTEEN STONE STATUETTES FROM THE LEO MILDENBERG COLLECTION THAT WERE DISPLAYED AT THE CLEVELAND MUSEUM OF ART. ALL THE STATUETTES WERE SIMPLY SHAPED ANIMAL FORMS (FIG. 1) WITH DEEPLY CARVED ENGRAVINGS ON THEIR UNDERSIDES; ALL HAVE SOME DRILLED PARTS, SUCH AS THE EYES AND NOSTRILS. KOZLOFF [5], WHILE SUGGESTING A LATE FOURTH MILLENNIUM B.C. NEOLITHIC DATE, COMMENTED THAT:

...THIS IS AN EXTRAORDINARY GROUP OF OBJECTS PROBABLY DERIVING FROM A CULTURE THAT IS NOT YET KNOWN. THE FACT THAT THESE SMALL STONE ANIMALS ARE DIFFICULT TO IDENTIFY AND INDEED, MAKE UP PART OF A UNIQUE FIND HAS CAUSED SOME SCHOLARS TO DOUBT THEIR AUTHENTICITY...(THEY ARE) SAID TO HAVE BEEN FOUND IN THE REGION NORTH OF LAKE VAN, TURKEY.

MICROSCOPIC AND MINERALOGICAL STUDIES

OUR INVESTIGATIVE APPROACH HAS BEEN A VARIANT OF AN INFLUENTIAL HYPOTHESIS FORMULATED BY SEMENOV [6]. IN THE STUDY OF ANCIENT LITHIC TOOLS, HE SUGGESTED THAT THEIR ORIGINAL

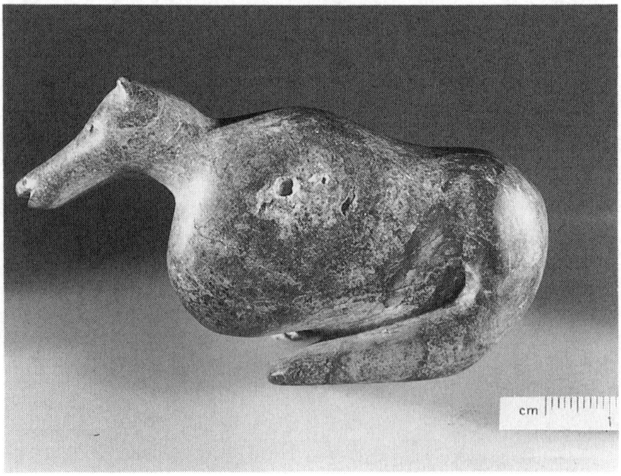

FIG. 1 GALLOPING FOX. MILDENBERG COLLECTION.

FUNCTION COULD BE ESTABLISHED IF FIRST ONE IDENTIFIED THE SURFACE MARKINGS ON THE TOOL, THEN SIMULATED THEM EXPERIMENTALLY ON A MODERN LITHIC OF SIMILAR STRUCTURE AND COMPOSITION. WE, HOWEVER, FOCUSSED OUR ATTENTION ON A PORTION OF AN ARTIFACT PRODUCED BY SUCH A TOOL, RATHER THAN THE TOOL ITSELF: SPECIFICALLY THE MORPHOLOGY OF THE SURFACE OF A DRILL-HOLE, RATHER THAN THE DRILL ITSELF.

THE MORPHOLOGY CAN BE SEEN WITH REMARKABLE CLARITY IN AN IMPRESSION OF THE DRILL-HOLE TAKEN NONDESTRUCTIVELY WITH A POLYVINYL SILOXANE (REPROSIL, L.D. CAULK, MILFORD, DE), HENCEFORTH REFERRED TO AS SILOXANE [7]. THE IMPRESSIONS ARE EXAMINED BY SCANNING ELECTRON MICROSCOPY (SEM). ONCE THE MATERIAL OF THE ARTIFACT HAS BEEN IDENTIFIED, AN ATTEMPT IS MADE TO DUPLICATE THE DRILLING CHARACTERISTICS EXPERIMENTALLY USING A VARIETY OF DRILL TYPES BOTH ANCIENT AND MODERN. THUS WE HAVE BEEN ABLE TO DEMONSTRATE A STEATITE STAMP SEAL OF THE JEMDT NASR PERIOD (CIRCA 3000 B.C.) WAS BORED WITH A WOODEN DRILL, AND THAT THE ANTERIOR TEETH OF THE ANCIENT MAYA, INLAID WITH JADE AND HEMATITE, WERE PROBABLY DRILLED USING A WOODEN DRILL AND A LOOSE ABRASIVE [8]. WE ALSO CONSLUDED THAT A MARBLE CYCLADIC STATUETTE, DATING CIRCA 2500 B.C., WAS DRILLED WITH FLINT [9] AND THAT HEMATITE SEALS OF THE OLD BABYLONIAN PERIOD IN MESOPOTAMIA WERE DRILLED WITH A COPPER ROD AND AN ABRASIVE [10].

THE STATUETTES IN THE MILDLENBERG COLLECTION WERE IDENTIFIED USING X-RAY DIFFRACTION AND EVALUATION OF REFRACTIVE INDEX AS DICKITE, A CLAY MINERAL OF THE KAOLIN FAMILY, MOHS 2-3. THIS IDENTIFICATION WAS CONFIRMED BY DIRECT ANALYTICAL COMPARISON OF A KNOWN DICKITE SPECIMEN FROM THE AMERICAN MUSEUM OF NATURAL HISTORY IN NEW YORK AND THAT SPECIMEN THEN ALSO SERVED AS ONE OF THE SUBSTRATES UPON WHICH WE UNDERTOOK OUR EXPERIMENTAL DRILLING. WE, HOWEVER, ALSO WERE ABLE TO OBTAIN A BIRD STATUETTE FROM THE NORBERT SCHIMMEL COLLECTION THAT HAD A

DAMAGED SURFACE UPON WHICH WE REPLICATED A SET OF DRILLINGS INTENDED TO MATCH MICROMORPHOLOGICALLY WITH THOSE IN THE SUSPECT SCULPTURES.

TO SIMULATE ANCIENT PROCEDURES, DRILLS OF FLINT, COPPER AND WOOD WERE USED IN A HAND-HELD CARPENTER'S BRACE THAT WAS OSCILLATED BACK AND FORTH IN A MANNER SIMULATING AN ANCIENT BOW-DRILL. MODERN DRILLING PROCEDURES WERE SIMULATED USING A MOTORIZED UNIT WITH A ROUND STEEL BURR AND AN ABRASIVE STONE OF ALUMINUM OXIDE SIMILAR TO THAT USED IN LAPIDARY. SILOXANE IMPRESSIONS WERE THEN TAKEN FROM TWO MILDENBERG ARTIFACTS AND FROM THE EXPERIMENTAL BORINGS, TO PROVIDE A COMPREHENSIVE DISPLAY OF SCANNING ELECTRON MICROGRAPHS. IT WAS CLEAR THAT THE DRILLING PATTERN ON THE ORIGINAL STATUETTES [11] WAS CLOSELY SIMULATED ONLY BY A FLINT DRILL (FIG. 2).

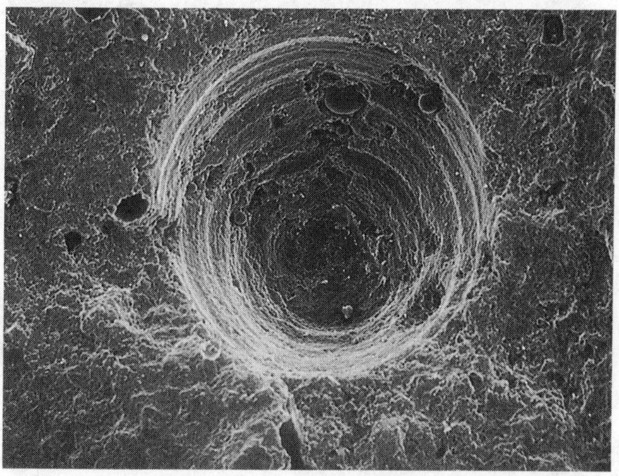

FIG. 2 PATTERN OF A DRILL HOLE IN A BIRD EYE WHICH WAS MATCHED EXPERIMENTALLY WHEN DRILLING WITH A CHIPPED FLINT DRILL. X 15.

EXPERIMENTAL WEATHERING

ANOTHER POTENTIAL CLUE AS TO THE AUTHENTICITY OF THESE STATUETTES COMES FROM THE PRESENCE OF FINE FRACTURE LINES ON THEIR SURFACES. THESE LINES SEEM TO BE A SPALLING PHENOMENON CAUSED BY REPEATED EXPANSION AND SHRINKAGE DURING CYCLES OF NATURAL WEATHERING. THE SURFACE OF EACH STATUETTE ALSO CONTAINED NUMEROUS POLISHING LINES RUNNING IN DIFFERENT DIRECTIONS IN A MANNER CHARACTERISTIC OF HAND POLISHING. WHEREVER POLISHING LINES WERE FOUND IN THE REGION OF A CRACK THEIR PATH WAS MISALIGNED (FIG. 3), THUS PROVING THAT THE FRACTURE OCCURRED AFTER THE POLISHING TREATMENT. WE CONFIRMED THIS CONCLUSION BY DIRECT EXPERIMENTATION ON THE METAMORPHASED CLAY AND STEATITE BLOCKS.

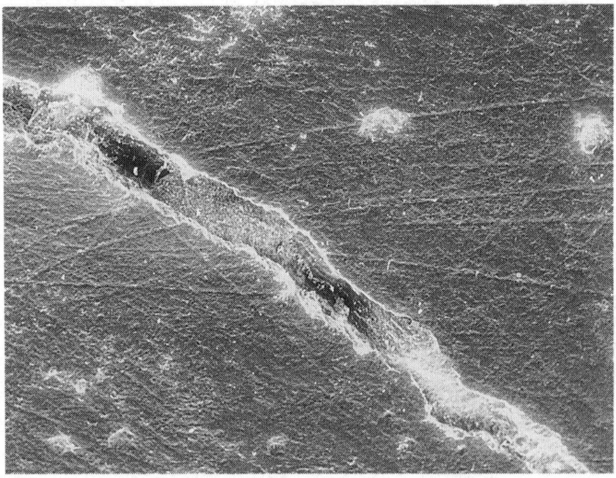

FIG. 3 MISALIGNMENT OF LINES FROM POLISHING ON THE SURFACE OF AN ARTIFACT WITH SUBSEQUENT SPALLING (CRACKING). X 25.

AS TO THE QUESTION OF WHETHER OR NOT A FORGER COULD INDUCE A SIMILAR SURFACE APPEARANCE ARTIFICIALLY, WE SOAKED PIECES OF DICKITE IN BOTH FRESH WATER AND SEA WATER FOR A MONTH, THEN FROZE AND THAWED THEM FOR A WEEK, THEN HEATED THEM AT 430°C IN AN OVEN, FOR THREE HOURS, AND RAPIDLY OVER AN OPEN BUNSEN FLAME. NO SPALLING OCCURRED UNDER THESE CONDITIONS.

THE EXPERIMENTAL APPROACH OF FUNCTIONAL ANALYSIS, DESCRIBED ABOVE, IS A VALUABLE APPROACH TO AUTHENTICITY PROBLEMS FOR STONE ARTIFACTS. IF, AFTER ALL, A PATTERN OF A MODERN TOOL MARK HAD BEEN DETECTED IN THIS STUDY, FORGERY OR REWORKING WOULD HAVE BEEN CONFIRMED WITH NO EQUIVOCATION. IN THIS STUDY, THE WORK PROVIDED ONLY EQUIVOCAL EVIDENCE FOR THE AUTHENTICITY OF THESE STATUETTES, BECAUSE THE TOOLMARKS WERE CONSISTENT WITH THOSE OF ANCIENT TECHNOLOGY, AND NO EVIDENCE OF FORGERY HAD BEEN FOUND.

SIMILAR CONCLUSIONS WERE ALSO DRAWN FROM A SECOND STUDY [12] OF THIRTEEN SIMILAR ARTIFACTS MADE FROM DICKITE.

SINCE THE RESULTS FROM THE TWO STUDIES WERE EQUIVOCAL, THE PROS AND CONS OF AUTHENTICITY CAN BE SUMMARIZED AS FOLLOWS:

1. THE MARKS IN THE DRILL HOLES WERE DUPLICATED USING A FLINT DRILL AND NO OTHERS; HOWEVER, THERE MAY BE OTHER HARD STONE TYPES THAT WE HAVE NOT TRIED.
2. A FLINT DRILL IS EVIDENCE OF ANCIENT MANUFACTURE, BUT A CONTEMPORARY FORGER COULD HAVE MADE A FLINT DRILL AND USED IT.
3. USING A FLINT DRILL AS AN AWL IS AN ANCIENT METHOD, BUT IT COULD ALSO BE USED BY A CONTEMPORARY CRAFTSMAN.
4. THE SHIFT IN LINEARITY OF FINE SURFACE ABRASION GROOVES AT CRACKS IN THE ARTIFACT IS DUE TO SPALLING, A LONG-TERM NATURAL WEATHERING PROCESS, BUT FORGERS COULD ARTIFICIALLY WEATHER STONE.

5. Our experimental stressing method did not produce spalling; but there may be other methods that we have not tried.
6. Dickite-rich stone comes from the Lake Van area in Turkey from which these artifacts supposedly came; but there is no definite proof as to where these dickite-rich stones came from or where the artifacts were made.
7. The fine abrasion grooves were produced by hand, which is characteristic of ancient manufacture; however, a contemporary craftsman also could have used hand finishing methods.
8. There is a consistency of technique and style typical of ancient culture; but this could be the consistency of a forger.
9. The initial stylistic information according to some scholars resembles artifacts from that region; however, the style is completely new and dissimilar to other styles of the area according to other scholars.
10. Where the drilling pierces completely through, as in the artifacts with handles, it is drilled from each end and has a biconical shape. This is typical of ancient drilling and consistent with the use of chipped stone drills. However, a knowledgeable forger could easily have duplicated this technique.
11. There are no other known artifacts made of this stone, which may or may not argue for the authenticity of the statuettes. However, a forger could have used this stone because of its availability and ease of working.

We concluded that while we had found evidence to suggest their authenticity and none for their forgery, two important equivocations were stressed. (1) A contemporary forger could have made a chipped stone drill and used it in the ancient manner, i.e., as a hand-held borer. (2) A contemporary forger may have had a way of artificial aging or stressing the completed artifact that we have not tried or tried sufficiently. Further research is needed for definitive evaluation of authenticity because more objects are becoming available for study.

The third study was concerned with a group of fifty artifacts that were donated to the Boston Museum of Fine Arts as part of a study collection (Fig. 4). They are bird forms similar in style to the previous examples analyzed and reported. One obvious difference is the larger size and the type of stone which was identified as serpentine and not dickite, (personal communication, communication, Dr. George Harlow, Curator, Mineralogy Dept., American Museum of Natural History and Deane K. Smith, Professor of Mineralogy, Pennsylvania State University). The basic method of investigation was the same as for the first two groups of artifacts; namely, that of a functional analysis involving siloxane impressions, scanning electron microscopy and experimental duplication.

Observations

Using a contemporary steel twist drill (Fig. 5), an identical match was produced to the tool marks found in the eyes of the artifact (Fig. 6 & 7).

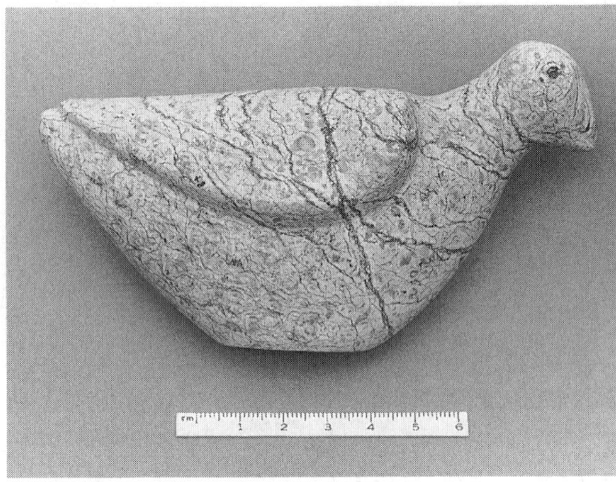

FIG. 4. BIRD FROM BOSTON FINE ARTS MUSEUM COLLECTION CARVED FROM SERPENTINE.

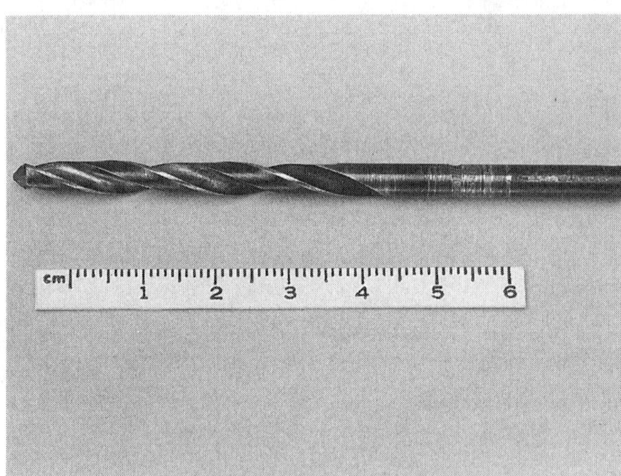

FIG. 5 MACROPHOTOGRAPH OF THE CONTEMPORARY TWIST DRILL USED IN THE FUNCTIONAL ANALYSIS.

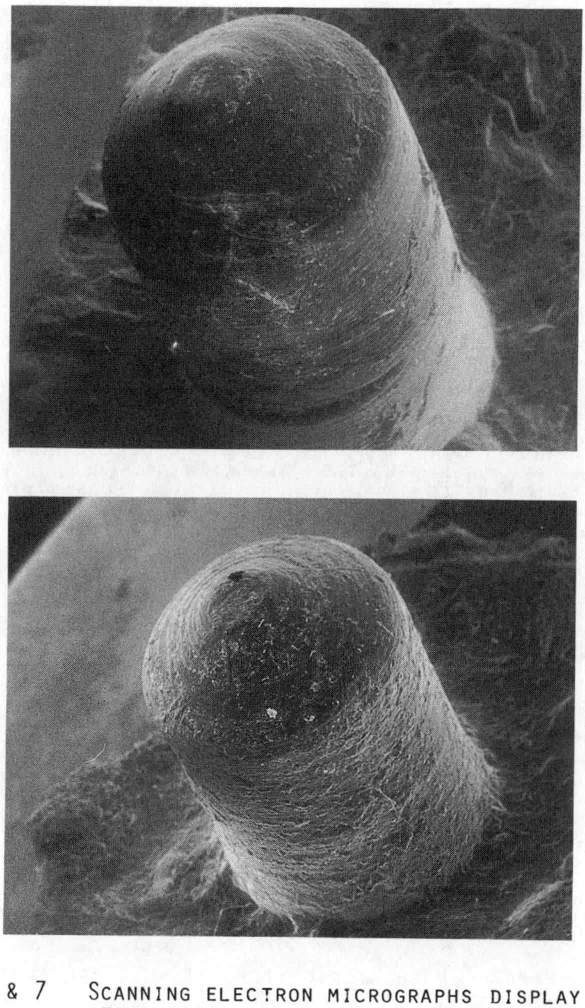

FIGS. 6 & 7 SCANNING ELECTRON MICROGRAPHS DISPLAY THE DETAILS OF THE DRILL HOLES IN THE ARTIFACT (FIG. 6) & THE EXPERIMENTAL SUBSTRATE (FIG. 7) IN WHICH A CLOSE RESEMBLANCE CAN BE SEEN FOR THE SHAPE, SIDE WALL AND LEADING EDGE CHARACTERISTICS. X 15.

The characteristics were a uniform parallel shape, concentric lines on the side walls, and a pyramidal leading edge. No other tool that we used, either contemporary or simulated ancient ones, produced a match experimentally. The contemporary drills used were; round and rectangular steel burs, diamond and green stone ball drills. The simulated ancient drills used were; chipped flint, copper rod and wooden rod with and without abrasives.

Conclusion

In the study of the artifacts from the Mildenberg collection, we found that we were able to obtain an experimental match using a simulated ancient tool; namely, a chipped flint drill and not by using a contemporary drill. We concluded that while there was no evidence of forgery, "a modern forger could also have made and used a flint drill...and that another type of drill, one not considered by us.......could produce the same pattern." In the present study, we conclude that definite unequivocal evidence for forgery exists. In this study, the substrate was serpentine unlike the dickite artifacts in the Mildenberg collection. The shape, sidewall and leading edge characteristics of drill holes in this latest study were quite definitely made with a modern drill.

It is evident that the method of investigation, namely the non-destructive recording of the structural characteristics of drill holes and observing them by scanning electron microscopy is useful. It may be limited to proving that an artifact was forged and not that it is genuine. For that, new methods of absolute and relative inquiry are needed. In other words, it would be highly desirable to be able to provide a data for stone in the manner that C^{14} provides for organic material.

References

1. P.R. Franke, Schwizer Munzblatter 30, 33, (1958).

2. O. Kurz, Fakes, (Dover, New York 1966) p. 309.

3. S.J. Fleming Authenticity in Art, (Crane, Russak, Inc., New York 1975) p. 73.

4. A. Reith Archaeological Fakes, (Barrie and Jenkins, London 1970) p. 68.

5. A.P. Kozloff Animals in Ancient Art, (Indiana University Press, Cleveland 1981) p. 9.

6. S.A. Semenov Prehistoric Technology, (Barnes and Noble, New York 1964) p. 2.

7. A.J. Gwinnett and L. Gorelick SEM, Inc. II 575-580 (1979).

8. A.J. Gwinnett and L. Gorelick Expedition 22 17-32 (1979).

9. A.J. Gwinnett and L. Gorelick J. Field Archaeology 10 378-384 (1983).

10. A.J. Gwinnett and L. Gorelick Expedition. 29 15-24 (1987).

11. A.J. Gwinnett and L. Gorelick MASCA 2, 88-90 (1982).

12. L. Gorelick and A.J. Gwinnett Archaeometry Proceedings 24th International Symposium. p. 63-72 (1986).

PART II C

Properties of Materials Used by the Artist, Artisan and Conservator

A GENERAL METHOD FOR DETERMINING THE MECHANICAL PROPERTIES NEEDED FOR THE COMPUTER ANALYSIS OF POLYMERIC STRUCTURES SUBJECTED TO CHANGES IN TEMPERATURE AND RELATIVE HUMIDITY

Marion F. Mecklenburg, Charles S. Tumosa and Mark H. McCormick-Goodhart
Conservation Analytical Laboratory
Smithsonian Institution, Washington

ABSTRACT

This study examines the methods for determining the effects of temperature and relative humidity (RH) on the dimensional and mechanical properties of artists' materials. Using both of these properties, typical cultural objects, such as paintings and photographs can be modeled on the computer to correlate the magnitude of developed stresses to environmental changes. Comparing these calculated stresses with the measured strength of the materials provides the opportunity to assess the potential risk from damage due to environmental change. Recognizing that the mechanical properties of polymeric materials are functions of the rate of the application of load, the time they are subjected to a load, the temperature and the relative humidity, it becomes a question of identifying those parameters that are the most important in determining the correct information needed for computer modeling.

INTRODUCTION

The stresses in a structure are usually determined with direct measurements using strain gages, stress coatings, or other systems that measure physical displacements of the structure.[1] These measurements and the mechanical properties of the materials used in its construction then help determine the stresses. Structures that are fully restrained and cooled do not exhibit these displacements and their material properties change. Paint film stresses, resulting from temperature or relative humidity fluctuations in paintings cannot be measured directly.[2] Traditional direct structural analysis requires displacements with moderately constant mechanical properties. Many structures, such as stretched paintings or photographic materials, exhibit little or no in-plane displacements and have highly variable mechanical properties. It then becomes necessary to find other techniques for analyzing such structures subjected to environmental changes.

One of the most powerful analytical tools for structural analysis is the digital computer with the technique of Finite Element Analysis (FEA).[3] FEA readily lends itself to complex structures composed of multiple materials. The fundamental principle of the method allows one to "sub-divide" a complicated structure into a "finite" number of simpler sub-structures called "elements". These elements are connected at points called "nodes". These simple elements are mathematically relatively easy to manipulate.

An example of a structural analysis using only a single element (Fig. 1) is a simple rod which is pulled with the force, F, at one end while the other end is stationary, or "fixed". The nodes, points 1 and 2, can be considered to be at the ends of the rod. In this analysis, the force, F, causes no displacement at node 1

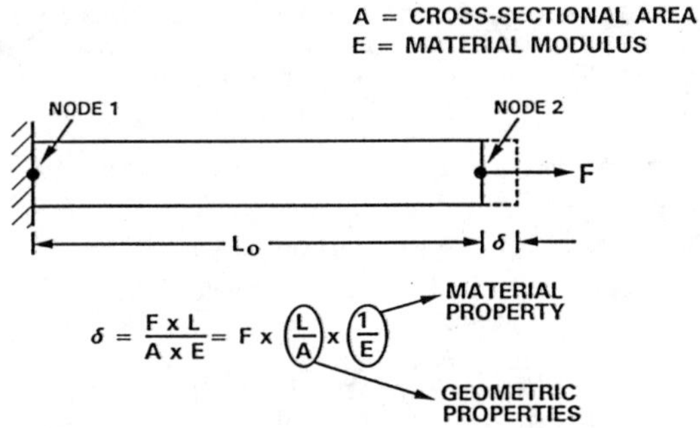

Figure 1. A simple structure modeled with a single element.

and a displacement, δ, at node 2. The magnitude of the displacement, δ, can be calculated as:

$\delta = (F \times l)/(A \times E)$ Eq. 1

where F is the applied force
l is the length of the structure, and in this case the length of the element,
A is the cross-sectional area of the structure (element) and
E is the modulus of elasticity, a material property.

The quantities, A and l, are geometric properties of the structure. In the example shown in figure 1, the geometry of the structure is quite simple and presents no mathematical complications. This is rarely the case in actual structures where the geometry can be complex (see fig. 2a). This complexity can be remedied by subdividing the structure into a "finite" number of "elements" as shown in figure 2b. These elements are geometrically simple and the structure is easily analyzed using a finite element program.

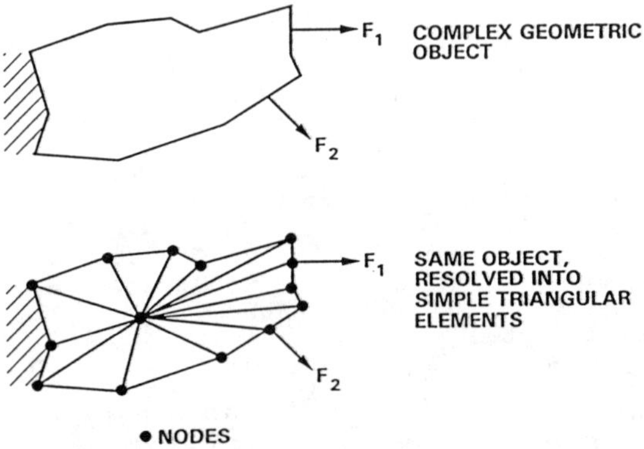

Figure 2. A complex structure modeled with several triangle elements.

In both the simple and complicated case, the mechanical and dimensional (physical) properties of the materials are of utmost importance, and are critical in correctly analyzing any structural problem. This paper concentrates on the mathematical and experimental procedures used in determining the physical and mechanical properties needed for the analysis of complex structures subjected to changes in temperature and relative humidity.

THE FUNDAMENTAL MECHANICAL PROPERTIES OF MATERIALS

From the tensile test of a material many fundamental properties of the material can be determined. The results of one of these tests is presented in figure 3. In this figure the different quantities of importance are labeled. Stress, σ, is equal to the applied force, F, divided by the test sample's cross-sectional area, A. Stated mathematically;

$$\sigma = F/A. \qquad \text{Eq. 1}$$

Engineering strain, ϵ, is the change in specimen length, δ, divided by the original length of the specimen, L_o, where;

$$\epsilon = \delta/L_o = (L_s - L_o)/L_o \qquad \text{Eq. 2}$$

and L_s is the "stretched" length of the specimen.

The strain times 100 is the percent elongation of the test specimen. The ultimate strength of a material, σ_{ult}, is the maximum stress the material can sustain, usually the point at which the material breaks. The modulus, E, is the measure of a material's ability to deform when subjected to stress, and may be viewed as a measure of the materials stiffness. Mathematically the modulus is:

$$E = \sigma/\epsilon. \qquad \text{Eq. 3}$$

The modulus is defined normally for the elastic region of a stress-strain plot where the material returns to its original

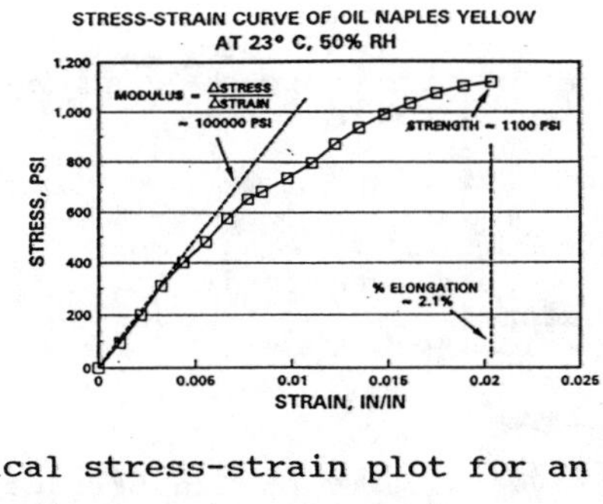

Figure 3. Typical stress-strain plot for an oil paint.

dimensions when the stress is removed. A material is said to exhibit plastic behavior if it remains permanently deformed after the removal of the stress. The yield point, σ_{yld}, is the stress where the material goes through a transition from elastic to plastic behavior. In many materials, this point is not clearly defined.

All of the properties defined above are altered with changes in temperature, relative humidity, and the speed with which the forces are applied.[4] In general, lowering the temperature or relative humidity, or increasing the speed of the applied force all increase the strength of the material, the stiffness of the material, E, and reduce the strain to failure. Raising the temperature, relative humidity, or slowing the speed of force application, lowers the strength and modulus, but increases the materials ability to deform, (strain to failure).

Because the time required for polymeric materials to equilibrate with new environments is relatively long, it is necessary to examine the mechanical properties of these materials under long-term or "equilibrium" conditions. The mechanical properties of painting materials are considerably different under these slow, long-term conditions than they are under rapid, dynamic conditions. Polymeric materials exhibit the ability to deform considerably when stresses are applied slowly. The maximum strengths attained are also quite low under these conditions when compared to rapid loading conditions.

STRESS DEVELOPMENT IN FULLY RESTRAINED MATERIALS RESULTING FROM CHANGES IN TEMPERATURE AND RELATIVE HUMIDITY

If any of the artists' materials are fully restrained and subjected to decreases in either temperature or relative humidity, they will experience an increase in tension. This results from the material's inability to contract while undergoing a loss of heat or moisture. The magnitude of the tensile stress the material experiences is a function of both the attempted shrinkage and the modulus of the material. The modulus in any of the artist's organic materials is dependent on both the temperature and relative humidity and must be taken into account in any calculation.

Temperature Effects

In general, if the thermal coefficient of expansion and the equilibrium modulus of a material are known, the stress levels resulting from cooling a restrained material can be predicted. The method of calculating this stress increase can be derived directly from the basic equation 3, repeated here;

$$E = \sigma/\epsilon \qquad \text{Eq. 3}$$

which can be rewritten as:

$$\sigma = E \times \epsilon. \qquad \text{Eq. 4}$$

To include the temperature effects, it must be recognized that both the modulus, E, and the strain, ϵ, are functions of temperature and therefore:

$$\sigma(T) = E(T) \times \epsilon(T). \qquad \text{Eq. 5}$$

Relative Humidity Effects

The effects of relative humidity on the artists' materials are analogous to those of temperature. Changes in relative humidity induce both dimensional and mechanical properties changes. In effect then, stress development in fully restrained materials subjected to changes in RH can be treated in a manner similar to restrained materials subjected to changes in temperature. The equation for relative humidity behavior would be:

$$\sigma(RH) = E(RH) \times \epsilon(RH) \qquad \text{Eq. 6}$$

STRAINS INDUCED BY TEMPERATURE OR RELATIVE HUMIDITY

Strains, as defined in Eq. 2, require a change in the length of the specimen. This presents an apparent paradox when discussing the strains developed in a material with fixed dimensions and subjected to either desiccation or cooling.

The strain, $\epsilon(T)$, as a function of temperature is derived by considering the shrinkage of the specimen as if it were free to do so. The specimen would contract upon cooling and it is effectively being "stretched" backed to its original restrained length. The "unstretched" length would be the free shrinkage length and the deformation would be the amount needed to stretch the contracted specimen to the restrained length. The amount of shrinkage is calculated from the thermal coefficient of expansion. Using the thermal coefficient of expansion, γ, the freely contracted length for the paint for any temperature is:

$$L_T = L_R + \gamma \times \Delta T \times L_R \quad \text{or}$$
$$L_T = L_R \times (1 + \gamma \times \Delta T) \qquad \text{Eq. 7}$$

where L_T is the free shrinkage length,
L_R is the restrained length,
ΔT is the change in temperature the material experiences,
γ is the thermal coefficient of the material, and for these calculations is considered to be a constant although that is not a requirement.

The strain at any temperature is now calculated as:

$$\epsilon(T) = (L_R - L_T)/L_T. \qquad \text{Eq. 8}$$

Equation 5 can also be written in terms of the thermal coefficient of expansion, temperature change, and the stretched length as:

$$\epsilon(T) = (L_R - L_R \times (1 + \gamma \times \Delta T))/(L_R \times (1 + \gamma \times \Delta T)) \qquad \text{Eq. 9}$$

or as

$$\epsilon(T) = -(\gamma \times \Delta T)/(1 + \gamma \times \Delta T). \qquad \text{Eq. 10}$$

The negative sign indicates that positive tensile strains are resulting from negative temperature changes, i.e. cooling. For changes in relative humidity, the equation for the strains developed in a fully restrained material and desiccated are:

$$\epsilon(RH) = -(\alpha \times \Delta RH)/(1 + \alpha \times \Delta RH) \qquad \text{Eq. 11}$$

and ΔRH is the change from the original RH to the new RH, with α the "moisture" coefficient of expansion.

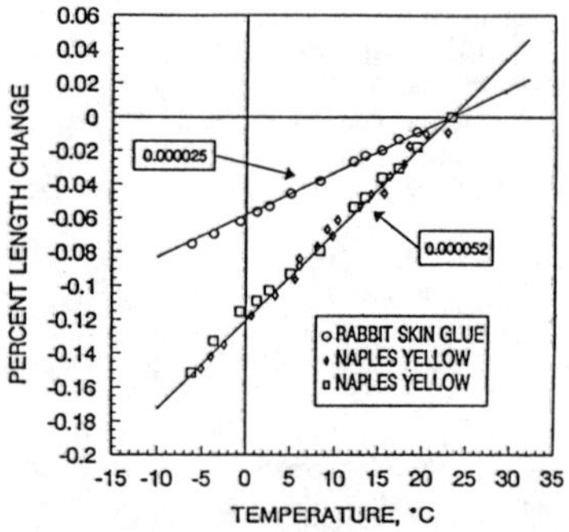

Figure 4. Plot of the percent length change versus temperature for rabbit skin glue and naples yellow oil paint. This data is used to determine the thermal coefficients of expansion for these materials.

Thermal Coefficients of Expansion

The thermal coefficient of expansion for thirteen year-old oil paint and hide glues were measured. The total change in the length of these materials is linear and quite small over the tested temperature range (see fig 4). The slope of these lines divided by one hundred will give the thermal coefficients (γ) of expansion for these materials. This is the change in strain per degree C. For the paint, the thermal coefficient, γ, is .000052 per degree C and for the rabbit skin glue, a thermal coefficient of about half that or .000025 per degree C. For comparison purposes, three common metals, copper, steel, and aluminum have coefficients of .000017, .000011, and .000024 per degree C respectively.

Moisture Coefficients of Expansion

Oil Paints

Most artists' materials will expand with an increase in moisture content and conversely shrink upon its loss. The driving mechanism for changes in the materials' moisture content is the ambient relative humidity. Oil paints, for example, have a highly non-linear variation in their dimensional response to relative humidity. Figure 5 illustrates a paint's dimensional response to RH. In this figure, flake white, ground in safflower oil, swells a maximum of .18% between 0% and 70% RH, with a larger increase, about .6%, occurring between 70% and 95% RH. This specimen, which was about .038 cm (.015 in) thick, took at least forty-eight hours to reach equilibrium with each environment at which the data were recorded.

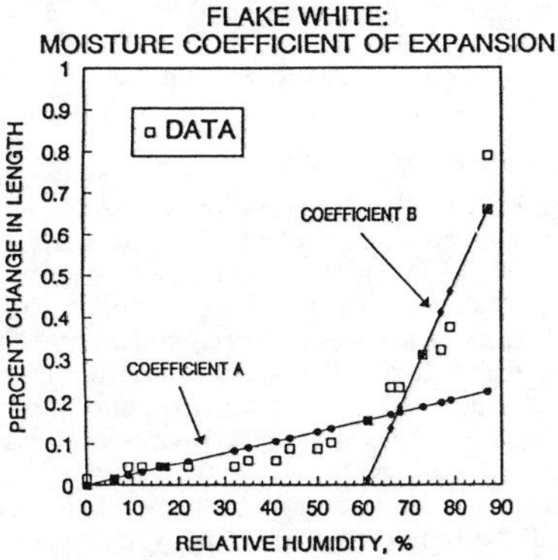

Figure 5. Plot of the percent change in length versus relative humidity for a typical oil paint. This data was used to determine the approximate "moisture" coefficients of expansion for this paint.

It is now possible to calculate two different "moisture" coefficients of expansion that linearly approximate the effect of moisture on the dimensional properties of the paint. These coefficients include the ranges from 0% to 70% RH and from 70% to 95% RH and can be calculated by using the percent length changes in each of the relative humidity ranges. For example the low relative humidity range, moisture coefficient, α_a, for the flake white is;

$\alpha_a = (.18/70)/100 = .0000257$, which is the change in strain per percent relative humidity. This coefficient will be used to calculate the stresses resulting from the desiccation of restrained paint. For the high relative humidity range;

$\alpha_b = (.62/25)/100 = .000248$, which is nearly ten times the low range coefficient. These are rarely going to be used since the equilibrium modulus of paints above 70% RH is nil. A non-linear treatment of the moisture coefficient is also possible.

Rabbit Skin Glue and Photographic Gelatin

One material, traditionally used in paintings, that is quite dimensionally responsive to relative humidity is rabbit skin glue (RSG). When further refined, hide glues are used as photographic gelatins found in the image and anti-curl layers of photographs.

As with the paints, only the low end moisture coefficients are useful. For the rabbit skin glue, α_a is .000264, over ten times that of the paints. For the more refined and processed photographic gelatin, α_a is .00044. This is significantly greater than the rabbit skin glue and remarkable considering that these are chemically very similar materials.

MECHANICAL PROPERTIES UNDER "EQUILIBRIUM" CONDITIONS

Temperature Effects

Depending on the material and the environment, artists' materials will "stress relax" over time. These times can be considerable.[5] For example, hide glues will stress relax to zero over a period of about 5 months at room temperature and low RH. Oil paints tested at low temperatures and moderate relative humidity give all indications of taking decades and even centuries to relax fully. On the other hand, the amount of time needed for these materials to equilibrate to a new environment is considerably shorter, for the paints, a matter of days and, for the glues, a matter of hours. Damage induced by environmental change occurs over these shorter time spans and the mechanical properties of the materials should be measured within similar time spans.

Direct Determination of E(T)

Several paints were mechanically tested under long-term loading conditions. Sample preparation and tests procedures are outlined in another paper.[6] The tests were conducted by applying a small strain to the test sample, about .007, and allowing it to stress relax fully. Once full stress relaxation was attained, a subsequent increment of strain was applied and the stress was again allowed to relax (Figure 6).

The time to relax fully the paints was typically between five and ten days, depending on the paint. This process was repeated until the specimen broke. Each test took several months to complete. The modulus was determined by taking the slope of the locus of relaxed points from the equilibrium test (see Figure 6). In this case, for Naples yellow at 23°C, 50% RH the modulus was 10,000 pounds per square inch (10 ksi), (68.9 Mega-pascals (MPa)).

The equilibrium modulus for the Naples yellow was measured at 23°C, 5% RH to be 47.5 ksi (327 MPa) and at -3°C, 5% RH was found to be 150 ksi (1,034 MPa). If it is assumed that E varies linearly with temperature (figure 7) then a linear function for E(T) can be fitted and stated as:

$$E(T) = 138,520 - 3770 \times T \qquad \text{Eq. 6}$$

where the units are psi for the modulus and the temperature is in degrees Celsius. That the function is linear is not unreasonable, since it has been shown that the thermal coefficient is linear over the temperature ranges considered here.

For the Naples yellow, the complete equation for a fully restrained sample of paint subjected to a drop in temperature at 5% RH is now:

$$\sigma(T) = -(138,520 - 3779 \times T) \times (.000052 \times \Delta T)/(1 + .000052 \times \Delta T).$$

where T is the current temperature and ΔT is the change from the original temperature to the current temperature.

Indirect determination of E(T)

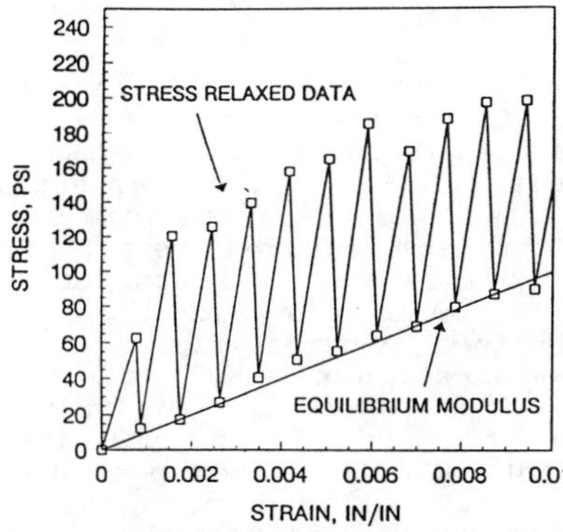

Figure 6. Direct determination the "equilibrium" stress-strain curve for the naples yellow paint by using the "stress-relaxation" method.

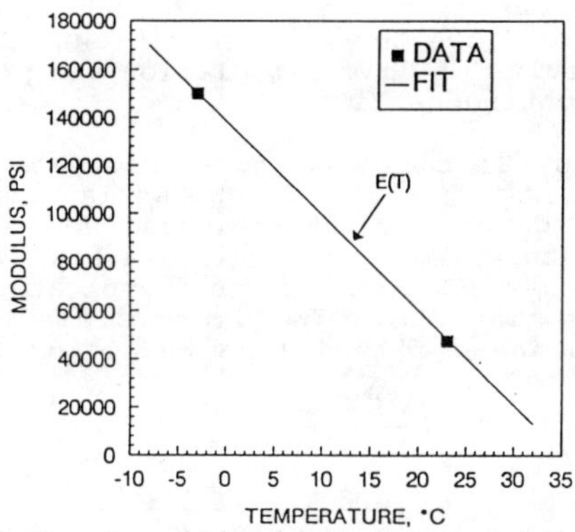

Figure 7. Plot of the "equilibrium" modulus data versus temperature, showing the fit used to determine E(T) for this oil paint.

Since the thermal coefficient was measured, it is possible to derive the equilibrium modulus from the restrained test itself. This can be done by rewriting Equation 5 as:

$$E(T) = \sigma(T)/\epsilon(T) \qquad \text{Eq. 7}$$

where $\sigma(T)$ is obtained from the restrained test and
$\epsilon(T)$ is obtained from the thermal coefficient of expansion.

At -3°C, the extrapolated value of the hide glue modulus is 705 ksi (4,860 MPa), which is only twelve percent less than the rapid loading modulus of 800 ksi (5,515 MPa). This is reasonable since the difference between the rapid loading modulus and the equilibrium modulus for the artists' materials diminishes with a reduction in temperature and relative humidity. E was measured directly at 23° C and 50% RH. Using these data points, E(T) was determined by assuming a linear fit.

Experimental verification

From an experimental point of view, it is not possible to measure easily the stress of a "fully" restrained specimen since the "load cell," the device that measures the stresses, is "compliant." This means the cell "gives" a bit as force is applied to it. This load cell compliance effectively relieves some of the restraint on the specimen and the stresses measured are actually less than if fully restrained. This compliance can, however, be included in the calculations that allow us to predict the behavior of the specimen subjected to restrained temperature changes. The compliance of the test device is a function of the total force and is measured during the test of the specimen. The value of the compliance, measured in units of length, is the raw data output, AO, times a compliance constant, KD. It is now possible to correct for the compliance of the load cell and calculate the expected stresses in the experimental restrained test where:

$$\sigma(T) = E(T) \times (\epsilon(T) - AO \times KD/L(T)) \qquad \text{Eq. 8}$$

This calculation was conducted for the Naples yellow using the measured modulus and the thermal coefficient of expansion of the material. The results of this calculation are presented in Figure 8 as the lower continuous line.

Also on that figure is the measured stress data (squares) for the Naples yellow paint. There is a substantial correlation between the predicted and the actual measured stress levels at the different temperatures while the relative humidity is held at 5%. This experiment was conducted with the temperature both increasing and decreasing. Also shown in figure 8 are the test data for rabbit skin glue and a fitted line resulting from the indirect determination of E(T).

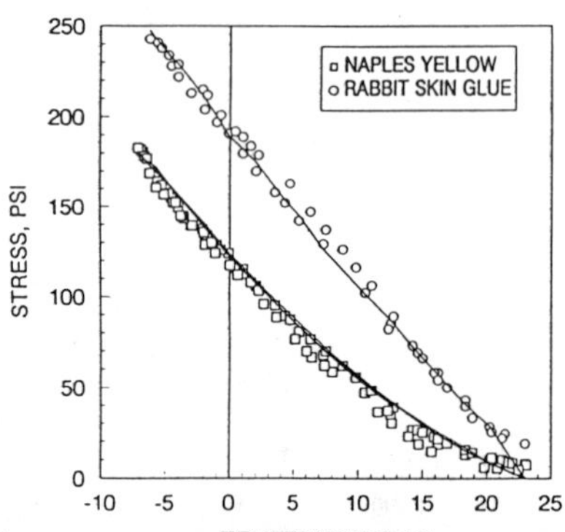

Figure 8. Stress versus temperature plot used to experimentally verify the mathematics developed computer modeling. The symbols are the measured data and the lines are the mathematical functions.

Calculation of the Fully Restrained Specimen

Using equation 5, it is now possible to calculate the stress development in Naples yellow and rabbit skin glue as if they were <u>fully restrained</u>. Figure 9 shows the expected stresses of these materials when fully restrained and chilled from 23°C to -6°C at 5% RH. The load cell compliance has a significant stress reduction effect since the fully restrained stresses are nearly twice for the glue and about fifteen percent higher for the paint.

The immediate conclusion to be drawn from Figures 8 and 9 is that while the coefficient of thermal expansion for hide glue is only one-half that of the Naples yellow paint, the hide glue still has a larger stress increase, not because of changes in dimension but because of a substantially higher modulus. It is important to recognize that both the modulus and the coefficient of expansion influence the magnitude of the resulting stresses in materials restrained and cooled.

Relative Humidity Effects

The effects of relative humidity on the artists' materials are analogous to those of temperature. Changes in relative humidity induce both dimensional and mechanical properties changes. In effect then, stress development in restrained materials subjected to desiccation can be treated in a manner similar to restrained materials subjected to decreases in temperature.

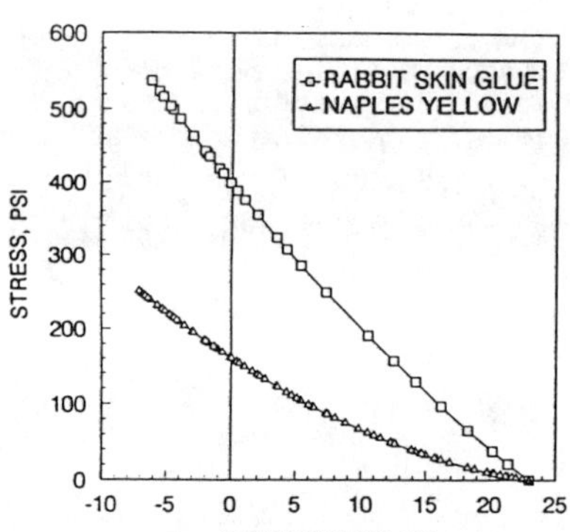

Figure 9. Predicted stress versus temperature for fully restrained rabbit skin glue and naples yellow oil paint.

Oil Paints

Two paints, Naples yellow and flake white, were restrained at 66% RH and 23°C, and desiccated to 58% RH where the stresses in the paints were .002 ksi (.014 MPa) and .004 ksi (.028 MPa) respectively. The paints were further desiccated to 5% RH and the stresses now reached were as high as .066 ksi (.455 MPa) and .092 ksi (.634 MPa) respectively (see Figure 10). The time for equilibration was forty-eight hours for .015 in (.0059 cm) thick films.

For these two oil paints, E(T) was determined directly from paint film testing. The equations for the modulus of the materials were: for Naples yellow, E(RH) = 51,800 - 840xRH and for flake white, E(RH) = 63825 - 840xRH, where the units are in psi and percent relative humidity. These two equations have the same slope and differ only in the intercepts. The difference in the intercept values is the difference in the modulus of each material at 50% RH. Using equation 6 (corrected for load cell compliance), and the expansion coefficient, it was possible to calculate the expected stresses. The value, α_a equal to .0000257, was used for both the Naples yellow and the flake white. The results of the calculations are shown in Figure 10 as the solid lines passing through the data points.

One interesting aspect of these calculations is the remarkable accuracy using linear approximations of the actual material behavior. Equally interesting, for the Naples yellow, is the magnitude of stress reached, (.066 ksi (.455 MPa)), at a desiccation from 66% RH to 5% RH, considerably less than that reached (.145 ksi (1.0 MPa)), when cooled from 23°C to -3°C at 5% RH.

Rabbit Skin Glue and Photographic Gelatin

Until now all of the discussions about the mechanical properties of the artists' materials and stress development in either cooled or desiccated materials have assumed that all behavior was linear. This assumption seems to be valid since all of the elastic modulus

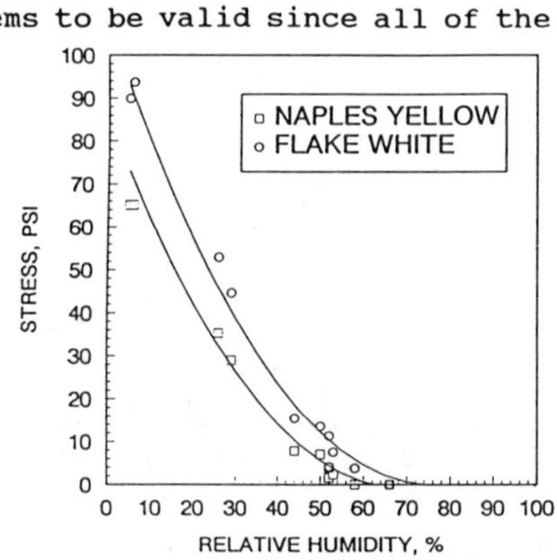

Figure 10. Paint stress versus relative humidity plots used to experimentally verify the mathematics developed for computer modeling. The symbols are the measured data and the lines are the mathematical functions.

equations result in successful stress predictions. For the rabbit skin glue, the behavior is almost never in the elastic range but exhibits a "quasi" plastic, non-linear behavior. The reason for this behavior is that the material wants to shrink extensively with desiccation. This results in strains, when the specimen is restrained and desiccated, that exceed the yield stress for any given relative humidity environment. Once these yield stresses are reached, the glue can have no further rise in stress no matter what strain level is attained. This presents the complication that the linear modulus is no longer valid, since any calculation using these values will result in stresses well in excess of the yield stresses.
What can be done is to determine an effective modulus, $E_c(RH)$, using Eq. 7 modified for RH, for all of the different relative humidity values. The effective modulus is for rabbit skin glue:

$$E_c(RH) = 200,000 + 2200 \times (75-RH_i) + (2200 \times RH)$$

where; RH_i is the value of the starting or initial relative humidity,
RH is the relative humidity at any time.
And a moisture coefficient of α_a = .000264 is as determined.
For the photographic gelatin:

$$E_c(RH) = 130,000 + 2500 \times (72-RH_i) + (1500 \times RH)$$

with moisture coefficient of .00044.

The results of the glue and gelatin modulus functions are compared to actual data in figures 11 and 12. The sample specimens were restrained at different relative humidities and desiccated. Both of the materials showed similar behavior and reach similar maximum stress values at slightly over 4 ksi (27.6 MPa). This is of particular interest since the coefficient of expansion of the glue is lower than the gelatin and the modulus of the glue is considerably higher than the gelatin.

COMPUTER ANALYSIS OF MUSEUM OBJECTS SUBJECTED TO CHANGES IN TEMPERATURE AND RELATIVE HUMIDITY.

Structural analysis was conducted using Finite Element Analysis (FEA) and the digital computer. The program used in the computer modeling in this paper was ANSYS®, Version 4.4, run on a 386, 33 megahertz desktop computer with 4 megabytes RAM.

Modeling The Desiccation of a Cibachrome Print.

There was interest in the stress development by desiccating the image layer of the photographic materials. Low relative humidity levels are currently recommended for long term preservation of photographic materials due to an incremental improvement in chemical stability. For example, 25 to 30% RH conditioning is recommended for color films prior to placing them in cold storage.[7] It is possible to model (using Ansys) a section of a cibachrome photograph. Cibachrome is manufactured by Ilford Photo Corporation and utilizes a unique silver dye bleach process. The finished print is structurally typical of a broad range of photographic materials coated on a polyester film base.[8] The image layer is gelatin and on the reverse there is an anti-curl layer which is gelatin with glass particles acting as a matting agent. The glass increases the stiffness of the material as well as reduces the moisture coefficient of expansion. So in effect we

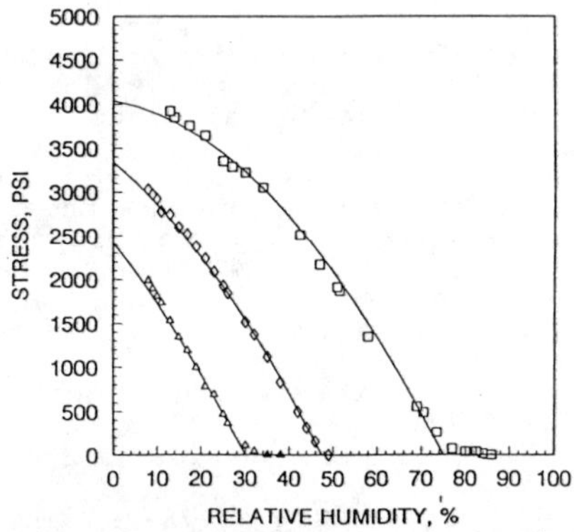

Figure 11. Rabbit skin glue stress versus relative humidity plots used to experimentally verify the mathematics developed for computer modeling. The symbols are the measured data and the lines are the mathematical functions.

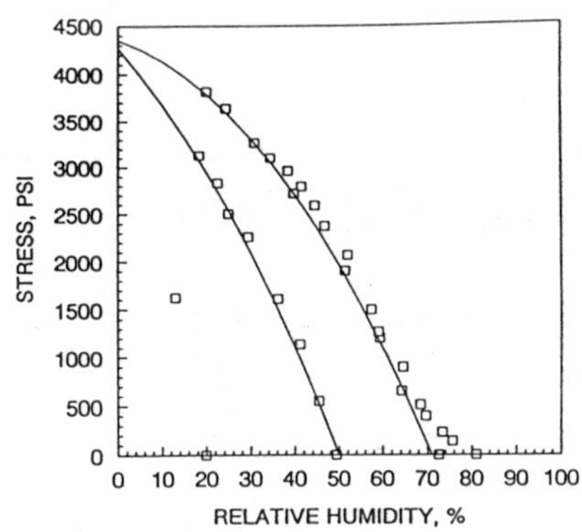

Figure 12. Photographic gelatin stress versus relative humidity plots used to experimentally verify the mathematics developed computer modeling. The symbols are the measured data and the lines are the mathematical functions.

have at least three material layers with distinct mechanical properties. The image layer is .000652 in. (.00166 cm) thick, the polyester base is .007 in. (.0178 cm) thick and the anti-curl layer is .00049 in. (.00125 cm) thick. The primary measurement was curling of the photograph as it dried out. If this calculation was right, then the rest of the stress analysis would be fairly reliable. In addition, it is possible to check the computed stress levels by comparing the results to the tested individual material layers. Figure 13 shows the dimensions of the model and the test specimen and the measured and calculated displacement. The actual test specimen was one inch wide and two inches long, held at the center. The test was started at 61.5% RH and deflections measured at 44.6%, 31%, 25% and 18.8% RH.

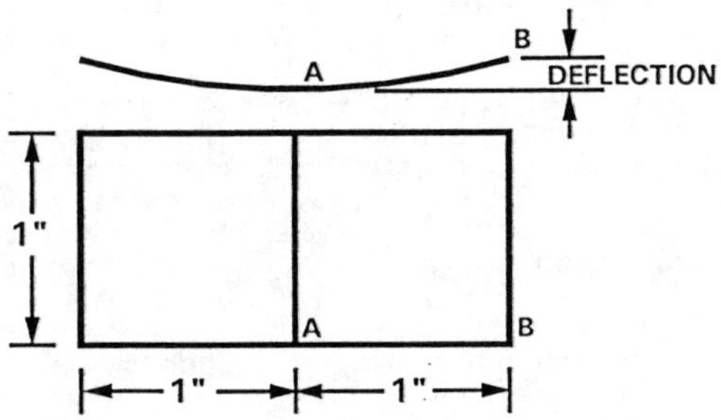

Figure 13. Dimensions of the Cibachrome experimental test specimen showing the displacement measured and computed.

The polyester layer was modeled with the measured moisture coefficient of expansion of .0000077 and a modulus that varied as a function of relative humidity (E(RH) = 695,000 - 1010xRH). Figure 14 shows the edge displacement results of the experimentally measured Cibachrome sample as well as the calculated results if the image layer and the anti-curl were the same gelatin material (labeled: model 2 material), and the calculated results if the model has an anti-curl layer as being stiffer and less responsive to dimensional change (labeled: model 3 materials). Notice that there is a rather distinctive plot for both the measured and correctly calculated displacements.

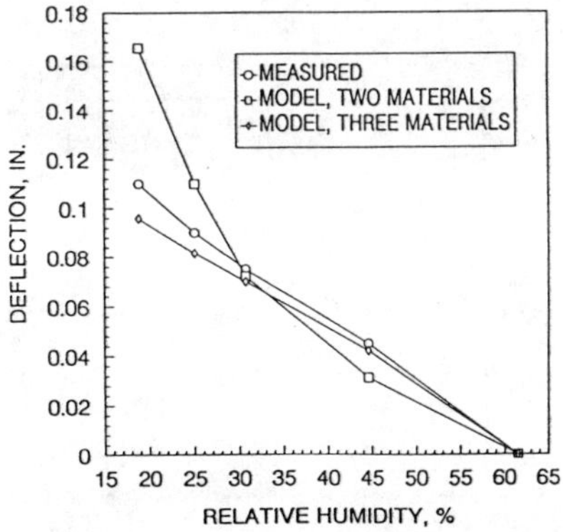

Figure 14. Computed and measured deflections of the Cibachrome computer model and test sample.

Figure 15 has the calculated maximum principal stress results which are as expected; higher than the simple experimental stress due to the biaxial stress field. From a conservation point of view, the stresses induced by cycling from commonly encountered environments to the presently recommended 25-30% RH level are undesirable. In the presence of defects such as cracks or at edges, the high stresses represent the source of delamination of the layers. Mechanical stability can now be quantified and should be considered along with improvements in chemical stability when storage environments are established for photographic materials.

The deflections in this test reach over ten times the thickness of the structure and over one tenth the gage length (1 inch.) of the specimen. This represents the successful modeling of a very large-displacement structure constructed essentially of polymers.

Modeling the Effects of Cooling a 30 in. x 40 in. (76 cm x 102 cm) Painting

A 30 in. x 40 in. (76 cm x 102 cm) painting was modeled and subjected to changes in temperature while the ambient relative humidity was held at 5%. The layer of the painting consisted of a linen .025 in. (.064 cm) thick, modeled as a .006 in. (.0152 cm) layer, a glue layer .002 in. (.00508 cm) thick and a Naples yellow (lead) oil paint layer .003 in. (.0076 cm) thick. The stretcher was not expanded. Each of the layers of the model painting were programmed with their respective material properties taken from the data presented earlier in the this paper. The glue layer was characterized using Equation 10 with a thermal coefficient of .000025 per degrees C and a changing modulus that varied linearly from 705 ksi (4,860 MPa) at -3°C to 550 ksi (3,791 MPa) at 23°C. This data was used to develop the equation:

$$E(T) = 689,875 - 7400xT \qquad \text{Eq. 16}$$

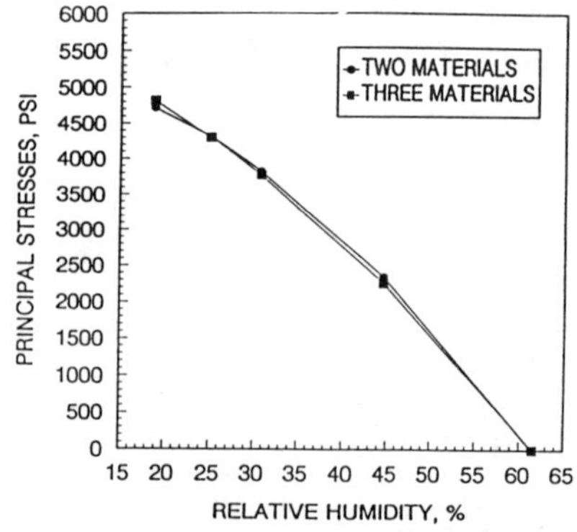

Figure 15. Calculated stresses in the image layer of the cibachrome computer model.

where T is any temperature. Equation 16 was the equation that allowed programming of the change of the rabbit skin glue's modulus with changing temperature at 5% RH. The fabric was programmed with an average constant modulus of 5 ksi (38.6 MPa) and a thermal coefficient of .00001 per degrees C. The distribution of the principal stresses in the paint layer, calculated by the computer is shown in Figure 16. This stress distribution is quite uniform, ranging from .303 ksi to .307 ksi (2.09 MPa to 2.11 MPa) and just reaches the measured breaking strength of Naples yellow paint at 5% RH and -2°C. This suggests that cracking of the paint layer is going to occur throughout the entire surface of the model painting when the temperature is decreased from 23°C to -3°C at 5% RH. It most certainly would have occurred if the starting environment was 23°C and 50% RH. This would be the consequence of the combined adverse effects of both desiccation and cooling.

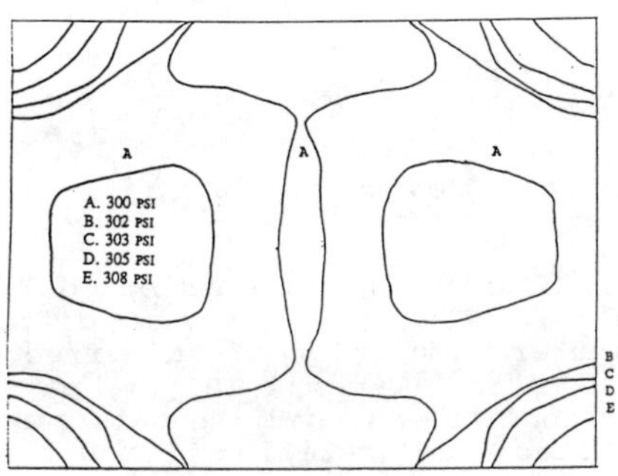

STRENGTH OF PAINT = 300 PSI

Figure 16. Computer calculated, maximum principal stress contours for a 30" x 40" traditional canvas supported oil painting subjected to cooling from 23$_{o}$C to -3$_{o}$C at low relative humidity.

The stresses have a directional bias as shown in Figure 17. In this figure, the principal stresses are indicated by the calculated directional vectors (arrows), and cracking that occurs will do so perpendicular to the vectors as shown by the continuous lines. This is a crack pattern that shows up frequently in paintings in North America, particularly those that are on strainers, i.e. stretchers with fixed, non-expandable corners. If the paint modeled was the flake white tested and reported in this paper, it would have also failed, but at a higher temperature, since this paint has a higher modulus but consistently less strength than the Naples yellow. The glue layer principal stresses reached levels ranging from 4.72 MPa to 5.13 MPa (.685 ksi to .745 ksi).

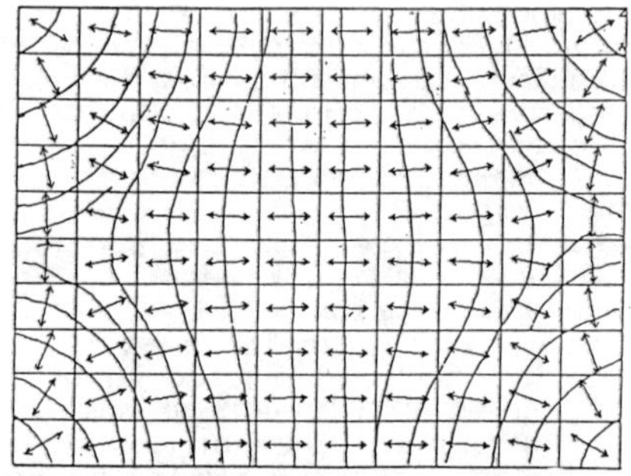

Figure 17. Principal stress directions and the projected crack pattern of a 30" x 40" oil painting subjected to cooling from 23_0C to -3_0C at low relative humidity.

Modeling the Effects of Desiccating a 30 in. x 40 in. (76 cm x 102 cm) in.) Paintings

While seemingly similar to the effects of cooling, reductions in ambient relative humidity have somewhat different effects. In starting the computer simulation of the effects of desiccation, the same 30 in. x 40 in. (76 cm x 102 cm) painting as described in the section on cooling was used. In this case the computer was programmed to alter the relative humidity instead of the temperature. Additionally, in order to obtain greater analytical detail, it was possible to take advantage of the painting's double symmetry and model only the upper right hand quadrant. This was done by programming the appropriate material properties-RH equations for each of the individual layers of the model painting as well as the proper boundary conditions. In this case, the paint layer again was Naples yellow. For the changes in relative humidity; E(RH) = 51,800 - 840xRH was the equation used. For the strain calculations, a moisture coefficient of.0000257 per percent relative humidity was used. For the glue layer, the modulus was obtained as discussed previously under RH and the strains were calculated using a moisture coefficient of.000264 per percent relative humidity. The fabric was assumed to be a minimally contributing material since glue sizing a stretched linen removes any residual stress in the fabric. In this case, the modulus was assigned a nominal constant 5 ksi (35 MPa) and the moisture coefficient was an average .0001 per percent relative humidity. The model painting was desiccated from 70% to 10% RH at 23°C. The model painting was modified to increase the glue thickness from .002 in. to .004 in. (.0051 cm to .01 cm). The results of the analysis showed the stresses in the paint layer ranging from .1

ksi (.698 MPa) to .28 ksi (1.93 MPa). Doubling the glue layer more than doubled the stresses in the paint layer. The distribution of the stresses are shown in figure 18. This distribution is not nearly as uniform as resulting from the cooling analysis. Cracking, if it occurs will initiate mainly in the corner regions of the painting as shown in figure 19. In this figure, both the direction vectors of the principal stresses and the possible crack pattern are shown. If the equilibrium breaking strength of the paint at 5% RH and 23 C was low, only .2 ksi (1.38 MPa), the cracks would run no farther than about 3 in. (7.6 cm) from the corner. Clearly it seems that low temperature has the greater potential for cracking an <u>undamaged</u> painting than low relative humidity. Changing the thickness of the glue layer influenced the stresses in the paint layer when desiccating the painting. This indicates that the very high glue layer stresses cause an interlayer interaction.

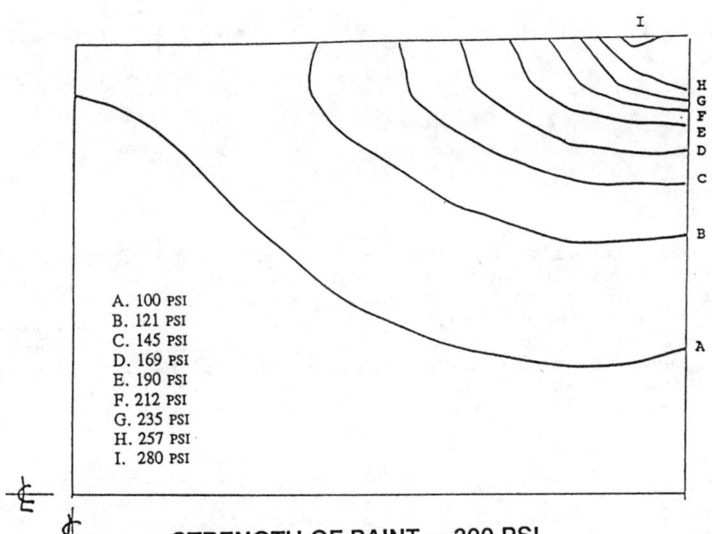

Figure 18. Computer calculated, maximum principal stress contours for a 30" x 40" traditional canvas supported oil painting subjected to desiccation from 70% to 10% relative humidity at 23_{o}C.

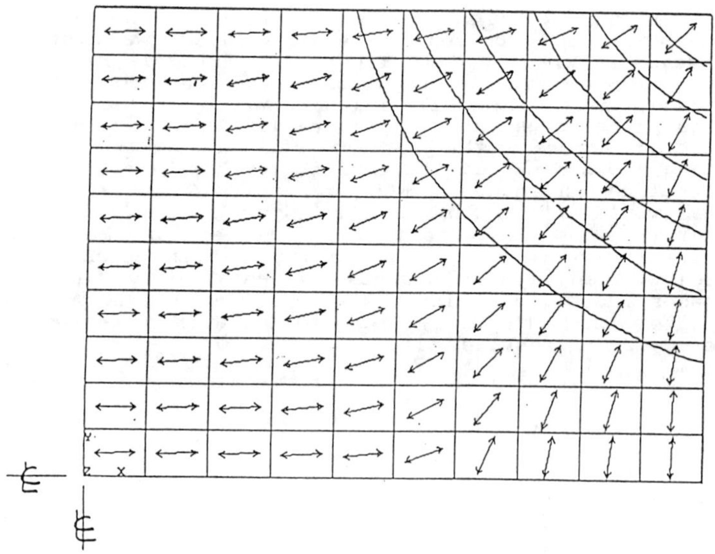

Figure 19. Principal stress directions and the projected crack pattern of a 30" x 40" oil painting subjected to desiccation from 70% to 10% relative humidity at 23_{o}C.

SUMMARY

In order to assess the risks to photographic prints and paintings associated with changes in temperature or relative humidity an accurate determination of the stresses in the different layers of a painting is necessary. In turn, those stresses must be compared to the strengths of the artists' materials at the specific environments in question and rates of loading encountered by the painting. Computer modeling in the form of finite element analysis, readily lends itself to performing the needed analysis. However, the ability to numerically model multiple layered structures such as paintings is extremely dependent on an accurate determination of the mechanical properties of the paintings constitutive materials. This includes assessing alterations to the mechanical properties brought about by time. This study has examined some of the materials typically found in a large proportion of canvas and panel paintings and concentrated on those paints such as the lead based oils that have shown to be the most brittle. Additionally the study confines the analysis to effects of temperature and relative humidity brought about over moderately long periods of time. In all of the analytical modeling, it was assumed that the materials were at equilibrium with the environment.

All of the materials respond dimensionally to the changes in temperature and relative humidity. The thermal coefficients of expansion for hide glue and lead paint were small, while their moisture coefficients were considerably higher. Restraining these materials caused stress development with desiccation and cooling, but the magnitude of the stress was a result of both the material's attempt to contract and its related modulus or stiffness.

Computer modelling of a typical canvas painting with glue size and lead oil paint layers showed that deep drops in temperature at low relative humidity (from 23° to -3°C, at 5% RH) resulted in both uniform and high stresses in the paint layer. These stresses equalled or exceeded the measured breaking strengths of thirteen year-old paint. It could be inferred that an actual painting, constructed with the materials modeled, would exhibit severe and extensive cracking over the entire surface of the painting. The same computer analysis also showed that the glue layer reaches stresses slightly more than twice those of the paint layer. It can also be concluded that severe cooling results in the independent response of the various layers of the painting with little interlayer interaction. Paint and glue tested at -3°C and 5% RH, showed extremely brittle behavior, shattering into multiple pieces when broken.

Computer modelling of the same painting, subjected to drops in relative humidity (from 70% to 10% RH), shows similar crack patterns as those found in cooling could develop but not nearly as extensive. In fact, it was not until an additional layer of glue was added to the model, that stresses in the paint layer were high enough to suggest cracking. Even then the analysis suggested that changes in RH would cause considerably less damage than cooling.

Computer modeling of the cibachrome print subjected to desiccation successfully calculated both the deflection and images layer stresses even though the deflection was large for such a thin film structure.

FOOTNOTES

1. A. Higdon, E.H. Ohlsen, W.B. Stiles, and J.A. Weese, Mechanics of Materials, Wiley & Sons, (1967) 58-59.

2. M.F. Mecklenburg, "Some Aspects of the Mechanical Behavior of Fabric Supported Paintings", Report to the Smithsonian Institution, (1982), 12-15.; M.F. Mecklenburg, "The Effects of Atmospheric Moisture on the Mechanical Properties of Collagen Under Equilibrium Conditions", AIC Preprints of papers presented at the 16th Annual Meeting, (1988), 231-244; M.F. Mecklenburg, "Applied Mechanics of Materials in Conservation Research", Materials Issues in Art and Archaeology II, Materials Research Society Symposium Proceedings, Vol 185, P.B. Vandiver, J. Druzik, and G.S. Wheeler, Eds., (1991), 105-124; G. Hedley, "Relative Humidity and the Stress/Strain Response of Canvas Paintings:

Uniaxial Measurements of Naturally Ages Samples", Studies In Conservation, Vol. 33, No. 3, (1988), 133-148; A. Karpowicz, "In-Plane Deformations of Films of Size on Paintings in the Glass Transition Region", Studies In Conservation, Vol. 34, No. 2, (1989), 67-74; A. Karpowicz, "A Study on Development of Cracks on Paintings", Journal of the American Institute for Conservation, Vol. 29, No. 2, (1990), 169-180.

3. R.D. Cook, Concepts and Applications of Finite Element Analysis, Wiley, (1974), 173-188; J.S. Przemienicki, Theory of Matrix Structural Analysis, McGraw-Hill, (1968).

4. M.F.Mecklenburg and C.S. Tumosa, "An Introduction Into the Mechanical Behavior of Paintings Under Rapid Loading Conditions", Art in Transit, M.F.Mecklenburg, Ed. The National Gallery of Art, Washington, D.C., (1991a) 137-172; M.F.Mecklenburg and C.S. Tumosa, "Mechanical Behavior of Paintings Subjected to Changes in Temperature and Relative Humidity", Art in Transit, M.F.Mecklenburg, Ed. The National Gallery of Art, Washington, D.C., (1991b) 172-214.

5. Mecklenburg and Tumosa, (1991b), 177-179.

6. Mecklenburg and Tumosa, (1991a), 137-139.

7. Eastman Kodak Co., Conservation of Photographs, Publication No. F-40, 1985.

8. Imaging Processes and Materials, Neblett's Eighth Edition, J.M. Sturge, V. Walworth, and A. Skepp, Eds., Van Nostrand Reinhold, New York, (1989) 129-130.

THE EFFECTS OF TEMPERATURE AND RELATIVE HUMIDITY ON THE MECHANICAL PROPERTIES OF MODERN PAINTING MATERIALS

Jonah D. Erlebacher, Eric Brown, Marion F. Mecklenburg*, Charles S. Tumosa
Conservation Analytical Laboratory, Smithsonian Institution, Washington, DC 20560
*To whom correspondence should be addressed.

ABSTRACT

The mechanical properties of strength, modulus, and elongation to break were studied for artists' acrylic and alkyd paints under varying conditions of temperature and relative humidity (RH). In the ambient environment, 23° C, 50% RH, acrylic paints are very flexible and are able to sustain large deformations (>50%). Alkyd paints are much stiffer and stronger, and they cannot sustain deformations nearly as dramatic as the acrylics. Acrylic paints at 5% RH are stiffer and stronger than at 50% RH and their ability to stretch is lessened. At temperatures below 15° C at 50% RH, the strength and stiffness of acrylic paints begin to rise rapidly. Some were found to be brittle at 5° C, and by -3° C, all were brittle. At a lower RH, some acrylic paints became brittle at a temperature near 11° C. These temperatures and relative humidities may be found in the transport environment of art objects, and may render them subject to possible damage.

INTRODUCTION

The measured mechanical properties of oil paint films in different relative humidities (RH) and temperatures have recently been used to predict potential damage to oil paintings during transit. Two sets of mechanical properties are useful, those under rapid loading conditions, such as shock and vibration, and those under slow-loading conditions, in which the paintings stay equilibrated to environmental conditions. By using the experimentally determined mechanical properties of traditional oil paint, a structural analysis computer program accurately predicted cracking patterns due to vibration, edge, and corner impacts of oil paintings.[1] Stress fields and crack patterns in oil paint films subjected to dry and cool environments were also determined.[2]

Experimental techniques are being extended to determine the mechanical properties of other artists' materials, in particular modern acrylic and alkyd paints. Acrylic paints are pigmented polymer emulsions in water. In general, they are more flexible than traditional oil paints. Alkyd paints, on the other hand, are oil paints modified so that there is much more cross-linking in the paint film when dry. As a result, alkyd paints tend to be stiffer and stronger than traditional oil paints.

In this paper, the mechanical properties under rapid loading conditions of acrylic and alkyd paints in different relative humidities and temperatures are reported. These properties will be used in further studies of the sources of cracking and damage in modern paintings. Other experiments are currently being conducted to determine the mechanical properties of acrylic and alkyd paints under equilibrium conditions and will be reported at a later date.

TERMS

Strength, stiffness and elongation are measurable engineering properties and are defined as follows:

Stress (σ) = Applied force/cross-sectional area of material, (F/A)
Strain (ϵ) = Change in material length, L, in direction of deformation/undeformed length = $\Delta L/L$.
ϵ x 100 = % **elongation** of material.
Modulus, E = Stress/strain (σ/ϵ). The modulus is a measure of a material's stiffness. The greater the modulus, the stiffer the material.
Strength = Maximum stress in a material before the material fails.

The method by which the mechanical properties of a typical paint are determined from its stress-strain curve is shown in Figure 1. The paint in this example is a sample of acrylic cerulean blue in an environment of 23° C, 50% RH.

TESTING EQUIPMENT

Stress-strain curves for the acrylic and alkyd paints were found by stretching samples on screw-driven tensile testers. Strain rates ranged from 0.001 in/in/sec to 0.01 in/in/sec. The testers were enclosed in plastic environmental chambers capable of maintaining constant temperature and relative humidity. Conditioned silica gel was used to maintain relative humidity. The tensile testers and environmental chambers were designed and constructed at the Conservation Analytical Laboratory (CAL) of the Smithsonian Institution.

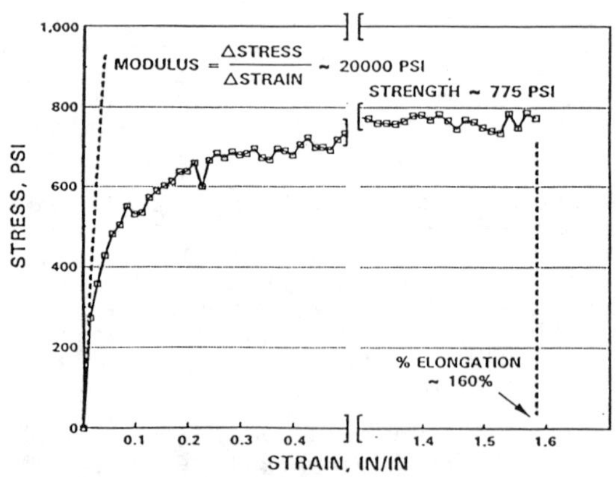

Figure 1. Stress-strain curve of acrylic cerulean blue at 23° C and 50% RH, showing its mechanical properties.

Test samples were taken from strips of paint cast between 1978 and 1982, making the youngest paint sample 10 years old. The samples were typical of commercially available acrylic paints. For testing purposes, each cast was cut into smaller strips approximately 0.2" wide and 3"-5" long. The different paint thicknesses varied from 0.005"-0.015".

MECHANICAL PROPERTIES OF ACRYLIC AND ALKYD PAINTS AT 23° C, 50% RH

Ten different acrylic paint samples from two different sources and eight different alkyd paints from a single source were tensile tested under ambient conditions, 23° C and 50% RH. Typical mechanical properties of these paints at this temperature and relative humidity are listed in Table 1.

Alkyd paints were found to be considerably stiffer and stronger than acrylic paints at 23° C, 50% RH. In this environment, the modulus of the alkyd paint samples ranged from 80,000 psi to 600,000 psi, while the modulus of acrylic paint samples ranged from only 10,000 psi to 45,000 psi. Alkyd paints were nearly twice as strong as acrylic paints. The strength of alkyd paints varied from 850 psi to 1600 psi, compared to a variation of strength in acrylic paints from 575 psi to 975 psi.

Table 1. Typical mechanical properties of acrylic and alkyd paint samples at 23° C, 50% RH

Paint Type	Modulus, psi	Strength,psi	Elongation to Break (%)
ACRYLIC PAINTS			
Burnt Sienna	40000	950	53%
Burnt Umber	11000	650	>293%
Cadmium Red	23000	800	174%
Cadmium Yellow	15000	625	195%
Cerulean Blue	20000	775	160%
Cobalt Blue	45000	975	54%
Iron Oxide Red	40000	900	60%
Ivory Black	12000	575	>363%
Raw Sienna	10000	865	279%
Titanium White	21000	819	255%
ALKYD PAINTS			
Alizarine Crimson	110000	850	2%
Burnt Umber	285000	1400	0.7%
Cadmium Yellow	190000	1000	2%
Iron Oxide Red	165000	1300	6%
Ivory Black	80000	1350	12%
Lead White	600000	1500	0.3%
Titanium White	275000	1600	1.4%
Yellow Ochre	140000	1050	3%

Interestingly, acrylic paints were able to withstand exceptionally large deformations at 23° C, 50% RH. All of the acrylic paint samples stretched at least 50% of their original length. Six of the samples stretched over 100%. The most flexible of the acrylic paints was a sample of ivory black, which stretched 363% of its original length before the test was stopped. The sample had reached the displacement limits of the tensile tester.

In the ambient environment, alkyd paints did not sustain elongations as large as the acrylics. An ivory black sample that stretched 12% was the only sample to stretch over 10% of its original length. The least flexible alkyd paint was a lead white, which stretched only 0.3% before breaking.

This same sample of lead white had the largest modulus (600,000 psi) of all the alkyd paints at 23° C, 50% RH. The acrylic ivory black sample that stretched 363% had the lowest modulus of the acrylic paints. In general, it was found that the higher the modulus, the less the acrylic and alkyd paints are able to stretch. Conversely, the lower the modulus, the more the paints are able to stretch.

The relative differences between the mechanical properties of acrylic and alkyd paints at 23° C, 50% RH indicate two different ways that these paints respond to shock. When force is applied, both kinds of paint initially deform and build stress. The rate at which stress builds with deformation during this period is determined by the modulus ($modulus*\Delta\epsilon=\Delta\sigma$). Once a maximum stress level is reached, acrylic paint films *plastically deform*. Rather than build more stress, they elongate. Alkyd paints plastically deform only slightly, if at all. Instead, they generally hold greater stresses. Some alkyd paints, in particular the lead white, did not plastically deform at all. This sample continued to build internal stress until it failed in a brittle manner, that is, without a plastic region in its stress-stain curve.

MECHANICAL PROPERTIES OF ACRYLIC AND ALKYD PAINTS AT 23° C, 5% RH

Nine types of acrylic paint and eight types of alkyd paints were tensile tested at 23° C, 5% RH. The mechanical properties of these paints at this temperature and relative humidity are listed in Table 2.

Desiccation to 5% RH at 23° C was found to stiffen all the acrylic and alkyd paint samples. In addition, their strengths rose and their ability to elongate decreased. The strengths and modulus of acrylic and alkyd paints at 5% RH and at 50% RH are compared in Figures 2 and 3.

The modulus increase of acrylic paints due to desiccation at 23° C is particularly dramatic. Titanium white, whose modulus rose from 21,000 psi to 140,000 psi had the largest relative stiffness increase of all the acrylic paint samples. The increase in modulus was accompanied by an increase in strength and a decrease in ability to stretch. Furthermore, when there were large increases in strength and modulus there were also large reductions in elongation ability. The maximum elongation of the acrylic titanium white sample mentioned above dropped from

Table 2. Typical mechanical properties of acrylic paint samples at 23° C, 5% RH

Paint Type	Modulus, psi	Strength,psi	Elongation to Break (%)
ACRYLIC PAINTS			
Burnt Sienna	150000	2000	14%
Burnt Umber	50000	1125	115%
Cadmium Yellow	68000	1075	80%
Cerulean Blue	38000	1140	113%
Cobalt Blue	100000	1600	17%
Iron Oxide Red	95000	1450	25%
Ivory Black	40000	950	>183%
Raw Sienna	32000	1200	>108%
Titanium White	140000	1900	40%
ALKYD PAINTS			
Alizarine Crimson	180000	1350	1.2%
Burnt Umber	400000	2400	0.7%
Cadmium Yellow	425000	1650	0.78%
Iron Oxide Red	300000	2000	2.8%
Ivory Black	180000	1930	7.5%
Lead White	850000	2400	0.3%
Titanium White	630000	2600	0.95%
Yellow Ochre	300000	2300	2.5%

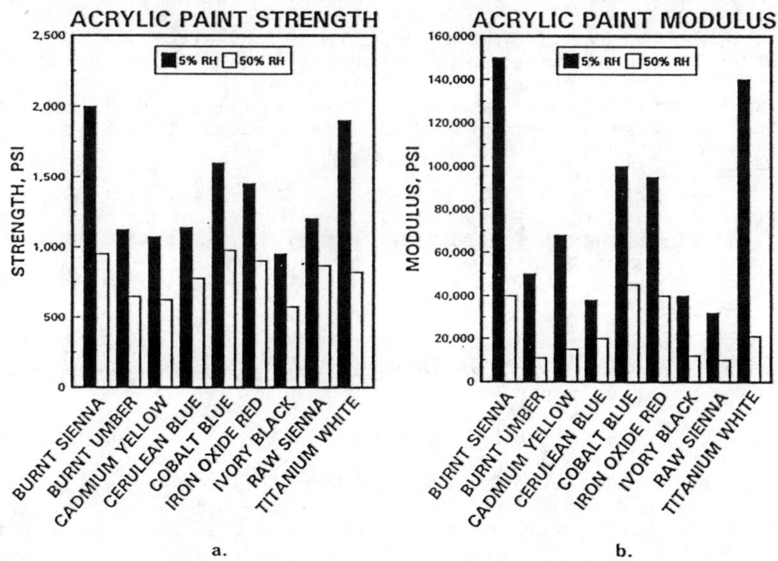

Figure 2. Strength (a) and Modulus (b) of acrylic paints at 23° C.

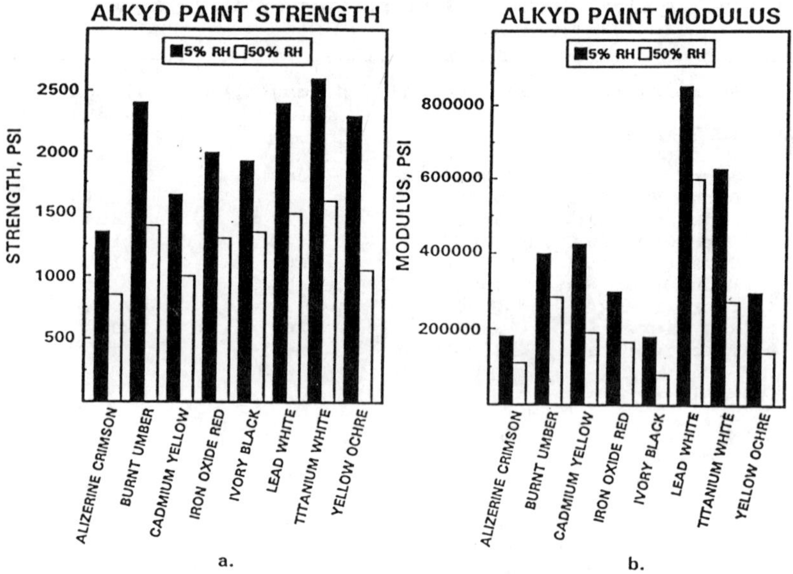

Figure 3. Strength (a) and Modulus (b) of alkyd paints at 23° C.

approximately 255% to 40%, which was also the largest relative drop in ability to sustain elongation seen in the acrylic paints.

Despite the large changes in strength, modulus, and ability to elongate during desiccation, acrylic paints still retained large abilities to stretch compared to alkyd paints. While at 23° C, 50% RH, no alkyd paint stretched over 12%, at 23° C, 5% RH all of the acrylic paints were still able to stretch at least 14% of their original length. Decreases in the ability of alkyd paints to elongate are losses of the ability to plastically deform when loaded. At 23° C, 50% RH, six of the eight alkyd paint samples stretched over 1% of their original length, but at 23° C, 5% RH only four of the samples were able to stretch that small amount. The loss of the ability to stretch in response to shock in desiccated environments at 23° C is more dangerous to alkyd paint films than to acrylic paint films.

MECHANICAL PROPERTIES OF ACRYLIC PAINTS AT DIFFERENT TEMPERATURES IN 40-50% RELATIVE HUMIDITY

The variation of strength, stiffness and ability to stretch versus temperature was studied for three acrylic paints in relative humidities of 40-50% RH. The paints were burnt umber, ivory black and cobalt blue. The cobalt blue used for testing was a relatively stiff acrylic paint, and the ivory black and burnt umber were relatively flexible acrylics. The strength and modulus of these acrylic paints at different temperatures in 40% to 50% RH is shown in Table 3. Table 4 shows the elongations to break of these paints at low temperatures.

Table 3. The mechanical properties of three acrylic paints at different temperatures, with relative humidity between 40% and 50%.

	Burnt Umber		**Ivory Black**		**Cobalt blue**	
Temp, °C	*Modulus (psi)*	*Strength (psi)*	*Modulus (psi)*	*Strength (psi)*	*Modulus (psi)*	*Strength (psi)*
-8.1	570000	3650	------	----	930000	4200
-4.3	485000	4050	450000	3850	720000	3200
-2.9	360000	2830	285000	2830	------	----
-1.4	340000	3740	275000	2700	560000	2700
3.4	260000	2730	260000	1950	360000	2550
5.6	225000	2250	210000	2200	290000	2020
8.5	200000	1450	80000	900	230000	2100
15.1	40000	620	13000	>220	58000	775
23	11000	650	12000	600	45000	950
33.4	2500	220	680	50	3300	>120

Figures 4 and 5 show the rise in the strength and the modulus of burnt umber, ivory black, and cobalt blue acrylic paints with decreasing temperature. Above 15° C, the modulus and also the strength of each acrylic paint is small, and the paints have a large ability to deform in response to shock. As the temperature dropped below approximately 15° C, the strength and modulus of all the acrylic paints rose rapidly. Associated with this rise was a similarly sharp decrease in the ability of these paints to stretch when rapidly loaded.

In 40% to 50% RH, all the acrylic paints were found to become extremely brittle at low temperatures. At -2.9° C, none of the tested paints exhibited the ability to deform plastically in response to rapidly applied force. Often, samples shattered when breaking. At temperatures below -2.9° C, samples of ivory black and burnt umber often broke eary, although their modulus generally kept increasing. This suggests that the paint films prematurely broke before a stress equal to their strength was developed in the entire sample. This may be due to very high stresses localized in areas of defects or impurities, a phenomenon called *fracture sensitivity*. It must be assumed that there will be defects present in actual paints and that this form of brittle behavior must be considered.

Table 4. The elongation to break of three acrylic paints at low temperatures and 40% to 50% RH

Paint Type	**Temperature, °C**						
	-8.1	-4.3	-2.9	-1.4	3.4	5.6	8.5
Burnt Umber	0.76%	1.28%	1.12%	1.86%	6.82%	>27.0%	>24.1%
Ivory Black	0.67%	1.63%	1.95%	7.35%	>30.2%	>50.0%	>16.7%
Cobalt Blue	0.52%	0.51%	-----	1.12%	1.43%	2.37%	6.46%

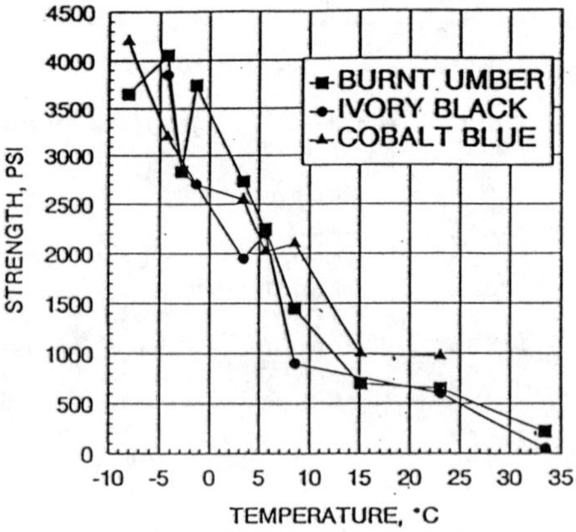

Figure 4. The variation of strength with temperature for three acrylic paints in environments near 50% RH.

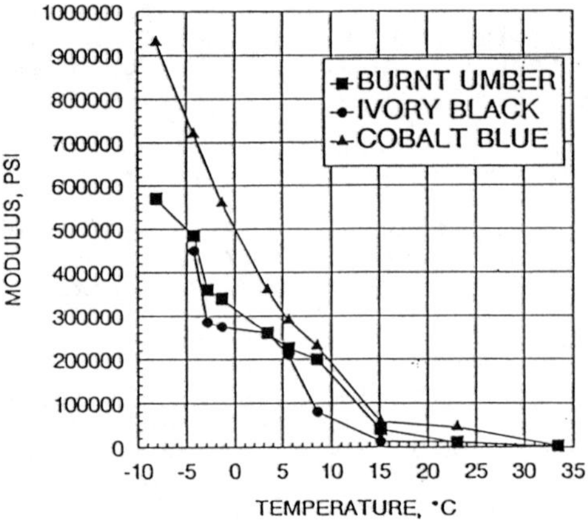

Figure 5. The variation of modulus with temperature for three acrylic paints in environments near 50% RH.

As temperatures increase from -2.9° C, with relative humidities between 40% and 50% RH, acrylic paints demonstrate the ability to plastically deform. More flexible acrylic paints have the ability to plastically deform at -1.4° C. For example, an ivory black sample stretched 7.35% at this temperature. However, stiff acrylics, like the cobalt blue, do not have a substantial ability to deform without shattering unless the temperature is over 5.6° C.

The bar charts in Figure 6 shows that the relative rise in strength and modulus in acrylic paints with desiccation to 5% RH at 23° C is much less severe than the strengthening and stiffening due to cooling to -2° C, maintaining 40% - 50% RH. If the reaction of these paints to rapid shock are compared, not only will stress build up faster in the cooled paint (because the modulus is higher), but the paint film will not hold the maximum stress through any plastic deformation. Once the maximum stress is reached, the paints will crack. At -2° C, all the paints tested exhibited brittle behavior. However, some acrylic paints, like the cobalt blue, exhibited brittle behavior at temperatures as high as 5° C.

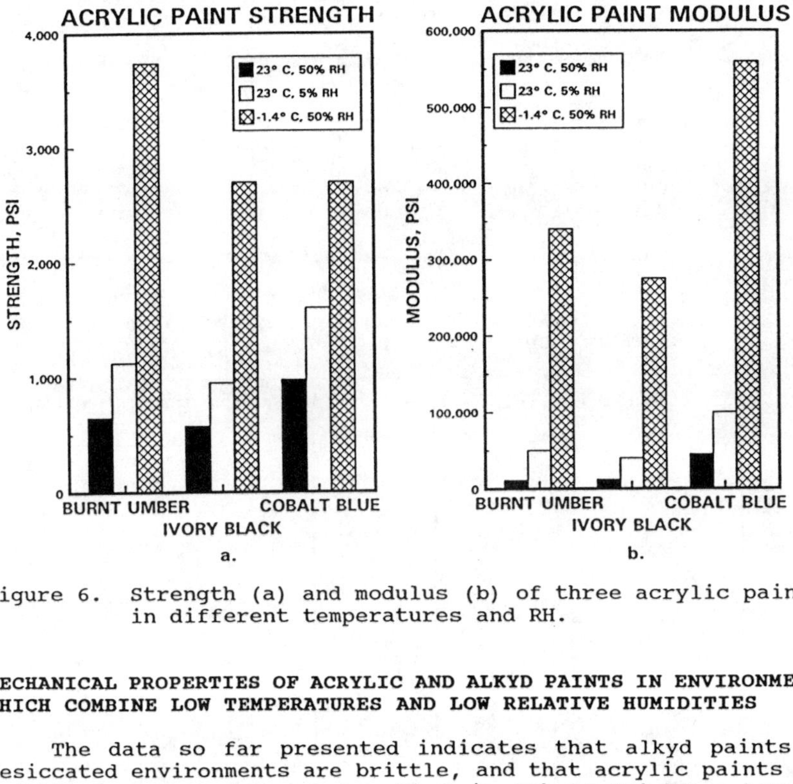

Figure 6. Strength (a) and modulus (b) of three acrylic paints in different temperatures and RH.

MECHANICAL PROPERTIES OF ACRYLIC AND ALKYD PAINTS IN ENVIRONMENTS WHICH COMBINE LOW TEMPERATURES AND LOW RELATIVE HUMIDITIES

The data so far presented indicates that alkyd paints in desiccated environments are brittle, and that acrylic paints are brittle in temperatures below 5° C in 40% to 50% RH. Acrylic paints in 5% RH at 23° C are stiffened, but they retain a substantial ability to stretch.

The combination of desiccation and cooling was found to embrittle severely acrylic and alkyd paints. Tables 5 and 6 list the mechanical properties of three acrylic and three alkyd paints in different temperatures at 5% RH. Figure 7 shows the modulus and strength of these three acrylic paints at 23° C 50% RH, 23° C 5% RH, 5° C 5% RH, and -6.5° C 5% RH.

Desiccation sharply raises the temperature at which acrylic paint exhibit brittle behavior. If the relative humidity is maintained between 40% and 50% RH, acrylic paints become brittle and fracture sensitive when their modulus nears or passes approximately 400,000 psi, usually under 5° C, as shown above. In 5% relative humidity, titanium white and burnt sienna samples each developed a modulus over 400,000 psi at 11° C. Furthermore, the burnt sienna sample was decidedly brittle at this temperature, breaking at an extension of only 1.44%. The titanium white sample showed a slight amount of plastic deformation.

Table 5. The mechanical properties of three acrylic paints at different temperatures, in 5% relative humidity.

Temp, °C	*Modulus (psi)*	*Strength (psi)*	*Modulus (psi)*	*Strength (psi)*	*Modulus (psi)*	*Strength (psi)*
ACRYLIC PAINTS	**Titanium White**		**Cadmium Yellow**		**Burnt Sienna**	
-6.5	1500000	5060	790000	4460	920000	3280
-5.2	1000000	3600	------	----	------	----
5.0	900000	6090	640000	3760	875000	4800
11.1	570000	3450	------	----	400000	2440
23.0	140000	1900	68000	1100	150000	1900
ALKYD PAINTS	**Alizarine Crimson**		**Lead White**		**Iron Oxide Red**	
-5.0	650000	1700	1250000	2500	1300000	4000
23.0	180000	1300	850000	2400	300000	2000

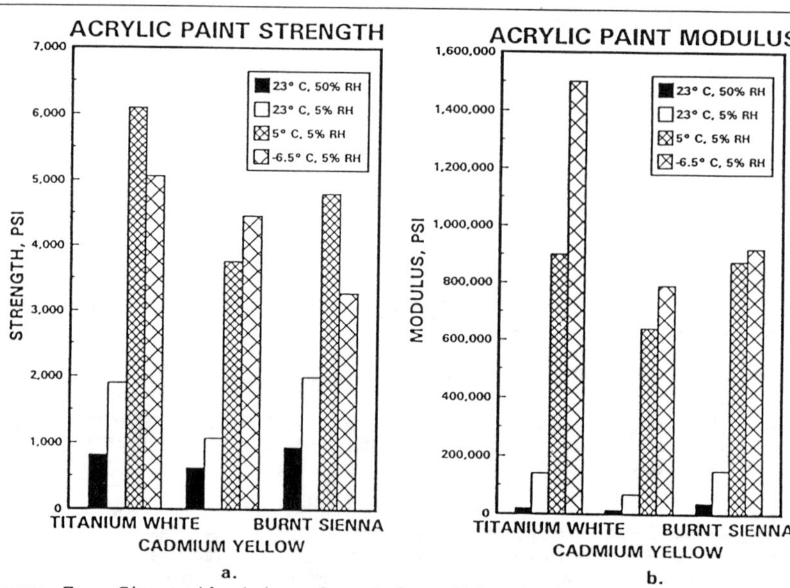

Figure 7. Strength (a) and modulus (b) of three acrylic paints at various temperatures and relative humidities.

Table 6. The elongation to break of three acrylic paints and three alkyd paints at 5% RH in different temperatures.

Paint Type	Temperature, °C					
	-6.5	-5.0	-5.2	5.0	11.1	23
ACRYLICS						
Titanium White	0.46%		0.38%	1.01%	4.78%	40.2%
Cadmium Yellow	0.63%		-----	0.63%	-----	76.0%
Burnt Sienna	0.38%		-----	0.61%	1.44%	13.2%
ALKYDS						
Alizarine Crimson		0.25%				12.0%
Lead White		0.15%				0.3%
Iron Oxide Red		0.5%				2.5%

At low temperatures, very brittle samples tested in 40%-50% RH tended to shatter prematurely. Extreme fracture sensitivity was also exhibited by acrylic paint samples tested at temperatures below -5.2° C in 5% RH, as shown in Figure 6(a). At 5° C, the acrylic paint samples all were extremely brittle, with high strength and modulus, but they did not seem to be fracture sensitive.

The combination of desiccation and cooling also greatly increased the strength and stiffness of the alkyd paint samples. However, the extraordinarily large rises in modulus and strength seen in acrylic paint samples in desiccated and cold environments were not seen in alkyd paint samples. Indeed, although alkyd paints are stiffer than acrylics at 23° C, 50% RH and at 23° C, 5% RH, acrylic paints are stiffer than alkyds at -5° C, 5% RH. The modulus of the titanium white at this temperature, 1,500,000 psi, was the largest modulus of any paint, including alkyds in all environments, seen in this study. Alkyd paints, however, still could not sustain as much elongation as acrylic paints even at -5° C, 5% RH. The alkyd lead white sample that failed after only 0.15% had the least amount of deformation before it shattered of all the tested samples.

CONCLUSION

The strength and stiffness of acrylic and alkyd paints become greater as relative humidity changes from approximately ambient RH to 5% RH. Acrylic paints were found to be extremely flexible at 23° C, 50% RH. At 5% RH, their ability to stretch is smaller, but remains substantial. Alkyd paints, in comparison, stretch significantly less than acrylic paints at 50% RH, and many can deform only small amounts at 5% RH.

As acrylic paints are cooled in 40-50% RH, their stiffness and strength rise as the temperature drops below 15° C. In this RH, some acrylic paints were brittle at 5° C. That is, they lost their ability to plastically deform after developing a maximum stress. At -3° C, many of the acrylic paints exhibited drops in strength although their modulus continued to increase. This may indicate fracture sensitivity in acrylic paints at very low temperatures.

Desiccation effectively increases the temperature at which acrylic paints become brittle when cooled. At 50% RH, some acrylics were brittle at 5° C, but at 5% relative humidity many were brittle at temperatures as high as 11° C. Alkyd paints are also very brittle at low temperatures and low relative humidities.

The experiments presented here indicate that cold transport environments are potentially dangerous to acrylic and alkyd paint films because their ability to respond to force by deforming is diminished, thus making them susceptible to shattering. Aircraft holds, for instance, generally have a temperature of approximately 10° C, although some holds have temperatures as low as 5° C.[3] Acrylic paint in 5% RH is brittle at both of these temperatures, but if its environment were buffered to a higher relative humidity the chance of it breaking due to shock or vibration would be lessened.

ACKNOWLEDGEMENTS

The authors wish to acknowledge the support for this research provided by the Scholarly Studies Program of the Smithsonian Institution.

REFERENCES

1. M. F. Mecklenburg and C. S. Tumosa, "An Introduction Into the Mechanical Behavior of Paintings Under Rapid Loading Conditions," *Art in Transit: Studies in the Transport of Paintings*, M. F. Mecklenburg, ed. (1991), 137-171.

2. M. F. Mecklenburg and C. S. Tumosa, "Mechanical Behavior of Paintings Subjected to Changes in Temperature and Relative Humidity," *Art in Transit: Studies in the Transport of Paintings*, M. F. Mecklenburg, ed. (1991), 173-216.

3. D. Saunders, "Temperature and Relative Humidity Conditions Encountered in Transportation," *Art in Transit: Studies in the Tranport of Paintings*, M. F. Mecklenburg, ed. (1991), 299-308.

DETECTION OF DELAMINATIONS IN ART OBJECTS USING AIR-COUPLED ULTRASOUND°

Alison Murray,* C.M. Fortunko,** Marion F. Mecklenburg,*** and Robert E. Green, Jr.*
* Department of Materials Science and Engineering and the Center for Nondestructive Evaluation, The Johns Hopkins University, Baltimore, MD, 21218, USA
** The National Institute for Standards and Technology (NIST), Boulder, CO, 80303, USA
*** Conservation Analytical Laboratory, Smithsonian Institution, Washington, D.C., 20560, USA

Abstract

The possibility of using air-coupled ultrasound to find delaminations between layers in paintings was investigated. Simulations of modern paintings were constructed with hardboard as the support layer, and a layer of gesso, an animal hide glue and chalk, as the upper layer. Delaminations were introduced between the two layers. Scanning the samples using air-coupled ultrasound clearly showed these flaws. A transmitting transducer, which was 25 mm in diameter and had a focal length of 51 mm, was placed 11 mm from the back surface of the sample and an identical receiving transducer was placed 51 mm from the front surface. The transducers were operated in a tone burst mode at a center frequency of 475 kHz. Signal enhancement techniques were used to improve the contrast of the data.

Introduction

The care of wooden art objects is rendered more difficult by the presence of cracks, voids, and delaminations. Computer analysis shows that cracks in wood promote severe stress concentrations that are aggravated by environmental and other stresses.[1] Unfortunately, many of these internal anomalies are not easily visible, making assessment of the various risks to art works difficult. In the case of a panel painting, its structural condition can be mapped using nondestructive testing techniques, in order to predict how it will respond mechanically over time in different environments. It can then be decided whether the object can be safely shipped to special exhibitions, and what plans can be made for proper display, storage, and conservation.

Various nondestructive methods for detecting voids, hidden cracks, and fine fractures in paintings have been investigated.[2] These methods have been incorporated into the repertoire of techniques from which conservators draw, to prevent the premature failure of objects.

Investigations of wooden panels and paintings have shown that xeroradiography has the advantage over x-ray radiography as xeroradiography's edge enhancement property enables flaws to be displayed prominently. It also uses equipment compatible with that typically found in conservation laboratories.

Research has shown that air-coupled ultrasound, a non-contact, nondestructive testing technique, can detect delaminations and cracks at certain angles that radiography cannot. Unlike radiographic techniques, ultrasound is not a health hazard for the user. The need for a non-contact ultrasonic method for use in the art conservation field has been discussed before.[3]

In working with ultrasound, industry has been able make use of traditional couplants, such as water, oil, and grease. Air coupling is a non-contact technique which may be faster and safer for examining museum objects. Air was not considered until recently because of the low signal-to-noise ratio caused by the high acoustic mismatch between solid materials and air. Air-coupled ultrasound has now matured into a viable method as the transducers and support electronics needed to overcome the signal-to-noise problem have become available.

There is a considerable body of literature on the subject of air-coupled ultrasound.[4-12] Historically, the main obstacle to the use of air-coupled transducers was the misconception that the attenuation in air is very high. In fact, it is only 20 dB/m at 600 kHz and 20°C. There is still, however, a substantial signal strength loss between the test material and the air, as the acoustic impedance of the material is five orders of magnitude larger. This is one of the major limitations of the air-coupled ultrasound technique. This limitation can now be overcome through the judicious selection of the critical components, including transducers, transmitting and receiving electronics, and appropriate signal processing methods.

The use of ultrasound in art conservation is not a new idea. Conservation scientists have performed ultrasonic tests on a variety of materials. For instance, ancient masonry buildings have been tested before and after repairs to determine the penetration and diffusion of grout, and to measure the degree of strengthening effected in masonries through repair.[13] In this work, correlations were found between the sound transmission data, aging tests, changes in the mechanical properties, and flaws.

The calcerous stone monument "St. George and the Dragon" was examined with ultrasound by Yakhont and Pimenov to determine the lithogical composition and other characteristics of the stone, as well as the existing damage.[14] The possibility of using ultrasound to test metal castings in order to show their condition and thickness has been discussed.[15] Measurements taken on wetted archaeological woods have shown the homogenization and strength of waterlogged wood.[16] Ultrasound was also used in conjunction with thermography to investigate whether Leonardo da Vinci's lost painting "The Battle of Anghiari" could be present beneath six large murals by Georgio Vasari.[17,18]

Inspection Problem

The objective of this work was to investigate whether air-coupled ultrasound was a feasible technique for finding delaminations in modern art paintings, between the support layer (hardboard) and the gesso layer. Such delaminations might be caused by environmental stress, imposed, for example, by changes in relative humidity, which ultimately affect the mechanical properties of the different layers. Another possible reason may be that the layers had never adhered properly.

Air-coupled ultrasound systems suffer from low signal to noise performance, which limits the flexibility of their use. For example, operation in the pulse-echo mode, where one transducer is used to send and receive the signal, can measure distances, but cannot perform analyses of subsurface features.

The safety of the object during examination is of utmost importance. Positioning the transmitting transducer close and unfocussed on the back surface of the panel, reduced the sound pressure and minimized the likelihood of damaging the object .

Experimental Configuration

The experimental set-up was reported in a previous paper.[20] One transducer sent the signal through the sample and another received the signal after it had travelled through the sample (Figure 1). Both transducers were 25 mm in diameter, had focal lengths of 51 mm, and had frequencies of 475 kHz. The transmitting sample was only 11 mm from the back surface of the hardboard panel, while the receiving transducer was 51 mm away from the gesso surface.

The transducers were held in a "U-shaped" yolk attached to a frame which could move along one axis (Figure 2). The apparatus was mechanized so that line-scans could be made of the samples. The stepping distance of the apparatus was 0.5 mm. In future work, the air-coupled ultrasonic system will be combined with c-scanning capabilities so that two dimensional images can be made. The stepping resolution will also be improved.

The resolution of the system was determined by scanning the signal from a small transducer (a "pinducer" of radius 0.8 mm) with a typical air-coupled transducer. At the focal distance, the receiving area is 3.2 mm in diameter. The actual resolution was calculated to be 1.6 mm, by subtracting two times the radius of the pinducer from the measured focal diameter. This is in excellent agreement with the theoretically predicted value (using the Rayleigh criterion) of 1.6 mm.

Other factors are important to consider when discussing the resolution of the system. The speed of sound in air is substantially lower than that in the hardboard. This means that the wavelength in air is shorter than that in hardboard. This mismatch caused severe refraction of the sound transmitted through the air-hardboard interface (Figure 3). The transducers are, therefore, not able to focus inside samples to find interior flaws.

Simulated panel paintings with anomalies were made by the Canadian Conservation Institute. Four panels of hardboard, 200 mm x 200 mm x 6 mm, were tested:

1. blank hardboard panel,
2. panel sanded overall to ensure good adhesion and to which a uniform layer of rabbit skin glue gesso of thickness of 0.6 mm,
3. same as panel 2 except that areas of failure were simulated by inserting thin sheets of plastic; the remaining surface was sanded to ensure good adhesion,
4. same as panel 2 except areas were left unsanded to promote adhesion failure and delamination.

The use of plastic sheet inserts (generic polyester) to simulate delaminations was developed by the aerospace industry. The method has been used on laminated composite panels both in industry and in art conservation.[21]

Panels 3 and 4 were also x-ray radiographed and xeroradiographed.

Results

The ultrasonic signal was able to traverse each panel. The experimental results from panel 3 are shown in Figure 4. The inset diagram shows where the line scan was made, and the main graph shows the amplitude of the received through-transmission signal as a function of position.

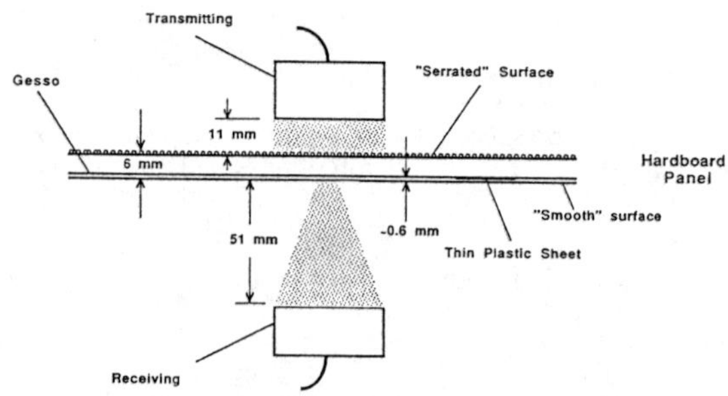

Figure 1 Experimental Configuration

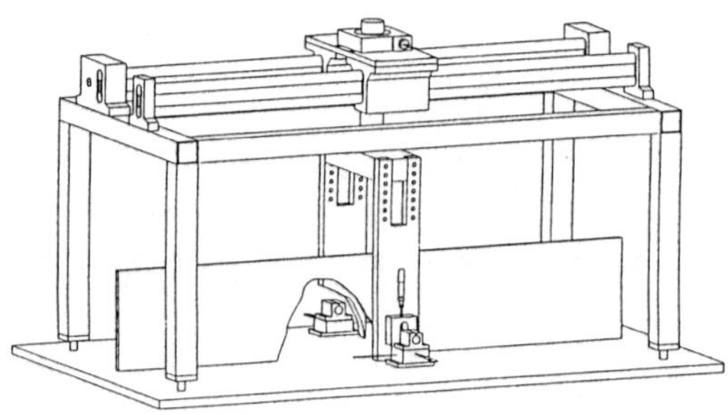

Figure 2 Fixture for Air-Coupled Ultrasonic Technique

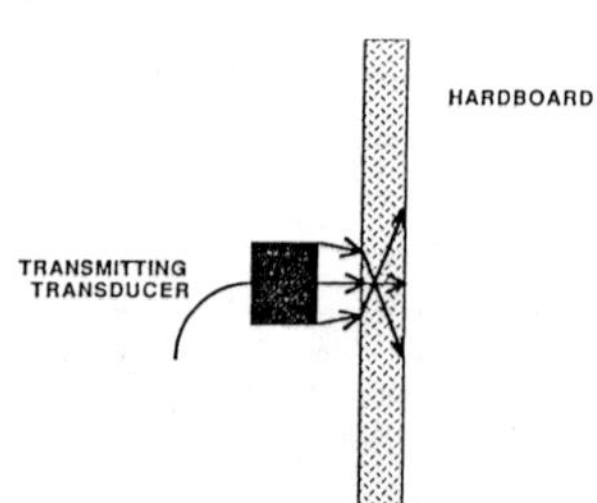

Figure 3 Ray Bending of Transmitted Sound

Two signal enhancement techniques were used to transform the ultrasonic data. "Running signal averaging" was used to obtain the mean value of five adjacent points which was used to remove rapid amplitude variations of the signal (dotted line in Figure 4). A simple transformation was then used to improve the contrast of the signal,

$$A= ((1-\langle T\rangle^2_N)/\langle T\rangle^2_{UNFLAWED})^{1/2}$$

where $\langle T\rangle_N$ is the running average of N adjacent amplitude points (N=5), and $\langle T\rangle_{UNFLAWED}$ is the average transmitted amplitude from the unflawed sample, panel 2, using all the data points. The enhanced ultrasound data from the hardboard panels 3 and 4 are shown in Figures 5 and 6. The results show the flawed regions in the panel with plastic inserts. The inserts were also visible with the radiographic techniques, although this was due to the increase in density from the inserts rather than from delaminations. Neither the ultrasound nor the radiographic techniques detected any delaminations from panel 4, where areas had been left unroughened to promote de-adhesion.

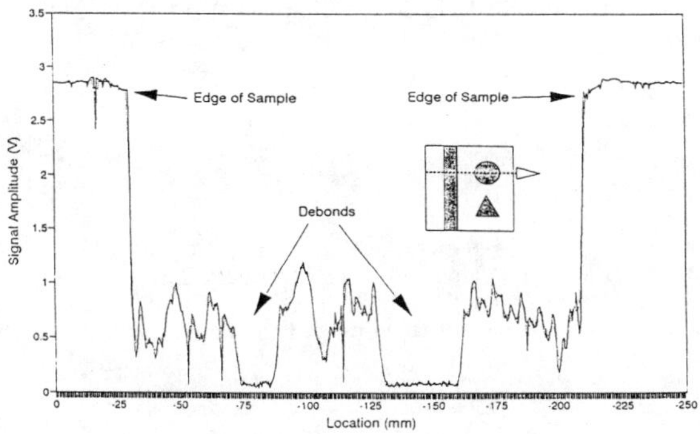

Figure 4 Ultrasonic Results for Line Scan of Panel 3

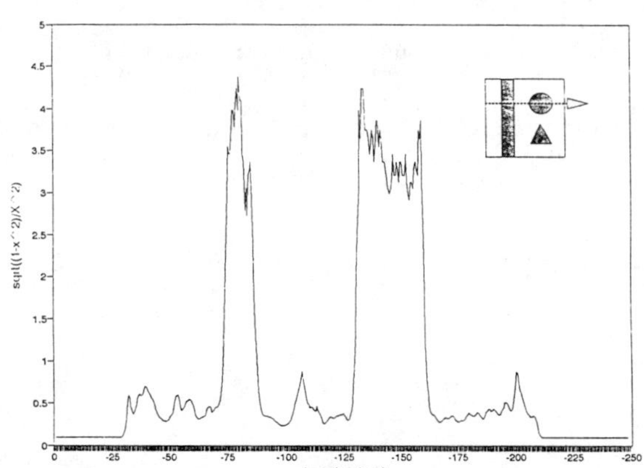

Figure 5 Ultrasonic Results for Line Scan of Panel 3 After Enhancement Techniques (the flawed areas contained plastic inserts, resulting in high contrast)

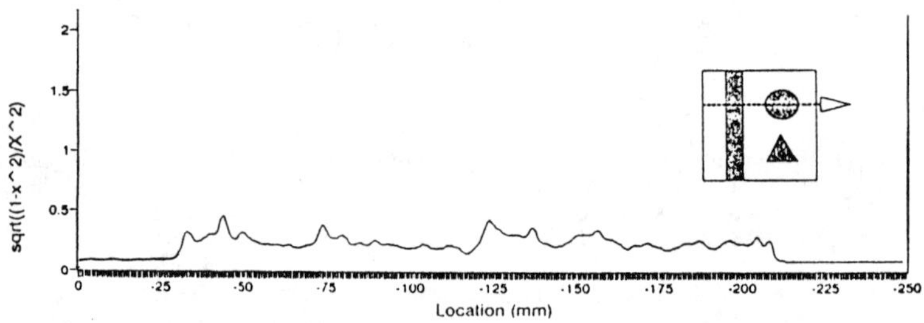

Fig. 6. Ultrasonic Results for Line Scan of Panel 4 After Enhancement Techniques (the flawed areas were left unroughened, resulting in low contrast)

Conclusions

Simulated delaminations have been reliably detected using air-coupled ultrasound. The areas of internal anomalies were enhanced using signal averaging and a transformation. It can be concluded that the air-coupled ultrasound technique shows great promise for the art conservation of panel paintings. Future work will include detection of delaminations on actual paintings.

Acknowledgements

This research has been supported by the Conservation Analytical Laboratory, the Smithsonian Institution; the National Institute of Standards and Technology (NIST), Boulder, Colorado; and the Center for Nondestructive Evaluation, the Johns Hopkins University. The authors thank Ian Wainwright and Helen McKay of the Canadian Conservation Institute, for lending the hardboard and gesso samples discussed in this paper; Dr. James W. Wagner, Dr. Dale E. Chimenti, and Robert D. Huber of the Johns Hopkins University, for their discussions and experimental work on the characterization of the air-coupled transducers; and Alan Price and Eric Jensen, at NIST, for their assistance with the ultrasound data collection and signal processing.

Bibliography

1. M.F. Mecklenburg and C.S. Tumosa, *Art in Transit - Studies in the Transport of Paintings* (National Gallery of Art, Washington D.C., 1991), pp. 173-216.
2. A. Murray, R.E. Green, M.F.Meckenburg, and C.M. Fortunko, *Art in Transit - Studies in the Transport of Paintings* (National Gallery of Art, Washington D.C., 1991), pp. 249-260.
3. J.F. Asmus and S.T. Pomeroy, *Preprints of papers presented at the 6th annual meeting for the American Institute for Conservation of Historic and Artistic Works, Fort Worth TX, June 1-4, 1978* (AIC, Washington D.C., 1978), pp. 12-20.
4. A.S. Birks, Forest Products Journal *22*, (6), 23-26 (1972).
5. C. Canali, G. de Cicco, B. Morten, M. Prudenziati, and A. Taroni, IEEE Transactions on Industrial Electronics *IE-29,* (4), 336-341 (1982).
6. B.T. Khuri-Yakub (Sec. 10), and A.S. Birks, J. Carlson, R. Harris, R.W. Smith, and R. Walters (Sec. 23) in *Nondestructive Testing Handbook, (2nd ed.) Vol. 7,* Sec. 10, Part 2, and Sec. 23, Part 1 (The American Society for Nondestructive Testing, 1991), pp. 320-325 and pp. 786-793.
7. B.T. Khuri-Yakub, J.H. Kim, C-H Chou, P. Parnet, and G.S. Kino, IEEE Ultrasonics Symposium *1988*, 503-506 .
8. M. Luukkala and P. Meriläinen, Ultrasonics, *22*, (5), 218-221 (1973).
9. M. Luukkala in *Review of Progress in Quantitative Nondestructive Evaluation, Vol. 9,* edited by D.O. Thompson and D.E. Chimenti (Plenum Press, New York, 1990) pp. 1987-1991.
10. A.J. Rogovsky, Materials Evaluation *1991*, 1491-1497
11. J. Szilard, in *Ultrasonic Testing - Non-conventional testing techniques,* edited by J. Szilard (John Wiley & Sons, 1982) pp. 381-409.
12. S.K. Yang, V.V. Varadan, and V.K. Varandan, Materials Evaluation *48*, 471-480 (1990).
13. M. Berra, L. Binda, G. Baronio, A. Fatticcioni in *2nd International Conference on Nondestructive Testing, microanalytical methods and environment evaluation for study and conservation of works of art, Perugia, Italy, April 17-20, 1988*, edited by M. Marabelli and P. Santopadre (Istituto Centrale per il Restauro and Associazione Italiana Prove non Distruttive, Italy, 1988) pp. I/10.1-19.
14. O. Yakhont and Y. Pimenov in *2nd International Conference on Nondestructive Testing, microanalytical methods and environment evaluation for study and conservation of works of art, Perugia, Italy, April 17-20, 1988*, edited by M. Marabelli and P. Santopadre (Istituto Centrale per il Restauro and Associazione Italiana Prove non Distruttive, Italy, 1988) p. IV/20.1.
15. Gilardoni S.p.A. Laboratori di Ricerca in *2nd International Conference on Nondestructive Testing, microanalytical methods and environment evaluation for study and conservation of works of art, Perugia, Italy, April 17-20, 1988,* edited by M. Marabelli and P. Santopadre (Istituto Centrale per il Restauro and Associazione Italiana Prove non Distruttive, Italy, 1988) pp. IV/10.1-8.
16. S. Miura, Hozon Kagaku *15*, 14-18 (1976) [Science for Conservation].
17. J.F. Asmus, H.T. Newton, Jr., S.T. Pomeroy, and M. Seracini, *Preprints of papers presented at the 5th annual meeting for the American Institute for Conservation of Historic and Artistic Works, Boston MA, May 30-June 2, 1977* (AIC, Washington D.C., 1977) pp. 12-17.
18. J.F. Asmus and S.T. Pomeroy, *Preprints of papers presented at the 6th annual meeting for the American Institute for Conservation of Historic and Artistic Works, Fort Worth TX, June 1-4, 1978* (AIC, Washington D.C., 1978) pp. 12-20.
19. J.C. Dainty, in *Laser Speckle and Related Phenomena*, edited by J.C. Dainty (Springer-Verlag, New York, 1975), pp. 1-7.
20. C.M. Fortunko, M.C. Renken, and A. Murray, IEEE Ultrasonics Symposium *1990*, 1099-1103.
21. S. Amadesi, F. Gori, R. Grella, G. Guattari, P. Pasquini, and G. Guattari in *Problemi di conservazione*, edited by G. Urbani (Editrice Compositori, Bologna, Italy, 1973) pp. 197-238.

MODERN TRANSPARENT PAPERS: MATERIALS, DEGRADATION, AND THE EFFECTS OF SOME CONSERVATION TREATMENTS

DIANNE VAN DER REYDEN, CHRISTA HOFMANN, MARY BAKER, AND MARION MECKLENBURG

Conservation Analytical Laboratory, Smithsonian Institution, Washington, D.C. 20560

ABSTRACT

Modern transparent papers, also known as tracing papers, require specific chemical and physical properties, such as translucency, smoothness, and whiteness, which are produced by specially formulated compositions and manufacturing procedures. Ironically, these special formulations may in some ways render such papers susceptible to problems from damage and degradation, while also making them particularly reactive to moisture and solvents that must be used in conservation treatments to correct these problems. In order to evaluate the effects of water and organic solvents on such papers, four research projects were designed to consider the variables of paper composition, properties, type of solvent, technique of application of solvent, and flattening. This paper summarizes findings on materials characterization, degradative effects of aging, and some effects of solvents used on transparent papers for stain reduction, humidification, and flattening.

INTRODUCTION

Most museums, libraries, and archives contain large numbers of a class of material generically known as transparent or tracing papers, which serve, among other things, as the support layer for technical and architectural drawings. Such drawings are used frequently by architects, engineers, historians and other scholars for many research purposes, and are often exhibited and reproduced. However, they can not be effectively replaced by photographs, facsimiles, or microfilm, since the design media may be difficult to decipher as it is often faded, smeared or smudged from handling. In addition, the drawings may have notations, added or erased, which indicate significant changes in designs important for an understanding of the history or evolution of architectural ideas and concepts, and these subtle but important changes may not be captured in reproductions, which also fail to accurately portray the variety of media, representing different elements and stages of design. Finally, reproductions occasionally crop writing scribbled in margins or on the reverse of the drawings. Consequently, scholars often require access to the original material. Unfortunately, the composition, size, age, past use, and fragility of these drawings, compounded by their frequent handling, makes them particularly vulnerable to damage. To prevent the total destruction of the original information, such drawings frequently require conservation treatment.

Conservation treatment of such artifacts is complicated not only by diverse and sensitive media, but also by of the very nature of the paper supports, which require specific chemical and physical properties, such as translucency, whiteness, and smoothness, produced by specially formulated compositions and manufacturing procedures. However, these special formulations and procedures may render such papers exceptionally susceptible to damage such as surface marring, embedded grime and stains, and degradation such as embrittlement and discoloration. Often the papers are extremely hygroscopic, which renders them particularly susceptible to cockling, planar distortions, and dimensional changes. Different manufacturing processes may also render the papers extremely sensitive to conservation treatments. This is especially true for treatments requiring solvents for washing or deacidification [16, 18, 48], lining with aqueous, solvent-activated or thermoplastic adhesives [3, 4, 9, 19, 21, 36, 39, 40, 48], consolidation [48], stain-removal [5, 17], or humidification and flattening [15]. The papers may also be sensitive to treatment application techniques, such as immersion or use of a suction disk or table [15, 22, 25, 42, 48].

RESEARCH DESIGN

Some conservation literature exists on the analysis and treatment evaluation of old transparent papers [15, 16, 29, 32], but less is available on modern transparent papers [21, 35, 39]. In 1987, the Conservation Analytical Laboratory began research characterizing several types of specialty papers, including coated and transparent papers [5], to enable conservators to 1) categorize the nature of many types of specialty papers in order to anticipate potential changes to properties from aging and solvent treatment, and 2) design treatments selecting solvents and application techniques appropriate to specific types of specialty papers. The ongoing work includes the four interrelated projects described below.

RESEARCH DESIGN FOR MODERN TRANSPARENT PAPERS PROJECT

PROJECT I. CHARACTERIZATION OF EXPERIMENTAL SAMPLES					
SELECTED SAMPLES FOR:	FIBER PROCESSING:	OVERBEATING:	NATURAL TRACING PAPER SAMPLE "G"		
	SHEET PROCESSING:	ACID IMMERSION:	GENUINE VEGETABLE PARCHMENT PAPER SAMPLE		
		CALENDERING:	IMITATION PARCHMENT PAPER SAMPLE		
		COATING AND/OR IMPREGNATING:	VELLUM PAPER SAMPLE "C"		
			PREPARED TRACING PAPER SAMPLE "A"		
IDENTIFICATION OF MATERIALS BY:	GENERAL OBSERVATIONS				
	MANUFACTURER'S INFORMATION				
	ANALYSIS:	SEM IMAGING	SEM/EDS	FTIR	GC/MS
MEASUREMENT OF PROPERTIES:	APPEARANCE:	COLOR	OPACITY	GLOSS	
	PHYSICAL:	STRENGTH	DIMENSIONAL STABILITY		
	CHEMICAL:	PH			

PROJECT II. EFFECTS OF ACCELERATED AGING

PROJECT III. EFFECTS OF SOLVENTS & APPLICATION			
WATER	ETHANOL	ACETONE	TOLUENE
APPLICATION TECHNIQUES:			
IMMERSION	POULTICE	SUCTION DISK	

PROJECT IV. EFFECTS OF HUMIDIFICATION & FLATTENING		
IMMERSION	HUMIDIFICATION CHAMBER	HUMIDIFICATION PACK
FLATTENING TECHNIQUES:		
AIR DRY	BLOTTER PRESS	SUCTION TABLE

Project I: Characterization: Samples of modern transparent papers were selected representing four primary categories (natural tracing papers, genuine vegetable parchment paper, imitation parchment paper or glassine, and vellum or prepared tracing papers) (Fig. 1). Characterization included identification of furnish material by fiber microscopy and staining, SEM/EDS, FTIR, and GC/MS, and of formation procedures by SEM imaging and UV microscopy. The samples were also characterized by measurement of chemical and physical properties including pH, color, gloss, opacity, transmission, mechanical strength, and dimensional stability.

Project II: Aging: Evaluation of the effect of accelerated aging on the properties of the selected transparent papers was done by measuring properties, before and after aging, of color, opacity, transmission, gloss, and tensile strength (Figs. 2, 3, & 4). Accelerated aged samples were also used to simulate aged or old papers for treatment and further testing.

Project III: Solvent application: Evaluation of the effects of four commonly used stain-removal solvents applied by three different treatment techniques to transparent papers has included measurement of opacity, gloss, and color (Fig. 5), and subjective observation of appearance (tidelines or ringing) in visible and ultraviolet light, and tracking dislocation of furnish materials by SEM and UV microscopy (Fig. 6) [42, 43].

Project IV: Humidification: Evaluation of the effects of three humidification techniques and two flattening techniques used for transparent papers included measurement of changes in dimensions, transmission and gloss (Figs. 7 & 8), tensile strength (Fig. 9), and SEM surface imaging (Fig. 10) [22]. Conditions of testing and analysis are listed in Appendix 1.

PROJECT I: CHARACTERIZATION

General terms used to describe transparent paper in the literature include tracing paper, oiled paper, onion skin paper, and waxed paper [1, 3, 4, 11, 35, 37, 47]. US Federal Specifications for Tracing Papers UU-P-561H 1972 currently recognizes four classes, with specific minimal requirements for fiber type (100% rag or chemical wood pulp), permanence, translucency, whiteness, and strength, and for properties of tear resistance, fold endurance and opacity, among other things [14, Appendix 2]. Properties may vary depending on intended use. For instance, surface finish might be rough for pencil work or smooth for ink illustrations. Manufacturers of modern transparent papers tend to categorize their products as a) natural tracing papers, b) vegetable parchment paper (genuine parchment paper or pergamet), c) imitation parchment (pergamyn or glassine), or d) vellum paper or prepared tracing papers. Each of these papers is made by either 1) processing the fiber stock by overbeating to insure that the fiber structure is broken down to eliminate air/fiber interfaces, and/or 2) processing the paper sheet to insure that the air pockets are eliminated or filled with a material having a refractive index similar to the paper fibers. A paper sheet can be made transparent by adding or applying a transparentizer (coating and/or impregnating agents), by immersion of the sheet in acid, and/or by calendering. Fiber processing is the primary procedure used to make a) natural tracing paper (highly overbeaten), while sheet processing is used to produce b) vegetable parchment paper or pergamet (immersed in acid), c) imitation parchment, pergamyn or glassine paper (highly calendered), or d) vellum or prepared tracing paper (both made with transparentizers).

Several papers were selected randomly to represent the aforementioned types of modern transparent papers. Glassine paper serves as a "transparent" control, since it has the fewest additives. The compositions of some of the papers are listed and described below; morphological structures have been documented by SEM imaging of surfaces and cross-sections [42, 43].

Fig. 1: SUMMARY OF GENERAL CHARACTERIZATION OF SELECTED MODERN TRACING PAPERS

		NATURAL SAMPLE "G"	VEGETABLE PARCHMENT	IMITATION PARCHMENT	VELLUM PAPER SAMPLE "C"	PREPARED TRACING SAMPLE "A"
MANUFACTURE INFO	ADDITIVES	sulphamic acid; artificial clay; rust inhibitors; defoaming agents; saponified oils; talc; nitrogen polymer or acrylonitrile/ butylacrylate copolymer	unfilled	unfilled	melamine formaldehyde, urea formaldehyde	silica, aromatic solvent
		surface size: styrene maleic anhydride compound, or quaternalized polymer; modified starch	unsized	unsized	internal and external size: starch	coating: styrene ester, cellulose ester
ANALYSIS	SEM IMAGING	fibers visible on surface but not in cross-section	fibers visible on surface & cross section	fibers pressed & visible in cross sect.	coated surface; impreg. cross.	heavily coated surf.; impreg. cross.
	SEM/EDS	S, Si, Ca, Hg traces: Al, Cl, Na	Si, Ca, Al, S	Al, Si traces: S, Cl, Na, Ca	Al, S; traces: Si, Cl, Na, K, Ca	Al, Si, Cl traces: S, Na, K, Ca
	FTIR	melamine (?) acrylate, oil	NA	NA	melamine, starch, syn. res.	styrene acrylate, alcohol polymer
	GC/MS	oil	NA	NA	oil	oil, resin

A highly overbeaten natural tracing paper sample, which we have designated "G", is made from a furnish of chemical wood pulp and numerous additives. It has the greatest degree of fibrillation of chemical wood pulp on the surface and virtually no fiber structure in cross-section. This is because the paper is produced in a large volume of water having only 6% fiber content while at a high temperature (80 degrees C), in order to soften the fibers and increase fibrillation. It is machine calendered. The density ranges between 1.1-1.3 grams/cubic cm.

The genuine vegetable parchment paper is made of chemical wood pulp and transparentized by a momentary immersion in sulfuric acid, which changes the wood fibers into an amyloid gel. The gel is then solidified by washing and neutralization, which bonds the fibers into a solvent resistant paper with high initial wet strength recommended by its manufacturer for off-set lithography and silk-screen printing. Testing has not been completed for this paper, so no findings are included in the current publication.

The calendared imitation parchment paper sample, or glassine control sample, shows in both surface and cross-section SEM photomicrographs highly fibrillated chemical wood pulp fibers, partially "gelatinized" by prolonged beating of the pulp, and highly compressed by calendaring. Glassine chemical woodpulp fibers are beaten in water with a 20-30% fiber content. The final sheet is dampened and then supercalendered, under high pressure and heat (approximately 2000+lbs/linear inch at 180-200 degrees C).

A vellum paper sample, "C", is made from 100% cotton fibers which do not fibrillate upon overbeating as well as chemical wood pulp. The twisted shape of cotton fibers prevents close conformation and traps air, leading to scattering of light at the fiber/air interface. Therefore, such paper requires transparentizers, made up of a coating and/or impregnant. This sample has an internal and external size of starch and melamine formaldehyde resin.

A prepared tracing paper sample, "A", is made of 100% "rag", and is technically in the same category as sample "C". However, it appears to be coated with a viscous material pitted by minute air pockets, as seen in SEM surface images. In SEM cross-section, this coating appears to impregnate the fibers, which are none-the-less distinct and appear less fibrillated than the other papers. It is transparentized by an aromatic solvent based synthetic resin and top coated by styrene and cellulose esters.

PROJECT II: DEGRADATION FROM AGING

The transparent papers in this study were artificially aged at 90 degrees C, 50%RH for four weeks which, it must be noted, substantially exceeds the conditions required by the Federal Specifications for Tracing Papers UU-P-561H 1972 (14, App. 2). To facilitate comparison with samples characteristic of more standard papermaking processes, measurements for untransparentized Whatman papers have been included in some graphs.

Colorimetry (Fig. 2): The transparent papers for the most part appeared to be about as stable as the glassine paper control, showing a slight increase in yellowing (b^*). The notable exception was sample "G", which showed considerable change in all values upon aging, darkening with a shift to red and yellow, and becoming a uniform brown color. This could be due to the presence of chromophores formed, as a result of our severe aging conditions, from the numerous additives in this particular paper. The manufacturer attributes the color change under our aging conditions to the presence of sulphamic acid (amido sulphuric acid) that is added to control sizing, flocculation and algae growth during processing, or to a modified starch added to surface sizing.

Opacity, Transmission, and Gloss (Fig. 3): Aging resulted in an increase in opacity that was visually apparent for all samples, although samples "G" and "C" underwent the greatest measurable increase. All the samples had a corresponding drop in percent transmission. Glassine was the only sample having high initial gloss, owing to supercalendering, and accelerated aging caused little change.

Strength (Fig. 4): The tensile strength of two tracing papers were measured and both had far greater strength than normally processed paper. This may reflect a combination of exceptionally strong sizes, impregnants and/or overbeating. Sample "G", which underwent pronounced overbeating to increase fibrillation and bonding, had substantially greater strength (ultimate stress) and stiffness, exceeding that of prepared tracing paper "A" (transparentized by coating and impregnation), or a heavily sized paper (W56). After accelerated aging, the papers tested underwent a reduction in strength, characteristic of embrittlement. This may be a result of a

Fig. 2: COLORIMETRY

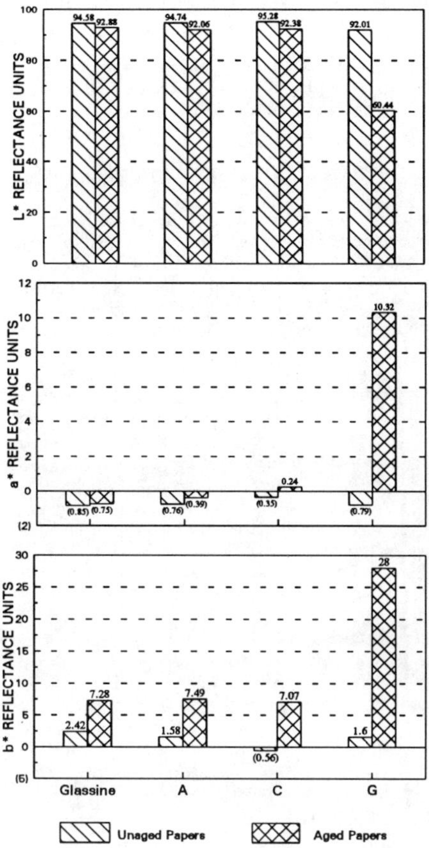

Fig. 3: OPACITY

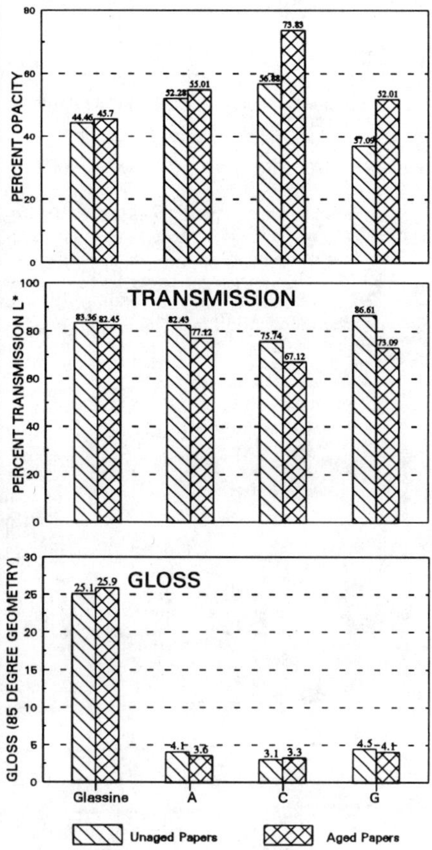

Fig. 4: TENSILE STRENGTH
TRANSPARENT AND WHATMAN PAPERS

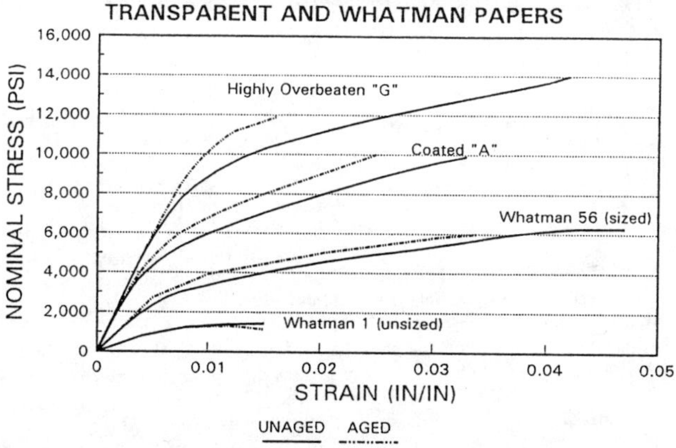

breakdown in sizing. In addition, overbeating, as in the case particularly of sample "G", may be responsible for substantial decreases in strength (ultimate stress) and elongation (ultimate strain), because it produces shorter fibers and weakens fibrillar attachment to fibers.

pH: Glassine and sample "A" had the highest initial surface pHs (6.05 and 5.8 respectively) which dropped slightly after aging (5.4 and 5.1). Sample "C" had a pH (4.3) which increased after aging (5.6). Sample "G" had a low initial pH (4.65) which dropped appreciably after aging (3.8).

PROJECT III: EFFECTS OF SOLVENT TREATMENTS ON TRANSPARENT PAPERS

Transparent papers, because of the special properties mentioned above, tend to be particularly susceptible to treatments with water or organic solvents. For the purpose of this project, four common stain removal solvents (water, ethanol, acetone, and toluene) were applied to samples of each transparent paper by three common conservation techniques (immersion, suction disk and poultice). Evaluation of changes in opacity, gloss, and planar uniformity caused by solvent type, application technique, and paper type indicates certain trends summarized below.

Fig. 5

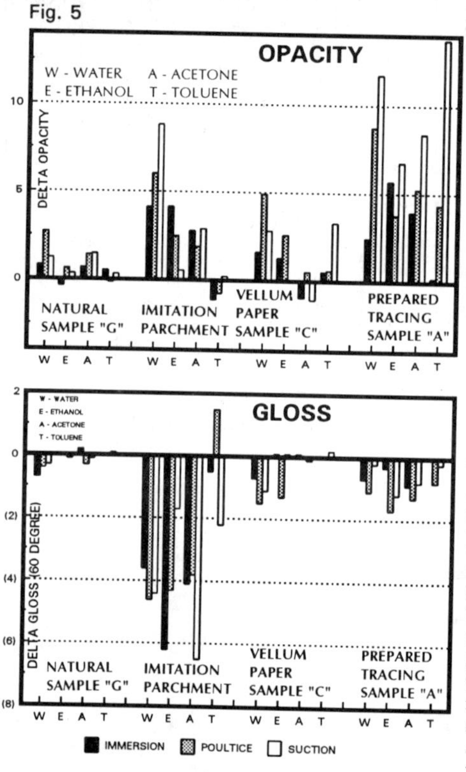

Solvent Evaluation: Water generally caused the greatest net changes of all solvents regardless of application technique or type of paper in this study, resulting in an increase in opacity and decrease in gloss (Fig. 5). Water was the only solvent that caused appreciable planar distortions, which were about the same regardless of application technique or type of paper. Acetone, ethanol and toluene generally caused less effect on opacity and gloss.

Application Technique Evaluation: Suction application frequently caused the greatest increase in opacity and poultice the greatest in gloss. Visual changes induced by poultice may result in part from residual poultice material. Suction application appears to have the greatest effect on cross-section morphology as seen with SEM imaging.

Paper Type: The slightly overbeaten and highly calendared glassine sample underwent the greatest loss of gloss following solvent treatment. The natural tracing sample "G" was the least affected by all treatment conditions. The heavily coated prepared tracing sample "A" had the greatest general reactivity to all solvents tested. Only immersion in toluene had no readily apparent affect on "A", whereas suction disk application of toluene caused an increase in opacity readily apparent in visible light. However, ultraviolet illumination revealed that areas treated by toluene applied by immersion and poulticing underwent reduced fluorescence and increased absorption, indicative of a break up of surface coating. As already suggested, this paper may be prepared with a toluene based resin. SEM imaging of this paper confirmed that each solvent affected the paper differently [41, 42, 43]. For instance, acetone applied by suction disk appeared to cause a loss of material (Fig. 6a). Immersion of the sample in ethanol appeared to cause a breakdown of material (Fig. 6b), precipitating small spheres which were identified by SEM/EDS as silicon based (Fig. 6c), concentrated in the areas of treatment as evidenced by SEM elemental dot mapping (Fig. 6d) of the same immersed area (Fig. 6b).

Fig. 6a Acetone applied by suction disk (right side of SEM photomicrograph)

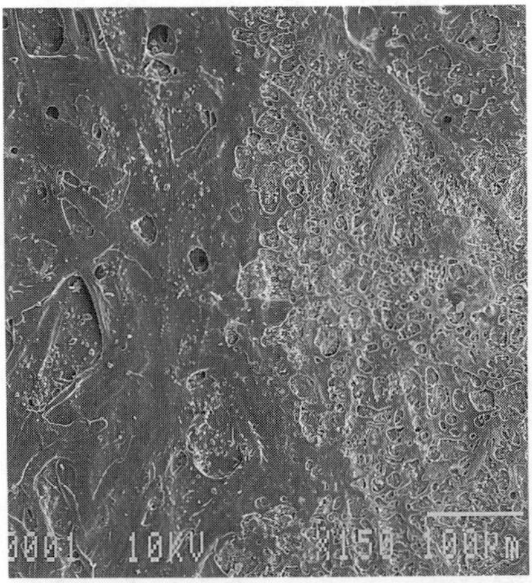

Fig. 6b Ethanol applied by immersion (the upper left)

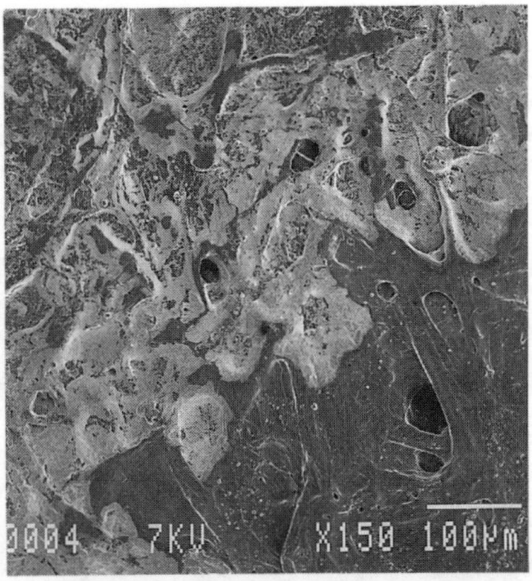

Fig. 6c Detail of 6b identified as silicon

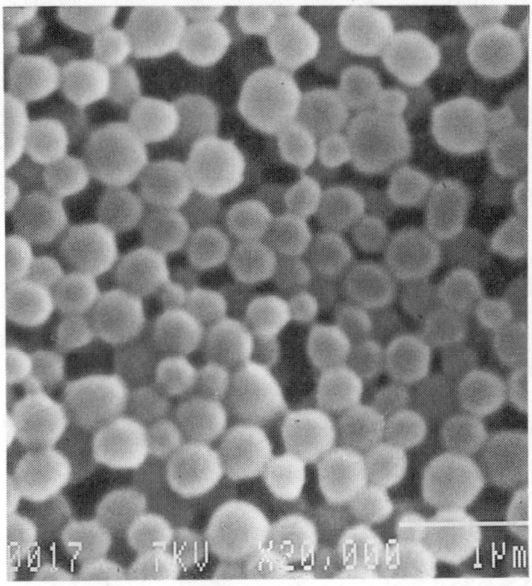

Fig. 6d Elemental dot map for Si in 6b

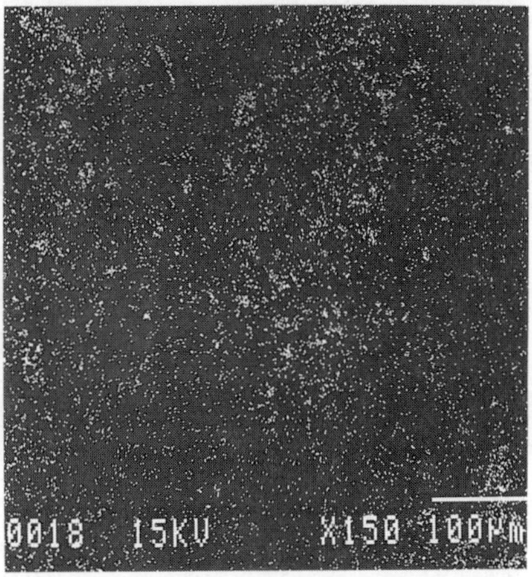

PROJECT IV: EFFECTS OF HUMIDIFICATION AND FLATTENING ON TRANSPARENT PAPERS

The fourth project investigated the effects of three humidification techniques (using immersion, an ultrasonic humidification chamber, and a humidification pack system) and two flattening techniques (on a suction table and in a traditional blotter press with c. 1PSI weight). Each technique was applied to the unaged transparent papers (to simulate the effect on "new" paper) and to aged paper (to simulate the effect on "old" paper). Evaluation of transmission, gloss, dimensional changes, and, in some cases, strength, indicate certain trends depending on humidification and flattening technique and paper type, as outlined in the following summary (Fig. 7 and Fig. 8).

Fig. 7 DELTA PROPERTIES AFTER HUMIDIFICATION AND FLATTENING OF UNAGED TRANSPARENT PAPERS ("NEW")

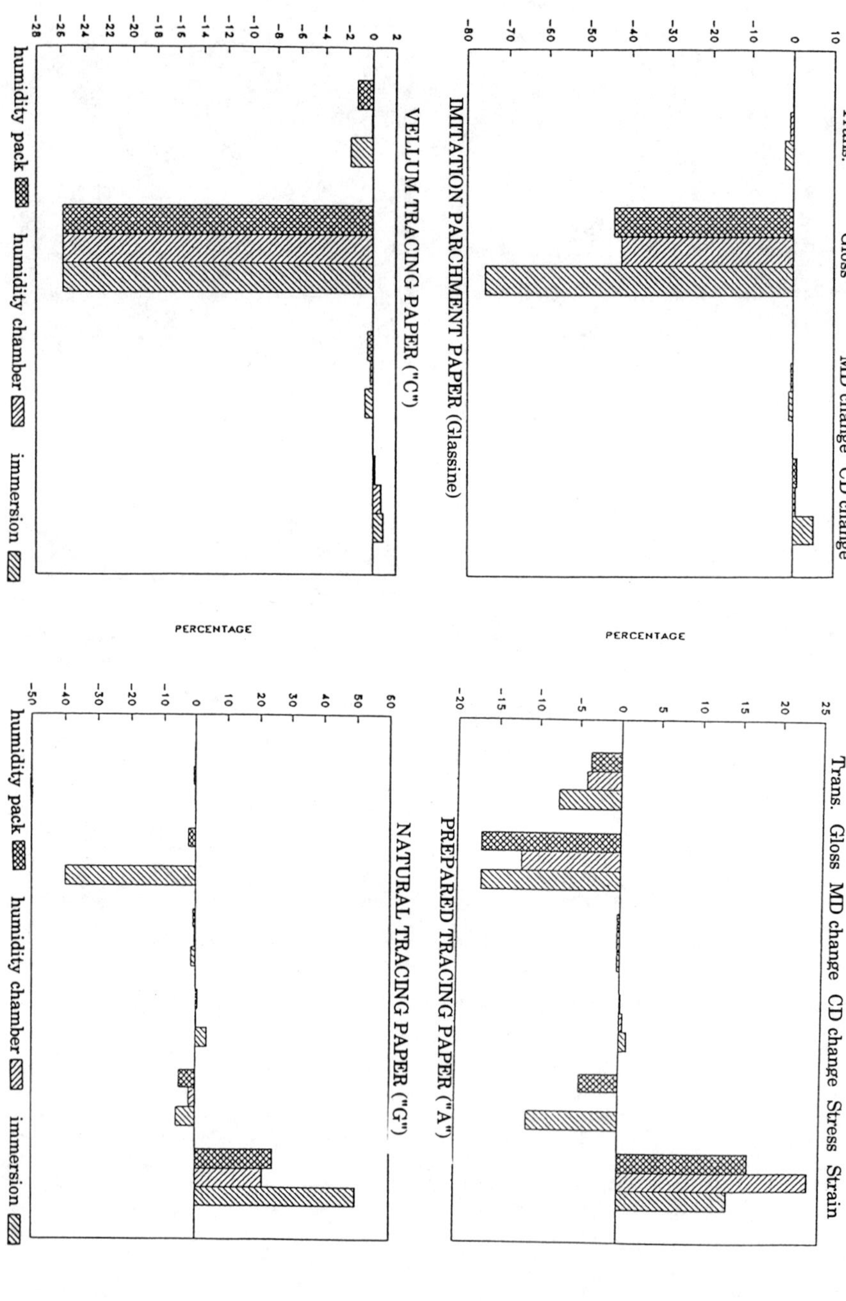

Fig. 8 DELTA PROPERTIES AFTER HUMIDIFICATION AND FLATTENING OF AGED TRANSPARENT PAPERS ("OLD")

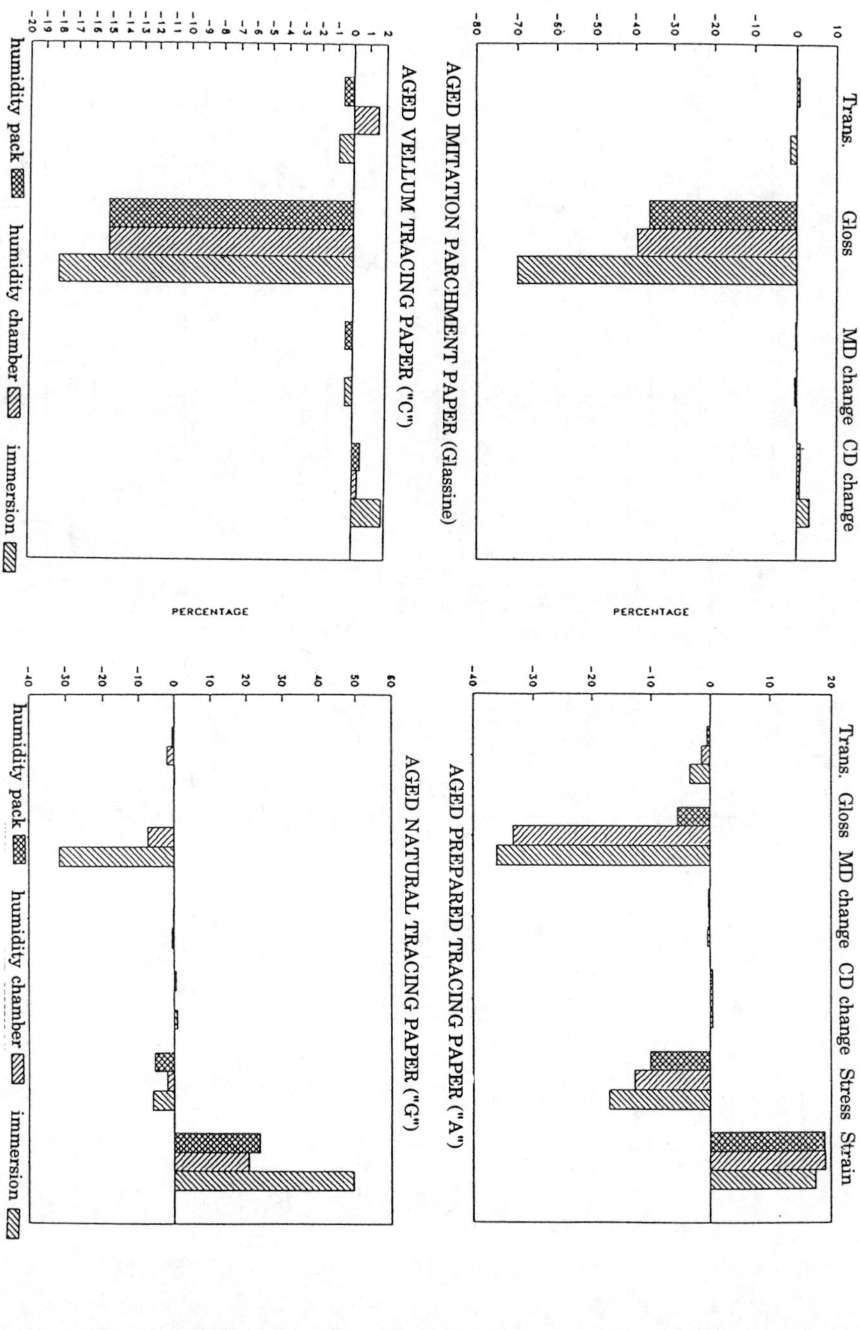

Fig. 9 TENSILE STRENGTH AFTER HUMIDIFICATION AND SUCTION TABLE DRYING OF UNAGED (NEW) AND AGED (OLD) TRANSPARENT PAPERS

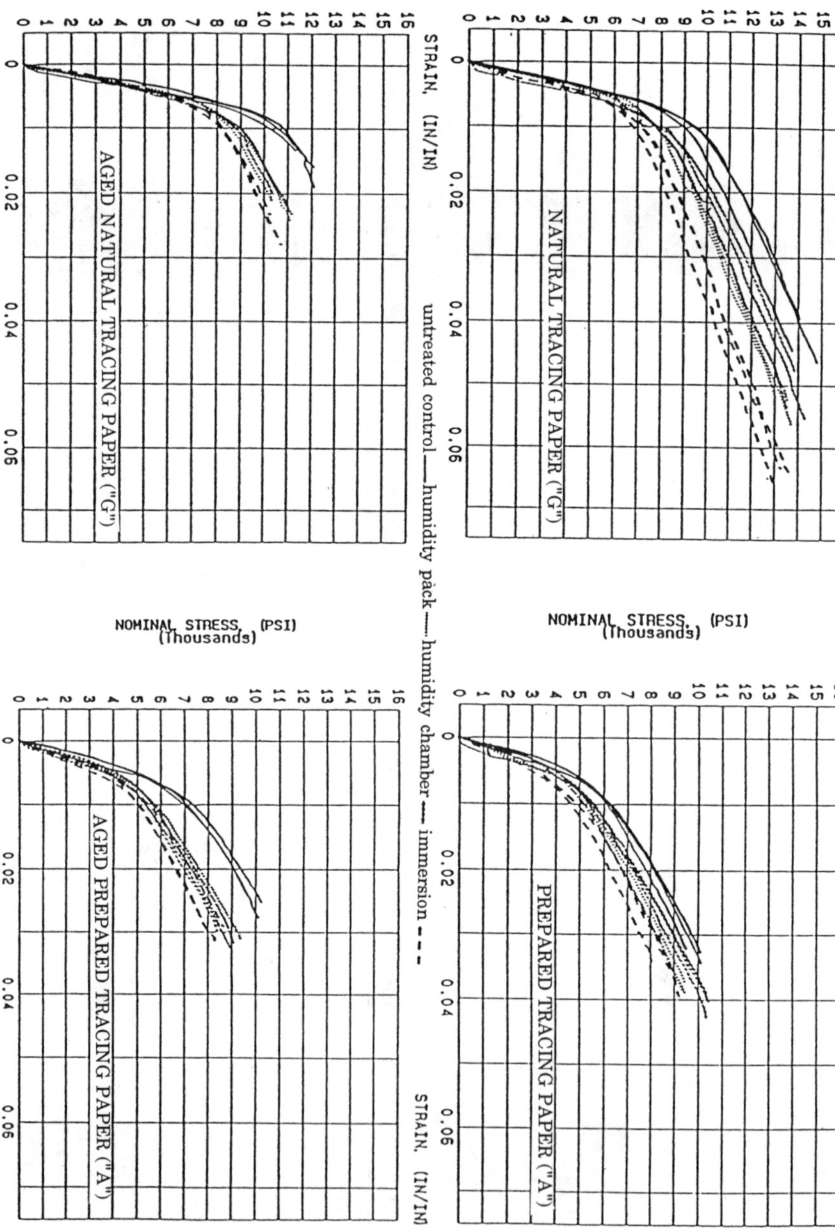

Water Application Technique Evaluation: The most severe planar distortion occurred with immersion, particularly for the overbeaten natural tracing "G", followed by humidification pack and chamber, which caused less planar distortion since both techniques effectively reduce water penetration. Immersion also caused the greatest dimensional changes (Figs. 7 & 8), and the greatest change in mechanical properties (Fig. 9), especially in comparison to humidity chamber. This was particularly true for the overbeaten natural tracing paper "G", for which immersion caused the greatest increase in strain to break, possibly from rebonding after the release during immersion of dried-in strain produced during manufacture. Immersion caused a greater decrease in the stress and strain to break for prepared tracing paper sample "A", possibly from the break up of the coating material.

Flattening Technique Evaluation: With respect to the two flattening techniques, compared to air dried controls, suction table drying of immersed samples caused severe distortion, especially for the overbeaten natural tracing "G", as compared with drying in a traditional blotter press. Dimensional changes in the cross-grain direction were greatest for the overbeaten and calendared papers, following immersion and drying on the suction table (Figs. 7 & 8). The coated papers were less affected dimensionally.

Paper Type: Heavily coated prepared tracing sample "A" incurred the greatest change in opacity following immersion, becoming mottled. This may result from a disruption of material which occurred to some degree for this paper regardless of humidification technique. As evidenced by SEM photomicrographs, immersion (Fig. 10a) and a humidification pack system (Fig. 10b) appeared to result in actual loss of the surface material in some areas.

Fig. 10a After immersion humidification

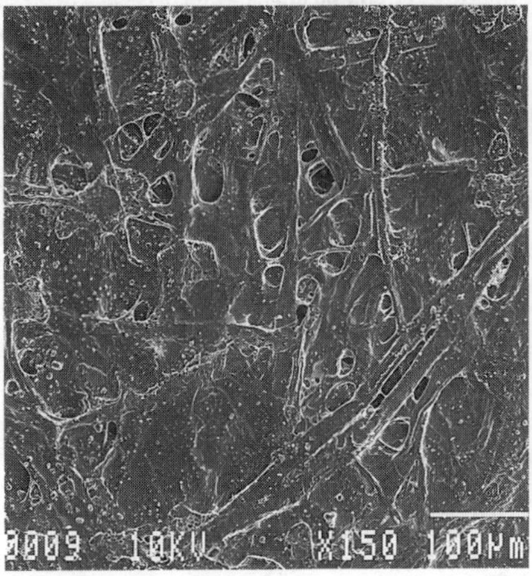

Fig. 10b After humidity pack

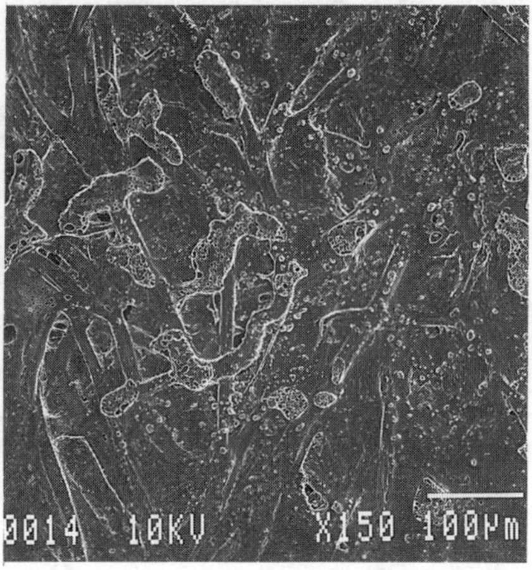

CONCLUSION

Some preliminary observations, which may aid conservators faced with the treatment of transparent papers, may be summarized for each project as follows:

Project I: Based on manufacturers' information, literature, and our own findings, there is a great deal of variation and overlap in the furnish and formation procedures for modern transparent papers, and this can make them difficult to distinguish. However, the furnish, formation and properties of modern transparent papers follow certain trends sufficiently different that conservators may be able to determine, with qualifications, whether a paper is characteristic of prepared tracing or vellum paper, a natural tracing paper, a vegetable parchment paper, or imitation parchment. This may be done, with varying degrees of accuracy, by several techniques, for instance: a) observing general appearance (high gloss for calendared imitation parchment, high translucency for natural tracing paper, high fluorescence for papers with transparentizers), planar and dimensional stability (greater for coated tracings), reactivity to water (greater for overbeaten natural or imitation parchment papers) or organic solvents (greater for heavily coated tracings), and strength (weaker for overbeaten natural tracings after aging); b) analyzing fiber content with polarized light microscopy (chemical wood pulp for natural tracings and parchment papers, cotton for tracings with transparentizers) or sheet morphology by SEM; and c) undertaking instrumental elemental analysis using SEM/EDS, FTIR, or GC/MS. In addition to the more frequently used GC/MS, microtome cross-sectioning of samples combined with SEM, FTIR, and UV microscopy appears to have great potential for providing detailed information on composition and distribution of furnish materials in samples.

Project II: Accelerated aging under the conditions outlined in this paper caused dramatic differences in the properties of all the samples, sometimes in excess of the standards considered acceptable by the US Federal Specifications [14, App. 2]. There are substantial differences in mechanical properties, before and after aging, depending on whether the samples were transparentized by overbeating or by transparentizing agents. In general, the heavily coated transparent paper sample seemed to have better retention of optical and strength properties after aging than the overbeaten natural tracing paper and imitation parchment paper samples.

Project III: The effects of solvents on the surface of transparent papers vary a great deal and seem to depend first on the composition of the paper, followed by the application technique and type of solvent. Of the papers studied here, the natural tracing paper was the least affected and the heavily coated paper was the most affected by the various solvents and application techniques. Water effected the greatest changes (increase in planar distortion and opacity and decrease in gloss) and toluene the least. The property most severely affected was translucency, which decreased in most cases. Prepared papers were the most sensitive to tidelines.

Project IV: Different techniques for humidifying and flattening transparent paper affect properties in different ways. Based on our findings, conservators who must immerse a transparent paper in water might find dimensional changes less severe if the paper is dried in a blotter press rather than on a suction table. Transparent papers with heavy coatings that might soften on exposure to moisture should be humidified in a humidity chamber rather than by prolonged contact with a humidification pack system. This latter technique, however, might be more appropriate for uncoated or natural transparent papers, since initial curling is prevented.

ACKNOWLEDGEMENTS

The authors would like to thank the many Conservation Analytical Laboratory interns who helped prepare and test samples for this project, in particular Jennifer Beers. Special thanks goes to SEM operator Melanie Feathers and Conservators Mel Wachowiak for his help with UV microscopy, and Ron Cunningham for his development of microtoming techniques for our transparent papers samples.

APPENDICES

1. Conditions of testing and analysis

2. Summary of US Federal Specifications for tracing papers (UU-P-561H 1972)

3. Bibliography

APPENDIX 1: Experimental procedures and instrumental analysis

pH measurements: The pH was measured with a Corning Model 12 Research pH meter with an Orion model No. 81-35 flat surface combination electrode. The rinsed electrode, with a pendent drop of deionized water, was lowered onto a square paper sample(1.5 x 1.5 cm) on a polyethylene bag padded with blotters. The pH was recorded after 5 min. The electrode was calibrated with pH 7 and pH 4 buffer solutions before each measurement session. The surface pH of the untreated new and aged controls of each paper type was measured (i.e. 1 measurement for each of 8 samples).

Microscopy: Microscopy was undertaken on a polarized light stage microscope, using transmitted and reflected visible and ultraviolet illumination. UV illumination employed a short pass (Kurz pass) KP 500 excitation filter and a chromatic beam splitter (TK510/K515).

Microchemical staining: For the iodine potassium iodide test for starch, 0.13 g iodine were dissolved in a solution of 2.6 g potassium iodide in 5 ml water. The solution was diluted to 100 ml (Browning 1977, 91).

SEM: SEM imaging and SEM analysis were carried out on a Jeol JXA - 840 A scanning electron microscope with Tracor Northern TN 5502 energy dispersive x-ray analysis system. For imaging the samples were mounted on aluminum stubs and gold coated. For elemental analysis the samples were mounted on carbon stubs and carbon coated.

FTIR: FTIR analysis was carried out on a Mattson Cygnus 100 Fourier Transform Infrared Spectrophotometer with a Spectratech IR-Plan Microscope. The surfaces of the bulk paper samples were analyzed by reflectance; for transmission spectra, fibers were pressed in a diamond anvil cell. To isolate coatings, samples of the papers were extracted with solvents; the solvents were evaporated and the residues analyzed by transmission in the diamond cell.

Gas chromatography: Samples of new papers C, A and G and of the aged paper G were hydrolysed in potassium hydroxide (10% in methanol) overnight, neutralized with 3M hydrochloric acid, extracted with diethyl ether, and then taken to dryness in a stream of nitrogen. The sample was redissolved in methylene chloride and then an equal volume of dimethyl formamide-dimethyl acetal was added to form methyl esters. The prepared sample was analyzed on a Carlo-Erba model 5360 gas chromatograph with a 30 m x 0.32 mm DB-1 column. One microliter of the sample plus a comparable volume of methylene chloride was injected. The injector temperature was 300°C. The initial column temperature of 50°C was immediately raised 10°C/min to 320°C. The effluent was detected with a flame ionization detector at 325°C. The chromatograms of the samples and that of a standard solution of methyl esters ("K101") were compared.

Artificial aging: The transparent paper samples were aged for four weeks in the dark at 90°C and 50% relative humidity in an Associated Environmental Systems HK-4116 Temperature/Humidity chamber. These conditions have been chosen as suitable for artificial aging studies [13]. A set of transparent papers was sewn with cotton thread into plexiglas frames so that all four corners were anchored and the samples did not touch one another. However the samples did vibrate in the oven draft.

Colorimetry, Transmission and Opacity: Color (specular reflectance included), total transmission (diffuse plus regular transmission), and opacity were measured with the HunterLab Ultrascan Spectrocolorimeter (D_{65}; 10° observer, diameter of area of view 1.2 in) using the CIE L*a*b* color notation, where L* represents the degree of brightness (100 white, 0 black), a* the degree of redness (positive numbers) or the degree of greenness (negative numbers) and b* the degree of yellowness (positive numbers) or the degree of blueness (negative numbers). Due to irregularities of the transparent papers, the standard deviation for the L* value of transmission was rather high, sometimes amounting to 0.5. For color and transmission five measurements were taken per sample and averaged. Opacity for the solvent treatment was measured by the ratio of reflection when samples were read against a white and black tile on the HunterLab Ultrascan Spectrocolorimeter.

Gloss: The gloss was measured with a Dr. Lange Labor-Reflektometer RL. Three measurements per sample were taken and averaged at the each of the following angles: 20°, 60° and 85°.

Tensile tests: The tensile properties were investigated using the Mecklenburg relaxation tensometer [28] with a horizontal load applied in the machine direction to the paper strips. Narrow strips of uniform width were cut with a mat cutter. After measurement of the paper thickness in five places with a micrometer, the paper strips were mounted horizontally in the apparatus exposed to laboratory atmosphere. After an initial equilibration period at a gauge length of 2.5 in, the strip was stretched 0.0025 inch and one minute later the stress sustained by the paper strip was recorded. This process was repeated once per minute until the paper strip broke. Measurements were made on three strips of each paper. From these data, nominal stress (force applied per cross-sectional area of the strip) and strain(change in length divided by gauge length) were computed. Nominal stress was plotted as a function of strain for each paper strip. The new and aged untreated controls and the new and aged treated samples of paper A and paper G that were dried on the suction table were measured three times each.

Dimensional changes: Dimensions of the samples were measured once per sample in mm (+/- mm) in cross and in machine direction before treatment, after humidification, after one day drying, after two weeks drying and after two weeks storage following drying. The dimensions of all 56 samples including new and aged untreated controls and new and aged treated samples were measured.

Planar distortion: New and aged untreated controls and new and aged treated samples were compared and subjectively ranked in daylight, raking light and transmitted light in a preliminary attempt to evaluate planar distortion. A total of 56 samples was ranked (ranging from 1 for least changed to 5 for most changed). A blind study is planned to provide more statistically accurate data.

Moisture: Moisture content was measured with a Sovereign electronic moisture meter, model 452 A. The samples humidified with Gore-tex were taken out of the sandwich and placed between polyester film supported by a blotter. The upper polyester film was removed and the surface of the paper measured. The same procedure was applied for the humidity chamber samples. The samples dried out very fast. The measurements had to be taken very quickly. The immersed samples were taken out of the bath and placed on polyester film supported by a blotter. One set of new (120 x 170 mm) and old (90 x 150 mm) untreated papers was cut for this purpose for each humidification technique. Four measurements were taken per paper sample and averaged. A total of 12 samples was measured.

Poultice: Diatomaceous earth (hydrated silica from diatom plant skeletons) was selected for its working properties since, unlike gel poultices (methylcellulose, agarose, starch paste, or hydroxypropylethylcellulose) it can be mixed with aqueous or non-aqueous solvents to form a plaster or paste that absorbs solutes as it dries to a powder, which can then be brushed off. It is more cohesive than fused silica. It is whiter than Fuller's earth, which is formed from hydrated silicates of magnesium, calcium, aluminum, or other metals. It is more controllable than organic solid poultices such as powdered cellulose, paper, or cotton. The diatomaceous earth was saturated with each solvent (approximately 1-2 ml solvent to 0.3 grams earth depending on solvent) and placed on the sample. Contrary to normal practice, the wet poultice was not surrounded by dry poultice which would reduce tideline formation.

Suction disk: Solvents were applied locally by dropper (6 drops of solvent, with drying between drops) on a 15 cm fritted glass bead disk (masked off with polyester film), which can reach a pressure of c. 25"Hg [45].

Suction table: The transparent papers were dried under low vacuum on a Nascor dual mode suction table, which can reach a pressure of c. 4.5"Hg [45].

Humidification pack system: The humidification pack system consisted of a damp blotter placed on a polytetrafluoroethylene membrane (a polyester felt laminate, 1/16" thick, produced by W.L. Gore & Associates, Inc.) which was held in contact with the transparent paper while under light weight (c. 1PSI).

APPENDIX 2: Federal Specifications

The US government has developed specific minimal standards for modern tracing papers. According to Federal Specifications (UU-P-561H 1972), modern tracing papers are classified as four types which are required to have certain chemical and physical properties as measured by particular tests. Briefly stated, the four types consist of three groups of 100% rag fiber papers and one group of chemical wood pulp fibers, each group with varying degrees of whiteness, translucency, strength, and permanence. The general requirements are that color does not change, according to subjective comparison, after accelerated aging (100 degrees C, 72 hours); that opacity averages, depending on classification type, from 28-70% and not increase more than 3-7% for rag papers or 9% for chemical woodpulp; and that fold endurance not drop by more than 50% after aging. Manufacturers have tried to meet these standards, despite inherent vagueness (i.e. specifying ill-defined 100% "rag" rather than cotton linters or alpha cellulose; subjective color "measurement") and the recent modification requiring minimum inclusions of 25-50% "recovered material" in rag and chemical wood pulps respectively.

APPENDIX 3: Selected Bibliography

1. Anderson, P. "Transparent Paper: An examination of its uses through the centuries, with a close investigation of a collection of early 19th century tracings." Unpublished Senior Essay, Yale College, 1990, History of Art Dept. pp. 1-47.

2. Andes, L.D. "Tracing and Copying Papers". In The Treatment of Paper for Special Purposes, Scott and Greenwood, London, 1907.

3. Bachmann, K. "Transparent Papers Before 1850: History and conservation Problems." In New Directions in Paper Conservation, Oxford, England: Preprints of the Institute of Paper Conservation 10th Anniversary Conference, 14-17 April 1983.

4. Bachmann, K. "The Treatment of Transparent Paper: A Review." The Book and Paper Group Annual, Vol. 2, 1983. pp. 3-13.

5. Baker, M., D. van der Reyden, and N. Ravennel. "FTIR Analysis of Coated Papers." The Book and Paper Group Annual, Vol. 8, 1989, pp. 1-12.

6. Bolam, F., ed. 1962. The Formation and Structure of Paper, transactions of the symposium held at Oxford September 1961. London: Technical section of the British Paper and Board Makers' Association (Inc.). Vol. 1 and 2.

7. Bolam, F., ed. 1976. The Fundamental Properties of Paper related to its Uses, transactions of the symposium held at Cambridge: September 1973. London: Technical section of the British Paper and Board Makers' Association (Inc.). Vol. 1 and 2.

8. Booker, P.J. A History of Engineering Drawing, Northgate Publishing Co. London, 1979.

9. Bush, A. "Conservation of 19th century tracing papers in the national Maritime Museum." In New Directions in Paper Conservation, Oxford, England: Preprints of the Institute of Paper Conservation 10th Anniversary Conference, 14-17 April 1986.

10. Cable, C. The Architectural Drawing: Its Development and History 1300-1950. Monticello, Illinois: Vance Bibliographies, 1978.

11. Casey, J., ed. 1980-83. Pulp and Paper: Chemistry and Chemical Technology. New York: Wiley-Interscience. 3rd Edition. 4 Vols.

12. Dawe, E.A. Paper and its Uses. Cosby, Lockwood and Son, London, 1914.

13. Erhardt, D., D. van Endt, W. Hopwood: "The Comparison of Accelerated Aging Conditions through the Analysis of Extracts of Artificially Aged Paper." AIC Book and Paper Preprints, 15th

annual meeting, AIC, Washington D.C., pp. 43-55.

14. Federal Specification UU-P-561H 1972

15. Flamm, V., C. Hofmann, S. Dobruskin, G. Banik. "Conservation of Tracing Papers." Preprints of the Triennial ICOM Meeting, Dresden 1990, pp. 463-467.

16. Flieder, F. Leclerc, C. Laroque, P. Richardin, B. Guineau. "Analyse et Restauration des Papiers Transparent Anciens." Les documents Graphiques et Photographiques, Analyse et Conservation 1986-1987 (Paris, Archives Nationales, 1988) pp. 93-139.

17. Flieder, F., Martine Leroy, Jose Carlos Andreoli, Francoise Leclerc. "Comparative Study of four Paper Bleaching Methods." IFLA Conference, 1991, pp.1-27.

18. Glaser, M.T. "Conservation of Drawings by Frank Lloyd Wright at the New England Conservation Center." AIC Preprints of the Annual Meeting at San Francisco, 1988, pp. 20-25.

19. Jirat-Wasiutynski, T. "Sprayed Poly (vinyl acetate) Heat Seal Adhesive Lining of Pen and Iron Gall Ink Drawings on Tracing Paper." JAIC 19(1980):96-102.

20. Higgins, H. and J. de Yong. "The Beating Process," in The Formation and Structure of Paper, edited by Francis Bolam, Vol.II, London 1962, pp. 651-695.

21. Hoffenk de Graaf, J., C.E.M. Wolff. "Eine Methode, um Transparentpapier zu Kaschieren." Maltechnik-Restauro 3 (1982): 198-203.

22. Hofmann, C., D. van der Reyden, and M. Baker. "The Effect of Three Humidification, Flattening, and Drying Techniques on the Optical and Mechanical Properties of New and Aged Modern Transparent Paper." Paper presented at IPC, Manchester, England, April 1992; forthcoming in The Paper Conservator, 1992.

23. Kouris, M. ed. Coating, Converting and Specialty Processes, Vol. 8., Pulp and Paper Manufacture Series, TAPPI Press. 1990.

24. Krill, J. English Artists Paper. London: Trefoil Publications Ltd., 1987.

25. McClintok, T. "The Drawings of Frank Lloyd Wright: Conservation of Works on Tracing Paper." In New Directions in Paper Conservation. Oxford, England: Preprints of the Institute of Paper Conservation 10th Anniversary Conference, 14-17 April 1986.

26. Maddox, H.A. Paper: Its History, Sources and Manufacture. London: Sir Isaac Pitman and Sons, Ltd., 1930.

27. Maroger, J. The Secret Formulas and Techniques of the Masters, trans. by Eleanor Beckham. NY. and London: the Studio Publications, Inc. 1948.

28. Mecklenburg, M. "The Role of Water on the Strength of Polymers and Adhesives." Doctoral dissertation, University of Maryland, College Park, MD, 1984.

29. Mills, J.S. "Analysis of Some 19th Century Tracing Paper Impregnants and 19th Century Globe Varnishes." In New Directions in Paper Conservation. Oxford, England: Preprints of the Institute of Paper Conservation 10th Anniversary Conference, 14-17 April 1986.

30. Mosher, R.H. ed. Specialty Papers, Their Properties and Applications. Remsen Press, Brooklyn, N.Y. 1950.

31. Orme, E. An Essay on Transparent Prints and on Transparencies in General, London: 1807.

32. Richardin, P., F. Flieder, S. Bonnaissies, C. Pepe. "Analyse par CG/SM des Produits d'Impregnation de Papier Calque Anciens." ICOM Preprints, Los Angeles 1990, pp. 482-488.

33. Robinson, J. "Fiber bonding," in Pulp and Paper Chemistry and Chemical Technology, J.P. Casey ed., third edition, Vol.III, New York 1980, pp. 915-961.

34. Roth, L., J. Weiner. Transparentising of Paper: Bibliographic Series No. 201. Institute of Paper Chemistry, Appleton, Wisconsin, 1963.

35. Rundle, C. "The Composition and Manufacture of Modern Tracing Papers." In New Directions in Paper Conservation. Oxford, England: Preprints of the Institute of Paper Conservation 10th Anniversary Conference, 14-17 April 1986.

36. Saucois, C. "La Restauration des Papiers Calques aux Archives Nationales." Paper presented at the Sixth Triennial Meeting, ICOM Committee for Conservation, Ottawa, September 1981.

37. Sindall, R.W. Paper Technology: An Elementary Manual on the Manufacture, Physical Qualities, and Chemical Constituents of Paper and of Paper-Making Fibres. London, Charles Griffin and Co. Limited, 1910.

38. Smart, H. "The Methods of Manufacture of Translucent coated Papers and their Conservation Problems." Unpublished 2nd year project, Camberwell School of Art.

39. Steinkellner, C. "Transparentpapier." Maltechnik Restauro, No. 1 & 2, 1979, No. 1, pp. 9-13, No. 2 pp. 113-117.

40. Stone, J. L. "Treatment of 19th Century Tracing Papers from the Frederick Law Olmsted Collection." Paper presented at the 10th Triennial Meeting, ICOM Committee for Conservation, Vol. II, 1987, pp. 731-738.

41. van der Reyden, D. "Recent Scientific Research in Paper Conservation," Journal of the American Institute for Conservation. 31 (1992):117-138.

42. van der Reyden, D., C. Hofmann, and M. Baker. "Effects of Aging and Solvent Treatments on Some Properties of Modern Tracing Papers," forthcoming in JAIC 1993.

43. van der Reyden, D., C. Hofmann, and M. Baker. "Some Effects of Solvents on Transparent Papers." Paper presented at IPC, Manchester, England, April 1992; forthcoming in The Paper Conservator, 1992.

44. Vaurio, P. "Transparentizing of Paper," TAPPI Journal Vol. 43, No.1, Jan. 1960.

45. Vincent, H. "The Materials Used in the Manufacture of Tracing Papers," Paper Trade Journal, TAPPI section, Dec. 1932, p.304-305.

46. Vitale, T. "Observations on the theory, use and fabrication of the fritted glass bead small suction disk device," The Paper Conservator, Vol. 12, 1988, pp. 47-67.

47. Watkins, S. "Chemical Watermarking of Paper," Journal of the American Institute for Conservation. 29 (1990) pp:117-131.

48. Yates, S. A. "The Conservation of Nineteenth-Century Tracing Paper." The Paper Conservator, Vol. 8, 1984, pp. 20-39.

48. Yates, M.I. Chartam Papers (Private correspondence) 1991-2.

EFFECTS OF WATER ON THE MECHANICAL PROPERTIES OF PAPER AND THEIR RELATIONSHIP TO THE TREATMENT OF PAPER

TIMOTHY VITALE
Conservation Analytical Laboratory, Museum support Center, Smithsonian Institution, Washington, DC. 20560

ABSTRACT

Through a series of experiments the mechanical properties of paper are explored. Hydrogen bonding is fundamental to the performance of paper and its disruption results in distinctive stress-strain behavior. Stress-strain curves were generated from which tensile strength, Young's modulus, percent stretch, and work (tensile energy absorption) were obtained.

It was found that the contribution of the fiber to the mechanical properties of paper is primarily elastic. Fibers are many times stronger than paper. Only fibers which have been severely deteriorated show measurable changes in stress-strain behavior. Fiber deterioration results in characteristically different stress-strain behavior than that which results from disruption of interfiber bonding.

Water immersion results in the disruption of interfiber bonds in paper, leaving only 2-3% of dry tensile strength. Interfiber bonds make a profound contribution to the mechanical properties of the paper. Aqueous treatment is shown to be a radical treatment, altering the original dried-in properties of the sheet. The release of structural bonds and dried-in strains during wetting and the subsequent reformation of interfiber bonds during drying are shown to be independent of water purity, be it ultrapure water, tap water, or water containing washing aids such as $Ca(OH)_2$, NaOH, $CaCO_3$ or Na_2CO_3.

The effects of immersion in organic solvents was explored. Solvents have effects on mechanical properties which are approximately proportional to the degree of swelling caused by the solvent. Water, the liquid which caused the greatest swelling of the liquids evaluated, is shown to be the most disruptive liquid followed by methanol and acetone; toluene caused virtually no change.

To explore the behavior of interfiber bonds paper was taken through a solvent exchange process. A sample was immersed in water and then taken through separate ethanol and acetone immersions to toluene, and dried. The result was a sheet with little bonding and decreases in all mechanical properties. To explore the surface tension and capillary action effects of water, the solvent-exchanged sheet was re-immersed in water. Upon drying, interfiber bonding was reintroduced which resulted in the regain of mechanical properties lost.

A paradigm for the mechanical behavior of paper is developed. Fibers contribute elastic behavior and interfiber bonds are a principal source of plastic behavior.

INTRODUCTION

Wetting with Liquids

Liquid water is an active solvent for paper. It affects paper in two distinct ways: (1) it is imbibed into the pores and (2) it swells and plasticizes fibers. Thus, within a sheet of paper there are two domains for the absorption of liquids: pores and fibers.

Cotton swells readily and becomes plasticized when humidified or wet.

Paper is commonly 25-50% voids (Casey, 1960). The decrease in percent solids after wetting suggests that the structure of the paper is disrupted by water. At the fiber saturation point the water content of a cotton fiber paper is approximately 22%; no free water is present, but the fiber is in equilibrium with 100% RH air. The fiber increases 25-30% in circumference, 10-12% in diameter, and 1-2% in length when wet at the fiber saturation point.

Hydrogen bonds in paper and fibers

The surface of any cellulose fiber is bristling with hydroxyl groups and oxygen containing groups. The fiber is a complex physical structure of fibrils which are composed of glucose polymer chains. There are 5 oxygens in one anhydroglucose ring in a cellobiose unit. Three are in hydroxyl groups at C_2, C_3 and C_6; there is a ring oxygen between C_5 and C_1, and a glucosidic bond between C_1 and C_4. These are all possible sites for hydrogen bonding. When hydroxyl groups come within roughly 3.6 Å of each other in cellulose, a hydrogen bond can form (Liang & Marchessault, 1959a; Marchessault & Liang, 1960; Blackwell & Marchessault, 1971.

Nissan, Byrd, Batten & Ogden (1985) consider paper an H-bond dominated solid and have established a force constant for stretch or compression of bonds using stress-cycling and 2 other methods. Nissan (1976) related Young's modulus to the force constant times the effective number of bonds per unit volume. Nissan (1977) presented an excellent review of hydrogen bonding in fibers and paper and related mechanical properties.

Within a chain three bonding sites are possible and between chains three more sites are possible, using the unit cell defined by Marchessault (1959). When polymer chains are aligned so that all 6 possible sites in each glucose unit are filled over a relatively large region (approximately 100-500 Å along the length), crystalline cellulose is formed. Amorphous cellulose has less than 6 of the possible sites filled. This is usually due to dislocations or other physical phenomena (Haigler, 1985). Cellulose with a lower degree of order is more accessible.

Interfiber bond sites are regions of less than crystalline bond density. The geometry of the polymer chains for the two interacting fibers is nominally random, rather than the regular biological, polymeric, and filamentary structure of the fibrils. These regions are therefore highly accessible to water.

The only restriction to disruption of glucose-OH...OH-glucose or glucose-OH...O-glucose hydrogen bonds is proximity of the water molecule to the bond. Water will not fit into crystalline regions. It can fit into intercrystalline cellulose (amorphous) and interfiber pores. Because the fiber is composed of regularly distributed crystallites in an accessible matrix the fiber is softened by water vapor or water.

Water and solvent interactions with cellulose

Common organic solvents swell cellulose less than water: methanol (swells 95% of water), ethanol (83%), acetone (63%), and toluene (2%). The degree of swelling depends on polarity and molecule size (Stamm, 1964). Only the metal complex solvents such as cupraammonium hydroxide, cupriethylenediamine hydroxide and ethylenediamine hydroxide will disrupt the crystalline structure causing the fibers to separate into relatively intact polymer chains.

Effects of water on mechanical properties

There is enough intercrystalline amorphous material in fibers (27% in cotton, up to 50% in other fibers) that they are readily plasticized by water (Treiber, 1975). A comparison of stress-strain curves in Morton (1975) shows that modulus for fibers decreases by 70% when wet.

The disruption of interfiber bonds and intercrystalline swelling are related. Table 1 shows Robertson's data (1970) on the effects of liquids on tensile strength, initial modulus, and toughness (work) correlated with Stamm's (1964) swelling data. Robertson reports his sample to be "standard blotter;" Stamm used wood coupons machined to a uniform fairly thick caliper which is required for measuring the small thickness increases due to swelling.

TABLE I: Decreases in mechanical properties due to immersion in various liquids. Mechanical strength data from: Robertson, A.A. (1970: Table I, p.1332) on a "standard" 0.02" blotter. Reported as percent change from control. Swelling data from Stamm, A.J. (1964: Table 15-2), for white pine and Sitka spruce cross-sections reported as percent change from control; water = 100% swelling.

	Tensile Strength	Youngs' Modulus	(Work) Toughness	Percent Swelling
WATER	95.7	95.7	92.4	100
DMF	90.6	90	85.4	
METHANOL	80.4	83.3	61.4	95
ETHANOL	66.3	68.1	40.4	83
PROPANOL	55	63.8	29.8	45
ACETONE	44.2	54.8	26.3	63
MEK	35	40.5	22.2	
BUTANOL	31	34.8	16.4	14
DIACETONE ALCOHOL	23	23.8	12.9	
TOLUENE	15	14.3	17.5	2
BENZENE	12	2.4	15.2	0
HEPTANE	8	8.6	2.9	
ISOOCTANE	5	4.3	8.2	
CONTROL	0	0	0	0

Physical structure in relation to mechanical properties

Fibers are more than simple rodlike structural elements in paper. Several workers have documented that paper fibers have crimps, curls, kinks and microcompressions. This is different from the relatively straight fibers found in wood and textile yarns. Page et al. (1985) reviews the sources and causes of curl, crimps, kinks and microcompressions in paper fibers. Page (1962) first observed microcompressions which are microscopic compressions in a fiber length where one fiber overlaps another fiber at (or approaching) right angles. The large shrinkage of fiber width during desorption, compared to length, causes the width of one fiber to compress the length of the fiber it is bonded to, by introducing microcompressions. It has been shown that sheets made with either excessive kinks (Page, 1966) or microcompressions have increased stretch behavior (Seth and Page, 1981).

In machine-made paper, fibers are oriented in a 3:1 ratio in the machine direction of the sheet. Fibers intersect at fairly small angles forming structural polygons which are hydrogen bonded at the intersections (Van den Akker, 1962; Mackeprang, 1980). During the early stage of loading the polygons and fiber segments deform in the plane of the strain. In addition, straining straightens the curls, crimps, kinks and microcompressions. Free fiber segments are twisted off axis, and both stress and shear stress is applied to the bonds (Van den Akker, 1962). As with all elastic behavior, when the stress is released the mechanical performance is reversed with no permanent deformation.

Hearle (1985) states that differences in fine structure among cellulose fibers (natural, regenerated, etc.) will have an effect on mechanical properties of the fibers. Zeronian (1976) shows, however, that within a single class of cellulose fiber only extensive changes in fine structure (conversion from Cellulose I to Cellulose II) have effects on mechanical properties. Cotton treated with 12% NaOH (3M) did not show increased break load as compared to cotton treated with 20% NaOH (5M) which did show an increase in strength. The latter causes mercerization of cotton, which is a change in unit cell structure from Cellulose I to Cellulose II.

EXPERIMENTAL DESIGN

Equilibrium stress-strain protocol

Stress-strain measurements were made on self contained stress-strain testbeds (Arch, McNamara, Hand & Mecklenberg, 1989) which employed an incremental increase of strain rather than load. The equipment is hand operated with digital strain gauge readout. Rate of elongation is very slow, in stepwise increments of 0.002 inches at one step per minute until the sample breaks. The strain is increased and then the resulting stress is read one minute later. This stress-relaxation method was developed by Mecklenburg (1988) and is referred to as equilibrium stress-straining. Equilibrium terminology is used because the sample is allowed to come to equilibrium with the strain before the stress is recorded. One minute relaxation, for the small strains applied, was found to be sufficient for the load to come to a reasonable equilibrium.

The stress-relaxation technique is used because low strain rates and one minute incremental relaxations of the material allows an equilibrium with the load to be reached; that is, the stress is not dropping dramatically. Most engineering work is optimized to duplicate manufacturing conditions where speed is desirable, but in the world of cultural materials rapid loading and straining generally do not occur. Temperature and humidity change over hours or days. Artifacts are generally handled with care to prevent fatigue or premature failure, in comparison to less precious commodities. The slow straining rates approach the "real" world for cultural materials. If equilibrium stress-straining data are compared to standard engineering data, the values will be lower in stress and modulus and higher in strain.

All measurements were conducted at 50% RH $\pm$2% and 70°F $\pm$2°F. In most cases a single sample was cut, measured for thickness and width, loaded into a tensile jig and held at equilibrium conditions in a sealed container for 1 hour prior to testing. Thickness was measured in 5 locations with a standard micrometer and averaged. In several cases (eg., controls) multiple samples were run and data averaged into master curves. Stress-strain master curves are created by mathematically averaging the stress and strain data independently at each incremental step for the multiple replicates of each sample. The final break load is calculated from the average of the last point from each of the runs. Curves with 2-6 runs are noted as n= after the data label. Machine direction (MD) samples took 15-45 minutes to run and cross machine direction (CD) took 60-150 minutes, depending on the paper or treatment.

There is often a slight toe in the stress-strain curves of paper. This was eliminated by applying a very slight initial strain (0.00004" -

0.0001") in the set-up phase. This procedure was introduced in later stages of the investigation; the change did not interfere with data evaluation or comparisons.

Materials

Five papers were used in the experiments. The principal experimental paper is an unsized (waterleaf) staple cotton fiber sheet. The waterleaf cotton and wood pulp analog, were both manufactured specifically for conservation research projects by Neenah Paper approximately 5 years ago. Two historic papers of different basic manufacture were included to provide relevancy to the paper conservation community. The machine-made historic paper was made by Strathmore approximately 75 years ago. The handmade historic paper is estimated to be 150-200 years old and was taken from the unused section of a ledger. A modern cotton filter paper, Whatman #1, was used in heat aged experiments (Erhardt, 1987 & 1988). However, its strength was less than a third of the staple cotton sheet so it was not used for the bulk of the study.

The two research papers from Neenah were made without sizing. Fiber analysis was done by the Institute of Paper Science and Technology (IPST, formerly Institute of Paper Chemistry). The waterleaf cotton paper is 91% cotton seed hair pulped from cotton knit cuttings in sodium hydroxide liquor with 2.2% softwood and 6.7% hardwood, both pulped using predominately the Kraft process. The waterleaf wood paper is 65% hardwood, 29% softwood and 6% cotton seed hair, free of lignin, and pulped with a variety of processes, but predominately Kraft.

The historic Strathmore paper was yellowed and felt slightly brittle. It was analyzed (IPST) to be 91% mixed softwood pulp and 9% cotton seed hair furnish, produced by either bleached Kraft or bleached sulfite processing with alum rosin sizing. The Conservation Analytical Laboratory received this paper in bulk as a gift. It was in an open package with the "Strathmore Quality" imprint on an incomplete label which was marked **Fairfiel(d) Japan**, 25" X 38," 70 lbs. It also has watercut deckled edges on the long sides. On the basis of the partial label the paper was dated by Strathmore to 1914-1919.

The historic ledger sheet is handmade wove and was lined for use in a ledger. The fiber content is a mixture of staple cotton and bast fibers which show unmistakable visual signs of degradation when magnified during fiber analysis, done at CAL. Sizing is alum and gelatin and was determined using standard spot tests from Browning (1969).

Whatman #1 filter paper is made from high alpha cellulose cotton linter pulp beaten and formed to 0.0069" thickness for an 11 micron particle capture size; no acid or alkaline wash or wet strength agents are added. Ash content is listed as 0.06%; cation content is 185 ppm Ca, 160 ppm Na, 7 ppm Mg, 5 ppm Fe; and anion content is 130 ppm Cl and 15 ppm S.

Principal experiments

Comparison of staple cotton fiber paper with virgin cotton fibers

To explore the relationship between the fiber and the paper into which it is made, data was extracted from Morton & Hearle (1975) for virgin cotton fibers. Because the Neenah waterleaf cotton paper was made from new textile cuttings it was as close to the virgin fiber as a machine made paper can be made. Figure 1 shows the virgin fiber plotted with the machine direction of the waterleaf staple cotton fiber paper made by Neenah.

Effects of severe fiber degradation of mechanical properties of paper

To explore the effects of fiber degradation on the mechanical properties of paper, degraded paper samples prepared by Erhardt (1987) were examined using the equilibrium stress-strain protocol. Erhardt's (1987, 1988) analysis of water extracted degradation by-products showed that dry higher temperature aging is different from humid aging. Glucose was associated with humid and moderate temperature dry aging and a five-membered-ring sugar is found in the extracts of the higher temperature dry aged samples. Only dry heat aged samples were used because the humid aged samples were folded into rectangles too small to for analysis. It can be seen in Figure 3, a photograph of Erhardt's (1987) entire sample population after pulping and extraction, that the humid aged samples (first row of samples) show the least degradation. Degradation is indicated by the presence of interfiber cross-linking, which is manifested as a fluffy appearance of the sample. In comparison, the heat aged sample series shows progressive and severe damage. The range of samples allowed specimens from slightly to severely degraded to be evaluated using mechanical analysis. Chemical evaluation of the intact polymers (degree of polymerization) was not done.

Wetting with pure water

The waterleaf cotton, waterleaf wood and Strathmore papers were immersed in deionized water for 60 minutes, drained for 3 minutes on a tilted and pre-cleaned polyester screen, and then dried between blotters at 0.01 PSI for 24 hours. The deionized water was ultrapure, processed through Hydroservices Corporation mixed-bed and activated charcoal columns. In-pipe water quality was 18.1 megohms at 20°C, after 30-60 minutes purity drops due to absorption of CO_2.

Wetting with water of lower purity

Samples were immersed in Suitland, MD, tap water. Suitland water is an imprecise mixture of predominately Potomac River water with Patuxet River water (Laurel, MD) from the regional water company. Yearly ionic averages are: cations, 35±6 ppm Ca, 9±3 ppm Na, 6±2 ppm Mg and anions, 29±6 ppm Cl, 27±11 sulfate; carbonate information not supplied. The immersion and drying protocol was the same as the ultrapure water method. The distilled water was the standard effluent of the Corning Megapure MP-3A still run at 30cc/min. The purity of the distilled water is often measured to be 1 Megohm, just after cooling. The waterleaf cotton, machine direction (MD), sample was not exposed to this portion of the experiment.

Wetting with water and washing aids

The following washing aids were individually added to the deionized water described above: pH 10.2 NaOH; pH 10.2 $Ca(OH)_2$; pH 10.15, 80ppm $Na_2CO_3.H_2O$ and pH 9.25, 80ppm $CaCO_3$. The calcium washing aids are those commonly in use by paper conservators. The sodium analogues were used in order to compare an ion more active with cellulose than calcium. Only 8 ppm calcium carbonate will dissolve in deionized water. The excess (80-8 = 72 ppm) calcium carbonate was added to make up for calcium that might be ion exchanged onto the cellulose or lost from solution by some other means. There were finely divided particles suspended in the liquid and on the bottom of the tray. The 80 ppm sodium carbonate completely dissolved, thus both the cation (Na^+) and CO_3^{-2} ions were in higher concentration than in the calcium carbonate solution. Treatment method was the same as the ultrapure water protocol.

Wetting with solvent

Solvent immersion was done in a manner similar to wetting with water. Several sheets of each paper type were immersed in baths of 1-2 liters liquid for approximately 60 minutes. Sheets were drained on edge for 10-15 seconds and laid horizontally on stretched polyester screen for approximately 3.5 hours. When almost dry, they were placed in a felt stack separated by two 1/2" felts and dried overnight at approximately 0.01 psi.

Solvent exchange method

To study the effects of paper with disrupted hydrogen bonds a solvent exchange process was used. Solvent exchange is known to cause disruption of hydrogen bonds in similar materials. The sheet was immersed in deionized water for approximately 60 minutes. The sheet was drained on edge for approximately a minute, then immersed in a liter of ethanol and allowed to stand for 15 minutes; it was than transferred to another ethanol bath for 15 minutes. Again, it was drained on edge for approximately 1 minute and immersed in a bath of acetone. After 15 minutes the paper was transferred to another acetone bath for 15 minutes. After acetone was drained from the sheet it was immersed in two successive one liter baths of toluene and dried on a polyester screen in a fumehood overnight. The preparation of individual samples was repeated several times. Each sample was mechanically evaluated against the others to be sure that maximum unbonding had been achieved.

Freeze drying method

In an attempt to achieve greater unbonding a cotton sheet was lyophilized following standard techniques using a Labconco Lymp-Loc 4.5 freeze dry unit. The sheet was immersed for approximately one hour in deionized water and placed in a jar soaking wet where it was fast frozen at -80 C°. The specimen was pumped down to 7.5 microns and held at -55 C° for 2 days.

RESULTS

Comparison of individual fibers and paper

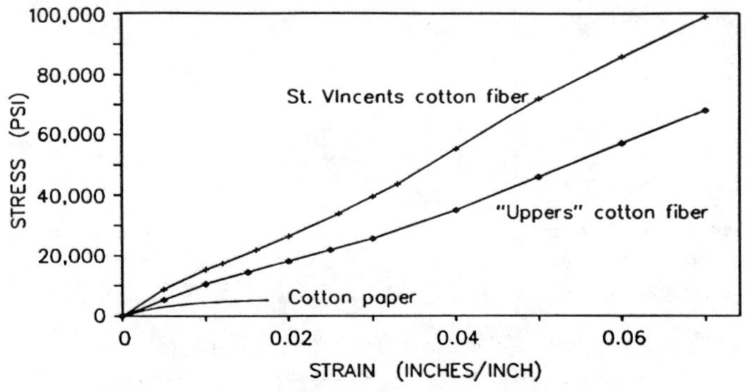

Figure 1: Comparison of cotton textile fibers to the waterleaf cotton paper MD. Source of the fiber data is Morton & Hearle (1975).

Using data from Morton (1975) on single cotton fibers and the waterleaf cotton sheet employed for most of the study, the strength of a fiber is compared to the strength of a network made from similar fibers. Figure 1 shows that the fiber network fails at 5-8% of the strength of the two staple fibers. The fiber's mechanical performance in the cotton waterleaf paper is from 0-5200 psi only. The stress-strain curves for the textile fibers show elastic behavior in the region where break load of the paper occurs; thus the contribution to the mechanical behavior of the paper is all recoverable.

The stiffness of the virgin fibers is approximately 1.1 Mpsi to 1.8 Mpsi compared to 0.69 Mpsi for the cotton paper. The fibers are substantially stiffer; the modulus ratio is as high as 3:1.

<u>The effects of severe fiber degradation on paper</u>

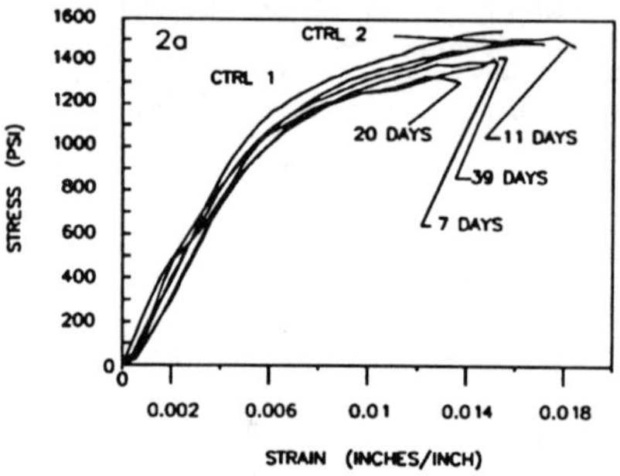

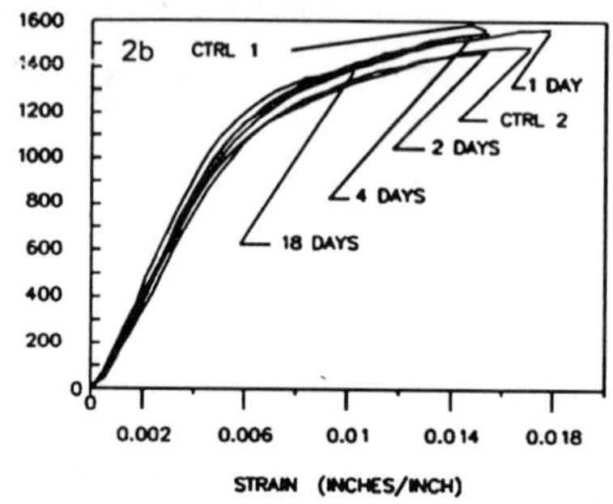

Figure 2a: Whatman #1 filter paper aged at 90°C in a dry oven for 7, 11, 20 & 39 days. Note that the samples fall within 15% of the controls and are randomly placed. Two controls were included to show typical scatter of data.

Figure 2b: Whatman #1 filter paper aged at 120°C dry for 1, 2, 4 & 18 days. Note that only the sample aged for 18 days is different from the controls.

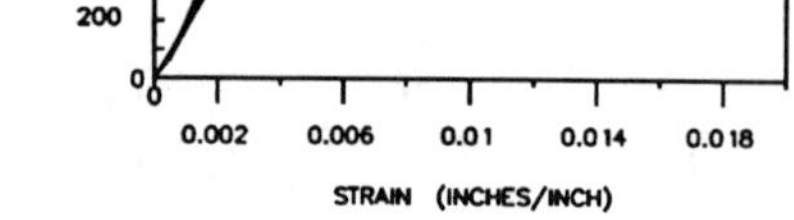

Figure 2c: Whatman #1 filter paper aged at 150°C dry for 1, 2, 4 & 8 days. Artifical aging has degraded the fiber, but not released the interfiber bonds. Degradation results in incremental losses of mechanical properties from those of the control's, along the stress/strain curve, rather than changes in all properties including modulus and yield point.

When a common cotton linter filter paper (Whatman #1) is degraded, only severe deterioration will result in a change in mechanical properties. When loss of mechanical properties does occur there is

decrease in strength

decrease in stretch

Figure 2a-c shows the effects of dry oven aging at 90°C, 120°C and 150°C on the mechanical properties of the cotton paper (Erhardt et al., 1987 & Erhardt, 1988). Samples in the dry aging series show the progressive effects of heat degradation only and were not intended to correspond to natural aging.

The samples degraded at 90°C dry show some variability of tensile strength and percent stretch but within 15% of the controls. In addition the properties do not change in predictable increments but rather are random. Variation of 15% from the mean within a group of samples is uncommon, but not rare; the aging influenced the paper slightly but the samples were influenced in a random manner. It is interesting to note that 39 days at 90°C (190°F) has no effect on the mechanical properties of paper.

Only one of the four samples dry oven aged at 120°C showed signs of degradation. After 18 days of aging the cotton linters lost some ability to stretch and hold a load. It is evident from Figure 1 that 90%+ of the fibers' strength must be degraded before the mechanical behavior of the paper will be influenced by the change. While 39 days of dry aging at 90°C did not alter the properties, 18 days at 120°C made a change in paper strength which is irrefutable and must have degraded the fiber by 90%+. The fact that there is an incremental progression from 1 day through 4 days is probably the result of chance because the first three samples are very similar to the controls and should not be considered different from the control.

Samples aged at 150°C show severe degradation of the fibers resulting in an incremental reduction of mechanical properties along the nominal stress/strain behavior of the control. Figure 2c shows the 4 levels of aging: 1, 2, 4 and 8 days; each with progressive loss in strength and percent stretch. With the possible exception of 1 day, the 150°C dry aging series exhibits incremental loss of mechanical properties from the control (see Figs. 2a & 2b) without changing either modulus (elastic behavior) or yield point (elastic limit). Degradation has not progressed far enough, therefore, to inhibit the reversible behavior of curls, crimps, kinks and microcompressions.

Visualization of degraded fibers

Figure 3 shows the samples after pulping and extraction. The information contained in the image of the pulped extracts provides valuable insight into the state of degradation in the paper. With increasing heat and time, samples cut into 1/4" squares and extracted, can not be broken down into pulp; they appear as a fluffy mass. The samples which formed disks of felted pulp indicate that the fibers could be successfully pulped. Samples showing a fluffy mass resisted pulping because they had wet strength. Wet strength is due to cellulose crosslinking at the interfiber bonds. In addition to wet strength, the last three samples (150°C dry; 2, 4 & 8 days) became increasingly reddish brown, indicating a further level of degradation.

Fiber rupture is not the principal mode of paper failure due to straining. Figure 4 a-f shows SEM micrographs of the Whatman #1 control and the 150°C dry aged for 8 days sample. Figures 4a & 4c show the pullout effect, and Fig 4e shows whole fibers which have been pulled free. Average fibers pull-out during straining; this demonstrates that interfiber bond rupture is the more common mode of paper failure. The fibrils do show some signs of rupture, but they are only filaments which have pealed away from the fiber body. Highly fibrillated fibers, created during pulping and beating, are generally fragments of whole fibers; thus they will rupture if their ultimate tensile strength is below the breaking load of the paper. Because the fibers are much stronger than the paper, whole fiber rupture occurs only if the fibers have been degraded by heat, light or chemical action.

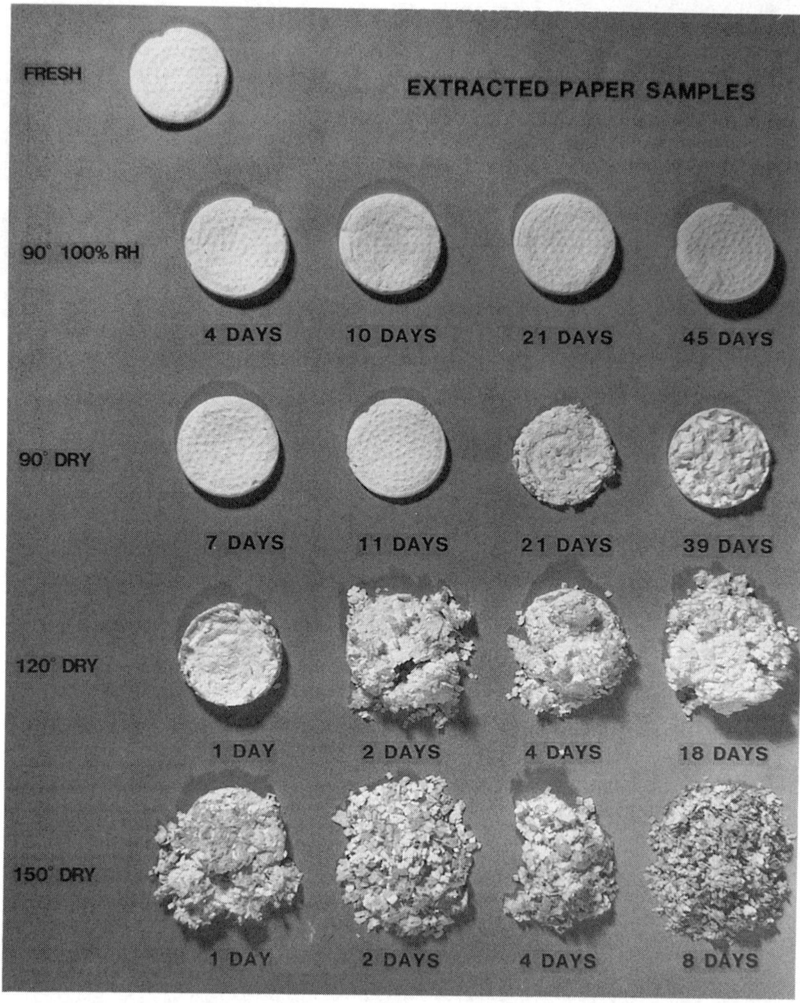

Figure 3: Whatman #1 filter paper control (upper left), aged at 90°C dry for 7, 11, 20 & 39 days (row 2); aged at 120°C dry for 1, 2, 4 & 18 days (row 3); and aged at 150°C dry for 1, 2, 4 & 8 days (row 4). The specimens in row one were not used in this study (see text). Note that some of the more degraded specimens have discolored as well as crosslinked. Samples were created, pulped, extracted and photographed by Erhardt (1987, 1988; photograph unpublished).

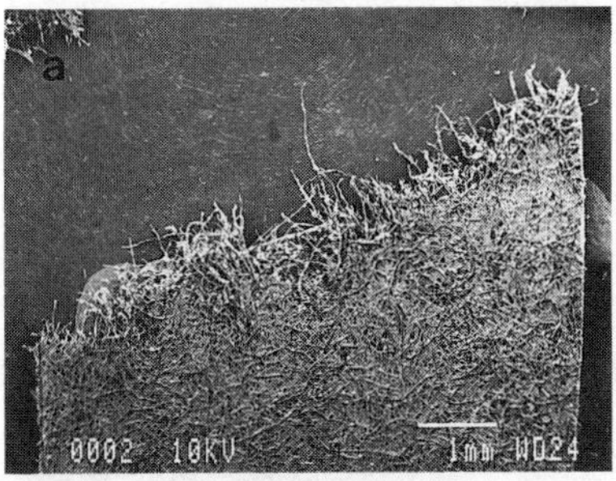

Figure 4a: SEM micrograph (15X) of Whatman #1 filter paper control at mechanical test rupture site.

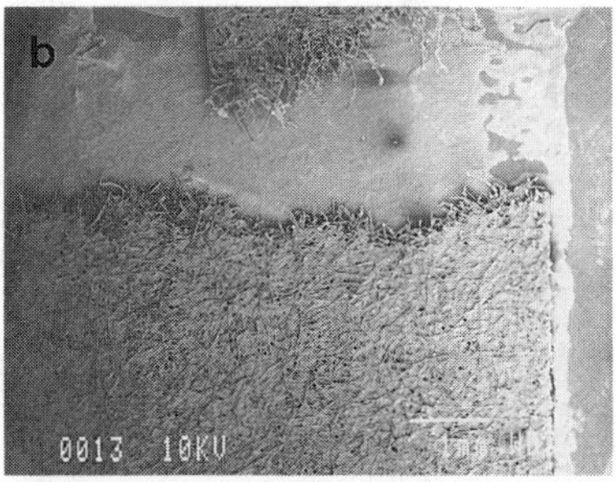

Figure 4b: Whatman #1 (15X) aged at at 150 C dry for 8 days at rupture site.

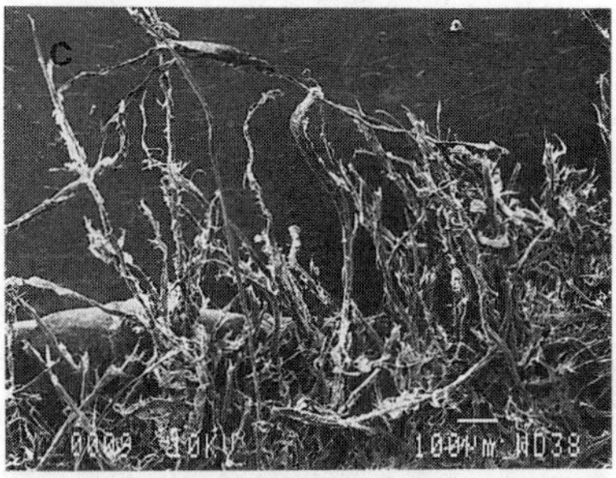

Figure 4c: Detail (75X) of Fig. 4a showing pulled-out fibers.

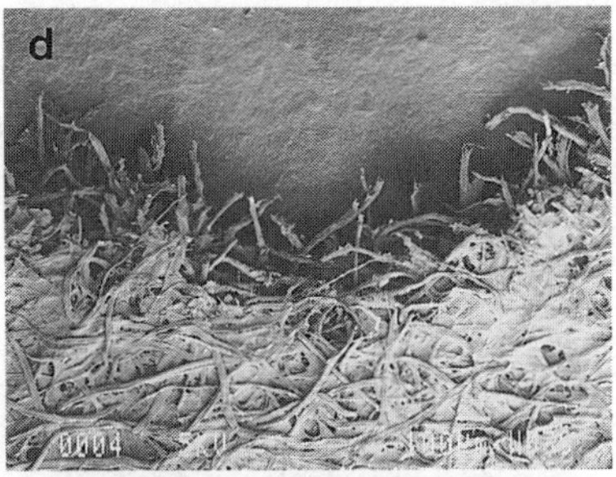

Figure 4d: Detail (75X) of Fig. 4b showing broken fibers.

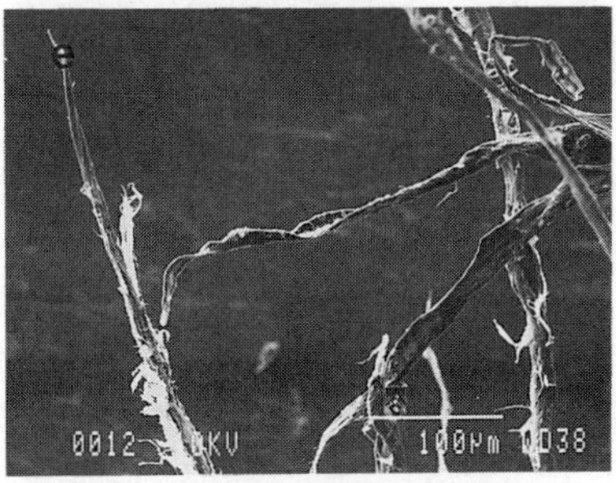

Figure 4e: Detail (250X) of Fig. 4c showing intact pulled-out fibers.

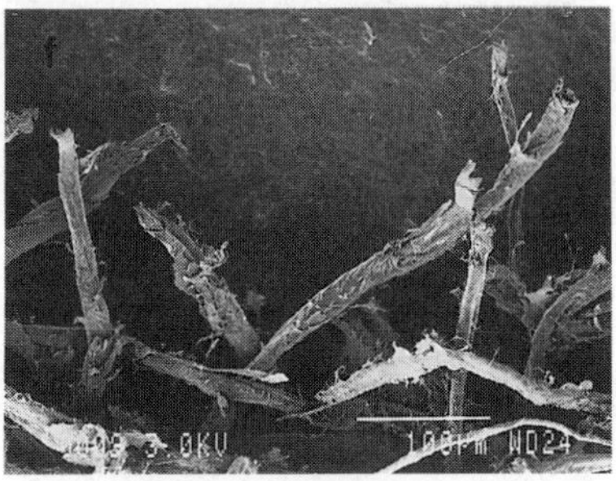

Figure 4f: Detail (250X) of Fig. 4d showing broken fibers.

Figure 4b,d & f shows the failure of paper with highly degraded fibers. Note that the fibers are stubs of those seen in Fig. 4a & 4b. Figure 4f shows whole fibers which are blunt broken stubs and the outer layers of fibrils which have broken away from the body of the fiber.

Wetting paper with water and wet-straining

Figure 5a-d shows the dramatic loss in all mechanical properties due to wetting. The remaining properties are due to fiber entanglement. The waterleaf cotton paper shows a 97% loss of its tensile strength; the Whatman #1 shows a 99% loss, and a "standard 0.02 blotting paper" reported by Robertson (1970: see Table I) has a 96% loss. The remaining strength is attributed to physical entanglement of the fibers. Careful analysis of the wet-strained examples (Figs. 5b & 5d) shows that failure begins at the maximum load the paper will hold and the sample does not fail catastrophically. In contrast, dry-straining has a clear catastrophic failure which is seen again and again.

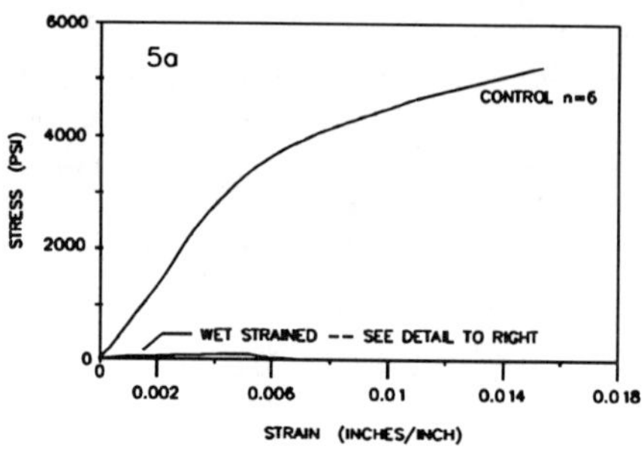

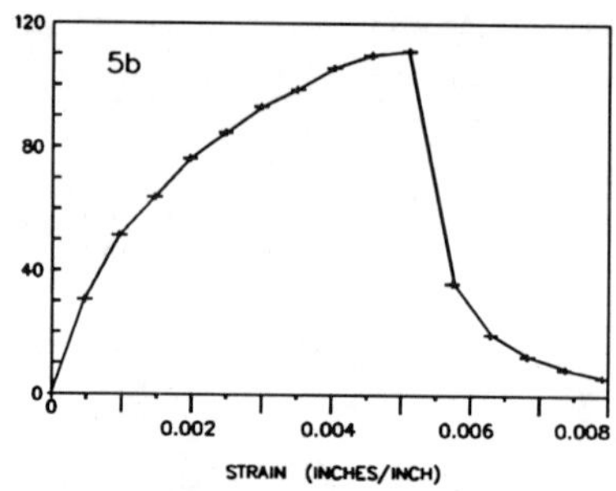

Figure 5a: Waterleaf cotton never-wet control (MD) and wet strained sample.
Figure 5b: Detail of the wet-strained sample from 5a; note slow rupture process.

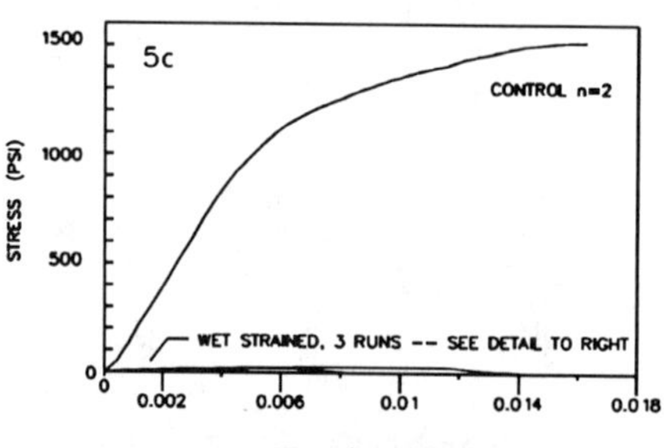

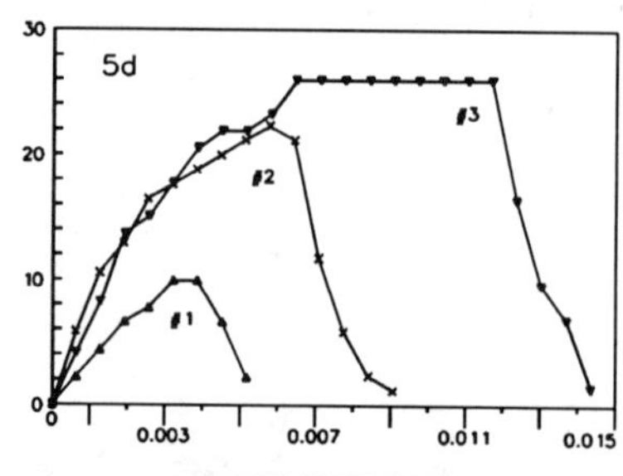

Figure 5c: Whatman #1 never-wet control (MD) and 3 wet strained samples
Figure 5d: Detail of 3 wet strained samples from 5c; note sustained load in #3.

Note that in Fig. 5b & 5d as the straining continues after maximum stress has been reached, a process of pulling the fibers past one another occurs. The resistance offered by physical entanglement holds a partial load. Eventually the stress decreases with increased straining, until failure occurs. Continuous application of strain with no catastrophic failure, along with no indication of decreased capacity to sustain a load for one of the Whatman #1 samples, followed by slow decay of load describes the pullout of long unattached components. Because paper is a network of felted fibers and wetting disrupts most of hydrogen bonds tying the network together, the fibers are allowed to be pulled past one another during wet straining.

Wet with pure water and dried

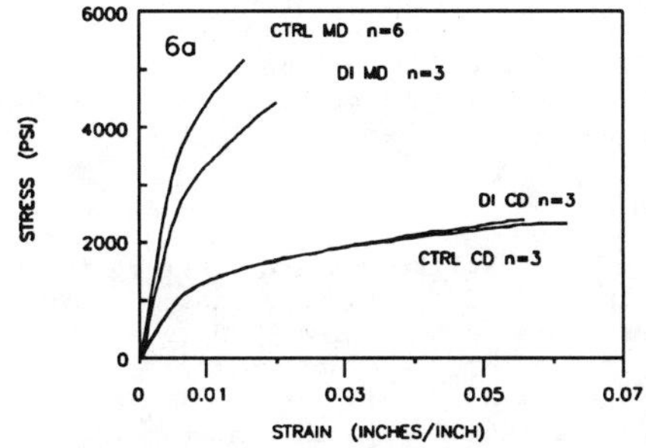

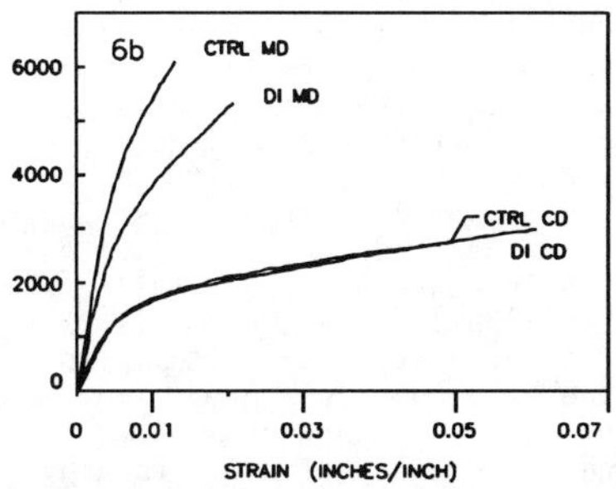

Figure 6a: Waterleaf cotton paper immersed in deionized water then dried (DI) and never wet controls (CTRL); both MD and CD.

Figure 6b: Waterleaf woodpulp paper immersed in deionized water then dried and never wet controls; both MD and CD.

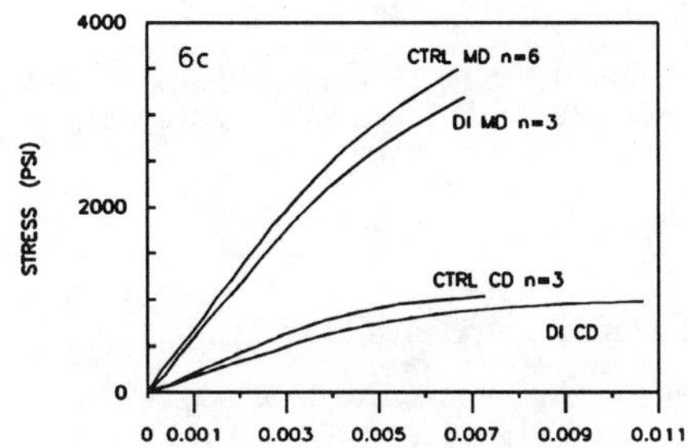

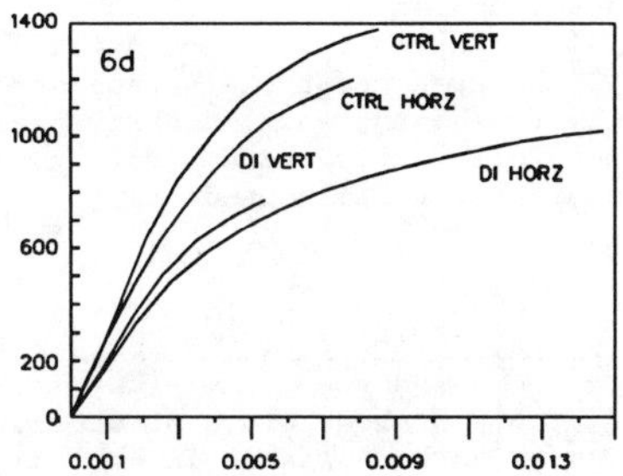

Figure 6c: Historic Strathmore alum–rosin sized paper, immersed in deionized water then dried, and never wet controls; both MD and CD.

Figure 6d: Historic handmade ledger paper, alum–gelatine sized, immersed in deionized water then dried and never wet controls; both horizontal and vertical directions.

Table II: Comparison of mechanical properties before and after washing in deionized water.

	PSI Tensile strength		PSI Modulus		g (within paper) Work		Inches/Inch Strain	
	MD	CD	MD	CD	MD	CD	MD	CD
waterleaf cotton CTRL	5166	2318	687,285	202,224	0.1858	0.332	0.0154	0.0619
waterleaf cotton DI	4396	2377	536,193	202,224	0.1986	0.374	0.0200	0.0558
percent difference	(15%)	3%*	(22%)	0%	6.8%*	13%	30%	10%
waterleaf wood CTRL	6102	2738	875,000	277,668	0.151	0.302	0.0131	0.0495
waterleaf wood DI	5326	2981	623,330	277,668	0.211	0.362	0.0209	0.0606
percent difference	(13%)	9%*	(29%)	0%	39%	20%	59%	22%
historic Strathmore CTRL	3487	1032	655,308	218,579	0.433	0.149	0.0067	0.0073
historic Strathmore DI	3187	980	571,429	163,934	0.405	0.231	0.0069	0.0107
percent difference	(9%)	(5%)*	(13%)	(25%)	(6.3%)*	55%	3%*	47%
historic ledger CTRL	1380	1198	321,700	259,100	0.201	0.155	0.0085	0.0079
historic ledger DI	772	1017	217,391	181,818	0.066	0.273	0.0054	0.0147
percent difference	(44%)	(15%)	(32%)	(30%)	(67%)	76%	(37%)	86%

* Differences less than 5-8% are within the scatter of high and low values on multiple-run samples.

When water is removed from the sheet, the network's strength returns. However, after wetting and drying there is a significant change in mechanical properties. Figure 6a-d shows this effect for the waterleaf cotton and wood, historic Strathmore and historic ledger papers in the machine direction (MD) and crossmachine direction (CD) or vertical and horizontal directions for the handmade paper. The trend is the same:

decrease in MD modulus
decrease in MD tensile strength
increase in MD & CD percent stretch

Changes in crossmachine direction strength and stiffness were observed for the handmade sized and naturally aged paper. The ledger paper has a 1.15 MD/CD tensile strength ratio which, when compared to ratios of 2.23 - 3.33 for the machine made papers, indicates that fibers were not as strongly oriented in the handmade sheet. Thus, wetting and drying after manufacture changes the stiffness and strength of both directions in the sheet. The machine direction results are most significant in machine made papers because the fibers are aligned in that direction.

Most of the samples show an increase in stretch in both directions. Furthermore, all of the papers shrank on drying; for example, the waterleaf cotton shrank 2.4% MD and 0.5% CD. The increase in the ability to stretch, in these samples, usually overwhelms the loss of tensile strength and modulus which results in an increase in work or tensile energy absorption (TEA), the area under the stress-strain curve. Work is defined here as the ability to absorb tensile energy (see Appendix). The historic ledger paper did not follow this trend; more than likely this is due to the difference in manufacture: it is handmade.

<u>Wetting and drying from tap and distilled water and salt solutions</u>

Figure 7 shows plots of stress-strain data for individual paper samples which were washed in various waters. Figure 8 shows the same for calcium washing aids and their sodium salt analogs. The stress-strain behavior of six control runs shown in Figure 9a when compared to the data for all waters and salt solutions suggests that all of the water types and salt solutions had a similar effect.

While much has been said and written about the effects of pure water on paper, the results shown in Fig. 6 show that wetting and drying with water alone has a larger effect on the mechanical properties of paper than effects due to changes in the content of the water. Thus, the temporary unbonding seen in Fig. 5a-d and the rebonding process depicted in Fig. 6a-d are not sufficiently affected by the inclusion of ions in the water, which could either be adsorbed onto the fibers or interfiber bonding sites, to influence mechanical properties.

Immersion in standard paper conservation washing aids (salt solutions) appears to have no additional effect on the mechanical properties of the papers explored. Figure 8 shows that immersion in (1) pH 10.2 CaOH, (2) pH 10.2 $Na(OH)_2$, (3) pH 9.25, 8/80ppm $CaCO_3$, and (4) pH 10.14, 80ppm $Na_2CO3 \cdot H_2O$ has no effect on mechanical properties when compared to the effects of immersion in ultrapure deionized water. Increased swelling, ionic attraction or inhibition, and changes in water surface tension are not sufficient to change mechanical properties of the waterleaf papers.

The effects of washing aids on the permanence of paper was not examined in this study.

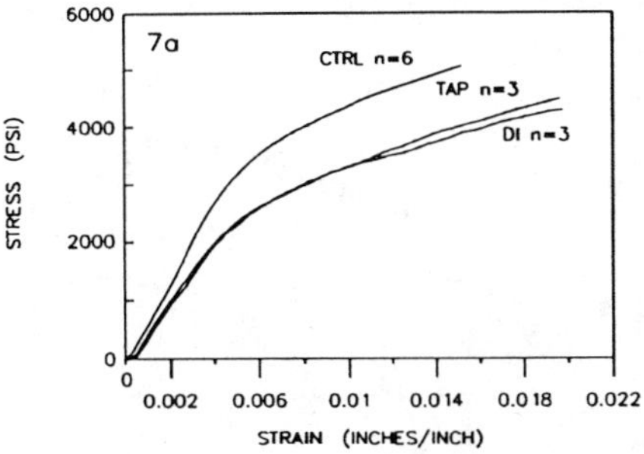

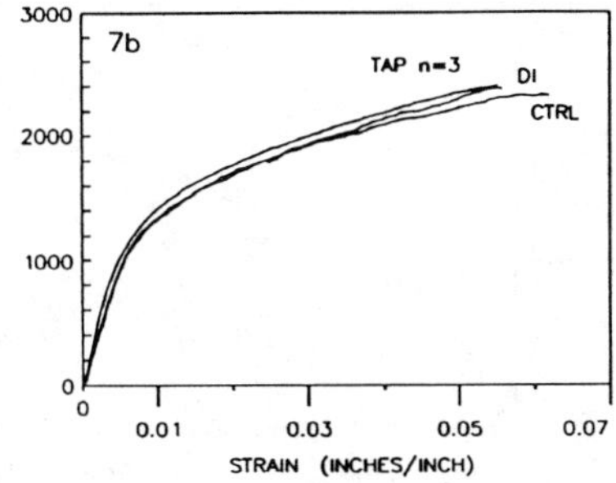

Figure 7a: Waterleaf cotton paper (MD) immersed and dried from deionized water and Suitland tap water; and never wet control.

Figure 7b: Waterleaf cotton paper (CD) immersed and dried from deionized water and Suitland tap water; and never wet control.

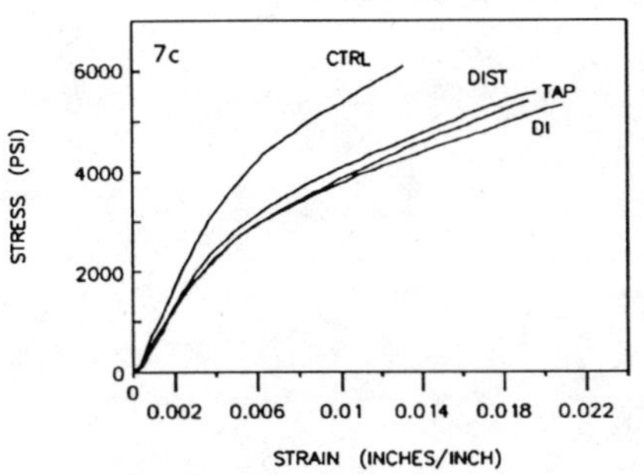

Figure 7c: Waterleaf wood paper (CD) immersed and dried from deionized water, distilled water and Suitland tap water; and never wet control.

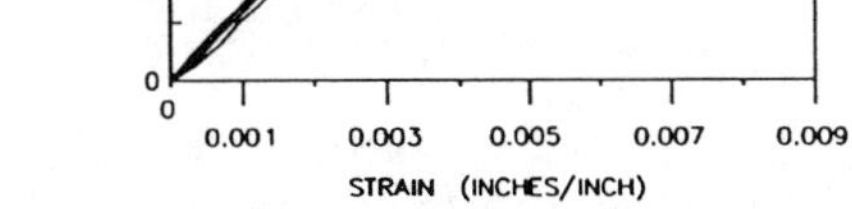

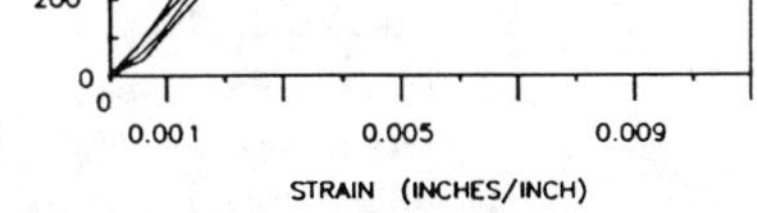

Figure 7d: Historic Strathmore paper (MD) immersed and dried from deionized water, Suitland tap water and distilled water; and never wet control.

Figure 7e: Historic Strathmore paper (CD) immersed and dried from deionized water, Suitland tao water and distilled water; and never wet control.

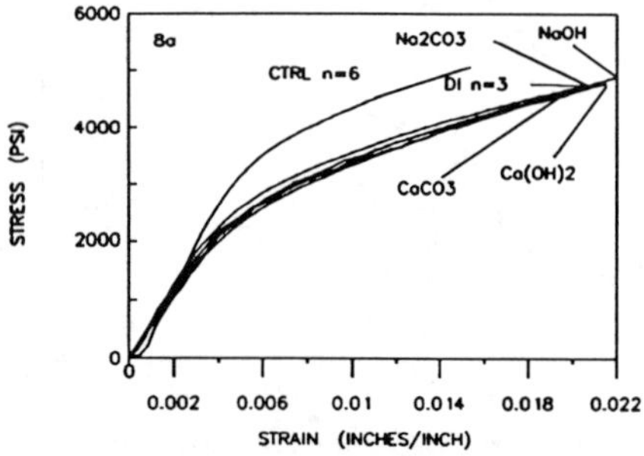

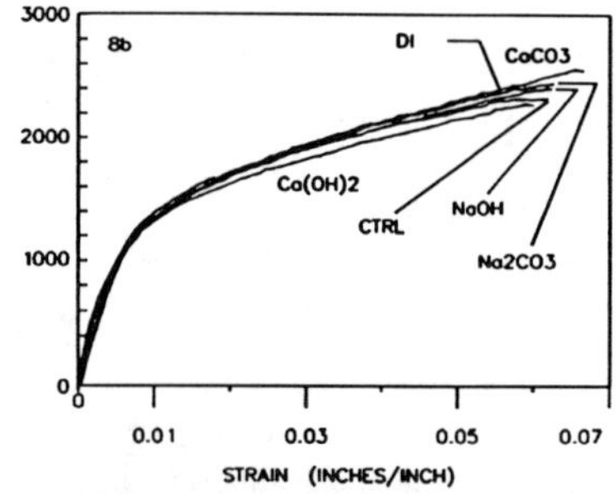

Figure 8a: Waterleaf cotton paper MD immersed in deionized water alone, with various washing aid solutions and the never wet control.

Figure 8b: Waterleaf cotton paper CD immersed in deionized water alone, with various washing aid solutions and the never wet control.

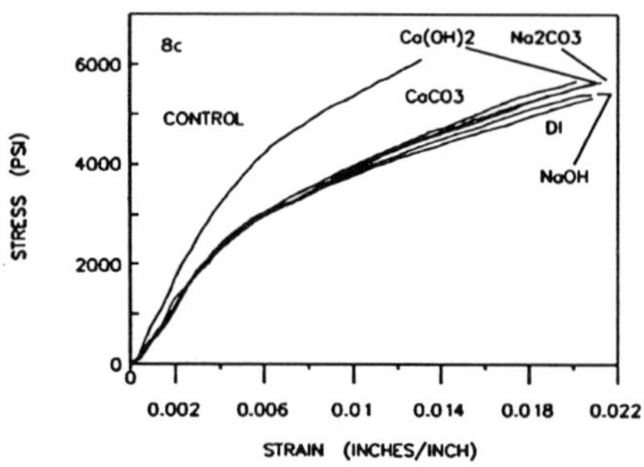

Figure 8c: Waterleaf wood paper MD immersed in deionized water alone, with various washing aid solutions and the never wet control.

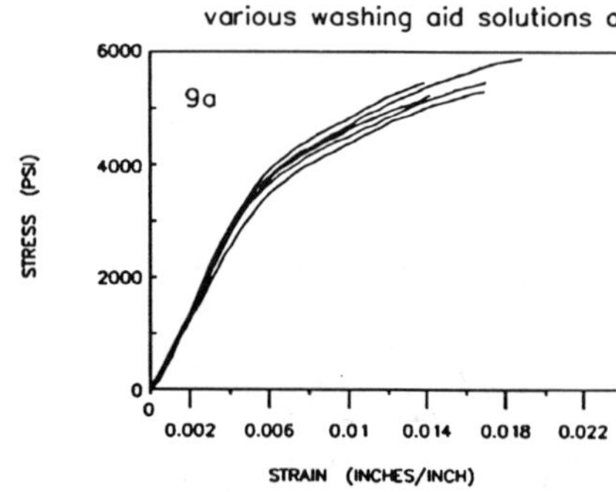

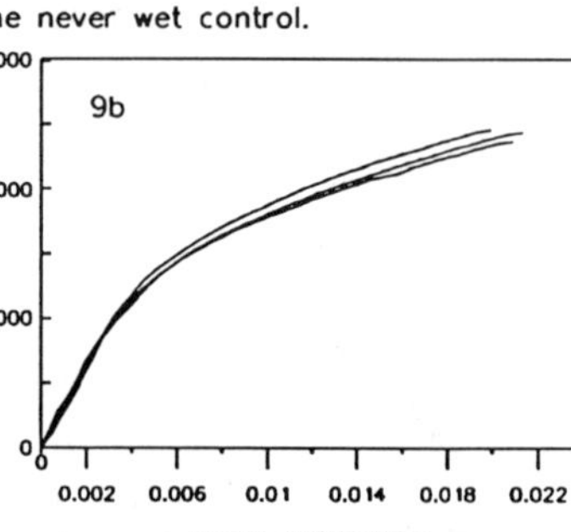

Figure 9a: Waterleaf cotton MD controls, six runs by two operators approximately one year apart (three each) and all on separate occasions. Shows scatter of data under worst possible conditions.

Figure 9b: Waterleaf cotton MD immersed in deionized water and dried, run by one operator at one sitting. Shows the least scatter of data under best conditions.

Wetting with solvents

Solvents have variable effects on paper. The data in Table I from Robertson (1970) and Stamm (1964) show that there is a relationship between swelling and release of bonds while immersed. In this investigation, solvents which cause marked swelling such as methanol and ethanol yielded paper with lower mechanical properties (see Fig. 10). Solvents which result in low swelling such as toluene caused almost no change in mechanical properties compared to the control. It has been shown that water causes a change in mechanical properties after immersion and drying. The differences in degree of change appear to be related to release of as-manufactured properties such as bonding and dried-in strains parallel to the process seen for water. If water causes the greatest change, then solvents which swell less should have an effect intermediate between the control and the water immersed and dried samples. Figure 10 shows those general results. The stress-strain behavior of acetone treated paper is slightly above that for water immersion. The toluene immersed sample is indistinguishable from the control. The methanol treated sample has a Youngs' modulus equal to that of the water sample, and a tensile strength 8% above that of water. However, this is close to the error of single measurements, so it does not provide conclusive evidence for incomplete release of interfiber bonds.

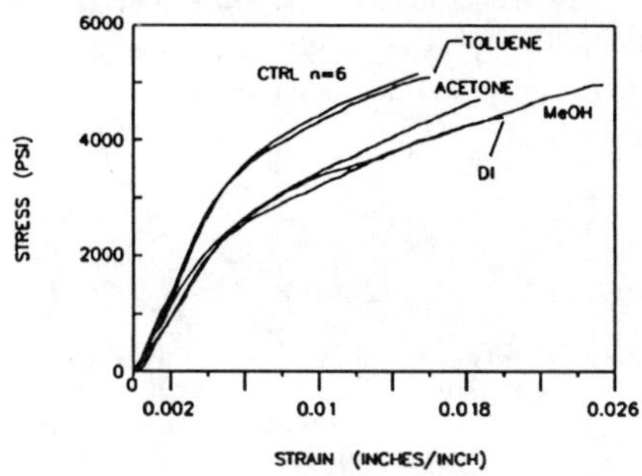

Figure 10: Waterleaf cotton paper (MD) untreated control and samples immersed in various solvents and dried in a felt stack at 0.01 psi.

Solvent Exchange and Freeze Drying

Figure 11 shows the effect of partial unbonding on the waterleaf cotton paper. The waterleaf wood and the historic Strathmore papers displayed the same behavior (data not shown). The trend was consistent:

- decrease in modulus
- decrease in tensile strength
- decrease in plastic yield point
- decrease in percent stretch (except freeze dried)
- decrease in work

The experiment confirms that bonding is fundamental to the mechanical properties of paper.

Further exploration of the effects of water on interfiber bonding was done by wetting a solvent exchanged sample and drying the sheet between blotters at 0.01 psi (see rewet & dried in Fig. 11b). The infusion of water into the unbonded sheet and subsequent drying largely reformed the pre-existing interfiber bonds. Drying the unbonded sheet from water recreated the network bonding commonly found in the sheet.

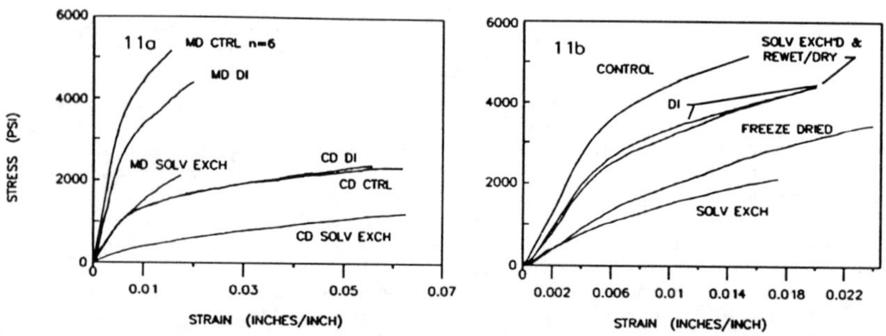

Figure 11a: Waterleaf cotton paper immersed in deionized water and dried, solvent exchanged to toluene and never wet control, both MD and CD.

Figure 11b: Waterleaf cotton MD immersed in deionized water and dried, solvent exchanged sample, freeze dried sample and a solvent exchanged sample which was wet with deionized water and dried at 0.01 psi. Note that wetting and drying an unbonded sheet cause it to perform mechanically as a simple wet and dried sheet.

DISCUSSION

Fiber modulus greater than sheet modulus

Seth and Page (1981) have shown that the relationship between the zero-span modulus and sheet modulus is 3:1. This correlates well with the observation made in regards to Figure 1, that fibers are stiffer than paper by a ratio of 3:1.

The yield point of the virgin cotton fibers is above the strength of the paper. Morton & Hearle (1975) state that there is no obvious yield point for cotton fibers, but if yield point is defined as a change in the initial slope of the stress/strain curve, and that definition is applied to Figure 1, then the yield point is above the strength of the paper. Jayne (1959) shows stress/strain curves for 3 wood pulps, all of which either have no yield point or one above 60,000 psi. This suggests that the fibers are always in their elastic regime when paper is strained.

Strength of individual fibers is well beyond the strength of paper. Study of the stress-strain behavior of individual fibers was undertaken by Webb & Richardson (1945), Grant et al. (1952), Meredith (1956), Jentzen (1964), Jayne (1959), Hardacker (1962 & 1969), Hardacker et al. (1973), Weiss et al. (1962) and reviewed by Mark (1967) and Hanby (1965) among others. The Hardacker et al. (1962) work included sample stress-strain curves of woodpulp fibers, which by their very source (logs) insures that they are thoroughly digested, cooked, pulped and often oxidatively bleached, resulting in a variety of stress/strain behaviors. Their strength ranged from 75,666 psi for the most processed (bleached western hemlock, sulfite processed) to 114,353 psi for the unbleached southern pine kraft processed fibers, the least processed example. Mark (1967) reviews 61 wood pulp types and several native cellulosic fibers including cotton. The strongest wood pulp fibers range from 132,277 psi to 210,505 psi. An average strength for the entire list of wood pulps, which included some very weak fibers, was roughly calculated to be 65,500 psi. Wood fibers are certainly many times stronger than the paper into which they are formed.

The literature has good agreement with the cotton fiber strengths reported in Figure 1. Meredith (1956) lists the strength of cotton fibers which range from 55,185 to 87,045 psi. Weiss et al. (1962) reports an average of 15 varieties of cotton at 63,860 psi. Webb & Richardson (1945) report the average of 129 cotton samples at 96,000 psi; minimum at 70,000 psi and maximum at 116,000 psi. Grant et al. (1952) corroborates Morton and Hearle (1975) quite well by reporting cotton strength from 57,000 psi to 107,000 psi.

Unique mechanical behavior exhibited by papers with degraded fibers

Fiber degradation results in unique mechanical behavior. Exposure to the very harsh artificial aging conditions resulted in a change in sample color and crosslinking which is evidence of severe degradation (see Fig. 3). When the mechanical behavior of paper with degraded fibers showed changes, it had an incremental loss of strength and percent stretch. It was shown that wetting and drying resulted in a decrease in strength and modulus and an increase in percent stretch. Degradation did not affect modulus. Wetting and drying increase percent stretch while degradation decreases stretch. Thus, the mechanical behavior of a paper with degraded fibers is unique within the other mechanical behaviors studied.

Evaluation of fiber degradation using stress-strain measurements

It has been shown that fibers are substantially stronger than paper. In the study of heat aged paper using stress-strain methodology, fiber degradation was not detected until severe degradation had occurred. It is evident that the fibers were degrading during heat aging. The strength of the paper, however, was so far below that of the fibers that changes in the mechanical strength of the fiber could not be detected due to the limits imposed by the mechanical measurement of paper. Therefore, the evaluation of the fiber degradation using mechanical properties is questionable.

Mechanical behavior exhibited by paper with bonding disrupted

Solvent Exchange vs freeze drying

The solvent exchange process was repeated on several replicates of the same sample. The final result always yielded a sheet that had more strength than expected. Eventually it was learned that solvent immersion would not strip all water molecules from the surface of the fiber. While rebonding was largely prevented by the nonpolar solvent, some interfiber bonding was facilitated by water "bridging" (from monomolecular bound water remaining on the fibers) during drying from toluene. Drying from boiling toluene may be a method to further reduce interfiber bonding. The hot liquid should cause the bound water to be forced free of the cellulose (Tappi, 1989).

Freeze drying was found to be a less effective debonding technology. This corroborates what was reported by Merchant (1957), who found the WAN-dried technique (water (W), alcohol (A) and nonpolar solvent (N)) superior for preventing bonding, and Broughton & Wang (1955) who found 63% decrease in strength from freeze drying and a 80% decrease when dried from toluene. The frozen water holds the cellulose surfaces apart, preventing complete reformation of interfiber bonds, as expected. The increased reformation of bonding, over the solvent exchange method, is attributed to the presence of free water during drying, albeit frozen. The bound water is available for interfiber bonding. Sublimation of the ice solid allows contact between fibers which results in interfiber bonds, in the presence of a source of water. Van Den Akker (1952), in a study to determine if the Campbell effect (Campbell, 1947 & 1959) was valid, explored freeze dying and found that by inhibiting capillary attraction, in the presence of water, bonding was decreased over normal drying, but not completely. Thus, bonding does occur, but cold fibers are stiffer than room temperature fibers; this prevents full scale bonding because the fibers can not conform to one another.

Shape of the stress-strain curve due to changes in bonding

Unbonded sheets (see Fig 11a-b) show loss of all mechanical properties. The change in the shape of the stress-strain curve is profound and completely different than that created from straining paper with severely degraded fibers.

When a sheet is wet and dried all mechanical properties are affected. The initial slope (modulus) decreases, and strength is decreased. It is assumed that the strength loss shows loss of bonding. While not as dramatic as unbonding the effects on the shape of the stress-strain curve are similar.

Unquestionably, changes in bonding can be effectively monitored through the use of the stress-strain methodology.

Source of paper strength

Fibers are much stronger than paper, thus they are not the strength limiting factor. Wet straining clearly shows that interfiber bonding is the major source of strength in paper (see Fig. 5a-d). The prolonged failure, after maximum load has been reached (and sustained in some examples), displays the effects of mechanical entanglement of the fibers. Thus, the sources of strength have been visualized.

Wetting and drying releases dried-in strains and decreases bonding

Decrease in dried-in strains

Green (1981 & 1983) demonstrated that there is stress relief upon wetting which is measured as permanent loss in dimension. All of the papers in this experiment showed shrinkage after wetting and drying; the waterleaf cotton had 2.4% (MD) and 0.4% (CD) shrinkage. Page and Tydeman (1962) show that unrestrained specimens dry with fibers that are kinked rather than relatively straight. Setterholm and Chilson (1965) show how lack of restraint in drying reduces modulus for 15 pulp types. Htun and de Ruvo (1978) have calculated internal stress and drying stress and found them to be equal and to exhibit no time dependency. That is, stress set into the sheet on drying does not diminish until it is released by water. They also found that drying restraint increases drying stresses.

Papers which have been wet and dried show an increase in ability to stretch. The release of strains created more kinks and curls by allowing the fibers to shrink and become less straight. It follows that straining a specimen with the dried-in strains released, would produce stress-strain behavior that shows increased ability to be stretched.

The loss of dried-in strains undoubtedly increases the kinks and curls in fibers. When the dried-in strains are recreated, paralleling the manufacture process, the kinks and curls are straightened resulting in the original stress-strain behavior of the paper (Vitale, 1992).

Decrease in bonding

Wetting and drying results in a decrease in tensile strength which is a decrease in ability to hold a load. This loss of strength can be interpreted as a change in the mechanism which is responsible for paper strength: the interfiber bonding. Tensile-induced shearing of one fiber from another dominates the plastic deformation portion of MD straining. Decreased bonding or decreased interfiber bonding area would lower the load required to peel away one fiber surface from another. In addition, loss of interfiber bonds or bond area would also legitimately lead to additional axial and torsional flexing of the somewhat longer free fiber segments during straining. It follows that this would result in additional stretch, which is the observation made from the data.

Simple model: fibers are elastic component; bonds are plastic component

Key literature

Most workers in the field of paper mechanics and physics agree that fibers are the source of **elastic** behavior in the sheet. Page (1969) and Seth and Page (1981), and many other workers show that **plastic** behavior is due to fiber physical properties (microcompressions) and the rupture of interfiber bonds during straining. Eberling (1970) and Eberling and Retulainen (1985), however, argue that plastic behavior is due to the rupture and reformation of interfiber bonds as the load is distributed to the sites. They propose that interfiber bonds are altered, but not ruptured until catastrophic failure occurs. This is mechanical behavior similar to the slip in aluminum during plastic deformation. Most workers (excellent review in Seth & Page, 1981) acknowledge that interfiber bonds are ultimately ruptured by the straining process.

Following the heat flow: heat dissipation equals bond rupture

Eberling's work using a microcalorimeter to follow heat flow during straining indicates two stages: (1) heat absorption, which is reversible, and (2) heat dissipation, which is irreversible. Heat absorption is due to Kelvin's thermoplastic effect where axial elongation results in a shape change which is accompanied by absorption of heat. Heat dissipation is attributed to irreversible rupture of hydrogen bonds. Unfortunately, because both intrafiber and interfiber bonds are hydrogen bonds, the source of the heat is yet to be localized. Eberling's finding that heat is given off during destraining suggests that ruptured bonds do not reform. The assertion that bonds are continually being ruptured and reformed during straining should also apply in the reverse, but this appears not to be the case.

Fibers contribute elastic behavior

It is shown that fibers have to be highly degraded for loss in mechanical properties to affect a sheet's mechanical performance (see Fig. 2a-c), and because the strength and yield point of fibers is greater than the strength of paper (see Fig. 1), the primary contribution of the fibers to paper is elastic behavior. Elastic behavior is reversible. Since fibers contribute only reversible behavior, the bonds between the fibers must be the site of irreversible heat generation during straining. Thus, Eberling's finding that heat dissipation is rupture of hydrogen bonds can be used to support the assertion that interfiber bonds are irreversibly rupturing during straining and that interfiber bonds are the source of plastic behavior in papers with strong intact fibers, such as cotton.

Figure 4a-e shows that papers with relatively intact fibers have fibers which are pulled out during straining. This verifies the assertion that the rupture of interfiber bonds is the major source of irreversible behavior. The rupture of fibrils does occur, but this irreversible failure of a component of a fiber is minor when SEM micrographs of rupture sites are examined (see Fig. 4c).

Mechanical behavior paradigm

Fibers are stronger than paper. Fibers degrade before the paper shows the effects; this indicates that factors other than fibers also influence the mechanical properties of paper. Loss of interfiber bonding results in the loss of all mechanical properties indicating that bonding is fundamental to the mechanical behavior of the paper. Fibers pull out rather than rupture, further indicating that fibers act largely as structural components and proves that interfiber bonds are the weak link. Fibers, if they do have a yield point, exhibit only elastic behavior in the relative low strength realm of the paper. Eberling's data show that plastic behavior in paper creates heat dissipation which is the rupture of interfiber bonds. The heat dissipation is not reversible on destraining; once broken the bonds do not reform. It is concluded: in paper with largely intact fibers, the fibers provide elastic behavior and the interfiber bonds are responsible for plastic behavior.

Application to conservation treatment

Respect for changes due to treatment

Wet handling

Conservation treatments which affect interfiber bonding must be undertaken with profound respect for changes which may occur. A wet sheet is very weak (see Fig 5a-d). The force needed to untangle and pull the surface of one fibre past another in a wet sheet is very small -- only 1-5% of the dry sheet strength (see Fig 5b & d). Fortunately very few papers weigh 20-100 pounds when wet. The break load of the wet Whatman #1 averages to approximately 19 psi and the waterleaf cotton to 100 psi. When two sets of clamped fingers are used to pull a wet sheet from a bath, the load distributed through the roughly 1 in^2 contact area is below the breaking strength of the wet paper. Obviously, the greater the surface area of the clamped fingers the better.

Evidence of manufacture

Wetting and drying results in decreases in the original strength and stiffness of the sheet. The manufactured-in data is lost from the record that was contained by the artifact.

The decreases in modulus and tensile strength may be detrimental, although the changes in mechanical properties are not profound. Change in as-manufactured properties is often considered an undesirable outcome of conservation treatment. The measurement of original mechanical properties is largely a destructive process because a sizable sample must be taken for tensile analysis. The loss of properties, which a conservator can not currently measure, may therefore be irrelevant.

There is evidence that original surface texture (Sugarman and Vitale, 1992), which is surely related to interfiber bonding and fiber kinks and curls, can not be completely recovered by current conservation drying procedures. Measurement of surface texture, however, can be done non-destructively by the trained conservator, and previous work (Sugarman and Vitale, 1992) has shown that drying treatments can be distinguished by trained observers. Surface texture is one of the connoisseur's tools for assessing the aesthetics and value of paper artifacts. Thus, properties related to changes in mechanical properties have value to the conservation community.

It has been shown that simply wetting with high swelling solvents (for cellulose) and drying can measurably decrease the fundamental mechanical properties of paper strength and stiffness (see Fig. 10). Alteration of the "as manufactured" properties should be weighed against the perceived gains of solvent treatments. But the effects are not a pronounced as with water.

Wetting and drying results in increased percent stretch. Increased stretch may be mechanically desirable in some situations. The grain of a book page is traditionally oriented so that the sheet flexure is along the CD direction. Increased stretch in the CD direction may enhance the ability of a page to withstand flexure. However, increased stretch under load may not correlate with page flexure. Deterring fiber failure from flexure induced fatigue may not be enhanced by an increase in percent stretch (CD) after wetting and drying.

Effects of ultrapure deionized water

Wetting with water containing salts as active as pH 10.2 sodium hydroxide is not different from wetting with deionized water when only the mechanical properties are evaluated (see Fig. 9a-c). Burgess (1991) shows that deionized water has little or no effect on the degree of polymerization (DP) of several historic papers. The effects of deionized water on paper appear to have been overstated in the past. As far as mechanical properties are concerned the effects of water on disruption of interfiber bonds are greater than the effects of the ions on the mechanical strength of the sheet. Burgess's findings regarding the limited change in DP of the cellulose in the fibers suggests that changes in fiber strength would be nil. Little or no damage to the polymers in the fibers confirms that the fibers did not degrade sufficiently (if at all) for there to be a change observed in the mechanical properties of the paper.

Presence or absence of ions has not effect on immediate mechanical or DP, but the lack of an ion such as calcium may alter the future of an artifact. Once that artificial aging conditions which parallel natural aging have been established (Erhardt et al, 1987) a study with normal concentrations of the calcium salts remaining in unsized cotton paper may show decreased deterioration, but that remains to be explored. It has been documented that calcium is not detrimental to paper. Barrow (1974) has reported that papers with high calcium content are in better condition than papers with little or no calcium content. Barrett (1989) confirms that "elements that show overall higher levels in papers in good condition include magnesium, calcium, and zinc."

Is there a danger of removing calcium from paper? It is often said that ultrapure deionized water is a "hungry solvent." Ultrapure deionized water has a purity of approximately 18 megohms (electrical resistivity). Distilled water commonly has a resistivity of 1.0 to 0.5 megohms. The difference between 18 and 1 megohm is 0.65 ppm of a generic ion. Distilled water is thought of as benign. When a historic paper is in ultrapure water for a very short time its collective ionic content will lower the resistivity of the water well below that of distilled water and approach that of tap water. Tap water commonly has a resistivity of 0.003 to 0.006 megohms, which is 100-200 ppm of a generic ion. Research into the leaching effects of any water is yet to be explored.

Solvent treatments

Wetting with common solvents is less drastic than wetting with water. Changes caused by methanol, the simplest alcohol, are similar to or slightly less than those caused by water. The tensile strength of the methanol sample is borderline between indistinguishable from the wet & dried sample and slightly stronger, suggesting that wetting with methanol may leave more interfiber bonding intact than water. Clearly dried-in strains have been lost from the methanol immersed sample, because the modulus decreased. It has been common conservation practice to follow solvent wetting with water wetting. In the case of methanol, it is probable that more loss of properties would result. If water wetting followed toluene then very substantial changes in mechanical properties would result.

Solvent exchange effects during treatment

It is unlikely that a treatment would progress, wet onto wet, from water immersion to ethanol immersion to acetone and then to toluene, but if for some reason it did, the sheet would be very weak as a result of the solvent exchange process. It is reasonable, however, to imagine that local stain removal on the suction table could follow some of the steps in the solvent exchange process. One might try to remove a stain with methanol, find that it didn't work, and then move to acetone. The mechanical properties of the treated region would be those of the methanol treatment, since methanol relieves more dried-in strains. If toluene treatment follows (wet onto wet) treatment with a polar solvent with which toluene is miscible, then the effects of solvent exchange are possible. The local region would have different mechanical properties, and the interfiber bonding would be decreased. In this case, wetting with water would restore lost interfiber bonding.

CONCLUSIONS

A simple paradigm for the stress-strain behavior of paper was developed. Fibers contribute elastic behavior; interfiber bonds contribute plastic behavior.

The cotton fibers are 12-20 times stronger and 3 times stiffer than the cotton paper. The low relative strength of the paper places the fiber's contribution to paper in the elastic portion of the fibers stress-strain curve. The low relative stiffness of the paper demonstrates that the elastic behavior of the paper is not due to the axial straining of the fiber.

Degradation of fibers in a sheet by dry heat aging eventually results in incremental loss of paper strength and stretch following the nominal stress/strain behavior of the sheet. Modulus and yield point are not affected because interfiber bonding is not disrupted.

Stress-strain behavior of paper with degraded fibers is unique and distinct from that caused by disruption, i.e., degradation of, interfiber bonding.

Major disruption in interfiber bonding, using the solvent exchange method, causes losses in all mechanical properties. Loss of bonding affects all mechanical properties. Interfiber bonding is the major feature affecting mechanical properties in paper, until 90% + of the fiber strength has been lost.

Disruption of interfiber bonds (solvent exchange unbonding) can be reversed by wetting with water and drying.

Wetting releases (1) interfiber bonding and (2) dried-in strains which were fixed at the time of manufacture. Organic solvents release bonds and dried-in strains proportionally to the degree of swelling they cause.

Water causes the greatest disruption of mechanical properties after drying of the liquids examined. Solvents have a parallel effect in proportion to the degree of swelling they cause. Common solvents can effect no greater change than that caused by water.

Disruption and recreation of interfiber bonding by water cause decreases in (1) MD strength and (2) MD modulus, and an increase in (3) MD stretch. There were indistinguishable changes in CD strength and CD modulus and a slight increase in CD stretch, with the exception of the handmade paper.

Increased ability to stretch allows the paper fiber network to absorb more work.

Disruption of interfiber bonding causes distinct stress-strain behavior.

Water alone, regardless of added salts or water hardness, has the most profound effect on the mechanical properties of paper.

Because fibers are stronger than the paper they are made into, monitoring fiber degradation by following mechanical properties is questionable.

Solvent exchange did not unbond a sheet completely. Lyophilization was less efficient at unbonding than the solvent exchange process.

APPENDIX

Definition of terms

Modulus of elasticity (or modulus) is the stiffness of a material in the elastic regime. **Young's Modulus** is modulus of elasticity in tension. Modulus is a ratio between the load and strain. It is reported as force per unit area pounds per square inch (PSI). An example: epoxy resin has high modulus, rubber has low modulus. Thus, stretching a film of epoxy a given amount develops a high load; stretching a film of rubber the same amount develops a low load.

Load at break is maximum amount of weight or load that a material will hold just prior to failure. Ultimate tensile strength is load at break divided by the original cross-sectional area; it is reported here as **tensile strength** in pounds per square inch (PSI).

Strain at break is the degree of stretch that a material can sustain before it fails. Percent stretch equals strain at break times 100. In the process of creating the stress-strain diagrams used here a constant rate of straining was applied. It is noteworthy that the rate of loading was constantly changing because a movable jaw was pulled away from a fixed jaw which contained a load cell. It is useful to make measurements at constant rate of strain, because a stress-stain curve can be developed which shows properties that can be compared to like or different materials.

Work is the integration of force per unit energy put into the system to create failure. Thus, work is the tensile energy absorbed by the sample. Clearly, the elastic component of the energy absorbed will be recovered when the sample fails. Paper technologists often use the term tensile energy absorption (TEA) for the area under the stress strain curve. Work for stress/strain curves [(psi)/(inches/inch)] are commonly reported as force per unit volume, footpound inches/cubic inch. Due to the difficulty of calculating work for the number of samples represented by the study, a simple cutout-and-weigh technique was used. Data is reported in g, grams of paper under the stress-strain curve (weights are consistent within the plot but not between plots). Many samples show an increase in work after treatment. Ability to absorb energy or work is often called toughness. Brittle materials have diminished ability to absorb work. Commonly, the material's ability to stretch is decreased while the tensile strength may remained the same or increase.

Yield point or plastic limit is the point at which the recoverable elastic behavior ends and the plastic behavior begins. The transition is signified by a change in the linearity of the stress-strain curve. Elastic behavior is linear; plastic behavior is non-linear. A decrease in yield point indicates that permanent deformation begins at lower loads and therefore some of the elastic component of the material has been lost.

BIBLIOGRAPHY

Arah, C.O., McNamara, D.K. Hand, H.M. and Mecklenburg, M.F. 1989. Techniques for Screening Adhesives for Structural Applications. **J. Adhesion Technol**. 3(4): 261-75.

Barrett, T.D. 1989. Early European Papers/Contemporary Conservation Papers: a Report on Research Undertaken from Fall 1984 Through Fall 1987. **The Paper Conservator**. 13: 4-108.

Barrow, W.J. 1974. Physical and Chemical Properties of Book Papers, 1507-1949. **Permanence/Durability of the Book - VII**. Richmond: W.J. Barrow Research Laboratory Inc.

Blackwell, J. & Marchessault R.H. 1971. Investigation of the Structure of Cellulose and its Derivatives, Ch. XIII in **Cellulose and Cellulose Derivatives**, editors Bikales N.M. & Segal, L., Pt. IV, pp. 1-37, New York: Wiley-Interscience.

Broughton, G. & Wang, J.P. 1955. The Mechanical Properties of Paper (3). **Tappi**. 38(7): 412-415.

Browning, B.L. 1969. **Analysis of Paper**. Marcel Dekker, New York.

Burgess, H.D. 1991. Investigation of the Effect of Alkali on Cellulosic Fibers, Part 1: Rag and Processed Wood Pulp Paper. In **Paper and Textile: the Common Ground**, Preprints of Scottish Society for Conservation & Restoration meeting Sept. 1991. 29-47.

Campbell, W.B. 1947. Physics of Water Removal. **Pulp and Paper Mag. of Canada**. 48(3): 103-109 & 122.

Campbell, W.B. 1959. The Mechanism of Bonding. **Tappi**. 42(12): 999-1001.

Casey, J.P.(ed). 1960. **Pulp and Paper Chemistry and Chemical Technology**, 2nd ed., 3 vol.s. New York: Interscience Publishers.

Casey, J.P.(ed). 1980. **Pulp and Paper Chemistry and Chemical Technology**, 3rd ed., 3 vols. New York: John Wiley & Sons.

Caulfield, D.F. 1978. The Effects of Cellulose on the Structure of Water- 2 in **Fiber-Water Interactions in Papermaking**, Vol 1, ed. Fundamental Research Committee. London: The British Paper and Board Industry Federation.

Ebeling, K. & Retulainen, E. 1985. **Physical Properties of paper: Mechanism of Deformation**. Proceedings of the International Symposium on Fiber Science Technology, Hakone, Japan. August, 1985.

Ebeling, K. 1970. **Distribution of Energy Consumption During Straining of Paper**, Doctoral Dissertation, Parts I & II, Appelton, WI: Institute of Paper Chemistry.

Erhardt, D. von Endt, D., Hopwood, W. 1987. The Comparison of Accelerated Aging Conditions Through The Analysis of Extracts of Artificially Aged paper. In **Preprints: AIC 1987 Annual Meeting**. Washington,D.C.: American Institute for Conservation. 43-55.

Erhardt, D. 1988. Paper Degradation: A comparison of Industrial and Archival Concerns. In **Tappi Proceedings: 1988 Paper Preservation Symposium**. Atlanta, GA: Tappi Press. 83-90.

Grant, J.M., Morlier, O.W. & Scott J.M. 1952. The Effects of Mechanical Processing of Cotton on the Physical Properties of Fibers. **Textile Research Journal**. 22(10): 682-687.

Green, C. 1981. **Dimensional Properties of Paper Structures**. Ind. Eng. Chem. Prod. Res. Dev. 20: 151-158.

Green, C. 1983. Relationship of Dimensional stability to Rheology of Paper Structure. In **Proceedings of the TAPPI 1983 International Paper Physics Conference**. Atlanta GA: Tappi Press. 165-171.

Green, C. 1987. Relationship of Dimensional Stability to Rheology of Paper Structure, **International Paper Physics Conference 1983**, Proceedings of TAPPI meeting Harwichport, MA. Sept. 1983. Reprinted: University Microfilms International Ann Arbor MI.

Haigler, C.H. 1985. The Functions and Biogenesis of Native Cellulose. In **Cellulose Chemistry and its Applications**. Ch. 2. eds T.P. Nevell & S.H. Zeronian. Chichester: Ellis Horwood Limited.

Hamby, Dame S. ed. 1965. **The American Cotton Handbook**. Vol 1. 3rd ed. New York: Interscience Publishers. 74-76.

Hardacker, K.W. 1962. The Automatic Recording of the Load-Elongation Characteristic of Single Papermaking Fibers. **Tappi**. 45(3): 237-246.

Hardacker, K.W. 1969. Cross-Sectional Area Measurement of Individual Wood Pulp Fibers By Lateral Compaction. **Tappi**. 52(9): 1742-1746.

Hardacker, K.W. & Brozinski, J.P. 1973. The Individual Fiber Properties of Commercial Pulps. **Tappi**. 56(4): 154-57.

Hearle, J.W.S. 1972. Mechanical Behavior of Nonwoven systems. In **Theory and**

Design of Wood and Fiber Composite Materials. Ch. 13. ed. B.A. Jayne. Syracuse University Press. 327-351.

Hearle, J.W.S. 1985. Mechanical Properties of Cellulosic Textile Fibers in **Cellulose Chemistry and its Applications**. Ch. 19. eds. T.P. Nevell & S.H. Zeronian. Chichester: Ellis Horwood Limited. 480-504.

Higgins, H.G. and Mckenzie, W. A. 1963. The Structure and Properties of Paper XIV. Effects of Drying on Cellulose Fibers and the Problem of Maintaining Pulp Strength. **Appita**. 16(6): 145-164.

Htun, M. & de Ruvo, A. 1978. Relationship Between Stresses and Internal Stresses and the Mechanical Properties of Paper. In **Fiber-Water Interactions In Paper Making**. Vol 1. eds. Fundamental Research Committee, London: Technical Division The British Paper and Board Industry Federation. 477-491.

Jayne, B.A. 1959. Mechanical Properties of Wood Fibers. **Tappi**. 42(6):461-467.

Jentzen, C.A. 1960. Some Mechanical Properties of Wood Fibers in Tension. **For. Prod.** J. 10(6): 316-322.

Jentzen, C.A. 1964. The Effects of Stress Applied During Drying on Some of the properties of Individual Fibers. **Tappi**. 47(7): 412-418.

Kallmes, O.J. 1972. A Comprehensive View of the Structure of Paper. In **Theory and Design of Wood and Fiber Composite Materials**, ed B.A. Jayne, Ch. 6. Syracuse: Syracuse University Press. 157-175.

Liang, C.Y. & Marchessault, R.H. 1959. Infrared Spectra of Crystalline Polysaccharides. I. Hydrogen Bonds in Native Cellulose. J. **Polymer Sci**. 37: 390. Designated: Liang (1959a)

Liang, C.Y. & Marchessault, R.H. 1959. Infrared Spectra of Crystalline Polysaccharides. II. Native Celluloses in the Region from 640 to 1700 cm^{-1}. **J. Polymer Sci**. 39: 269. Designated: Liang (1959b)

Lobben, T.H. 1976. The Tensile Stiffness of Paper (2) Activation Studied by Freeze Drying. **Norsk Skogindustri**. 30(3): 43-48.

Marchessault, R.H. & Liang, C.Y. 1960. Infrared Spectra of Crystalline Polysaccharides. III. Mercerized Cellulose. **J. Polymer Sci**. 43: 79.

Mackeprang, M. 1980. Nonwovens, in **Mechanics of Flexible Fiber Assemblies**. Tutorial Discussion II. eds. J.W.S. Hearle, J.J. Thwaites, J. Amirbayat. Netherlands: Sijthoff & Noordhoff. 543-547.

Mark, R.E. 1967. **Cell Wall Mechanics of Tracheids**. New Haven: Yale University Press.

Mecklenburg M.F. 1988. The Effects of Atmospheric Moisture on the Mechanical Properties of Collagen Under Equilibrium Conditions. **JAIC**. Preprints 1988 AIC meeting New Orleans, Louisiana. 231-244.

Merchant, M.V. 1957. A Study of Water-Swollen Cellulose Fibers Which have been Liquid-Exchanged and Dried from Hydrocarbons. **Tappi**. 40(9): 771-781.

Meredith, R. 1956. Fiber Strength. In **The Mechanical Properties of Textile Fibers**. Ch. VII. Amsterdam: North-Holland Publishing Company. 129-

150.

Morosoff, N. 1974. Never Dried Cotton Fibers. III. Crystallinity and Crystallite size. **J. Appl. Poly Sci**. 18: 1837-54.

Morton, W.E. & Hearle, J.W.S. 1975. **Physical Properties of Textile Fibers**. London: Heinemann [for] The Textile Institute.

Nissan, A.H. 1976. H-Bond Dissociation in Hydrogen Bond Dominated Solids. **Macromolecules**. 9: 840-850.

Nissan, A.H., 1977. **Lectures on Fiber Science in Paper**, Joint Textbook Committee of the Paper Industry, ed. W.C. Walker. Atlanta, GA: Tappi Press.

Nissan, A.H., 1978. Water Effects on Young's Modulus of H-Bonded Solids. In **Fiber-Water Interactions in Papermaking**. eds. Fundamental Research Committee. Vol 1. London: Technical Division The British Paper and Board Industry Federation. 609-640.

Nissan, A.H., Byrd, V.L., Batten Jr., G.L. & Ogden, R.W. 1985. Paper as a H-Bond Dominated Solid in the Elastic and Plastic Regimes. **Tappi Journal**. 68(9): 118-124.

Page, D.H. 1966. The Axial Compression of Fibers -- A Newly Discovered Beating Action. **Pulp and Paper Mag. of Canada** 67(1): T2-T12.

Page, D.H. 1969. A Theory for the Tensile Strength of Paper. **Tappi**. 52(4): 674-681.

Page, D.H., Seth, R.S., Jordan, B.D. Barbe, M.C. 1985. Curl Crimps, Kinks and Microcompressions in Pulp Fibers - their Origin, Measurement and Significance. In **Papermaking Raw Materials**, ed. Punton, Vol 1, London: Mechanical Eng. Pub. Ltd. 77-91.

Page, D.H. & Tydeman, P.A. 1962. A New Theory of the Shrinkage Structure and Properties of Paper. In **The Formation and Structure of Paper**. ed. F. Bolam. Vol 1. London: Technical Section of the British Paper and Board Makers' Association (Inc.). 397-425.

Robertson, A.A. 1966. Measurement and Significance of the Water Retention Properties of Papermaking Fibers in **Consolidation of the Paper Web**. ed. Bolam, F. London: Technical Section of the British Paper and Board Makers' Association. 90-117.

Rowland, S. 1977. Cellulose: Pores, Internal Surface, and the Water Interface. In **Textile and Paper Chemistry and Technology**, ACS Symposium Series 49. ed. Arthur, J.C. Wash. DC: American Chemical Society.

Rowland, S. 1979. Solid-Liquid Interactions: Inter- and Intracrystalline Reactions in Cellulose Fibers, Ch. 7. In **Applied Fiber Science**, Vol. 2. ed. Happey, F. London: Academic Press. 205-237.

Schniewind, A.P. 1964. Fibers and Pulp Properties: Shear strength of Single Fiber Crossings. **Tappi**. 47(4): 244-248.

Seth, R.S. & Page D.H. 1981. **The stress-strain Curve of Paper**. In **Role of Fundamental Research in Papermaking**, Vol 1. ed. J. Brander. London: Mechanical Engineering Publications Ltd. 421-452.

Setterholm, V.C. & Chilson, W.A. 1965. Drying Restraint: Its Effect on the Tensile Properties of 15 Different Pulps. **Tappi**. 48(11): 634-40.

Stamm, A.J. 1964. **Wood and Cellulose Science**. New York: Ronald Press Co.

Sugarman, J. & Vitale, T.J. 1992. Observations on the Drying of Paper: Part I, Five Drying Methods and The Drying Process. **JAIC**: In press, Winter, 1992.

Tappi. 1989. **Tappi Test Methods**. Vol. 1. Atlanta, Ga: Tappi. T 208 om-84.

Treiber, E. 1975. Cellulosics. in **Polymer Handbook**, eds. Bandrup, J. & Immergut, E. H.. New York: Wiley & sons. V87-121.

Van Den Akker, J.A. 1952. A Note on the Theory of Fiber-Fiber Bonding in Paper, The Influence on Paper Strength of Drying by Sublimation. **Tappi**. 35(1): 13-15.

Vitale, T.J. 1988. Observations on the Theory, Use and Fabrication of The Fritted Glass Bead Suction Disk Device. **The Paper Conservator**. 12: 47-76.

Vitale, T.J. 1992. Effects of Drying on the Mechanical Properties of Paper and their Relationship to the Treatment of Paper. These Proceedings.

Webb, R.W. & Richardson, H.B. 1945. **Comparative Significance of Alternative Cotton Fiber Length and Strength Measures in Relation to Yarn Strength**. Washington, DC: U.S. Department of Agriculture, Production and Marketing Administration.

Weiss, L.C., Orr, R.S., Redmann, J.J. & Grant, J.N. 1961. Moduli of Cotton Fibers and Yarns in Relation to X-ray Angles. **Text. Res**. J. 31(9): 787-793.

Zeronian, S.H. 1985. Intercrystalline Swelling of Cellulose, Ch. 5, in **Cellulose Chemistry and its Applications**, eds. T.P. Nevell & S.H. Zeronian, Chichester: Ellis Horwood.

Zeronian, S.H. 1985. Intracrystalline Swelling of Cellulose, Ch. 6, in **Cellulose Chemistry and its Applications**, eds. T.P. Nevell & S.H. Zeronian, Chichester: Ellis Horwood.

Zeronian, S.H. & Cabradilla, K.E. 1972. Action of Alkali Metal Hydroxides on Cotton. J. **Appl. Polym. Sci**. 16: 113-128.

ACKNOWLEDGEMENTS

I am indebted to Marion Mecklenburg, Assistant Director for Conservation Research at CAL, for extensive consultations on mechanics and introduction to the technique of equilibrium stress-straining. Without the valuable assistance of John Childs, formerly a CAL Paper Lab intern and now a practicing furniture conservator, who performed much of the stress-strain evaluations, this work would not have explored as many possibilities with such aplomb. John Vitagliano, also formerly a CAL Paper Lab intern and now a practicing paintings conservator, also performed a portion of the stress-strain evaluations in this work, for which I am grateful. W. David Erhardt, Senior Research Organic Chemist at CAL, provided portions of his heat aged Whatman #1 samples for analysis and a photograph of the pulped samples.

Noreen Tuross, Research Archeo-biologist at CAL, generously performed the freeze drying of the waterleaf cotton paper. James Murphy of Strathmore Paper provided the historical data and interpretation of the Strathmore paper package label. Jeff Geiger of Whatman provided the data on the Whatman #1 filter paper. Chandru Shahani, Research Officer of the Preservation Directorate, Research & Testing Office of the Library of Congress, very generously gave CAL the waterleaf cotton and waterleaf wood pulp papers used in this experiment; this work would not have been possible without his assistance. Finally, I am extremely grateful to Duane Chartier, formerly a Smithsonian Research Fellow and now a Contract Research Scientist in Los Angeles, CA., who offered critical insight into behavior of solvent exchanged and freeze dried samples.

Many thanks to Carol Grissom, Chief of Objects Conservation, and David Erhardt, Senior Research Chemist, of CAL who served as readers and critics.

EFFECTS OF DRYING ON THE MECHANICAL PROPERTIES OF PAPER AND THEIR RELATIONSHIP TO THE TREATMENT OF PAPER.

Timothy Vitale, Conservation Analytical Laboratory, Smithsonian Institution, Museum support Center, Washington DC 20560.

ABSTRACT

The effects of drying on paper after water immersion are explored by evaluating mechanical properties. When paper is wet, structural interfiber hydrogen bonds are disrupted; drying reverses this process. However, some "as manufactured" mechanical properties are lost after wetting and drying. The method of drying was found to be critical for regaining the mechanical properties of a dried sheet.

The effects of (1) wet-pressing (weight on paper during drying) and (2) wet-straining during drying are assessed. Application of incremental increases of the two variables results in changes of the mechanical properties which eventally reach and exceed the original mechanical properties which are lost during standard water immersion and drying treatments.

The treatment method which most closely reproduces the "as manufactured" properties involves wet-pressing and wet-stretching. In particular, **friction stretch-drying in an air-bag press** can approximate original drying conditions. However, alterations of optical properties of the paper and design media due to elevated pressure prevents endorsement of the methodology.

INTRODUCTION

Conservation drying

Drying involves restoring the interfiber bond network to paper, lost when a sheet is wet. It was shown in a previous paper (Vitale, 1992) that some strength and stiffness is lost due to simple wetting and drying. The procedure generally used by conservators to dry paper involves drying in the open air, on a screen, or in a stack of blotters with very little weight applied to the blotters. This technique does not recreate the "as manufactured" mechanical properties.

Manufacture of paper

When paper is made, wet pressing and wet stretching are used in the manufacture process. Hand and machine papermaking both use presses to squeeze water from the sheet. Pressing removes water which increases the solids content and brings the fibers closer together so that more complete interfiber bonding will occur. Robertson (1959) showed the correlation between percent solids and wet web strength. At 47-50% solids interfiber bonding begins and continues at approximately the same rate until 80% solids. At approximately 80% solids the fibers are still fully saturated (equal to 100% RH) and pliable. This state is often referred to as the fiber saturation point. At the fiber saturation point the paper is at 90% of its ultimate strength. As the fibers dry to equilibrium room conditions

(50% RH equals approximately 6-7% moisture content or 92-95% solids) the increase in their stiffness does not add significantly to the paper's strength. The final stages of drying, which take longer because water must transfer from inside the fibers, contribute less to the overall mechanical properties of the sheet.

Wet stretching is the process of applying an in-plane tension to the sheet such that the fibers are not pulled past one another, but are straightened somewhat. The papermaking machine provides wet stretch as a consiquence of its operation parameters. In handmade paper making, wet stretching occurs as a result of being held in an expanded state during a lengthy water reduction process (pressing). Handmade papers are wet stretched to a much smaller extent than machine made paper.

Handmade paper is created on a wire mesh screen in a wood frame with a deckle removable frame which is dipped into a vat of fiber pulp. The frame traps a fixed amount of pulp. When sufficient water has drained away from the pulp on the screen the frame is removed and the wet mat is couched onto a felt. When a sizable number of sheets, on felts, have been amassed into a post (144 sheets), they are placed in a press and water is expressed from the sheets. Under fairly high pressure, dewatering reduces the thickness by approximately 75% and weight by about 90% (Hunter, 1978). During the first pressing the sheets are held in an expanded state while some of the network bonds are formed. This locks some wet straining into the sheet. The post is taken from the press and the wet sheets separated from the felts. The damp sheets are stacked on one another and subjected to further pressing where less pressure is used because there are no felt separators. While less water is removed in the second pressing, more bonding and wet strains are created. The sheets are peeled apart and hung on horsehair rope lines in a drying attic. It is important to note that multiple pressings are used and that press-and-hold method fixes dried-in strains into the sheet.

Barcham Green estimated that an old screw press could achieve 150-250 psi (Vitale, 1988). Using the compressive strength of maple (7650 psi) to model a press pictured in Hunter (1978) with a 5" diameter wood screw and a 15" headblock, the press could handle the 250 psi load suggested by Green, over the area of a standard sized sheet (foolscap size: 13.5" X 17"), and substantially more. Hunter (1978) shows several press designs including simple weighted levers, screw presses and the "Sampson," a six person lever and windlass press. It is important to note that the more water that was removed in pressing the faster the sheet would dry, and the better the economics would be, but the post was held in the press for some time as the hand papermakers amassed the next post of sheets for pressing.

Machine-made paper is formed on a moving wire screen. After draining in the formation section of the machine, the moving wet web is couched onto a felt web in the first stage of the drying section of the paper machine. Water is expressed from the web as it is momentarily nipped at elevated pressure between two opposed rollers. Generally, the web is nip-pressed two or more times in the drying section. When the mat moves to the second nip press it, generally, has enough strength to be free of the felt couching web. Between each stage or set of independent nip rollers the machine's speed is increased slightly. The effect is to put several slight stretchs on the drying fiber web. Thus between the first and second stage of the drying section the sheet is stretched slightly. Again, between the first second nips in the secondstage of the drying section the sheet is stretched. In the final drying stage, prolonged moderate pressure from highly tensioned felts over large diameter (often heated) drying cylinders completes the process. There are many variations on the machine drying process, but the principles are the same: (1) opposed rollers applying pressure to squeeze water from the wet web and (2) multiple nip-presses running slightly faster than the previous one.

Drying pulp into paper

Pye, Washburn & Buchanan (1966) clearly define the stages of pressing and drying and the effects on the surface of the paper. Kershaw (1980) documents a common wet stretch applied to the web: 2% before major hydrogen bonding begins and 4% between the first press and the dryers. The 2% stretch before bonding merely separates and aligns the fibers; the 4% wet stretch is set into the sheet by interfiber bonding during drying.

Robertson (1959, 1963, 1966), Rance (1954, 1980) and Campbell (1933) have produced definitive work on the formation and consolidation of the wet web into paper. The primary observations are that (1) major structural hydrogen bonding begins at 47-50% solids in papermaking and (2) the slope of the strength increase vs percent solids curve changes (decreases) when the fiber saturation point is reached (no free water in the network at 100% saturated fiber). The average fiber, approximately 20 microns in diameter, develops 540 PSI of force pulling the two surface together when a water capillary is formed between the the two cylinders (Campbell, 1933). Fibrils, which can be 2 microns in diameter, develop 2480 PSI during bonding due to capillary action and surface tension of water. This demonstrates why, with no outside assistance, fiber networks bond together when dried from water.

Many workers have explored ways to increase interfiber bonding and strength, as well as sheet homogeneity and surface texture uniformity on pulps formed into sheets. MacLaurin & Whalen (1954) explored the effects of multiple wet pressing and found that increasing wet pressing pressure from 50 to 150 psi resulted in an 8% increase in tensile strength. Schultz (1961) explored most of the ramifications of wet-straining: increases in tensile strength, modulus and light scattering coefficient and decreases in elongation, work and creep. He concluded that wet-straining causes a more uniform distribution of stress under load. Fahey & Chilson (1963) showed that wet straining decreased expansion of a sheet when subsequently exposed to moisture and was the primary factor controlling dimensional stability. Stetterholm & Chilson (1965) reported that prevention of shrinkage and wet straining of 15 different pulps formed into sheets has substantial effect on tensile properties. Parsons (1972) documented that tensile strength increased due to wet-straining and drying with restraint. Byrd (1981) reported that 1.5% wet-straining (not including 1% shrinkage) increases modulus and tensile strength in the direction of strain. Wet straining had little effect on cross-direction (CD) tensile strength or modulus. Baum, et al. (1984) showed that wet straining increases stiffness and that fibers are straightened because thickness is increased. The straightening results in less conformation of one fiber to its neighbors which causes an increase in thickness. Many workers also reported data that showed there was increase in sheet homogeneity and surface uniformity. Because it is not necessary for the paper industry to change the properties of paper after it has been made, there appears to be no literature on the subject of wet straining pre-dried sheets.

EXPERIMENTAL DESIGN

Equilibrium stress-strain protocol

The stress-strain measurements were made using equipment described in earlier work (Vitale, 1992). It employs an incremental strain increase, rather than loading. The equipment is hand operated with a digital strain gage readout (Arch, McNamara, Hand & Mecklenburg, 1989). Rate of elongation is very slow: stepwise increments of 0.002 inches per minute. The stress-relaxation method was developed by Mecklenburg (1988). At low strain rates and one minute incremental relaxations, the material is allowed to come to equilibrium with the load after each increment in strain. Most engineering test conditions are chosen to duplicate manufacturing conditions where speed is desirable for reasons of economy, but in the realm of cultural materials rapid loading and straining generally do not occur.

All measurements were done at 50% RH +/- 2% RH and 70°F +/- 2°F. In most cases three samples were cut. Each sample was measured for width and thickness (using a standard micrometer in five places along the strip) and loaded into tensile straining jig and held at equilibrium conditions in a sealed container for 1 hour prior to testing. In most instances multiple runs (n=3) were done. Stress-strain curves of averaged values were created by mathematically averaging the stress and strain data independently at each incremental step for the multiple replicates of each sample. The final break load and strain was calculated from the average of the last stress and strain value from each of the runs.

Materials

Primarily one paper was used for these experiments. It was manufactured in large quantity specifically for the Conservation Analytical Laboratory's research projects by Neenah Paper. Fiber analysis was done at CAL following Institute of Paper Chemistry (now known as Institute of Paper Science and Technology) methodology. This waterleaf cotton was substantially the same as the material use in the previous experiments (Vitale, 1992): 89.5% soda cooked staple cotton seed hair pulped from knit cuttings, 2.2% softwood and 8.4% hardwood (wood pulps were predominately Kraft processed).

In the experiments reported in Figures 2a & 2b the LC waterleaf cotton paper (made for the Library of Congress by Neenah Paper) was used. With the exception of the comparison to the CAL cotton in Figure 1 this was the only use of the LC cotton paper. The LC cotton waterleaf paper was used in the liquid wetting experiments reported on previously (Vitale, 1992). The LC waterleaf cotton paper is 91% cotton seed hair, pulped from cotton knit cuttings, in sodium hydroxide (soda process) liquor, 2.2% softwood, 6.7% hardwood, cooked/pulped using predominately the Kraft process.

Experiments

Wetting and drying

The samples were immersed in deionized water (18.1 megohm at 20.5 °C) for approximately 1 hour. The sheet was drained vertically for approximately 30 to 60 seconds and blotted-off between 2 fresh blotters for 5 seconds. The saturated specimen was placed in a stack of blotters with a plexiglas platten and held at 0.01 psi (480 in^2 with 5 lbs of blotters and platten) for a minimum of 24 hours. Extensive experiments comparing air drying on a screen and 0.01 psi low restraint showed no difference in mechanical properties (see Fig. 1c). The primary difference was that the air dried samples were highly cockled; this made the initial set-up in the tensile strain jig problematic. All samples marked WET & DRY were dried at 0.01 psi which was considered equivalent to drying with no restraint.

Wet pressing

Wet pressing explored the effects of increased pressure on the bonding properties. Following immersion for one hour in deionized water (18.1

megohm at 20.5 °C), the 8.5" X 8.5" sample was drained for 30 to 60 seconds and then blotted-off for 5 seconds between fresh dry blotters in preparation for wet pressing. The sheet was placed between a 1" stack of slightly larger (1/4" - 3/8" on all sides) blotters and sandwiched between two 3/4" high density particle board plattens, then loaded into a hydraulic press and taken to high pressure. The experiments shown in Figure 2 were pressed at 100 psi or 120 psi nip for 5 minutes and designated 100 NIP or 120 NIP. In later experiments pressure was applied for prolonged periods, i.e., overnight. The samples designated 100 DRY or 120 DRY were held for 18 hours at elevated pressure. The initial loss of pressure due to compaction of blotters was monitored and corrected for 60 minutes, until the sheet was substantially dry (determined in advance); then the sample was left overnight. After approximately 12-18 hours all of the samples were at about 50 psi when removed from the press. Due to the extensive inter-sheet hydrogen bonding, one blotter had to be peeled away from the sample sandwich, then the sample peeled from the second blotter.

Wet straining

Wet straining was used to explore the effects of drying under tension on mechanical properties. Wet straining was done in coordination with wet pressing or the sheet would have pulled itself apart. A single axis wet straining frame was constructed from 0.5" plexiglas. Two 7" X 1" jaws with seven evenly spaced screws held the pre-dried edges at a starting distance of 6". One jaw was held fixed while the other was pulled towards the frame by tightening three machine screws. This press & stretch technique started with an initial 100 psi nip (see above) for approximately 5 seconds. The slightly dryer 8" X 6" sheet was removed from full size blotters and transferred to a jig which rigidly held and protected the 6" X 6" section, while a 1" clamping margin with pre-punched holes was dried with a low power hair drier. These edges were then clamped into the frame and an equilibrium tensioning step followed. Depending upon the final dry stretch intended, the screws were turned incrementally a half or full turn, followed by a 5 second nip at 150 psi; the process continued until the desired stretch had been achieved. The least number of stretch and press steps was used. The final pressing was held for 12-18 hours, i.e., overnight. The hydraulic pressure trailed off to about 100 psi after 1-2 hours and was approximately 75 psi the following day. The 150 psi nip was done using two 1" thick stacks of blotters 6" X 6.25". The edges of the frame rested horizontally on foam strips which prevented the weight of the frame from ripping the wet section of the sample. The process proved to be equal parts of applied technology and craftsmanship.

The CAL waterleaf cotton sheet shrinks 0.5% in the machine direction (MD) and 2.5% in the cross direction (CD) when free dried from the manufactured state; thus, simply preventing the sheet from shrinking stretches the sheet 0.5% in the MD. One full turn of the wet straining jig produced 0.7% wet stretch. Thus, the first press & stretch increment was 0.85% (0.5% for non-shrinkage plus a half turn, 0.35%). The final stage was 3.3% (2.8% wet-strain plus 0.5% non-shrinkage). A 4.0% press & stretch was attempted, but the sample ripped sometime during the overnight pressing.

Friction drying

In 1990, Ball et al. reported on an investigation of the effects of friction drying conducted at Queens University. Subsaquently she recreate her friction drying experiments on the CAL cotton paper and evaluated the results using the equilibrium stress-strain methodology. Friction drying is done to impart stretch to a sheet during drying. A Japanese paper "friction sheet" is adhered to a specimen sheet with water (so-called friction). The capillary effect, with water's high surface tension, provides some adhesion when free water is present and leads to inter-sheet bonding as the sheet dries. Pressure is applied to the sandwich through the Japanese paper "friction" sheet. Two methods were employed: pounding and smoothing. This causes a decrease in the capillary size and thus increases adhesion. The friction sheet is pasted at the edges only, to a rigid planar surface such as coated plywood, jointed wood plank, laminate-covered plywood or a kari-bari (Japanese drying screen). Often in conservation practice, the sandwich is air dried; in general the treatment produces a flat sheet with no cockling. For these experiments, the edges were pasted down to a jointed wood panel, and the sandwich was either air dried or pressed until dry, usually a day later. The specimen sheet easily separated from its friction mount.

The friction drying experiments were all done on the CAL waterleaf cotton paper. Two experimental segments were explored: air drying, and pressure drying. The friction sheet was the Japanese paper Goyu (purchased from TALAS in December 1985) it is a bleached, .005" thick (moderately thick) paper which shrinks 0.25% MD and 0.5% CD from the wet state.

The air drying segment of the experiment was done in two ways. Placed on an uncoated wood drawing board, the edge-pasted Goyu, over the waterleaf cotton specimen, was pounded with the flat 3" X 6" face of the traditional Japanese brush using the traditional rhythmic overlapping and crosswise repeat technique on an uncoated jointed wood drawing board. The second method followed the same general protocol but a traditional Uwabake Japanese smoothing brush was used to apply pressure which increased adhesion. Both samples were air dried.

The pressure dried experiments used the smoothing brush method described above. The Goyu and specimen sandwich were brushed to improve capillary adhesion, then the Goyu's edges were pasted to the wood panel. The sandwich was dried with pressure, using three methods: with felts, blotters/weight and an air bag press with blotters. The felt method used a stack of soft thick (1/2") felts in contact with the Goyu "friction" sheet and topped with a 1/2" plexiglas platten for a total of 0.026 psi pressure. The felt in contact with the friction sandwich was changed 3 times within the first 30 minutes. For the weighted blotters experiment, the blotters were stacked on the Goyu friction sheet, as before, but with a 1/2" plexiglas platten and a bag of shot weight for a total of 0.11 psi pressure. The blotters were not changed. The final method involved placing the joined wood plank with a stack of blotters over the Goyu, as before, in a specially built air-bag press. The bag was inflated to 20 psi, the blotters were changed twice in the first 30 minutes. Of all the samples, the latter showed the greatest inter-sheet adhesion after drying.

RESULTS

Effects of water on mechanical properties: previous work reviewed

It has been shown previously (Vitale, 1992) that water immersion of a cotton waterleaf paper releases substantially all interfiber bonding resulting in a 95-97% loss of tensile strength when the paper is wet. When the sheet is then dried, a portion of the original bonding and dried-in strains are reduced so that stiffness and strength were lowered. Figures 1a & 1b show the effects of wetting and drying and immersion (only) on the cotton LC cotton waterleaf control. These can be compared to the CAL cotton paper in Figure 1c; the primary paper in this experimental series. The two sheets are similar: they are both waterleaf (unsized) cotton staple fiber papers manufactured by Neenah Paper for experimental use. They are different in that (1) the tensile strength of the CAL cotton is 11% higher than LC cotton sheet, (2) the tensile strength of CAL cotton wet & dried sample is 7.6% less than the control as compared to 15% for the LC cotton wet & dried sample and (3) the increase in percent stretch between wet & dried and control is 52% as compared to 30% for the LC cotton sheet. These differences are minor in comparison to the range possible; the papers have the same qualitative behavior after wetting and drying.

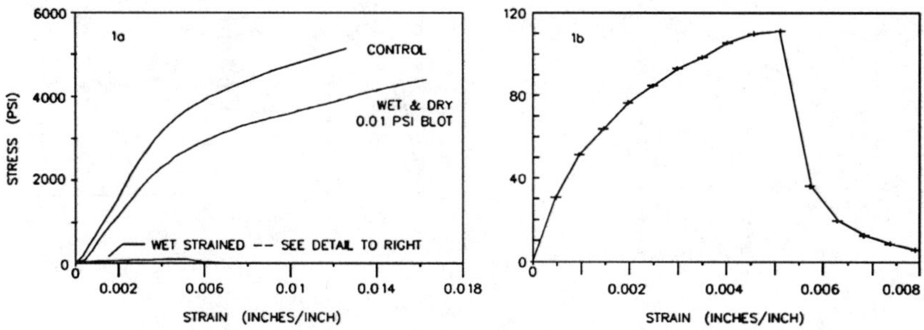

Figure 1a: LC waterleaf cotton untreated control (MD), wet and dried at 0.1 psi sample and wet strained sample.

Figure 1b: Detail of wet strained LC waterleaf cotton (MD) sample, note slow rupture process.

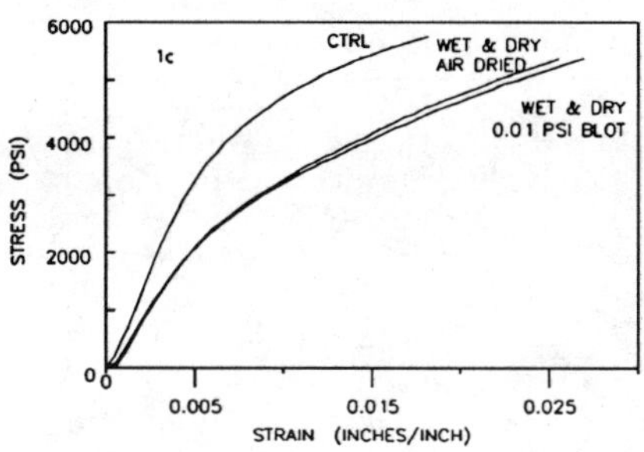

Figure 1c: CAL waterleaf cotton (MD) untreated control sample, a sample wet and air dried on a screen, and a sample wet and dried in a stack of blotters at 0.01 psi.

Wet pressing

Wet pressed samples show improvements in mechanical properties over the wet & dried low restraint drying, but the results are not consistent. The original tensile strength of the control is regained with prolonged wet pressing, but not with the simple wet nipped samples. Percent stretch shows a possible increase overall. Work increased only in the prolonged wet pressed samples. Stiffness increased for one of the prolonged wet pressed sample, but not for the other.

Figure 2a shows the effects of pressing the LC cotton paper at 120 psi for either a 5 minute nip (120 NIP) or sustained until dry (120 DRY). The 120 NIP sample shows slight increases in percent stretch (8.5%) and an increase in tensile strength (7%) over the WET & DRY sample. The 120 DRY sample shows a return to original tensile strength of the control and an additional increase in stretch (6.5%) over the 5 minute nip pressed sample (total 15%). Thus, prolonged contact with blotters under pressure has

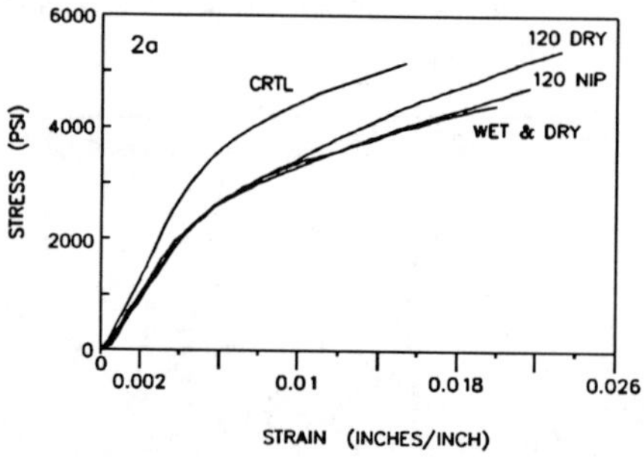

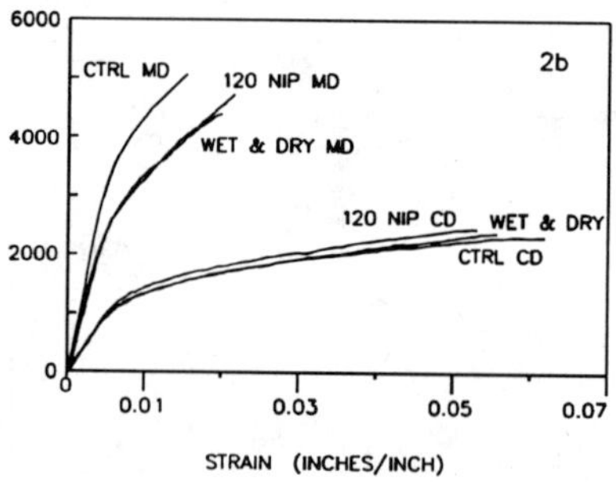

Figure 2a: LC waterleaf cotton paper (MD) untreated control, wet and dried at 0.01 psi, nip pressed at 120 psi for 5 minute then in blotters at 0.01 psi until dry, and pressed at 120 psi until dry. See Vitale (1992).

Figure 2b: LC waterleaf cotton MD & CD untreated controls, wet and dried at 0.01 psi, and nip pressed at 120 psi for 5 minutes then dried at 0.01 psi in blotters. See Vitale (1992).

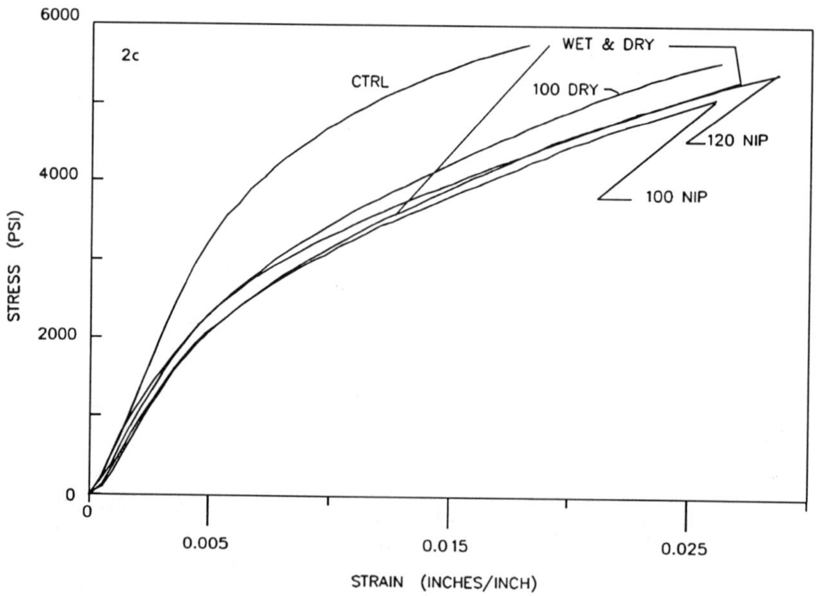

Figure 2c: CAL cotton (1) untreated control and simple (2) wet & dried samples with 3 press drying experiments: (3) nip pressed at 120 psi and (4) 100 psi (independently) for 5 minutes then into blotters at 0.01 psi until dry, and (5) pressed at 100 psi until dry.

increased tensile strength and toughness. Figure 2b shows the effects of the 120 psi nip drying in both MD and CD. There are no significant differences in the CD samples.

Figure 2c shows the results for the CAL cotton paper treated as the 120 NIP sample, similar to Fig. 2a, and another nip-pressed experiment at 100 psi (100 NIP) and a prolonged pressing sample, 100 DRY. Prolonged pressing resulted in the sheet being stretched most of the 0.5% it would normally have shrunk if left to dry with little or no restraint. Wet pressing with no stress release therefore creates some wet straining which is manifested as an increase in modulus (9%) and strength (4%) over the WET & DRY sample.

The 120 NIP CAL cotton sample exhibits behavior inconsistent with the other wet pressed samples. There is an indistinguishable difference in tensile strength, a 6% increase in percent stretch and a very large 21% increase in stiffness over the wet & dried sample. This indicates that something out of the ordinary occurred during this particular 5 minute wet-pressing process.

The nip dried samples are different from the prolonged dried samples for both LC and CAL cotton papers. While percent solids was not determined on the samples upon removal from the press, the nipped sheets had clearly gone from translucent to white/opaque indicating significant water loss. The remainder of the nipped sheet's drying occurred at 0.01 psi, which imparted force sufficient to prevent the sheet from cockling and rippling but did not enhance fiber to fiber contact. Both the LC and CAL cotton sheets shrink 0.5% (MD) during free drying. These sheets shrunk almost this amount due to the virtually nonexistent drying pressure.

Both the LC cotton and CAL cotton samples show an increase in slope of the plastic portion of the stress-strain curve, when compared to the WET & DRY samples. The slope of the 120 DRY change is more dramatic than the 100 DRY sample.

The 100 DRY sample shows a 19% increase in stiffness over the wet & dried sample. The major difference between the 100 NIP and 100 DRY was that strain was dried into the sheet because the normal 0.5% sheet shrinkage was not allowed to occur in the 100 DRY sample. The additional work to failure in the 100 DRY sample suggests that load is being distributed more widely within the sheet during straining.

Wet straining

Wet straining imparts dried-in strains to paper. When a sheet is wet and dried, it was found that dried-in strains are lost (Green, 1981 & 1987; Vitale, 1992). It is apparent in Figure 3 that for CAL waterleaf cotton paper the application of 2.6% wet stretch with wet pressing (150 psi) restores the mechanical behavior lost during simulated conservation treatment. The literature commonly reports that papers are wet strained several percent on the papermaking machine. Recreation of the original mechanical properties of the control, therefore, was accomplished by imitating the papermaking process.

Figure 3 shows that by introducing wet straining, modulus and yield point are affected. Wet pressing shows, Fig 2a-c, a trend of increases tensile strength with an increase in percent stretch. Wet straining with wet pressing causes tensile strength to increase while percent stretch decreases. The more strain that is dried into the sheet, the less the sheet stretched during tensile testing. The change in modulus and yield point, while not continuous, shows the trend of an increase with increasing wet straining. Increase in work was also observed, suggesting that the load is distributed more efficiently into to the network, i.e., more load-elongation (energy) is absorbed before failure, with wet straining.

Wet straining decreases the ability of the sample to be stretched when dry. Half of the ability of the dry paper to stretch is lost between 0.85% and 3.3% stretch. At 4% wet straining the sample ripped in the experimental jig; obviously 4% is very close to the limit that paper can sustain at 150 psi wet pressing.

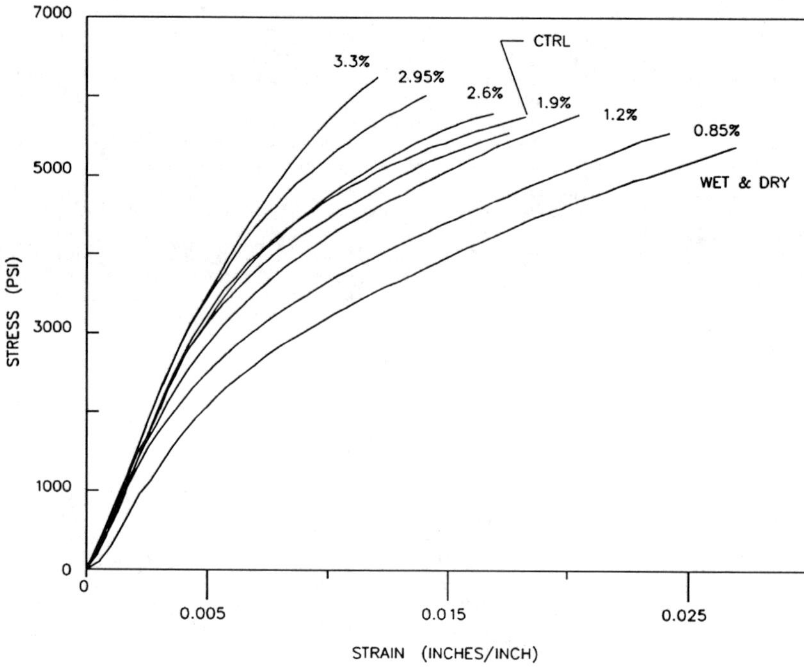

Figure 3: CAL waterleaf cotton paper (MD) untreated control, simple wet & dried sample, and sheets which were wet then wet pressed (150 psi) and wet stretched 0.85%, 1.2%, 1.9%, 2.6%, 2.95% and 3.3% (including 0.5% non-permitted skrinkage).

Friction drying

In comparison to standard conservation drying procedures, conventional friction drying procedures appear to have no effect on mechanical properties. Figure 4 shows that samples (1) pounded & air dried, (2) smoothed & air dried, (3) smoothed and felt dried at 0.026 psi and (4) smoothed and blotter dried at 0.11 psi have indistinguishable mechanical behavior.

The use of the friction drying procedure modified by application of 20 psi wet pressing, increases the modulus slightly, 6.8% (from 515,000 psi to 550,000 psi) with virtually no change in yield point. The changes appear to be similar to those shown in Fig. 2c for the 100 DRY sample; the procedures are parallel. In addition, the performance of the 20 psi friction-dried sample parallels the behavior of the 0.85% wet straining sample in Fig. 3. Two runs of the 20 psi friction dried procedure were run, designate #5 and #6 in Fig. 4. The variation in strength is insignificant, the shape of the stress-strain curves are almost identical. The approximate 12% difference in precent stretch is most likely due to vicissitude of implementation of the press-friction drying procedure. The replication of the 20 psi friction dried samples generally shows that the drying procedure is valid and repeatable.

The samples dried with light pressure, FELTS AT 0.026 PSI and BLOTTERS AT 0.11 PSI, have the same behavior as the air-dried samples. Thus, the slight increase in pressure is insufficient to prevent shrinkage.

When pressure of 20 psi (limit of our air bag press) is used to prolong contact between the fixed friction sheet and the specimen, a portion of the normal shrinkage is prevented. The small amount of dried-in strain imparted to the sheet is enough to increase modulus, tensile strength and work. It appears that periodic release of pressure to change blotters did not affect the ability of the system to impart dried-in strain to the specimen sheet.

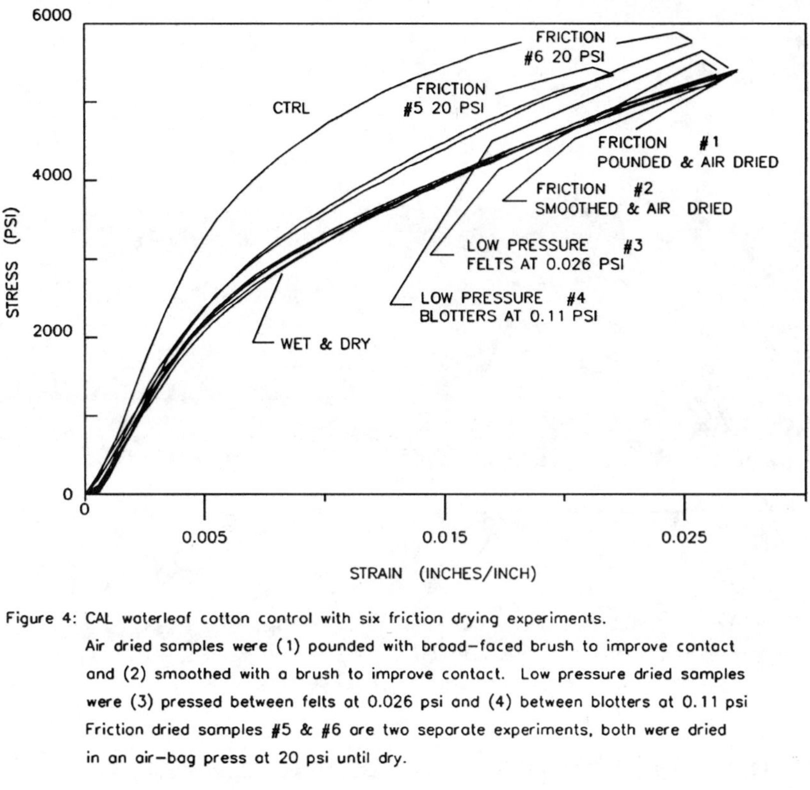

Figure 4: CAL waterleaf cotton control with six friction drying experiments. Air dried samples were (1) pounded with broad-faced brush to improve contact and (2) smoothed with a brush to improve contact. Low pressure dried samples were (3) pressed between felts at 0.026 psi and (4) between blotters at 0.11 psi Friction dried samples #5 & #6 are two separate experiments, both were dried in an air-bag press at 20 psi until dry.

DISCUSSION

Wet pressing

It has been shown that wet pressing increases bonding (Campbell, 1947; Casey, 1960; Schultz, 1961; Stetterholm & Chilson, 1965; Parsons, 1972). The trend in these experiments is for the increased bonding to result in increased tensile strength and percent stretch.

The wet pressing experiments showed inconsistent results due to (1) the small changes cause by the experiments and (2) the differences in the two papers used in the experiments. In summary, the trend is for wet pressing alone (nipped samples) to produce increased tensile strength and percent stretch and for prolonged wet pressing samples to produce increased tensile strength and stiffness.

In the wet pressing experiments, where 20 psi to 100-120 psi was used to enhance bonding, the pressure can be compared to the 540 psi capillary forces (for 20 micron fibers) which Campbell (1933) defined as the pressure pulling two fiber cylinders together (specific length) due to the capillary/surface tension forces during drying. Wet pressing will push fibers into better contact which will improve conformation of one fiber over another and thus increase bonded area, but they do not appreciably enhance the capillary forces, e.g. 540 psi + 120 psi, which pull the fiber surfaces near enough, 3.6 Å, for interfiber bonds to form (Liang & Marchessault, 1959).

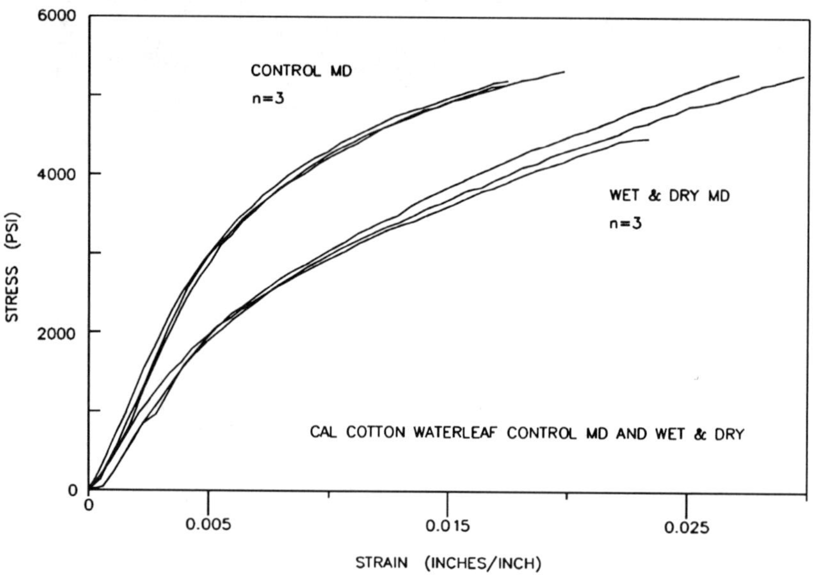

Figure 5: CAL cotton waterleaf MD, control and immersed in deionized water and dried at 0.01 psi. Three repetiotions of each sample shows the typical scatter for the data. Note that the control displays more homogeneous behavior than the wet & dried sample.

Wet straining enhances distribution of load to increase strength

Wet-pressing the waterleaf cotton sheets tends to increase tensile strength. It is shown that wet straining also increases tensile strength, modulus and yield point. From these data and the literature, it is concluded that wet-pressing increases bonding and that wet straining reduces elasticity of the network by pre-straightening the free-fiber segments. In addition, reduction of fiber curls, crimps, kinks and microcompressions (Page et al., 1985) results in an increase in modulus and elastic yield point, which demonstrates that the elastic properties of the sheet have been modified.

The application of wet straining decreases the ability of the sheet to stretch and at the same time increases the stiffness. Wet pressing introduces dried-in strains. The direct ramification is for free fiber segments between interfiber bonding sites to be straightened. It has been reported (Schultz, 1961) that wet pressing created better load distribution. It is reasonable to assume, therefore, that wet straining disrupts or prevents inherently weak bonding sites and straightens fibers so that (1) better and more complete contact can be made at bonding sites and (2) the load is passed from one site to the next through strong interfiber bonds.

Tensile strength increases with increased wet straining at a constant wet pressing pressure; this suggests that the load is being distributed more efficiently to the interfiber bonding sites. Wet-straining, therefore, increases strength by rupturing weak bonds during the pressing process. Pre-aligning or straightening of the free fiber segments between the interfiber nodes appears to allow more efficient transmission of load.

Tensile strength is increased by both wet pressing and wet straining. It has been shown that the strength of the fiber is greater than the strength of the paper (Vitale, 1992), thus, increase in tensile strength, is the ultimate measure of the interfiber bond efficiency -- the better the interfiber bonding the greater the strength of the sheet.

Application to conservation

Mechanical properties lost during treatment

Mechanical properties manufactured into a sheet are lost upon wetting and drying. Air drying and drying with 0.01 psi, 5 pounds of platten and blotters for a 20" X 24" blotter stack, are not mechanically different. Pressing at 0.1 psi or with 1 psi (50 pound of lead shot on a 20" X 24" blotter stack, a common conservation practice) has no effect either (unpublished work). Increasing the wet pressing pressure to 100 psi for a 5 minute nip has indistinguishable effects on mechanical properties. Pressing at 100 psi overnight, which would be the equivalent of 48,000 pounds resting on a 20" X 24" platten increases the tensile strength, over the wet & dry sample, sometimes.

Wet straining with wet pressing has been found to restore mechanical properties lost from the conservation procedure of washing. It is commonly reported in the literature that 2-6% wet straining of the wet web imparts desirable mechanical properties to a finished machine-made sheet. These results show that 2.6% wet straining (including the 0.5% shrinkage, which was prevented) matchs the as-manufactured mechanical properties of the Neenah sheet.

Effects on surface texture

The sheet dried with 0.01 psi pressure had no cockles, while the air dried sheet was very cockled. While a small amount of weight does not influence mechanical properties increasing weight on the blotters from 5 pounds to 50 pounds does alter the appearance (unpublished data).

The effects on surface texture from wet pressing are not likely to be acceptable. In an informal visual evaluation of surface texture by paper conservators in a CAL mid-career development seminar (Drying and Flattening) the CAL waterleaf cotton paper's surface texture was found to be satisfactorily recreated when friction-press dried at 20 psi (the limit of the air-bag press used). The CAL cotton sheet was manufactured to be quite uniform and smooth, thus pressure recreated the original method of manufacture (however, not the force).

Another paper used in the seminar was an internally sized handmade linen paper, manufactured as an art and conservation paper and given a common handmade surface texture. The paper was judged by means of scaled visual analysis to be too smooth when dried using the 20 psi friction-press method. Thus, press drying at 20 psi with blotters as the texture "molding" surface is not an ideal conservation procedure when a sheet has significant surface texture. If pressing a smooth bond, tracing or etching paper, however, 20 psi might be an acceptable pressure for recreating original surface texture.

Recreating original surface texture was found not to be possible on 12 historic papers using conservation drying procedures (0.001- 0.1 psi) in previous work (Sugarman & Vitale, 1992). In the study a best method was identified, which was quite good. However, the comments of the visual evaluators (paper conservators) noted that the surface texture was expanded compared to that of the controls and the visual scaling values of the treated samples was always higher than the corresponding controls. As is standard conservation practice, very little pressure was used in those experiments. Thus, it would appear that application of pressure may be desirable for some papers. Considering that substantial pressure was used in the manufacture of both handmade and machine-made paper, application of drying pressure is within the experience of the material. Defining the lower limits of the pressure required to recreate original surface texture is not yet resolved. Whether it is important to recreate mechanical properties lost during washing and drying is an issue for discussion. Recreation of original texture, however, is always desirable.

Friction drying

In the standard friction drying methods, the Goyu friction sheet holds the specimen expanded through inter-sheet hydrogen bonding during drying, up to a point. As the free water declines the capillary forces build, and then wane, when the free water is gone. The inter-sheet hydrogen bonding is formed through intimate contact with the host. Eventually the bonding fails as the specimen sheet shrinks during the later stages of drying in relation to the Goyu which does not. This procedure, thus, allows the specimen sheet to dry with its normal shrinkage. The wet pressed friction dried sample, on the other hand, holds the specimen in contact with the host Goyu sheet, with applied pressure, preventing the specimen from shrinking. The lower limits of pressure needed to prevent shrinkage of the specimen are not yet defined.

CONCLUSIONS

Wet pressing tends to increase tensile strength and percent stretch.

Wet straining, with wet pressing, increases tensile strength, Young's modulus and yield point. Percent stretch at break decreases due to the dried-in strains which were set into the sheet as the result of treatment. Increase in work for low percentage wet straining suggests better distribution of load during straining. At higher percentages of wet straining the decrease in work suggests that curls, crimps and kinks in free fiber segments are the source of energy absorption.

Standard friction drying procedures produce the same mechanical results as standard blotter drying conservation procedures.

Friction drying with wet pressing can be used to regain a portion of the mechanical properties lost due to wetting and drying.

Mechanical properties lost during the conservation treatments of washing and drying can be recovered using wet straining with wet pressing.

Tensile strength is the ultimate measure of interfiber bonding efficiency.

BIBLIOGRAPHY

Arah, C.O., McNamara, D.K., Hand, H.M. and Mecklenburg, M.F. 1989. Techniques for Screening Adhesives for Structural Applications. **J. Adhesion Technol**. 3(4): 261-75.

Baum, G.A. Pers, K., Shepard, D.R. & Ave'Lallemant, T.R. 1984. Wet Straining of Paper. **Tappi Journal**, 67(5): 100-104.

Ball, C. & Hanlan, J. 1990. **The Effects of Drying Method on Subsequent Properties of Washed Paper**. Presented at Drying and Flattening Seminar, Nov. 28-30,1990 at the Conservation Analytical Laboratory, Smithsonian Institution: available through BCIN data base as a Research Project Qualification Paper for Art Conservation Degree at Queens University.

Byrd, V.L. 1981. Fiber Orientation and Drying Restraint: Effects on Interlaminer Shear and Other Paper Properties. **Svensk Papperstidning**, Research Section, no. 15. R 105-09.

Campbell, W.B. 1933. The Cellulose-Water Relationship in Papermaking. **Bull. For. Serv. Dept. Int. Canada**. 84: 5-52.

Campbell, W.B. 1947. Physics of Water Removal. **Pulp and Paper Mag. of Canada**. 48(3): 103-109 & 122.

Campbell, W.B. 1959. The Mechanism of Bonding. **Tappi**. 42(12): 999-1001.

Casey, J.P.(ed). 1960. **Pulp and Paper Chemistry and Chemical Technology**, 2nd ed., 3 vol.s. Interscience Publishers, New York.

Fahey D.J. & Chilson, W.A. 1963. Mechanical treatment for improving Dimensional Stability of Paper. **Tappi**. 46(7): 393-99.

Green, C. 1981. **Dimensional Properties of Paper Structures**. Ind. Eng. Chem. Prod. Res. Dev. 20: 151-58.

Green, C. 1983. Relationship of Dimensional stability to Rheology of Paper Structure. In **Proceedings of the TAPPI 1983 International Paper Physics Conference**. Atlanta GA: Tappi Press. 165-71.

Green, C. 1987. Relationship of Dimensional Stability to Rheology of Paper Structure, **International Paper Physics Conference 1983**, Proceedings of TAPPI meeting Harwichport, MA. Sept. 1983. Reprinted: University Microfilms International Ann Arbor MI.

Htun, M. 1986. The Control of Mechanical Properties by Drying Restraints. In **Paper Structure and Properties**, eds J.A. Bristow & P. Kolseth. Ch. 15. New York: Marcel Dekker Inc. 311-26.

Hunter,D. 1978. **Papermaking, The History and Technique of an Ancient Craft**, New York: Dover.

Kershaw, T.N. 1980. Sheet Formation and Drying. In **Pulp and Paper Chemistry and Chemical Technology**. 3rd ed. Vol II, ed J.P. Casey. New

York: Wiley-Interscience.

Liang, C.Y. & Marchessault, R.H. 1959. Infrared Spectra of Crystalline Polysaccharides. I. Hydrogen Bonds in Native Cellulose. J. **Polymer Sci**. 37: 390.

MacLaurin, D.J. & Whalen, J.F. 1954. Effects of Wet Pressing on Sheet Properties. **Tappi**. 37(12): 608-13.

Page, D.H., Seth, R.S., Jordan, B.D. Barbe, M.C. 1985. Curl, Crimps, Kinks and Microcompressions in Pulp Fibers - their Origin, Measurement and Significance. In **Papermaking Raw Materials**, ed. Punton, Vol 1, London: Mechanical Eng. Pub. Ltd. 77-91.

Parsons, S.R. 1972. Effects of Drying Restraint on Handsheet Properties. **Tappi**. 55(10): 1516-21.

Pye,I.T., Washburn, O.V. & Buchanan, J.G. 1966. Structural Changes in Paper on Pressing and Drying,. In **Consolidation of the Paper Web**. ed. Bolam, F. London: The Technical Section of the British Paper and Board Makers' Association (Inc.).

Rance, H.F. 1980. Consolidation of the Paper Web. In Handbook of Paper Science, Vol 1 The Raw Materials and Processing of Papermaking. ed. Rance, H.F. Amsterdam: Elsevier.

Rance, H.F. 1954. Effect of water Removal on Sheet Properties. **Tappi**. 37(12): 640-49.

Robertson, A.A. 1959. The Physical Properties of Wet Webs 1. Fiber-water Association and Wet-Web Behavior. **Tappi**. 42(12): 969-78.

Robertson, A.A. 1963. The Physical Properties of Wet Webs 2. Fiber Properties and Wet Web Behavior. **Svensk Pappirtidning**. 66(12): 477-97.

Robertson, A.A. 1966. Measurement and Significance of the Water Retention Properties of Papermaking Fibers in **Consolidation of the Paper Web**. ed. Bolam, F. London: Technical Section of the British Paper and Board Makers' Association. 90-117.

Schultz J.H. 1961. The Effects of Straining During Drying on the Mechanical and Viscoelastic Behavior of Paper. **Tappi**. 44(10): 736-44.

Stetterholm, V.C. & Chilson, W.A. 1965. Drying Restraint: Its Effects on the Tensile Properties of 15 Different Pulps. **Tappi**. 48(11): 634-40.

Sugarman, J.S. & Vitale T.J. 1992. Observations on the drying of paper: Part I, Five Drying Methods and the Drying Process. in press, **JAIC**, Summer 1992.

Vitale, T.J. 1992. Effects of Washing on the Mechanical Properties of Paper and their Relationship to the Treatment of Paper. These Proceedings.

Vitale, T.J. 1988. Observations on the Theory, Use, and Fabrication of the Fritted Glass Bead, Small Suction Table. **The Paper Conservator**, 12: 47-67.

ACKNOWLEDGEMENTS

I am indebted to **Marion Mecklenburg**, Assistant Director for Conservation Research at CAL, for extensive consultations on mechanics and introduction to the technique of equilibrium stress-straining. Without the valuable assistance of **John Childs**, formerly a CAL Paper Lab intern and now a practicing furniture conservator, who performed much of the stress-strain evaluations, this work would not have explored as many possibilities with such elegance. **C. Mei-An Tsu**, an intern in the CAL Paper and Furniture Conservation Labs and currently an Art Conservation graduate student, performed some of the final mechanical strength measurements; her work is greatly appreciated. **Cynthia Ball** was a visiting researcher for one week at CAL to recreate her work on friction drying with the CAL cotton experimental paper and evaluate the results using the equilibrium stress-straining technology. I am grateful she has agreed that the data should be report here. Finally, I am very thankful to **Chandru Shahani**, Research Officer of the Preservation Directorate, Research Office at the Library of Congress for the gift of the LC waterleaf cotton paper.

Many thanks to Carol Grissom, Chief of Objects Conservation, and David Erhardt, Senior Research Chemist, of CAL who served as readers and critics.

PROTECTION OF ARCHIVAL RECORD FROM POLLUTANTS THE DIFFUSION OF SO_2 THROUGH ARCHIVAL BOXBOARD

Charles M. Guttman and Kenneth L. Jewett
Polymers Division, NIST, Gaithersburg, MD 20899

ABSTRACT

The diffusion constants of sulfur dioxide in boxboards, commonly used to store archival documents, have been measured. For the most common boxes used by National Archives and Record Administration a diffusion constant of about 0.001 cm^2/sec was determined independent of gas concentration. These results are discussed in terms of a model of the microenvironment provided by these boxboards as used in archival storage.

INTRODUCTION

Archives, libraries, and museums are involved in long term storage of records. Protecting these documents from degradation is a matter of great concern. Because temperature and atmospheric conditions in storage areas are important in mediating the rate of degradation of the stored records, temperature and humidity are controlled. However, atmospheric pollutants such as particulate matter, various oxides of nitrogen, sulfur dioxide (SO_2) and ozone cause degradation of records [1]. In some cases the removal of air pollutants from the external air by the air conditioning system is ineffective.

In storage areas such as stacks, many records are often kept in containers, and not exposed directly to the stack environment. These records experience a "microenvironment" which may be different from the macroenvironment of the stacks. If this microenvironment in fact is, or could be made to be less aggressive, then expensive changes in the air conditioning system might be obviated.

Recently, Passaglia [2] studied this microenvironment with specific reference to storage in the National Archives and Record Administration (NARA). He suggested that a study of the protection mechanism provided by the container, in particular the cardboard boxes used to safeguard a large quantity of archival material, would be valuable. Passaglia presented models and calculations to estimate the effectiveness of various containers for protection of archival materials from environmental pollutants. However, in order to utilize these models, values of the absorption and diffusion constants of the pollutants in archival boxboard are needed.

Herein, we report measurement of the diffusion constant of sulfur dioxide in various boxboard materials commonly used to make storage boxes. By incorporating these data into Passaglia's model we are able to ascertain the effectiveness of the archival boxes in providing a microenvironment that is different from the stacks.

EXPERIMENTAL

Details of the apparatus and materials are given in an earlier report [3]

Paper and boxboard are porous materials. The fraction of the volume of the paper or boxboard which is in pores affects the diffusion rate through the boxboard. The volume fraction of pores in boxboard was estimated by measuring the volume uptake of a low surface tension non-volatile liquid, hexadecane [3]. Data from this measurement are given in table 1.

Sulfur dioxide diffused through boxboards G, H, and J so slowly that a diffusion constant could not be determined in our apparatus. However, the smallest concentration we could have measured may be estimated by using the gas chromatography flame photometric detector (GC-FPD) calibration curve for SO_2. From this we estimated that the smallest diffusion constant we can determine by our measurement technique is 2×10^{-5} cm^2/sec[3]. The diffusion constants of boxboards G, H, and J must be less than this value.

TABLE 1

BOXBOARD	POROSITY CONSTANT	DIFFUSION $cm^2/s \times 10^5$	DESCRIPTION OF BOXBOARD+	ID
A	0.55	110.	60 pt, composite, acid free, cut from NARA box	
B	0.53	180.	60 pt, composite, low-lignin, cut from NARA box	
C	0.51	40.	60 pt, composite low-lignin	
D	0.47	30.	20 pt, lignin-free, high-density	
E	0.38	59.	10 pt, lignin-free, high-density	
F	0.45	310.	20 pt, lignin-free, low-density	
G	0.46	< 2	40 pt, composite, two pieces of D glued together	
H	0.44	< 2	60 pt, composite, three pieces of F glued together	
J	0.50	< 2	60 pt, composite, two E on outside glued to two F on inside	

\+ Boxboards identified as composite are made by gluing thinner boards together. All boards have 3% $CaCO_3$ reserve. All boards except boxboard A are pH 8.5.

Since SO_2 is absorbed by paper [4], one might be concerned that no SO_2 was observed through boxboards G, H, and J due to the massive absorption of SO_2. Concentration of the gas before it impinged on the boxboard (C_B) was routinely measured. This measurement was made to check the calibration on the GC-FPD and the mass balance in the system. During the measurement of boxboards G, H and J, the concentration of analyte entering and leaving the top chamber was not significantly different from C_B. If significant absorption of SO_2 had occurred, a large concentration drop would have been expected.

All diffusion constants are obtained after steady state has been reached. Normally the experiment is started approximately 24 hours before data collection is initiated. Points are then taken regularly two or three times daily. Steady state is assumed when the value of the diffusion constant shows no drift with time for at least 36 hours. Measurements obtained under these conditions eliminate all absorption interferences.

RESULTS AND DISCUSSION

Effect of SO_2 concentration on diffusion constant.

No concentration dependence of the diffusion constants was found for boxboards A and B, the two NARA boxboards, when measured at SO_2 concentrations of 147 ppm, 60 ppm and 10 ppm. These results suggest that the limiting low concentration diffusion constant of SO_2 in these paperboards has been measured.

Porosity effects on diffusion constants

Data from porosity experiments are summarized in table 1. Whereas significant sample weight gains were measured, the dimensions of paperboard samples remained unchanged. Therefore, it appears that the hexadecane did not swell boxboard samples, but only entered the pores of samples.

Although the chemical composition of papers or boxboards may vary from one grade to another, the physical structure is always porous. Literature suggests that papers have pores from between 0.1 and 100 micrometers [5].

Porous catalysis is used in several chemical manufacturing processes, and diffusion in these matrices have been studied in great detail [6]. In models of diffusion of gases through porous structures, diffusion of the gas through porous media is dominated by the diffusion of that gas through pores. Thus, for the random pore model [6], the diffusion constant of the pollutant in porous media is decreased by the volume available to the gas (i.e. the fraction of actual volume in the pores), and a tortuosity factor which describes increase path length the gas must travel. These ideas are expressed as

$$D_p = D_g \, f_p/t_p, \qquad (1)$$

where D_p is the diffusion constant of the pollutant gas in porous media, D_g is the diffusion constant of the gas in air, f_p is the fraction of volume of the paperboard in pores, and t_p is the tortuosity factor for the material.

According to the manufacturer, the boxboards D, E and F have no glue layer. They have porosities similar to those of the composite boards G, H, and J which are made by gluing various combinations of D, E, and F together. These composite boxboards have a much lower diffusion constant than those of the boxboards they are made from. This suggests that the glue may have an effect on the measured diffusion constant.

Effect of composite nature of boxboard on diffusion constant.

Many of the boxboards used to make boxes for archival storage are composite boards, made up from thinner boards that are glued together. This glue layer may effect SO_2 permeability through the boxboard. Two extreme cases may be considered. In one case the glue attaches to the matrix and does not fill or affect the pores. In the other case the glue creates a well defined intact polymer film with no holes between the two paperboards. In the first case, SO_2 flux through the boxboard is not affected by the presence of the glue. In the latter case, the glue may completely control the flux.

Boxboard H is a composite board manufactured by gluing three layers of boxboard F together. The manufacturer of the boxboard claims that there is about 4.5 kilograms of dry adhesive left per 280 square meter of boxboard laminated. Assuming the density of the dry adhesive is one gm/cm^3, we estimate a total thickness of adhesive of 0.0016 cm for the two glue films between the three pieces of boxboard.

For a composite board the diffusion constant may be estimated from the diffusion constant of its components as [7]:

$$L/D = \Sigma l_i/D_i = l_f/D_f + l_b/D_b \, , \qquad (2)$$

where the sum is over the constituents of the composite. l_f is the total thickness of the film (0.0016 cm) and l_b is the total thickness of boxboard

(0.16 cm). From data reported herein, the diffusion constant of boxboard F, D_b, is 0.003 cm^2/sec. For D_f, the diffusion constant in the film, we assign a value of 1.0 10^{-8} cm^2/sec. This value is estimated from a literature value for various gases in PVA films [7]. Then from equation 2, the diffusion constant for the composite board, boxboard H, is estimated to be 1.0×10^{-6}. This value is within our estimated upper bound for the diffusion constant of boxboard H given in table 1.

If the glue film is not intact, the value of l_f/D_f could change by many orders of magnitude. This loss of total coverage by a film could arise in the original laying down and drying of the film or could result from the aging of the film. This may explain why some composite boards give the high diffusion constant, boxboards A, B, and C, and some very low, boxboards G, H, and J. Thus, the properties of the glue used to make up the boxboard must be considered when estimating the overall flux of pollutants through a boxboard whose diffusion constant has not been measured.

Comparison of diffusion through the walls and gaps of boxes.

Passaglia [2] studied the microenvironment provided by archival boxes in protecting their contents from atmospheric pollutants. In that study he points out the importance of diffusion through gaps as a mode to bring stack pollutants into the archival box microenvironment. In his study, Passaglia assumed the boxboard had a diffusion constant of 10^{-8} cm^2/s. He concludes that the flux brought about by diffusion through gaps is much higher than the flux resulting from diffusion through the paperboard itself. This suggests that closing the gaps will result in better controlling the microenvironment.

Although his models are essentially correct, Passaglia had no measured values of the diffusion constants of pollutants through boxboard available to him. From the measurements provided in this work we find diffusion constants ranging from 10^{-3} to 10^{-6} cm^2/s for the various measured boxboards.

With the value of the diffusion constant for boxboard in the range of the NARA boxboard value, we find that diffusion through the boxboard is faster than through gaps. See figure 1. Thus, the use of boxes without gaps is of no importance for these boards. However, for the boxboards with much smaller diffusion constants like those values found for boxboards of G, H, and J, there may be some advantage to sealing the gaps.

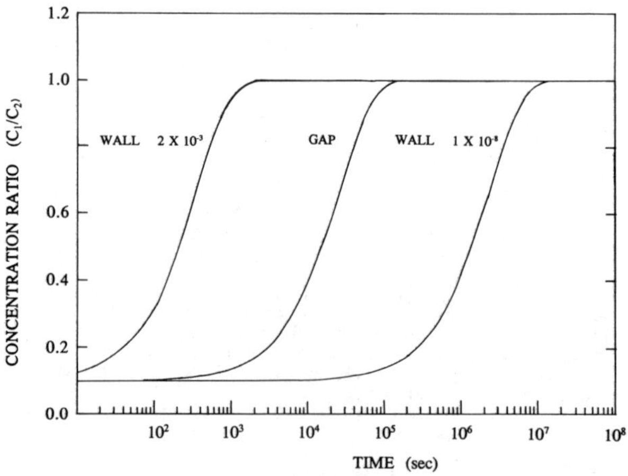

Figure 1. Ratio of the interior concentration, C_1, of an archival box to the concentration outside the box, C_2, as a function of time for a empty box for flow through gaps, for flow through box walls where the boxboard has a diffusion constant of 10^{-8} cm^2/s or for flow through boxboard where the boxboard has a diffusion constant of 2×10^{-3} cm^2/s.

CONCLUSIONS

The diffusion constant of the common atmospheric pollutant, SO_2, in the boxboards used to make up some archival boxes was determined to be 0.004 cm^2/s and less. We suggest that although the diffusion in these boxboards appears to be close to that in a porous medium, the effect of glue on the diffusion constant is important. This layer of glue may be controlling the diffusion in some boxboards. We suggest that for those boxboards with diffusion constants in the range of .004 cm^2/s that the total flux through a normal archival box is controlled by the diffusion though the boxboard, while for some of the boxboards with diffusion constants less than 10^{-6} cm^2/sec the diffusion through the gaps dominate the flux through the walls of the box.

REFERENCES

1. N.S. Baer and P. N. Banks, Int. J. Museum Mgmnt. and Curatorship 4, 9-20 (1985).
2. E. Passaglia, The Characterization of Microenvironments of Archival Records: A Research Program, 1987 NBSIR 87-3635, National Bureau of Standards, Gaithersburg, MD.
3. C. M. Guttman and K.L. Jewett, NISTIR 4719.
4. F. Lyth Hudson, R. L. Grant and J. A. Hockey, J. Appl. Chem. 14, 444-447 (1964),
5. J. P. Casey, Pulp and Paper Chemistry and Chemical Technology, Second Edition, Interscience Publishers, Inc., New York, Ch. XVII, (1961).
6. J. M. Smith, Chemical Engineering Kinetics, Third Edition, McGraw Hill, New York 1981.
7. J. Crank, The Mathematics of Diffusion, 2nd Edition, Clarendon Press, Oxford, 1975.
8. J. Brandrup and E. H. Immergut, Polymer Handbook, Third Edition, John Wiley and Sons, Inc. (1989).

Acknowledgments. The support for this project came from a grant from NARA. The authors would also like to thank W. Hollinger of Conservation Resources for providing samples and discussions on the boxboards and manufacturing process. The authors also appreciate discussion with E. Passaglia of NIST (retired) and W. K. Wilson of NIST (retired) and NARA.

THE MOISTURE BUFFERING CAPABILITY OF MUSEUM CASES

Vinod Daniel* and Shin Maekawa*
* The Getty Conservation Institute, 4503 Glencoe avenue, Marina Del Rey, CA 90292

1. ABSTRACT

Hygrometric half-lives and air exchange rates of four commonly used types of museum cases: Solander boxes, portfolio boxes, music boxes, and document boxes were measured. This paper will present the experimental results and illustrate the effect of the construction materials and closure characteristics on the moisture buffering capability of museum cases.

2. INTRODUCTION

Museum cases provide a less aggressive micro-environment for valuable photographs, documents and paintings than the macro-environment in the storage room. Relative humidity along with temperature and pollutants are buffered by these cases.

Maintenance of a stable relative humidity is essential for the preservation of stored paintings and documents. Since the storage area often does not have relative humidity control, it is important museum storage cases buffer sudden changes in relative humidity as during a Santa Ana condition in Southern California when the relative humidity stays very low for prolonged periods of time, or a wet rainy day when the relative humidity is very high.

The buffering capability of museum storage cases depends on the hygroscopicity and hysteresis of the case materials and their closure characteristics. These properties will be illustrated based on the results of our tests on four different types of museum cases: Solander boxes, portfolio boxes, music boxes, and document boxes.

3. THEORY

3.1 Hygrometric half-life

The rate of moisture transfer between the micro-environment inside the museum case and the macro-environment in the storage room, under isothermal conditions can be approximated as a first order differential equation

$$d(RH)/d\tau = k(RH_{env} - RH) \qquad (1)$$

where "RH", the relative humidity inside the museum case; "RH_{env}", the relative humidity of the macro-environment outside the case;

"τ", the time; and "k", the proportionality constant incorporating all modes of moisture transfer i.e diffusion through the walls, diffusion through the openings, and adsorption/absorption and subsequent desorption through the walls [1,2].

Integration of equation (1) between the initial condition of relative humidity "RH_o" and time "t_o" and final condition of relative humidity "RH_f" and time "t_f", and rearranging gives

$$k = \frac{\ln[(RH_{env} - RH_o)/(RH_{env}-RH_f)]}{(t_f-t_o)} \qquad (2)$$

And the hygrometric half-life "$T_{1/2}$can be evaluated as

$$T_{1/2} = \frac{Ln(2)}{k} \qquad (3)$$

The hygrometric half-life is the time required for the relative humidity inside a case to change to half the relative humidity difference between the micro environment inside the museum case and the macro environment in the storage room. The moisture buffering capability of the museum case can be measured by its hygrometric half-life: the greater the hygrometric half-life, the better the moisture buffering capability.

3.2 Air exchange rate

The rate of air exchange between the museum case and the macro-environment outside can be expressed by the first order relationship

$$d[O_2]/d\tau = k(20.9 - [O_2]) \qquad (4)$$

where "$[O_2]$", the oxygen concentration in percent inside the museum case; "τ", the time; and "k", the proportionality constant. Knowing the initial oxygen concentration "$[O_2]_0$", time "t_0", final oxygen concentration "$[O_2]_f$", and final time "t_f", the rate constant "k" can be evaluated as

$$k = \frac{Ln[(20.9-[O_2]_0)/(20.9-[O_2]_f)]}{(t_f-t_0)} \qquad (5)$$

and hence the half-life "$T_{1/2air}$" for air exchange can be evaluated as

$$T_{1/2air} = \frac{Ln(2)}{k} \qquad (6)$$

4. EXPERIMENTAL

4.1 Description of museum cases

The hygrometric half-life as well as the air exchange rate of four commonly used types of museum cases, differing in construction materials as well as closure characteristics were tested. A brief description of the tested cases are given below.

Solander box (16.5" x 20.5" x 2.5"): Made of basswood on the sides and .098 binder's board on the top and bottom with Truecore card stock lining and a book cloth fabric. The boxes have a lip arrangement for easy opening [3].

Portfolio boxes (14" x 18" x 2"): Made of 0.100 thick heavy weight boxboard lined with acid free Tyvek, the outside of the box is covered with black book cloth [4].

Document cases (15.25" x 10.25" x 5"): Made of 0.060" thick grey board, the box stands vertically and the top fits very loosely over the sides [5].

Music boxes (15" x 11.5" x 2.75"): Made of 0.060" thick lignin free paper, they feature a two-piece design with a drop front for easy access [5].

4.2 Hygrometric half-life

Five museum cases of each type conditioned at 48% ± 2% RH was placed in the environmental chamber. The relative humidity of the chamber was increased to 82% ± 2% RH to simulate a wet rainy day, or decreased to 32% ± 2% RH to simulate a dry day. The temperature and relative humidity inside the museum case as well as in the environmental chamber was monitored using a Shinyei resistive bulk polymer relative humidity sensor and thermistor. Data was collected every second, averaged over a 15 minute period and stored in a Campbell Scientific, Inc. CR10 data logger. The data were transferred to a personal computer and analyzed. The schematics of the experimental setup are shown in figure 1.

4.3 Air exchange rate

Three museum cases of each type were flushed with dry pre-purified grade nitrogen gas and the oxygen concentration inside the museum case was monitored using a GC Industries, model 501, 0-25% oxygen monitor. For the experiments with a sealed case, Solander and portfolio boxes were sealed with silicone grease and duct tape. Schematics of the experimental setup are shown in Figure 2.

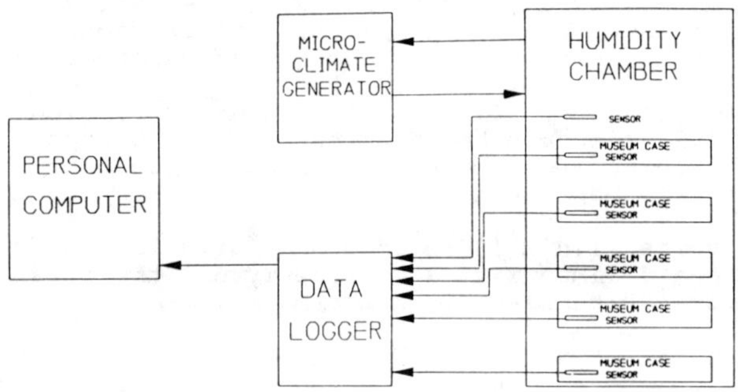

FIG 1. MEASUREMENT OF HYGROMETRIC HALF-LIFE

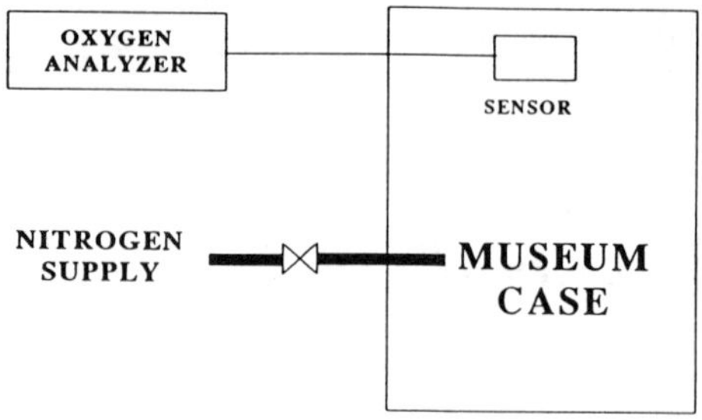

FIG 2. MEASUREMENT OF AIR EXCHANGE RATE

5. DATA, RESULTS AND DISCUSSION

The hygrometric half-lives expressed as mean ± one standard deviation of the four types of museum cases against both high and low environmental relative humidities are given in Table I [6].

Table I. Hygrometric half-lives of museum cases

TYPE OF CASE	ENVIRONMENTAL RELATIVE HUMIDITY	HALF-LIFE IN DAYS	HALF-LIFE RANGE IN DAYS
SOLANDER	HIGH	4.3 ± 1.0	3.0 - 6.3
SOLANDER	LOW	6.7 ± 2.4	3.7 - 10.7
PORTFOLIO	HIGH	2.9 ± 0.4	2.5 - 3.4
PORTFOLIO	LOW	5.4 ± 0.9	4.3 - 6.4
MUSIC	HIGH	0.5 ± 0.1	0.4 - 0.6
MUSIC	LOW	1.1 ± 0.2	0.8 - 1.4
DOCUMENT	HIGH	0.4 ± 0.1	0.3 - 0.4
DOCUMENT	LOW	1.5 ± 0.5	0.8 - 2.1

The hygrometric half-lives for all museum cases was better at low relative humidity than at high relative humidity, illustrating the hysteresis of the construction materials, i.e rate of moisture absorption/adsorption being faster than rate of moisture desorption.

The hygrometric half-lives of different museum cases [6] in Table I show a wide variation of values. The closure characteristics of museum cases are different, as are hysteresis characteristics of construction materials, hence this wide range of hygrometric half-lives observed. The Solander boxes showed a wider variation of half-lives than other types of cases since they are individually made unlike the other cases which are machine fabricated products.
The half-lives for air exchange for both sealed as well as unsealed museum cases are given in Table II. The values are expressed as mean ± one standard deviation.

Table II. Half-life for air exchange of museum cases

TYPE OF CASE	CLOSURE CONDITION	HALF-LIFE FOR AIR EXCHANGE IN MINUTES
SOLANDER	NOT SEALED	34 ± 5
SOLANDER	SEALED	281 ± 53
PORTFOLIO	NOT SEALED	144 ± 23
PORTFOLIO	SEALED	723 ± 88
MUSIC	NOT SEALED	56 ± 3
DOCUMENT	NOT SEALED	9 ± 1

As would be expected the half-life for air exchange was

significantly better for the sealed cases than for non sealed cases. Hence any improvement of the closure in a museum case will improve its moisture buffering capability.

Air exchange rates of Solander boxes were significantly lower than for portfolio boxes as illustrated in Table II, undoubtedly indicating that the closure characteristics of portfolio boxes are better. Since the hygrometric half-life of Solander box is longer than that of a portfolio box, we infer that the materials used in a Solander box absorbed/adsorbed more moisture than the materials of a portfolio box.

Music boxes had a higher half-life for air exchange than document cases, but their hygrometric half-lives were nearly the same. This (along with the fact that Solander boxes have a higher hygrometric half-life than portfolio boxes) illustrates that the hygroscopicity of the construction materials is more significant than air-exchange rate for buffering moisture changes.

The air exchange rate for document boxes are significantly lower than all the other museum cases since they had sufficient openings on the sides, as well as a loose fitted top. Also, because the construction materials of document boxes cannot retain much moisture, their hygrometric half-life was low.

Mat boards absorb significant moisture and hence are able to improve the moisture buffering capability of a museum case. When 30 matboards (100% cotton rag) were placed in a Solander box, and the relative humidity monitored with a sensor placed on top the half-life increased from 4.3 ± 1.0 days to 9.1 ± 0.7 days under high environmental relative humidity, and from 6.7 ± 2.4 days to 19.5 ± 9.6 days under low environmental relative humidity

7. CONCLUSIONS

Solander boxes and portfolio boxes significantly buffer changes in relative humidity. The moisture buffering capability is proportional to the hygroscopicity of museum case materials; for this reason matboards improve the hygrometric half-life and hence the moisture buffering capability. Closure characteristics are clearly important for each particular type of museum case.

REFERENCES

1) Vinod Daniel, Shin Maekawa, The Moisture Buffering Capability of Solander Boxes, Final report to Getty Conservation Institute.
2) Vinod Daniel and Shin Maekawa The Moisture Buffering Capability of Solander boxes, WAAC Annual Meeting, Seattle, WA, Sep 29-Oct 2, 1991 (Unpublished).
3) Catalog from Light Impressions Archival Supplies, Rochester New York-14607
4) Catalog from University Products, Holyoke, Massachusetts-01041
5) Catalog from Conservation Resources, Springfield, Virginia-22151
6) Vinod Daniel, Shin Maekawa, "Hygrometric Half-life of Museum Cases" (Unpublished, in review)

CHANGES IN THE INFRARED ABSORBANCE AND COLOR OF AGED CELLULOSE FILM

FRANCES WHITAKER MAYHEW* AND IRA BLOCK**
*University of Delaware, 201 Alison Annex, Newark, DE 19808
**University of Maryland, 2100 Marie Mount Hall, College Park, MD,20742

ABSTRACT

The accelerated aging of unplasticized Cellophane film specimens was conducted at temperatures ranging from 80 to 150 °C in a dry, forced-air convection oven for various periods of time. Colorimetric evaluation showed increasing discoloration which could be partially reversed by rinsing in deionized water. The infrared spectra of samples aged within this temperature range show the same changes, dependent upon aging time. This indicates that the species formed in the films at temperatures above and below the boiling point of water are the same, and that the moisture content of the films may affect the reaction rate, but not the mechanisms. The observed data in the visible region fit the standard pseudo-first-order kinetics model and are in excellent agreement with those determined for both the color change of cotton and the UV absorbance of Cellophane under similar conditions.

INTRODUCTION

The accelerated thermal oxidation produced by baking cellulosic materials in a convection oven at temperatures well above ambient conditions has been utilized by researchers in textile and paper conservation to simulate the effects of natural aging. To date, a consensus on the appropriate conditions for accelerated aging has not developed. Questions have been raised regarding (1) the appropriate temperature to be used and (2) whether the presence of moisture during baking alters the chemical reactions.

Block [1] has argued that, since the degradation rate of cellulosic materials follows pseudo-first-order kinetics and fits the Arrhenius model for temperatures between 60 and 150 °C, moisture acts only as an accelerant and does not change the mechanisms of the reactions. Therefore, the results of accelerated aging in a dry oven at temperatures above 100 °C should be valid predictors of natural aging. Others have claimed that moisture is necessary [2] and have argued that "dry-oven aging has little predictive value for oxidative degradation" [3].

In this work, the infrared spectra and the color change of Cellophane film samples baked at temperatures above and below the boiling point of water were studied in an attempt to clarify these issues. Cellophane film was chosen for this work because it is regenerated cellulose, it is transparent and it is manufactured to very close tolerances. The transparent quality eliminates the necessity to destroy the sample in order to obtain an infrared spectrum, so that serial evaluation of the same sample before and after treatment is possible. The precision of manufacture of the film reduces the need for special purification procedures and allows semi-quantitative comparisons of different samples.

Oxidation of Cellulose

Cellulose is a linear polymer, the repeat unit of which is (beta)-D-1,4 glucopyranose [4]. This configuration produces a well ordered and highly crystalline fiber fine structure [5]. The cellulose macromolecule can be degraded by four general mechanisms: oxidation, hydrolysis, microbial attack and mechanical stress. In the museum setting, mechanical stress and microbial attack can be reduced by proper handling and storage, and by good housekeeping. Hydrolytic degradation requires a rather high level of moisture to provide the appropriate pH condition. However, oxidative attack occurs under all conditions where textile materials are exposed to the atmosphere. This work focusses on oxidative degradation.

Oxidation of cellulose by molecular oxygen occurs readily, and is catalyzed either photochemically or by alkali. The oxidation of cellulose produces aldehyde, ketone, and carboxyl groups as well as hemiacetal end structures [6,7]. Most oxidants are nonspecific in their mode of attack, so that oxidation may occur simultaneously at many sites along the molecule. Some of the possible primary oxidation states are illustrated in Figure 1. Some of these oxidation reactions are produced by accepting a molecule of water, and others are produced through the release of a molecule of water. However, each of these primary oxidation reactions leads to ring scission or chain scission through beta-alkoxy elimination [4] or by weakening the 1-4 glycosidic ether link [8].

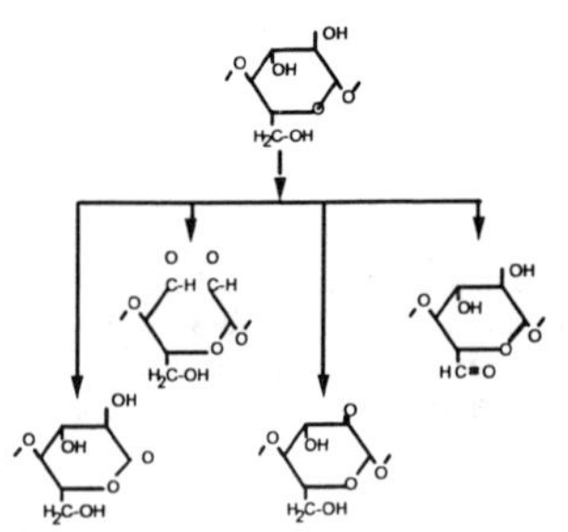

Figure 1. Various paths for cellulose oxidation

Studies of the effects of heating on the physical properties of cellulose indicate that these properties remain relatively constant for samples heated to temperatures below 160 °C [9]. As has been noted, "At the lower temperatures, it is difficult to distinguish between the thermal degradation and the normal aging of cellulose (which is accelerated by heating) "[10].

Infrared Spectrophotometry

Spectrophotometry is the measurement and interpretation of the electromagnetic radiation absorbed or emitted when molecules, atoms or ions of a material move from one allowed energy level to another [11]. The infrared (IR) region of the electromagnetic spectrum occurs within the range of 4000 -400 wavenumbers (1/cm), corresponding to wavelengths of 2.5 - 25 micrometers. The discreet absorption bands, caused by allowed quantum vibrational transitions, produced in this region indicate the presence of certain atomic structures or bond types. Major absorption bands characteristic of cellulose moieties are listed in Table I.

Table I Assignment of Absorption Bands in the IR Spectrum of Cellulose [13,14,15,16]

Frequency (1/cm)	Band Assignment
890	$-CH_2-$ Deformation
1200 - 1430	C-H or C-OH Deformation
1620 - 1640	H-O-H Stretching
1700 - 1725	-COOH Stretching
1720 - 1740	>C=O Stretching
2880	C-H Stretching
3160 - 3520	-O-H Stretching

In this study, the appearance of carbonyl (>C=O) and/or carboxyl (-COOH) species that signify the onset of cellulose degradation are of specific interest. Evidence of these species occurs at wavenumber bands of 1705-1725 for acyclic ketones, 1720-1740 for aldehydes and 1700-1725 for organic acids [13,14]. Infrared spectra can be effectively utilized for qualitative analysis, but are of lesser value in quantitative analyses [12].

Colorimetry

In the visible region of the spectrum, 400 - 700 nanometers, the allowed quantum transitions are very close together, so that visible spectra, unlike the discreet IR absorption peaks, are characterized by broad absorption bands. The effect of oxidation on cellulose is to produce a broad absorption band in the visible region that causes the material to take on a yellow-brown color. The depth of this color is a measure of the extent of conversion of the material to species of oxidized cellulose.

The Lab color analysis system incorporates two opponent chromaticity scales to measure hue, the "a" scale for red(+) and green(-) and the "b" scale for yellow(+) and blue(-). An "L" scale measures value from black (0) to white (100) [17]. The overall color change is a calculated value designated as delta E. The Hunterlab Color Difference Meter and Lab color analysis system was used in this study.

Reaction Kinetics

Since the thermal oxidation of cellulose has been shown to follow pseudo-first-order kinetics [18], and since the oxygen concentration remains constant in open systems, it can be shown that for the process

$$\text{Cellulose + Oxygen} \xrightarrow{k} \text{Oxidized cellulose species,}$$

$$\ln\{[\text{Cell}]_o/[\text{Cell}]_t\} = kt \qquad (1)$$

where $[Cell]_o$ is the initial molar concentration of cellulose, $[Cell]_t$ is the molar concentration at time "t" and k is the reaction rate constant. The appearance of oxidized cellulose [Oxy] can be traced by finding the difference between the original concentration and the concentration at any point in time. Thus,

$$[Oxy] = [Cell]_o - [Cell]_t, \text{ therefore}$$

$$[Cell]_t = [Cell]_o - [Oxy]. \tag{2}$$

Substitution of the value of $[Cell]_t$ from Equation 2 into Equation 1 gives:

$$\ln\{[Cell]_o/([Cell]_o - [Oxy])\} = kt \tag{3}$$

which produces a log/linear relationship that can be analyzed by simple linear regression techniques.

Unless there is a chemical or physical change which affects the reaction mechanisms, the rate at which a chemical reaction proceeds will increase with increasing temperature according to the Arrhenius relationship:

$$\ln k = -(E_a/R) * 1/T + \ln A \tag{4}$$

where E_a = the activation energy, R = the gas constant, T = absolute temperature and A is a constant of integration [11].

Photosensitive materials can interact with electromagnetic radiation. Transmittance is the fraction of incident radiation transmitted by such an analyte. Transmittance is expressed as:

$$T = P/P_o$$

where P = the power of the transmitted radiation and P_o = the power of the incident radiation. Absorbance is the negative logarithm of transmittance. It is expressed as:

$$A = -\log_{10} T = \log_{10} P_o/P \quad [12].$$

The Beer-Lambert Law states that the absorbance of a photosensitive analyte is directly proportional to its concentration through the expression:

$$A = \epsilon * l * C \tag{5}$$

where A is the absorbance value, C the concentration of the analyte, l = the length of the light path (which is constant for each analytical instrument), and ϵ is the light absorbing capacity of the analyte [12]. Therefore, the change in the concentration of oxidized cellulose can be traced by a change in the absorbance value. If the concentration of the analyte (oxidized cellulose) in Equation 3 is replaced by the infrared or visible absorbance, the expression becomes:

$$\ln[A]_\infty/([A]_\infty - [A])) = kt \tag{6}$$

where $[A]_\infty$ is the absorbance or the color change of an aged cellulose sample at "infinite time".

When substituting overall color change (delta E) for absorbance (A), Equation 6 becomes:

$$\ln([E]_\infty/([E]_\infty - [E])) = kt. \tag{7}$$

EXPERIMENTAL PROCEDURES

Materials

Unplasticized Cellophane film was supplied by the E. I. duPont de Nemours Company of Wilmington, Delaware and used without further treatment.

Equipment

A Blue M STABIL-THERM Constant Temperature Cabinet was used for accelerated aging; a Perkin-Elmer Mod. No. 281B Infrared Spectrophotometer was used to record the IR spectra, and a Hunterlab Model D25B Color/Difference Meter was used to measure changes in color of the aged films.

Procedures

All prepared specimens were stored in a glass desiccator prior to and following treatment. Twelve 3 1/2 " square specimens were used to evaluate color change and changes in the IR spectrum at each temperature. Pairs of specimens in each treatment group were identified by marking with a pattern of holes to prevent contamination by foreign materials such as ink. One 3 1/2" specimen was selected as a reference. This specimen received no treatment and was removed from the desiccator only when needed for comparison to the treated specimens.

Experimental runs were performed at 150, 140, 120, 100 and 80 °C. At each temperature the samples were baked for periods of 2, 5, 10, 20, 50 and 100 hours. Each specimen was placed on the oven rack and weighted with glass microscope plates to prevent curling during accelerated aging. Following aging, each specimen was placed in the desiccator to cool. Aged specimens were then appraised with the IR spectrometer and the color/difference meter. For IR measurements, the films were mounted in a sample holder and placed in the sample beam of the spectrophotometer. Air was used as the reference. For color measurements, the films were backed by a white standard.

After the initial colorimetric evaluation, one of the specimens from each pair was soaked for several hours in deionized water and allowed to dry between glass plates. When completely dry, the rinsed specimen was again colorimetrically evaluated and returned to the desiccator.

RESULTS AND DISCUSSION

Infrared Spectroscopy

The IR spectra of cellophane samples aged at temperatures between 80 and 150 °C and that of the unaged control sample are shown in Figure 2. It should be noted that interference fringes created by the ricochet of radiation through the film specimen occur in the region 1850-2500 cm^{-1}. The spectra in the 1400 - 1800 wavenumber region are shown, enlarged, in Figure 3. Each line represents the spectrum of a cellophane sample. The initial pen position for each successive spectrum was lowered to separate the tracings and facilitate comparison between them. The line designated by the O represents the untreated Control, line A was created by a sample baked for 100 hr at 80° C, line B was created by a sample baked 100 hr at 100° C, C was produced by a sample baked 50 hr at 120° C, D was produced by a sample baked 10 hr at 140° C, and E is the spectrum of a sample baked 10 hr at 150° C.

Examination of Figures 2 and 3 shows that the major spectral changes are (1) a decrease in %-transmission between 1630 and 1640 cm^{-1}, indicating an increased propensity to absorb water after aging; and (2) a monotonic decrease in %-transmission between 1725 and 1740 wavenumbers, indicating the formation of carbonyl species (aldehydes and ketones).

It should be noted that carboxylic acid species, if present, would cause an absorption peak between 1700 and 1725 cm^{-1}. In addition, the apex of the carbonyl absorption band is affected by the fiber fine structure and can broaden into the -COOH region [15,19]. Thus, the results of IR analysis cannot definitly confirm or deny the presence of carboxylic acid species. However, the work of DeSimone and Block [20] on Cellophane samples baked at 150 °C for up to 500 hr showed no evidence of acid formation.

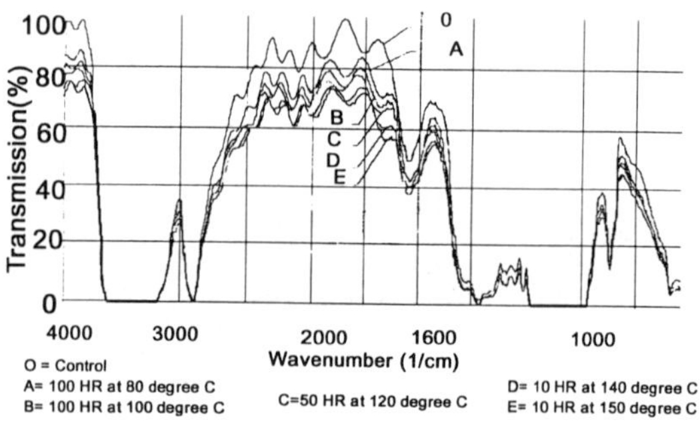

Figure 2. Infrared spectra of cellulose film aged at different temperatures. O = Control, A = 100 hr at 80° C, B = 100 hr at 100° C, C = 50 hr at 120° C, D = 10 hr at 140° C, E =10 hr at 150° C.

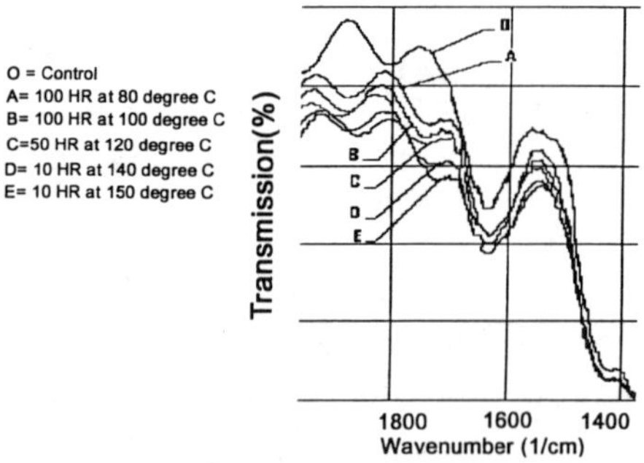

Figure 3. Infrared spectra of aged cellulose film in the 1400 - 1900 cm^{-1} region. O = Control, A = 100 hr at 80° C, B = 100 hr at 100° C, C = 50 hr at 120° C, D = 10 hr at 140° C, E =10 hr at 150° C.

Insofar as the absorption peak in the >C=O region is produced more rapidly at higher temperatures while the shapes of the spectra remain the same, the present work suggests that the oxidation reaction, as indicated by the creation of carbonyl species, changes in rate, but not mechanism, within the stated temperature range.

Colorimetry

A typical plot of color change (Delta E) vs time is shown in Figure 4. It should be noted that the curve cannot be fitted by a single linear equation , but comprises both a short-term (up to about 10 hr) and a long-term (after about 20 hr) portion, each of which does fit the linear form. This phenomenon holds true for all temperatures studied.

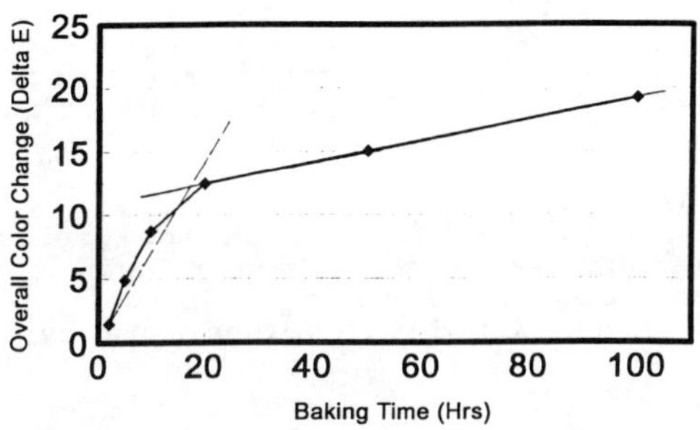

Figure 4. Color change vs time for Cellulose film aged at 150 °C

In order to determine the effect of temperature on the rate of oxidation by use of Equation 7, it is necessary to determine the value of $[E]_\infty$. The procedure is illustrated in Figure 5, in which the inverse color change (1/Delta E) is plotted vs time. In this form, the curve for the long term oxidation (beyond 20 hr) is effectively linear, and the slope and intercept can be found by simple linear regression techniques. By inserting that slope and intercept into the generic linear equation ($y = mx + b$) and extending the resulting line, the approximate value for delta E at "infinite time" is found to be 24 color units. The value of $[E]_\infty$ in Equation 7 was, therefore, set at 24. The estimate "24" is the lowest value for $[E]_\infty$ that will not appreciably affect the result. Estimates as high as 40 could be used without significant change in the reaction rate.

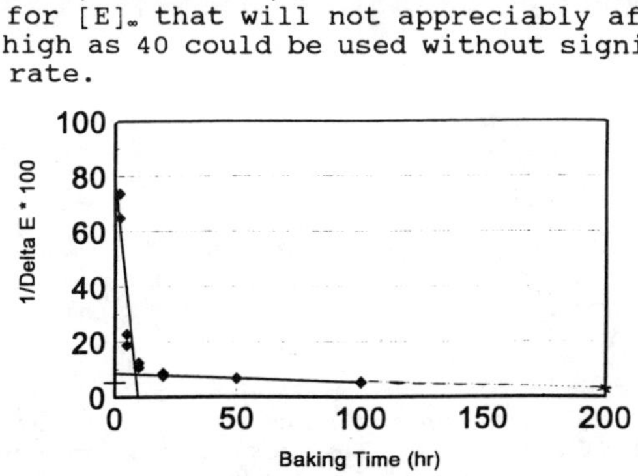

Figure 5. Inverse color change vs time for Cellophane film aged at 150° C.

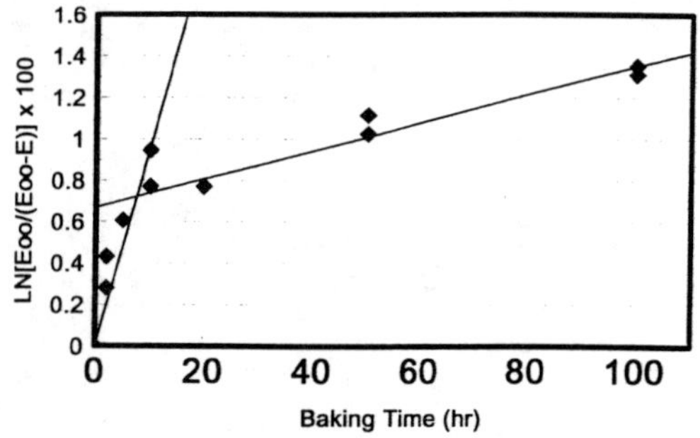

Figure 6. Method of determining reaction rates from color change for films aged at 150° C. Solid lines are best fit from linear regression.

Setting the value of $[E]_\infty$ at 24 in Equation 7 yields the type of curve shown in Figure 6, from which the pseudo first-order reaction rate constants (k), for both short and long-term oxidation, may be determined as a function of temperature. These reaction rates are given, along with values found by other authors, in Table II.

Table II Reaction Rate Constants for Color Change and Ultraviolet Absorption for Cellophane Film and Cotton Cloth

Temperature (°C)	Reaction Rate Constant $\times 10^{-4}$/hr Short Term	Long Term
80	2.6	0.2
90*	3.0	0.4
100	9.6	0.7
100+	---	0.7
110*	9.0	1.5
120	19.2	3.4
120+	---	3.7
140	80.2	17.0
140*	52.0	19.9
140+	---	16.3
150	247.0	34.0
150+	---	26.6

* Data from Reference 20 on Cellophane film
\+ Data from Reference 21 on cotton cloth

Using the Arrhenius Equation (Equation 4) to plot these values, permits calculation of the activation energy (E_a) by linear regression (Figure 7). The activation energy values are given in Table III.

Table III Activation Energies from Color Change and Ultra-Violet Absorption for Cellophane Film and Cotton Cloth

	Activation Energy (Kcal/mole) Short Term	Long Term
Color Change (Film)	22.1	24.3
UV Absorption (Film)*	23.0	23.6
Color Change (Cotton)+	----	23.5

* Data from Reference 20 on Cellophane film
\+ Data from Reference 21 on cotton cloth

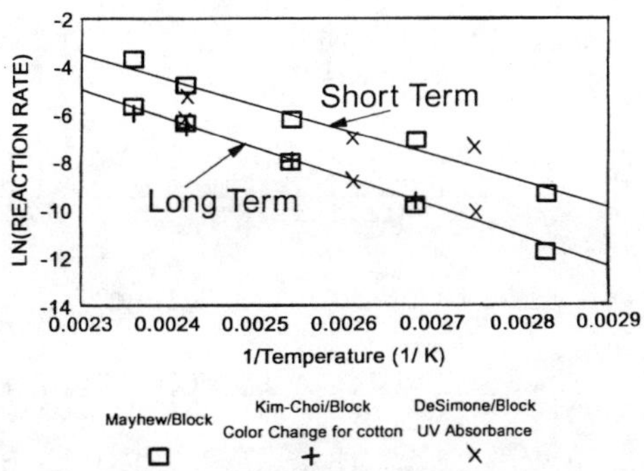

Figure 7. Arrhenius plot of color change and Ultra Violet absorption in cellophane film and cotton cloth. Square = this work. x = Ref.20, + = Ref. 21. - - - = long term, ---- = short term.

It should be noted that (1) the reaction rates for color change of both Cellophane film and cotton cloth are essentially the same (2) the reaction rate constants for UV absorption and for color change in Cellophane film are essentially the same, and (3) the activation energies calculated for these changes show no statistically significant difference at the 0.5 confidence level.

Furthermore, these results are in excellent agreement with those of Kim-Choi and Block [21] who also measured the tear strength of cotton fabric and with Roy [22] who examined the change in degree-of-polymerization of cotton fabric.

It should be noted that rinsing baked samples in distilled water reduced the overall color change. However, IR scans of selected rinsed samples that had all discernable color removed continued to show the presence of carbonyl species. In addition, the rinse solution had a noticeably sweet fragrance, indicating the presence of saccharides.

CONCLUSIONS

The Infrared spectra of Cellophane films baked at temperatures from 80 to 150 °C are qualitatively identical, demonstrating that the reaction mechanisms do not change over this temperature range.

The kinetics of color change of these films fit the standard first-order model, except that there are definite short-term and long-term reaction rates. This finding is in agreement with other reports on the behavior of cellulosic materials.

The reaction rates determined in this work for the color change of Cellophane film are in excellent agreement with those determined for both the color change of cotton and the UV absorbance of Cellophane under similar conditions.

The activation energies determined in this work are in good agreement with those obtained from studies of other chemical and physical properties of cellulosic materials.

REFERENCES

1. I. Block, ICOM Committee for Conservation, Copenhagen (1984) 84.9.7-84.9.11.

2. E. L. Graminski, E. J. Parks, and E. E. Toth, E. E. in Durability of Macromolecular Materials. (Am. Chem. Soc., Washington, D. C. 1979) 341-355.

3. J. C. Williams, C. S. Fowler, M. S. Lyon, and T. L. Merrill, in Preservation of Paper and Textiles of Historic and Artistic Value. edited by J. C. Williams (Am. Chem. Soc. Washington, D. C., 1977) 37-61.

4. R. H. Peters, Textile Chemistry, Volume I: The Chemistry of Fibres. (Elsevier Publishing Company, New York, 1963) 180.

5. S. P. Roland, in Applied Fibre Science, Volume 2. edited by F. Happey (Academic Press, New York, 1979) 207.

6. L. F. Mc Burney, in Cellulose and Cellulose Derivatives, 2nd edition. edited by E. Ott, H. M. Spurlin, and M. W. Grafflin (Interscience Publishers, Inc., New York, 1954).

7. T. P. Nevell, in Cellulose Chemistry and Its Applications. edited by T. P. Nevell, and S. H. Zeronian (John Wiley and Sons, New York, 1985).

8. A. Sharples, in Cellulose and Cellulose Derivatives, Part V. edited by N. M. Bikales, and L. Segal (John Wiley and Sons, Inc, New York, 1971).

9. S. H. Zeronian, in Cellulose Chemistry and Technology. edited by J. C. Arthur (Am. Chem. Soc., Washington, D. C., 1977).

10. F. Shafizadeh, W. DeGroot, and G. McGinnis, in Cellulose Chemistry and Its Applications. edited by T. P. Nevell, and S. H. Zeronian (John Wiley and Sons, New York, 1985) 266-289.

11. G, M. Barrow, Physical Chemistry, 4th Edition. (McGraw-Hill Book Co., New York, 1979).

12. D. A. Skoog, Principles of Instrumental Analysis, 3rd. Edition. (Saunders College Publishing, Philadelphia, 1985).

13. C. D. Craver, The Coblentz Society Desk Book of Infrared Spectra. (The Coblentz Society, Kirkwood, Mo, 1977).

14. R. M. Silverstein, G. C. Bassler, and T. C. Morrill, Spectrometric Identification of Organic Compounds, 4th Edition. (John Wiley and Sons, New York, 1981).

15. R. J. Berni, and N. C. Morris, in Analytical Methods for a Textile Laboratory, 2nd Edition. edited by J. W. Weaver, (Amer. Asso. of Tex. Chem. and Color., Research Triangle Park, NC 1968).

16. D.L. Pavia, G. M. Lampman, and G. S. Kriz, Introduction to Spectroscopy. (Saunders College Publishing, Philadelphia, 1979).

17. K. McLaren, in The Color Science of Dyes and Pigments. (Adam Hilger Ltd., Bristol, 1983) 142-144.

18. B. L. Browning, and W. A. Wink, TAPPI 51(4) (1968) 156-163.

19. R. G. Zhbankov, Infrared Spectra of Cellulose and Its Derivatives. Translated by A. B. Desham, (Consultant's Bureau, New York, 1966).

20. L. DeSimone, and I.Block, in Historic Textile and Paper Materials II. (Amer. Chem. Soc., Washington, D.C., 1989).

21. H. K. Kim-Choi, and I.Block, in Historic Textile and Paper Materials: Conservation and Characterization. (Amer. Chem. Soc., Washington, D.C. 1986) 411-426.

22. Roy, A. M. MS thesis, University of Maryland, 1985.

Editor's Note: Because the experimental work was conducted at very low RH (2% or less), several reviewers have cautioned against applying these results to aging which occurs at higher RH values.

PART III

Ancient Technology, Craft and the Role of Technology in Material Culture

PART III A

Craft Reconstruction From Evidence of Workshops, Industrial Debris and Artifacts

1. Production Events, Tools and Techniques

ANALYSIS OF USE WEAR ON CERAMIC POTTER'S TOOLS

ROBERT C. HENRICKSON
Conservation Analytical Laboratory, Smithsonian Institution, Washington, D.C. 20560

ABSTRACT

Analyses of the distribution and nature of abrasion on ceramic rings from the third millennium B.C. portion of the Godin III pottery assemblage in western Iran (2600-1400 B.C.) indicate that it represents use wear. Residual scars on pottery vessels and reconstruction of vessel forming sequences support identification of these rings as potter's tools. They provide interconnections among varied sizes of Buff Common ware vessels made using different forming and finishing methods. This strengthens the inference of workshop production by specialist potters who make the full range of Buff Common ware vessels.

INTRODUCTION

Analysis of abrasion traces on cylindrical ceramic rings (D=6-10 cm; H=1.5-3 cm) from the early phases of the Godin III assemblage of western Iran (2600-1400 BC) indicate that these represent use wear.[4-6, 8] This consists of scoring, extensive abrasion, and polishing, all restricted to the relatively sharp upper edge and adjacent interior surface of the profile. Macroscopic and low-power microscopic examination of the nature, distribution, and orientation of the use wear, and details of the edge bevelling clarify the probable methods of use. Residual scars on Godin III Buff Common ware vessels, together with reconstruction of forming and finishing sequences, support the inference that ceramic rings were potters' tools, one type in a range of scrapers used in pottery production. Scraper rings provide a technological link between the smaller wheelmade and larger handbuilt Buff Common ware vessels of the Godin III assemblage, and support inference of workshop production.

Eight fragmentary ceramic rings were recovered from well-stratified Godin III period contexts (2600-1400 B.C.) at the site of Godin Tepe in the Kangavar Valley of the central Zagros highlands of Iran (Table I).[4,5] Seven of the eight come from Godin III:6 (2600-2300 B.C.) contexts.

DESCRIPTION OF THE GODIN III CERAMIC RINGS (Figures 1-3; Table I)

The Godin III ceramic rings from Godin Tepe are 7-10 cm in diameter, with a height of 2.5-4.0 cm (Table I, Figures 1-2). The edge with the lesser diameter is relatively thin and pointed and usually shows evidence of abrasion; the other edge is thickened and blunt (Figure 3). The production sequence and method of use define the 'upper' and 'lower' edges of these rings. In addition to surface marks ('wheelmarks' or 'throwing marks'), the diagonal orientation of inclusions and voids within the fabric shown by xeroradiography indicate that rings were thrown on a potter's wheel (Figures 1-2).[9]

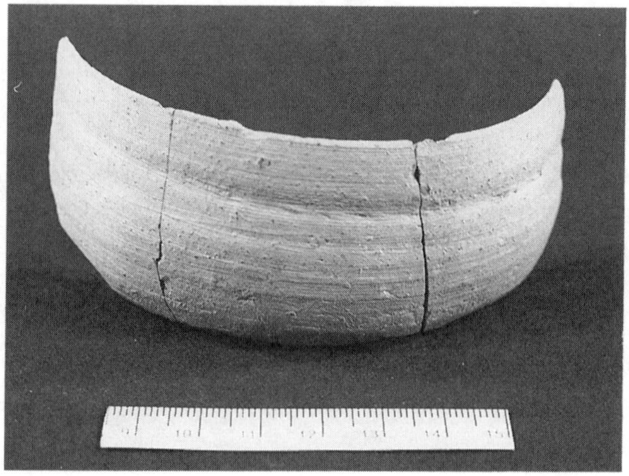

Figure 1a. Exterior view of Ring 71-2056 (Type 1).

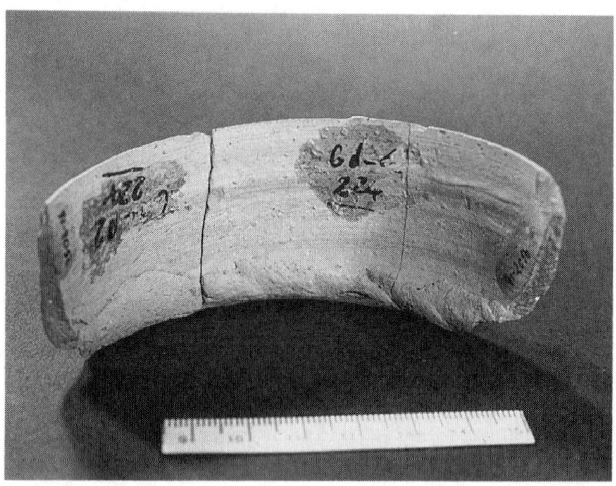

Figure 1b. Interior view of Ring 71-2056 (Type 1).

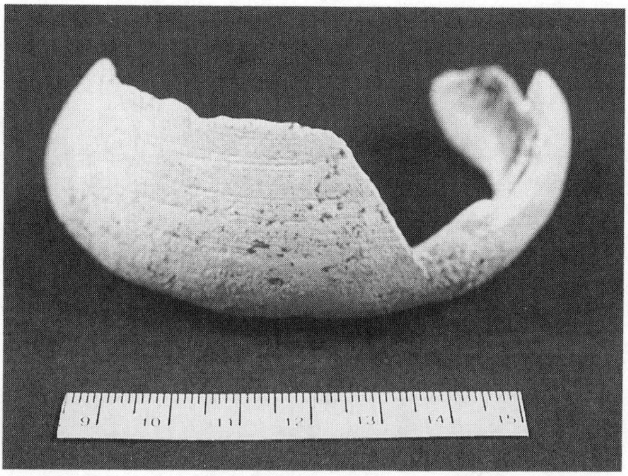

Figure 2a. Exterior view of Ring 71-2330 (Type 2).

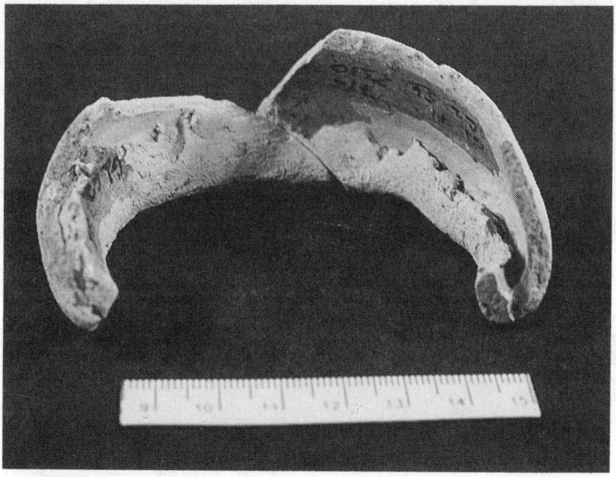

Figure 2b. Interior view of Ring 71-2330 (Type 2).

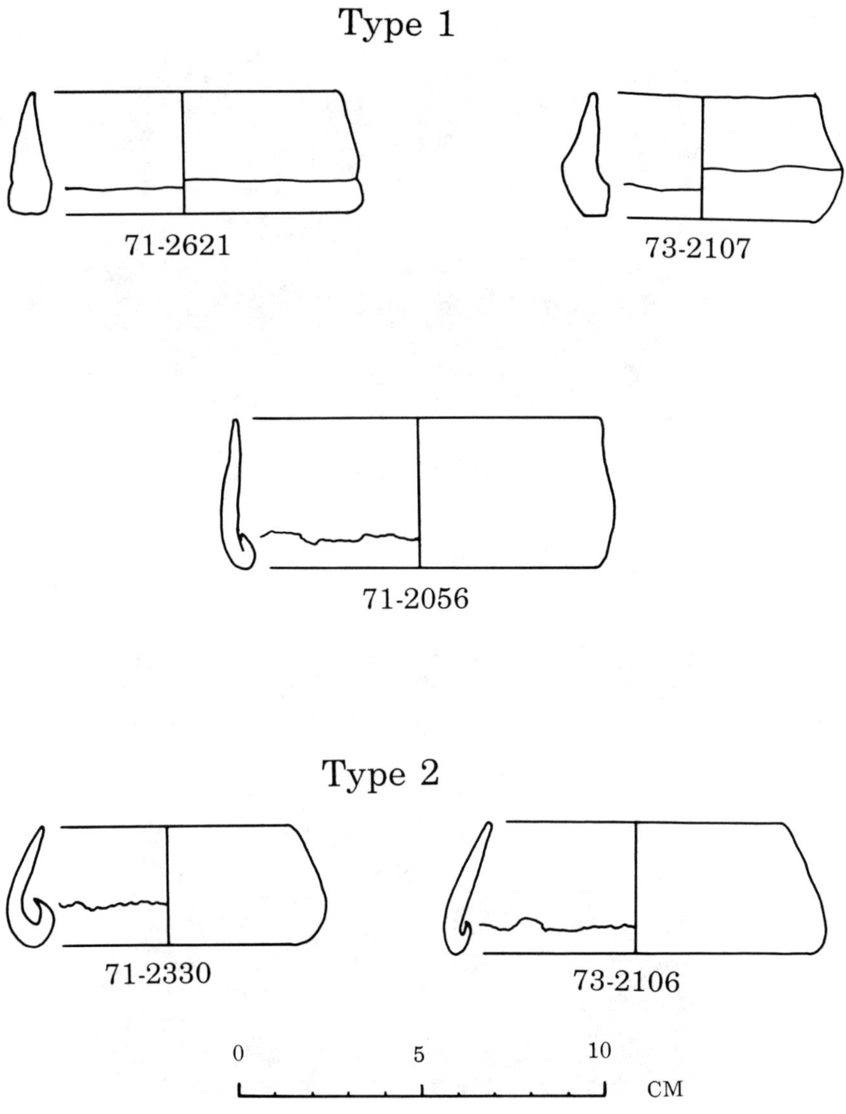

Figure 3. Drawings of Type 1 and Type 2 Rings.

Table I: Descriptions, Use Wear, and Proveniences of Ceramic Ring Scrapers from Godin Tepe

Godin No.	Type	Upper Diam.	Lower Diam.	Ht.	% of Circ	Location and Nature of Use Wear	Level	Area	Context
71-2056	1	10	10	3.8	50%	UPPER: Bevelled, radial scoring INTERIOR: Polished, vertical scoring	III:6-III:5	6168/5037	Wash
73-2107	1	6	8	3.5	25%	UPPER: Bevelled, radial scoring, chipped INTERIOR: Chipped, polished	III:6	6051	Collapse /wash
71-2621	1	8	9	3.4	15%	UPPER: Radial scoring INTERIOR: Heavily chipped	III:6	6093	Wall
69-2585	2	7	9	2.2	20%	UPPER: Light abrasion INTERIOR: Light polish	III:4-III:3	4048/3010	Trash pit
71-2330	2	7	9	3.3	75%	UPPER: Worn INTERIOR: Chipped, worn	III:6/III:4	6166/4056	Wash/floor
71-2405	2	5	8	2.8	--	(Sherd unavailable)	III:6	6018	Trash/collapse
71-2702	2	±6	9	2.5	15%	UPPER: Edge broken away	III:6	6085	Wash/floor
73-2106	2	8	10	3.5	33%	No wear identifiable; overfired	III:6	6033	Collapse /wash

Basic forming begins with throwing a hollow cylinder or truncated cone 7-10 cm in diameter and at least 4-5 cm high. Two general types of rings may be distinguished as extremes of a typological continuum. *Type 1* (Figures 1, 3) is made by cutting off a vertical-sided ring, 2.5 to 4 cm tall, whose wall thickness increases below a pointed vertical rim, yielding a roughly triangular cross-section. The lower portion of the ring is then trimmed or squared with a sharp straight-edged tool. *Type 2* (Figures 2, 3) has relatively thin walls of uniform thickness which curve inward markedly toward the sharp upper edge. A ring segment 3 to 4 cm high is cut off and the lower cut edge folded inward to strengthen the base of the ring. The paste varies from having no visible temper to medium grit tempered (grain diameter $\leq$ 0.1-0.2 mm), similar in range to clay used for producing Godin III Buff Common ware vessels. The rings range from well-fired to semi-vitrified, representing firing temperatures of 800-1100°C, comparable to those for Buff Common ware vessels and suggesting kiln firing.[9]

The typological range of Godin III ceramic rings is similar to that of earlier rings, which date primarily to the Late Uruk (later fourth millennium B.C.) and continue into the first half of the third millennium B.C.[3,9] Alden has identified the Late Uruk rings as potters' scraping tools used in mass-production of pottery.[1-3]

ABRASION (USE WEAR) ON GODIN III CERAMIC RINGS (Figures 4-5; Table I)

Abrasion is restricted to the upper edge itself and the adjacent inner face of the Godin III rings. Its consistent location and nature suggest it is an integral byproduct of their use.[15] This is the primary evidence for their original purpose and method of use. Alden uses the wear on the upper edge of Late Uruk rings (late fourth millennium BC) as his principal evidence for identifying them as scrapers.[3]

The two types of rings defined above have consistent but somewhat individual patterns of abrasion (Table I). The lack of well-developed abrasion on the Type 2 rings may produce some of the ambiguity and apparent differences in use. Nonetheless, the differences in profile (angle of wall relative to ring axis) and the relation of the abraded working edge to the plane of the upper edge (bevelled vs flat) suggest that Type 1 and Type 2 rings may have had somewhat different types or loci of use.

Type 1 Rings (Figures 1, 3, 4-5)

The best preserved ring from Godin Tepe has extensive and well-developed abrasion on and adjacent to the upper edge (Ring Gd 71-2056; Figures 1, 4-5). It combines elements of both Types 1 and 2, but the vertical profile and pattern of use wear correspond better to Type 1. Abrasion has bevelled the upper edge at a slight angle (approximately 10-15 degrees) to the plane defined by the upper edge of the ring (Figures 4a, 5a). The *inner* face just below is scored and polished (Figures 4b, 5b); the outer face shows no abrasion. This combination of probable use wear indicates that this is the working edge.

On the top of the working edge itself, fine scratches and pedestalled harder inclusions protecting their elongated 'shadows' or 'drumlins'[17] of softer clay matrix are all essentially perpendicular to the curve of the working edge (Figures 4b, 5b). This indicates radial movement of clay across the cutting edge.[3,17] Since the radial scores on the working surface itself overlap one another at small angles (Figure 4a,

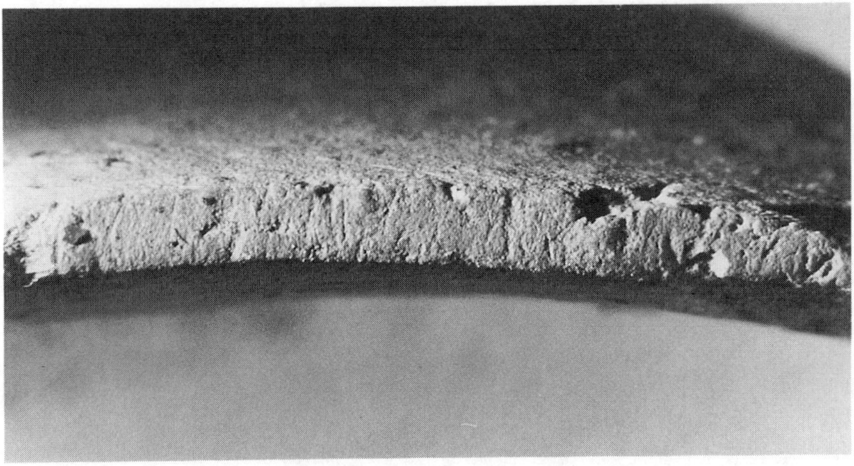

Figure 4a. Use wear on Ring 71-2056:
Top view of radial scoring.

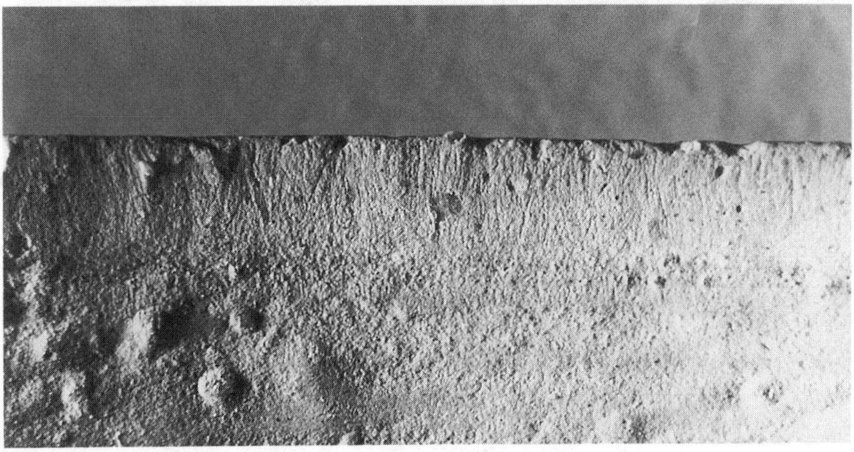

Figure 4b. Use wear on Ring 71-2056:
Scoring and 'drumlins' on interior face at upper edge.

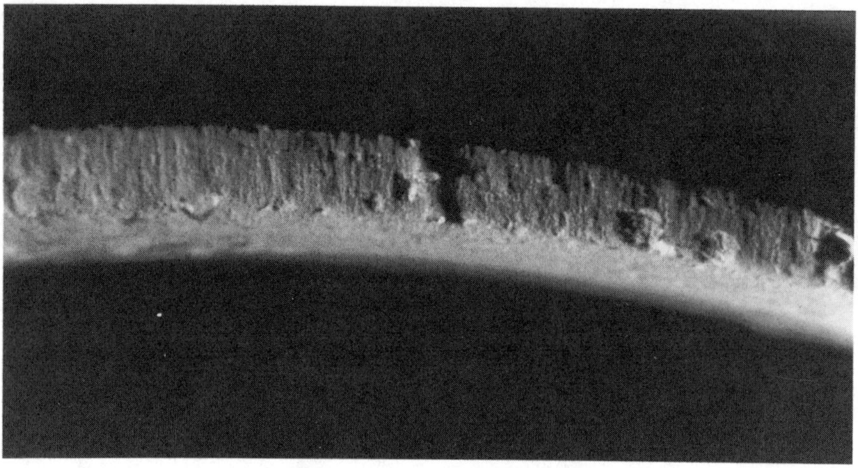

Figure 5a. Use wear on Ring 71-2056:
Top view of radial scoring and 'drumlins'.

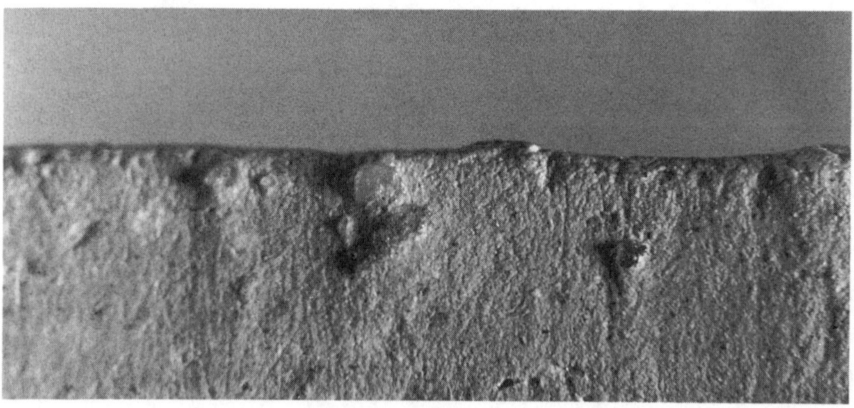

Figure 5b. Use wear on Ring 71-2056:
Scoring and 'drumlins' with inclusions
on interior face at upper edge.

5a), even in small segments of the working edge, they suggest repeated yet episodic uses of the scraper.

Judging from the bevelling of the edge, the ring was held with the *plane of the working edge* at a shallow angle, perhaps 10-15 degrees, to the surface being scraped and pulled *unidirectionally* toward the user. The axis of the ring was thus not quite perpendicular to the surface scraped. Only this stance would yield the bevelling found around the entire preserved circumference of the ring. Its uniformity around the entire (preserved) circumference must result from episodic uses of the ring, with slightly different but overlapping segments of the ring edge being used for scraping each time. This would also yield the observed criss-cross of radial scoring. Consistent use of only part of the circumference would result in differential wear, either yielding subtle scalloping or overall inclination of the working edge relative to the base (cf. Figures 6-7).

Polishing and scoring is found *only* on the 2-3 mm of the *inner* face *immediately* below the working edge. This provides further evidence for the ring having been held at an angle to the worked surface and drawn unidirectionally. Linear scratches and the elongated axes of the clay matrix 'drumlins' around the hard inclusions are all *perpendicular* to the plane of the working edge (Figures 4b, 5b). These result from the movement of waste clay away from the cutting edge along the inner surface of the ring. A canted ring stance would encourage the waste shaved clay to slide upward and away from the working edge, abrading the inner surface adjacent to the cutting edge as it passed.

The absence of use wear on the *exterior* surface of the ring at or near the working edge also indicates *unidirectional* scraping, with the ring held at an angle to the surface and *drawn toward the user*. The curve of the cutting edge opens toward the direction of pull. Any *pushing* would yield abrasion or polish on the *exterior* surface adjacent to the cutting edge, as clay was bulldozed ahead of the ring.

Reciprocal motion, with the ring held flat to the worked surface and moved back and forth, would leave abrasion in two diagnostic loci. The cutting edge would be flattened in the plane of the ring's upper surface, not bevelled at an angle. Polishing and scoring would occur on both the interior *and* exterior faces of the ring adjacent to the working edge. Neither is found, so again a unidirectional pulling motion is likely.

Other Type 1 rings are very fragmentary but have the same general pattern of abrasion: rounding or double bevelling on the edge and polishing on the interior surface (Table I; Figure 3).

The use wear on Type 1 vertical-sided scraper rings indicates that the tool was held with the working edge at a relatively shallow angle to the surface scraped and drawn unidirectionally toward the user. Holding a *Type 1* ring at a shallow angle would both *increase* the width while slightly *reducing the depth* of the working edge's bite, thereby making a wider yet thinner cut of more uniform thickness. This effect is more pronounced on *concave* rather than flat or convex surfaces. Convex, flat, and concave surfaces, and the radius of the curve relative to that of the ring, would yield differing types of use wear (see below and Figures 6-7).

Type 2 Rings

Type 2 rings, in contrast, have a strongly incurved profile and a somewhat different pattern of use wear. This profile would provide a relatively shallow bite on a concave (interior) surface, but this might be desirable on medium-sized vessels (see below). This also lessens the need for an angled stance during use in order to allow movement of waste clay away from the working edge, in contrast to the straight-walled Type 1. The limited use wear on well-preserved on Type 2 rings, and the small fragments with well-developed and uneven use wear, make it difficult to determine whether the use wear forms a bevel or lies in the plane of the upper edge. If the use wear indeed lies in the plane of the upper edge, the Type 2 rings might have been used on the convex exterior surfaces of vessel, but there is no direct surviving evidence on the exterior of vessels for such a use. The additional absence of well-developed abrasion (use wear) on either the exterior or interior surfaces of the Type 2 rings adjacent to the working edge leaves some question as to their method of use.

Type 2 rings are similar in both form and general patterning of use wear to the most of the Late Uruk scrapers which Alden argued were used on "the outsides of leather-hard pots",[3] vessels of sizes comparable to those scraped in the Godin III assemblage.

CERAMIC RINGS AND THEIR USE IN THE GODIN III TRADITION (Table II; Figures 6-10)

The varied residual traces and scars left on both the interior and exterior surfaces of Godin III pottery vessels by forming and finishing processes clearly demonstrate use of a wide variety of scrapers and other tools in the course of pottery production. Among the varied surviving scraping scars and traces,[4,8] some may reasonably be associated with ring scrapers.

Table II: Surface Markings on Interiors of Godin III Buff Common Ware Vessels

	VESSEL SIZE			
PROFILE	SMALL < 16-18 cm	MEDIUM 18-35 cm	LARGE 35-55 cm	OVERSIZE > 55 cm
Rim	Wheelmarks	Wheelfinish	Wheelfinish	Wheelfinish
Neck	Wheelmarks	Wheelfinish	Wheelfinish	Wheelfinish
Shoulder	Wheelmarks	Wheelfinish	Wheelfinish	Wheelfinish
Lower Body	Wheelmarks	**Scraped**	**Scraped**	**Scraped**
Base	Wheelmarks	**Scraped**	**Scraped**	**Scraped**

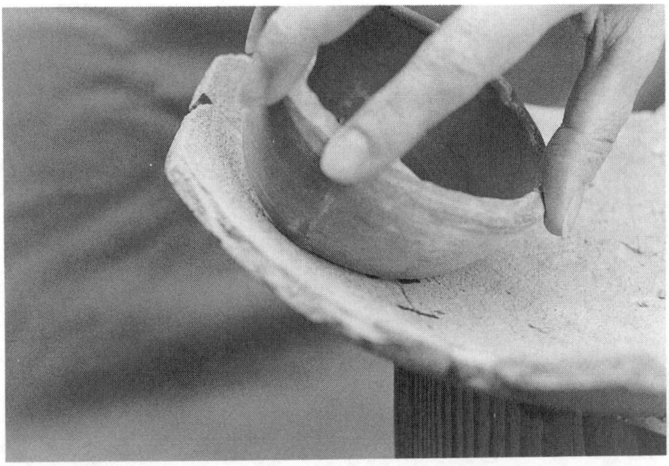

Figure 6a. Replicated ring, with an upper (working) edge diameter of 7 cm, held at a shallow angle to the *concave interior* lower body surface of a MEDIUM vessel sherd (diameter 30-35 cm), showing reconstructed use of a ring scraper.

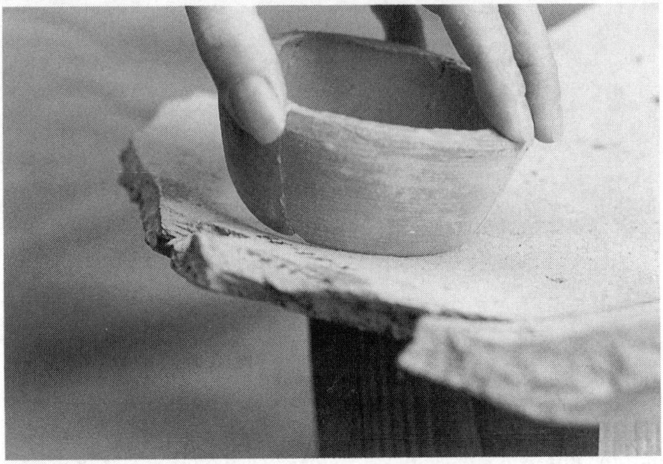

Figure 6b. Replicated ring, with an upper (working) edge diameter of 7 cm, held at a shallow angle to the *concave interior* lower body surface of a LARGE vessel sherd (diameter ca. 50 cm), showing reconstructed use of a ring scraper.

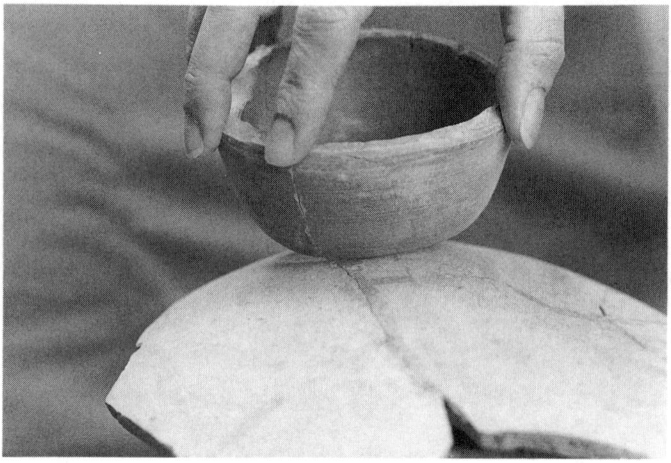

Figure 7a. Replicated ring, with an upper (working) edge diameter of 7 cm, held at a shallow angle to the *convex exterior* lower body surface of a MEDIUM vessel sherd (diameter 30-35 cm), showing limited contact and concentration of potential use wear.

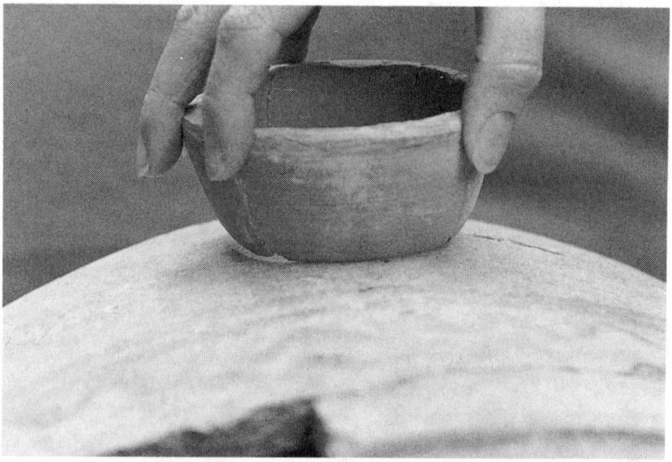

Figure 7b. Replicated ring, with an upper (working) edge diameter of 7 cm, held at a shallow angle to the *convex exterior* lower body surface of a LARGE vessel sherd (diameter ca. 50 cm), show potential degree of contact for use wear. No evidence of such exterior scraping is found on any large or oversize vessel.

Scraping scars and traces occur predominantly on the *interior* of the lower body of *medium-sized and larger* Godin III painted Buff Common ware vessels (Table II). Basic Buff Common ware vessel forms are few in number but have a wide range of sizes.[4,5] The predominant form is a carinated restricted vessel with varied profiles and proportions (Figures 8-9). The other relatively common vessel form comprises bowls with a rounded to carinated profile (Figure 8). Relatively uncommon forms, such as cylindrical beakers, also were scraped.[5] In Godin III:6-5, Buff Common ware vessel types included oversize ovoid storage jars with 'cable' decoration (Figure 10). Through the third millennium B.C. (Godin III:6-4 [2600-1900 B.C.]) Buff Common ware constituted a major though decreasing proportion of the assemblage (68% - 30%). No other contemporary ware (Buff Coarse, Grey-black, Red Slipped) preserves evidence for scraping.[5-7]

Scars preserve the curvature of the scraper on a limited number of Buff Common (painted) ware vessels from Godin III:6-2. Fitting trial templates of differing radii held at various angles to these scars yields a *4-6 cm radius of curvature, identical to that of the cutting edge of the ceramic rings.* Taken together with the evidence of edge wear on the rings themselves, the evidence for the rings' use as scrapers is strong.

Given that all of the abundant residual traces left by scraping are on the concave interior of the lower body of Godin III Buff Common ware vessels, a scraper would not only have to have a curved cutting edge with a radius smaller than that of the vessel wall but also be held at an angle (Figures 6-7). This would:

1) yield the relatively wide and shallow cuts seen on vessel interiors (Figures 9-10),
2) tend to bevel the ring's cutting edge, as is seen on at least Type 1 rings (Figures 1, 4-5), and
3) produce the use wear (radial scoring and drumlins) observed on the cutting edge and adjacent interior surface of the rings (Figures 4-5).

Use on an *concave* (curved or semi-spherical) interior surface has two advantages. First, holding a ring at a slight angle to the surface matches the ring's curve more closely to the larger curve of the vessel, yielding a relatively broad and shallow cut rather than a gouge (Figure 6-7). The ring could not act as scraper if its axis were perpendicular to a (concave) curved vessel wall; the sides rather than the cutting curve would touch the vessel wall. Second, interior scraping scars may be left unfinished without aesthetic problems, particularly on restricted mouth vessels (the most common form), yielding a small production efficiency. Indeed, scraping scars were left relatively unfinished even on the interior of Godin III Buff Common ware bowls (Figure 8).

Scrapers were clearly *not* used on the exterior of vessels whose interior was scraped, except perhaps at the base (Henrickson 1991). The exterior surface of the body of medium and larger vessels, from shoulder to near the base, often retains light linear impressions of string(?) bindings used to strengthen and support the leather-hard pots as they dried.[4,9] Since these imprints survive, no scraping was done on that part of the vessel. The copiously documented interior scraping obviates the need for further external scraping to reduce wall thickness. The *exterior* of the base of Buff Common ware vessels, however, may have been reworked slightly after primary vessel forming, to finish shaping the base. Interior scraping took place after the base profile was adjusted, since the interior scraped surfaces have continuous and undistorted curves. Although no trace of scraping remains on the outside of the base on any Buff Common ware vessel to confirm such a process, a smoothed finish is desirable there.

Ceramic rings were only one of a wide variety of scrapers used in the Godin III and other assemblages, including ones made of various materials with differing profiles, edges, and methods of use, such as worked sherds.[13,14,16] In the more than a millennium-long Godin III tradition the types and details of such tools may well have changed with time.

IMPLICATIONS FOR ORGANIZATION OF GODIN III POTTERY PRODUCTION

The use of scrapers, particularly ceramic rings, offers new insight into the organization of Godin III Buff Common ware production and contributes to definition of its technological style.[12] The rings are a wheel-thrown tool used exclusively in the latter stages of forming and finishing hand-formed vessels. Rings thus provide support for production of the full size ranges of all Buff Common ware vessel types by the same potters. This reinforces the inference of specialist production first based on stylistic analysis of the vessel decoration and shape[4,5], and later on neutron activation analysis.[10]

The basic vessel forms in Buff Common ware have consistent interrelationships among vessel size, method of forming, and presence/absence of scraping. Each vessel type and size has a characteristic forming and finishing sequence[8,10]; bowls are made like the lower bodies of equivalent size carinated restricted vessels.[7] Scraping scars are found exclusively on the interior of medium and larger sized Buff Common ware vessels (maximum diameter of $\geq$16-20 cm). This size-dependent dichotomy corresponds to the technological breakpoint separating true wheelmade (thrown) from hand-formed but wheel-finished Godin III vessels.[4,7-9] As discussed above, rings are thrown on a potter's wheel, using paste comparable to that used for all sizes of Buff Common ware vessels. Ceramic ring scrapers thus link pottery production based on true throwing on a potter's wheel with that based on wheel-finishing of handformed vessels.

Small Buff Common ware vessels (maximum diameter and height <20 cm) of all types are thrown on a true potter's wheel. With few exceptions, these are not scraped or smoothed internally, so their relatively thin walls retain the interior wheelmarks left by throwing.[4,8] The rings' diameters of 8-11 cm fall in the mid to upper range of neck and rim diameters of Godin III vessels thrown on a potter's wheel.[4,5,8,10]

Medium and larger vessels (maximum diameter >20-25 cm) in Buff Common ware are hand-formed (coiled, coiled and drawn, perhaps slab-built, or some combination) and wheel-finished.[4,7,8,10,11,13] For restricted vessels (rim and neck less than carination or maximum diameter), a minimum neck inner diameter >12 cm is large enough to admit a ring, especially when allowance is made for subsequent shrinkage during the drying of the vessels; scraping a bowl interior offers no problems of access. Scraping strokes on the interior of medium vessels tend to be roughly horizontal; on large and oversize vessels they tend to be vertical (Figures 9-10).[4,8]

The medium and larger vessels, whether carinated pots or bowls, would need a relatively thick-walled lower body to resist stresses during forming and to support the weight of the vessel's upper body until drying to leather hardness stiffened the

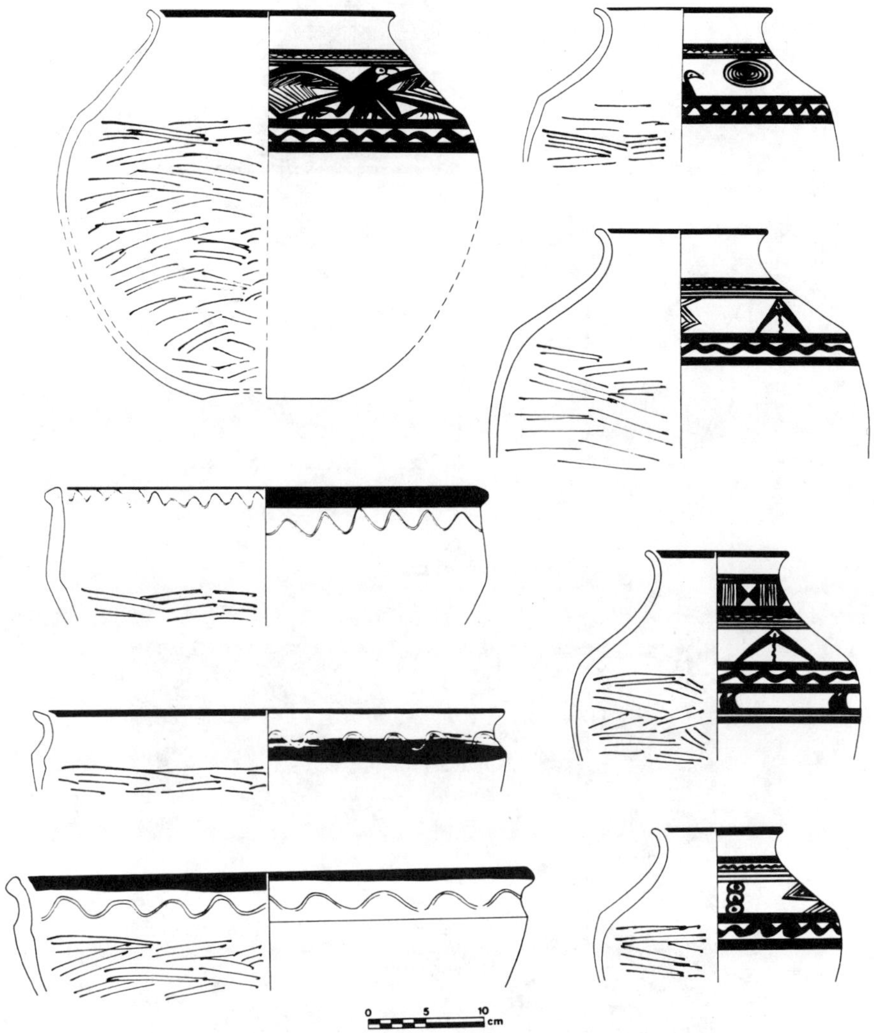

Figure 8. Medium Godin III:6-5 Buff Common ware carinated restricted vessels and bowls, with area of scraping on medium vessel lower interiors marked (note thinning of profile).

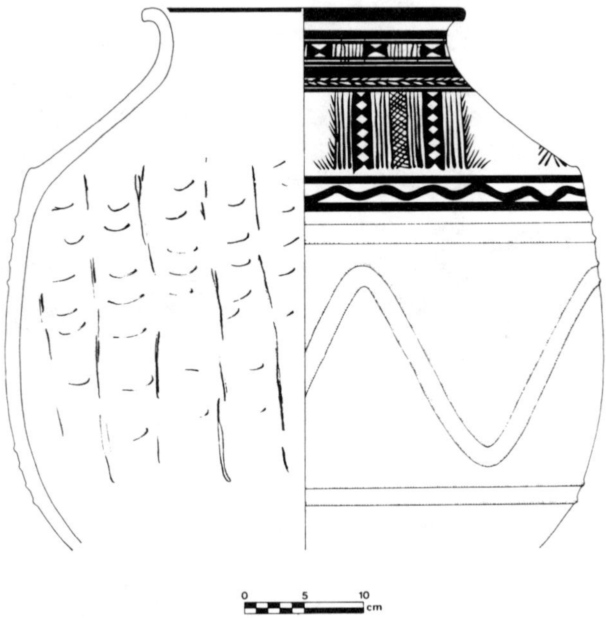

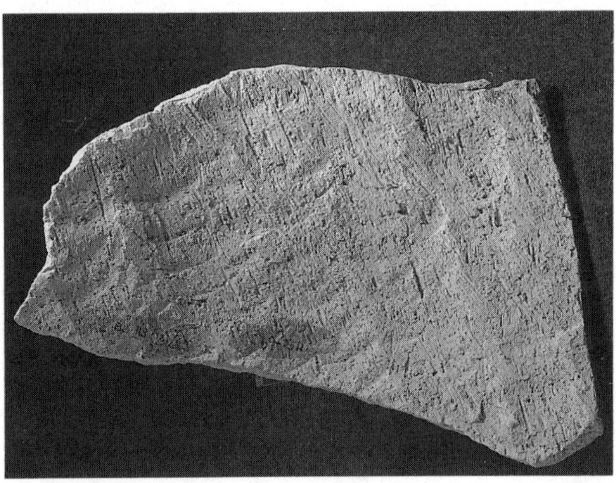

Figure 9. Drawing of a large Godin III:4 Buff Common ware vessel (Gd 69-2511) and photographs of portions of the interior of the vessel with probable ring scraper scars just below the carination and on the lower body.

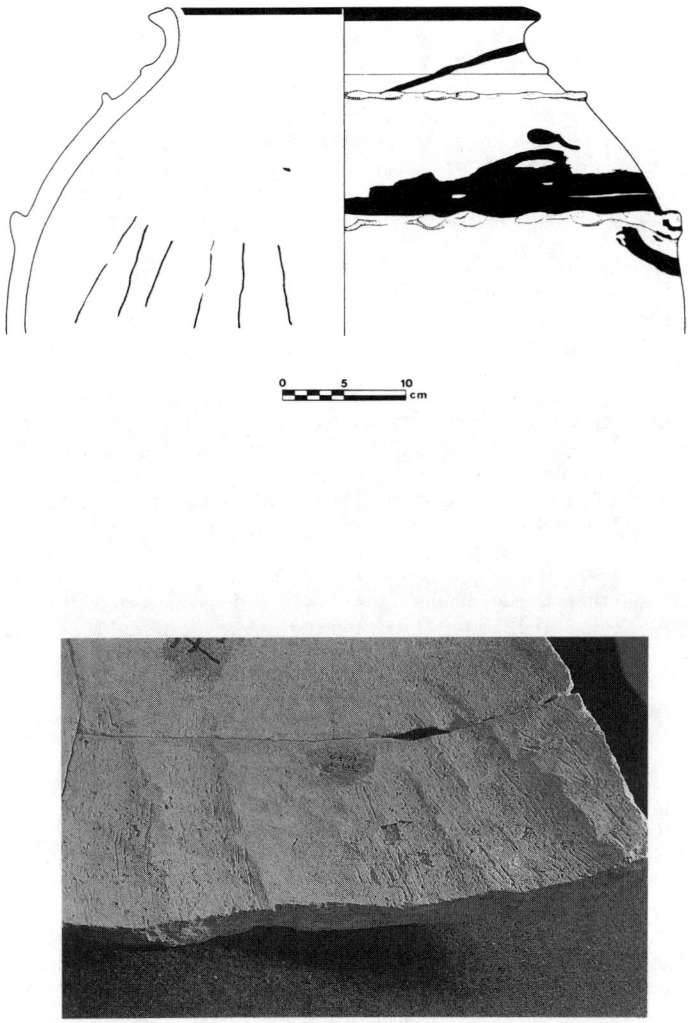

Figure 10. Drawing of an oversize Godin III:6 ovoid storage jar with 'cable' decoration (Gd 73-2018) and a photograph of interior scraping scars on the lower shoulder.

clay. After that, thick vessel walls become more a liability. Scraping to reduce lower body wall thickness would facilitate final drying and firing, and reduce vessel weight.

Varied interior scraping scars (relatively smooth to facetted or gouged) extend from the base to the lower edge of the shoulder on medium and larger carinated vessels and to the reversal of curve on bowls (Figures 8-10).[8,16] Since preserved Godin III scraping scars are on the *interior*, scraping was not used to modify gross vessel shape, except perhaps to adjust the exterior base profile (see above).[13] This interior scraping sometimes results in excessive thinning of the lower body vessel wall. In vessels whose maximum diameter is approximately 30 cm, the lower body may have a thickness of only 0.3-0.4 cm or less, while the upper body (above the carination, whether pot, jar, or bowl) is usually at least 1.0 cm thick and the intermediate carination 1.0-1.5 cm or more. Such thinning would not contribute to vessel durability.

The *formal* style of Godin III Buff Common ware painted decoration (essentially all Buff Common ware has painted decoration)[5,7], with its persistent coherent structure related to vessel form and its distinctive component motifs, provides interconnections among the small, medium, large, and oversize carinated vessels. The fundamental structure of the decoration, the motif combinations used in secondary zones, and many of focal motifs and motif sequences in the main zone of decoration crosscut size groups and associated forming methods.[4-6]

Neutron activation study of Buff Common from Godin III:5 (2300-2100 BC) and III:4 (2100-1900 BC) indicates that paste groups also crosscut vessel size groups and their distinctive forming methods. The use of very similar pastes for vessels of varied sizes and forming methods suggests production within a single workshop, replicating the stylistic results which also crosscut size and forming groups; even the pastes of oversize ovoid storage jars (e.g., Figure 10) are included.[10]

Taken together, these various lines of evidence all support the inference that a relatively few (specialist) potters produced the *full range* of vessel sizes and types in Godin III Buff Common ware.[4,8]

SUMMARY

Analysis of use wear on ceramic rings, residual scars on pottery vessels, and reconstruction of vessel production sequences together demonstrate that ceramic rings were potters' tools. Scraping is an integral part of the technological style of the Godin III (painted) Buff Common ware production tradition. Small Godin III Buff Common ware vessels and ceramic rings are thrown but not scraped. Larger vessels, however, are hand-formed and wheel-finished. Reducing lower body wall thickness by scraping was an essential step in the later stages of their forming and finishing. The varied types of residual interior scars include one variety with a radius of curvature similar to that of the ceramic rings. Ring scrapers and their use wear are appropriate for use on concave (i.e., interior) surfaces. The wheel-thrown ring scrapers, and scraping in general, provide technological links among the various types and sizes of Godin III Buff Common painted ware vessels made using combinations of hand-forming and wheel-based techniques. Combined with the interconnections based on stylistic analysis of painted decoration and the paste characterization by neutron activation analysis, the proposal of specialist production of the full range of Buff Common ware pottery in workshops is strengthened.

ACKNOWLEDGEMENTS

Particular thanks are due to T. Cuyler Young, Jr. (Royal Ontario Museum, Toronto [ROM]), for continuing support in my work on Godin III, and to W. Pratt (ROM) for aid in long-distance access to collections. P. B. Vandiver, M. J. Blackman, M. Aronson, W. M. Sumner, M. Hegmon, and E. F. Henrickson have commented on various portions or drafts of this paper. Some of the research and writing was done while I was a Post-doctoral Research Fellow in Materials Analysis at the Conservation Analytical Laboratory (CAL) of the Smithsonian Institution, whose facilities and staff have aided this study. P. B. Vandiver (CAL) threw the ring used in Figures 6-7. C. Tumosa (CAL) did the xeroradiography on the Godin ceramic rings. All interpretation and errors remain solely my responsibility.

REFERENCES

[1] J. R. Alden, PhD dissertation, University of Michigan, 1979.

[2] J. R. Alden, in Contexts for Prehistoric Exchange, edited by T. Earle and J. Ericson (Academic Press, New York, 1982), pp. 83-101.

[3] J. R. Alden, Paléorient 14(1): 143-150 (1988).

[4] R. C. Henrickson, in Technology and Style, edited by D. Kingery (Ceramics and Civilization II, Amer. Cer. Soc., Columbus, OH, 1986), pp. 53-85.

[5] R. C. Henrickson, Iranica Antiqua 22, 33-115 (1987).

[6] R. C. Henrickson, in The Archaeology of Western Iran: Settlement and Society from Prehistory to the Islamic Conquest, edited by F. Hole (Smithsonian Institution Press, Washington, 1987), pp. 205-227.

[7] R. C. Henrickson, in Cross-Craft and Cross-Cultural Interaction in Ceramics, edited by M. R. Notis and P. E. McGovern (Ceramics and Civilization IV, Amer. Cer. Soc., Westerville, OH, 1989), pp. 81-146.

[8] R. C. Henrickson, in Materials Issues in Art and Archaeology II, edited by P.B. Vandiver, J.R. Druzik, and G. Wheeler (Mater. Res. Soc. Proc. 185, Pittsburgh, 1991), pp. 523-541.

[9] R. C. Henrickson, manuscript in preparation.

[10] R. C. Henrickson and M. J. Blackman, in Chemical Characterization of Ceramic Pastes in Archaeology, edited by H. Neff (Prehistory Press, Madison, WI, 1992), pp. 125-144.

[11] R. Johnston, in Material Culture: Styles, Organization, and Dynamics of Technology, edited by H. Lechtman and R. S. Merrill (1975 Proc. of the Amer. Ethnological Soc., 1977), pp. 169-210.

[12] H. Lechtman, in Material Culture: Styles, Organization, and Dynamics of Technology, edited by H. Lechtman and R. S. Merrill. (1975 Proc. of the Amer. Ethnological Soc., 1977), pp. 3-20.

[13] O. S. Rye, Pottery Technology: Principles and Reconstruction (Taraxacum, Washington, 1981).

[14] A. O. Shepard, Ceramics for the Archaeologist Sixth Printing (Carnegie Institution of Washington, 1968).

[15] M. B. Schiffer and J. M. Skibo, American Anthropologist 91, 101-115 (1989).

[16] A. van As, in Pottery Technology: Ideas and Approaches, edited by G. Bronitsky (Westview, Boulder, CO, 1989), pp. 41-79.

[17] M. R. Waterworth and E. Blinman, Pottery Southwest 13(2), 4-7 (1986).

A new look at stone drills of the Indus Valley Tradition

J. Mark Kenoyer* and Massimo Vidale**

*J. Mark Kenoyer
Department of Anthropology, University of Wisconsin, Madison
**Massimo Vidale
Is. M. E. O. Rome, and Smithsonian Institution, Washington, D. C.

Abstract

Drilling technology of the Indus Valley Tradition was highly specialized and various types of chert and jasper were used to drill different types of materials. Earlier studies used primarily macroscopic observations to define features such as the manufacturing technique of drills, the raw materials and the mechanics of drilling. These generalizations can be revised given the discovery of important workshop areas and the availability of SEM, XRD and electron microprobe analysis. This paper will summarize the current state of drilling research and define two categories of drills that were used in antiquity; *tapered cylindrical drills* and *constricted cylindrical drills.* Directions for future research on the relationship between drilling and other contemporaneous technologies are discussed.

Introduction

Drilling technologies in general and, more specifically, the perforation of hard stones have been a major topic of research in South and West Asia since the discovery of the ancient Indus Valley Civilization in the 1920s and 30s (Figure 1). The urban phase of this culture dates from between 2600 B. C. to 1900 B.C., and has technological roots that can be traced back to 6500 B. C. in the early Neolithic period [1, 2]. Because of the fact that most of the tools, raw materials and manufacturing residues of ancient bead makers are preserved archaeologically, this craft may be efficiently used by archaeologists for reconstructing important aspects of the ancient organization of production [3].

In this article we will examine some of the previous research on prehistoric bead drilling in West and South Asia and propose new terminologies for discussing types of drills used for perforating stone beads during the long history of the Indus Valley Tradition (+6500 B. C. to 1300 B.C.). Two new terms are introduced to differentiate specific varieties of cylindrical drills with dimple impressions at the distal tip. The term *tapered cylindrical drill* refers to drills made on blades that were generally used for the decoration of soft stones and the perforation of short beads made from harder stone (Figure 2). Tapered cylindrical drills have a long historical use and are found in a wide range of cultural contexts spanning the entire region from the Mediterranean to the Indus Valley. In contrast, *constricted cylindrical drills* (Figure 3) are a unique form of standardized and specialized tool

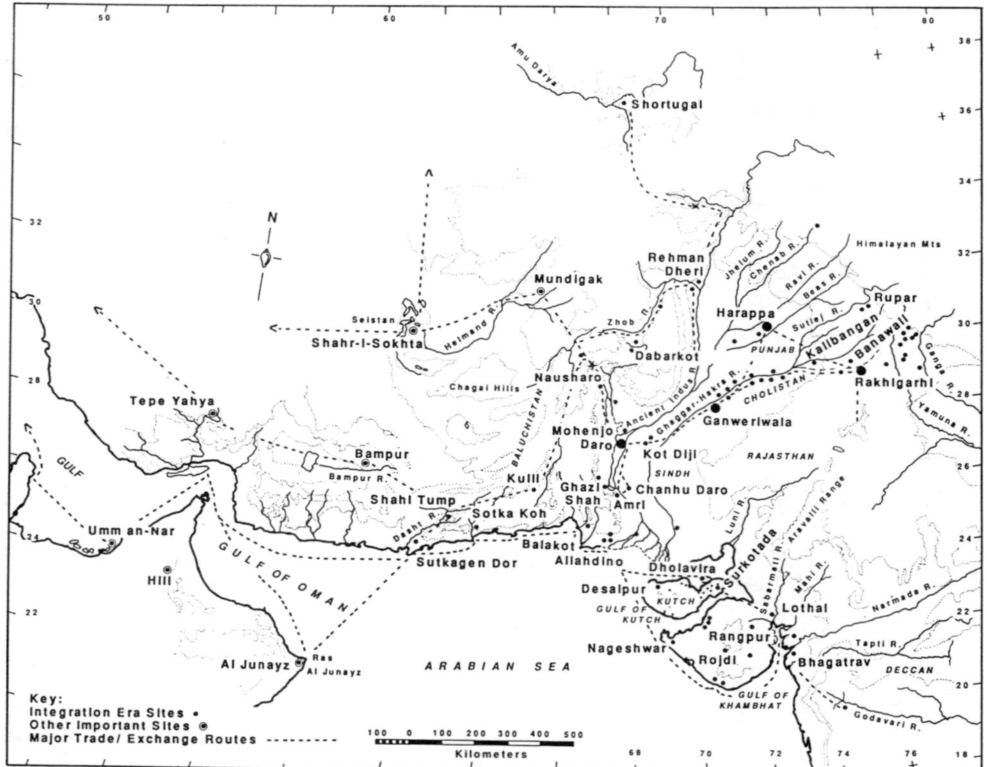

Figure 1. Major Sites of the Indus Valley Tradition, Integration Era, Harappan Phase.

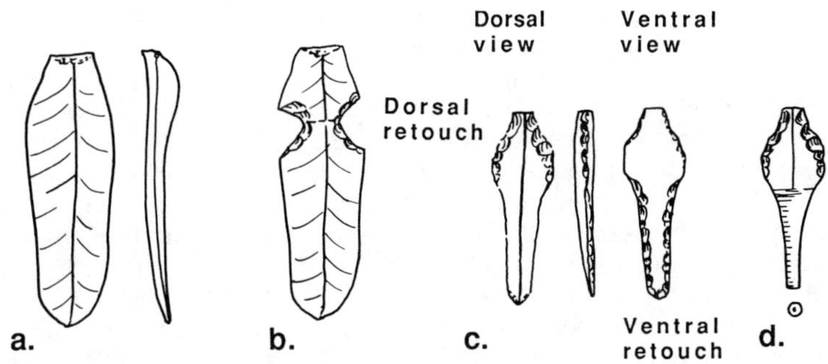

Figure 2. Tapered Cylindrical Drill Manufacture. **a.** unmodified blade; **b.** notched blade; **c.** finished drill; **d.** used drill with tapered cylindrical form.

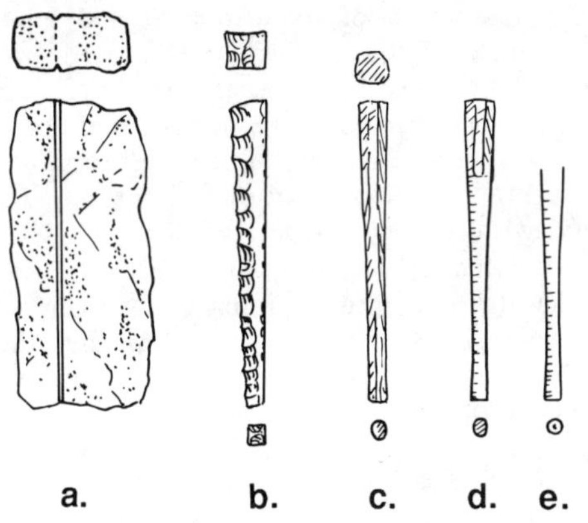

Figure 3. Constricted Cylindrical Drill Manufacture. **a.** grooved or sawn blocklet; **b.**chipped drill rough-out; **c.** longitudinally ground and faceted drill blank; **d.** finished drill with tapered midsection; **e.** used drill with distal depression.

developed by artisans in the cities of the Indus Valley region for the perforation of long beads made from agate/carnelian and jasper.

The cultural implications of these two types of drills are significantly different. Tapered cylindrical drills represent a common technology used in a wide range of production contexts. These drills were made from a variety of materials and were used to perforate different types of materials. On the other hand, the use of constricted cylindrical drills involved a more complex series of procurement and manufacturing processes. These drills were used primarily for the perforation of long beads, which were not only much harder to drill, but appear to have been considerably more valuable than the types of beads commonly manufactured with the tapered cylindrical drills. The use of constricted cylindrical drills is clearly associated with specialized production in the context of a state level society and may represent a technology that was restricted or controlled. This control could have been achieved through the restricted acquisition of the drill and bead raw materials, the skills needed to manufacture the drill bits, and the techniques used for the drilling of long beads.

The constricted drills also appear to have been the prototype for later diamond drilling technology that developed in Western India during the first millennium B.C. The connection between state controlled production, constricted stone drills and the development of diamond drilling are not the subject of this paper, but will be discussed briefly in the conclusion.

Current research by the authors involves the examination of drills and drilled surfaces using a combination of experimental and analytical techniques to understand the nature of the raw materials used to manufacture drills and the effects of the drills on the material being drilled. These techniques include experimental reconstructions of drills and drilling techniques along with the analysis of both experimental and archaeological materials using scanning electron microscopy (SEM), scanning electron microscopy with energy dispersive x-ray spectrometry (SEM EDS), x-ray diffraction (XRD) and electron microprobe x-ray analysis. The drills and drill impressions were examined at many different magnifications under the SEM, but the magnifications of 100x to 300x appear to be optimal for determining the different types of abrasive properties and effects.

Background and Previous Research

The first major research on drilling technology of the greater Indus region focused on bead drilling in Sindh and was carried out by E. J. H. Mackay [4-6]. At both Mohenjo-daro and Chanhu-daro (Figure 1) Mackay discovered various types of tiny stone drills that were used to perforate stone beads. Unfortunately he did not differentiate between the two types discussed above and he was not conclusive in his published interpretations of the role of these stone drills in the perforation of long carnelian beads. At the site of Chanhu-daro Mackay discovered large quantities of unfinished and finished drills that were directly associated with long carnelian bead production, but these were only briefly reported in his publication and he died before he could complete their study. He suggested that instead of stone drills, copper drills used with abrasives might have been utilized to perforate long beads made from carnelian and jasper [4-6].

The confusion regarding the role of stone drills was next addressed in the course of research by Piperno and Tosi on the stone bead drilling represented at Shahr-i Sokhta, Iran (Figure 1)[7, 8]. These studies were a part of a larger research on stone bead making industries of the Third Millennium B.C., carried out at sites in eastern Iran such as Tappeh Hesar and Shahdad, and at various sites in Turkmenia (for bibliography see [9]). Most of these studies were conducted using macroscopic or low magnification microscopic observations and demonstrated that drills made from stone blades were used to perforate lapis lazuli and turquoise beads.

In the workshop areas of Shahr-i Sokhta (dated about to 2700 BC) actual stone drills have been recovered with lapis lazuli dust adhering to the surface, but all of the drill bits are larger than 1 mm in diameter [10]. These stone drills were undoubtedly used for drilling lapis lazuli beads and may have been used for drilling other materials. There were however, beads with tiny perforations less than 1 mm in diameter, and smaller than any of the stone drills recovered from the excavations. The lack of tiny chert drills that would match the size of these holes has led Piperno and others to suggest that the tiny drill holes were made by copper drills and abrasives [7, 8]. Here again we see the proposition that metal drills with abrasives were used by the protohistoric artisans of South Asia. On the other hand, the tiny stone drills needed to perforate such beads may have been discarded in areas of the site that were not excavated or blown away by the strong winds that deflated many areas of the site.

Another site where there is evidence for the use of stone drills is at the site of Shahdad in Eastern Iran. During the protohistoric period, different colors of semi translucent cherts and jaspers were used at Shahdad to perforate large amounts of small carnelian beads (short truncated bicones). Most of these drills appear to have been made from burin spalls, and are comparable with drill bits from Tappeh Hesar, while others were made from tiny blades [11]. On the whole, the picture we get from Eastern Iran and Central Asia is of well established regional technological traditions capable of coping with the material properties and marketing constraints of locally exploited semiprecious stones.

Another variety of drill found both at Shahr-i Sokhta, Iran and Mehrgarh, Pakistan [1] was produced from a fine grained jasper-like rock with conchoidal fracture and good chipping properties, distinguished by a uniform light greenish color. It was originally called *pthanite* [7] but this term should be discarded because it is not an officially recognized scientific term. The precise characterization of this green rock is still incomplete, but preliminary X-ray Diffraction of a sample from Shahr-i-Sokhta suggests that it is some form of iron magnesium silicate. More precise identification must await detailed petrographic analysis.

All of the drills mentioned above have generally been described as being cylindrical in shape with a concave dimple at the tip and have been compared with those found by Mackay at Chanhu-daro and Mohenjo-daro. While the distal tip is cylindrical and has a dimple impression, the drill bit tang usually has an irregular section, triangular or trapezoidal due to the fact that these drill bits were made from blades (Figure 2). It is quite clear that these drills are not the same as those reported by Mackay from Chanhu-daro and should be identified as tapered cylindrical drills (see below).

Other studies on West and South Asian material have been made using potentially more powerful instruments, such as SEM on silicone impressions of bead drill holes [12-16]. Unfortunately these studies did not involve primary analysis of the types of stone drills that are the focus of our paper. Furthermore, the interpretations of drilling technology made by the authors have been based primarily on the examination of silicone impressions at a very low magnification (around 10X to 15X) and therefore cannot be adequately compared with the results of our studies.

Our studies of drill hole impressions indicate that at 15X it is often difficult to differentiate the type of abrasive action and surface modification resulting from different types of drills (Figure 4). Drill hole impressions were made from a carnelian bead perforated by a double diamond tipped drill in Khambhat, India [17, 18], a carnelian bead perforated by a Harappan stone drill and a pre-Columbian quartz bead drilled with an unknown abrasive. When these same drill hole impressions are viewed at 100X to 300X the differences in drill cutting action and surface modification become clearly apparent (Figure 5).

In summary previous studies have confused two or more different types of drills and their technological performance. In the following sections we will outline the distinctions between two major types of drills and present the results of our recent studies of constricted cylindrical drills from major sites of the greater Indus Valley.

Tapered Cylindrical Drills

The important morphological feature of this type of drill is that the drilling shaft portion is tapered, with the distal end being significantly smaller than the medial portion of the drill bit. Because of this feature we suggest that these drill bits should be called tapered cylindrical drills in order to distinguish them from another variety of drill bit found at the major settlements of the Indus Valley Civilization.

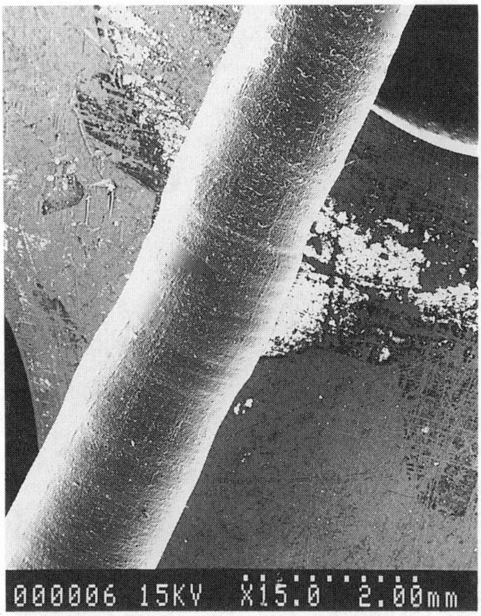

a.

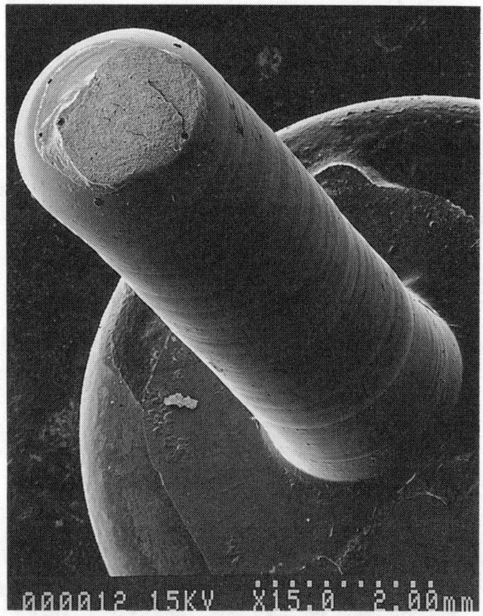

b.

Figure 4. SEM photographs of drill hole silicone impressions at 15X. **a.** drill hole impression of a carnelian bead from Khambhat, India, perforated with a double diamond drill; **b.** drill hole impression of a carnelian bead from Harappa, Pakistan probably perforated with a constricted cylindrical stone drill; **c.** drill hole impression of a Pre-Columbian quartz bead from South America (possibly Tirona Culture), perforated with an unidentified drill and abrasive.

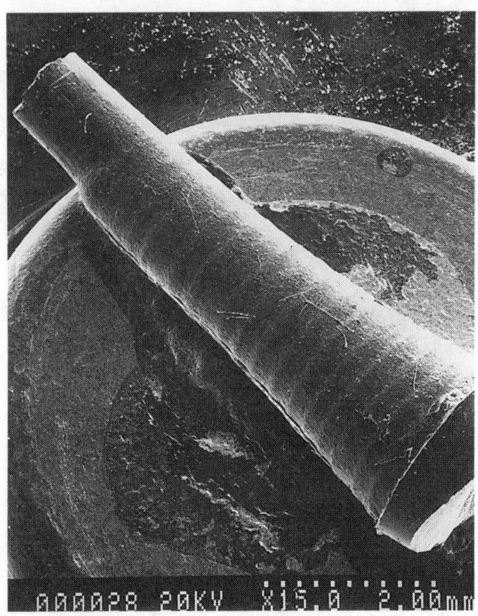

c.

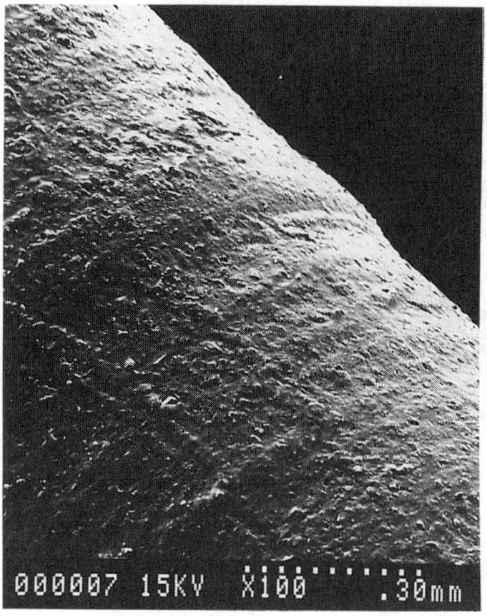

a.

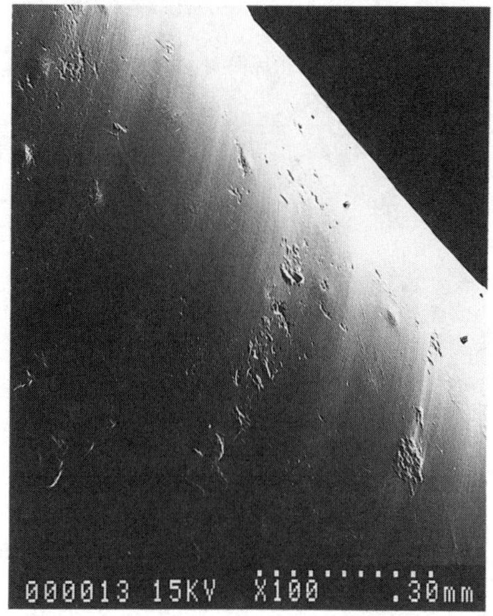

b.

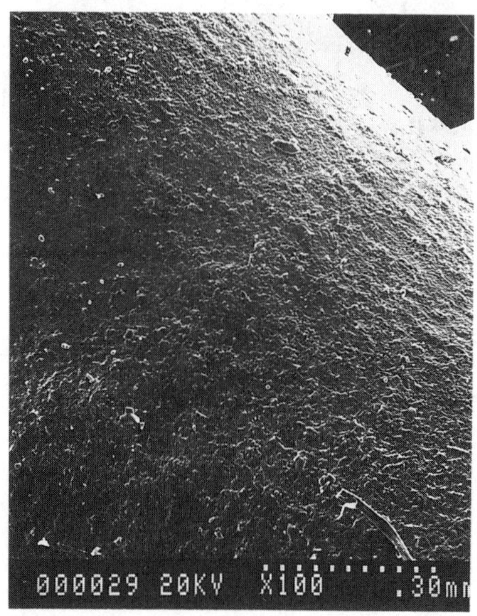

c.

Figure 5. SEM photographs of drill hole silicone impressions at 100X. **a.** drill hole impression of a carnelian bead from Khambhat, India, perforated with a double diamond drill; **b.** drill hole impression of a carnelian bead from Harappa, Pakistan probably perforated with a constricted cylindrical stone drill; **c.** drill hole impression of a Pre-Columbian quartz bead from South America (possibly Tirona Culture), perforated with an unidentified drill and abrasive.

At Indus Valley sites, the raw materials used to make this type of drill bit include both fine grained greenish silicates (similar in appearance to those found at Shahr-i-Sokhta), as well as grey-black jaspers, and colored cherts. No petrographic or mineralogical analyses have been reported on these materials, however we have been able to analyze two samples of green stone from Mehrgarh using X-ray diffraction. One sample corresponds to the pattern available for vesuvianite (calcium magnesium aluminum silicate hydroxide) while another sample is more closely comparable to the iron magnesium silicate from Shahr-i-Sokhta mentioned above. While these identifications are still inconclusive, it is interesting to note that more than one variety of greenish stone was being used at Mehrgarh to manufacture of such drills.

Tapered cylindrical drills are usually made from a thin flake or parallel sided blade that has a single ridge which is used to center the drill. Occasionally a double ridged blade may be used if the ridges are close together. This usually results in a very sturdy and thick drill. These drill bits are not very long as the raw material is brittle and tends to flake very easily and snap. The drilling portion of the drill bit usually falls between 4 to 7 mm although longer examples are on record (for example from Shahr-i-Sokhta).

The blade is flaked with different techniques to produce a pointed distal end and a thick proximal end that is used for hafting. Usually the distal end is modified with inverse backing retouch (flakes being removed from the ventral surface of the blade), but occasionally the tip is bevelled with the backing retouch being normal on one edge (i.e. on the dorsal surface) and inverse on the other (Figure 2). The proximal end or tang of the drill bit is usually shaped with bidirectional backing retouch to form a trapezoidal or triangular section. Through repeated use in drilling a hard material, the angular or bevelled distal portion of the drill becomes chipped and abraded until a tapered cylindrical shape is formed. Archaeological examples of tapered cylindrical drills usually have a concave depression at the tip.

Drills that can be classified as tapered cylindrical in shape have been reported from sites in Mesopotamia [19] and Eastern Iran [11, 20, 21]. In the greater Indus Valley these types of drills are found at the sites of Mehrgarh [22] (Figure 6), Ghazi Shah [23], Rehmandheri [24] and Nagwada [25].

Experimental Studies and Analysis of Drilling

The authors have conducted replicative experiments using tapered cylindrical drills made from flakes of greenish rock recovered from Shahr-i-Sokhta and Mehrgarh, as well as with similar greenish rock collected in Khambhat, Gujarat. The drills were chipped and shaped in order to replicate the archaeological examples. After chipping, they

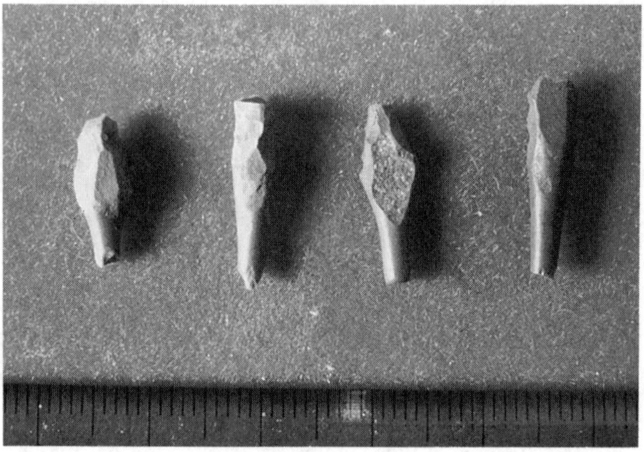

Figure 6. Tapered Cylindrical Drills, Mehrgarh, Pakistan, Period III (circa 3500 B.C.).
(Courtesy J.-F. Jarrige, Photo by M. Tosi).

were hafted in wooden shafts using lac (insect resin), a material that would have been available to the ancient Harappan artisans. The bead blank to be drilled was held in a wooden vise and drilled using a hand powered bow drill (Figure 7). A continuous stream of water was used to cool the drill and wash out the abrading materials. The tapered drill is efficient for making shallow holes in wood, shell, pottery, bone and rocks such as lapis lazuli, steatite, hematite, limestone and sandstone. When the drill tip becomes totally cylindrical it is no longer suitable for drilling softer substances, however it may still be used effectively on harder materials.

The perforation of the softer materials is achieved by the gouging or cutting out of the material. Silicone molds of the drill hole show that the drilling striae are irregular in depth and fairly parallel. The best SEM magnification for observing these striae is between 100 and 300 x. The pitted surface is not polished in the depths of the grooves, but often the tops of the grooves are abraded and polished.

When green stone drills are used on a hard material such as agate, the drill and the drill hole become highly polished. Due to the tapered form of the drill bit, the abraded surface is not confined to the tip, but

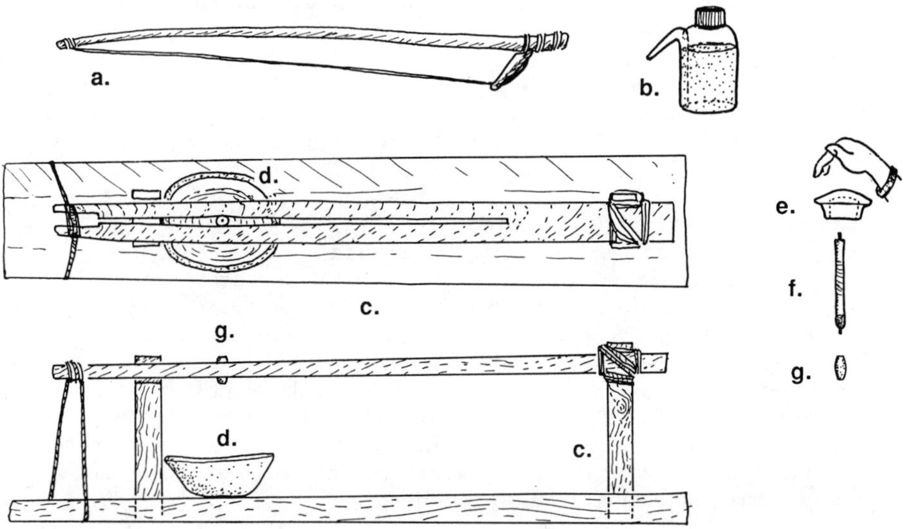

Figure 7. Experimental Drilling apparatus.
a. bow; **b.** squirt bottle; **c.** wooden vise; **d.** and container for water; **e.** drill back; **f.** drill; **g.** bead.

includes the entire tapered shaft of the drill bit. This increased area of abrasion and friction slows down the drilling process and the drill bit is difficult to cool due to the lack of space for water to flow between the drill shaft and the bead surface. The fact that there is no room for lateral movement as the drill is being rotated and the difficulties in cooling the drill bit result in considerable breakage.

Initially the distal tip of the drill is jagged but through repeated use on hard materials, such as carnelian, it becomes rounded. Through continued drilling without any further modification of the tip we found that the tip remained rounded and lost its cutting power. Furthermore, the tip did not form a concave depression as has been seen on archaeological drill bits. In order to increase the cutting performance of the drill bit we began to repeatedly flatten the tip by grinding on a silicon carbide grinding stone. The drill tip was carefully cleaned before resuming the drilling. Through this process of flattening and drilling, the drill gradually developed a concave depression at the tip. This experimental simulation suggests that the concave depression is not simply the result of normal drilling dynamics as previously proposed by Piperno [7], but results from a combination of drilling dynamics and intentional modification.

The rate of perforation using a bow drill was difficult to measure because of repeated interruptions to cool the hot drill back (Figure 7), wash out the drill hole, and to grind the drill tip. However, these same factors are also present in drilling with a constricted cylindrical drill or a double diamond drill and therefore the rates of drilling between these different types of drill can be compared and will be discussed in more detail below.

Constricted cylindrical drills

The disadvantages present in the tapered cylindrical drill seem to have been addressed by artisans of the Indus Valley Tradition at the sites of Mohenjo-daro, Chanhu-daro and Harappa. Although they still continued to use tapered cylindrical drills for some perforation activities, they also invented a new form of drill which we will call the constricted cylindrical drill. The constricted cylindrical drill has a long cylindrical shape that is wide at the tip and constricted in the midsection (Figure 3). This type of drill was made from a single type of raw material that was used to produce standardized bead drills having a specialized morphology and size grading.

Raw materials

The raw material used for making constricted cylindrical drills has been found as angular blocks at the sites of Mohenjo-daro, Chanhu-daro and Harappa (Figure 8). Visually the rock is composed of a mottled greyish-green to yellow-brown matrix with irregular dark brown to black patches or dendritic formations. Fault lines in the stone are visible and often stained with reddish color, possibly due to iron oxide. These fault lines are strongly bonded and cleavage of the block often cuts across visible fault lines. There is no pattern of large scale conchoidal fracture as normally seen in chert or jaspers, but within its irregular cleavage planes the stone, although very hard, has excellent chipping properties. The stone feels quite heavy to the touch, but the specific gravity is difficult to calculate due to the fact that it is composed of different minerals. Two samples of combined dark brown -black and yellow-brown matrix had a specific gravity of 2.26 to 2.35. Sillimanite ranges from 6 to 7 1/2 hardness on the Mohs' scale and has a specific gravity of 3.25. The sample rock can easily scratch quartz but is unable to scratch beryl and therefore falls between 7 and 8 on the Moh's scale of hardness. Nevertheless, the rough surface of the rock can be grooved by a chert blade, and chert is generally between 6 1/2 and 7 hardness.

In order to determine the characteristics of this rock, a sample from Mohenjo-daro was selected for x-ray diffraction and electron microprobe analysis. Samples of the yellow-brown matrix and the brown-black portions were analyzed separately. Based on the XRD spectrum and the calculation of the unit cell values, the yellow brown matrix is made up primarily of quartz and sillimanite [26]. (XRD interpretation was made by Dr. S. Bailey, Department of Geology,

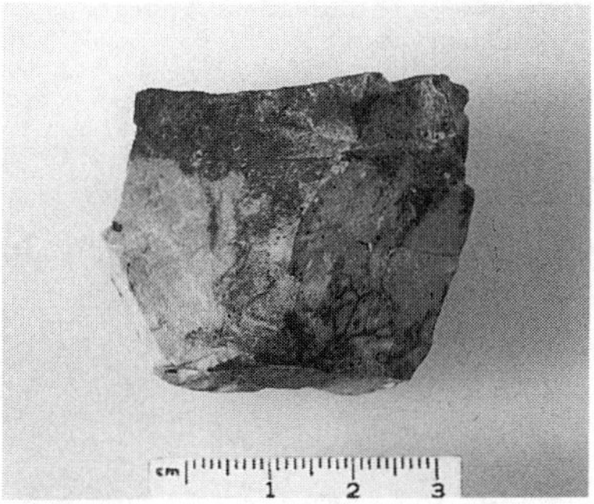

Figure 8. "Ernestite" sample from Mohenjo-daro.

University of Wisconsin, Madison). However, there are some peaks in the XRD spectrum that may represent the presence of mullite or an intermediate phase that has not been identified. The brown-black portion appears to be primarily quartz with hematite and some sillimanite/mullite. It is important to note that the sample from Mohenjo-daro has no trace of cordierite or corundum which are often associated with mullite and are common in rocks called porcellanite [27, I. Freestone, personal comunication].

The electron microprobe x-ray analysis of the yellow brown and the brown-black components revealed a matrix of quartz with concentrations of iron/titanium oxide phase. The concentrations of hematite and iron-titanium oxides endow the rock with cutting and polishing properties, while the matrix of quartz (probably as quartzite) and sillimanite results in a strongly bonded structure capable of withstanding the stresses of drilling. The abrading surface of the drill retains a rough irregular surface (Figure 9) that helps in cutting the carnelian, which is not as strongly bonded. Abrasion is probably attained because of the inherent tenacity or toughness of the sillimanite matrix and through the concentration of iron-titanium oxides in the rough surface of the drill bit as these minerals gradually erode from the drill itself.

Although we have been able to identify the major components of this sample, it has not been possible to come to a final conclusion regarding its classification. We propose "Ernestite" (after Ernest J. H. Mackay) as a temporary name for this material until further petrographic studies can complete the characterization. We can say that it is a fine-grained metamorphic rock composed primarily of quartz, sillimanite, mullite, hematite and titanium-oxide phases. The samples of "Ernestite" from Mohenjo-daro, Chanhu-daro and Harappa that we have been able to examine are clearly tougher than normal sillimanite and and were used quite effectively for drilling carnelian, agate, jasper and other ornamental rocks.

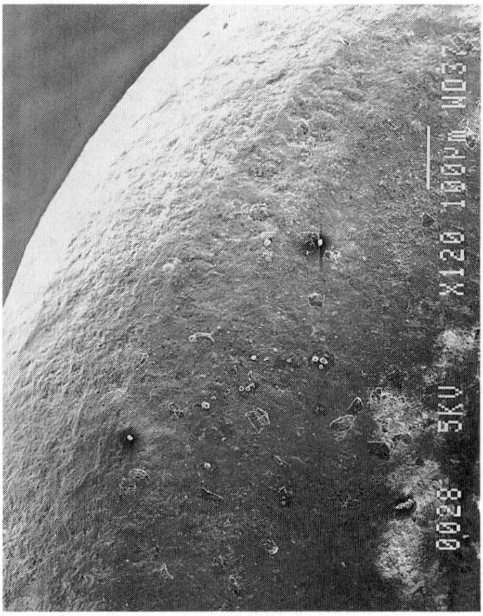

Figure 9. Detail of drill tip at 120X showing rough surface after use. Archaeological sample from Mohenjo-daro.

Drill Morphology

The distinctive morphology of the constricted cylindrical drills can be divided into three components for discussion purposes: 1) the distal tip, 2) the constricted medial portion and 3) the proximal tang portion.

1. The distal tip of this drill is wider than the medial shaft and ensures that the entire cutting pressure is focused on the area of the rotating circular tip. The tip begins as a flattened oval that is gradually rounded into a circular ring with a concave dimple depression in the center (Figure 10a). At this point the cutting pressure becomes even more concentrated, being primarily the circular ring shaped cutting edge. Due to the fact that a circular ring does not have a leading edge, the edges of the circular tip of the drill were repeatedly ground flat on two faces, essentially creating two facets. These facets result in two leading edges that increase the cutting ability of the drill tip, exposing new areas of abrasive raw material of which the drill itself is composed.

After each phase of resharpening or faceting, the drill tip becomes smaller and requires additional thinning of the medial portion. This results in progressively thinner and short drills that must eventually be discarded or used for drilling smaller sizes of beads.

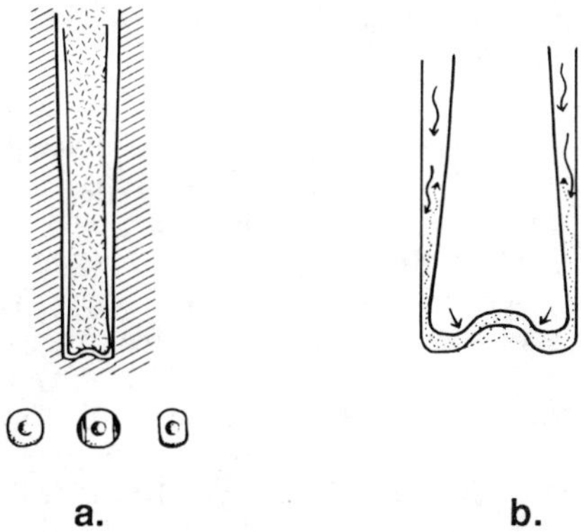

Figure 10. Detail of Constricted Cylindrical Drill Tip. **a.** round and faceted drill tip with concave depression. **b.** schematic section of drilling to show movement of abraded particles and water flow.

Once formed, the concave dimple depression appears to have been an important component of the drilling dynamics. The concave dimple of the drill tip leaves a convex projecting "nipple" at the center of the drill hole. The lateral movement of the drill itself results in pressure on the nipple such that it becomes chipped and broken quite rapidly whenever it begins to form.

The tiny chips of broken nipple become ground into a slurry beneath the spinning drill tip or are sucked out through the plunger motion of the drill bit at it is moved up and down in the drill hole (Figure 10b).

2. The drilling mechanism of the tip is closely associated with the constricted medial portion. The space between the drill and the drill-hole wall was increased by making the drill itself narrower in the medial portion behind the tip. This feature reduced the amount of friction between the drill and the bead and it allowed minimal lateral movement without snapping of the drill bit or the bead. Furthermore, it provided space for the exit of the drilling slurry and entrance for cooling water through a vertical plunging motion of the drill bit in the perforation. Most of the long carnelian beads broken in manufacture are snapped at the middle, where heat from drilling friction would have been difficult to control and where the maximum effect of lateral movement was present.

3. The constricted cylindrical drills involved considerable labour for the preparation of a regular and symmetrical tang. This tang was probably wrapped with a cord and hafted by pressure into a specially designed drill shaft. No traces of adhesive had been found on the

specimens observed so far, either macroscopically or with SEM. In contrast, adhesive residues are relatively common on other types of stone drills, both in the greater Indus Valley and Eastern Iran. The careful shaping of symmetric tangs optimized the centering of the drill bits into the wooden shaft, and this could reduce the danger of breaking expensive beads because of lateral movements and eccentric pressure.

After the drill bits were broken or worn down, the tangs were often reused as pivots on the back of the drill shaft. This form of industrial recycling has been reported from the Moneer site at Mohenjo-daro, on the basis of macroscopic and SEM observations of distinctive wear traces on the back of the tang [9]. This evidence is an additional indicator of the high levels of efficiency which characterize the making and use of the constricted cylindrical drills.

Constricted Cylindrical Drill Manufacture

The recovery of constricted cylindrical drill making debitage at Mohenjo-daro, Harappa and Chanhu-daro allows a preliminary reconstruction of their manufacturing sequence. The manufacture begins with the cleavage of a large lump into small angular blocklets (Figure 3a). These blocklets were then scored with longitudinal grooves, apparently made with chert blades. Our experimental studies show that chert blades can incise a shallow groove in the blocklets. With the assistance of these guiding grooves which encircled the blocklet, a small square sectioned drill rough-out was detached. In the Third millennium B.C., grooving and splitting was a common technique for cutting relatively soft stones such as limestone, lapis lazuli or steatite, but craftsmen of the Indus Valley were able to apply it to a much harder and heterogeneous material. The drill rough-outs are thicker at one extremity, which was to become the proximal end or tang of the drill bit (Figure 3b).

The square sectioned drill rough-out was further reduced to a slender square column (Figure 3b), using primarily pressure flaking, although inverse indirect percussion appears to have been used in some instances. The micro-debitage from this operation (in the form of tiny, scale-like flakes) has been found in the lapidary activity area of the Moneer site in Mohenjo-daro, together with finished and unfinished constricted cylindrical drills [9].

In the subsequent stage of manufacture, the slender, square column was ground longitudinally and at an oblique angle to form a faceted polyhedron (Figure 3c and Figure 12) [28]. In the final stages of grinding, only the distal portion was modified and this was done by grinding the polyhedron into first a cylindrical shape and then a constricted cylinder (Figure 3d). The constriction resulted in the distal point of the drill bit being greater in diameter or width, than the medial portion of the drill bit.

Grinding stones used to shape the drill bits were made from very hard and fine grained sandstones or quartzite. These grinding stones are found in different shapes, ranging from small tabular pieces to irregular columnar shapes. Their use for shaping drills can be defined on the basis of association (as at Chanhu-daro) and by the nature of the grinding facets and striae.

Prior to actual drilling, the distal tip of the drill was ground flat and the edges of the tip were faceted, resulting in a oval shaped distal end (Figure 10a). Through repeated drilling the flattened oval form of the distal end becomes rounded and develops a concave dimple in the center (see above for discussion). In order to resharpen the drill bit, the edges were repeatedly faceted to provide additional cutting potential.

Drill Sizes and Use

In order to drill carnelian beads that were between 70 and 100 mm long, the ancient artisans appear to have used several sizes of drills. Most of the long carnelian beads made during the Harappan Phase of the Indus Valley Tradition reveal that three or four different sizes of drills were used to perforate one half of the bead (Figure 11). The bead was then turned over and the opposite side was perforated. In most cases the bipolar perforations meet precisely in the center of the bead. The use of drills of different sizes is often clearly visible in drill sections of translucent beads as well as in the silicone impressions of their perforations.

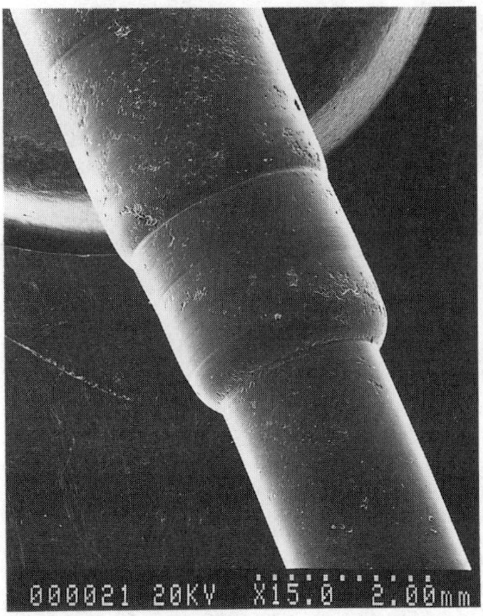

Figure 11. Detail of silicone impression of carnelian bead drill hole showing the use of different drill sizes. Archaeological bead from Chanhu-daro (Museum of Fine Arts, Boston).

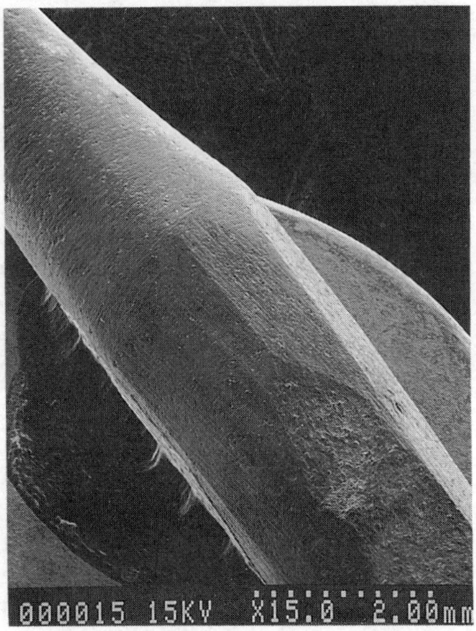

Figure 12. Drill tang portion showing faceted polyhedral section. Archaeological drill from Mohenjo-daro.

On the basis of experimental reconstructions and ethnographic observations of traditional bead drilling, the drilling was most probably accomplished with a bow drill cooled with dripping water or by keeping the bead and drill totally immersed in water [17, 18]. In the drilling process, the drill would have been held very rigid to avoid excessive lateral movement as this would snap the delicate drill bit. On the other hand there would have been repeated vertical movement, moving the drill bit up and down inside the perforation like a plunger. This movement would remove particles of chalcedony chips and powder and introduce cool water that was essential in order to avoid thermal fracture of the carnelian and the drill point (Figure 10b).

Experimental drilling using different drills made from the same blocklet of "Ernestite" (collected from Chanhu-daro) resulted in slightly different rates of abrasion (Table 2). All of the drilling was conducted on similar qualities of heat treated carnelian and the drilling itself was done by Kenoyer to maintain a standard level of technical expertise. Initial experiments in 1991 resulted in a drilling rate of 1 mm in 15 to 20 minutes. Experiments in 1992, using several different drills and bead blanks made from similar colors of carnelian. averaged 2.37 mm per hour. These results can be compared with the green jasper-like rock (discussed above) which has a drilling rate of approximately 0.83 mm per hour.

These preliminary results show that the drills made from "Ernestite" are more than twice as efficient as those made from the green rock. Experimental drilling with copper drills using ground silicon carbide (220 mesh) and ground corundum (ruby) had rates of drilling at 2.62 mm per hour and 1.26 mm per hour respectively.

It appears that "Ernestite" drills are much more efficient than copper with corundum when using a bow drill and manually applied abrasive. This may explain why the use of abrasives in drilling carnelian, agate or jasper beads has not been documented from any Indus Valley site.

A drill made with the double diamond mounting [17] had a rate of 81.16 mm per hour when used by Kenoyer (Table 1). In stark contrast, the professional bead drillers in Khambhat have an average rate of 536.45 mm per hour (n=102). When drilling beads from 73 to 82 mm the rate drops to 500 mm per hour (n=5) and for beads between 90 and 100 mm the rate is even lower, at 388 mm per hour (n=7). There is no doubt that the ancient Harappan bead drillers would have drilled much faster than in our experimental reconstruction, but given the reduced cutting ability of the "Ernestite" drills, it is unlikely that they would have been six times faster as is the case with the average for diamond drilling.

Table 1. Experimental Drilling Rates

Drill Type	mm per hour
"Ernestite" (constricted cylindrical drill)	
Average Drilling Rate	
(n=5 drilling experiments)	2.37
Green Jasper (tapered cylindrical drill)	
Drilling Rate	0.83
Copper drill with corundum	1.26
Copper drill with commercial silcon carbide	2.62
Double diamond drill	
(drilled by Kenoyer, n = 1)	81.16

Conclusions

On a superficial level, the tapered cylindrical drills of the protohistoric period in West and South Asia are easily confused with the constricted cylindrical drills of the Harappan phase of the Indus Tradition. The visual differences between these two types of drills may seem very insignificant, but in fact they involve two very different types of drilling mechanics and represent two contrasting technological concepts. In the tapered specimens, the general cylindrical shape of the bit is a by-product of the friction between the drill and the perforation; in the constricted drills, the general cylindrical shape is due to an intentional choice involving the delicate shaping of the bit and having substantial functional implications. The combination of the features of drill tip, medial portion and tang make constricted cylindrical drills the most complex and refined perforation tools of protohistoric West and South Asia (Figure 13).

Our studies on agate bead making suggest that, during the Third millennium B.C., the focus of the production of extremely long beads, particularly of carnelian resulted in the specialization in manufacture of long cylindrical bead drills. The demand for the long beads required by the elites of the cities of the Indus valley lead the artisans to discover and systematically exploit a new rock containing minerals particularly suitable for the drills. Specific manufacturing techniques were developed for making these drills and they were probably transmitted through a closely monitored apprenticeship. The need to control the drilling process more and more efficiently resulted in significant innovations involving the tang and the medial section of the drills.

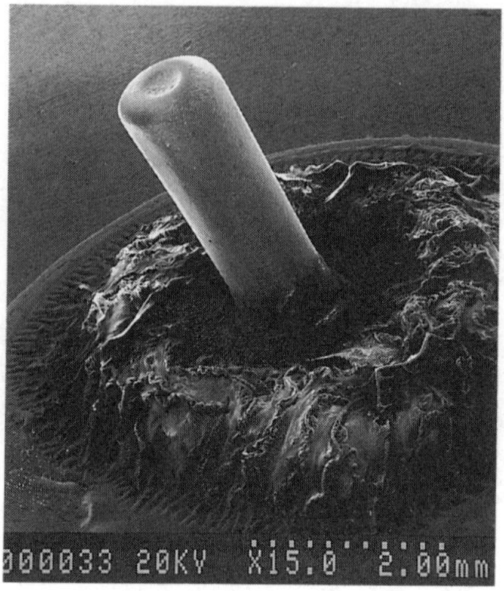

Figure 13. SEM of Constricted cylindrical drill, Mohenjo-daro. Side view showing the constricted medial portion of the drill.

Furthermore, the artisans learned to resharpen the rounded tips of the constricted drills to improve their cutting potential.

The distinctive constricted form of the stone drill shaft and the oval shaped tip with faceted edges may have provided a template for the later invention of double diamond drilling in the early historic period [17]. The double diamond drills still used in the agate workshops of Khambhat are constructed with a metal (iron) shaft that is constricted behind the drilling tip (Figure 14a). The tip itself is set with two tiny rounded diamond chips resulting in an oval shaped distal tip and a depression forms in the center of the tip during the drilling process (Figure 14b).

While it is not impossible that the similarities in shape are simply coincidental, our current research indicates that the constricted cylindrical stone drills and the double diamond drill technique are both unique to the Indian subcontinent. The invention of the double diamond drill technique is still not well documented but current research by the authors on carnelian beads from the site of Nagra (near Khambhat) has determined that it was present sometime prior to 600 B.C. in western India [29]. Although there has been considerable scope for technological diffusion, there is no evidence for the use of these techniques in the regions to the west, and even today, only single diamond drilling is practiced in Afghanistan.

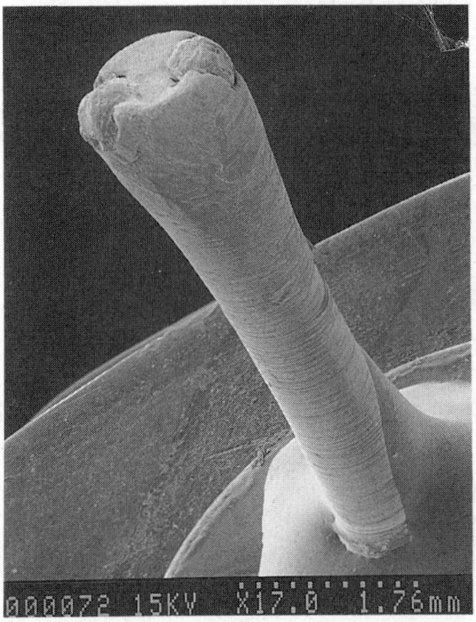

a.

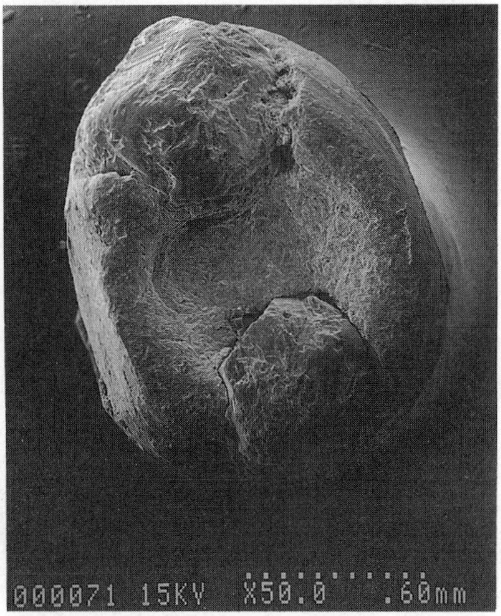

b.

Figure 14. SEM of Double Diamond Drill from Khambhat **a.** side view at 15X showing the constricted medial portion of the drill; **b.** plan view at 50X showing the oval shape of the distal tip with a concave depression in the center.

The connection between constricted cylindrical stone drills and the constricted cylindrical double diamond drills raises the question of the cultural continuity or discontinuity between the Harappan phase of the Indus tradition and the early historic period in western South Asia. This question has been debated at length by several scholars, with little consensus. In this regard, the detailed study of protohistoric technologies and the evaluation of their social implications is a rewarding field of enquiry. For example, the reconstruction of the Harappan stoneware bangles manufacture [30, 31] shows that this peculiar technology developed out of a technological know-how widely spread in the Subcontinent from very ancient times , i.e. ceramic firing at high temperatures in reduced conditions within closed containers. However, at the end of the Harappan phase, the sophisticated stoneware technology disappeared along with the social context of production and use.

While the disappearance of the Harappan stoneware bangle technology would point towards discontinuity, the evolution of carnelian drilling techniques beginning with stone drills and then diamond tipped metal drills, suggests a specific process of cultural continuity. The continued use of carnelian, agate and jasper beads by the elites of the subcontinent provided the need for continuity in drilling technology. It is also possible that an increased demand and a larger market spurred the experimentation with different materials to find a more efficient cutting tool. Corundum is commonly available in northern Pakistan and western India, but diamonds are found primarily in the central peninsular subcontinent [32]. It is not unlikely that the expansion of the Early Historic states into peninsular India was stimulated by the promise of new resources and the need to produce elite commodities more efficiently. On the basis of Early Historic texts [33], agate and carnelian bead production were an important commodity for the state and therefore needed to be properly controlled.

The important connections between resource availability, socio-political needs and technology can be addressed and it is important not to loose sight of these issues when one becomes engulfed with the details of modern archaeometric techniques. We are only beginning to be able to articulate these complex processes thanks to the collection of extensive, well documented bodies of archaeological data and new archaeometric research methods. Through continued research and dialogue with we hope to develop more precise interpretations of these cultural and technological processes.

Acknowledgements

Our research on drilling technology and bead manufacture has been supported by grants from the Smithsonian Institution Foreign Currency Program and Short Term Fellowships at the Conservation Analytical Laboratory. J. M. Kenoyer has carried out much of his research on ancient technologies through a National Science Foundation Presidential Young Investigator Award and M. Vidale has received support through a NATO Fellowship. Additional support has been provided by our respective institutions.

Special thanks to Dr. S. Bailey and Dr. E. Glover, U.W. Madison, Department of Geology, for their assistance in the use of the X-Ray Diffraction and Electron Microprobe equipment. We would like to thank Nancy Lawrence, SEM Lab, Department of Entomology, U.W. Madison for her patience and expertise in taking the SEM photographs.

We would also like to thank the editors and reviewers whose comments were very constructive and helped to refine the presentation of this paper considerably.

References

1. J. F. Jarrige, in Harappan Civilization, edited by G. L. Possehl (Oxford and IBH Publishing Co., New Delhi, 1982), p. 79-84.
2. J. F. Jarrige and R. Meadow, Scientific American 243 (2), 122-133 (1980).
3. J. M. Kenoyer, in Old Problems and New Perspectives in the Archaeology of South Asia, edited by J. M. Kenoyer (Dept. of Anthropology, U.W.-Madison, Madison, WI, 1989), p. 183-192.
4. E. J. Mackay, Journal of the American Oriental Society 57 1-15 (1937).
5. E. J. H. Mackay, Further Excavations at Mohenjo-daro, (Government of India, New Delhi, 1938).
6. E. J. H. Mackay, Chanhu-Daro Excavations 1935-36, (American Oriental Society, New Haven, CN, 1943).
7. M. Piperno, in South Asian Archaeology 1971, edited by N. Hammond (Duckworth, London, 1973), p. 119-130.
8. M. Piperno, East and West 23 (1-2), (1973).
9. M. Vidale, in Interim Reports Vol. 2, edited by M. Jansen and G. Urban (IsMEO-Aachen, Aachen, 1987), p. 113-150.
10. M. Tosi and M. Piperno, Expedition 16 (1), 15-23 (1973).
11. S. Salvatori and M. Vidale, Rivista di Archeologia 6 5-10 (1982).
12. L. Gorelick and A. J. Gwinnett, Expedition 23 (4), 17-30 (1981).
13. L. Gorelick and A. J. Gwinnett, Archeomaterials 3 39-46 (1989).
14. L. Gorelick and A. J. Gwinnett, Expedition 25 (3), 40-47 (1983).
15. A. J. Gwinnett and L. Gorelick, Expedition 22 (1), 17-32 (1979).
16. A. J. Gwinnett and L. Gorelick, Expedition 24 (1), 10-23 (1981).
17. J. M. Kenoyer, M. Vidale and K. K. Bhan, World Archaeology 23 (1), 44-63 (1991).
18. J. M. Kenoyer, Ornament 10 (1), 18-23 (1986).
19. E. J. H. Mackay, in Anthropology Memorial (Field Museum of Natural History, New York, 1929).
20. G. M. Bulgarelli, in South Asian Archaeology 1977, edited by M. Taddei (Istituto Universitario Orientale, Naples, 1979), p. 39-54.
21. M. Tosi, in Tappeh Hesar: Reports of the Restudy Project, 1976, edited by R. H. Dyson and S. M. Howard (Casa Editrice Le Lettre, Firenze, 1989), p. 13-24.
22. J. F. Jarrige, in South Asian Archaeology 1979, edited by H. Härtel (Dietrich Reimer, Berlin, 1981), p. 93-114.
23. L. Flam, in Harappan Civilization (2nd Edition), edited by G. L. Possehl (Oxford and IBH, New Delhi, 1992 in press)
24. F. A. Durrani, Ancient Pakistan 6 1-232 (1988).
25. K. T. M. Hegde, V. H. Sonawane, D. R. Shah, K. K. Bhan, Ajitprasad, K. Krishnan and S. P. Chardnan, Man and Environment 12 55-65 (1988).

26. W. L. d. Keyser, Transactions of the British Ceramic Society 50 349-365 (1951).
27. S. O. Agrell and J. M. Langley, Proceedings of the Royal Irish Academy 59 (B-7), 95-110 (1958).
28. M. Vidale, East and West 40 (1-4), 301-314 (1990).
29. J. M. Kenoyer, K. K. Bhan and M. Vidale, Agate Beadmaking: An Ethnoarchaeological Study, (In Preparation, 1993).
30. M. A. Halim and M. Vidale, in Interim Reports Vol. 1, edited by M. Jansen and G. Urban (RWTH-IsMEO, Aachen, 1984), p. 63-97.
31. M. Vidale, in Old Problems and New Perspectives in the Archaeology of South Asia, edited by J. M. Kenoyer (Wisconsin Archaeological Reports, Madison, WI, 1989), p. 171-182.
32. O. H. K. Spate, India and Pakistan: A General and Regional Geography, (Methuen and Co. Ltd., New London, 1963 reprinted).
33. K. Prasad, Cities, Crafts and Commerce under the Kusanas, (Agam Kala Prakashan, Delhi, 1984).

A THIRD MILLENNIUM B.C. GLAZED QUARTZ BEAD FROM TELL ES-SWEYHAT, SYRIA

PAMELA B. VANDIVER*, MARK FENN* AND T.A. HOLLAND**
*Conservation Analytical Lab, Smithsonian Institution, Washington, D.C. 20560 USA
**Oriental Institute, University of Chicago, Chicago, IL. USA

ABSTRACT

The probable composition of a weathered, blue-green glaze on a ground quartz bead was reconstructed. The bead was excavated from Operation 5 at Tell es-Sweyhat, Syria, in 1991 and is dated to the last quarter of the third millennium B.C. Special sample handling was required by the poor state of preservation of the glaze. Microprobe analyses and replicate melts showed that the composition was 60% SiO_2, 20% CuO and 20% flux, probably as soda, potassia or a combination. This composition is unusual for ancient glasses and glazes because of its high copper oxide content, which may indicate a link with copper or malachite technology. Very few glass and glaze compositions are known from the third millennium B.C., and none of these has such a high copper concentration, nor are there examples reported from the mid-second millennium B.C. with one exception of glazed tiles from Kerma in Sudan. This rare and unusual composition suggests that there may have been several paths to the development of glasses and glazes. Some of these paths were successful, and some others, as this one, did not continue and were "dead-end" technologies.

INTRODUCTION

During the 1991 excavation season of the Tell es-Sweyhat Expedition, Syria, sponsored by the Oriental Institute of the University of Chicago and the University Museum of the University of Pennsylvania, a glazed rock crystal bead was recovered. Examination and analytical testing of the bead were carried out at the Conservation Analytical Laboratory of the Smithsonian Institution to determine the extent of weathering and the composition and microstructure of the glaze. We had planned to melt a glass of similar composition in order to determine its working characteristics, until we realized the extent of corrosion of the glaze.

ARCHAEOLOGICAL CONTEXT

Tell es-Sweyhat is located about 3 km east of the Euphrates River in north central Syria (36° 17' north, 38° 15' east), approximately 64 kilometers south of Carchemish. The site covers roughly 30 hectares. Carbon-14 dating and pottery analysis indicate that the main occupation levels of the site date to the late third millennium. The 1991 season was the fifth at the site. Several preliminary reports have been published (1,2,3,4).

The bead, batch number 2604, is from locus 17/3 of operation 5, which lies in grid squares H/J-7, just north of the 1970's excavations in area I^4 (5). At the present stage of excavation, locus 17/3 appears to be a pit, the top of which is at the level of the foundations of wall 18/1 that is clearly

Bronze Age in date. Inspection of the pottery in-situ in this locus places it in the last quarter of the third millennium. A large jar stand from locus 17/1, just above the pit, is typically third millennium in form and decoration. In addition to the bead and the pottery, the pit also contained ocher, ground stone, shell, a figurine fragment, a possible vitrified brick, carbon and a piece of worked bone, as well as some unworked bone.

EXAMINATION OF THE BEAD

Only about half of the bead remains. It is approximately spherical, 6.1 mm long and 8.6 mm in diameter (Fig. 1). It is broken along the plane of the stringing hole, the walls of which are smooth and parallel throughout, suggesting it was drilled from only one direction. The exterior edges of the hole are smoothly rounded.

The body of the bead is colorless and varies from opaque to transparent. The broken surfaces exhibit conchoidal fractures. The glaze is bright medium green and is found in small patches on the outer surfaces of the bead (dark areas in Fig. 1 are glaze). Somewhat more glaze remains lining the stringing hole, including an area of blue-green glaze at one end. This thicker and brighter colored glaze is the least weathered. Shortly after excavation much of the green glaze layer coating the bead spontaneously exfoliated. This phenomenon has been observed and reported at several other excavations, including Nuzi and Nippur in northern and southern Iraq, respectively. Observation of this glaze in the lab and extended observation of glazes and glasses in the field at Nippur suggests that an increase in temperature and decrease in humidity from that found in-situ in the ground caused shrinkage of the glaze, followed by cracking and delamination. It is likely that only the least weathered parts of the glaze remained attached to the bead.

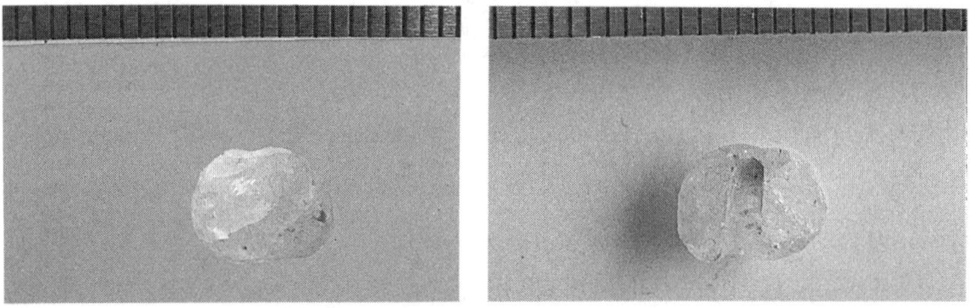

Fig. 1 Exterior (left) and fractured interior (right) of the glazed quartz bead. Dark areas are glaze. Scale is in millimeters.

A TECHNICAL EXAMINATION

A small sample of the body of the bead was subjected to X-ray diffraction using a Gandolfi camera and identified as quartz.

The glaze appeared friable and soapy under a microscope and had a Mohs hardness of 1, similar to talc. When touched with a needle, the glaze crushed; there was no evidence of any possibility of a conchoidal fracture, typical of unweathered glasses. The surface was layered and pitted with products of network break-down, similar to those shown in the micrograph in the paper by Schulz, Clark and Folz later in this volume.

As soon as possible after excavation, four pieces of the exfoliated glaze, measuring 0.1 to 0.5 mm in maximum dimension, were mounted as cross sections in epoxy and polished for wavelength dispersive microprobe analysis, as described below. This procedure incarcerated part of the sample, and, it was hoped, prevented further changes to any pristine glass which might remain below the weathered surface.

Another piece was mounted in cross section on a stub, coated with unusually heavy layers of carbon and gold and placed in a JEOL 840-II scanning electron microscope with a Tracor-Northern 1700 energy dispersive x-ray analyzer. At 3 kV, the glaze fragment charged and would not hold still beneath the beam; it contained sufficient water that imaging was not possible (thus, no micrograph of the sample is included). At 25 kV, the only elements present were Si and Cu. Surprisingly, no fluxes, that is no Na, K, Ca or Mg, were found; neither was there any indication of arsenic or tin. Our system can detect 1 wt%, with questionable peaks close to background which usually measured about 0.5 wt% when later subjected to microprobe analysis. The samples were not large enough for x-ray fluorescence which has a more sensitive detection limit of parts per hundred.

Because of the extensive weathering, the cross sectional samples embedded in epoxy were ground and polished parallel to the long direction of the mounted samples for the least possible time, 15 sec. with 600-grit (or 25 micron) SiC, 15 sec. with 0.3 micron alumina, and 15 sec. with 0.05 micron alumina. The inner parts of two of the samples did not dissolve during the 45 seconds of sample preparation. Microprobe analysis was conducted on a 9-spectrometer ARL wavelength dispersive unit at 15 kV for 10 seconds by moving the sample and stage beneath a beam defocused to 20 microns. Ten analyses gave the following results:

	SiO_2	CuO
Average	60.0%	20.0%
Range	59.5-60.6	19.2-20.4

The other seven elements analyzed were Al_2O_3, CaO, MgO, K_2O, Na_2O, FeO, P_2O_5. The values obtained were all at or below detection limits near 0.06%. The mean soda value was 0.05, and the mean potassia value was 0.04 (i.e., both below detection). The missing 20% was not expected. Low microprobe totals usually are ascribed to the presence of elements which were not standardized for analysis, to porosity, water or other organics, or to surface roughness. EDS analyses showed no elements were missed; the part of the sample analyzed was smooth, nonporous and contained no epoxy mounting medium. Moving the stage should not have caused a problem because the beam was already defocused to 20 microns. The low totals are probably due to such weathering reactions as the sodium and silicon hydroxide exchange reactions, other reactions involving alkali loss and both adsorbed and absorbed water.

Interpretation of this composition involves reference to the SiO_2-CuO phase diagram (6). The above composition melts at about 1690°C, far above the temperature expected for second,

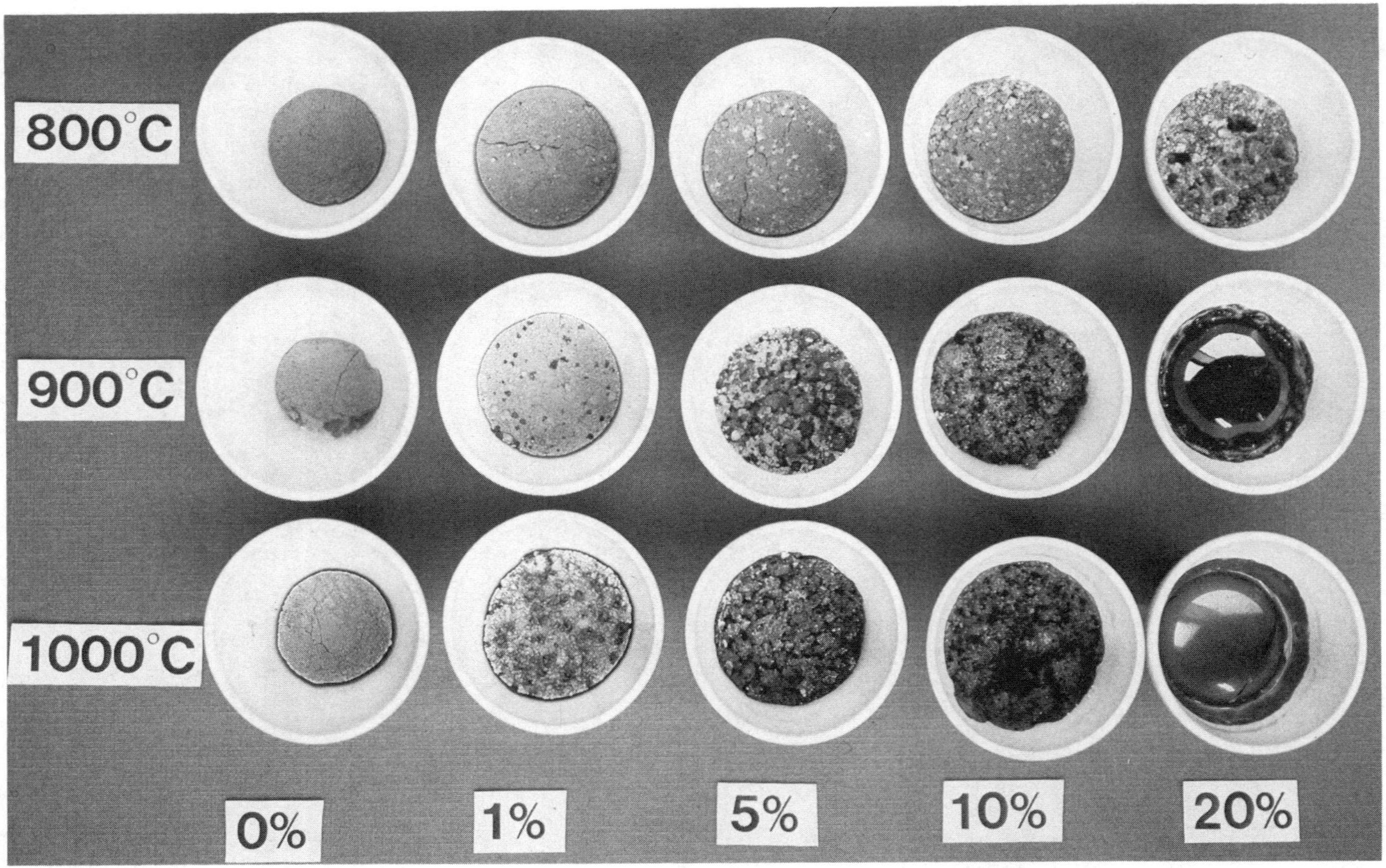

(See Plate 1 on page v in the front of this book)

Figure 2. Replicate melts showing variation in content of copper oxide and relation to incremental temperature variation. Note the only glasses that formed contain 20% CuO and were third or fourth millennium B.C. soda-lime-silicate technologies that tend to involve the temperature range from 800-1000°C (7,8,9,10). However, when the concentration of copper oxide drops to about 8%, the liquidus decreases to a eutectic of about 1060°C, which may vary a little depending on the oxidation state of the copper. This temperature is still above that found in most early soda-lime-silicate technologies.

A REPLICATIVE MELT PROJECT

In order to determine whether the missing 20% was composed primarily of flux, several possible compositions were melted to test whether glass would form. We hypothesized that the Cu^{2+} cation, having the same valence and nearly the same ionic radius, may act as a network modifier to stabilize the glass structure, similarly to the Ca^{2+} ion (We are currently testing the validity of this conjecture using electron spin resonance). In this case the Cu^{2+} ion would act to stabilize the glass structure by providing a Cu-O bond which is intermediate in strength between Si-O with a higher bond strength and Na-O or K-O with lower bond strengths. We found no references to Cu-O bond strength calculations. However, the lattice energies of CuO and Cu_2O are similar to those of the alkaline earths, CaO and MgO, and on the order of 3500 KJ/mole and intermediate between the glass formers and alkalies (about 2300 KJ/mole for K_2O or Na_2O and above 13,000 KJ/mole for SiO_2 and Al_2O_3). However, the enthalpies of formation of the copper oxides are about one-third those for the alkaline earths. Adding further credibility to this conjecture is evidence that the durability of an alkali-copper-silicate using the standard A.S.T.M. titration test is similar to that for a soda-lime-silicate.

The experimental part of this study consisted of melting four glasses and one control at three different temperatures, 800, 900 and 1000°C for 18 hours. The model glasses had compositions of 60% SiO_2 (200-mesh silica flour) and 20% CuO (introduced as $CuCO_3$), with additions of no alkali, 1%, 5%, 10% and 20% Na_2O (added as Na_2CO_3). Soda was chosen, rather than potassia, because it is the more active flux. The batch was mixed dry in a Patterson-Kelly V-blender for 30 minutes and slowly brought up to temperature over a period of 30 minutes, and then melted for 18 hours in alumina crucibles in a Thermolyne muffle furnace in an oxidizing atmosphere.

The results show that only 20% Na_2O at 900° and 1000°C produced a fully melted glass (Fig. 2 and color plate), although several of the other melts evidenced complex and often localized reactions, especially when compared with the controls.

SIGNIFICANCE OF OBJECT AND CONCLUSION

The composition of the glaze has an unusually high concentration of copper oxide. About 1 to 3% is required for an intense blue color. Tite (11) has shown that for green coloration to occur, a common solution was to add calcium oxide. Since no calcium oxide was found in this glaze and since we know that it migrates very little during weathering, we presume that the original color was a strong, dark blue which would have been translucent over the nearly transparent quartz bead. Internal cracks have caused light to scatter and the color to appear green. The presence of 20% copper oxide with no tin or arsenic over about 0.5-1% suggests that unalloyed copper, copper oxide or copper carbonate was added to fine quartz powder to make the glaze, which would suggest a link with copper-smelting or -using technology or with a malachite technology.

Very few glass and glaze compositions are known from the third millennium B.C. (12); and none of these third millennium examples has such a high copper concentration. Only one exception of tiles used to line tombs from the mid-second millennium B.C. site of Kerma in the Sudan has glazes on so-called Egyptian faience which contain such high copper oxide values (13). One idea which Cyril Smith often set forth was that many pyrotechnologies may have derived from the making of beads, protective charms and other small decorative objects; such tasks essentially involved experimentation with materials over a fire or partial enclosure. This unusual composition indicates that there may have been several paths to the development of glasses and glazes which by the mid-second millennium B.C. became well-established technologies with more limited ranges of composition and processing. Some of these paths were successful, and some, such as this one, were unsuccessful and did not continue. We only find rare remnants of a such a "dead-end" technology.

Such studies as the one reported here are relatively simple to do, certainly suitable as a project for an undergraduate student in a Department of Materials Science. They are not path-breaking studies, but require recognition and definition of a problem, special handling of a particular sample or samples to solve that problem, a plan of solution, data and a considerable amount of interpretation. Certainly if enough are carried out, we will know much more about materials processing and development, and many undergraduate students will have had direct experience with the history of their discipline.

REFERENCES

1. Holland, T.A., "Preliminary Report on Excavations at Tell es-Sweyhat, Syria, 1973-4", Levant 8 (1976) 36-70.

2. Holland, T.A., "Preliminary Report on Excavations at Tell es-Sweyhat, Syria, 1975", Levant 11 (1977) 36-65.

3. Weiss, H., "Archaeology in Syria", American Jour. of Archaeology 95 (1991) 683 ff, contains a short note by T.A. Holland and R.L. Zettler, pp. 717-719.

4. Holland, T.A., "Excavations at Tell es-Sweyhat in Syria, 1989 and 1991", The Oriental Institute Notes and News no. 134 (Summer 1992) 1-6.

5. See contour plan, fig. 1, in Oriental Institute News and Notes, no. 133 (Spring 1992) 1.

6. Levin, E.M., C.R. Robbins, H.F. McMurdie, Phase Diagrams for Ceramists, vol. 1 (1964) no. 164 and vol. 2 (1969) no. 2142-2144, American Ceramic Society, Westerville, OH.

7. Vandiver, P.B., "Appendix A: The Manufacture of Faience", in A. Kaczmarczyk and R.E.M. Hedges, Ancient Egyptian Faience, Aris and Phillips, Warminster, England, 1983, pp. A1-A123, and

summarized in P.B. Vandiver, "Technological Change in Egyptian Faience", in J.S. Olin and A.D. Franklin, eds., *Archaeological Ceramics*, C.A.L., Smithsonian Institution Press, Washington, D.C., pp.167-179.

8. Vandiver, P.B., in *Early Pyrotechnology*, ed. T.A. and S.F. Wertime, Smithsonian Institution, 72-92; and P.B. Vandiver, in *Jour. of Glass Studies* 26 (1983) 238-248.

9. Vandiver, P.B., Mc. Gibson, and A. McMahon, submitted to *Science*.

10. Vandiver, P.B., C.S. Swann and D. Cranmer, "A Review of Mid-Second Millennium B.C. Egyptian Glass Technology at Tell el-Amarna", *Materials Issues in Art and Archaeology II*, MRS Symp. Proc. 185 (1991), pp. 609-616, and B. McCarthy and P.B. Vandiver, "Ancient High-Strength Ceramics: Fritted Faience Bracelet Manufacture at Harappa (Pakistan), ca. 2300-1800 B.C., *Ibid.*, pp. 495-510.

11. Tite, M.S., paper presented at the Archaeometry Conference, Los Angeles, CA, April 1992.

12. Moorey, P.R.S., *Materials and Manufacture in Ancient Mesopotamia: The Evidence of Archaeology and Art*, BAR Int'l Series 237 (1985) 194-222.

13. Analyses of tiles from Kerma, in Vandiver, "Appendix", *op.cit.*, p. A50, and A. Kaczmarczyk et al., in prep.

PART III A

Craft Reconstruction From Evidence of Workshops, Industrial Debris and Artifacts

2. Production Sites and Process Reconstruction

THE WORK OF THE POTTER IN ANCIENT MESOPOTAMIA DURING THE SECOND MILLENNIUM B.C.

ABRAHAM VAN AS AND LOE JACOBS
Department of Pottery Technology, Archaeological Centre, University of Leiden, Reuvensplaats 4, P.O.B. 9515, 2300 RA Leiden, The Netherlands

INTRODUCTION

In the beginning of the second millennium B.C. Babylon became the centre of power in Mesopotamia. Hammurapi (1792-1750 B.C.) was one of the most important kings of the First Dynasty of Babylon. He is above all known for his law code (*Codex Hammurapi)*. At the height of his power the Old Babylonian Empire extended as far as Sumer in the south and to Nineveh in the north. After the Old Babylonian times a dark period followed in the history of Mesopotamia. The conquest of Babylon in 1595 B.C. by the Hittite king Mursilis I ended the First Dynasty of Babylon. His allies, the Kassites from the Zagros Mountains, occupied Babylon without breaking the Babylonian traditions. Dur Kurigalzu became their capital. In 1157 B.C. the Kassite Dynasty was attacked from Elam (southwestern Iran) and came to an end.

This article is concerned with the technological characteristics of the pottery made according to traditional methods in the second millennium B.C. (Old Babylonian and Kassite periods), which was found at Tell ed-Dēr [1], Sippar [2], Isin [3], and Nippur [4] (Fig. 1). This pottery is not a highly sophisticated product; it has all the characteristics of fast-produced, standardized and wheelmade consumer ware of small and large vessels consisting of plates and bowls, cups, goblets and jars (Figs. 2 and 3). It is in general undecorated and of a relatively uniform buff colour. We will try to present an idea of the work of the Mesopotamian potters who made this ware, from digging, preparing and shaping the clay to firing the pottery.

The technological analysis was carried out as part of the preparation of a Corpus of Mesopotamian Pottery by a number of archaeologists and the Department of Pottery Technology of the University of Leiden united in the Working Group on Mesopotamian Pottery. This pottery corpus should be a reference system of pottery shapes for dating and comparative purposes. It should also present the technological information and its relationship to the morphological data.

MATERIAL AND METHODS

The technological investigation of the pottery was carried out between 1985-1990 during excavation campaigns at the archaeological sites Tell ed-Dēr, Isin and Nippur. Moreover, the study includes pottery from Nippur, stored at the Oriental Institute (Chicago) and at the Dickson Mounds Museum (Lewistown/USA). Complete pots and sherds were available for research. No statistical formula, as done by H.J. Franken in his study of Iron Age pottery from Jerusalem [5,p.69-71], was used in order to establish the minimal amount of pots and sherds necessary to get a reliable answer to the research question. Almost all the second millennium B.C. pottery from Tell ed-Dēr and Nippur has been

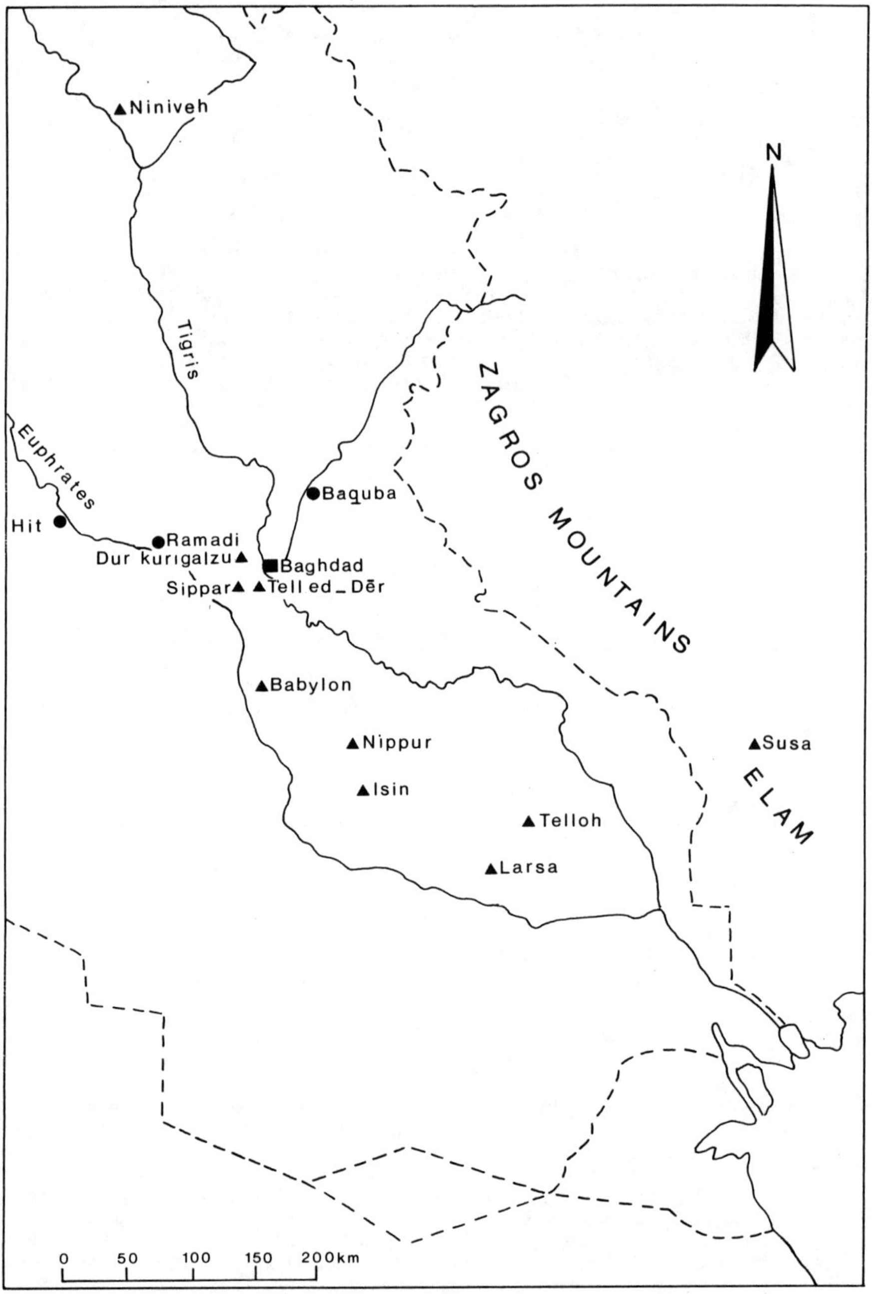

Fig. 1. Map of Mesopotamia.

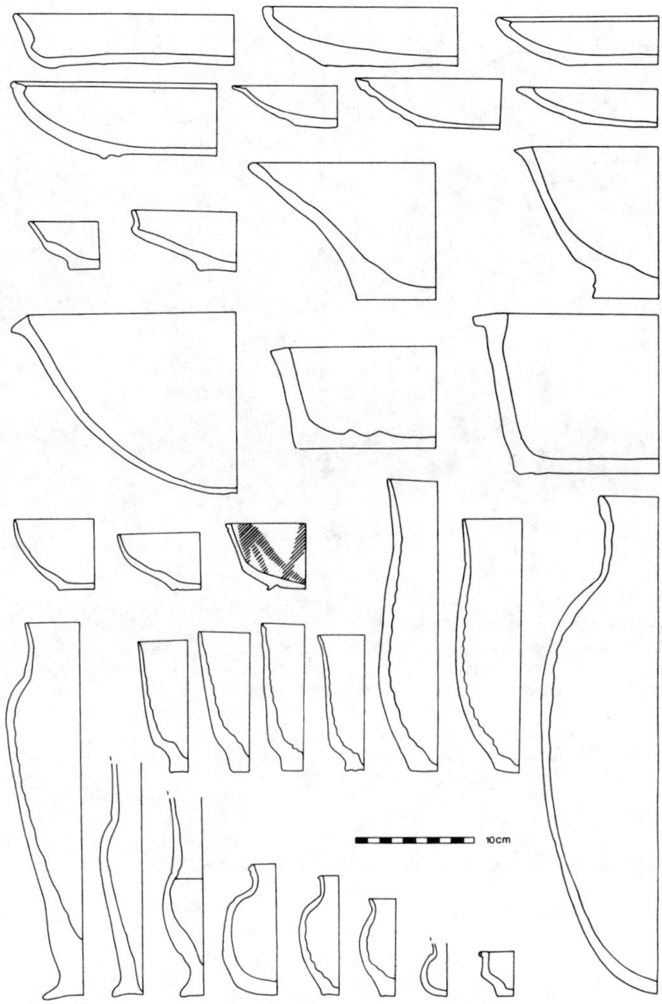

Fig. 2. A selection of pottery from the second millennium B.C. (Old Babylonian and Kassite periods) from Tell ed-Dēr.

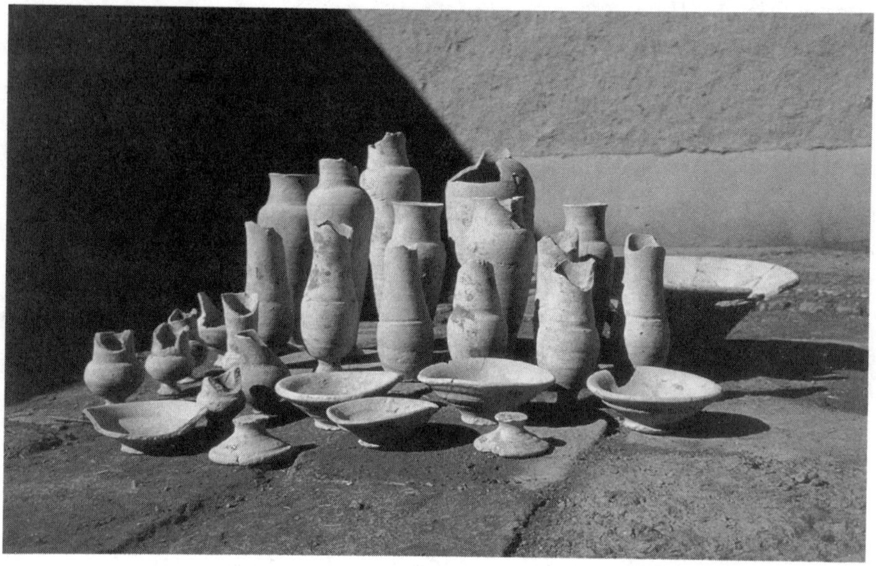

Fig. 3. A selection of pottery from the second millennium B.C. (Old Babylonian and Kassite periods) from Nippur.

examined. According to the excavators the sample of pottery taken from Sippar and Isin is representative of the total repertoire of these sites. To complete our knowledge of the pottery from the sites mentioned above and for comparative needs pottery material from Susa (Iran), excavated by R. Ghirshman, was studied in the Louvre in Paris. A small collection of pottery from Telloh and Larsa was also inspected towards this goal.

The technological research-work was executed according to methods currently used by the Institute of Pottery Technology at the University of Leiden [6,7,8,9,10]. Summarizing, an ideal technological research-work includes the following aspects:

1. *Analysis of the excavated pottery.*

1.1. Shaping technique.

1.1.1. Reconstruction of the techniques by the analysis and interpretation of technical characteristics which have been left in the pottery as traces of the techniques used.

1.1.2. Verification of the reconstructed techniques by practical experimentation with clays of a quality which is comparable with that used in antiquity.

1.2. Firing technique.

1.2.1. Reconstruction of the firing atmosphere by interpretation of the colours of the sherds (inside, outside, core) before and after refiring under standardized conditions.

1.2.2. Reconstruction of the firing temperature by measurement of the shrinkage behaviour of the sherds after refiring in steps of increasing temperature.

1.3. Non-plastics.

1.3.1. Analysis of the quantity (in percentages of volume) and grain size of the non-plastics.

1.3.2. Qualitative analysis of the non-plastics in fresh and polished breaks of the sherds, as far as possible by using a stereomicroscope (40 x magnification).

1.3.3. Qualitative analysis of the non-plastics by means of thin-sections under a polarization microscope.

2. *Study of clay samples from near the archaeological sites.*

2.1. Workability tests.
Experiments to determine the behaviour of clays in relation to the potter's craft, e.g. the plasticity, the measure of shrinkage when dry or fired, the firing colour, porosity, etc.

2.2. Quantitative and qualitative analysis of the non-plastics in the clay (see 1.3) and comparison of the results with those of the analysis of the non-plastics in the sherds.

3. *Study of a local traditional type of working potters in the area of archaeological research.*
Local potters can offer information on the possibilities of local materials for their craft as none other.

It is not always possible to carry out all the above-mentioned aspects. Restrictions may be imposed by factors such as the character of available pottery excavated (complete pots and/or sherds); the possibility to make thin-sections to a sufficient extent; the ability to execute refiring tests; and the presence or absence of local potters.

In the case of our technological research-work in Iraq on the second millennium B.C. pottery from the above-mentioned sites the conditions were rather positive. All aspects of a ceramological investigation could be executed to various degrees. In the absence of electricity, an electric kiln could not be used for the refiring tests or the firing of the test bars made of the clay samples. The results in our gas-fired test kiln for use in the field equipped with a pyrometer and oxyprobe (atmosphere control) were less exact (Fig. 4). The number of samples for thin-sections had to be limited because of the restrictions in the Iraqi law with respect to the export of antiquities.

THE DIGGING AND PREPARATION OF THE CLAY

The Mesopotamian alluvial plain consists of secondary clay which was deposited by the Euphrates and Tigris rivers. For this reason, it may be assumed that the Mesopotamian potters in the second millennium B.C. collected their clay from largely available local deposits. At present, the potters who work northeast of Baghdad, along the road to Baquba, also use local clay. They collect the clay in a dry state, and soak it for twenty-four hours in a shallow pit to which well water is supplied through a gully (Figs. 5 and 6). In antiquity the clay was also soaked for a short time. This is indicated by undissolved lumps of clay which are often found in the sherds. They can be seen in fresh breaks and appear as conglomerates with a structure which is

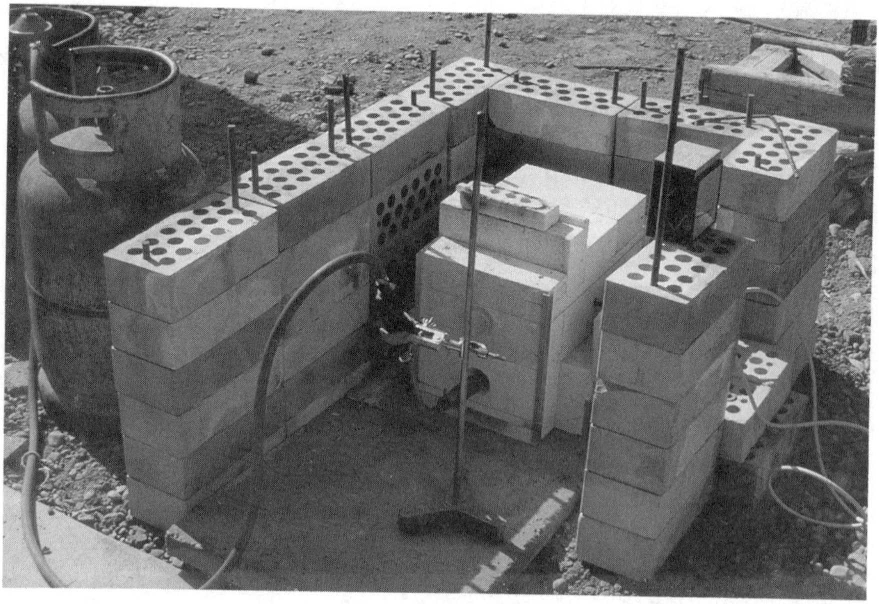

Fig. 4. Gas-fired test kiln for use in the field.

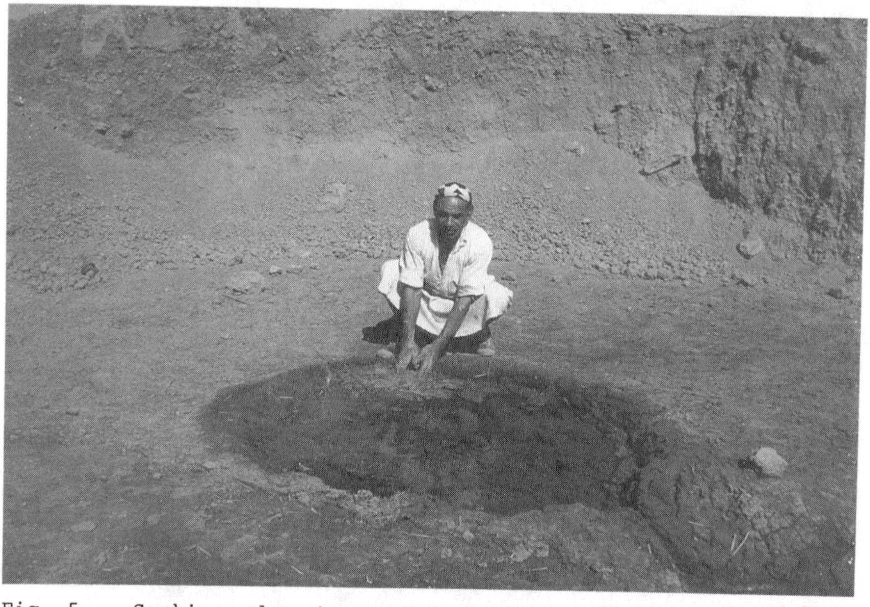

Fig. 5. Soaking clay in a shallow pit (a potter northeast of Baghdad).

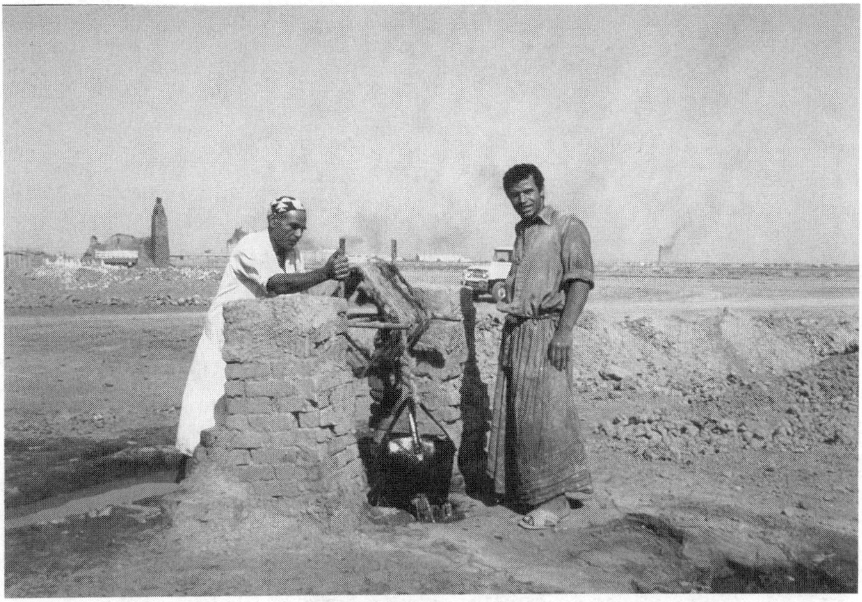

Fig. 6. Well used by the potters northeast of Baghdad.

identical to that of the matrix.

On comparison of the quantitative analysis of the non-plastics in the pottery and in the clay samples, it is unlikely that the clay was levigated, as to remove the coarser particles. Besides this, the fine clay necessary to produce a more delicate pottery was available in sufficient quantities by nature. It is also unlikely that non-plastic temper other than organic material was added to the clay during soaking and kneading processes. In most cases the clays could be used as they were found. The natural clay had little plasticity and strength (a short or lean clay).

One can conclude from the workability properties of the clay samples, that the potters did not find it necessary to reduce the plasticity of the natural clay or its resistance to deformation. These properties were already present as required. It became also evident by experiments that the clays did not show more than usual shrinkage. A decrease of shrinkage by adding mineral temper was therefore not necessary. It would only have increased the lack of coherence of the clay. From studying the pottery it appeared that the occurrence of stretching traces in the wall due to lack of coherence of the clay was indeed a problem to the potters of the second millennium B.C. (Fig. 7). Nowadays, the potters northeast of Baghdad still meet the same problems with their clays (Fig. 8). Such stretching traces are in fact cracks which were beginning to occur when a poor clay is being shaped outwardly during the throwing process or when the potter pulls the wall up too much. The cracks which occur during drying are not caused by excessive shrinkage as is the case with very plastic clays. They are the result of lack of coherence in the clay. To enhance the coherence of the clay and its resistan-

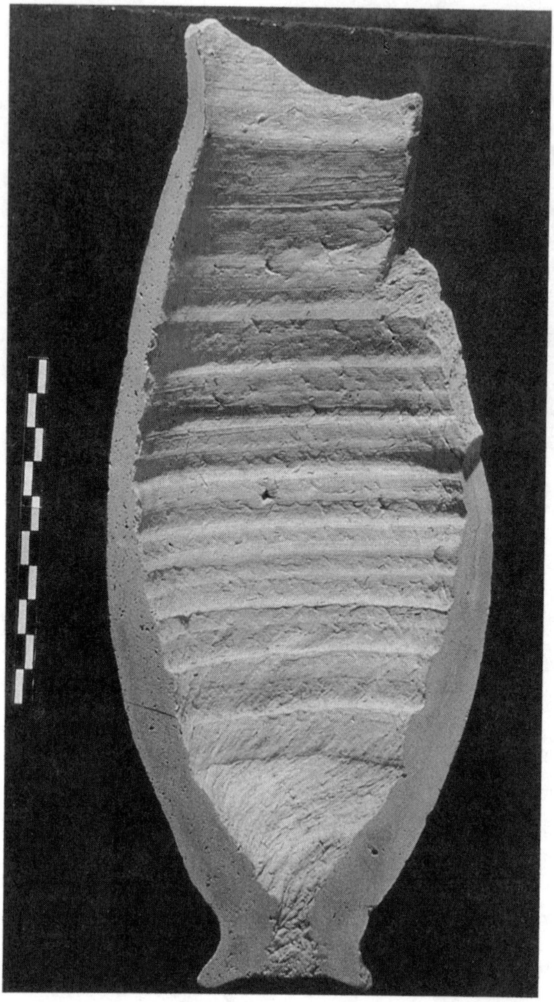

Fig. 7.
Stretching traces visible on a goblet from the Kassite period.

ce to tear, the potters added fibrous material to the clay. This way the structure of the substance was reinforced. By adding organic material the drying process was slightly slowed down, because organic material contains water for a longer period than clay. For the taller vessels chopped straw of a particular coarseness was used. This may well have been remainders from the harvest. In general, we found organic temper, in varying quantities and sizes, in the handmade and wheelmade pottery. It is possible that the straw in the clay used on a wheel was added in the form of manure. As a result, not only did the coherence of the clay increase, but also the plasticity. Current potters northeast of Baghdad use *Typha latifolia* (Fig. 9). By adding this fine fibrous tempering material, the cohesion and resistance

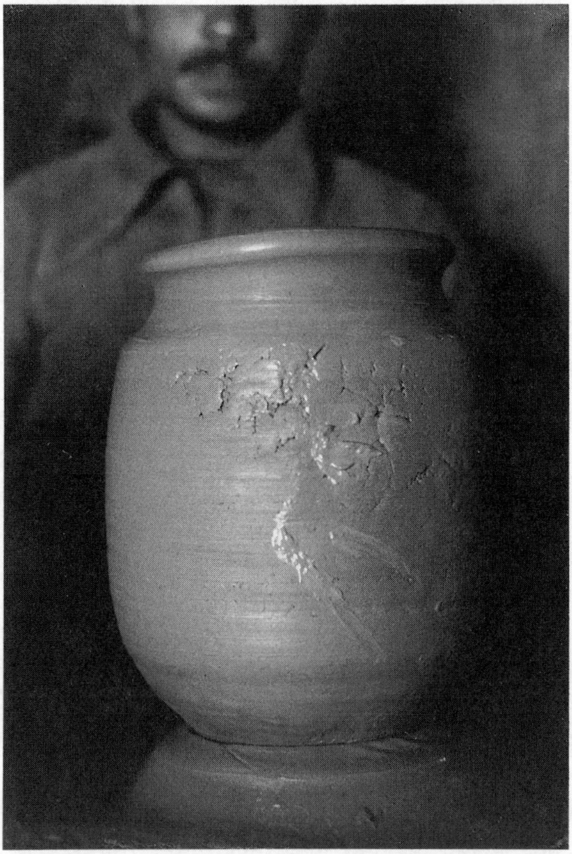

Fig. 8. Stretching traces visible on a pot made by a potter northeast of Baghdad.

to tear, which are both important during throwing and drying, are improved.

Finally, as elaborated below, a clay body tempered with an extra amount of organic material was used in order to repair or to prevent cracks during drying and firing.

THE SHAPING TECHNIQUES

Most of the pottery was thrown on a fast potter's wheel. The techniques applied to make the thrown pottery can be differentiated in four categories: (1) thrown from the cone; (2) thrown out of one piece of clay; (3) thrown in coils; and (4) thrown in cylinder parts. The use of these specific techniques was directly related with the shape, size and function of the pottery: trays, bowls, goblets, large jars etc. Several technical variants were observed within each technique such as the various techniques of finishing the rims (thickening by pushing up or pressing down the clay, and the various techniques of scraping the bases of the wheelmade trays (by hand in various

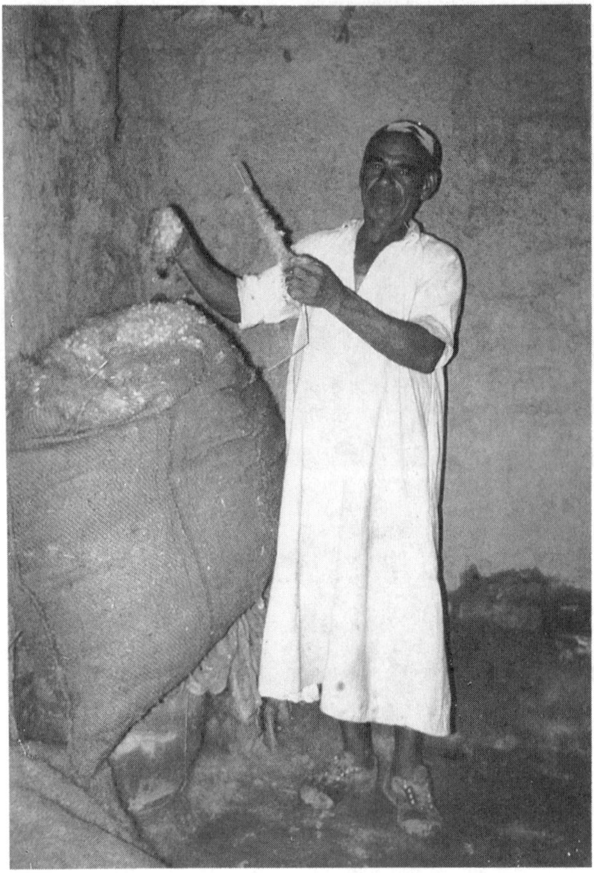

Fig. 9.
Fine fibrous tempering material (*Typha latifolia*).

directions or on a fast spinning wheel).

On the basis of the cut-off traces visible on the bottom side of the pot, the shape and the limited size of the pottery, it can be deduced that the greater part of the small and medium sized pottery was thrown from the cone. This technique was especially suitable for the production of large amounts of pottery. It seems that with this mass production a careful finish was not regarded as necessary.

Verification experiments made it clear that it was not easy while producing the goblets to create a base of the correct thickness [11] (Fig. 10). The thickness of the base was determined by the height of the cut with regard to the level of the base on the inside. A thick base was created when the goblets were cut off at a low height (as regards the level of the base on the inside). Such a base could not dry in an equal manner. This created shrinkage tensions, whereby cracks could appear. When the goblets were cut off at a high level, a very thin base was made. Such a weak base could easily crack, even with the

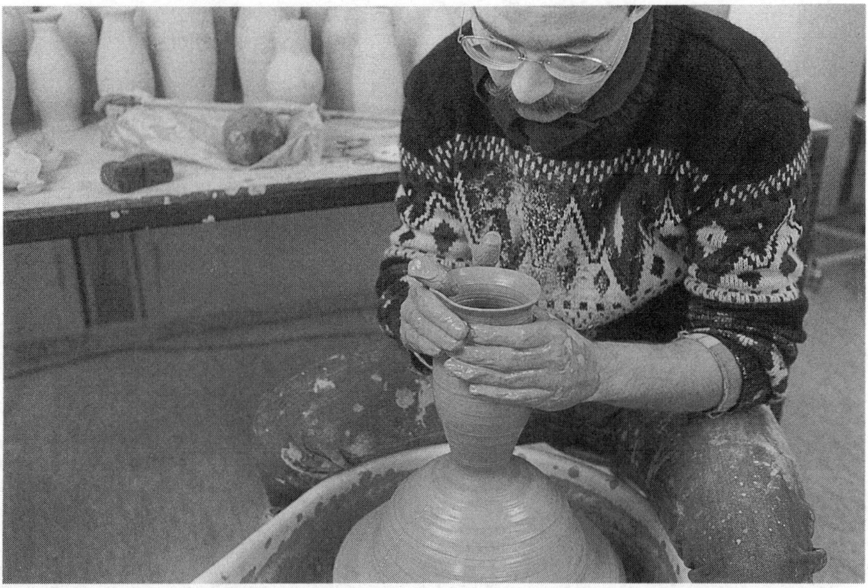

Fig. 10. Loe Jacobs testing the reconstructed shaping technique of an Old Babylonian goblet with clay from Iraq.

smallest shrinkage tensions which always occur. Already soon the Old Babylonian potters started to repair drying cracks as soon as they appeared, by filling them up with clay tempered with an extra amount of organic material. Because the organic fibers keep the clay together, the resistance against cracking at that place was strengthened. This method was not always very success-full. Very often the drying cracks partly redeveloped when the drying process was continued. This could happen because the extra organically tempered clay filling was not very well integrated with the clay of the goblets. For this reason, at the end of the Old Babylonian period, the potters started to provide the base part of their pottery with extra organically tempered clay before cracks even appeared. It was added to the clay of the base before raising the walls. This method became the prevailing technique during the Kassite period. From the profiles of the cut-through goblets can be seen which of both techniques (repairing or preventing) was used (Figs. 7 and 11).

In view of the limited plasticity of the clays it was necessary to make large pots from coils or cylinder parts. The sherds from pots of this category showed traces of the joints of coils. During the production of a particular category of high beakers, pre-fabricated cylindrical parts were joined together. The wall was thrown again after attaching the cylinders to each other. As a result of the joining traces of these parts one may conclude that the beakers depending on their size were formed from about four to seven of these elements.

In addition to the large collection of thrown pottery, there is also a small group of handmade pottery. This group includes miniature vessels and a category of large bowl-like vessels. The latter were formed by means of coiling, pinching and scraping.

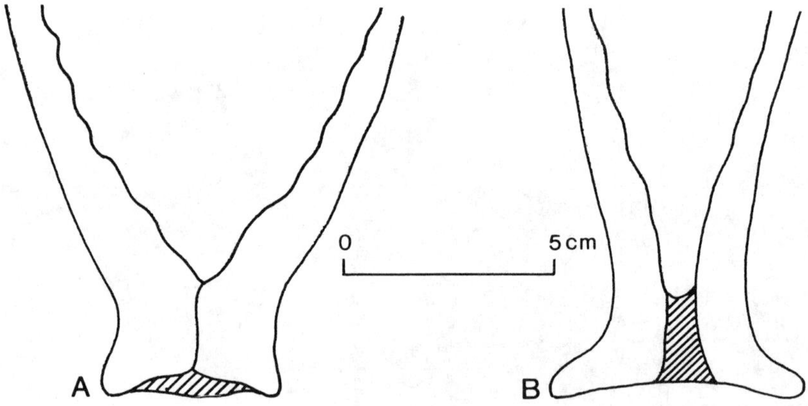

Fig. 11. Profiles of cut-through bases of goblets: the use of clay tempered with an extra amount of organic material. (A) Reparation; (B) Prevention.

The outside was smoothed by a hand covered with slip. Finally the vessels were sometimes decorated by fixing coils of clay to the wall. In the same way the rim was thickened and a ring to stand on was attached. It is remarkable to see the occurrence of bitumen on the inside of this pottery. It was probably applied to decrease the permeability of the wall. Bitumen was also used to repair broken pottery. Certain trays show also indications that they were handmade, although with the use of a turn table, as opposed to other wheelmade trays. The handmade trays have an irregular shape whose base is not smooth and level. The traces of stretching on the underside of the bottom are a result of pushing the clay on an uneven and dry surface. The base and the low raised rim were joined together in different ways. In most cases a coil of clay, which was later remodelled into the low raised rim, was fixed to the clay which had been flattened to form the base. Sometimes, however, the flattened piece of clay was put inside a coil which had been made into a ring and then fixed onto the base.

Finally, mention should be made of a particular type of jars occuring within the repertoire of which the upper section (upper part of the wall, shoulder, neck and rim) shows traces of throwing while the lower section (lower part of the wall and the base) shows traces of pinching, fixed coils, as well as traces of fixed flattened pieces of clay. Traces of wringing are visible on the inside at the centre of the base. One may conclude from this that while the upper section was thrown the lower part of it was at first left thick. After a drying phase this lower part was pressed and consequently gradually built up either by fixing coils and pinching or, alternatively, by using flat pieces of clay. Towards the end of the man's work, the jar was almost closed again. Next, if necessary the outside was corrected by tapping and finally it was smoothed by wiping it with a hand dipped in slip. In some cases an extra ringstand was attached. The technique may indicate specific work assignments. One potter probably made the top section while another completed the pots by coiling, pinching and tapping.

FIRING TECHNIQUE

On the basis of colour interpretation, refiring tests, and experiments with clay samples it can be stated that the pottery was in general fired in a potters kiln in a neutral to lightly oxydizing atmosphere, at temperatures varying between 800°C and 1000°C.

"In ancient Mesopotamia thorny desert shrubs called 'ašâgu' were collected in considerable quantities and the farm waste such as chopped straw and reeds and the cameldung still used as cheap fuel by potters and lime burners must have figured largely" [12,p.15]. Elsewhere these and other dried plant materials are and were also used for firing potters kilns [13,p.241; 14,p.39; 15,p.72; 16,p.6; 17,p.17; 18,p.77]. One needs a lot of organic fuel to bring a rather large potters kiln during some considerable time on 800°C-1000°C. Although ancient Mesopotamia was greener than present day Iraq, it is hardly unthinkable that firing of such masses of pottery as in the second millennium B.C. should not have led to a lack of fuel [19,p.35; 20,p.210]. Oil has twice as much caloric value than wood [21,p.251]. For this reason it should be plausible to find that in a region like the area near Hit, where oil is under such a pressure that it comes up to the the earth's surface by nature, crude oil was used as fuel for potters kilns [19,p.16,24,33,35; 22]. Especially the light oil (petrol) coming up to the surface from certain sources was very suitable for fuel. Laboratory research made it clear that the ancient oillamps from this region were burnt with the lighter fractions of crude oil, not with vegetable oils [23, p.90]. The use of oil products was known to Mesopotamian potters in the second millennium B.C. as indicated by the bitumen paint on the surface of a small part of the pottery. Large jars and pots were also found totally covered on the inside with bituminous material. It is plausible that it was a means of limiting the permeability. R.J. Forbes [19,p.64] has suggested that it is not impossible that some of the bitumen-lined jars are really the actual melting pots of mastic, used as mortar in architecture. Especially the heavy crude oil was used for the production of bitumen. If necessary, this crude oil was condensed by means of evaporation of the lighter fractions. Assuming that crude oil was used as fuel for potters kilns there must have been a system of transport between the oil sources and the potters workshops. The bitumen-lined jars could have been used for transportation of the crude oil.

Finally, we have another indication for the use of crude oil as fuel for potters kilns in antiquity in this region. In the present-day potters workshops northeast of Baghdad, where the potters produce their pottery in a traditional way, there is no question of any modernization at all, although at first glance the use of crude oil as fuel for the kilns by these potters looks like a modernization (Fig. 12). The primitive technology, however, makes it acceptable that the firing technique could have been known in the past. No installation with burners or oil pumps was used. The crude oil, poured into a hole in front of the firing chamber, was continuously scooped from the pit into the fire with help of a bowl or soup ladle (Fig. 13). This was to spread the oil. The result was a roaring fire, filling the wide low chamber, more than three meters in section. It is necessary to transform the oil from a liquid into a mist or vapor to make it burn effectively [21,p.158; 16,p.13]. This simple way could have been applied in antiquity. The colour of the pottery fired in oil-burnt kilns does not differ from the colour of the pottery from the

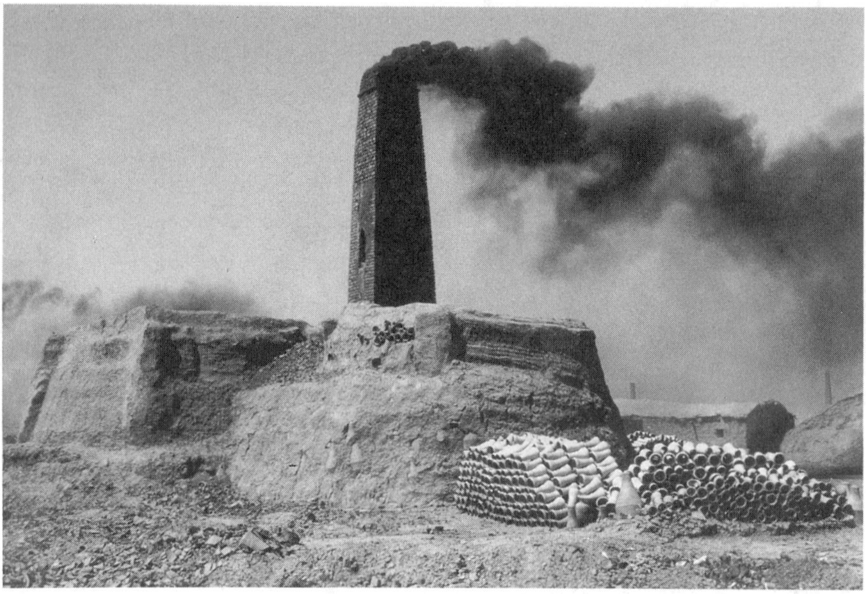

Fig. 12. Oil-burnt pottery kiln northeast of Baghdad.

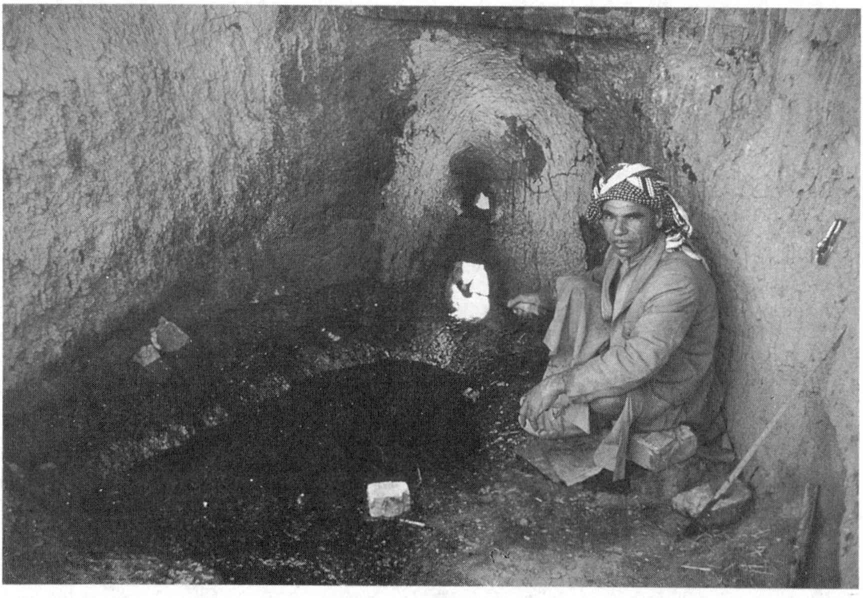

Fig. 13. Scooping crude oil into the fire of the kiln.

second millennium B.C. If it should be possible to find remnants of possible combustion products such as sulphur combinations we should have a stronger indication of the use of crude oil as fuel. The crude oil from the country around Hit and Ramadi on the south bank of the Euphrates has a very high sulphur content [19,p.32].

SUMMARY AND CONCLUSION

We have tried to give an idea of the work of the Mesopotamian potter in the second millennium B.C., especially the Old Babylonian and Kassite periods. To that end the shaping and firing technique of the excavated pottery from Tell ed-Dēr, Sippar, Isin and Nippur was analysed on the basis of relevant characteristics. The reconstructed shaping techniques were subsequently tested with appropriate clays. Macro- and microscopic studies were carried out in order to determine the basic materials used for the pottery production. Clay samples were collected from near the archaeological sites and these clay samples were used for workability tests. Attention was also paid to current potters' techniques in the area of research.

The technological analysis of the pottery has shown that the basic raw materials (Euphrates clays) and shaping techniques which were applied in the prevalent pottery from Tell ed-Dēr, Sippar, Isin, and Nippur and the pottery which was taken from the other sites for comparison do not differ from each other. In the course of the second millennium B.C. minor changes in the shape of the pottery can be noticed. There are, however, no important big technological changes after the Old Babylonian period. In the Kassite period the potters did not break with the Old Babylonian traditions. The only important difference lies in a technological advance of the Kassite ceramics, for during that period the potters had learned how to prevent cracks in the base part of the pots. This was a problem which the potters in the Old Babylonian period were unable to resolve.

References

1. H. Gasche, Dossiers Histoire et Archéologie 103, 56-58 (1986).
2. W. Al Jadir, Dossiers Histoire et Archéologie 103, 52-54 (1986).
3. B. Hrouda, Dossiers Histoire et Archéologie 103, 72-75 (1986).
4. McG. Gibson, Dossiers Histoire et Archéologie 103, 68-71 (1986).
5. H.J. Franken, and M.L. Steiner, Excavations in Jerusalem 1961-1962 Vol II. The Iron Age Extramural Quarter in the South-east Hill, (Oxford University Press, Oxford, 1990).
6. H.J. Franken, Newsletter Department of Pottery Technoloy Leiden 1, 1-4 (1983).
7. L. Jacobs, Newsletter Department of Pottery Technology 1, 34-35 (1983).
8. A. van As, in The many dimensions of pottery; ceramics in archaeology and anthropology, edited by S.E. van der Leeuw and A.C. Pritchard (Amsterdam, 1984) pp. 131-164.
9. P. Stienstra, Newsletter Department of Pottery Technology Leiden 4, 29-48 (1986).

10. A. van As, presented at the 1992 Conference on the History and Archaeology of Jordan, Irbid (in press).
11. A. van As and L. Jacobs, Newsletter Department of Pottery Technology Leiden 5, 39-53 (1987).
12. R. Forbes, Studies in ancient technology VI (Brill, Leiden, 1966).
13. M.B. Annis, World Archaeology 17(2), 240-255 (1985).
14. A. van As, L. Jacobs, and M.-H. Wijnen, Newsletter Department of Pottery Technology Leiden 6, 35-40 (1988).
15. A. van As and M.-H. Wijnen, Newsletter Department of Pottery Technology Leiden 7/8, 69-74 (1989/90).
16. J. Gregory, Kiln building, (London, 1977).
17. R. Hampe, and A. Winter, Bei Töpfern und Zieglern in Süd-Italien, Sizilien und Griechenland, (Mainz, 1966).
18. P.T. Nicholson, and H.L. Patterson, World Archaeology 21(1), 71-87 (1989).
19. R. Forbes, Studies in ancient technology I, (Brill, Leiden, 1964).
20. F.R. Matson, Ceramics and Man, (London, 1966).
21. R. Fournier, Illustrated dictionary of practical pottery, (London, New York, 1977).
22. L. Jacobs, Profiel, Contactorgaan historische Leidse archeologieën Rijksuniversiteit Leiden 2, 45-56 (1991).
23. H. Gasche, Mesopotamian history and environment I, (Gent, 1989).

* We would like to thank the other participants in the Working Group on Mesopotamian Pottery for their amicable cooperation. Our gratitude also goes to the Department of Antiquities, Baghdad. Financial support for our research-work has provided by the University of Leiden; the Netherlands Organization of Scientific Research (NWO); the Belgian Archaeological Expedition to Iraq; the German Archaeological Mission to Isin (University of Munich); and the Oriental Institute Expedition to Nippur (The University of Chicago). We thank Gloria London for carefully reading the manuscript.

THIRD MILLENNIUM B.C. TIN PROCESSING DEBRIS FROM GOLTEPE (ANATOLIA)

Pamela Vandiver*, K. Aslihan Yener* and Leopold May**

*Conservation Analytical Lab, Smithsonian Institution, Washington, D.C. 20560, USA
**Chemistry Department, The Catholic University of America, Washington, D.C. 20064, USA

ABSTRACT

Ceramic remains of fragmentary vessels, believed to be smelting crucibles based on microscopic field investigation, were analyzed using a variety of techniques in order identify the metal being processed and to reconstruct the ancient processing, especially ranges of composition, temperature, atmosphere, and the constraints on processing which each of these processing variables would have imposed. The identification of the metal being processed in the crucible surface residues and the slags provide data to resolve a dispute over the nature of the ancient metallurgical processing at Goltepe; gold and iron have been suggested as alternate metals but were not found. Intensive study of the ceramic and slag microstructure, composition and thermal history was required to establish the processing constraints which will be used to aid future replicative smelting experiments at the site.

INTRODUCTION AND PREVIOUS STUDIES

The analyses of slags, surface residues and earthenware refractories from the mountainous site of Goltepe in south central Turkey provide evidence of tin metal processing in the mid-third millennium B.C. This is the earliest evidence of tin processing in the ancient world. Tin was an important metal because, when it is alloyed with copper, bronze is formed. This important metal was formed into objects of value and traded throughout centers of power as a prestige good, and in fact, the Bronze Age is the name of the period in which this sophisticated technology is found. Bronzes are alloys of about 90 to 95% copper with 5 to 10% tin and contain impurities of about a percent or less of iron, arsenic and manganese, and even less of other elements(1). A typical range of ratios of iron to tin in bronzes of this period is 1 part Fe to 4-10 parts Sn. The tin component can be added to copper in the form of the metal, the oxide or a mixture of, for instance, iron and tin oxides, so that many possibilities for various usable products of tin extraction exist(2).

Veins in the Kestel mine a short distance from the habitation and industrial processing site of Goltepe analyzed by x-ray diffraction and atomic absorbtion have been shown to contain quartz, calcium carbonate, magnetite and tin oxide as SnO_2, cassiterite, originally in a concentration of 0.7%, but more recently assays have shown 1.5% (3). Another source of tin ore, which would be expected based on geological evidence, is alluvial sands and gravels below the mine. However, geological survey has shown this evidence is buried by recent sedimentation to a depth of about 15 meters. Submillimeter particles of cassiterite from the floor of the mine have been isolated visually and sensually by their brown to purple color, translucency, rhombohedral cleavage and relatively high density(3).

ARCHAEOLOGICAL CONTEXT OF THE TECHNOLOGY

In front of and below the mine of Kestel were found many ground

Mat. Res. Soc. Symp. Proc. Vol. 267.

stone tools, probably used in ore dressing, beneficiation and other grinding operations. Below this activity site is a stream which may have been used to separate the less dense limestone with a specific gravity of 2.71 and quartz (sp. gr. of 2.65) from the more dense magnetite (5.18) and cassiterite (6.95). Further grinding may have been required, because the similarity of the densities of magnetite and cassiterite makes them difficult to separate. However, the Burgher's vector of cassiterite is less than that of magnetite, such that an equivalent amount of grinding would reduce cassiterite to a finer particle size than magnetite.

On top of a hill across the stream from the mine is the site of Goltepe. It is far removed from the major north-south trade routes, but commands a view of the access route from the lower valley. The top of the hill, where the largest residual hearth area in which bright fire reddened earth was found, is exposed to considerable wind, which may have been channelled as a natural draft into furnaces. Pit houses partially dug into rock were excavated on more sheltered side slopes. In addition, part of what is probably a circuit wall was excavated, implying there was a need to expend effort for seclusion or protection. The artifact assemblage excavated from this workshop/habitation site includes grindstones, multifaceted molds, ore nodules, slag and metal fragments. From the small size of the grindstones and crucibles with diameters of 30 to 40 cm, and from the presence of a wall around the village, we surmise that tin was a precious commodity which was processed in small quantities, and one which required a protected production site to insure success, and perhaps secrecy.

THE FRAGMENTARY CERAMIC VESSELS IDENTIFIED AS CRUCIBLES

Twenty-four ceramic fragments from bowl-shaped crucibles, or possibly as partial covers, were analyzed as a representative ten percent sample from the 250 excavated in the 1990 season (Fig. 1). Most of the vessel fragments were found in secondary contexts, some on floors, as reported elsewhere(4), and were dated directly by thermoluminescence and the associated charcoal was dated by radiocarbon methods. In addition, considerable Early Bronze Age pottery at the site provides a baseline for chronology(4).

Most of the ceramics look like vegetal-tempered cookpots, except that they are blackened on the insides (Fig. 2). Low-power optical microscopy shows that adhering to the inner surfaces are small, submillimeter droplets and rarely elongated driplets of a nonwetting, black glassy slag. There are also patchy areas of surface with a layer having a brown to gray porous, compact powdery residue which in a few areas is sufficiently hard to be considered a layer or concretion. Some of the inner surfaces are somewhat bloated to a depth of less than a millimeter.

Considerable variability in fabric texture is found with sand tempered, slag resistant layers on the interiors of some of the crucibles and vegetal fiber-tempered backing or exterior layers. These exterior layers have considerable closed porosity, sometimes up to 40vol%, and they have superior insulation and better thermal shock properties (Fig. 2). The method of manufacture was consistent and involved slab building in pancake-like pieces, often with a strip or coil added to form the rim. Because of the small size and because there are no butt joints commonly found as where two sections are joined, each crucible was probably formed at one sitting. No evidence of differential drying shrinkage or delamination occurs between layers

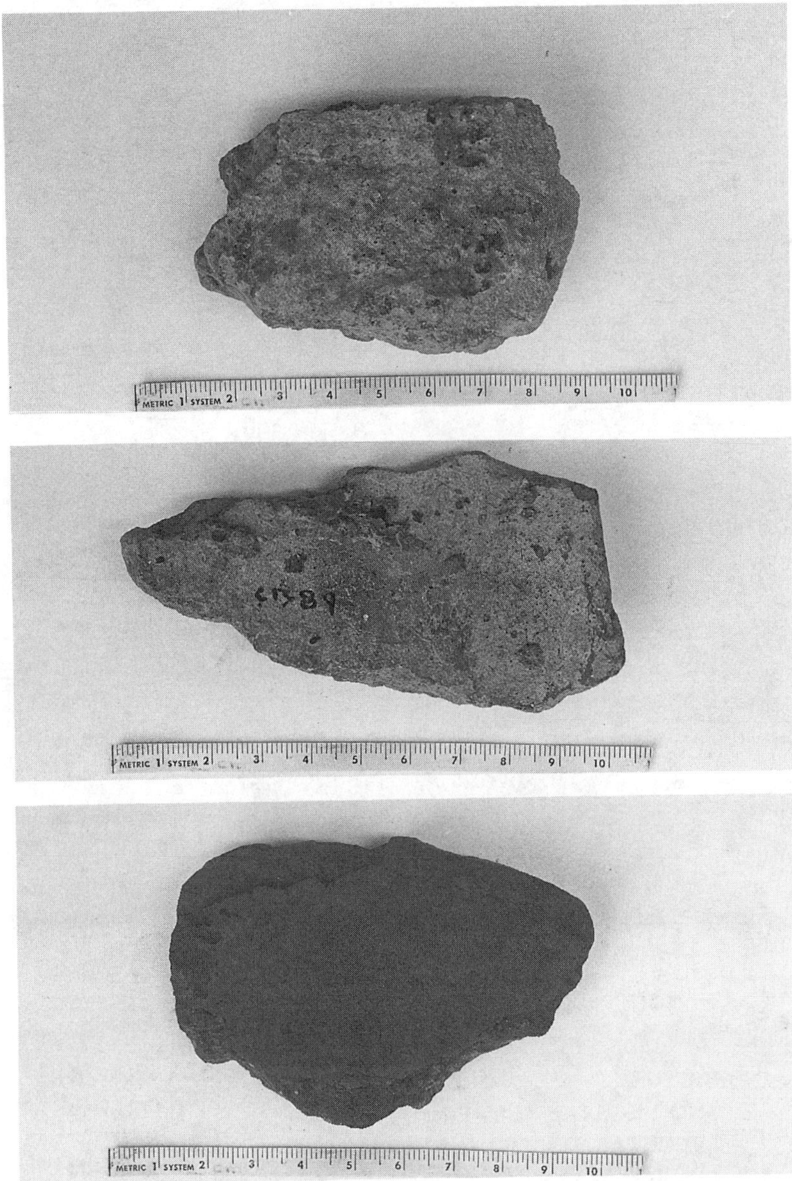

Fig. 1. Representative ceramic crucible fragments from Goltepe. (Top) Crucible 4, top view, showing inner scoriaceous surface, (Middle) Crucible 23, thin-walled, with powdery surface containing tin and other elements, (Bottom) atypical Crucible 24 with brown, powdery exfoliating surface containing elemental arsenic.

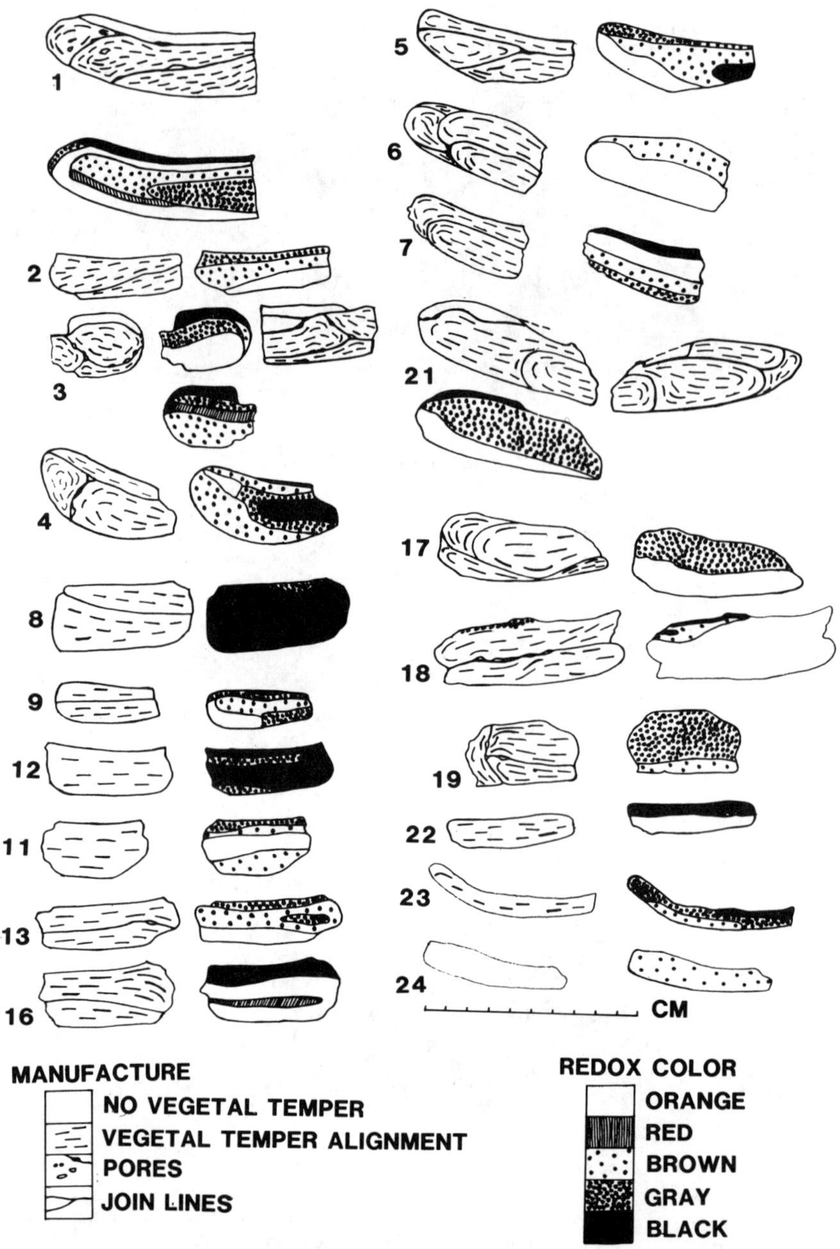

Fig. 2. Cross sectional drawings of Goltepe crucibles showing slab construction and colors caused by oxidation (red and orange), mixed redox state (brown) and reduction (grey and black).

to indicate that the crucibles were relined or reused.

The clay body, as shown by x-ray diffraction, consists of an illitic clay, albite, quartz and, by optical microscopy, quartzite and a variable amount of chopped, fibrous vegetal temper for which no diffraction pattern of a carbon-containing species was found in lower fired examples. The illite lines were identified on a low-fired, exterior fragment of one of the crucibles. A platey morphology, particle size of 1-3 microns viewed by SEM and an EDS pattern showing potassium are consistent with this x-ray identification of illite. The fragments probably functioned as crucibles, especially those where bloated, reduced, scoriaceous surfaces are present, or as partial lids, especially where brown, casiterite-containing surface residues and small droplets of slag remain. Some sort of lid may have been used to partially cover the charge to kept out oxygen during the early stages of sintering, although no in situ features of furnaces have been found. However, the lack of a greater quantity of slag also suggests that a layer near the crucible might have consisted mainly of charcoal. Considerably more reduction is required to achieve the highly reduced atmosphere required to smelt tin than is required to smelt copper ores.

The compositions of four of the crucibles were analyzed with an ARL wavelength dispersive electron beam microprobe defocused to 15 microns and using a finely powdered 200 mg. sample previously fused in a carbon arc furnace in a tungsten boat. Three of the higher fired crucible fragments, prior to fusion, were red or grey in color and quite homogeneous in bulk composition and mineral assemblage as identified by x-ray diffraction (albite, quartz, illite and glass).

Sample Number	SiO_2	Al_2O_3	FeO	MgO	CaO	K_2O	Na_2O	TiO_2	MnO	Total
	Weight Percent									
No. 12	63.0	20.8	1.27	0.67	0.56	4.52	6.64	0.04	0.06	97.59
No. 8	64.0	19.7	0.54	0.19	0.34	3.71	7.64	0.01	0.03	96.21
No. 1	61.8	20.1	2.29	1.23	0.82	4.43	5.74	0.24	0.08	96.28
	Standard Deviation									
No. 12	4.8	1.7	0.9	0.7	0.3	3.3	3.2	0.06	0.08	(n=11)
No. 8	4.1	1.4	0.4	0.17	0.21	2.9	4.1	0.01	0.01	(n=10)
No. 1	10.9	3.6	2.0	3.1	0.54	2.2	3.2	0.71	0.12	(n=18)

Table I. Microprobe Analyses of Ceramic Crucible Fragments.

These compositions are unusual for refractories and lie in the least stable, flux-rich corner of a plot of ancient refractory compositions measured by Freestone(5). EDS analyses reveal no sulfur or chlorine peaks which would imply that the ceramic has been contaminated with salts. None of the excavated ceramics were coated with salts, other than a minor amount of calcium carbonate. The site is a well-drained hill with very slightly acidic soil. Much of the region consists of uplifted granitic formations, with highly alkaline, igneous rocks of composition similar to those in Table I and published in Hughes(6). Thus, these compositions which are so high in soda and potassia are not surprising. The crucible compositions lie in the compatibility triangle of albite, nepheline and sodium-silicate in the Na_2O-Al_2O_3-SiO_2 phase diagram(7). The first two phases have been shown to be very difficult to precipitate from glass melts, and the third is not stable in a moist atmosphere. The equilibrium temperature at which the first liquid starts to form is 732°C(7), although the ceramic is not molten enough for viscous flow until about 1150°C, as tests refiring 1 cm cubes of the crucible fragments have shown.

The fourth crucible, no. 24, is visually and compositionally atypical. The color is a bluer brown than any other, and texture is coarse, less varied in size, friable and layered, as if fired many times or reacted with a gaseous and corrosive composition. The composition of the surfaces was also atypical, as identified by XRF (see Table 3 below) in that considerable elemental arsenic and some copper were identified. This is the only crucible fragment with copper. The x-ray diffraction pattern of a half millimeter deep surface layer showed that quartz, hematite, illite, cuprite and possibly arsenic oxides are present. The bulk composition shows this crucible to be an outlier; especially note the low alkali values and almost inverse ratio of iron oxide to alumina.

	SiO_2	Al_2O_3	FeO	MgO	CaO	K_2O	Na_2O	TiO_2	MnO	Total
Weight Percent	57.6	2.3	20.4	0.2	0.5	1.1	0.04	0.02	0.09	79.9
Standard Deviation	7.2	2.5	1.1	0.1	0.1	0.1	0.01	0.02	0.02	(n=20)

Table 2. Unusual Composition of Crucible no. 24.

This large difference in composition and mineral assemblage points to a different function and production event. We hypothesize that this crucible was used in the intentional production of arsenic oxide, which will be discussed below as a possible additive to the ore charge.

The firing temperature range of the ceramics is 800-950°C, or perhaps 1000°C, as determined according to the methods of Kingery and Tite(8) as shown in Figs. 3 and 4. Tite's methods tend to emphasize lower magnification (generally below 1000x) polished sections in which it is easy to demonstrate that the degree of vitrification is representative of the ceramic and in which major morphological changes are readily apparent to a nonspecialist. Kingery's methods tend to emphasize higher magnification (generally 1000-5000x) fractured sections in which observation of submicron clay particle morphology and fine changes in sintering behavior are more easily studied to determine fine morphological changes. To insure representative areas are studied, composition was held constant using EDS, and both ranges of magnification and types of samples were evaluated. Fig. 3(top) shows a low-fired exterior cross section which contains irregular-shaped pores and which retains angular granular structures at 850x which are characteristic of poorly sintered raw materials. Fig. 3(bottom) shows a more vitrified structure at 750x in which more glass has melted to fuse the particles together and in which the pores have begun to round as flow has occurred. Fig. 4 shows the low-fired outer surface of crucible 3 which was divided into pieces and refired to 700, 800, 900, 1000, 1100°C, freshly fractured in cross section to insure similar orientation and studied at 1000x and 5000x for changes in microstructure with increasing temperature. This is one of five different crucible fragments which was refired and used as a standard indicator of temperature. The sample at 800x is somewhat sintered as the particles are subangular, have distinctive particulate surface features and are poorly joined to one another with only minor amounts of glass. The sample at 900x is well sintered as the particles are rounded, have smoother surface features, and the pores are well rounded with lots of glass coating the particles and insides of the pores. The sample at 1000x is overfired, as the pores have begun to enlarge and bloat, particles are almost spherodized and considerable glass flow is beginning to occur. The original samples which were not refired match the incremental range from 800°C to just short of the

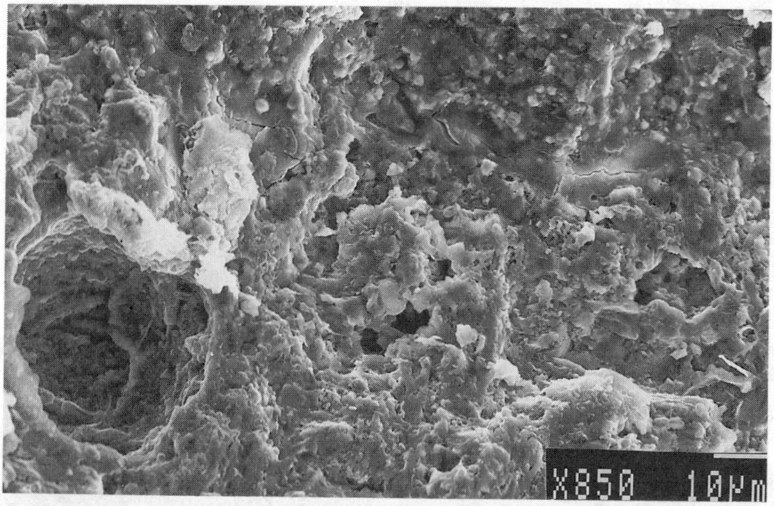

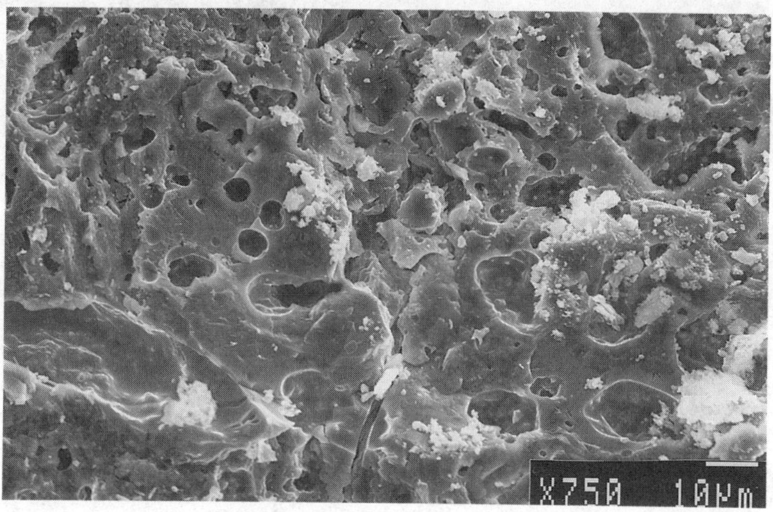

Fig. 3 (top). Fractured cross section of porous, friable, low-fired, red exterior showing fine submicron particles of clay still clearly visible although somewhat joined together and pore (lower left) with rough, irregular inner surface, all indicative of low firing. Prior to firing the clay particles would have appeared as submicron angular platelets; (bottom) reduced, blackened inner layer shows a more glassy appearance with rounded, well fused pores.

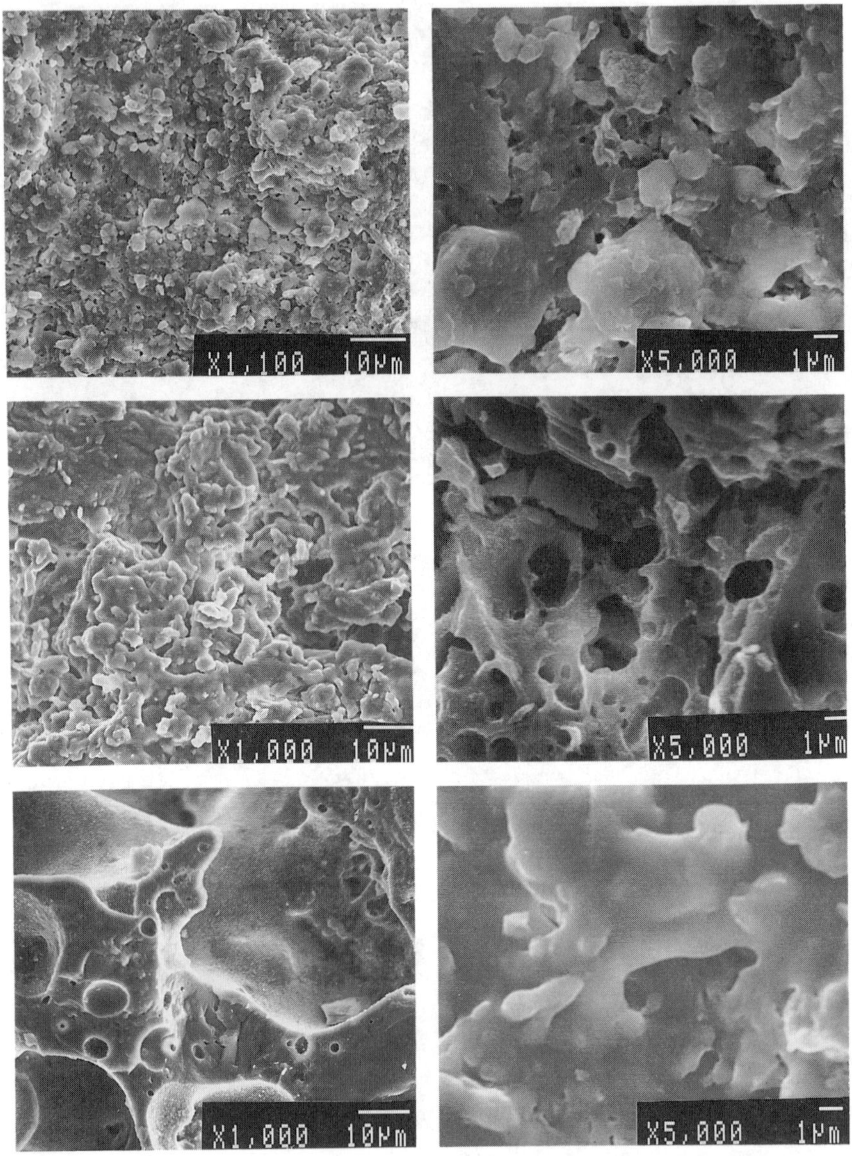

Fig. 4. Firing temperature range of the crucibles is shows by matching samples of theoriginal crucible microstructures with these samples refired to 800°C (top), 900° (middle) and 1000° (bottom) for a 5 minute ramp and 15 minute soak at each temperature. Each temperature is shows at two magnifications, 1000x (left) and 5000x (right).

Fig. 5. Schematic drawing of lower part of tin smelting furnace; possible furnace superstructure and mechanism for flow of gas (draft) cannot be reconstructed on present evidence.

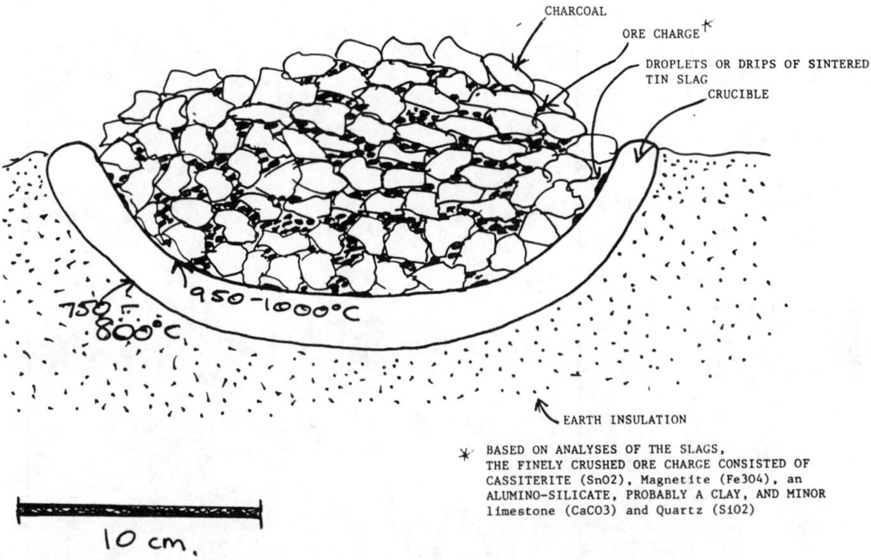

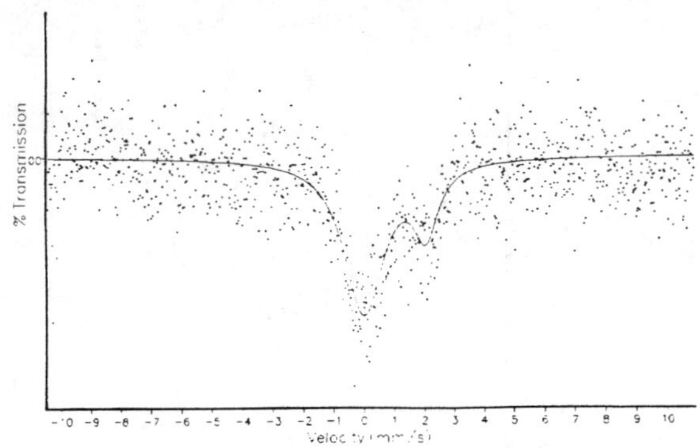

Crucible 8

Mossbauer Spectrum

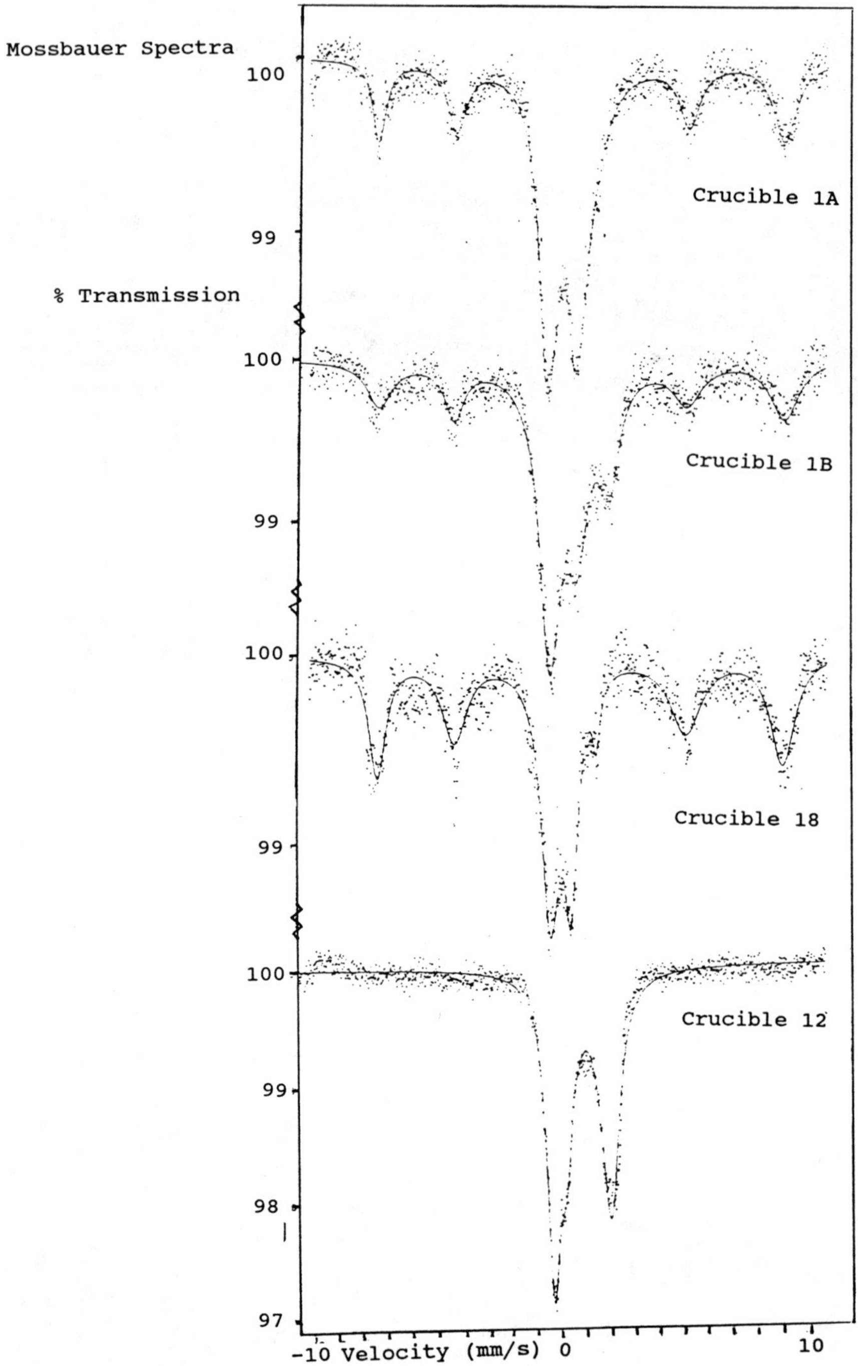
Mossbauer Spectra
% Transmission
100
99
Crucible 1A
100
99
Crucible 1B
100
99
Crucible 18
100
99
98
97
Crucible 12
-10 Velocity (mm/s) 0
10

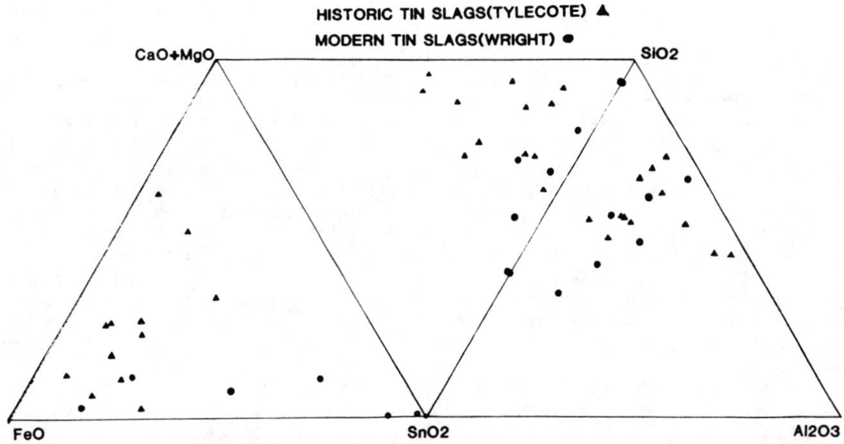

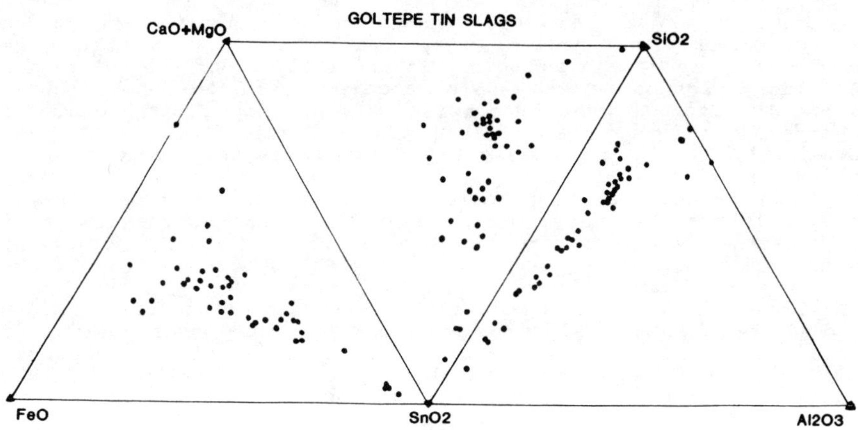

Fig. 6. Summary of current data on the compositions of tin slags from historic periods based on Tylecote (triangles) and modern tin slags from Wright (dots).

1000°C refired temperature.

There is an envelope of processing in a plot of time versus temperature for a firing in which the microstructure produced may be the result of a longer time at a lower temperature, or visa versa. Because these crucibles contain such thin inner layers of high-fired ceramic (0.6 mm being the thickest) and because of the presence of so many colors which indicates mixed oxidation, we conjecture that they were fired for only a short time and did not have time to equalize either temperature or atmosphere throughout the cross section. They may have been fired for a half hour. We further conjecture that the crucibles were set in the ground during the smelting operation, as schematized in Fig. 5, in order to account for the lower temperature and oxidized exteriors. It is possible that the crucibles were not prefired as ceramics, but rather, they were fired for the first time with the tin ore charge in place. It is also possible that the temperature of the ore charge may have been higher than indicated by the refiring tests, and that charcoal next to the crucible may have insulated the refractory.

Mossbauer spectroscopy was used to study the variation in the iron species in relation to the color in the polished surfaces of 4 crucible fragments (Figs. 2 and 6). The spectra were obtained on the Mossbauer spectrometer model MS-900 (Ranger Scientific) in the acceleration mode with moving source geometry at room temperature using 700 mg of sample in a polethylene cell. The resultant spectra were analyzed by a constraint, least-square fit to Lorentzian-shaped lines. The parameters are relative to the center of the iron foil spectrum. The samples were crushed and ground from an orange, presumably oxidized region (1A), from a brown, presumably less oxidized, region (1B) of crucible fragment 1, orange area of fragment 18, black, presumably reduced, area of fragment 12. The spectra of 1A and 18 (Fig. 6a and 6d) consist of two subspectra; magnetic and superparamagnetic hematite of differing proportions (1A- 38.6%, 61.4%; 18- 59.5%, 40.5%, respectively). The spectrum of 1B (Fig. 6b) includes a third subspectrum due to ferrous iron with 44.1%, 47.4% and 18.5% for magnetic and superparamagnetic hematite and ferrous iron. This indicates that the iron in 1B is less oxidized than in 1A with 18.5% of the iron in 1B reduced. The sample no. 12 contains only non-magnetic ferric (85.2%) and ferrous (14.8%) species. The differing amounts of hematite and ferrous iron in these samples account for the different colors of the samples. The blackest of the crucibles, no. 8, had very little signal compared with the other samples and contained very little iron (Fig. 6 and Table 1). The intense black color is probably from reduced carbon, indicating heavy reduction; and this was confirmed by C-N-O analysis showing higher amounts of C in the crucible 8 compared with the others.

Crucibles analyzed by Mössbauer spectroscopy

Run	Sample	Color	% Ferric		% Ferrous
			%Fe_2O_3	% Super*	
1762	1A OR (A) OXID	Brown	38.64	61.36	0
1763	1B O-R (B) RED	Greenish?	44.10	47.36	18.54
1765	12 RH	Black	0	85.18	14.82
1764	18 O-S	Brown	59.52	40.48	0

Mössbauer parameters**

Run	Sample	H (Fe_2O_3)	Super* QS	Super* IS	Ferrous QS	Ferrous IS
1762	1A OR (A) OXID	512.5	1.13(21)	0.16(5)		
1763	1B O-R (B) RED	511.1	1.05(22)	0.18(5)	2.39(22)	0.93(5)
1765	12 RH	-	0.67(21)	-0.01(5)	2.16(22)	0.91(5)
1764	18 O-S	509.4	0.79(21)	0.07(5)		

* Super = % Supermagnetic part or ferric iron.

** All parameters in mm/s relative to iron foil. QS = quadrupole splitting, and IS = isomer shift.

SURFACE RESIDUES

The interior and exterior of each crucible fragment and those interiors with large areas of slag or brown concreted powder were analyzed using nondestructive x-ray fluorescence (United Scientific Dubois Object Analyzer with Tracor Northern 5502 Analysis System), shown in Table 3. Elemental tin was detected on the interior surfaces of 21 of the crucibles, but not on any of the exterior surfaces. No tin was found on one thin-walled (no. 22) and two thick-walled crucibles (nos. 20 nd 21). Copper was found only on the atypical crucible no. 24. These analyses indicate that tin was processed on the interiors of the crucibles and that it was processed separately from copper.

Arsenic was found on half of the crucibles in concentrations of parts per hundred, with the exception of no. 24 had areas which contained about 2-5% arsenic oxides. On three examples arsenic was found on the outsides, but not on the insides of the crucibles; on one, it was found both on the outside and inside. On four it was found on the interiors only. We conclude that variable amounts of arsenic were present, but also that it was not always present. No ore deposits or minerals containing both tin and arsenic oxides are reported; and none are present in the reference collections of the Museum of Natural History (Russell Feather, personal communication). Because they do not occur in nature together, we suggest that the arsenic and tin oxides were added together intentionally by a human agent. Arsenic is not present in the ore from Kestel, although separate sources are found in the region.

We suggest that tin oxide, probably as SnO, and arsenic oxides may have deposited from the vapor phase. The rounded particle morphology, typical of precipitation from the vapor phase, of the tin-containing particles in a brownish inner crucible surface is viewed using SEM and EDS (Fig. 7). The high vapor pressure and low vaporization temperature of arsenic oxide may acount for the variable amount of arsenic oxide present.

Table 3. Elemental Identification of the Interior and Exterior Crucible Surfaces by Nondestructive X-Ray Fluorescence (no gold, silver or lead was identified). Peaks are listed in order of decreasing intensities. Minor peaks are 10% or less of the height of the highest major peak.

Crucible No., Locus and Type of Area Sampled		Interior Surface Major	Interior Surface Minor	Exterior Surface Major	Exterior Surface Minor
THICK-WALLED CRUCIBLES CONTAINING TIN					
1 MRN325	Ceramic	Fe,Ca	Sr,K,Ti,Mn	Fe	K,Ca,Ti,Sr
	Slag Drip	Sn,Fe,Ca	Ti,Mn,Sr,Rb		
	Slag Drop	Fe,Sn	Ca,Ti,Mn,Sr		
	Residue	Sn,Fe,Ca	Sr,Ti	Fe,Ca	K,Ti,Mn,S,Rb
2 MRN325	Ceramic	Fe,Ca	K,Ti,Mn	Fe,Ca	K,Ti,Mn,Sr,Rb
	Slag Drop	Fe,Ca,Sn	Ti,Mn,Sr,Rb		
3 MRN325	Scoria	Fe,Sn,Ca	Ti,Mn,Sr,Rb	Ca,Fe	K,Rb,Sr
4 MRN325	Scoria	Fe,Ca	Sn,K,Ti,Sr,Rb	Ca,Fe	K,Ti,Mn,Sr
6 MRN325	Scoria	Fe,Sn	Ca,K,Mn,Sr,Rb	Fe	Ca,K,Ti,Mn,Sr
8 MRS325	Scoria	Fe,Sn	Ca,Sr	Fe	Ca,K,Sr
9 MRN325	Slag Drip	Fe,Sn	Ca,Sr	Fe	Ca,K,Ti,Sr,Rb
10 MRN325	Scoria	Fe,Sn,Ca	Sr,Rb	Fe	K,Ca,Ti,Sr,Rb
14 MRN325	Scoria	Fe,Sn	Ca,Rb,Sr	Fe,Ca	Sr,Rb,K,Ti
15 MRN325	Scoria	Fe,Ca,Sn	Sr	Fe,Ca	Sr
16 MRN636	Ceramic	Fe,Ca	K,Mn,Ti,Sr,Rb	Fe	Ca,K,Ti,Mn,Sr
	Slag	Fe,Sn	Sr		
17 MRN347	Scoria	Fe	Sn,Ca,K,Sr,Ti,K	Fe	Ca,K,Ti,Sr
THICK-WALLED CRUCIBLES CONTAINING TIN AND ARSENIC					
5 MRN325	Scoria	Fe,Sn	Ca,Sr,Mn,Ti	Fe	Ca,K,Ti,As,Sr
7 MRN325	Scoria	Fe,Sn	Ca,Sr,As,Rb	Fe	Ca,K,Ti,Mn,Sr,Rb
11 MRN325	Ceramic	Fe,Ca	Sr,As,Sn,Ti,K,Mn	Fe	Ca,Sr,K,As
	Residue	Fe,Ca,As	Sr,Sn,Ti,K	Fe,Ca,As	Sr,Sn,K,Ti,Mn
12 MRN325	Scoria	Fe,Ca	Sr,Ti,Mn,As	Fe	Ca,K,Ti,Mn,As,Sr
18 MRN1401	Scoria	Fe,Ca	Ti,As,Sn,Sr	Fe,Ca	Ti,As,Sr
19 MRN1401	Scoria	Fe,Ca,As	Fe,Sn,Ti,K	Fe,Ca,As	Sr,Sn,K,Ti,Mn
THICK-WALLED CRUCIBLES CONTAINING ARSENIC					
20 MRN1359	Ceramic	Fe,Ca	As,Sr,K,Ti,Mn		
21 MRN409	Residue	Fe	Ca,Ti,K,As,Mn,Sr		
THIN-WALLED CRUCIBLE CONTAINING ARSENIC					
22 MRN38	Residue	Fe,Ca	As,K,Ti,Sr		
THIN-WALLED CRUCIBLES CONTAINING TIN AND ARSENIC					
23 C1389	Slag Drip	Fe,Sn	Ca,Sr	Ca,Fe	Sr,As
24 MRN409	Residue	Fe,As	K,Ca,Sn,Sr,Sb,Cu	Fe,As	Ca,K
	Slag	Fe,As	K		

SLAGS

Compositional analysis by x-ray diffraction indicates quartz, calcium carbonate, a ferroaugite (faissaite), glass, cassiterite and possibly cristobalite and tin metal. Of the 128 diffraction peaks, no match was found for 31, and only one tin metal peak was found with

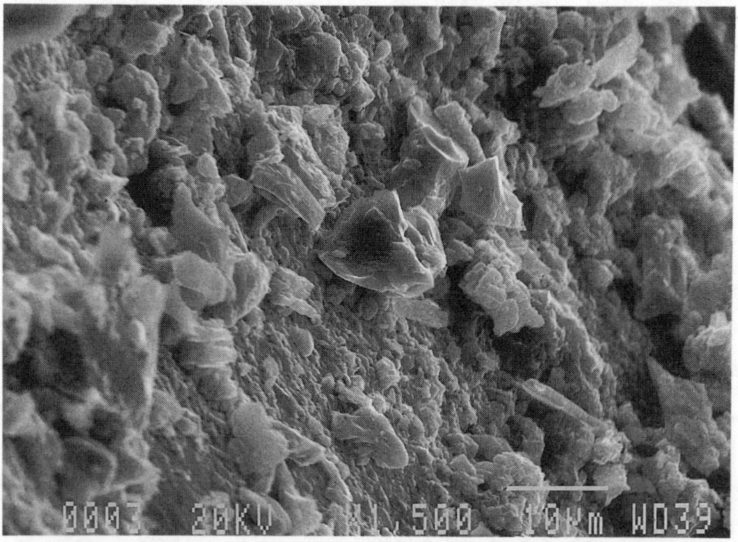

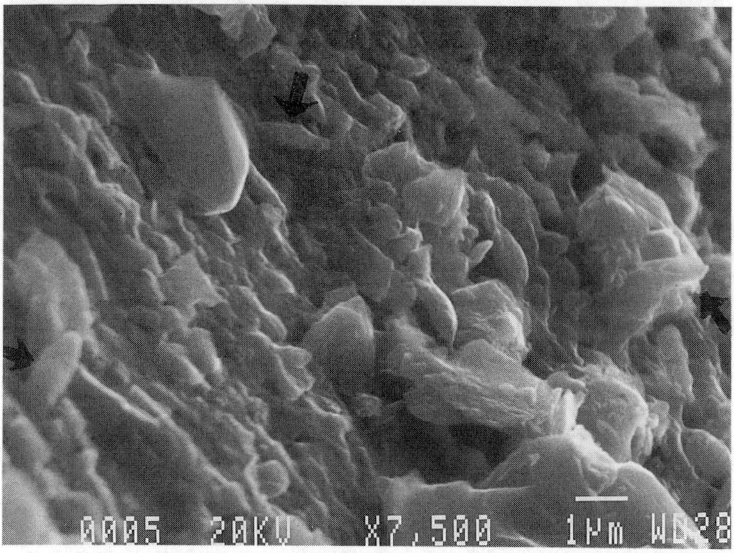

Fig. 7. Inner surfaces of crucibles show a powder well-adhered to the surface (upper) and composed of fine subangular to rounded particles which contain tin (lower). The rounded and acicular (arrow) particles measure from 0.1 or less to 3 microns and are evidence for precipitation from a gaseous phase. Contaminants and weathering products are also present on this surface, such as the 10-micron quartz particle (top, center).

another possible. Faissaite is a ferroaugite of the clinopyroxene family, $((Fe,Ca,Mg,Al,Ti)_2(Si,Al)_2\ O_6)$ with a composition of approximately 9.7% FeO, 24.5% CaO, 7.4% MgO, 19% Al_2O_3 and 38% SiO_2. In this mineral iron is present mostly in the reduced Fe2+ state. This mineral is not present in the Kestel mine ore and would have precipitated during heat treatment. The presence of three of the phases present in the ore (limestone, quartz, magnetite) also in the slag is another indication of incomplete equilibration of the raw materials in the slag, and thus of a short period of time at a low temperature compared to the glass melting temperature of the slag. The probable presence of cristobalite would argue for resmelting of at least part of the slag. The probable presence of modern tin metal may be the result of inclusion of metallic tin in the glass phase such that diffusion of oxygen was slowed to diffusion coefficient values on the order of 10^{-23}
or geologic time.

Further testing of the viscosity of the slag was conducted. Differential thermal analysis (Perkin Elmer automated DTA-1700) of a replicate melt as well as of 30 mg of slag was conducted. By running each sample twice from room temperature to 1100°C at 10oC/min in air, such that most of the crystals were melted with the exception of the quartz, a glass transition temperature range, or T_g, was determined having a mean value of 860+-25°C. This value is high for common earthenware and stoneware glazes. Fluid flow usually occurs about 200° above T_g, so the slag would not have been fluid enough to flow at 950°C.

Backscattered SEM images of the microstructure of a representative droplet of slag are shown in polished cross section in Fig. 8. The upper micrographs at 250 and 500x show the structure of a calcium-containing particle (upper left), pores (dark grey), and mixed phases which have sintered together. Cassiterite (white particles) have precipitated in greatest concentration at the grain boundaries between the fused grains. The lower two micrographs at 3500x and 8500x show the cassiterite needles at closer range. Note that many of the needles are seen on-end which makes them appear as equiaxed particles (The shape of the particles, dendrites, was confirmed by etching in HF). The cassiterite particles measure about 20 microns maximum length by about 0.7 microns in maximum diameter. Most dendrites are 5 to 8 microns long by 0.5 microns in diameter. These are similar to cassiterite particles found as defects in modern float glass melts, but they are not as large as those in historic tin slags which measured 150-200 microns, as found by Tylecote and cited by Bachmann(9). These Goltepe cassiterite particles are barely at the limit of resolution of the naked eye, such that processing would have involved a grinding step in which the required powder was below or just at the limit of visual recognition. However, it is likely that the slag droplets on the crucible fragments were considered discard, and thus selection of tin-bearing slags would have concentrated on larger cassiterite particles.

The large, grey rhombohedral phase is rich in iron and may well be the faissaite phase. The background glass contains a slightly less grey, lathlike phase with particles averaging about 5 microns. Being less dark, we know it has a lower atomic number than the glassy matrix, but the volumes are sufficiently small that the EDS spectra for the two areas are the same and consist of peaks in order of descending peak height for silica, calcia, iron oxide, alumina, potassia, soda. These laths may be cristobalite or a siliceous phase. However, the detectors were set for best resolution which is the differentiation of an average atomic number of 2, so that the laths may represent another solid solution augite phase which was not recognized by diffraction.

Some examples of well-precipitated, heterogeneous slag compositions are given below in Table 4.

Table 4. Analyses of Slags by Electron Beam Microprobe Analysis

SiO2	Al2O3	FeO	MgO	CaO	K2O	Na2O	TiO2	SnO2
Contiguous 7-micron diameter points taken 10 microns apart:								
47.96	12.19	12.31	2.30	14.29	4.21	1.68	0.45	1.42
0.50	0.37	1.66	0.19	0.61	0.32	0.14	0.38	94.23
55.38	13.49	10.83	2.55	14.38	3.88	1.58	0.44	1.04
50.80	11.79	12.71	3.09	16.31	3.00	1.35	0.44	1.58
30.45	7.18	6.95	1.27	8.78	2.33	0.89	0.43	38.07
7.17	2.13	2.85	0.55	2.22	0.71	0.36	0.40	84.1
58.18	12.29	10.83	3.02	14.95	3.49	1.37	0.39	1.49
58.03	14.53	8.52	2.21	12.57	4.48	1.79	0.28	1.05
0.73	0.85	1.74	0.16	0.69	0.25	0.08	0.42	95.42
8.17	2.22	2.87	0.62	2.24	0.83	0.19	0.41	82.19
High Iron-containing areas:								
30.94	3.79	36.58	0.68	11.39	1.65	0.58	0.11	5.67
36.10	5.88	34.53	0.53	4.04	2.83	0.85	0.15	9.48

Unfortunately, the microprobe does not have as high resolution a backscattered detector as the SEM, nor are the probe optics configured for optimal imaging, so that we cannot relate the probe compositions directly to the micrographs.

Surprisingly few analyses of tin slags have been published. Compositional data for historic and modern tin slags derived, respectively, from Tylecote and Wright(10) are plotted in three ternary diagrams showing the variation of tin oxide with iron oxide, calcium plus magnesium oxides and silica and alumina (Fig. 9). Their analyses tend to be higher in iron, alumina and silica than the slags from Goltepe. The Goltepe slag compositions represent "lineblends" between tin in the center and a limited ratio of the components represented on the other sides of the triangles. If the tin in the center is 100% pure, then each 10% increment in distance from the center represents a 10% drop in tin concentration until one reaches the outer lines from which one may derive a ratio of the components in the two outer corners by taking the ratio of the outer line lengths (the line length opposite alumina for alumina, and visa versa). These narrow ranges of composition indicate that precursors of specified composition were utilized in a limited range of proportions in batching of raw materials into the crucibles. Precipitated phases, such as faissaite and tin oxide, are represented as well as the ratios of the other components which combined to form the major fraction, a ferrugenous-calcia-alumina-silicate glass. The average of 856 probe analyses of the various slag droplets is 20 wt% of SnO_2. Based on the probe compositions and x-ray data, the raw materials for a charge must have consisted of 1 part quartz, 1 part calcium carbonate, 1 part magnetite and 1 part cassiterite and 1 part of a clay or other alumino-silicate rich in soda (3-8%) and potassia (3-5%). Perhaps a small amount of potassia or soda-rich ash was added to increase the alkali total.

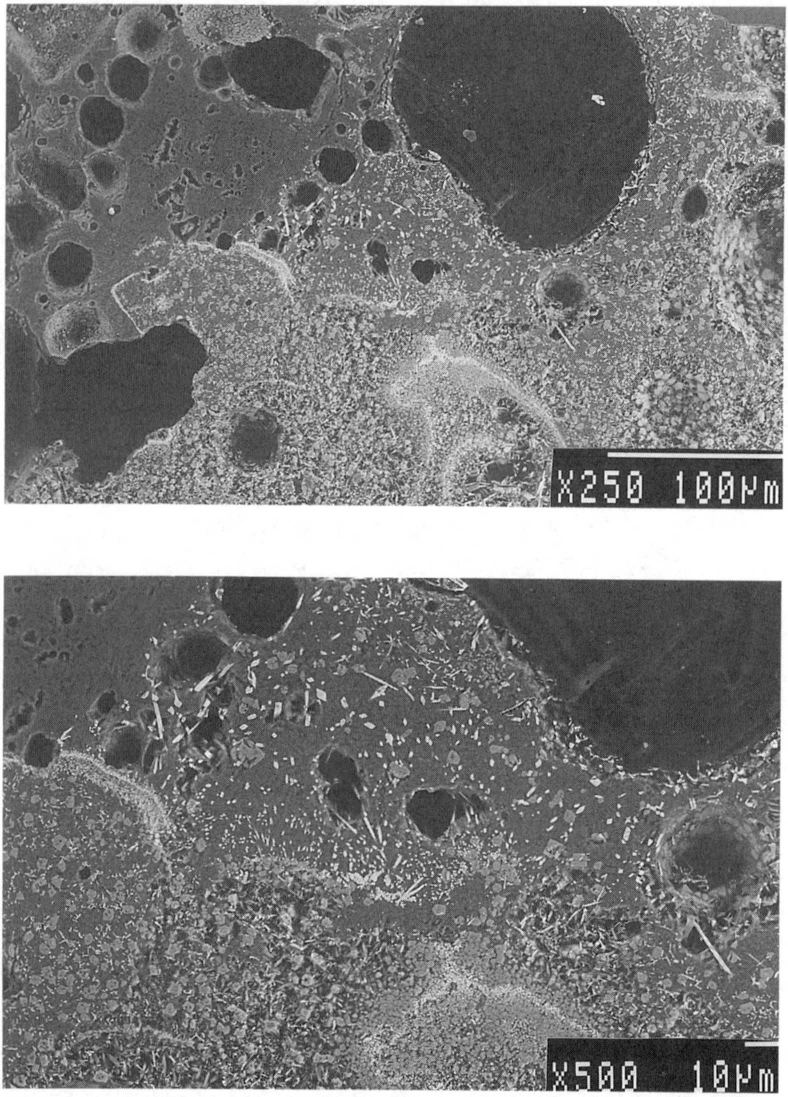

Fig. 8. Sequential enlargements of a backscattered SEM image of the microstructure of a droplet of tin slag in cross section showing sintered structure of calcium-containing particle (upper left) and sintered grains of mixed phases. Rounded black areas are pores.

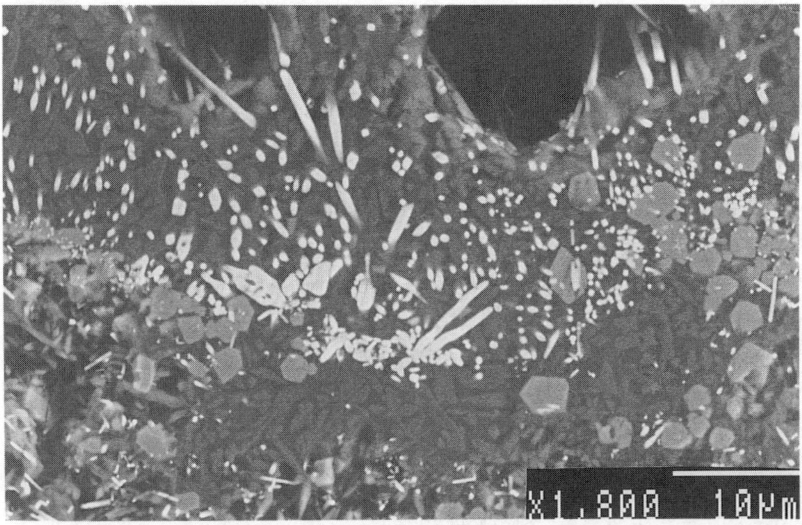

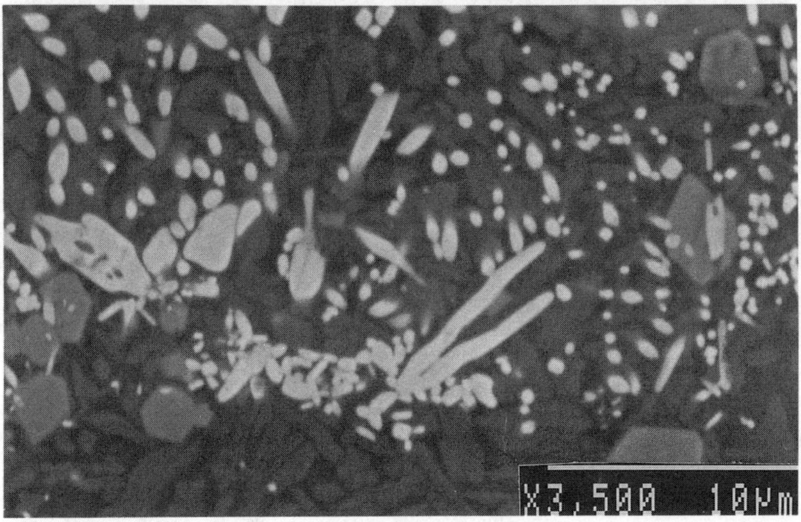

Fig. 8, cont. White dendrites are the tin-rich phase; grey, angular, equiaxed particles rich in iron are beliefed to be faissaite. Matrix consists of grey laths in a black ground of glass and has an EDS spectrum of Si, Ca, Fe, Al, K, Na. Because of their fine size, the two phases in the matrix cannot be distinguished. The backscattered system is differentiating two phases which differ in average atomic number by slightly over 2.

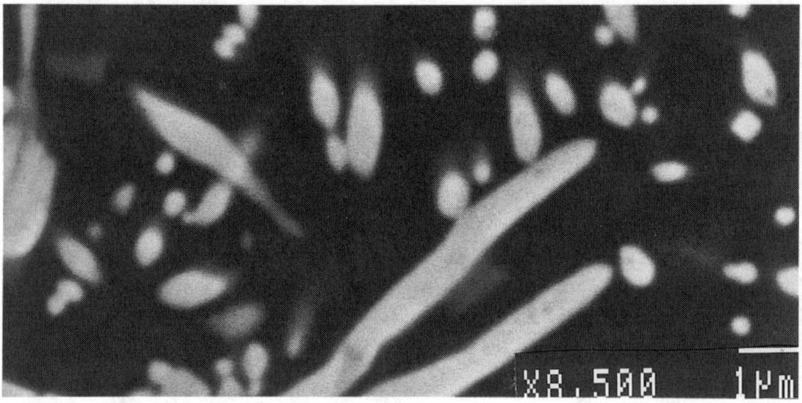

Fig. 8, cont.

PROCESSING SEQUENCE AND PARAMETERS

Activities demonstrated by analysis include a multiple-step, low-temperature process carried out between 800 and 950°C for a relatively short duration, perhaps of a half hour to a few hours. Processing involved intentionally producing tin metal by reduction firing of tin oxide in crucibles, interspersed with one or probably more intermediate grinding and/or density separation techniques, such as sedimentation, vanning or winnowing. Arsenic oxide does not occur naturally in the ore from Kestel and is not known in nature in conjunction with tin oxide. Arsenic was an intentional addition, and it acted as an internal chemical-reducing agent in the smelting process. Whether or not this was the reason it was added, we, of course, cannot determine. The raw materials being processed in the crucibles probably consisted of tin oxide (cassiterite, SnO_2) with no copper ores present, along with calcium carbonate ($CaCO_3$), a high-iron component (probably magnetite, Fe_3O_4) containing minor amounts of magnesia and titania, probably an alumino-silicate containing more than 12% potassia and soda, and a carbonaceous material, such as charcoal, as a physical-reduction agent.

The heat treatment most closely follows a sintering model in which particulate materials partially fused together forming a high viscosity glass into which crystals nucleated and grew mainly at grain-grain, grain-glass or grain-pore boundaries. If these were the crucibles in which smelting occurred, then the glass was of such a high viscosity that it could not have flowed out of the crucible by pouring or tapping. The tin metal was present in sufficiently small grains and was sufficiently divided without much interconnectivity such that gravity separation in the melt was not possible. The glass melt did not equilibrate the raw materials into the melt, not did the temperature of the crucibles equilibrate through the crucible wall thickness. However, sufficient time was allowed for the precipitation of very small cassiterite and faissaite crystals on the order of a few microns. Removal of the metal prills would have required chipping, grinding and density separation as well as reheating at lower temperature.

Fig. 9. Schematic diagram of tin processing at Goltepe with excavated material correlates in parentheses. Question marks occur where there is no archaeological evidence or material correlate.

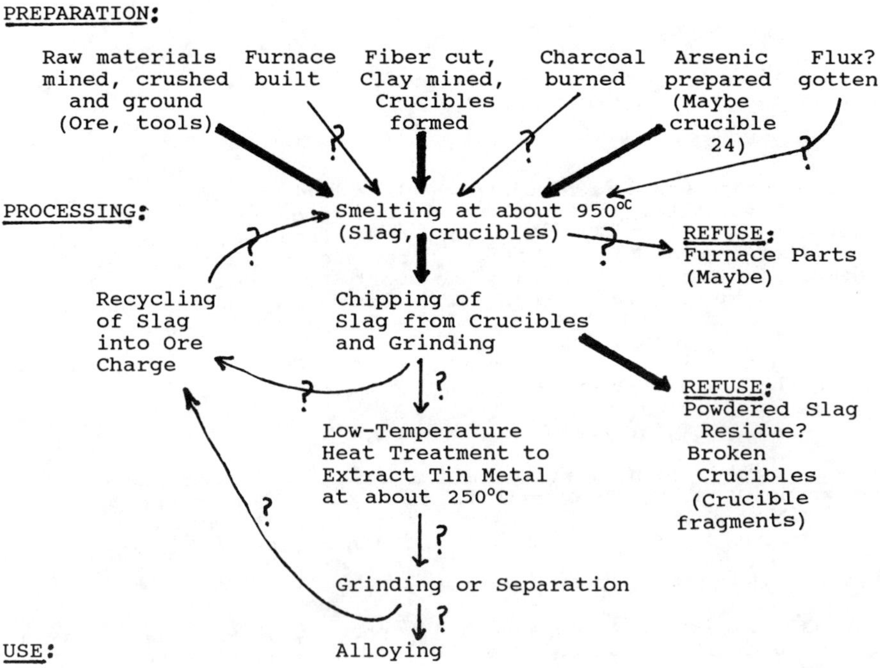

Modern tin processing is a 3-step process involving: 1) primary smelting of the ground and separated ore, a glass formation step at 1300-1400°C in which metallic tin and tin silicates are formed, followed by grinding; 2) retreatment to recover tin from its complexes; and 3) refining of the resultant tin(11). The evidence from Goltepe consists of the first step in the modern process, but carried out at a much lower temperature. Earl has summarized evidence for the gradual evolution of these processes during the Industrial Revolution in England(12). Some descriptions from the last century tell of a process in which a high temperature smelting step is followed by a low temperature heat treatment carried out at about 232-300°C involving "sweating out" of the metallic tin which is exposed on ground surfaces of the slag as molten metal droplets which wet one another and pull away from the residuum(13). Then the residuum is reintroduced into the smelting step because addition of a small amount of glassy material, if well mixed, aids in the sintering and formation of more glass. In modern glass practice, batches melt more rapidly and with less fuel, if 20 to 33% of crushed scrap glass is added or, in modern parlance, recycled.

The low temperature process we conjecture above is similar to the nucleation and precipitation of metallic crystals in highly viscous glasses or glazes. When compared to a glass or glaze processing model occuring at about 1000°C, the process shown in Fig. 6 is feasible. In addition, experiments by Parker(14) demonstrate the efficacy of this process at the low temperature proposed. As part of his thesis for which the intent was to save fuel, Parker was able to smelt tin at low temperatures in a manner similar to that conjectured here.

We know from Ellingham diagrams that for the reaction,

$SnO_2 \quad Sn + O_2$,

a temperature of 950°C and a highly reducing atmosphere of $p_{o2} = 10^{-14}$ atm is required. For the reaction,

$2SnO \quad 2Sn + O_2$,

a range of temperature is required which is higher than that found at Goltepe, 1250-1540°C, although a lesser reducing atmosphere is required, $p_{o2} = 10^{(-6)}$ to $10^{(-12)}$ atm(8b). A bonfire can reach $10^{(-4)}$, and copper is smelted at $10^{(-6)}$; thus $10^{(-14)}$ represents a considerable feat of reduction. Our high viscosity and poorly homogenized melt at 950-1000°C could use charcoal as a physical reducing agent until the glass began to form; then a locally acting chemical reducing agent would be of benefit.

We suggest that the role of As_2O_3 would provide such a reducing agent in the initial stages of glass formation, because

$As_2O_3 + O_2 \quad As_2O_5$, at about 650°C,

or in other words acts as an oxygen scavenger until about 800°C (15). At or above 900°C the As_2O_5 reverts to As_2O_3 and the extra O_2 forms bubbles, which is the usually described function of arsenic oxides as oxygen liberator in fining (or debubblizing) glass melts(16).

TIN PROCESSING TECHNOLOGY AT GOLTEPE

Cassiterite is present as a major phase of the slags and as a constituent of the powdered layers on the interiors of the ceramic fragments. The raw materials and resultant phases have been identified. The raw materials were quartz, calcite, magnetite, cassiterite and a potassium-rich clay or other alkali-rich alumino-silicate, although we cannot disprove inclusion of ash. The resultant phases were tin metal, cassiterite, faissaite, cristobalite and glass. We can positively identify all but the tin metal phase.

The understanding of ceramic technology demonstrated in the fabrication and usage of the crucibles indicates a fairly sophisticated understanding of refractory behavior, even though the composition of the clay body chosen for the refractory was at the lower limit of stability for even an ancient refractory because of the high combined total of fluxes, soda and potassia, and because of the low silica. Even though the ceramic technology is quite advanced, the glass melting and metallurgical processing does not display the same in-depth understanding. The smelting process was probably not a very efficient process because of the high viscosity of the slag in which the nucleation of tin crystals probably would have occurred fairly readily as indicated by their large number, but the growth and agglomeration of these crystals in such a viscous glass would probably have been the rate limiting process as indicated by the small size and lack of aggregation of the crystals. Further studies should investigate the change in the process of tin smelting with time using TL-dating as a way of establishing temporal context, or relative date.

Although this third millennium tin processing debris is the earliest tin slag known or analyzed, phase characterization of 25 ceramic fragments and slags demonstrates that the technology being practiced has the consistent, goal-oriented practice of a well-established craft and is not at all consistent with the heterogeneity expected of an experimental trial-and-error process. This technology is characterized by:

Production events with limited range of variability,
Evidence of working within constraints of raw materials, achieveable temperatures and physical chemistry,
Careful observation and use of the geological resource base,
Complex, multiple-stage technology, not all of which is directly observeable, so it is based on considerable experience,
Combination of materials which are not in immediate juxtaposition in nature,
Identification of a reproduceable, rate-limiting step, and
Pyrotechnology which involved considerable control of process.

The technology suggests interactions with the geological surroundings such that fine distinctions in the environment were recognized, experimented with and incorporated into patterns of activities which, based on the extensive effort expended, involved making useful things. Goltepe is a difficult to reach, walled site high in the mountains, which is distant from main trade routes. Some special features of the technology are indicated by the small size of grindstones and crucibles, by the fineness of sintered grains which is an indication of the fineness of grinding of the original charge and by the small amounts of material processed and left as refuse at the site. The technology was special, secretive and directed to producing a precious commodity.

ACKNOWLEDGEMENTS

The excavations at Goltepe are directed by Aslihan Yener and funded by the National Endowment for the Humanities, the National Geographic Society and the Institute for Aegean Prehistory. They are under the augpices of the Nigde Archaeological Museum, the Ministry of Culture and the General Directorate of Monuments and Museums of Turkey. Dating to the mid-third millennium B.C. was conducted by radiocarbon measurements on associated charcoal by Beta Analytic and by thermoluminescence technique directly on the fine fraction of the ceramic fragments at the Archaeometry Lab of the University of Maryland Physics Dept., directed by Prof. William Hornyak.

We wish to thank E.V. Sayre for suggesting this collaboration, Melanie Feather of CAL who aided the XRF analyses, and Eugene Jarosewich and Joseph Neelen, Dept. of Mineral Sciences, National Museum of Natural History. We are most grateful to such helpful advisors and discussants as Martha Goodway, Ian Freestone, Cyril Smith, Arthur Steinberg, Richard Bradt, Ramana Reddy, John F. Elliot, Brian Earl, David Killick, Roger Moorey, C.C. Lamberg-Karlovsky, Liz Penisi, John Travis, David Sours, Russell Feather and Jim Feathers.

REFERENCES

1. P.R.S. Moorey, Materials and Manufacture in Ancient Mesopotamia: BAR Int'l 237 (1985) 17-20, 51-68, and E.S. Hedges, Tin and Its Alloys, Arnold (London 1960).

2. M. Goodway, personal communication.

3. K.A. Yener, H. Ozbal, E. Kaptan, A.N. Pehlivan, M. Goodway, "Kestel: An Early Bronze Age Source of Tin in the Taurus Mountains, Turkey," Science 244 (1989) 200-203.

4. K.A. Yener, P.B. Vandiver, "Tin Processing at Goltepe, an Early Bronze Age Site in Anatolia," Am. Jour. Archaeology, in press; W. Hornyak et al., "TL Dating of a Crucible Fragment from an Early Tin Processing Site in Turkey," in preparation.

5. I.C.Freestone, "Refractories in the Ancient and Preindustrial World," in W.D. Kingery, ed., High-Technology Ceramics: Past, Present and Future, III, A. Cer. S. (Westerville, OH, 1986 35-64, and I.C. Freestone, "Refractory Materials and their Procurement," in Old World Archaeometallurgy, R. Maddin, ed., M.I.T. Press (Cambridge 1991) 151-162.

6. C.J. Hughes, Igneous Petrology, Elsevier (Amsterdam, 1982) 92-93.

7. E.M. Levin, C.R. Robbins and H.F. McMurdie, Phase Diagrams for Ceramists, I, A.Cer.S. (Westerville, OH, 1964) 181 (fig. 501).

9. W.D. Kingery and J.D. Frierman, "The Firing Temperature of a Karanova Sherd and Inferences about South-east European Chalcolithic Refractory Technology," Proc. Prehistoric Soc. 40 (1974) 204-205, and M.S. Tite et al., "Technological Studies of Ancient Ceramics," in T.A. and S.F. Wertime, eds. Early Pyrotechnology (Washington, D.C., 1982) 61-72.

9. C. Clark-Monks and J.M. Parker, Stones and Cord in Glass (Sheffield 1980) 95, 185; E.R. Begley, Guide to Refractory and Glass Reactions (Boston 1970) 101, and F.H. Norton, Refractories (New York 1968) 369; H.G. Bachmann, Identification of Slags from Archaeological Sites, Inst. of Archaeology Occ. Publ. 6 (London 1982) 27.

10. R.F. Tylecote, Early History of Metallurgy in Europe (London, 1987) and P.A. Wright, The Extractive Metallurgy of Tin (New York, 1966) 87-89.

11. J.M. Floyd, "The Physical Chemistry of Tin Smelting," in J.M. Cigan, T.S. Mackay and T.J. O'Keefe, eds., Lead-Zinc-Tin '80 (Washington, D.C., 1980) 508-531, and B.T.K. Barry and C.J. Thwaites, Tin and Its Alloys and Compounds, Halsted (New York 1983).

12. Bryan Earl, "Tin Preparation and Smelting," in J. Day and R.F. Tylecote, eds., The Industrial Revolution in Metals, The Institute of Metals (London 1991) 37-83.

13. M.J. Errington de La Crois, Les Mines d'Etain de Perak Presqu'Ile de Malacca (Paris 1882); L. Wray, Jr., Alluvial Tin Prospecting (Taiping, Malaya, 1893) 1-5; G.R. Lewis, A Study of the Medieval Tin Miners of Cornwall and Devon, Barton (Truro 1908) 10-19.

14. R.H. Parker, A.R. Mitchell and F.H. Lai, "Reduction of Cassiterite Concentrated by Solid Carbon in Reverberatory Tin Smelting," Trans. Instn. Min. Metall. (Sect. C: Mineral Process. Extr. Metall.) 9 (May-Aug. 1990) C93-C99.

15. M. Cable, "Principles of Glass Melting," in D.R. Uhlmann and N.J. Kreidl, eds., Glass: Science and Technology, vol II, Academic Press (Orlando 1984) 19-21, and N.J. Kreidl, "Inorganic Glass-Forming Systems," in D.R. Uhlmann and N.J. Kreidl, eds., Glass; Science and

Technology, vol I, Academic Press (Orlando 1983) 151-153.

16. H. Mulfinger, J. A.Cer.S., 49 (1966) 462-467; M. Cable and M. Haroon, J. Glass Technol. 11 (1970) 48-53, L. Nemec, J. Glass Tech., 15 (1974) 153-161, L. Nemec, J. A.Cer.S. 69 (1977) 436-440; H. Verweij, J. A.Cer.S. 62 (1979) 450-455; L. Nemec, Jour. Glass Tech., 21 (1980) 134-144.

FURTHER STUDY OF THE PROCESS OF TECHNOLOGICAL EVOLUTION OF ANCIENT CHINESE POTTERY AND PORCELAIN

LI JIAZHI*, LUO HONGJIE* AND GAO LIMING**

*Shanghai Institute of Ceramics, Chinese Academy of Sciences, 1295 Ding Xi Road, Shanghai 200050, China
**Northwest Institute of Light Industry, Ceramics Engineering Department, Xiangyang, Shaanxi, P.R.C.

ABSTRACT

The chemical compositions and firing temperatures of seventy-four stamped pottery and proto-porcelain fragments which were excavated from Jiangxi, Zhejiang, Henan and Fujian were studied. Their chemical compositions, together with a database of more than 600 pottery and porcelain bodies and 100 proto-porcelain and porcelain glazes previously investigated, were processed and evaluated by correspondence analysis. Their firing temperatures and those of more than 200 pottery and porcelain samples were compared as well. The results of this investigation yield a more objective and comprehensive view of the process of technological evolution of ancient Chinese pottery and porcelain than heretofore possible. The place of stamped pottery and proto-porcelain in the evolution process is better understood, and the differences between the patterns of continuity and change in the north and the south of China is further discussed in this paper.

INTRODUCTION

In China pottery and porcelain have been manufactured over a widespread area for more than 10,000 years. The process of technological evolution has been an attractive research topic for scholars devoted to the study of ancient pottery and porcelain. This study has a long and distinguished history. Since the early fifties Zhou Ren, Li Jiazhi, Zhang Fukang and others have studied pottery, proto-porcelain and porcelain fragments from excavations at archaeological sites, graves in the Huanghe and Changjiang River valleys, and representative samples of many famous porcelains to determine the nature and variation of manufacturing technology (1-8). An outline of the process of technological evolution was drawn from their work, and major technological differences were found between the northern and southern river valleys of China as might well be expected of geologically distinct areas.

One significant aspect which heretofore has been known only incompletely is the relationship between these two types of pottery, stamped pottery and proto-porcelain, and the role their development played in the evolution of Chinese ceramic technology. Only a few examples had been studied prior to this study, and these were insufficient to cover the different periods of manufacture. In order to overcome these shortcomings, more stamped pottery, proto-porcelain and porcelain samples were chosen and collected for study to complement the previous work.

With the use of a computer, multivariate statistical analysis was applied to the study of ancient pottery and porcelain to demonstrate relationships of composition in multi-dimensional space. Using correspondence analysis we have achieved more objective results with better separation of results and a better understanding of the pattern of overlap than previous studies.

EXPERIMENTAL PROCEDURE

Seventy-four fragments of pottery and proto-porcelain were collected from the provinces of Zhejiang, Fujian, Jiangxi and Henan. The chemical compositions of each body and glaze were determined by x-ray fluorescence (XRF) and quantitative energy-dispersive x-ray analysis (EDAX) and are listed in Tables 1 and 2, respectively. Their firing temperatures were measured by dilatometer when sufficient sample was available and are given in Table 3. Dilatometer measurements are denoted by the symbol "*", shown to the right of the data given in Table 3. Other firing temperatures were based on information given in references 1-8.

Using correspondence analysis (9) to process the chemical compositions for 74 samples analyzed in this study as well as those for 609 pottery and porcelain bodies, now indexed in the Chemical Composition Database for Ancient Chinese Pottery and Porcelain (CCDACPP), three factors (F1, F2 and F3) were selected in which the value of accumulated variances is larger than 80%. The F1 and F2 and the F1 and F3 factor loading diagrams are shown in Figures 1 and 2, respectively.

Moreover, using correspondence analysis to analyze the chemical compositions for 12 proto-porcelain glazes studied for this paper and those for 96 pottery coatings, slip glazes and glazes before the Five Dynasties, which are also indexed in the CCDACPP, two factors were found in which the value of accumulated variances is larger than 70%. The F1 and F2 factor loading diagram is shown in Figure 3.

DISCUSSION

The Evolutionary Process of Chemical Compositions for Northern and Southern Pottery and Porcelain

As shown in Figs. 1 and 2, the process of evolution in chemical composition for northern and southern pottery and porcelain bodies is very obvious. As the ceramic types were changing from pottery to stamped pottery, and then to proto-porcelain, the SiO_2 contents in the bodies were increased step by step, meanwhile the fluxing agents (such as CaO, Fe_2O_3 and MgO) were decreased correspondingly. However, Al_2O_3 content seemed to change very little until the Song Dynasty. In addition, the regions of chemical composition are connected and overlap from pottery to stamped pottery, proto-porcelain and then porcelain. However, for the northern pottery and porcelain bodies, the process of evolution is different. The pottery bodies are located in an area with a high content of fluxing agents and a low content of SiO_2. Compared with the northern pottery bodies, the northern porcelain bodies are located in an

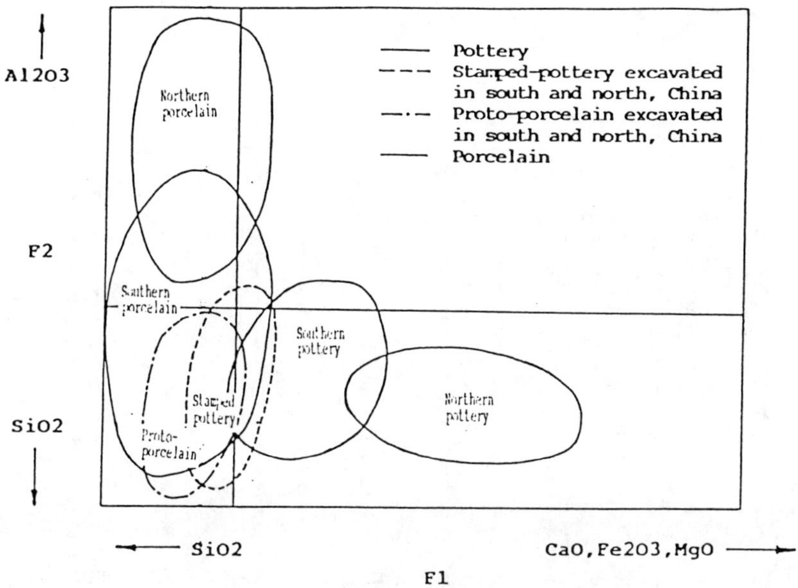

Fig. 1. F1 and F2 Factor Loading Diagram for Ancient Chinese Pottery and Porcelain Bodies Based on Chemical Composition Variations

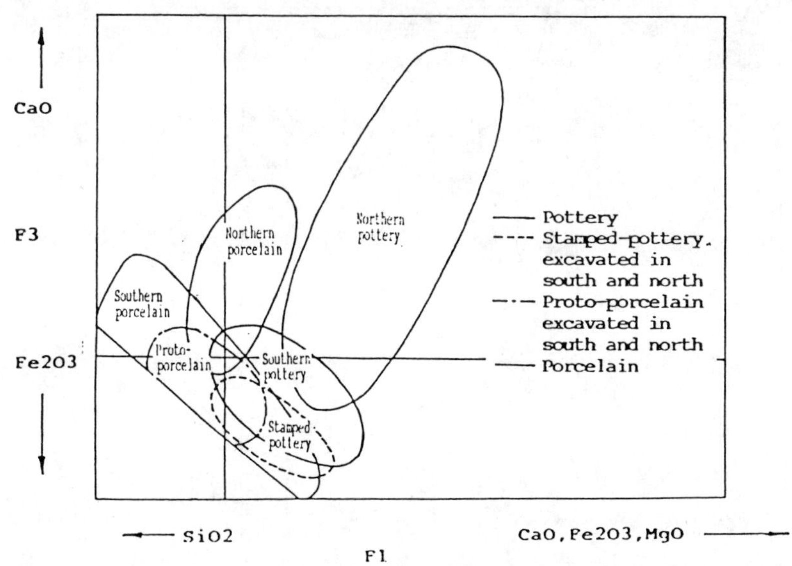

Fig. 2. F2 and F3 Factor Loading Diagram for Ancient Chinese Pottery and Porcelain Bodies Based on Variation in Chemical Composition

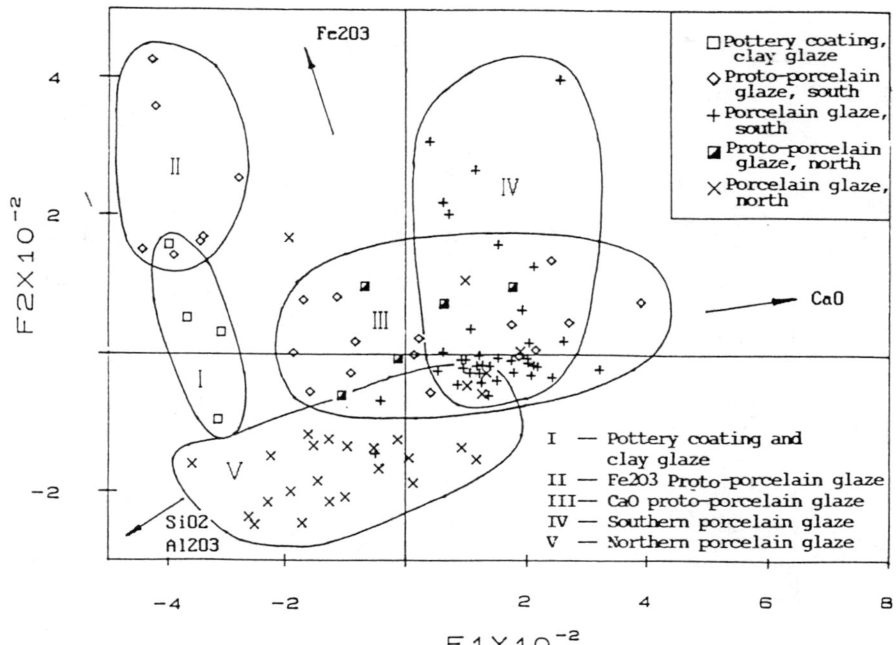

Fig. 3. F1 and F2 Factor Loading Diagram of Pottery Coatings and Porcelain Glazes before Five Dynasties Based on Variations in Chemical Composition

Table I. The Chemical Compositions of Stamped Pottery and Proto-Porcelain Bodies (wt%).

Code	Name	Manufactured period	Manufactured site	SiO2	Al2O3	Fe2O3	TiO2	CaO	MgO	K2O	Na2O	MnO	P2O5	I.L.	Total
JYJ2	S.P.*	The late Shang Dynasty.	Jiaoshan, Yingtan, Jiangxi.	68.02	21.62	3.85	0.96	0.37	1.19	1.55	0.60	0.031	0.062	2.11	100.36
JYJ3	S.P.	The late Shang Dynasty.	Jiaoshan, Yingtan, Jiangxi.	66.88	20.76	5.52	1.00	0.44	1.50	1.84	0.51	0.035	0.032	1.80	100.32
FF01	S.P.	The late New Stone Age.	Tanshishan, Fuzhou, Fujian.	65.09	22.67	5.37	0.90	0.48	0.69	2.51	0.46	0.055	0.200	2.06	100.49
FF02	S.P.	The late New Stone Age.	Tanshishan, Fuzhou, Fujian.	66.13	21.83	3.66	0.95	0.35	0.99	3.45	0.67	0.060	0.140	1.56	99.79
FQ01	S.P.	The late New Stone Age.	Nanan, Quanzhou, Fujian.	66.32	21.42	5.87	0.86	0.25	0.74	2.88	0.68	0.041	0.046	0.79	99.90
FQ02	S.P.	The Shang Dynasty.	Nanan, Quanzhou, Fujian.	63.01	27.66	3.73	1.08	0.33	0.69	2.66	0.44	0.040	0.048	0.68	100.37
FQ03	S.P.	The late New Stone Age.	Quanzhou, Fujian.	57.50	29.02	8.56	0.81	0.37	0.68	1.41	0.41	0.036	0.130	0.91	99.84
FQ04	S.P.	The late New Stone Age.	Quanzhou, Fujian.	75.24	16.51	1.50	0.62	0.21	0.47	2.98	0.35	0.040	0.064	1.52	99.50
HXZ05	S.P.	The Warring States.	Xinzheng, Henan.	66.95	18.78	6.01	0.87	0.72	1.28	2.37	1.31	0.100	0.066	1.30	99.76
JFL1	S.P.	The Shang Dynasty.	Leipishi, Fuzhou, Jiangxi.	64.78	24.12	4.15	1.08	0.32	1.03	2.77	0.75	0.075	0.110	0.60	99.78
JFQ1	S.P.	The Shang Dynasty.	Qipanlao, Fuzhou, Jiangxi.	62.11	27.51	3.85	1.16	0.20	0.89	2.86	0.33	0.021	0.076	0.59	99.60
JFQ2	S.P.	The Shang Dynasty.	Qipanlao, Fuzhou, Jiangxi.	64.22	25.33	4.31	1.15	0.18	0.87	3.13	0.45	0.027	0.066	0.15	99.88
JNX1	S.P.	The Shang Dynasty.	Xingizhou, Nanchang, Jiangxi.	66.88	23.12	3.48	1.03	0.41	0.92	2.95	0.56	0.028	0.044	0.67	100.09
JNX2	S.P.	The Shang Dynasty.	Xingizhou, Nanchang, Jiangxi.	67.79	19.28	6.34	1.21	0.16	0.43	0.40	0.18	0.024	0.440	3.87	100.12
JQF4	S.P.	The West Zhou Dynasty.	Fanchengdui, Qingjiang, Jiangxi.	65.52	23.64	4.14	1.04	0.16	1.14	2.97	0.49	0.024	0.070	0.70	99.89
JQS1	S.P.	The middle West Zhou to the middle Spring and Autumn.	Wucheng, Qingjiang, Jiangxi.	66.98	20.64	5.82	0.98	0.37	1.43	2.46	0.53	0.067	0.088	0.19	99.56
JQS10	S.P.	The late Shang Dynasty.	Wucheng, Qingjiang, Jiangxi.	67.64	22.04	3.71	0.99	0.37	1.63	2.50	0.71	0.054	0.120	0.46	100.22
JQS14	S.P.	The late Shang Dynasty.	Wucheng, Qingjiang, Jiangxi.	80.08	11.82	4.38	1.14	0.17	0.89	0.41	0.07	0.030	0.160	0.65	99.80
JQS17	S.P.	The Shang Dynasty.	Wucheng, Qingjiang, Jiangxi.	69.37	21.72	2.85	1.26	0.41	1.09	2.44	0.50	0.038	0.070	0.46	100.21
JQS2	S.P.	The middle West Zhou to the middle Spring and Autumn.	Wucheng, Qingjiang, Jiangxi.	74.45	15.62	4.05	1.13	0.32	0.98	1.57	0.28	0.056	0.160	1.47	100.09
JQS3	S.P.	The middle West Zhou to the middle Spring and Autumn.	Wucheng, Qingjiang, Jiangxi.	66.48	23.09	3.83	1.03	0.17	1.06	3.17	0.41	0.031	0.080	0.68	100.03
JQS5	S.P.	The late Shang Dynasty.	Wucheng, Qingjiang, Jiangxi.	70.03	19.26	6.02	1.18	0.36	0.63	1.10	0.24	0.032	0.380	1.29	100.52
JQS6	S.P.	The late Shang Dynasty.	Wucheng, Qingjiang, Jiangxi.	77.80	14.31	2.23	1.20	0.30	0.64	1.68	0.46	0.030	0.072	1.02	99.74
JQS7	S.P.	The late Shang Dynasty.	Wucheng, Qingjiang, Jiangxi.	79.24	11.82	4.15	1.19	0.26	0.42	0.61	0.18	0.029	0.430	1.48	99.81
JQS8	S.P.	The Late Shang Dynasty.	Wucheng, Qingjiang, Jiangxi.	75.18	16.79	2.52	1.02	0.28	0.72	2.10	0.45	0.026	0.140	0.95	100.18

Table I, continued.

Code	Name	Manufactured period	Manufactured site	SiO_2	Al_2O_3	Fe_2O_3	TiO_2	CaO	MgO	K_2O	Na_2O	MnO	P_2O_5	I.L.	Total
JQZ3	S.P.	The West Zhou Dynasty.	Zhuweicheng, Qingjiang, Jiangxi.	69.12	20.36	4.72	1.10	0.28	0.96	1.34	0.52	0.032	0.410	0.94	99.78
JQZ4	S.P.	The West Zhou Dynasty.	Zhuweicheng, Qingjiang, Jiangxi.	70.29	18.07	5.98	1.03	0.44	0.85	1.44	0.44	0.059	0.430	0.93	99.96
JQZ5	S.P.	The West Zhou Dynasty.	Zhuweicheng, Qingjiang, Jiangxi.	61.63	26.28	5.22	1.21	0.26	1.24	1.77	0.48	0.053	0.420	1.21	99.77
JWG1	S.P.	The West Zhou Dynasty.	Shizhinao, Wanzai, Jiangxi.	60.28	28.46	2.90	1.26	0.21	0.80	2.61	0.34	0.031	0.960	2.25	100.10
JWG2	S.P.	The West Zhou Dynasty.	Shizhinao, Wanzai, Jiangxi.	70.33	21.02	2.75	0.93	0.13	0.53	2.39	0.25	0.036	0.150	1.25	99.77
JYB1	S.P.	The late Shang Dynasty.	Banlishan, Yingtan, Jiangxi.	68.35	20.10	5.36	0.94	0.43	0.98	1.58	0.51	0.061	0.350	1.11	99.77
JYB2	S.P.	The late Shang Dynasty.	Banlishan, Yingtan, JIangxi.	67.47	21.14	5.19	1.17	0.26	0.78	1.43	0.51	0.016	0.120	1.79	99.88
JYJ1	S.P.	The late Shang Dynasty.	Jiaoshan, Yingtan, JIangxi.	63.53	22.52	6.75	1.08	0.54	1.56	1.95	0.77	0.064	0.056	0.99	99.81
JYJ10	S.P.	The late Shang Dynasty.	Jiaoshan, Yingtan, JIangxi.	64.96	24.42	4.23	1.04	0.85	1.37	1.67	0.40	0.031	0.032	0.69	99.69
JYJ11	S.P.	The late Shang Dynasty.	Jiaoshan, Yingtan, JIangxi.	68.82	19.61	4.71	1.18	0.25	0.59	0.90	0.31	0.017	0.051	3.13	99.57
JYJ4	S.P.	The late Shang Dynasty.	Jiaoshan, Yingtan, Jiangxi.	62.39	23.66	7.62	1.06	0.81	1.28	1.59	0.58	0.063	0.120	0.61	99.78
JYJ5	S.P.	The late Shang Dynasty.	Jiaoshan, Yingtan, Jiangxi.	70.37	18.25	4.86	0.96	0.48	0.87	1.61	0.39	0.072	0.074	1.98	99.92
JYJ9	S.P.	The late Shang Dynasty.	jiaoshan, Yingtan, Jiangxi.	66.01	21.20	6.19	1.00	0.58	1.22	1.59	0.41	0.061	0.037	1.30	99.60
JYT1	S.P.	The West Zhou Dynasty.	Taipinggang, Yifeng, Jiangxi.	62.15	27.57	2.99	0.94	0.19	0.60	3.76	0.44	0.052	0.091	0.62	99.40
JYT2	S.P.	The West Zhou Dynasty.	Taipinggang, Yifeng, Jiangxi.	60.67	29.52	2.76	0.93	0.17	0.69	3.42	0.56	0.053	0.091	0.93	99.79
JZS01	S.P.	The late Shang Dynasty.	Zaishan, Jiangxi.	66.78	22.13	5.59	0.92	0.33	0.96	2.46	0.18	0.019	0.044	0.25	99.66
JZS02	S.P.	The late Shang Dynasty.	Zaishan, Jiangxi.	68.64	17.57	7.73	0.88	0.44	1.27	2.05	0.22	0.180	0.082	0.67	99.73
JZS03	S.P.	The late Shang Dynasty.	Zaishan, Jiangxi.	64.64	23.06	6.00	0.96	0.38	1.03	2.69	0.40	0.027	0.058	0.34	99.59
ZSD1	S.P.	The Spring and Autumn Dynasty.	Dongbao, Shaoxing, Jiangxi.	69.54	17.57	5.76	1.16	0.46	0.97	1.82	1.37	0.054	0.084	1.07	99.86
ZSD2	S.P.	The Spring and Autumn Dynasty.	Dongbao, Shaoxing, Jiangxi.	62.95	21.10	7.81	0.89	0.64	1.10	2.44	1.85	0.064	0.093	0.97	99.91
ZSD3H	S.P.	The Spring and Autumn Dynasty.	Dongbao, Shaoxing, Jiangxi.	64.88	17.79	10.28	1.09	0.42	0.94	2.09	0.67	0.100	0.150	1.40	99.81
ZSD4	S.P.	The Spring and Autumn Dynasty.	Dongbao, Shaoxing, Jiangxi.	67.20	19.59	6.91	0.91	0.49	1.16	2.39	0.82	0.096	0.098	0.52	100.18
ZSD5	S.P.	The Spring and Autumn Dynasty.	Dongbao, Shaoxing, Jiangxi.	64.81	19.83	8.79	1.04	0.30	0.99	2.17	0.71	0.057	0.086	1.25	100.03
ZSD6	S.P.	The Spring and Autumn Dynasty.	Dongbao, Shaoxing, Jiangxi.	69.04	18.97	5.86	1.01	0.45	1.03	2.14	0.78	0.034	0.130	0.97	100.41
HZZ02	H.P.**	The middle Shang Dynasty.	Zhengzhou, Henan.	70.05	19.89	3.93	0.98	0.22	0.94	2.38	0.39	0.035	0.087	0.61	99.51
JQZ1	H.P.	The East Zhou Dynasty.	Zhuweicheng, Qingjiang, Jiangxi.	64.78	24.21	4.00	1.16	0.34	1.42	2.95	0.43	0.039	0.120	0.13	99.58

Table I, continued.

Code	Name	Manufactured period	Manufactured site	SiO_2	Al_2O_3	Fe_2O_3	TiO_2	CaO	MgO	K_2O	Na_2O	MnO	P_2O_5	I.L.	Total
JQZ2	H.P.	The Zhou Dynasty.	Zhuweicheng, Qingjiang, Jiangxi.	77.23	14.47	2.18	1.26	0.21	0.56	1.80	0.43	0.030	0.093	1.28	99.54
JW01	H.P.	The middle Shang Dynasty	Wucheng, Qingjiang, Jiangxi.	73.87	18.06	2.61	1.17	0.20	0.65	1.70	0.33	0.027	0.170	0.97	99.76
JYB3	H.P.	The late Shang Dynasty.	Banlishan, Yingtan, Jiangxi.	70.99	17.02	5.90	1.18	0.21	0.79	1.04	0.21	0.047	0.480	1.68	99.55
HXZ01	P.P.***	The West Han Dynasty.	Xinzheng, Henan.	72.87	18.95	3.38	1.03	0.25	0.87	2.07	0.92	0.047	0.072	0.08	100.54
HXZ02	P.P.	The West Han Dynasty.	Xinzheng, Henan.	72.37	18.77	2.52	0.97	0.37	0.77	2.61	1.16	0.058	0.063	0.15	99.81
HXZ03	P.P.	The West Han Dynasty.	Xinzheng, Henan.	71.59	18.95	2.49	0.93	0.39	0.75	2.72	1.18	0.058	0.060	0.66	99.78
HXZ04	P.P.	The Shang Dynasty.	Xinzheng, Henan.	75.72	16.02	3.50	1.07	0.32	0.87	1.50	0.70	0.056	0.097	0.04	99.89
HZZ01	P.P.	The middle Shang Dynasty.	Zhengzhou, Henan.	77.58	15.66	1.55	0.97	0.25	0.65	1.78	0.41	0.032	0.062	0.85	99.79
JL01	P.P.	The middle to the late West Zhou D.	Taishan, Anyi, Jiangxi.	76.50	16.49	1.20	0.96	0.10	0.51	2.35	0.60	0.011	0.150	1.04	99.91
JQF1	P.P.	The Shang to the West Zhou Dynasty.	Fanchengdui, Qingjiang, Jiangxi.	81.69	10.94	2.16	1.11	0.22	0.57	1.29	0.25	0.034	0.260	1.16	99.68
JQF2	P.P.	The Shang to the West Zhou Dynasty.	Fanchengdui, Qingjiang, Jiangxi.	74.29	14.77	4.86	1.16	0.34	0.87	1.60	0.40	0.047	0.340	0.72	99.40
JQF3	P.P.	The Shang to the West Zhou Dynasty.	Fanchengdui, Qingjiang, Jiangxi.	77.31	13.43	3.81	1.24	0.28	0.68	0.87	0.28	0.031	0.610	1.23	99.77
JQS11	P.P.	The late Shang Dynasty.	Wucheng, Qingjiang, Jiangxi.	72.54	18.29	3.16	1.49	0.23	0.86	1.99	0.33	0.053	0.097	1.45	100.49
JQS13	P.P.	The late Shang Dynasty.	Wucheng, Qingjiang, Jiangxi.	74.18	17.14	3.22	1.16	0.33	0.74	1.45	0.49	0.042	0.160	1.04	99.95
JQS15	P.P.	The late Shang Dynasty.	Wucheng, Qingjiang, Jiangxi.	72.16	18.63	3.76	1.12	0.56	1.01	1.77	0.47	0.041	0.240	0.64	100.40
JQS16	P.P.	The early Shang Dynasty.	Wucheng, Qingjiang, Jiangxi.	79.40	14.19	1.95	1.12	0.23	0.57	0.91	0.31	0.024	0.250	1.15	100.10
JQS4	P.P.	The middle West Zhou to the Spring and Autumn Dynasty	Wucheng, Qinhjiang, Jiangxi.	83.37	10.14	2.26	0.83	0.29	0.56	1.19	0.24	0.049	0.270	1.18	100.38
JQS9	P.P.	The late Shang Dynasty.	Wucheng, Qingjiang, Jiangxi.	69.23	21.43	3.05	0.96	0.29	0.56	2.92	0.52	0.068	0.054	0.47	99.55
JS01	P.P.	The late West Zhou to the early Spring and Autumn.	Shengdui, Jiujiang, Jiangxi.	76.46	16.52	1.18	0.96	0.10	0.49	2.36	0.65	0.007	0.160	1.04	99.93
JS02	P.P.	The late West Zhou to the early Spring and Autumn.	Shengdui, jiujiang, JIangxi.	71.34	20.17	1.56	0.86	0.39	0.42	3.45	0.85	0.025	0.074	0.77	99.91
JYJ6	P.P.	The late Shang Dynasty.	Jiaoshan, Yingtan, Jiangxi.	69.30	19.08	4.78	1.39	0.35	0.86	1.60	0.40	0.028	0.070	1.84	99.70
JYJ7	P.P.	The late Shang Dynasty	Jiaoshan, Yingtan, Jiangxi.	69.18	19.02	5.32	1.04	0.39	1.01	1.91	0.52	0.067	0.063	1.32	99.84
JYJ8	P.P.	The late Shang Dynasty	Jiaoshan, Yingtan, Jiangxi.	70.51	18.00	4.67	1.04	0.42	0.98	2.09	0.48	0.052	0.059	1.62	99.92

* S.P. ---- Stamped Pottery; ** H.P. ---- Hard Pottery; *** P.P. ---- Proto-Porcelain.

Table II. The Chemical Composition of Proto-Porcelain Glaze in Zhejiang Province (wt%).

Code	Manufactured period	Manufactured Site	SiO_2	Al_2O_3	Fe_2O_3	TiO_2	CaO	MgO	K_2O	Na_2O	MnO	P_2O_5	Total
JL01	The middle or late Zhou Dynasty.	Taishan, Anyi.	50.00	12.07	1.74	0.47	25.20	3.52	3.21	0.53	0.730	2.030	99.50
JQS11	The late Shang Dynasty.	Wucheng, Qingjiang.	60.09	18.12	3.51	1.86	9.68	2.26	3.64	0.36	0.160	0.310	99.99
JQS13	The late Shang Dynasty.	Wucheng, Qingjiang.	58.00	16.93	2.44	1.01	13.66	3.15	3.56	0.50	0.160	0.570	99.98
JQS15	The late Shang Dynasty.	Wucheng, Qingjiang.	60.98	17.70	6.90	1.32	4.61	2.18	4.34	0.50	0.880	0.580	99.99
JS01	The late West Zhou to the early Spring and Autumn.	Shendui, Jiujiang.	56.11	13.88	1.69	0.72	18.93	3.41	2.39	0.47	0.520	1.370	99.49
JS02	The late West Zhou to the early Spring and Autumn.	Shendui, Jiujiang.	51.85	14.00	3.56	0.67	21.23	2.88	3.04	0.64	0.160	1.370	99.40
JYJ6	The late Shang Dynasty.	Jiaoshan, Yingtan.	61.81	17.50	3.24	1.92	8.45	1.69	4.63	0.34	0.090	0.320	99.99
JYJ7	The late Shang Dynasty.	Jiaoshan, Yingtan.	61.55	16.78	10.11	1.25	1.67	1.86	5.68	0.64	0.210	0.230	99.98
JYJ8	The late Shang Dynasty.	Jiaoshan, Yingtan.	61.69	17.97	5.00	0.96	4.49	1.72	7.43	0.47	0.050	0.220	100.00
JQF-1	The Shang to the West Zhou Dynasty.	Fanchengdui, Qingjiang.	72.67	8.57	4.24	0.34	3.65	0.68	8.99	1.27	0.000	0.000	100.41
JQF-3	The Shang to the middle Spring and Autumn Dynasty.	Fanchengdui, Qingjiang.	68.49	12.16	8.97	1.25	0.91	1.76	5.09	0.77	0.480	0.000	99.88
JQS-4	The middle West Zhou to the Spring and Autumn Dynasty	Wucheng, Qingjiang.	67.36	13.90	5.47	1.36	3.18	1.88	5.31	1.55	0.000	0.000	100.01

Table III. The Firing Temperature of Ancient Chinese Pottery and Porcelain.

Code	Name	Manufactured or excavated site	Manufactured period	T (℃)	
46A	Black pottery	Chengziyai, Shandong	The New Stone Age	1000	
64A	Red pottery	Shandong	The New Stone Age	950±20	
60C	Red pottery	Changan, Shaanxi	The New Stone Age	1000±50	
10	Red-Grey pottery	Changan, Shaanxi	The New Stone Age	1000±20	
YMT1	Black pottery	Hemudu, Zhejiang	The New Stone Age	850-900	
YMT10	Grey pottery	Hemudu, Zhejiang	The New Stone Age	950-1000	
YMT12	Red pottery with sand	Hemudu, Zhejiang	The New Stone Age	900-950	
YMT4	Black pottery	Hemudu, Zhejiang	The New Stone Age	830-870	
YMT5	Grey pottery with sand	Hemudu, Zhejiang	The New Stone Age	800-850	
YMT7	Red pottery	Hemudu, Zhejiang	The New Stone Age	800-850	
YMT8	Red-grey pottery	Hemudu, Zhejiang	The New Stone Age	800-850	
YMT2	Black pottery	Hemudu, Zhejiang	The New Stone Age	800-850	
YMT3	Black pottery with sand	Hemudu, Zhejiang	The New Stone Age	880-930	
25	Grey pottery with sand	Ruixian, Henan	Yin Dynasty	950	
28	Red pottery	Zhengzhou, Henan	Yin Dynasty	1000±20	
ZSD1	Stamped pottery	Dongbao, Shaoxing, Zhejiang	The Spring and Autumn	1120±20	*
ZSD2	Stamped pottery	Dongbao, Shaoxing, Zhejiang	The Spring and Autumn	960±20	*
ZSD3	Stamped pottery	Dongbao, Shaoxing, Zhejiang	The Spring and Autumn	1190±20	*
ZSD4	Stamped pottery	Dongbao, Shaoxing, Zhejiang	The Spring and Autumn	1190±20	*
ZSD5	Stamped pottery	Dongbao, Shaoxing, Zhejiang	The Spring and Autumn	1120±20	*
ZSD6	Stamped pottery	Dongbao, Shaoxing, Zhejiang	The Spring and Autumn	1050±20	*
JQZ1	Hard pottery	Zhuwucheng, Qingjiang, Jiangxi	The East Zhou Dynasty	1050±20	*
JQS17	Stamped pottery	Wucheng, Qingjiang, Zhejiang	The Shang Dynasty	1060±20	*
JFL1	Stamped pottery	Leipishi, Fuzhou, Jiangxi	The Shang Dynasty	1090±20	*
JFQ1	Stamped pottery	Qipanlao, Fuzhou, Jiangxi	The Shang Dynasty	1170±20	*
JFQ2	Stamped pottery	Qipanlao, Fuzhou, JIangxi	The Shang Dynasty	1050±20	*
JNX1	Stamped pottery	Xingizhou, Nanchang, Jiangxi	The Shang Dynasty	1130±20	*
JQS10	Stamped pottery	Wucheng, Qingjiang, Zhejiang	The late Shang Dynasty	990±20	*
JQS5	Stamped pottery	Wucheng, Qingjiang, Zhejiang	The late Shang Dynasty	1080±20	*
JQS6	Stamped pottery	Wucheng, Qingjiang, Zhejiang	The late Shang Dynasty	1190±20	*
JQS7	Stamped pottery	Wucheng, Qingjiang, Zhejiang	The late Shang Dynasty	1160±20	*
JQS8	Stamped pottery	Wucheng, Qingjiang, Zhejiang	The late Shang Dynasty	1130±20	*
JYB1	Stamped pottery	Banlishan, Yingtan, Jiangxi	The late Shang Dynasty	1130±20	*
JYB2	Stamped pottery	Banlishan, Yingtan, Jiangxi	The late Shang Dynasty	1100±20	*
JYB3	Hard pottery	Banlishan, Yingtan, Jiangxi	The late Shang Dynasty	1100±20	*
JYJ1	Stamped pottery	Jiaoshan, Yingtan, Jiangxi	The late Shang Dynasty	1030±20	*
JYJ10	Stamped pottery	Jiaoshan, Yingtan, Jiangxi	The late Shang Dynasty	1030±20	*
JYJ11	Stamped pottery	Jiaoshan, Yingtan, Jiangxi	The late Shang Dynasty	1130±20	*
JYJ2	Stamped pottery	Jiaoshan, Yingtan, Jiangxi	The late Shang Dynasty	1010±20	*
JYJ3	Stamped pottery	Jiaoshan, Yingtan, Jiangxi	The late Shang Dynasty	960±20	*
JYJ4	Stamped pottery	Jiaoshan, Yingtan, Jiangxi	The late Shang Dynasty	1080±20	*
JYJ5	Stamped pottery	Jiaoshan, Yingtan, Jiangxi	The late Shang Dynasty	1170±20	*
JZS01	Stamped pottery	Zaishan, Jiangxi	The late Shang Dynasty	1070±20	*
JZS03	Stamped pottery	Zaishan, Jiangxi	The late Shang Dynasty	1140±20	*
ZHJ1(1)	Black pottery with clay glaze	Jiangshan, Zhejiang	The early Shang Dynasty	1050	
ZHJ1(2)	Black pottery with clay glaze	Jiangshan, Zhejiang	The early Shang Dynasty	1170	
ZHJ1(3)	Black pottery with clay glaze	Jiangshan, Zhejiang	The early Shang Dynasty	1150	
ZHJ1(4)	Black pottery with clay glaze	Jiangshan, Zhejiang	The early Shang Dynasty	1200	
ZHJ2(1)	Black pottery with clay glaze	Jiangshan, Zhejiang	The early Shang Dynasty	1190	
ZHJ2(2)	Black pottery with clay glaze	Jiangshan, Zhejiang	The early Shang Dynasty	1180	
HXZ01	Hard pottery	Xingzhen, Henan	The West Han Dynasty	950±20	*
JQF4	Stamped pottery	Fanchengdui, Jiangxi	The West Zhou Dynasty	1140±20	*

Table III, continued.

Code	Name	Manufactured or excavated site	Manufactured period	T (°C)	
JQZ3	Stamped pottery	Zhuwucheng, Qingjiang, Jiangxi	The early West Zhou Dynasty	990±20	*
JQZ4	Stamped pottery	Zhuwucheng, Qingjiang, Jiangxi	The early West Zhou Dynasty	1160±20	*
JQZ5	Stamped pottery	Zhuwucheng, Qingjiang, Jiangxi	The early West Zhou Dynasty	1020±20	*
JWG1	Stamped pottery	Shizilao, Wanzai, Jiangxi	The early West Zhou Dynasty	1060±20	*
JWG2	Stamped pottery	Shizilao, Wanzai, Jiangxi	The early West Zhou Dynasty	1100±20	*
JYT1	Stamped pottery	Taipinggang, Yifeng, Jiangxi	The early West Zhou Dynasty	1100±20	*
JYT2	Stamped pottery	Taipinggang, Yifeng, Jiangxi	The early West Zhou Dynasty	1110±20	*
JQS1	Stamped pottery	Wucheng, Qingjiang, Zhejiang	The middle West Zhou to the middle Spring and Autumn	1140±20	*
JQS2	Stamped pottery	Wucheng, Qingjiang, Zhejiang	The middle West Zhou to the middle Spring and Autumn	980±20	*
JQS3	Stamped pottery	Wucheng, Qingjiang, Zhejiang	The middle West Zhou to the middle Spring and Autumn	1100±20	*
FF01	Stamped pottery	Tanshishan, Fuzhou, Fujian	The late New Stone Age	1000±20	*
FF02	Stamped pottery	Tanshishan, Fuzhou, Fujian	The late New Stone Age	970±20	*
FQ03	Stamped pottery	Quanzhou, Fujian	The late New Stone Age	1030±20	*
FQ01	Stamped pottery	Nanan, Quanzhou, Fujian	The late New Stone Age	1010±20	*
FQ02	Stamped pottery	Nanan, Quanzhou, Fujian	The late New Stone Age	1080±20	*
41	Hard pottery	Zhengzhou, Henan	Yin Dynasty	1180±20	
HXZ05	Stamped pottery	Xingzhen, Henan	The Warring States	1040±20	*
JQZ2	Hard pottery	Zhuwucheng, Qingjiang, Jiangxi	The Zhou Dynasty	1070±20	*
W1	Proto-porcelain	Jingxian, Hebei	The Northern Dynasties	1200±20	
ZSH1	Proto-porcelain	Houshan, Shaoxing, Zhejiang	The late Spring and Autumn to the early Warring States	970±20	*
ZSH2	Proto-porcelain	Houshan, Shaoxing, Zhejiang	The late Spring and Autumn to the early Warring States	980±20	*
ZSH3	Proto-porcelain	Houshan, Shaoxing, Zhejiang	The late Spring and Autumn to the early Warring States	1000±20	*
ZSH4	Proto-porcelain	Houshan, Shaoxing, Zhejiang	The late Spring and Autumn to the early Warring States	980±20	*
ZSH5	Proto-porcelain	Houshan, Shaoxing, Zhejiang	The late Spring and Autumn to the early Warring States	1000±20	*
ZSH6	Proto-porcelain	Houshan, Shaoxing, Zhejiang	The late Spring and Autumn to the early Warring States	920±20	*
ZSH7	Proto-porcelain	Houshan, Shaoxing, Zhejiang	The late Spring and Autumn to the early Warring States	1000±20	*
ZH-7	Proto-porcelain	Yingxing, Jiangsu	The East Zhou Dynasty	1140±20	
33	Proto-porcelain	Houma, Shanxi	The East Zhou Dynasty	1230±20	
H2	Proto-porcelain	Zhenjiang, Jiangsu	Tha Han Dynasty	1130±20	
J1	Proto-porcelain	Yingxing, Jiangsu	The Jin Dynasty	1260±20	
ZHJ3	Proto-porcelain	Jiangshan, Zhejiang	Tha Shang Dynasty	1220	
JQS11	Proto-porcelain	Wucheng, Qingjiang, Zhejiang	The late Shang Dynasty	1040±20	*
JQS13	Proto-porcelain	Wucheng, Qingjiang, Zhejiang	The late Shang Dynasty	1040±20	*
JQS15	Proto-porcelain	Wucheng, Qingjiang, Zhejiang	The late Shang Dynasty	1250±20	*
JQS9	Proto-porcelain	Wucheng, Qingjiang, Zhejiang	The late Shang Dynasty	1100±20	*
JYJ7	Proto-porcelain	Jiaoshan, Yingtan, Jiangxi	The late Shang Dynasty	1120±20	*
JYJ8	Proto-porcelain	Jiaoshan, Yingtan, Jiangxi	The late Shang Dynasty	1130±20	*
JQS16	Proto-porcelain	Wucheng, Qingjiang, Zhejiang	The early Shang Dynasty	960±20	*
HZZ01	Proto-porcelain	Zhengzhou, Henan	The middle Shang Dynasty	1190±20	*
JQF1	Proto-porcelain	Fanchengdui, Jiangxi	The Shang to the West Zhou D.	1150±20	*
JQF2	Proto-porcelain	Fanchengdui, Jiangxi	The Shang to the West Zhou D.	920±20	*
JQF3	Proto-porcelain	Fanchengdui, Jiangxi	The Shang to the West Zhou D.	1170±20	*
H3	Proto-porcelain		The West Han Dynasty	1270±20	

Table III, continued.

Code	Name	Manufactured or excavated site	Manufactured period	T (°C)	
HXZ03	Proto-porcelain	Xingzhen, Henan	The West Han Dynasty	1055±20	*
ZH-9	Proto-porcelain	Fufeng, Shaanxi	The West Zhou Dynasty	1280±20	
M-198	Proto-porcelain	Luoyang, Henan	The early West Zhou Dynasty	1220±20	
M-250	Proto-porcelain	Luoyang, Henan	The early West Zhou Dynasty	1150±20	
M-32:5	Proto-porcelain	Luoyang, Henan	The early West Zhou Dynasty	1220±20	
M-37	Proto-porcelain	Luoyang, Henan	The early West Zhou Dynasty	1130±20	
M-668	Proto-porcelain	Luoyang, Henan	The early West Zhou Dynasty	1120±20	
ZHJ5	Proto-porcelain	Jiangshan, Zhejiang	The late West Zhou Dynasty	1250	
ZHJ4(1)	Proto-porcelain	Jiangshan, Zhejiang	The early West Zhou Dynasty	1200	
ZHJ4(2)	Proto-porcelain	Jiangshan, Zhejiang	The early West Zhou Dynasty	1200	
JQS4	Proto-porcelain	Wucheng, Qingjiang, Zhejiang	The middle West Zhou to the middle Spring and Autumn D.	1150±20	*
JL01	Proto-porcelain	Anyi, Jiangxi	The middle West Zhou Dynasty	1250±20	*
	White porcelain	Ding kiln, Hebei		1300	
	Celadon	Julou, Hebei		1165-1180	
	Celadon	Qinghe, Hebei		1250-1260	
193	White porcelain	Dehua, Fujian	The Northern Song Dynasty	1210±20	
NST2(2)	White porcelain	Dehua, Fujian	The Northern Song Dynasty	1260±20	
NST3(2)	White porcelain	Dehua, Fujian	The Northern Song Dynasty	1260±20	
192	Black porcelain	Jian kiln, Fujian	The Northern Song Dynasty	1330±20	
	White porcelain	Ding kiln, Hebei	The Northern Song Dynasty	1320±20	
185	Celadon		The Northern Song Dynasty	1260±20	
150	Black porcelain		The Northern Song Dynasty	1270±20	
181	Celadon		The Northern Song Dynasty	1270±20	
NSL-1	Celadon	Longquan, Zhejiang	The Northern Song Dynasty	1160±20	
1	Black porcelain	Shangyu, Zhejiang	The East Han Dynasty	1220±20	
2	Black porcelain	Shangyu, Zhejiang	The East Han Dynasty	1200±20	
3	Black porcelain	Shangyu, Zhejiang	The East Han Dynasty	1240±20	
H4	Celadon	Shangyu, Zhejiang	The East Han Dynasty	1160±20	
H5	Celadon	Shangyu, Zhejiang	The East Han Dynasty	1310±20	
SY8-5	Celadon	Shangyu, Zhejiang	The East Han Dynasty	1260±20	
SY8-7	Celadon	Shangyu, Zhejiang	The East Han Dynasty	1270±20	
141	Black porcelain	Deqing, Zhejiang	The East Jin Dynasty	1150±20	
J7	Celadon	Jinhua, Zhejiang	The East Jin Dynasty	1180±20	
J6	Celadon	Shaoxing, Zhejian	The East Jin Dynasty	1270±20	
142	Black porcelain	Yuhang, Zhejiang	The East Jin Dynasty	1130±20	
227-1	White porcelain		The Jin Dynasty	1190±20	
	White porcelain	Ding kiln, Hebei	The Jin Dynasty	1250±20	
105	White porcelain		The Jin Dynasty	1210±20	
106	White porcelain		The Jin Dynasty	1230±20	
243	Celadon	Xunyi, Shanxi	The Jin Dynasty	1260±20	
277	Black porcelain		The Jin Dynasty	1230±20	
LB-A00	White porcelain	Liaoning	The Liao Dynasty	1280±20	
LB-A002	White porcelain	Liaoning	The Liao Dynasty	1270±20	
LB-B002	White porcelain	Liaoning	The Liao Dynasty	1230±20	
LB-C001	White porcelain	Liaoning	The Liao Dynasty	1270±20	
LB-D001	White porcelain	Liaoning	The Liao Dynasty	1200±20	
LB-F	White porcelain	Liaoning	The Liao Dynasty	1200±20	
LB-G	White porcelain	Liaoning	The Liao Dynasty	1270±20	
LB-GVAN	White porcelain	Liaoning	The Liao Dynasty	1260±20	
LB-H00	White porcelain	Liaoning	The Liao Dynasty	1280±20	
LB-Z001	White porcelain	Liaoning	The Liao Dynasty	1240±20	
M-1	Blue-and-white porcelain		The Ming Dynasty	1280±20	

Table III, continued.

Code	Name	Manufactured or excavated site	Manufactured period	T (°C)	
129	White porcelain	Pengxian, Sichuan	The Song Dynasty	1250±20	
61	Celadon		The Song Dynasty	1280±20	
S6-2	Celadon	Yue kiln, Zhejiang	The Song Dynasty	1050±20	
117	White porcelain	Anyang, Henan	The Sui Dynasty	1230±20	
118	Celadon	Anyang, Henan	The Sui Dynasty	1150±20	
119	White porcelain	Anyang, Henan	The Sui Dynasty	1170±20	
135	Celadon	Wuchang, Hebei	The Sui Dynasty	1230±20	
HG1	White porcelain	Gongxian, Henan	The Sui Dynasty	1290±20	
HG2	White porcelain	Gongxian, Henan	The Sui Dynasty	1260±20	
HG3	White porcelain	Gongxian, Henan	The Sui Dynasty	1380±20	
HG4	White porcelain	Gongxian, Henan	The Sui Dynasty	1290±20	
HG5	White porcelain	Gongxian, Henan	The Sui Dynasty	1340±20	
HG6	White porcelain	Gongxian, Henan	The Sui Dynasty	1290±20	
HN1	White porcelain	Xing kiln, Hebei	The Tang Dynasty	1370±20	
HN4	White porcelain	Xing kiln, Hebei	The Tang Dynasty	1260±20	
HN5	White porcelain	Xing kiln, Hebei	The Tang Dynasty	1340±20	
57	Black porcelain	Gongxian, Henan	The Tang Dynasty	1290±20	
T1	White porcelain	Gongxian, Henan	The Tang Dynasty	1150±20	
	Celadon		The Tang Dynasty	1120±20	
77	Black porcelain		The Tang Dynasty	1320±20	
77	White porcelain		The Tang Dynasty	1240±20	
149	Celadon	Yixing, Jiangsu	The Tang Dynasty	1260±20	
220	Celadon	Yixing, Jiangsu	The Tang Dynasty	1220±20	
89	Celadon	Tongchuan, Shaanxi	The Tang Dynasty	1230±20	
133	Celadon		The Tang Dynasty	1240±20	
144	Celadon	Wunzhou, Zhejiang	The Tang Dynasty	1270±20	
92	Celadon	Xiangshan, Zhejiang	The Tang Dynasty	1200±20	
143	Celadon	Yuyao, Zhejiang	The Tang Dynasty	1240±20	
	White porcelain	Ding kiln, Hebei	The late Tang Dynasty	1300±20	
ZSS2	Celadon	Shaoxing, Zhejiang	The late Tang to Northern Song	1120±20	*
ZSS5	Celadon	Shaoxing, Zhejiang	The Late Tang to Northern Song	1130±20	*
	White porcelain	Ding kiln, Hebei	The Five Dynasties	1300±20	
150	Celadon		The Five Dynasties	1260±20	
151	White porcelain	Jingdezhen, Jiangxi	The Five Dynasties	1180±20	
T2-1	White porcelain	Jingdezhen, Jiangxi	The Five Dynasties	1150-1200	
147	Celadon		The Five Dynasties	1220±20	
145	Celadon		The Five Dynasties	1220±20	
J4	Celadon	Shangyu, Zhejiang	The West Jin Dynasty	1300±20	
J5	Celadon	Shangyu, Zhejiang	The West Jin Dynasty	1180±20	
SY-16	Celadon	Shangyu, Zhejiang	The West Jin Dynasty	1220±20	
Y-1	Blue-and-white porcelain		The Yuan Dynasty	1280±20	
Y-2	Blue-and-white porcelain		The Yuan Dynasty	1290±20	
Y-3	Blue-and-white porcelain		The Yuan Dynasty	1240±20	
Y-5	Blue-and-white porcelain		The Yuan Dynasty	1280±20	
Y2	Blue-and-white porcelain		The Yuan Dynasty	1100-1150	
YG'(2)	White porcelain	Dehua, Fujian	The Yuan Dynasty	1260±20	
YG(2)	White porcelain	Dehua, Fujian	The Yuan Dynasty	1270±20	
YT12	White porcelain	Dehua, Fujian	The Yuan Dynasty	1260±20	
YT15	White porcelain	Dehua, Fujian	The Yuan Dynasty	1270±20	
YT7(1)	White porcelain	Dehua, Fujian	The Yuan Dynasty	1280±20	
81	White porcelain		The Yuan Dynasty	1280±20	
158	Black porcelain	Jingdezhen, Jiangxi	The Yuan Dynasty	1250±20	
79	Black porcelain	Linfeng, Shanxi	The Yuan Dynasty	>1300	

Table III, continued.

Code	Name	Manufactured or excavated site	Manufactured period	T (°C)
M-4	Blue-and-white porcelain		The Ming Dynasty	1240±20
M1			The Ming Dynasty	1200±20
M2	White porcelain		The Ming Dynasty	1200±20
M3	Blue-and-white porcelain		The Ming Dynasty	1200±20
199	White porcelain	Dehua, Fujian	The Ming Dynasty	1200±20
MF1	White porcelain	Dehua, Fujian	The Ming Dynasty	1200±20
MTB(1)	White porcelain	Dehua, Fujian	The Ming Dynasty	1270±20
ML1	Celadon	Longquan, Zhejiang	The Ming Dynasty	1200±20
HB4	Celadon	Shangyu, Zhejiang	The Southern Dynasty	1190±20
SST1(1)	White porcelain	Dehua, Fujian	The Southern Song Dynasty	1270±20
SST3(1)	White porcelain	Dehua, Fujian	The Southern Song Dynasty	1250±20
S5	Celadon	Southern Song offical kiln	The Southern Song Dynasty	1150±20
S3-1	Celadon	Longquan, Zhejiang	The Southern Song Dynasty	1200±20
SSL-1	Celadon	Longquan, Zhejiang	The Southern Song Dynasty	1230±20
SSL-7	Celadon	Longquan, Zhejiang	The Southern Song Dynasty	1250-1280
98	Celadon		The Qing Dynasty	1280±20
C-1	Blue-and-white porcelain		The Qing Dynasty	1320±20
C-2	Blue-and-white porcelain		The Qing Dynasty	1300±20
C13	White porcelain		The Qing Dynasty	1300±20
C14	White porcelain		The Qing Dynasty	1300±20
C14/58	White porcelain	Dehua, Fujian	The Qing Dynasty	1270±20
C74	White porcelain	Dehua, Fujian	The Qing Dynasty	1270±20
SHT1(2)	Celadon	Shangyu, Zhejiang	The Three Kingdoms	1240±20
90	Black porcelain	Ankou, Gansu	The Song Dynasty	1330±20
87	Black porcelain	Cixian, Hebei	The Song Dynasty	1310±20
S4	White porcelain	Ding kiln, Hebei	The Song Dynasty	1150±20
84	White porcelain	Qiuyang, Hebei	The Song Dynasty	1270±20
85	Black porcelain	Qiuyang, Hebei	The Song Dynasty	1350±20
86	Black porcelain	Qiuyang, Heibei	The Song Dynasty	>1300
68	Black porcelain	Baofeng, Henan	The Song Dynasty	1320±20
69	Celadon	Baofeng, Henan	The Song Dynasty	1240±20
	Celadon	Ru kiln, Henan	The Song Dynasty	1200±20
	Celadon	Jun kiln, Henan	The Song Dynasty	1270
	Celadon	Jun kiln, Henan	The Song Dynasty	1270
	Celadon	Jun kiln, Henan	The Song Dynasty	1250
	Celadon	Jun kiln, Henan	The Song Dynasty	1270
	Celadon	Jun kiln, Henan	The Song Dynasty	1250
S2-1	Celadon	Jun kiln, Henan	The Song Dynasty	1150±20
	Celadon	Linru, Henan	The Song Dynasty	1270±20
72	Celadon	Linru, Henan	The Song Dynasty	1270±20
	Celadon	Ru kiln, Henan	The Song Dynasty	1240±20
S1-1	Celadon	Ru kiln, Henan	The Song Dynasty	1150±20
172	Black porcelain	Jizhou, Jiangxi	The Song Dynasty	1270±20
S10-1	Celadon	Jingdezhen, Jiangxi	The Song Dynasty	1250±20
S9-2	Celadon	Jingdezhen, Jiangxi	The Song Dynasty	1100-1150
62	Celadon		The Song Dynasty	1270±20
76	White porcelain	Jiexiu, Shanxi	The Song Dynasty	1280±20
4	Black porcelain	Linfen, Shanxi	The Song Dynasty	1260±20
247	Celadon	Huangpu, Shaanxi	The Song Dynasty	1230±20
	Celadon	Yaozhou kiln, Shaanxi	The Song Dynasty	1300±20
S7-1	Celadon	Yaozhou kiln, Shaanxi	The Song Dynasty	1280
宋瓷-1	Celadon	Yaozhou kiln, Shaanxi	The Song Dynasty	1310
130	Black porcelain	Guangyuan, Sichuan	The Song Dynasty	1290±20

area with a high content of Al_2O_3 and low amount of fluxes. The chemical composition areas for the northern pottery and porcelain bodies are fully separated from each other, and no connected parts are found between them (Fig. 1). A few examples of stamped pottery and proto-porcelain which were excavated in the north do not form the transitional area from pottery to porcelain as the southern samples do. No obvious relationship in chemical composition can be found among the northern pottery, stamped pottery, proto-porcelain and porcelain. Therefore, it can be seen that the three stages of development (pottery, proto-porcelain and porcelain) which we presented previously (10) is only suitable for the southern samples, but not for the northern development. The southern technological development represents a gradual and connected compositional change with large fields of variation in raw materials which are suitable for each pottery and porcelain while an intermediate and restricted range is found during the transition. The northern tradition of body composition has two distinct pottery and porcelain fields and may suggest separate raw material types rather than a continuum, or that the technologies developed separately.

The above study helps us to understand what kind of roles the stamped pottery plays in the developmental process. When the technique of producing pottery had developed to a level where the pottery is quite hard, a category which includes mainly stamped pottery and not the proto-porcelain, the content of SiO_2 for the stamped pottery is higher than previous pottery bodies but lower than proto-porcelain bodies. The excavated ceramics also show that the periods of manufacture of stamped pottery were earlier than than of proto-porcelain, although some stamped pottery and proto-porcelain fragments were found as wasters in the same kilns. The above statistical results were obtained from a larger number of samples, and the excavated materials all show that the stamped pottery had established a strong foundation of manufacturing technique upon which the appearance of much early proto-porcelain is based. This result is in accord with the opinions of the community of archaeologists excavating these sites.

As shown in Fig. 3, the proto-porcelain glazes are located in two areas, Area II and Area III, which are separated. The samples in Areas II and III belong to the glazes containing high contents of Fe_3O_3 and CaO, respectively, which we now call the Fe_2O_3-rich proto-porcelain glaze and the CaO-rich proto-porcelain glaze. These two terms will be used below. The pottery coatings and clay-based slip glazes are all located in Area I. They have a higher content of SiO_2 and are lower in Fe_2O_3 and CaO. In discussing the relationships in chemical compositions of these three areas, it can be seen that the Fe_2O_3 proto-porcelain glazes have a much closer relationship with the pottery coatings and clay-based slip glazes. It is possible that the Fe_2O_3 proto-porcelain glazes were developed from the pottery coatings or clay glazes by decreasing the content of SiO_2, and increasing the content of Fe_2O_3. Figure 4 also shows that by increasing the CaO content, the Fe_2O_3 glaze will change to the late black glazes, including Jizhou, Jian and other famous black glazes. In addition, the CaO proto-porcelain glazes in the Area III seem to be developed from the pottery coatings and clay-based slip glazes by increasing the CaO content. This direction of technological change laid a good foundation for development of the early southern celadon or green glazes. The technique of

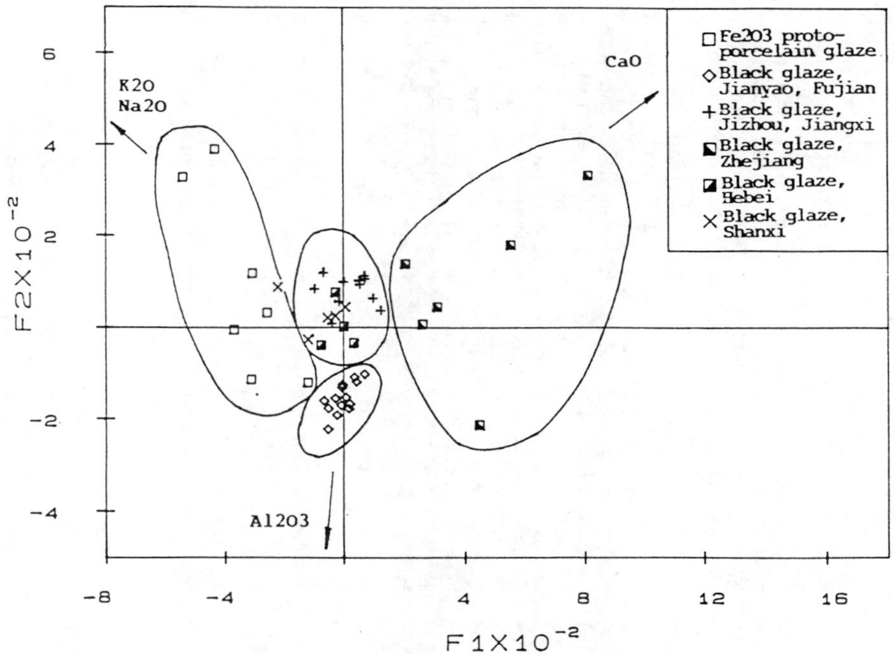

Fig. 4. F1 and F2 Factor Loading Diagram of Fe_2O_3 Proto-Porcelain Glaze and Black Glaze in Southern and Northern China

manufacture for the Fe_2O_3 and CaO proto-porcelain glazes were discussed in detail in references 1-8 and will not be repeated here. From the above discussion, we conclude that the process of gradual compositional evolution from ancient pottery coatings to porcelain glazes as presented in reference 6 was correct for the southern style of technology and that a northern technological development can be isolated and further characterized.

According to the above data and its evaluation, the developmental process for ancient Chinese pottery and porcelain can be summarized in the chart presented as Fig. 5. This process is divided into four successive periods based on analysis of the bodies: pottery of the New Stone Age, stamped pottery and proto-porcelain of the later part of the New Stone Age to the Shang and Zhou dynasties, the early celadon of the Han to Five dynasties, and the late celadon and white porcelain of the Song to Qing dynasties. For the glazes, the four periods are: pottery coating and clay-based glazes of the New Stone Age, the Fe_2O_3 and CaO proto-porcelain glazes of the Shang and Zhou dynasties, the glazes of the Han to Five dynasties, which include the black glaze, CaO glaze of the early southern celadon (yue) and CaO-alkali glaze of the early northern white porcelain glazes, and the CaO-rich glaze together with the CaO-alkali glaze and alkali-CaO glaze of the Song to Qing dynasties.

It is worth pointing out that although the gradual evolutionary process of ancient Chinese pottery and porcelain is mainly reflected in the southern technology, the white porcelain and the CaO-alkali (and even alkali-CaO) glazes appeared at northern kilns sites in the Sui and Tang periods. However, in the south, the white porcelain was not manufactured until the Five Dynasties, and the CaO-alkali glaze did not appear until the Song Dynasties at southern kiln sites.

The Firing Temperature Evolution Process for Pottery and Porcelain from the North and South

From Table 3 the characteristics of the various firing temperatures for different kinds of pottery and porcelain are summarized as follows:

Classification	Sample Numbers	The Highest Firing Temperature	The Mean Firing Temperature (°C)
Pottery	15	1000	920
Stamped Pottery	55	1200	1180
Proto-Porcelain	37	1280	1120
Porcelain	146	1350	1245

It is obvious that the firing temperature of the stamped pottery is as high as 1080°C, which is 160°C higher than that of the pottery. To increase the mean firing temperature of the stamped pottery by 160 degrees is a very difficult task, especialy in ancient times, and it is an amazing achievement in high temperature technology. The mean firing temperature of the proto-porcelain is only 40 degrees higher than that of the stamped pottery. Moreover, some proto-porcelain and stamped pottery were fired in the same kilns and have the same firing temperatures, as shown in Table 2. Therefore, the manufacture of the stamped pottery is a transitional stage but one in which control of high temperature was practiced and which prepared a

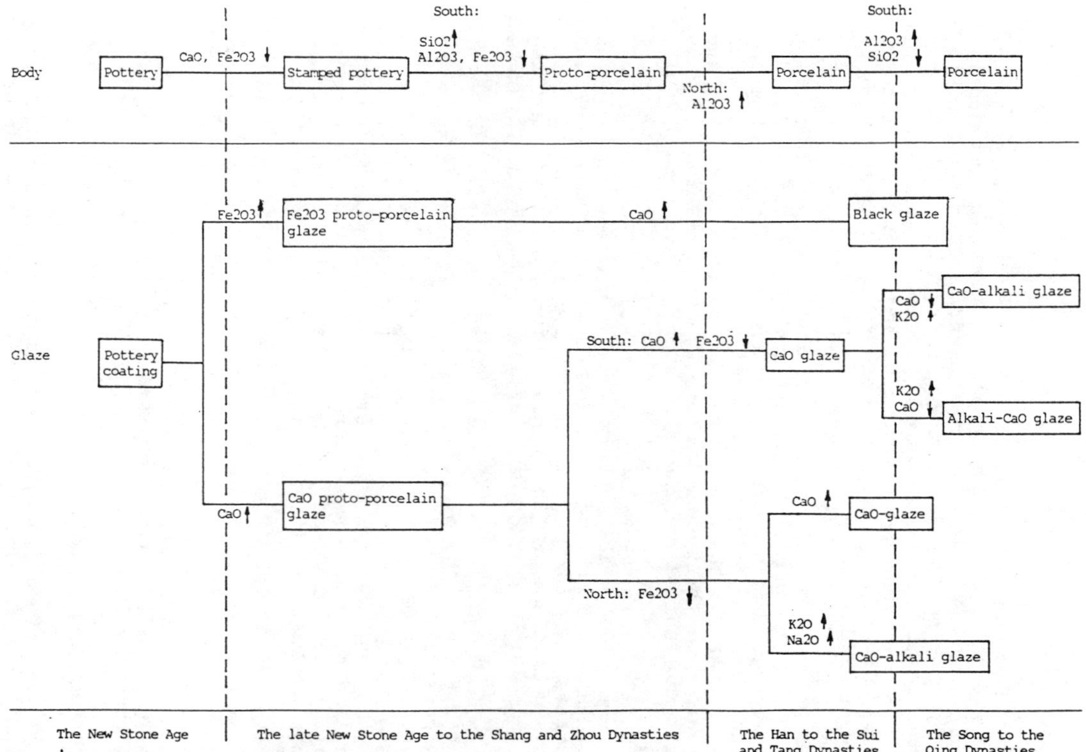

Note: ↓ refers decrease in content of some oxides.
↑ refers increase in content of some oxides.

Fig. 5. Chart Showing Sequences of Development of Body and Glaze Technologies with Time as well as Variations between North and South China

technological basis for further developments of proto-porcelain and then porcelain. From the above discussion it can be seen that the stamped pottery and proto-porcelain acted like a bridge to connect the pottery and porcelain in controlling high temperature to fire these ancient ceramic wares.

CONCLUSION

With the aid of multivariate correspondence analysis, the pattern of evolution of ancient Chinese pottery and porcelain is elucidated, and the roles which stamped pottery and proto-porcelain played in the development of northern and southern ceramic technology are discovered. From a vast amount of scientific data on chemical composition and firing temperature, the basis of such developments are isolated. The results may be summarized as follows:

(1) The process of technological development for ancient Chinese pottery and porcelain bodies and glazes can be divided into four periods according to the criteria provided by chemical composition. For the bodies these periods are in chronological order: the pottery of the New Stone Age, the stamped pottery and proto-porcelain of the late New Stone Age to the Shang and Zhou dynasties, the early southern celadon and northern white porcelain of the Han to Five dynasties, and the celadon and white porcelain of the Song to Qing dynasties. For the glazes, they are successively: the pottery coating and clay-based glazes of the New Stone Age, the proto-porcelain glaze of the Shang and Zhou dynasties, the early southern celadon glazes and northern white porcelain glaze of the Han to Five dynasties, and the celadon and white porcelain glazes of the Song to Qing dynasties.

(2) For different kinds of pottery and porcelain wares, their firing temperatures vary as follows: for pottery a mean firing temperature of 920°C (with the highest firing temperature at 1000°C), for stamped pottery a mean firing temperature of 1080°C (with the highest firing temperature at 1200°C), for proto-porcelain a mean firing temperature at 1120°C (with the highest firing temperature at 1280°C), and for porcelain the mean firing temperature at 1240°C (with the highest firing temperature at 1380°C).

(3) The manufacturing process of stamped pottery and proto-porcelain shows an important breakthrough not only in choosing of raw materials to be used in the batch composition, but also in mastering the high temperature technique of firing. They were developed on a foundation of manufacturing pottery, and meanwhile strongly influenced the manufacture of porcelain, thus playing very important roles in the development of ancient Chinese pottery and porcelain.

(4) The process of technological evolution for ancient Chinese pottery and porcelain continued through 8,000 years without a break, and perhaps throughout a period of 10,000 years. The technological characteristics of each developmental stage have been demonstrated as it was mainly reflected in the tradition of southern pottery and porcelain. Because there was no experience in the manufacture of stamped pottery and proto-porcelain in the north, the process of change for the northern pottery and

porcelain was not continuous and developing the difficult-to-find connections which will complete the picture of its pattern of evolution remain for future work by collaborating archaeologists and scientists.

ACKNOWLEDGEMENT

This work has been supported by the National Science Foundation of China.

REFERENCES

1. Zhou Jen, Lee Chia Chih (now Li Jiazhi), "An Investigation on the Technical Aspects of Chinese Ancient Ceramics, The Chinese Journal of Archaeology 1 (1960) 89ff.

2. Zhou Jen, Lee Chia Chih (Li Jiazhi), "Researches on the Body and Glaze of Porcelain of Different Periods of Jingdezhen", Silicate 4, no. 2 (1960) 49ff.

3. Zhou Jen, Chang Fu-kang, Cheng Yang-fu, "Technical Studies on the Lungchuan Celadons of Successive Dynasties", The Chinese Journal of Archaeology 38 (1973) 131ff.

4. Li Jiazhi, Chen Xianqiu, Deng Zequn et al., "A Study of the Pottery Unearthed at Homudu', Journal of the Chinese Ceramic Society 7, no. 2 (1979) 105ff.

5. Li Jiazhi, "Formation and Development of the Green Glaze in Zhejiang Province, Journal of the Chinese Ceramic Society 11, no. 1 (1983) 1ff.

6. Li Jiazhi, "A Study on the Emergence of Porcelain in Ancient China, Journal of the Chinese Ceramic Society 6, no. 3 (1978) 190ff.

7. Li Jiazhi, Deng Zequn, Zhang Zhigang et al., "Studies on the Black Pottery with Clay Glaze and Proto-porcelain in Jiangshan County, Zhejiang Province, Scientific and Technological Insights on Ancient Chinese Pottery and Porcelain, edited by the Shanghai Institute of Ceramics, Science Press, Beijing, 1986.

8. Li Jiazhi, "The Evolution of Chinese Pottery and Porcelain Technology", Ceramics and Civilization, vol. 1, edited by W.D. Kingery, The American Ceramic Society, Westerville, Ohio, 1983, 135ff.

9. Yu Chongwen, et al., The Methods and Applications of Geomathematics, Metallurgical Industry Press, Beijing, China, 1980, pp. 231-236 (in Chinese).

10. Luo Hongjie, Li Jiazhi and Gao Liming, "The Chemical Composition and Microstructure Study for the Proto-Porcelain Glaze", in preparation.

THERMAL ANALYSIS OF EARLY NEOLITHIC POTTERY FROM TEPE GANJ DAREH, IRAN

A. YELON, A. SAUCIER[1], J.-P. LAROCQUE[2],
P.E.L. SMITH[3], AND P. VANDIVER[4]
Département de Génie Physique, École Polytechnique,
P.O.Box 6079, Station A, Montréal (Québec) H3C 3A7, CANADA

1 Present address: Institute for Energy Technology, Box 40, 2007 Kjellar, Norway
2 Present address: CN Rail, 935 Lagauchetière W., Montréal
3 Département d'Anthropologie, Université de Montréal, Montréal (Québec) H3C 3J7
4 Conservation Analytical Laboratory, Smithsonian Institution, Washington, DC 20560

ABSTRACT

A series of pottery fragments from an early Neolithic site in Iran is analyzed to determine the original baking temperatures. Two magnetic techniques are used: remanent magnetization decay, and hysteresis loop analysis. The results are complementary, although some are, at first sight, contradictory. The causes of these apparent contradictions are explained. The analyses confirm that the pottery was baked under highly variable conditions, probably over open fires. Control of baking was poor, and there was considerable variability not only from one vessel to another but even from one zone to another on the same vessel. Our results suggest the need for caution in interpreting often ambiguous findings in this field of ancient technology.

INTRODUCTION

The possibility of determining the thermal history of early Neolithic pottery is of interest from several points of view. First, it provides information on the technology of its preparation, illuminating the early histoty of pyrotechnology, and its condition in a particular culture. Further, information as to whether the pottery received secondary heating may cast light on its use, and consequently on other aspects of the culture under study. A number of techniques have been developed for obtaining such information (see, for example, Goodyear, 1971, Heimann and Franklin, 1981, Matson, 1981, Périnet and Courtois, 1983). Among the most interesting of these are two magnetic techniques, the Thellier and Thellier (1959) method of remanent magnetization (RM) decay (Schwarz and Christie, 1967, Barbetti et al, 1977, Arbour and Schwarz, 1982) and the hysteresis loop method (Coey et al, 1979). We have applied these two tchniques to an investigation of very early Neolithic pottery from the site of Tepe Ganj Dareh, in Iran.

As we shall show, the two techniques provide complementary information. As we shall also show, some of the results appeared to be contradictory at first. The simple technique of placing the pottery in boiling water, to determine if it had been heated above its sintering temperature, was therefore applied to some of the samples. This helped to resolve the apparent contradictions. Our present understanding of the thermal history of our samples casts some light both on early pyrotechnology and on the uses to which the pottery was put. In addition, the ambiguities initially presented by our measurements serve to illustrate problems which may be encountered with such samples, which should be considered in future studies.

THE ARCHAEOLOGICAL PROBLEM

The small mound or *tepe* of Ganj Dareh is located in the Zagros Mountains of western Iran. It was excavated between 1967 and 1974 (Smith 1978, 1983), and the results are now being analyzed and published. In its approximately 7 meters of deposits the mound reveals five principal occupation levels: A, B, C, D, and E, reading from the top down. The time span involved seems concentrated in the 8th millennium B.C. for the most part. The later levels (A,B) may extend into the early 7th, while the earliest occupation, level E, whose radiocarbon determinations are inconsistent, is probably no younger than ca. 7500 B.C. There are some traces of Iron Age presence (mainly potsherds) on the surface of the mound.

The strata (except for level E, situated on virgin soil and without solid architecture) contain rectangular buildings of mud and brick, some with several stories. Study of the faunal remains have now shown that herds of morphologically wild goats were being selectively culled by the inhabitants from level D onward, that is, were under some degree of cultural control which presumably anticipated full domestication (Hesse 1982). Domesticated barley was present (and, we presume, under local cultivation) in all levels (van Zeist et al, 1986). Thus Ganj Dareh seems to exhibit a series of small communities very near the beginning of food production in this part of the Near East, although domesticates still accounted for only a small proportion of the total diet of the inhabitants.

Another interesting feature is a small quantity of rather simple pottery vessels. Only one sherd occurs in level E, and the pottery is uncommon in levels C, B and A. In level D, however, the pottery is much more abundant, thanks in part to an intense fire which hardened and preserved the clay objects in this level: architecture, figurines, various other artifacts, and the pottery. The vessels occur in two forms: very large jars of thick, coarse clay with capacities of up to several hundred liters which were probably storage vessels; and small, even miniature vases, pots and saucer-like items whose functions are problematic. None of the pottery seems to have been suitable for cooking purposes, judging by their sizes and forms. None was painted, and decoration was restricted to occasional simple impressions.

This pottery is among the earliest known in the Near East, and the large jars may represent the earliest case of pottery designed for purely utilitarian use in this region. It therefore presents a valuable opportunity for illuminating the early history of this important technological innovation. We already know, from the well-preserved remains of level D, that the inhabitants of Ganj Dareh were highly accomplished craftsmen in many types of ceramic technology. We also know from the presence of several fairly sophisticated lime kilns built of bricks and stones, and the occurrence of much lime and plaster, that this branch of pyrotechnology was reasonably advanced. How elaborate and precise, therefore, were the thermal techniques used in manufacturing the pottery?

EXPERIMENTAL METHOD

Magnetic measurements on pottery are based on the fact that all clays contain iron oxides, with typical concentrations of iron between 1 and 3%. Thus, the natural remanent magnetization (NRM) observed in a Thellier measurement will be that acquired by these oxides when cooled from temperatures at which they are demagnetized. The hysteresis loop measurement determines properties of these oxides, which will have been altered in the past by the high temperature chemical reactions to which they have been subjected.

The Thellier and Thellier method (1959) is based on the additivity of partial thermoremanent magnetization (TRM) acquired by cooling in a constant field over successive temperature intervals. Even though the susceptibility may vary in a complicated way, the variation of remanent magnetization with maximum temperature will be monotonic. The RM is expected to be proportional to the intensity of the field in which it was kept during cooling. The magnetization acquired by cooling in the laboratory should be the same as that which is lost by heating to that temperature if the ancient and modern fields are the same. Thus, if a sample has been heated above the highest blocking temperature of the iron oxides (typically 500-650°C), a plot of NRM versus TRM, with temperature as a parameter should be a straight line. The slope of this line will indicate the ratio of the ancient magnetic field to the laboratory field in which the TRM is measured. If a piece of pottery has never been heated as high as the highest blocking temperature of its oxides, we expect to observe an anomaly in the NRM versus TRM curve at the highest temperature to which it was heated. Thus, we expect to be able to determine this temperature by this method.

Samples for this measurement consisted of sherds embedded in a 1" cube of plaster. These were demagnetized by heating in a field-free volume, and remagnetized by cooling in a field of 0.5 Oe. Using a spinner magnetometer, the magnitude of the RM could be determined in each case. In principle, we could have also determined its direction, but this was not done.

The hysteresis loop method is based on the fact that the granularity of the clay and the composition of the oxides it contains are strongly affected by the firing temperature and atmosphere, and that these strongly affect the hysteretic magnetic properties. Thus, if one knows how these properties evolve with temperature for likely atmospheric conditions, one may put these parameters on master curves. If the values of these properties for an unknown sample then correspond to a temperature program and atmospheric condition, it is then a reasonable hypothesis that the sample was fired under these conditions.

Coey et al (1979), who were the first to apply the technique, studied the evolution of three parameters, saturation magnetization, M_s, remanent magnetization, M_r, and coercive force, H_c, as a function of firing temperature under oxidizing and reducing atmospheres. They were able to demonstrate that the red, buff, and grey ware of the Tureng Tepe culture of Iran (c 3000-c 1650 B.C.) were produced under different atmospheres. They were able to show that the introduction of firing under reducing atmosphere constituted a useful technological innovation.

In order to apply this method, it is necessary to have a sample of clay similar to that from which the pottery was made, and which can be fired in a systematic way, in order to establish master curves of the hysteretic properties as a function of firing conditions. In addition, since the iron content will vary from one sample to another, even if they came from close proximity, it is necessary to determine the iron content of each sample measured. This can be done by a number of methods (Brissaud et al, 1985). In the present study, we have employed neutron activation. The M_s and M_r for both the master curve and the samples are normalized to the iron content.

Since the hysteresis loop changes continue up to above 1000°C, Coey et al (1979) were able to explore temperatures inaccessible to the Thellier technique, which is limited by the blocking temperature, as well as atmosphere effects. However, since the changes they observed were very small below 400°C, had they had pots fired below that temperature, they would only have been able to give an upper limit to that temperature. Thus the two techniques may be seen as complementary, confirming each other in some cases, and yielding different information in others. This promised to be particularly useful in our study of pottery which was much older than that of Tureng Tepe and which was rather soft, suggesting that the firing temperatures might have been quite low. Because of the greater age of our samples, we were confident that we could limit ourselves to oxidizing atmospheres.

For this measurement, pieces of pottery were ground into powder, and measured in a vibrating sample magnetometer. Standard samples were prepared by grinding clay in a mortar, and removing particles. The sample was heated on a slab of ceramic, after which a sample was prepared for measurement as in the case of the pottery sherds.

In the face of results of our magnetic measurements which appeared anomalous or contradictory in a number of cases, we applied the simple, but relatively unambiguous, technique of boiling samples in water. This serves to distinguish between samples which have been heated, at least once, to a temperature at which they have at least begun to be sintered from those which have never been heated to such a temperature. The former will remain intact, while the latter will disintegrate within a few minutes. Without much additional study, this temperature is very ill defined, as some samples may begin to sinter in the range 400-450°C, whereas others may not begin until several hundred degrees higher, and "strong" sintering does not take place until about 800°C (Heimann and Franklin, 1981). Despite these limitations, this approach helped considerably in clarifying the situation and permitted us to find a consistent interpretation of our results.

HYSTERESIS LOOP MEASUREMENTS: RESULTS AND INTERPRETATION

All of our samples were studied using the hysteresis loop measurement technique, while smaller subsets of these were investigated by the Thellier and boiling water techniques. Of course, the precise piece which is employed for one of these measurements cannot be re-employed for either of the other two. This technique requires the establishment of a master curve, using clay similar to that employed in the manufacture of the pottery to be studied. We assumed that any early pottery would have been made with material very similar to that originally found in the vicinity of the site. This was confirmed by the determination that clay from virgin soil beneath level E of Ganj Dareh contained typically 3.2-3.5% iron, by weight, compared with values of 2.5-3.8% for pottery from levels A to D. In comparison, Iron Age pottery from the site contained 3.9% Fe, whereas imported Iron Age pottery contained 5.2% Fe, and a sample of modern soil contained 4%. Our hypothesis was further supported by the fact that most (but not all) pottery samples from levels A to D could be placed in a relationship with the master curve we eventually obtained, whereas both the present day soil and the imported Iron Age pottery were very far from this curve. Other measurements, such as trace element analysis, were not attempted, as they were unlikely to either strengthen or weaken our conclusions.

Clay from virgin soil beneath level E was fired for the establishment of master surves for the hysteresis loop method. We assumed that firing in open air would most closely approximate the conditions under which the pottery was produced, so the master samples were prepared in a tube furnace with the tube open to the air. In establishing the master curves, one does not know a priori whether the comparison should be made with a standard which was fired for a few minutes at the highest temperature or with one which was fired for longer times, and whether the appropriate standard would be one heated directly from room temperature or one which was fired at the next lowest temperature and then refired. In addition, two samples of clay will give somewhat different results. Fortunately, two different curves produced under quite different conditions showed differences which were relatively small, except at the very highest temperatures (A. Saucier, unpublished report). As a result, we expect qualitative, but not quantitative, agreement between the parameters of the master curve and those of the samples. In a test of three samples which were heated to known temperature and compared with the master surve for determination of the temperature, it was found that H_c gave the most reliable comparison.

In Table I and in Figs. 1-3, we present the results for a sample containing 3.2% Fe. In order to produce the master curve, this sample was fired for three hours at each temperature. A piece was then removed, to be ground for measurement, and the rest was returned to be heated to the next higher temperature. M_s is taken as the magnetization at 9000 Oe. As expected from the work of Coey et al (1979) the changes are relatively small until about 500°C, although the change in H_c is large enough to offer hope that we might be able to determine low firing temperatures from this parameter.

The samples studied by this method, some of which were also studied by other methods, are presented in Table II, in which we give the catalog number, the level in which it was found, and comments on the sample and the situation in which it was found. The procedure we used to attempt to estimate the firing temperature of these sherds, based upon the fact that H_c gave the most reliable values, was to determine a range of possible temperatures from the two magnetization values, and then to take the estimated temperature (or temperatures) from the corresponding H_c values in this range.

The procedure may be illustrated with one of the known samples, fired at 830°C. The value of M_r suggests a firing temperature of 800°C or below. The value of M_s suggests a value of 830°C or below. The value of H_c would permit values of 820, 860, 960, or 1025°C. The only one of these which is in reasonable agreement with the limits imposed, is 820°C, in excellent agreement with the true value. The other two known samples also give values within 10° of the correct value using the same procedure.

In Table III, we present for each sample, including the sample for which the firing conditions were known, the percentage of Fe, and the hysteresis loop parameters, M_s, M_r, and H_c, as well as the results of this analysis. We find that the samples may be divided into four categories. Three of these are temperature ranges; we find that most of the samples may be grouped as being near 150°, near 600°, or in the 900-1150° range.

For some of the samples the agreement is excellent. For example, the samples in the 1000°C category marked with an asterisk are those for which all three parameters are consistent with a temperature within 60° of the value given as the "baking temperature". However, we note that other samples agree less well. This disagreement might be attributed to mineralogical differences or to differences in the firing atmosphere, when compared with the standard. In the anomalous category were samples for which the three different parameters indicate different temperature ranges, or for which the measured parameters are completely outside of the range of parameters of the master curve.

Some of the results of Table III inspire confidence in the method. The two pieces of sample 31007 gave similar results. The only sample in the 150° category is the sample from level E, the oldest sample from the site. All of the samples from level D, which shows

TABLE I

Magnetic Properties of a Soil Sample Heated in Air at Various Temperatures

Temperature (°C)	M_r (emu g^{-1})	M_s (emu g^{-1})	H_c (Oe)
virgin	0.21	0.92	87
48	0.21	1.19	87
108	0.20	1.28	82
174	0.16	1.15	71
222	0.16	1.14	67
270	0.15	1.00	68
330	0.16	1.04	66
380	0.15	0.93	63
430	0.14	0.86	59
480	0.16	1.13	59
540	0.15	0.82	65
585	0.15	0.69	70
624	0.15	0.56	94
675	0.11	0.43	87
730	0.08	0.39	64
787	0.05	0.45	33
845	0.71	1.23	13
900	0.89	4.53	42
945	0.48	3.26	27
990	0.40	2.59	32
1045	0.50	2.18	66
1100	0.77	3.09	75
1150	0.15	1.22	78
1200	vitrified, unmeasurable	vitrified, unmeasurable	

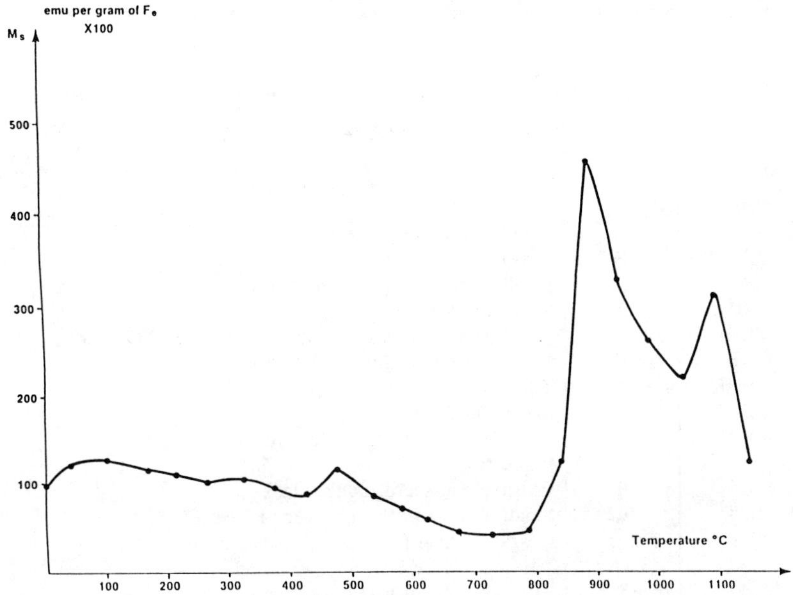

Fig. 1: Remanent moment, M_r, versus baking temperature for a clay sample.

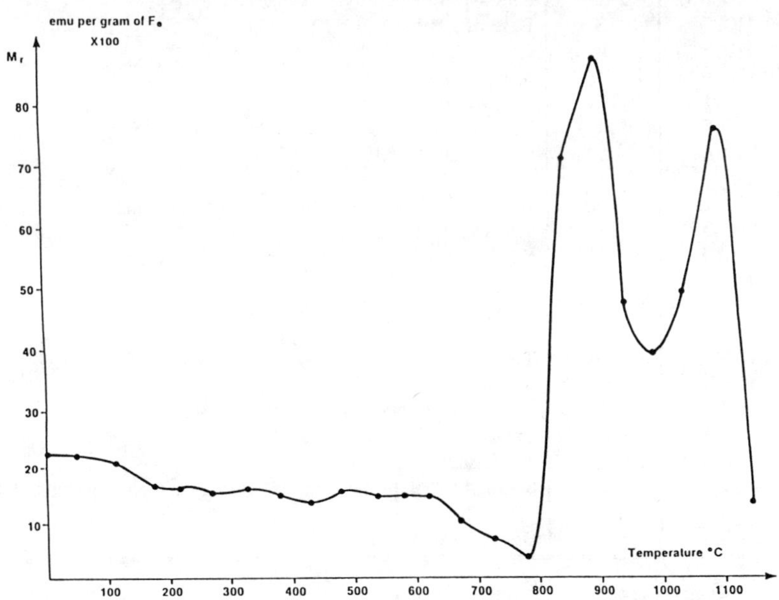

Fig. 2: Saturation moment, M_s, versus baking temperature for a clay sample.

TABLE II

Characteristics of Specimens Studied

Catalog Number	Level	Description
4001	AB	Small sherd with fingernail (?) impressions.
4003	B	Possible rim-sherd, thin.
4004	A	Small sherd with fingernail (?) impressions.
4005	A	Small vase (?) fragment.
4015	A	Small grey-black sherd with fingernail (?) impressions.
4016	A	Grey sherd with fingernail (?) impressions.
4018	A	Thick sherd, blackened, plain.
4020	A	Sherd, hard, plain, blackened exterior.
4024	AB	Small vessel (?) fragment, hard, with fingernail (?) impressions.
4026	B	Possible rim-sherd, thick, black.
4027	BC	Small buff sherd from saucer-like vessel.
4028	BC	Coarse reddish fragment, possible sherd.
4037	AB	Base of small crude pot, grey, soft.
4042	ABC	Rim-sherd (?) of small pot, plain, buff.
4046	D	Small intact pot, buff, globular.
31006	B	Rim-sherd of bowl (?), coarse.
31007	B	Base of small vase, buff.
31008	B	Clay fragment, buff.
31010	Iron Age	Fine, thin sherd, probably imported.
31011	BD	Base of small crude vessel, buff.
31015	BC	Grey sherd, coarse.
31024	B	Base of small vase (?), buff.
46002	A	Base of small vase (?), hard, brown. Recent?
46005	A	Thick plain sherd, buff.
46010	A	Small sherd (?), possibly impressed or notched.
46016	B	Base of small vase, buff.
46018	B	Sherd, soft, buff, fingernail (?) impressions.
46019	B	Base of small vessel (?), grey.
46041	Iron Age	Thick, coarse sherd.
46070	C	Fragment of possible bowl, thick, coarse, buff.
46106	CD	Fragment of small vase, thick, coarse.
46119	D	Fragment of large thick-walled pot ("storage jar").
46129	Iron Age	Thick sherd.
46215a	E	Small sherd with punctations (only specimen from this level).
52046	D	Brick, hard, burnt

TABLE III

Hysteresis Loop Properties and Estimated Baking Temperatures of our Samples

Catalog Number	Fe Content (%)	M_r (emu/g) x 100	M_s (emu/g) x 100	H_c (Oe)	Apparent Baking Temp. (°C)	Category
4005	3.3	178	859	78		Anomalous
4015	2.8	322	1190	100		
4016	3.2	144	678	68		
31010	5.2	160	735	75		
31015	2.9	203	907	78		
46002	2.9	36	174	88		
46010	3.2	59	218	102		
46041	3.0	293	1125	86		
46129	3.9	170	900	69		
46125a	3.4	19	126	80	150	150°C
Known 620°	3.2	19	74	109	625	600°C
4024	3.2	25	110	88	600	
4026	3.0	20	80	84	100/600	
4037	3.1	16	72	101	630	
31007a	2.9	15	89	90	600	
31007b	2.8	20	111	93	650	
31011	2.9	19	100	92	630	
46018	3.0	19	105	95	630	
Known 830°	3.2	6	81	23	820	800°C
Known 1045°	3.2	90	416	55	1040	1000°C
4001*	3.0	21	163	53	1060	
4003	3.2	20	481	50	1040?	
4004*	3.8	86	439	40	900	
4018	2.6	24	119	78	100/600/1150	
4020*	2.7	40	208	69	1060	
4027*	3.2	33	250	72	1075	
4028*	2.8	29	130	81	1150	
4042*	3.3	31	231	62	1040	
4046*	3.3	51	297	68	1050	
31006*	2.6	24	191	71	1075	
31008*	2.9	24	188	77	1100	
31024	2.5	26	454	76	1125?	
46005*	3.8	33	210	71	1075	
46016	3.0	44	367	60	1040	
46019*	3.2	31	212	70	1075	
46070	2.6	25	134	88	1150	
46106*	3.0	23	214	68	1050	
46119	3.6	77	483	65	900/1050	
52046	3.2	91	575	65	1050	

abundant evidence for a fire in which the village was destroyed, are in the high temperature category, as we would expect. The Iron Age samples and the sample which was suspected of being a recent intrusion are all anomalous, as we would expect for material made from different clays and using a technology different from that for which the master curve was established. However, two factors cast doubt on the reliability of the method. First, samples from levels A, B, and C are to be found both in the 600° and 1100° categories. This might reflect the reality of a technology which produced widely varying results, or it might reflect a difficulty with the method. More disturbing are the samples from those same levels in the anomalous category, strongly suggesting a problem with the method, or with our assumptions. As we shall show, as a result of our other measurements we are now able to propose explanations for both of these observations.

THERMOREMANENCE MEASUREMENTS: RESULTS AND INTERPRETATION

Remanence measurements are more difficult to perform than the hysteresis loop measurements, first because they require between 0.5 and 1 g of material for the former, compared with less than 0.1 g for the latter. This is a serious problem since, as indicated above, samples are uncommon. That is, our samples are few and small. Second, remanence measurements are far more demanding of equipment and time than hysteresis measurements, requiring two heatings for each experimental point. In an attempt to confirm the results of the hysteresis loop measurements, and to clarify the questions arising from them, we have performed remanence measurements on some of the same sherds.

The samples studied are presented in Table IV. Some NRM vs TRM curves are shown in Figs. 4-7; these illustrate the type of behavior we have observed. Fig. 4 shows the results for a sample which we had prepared and heated to 300°C under a field of 0.5 Oe. An anomaly is clearly visible close to the firing temperature, as expected. In Fig. 5, we show the results for a sample which seems clearly to have been fired at a temperature near 420°C or higher. The curve of Fig. 6 seems to indicate that the sample had been fired at some temperature, perhaps 420°, perhaps considerably higher, and then reheated in the 200-360° range one or more times. Reheating, rather than chemical alteration seems to be the most likely explanation of the maxima and minima observed. Similar curves were observed on several samples. Finally, Fig. 7 illustrates the behavior of a sample for which we believe we can discern both a firing near 400° and a second heating near 300°. We also have more than one example of this type of behavior. The hypothesis of multiple heatings is also supported by the fact that NRM is far smaller than the TRM in most of the samples, as indicated in Table IV, where we show the ratio of the maximum NRM to that of the maximum TRM, for a number of samples. The very small values for most of these suggest the presence of many vector components of magnetization, which tend to cancel. We note that sample 4024, for which the straight line NRM vs TRM suggested that it had not been reheated, has a ratio close to 1.

In Table IV we present the results of the remanence measurements, comments on the behavior observed, and, for comparison, the results of the hysteresis measurements on the same samples. The most striking fact concerning these samples is the large fraction which have been, or appear to have been, reheated. Visual observation suggests that this reheating was quite local. As already mentioned, the non-storage vessels from Ganj Dareh were probably far too small to have been suitable for cooking over a fire. The best preserved, a spherical narrow-mouthed pot, has a capacity of only ca. 160 - 170 ml (Smith and Crépeau 1983), and the others probably have even less. Just what they were meant to contain is in fact a problem for which

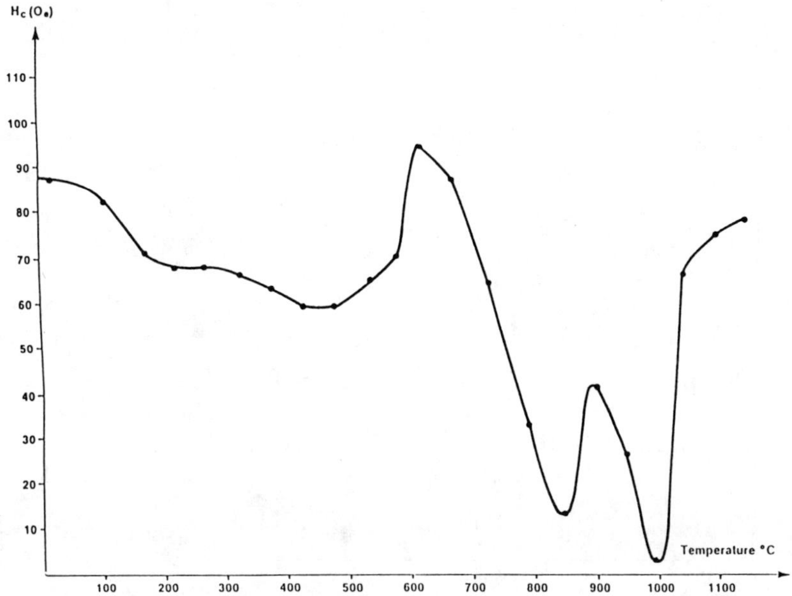

Fig. 3: Coercive force, H_c, versus baking temperature for a clay sample.

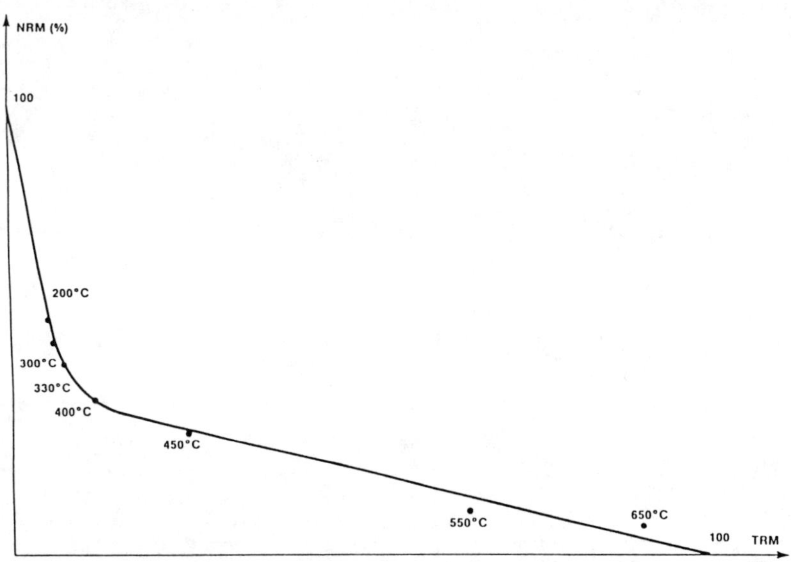

Fig. 4: Natural remanent moment (NRM) versus partial thermoremanent moment. (TRM) for a clay sample baked at 300°C.

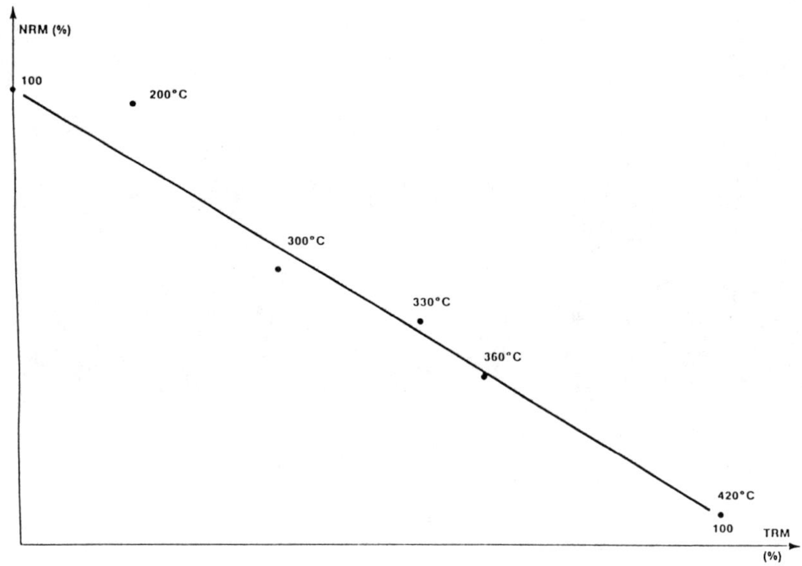

Fig. 5: NRM versus TRM for sample 4037.

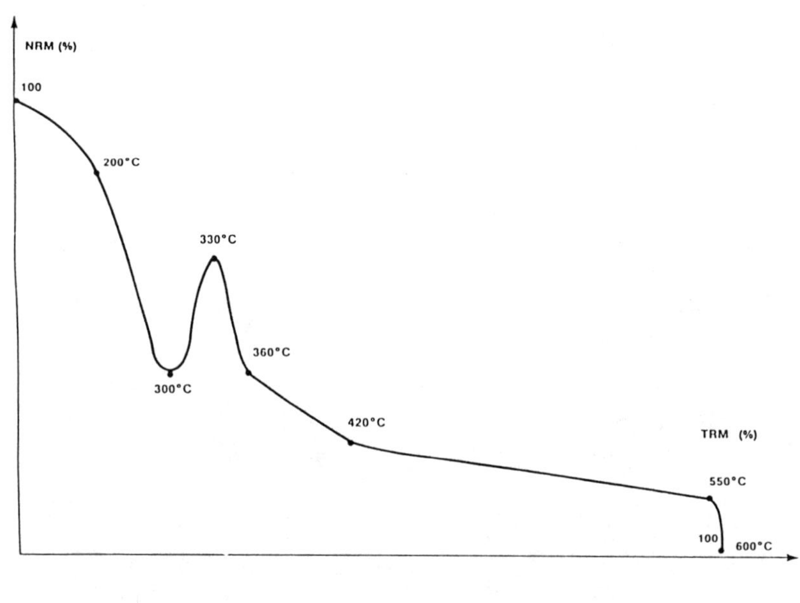

Fig. 6: NRM versus TRM for sample 31007b.

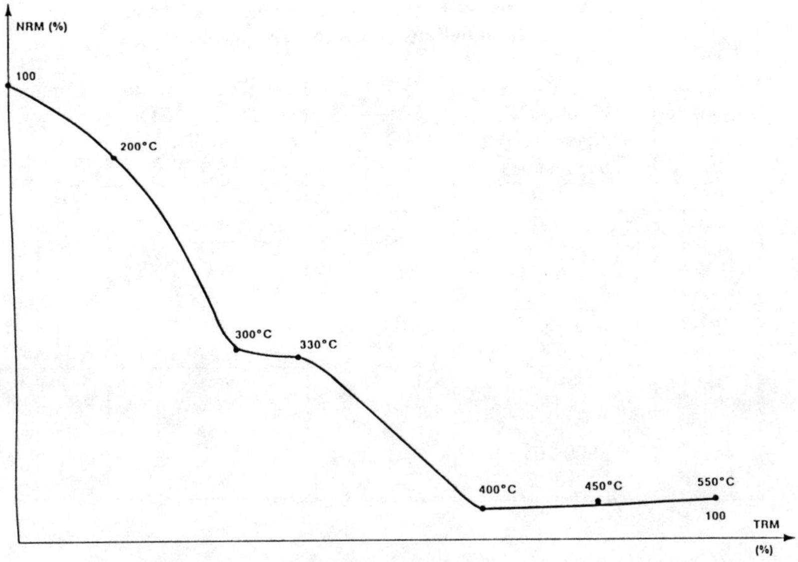

Fig. 7: NRM versus TRM for sample 31011.

TABLE IV

Comparison of Hysteresis and Thermoremanence Determinations of Baking Temperature

Catalog Number	Apparent Baking Temp. (°C) (hysteresis)	NRM_{max} / TRM_{max}	Apparent Baking Temp. (°C) (Thellier)	Comments
46129	Anomalous		>570	Iron Age
46215	150	0.61	330?	Level E
4024	600	0.69	600?	Reheated?
4037	630	0.96	420	Soft, grey
31007a	600		>570?	Reheated
31007b	650	0.12	>550?	Reheated
31011	630	0.053	400	Reheated?
46018	630	0.19	500	Reheated?
4042	1040	0.11	>650	Reheated
46005	1075		>570	Reheated?
46106	1050	0.035	>650?	Reheated
46119	900/1050		>570	Pottery D
52046	1050		>570	Brick D

TABLE V

Samples Tested in Boiling Water

Catalog Number	Apparent Baking Temp. (°C) (hysteresis)	Apparent Baking Temp. (°C) (Thellier)	Effect of Boiling Water
4016	Anomalous		Stayed together.
31015	Anomalous		Stayed together.
46010	Anomalous		Stayed together.
4026	100/600	420	Stayed together.
4037	630		Fell apart immediately.
4001	1050		Fragmented within 5 minutes.
4003	1040?		Fell apart immediately.
4020	1060		Stayed together.
4028	1150		Interior fragmented within 5 minuted. Thin exterior layers stayed together.
4046	1150		Stayed together.

we have at present no solution: perhaps some valuable substances, whether liquid or solid, that were not directly related to subsistence (see also Cauvin (1974) for a comparable interpretation of rather analogous pottery from the early 8th millenium B.C. at the site of Mureybat in Syria).

We may note several other points from the Table. All of the pots which were believed, from hysteresis measurements, to have been fired at 900° or above, and the Iron Age pottery, which was anomalous in the hysteresis results, yield demagnetization curves compatible with high temperature firing. Those samples which hysteresis measurements had suggested were fired in the vicinity of 600° show less consistent behavior. One seems to have been fired above its blocking temperature, while the others indicate maximum temperatures between 400 and 600°C. Thus, the results qualitatively confirm that these samples were fired at relatively lower temperatures, but there are discrepancies in the values, which may range up to 200° lower for the remanence method. Finally, the sample from level E yields the lowest value for the firing temperature, as it did in the hysteresis measurement. But there is a difference in values of 150°, with the remanence value being the higher this time. Thus, our previous results are confirmed qualitatively, but the quantitative agreement is poor. In addition, we have not accounted for most of the anomalous results of the hysteresis loop measurements.

SINTERING MEASUREMENTS: RESULTS AND INTERPRETATION

In an attempt to verify the results presented above, and with the hope of clarifying the source of the anomalies in the hysteresis loop measurements, we boiled several samples in water, as described above, to determine whether they had been sintered. The samples studied are presented in Table IV, in which we present the catalog number, the results of the two previous measurements (where they exist), and the results of this measurement. In general, we confirm the previous conclusion that the firing conditions of the samples in question were quite variable, a few samples having been fired below the sintering temperature, some (the 5 samples having been fired below the sintering temperature, some (the 5 minute group), close to that temperature, and the majority, above.

Most striking, however, is the result for sample 4028. Another piece of this sample had been found to be in the 100° range. The fragment of this sample used for the boiling water measurement was found, by visual inspection, to be hard on the surface and crumbly in the interior. When placed in boiling water, most of the sample dissolved, leaving two surface layers more or less intact. It is almost impossible to imagine this result as arising from post-firing chemical alteration. We conclude that the exterior of the fragment had reached sintering temperature, whereas the interior never did.

The most reasonable explanation of this observation is that the sample was baked over an open fire. Where the flames touched it, it became very hot; where the flames did not reach, it remained relatively cool. The fragment studied here would have been touched by the flames long enough to bring the surface to a high temperature, but not enough to conduct heat to the interior. Assuming that this was a relatively frequent occurrence will permit us to propose a simple explanation of all of the results reported here.

CONCLUSIONS

Early Pyrotechnology

We have shown that during the time of layers C to A in Tepe Ganj Dareh pottery was baked, probably over open fires. This resulted in a product which was relatively poorly controlled, with firing temperatures varying from one pot to another, from one place to another on the same pot, and from the exterior to the interior of the same piece. After they were prepared, they were again heated locally, perhaps for heating some valued liquid or solid, as discussed below.

The data we have obtained here do not permit us to draw any conclusions as to whether the technology evolved between level C and level A. This would require a large scale statistical study, based on small scale, local evaluation of firing temperatures. We cannot draw any sweeping conclusions on the basis of the one sample from level E but the fact that this one sample yielded a firing temperature lower than that obtained on any of the samples from levels C to A suggests that there may have been an evolution in the technology between the time of level E and that of level C. Unfortunately, the level D pottery, because it was secondarily baked in the conflagration, cannot be used to document such an evolution.

Since Tepe Ganj Dareh contained several well made structures which have been identified as kilns for the preparation of lime, we are drawn to the conclusion that lime-making technology was more advanced than was pottery making. This is in agreement with the findings for other sites in the Near East (Frierman, 1971, Gourdin and Kingery, 1975), where the authors concluded that plaster making technology had preceded the technology of firing pottery. This suggests that pottery manufacture only made its next important advance - the production of hard-baked, durable, vessels destined for cooking and other purposes - when techniques developed in another field of pyrotechnology were transferred to it.

Experimental Methods

We have seen that each of the techniques employed in this study provided useful information concerning the thermal history of the samples under investigation. We have also seen that the magnetic techniques need to be applied with considerable prudence, if they are to yield information rather than confusion. We were fortunate to have used that fragment of sample 4028 in the sintering test in that measurement, rather than in the hysteresis loop measurement. Had we ground this fragment to powder and measured its hysteretic properties, the mixture of material with two (or a range of) firing temperatures would surely have yielded anomalous results. This is clear, since the hysteretic properties of a powder consisting of fraction f fired at T_a and of (1-f) fired at T_b will not, in general, correspond to T_a, T_b nor to some average of the two. We assume that the other anomalous measurements on samples produced at Ganj Dareh in levels C to A (as contrasted to samples made elsewhere, or later, which we expect to be anomalous) may be explained in a similar fashion. The search for a unique firing temperature of a sample which has been baked non-uniformly, and the possibly reheated, is obviously quixotic.

In order to apply the magnetic techniques in a systematic way, we would need to follow one of two possible procedures. The samples might be screened by a technique which assured they were uniform in some property related to firing over the volume to be studied. This technique would not have to yield a value of firing temperature itself. Alternatively, we might imagine increasing the sensitivity of the magnetic techniques, so that we could be certain that an entire specimen is uniform. This could be done, using SQUID magnetometers, for example (see Barone and Paterno (1982)).

The study described here has enabled us to understand a little better the procedures used by one group of early ceramicists in one part of the Old World. We hesitate to extrapolate from their experience to other sites and regions: early pottery making, like early food production, was probably an enormously varied set of operations and decisions whose complexity we have only begun to appreciate. The results of our own analytical procedures may serve, however, as a cautionary tale to other colleagues regardless of their region of study.

ACKNOWLEDGEMENTS

The authors are grateful to the late Dr. E.J. Schwarz of the Geological Survey of Canada, who lent his spinner magnetometer for remanence measurements and who provided valuable guidance and suggestions for their interpretation. They are also thankful to Dr. G. Kennedy of Ecole Polytechnique, who performed the neutron activation analyses and to Prof. J.M.D. Coey of Trinity College, Dublin, for many helpful discussions. Prof. W.D. Kingery, then of M.I.T., was kind enough to offer very helpful comments on the manuscript. This work was supported in part by the Social Sciences and Humanities Research Council of Canada, and by the Natural Sciences and Engineering Research Council of Canada.

REFERENCES

Arbour, G. and Schwarz, E.J., Archeomagnetic Intensity Study of Indian Potshards from Quebec, Canada, J. Geomag. Geoelectr., **34**, 129-136 (1982).

Barbetti, M.F., McElhinny, M.W., Edwards, D.J. and Schmidt, P.W., Weathering Processes in Baked Sediments and their Effect on Archaeomagnetic Field-Intensity Measurements, Physics of the Earth and Planetary Interiors, **13**, 346-354 (1977).

Barone, A. and Paterno, G., Physics and Applications of the Josephson Effect, Wiley, N.Y., 1982, Ch. 13.

Brisaud, I., Houdayer, A., Jehanno, C. and Sabir, A., Pottery Analysis using PIXE, Neutron Activation, Scanning Electron MIcroscopy, and X-Ray Fluorescence Techniques, J. Radioanalytical and Nuclear Chemistry, Articles **89**, 473-486 (1985).

Cauvin, J., Les débuts de la céramique sur le Moyen-Euphrate: nouveaux documents, Paléorient, **2**, 199-205 (1974).

Coey, J.M.D., Bouchez, R. and Dang, N.V., Ancient techniques, J. Appl. Phys. **50**, 7772-7777 (1979).

Frierman, J.D., Lime Burning as the Precursor of Fired Ceramics, Israel Exploration Journal **21**, 212-216 (1971).

Goodyear, F.H., Initial firing temperature, composition and provenance of Pottery. Science and Archaeology, No. 6, 12-14 (1971).

Gourdin, W.H. and KIngery, W.D., The Beginnings of Pytotechnology: Neolithic and Egyptian Lime Plaster, J. Field Archaeology **2**, 133-150 (1975).

Heimann, R. and Franklin, U.M., Archaeo-thermometry: The Assessment of Firing Temperatures of Ancient Ceramics. J. Internatl. Inst. Conservation-Canadian Group **4**, 23-45 (1981).

Hesse, B., Slaughter Patterns and Domestication: The Beginnings of Pastoralism in Western Iran, Man (N.S.) **17**, 403-417 (1982).

Matson, F.R., Archaeological Ceramics and the Physical Sciences: Problem Definition and Results. J Field Archaeology, **8**, 447-456 (1981).
Périnet, G. and Courtois, L., Évaluation des températures de cuisson de céramiques et de vaisselles blanches néolithiques de Syrie, Bull. Soc. Préhistorique Française, **80** (5), 157-160 (1983).
Schwarz, E.J. and Christie, K.W., Original Remanent Magnetization of Ontario Potsherds, J. Geophys. Res. **72**, 3263-3269 (1967).
Smith, P.E.L., An Interim Report on Ganj Dareh Tepe, Iran, American Journal of Archaeology **82**, 538-540 (1978).
Smith, P.E.L., Ganj Dareh: An Early Neolithic Site in Iran. Archiv fur Orientforschung **29**, 300-302 (1983).
Smith, P.E.L. and Crépeau, R., Fabrication expérimentale de répliques d'un vase néolithique du site de Ganj Dareh, Iran: recherche technologique. Paléorient **9**, 55-62 (1983).
Thellier, E. and Thellier, O., Sur l'intensité du champ magnétique terrestre dans le passé historique et géologique, Ann. Géophys. **15**, 285-377 (1959).
Van Zeist, W., Smith, P.E.L., Palfenier-Vegter, R.M., Suwijn, M. and Casparie, W.A., An Archaeobotanical Study of Ganj Dareh Tepe, Iran, Palaeohistoria **26**, 201-224 (1984 (1986)).

Editor's Note: The pottery from Ganj Dareh is some of the earliest in the Southwest Asia, but the real significance of this article resides in the determination that after an initial firing to harden the clay into a ceramic, secondary usage at very much lower temperatures can be determined. This paper leaves some questions about methodology open, but we believe that the findings are of sufficient merit as to warrant publication.

Several reviewers requested more information regarding the pottery and other ceramics from Ganj Dareh. There is a considerable diversity in fabric type, object type and manufacturing technology as well as range of firing temperature. In general, most of the pottery is tempered with an addition of fibrous vegetal material, probably for strength during building and to prevent shrinkage of the impure seabed clays common at the site. Large storage pots were made by sequential slab construction from this fabric. In addition, some very small (less than 5 cm) bowls were made of partly dried clay which was ground or broken into 2 to 5 mm pellets and then wetted, but without being wetted throughout. The pores have unusual, but typical, shapes having multiple points, rather that being rounded. This type of poorly plastic fabric is very common among pottery students who are unfamiliar with common clay preparation methods. Another type of fabric is common among the figurines and tokens which is without vegetal temper but completely wetted through and thus a workable, plastic clay. A chapter on this pottery is found in P. Vandiver's Ph.D. thesis from M.I.T., and eventually a proper report will appear.

THE TECHNICAL EXAMINATION OF SOME NEOLITHIC CHINESE LIANGZHU CERAMICS IN THE HARVARD UNIVERSITY ART MUSEUMS COLLECTION

ANDRZEJ DAJNOWSKI*, EUGENE FARRELL**, AND PAMELA VANDIVER***
*Chicago Park District, Research & Planning, Chicago IL
**Harvard University Art Museums, Cambridge MA
***Conservation Analytical Laboratory, Smithsonian Institute, Washington DC

ABSTRACT

At the Neolithic archaeological site at Liangzhu in China, located 16 kilometers northwest of Hangchou in Cheking province, many objects of stone, jade, and black pottery were discovered between 1936 and 1939. While the exact date of the Liangzhu culture is uncertain, it is considered to be between 3500 and 2000 BC. A characteristic feature of the Liangzhu-culture pottery is a thin black finish and a layered structure of the body consisting of a black core sharply changing to a gray or red band ending in the black surface.

The focus of this paper is to explain why the cores of the vessels are black and why the colored bands occur and what they indicate about the firing conditions of the pottery. Analyses were carried out using SEM, electron-beam microprobe, X-ray diffraction, photo-electron spectroscopy, and polarizing microscopy. From this evidence, it is clear that a clay vessel containing abundant charcoal was thrown on the wheel and then fired under reducing conditions so that charcoal was retained and iron reduced. Just before completion of the firing or during the cooling phase, air was allowed to enter the kiln and a limited thickness of the outer core was oxidized. Then, for decorative purposes, the surface was heavily smoked, and/or a thin iron-containing slip layer was reduced in smoke and burnished.

INTRODUCTION

The ceramic objects studied in this paper came from the Neolithic site of Liangzhu, located sixteen miles northwest of Hangchou in the province of Cheking, China, close to the coast.

In 1936, a group of Chinese archaeologists from the West Lake Museum discovered the site and named it Liangzhu, after a nearby town. A stratum two meters thick yielded stone, jade, and black pottery of various shapes. Soon after the excavation of Liangzhu had begun, the Japanese invasion of China prevented further digging. In 1943, many of the excavated pieces left China and entered a private collection, eventually to become a part of the collections of the Harvard University Art Museums [1, 2, 3].

The exact dating of the culture is still a controversial issue. Liangzhu existed somewhere between 5500 and 4000 years ago, as suggested by Professor K. C. Chang of Harvard University [4]. Thermoluminescence dating was done on two of the objects studied in this paper. The date of one established the last firing between 3100 and 4800 years ago; and the second, 3600 to 5500 years ago, which is reasonable for the Liangzhu period.

We have examined approximately 150 pots and shards. In the process we have found that characteristic features of the Liangzhu-culture pottery are a black finish and a layered body when seen in cross section. It consists of a black core, with gray or red layers on both sides of the core. Fig. 1 shows a pedestal stand used to support an attached cup. The stand is decorated with horizontal bands and crosshatching lines at the bottom. It also has a layered character in cross section. The vessels may have been used for ritual wine service. A similar piece was examined that had a very fine polished surface decorated with six double lines of bands on which a bowl could have been placed. Some of these objects have been found intact, with the bowls and bases together.

Not all the pottery found at Liangzhu is of this type. Many different types were found, but they almost all have the characteristic black cores with gray or red side bands.

Our research has focused on trying to explain how the layers were formed. They could have been produced by the application of slip or a lamination process or by reduction oxidation during firing [5]. A slip or lamination would involve similar-appearing clay layers of different composition, and reduction oxidation would produce a change in iron from Fe_2O_3 to Fe_3O_4 or FeO in a homogeneous clay body. The black and gray layers were not a part of a decorative scheme; they are, however, an important factor in understanding the manufacture of the objects.

Fig. 2 is an X-ray of three pedestal stands. The middle piece shows the stem from the upper element that was used for attachment. There is no indication of joints, which would suggest that the pots were built with coils of clay. On the contrary, nearly all of the pieces that we have examined show evidence of having been thrown on a potter's wheel. All show a very fine polished exterior surface and unfinished interior.

The spiral effect seen in Fig. 3 suggests that a wheel was used in its manufacture. Other pieces show evenly formed bands with uniform distances between them and a very fine polished finish of the exterior, further evidence of using a wheel.

ANALYSES

In order to understand the internal layering of the pottery, we have used scanning electron microscopy (SEM), wavelength dispersive electron beam microprobe analyses, X-ray photo-electron spectroscopy (XPS), and X-ray powder diffraction analyses on representative pots from the collection. Thin sections and cross sections were prepared and studied.

A survey analysis with SEM gave us an initial idea of the chemical composition. Polarizing microscopy was used to study the structural characteristics of the samples.

In Fig. 4 one can see a sample cut from the pedestal stand shown in Fig. 5. The layering and the porosity are obvious. We have done X-ray powder diffraction analyses of the surface, the gray, and the black layers. In each of them we have found quartz, and illite or muscovite [6]. X-ray diffraction analyses done at the Conservation Analytical Laboratory showed that some samples were a mixture of different clay minerals, micas, and feldspars, consistent with the lowland environment from which the pottery was derived.

Thin sections show that the area of the color change between the black and the gray is a diffused transition zone consistent with oxidation-reduction. If the transition were the result of lamination or the application of slip, one would expect to see a very sharp line of demarcation between the two bands. More carbon was found in the core, giving its characteristic black color, although carbon particles were found in all layers.

It is known that the Chinese used straw, reeds, seeds, or other organic materials for tempering clay. Charcoal was found in some of our samples. SEM photomicrographs show both the cellular structure and the clay fabric (Fig. 6 and Fig. 7).

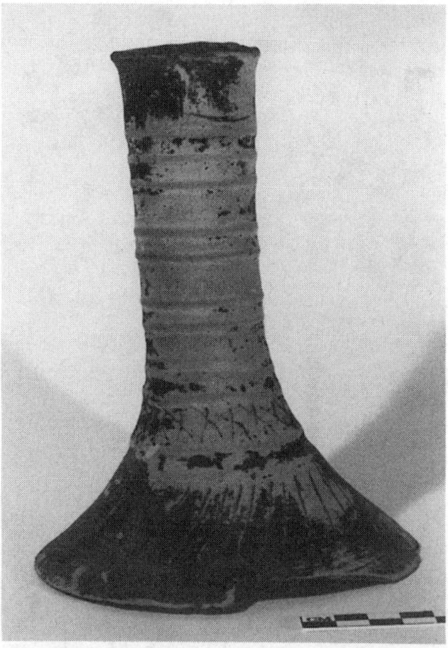

Fig. 1 A typical pedestal stand used to support an attached cup. Not all the Liangzhu ceramics are this type; their shapes and use varied widely. The scale is in centimeters.

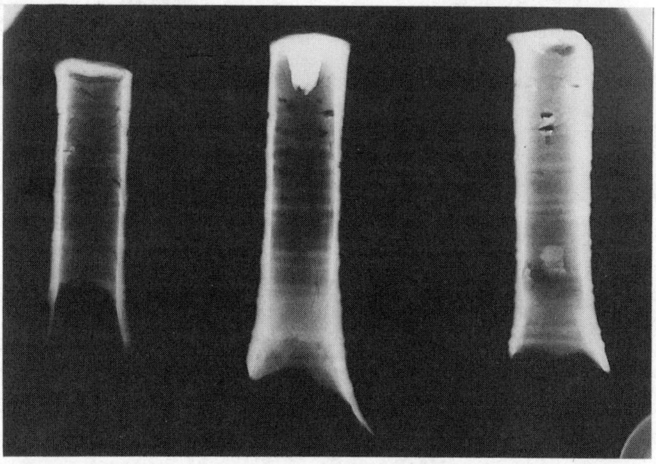

Fig. 2 An X-ray of three pedestal stands. The center stand contains the remnants of an insert designed to attach a cup to the hollow center of the base. The X-rays do not show any evidence of joints which would suggest that the bases were built up from coils of clay.

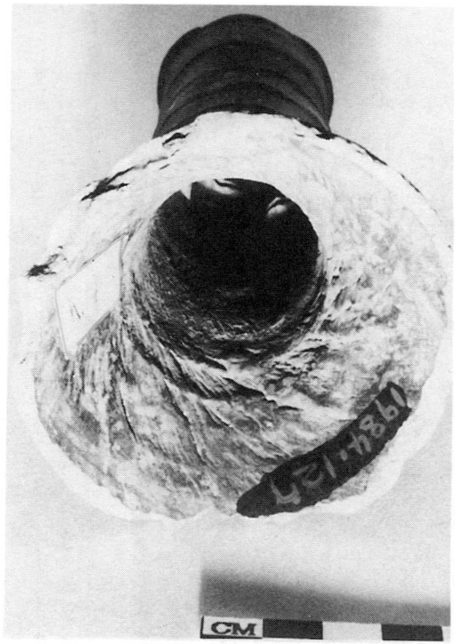

Fig. 3 The spiral seen on the inside of this pedestal base was formed by a slurry and is evidence that it was turned on a wheel.

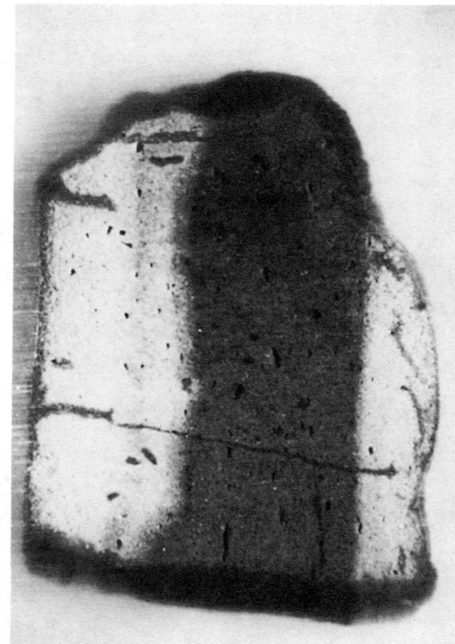

Fig. 4 A cross section taken from the pedestal stand seen in Fig. 5. Note the gradual change from the black core to the gray bands along the edges. The width of the section is approximately 0.5 cm.

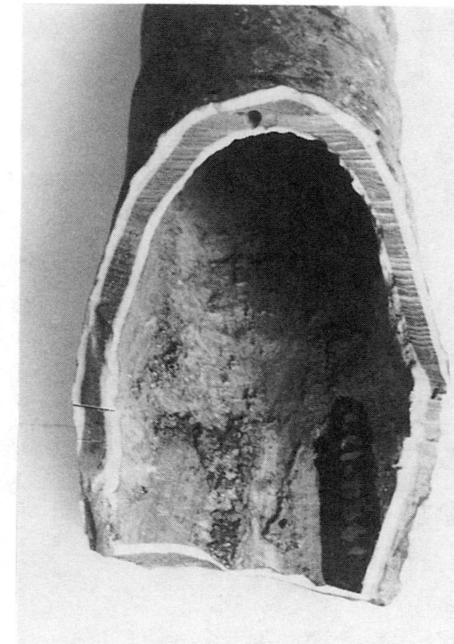

Fig. 5 Pedestal stand number 1984.122 showing the alternating black and gray layers of the wall structure. The section shown in Fig. 5 was taken from the upper part of the wall as seen in the photograph.

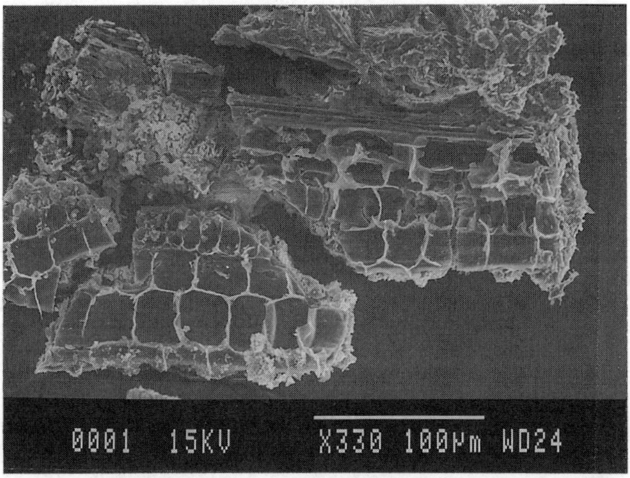

Fig. 6 Photomicrograph of a charcoal inclusion taken in a scanning electron microscope at 330 x magnification. The cellular structure of the plant material inclusion has survived the kiln temperature of 700-800 degrees C.

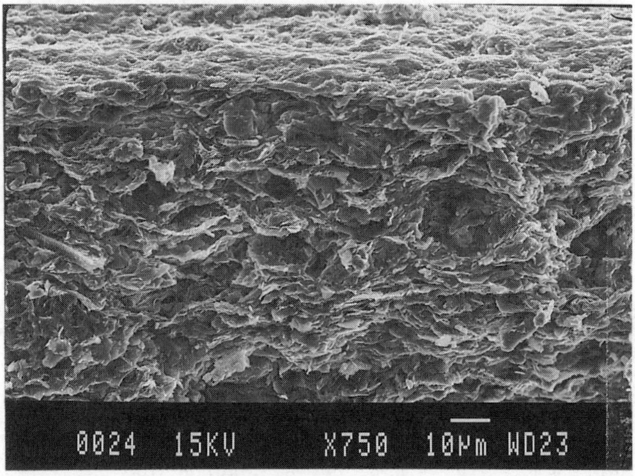

Fig. 7 Well preserved platty mineral material of a cross section removed from a pedestal stand seen at 750x magnification in a photomicrograph taken with a scanning electron microscope.

We have summarized the analytical data graphically. Microprobe analyses were used to determine the weight percentages of the elements present. The totals were often less than 100% and were consequently normalized. The standard of deviation was around 1.5 to 4.0 percent of stated values for major elements and 3.0 to 10.0 percent for minor elements. The XPS data is in atomic percent. XPS was used to detect and measure organics including carbon which could not be measured on the microprobe. The standard of deviation was 4% for survey analyses and 3% for high-resolution analyses–standard values determined for the instrument. XPS analyzes a surface of the sample 10 to 50Å in depth, while the electron beam of the microprobe penetrates 1 to 2 microns.

The bar graph in Fig. 8 shows the results of analyses done on the microprobe and Fig. 9 shows the corresponding XPS analyses. As may be seen from the probe results, the two layers are identical in composition. The XPS analysis shows that there was twice as much carbon in the black layer as in the gray and that the black layer had less oxygen than the gray.

Samples with black cores and red side bands were also found to be identical in composition, except for the high carbon in the core. In one case we sectioned the join between a cup and a base where we were sure that two layers had been joined together. The chemical analyses seen in Fig. 10 for microprobe and Fig. 11 for XPS reflects the difference between the black layer from the base and the red-and-gray layer to which it had been fused.

In an exceptional case that proves the rule, one pot was in fact found to be made in part by a slip or lamination process. This was obvious from visual inspection and confirmed by analysis. Bar graphs for the microprobe results in Fig. 12 and XPS analyses in Fig. 13 demonstrate the difference in the composition of the clay in each layer, as is easily visible on the diagrams. Differences in silica, iron, aluminum, and a surprisingly high phosphorus content of the black layer (6.27%) demonstrate that there are three different clay compositions. XPS supports this observation by indicating a difference in oxygen, carbon, and other elements in the layers. The high phosphorus content may have found its way to the sediments of which the pottery is made from the local rocks which are known to contain phosphates [7].

FIRING TEMPERATURE

Evidence for the firing temperature came from several sources. Hsin-yuan Tu, a scientist from Taiwan, did refiring tests and established a firing temperature for three Lungshanoid black pots in the range of 600-620°C [8]. From the evidence, our samples also appear to have been fired at a low temperature. Refiring tests carried out at the Conservation Analytical Laboratory indicate that the firing temperature was between 700-800°C. Furthermore the structure of the mica is preserved in X-ray diffraction patterns of our samples, and micas do not break down much under 900°C. Also the body is very soft and can be easily scratched with a fingernail, and carbonaceous inclusions did not burn off completely. Carbon is known to burn off at around 900-950°C [9].

SUMMARY AND CONCLUSIONS

In summary, our observations lead us to believe that the Liangzhu potters were very technologically advanced. They were able to control the kiln environment and composition of the clay in order to achieve the effects they sought. They were also capable of skillful usage of some kind of a wheel for producing the shape pottery they desired.

In general, the analyses have confirmed what was surmised from visual

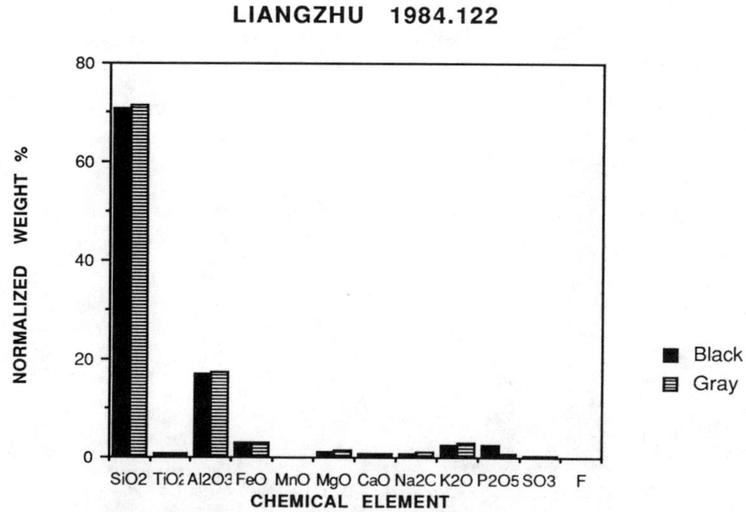

Fig. 8 Bar graph of microprobe analysis of sample 1984.122

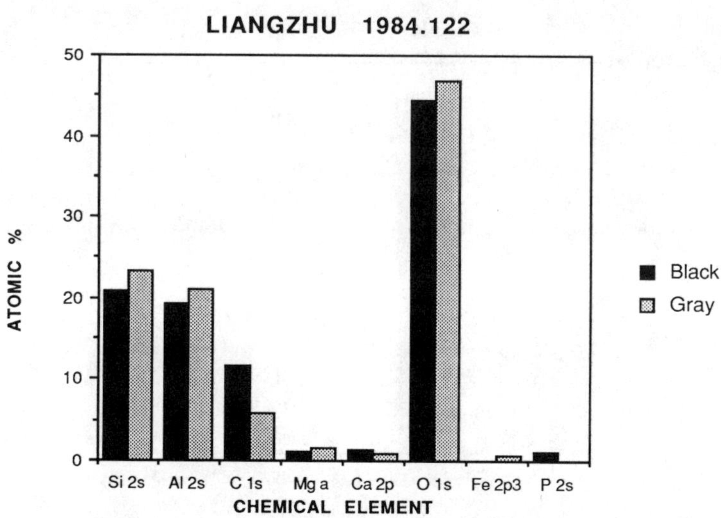

Fig. 9 Bar graph of XPS analysis, sample 1984.122

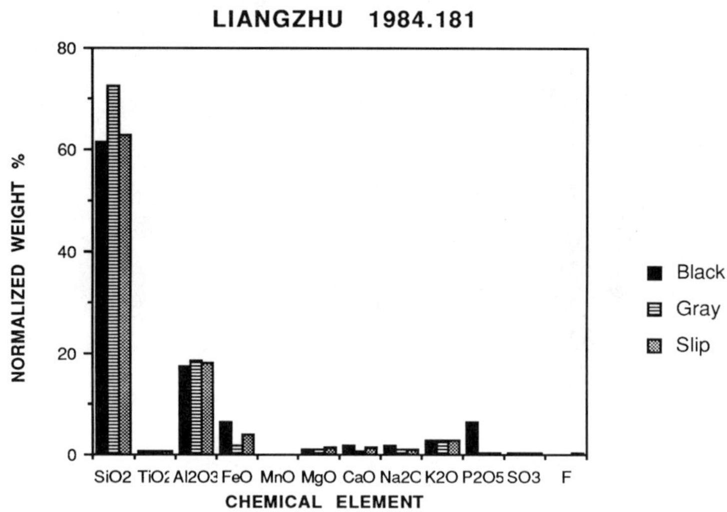

FIG. 10 Bar graph of microprobe analysis of sample with a slip; 1984.181

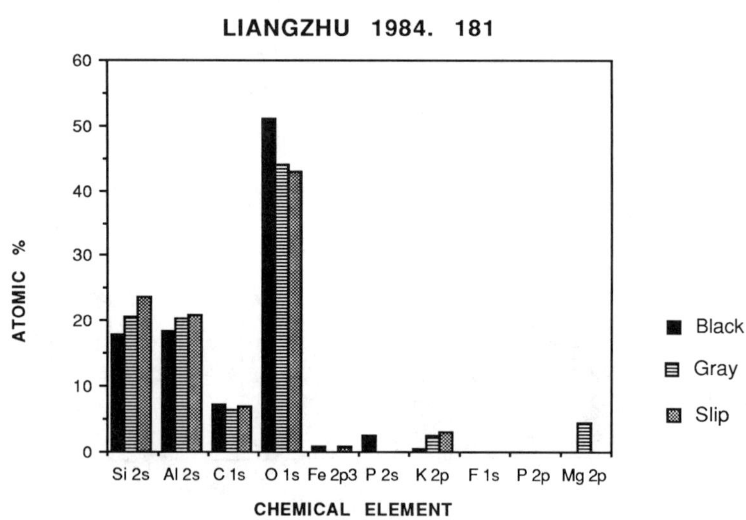

FIG.11 Bar graph of XPS analysis of sample 1984.181

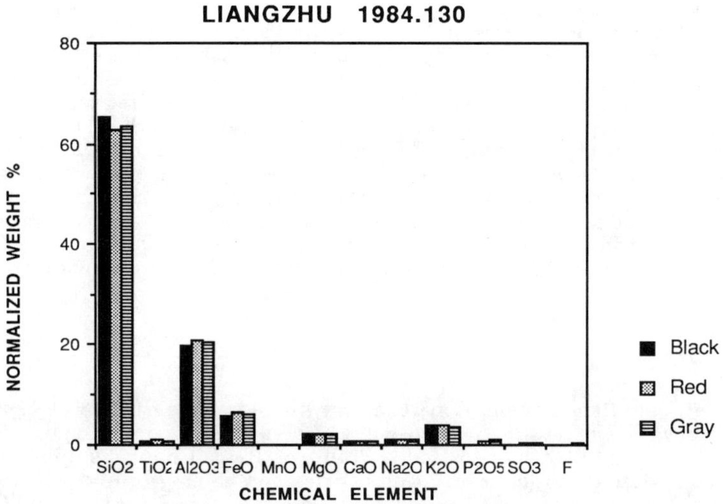

Fig. 12 Microprobe analysis of laminar sample 1984.130

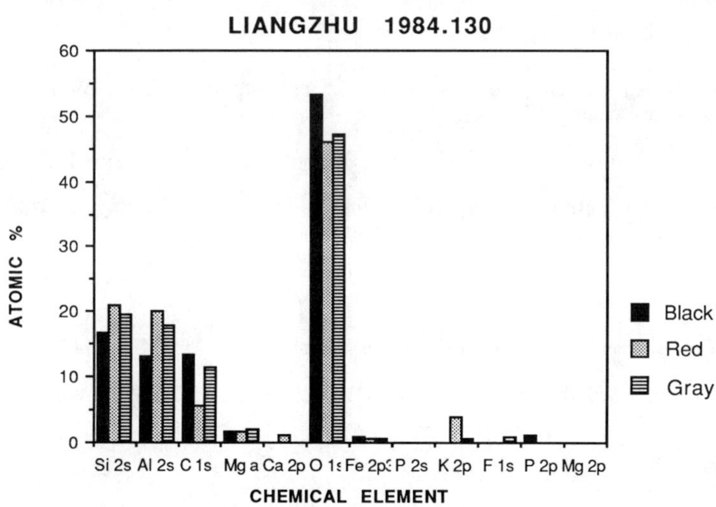

Fig. 13 XPS analysis of sample 1984.130

observation. Microscopic examination of thin sections shows a high carbon content in the black bands and in the burnished surface. In some cases, isolated pieces of charcoal can easily be identified in thin section. Scanning electron microscopy has revealed that the cellular character of the charcoal has survived the firing. The presence of a high percentage of carbon in the black core and less in the outer gray-white layers was verified using X-ray photo-electron spectroscopy.

From this evidence it is clear that a clay vessel containing abundant charcoal or plant material was thrown on the wheel and then fired under reducing conditions, so that charcoal was retained and iron reduced in the clay body. Just before completion of the firing or during the cooling phase, air was allowed to enter the kiln and a limited thickness of the outer core was oxidized. Then, for decorative purposes, the surface was heavily smoked, and/or a thin iron-containing slip layer was reduced in smoke and burnished. Despite the fact that most of the Liangzhu pots had been buried for approximately 5000 years and are so soft that they can be easily scratched with a fingernail, many of them are in surprisingly good condition.

ACKNOWLEDGMENTS

We would like to thank Robert Mowry, the Associate Curator of Asian Art at the Harvard University Art Museums, for suggesting and supporting this project. We would also like to thank Professor K. C. Chang for his valuable suggestions, Stella Kao for translating articles and helping to find information, and Henry Lie, Director, and Cherrie Corey, Administrative Director of the Center for Conservation and Technical Studies, Harvard University Art Museums for their support. We are especially grateful to the National Endowment for the Arts and the Getty Grant Program for their support of the Advanced-Level Training Program at the Center for Conservation and Technical Studies which made this research possible.

REFERENCES

1. Cheng Te-Kun, *Archaeology in China*, vol. 1, Prehistoric China (W. Heffer & Sons Ltd., Cambridge, 1959).

2. Sterling S. Beath, *The China Journal*, December, vol 6 (1939), pp 262-266.

3. Ho Tien-shin, Volume 1, The Society for the Study of the History and Geography of Kiangsu and Cheking, Shanghai (1937).

4. K. C. Chang,*The Archaeology of Ancient China* (Yale University Press, New Haven and London, 1986).

5. S. Kaplan, Archives of the Chinese Art Society of America, III (1948-1949).

6. R. E. Grim, *Clay Mineralogy* (McGraw-Hill Book Company, New York, 1968).

7. Yang Zunyi, Cheng Yuqi, and Wang Hongzhen, *The Geology of China* (Clarendon Press, Oxford, 1986).

8. K. C. Chang, Department. of Anthropology (Yale University Press, New Haven and London, 1969).

9. J. V. Noble, *The Techniques of Painted Attic Pottery*, (Watson-Guptill Publications, New York, 1965), p. 32.

Editor's Note: The explanation of firing is very complex. If you fire a pot in a bonfire, you get a black core and a red surface. The carbon is not burned out of the core because of the short time of the firing and the poorly oxidizing conditions. The fire burns down, the pot remains hot in the ashes, the surfaces oxidize to a depth of a millimeter or a few millimeters. If the hot pot is taken out of the fire with a stick and buried in a heap of green vegetation, the result is a bluish shiny surface. No need for kilns or special reducing atmospheres. The fabrics and finishes are typical of much prehistoric pottery in western Europe. They had no kilns, and the P_2O_5 is absorbed from the soil (see Freestone et al., 1985, Archaeometry).

Another Editor's Note: This editor has made over a hundred firings of pottery in open pits to produce both oxidized and reduced colors. Your explanation is excellent for one scenario of firing, but our problem with Liangzhu is to try to determine if a kilns or open pits were used. Kilns had been used in China for oxidizing and reducing firings for a considerable length of time of about two to three millennia prior to the manufacture of this ware. We know of no criteria based on artifact analysis, rather than finding and excavating a kiln, which will allow us differentiate kiln-fired from open-pit fired ceramics.

TERRACOTTAS AROUND CLODION : CHEMICAL AND MINERALOGICAL STUDIES AND THERMOLUMINESCENCE ANALYSIS.

Frédéric Saltron, Anne Bouquillon and Guirec Querré
Laboratoire de Recherche des Musées de France, Palais du Louvre 75041 Paris cedex 01.

ABSTRACT

Forty one terracottas by Clodion, Marin and P. Michel, all french sculptors of the XVIIIth century, have been analyzed in the L.R.M.F.

Two different kinds of analysis have been performed.

The results obtained by thermoluminescence allowed to partition the works between one group contemporaneous with the Clodion period and the other one consisting of posterior works. Although the lowest limit in the use of thermoluminescence seems to be reached, the conclusions are, more often than not, in agreement with the hypotheses of the curator.

X-Ray diffractometry, microprobe analysis and ion-beam methods (PIXE) have led to distinguish between siliceous pastes corresponding to artifacts created in Paris and carbonated pastes gathering terracottas made during the italian stays of the artists.

INTRODUCTION

Few archaeometric papers deal with ceramological studies concerning recent artefacts. The great exhibition "Clodion" in the Louvre museum (march-jun 1992) was the occasion for the LRMF to conduct an in-depth study of a series of terracotta sculptures of th French XVIIIth century [1,2,3].

The artefacts are attributed to Clodion, to his brothers Pierre and Sigisbert Michel and to J.C. Marin, a pupil of Clodion.

This work is aimed at answering two main problems: (*i*) distinguish copies from fakes of original sculptures of the XVIIIth century.

(*ii*) determine for authentic objects, where they were created: in Rome or in Paris.

Different analytical techniques have been used for this study: thermoluminescence for authentication, X-ray diffraction, microprobe and PIXE analysis for mineralogical and chemical paste caracterization.

I - GENERAL FEATURES.

A - Clodion and the contemporary sculptors, a brief historical reminder.

Claude Michel nicknamed Clodion (1738-1814) has learned the technique of sculpture in Paris between 1754 and 1762, first at his uncle Adam's workshop, then in the Ecole des Elèves Protégés. From 1762 to 1771, he exercises his art in the French Academy in Rome. After that period, he establishes his own workshop in Paris where his brothers

Sigisbert, Pierre and Nicolas work with him, along with pupils, especially J.C. Marin (1759-1834). During this second french period, Clodion returned to Italy only for a short nine months stay (1773-1774). The production of Clodion and his workshop is very diversified going from official monumental sculptures to small decorative objects. Most famous are the neoclassical terracottas, highly valued from the XVIIIth to the XXth century.

B - Analysed sculptures.

41 artefacts have been studied from which 31 are attributed to Clodion, 6 to Marin, 3 to Pierre Michel and 1 to Sigisbert Michel: statues, bas-reliefs, medallions of very different themes, most often small sized, modelled in fine clay without any apparent temper, coating or polychromy. The names of all the sculptures are given in the table.

II - AUTHENTICITY.

A - Thermoluminescence (TL) Technique.

With this method, it is possible to evaluate the age of terracottas by measurement of the total irradiation dose absorbed by the artefact from its latest firing (equivalent dose, E.D.) [4,5].

Conscious of the fact that the TL method reaches its extreme limits of application because the artefacts are recent and because a short period separates the authentic sculptures from possible copies and fakes of the XIX-XXth centuries, we have multiplied the methods of evaluation and attempted to be very careful in the interpretation of our results.

Most of the 41 terracottas have been analysed by three different TL methods on 40mg sample:

- "fine-grain" method [6]. This method gives the most accurate data.
- Predose method [7] on single disc. For nine samples, the equivalent dose could not be measured.
- Test on one disc (NTL, ATL, BC) gives an approximative value of the total dose absorbed (ED). By using a 0.62Gy standard beta irradiation for each sample, we made a comparative study on glow-curves and dose value. All the 41 sculptures yield a result.

We established our hypothesis only on the basis of the equivalent dose. It is not possible to know the external dose received by the materials and, for technical reasons, we did not measure the internal activity. We have the advantage to have on hand reference sculptures to calibrate the TL data and to work by analogy with studies of Fleming, Aitken and Stoneham [8,9,10].

B - Results and discussion.

Five sculptures are well known , authentic, signed and dated pieces, they are used as references: three by Clodion, "la Pleureuse" n°24 (In Roma, 1766), "la Pleureuse" n°23 (1766), "la Comtesse d'Orsay" n°30 (1773), 2 by Marin, "la Jeune Fille à la Colombe" (1791), "la Bacchante Couchée"

STYLE	MEAN E. D.			PASTE	SCULPTURE (ANALYSE NUMBER)	
	> 0.7Gy	0.5 - 0.7	<0.5 Gy			
+	+			S	38 - Triomphe d'Ariane	PARIS
+	+			S	6 - Bacchante endormie (Pierre Michel	
+	+			S	30 - La comtesse d'Orsay expirant	
+	+			S	7 - Jeune femme allongée et lisant (P. Michel)	
+	+			S	34 - Buste de jeune fille	
+	+			S	33 - Buste de jeune fille	
+	+			S	36 - Triomphe de Galatée	
+		+		S	3 - Monument à Fifi	
+		+		S	41- Montesquieu	
+		+		S	20 - Bacchant assis tenant une grappe de raisin	
+		+		S	22 - Jeune fille à la colombe (Marin)	
+		+		S	26 - Egyptienne au Naos	
+		+		S	19 - Jeune fille lisant (P. Michel)	
+		+		S	18 - Monument à un chien	
+	+			C	23 - Pleureuse	ROME
+	+			C	31- Allégorie en l'honneur d'un sculpteur (Marin)	
+	+			C	24 - Pleureuse	
+	+			I	12 - Bacchante et enfant (B. Michel)	ND
+		+		I	35 - Buste de jeune fille (Marin)	
+		+		I	17 - Bacchante courant	
+		+		ND	37 - Bacchante couchée avec 5 enfants (Marin)	
-			+	C	28 - Satyre couronnant une bacchante	
-			+	I	29 - Deux satyres et une bacchante	
-			+	C	14 - Baigneuse debout (Marin)	
-			+	S	10 - Satyresse enfant au nid	
-			+	S	5 - Flore	
-			+	S	39 - Bacchant et bacchante debouts à la grappe	
-			+	C	21 - Sainte Bibienne (d'après Bernin)	
-			+	S	9 - Satyre enfant au hibou	
-			+	S	13 - Bacchante assise à la coupe et enfant	
-			+	C	27 - Médaillon - Satyresse avec un enfant	
-	+			S	2 - Satyresse au tambourin	
-	+			S	8 - Triomphe de Bacchus	
-		+		S	32 - Deux bacchantes et un bacchant	
+			+	S	40 - Erigone	
+			+	S	1- Satyre flûteur	
-	0			I	11 - Bélier, brebis et agneau	
	0			S	4 - Pomone	
	0			S	16 - Bacchante couchée avec enfants (Marin)	
	0			S	15 - La gimblette	
	0			S	25 - Sainte Cécile	

TABLE
STYLE: hypothesis of the curator on the authenticity:
+ : XVIIIth c. - : XIXth-XXth c.
E.D.: Equivalent dose, Thermoluminescence.
PASTE: groups of composition S: siliceous; C: carbonated; I: Intermediate; ND: no determine.

(Marin fecit 1793). All the artefacts analysed by the three TL methods give an annual dose close to 0.3 Gy. On the basis of this annual dose value using the equation for TL age and taking the error ranges into account, we propose the following. Sculptures for which the equivalent dose is greater than 0.7 Gy can be considered, in all likelihood, to belong to the Clodion activity period. Sculptures with an equivalent dose lower than 0.5 Gy have probably been created recently and sculptures with intermediate equivalent dose, between 0.7 and 0.5 Gy, cannot be reasonably attributed to any period.

On the basis of all TL data, we distinguish three cases (see table):

1- The samples for which the three TL results (fine-grain, predose, test) are extremly close.

* from 1.5 to 0.7 Gy: six art pieces to be considered as contemporaneous of Clodion,
* from 0.7 to 0.5 Gy: six sculptures without precision on age but possibly created during the XVIIIth century,
* from 0.5 to 0.2 Gy:seven terracottas fired during the XIXth or XXth century.

2- The samples for which the three TL methods cannot be applied or when the different values of the equivalent dose are scattered but yet clearly define a general tendency. This is the case for seven terracottas attributed to the XVIIIth century, for five sculptures in the 0.5-0.7 Gy range and for five other pieces under 0.5 Gy which have probably been created after Clodion's death.

3- The last group consists of five sculptures showing unexploitable results (n°4,15,16,25) or queer equivalent dose for the sculpture "Bélier, Brebis, Agneau". This high value is due to an important amount of radiogenic element (uranium) in the paste that has been measured by chemical analysis.

C - Confrontation with historical and stylistic data.

Except for the reference artefacts, we made the analysis without the knowledge of the certainties or intuitions of the curator about the different pieces. TL results and stylistic appreciations are really in a very good agreement as may be seen in the table. On the 41 analyzed terracottas, the two approaches give a different answer only for five sculptures, for instance the "Satyresse au tambourin".

III - ORIGIN OF THE SCULPTURES.

The LRMF operates a particle accelerator which allows to perform all the ion beam methods of analysis. The program "Clodion" was the occasion to apply for the first time in the LRMF, the PIXE method (proton Induced X-Ray Emission) to terracottas. We have established the analysis requirements which are summarized below.

SAMPLING: 40mg powder sampled with a diamond drill

INSTRUMENTATION: pellets stuck with carbon glue on a plexiglas sample carrier, carbon metallization.

ANALYSIS: for major and minor elements: beam energy=1MeV, filter 13µm kapton; for trace elements: beam energy 3MeV, filter 100µm aluminium - Analysis under vacuum.
INTERPRETATION of the spectras: PIXAN soft.

To verify the validity of our measurements, we have simultaneously analyzed three standards, River sediments RS-NBS, granite Ga and diorite Dr-N CRPG. In a second stage, we have compared the results obtained on the same pellets, but with an electronic probe coupled with a SEM. The results were very similar. For the determination of the mineralogical composition, X-Ray diffraction has been performed on 14 selected samples.

The 41 pieces have been analyzed. All the results are detailed in the table. Nevertheless, to assess the origin of the sculptures, it was necessary to set aside all the recent or problematic artefacts. So, only 25 terracottas are taken into consideration in this chapter.

Two main homogeneous groups are identified. Their mean chemical composition (major and minor elements only), is given below.

Groupe		SiO_2	Al_2O_3	Fe_2O_3	MnO	MgO	CaO	Na_2O	K_2O	TiO_2	SO_3
S	X (%)	67.4	23.8	2.7	tr.	0.8	1.1	0.2	0.8	1.6	1.4
	(sigma)	(2.6)	(2.2)	(0.6)	-	(0.3)	(0.6)	(0.2)	(0.3)	(0.4)	(0.9)
C	X (%)	47.5	14.3	6.0	0.09	0.35	22.3	0.83	3.1	0.56	1.59
	(sigma)	(1.2)	(0.77)	(0.8)	(0.03)	(0.35)	(1.85)	(0.83)	(0.74)	(0.06)	(1.1)

* S. GROUP: siliceous terracottas (18 s.) characterized by high amounts of silica and alumina (Si02 + Al2O3 >90%) and by a simple X-Ray diffractogram where the quartz alone is identified. All the TL glow-curves show the same profile with a well developped low temperature peak at 110°C.

This group contains 9 attested and signed sculptures known for their parisian origin (3,18,25,26,30,33,34,38,41).

* C Group : carbonated pastes (5 s.) containing more than 20% CaO and showing a slight enrichment in alcaline elements and iron.

The mineralogical composition is more diversified: quartz, calcite, feldspars and some calcium-silicates (diopside, gehlenite). This mixing allows to obtain some information about the firing temperatures. Indeed, according to theoretical studies on marls published by Magetti and al. [11] and Echallier and Mery [12], some minerals or more precisely, some mineral associations act as thermometers. Depending upon the different sculptures of this group, we deduce that the firing temperatures are between 850 and 950°C. This group includes the two "Pleureuses" signed "in Roma, 1773".

By using statistical procedures, particularly principal components analysis, we have proposed some assumptions about the origin of 11 of the terracottas. We have supposed that two close points will belong to the same group and will correspond to a same origin for the raw material.

So 10 artefacts are linked to the parisian siliceous group, for example "la Bacchante endormie", "Deux bacchantes et un bacchant", "le Triomphe de Galatée"... and only one is tied to the italian carbonated group, "l'allégorie en l'honneur d'un sculpteur".

Three other sculptures are characterized by heterogeneous intermediate silico-calcic compositions. As there is no historical reference for these artefacts, nothing can be said about their origin.

It is interesting to present the results concerning one of the two problematic pieces: the "Bélier, brebis et agneau" which has an uranium-enriched paste; this unusual fact explains the high results obtained by TL.

To summarize, 18 sculptures would be parisian, 3 italian and 3 have an undetermined origin. These results have to be considered only as working assumptions by the curators, because they are only supported by scientific arguments without any stylistic or historical considerations.

In addition, according to the small number of italian works analyzed, it would be suitable to increase the italian corpus to verify and to test our propositions.

CONCLUSION

With this study, we tried to show the interest of scientific investigations for the conservation and the knowledge of the recent terracottas.

Thermoluminescence is useful to complete stylistic points of view, even for so recent periods, on the condition that such a study is performed on a representative number of pieces, with some attested artefacts, and that all the results are carefully, critically used and interpreted.

Results obtained with the same thermoluminescence techniques, but on isolated samples, would be, in our opinion, dangerous and without any credibility.

All the data will constitute the first stages of a data bank concerning terracottas of the 18th century.

1. F. Saltron, report n° 1806 LRMF, (1991).
2. A.L. Poulet and G. Scherf, Clodion (Réunion des Musées Nationaux, Paris, 1992), p. 489.
3. F. Saltron, A. Bouquillon and G. Querré, in ibid. pp.415-419.
4. M.J. Aitken, Thermoluminescence Dating (Academic Press, London, 1985), p.359.
5 M.J. Aitken, Science-based Dating in Archaeology (Longman, London, 1990), p.274.
6 D.W. Zimmerman, Archaeometry, 13, 1, (1971).
7. S. Fleming, PACT, 3, (1979).
8. S. Fleming and D. Stoneham, Archaeometry, 15, 2, (1973).
9. S. Fleming, Thermoluminescence Techniques in Archaeology, (Clarendon press, Oxford, 1979), p.233.
10 M.J. Aitken and S.J. Fleming, Topics in Radiation Dosimetry, Sup. 1, (1972).
11. M. Maggeti, H. Westley and J. Olin, in Archaeological Chemistry (Am. Chem. Soc., 1984).
12. J.C. Echallier and S. Méry, internal report IGAL, (1989).

THE ROMAN CERAMICS OF SALOBREÑA (GRANADA-SPAIN) AND THEIR RAW MATERIALS.

ANA Mª DE ANDRES * AND ISABEL MUÑOZ **
* Inst. of Materials Science, sede D (C.S.I.C.) Serrano 113, E-28006 Madrid.
** Dept. of Geology, Alcalá University, Nal. II, km. 33.600, E-28871 Madrid.

ABSTRACT

Nineteen roman ceramic sherds found near Salobreña (Granada, Spain), in the western coast of the Mediterranean Sea, as well as different ceramic clays from the surroundings are studied. Both clays and ceramic sherds are characterizad by X-ray diffraction and spectrometry, differential thermal analysis, and scanning electron and optical microscopies. A good number of the ceramic pieces, among which some "Terrae Sigillatae", have a composition similar to that of the local clays and, thus, have been probably manufactured at Salobreña. Only a few of them have a foreign origin. For most of them, the firing temperature was about 800-850 °C, although some have been produced at 900-1000 °C, and some others at 1000-1100 °C. It is concluded that Salobreña appears to have been an important settlement just in the third century of the Christian era.

INTRODUCTION

The region of Salobreña shows a two-fold interest, geological and archaeological, as the result of an important evolution of the coast line which took place 6.000-4.000 years ago (Fig. 1). During this period there was a first sedimentation from the river Guadalfeo, and more recently, 500 years ago, a second deposition that continues up to the present (1, 2). This suggests that the site of Salobreña become strategic for the man of the second century B. C. At that time the settelments were in the slope of the mountains near the sea by reasons of defense and economic control. This has been shown by German and Spanish archaeologists, who study the commercial and cultural influence of this settlement in other colonies of the Mediterranean coasts (1, 3, 4). The authors have chosen Salobreña after founding data on the large development of this site during the Roman period, specially along the second and third centuries of our era. In those days cultural exchanges and international trade by sea were favoured by being Salobreña settled at the coast. It was during the Roman period when a flourishing pottery industry took place. Its zenith came with the elaboration of the famous vases "Terrae Sigillatae". In this way, our interest in this study is to know how the man's technology at this period was (5).

MATERIALS AND METHODS

A selection among several archaeological wares belonging to the Archaeological Museum of Salobreña was performed. The most relevant of the Roman period were chosen in order to know their origin and characteristics of manufacture. The study comprises nineteen of these ceramic pieces: five pots (Nos. 1 to 5), three tiles (Nos. 6 to 8), six amphoras (Nos. 9 to 14) and five "Terrae Sigillatae" (Nos. 15 to 19).

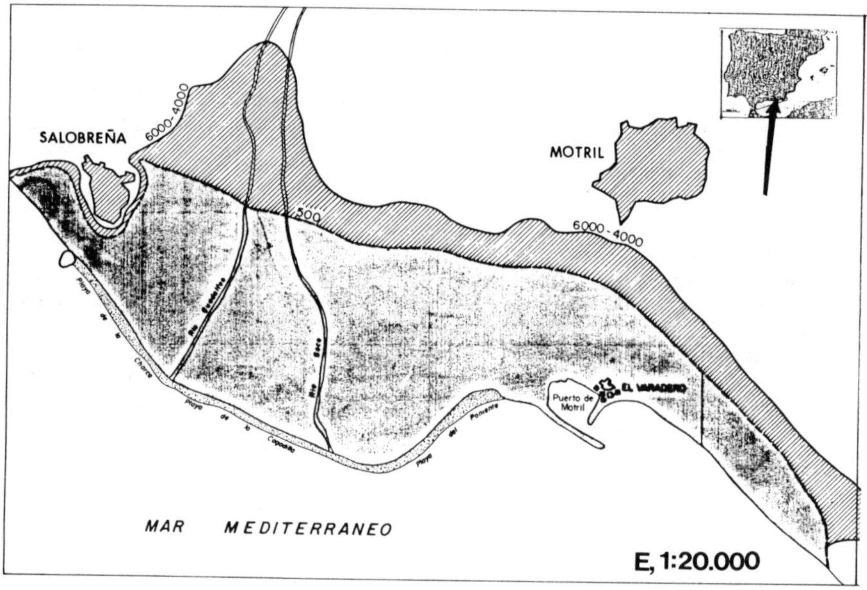

Fig. 1.- Evolution of the Mediterranean coast since 6.000 - 4.000 years up to the present (1).

Two outcrops of ceramic clays with a traditional use in the manufacture of pottery and even employed nowadays, were chosen. In the first outcrop, located 1 km northwest Salobreña, sherds from three differents levels, separated 20-30 m. each other, labelled Sal 1, Sal 2 and Sal 3, were taken. The second outcrop, 6 km at the southeastern of Motril, is labelled P.N. (6). Both ceramic clays were chosen because they were onto an emerged area in the Roman Period.

Chemical and mineralogical characterizations of these pieces as well as of some clays from Salobreña, have been performed. In this way, by comparison between these raw materials and the bulk of the archaeological ceramics we try to elucidate their local or foreign origin, and their behaviour during firing. In order to obtain this information, different physical and chemical techniques, such as spectroscopy of dispersive X-ray energy along with scanning electron microscopy (SEM-EDX) and X-ray diffraction (XRD) have been employed (7).

RESULTS AND DISCUSSION

These local clays are non calcarea, rich in alumina, silica and potasa (Table I); and are constituted by quartz, ortoclase and albite as feldspars, and illite, kaolinite and smectites as clay minerals (Table II). On the other hand, these raw materials are a good ceramic support and stoneware at 1100 °C, as shown by their ceramic properties: apparent density - open porosity and linear contraction - water sorption capacity, and behaviour during firing.

Table I. Chemical Analysis of the local clays and archaeological potsherds by EDX.

SAMPLES	Al_2O_3	SiO_2	K_2O	CaO	TiO_2	Fe_2O_3
P.N.	29.1	55.4	4.2	0.9	0.8	9.6
Sal 1	18.1	65.8	5.1	1.4	1.0	8.6
Sal 2	19.7	65.0	4.8	0.9	0.9	8.7
Sal 3	16.9	68.2	4.7	0.6	0.8	8.8
1	14.4	61.4	3.3	12.4	0.8	7.7
2	14.4	69.0	4.5	2.3	1.1	8.7
3	19.4	61.7	4.5	2.8	0.9	10.8
4	15.3	69.9	4.1	1.0	2.5	7.1
5	14.5	62.2	4.0	8.2	1.1	9.9
6	18.6	59.6	4.6	7.4	0.6	9.2
7	15.2	57.8	3.1	16.4	0.8	6.8
8	15.6	57.7	4.3	3.6	1.0	14.8
9	17.1	62.8	4.6	6.9	0.5	8.2
10	18.2	56.7	5.2	8.9	1.1	9.9
11	20.1	63.1	4.0	1.8	0.9	10.1
12	16.9	58.7	4.2	9.0	1.4	9.9
13	17.9	63.1	4.2	2.1	1.3	11.3
14	15.3	64.2	4.8	4.1	1.3	10.4
15	15.8	69.0	3.6	1.4	2.0	8.3
16	20.2	65.2	3.1	2.3	1.4	7.8
17	19.5	63.1	3.7	2.3	0.8	10.6
18	21.5	63.4	2.9	1.6	1.8	8.8
19	14.3	68.6	4.1	3.5	1.6	8.0

The morphological characteristics of the archaeological ceramics such as surface and core colours, presence of slip, size and quantity of tempers and even the thickness of the walls are included in Table III (8). Only the pot No. 1 and the five "Terrae Sigillatae" present slip. The whole of them contain feldspars-plagioclases and quartz as a natural tempering material, except apparently Nos. 1, 13 and 18. Mica grains have the same size of particle (0.12 mm). The quantity of tempers, as one could expect, depend on the use of the ceramics, that is, tiles and amphoras are done as container or supporter, so there is a great quantity of them with a big wall, but the luxe pots, such as the "Terrae Sigillatae", show only a few of them, and even a thin wall.

From data of Table I it can be observed that only a small number of ceramics are elaborated with calcareous clay (CaO> 6.4%) (9). These are pots Nos. 1 and 5, tiles Nos. 6 and 7 and amphoras Nos. 9, 10 and 12. Silica use to be related to the presence of quartz. The percentage of alumina and potasa is related to the presence of illite, micas and feldspars. TiO_2 and Fe_2O_3 are usually associated to clay minerals and play a prior rule as colorant agents; the first one, used as a colour of under cover, produces yellow shades; the second, gives a black or red coloring in a reductive or oxidative atmosphere respectively. These hetereogenities could indicate a different procedence, or several criteria of elaboration.

Tabla II. Mineralogical composition of the local clays

	Q	F	M	Fe	K	I	S	Am	A	Ha
Sal 1	x	++++	+++	+++	++	++	+++	-	++	++
Sal 2	x	+++	++++	+++	++	+++	++	-	++	++
Sal 3	x	++++	++++	+++	++	++	+++	+++	++	+
PN	x	++	++	++	++	++	+++	++	++	++

Key: + = 0.5-2%, ++ = 2-10%, +++ = 10-25%,
++++ = 25-95%, x = 100%
Q = quartz, F = feldespars, M = micas,
Fe = iron oxides, K = kaolinite, I = illite,
S = smectite, Am = amphibol, A = anatase,
Ha = halite.

The percentage of K_2O except for pot No. 1 and tile No. 7, are very close, and comparable with the results of the local clays obtained by EDX. The "Terrae Sigillatae" show a similar percentage of potassium and titanium but minor than the local clays, meanwhile the value of iron oxide is the usual of the rest of ceramics, as well as silica, calcium and alumina, whose percentages show a big similarity to those of the local clays. This seems to indicate that these vases have been made in potteries of the region.

Table III. Morphological characteristics of the archaeological ceramics from Salobreña (Spain).

WARE	COLOUR s.	COLOUR c.	COLOUR sl.	TEMPERS quartz (size in mm.)	TEMPERS mica (size in mm.)	TEMPERS others (size in mm.)	WALL THICKNESS (mm.)
1	b	o	b	-	0.12	0.12	9
2	b	b	-	0.74-3.72	0.12	0.12	10
3	g	g	-	0.24-0.74	0.12	0.24-0.37	4
4	g	g	-	0.24-0.74	0.12	-	6
5	g	g	-	0.25-1.24	-	-	16-18
6	r	m	-	0.25-0.87	-	0.25-0.62	33
7	o	o	-	0.25-2.48	-	0.37-1.86	23
8	r	b	-	0.37-1.49	-	0.25-0.62	20
9	g	g	-	0.25-1.24	-	0.25-0.99	-
10	g	g	-	0.25-1.24	0.12	0.25-0.62	18
11	g	r	-	0.25-2.11	-	0.12-0.49	14
12	b	b	-	0.25-1.24	0.12	0.62	7
13	o	o	-	-	0.12	0.12-0.37	10
14	o	o	-	0.25-0.62	-	0.37-1.24	12
15	r	r	r	0.25-0.62	0.12	0.12-0.25	4
16	r	r	r	0.25-0.62	0.12	0.12-0.49	5
17	r	g	r	0.25-0.62	0.12	0.12-0.49	4
18	r	r	r	-	0.12	0.12-0.25	4
19	r	r	r	0.25-0.62	0.12	0.12-0.37	4

Key: g = grey, m= maroon, b = black, o = ochre, r = red.
s. = surface, c. = core, sl. = slip.

And finally, by XRD, it can be elucidated which the tecnology of the Roman settelments was, comparing raw materials of the bulk with the local clays, and trying to determinate at what temperature they were fired (10). A great quantity of quartz has been detected (Table IV), what is coincident with the high value observed by EDX. The presence of carbonates, in pot No. 5, tiles Nos. 6 and 7, and amphoras Nos. 9, 10 and 12, indicates a maximum of 850 °C of firing. The existence of calcite in pieces Nos. 3, 15 and 16 has to be attributed to the burial conditions because minerals of high temperature have also been detected (9). Pot No. 1 and "Terrae Sigillatae" have been elaborated at 950-1100 °C, because minerals of neoformation, such as gehlenite, wollastonite and diopside, have been found. In pots Nos. 2 and 4, others minerals of high temperature, such as mullite and cristobalite syntesized from kaolinite have also been observed. The other ceramics seem to be elaborated at a temperature under 850 °C.

As orthoclase was detected not only in the local clays, but also in the pots, tiles No. 6, and amphoras (except Nos. 13 and 14) a local origin can be deduced for them. In the "Terrae Sigillata" the presence of sanidine, produced from orthoclase during heating at 900 °C (11), could indicate a local origin. The rest of the tiles, pieces Nos. 7 and 8, and amphoras No. 13 and 14 have plagioclases as raw materials and, then, a foreign origin.

Table IV. Mineralogical identification of the cermics wares, by XRD.

	Q	F-P	M	I	K	Fe	IL	A	C	G	D-W	Mu	Cr
1	X	++++	+	-	-	++	-	+	-	++	+++	+	-
2	X	++	+	+	-	++	-	-	-	-	-	-	-
3	X	+	+	+	-	++	-	+	+	-	-	++	++
4	X	++	-	-	-	+	-	+	-	-	-	-	-
5	X	+++	+++	-	-	++	-	+	++	-	-	-	-
6	X	++++	+++	++	-	++	+	+	++	-	-	-	-
7	X	+++	-	-	-	++	-	+	+++	-	-	-	-
8	x	++	++	++	-	+	-	+	-	-	-	-	-
9	x	++	++++	++	-	+	++	+	++	-	-	-	-
10	x	++++	+	++	-	+++	++	+	++	-	-	-	-
11	x	+	-	-	++	+	++	+	-	-	-	-	-
12	x	+++	-	+	-	++	-	+	++	-	-	-	-
13	x	++	++++	+++	++	++	+	++	-	-	-	-	-
14	x	+++	+	++	-	++	+	+	-	-	-	-	-
15	x	++	-	+	-	++	-	+	+	+	++	-	-
16	x	+	+	+	-	++	-	+	+	+	+	-	-
17	x	+	-	+	-	+	-	+	-	++	+	-	+
18	x	+	-	+	-	++	-	+	-	+	+	-	-
19	x	++	-	+	-	++	-	+	-	++	+	-	+

Key: + = 0.5-2%, ++ = 2-10%, +++ = 10-25%, ++++ = 25-95%, x = 100%
Q = quartz, F-P = feldespars - plagioclases, M = micas, I = illite, K = kaolinite, Fe = iron oxides, Il = ilmenita, A = anatase, C = carbonates, G = gehlenite, D-W = diopside and wollastonite, Mu = mullite, Cr = cristobalite.

CONCLUSIONS

We have found local ceramic wares with different types of manufacture and elaborated at various temperatures. These temperatures have not been chosen without any sense, but according to the function of the ceramics. Those elaborated as containers or supportes, as amphoras and tiles, present a temperature of firing less than 850 °C, thick walls and a very high quantity of temper. Those for a domestic use have been heated at 900-1000°C, in order to be resistent to firing. And, lux ceramics for decoration show as characteristics softness at touching (slip), thin walls (no impurities, less temper), and more elaboration (a higher temperature, 1000-1100 °C), all these conditions leading to a major development in the technology of manufacture. In this way, Salobreña seems to have had an important local pottery during the Roman Period, capable of producing pots, tiles, amphoras and even luxes ceramics like the "Terrae Sigillata". Moreover, in those days, the presence there of foreign pottery is a sign of its commercial and cultural exchange with other villages.

REFERENCES

1. O. Arteaga, G. Hoffmann, H. Schubart, H.D. Schultz, in Forschungen zur Archäeologie und Geologie im Raum von Torre del Mar 1983/84, (edited by Verlag Philip von Zabern, Madrid, 1988) pp. 107-126.

2. G. Hoffmann, PhD thesis, Bremen University, 1988.

3. M. Pellicer, Excavaciones en la Necrópolisis Púnica "Laurita", del Cerro S. Cristóbal, Almuñecar (Granada), (Exc. Arq. Esp., 17, Madrid, 1983).

4. F. Barreca, L'Espansione Fenicia nel Mediterraneo. (Edited by CNR, Roma, 1971).

5. G.M.A. Richter, in A History of Technology, (edited by Oxford at the Clarendon Press, vol. II, Great Britain, 1979), pp. 284-310.

6. Geological Maps of Motril (no. 1055) and Albuñol (no. 1056). E, 1:50.000 (edited by Instituto Geológico y Minero de España, Madrid, 1981).

7. N. Cuomo, La Ceramica in Archaeologia. Antiche Tecniche di Laborazione e Moderni Metodi d'Indagine. (Edited by La Fenice, Roma, 1988).

8. A.O. Shepard, Ceramics for the Archaeologist. (Carnegie Inst. of Washington. Washington D.C., 1976).

9. R.B. Heimann, in Archaeological Ceramics. (Smithsonian Inst. Press, U.S.A., 1982) pp. 89-96; M. Maggetti, ibid. pp. 97-109.

10. I. Muñoz, Thesis of Graduation. Alcalá University, Madrid, 1991.

11. J. Borg and S. Smith. Amer. Miner. 54, 163-181 (1969).

A PRELIMINARY SCIENTIFIC STUDY OF TRADITIONAL KOREAN CELADONS AND THEIR MODERN DEVELOPMENTS.

C.K.KOH CHOO
Department of Chemistry, Chung-Ang University, Seoul, Korea

ABSTRACT

In studying celadons culture of Korea scientifically 81 sherds from nine different kiln sites of different periods are analyzed for their macroscopic, compositional and microstructural characteristics. The raw materials used for modern reproduction are analyzed for their chemical and mineralogical composition and compared with those of traditional pieces.

INTRODUCTION

The ceramics culture of traditional Korea has a long and rich history of various interactions with China and Japan on one hand and indigeneous creative forces on the other. Its artistic achievements, especially of the Pisaek 翡色(jadelike color) and inlaid celadons, together with Korean technological achievements of world's earliest observatory of 7th century, metal movable type printing some 200 years before Gutenberg and invention of iron-clad submarine in 16th century, assume the position of highest pride in Korean people's heart. However, it is only natural that these achievements are not at all recognized abroad today, because they never really developed to their full possibility. Such ironic character of Korea's achievements, indicative of highest level of creativeness and inventiveness and simultaneously an extreme case of intellectual apathy to ignore these possibilities once the immediate need or interest for them pass, is very much present in Korean ceramics heritage. Korean potters left a legacy of simple natural beauty behind them, but they suffered as people of lowest class and means. Their technology was always ignored by learned circle of influencial scholars and officials. There is hardly any record of techniques and processes in literature, and actually most had to be relearned in recent times in reproducing the traditional ceramics.

Various types of potteries found mostly in the west coast and in southeastern part of Korea are believed to be as old as seven to eight thousand years[1]. The extent of ceramics production in the Three Kingdoms period and Unified Shilla period(668-935) can be seen in the enormous amount of ceramics found in tombs of that period. These dark grey unglazed stonewares of extreme hardness give bell-like resonance when tapped, and their firing temperature is supposed to be as high as 1200℃. This indigeneous technology and the influence from Yue in 9th century are seen as the beginning forces of celadon production of Koryo period(918-1392). While much similarity in shape and decoration with Yaozhou, Ru, Ding, Longquan, and Jindezhen wares are found in early celadon wares, those made from the peak stage of 12 and 13th century are unique in color, form, inlay and copper-red technique. With the change of dynasty from Koryo to Chosŏn(1392-1910), the difficult inlay technique was changed into easier freer slip decorations of various types, named as Punch'ong (powder-decorated grey blue) wares. The other kind of Chosŏn ceramics the white and the blue and white wares produced much in government-sponsored kilns strived for the purity of white color of Ming dynasty, but most are gray-tinted white giving the feeling of softness.

The Korean ceramics have been often characterized with qualities of vitality, freedom, simplicity, spontaneity, and playful humor[2]. In many pieces of same shape and type that are remaining, the perfection for exactness is not seen but slight irregularities result in the divergence. All of the representative types are all simply decorated and glazed in subdue colors of gray, gray-blue, and white, and various shades of brown and black. The most colorful decoration is done with copper red and cobalt blue, but they are sparingly used all as underglaze decoration. The

Table I. Locations and characteristics of traditional Korean ceramic sherds.

No	Location of kiln sites	Date	No. of sample	Cra-ckled	Thickness, sample, cm	Bulk density	Glaze Colors[1]	Thickness Glaze, μm[2]
I	Kangjin	Mid 9th						
	Yongunri	-E 10th	4	3	0.1-1.0	2.18-2.33	YL - GN	75-400
	Yongunri	10	5	2	0.3-0.7	2.01-2.28	GY - BL	
	Sadangri	11	5	4	0.3-1.2	2.01-2.24	YL-GN, BL-GY	250-350
	Sadangri	E 12	3	2	0.2-0.4	2.24-2.28	GY - BL	300-650
	Sadangri	E,M 13	2	1	0.3-1.3	2.27-2.28	YL-GY-BL	
	Sadangri	M 14	3	2	0.2-1.3	1.99-2.22	GY - BL WH-GY-BL	150-550
II	Kohŭng Undaeri	L 10 -E 11	10	4	0.3-1.8	1.91-2.33	GY - BL GY-GN, BK	125-600
III	Puan Yuchŏlri	12	6	4	0.2-0.5	2.14-2.32	GY-BL, GY-GN	250-500
IV	Sŏsan Osari	L10-E11	7	2	0.2-1.6	2.08-2.30	GY-GN, BN-GN	75-275
V	Kongju Sinyŏngri	11	7	7	0.3-1.5	2.07-2.28	YL-GN, BN-GN	200-450
VI	〃 Chunghungri	15	10	1	0.5-1.9	1.95-2.29	GY, BL-GY	75-350
VII	Koyang Wŏndangri	10	4	1	0.5-1.8	2.04-2.17	BN-GN, YL-GN	50-200
VIII	Yongin Sŏri	E 11	5	4	0.3-1.5	2.01-2.25	YL-WH, GY-GN	
IX	Ich'ŏn	Modern	10	6	0.2-0.7	2.17-2.44	GY-GN, GY-BL	200-500

1 : abbreviations for glaze color
BK:Black, Bl:Blue, BN:Brown, GN:Green, GY:Gray, YL:Yellow, WH:White
2 : The thickness is measured from pictures of optical microscope.

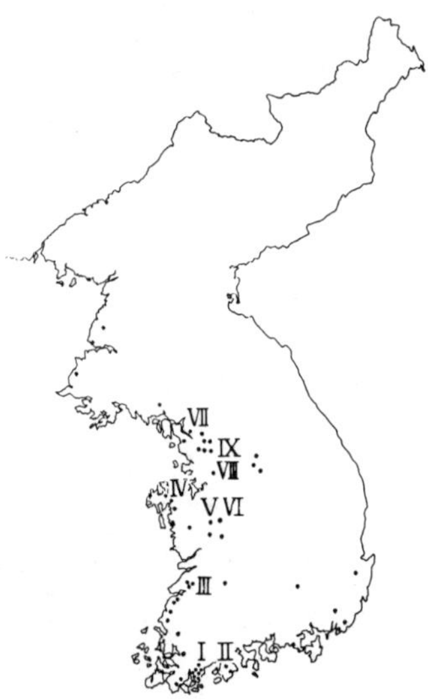

Figure 1. Map of kiln sites covered in this study.

colorful Tang sancai and overglaze decoration never developed in Korea, and the beauty was expressed more in elegant contour lines and often quickly applied abstract decoration. The rather low reflectivity result in much less shiny and bright surface than Chinese ceramics, drawing one's attention inwardly into the glaze and body. With such lack of colorfulness and brilliance Korean ceramics succeeds unexpectedly in drawing attention with "a subtle technique of silence as a form of communication.....Korean ceramics sit quietly and wait to be recognized in a silent sort of way, not demanding attention with brilliant designs"[3].

Such persistent almost stubborn adherence to simple form and decoration makes the study of korean ceramics more interesting.Were the Korean potters limited technologically and in availability of raw materials, or were their artistic and spiritual inclinations together with the users' tastes more responsible? What social,economic,religious,and political background is responsible for the development of Korean and in general Oriental reference for celadons and less colorful ceramics which contrasts sharply with Western taste for overglaze Chinese porcelains? In due time the technological and scientific study of Korean ceramics should be combined with these other methods of inquiry for a comprehensive understanding of ceramics

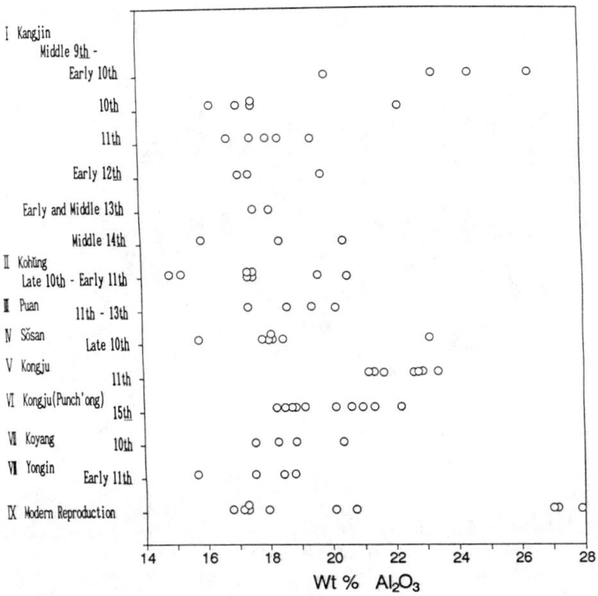

Figure 2. Al_2O_3 content of traditional Korean ceramics body.

heritage. Even from the start of such an inquiry, the dual reality of exceptional natural beauty providing a source of comfort and encouragement and of the harsh political environment of constant wars and human exploitation, that Koreans faced throughout their history should be kept in a close perspective for understanding the ironic nature of Korea's greatest achievements.

MACROSCOPIC OBSERVATION

Altogether 81 pieces of ceramic sherds were analyzed for their macroscopic and microscopic characteristics. As shown in table I, they are divided into 9 groups according to the kiln sites they originated from. The kiln sites of the first group Kangjin was by far the most important center of Korean celadons as can be seen in the map of figure 1. It is located in the southwest and the groups are numbered consecutively from it upward to Kyonggido, a province where the government kilns for white porcelain was located in Chosŏn period and now the modern reprodution studios are concentrated. The nineth group is from two modern studios of this area. They are all celadon kilns, except the groups VI and VII. VI is an early Punch'ong and VII is a rare Koryo kiln for white ceramics. In group II is included 5 pieces of black-glazed sherds. Puan kiln of group III was second only to Kangjin for both the quantity and quality of celadons.

The sample thickness indicates the part of ceramic ware the sherds come from, thin ones being usually the upper side and the thicker ones from the bottom and the foot of a dish or bowl. The bulk density ranges from 1.95 to 2.44, majority of good quality being between 2.2 and 2.3. About half (43 of 81) of the sherds had crackled glaze surface. Only the group V pieces were all crackled, while most of VI and VII were uncrackled. The glazed colors were all shades of gray, blue, green, often tinted with yellow and brown. The Pisaek can best be called a gray-blue. The glaze thickness ranges from 50 to 650μm.

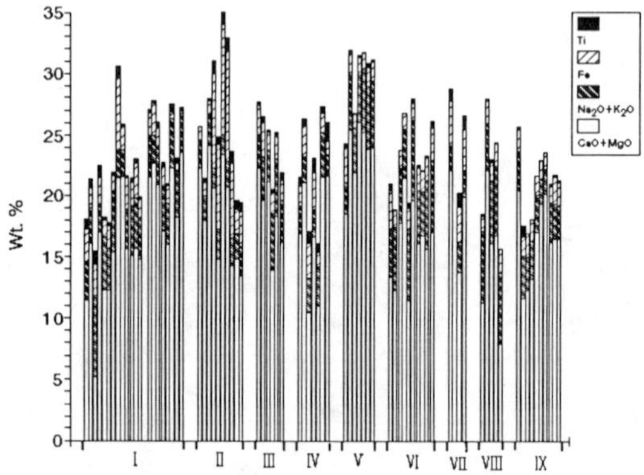

Figure 3. Flux, iron, and titanium contents of Korean ceramics glaze.

COMPOSITIONAL ANALYSIS

Cross-section of the sherds embedded in epoxy resin were prepared for compositional and microstructural analysis. Link energy dispersive X-ray analyzer (AN 10/85s) with scanning electron microscope (Akashi DS-130s) was used both for glaze and body compositions. Three different measurement of an area each 180x240 $(\mu m)^2$ for glaze and 300x400$(\mu m)^2$ for body were averaged. Powder body samples of about 10 sherds were analyzed by classical wet method, and the results agreed with those of cross-section EDX method within scatter of the data. Because of space limitation the raw data of these analyses is not shown here.

In figure 2 is shown the Al_2O_3 content of the body composition. About 80% of the sherds are found in the range between 16 and 21% with largest concentration at 17-18%. The sherds with greater than 21% occur only in the earliest Kangjin samples, two Kongju kilns, and modern replica group. Kongju kilns in general seem to represent higher Al_2O_3 content since all the sherds of both Kongju celadons and Punch'ong group lie above 18%. Such values of Al_2O_3 are closer to more siliceous celadons from southern Chinese kilns of Yue, Hongzhou and Longquan and Jindezhen than to the northern kilns of Ru and Yaozhou. The flux(CaO, MgO, K_2O and Na_2O), iron and titanium content is marvellously constant. The majority have flux values between 3 and 5% with K_2O content being the largest between 2.5 to 4%, much like Chinese celadons. The TiO_2 level is similar to Chinese, too, with values between 0.3 and 1.6. The iron content of most 2 and 3% seems slightly higher than Chinese celadons, which are usually lower than 2%.[4-6] The black-glazed wares found in group II have high values of iron between 4.8 and 7.5%, while Koyang sherds of group VII have high values between 4.0 and 4.9. As expected, the Yongin white ware has different values from the celadon. The flux content is increased to 7.5% with K_2O content being 4.1 to 7.1%, iron content all less than 2.0, and titanium content less than 0.3. These values are comparable to those reported for 8 white porcelain sherds of 18th century[7].

Figure 3 shows the flux, iron and titianium content of glaze. Calcium content varies widely from 4.2 to 24.8%, although a value lower than 10% is found only in 10samples. Such a wide variance occurs even within a group. The average value is 16.1% for group I, 17.5% for II, 17.3 for III, 15.4 for IV, 21.8 for V, 15.1 for VI, 16.7 for VII, 12.9 for VIII, and 15.0 for IX. The Kongju celadon with the highest average is the only group where all seven sherds have consistently high values. In previous works on Korean celadon glazes, there has been a conflicting report about whether the calcium oxide level is different from kiln to kiln or not [8-10].
Although more work is needed for any conclusive results, Kongju celadon of this study indicates a certain possibility for diffent kinds of lime and lime-alkali glaze

Table II. Chemical composition and mineral content of Kangjin unglazed celadon bodies and raw materials used for modern reproduction.

Sample	SiO_2	Al_2O_3	Fe_2O_3	MgO	CaO	Na_2O	K_2O	TiO_2	P min[5]	M min[5]	T min[5]
HOII-3(9,10th)[1]	68.5	24.5	3.0		0.2		2.4	1.2	q	a	o,m
HMOI-4(10th)[1]	69.9	21.2	2.9		0.5	1.2	3.4	1.0	q	o	a,m
CJ12-1(12th)[1]	72.3	21.3	1.9		0.3		3.4	0.9	q		o,a,m
CJ12-2(12th)[1]	71.7	21.2	2.6		0.4		3.3	0.8	q		m,a
Pottery stone[2]	78.9	15.0	0.7			2.9	2.5		q	a	o,m
Multo[2]	75.3	18.0				2.0	4.7		q	a	o,m
Celadon body[3]	57.8	31.9	3.2	0.7	2.6	1.3	2.3	0.3	q	a	k,m
Celadon body[4]	65.7	25.7	3.2		0.4		3.7	1.0	q	h, m	o,a,k
Celadon glaze[3]	62.8	16.2	3.4		11.7	1.9	3.4	0.6	c, q	a	
Celadon glaze[4]	61.6	16.3	1.3		13.4	1.4	5.4	0.6	q	c, o	a

1 Celadon body sherds of the first firing found in Kangjin kiln site.
2 Raw materials being used in a reproduction studio in Ich'ŏn, Kyŏnggido near Seoul.
3 Raw materials for traditional ceramics being sold in Yŏju, Kyŏnggdo near Seoul.
4 Raw materials used in Tonggukyŏ, a celadon reproduction studio in Ich'ŏn.
5 P min(pricipal mineral): Mineral with higher than 30% content in X-ray diffraction
M min(minor mineral) : 10-30% ;T min(trace mineral): Less than 10%
a : albite c : calcite h : halloysite
k : kaolinite m : muscovite o : orthoclase q : quartz

developed in different kilns and in different periods.

As with the previous works, only with small number of exceptions the range of SiO_2 is found to be 55 to 65% and of Al_2O_3 10-15%. Iron level is 6.1 to 9.1 for blackglazed. Its level in Yongin white ware is rather low between 1.3 and 1.9, while titanium level of 0.1-0.3 is also lower than other glazes. Manganese, reported to be responsible for more gray tone of Korean celadons compared to Chinese is found from 0.1 to 0.8%. In 8 of 9 modern replicas manganese is not found at all, and instead BaO is found in 5 samples. P_2O_5 is found in certain groups only, Kangjin group, Kongju celadon, and two Kyonggido groups of VII and VIII.

MICROSTRUCTURE

Only preliminary study of microstructure has been carried out with optical microscopy and scanning electron microscopy and X-ray diffraction works. Vandiver et al.[9] noted in glaze bubbles, undissolved quartz and black particles, but no anorthites. Newman [10] did not see any magnetite black particles but noted a case of extensive anorthite in glaze. In this study, bubble, quartz, and some other crystals were observed in microscopic pictures. Most of X-ray diffraction of glaze show well vitrified glass pattern only, but often anorthite peaks as well as quartz peaks were faintly, but quite clearly observed. There is some possibility that wollastonite crystals are found, too. Bubble and anorthite crystals are more extensively observed in interface. In body quartz crystals of various sizes up to 100μm are observed along with pores of various sizes and particles of pure titanium oxide and mixture of titanium and iron. In X-ray diffration, quartz and mullite peaks are always observed, whereas tridymite peaks are observed only in more vitrified bodies. When sherds of Kongju celadon and Punch'ong(group V and VI) were refired to 1250℃ in order to estimate original firing temperature vitrification occurred more extensively and tridymite peaks were newly developed. A sample refired to 1100℃ had similar X-ray pattern as the original sherd, showing that original firing temperature for these samples was between 1100 and 1250℃. More work on firing temperature is needed in the future.

MODERN DEVELOPMENTS

As mentioned in introduction, the techniques of making celadon had to be "rediscovered" in this century. Like their ancestors the pioneer potters suffered much from poverty and indifference of the public. However, as the economy and the conscious desire to learn more about their cultural heritage grew in 1970's and much of 1980's, the pottery and porcelain making became quite a booming enterprise.

A quick comparison between the old and new technology can be made from table Ⅱ. The X-ray patterns for Kangjin unglazed bodies are almost the same as the patterns for the raw materials of pottery stone and multo, which can be translated literally as water clay. These raw materials sold in a wholesale shop can be made into ceramics bodies singly as they are. Such a practice of using a single raw material for a ceramics body is called Tanmi, 單味 meaning a single flavor. These raw materials can be inferred to be similar to the Chinese porcelain stones which provide both plasticity and flux in form of sericite. In fact the chemical composition of the sherd bodies shows that much of traditional celadons was made using one such material.

The shop also carried materials composed of several different components, which were prepared for direct use as celadon body or glaze. The chemical and mineral composition of this celadon body material and another such material prepared at a nearby studio are quite different from Kangjin samples. Aluminum content is much higher and clay in form of kaolinite and halloysite are present. The modern firing method and the kiln itself are very different from traditional hill-climbing wood-fired kiln. Much better control is obtained in a factory-styled kiln fueled with gas. The products are understandably much cleaner and shinier, but lacks that essential quiet feeling of softness and depth one receives from traditional Korean celadons. A deeper understanding of the traditonal craft based upon scientific and technological analysis with interdisciplinary approach can make the successful reproduction and further development possible.

REFERENCES

1. W. Y. Kim, *Art and Archaeology of Ancient Korea*, 80-93 (The Taekwang Publishing Co., Seoul, 1986)
2. Y. S. Koh, *Essays on History of Korean Art and Esthetics*, 3-13 (T'ongmungwan, 1963, in Korean)
3. E. B. Adams, *Korean Pottery Heritage*, Vol 1, 6, 21-22 (Samhwa Printing Co. Ltd, 1986)
4. C. Xianqiu, C. Shiping, T. Wuyang and Z. Xuelin, *Scientific and Technological Insights on Ancient Chinese Pottery and Porcelain*, 148-152 (Science Press, Beijing, 1986)
5. G. Yenyi and L.Guozhen, *Scientific And Technological Insights on Ancient Chinese Pottery and Porcelain*,153-160 (Science Press, Beijing, 1986)
6. L. Guozhen and G. Lingxiang, W. D. Kingery(ed.) *Ceramics and Civilization*, Vol Ⅲ, 129-151 (The American Ceramic Society, Inc., 1986)
7. M. Tregear, M. A. Pollard, H. Hatcher, and N. D. Wood, Presented at the Second International Conference on Ancient Chinese Pottery and Porcelain, (Bejing, 1985)
8. S.Choi-Bai, *Catalogue of the Collection at the Museum of East Asian Art, City of Cologue*, 233-241 (Museum für Ostasiatische Kunst, 1984)
9. P. B. Vandiver, L. A. Cort and C. A. Handwerkes, W. D. Kingery(ed.) *Ceramics and Civilization*, Vol Ⅳ, 365-388 (The American Ceramic Society, 1989)
10. R. Newman, 429-434, *Materials Research Society Symposium Proceeding*, Vol 185 (Materials Research Society, 1991)

WHITE INCRUSTATION ON POTTERY FROM NEOLITHIC TELL SAMOVODENE, BULGARIA

JULIA GENCHEVA

National Scientific Centre of Monumental Culture
Bulgarian Academy of Sciences
2 Str. King Ivan Alexander
5000 Veliko Turnovo, Bulgaria

ABSTRACT

Twelve samples of lime plaster were examined, including a white incrustation on vessels and religious tables and the contents of a large vessel filled with the white substance. These materials are from Tell Samovodene, Bulgaria which was inhabited during the Neolithic period between 5750 and 5400 B.C. Crystal-optical analysis, X-ray diffraction, emission spectroscopy and scanning electron microscopy (SEM) were used in the analysis. It was established that the white contents of the vessel consisted of slaked lime which had been carbonized in antiquity. The white incrustation was shown to be a lime plaster, composed of slaked lime and quartz. We are continuing investigation of the kiln, found on the periphery of the tell; preliminary data suggest the kiln was used for lime burning.

INTRODUCTION

The Neolithic period Tell Samovedene is one of the largest researched archaeological sites in northeastern Bulgaria in the basin of the Yantra River. The stratigraphy represents a standard for the chronology of the Neolithic period in this part of Bulgaria and in the area of the Lower Danube (1). There are five cultural layers and eleven building horizons which date from the end of the Early Neolithic to the Middle Neolithic, a period between 5750 and 5400 B.C. A religious complex was discovered with a shaft which is the earliest known in southeastern Europe. A fortification system was also excavated which consists of a ditch, rampart and palisade. This system of fortification is the earliest in the Balkan Peninsula, and the only parallel is found in Oltenia, Rumania (2). On the periphery of the tell was found a complex of pottery kilns.

Pottery is the most widespread material remaining at Tell Samovodene in the Neolithic period. One's impression is that part of the vessels, religious tables and anthromorphic and zoomorphic figurines are incrusted with a white substance. Some pottery fragments with the white substance infilling incised areas are shown in Figures 1 and 2.

Some Bulgarian specialists have described the white substance as a paint (3), ochre (4), white clay (5), matter (6) in addition to others. The first attempt to research the nature of the white substance from prehistoric settlements in south Bulgaria was made in 1971. Spectroscopy was used to conclude that the white substance was gypsum.

When the large vessel filled with a similar white substance was found during archaeological excavation at Tell Samovodene, it was decided to examine whether the white substance was gypsum or not and to compare it with other samples of white substance or substances from the site. Such white materials are a common occurrence in some cultural layers, in floors of houses and other examples which resemble intentional plaster usage.

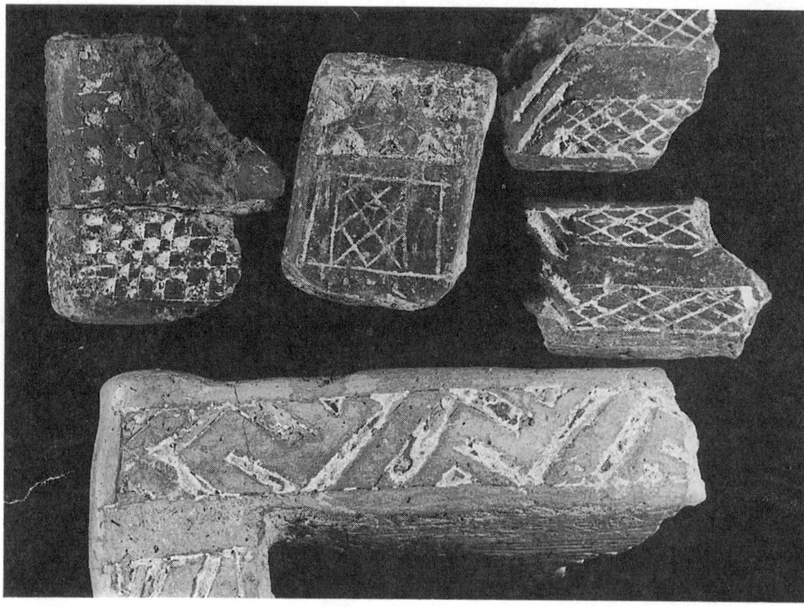

Fig. 1. Pottery fragments with white decorative incrustations from Neolithic Tell Samovodene.

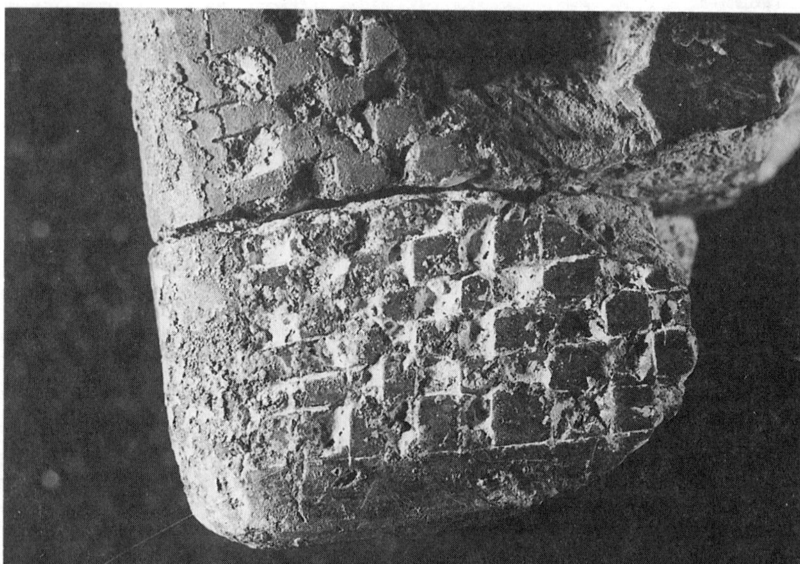

Fig. 2. Pottery fragments with white incrustations shown with slight increase in magnification.

SAMPLING

The large vessel filled with a white substance (Fig. 3, arrow) relates to actitivies at the end of the Early Neolithic period. The diameter of the mouth is 40 cm, the base diameter is 47 cm, and the height is 53 cm. After appropriate conservation, the vessel with its contents was exhibited in the exposition of the prehistoric hall at the Museum of History entitled "Veliko Tarnovo - A Capital of Bulgaria in the XII-XIV Centuries". A sample was taken from the contents of the large vessel, in addition to the incrustation of the pottery vessels, religious tables, the cultural layer which had a large quantity of the white substance. Other samples were taken from the floor of a house, plaster from a house wall and from supposed raw material. The samples were taken very carefully, under a microscope. The aim was to choose pure substance under the uppermost, superficial layer.

Sample 1, white substance from the large vessel, sector III, the end of the Early Neolithic.
Sample 2, white substance from a plaster floor of house 2, sector VI, the end of the Early Neolithic.
Sample 3. white substance from a cultural layer, sector III, the end of the Early Neolithic.
Sample 4, white substance from the place named "White Earth", in the contemporary village, Samovodene.
Sample 5, limestone from the place "Ostra Mogila" near the village Samovodene.
Sample 6, limestone from a lime-pit over the lime-kilns near the village Samovodene.
Sample 7, limestone from the Neolithic Tell Samovodene.
Sample 8, white substance from incrustation of a pottery fragment, square VI^{a}, the transition period from Early to Middle Neolithic.
Sample 9, white substance from incrustation of a pottery fragment, sector V, the end of the Early Neolithic.
Sample 10, white substance from incrustation of a pottery fragment, sector III, the end of the Early Neolithic.
Sample II, white substance from incrustation of a religious table, sector III, the end of the Early Neolithic.
Sample 12, white substance from a lime plaster of a wall of a house, sector III, the end of the Early Neolithic.

All samples were examined with crystal-optical analysis and semi-quantitative emission spectroscopy (8). Some samples were subjected to X-ray diffraction and scanning electron microscopy (SEM).

RESULTS AND DISCUSSION

The crystal-optical analysis showed that all the samples contain calcite as a main or secondary mineral (Table 1). The clay minerals in the table were examined with X-ray diffraction. The sequence of the mineralogical composition in Table 1 is shown in decreasing quantity. Samples 5, 6 and 7 are limestone consisting of coarse-grained calcite. Because these raw materials are near the Neolithic Tell Samovodene, the decision was made to examine these limestone for the supposed origin of the white substance. Samples 1, 2, 3, 9, 10 and 11 consist mainly of calcite which is micro-grained, except sample 1 which

Table 1

Crystal-optical analysis of the samples

Sample	Mineralogical composition
1[a]	close-grained calcite, quartz
2[a]	micro-grained calcite, quartz, mica, chlorite
3[a]	micro-grained calcite, quartz, mica, chlorite
4[a]	mica, chlorite, kaolinite, quartz, micro-grained calcite
5[a]	limestone built of coarse-grained calcite
6[a]	limestone built of coarse-grained calcite
7[a]	limestone built of coarse-grained calcite
8[a]	micro-grained calcite + dolomite, quartz, mica, chlorite
9[a]	micro-grained calcite, quartz, mica, chlorite
10[a]	micro-grained calcite, quartz, mica, chlorite
11[a]	micro-grained calcite, quartz, mica, chlorite
12[a]	quartz + micro-grained calcite, clay minerals

Table 1. Summary of results by crystal-optical analyses of the samples.

Sample	Si	Mg	Al	Fe	Ca	Pb	Sn	Be	Cr	Ni	Na	Cu	Zn	Mn	Co	Ag
13[a]	over 1%	over 1%	over 1%	over 1%	over 1%	10 20	-	10 20	20 50	5 10	over 1%	20	-	100 200	2 5	0,5 1
14[a]	over 1%	over 1%	over 1%	over 1%	over 1%	20 50	-	10	20	10 20	over 1%	10 20	-	1000 2000	2 5	0,5 1
15[a]	over 1%	over 1%	over 1%	over 1%	over 1%	20 50	-	-	10	5	over 1%	10	-	200 500	2	0,5
There are not discovered: S8, 6e, Sn, Zn, As, Te, W, Bi, Mo, V																

Table 1, continued. Results of crystal-optical analysis.

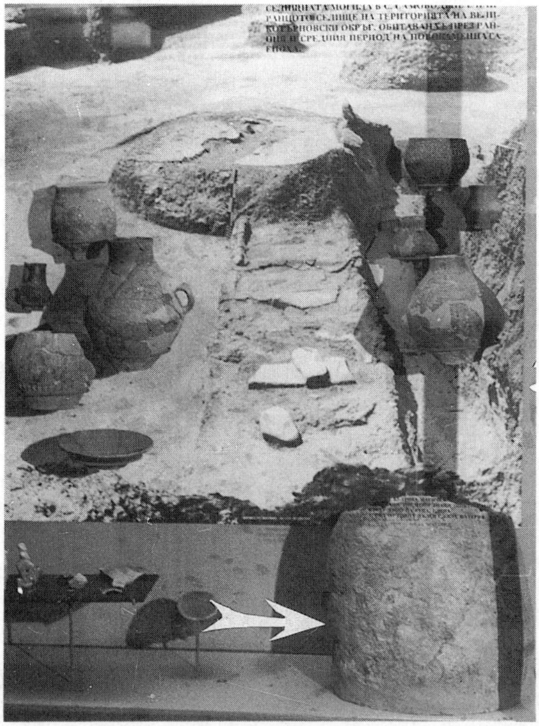

Fig. 3. Pottery vessel filled-up with white substance (arrow).

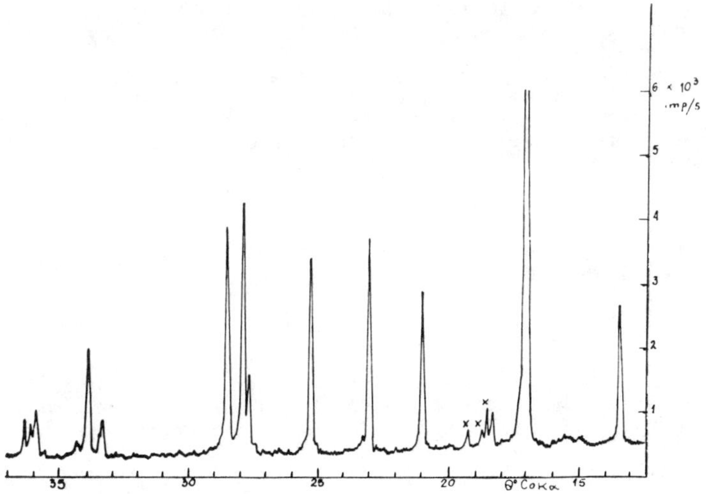

Fig. 4. Results of X-ray diffraction for sample 1.

is close-grained. Sample 8 consists of calcite and dolomite, both micro-grained. Sample 12 is composed mainly of quartz and micro-grained calcite. Sample 4 consists mainly of the clay minerals, mica, chlorite and kaolinite. The samples consist of little other minerals. The presence of very minor amounts of other minerals are: for sample 1, very little quartz, for samples 2, 3, 4, 8, 9, 10 and 11, little quartz, mica and chlorite, and for sample 12, the clay minerals cited above.

The crystal-optical analysis proves that the white substance of incrustation from the pottery fragments and the contents of the large vessel is mainly calcite. For the pottery incrustation of sample 8, it is calcite and dolomite.

Samples 1, 8 and 10 were examined by X-ray diffraction (Figures 4, 5 and 6). Sample 1 contains calcite as the main component and very little quartz, perhaps about 1%. Sample 8 consists of calcite and dolomite as basic components with a little quartz, about 5%, as well as clay minerals under 1%. Sample 10 is composed of mostly calcite, about 95%, about 5% quartz and less than 1% clay minerals.

The results of emission spectroscopy for all the samples are shown in Table 2. Sample 9 is distinguishable from all the others. It does not consist of Al, Pb, Ga, Cr, Ni or Zn. The element Mn is only present in trace amount. Manganese is a typical element in materials from Neolithic Tell Samovodene. Maybe this vessel may have been carried from another place during the Neolithic period. The quantity of the element Si is high in sample 4, a possible raw material from the place in the modern village called "white earth". This high silica content differentiates it from all the other samples. Sample 4 is composed mainly of clay minerals. It was interesting to determine whether the name "white earth" meant that the substance might have had a composition in common with the incrustation; obviously these are different substances. The results of the modern limestone analyses when compared with the ancient samples show that not one of them was used as a raw material. Sample 10 is closest to the limestone from the lime-pit over the lime-kilns near the contemporary village of Samovodene.

The microphoto of sample 1 taken in polarized light shows that the crystals of calcite have regularly shaped rhombs (Fig. 7), indicating secondary crystallization of the carbonate. The microstructure seen with scanning electron microscopy reveals rhomboidal and spherical particles of calcite in samples 1, 8 and 10 (Figs. 8, 9 and 10, scale in microns).

The white substance in the large vessel is mainly calcite with under 1% quartz and probably clay minerals. The impurities of quartz and clay in samples 8 and 10 are greater in quantity than samples 1, but not more than 5%.

CONCLUSIONS

The results provide an opportunity to express the opinion that the white substance inside the vessel and from the incrustations on pottery are slaked lime which recarbonated in antiquity. The chemical reactions can be expressed as follows:

$$CaCO_3 \quad = \quad CaO + CO_2 \qquad -12.5 \text{ kcal}$$

$$\text{at } 500^\circ \text{ to } 900^\circ\text{C}$$

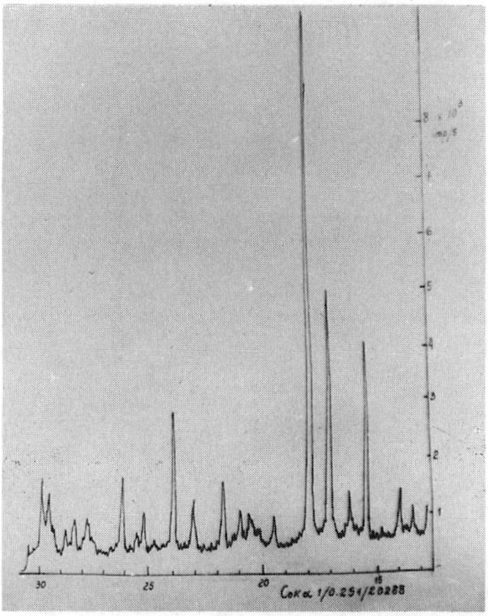

Fig. 5. Results of X-ray diffraction for sample 8.

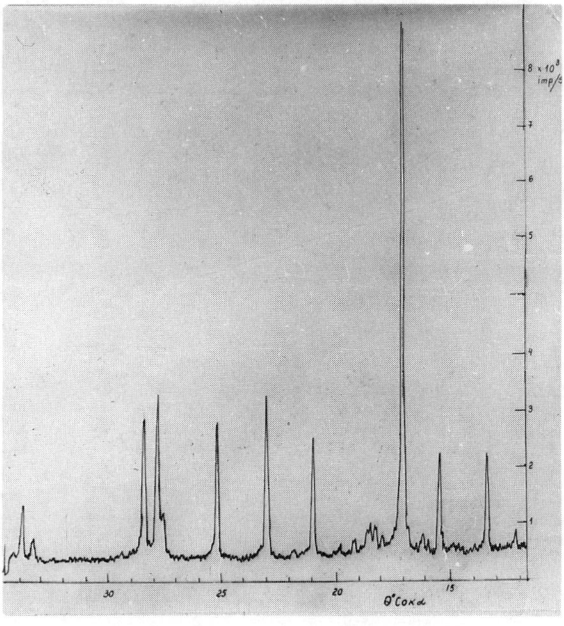

Fig. 6. Results of X-ray diffraction for sample 10.

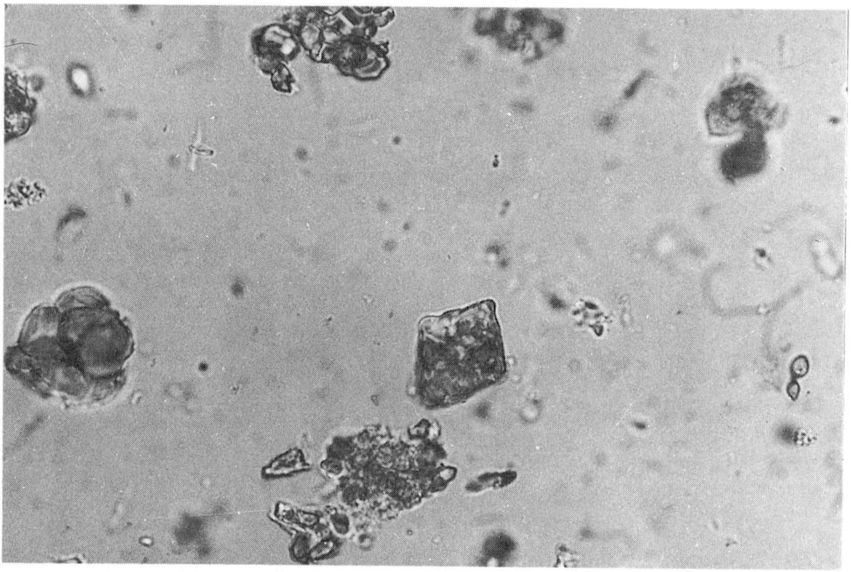

Fig. 7. Microphoto of sample 1 in polarized light, 1 Nicol.

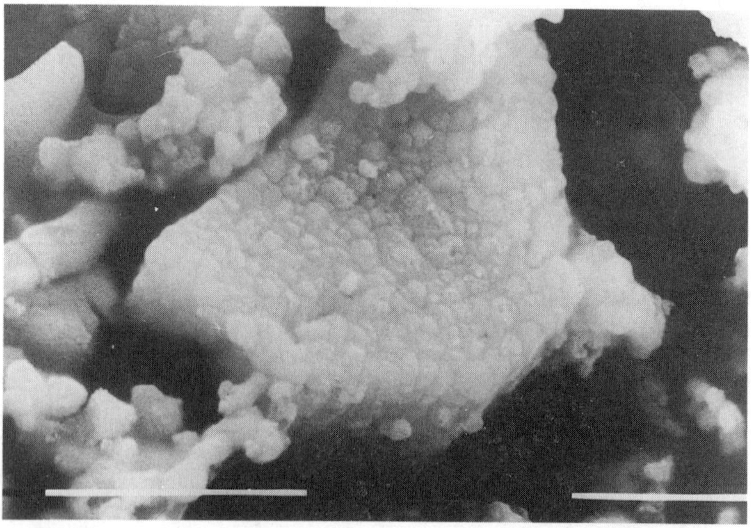

Fig. 8. Microstructure of sample 1, SEM at 4,000x.

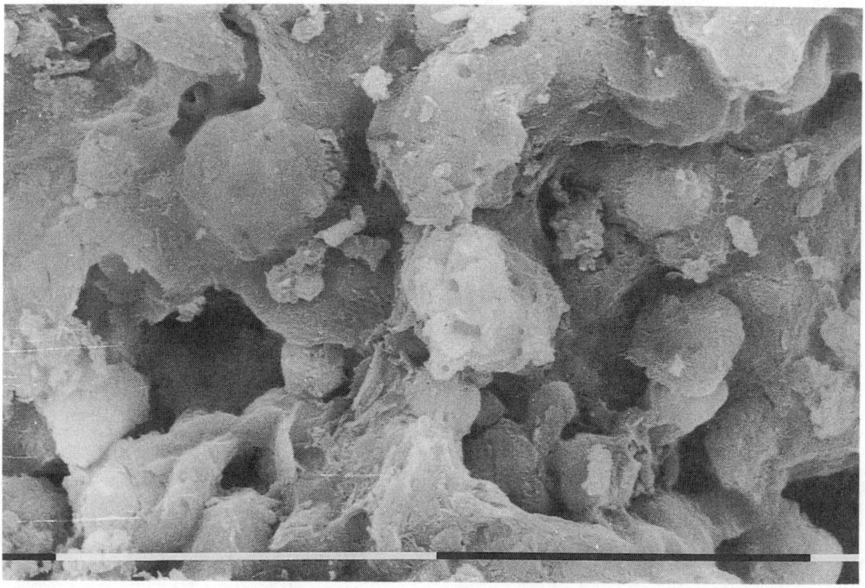

Fig. 9. Microstructure of sample 8, SEM at 5,000x.

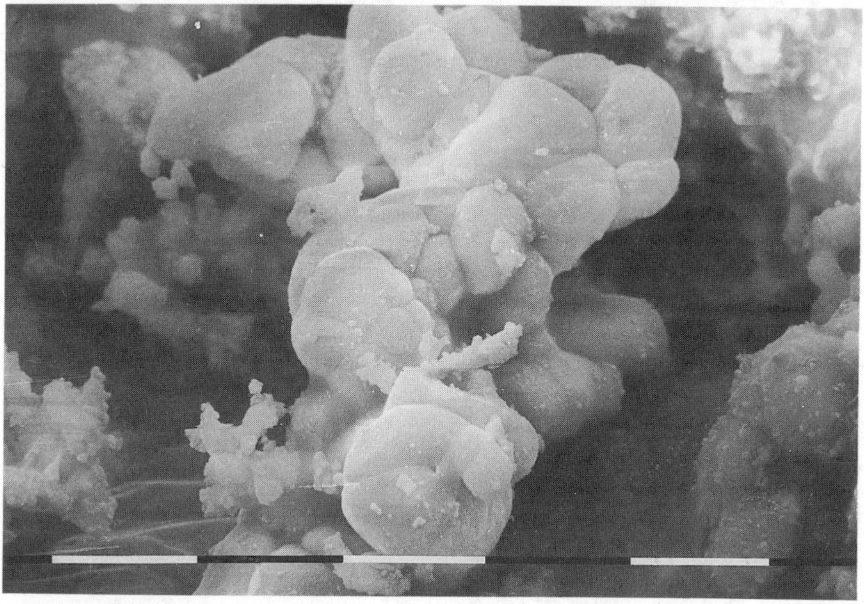

Fig. 10. Microstructure of sample 10, SEM at 2,000x.

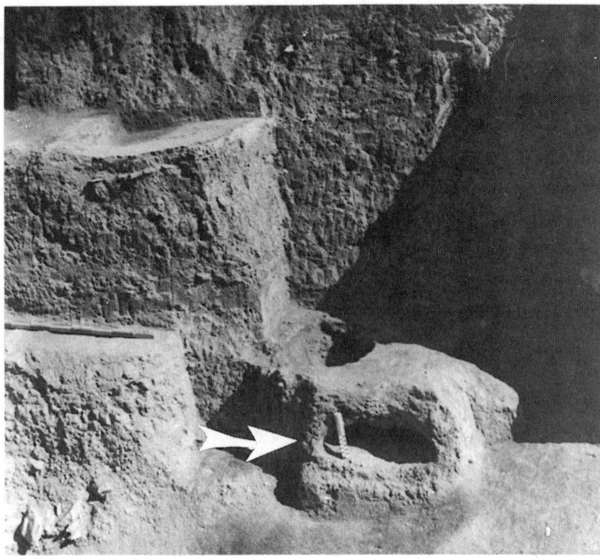

Fig. 11. Neolithic kiln presumably for lime preparation.

Table 2

Results of emission spectrography

Sample	Si	Mg	Al	Fe	Ca	Pb	Sn	Ga	Cr	Ni	V	Cu	Zn	Mn	Co	Ag
1ª	+	+++	+	+	+++	10	2	10 30	1 3	2 5	–	20	10 20	500 1000	2	–
2ª	++	+++	++	+	+++	5	tr.	10	1	2	–	10	–	500	1 2	–
3ª	+	+++	+	++	+++	10	–	10	1 3	5	–	20	20	1000	2	–
4ª	++++	+++	++	+++	++	10 20	2 5	30	20	10 20	50	20	20	200 500	2 10	–
5ª	+	+++	tr.	+	+++	10	–	5	1 3	2	–	10	10	2000	–	–
6ª	+	+++	tr.	+	+++	10	2	tr.	–	2	–	10	–	100 200	–	–
7ª	+	+++	tr.	+	+++	5 10	–	–	–	1 2	–	10	–	100 200	–	–
8ª	++	+++	+	+	++	10 20	–	10 20	–	–	2	10	50	200 500	–	50 100
9ª	+	tr.	–	tr.	+	–	–	–	–	–	–	tr.	–	tr.	–	–
10ª	++	++	tr.	++	+++	20 50	2 5	5 10	–	5 10	tr.	20 50	200	5000 1%	–	–
11ª	+++	+++	++	++	++++	tr.	–	tr.	–	2	–	10 20	–	100 200	–	–
12ª	++	+++	++	++	+++	10	2	10 20	1 3	5	–	10 20	10	500 1000	1 2	–

There are not discovered: As, Te, Sb, Be, Ge, W, Bi, Mo

Table 2. Results of emission spectroscopy.

$$CaO + H_2O = Ca(OH)_2 \qquad +15.5 \text{ kcal}$$

$$Ca(OH)_2 + CO_2 = CaCO_3 + H_2O \quad +17 \text{ kcal},$$

or in the presence of water:

$$Ca(OH)_2 + H_2CO_3 = CaCO_3 + 2\ H_2O$$

The content of the large vessel is slaked lime which was probably used for preparation of lime plaster. The white incrustation of the pottery is lime plaster. The lime plaster is composed of slaked lime and quartz. The clay minerals are impurities in the lime plaster.

The use of lime plaster has been known since the Pre-Pottery Neolithic period (9). The raw material which was used to make lime plaster at Neolithic Samovodene was different from archaeological sites.

The region of the Neolithic Tell of Samovodene is rich in limestone. It is neither accidental that a lime-pit is located at one end of the contemporary village of Samovodene, nor is the ancient use of limestone out of place in such a resource rich area. An interesting kiln was found at the northwest periphery of the tell (Fig. 11) which has neither evidence of being used for food nor pottery. Further investigation will show if the manufacture of slaked lime was carried out in such a kiln at the Neolithic Tell Samovodene at 5750 B.C.

REFERENCES

1. P. Stanev, "The Results and Perspectives of the Prehistorical Investigation in the Region of Veliko Tarnovo", Bulletin VI, Museum of History, Veliko Tarnovo, 1991, pp. 99-107.

2. M. Nica, "Nouvelles Donnees Le Neolithiques Ancien d'Oltene", DICIA 23 (1977) p. 21, fig. 14.

3. V. Nikilov, "Ornament of Early Neolithic Drawn Pottery: Systemization and Characterization", Archaeology 1-2 (Sofia, 1983), p. 29.

4. P. Detev, "The Excavation of the Settlement Yasa-Tepe in Plovdiv, 1959", ACAMP 6 (1960), p. 29.

5. R. Popov, "The Tell Deve-Bargan", AHM (1922-1925) p. 102.

6. St. Vaklinov, Archaeology I (Sofia, 1973), p. 58.

7. A. Pejnov, K. Lautlieva, "To the White Substance of the Neolithic and Eneolithic Pottery", Archaeology 3 (Sofia, 1971), pp. 58-66.

8. I would like to thank Prof. B. Alexiev, Department of Petrology and Lithology at Sofia University, for help during this investigation.

9. W.D. Kingery, P.B. Vandiver and T. Noy, "An 8500-Year-Old Sculpted Head from Jericho, Israel", MRS Bulletin 17, 1 (1992) Art and Technology, pp. 46-52.

METALWORKING AT HENGISTBURY HEAD, DORSET AND THE DUROTRIGAN COINAGE: A REINTERPRETATION OF AN IRON AGE AND ROMAN INDUSTRIAL SITE

CHRIS J. SALTER* AND J. PETER NORTHOVER*
*Department of Materials, University of Oxford, Parks Road, OXFORD, OX1 3PH, United Kingdom

ABSTRACT

The excavation of the metalworking areas of the late Iron Age and early Roman port of Hengistbury Head in southern England revealed evidence of a wide range of processes. These involved gold, silver and copper alloys and their connections with the local Iron Age coinage as well as the casting of bronze artefacts and ironworking. Since publication in 1987 re-analysis of some material coupled with an extensive analysis of the associated coinage has led to a re-interpretation of the material. This paper presents the new perspective on the site as a metallurgical centre that has emerged as a result of our analyses.

INTRODUCTION

Hengistbury Head is a promontory on the south coast of England to the west of the Isle of Wight. It protects a natural harbour formed by the estuaries of the Avon and Stour from the open sea. (A full description of the site's topography is given the excavation reports [1,2]. Iron Age and early Roman occupation was mainly along the north side of the headland. Salt was produced locally and the headland, formed of tertiary sands and clays, contains high grade iron ore in the form of ironstone nodules or 'doggers'. These were mined in the nineteenth century; the removal of banks of doggers from the base of the seaward cliffs led to the massive erosion and partial destruction of the headland. The head is cut off by a massive double rampart, possibly built in the 5th-3rd centuries BC.

Several areas of the headland have been excavated, principally by Bushe-Fox in 1911-12 [1], St. George-Gray in 1918-24 and Cunliffe in 1979-84 [2]. Unusually for its date Bushe-Fox's report contains an extensive metallurgical report, by Prof. W. Gowland, a remarkable pioneer of archaeometallurgy. The metallurgical material from subsequent excavations was studied by the present authors [4,5]. These reports between them suggested that the following processes could have been carried out in the later Iron Age and early Roman period, say 100BC-AD100:-

1) Extractive processes: possible smelting of imported copper ores or matte and extraction of silver from silver-bearing matte; possible smelting of local iron ore.

2) Hearth-based processes: cupellation of silver alloys; possible minting activities implying hot-striking of coins; iron smithing.

3) Crucible processes: possible refining of copper; making and casting of bronze alloys; melting of silver alloys, possibly in relation to the coinage.

THE COINAGE

Very many pre-Roman coins were found at Hengistbury, most from an area (site 33) postulated as a mint by Bushe-Fox. The majority comprised around 1700 cast bronze and 1475 struck base silver or copper coins of the Durotriges, the Celtic tribe whose territory was served by the harbour. There were about 50 gold and finer silver Durotrigan coins and a handful of gold and silver coins from other tribes, mainly in southern England. The most significant group is of 19 base silver coins from the Coriosolites, the tribe's trading partners across the sea in Brittany, northwest France [6-8].

The evolution of the Durotrigan coinage is of great importance to our interpretation of the metallurgy at Hengistbury so a brief review is in order. The first Celtic gold coins reached Britain from areas of northern France in the second half of the 2nd century BC. These were large heavy coins made initially from an unrefined natural gold with about 25% silver and only a small addition of copper. The gold coins from this area supplied the bulk of the gold circulated and later coined in southern England until the end of the Gallic War around 50BC [9-12]. By the late 50s BC the weight of the coins on both sides of the Channel had fallen by 25% and the gold content had dropped to about 50%. The debasement policy was consistent, the same near eutectic silver-copper alloy always being used, a policy which tended to minimise colour change. The first British-struck coinage to circulate in the Durotrigan area derived from this trend but then the alloy diverged. This may have been due to the use of bullion supplied to a different standard by another Gallic tribe, (British gold sources were inactive at this time), but any subsequent debasement seems to have been with silver rather than a silver-copper alloy and the coins become proportionately richer in silver. The Roman subjugation of France cut this supply as the Gallic gold coinage was suppressed. However, local silver and bronze issues continued to circulate in some areas, an important point for Hengistbury Head.

During the consequent period of restricted gold supply in the late 50s and 40s BC the Durotriges moved to a white gold alloy with around 15% Au, 55-75% Ag, so silver was clearly still, available. The coins were of good weight, around 6.0g, and some are still in excellent condition. This phase appears to have been short-lived: the gold content dropped sharply and then disappeared except as an impurity. Most other coin issuing tribes in southern Britain managed to re-establish a stable gold currency on a new alloy standard and develop bi- or even tri-metallic currencies. For some reason the Durotrigans could not or did not achieve this. There is good evidence that there was a strong barrier to trade with their wealthier neighbours to the east, the Atrebates, who could have supplied them with gold, and they may have been in political difficulties with the Romans [9]. Their northern neighbours, the Dobunni, with whom they did trade, probably did not have enough gold to spare. Their Coriosolite partners could only supply silver alloys, and debased ones at that [13-14] and any connections with free Europe would have been tenuous at best.

When the Durotriges lapsed from white gold to silver the sorry story of debasement and weight reduction was repeated: successive 'standards' were around 75-80% Ag/5.0-5.4g; 40-50% Ag/4.5-4.8g and 25-30% Ag/4.0-4.3g; fractional coins also exits. When the silver content fell below 70%, tin became an additional element in the alloy, suggesting a change in either policy or supply [12]. Eventually even the lowest standard collapsed and there are many coins with only 10% silver or less. This sequence is difficult to date but may have come to an end by the beginning of the 1st century AD. The end point was a struck copper or low tin bronze coinage with up to 1-2% tin, a silver impurity, and a weight of about 3.4g. Thereafter the struck coinage was abandoned. A cast bronze coinage was then produced with what is interpreted as a very basic version of the struck design. This too seems to have become 'debased' with concurrent reductions in weight and tin content. However, it is virtually impossible to develop a chronology for this coinage so it is not at all certain that the variation of weight and tin content was in fact time dependent. The bronze is leaded and has a zinc impurity; it seems characteristic of British metallurgy in the years around the Roman invasion [12]. Production may have continued after the occupation for circulation in the immediate Hengistbury area [7-8].

The material from the Hengistbury excavations is now considered alloy type by alloy type.

GOLD ALLOYS

Three gold objects have been found at Hengistbury that definitely date to our period of interest: a bracelet and two torc fragments. The bracelet imitates twisted and plaited wire torcs from other finds in Britain, notably Ulceby and Snettisham [15-17] and contains 78.4% Au, 18.9% Ag, 2.6% Cu, matching the earliest imported Gallic gold coins and not far from the earliest coins to circulate in the Durotrigan area [10-11]. The fragments of a button-terminal torc is parallelled by mercury-gilded examples at Snettisham [12] but contains 44.3% Au, 46.3% Ag, 9.0% Cu, a composition which is repeated in gold objects from Snettisham and in coinage from the Durotrigan area. The third object is a white gold ring torc terminal, now much corroded; analysis gave 28.3% Au, 70.1% Ag, 0.95% Cu; a metallographic examination suggests that corrosion has resulted in an enrichment of the gold content and a depletion of copper. This white gold is similar to that in the Durotrigan white gold coinage but a comparison is hindered by the corroded state of the torc terminal.

The button terminal torc and the bracelet, together with a small gold strip, were found crumpled together and must be seen as a small hoard of scrap; the other torc terminal was found nearby. They represent a microcosm of the Snettisham treasure, especially the scrap component of it, and were probably, although not necessarily deposited at the same time in the mid-1st century BC. Analysis has demonstrated close links between the metal of the torcs and of contemporary Celtic coins. It is highly probable that the coins were used as the bullion supply for the torc workshop and at least possible that the reverse process took place at Hengistbury Head with scrap torcs being re-cycled into the Durotrigan coinage, perhaps at a time when other gold supplies were drying up. Although a touchstone [1] suggests at least the possibility that gold alloys were being prepared and tested on the site. However there is no evidence of the striking of Durotrigan gold or white gold coins on the site: numismatists suggest that the mint was some kilometres inland at Badbury Rings, a massive fortified settlement [8].

SILVER ALLOYS

In contrast there was literally massive evidence for the working of silver alloys at Hengistbury. Bushe-Fox's Site 33 yielded a block of silver alloy weighing 8.8kg. Gowland's analysis gave 46.43% Ag, 1.2% Au with no tin; unfortunately the piece has now disappeared. Site 33 also produced other debris from the melting of silver alloys. Microprobe analysis of part of a corroded fragment from the recent excavations gave 52.02% Ag, 1.29% Au, 5.02% Sn and 0.66% Sb, these values enhanced by corrosion; the balance consists of copper oxide corrosion products. The fragment was unfortunately unstratified but the state of corrosion is compatible with very long term burial and the assumption must be that it was of either Iron Age or Roman date. A metallographic analysis of the fragment suggests that the overall silver content probably fell within the 40-50% 'standard' band as does the large block.

Key points in assessing the relationship of this material to the production of Durotrigan coinage are the tin and antimony contents. Early silver coins from most British tribal areas show a variety of alloys some with tin, some without: as many of the coin types are newly identified few conclusions can be drawn as yet. Analysed post-Gallic War Celtic silver coins from Britain fall into two main categories [12]. Those tribes in the south and east of England which had the closest relationship with the Romans produced silver 'units' of good silver, normally better than 95%, and fractions which were baser, generally with 80-90% silver, but sometimes less. These compositions parallel Roman denarii and quinarii and it is assumed that Roman silver was being re-cycled. Besides being of good silver these coins generally do not contain tin or antimony. Around the periphery of this zone the silver coins of the Iceni, the Dobunni and the Durotriges are generally debased although only with the Durotriges did this reach an extreme. With all three tribes the debased silver contains antimony and tin. Except for some specific coin types these elements remain at impurity levels for the

Iceni and Dobunni but in the Durotrigan silver when it falls below 70% Ag the tin must be regarded as originally an alloying addition at 1-5% and the presence of the antimony impurity is much more consistent and at a higher level.

Cunliffe ([2], p342-3) suggests that the Durotriges obtained silver from silver-rich lead ores in the Mendips to their northwest; however there is no independent evidence that the such ores were being exploited during the 1st century BC. It is, of course, also possible that Roman silver was recycled into the Durotrigan coinage and debased with a bronze with an antimony impurity but another proposition is more attractive, that they were re-cycling the coinage of their nearest Continental partners the Coriosolites. Their coinage has been extensively studied [13-14] and shown to be made of a series of base silver (billon) alloys containing added tin. Analysis of their impurity patterns [12] shows close agreement with the Durotrigan coinage below 70% Ag, and with the analysed waste fragment. However, although a connection with the coinage seems certain, we cannot say whether this single fragment relates to the manufacture or destruction of that coinage.

On the other hand, the Bushe-Fox block contains no tin or antimony and so cannot be linked with later stages of the Durotrigan silver coinage. It is also base and would require extensive refining to bring it to the standard of the earlier coinage and is probably not connected with it. It may however relate to the other evidence for silver refining at Hengistbury Head described in later parts of this paper. As we shall see, there are unlikely to be contemporary with the Durotrigan silver coinage but rather date to the period of the Roman conquest and the decades following.

Further direct evidence for the working of silver comes from the remains of cupellation hearths excavated by Bushe-Fox and analysed by Gowland. They consist of complete circles of segments formed by scraping the material from the surface of the melt in the hearth to its circumference [3]. Gowland found lead and copper oxides, small amounts of silver (3.15%, 7.86%) and traces of gold together with bone-ash and slaggy material. One piece survives and is currently being analysed. It appears to be atypical compared with the samples analysed by Gowland in having a very low bone-ash content. Its microstructure is hard to interpret as we are not sure of the correct orientation of the fragment we are studying. Metal prills are concentrated towards what appears to be the lower surface: these are formed of silver-copper alloys from the eutectic down to nearly pure copper. Some of the prills are highly corroded. The material between the prills is mainly the stoichiometric mixed copper-lead oxide Cu2O.PbO [18]. Towards the top of the section the density of metal inclusions is much lower and there is some separation of lead and copper oxides interspersed with what appears to be sintered bone-ash.

Although these hearths were reported as coming from areas that included pre-Roman metallurgical activity, it is more probable that they derive from the Early Roman period (i.e. after AD43-44). It has yet to be demonstrated that this process for silver-refining was known to the Celtic gold- and silversmiths and moneyers in Britain. As far as we can see their work was dependent on imported bullion, adjusting compositions with whatever material was to hand. The earliest clear evidence for cupels comes from the mid-1st century AD, i.e. the years of the Roman conquest. The direct parallels for the Hengistbury cupellation hearths come from Roman urban sites. One of these, Wroxeter, was established on a greenfield site in the second half of the 1st century AD and so must be purely Roman. What was being refined in this way and why is another question. Roman silver, either coin or plate, in the 1st century AD was of a high fineness, and would not need extensive refining for re-use. The Bushe-Fox block is the sort of material that should probably be associated with the hearths and, as suggested above, that too perhaps ought to be given a Roman date. Other material being recycled might have been base silver Celtic metalwork and coins, such as the Site 33 hoard; however these contain tin which did not show up in either the Bushe-Fox block or the hearth residues.

BRONZE CASTING

Bronze droplets, oxidised bronze and numerous crucible fragments provide abundant evidence for bronze casting. Most seem to date to the period 50BC-AD50 although there may be some extension on either side of this time-span. There is also one bowl crucible of early Roman type, probably from the 1918-24 excavations but not included in any subsequent publication, that was used for melting brass. The majority of identifiable crucible fragments are from the triangular type common from the 3rd to the 1st century BC in southern Britain and continuing in use in more conservative areas until the 1st century AD. There is not yet sufficient dated material to say to which group the Durotriges belonged; probably they were a little more conservative than the rest of southern Britain. These crucibles were used in a most inefficient way with a strong blast directed across the rim causing excessive oxidation of the bronze and severe slagging or even collapse of the crucible wall. This intense oxidation can make matching impurity patterns between waste and artefacts very difficult but at Hengistbury Head sufficient bronze survives as waste or attached to crucibles to allow some conclusions to be drawn.

Work on other sites in southern and western Britain [19] has shown that the main bronze type in use on metalworking sites in the region from the 5th until the mid-1st century BC is characterised by arsenic, cobalt and nickel as impurities with Co> Ni; antimony is at trace levels only. This metal is absent from the metalworking waste at Hengistbury and there is only one potential example among the artefacts. Although not conclusive this supports the contention that bronzeworking is concentrated in the period after 50BC. Within this period the waste could relate either to artefacts or to the struck copper or cast bronze coinages. The answer is simply found: the cast bronze coins are alloyed with lead, mainly in the range 5-15%, and have a zinc impurity up to 1%, while all the waste analysed is free of zinc and generally has much lower lead than in the coins. Thus the crucibles and droplets relate to the casting of bronze artefacts; the types produced are not known as there are no moulds or large mis-casts to guide us. The impurity patterns of the waste, mainly with As/Sb/Ag, have many parallels in the 1st century BC/1st century AD, not least among the artefacts at Hengistbury Head itself.

Some of the crucible residues contain copper-tin intermetallics and high densities of cassiterite crystals. Although the melting of bronze under oxidising conditions can produce cassiterite in some quantity and leave prills sufficiently enriched in tin to form the ϵ phase, some of the Hengistbury examples seem to be different and hint at the melting of tin, perhaps to make specific bronze alloys; tinning was also a technique in vogue at this time. Contrary to our earlier published remarks there is no evidence for the making of bronze by the crucible reduction of tin in contact with copper.

The intense concentration of the cast bronze coins on the headland, and in particular at Site 33, suggests that their manufacture must have taken place close by. The overall distributional evidence for the coinage underlines this connection with Hengistbury Head with all but four of the 1700 or so known coins being found within 8km of the site. It is of course possible that the large accumulation around Site 33 represents metal collected for recycling by the Romans after the conquest, as we have suggested might be the case with the silver. As there are no residues associated with the melting of the coinage alloy we cannot come to any conclusion. It should be pointed out, however, that there is numismatic evidence for the continued, if very localised circulation of these coins well into the Roman period, possibly into the 2nd century AD.

OTHER PROCESSES INVOLVING COPPER

The coins that were previously regarded as struck bronze have now been identified as very base silver or plain copper, sometimes with 1-2% tin; struck copper coins were also issued elsewhere but there seems no connection between the two groups. Too few coins have actually been analysed to divide the 1400+ relevant coins at Hengistbury Head between base silver and struck copper but the production of the latter certainly seems to have been large requiring the processing of a significant amount of copper.

Again the presence of such a large hoard at Hengistbury may indicate no more than the accumulation of metal for recycling and refining, but there is a variety of copper processing waste to be described of which at least part may relate to the coinage:-

1) Bushe-Fox and Gowland [3] record the discovery of ingot copper; Gowland's analyses show a significant sulphur content hinting at the use of sulphide ores. So far in southern Britain ingot copper has only been recorded in Late Bronze Age and Roman contexts, with none of confirmed pre-Roman Iron Age date. Late Bronze Age ingots are almost all characterised by high sulphur contents (up to 2%) but Roman ingots are more varied, there being examples with oxide rather than sulphide inclusions [20]. Given the small amount of Late Bronze Age material at Hengistbury Head [2] the balance of probabilities must be that the ingot copper there is Roman in date.

2) A hearth bottom recovered during the 1979-84 excavations had clearly been used for the melting of a low tin bronze with significant arsenic, antimony and silver contents, and a copper containing just arsenic and silver (sulphide inclusions demonstrate locally reducing conditions so that these impurity patterns are representative). These impurities, as well as the low tin contents, characterise the struck copper coinage.

3) Two cakes of matte; both would have been over 60mm in diameter when complete and at least 15mm thick. These consist of a matrix of solidified copper sulphide with silicate slag inclusions indicating that it is truly a smelting product and not an artificially created item such as niello. The matte also contains a few percent of lead as small droplets, which often contain particles of silver. Gravity segregation has resulted in the concentration of these metals towards the base of the matte cake. The nearest copper ore deposits with an appreciable silver content are in Devon, 150km to the west of Hengistbury, with poorer sources across the sea in Brittany. If the matter were further smelted a silver-containing copper would result which could match some of the struck coins. On the other hand the matte could represent waste from a process to extract silver from a silver-rich copper ore by allowing it to partition to lead, which could subsequently be separated from the matte by gravity segregation. The matte must either have been produced on the site from imported ore, or brought to the site in its present form for further processing. Unfortunately, the matte cannot be closely dated here but matte, albeit of other types, has been found on very late Iron Age sites elsewhere in southern England [12].

4) Further support for the suggestion that copper smelting was carried out at Hengistbury comes from 'iron' slags. Although most of the material examined was probably produced by iron smithing activity, a number of fragments contained copper and copper sulphide inclusions. It is not unusual for some iron-working slags of this period to contain copper when copper and iron have been worked in the same hearth. However, the presence here of copper sulphide inclusions is extremely rare (the only case seen during the examination of at least 500 ancient British 'iron-working slags'), as is the deliberate addition of quartz and chert crushed to a 3-5mm particle size [5]. Therefore, although the quantity of material is small, it does point to the active processing of semi-smelted copper matte and, possibly, the refining of very impure crude copper being carried out at Hengistbury Head. More importantly, it is clear that slags containing traces of copper and/or crushed silica cannot be instantly dismissed as iron or mixed ferrous/non-ferrous forging slags, even when there is no obvious nearby source of copper ores.

IRON

Given the small amount of slag recovered (9kg), and that some of this has been identified as relating to copper, it remains for us to consider what evidence there was for the use of the high grade iron ore available locally. Traces of iron smelting on top of the headland which have been reported in association with the iron ore doggers have largely been destroyed by quarrying and erosion and cannot be dated. Evidence of iron smelting apart from that is extremely limited as is the quantity of iron smithing debris, and both are commensurate with local use only. One further intriguing item is a piece of white cast iron recovered by Bushe-Fox. At present it seems clear enough that Iron Age and Romano-British smiths could not decarburise this accidentally produced iron and so it was discarded as unforgeable. Pieces have been found on other Romano-British sites [5].

CONCLUSIONS

The re-examination of debris from Hengistbury Head is still going on but the picture of metallurgical activity is already clearer with the various processes concentrated in a small number of functional groups:-

1) The evidence of scrap Snettisham-style gold alloy objects and a touchstone suggests the recycling of gold alloys: the most probable context is some association with the production of the Durotrigan gold/white gold coinage but this is far from proven.

2) The casting of bronze artefacts, most probably in the last half of the 1st century AD but potentially up to the Roman conquest. There is undated evidence for the melting of brass.

3) Some processing of copper, perhaps in relation to the Durotrigan coinage.

4) A complex series of operations involving the refining and, possibly, even the extraction of silver. This could be connected with the matte, the Bushe-Fox silver alloy block and the cupellation residues. A very late Iron Age or, more probably, a Roman date should be assigned.

5) Smelting and smithing of iron mainly for local use.

We have then a picture of Hengistbury Head as an industrial zone where imported materials (copper, matte, silver and gold alloys) are processed and local resources (iron and salt) exploited. We also know that glass was worked and shale carved [2]. In attempting to move beyond a simple identification of processes to an understanding of their mechanisms we have discovered that there has been remarkably little detailed analysis of both mattes and cupellation residues in an archaeological context. Our analysis of these materials is therefore being pursued further. We are also looking for other examples for the long-distance transport of ores and semi-smelted products in the ancient world.

REFERENCES

[1] J.P. Bushe-Fox, Excavations at Hengistbury Head, Hampshire, in 1911-12, Soc. of Antiquaries Res. Rep., 3, 1915.
[2] B.W. Cunliffe, Hengistbury Head, Dorset, Vol. 1, Oxford Univ. Committee for Archaeology Monograph, 13, 1987.
[3] W. Gowland, Report on the metals and metallurgical remains from the excavations at Hengistbury Head, in [1], 72-83.
[4] J.P. Northover, Non-ferrous metallurgy in [2] 186-196.
[5] C.J. Salter, Ferrous metallurgy and other slags, in [2] 197-205.
[6] L. Sellwood, The non-Durotrigan coins, in [2], 138-140.
[7] M. Mays, The Durotrigan coins, in [2], 140-141.
[8] M. Mays, The coinage of the Durotriges, unpublished DPhil thesis, University of Oxford, 1985.

[9] R. van Arsdell, Celtic coinage in Britain, Spink, 1989.
[10] M.R. Cowell, W.A. Oddy and A.M. Burnett, Brit. Num. Journ., 57, 1988, 1-23.
[11] M.R. Cowell, An analytical survey of the British Celtic gold coinage, in M. Mays, ed., Celtic coinage: Britain and beyond, BAR British Series, 222, 207-234.
[12] J.P. Northover, Materials issues in the Celtic coinage, in M. Mays, ed., Celtic coinage: Britain and beyond, BAR British Series, 222, 207-234.
[13] K. Gruel, Le Trésor de Trebry, Études Numismatiques Celtiques, 1.
[14] K. Gruel and A. Taccoen, Petit numéraire de billon émis durant et après la conquête romaine dans l'Ouest de la Gaule, in M. Mays, ed., Celtic coinage: Britain and beyond, BAR British Series, 222, 207-234.
[15] L. Sellwood, Objects of gold and silver, in [2], 156-7.
[16] R.R. Clarke, The Early Iron Age treasure from Snettisham, Norfolk, Proc. Prehist. Soc. 20, 1954, 27-86.
[17] I.M. Stead, The Snettisham treasure, excavations in 1990, Antiquity, 65, 1991, 447-65.
[18] E.M. Levin, C.R. Robbins and H.T. McMurdo, Phase diagrams for ceramists, American Ceramic Society Inc., 1964, 86, Fig. 163.
[19] J.P. Northover, Non-ferrous metalwork and metallurgy, in N.M. Sharples, Maiden Castle, 1985-6, English Heritage, Arch. Rep., 19, 156-161.
[20] J.P. Northover, unpublished data.

PART III B

Pattern of Production, Raw Material Constraints and Style of Technology

A Technological Marker of the Penetration into North America : Pressure Microblade Debitage, Its Origin in the Paleolithic of North Asia and its Diffusion

M-L. INIZAN, M. LECHEVALLIER and P. PLUMET

Until recently[1] it was admitted that the pressure debitage technique had been invented at the end of the Upper Paleolithic, and its presence in different geographical zones was difficult to explain.

Now that we are able to recognize it thanks to advances in experimentation, its invention seems to have been made some 20,000 years ago in the vast area where the Mongoloid people, who were to occupy the Far East, the Beringian zone and America, first appeared.

We are going to describe the methodological approach which allowed us to imply pressure debitage as a cultural marker in the history of prehistoric penetration into North America and to show the specificity of this technique in the Paleolithic of North Asia. Our proposal is a development of an unpublished paper presented at the Novosibirsk Symposium in 1990[2]. The opportunity we had then to examine lithic industries from Siberia convinced us of the early use of the pressure technique.

1. Technology, a conceptual approach of material culture

This approach is based upon the study of technical activities, as proposed by A. Leroi-Gourhan who wrote the word **Technology** with a capital T, to avoid any confusion with **technology** (know-how) (Leroi-Gourhan, 1943). Technology can be considered as a **system** in which techniques and socio-economic phenomena constitute the material basis of societies. Thus techno-economy, socio-economy and technology (know-how) cannot be dissociated. If system theory is recent (the fifties), in fact, this concept was adopted by Leroi-Gourhan a long time before its formalization.

Technology aims at defining prehistoric societies by the analysis of the relationship between the technical system and the socio-economic field. On the other hand, typology, which has been for long the principal approach to study lithics, is used to characterize cultures within a chronological frame. Thus Technology and typology are different and their fields do not overlap.

In the technological approach all techniques are considered an integral part of the **technical system.** In stone-knapping viewed as a sub-system, the various operations involved should be taken into account as well as their interdependance. For this purpose the concept of "**chaîne opératoire**" [3] which is also one of the fundamental aspects of Leroi-Gourhan's work, is used : it is a tool with multiple applications, quite versatile, with which it is possible to analyze technical processes in all their detail (Balfet, éd., 1991).

The study of a *chaîne opératoire* in lithics consists in reconstructing all the processes having affected the raw material from its search, its transformation until all the elements

[1] cf. infra Historique 2.2

[2] International Symposium on 'Chonostratigraphy of Paleolithic of North, Central, Eat Asia and America (the paleoecological aspect)' held at Novosibirsk, U.S.S.R., july 23-August 11, 1990

[3] This French expression seems to be increasingly used by English-speaking technologists, as shown by the volume 9/1 of the Cambridge Archaeological Review which deals with *Technology in the Humanities*. It will therefore be left untranslated throughout this work.

which have been extracted from it, whether tools, weapons or waste, are discarded. This *chaîne opératoire* can be segmented, according to the specific activities with which one is concerned, into various fields such as : raw material procurement, debitage, tool making, etc.., keeping in mind the fact that lithics imply precise intentions concerning their use in different activities.

In this way we propose to study a blade debitage technique, the pressure debitage, through a combination of identifiable elements such as raw material, gestures, tools and know-how. First we should measure the part of determinism linked to these four components and the universal constraints of physical (raw material) or biological order (psycho-motricity, training...) to be able to turn this technical fact into an attribute of a culture and eventually to follow its diffusion.

2. Pressure debitage : a technique and different methods of blade debitage

If blade debitage by percussion is present sporadically in the Lower Paleolithic in some Acheulean industries of Europe, of the Near East, etc..., it is used by many cultures of the Upper Paleolithic. The production of bladelets and microblades which concerns microlithic industries is also a characteristic of the Upper Paleolithic. It must be stressed that pressure debitage applies only to blade debitage (blades, bladelets or microblades).

2.1 Definition of the blade debitage

Blade debitage, whatever the size of the products (blade, bladelet, microblade) and the technique used for knapping, is a preconceived procedure organized for the production of a large number of blanks on one core. Standardized by the presence of arrises, blade products are flakes which are at least twice longer than wide, according to a widely adopted convention. Blades are blanks over 5 cm long and up to 50 cm. Bladelets or microblades are under 5 cm long, eventually reduced to 1 cm with a 3 or 4 mm width.

To produce blades at will can rarely be done with an unmodified piece of raw material. A preparation of the knapping surfaces by creating guiding ridges and of the striking or pressure platform(s) is necessary. Thus blade debitage corresponds to a specific idea of the shape of the tool blanks.

Any debitage can be done not only by different **techniques** : direct percussion (with a hammer in stone, wood, horn, bone or metal), indirect percussion or pressure, but also by using different **methods**. The method is conceptual and rational. A knapping method is a succession of acts in which several techniques may be used to perform a design.

Thus traditions may be defined by identifying the technique(s) on the one hand, the debitage methods on the other. These methods are determined by the use of distinct knapping schemes with different levels of know-how, transmitted through training.

A problem of dimensions

The knapping of long blanks requires appropriate raw material (in quality and size). The longer the products, the more difficult it is to achieve. On the contrary, small-sized blade products require a less elaborate technical skill, and it is easier to find a good quality raw material in small-sized nodules or tabular flint. On the other hand, the presence of microblades, retouched or not, imply their use as inserts which refer to another technical sub-system, quite complex with such problems as shaft making and the use of glue.

2. 2 History of pressure debitage

D.E. Crabtree in 1964 interpreted ethnographic documents on obsidian knapping by the Aztec and was the first to prove the existence of pressure debitage and to replicate it (Crabtree, 1968). This technique remained for a long time considered as a particular technique, specific of the recent cultures of Mesoamerica. But J. Tixier, taught by Crabtree, later identified it in the Late Capsian of North Africa ; he then described the knapping scheme and the characteristic features of the technique (Tixier, 1976)[4].

The first references to pressure debitage in the last two decades have been sporadic and isolated : in Afghanistan (Davis and Dupree, 1972), in Japan (Tozawa, 1974). It has been recently recognized in Mesopotamia (Inizan and Tixier, 1983 ; Inizan, 1985), in the Zagros (Dittemore, 1983), in northern Iraq (Kozlowski, 1989), in eastern Anatolia (Calley, 1985). These examples from the Near and Middle East all concerned first farming communities of the 8th-7th millennia, the prehistoric industries of the Levant (Syria, Palestine) and Arabia not being concerned. It has also been identified in Pakistan in the Neolithic of Mehrgarh (Inizan and Lechevallier, 1985), and in India at Bagor and Navdatoli (Inizan and Lechevallier, 1990). It has been observed in Egypt at the Archaic Period (Midant-Reynes, 1983) and in neolithic industries in western and central Europe (Perlès, 1984 ; Binder, 1984 ; Kozlowski, 1987). Recently, J.G. Flenniken identified its use in paleolithic industries of the Dyuktai culture in Siberia (Flenniken, 1987).

The pressure debitage technique is not often mentioned in North Asia and in North America. However the originality of what we call the "knapping schemes" in the debitage of small-sized bladelets in paleolithic industries (Gobi cores, wedge-shaped cores, Yubetsu technique, etc.) has been noticed and studied by many prehistorians for some thirty years now in Siberia, China, Tibet, Mongolia, Korea and Japan. What we need to study now is the relationship between technique, raw material and knapping schemes.

2.3 Technique of pressure blade debitage

It is no longer a stroke (percussion) that produces the fracture ripple but a pressure transmitted by a pointed tool in wood, horn or bone and later in metal. The strong pressure made with this point on the edge of a preformed core makes the controlled detachment of blades and bladelets possible (pressure retouching in many bifacial industries uses the same technique but there is no obvious link so far with the debitage).

Criteria of identification of this technique

Due to the advances in experimentation, mainly by American and French knappers [5], it is now possible to analyze the debitage products and to differentiate pressure debitage from the punch or indirect percussion which can be very close on certain points :

- On cores :

arrises are quite rectilinear and parallel (fig. 1); they differ from those produced on cores knapped by indirect percussion.

The morphology of a core can vary : it depends mainly of the technical scheme wanted by the knapper. The various methods adopted will be analyzed through the observation of the way the core was shaped out. Some of these knapping schemes have been quite well analyzed (Yubetsu, Araya, Dyuktai, etc., cf. Morlan, 1967,1970 ; Hayashi, 1968 ; Smith, 1974 ; Kobayashi, 1970 ; West, 1982 ; Anderson, 1970, etc.).

The pressure platform (as well as the striking platforms) can be smooth or facetted.

- On full debitage blades and bladelets (fig.1),

edges and arrises tend to be rectilinear;
the mesial part is thin;
the lower face has no marked ripples;
the butt is always narrower than wide
the blade reaches its maximum width close to the butt .

Butts are good indicators of pressure platforms and of the technical procedure (abrasion or non abrasion of overhangs). They are very small, even punctiform.

---

[4]In 1982 J. Tixier organized a meeting on blade debitage, Half of it was dedicated to pressure debitage (*Economie du débitage laminaire, 1984*)

[5]American experiments have been mainly dedicated to the replication of the obsidian blades produced by the Aztec knappers (Clarke, 1982). In France, the knappers of a Technology Department of the CNRS (ERA 28), using equipments compatible with prehistoric cultures, are guided in their experimental tests on different raw materials by the study of archaeological assemblages

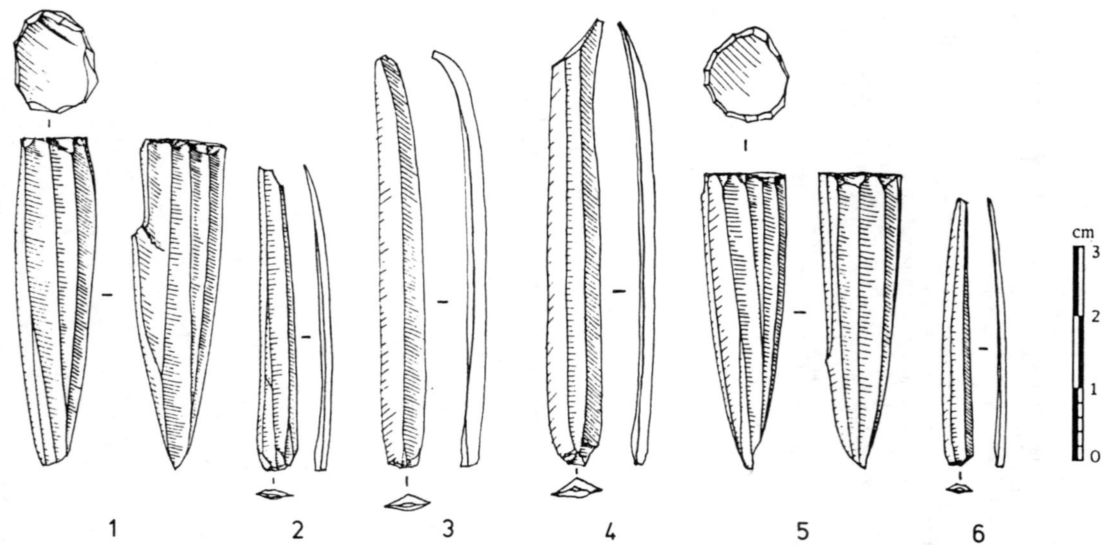

Fig.1 Core and bladelets (Merhgarh, Pakistan) : 1, 2, 3, 4; experimental core and bladelet by J. Pelegrin : 5, 6.

The directions of the arrises are sound indicators of the shape and size of cores from which the blades proceeded.

The reconnaissance of pressure is well-founded when it is based upon large assemblages. However a single core, such as a "bullet-core", of very small dimensions, showing the criteria described above, will be a safe indicator of the use of pressure, by exclusion of percussion, and impossible to use on very small-sized products. This is due to the constraints linked to raw material inertia.

Pressure (debitage and retouch) and heat treatment

Flint heating, invented some 18,000 years ago by the Solutreans to improve pressure retouching (Bordes, 1969), is a technique which has been commonly used for the bifacial retouching of many types of points in neolithic cultures. It has also been observed on pressure made blanks. Though it does not seem to have been used systematically, its presence shows however the complexity of the technical procedure involved (Binder and Gassin, 1989 ; Inizan and Lechevallier, 1985 ; Flenniken, 1987). It has been possible through experimentation to identify this technique and to appreciate its advantages, as well as the technical knowledge involved (pyrotechnics).

Advantages of this technique

Pressure debitage offers the possibility to spare raw material because cores can be worked further than with other techniques, as demonstrated by the presence of minute cores in archaeological assemblages. In a same volume of raw material, up to four or five times more blanks can be obtained by pressure, compared to percussion[6]. Its invention may have played a decisive part in the use of obsidian which is a material specially well-adapted for that technique.

The precise control of the debitage process helps to produce standardized blanks ; the development of geometric microliths (which require this type of regular blades for their production) is often related to the pressure technique. However there are also many industries with geometric microliths made on percussion-produced blanks.

Limitations of this technique

Experimentation has shown the constraints of the technique concerning raw material, specific equipment, technical skill and psychomotor and cognitive processes (training).

Pressure debitage requires a homogeneous raw material. Finer the grain is, easier the knapping will be (as is the case with obsidian). A sophisticated preparation of the core is needed to knapp efficiently without wasting too much raw material.

Experimentation has proved the need of a specific technical equipment, adapted to the various dimensions of the blanks wanted. The physical strength of the knapper limits the blanks' size. According to J. Pelegrin (1988), bladelets under 5-6 cm long can be obtained with the core held or fixed in the hand (fig.2 : a). In this case pressure is applied with a short point in the other hand or with a longer point whose proximal end is maintained by the shoulder or under the arm. A breast or ventral clutch will be used for longer blades, up to 15-18 cm (fig.2 : b). For larger blanks showing the characteristic features of pressure, the use of a lever is probably involved. The replication of these large blades, such as those found at Varna, Bulgaria, or in the Uruk culture (28 cm long blades), has not yet been realized.

Taken all together these requirements imply a direct transmission of the knowledge and a long training.

Microlithism and pressure

The term of "microlithism" seems to have different meanings according to the authors. We object to the notion of microlith being restricted to geometric microliths as it depends upon a narrow conception of typology in which microliths equal culture. Microliths whether geometric or not are tools or rather tool elements of small size. They represent often only a part (sometimes very small) of the tool kit. In fact all microlithic industries

---

[6]According to the experimental data given to us by J. Pelegrin

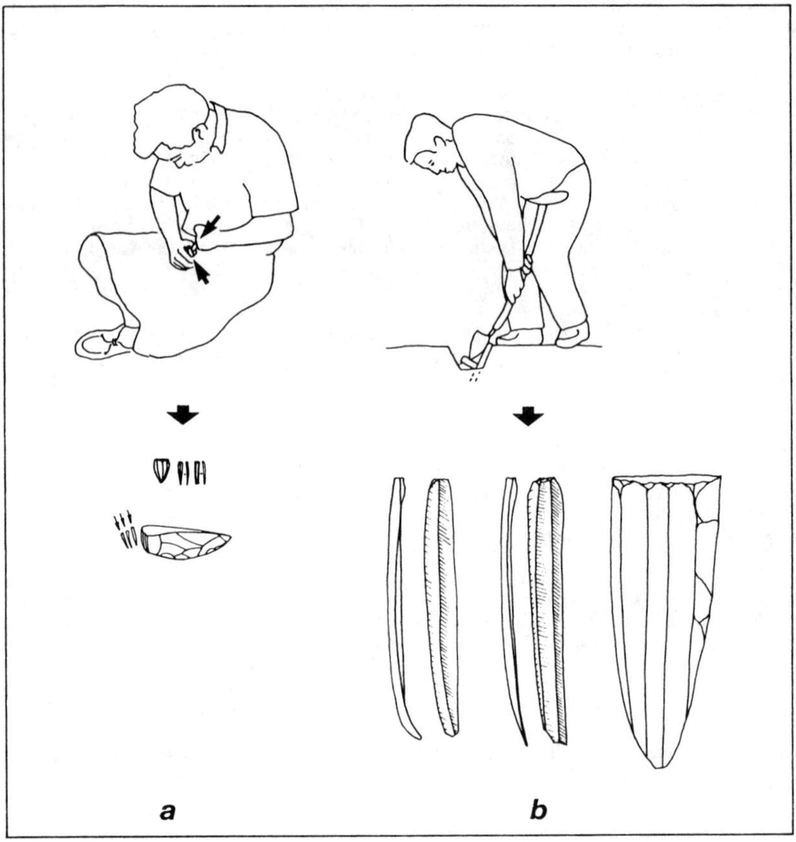

Fig 2 Equipment and positions used by J. Pelegrin in microblade (a) and large blade (b) production

have also tools of large dimensions, often made with other kinds of raw material. For example, the preparation of skins requires long and efficient cutting edges with the possibility to resharpen them, something that cannot be realized with microliths.

We agree (Keally, 1990) that the expression "microblade industry" corresponds to a technical diagnostic -most probably pressure with specific preparation of cores- and not to dimensions of the products which can be very small bladelets as well as large blades.

Microlithism can also be linked to a problem of raw material procurement or size. It is also possible to make microblades by percussion but when the core becomes too small, the knapper will not be able to go on with the pressure technique.

2.4 Debitage methods

If, as we have just seen, it is possible to identify the pressure technique, it must also be said that there are different knapping methods, with variants, as described by many authors (often without refering to pressure debitage). For us these variants may be grouped into two methods, the first of which concerns the vast geographic region of north-eastern Asia and North America for microblade debitage only. These methods emphasize two different options in the core volume treatment and lead to configurations specific to blade products :

The Yubetsu method (Shirataki, etc...) (fig. 3)

We propose the term of "Yubetsu", without refering to its regional meaning, simply because it is in Japan that this method has been described for the first time as a pressure debitage method (Tozawa, 1974). The main steps of the sequence are as follows :

- the raw material is shaped by bifacial flaking by percussion, to obtain a more or less regular biface, often asymetrical ;
- one of the ridges of the biface is then taken off, usually by successive blade detachments (named ski spalls). The first ski spall has a triangular section and the following ones have trapezoidal sections. These blades show characteristic features by which it is possible to reconstruct the knapping scheme, even without refittings ;
- the negative left on the core by these characteristic blades is used as pressure platform for microblade detachment ;
- the first microblade obtained shows the negatives of bifacial flaking of the other ridge of the biface ; it is a crest as defined in all blade debitages.
- the narrow section of the biface offers a reduced surface of debitage, producing the specific shape of the core, called "wedge-shaped" core for that reason.

The debitage being practiced in the width of the biface, with a debitage surface reduced to the core thickness, the laminar blanks have a relatively constant length and small size (Aita, Kato and Yamanaka, 1991). This method is efficient ; however, a lot of raw material is wasted in the shaping of the bifacial preform and in the pressure platform preparation.

The cores named "Gobi" which are characteristic of many industries of north-eastern Asia proceed from the use of this method. However it would be necessary to know if it is the case with all the so-called Gobi cores and if the pressure technique has been really utilized.

The "classical" method

We call it classical because it is similar to many percussion blade debitage methods used in the Upper Paleolithic. In contrast with the former method, the shaping of the core volume and the exploitation of a much larger debitage surface allow the production not only of more small-sized blanks but also of much larger blanks (raw material permitting).

- The shaping of the core, depending upon the original form of the raw material, is done most often by percussion with at least one crested ridge to guide the debitage,
- The creation of a pressure platform, with technical variants, can be done before the preparation of the debitage surface. The pressure platform may be smooth or facetted by centripetal flaking. Overhangs are abraded or not, depending upon the nature of the tool used for pressing (bone, ivory or metal).

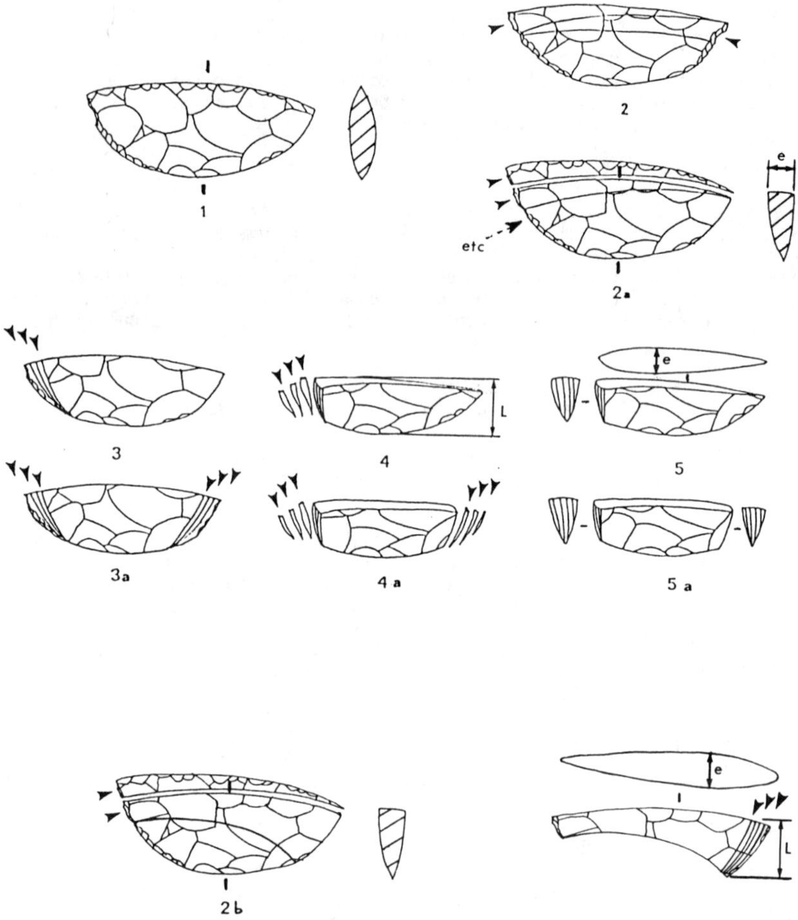

Fig 3 Reduction sequence of the Yubetsu method. Bifacial preparation by percussion: 1. Unidirectional or opposed (one or more) flaking along a ridge of the biface to create a pressure platform : 2, 2 a. Bladelet detachment along the other ridge : 3-5. Variant with two detachement fronts : 3a-5a

The shape of the cores depends upon the debitage surface ; they may be conical, cylindrical. If the shaping involved only part of the raw material as is often the case with pebbles, the knapped surfaces are rounded or semi-circular. Cores with a single flat surface are also observed.

Other methods of microblade debitage

Microblade production can also be realized on flake cores. Another tradition is that of the burin blow technique according to which a thick flake is truncated (hard hammer) and bladelets are struck off from the truncation. We have observed this method on the assemblage of Makarovo IV (Baikal region), which is dated from over 30,000 B.P. at the beginning of the Upper Paleolithic.

Using the assemblages of two sites where the Yubetsu method for the first one (Xiachuan, China) and the classical method for the second (Mehrgarh, Pakistan) have been observed, we propose to make a technological interpretation of their lithic industries.

The example of Xiachuan, Shanxi, China

This site, located in the Zhongtiao mountains, is dated between 24,000 and 14,000 B.P. It is part of a series of localities with similar industries (Wang *et al.*, 1978). Using the techno-typological description of the assemblage published by Chen Chun and Wang Xiang-Qian (1989), we try here to reconstitute the production sequence.

- Raw material economy

Various rock materials were exploited, mainly sandstone and quartzite of local origin and exogenous flint (one or two days walk) ; there is no information upon the size, nature, quality of these materials.

- Production sequence

Two different *chaînes opératoires*, according to two main categories of raw material (raw material economy), can be distinguished : that on sandstone-quartzite for the production of large tools and that on flint for producing flakes, blades and microblades. The reduction sequences proposed by the authors for the microblade cores is based upon their typological classification but the possible link between the various shapes is not exploited.

Only the microblade production is relevant from the use of pressure. As far as we can judge from the drawings, the wedge-shaped, naviform and funnel-shaped cores proceed from the shaping of bifacial preforms in which the single pressure platform is made by a longitudinal blade detachment (ski spall). The conical and cylindrical cores could represent the final phases of use of the former cores. The knapping experiments show that this morphological transformation of the cores is a normal process when the raw material is exploited to the final possible stages. Their presence would thus mean an exhaustive utilization of the raw material.

- Tool management

Besides the large tools like adzes or choppers, the tools on flakes and blades include an important production of scrapers, borers and burins. There are points on blades or bifacial artefacts, the most regular and smallest of which could be primitive arrowheads, as well as backed blades. The microblade production which was no doubt abundant is not justified by their transformation into tools ; at least no mention of them is made.

The example of Mehrgarh, Baluchistan, Pakistan.

Mehrgarh, a site occupied from the 7th to the middle of the 3rd millennium B.C. (Jarrige and Meadow, 1985), gives the opportunity to follow the development of techniques during a long sequence. A technological study of the debitage, including the search for the origin of the raw material, is in process (Inizan and Lechevallier, 1985, 1990; Lechevallier, in prep.). Lithics played an important part in the Neolithic but less and less during the Chalcolithic (Vth millennium) and the Bronze Age (IVth-IIIrd millennia). However this loss in quantity is compensated by a gain in quality (large standardized products).

The type of production in the Neolithic is a blade-bladelet pressure-flaked industry produced by the "classical" method. Percussion is used for the roughing out of preforms and indirect percussion for the production of the first blades.

- Raw material management

It is a homogeneous, light-coloured flint of undetermined origin. Locally flint pebbles are quite rare in the river bed but it is available at about one or two days' walk in the mountains (Aubry *et al.*, 1988). There is no exchange system. It will be imported later, in the Bronze Age, from a long distance (Rohri flint some hundred km away). Heat treatment is evidenced on a few pieces ; its use does not seem to have been systematic.

- Production sequence

The testing of raw material and the first roughing out were done on the extraction or collection sites, as shown by the rarity of cortical flakes. The core roughing out was done by percussion flaking. The first blades issued from large-sized cores (up to 90 mm height) were obtained by indirect percussion ; then (below 50-60 cm) pressure only was used.

The cores are flat (sometimes with a cortical back) or subconical (true bullet cores are exceptional). There are no wedge-shaped or Gobi cores. The pressure platforms are unfacetted, with a 70° or 80° angle ; abrasion was used to get rid of overhangs until the Chalcolithic. The negative marks of crested blades used for core preparation are visible at the negatives of crossed transversal flakes. Through the study of crested blades it is possible to reconstruct the reduction sequence.

Blades and bladelets (the distinction between the two is a 10 mm width) are thin and regular, with parallel edges and ridges and a constant thickness. Their mean length increases regularly : 37 mm in the Neolithic, 42 mm in the Chalcolithic and 50 mm in the Bronze Age. Their section is most often trapezoidal. The butt is punctiform, the bulb short but well-marked. The dimensions are a sign of continuous production from the largest blades by indirect percussion to middle-sized blades, bladelets and microblades by pressure. If we refer to J. Pelegrin's knapping experiments, a debitage in the hand is implied for the smallest blanks (up to 3-4 cm long) and the use of a fixation device for the core and of a shoulder clutch to produce the longest blades obtained by pressure (Pelegrin, 1988). Pressure with metal points appears first in the Chalcolithic. The overhangs are no longer abraded, the blade butts are wider. The pressure obtained largest blades (about 16 cm) are known only at the Bronze Age.

- Finally. Tool management in the Neolithic.

Blades and bladelets are tool blanks : whole pieces were chosen for end-scrapers, burins, borers, retouched blades and fragments for geometric microliths (lunates, trapezes and triangles). The assemblage is not "microlithic" but contains a low percentage of geometric microliths (4.6 % of tools), used as composite elements hafted in weapons or sickle shafts. Flakes are utilized without further retouching. Numerical data (not final) give a good idea of the production intentions. Cores (411) represent only 2.5.% of the total number of artifacts (20,753) ; bladelets (11,296) - four times more in number than blades (3,688) - constitute half of the assemblage. If tools (4,076), mostly on blades or bladelets, are added, one gets a blade production of about 19,000 pieces, that is about 46 blade blanks per core.

- Evolution

This procedure is stable during the whole Neolithic. It changes in the Chalcolithic, from a technological standpoint (use of metal for pressure point) and an economic standpoint (disappearance of microliths ; sickle elements are made of unretouched blade or bladelet fragments). In the Bronze Age there is a drastic change : large blades (up to 15 cm long), which are few in number, are imported since they are made of exogenous flint and since the characteristic debitage waste is not present. A production of geometric microliths on wide blade fragments reappears at the same time.

3 - The archaeological sources of pressure debitage

3-1 The oldest manifestations of pressure debitage

The formal recognition of pressure debitage as such in the Sino-Siberian area is still exceptional. However, the sites concerned point to an area centered upon the Baikal region, the Aldan Basin in Siberia, Mongolia and North China as the one where pressure debitage first appeared in the world.

In Siberia the assemblages of Dyuktai, Ust'Mil', Ezhantsy, Ilkhine, Verkhne Troitskaia in the Aldan region, Makarovo II, Kurla, Verkholenskaia Gora west of the Baikal Lake were described by Flenniken as being pressure-flaked (Flenniken, 1987). He is the only one who demonstrated the specific reduction sequences used in this region and who replicated them. He also mentioned the use of heat treatment. The C14 dates which range between 35,000 BP and 11,400 BP are widely considered as being not older than 20,000 BP (Yi and Clark, 1985 ; Chen and Wang, 1989).

For Gai Pei, as well as for many authors (Wang *et al.*, 1978), there is a local evolution in North China leading to microblade production (the techniques involved are not described). For him some of the conditions required for its appearance would be present as early as Zhoukoudian Locality 15 (100,000-200,000 BP) and much later at Shiyu, Shanxi (28.000 BP), and at Salawusu, Inner Mongolia : the small size of cores and blanks, the bifacial preparation of tools which could be applied to shape out core preforms (Gai Pei, 1985 and 1991). Wang Yuping and J. Olsen (1985 : 255) mention that the Salawusu blanks are produced by direct percussion and by pressure. A single C14 date of 35,340 BP and Uranium determinations between 37,000 and 50,000 B.P. have been obtained for these assemblages (Gai Pei, 1985 : 232).

From the assemblages with cores characteristic of the Yubetsu method (Gobi cores, some of the wedge-shaped cores), showing the same knapping procedures as in the Dyuktai tradition, it seems that pressure debitage was present in the Upper Palaeolithic from Central Siberia to the Pacific Coast. These cores, associated to a microblade debitage, are present from Central Siberia to the Pacific Coast in the Upper Paleolithic and, under their characteristic form of "true microblade cores", have been used as markers (Vasilevsky, 1976 ; Larichev *et al.*, 1990). They are mentioned in the Ienissei basin at Afontova Gora (the date of 20,900 BP obtained for the oldest level, considered too old, would be around 16,000-17,000 BP) and at Kokorevo I et II (ca. 13,300 BP), at Tachtyk (upper level : 12,690 BP), along with the use of a typical Upper Palaeolithic large blade technology by indirect percussion (Abramova, 1983 ; Michael, 1984 : 30).

In Mongolia, the same cores are found at Khere-uul and Moil'tyn-am (Vasilevsky, 1976 ; Smith, 1974). In Tibet, as well as in Xinqiang, no proper dates have yet been obtained. In North China pressure debitage is fully used at dates between 24,000 and 14,000, at Xiachuan, Shanxi (Chen and Wang, 1989). In Japan, most authors agree that microblade debitage has a continental origin ; its oldest manifestation would be at Hokkaido around 15,000 BP or even before (Shirataki, Okedo-Azumi)(Morlan, 1967 ; Kobayashi, 1970)[7]. In the North Pacific area, the oldest site with Dyuktai debitage would be Ushki in Kamchatka (level V : 10,860 BP).

It is thus reasonable to locate the origin of the technique in the vast area of North East Asia in the Upper Paleolithic. After Mochanov and many others we can emphasize the fact that the oldest dates come from the nuclear area at the centre of a vast region, limited to the west at the Ienissei and to the east by Japan, Kamtchatka, and North America where it was present around the end of the Upper Paleolithic.

3-2 Pressure debitage in Europe, the Middle East and South Asia

Beyond the western border of the Yubetsu method area, pressure debitage is to be found much later in Europe[8], Central Asia and the Middle East. In account of these geographical and chronological gaps, the problems of diffusion - that is of techniques

---

[7]cf Regional Focus-Hokkaido by Keally, Chiba, Goto and Tomikawa, Hata, Matsuya, Sato and Kitazawa, Nishida and Ishikawa, Sugiura, Suzuki in : Current Researchin the Pleistocene (University of Maine), 7 : 143-164

[8] The illustrations in the book on *The Mesolithic of the SSSR* show the wide part of the pressure technique in the blade production (Rybakova (ed.) 1982).

transfer and evolution (in these regions, the "classical method" only is present) - have to be raised.

Pressure debitage has been recognized in various mesolithic and neolithic industries in Eastern Europe, in the Balkan area as well as in the Neolithic of Western Europe between 8,000 and 6,000 B.P. Some authors have noted the appearance of a standardized blade debitage - often of excellent quality - that nothing before let us foresee, and which co-existed with the local traditional industries. We refer to the 1985 Warsaw meeting on *Chipped Stone Industries of the early farming cultures in Europe*, where Kozlowski (1987), Perlès (1987), Cahen *et al.* (1987) considered the pressure technique as being borrowed.

The use of pressure debitage is more ancient in the Middle East where it is found on flint and obsidian assemblages of the Zagros area at about 9,000 BP ; the oldest well-dated site so far is Nemriq, Iraq, at about 10,000 BP (Kozlowski, 1989). Pressure use is well documented in the Indus Valley at Mehrgarh, in Pakistani Baluchistan, from 9,000 BP on as we have seen already, in India at Bagor, Rajasthan at about 6-7,000 BP (Inizan and Lechevallier, 1985 and 1990). It is present in the Ganges basin at Ghagharia, an undated mesolithic site (Sharma and Clark, 1983).

It should be noted that the use of pressure technique is totally absent from the tropical zone of South India, South East Asia and South China where there is no blade tradition.

3-3 North America

Recognition of the pressure debitage technique as such in North America is still exceptional (West, 1980). However the many core reduction sequences described and their likeliness to those found in N.E. Asia allow to acknowledge the use of pressure in N.W. America (Chen and Wang, 1989). The use of various methods has been identified in many blade industries which have been very early associated with a Siberian origin, as at the Campus site, near Fairbanks, Alaska, in 1933 (Nelson, 1937). This site has been considered the oldest evidence in North America of an Asiatic influence until the recent study by Mobley (1991) who proved the site as being not older than 3,500 B.P. (3,500 +/- 140 BP and 2,727 +/- 125 BP). The multiplication, in Alaska and Yukon as well as in Siberia, of sites containing cores, microblades and their by-products led to the distinction of many archaeological complexes and traditions that Chen and Wang (1989) group in three traditions.

The oldest one is the *American Paleoartic tradition*, parallel to the Siberian Paleoarctic, present between 15,000 and 13,000 BP in the Akmakian level of Onion Portage, Alaska, and at Blue Fish cave, Yukon ; many sites are dated between 11,000 and 10,000 BP : 12,900 BP at Blue Fish Cave (Morlan and Cinq-Mars, 1982 ; Cinq-Mars, 1990), 11,470 BP at Putu (Alexander, 1987), 10,075 BP at Hidden Falls, 10,790 BP at Dry Creek, 10,540 BP at Gallagher (Dixon, 1975). They are found mostly inland as at the Healy and Denali lakes, refuge area not covered over by glaciers at the end of the Wisconsinian.

The second one, the *Arctic Small Tool tradition* (Irving, 1964), appeared between 5.000 and 4.000 BP in the whole Arctic, from Alaska with the Denbigh Flint Complex (Giddings, 1964, 1967) or the proto-Denbigh Complex (Anderson, 1988) to Greenland with the Independence culture (Knuth, 1967) and Newfoundland and North shore of the lower Saint-Lawrence River to the South with the Groswater and the Dorset cultures (Memorial University of Newfoundland, 1986 ; Plumet *et al.*, 1992). The Artic Small <Tool tradition followed an evolution through the eastern Paleoeskimo until 500 B.P. but disappeared in Alaska as early as 3,000 B.P. (Anderson, 1988; Dumond, 1987). These two traditions in which pressure microblade debitage is well documented are an evidence of the expansion of hunters-gatherers' bands in the wide coastal and inland territories of Beringian and arctic tundra.

Finally, a microblade component, associated with various typological traditions (tanged or notched points for example), is active until about 500 BP inside Alaska and the N.W. coast, in southern and central British Columbia and even in Alberta. One of the best known is the *Plateau Microblade tradition* as found in the Nesikep tradition (Sanger, 1968 ;

Fladmark, 1988). In some of these traditions, the pressure technique co-exists with percussion, in proportions which are not known.

4 - Interpretation : Origin and expansion of pressure debitage

4-1. Geographical origin and modes of appearance (fig.4)

The early appearance, at the end of the Pleistocene, inside the South Siberian and Mongolian area, of microblade industries among hunter-gatherer societies is not unique. Microblade industries are known in South Africa which date from about 80,000 B.P. (Clark, 1982).

However the specificity of North East Asia lies in the emergence, 20,000 years ago, of a true laminar production of bladelets or microblades with the use of pressure. The production of blades by percussion never developed there as in western Eurasia where sophisticated techniques implying direct and indirect percussion were elaborated.

Although this technique alone cannot define a culture, its invention and adoption is far from being subsidiary. The presence together of a technique unknown elsewhere and of an original knapping scheme as the Yubetsu model in the sibero-sino-mongol area at the earliest known dates, leads us to think that there is a unique hearth of invention for the pressure debitage technique. However its date of appearance around 20,000 B.P. remains now to be precised.

The processes of appearance concern a production of lamellar products, even micro-lamellar, which remained stable during thousands of years. This technical stability also shows the limits of the technical know-how and equipment used. According to what we know only small-sized products can be obtained with a simple fixation of the core and the strength applied with a hand-held bone or ivory point. These microblades were used unmodified as inserts, as indicated by the bone artifacts found associated with them in some sites of these regions.

When and where did the "classical" knapping scheme appear? Was it really in competition with the "Yubetsu" procedure? Or did it develop only in the Neolithic with the need of a blade production?

4-2 Microblade debitage in North America and pressure technique

J. Kozlowski and H.G. Bandi (1984) linked the presence of a blade production in Siberia and North America to an extension eastwards of an *Epigravettian* issued from Europe and H. Müller-Beck (1965) to an Aurignacoid wave coming from Europe. In 1974, J.W. Smith, after N.C. Nelson, proposed to group the archeological formations of Siberia and N.W. America in a *Northeast Asian-Northwest American Microblade tradition* or NANAMT. In 1976, R.S. Vasilevsky proposed N.E. Asia as the origin of the Yubetsu-like debitage at the end of the Paleolithic, the distribution of this technique from Japan to North America indicating common roots. J. Mochanov (1978, 1980) clearly distinguished two large ethnocultural areas, that of Malta-Afontova Gora to the West, with European affinities, and that of Dyuktai to the East, close to the North American industries. In 1980, F.H. West, comparing the Denali industries with those of Ushki and Dyuktai, also grouped the microblade formations under the concept of *paleoartic tradition*, occurring from the Lena to the Yukon. Smith and West thus disagreed with R.G. Hopkins (1987), D.M. Klein (1971) and W.R. Powers (1973) who understated the affinities between Siberian and Alaskan industries, but stressed the important differences between the European Upper Paleolithic industries and those to the East of the Ienissei. West is one of the few prehistorians who clearly identified pressure blade debitage in some Beringian industries, although without pointing out its importance as a technical marker.

Smith (1974 : 363), after A.P. Okladnikov (1974), also located the origin of microblade industries of eastern Siberia in central Siberia or in northern Mongolia while Mochanov placed its origin in China (Mochanov, 1969 a,b).

In fact, if the pressure blade debitage first appeared, as we have said before, in N.E. Asia around 20,000 B.P., it coincides with the territory occupied today by the Mongoloid people. The definition of its expansion territory should allow us to confirm the emergence,

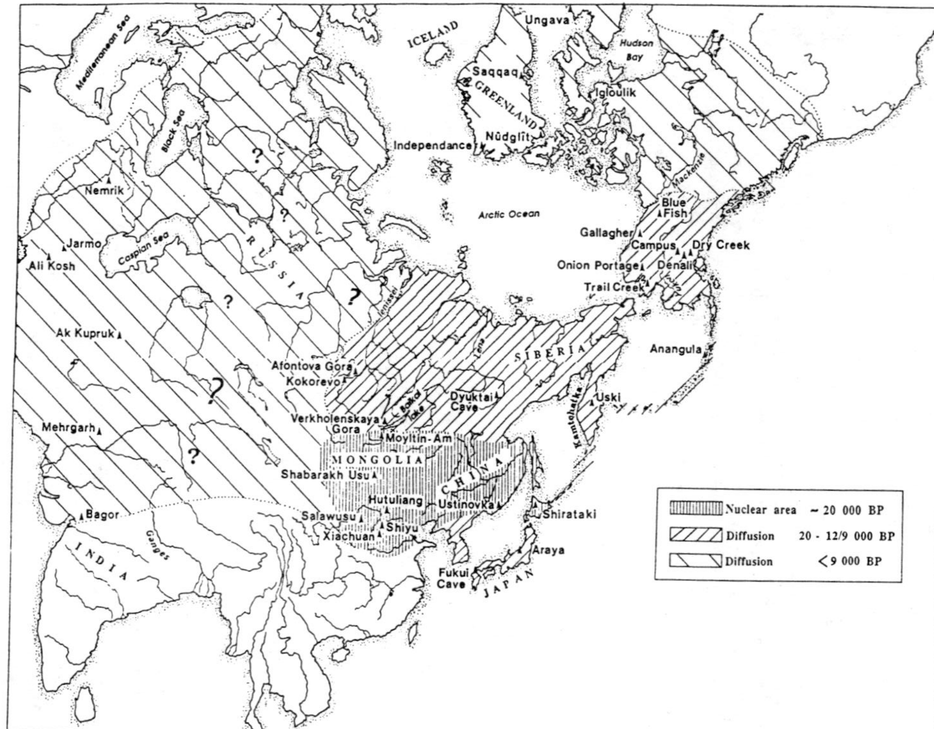
Mediterranean Sea
Black Sea
Caspian Sea
Nemrik
Jarmo
Ali Kosh
Ak Kupruk
Mehrgarh
Bagor
INDIA
Ganges
RUSSIA
ICELAND
Ungava
Saqqaq
GREENLAND
Nûdglît
Independance
Hudson Bay
Igloulik
Mackenzie
Arctic Ocean
Blue Fish
Gallagher
Campus
Dry Creek
Onion Portage
Denali
Trail Creek
Anangula
SIBERIA
Lena
Jenissei
Afontova Gora
Kokorevo
Baikal lake
Dyuktai Cave
Verkholenskaya Gora
Moyltin-Am
MONGOLIA
Shabarakh Usu
CHINA
Hutuliang
Ustinovka
Salawusu
Shiyu
Xiachuan
Kamtchatka
Uski
Shirataki
Araya
Fukui Cave
JAPAN
Nuclear area ~ 20 000 BP
Diffusion 20 - 12/9 000 BP
Diffusion <9 000 BP

Fig.4 Map

at the end of the Pleistocene, of a vast cultural unit distinct from the paleolithic cultures west of the Ienissei (Abramova, 1989). That Beringia, now under water, may have been its most populated area, as F.H. West suggested, is something which cannot be verified, but which is not essential, as Alaska was then the eastern frontier of Asia and not the gateway to America. At that time America was restricted to the south of the inland sea and opened to the North only later by annexing a piece of land detached from far-eastern Asia. It is possible that it had already been occupied but this is another problem concerning another geographic area. The microblade industries which must derive from the Beringian industries maintained themselves there quite a long time. They are found within the limits of the territories occupied before by the glaciers.

The coastal cultures depending upon arctic resources are more recent than those from the interior, even though Anangula dates back to 8,000 BP ; they are distinct from the Beringian formations gathered under the name of Paleoarctic tradition. At Anangula, in the Aleoutian area, as at Gallagher, in northern Alaska, blades are in large number. It is thus likely that within Beringia pressure technology was not dominant. It is found in association with bifacial or specialized points, even in some cases with fluted points.

A systematic technological study of the microblade industries of Siberia, North America and the Arctic would help determine the various streams of penetration into America. The hypothesis of F.H. West, who sees the origin of Clovis points in the blade knapping techniques of the Paleoarctic, could thus be verified, taking into account the possible role played by pressure in fluting and in the microblade production in Alaskan and Yukon industries to which these points are associated. In the same perspective, the splendid obsidian blade production of the Aztec may be the ultimate legacy of the Siberian hunters going to conquer the New World.

4-3 Pressure among the early farmers

The appearance of pressure debitage in the context of early farming cultures between the Euphrates and Indus rivers is a phenomenon which, as far we know, appeared suddenly. There are no antecedants to it in the epipaleolithic industries with geometric microliths of the Near East. If in the Indus valley nothing is known of what occurred before the Neolithic (cf. Mehrgarh), the situation is different in the Middle East. There, in spite of the few data at our disposal, blade industries are known in the Upper Paleolithic and in the Epipaleolithic. But in the Zarzian (Zagros) epipaleolithic industries percussion appears to be the only technique in use (Smith, 1986). The sudden occurrence in the Neolithic of pressure debitage in these regions is thus striking, especially as the method used is the "classical" one : it is used for the production of medium-sized blades and not only of microblades. And, as we have seen, the production of blades, longer than 6 cm, needs a complex procedure. Its elaboration certainly derived from a process using simple means, which is the production of microblades pressed in the hand.

It is clear to us that only the "classical" debitage is fit for the exploitation of large modules with a good raw material reliability. This is perhaps one of the reasons for the abandonment of the Yubetsu method.

4-4. Technological perspectives

If we adopt the concepts of "tendency" and "fact" as proposed by A. Leroi-Gourhan, percussion debitage which has an unavoidable character is inscribed in the logical order of human activities. Pressure debitage on the other hand belongs to a different order : the "fact" unpredictable and singular,which is an "unstable compromise between tendencies and the context" (Leroi-Gourhan, 1943 : 25).

The invention and adoption of the technique in the Upper Paleolithic by hunters in the Sibero-Mongolian area can be questioned in terms of raw material and tool management. Between 20,000 and 10,000 BP, raw material procurement on these vast icy territories implied a seasonal cycle of activities. The hunting parties required a light equipment which was partly provided by the efficient production of microblades. However their use is directly linked to the need of knowing how to haft them.

The association of the pressure technique and of the particular core shaping in the Yubetsu method leads us to look for the origin area of this debitage but also for the time and the conditions of its abandonment. Was this knapping scheme abandoned because there was a need of larger blade products, thus creating the necessity to spare the raw material ? Was a techno-economic change observed at the time of the shift toward the classical method ?

The adoption of the pressure technique, with the classical method, in part of the oriental and European world among farmers (one still needs however to evaluate the extension of this phenomenon) is part of the great changes in the technological system which had to address new needs but also new requirements in a new economic system. The increasing production of larger and larger blades by pressure debitage, in the same time as the quantitative decrease of blank production, implies the interference of highly qualified specialists. It is also in this economic system that long-distance trade developed to obtain raw material, such as obsidian which was needed for the specific use of that debitage technique.

We are still far from knowing exactly how the technique itself developed. There are as usual more questions than answers. New sites need to be discovered and well-dated lithic assemblages to be studied in a technological perspective....

Marie-Louise INIZAN
ERA 28 - CNRS
1, place Aristide-Briand
92195 Meudon Cedex, France

Monique LECHEVALLIER
UPR 316 - CNRS
Musée Guimet
6, place d'Iéna
75116 Paris, France

Patrick PLUMET
Laboratoire d'archéologie
Département des sciences de la terre
Université du Québec à Montréal (Québec)

Bibliography

ABRAMOVA Z.A.
1979 *Le Paléolithique de l'Iénisséi*. Novosibirsk
1983 Problèmes relatifs à l'étude du Paléolithique dans la région de l'Iénisséi (Sibérie). *L'Anthropologie*, 87,2 : 207-213
1989 Two models of cultural adaptation in Late Palaeolithic Eurasia. *In* : PRASLOV N.D. *et al.* The Steppes in the Late Palaeolithic. *Antiquity*, 63 : 784-792
AITA Y., KATO M.& YAMANAKA I.
1991 Le remontage des pièces lithiques : une illustration des différentes techniques du Paléolithique supérieur au Japon. *In* : *25 ans d'études technologiques en Préhistoire. Bilan et perspectives* : XIè rencontres internationales d'archéologie et d'histoire d'Antibes. Actes des rencontres 18-19-20 octobre 1990. Juan-les-pins, A.P.D.C.A. : 255-262
ALEXANDER H. L.
1987 Putu, a fluted point site in Alaska. *Simon Fraser University, coll. Publ.* 17. Vancouver, Dept. of Archaeology
ANDERSON D.D.
1970 Microblade traditions in northwestern Alaska. *Arctic Anthropology*, 6 : 2-16
1988 Onion Portage, the archaeology of a stratified site from the Kobuk River, northwest Alaska. *Anthropological Papers of the University of Alaska*, 22,1 : 163
AUBRY T., AHMED Z. & BALOCH M.K.
1988 A preliminary study of cherts from Balochistan. *Acta Mineralogica Pakistana*, 4 : 102-113
BALFET. H. (ed.)
1991 *Observer l'action technique. Des chaînes opératoires, pour quoi faire* ? Paris : Ed. du C.N.R.S.

BANDI H.G.
1969 *Eskimo Prehistory* (3rd ed. 1972). U. of Alaska Press
BINDER D.
1987 Le Néolithique ancien provençal. Typologie et technologie des outillages lithiques. 24ème Supplément à : *Gallia Préhistoire*. Paris : Ed. du CNRS
BINDER D. & GASSIN B.
1988 Industries lithiques. Tracéologie et technologie. *BAR. Int. Ser.* 411 (i) : 93-125. Oxford
BORDEN C.E.
1975 Origins and Development of early Northwest Coast Cultures to about 3.000 BC. *Collection Mercure*, dossier n° 45. Commission archéologique du Canada, Musée national de l'Homme.
1979 Peopling and early Cultures of the Pacific Northwest. *Science*, 203,4384 : 391-394
BORDES F.
1969 Traitement thermique du silex au Solutréen. *C.R.S.M. de la Société Préhistorique française*, 7 : 12-25
BUTZER K.W.
1991 An Old World perspective on potential Mid-Wisconsinian settlement of the Americas. *In* : DILLEHAY T.D. & MELTZER D.J. (eds).*The first Americans : search and research.*: 137-156. II Ann Arbor, Boston, London : CRC Press
CAHEN D., GASPAR J.P. & OTTE M.
1987 Industries lithiques danubiennes de Belgique. *In* : KOZLOWSKI J.K. & KOZLOWSKI S.K (eds). *Chipped Stone industries of the early farming cultures in Europe* : 247-330. Cracow
CALLEY S.
1985 Les nucléus en obsidienne de Cafer Hüyük (Turquie). Etude préliminaire sur les techniques de taille.*Cahiers de l'Euphrate*, 4 : 87-108. Paris : Editions Recherche sur les Civilisations
CHEN CHUN & WANG XIANG-QIAN
1989 Upper Paleolithic Microblade industries in North China and their relationships with Northeast Asia and North America. *Arctic Anthropology* 26, 2 : 127-156
CINQ-MARS J.
1990 La place des grottes du Poisson-Bleu dans la préhistoire béringienne. *Revista de Arqueologia americana* (Washington), 1 : 9-32
CLARK J.D.
1985 The microlithic industries of Africa : their antiquity and possible economic implications. *In* : MISRA V.N. & BELLWOOD P. (eds) *Recent Advances in Indo-Pacific Prehistory* : 95-103. New Delhi : Oxford & IBH Publishing Co.
CLARKE J.E.
1982 Manufacture of Mesoamerican prismatic blades : an alternative technique. *American Antiquity*, 47, 2 : 355-376
CRABTREE D.E.
1968 Mesoamerican polyhedral cores and prismatic blades. *American Antiquity*, 33, 4 : 446-478
Current Research in the Pleistocene (University of Maine), 7
1990 143-164
DIKOV N.N.
1968 The discovery of the Paleolithic in Kamchatka and the problem of the initial occupation of America. *Arctic Anthropology*, 5, 1 : 19- 203
DIXON J.E.
1975 The Gallagher Flint Station, an early man site on the north slope, Arctic Alaska, and its role in relation to the Bering Land Bridge. *Arctic Anthropology*, 12,1 : 68-75
DITTEMORE M.
1983 The soundings at M'lefaat. *In* : BRAIDWOOD L.S. *et al.* (eds). Prehistoric investigations along the Zagros. *O.I.P.*, 105 : 671-703. Chicago : University of Chicago Press

DUMOND D.E.
1977 *The Eskimos and Aleuts*. New York, London : Thames & Hudson
DUPREE L. & DAVIS R.S.
1972 Prehistoric research in Afghanistan. *Transactions of the American Philosophical Society* 62 : 14-32
FLADMARK K.R.
1988 Préhistoire de la Colombie britannique. *La Préhistoire du Canada*. Montréal, Ottawa, Fides : Musée National de l'Homme
FLENNIKEN J.J.
1987 The paleolithic Dyuktai pressure blade technique of Siberia. *Arctic Anthropology*, 24,2 : 117-132
GAI PEI
1985 Microlithic industries in China. *In*: WU RUKANG & OLSEN J.W. (eds). *Palaeoanthropology and Palaeolithic Archaeology in the People's Republic of China.* : 225-241. Orlando : Academic Press
1991 Microblade tradition around the northern Pacific rim : a Chinese Perspective. *Paper presented at the International Congress for Quaternary Research XIII International Congress, Beijing. August 1-9, 1991*
Giddings J.L.
1964 *The Archeology of Cape Denbigh*. Providence : Brown University Press
GIDDINGS J.L.
1967 *Ancient men of the Arctic*. New York : Alfred knof
GIDDINGS J.L. & ANDERSON D.D.
1986 *Beach ridge archaeology of Cape Krusenstern-Eskimo and pre-Eskimo settlements around Kotzebue Sound, Alaska*. Washington : US Dept of the Interior National Park Service
HAYASHI K.
1968 The Fukui microblade technology and its relationships in Northeast Asia and North America. *ArcticAnthropology*, 5 : 128-190
HOLE F., FLANNERY K.V. & NEELY J.A.
1969 Prehistory and human ecology of the Deh Luran plain. An early village sequence from Khuzistan, Iran. *Memoirs of the Museum of Anthropology*, University of Michigan, Ann Arbor
HOPKINS D.M.
1967 *The Bering Land Bridge*. Stanford : Stanford University Press
INIZAN M.-L.
1985 Le débitage par pression au Moyen-Orient. Premières observations. *In* : *Recueil Jean Deshayes* : 43-54. Paris : Ed. Recherche sur les Civilisations
INIZAN M.-L. & LECHEVALLIER M.
1985 La taille du silex par pression à Mehrgarh, Pakistan. La tombe d'un tailleur ? *Paléorient*, 11,1 : 111-118
1990 A techno-economic approach to lithics : some examples of blade pressure debitage in the Indo-Pakistani sub-continent. *In* : TADDEI M. (ed.). *South Asian Archaeology* : 43-59. Venice, 1987
In press Pressure debitage and the microlithic assemblage at Bagor, North-West India. *B.K. Thapar Felicitation volume*. New Delhi
IRVING W.N.
1964 *Punyik Point and the Arctic Small Tool tradition*. PhD thesis. Ann Arbor :University of Wisconsis
JARRIGE J.-F. & MEADOW R.H.
1985 The antecedants of Civilization in the Indus Valley. *Scientific American*, 243,2 : 122-133
KEALLY C.T.
1990 Hokkaido Pleistocene Archaeology and the Earliest Americans : *Some Comments. Current Research in the Pleistocene* (University of Maine, Orono), 7 : 143-146
KLEIN R.G.

1975 The relevance of old world archaeology to the first entry of man into the new world. *Quaternary Research*, 5,2 : 391-394. New York and London : Academic Press

KNUTH E.

1967 Archaeology of the Musk-ox way. *Contribution du Centre d'Etudes arctiques et Finno-Scandinaves*, 5. Paris : Ecole Pratique des Hautes études, 6° Section

KOBAYASHI T.

1970 Microblade industries in the Japanese archipelago. *Arctic Anthropology*, 7 : 38-58

KOZLOWSKI J. & BANDI H.G.

1984 The paleohistory of circumpolar Arctic colonization. *Arctic*, 37,4 : 359-372. Calgary

KOZLOWSKI S.K.

1987 The pre-neolithic base of the early neolithic stone industries in Europe. *In* : KOZLOWSKI J.K. & KOZLOWSKI S.K. (eds.).*Chipped Stone Industries of the early farming cultures in Europe* : 9-18. Krakow

1989 The neolithization of South-East Europe. An alternative approach. *Varia Archaeologica Hungarica*, II. Budapest

LARICHEV V., KHOL'USHKIN U. & LARICHEVA I.

1990 The Upper Paleolithic of Northern Asia : Achievements, Problems and Perspectives. II. Central and Eastern Siberia. *Journal of World Prehistory*, 4,3 : 347-385

LEROI-GOURHAN A.

1943 *Evolution et techniques. I : L'homme et la matière.* II *: Milieu et Techniques.* Paris : A Michel

MEMORIAL UNIVERSITY OF NEWFOUNDLAND

1986 Palaeo-Eskimo Cultures in Newfoundland, Labrador and Ungava. Saint-Jean. *RA : Reports in Archaeology*. MEMOR : Terre-Neuve Memorial University of Newfoundland

MICHAEL H.N.

1984 Absolute chronologies of late pleistocene and early holocene cultures of northeastern Asia. *Arctic Anthropology*, 21,2 : 1-68

MIDANT-REYNES B.

1983 Le débitage de lames de silex par pression : les éléments de faucilles de 'Ayn-Asil (Oasis de Dakhla). *Bulletin de l'Institut français d'Archéologie orientale*, 83 : 257-262

MOBLEY C.M.

1991 *The Campus site : a prehistoric camp site at Fairbanks, Alaska.* Fairbanks : University of Alaska Press

MOCHANOV J. A.

1969a The early Neolithic of the Aldan. *Arctic Anthropology*, 6 : 95-103

1969b The Ymyia Khtakh Late Neolithic Culture. *Arctic Anthropology*, 6 :115-118

1978 Stratigraphy and Absolute Chronology of the Paleolithic of Northeast Asia, according to the work of 1963-1973. The Palaeolithic of Northeast Asia and the problem of the peopling of America. *In* : BRYAN A.L.(ed.), Early Man in America : 54-66. *Occasional Papers of the Dept. of Anthropology*, U. of Alberta,1

1980 Early migrations to America in the light of a study of the Dyuktai Paleolithic Culture in Northeast Asia. *In* : BROWMAN D. (ed.), *Early Native Americans* : 119-131. The Hague : Mouton

MORLAN R.E.

1967 The preceramic period of Hokkaido : an outline. *Arctic Anthropology*, 4 : 164-220

1970 Wedge-shaped core technology in northern North America. *Arctic Anthropology*, 7 : 17-37

MORLAN R.E. & CINQ-MARS J.

1982 Ancient Beringians : Human occupation in the late Pleistocene of Alaska and the Yukon Territory. *In* : HOPKINS D.M, MATTHEWS J.V.Jr., SCHWEGER

C.E. & YOUNG S.B. (eds). *Paleoecoloy of Beringia* : 353-381. New York and London : Academic Press
MULLER-BECK H.
1965 Non-projectile-point stone industries in America and the problem of their Eurasiatic origins (roots). *Abstracts 7th Congress INQUA*, USA.
NELSON N.C.
1937 Notes on cultural relations between Asia and America. *American Antiquity*, 2,4 : 350-372
OKLADNIKOV A.P.
1974 A Stone Age settlement on the Khere-ul mountain (Eastern Mongolia) and preceramic cultures of Japan. Investigations in history and philology. *Articles collected in the honour of Academician N.I. Conrad.* : 322-338. Moscow (in russian)
PELEGRIN J.
1988 Débitage expérimental par pression : "Du plus petit au plus grand". Technologie préhistorique. *Notes et Monographies techniques*, n° 25. Paris : Ed. du CNRS
PERLES C.
1984 Débitage laminaire de l'obsidienne dans le Néolithique de Franchthi (Grèce) : techniques et place dans l'économie de l'industrie lithique. *Préhistoire de la pierre taillée*, 2 : 129-137. Paris : C.R.E.P.
1987 Les industries du Néolithique précéramique de Grèce : nouvelles études, nouvelles interprétations. *In* : KOZLOWSKI J.K. & KOZLOWSKI S.K.(eds). *Chipped Stone Industries of the early farming cultures in Europe* : 19-39. Cracow
POWERS W.R.
1973 Paleolithic Man in Northeast Asia. *Arctic Anthropology*, 10,2 : 1-106
PLUMET P. *et al.*
1992 *Paléoesquimau et Amérindien : Recherches préliminaires dans la région de Bland-Sablon, Basse Côte-Nord*, Québec. Laboratoire d'archéologie, Université du Québec à Montréal.
Préhistoire de la Pierre Taillée, vol. 2. Economie du débitage laminaire
1984 Paris : Cercle de Recherches et d'Etudes Préhistoriques
RYBAKOVA B.A. (ed.)
1989 *Le Mésolithique de l'URSS*. Moscou : Ed. Nauka
SHARMA G.R. & CLARK J.D. (eds)
1983 Paleoenvironments and Prehistory in the Middle Son Valley. *Archaeology of the Vindhyas and the Ganga Valley*, 7. Allahabad : University of Allahabad
SMITH J.W.
1974 The Northeast Asian-Northwest American microblade tradition (NANAMT). *Journal of Field Archaeology*, 1 : 350-364
SMITH P.E.L.
1986 *Palaeolithic Archaeology in Iran*. Philadelphia : The University Museum
TIXIER J.
1976 L'industrie lithique capsienne de l'Ain Dokkara, région de Tébessa. Fouilles de L. Balout. *Libyca*, 24 : 21-54
TOZAWA M.
1974 *The stone implements in the preceramic age in Japan.* Archaeological Institute of Meiji University, Kanda, Surugadai
VASILEVSKY R.S.
1976 The Pacific microlithic traditions and their Central-Asian roots. *Colloque VII* : 131-153. U.I.S.P.P., Nice
VINOGRADOV A.V.
1972 *Résultats de l'archéologie de l'Asie centrale, I. Sur l'expansion des complexes du Néolithique ancien du type de Darya Say (d'après les travaux de 1970)* : 32-34
1979 Recherche des gisements datant de l'âge de la pierre en Afghanistan du Nord (la région d'Andkoi-Akcha, en 1969 et en 1976). *In* : *La Bactriane Ancienne. Matériaux de l'expédition afghano-soviétique*, II. Paris : Ed. du C.N.R.S.

WANG YUPIN & OLSEN J.W.
1985 Aspects of the Inner Mongolian Palaeolithic. *In*: WU RUKANG & OLSEN J.W. (eds). *Palaeoanthropology and Palaeolithic Archaeology in the People's Republic of China.* New York and London : Academic Press
WANG, JIAN, WANG XIANG-QIAN, & CHEN ZHE-YING
1978 Xiachuan Culture. Kaogu Xuebao (Chinese Journal of Archaeology) 3 : 259-288
WEST F. H.
1980 *The Archaeology of Beringia.* New York : Columbia University Press
YI S. & CLARK G.
1985 The "Dyuktai Culture" and New World Origins. *Current Anthropology*, 26 : 1-20

ARCHAEOLOGICAL BITUMEN: INDENTIFICATION, ORIGINS AND USES OF AN ANCIENT NEAR EASTERN MATERIAL

JACQUES CONNAN* AND ODILE DESCHESNE**

*Elf-Aquitaine, Direction Exploration - Centre Scientifique et Technique Jean Feger, Paris, France
**Department of Oriental Antiquities, Louvre Museum, 34-36 Quai du Louvre, 75058 Paris, Cedex 01, France

ABSTRACT

Prior to their modern invention, the use of petroleum products (bitumen, asphalt, etc.) was common in the prehistory of the Near East. In Mesopotamia where they are plentiful, often occurring at the surface of the ground, these products have been utilized in diverse ways: for construction and caulking of buildings, as a waterproofing agent, and even as a medicine to remedy coughing. Archaeologists have questioned the location of principal sites exploited in antiquity and the ways that these products have been used. As the result of a close collaboration between petroleum specialists and archaeologists, this article explains how geochemical analysis has enabled identification of the provenance of bitumen utilized at Babylon in Iraq, how the inhabitants of Susa in Iran used bitumen, its preparation and also its fabrication as an artificial material which has proven to be as durable as stone.

INTRODUCTION

Today the word "bitumen" evokes a petroleum product or a coating on roadways, but one knows less about its usage further back in antiquity and that it was exploited in the Indus valley, in Mesopotamia and also in Egypt. Cuneiform texts on clay tablets inscribed at the end of the third millennia B.C., the accounts of Greek and Roman historians, the results of archaeological research during the last decades, all combine to establish that bitumin, utilized for its property of impermeability to water, was a material of construction, an adhesive and also as a remedy for sore throat. bitumen represented a not negligible part of exchanges in the economy of the ancient Near East similar in effect to trade goods such as copper, diorite or lapis lazuli.

Although commonly found on archaeological sites, bitumen is, paradoxically, badly known and usually poorly characterized. Archaeologists have agreed on the name, "archaeological bitumin", as a way of identifying archaeological samples which contain petroleum products generally called bitumen or asphalt, for these products have a dark color and a composition which appears to be dense and viscous and which seems solid. The unprocessed materials (asphaltum, oil, bituminous rock, etc.) that serve as raw materials for the elaboration of archaeological bitumen sometimes originate from the immediate environment of various sites, as the Susiana region in Iran, but they can also be exotic materials, imported from afar and therefore rare. The maritime or riverine transport of bitumen has been suspected to operate over considerable distances. Archaeologists, therefore, have posed questions regarding the treatment and conditioning of the raw materials, the existence of different recipes or mixtures, and also the geographical origin of archaeological bitumen. Only researchers working on petroleum products have within their power to ability to satisfy this curiosity. For a petroleum chemist, in effect, archaeological bitumen is amenable to analysis, as it does not matter which petroleum product is present, and the analyses can permit the localization of its geological origin and the characterization of that part of the raw materials (asphalt, bituminous rock, oil, etc.) from which it has been prepared.

THE USES OF BITUMEN IN ANTIQUITY AND ITS TRADE

The first analyses of this type have been realized by R.J. Forbes of Shell between 1936 and 1969 (1). Next in the year 1975, G. Hansen of the German petroleum company Wintershall has retraced the story of petroleum across the ages (2), while R.F. Marschner of the company Amoco, associated with Henry T. Wright, American archaeologist at the University of Michigan at Ann Arbor, undertook in 1976 the geochemical study of archaeological bitumen from the Near East to determine the source of supply and to define the nature of the raw materials (3). Some timely analyses have equally been realized by M.I. Venkatesan of the University of California at Los Angeles on the bitumens from the site of Terqa in Syria excavated from 1975 by a group of researchers at the University of California and undertaken since 1986 by a French mission under the direction of O. Roualt of the College de France (4).

A systematic study does not exist on this question. To respond to the expectation of French archaeologists working in the Near East, we have begun to claim this domain and to systematically study the archaeological bitumins, enumerating the many uses which have been found and proceeding to the physico-chemical analyses of bitumins unearthed by recent excavations where samples were taken from objects chosen from the collections of the Department of Antiquities Orientales of the Louvre Museum. The synthesis of our first research shows at what point bitumen became important and attempts to find out if one can qualify the civilizations of the Near East as "civilizations of bitumen".

As early as the Neolithic period, people discovered the intrinsic qualities of bitumen: their finding of its impermeability to water, its adhesive capabilities, its formability or malleability, and its capacity to mix with other materials. These properties have found a multitude of applications; certain have served until our days; others, more specific in antiquity, have fallen in disuse or have been replaced by more efficient products; some others at last, surprisingly, reveal more about beliefs than rational choices.

A first constraint imposes itself; the bitumen has not been employed by all in the Near East with the same intensity. Its utilization varied according to the epochs in relation to the availability of natural resources in the countryside concerned. It is in central Mesopotamia where sites are most important and in southern Mesopotamia where supply is facilitated by the presence of the Tigris, of the Euphrates and of an important network of canals that it has been most utilized, notably in the domain of brick construction: for instance in the processional way at Babylon where it has served as a mortar in the construction of the Ishtar gate. As early as the fifth millenium, it entered into the composition of bituminous mortars at Eridu in Mesopotamia and at Susa in Susiana (5). This usage intensified until the third millennium. However, this mortar was not utilized everywhere. As bitumen remains an expensive material and is found difficult to work in some areas where wood is rare--it was necessary to apply heat to render it fluid before mixing with other varieties of bitumen or with some mineral or vegetal additives--one employs it therefore essentially there where it will serve at the time for impermeability and as a binding agent. The waterproofing is obtained by applying with a trowel some layers more or less full of bitumen. This is just as one has proceeded for all the terraces of Babylon (6) (the famous "Hanging Gardens"), for the ground courses of the palace of Larsa (7), excavated by Jean-Louis Huot of the Univesity of Paris-I or those of rooms evident at the palace of Mari (8) excavated by Andre Parrot then by Jean-Claude Margueron of the University of Strasbourg. The alters of sacrifice, the reserves of lustral water at Mari or at Tell Asmar (9), the cisterns of the palace of Darius at Susa, the wells as those of Entemena at Tello (10), just as the greatest part of the bathrooms and toilets of Mari have equally been made impermeable in this manner.

The roadways did not exist, and bitumen has not served in roadway coatings, with the exception however of the processional way of Babylon and of the passages of circulation delimited by the courses of the palace at Mari.

Since the Neolithic, bitumen served also to waterproof containers. Storage pits, earthenware jars, baskets and ceramics were coated with bitumen on the interior or exterior in order to effect the protection of the contents from air, water, micro-organisms and moisture. In this spirit, one envelopped the dead in a bituminous coating or one sealed with bitumen the covers of funerary jars and sarcophaguses sometime with bitumen on the interior.

Of course, naval construction has, also, much use of the impermeable qualities of bitumen. Boats of plaited or woven reeds have been coated on the interior and exterior, all as were still the "guffahs" which navigated on the Tigris at the beginning of the century. The texts, which tell the legend of the voyage of the god Nanna to Nippur (11), the epic of Gilgamesh (12), or the Bible when the construction of the arc of Noah or the boats improvised by Moses to traverse the Nile, provide abundant references to this technique. But recently, the discovery at Ra's al-Junayz in the sultanate of Oman of important samples of bitumen, when the excavations were directed jointly by Serge Cleuziou of the University of Paris-I and Maurizio Tosi of the Istituto Studii Medio e Estremo Oriente (I.S.M.E.O.) in Rome, have permitted the reconstruction of the techniques of fabrication of these boats (13).

Effectively an adhesive, bitumen has equally served to haft very diverse tools. Some blades of neolithic flint frequently carry traces of bitumen where one can distinguish in negative the imprints of a piece of wood, bone or terra cotta, and the diverse tools, including hachets, hoes and points, were also hafted. This technique has been carried out at least since the end of the second millennium at Susa. Some ceramics, stone vases and statuettes have been repaired with bitumen, whereas sculptors have used it to fix the eyes of shell and stone, the hornes in gold or in silver, some details of clothes, on the statues. On can in a like manner realize some beautiful compositions in mosaic of rose calcite, lapis lazuli, shell and mother of pearl glued on supports of wood; the most celebrated are the standards of Ur and of Mari and the lyres and the games found in the royal tombs at Ur.

The artists of antiquity have also been inspired by the plastic and decorative qualities of bitumin. Modeled then surfaced with a layer of gold, silver or bronze, it has served to as a substrate to make beads, rings, pendants or appliques such as those which were found on Iranian sites of the thirteenth century B.C. Utilized, as painting, it has served to decorate the temple at Uruk, the beards and the hair of statues or to color skin and leather. Applied to the base of numerous walls, in the fashion of a panelling, it maintained its waterproofing property and also served as decoration.

In antiquity, bitumen was equally a universal remedy. Pure or mixed ingredients quite diverse and very unexpected, one has utilized them in potions, fumigations, poltices or in cream. A disinfectant, it also calmed fever, asthma, toothache, sore-throat and rheumatism. The medical treatises prescribed some bitumen also for standing sickness! One utilized it still for tending to animals and as an insecticide. The Egyptians, who imported it from the Dead Sea, pulled a profit from its antioxidizing qualities to embalm corpses (14).

Some other uses of bitumen we have brought out appear much more surprising. Difficult in effect to understand is why the inhabitants of Susa, who possessed calcite in abundance, were ingeneous to compose a material artificial from partly bitumin, bituminous calcite and crushed or ground calcite! Hard, of aspect enough homogeneous, of grey-black color and totally all very nearly stone, this material has been sculpted principally in the middle of the third millennium and then at the beginning of the second millennium, as well the production was continued afterward (Fig. 1). More than 300 objects, remarkable by their diversity and for certain by their artistic quality, have been discovered in the course of excavations organized by Jacques de Morgan. He was concerned with bowl-like vessels, cult objects, ornaments (15) or sculpted plaques in bas-relief. Certain objects have a polished surface, some others carry traces of the chisel of the sculptor or of an incomplete polishing.

Other usage less unusual if that of magic. At one time for benefit or at another time for bad effect, bitumen was often in magical formulas. A statuette fashioned in bitumen, mutilated or broken, was substituted for that on which was given the malediction, a usage which was still perpetuated in Medieval times. At last, justice itself utilized bitumen as one could in the guise of punishment smear the head of a offender.

In the Sumerian and Akkadian texts of the third and second millennia, which were commercial, historical or poetic, the abundance of terms designating bitumen attest equally to the variety of its utilizations. In effect, as well as not only giving few elements on the techniques of the epoch, several words served to define the quality of bitumin, liquid or solid, mixtures, etc. In the Sumerian texts, 15 terms at least served to designate bitumen. Transported by boat in variable quantities, bitumen was worked in skin containers or amphoras and its price appeared to vary only as the quality of transported product and as a function of the importance that was accorded to it by each epoch. In a general manner, the mixtures ready to employ were more expensive. Unfortunately,

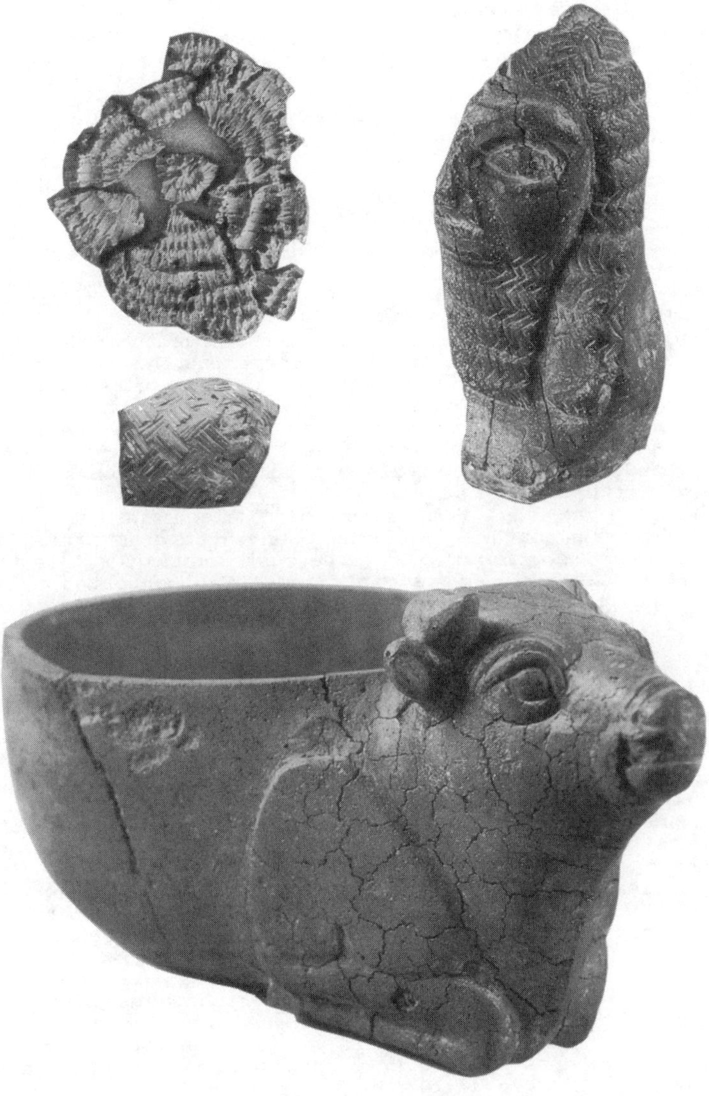

Fig. 1. Bitumen-containing objects: (a) Coated basket from Susa, Iran, and (b) sculpted figurine and (c) tomb figurine made of artificial mixture of bitumen, bituminous calcite and ground calcite (J.-Y. Chambault, coll. Musee du Luvre).

the indications concerning the localization, the mode of exploitation and of treatment of poor quality bitumen, the same as that which could inform us on the nature of mixtures and the techniques given in work. Very happily, the analyses that we have carried out on archaeological bitumen permit us little by little to fill these gaps. The first stage of our research has been to interest ourselves in the origin of archaeological bitumen.

The ancient texts, which date to the end of the third millennium, which concern accounts of Greek and Roman authors like Xenophon, Ptolemy, Herodotus, Plutarch, Strabo and Pliny, or from the Bible, accord all on the existence of three sources of bitumin: the country of Madga in the actual region of Kirkuk, Hit in Mesopotamia on the right bank of the Euphrates and Tilman or Dilmun, which provided commercialized bitumen to Lagash (Fig. 2).

Mesopotamia has been by far the largest producer of bitumen. It is therefore logical to see there the head of some utilizer regions. The principal sites with known asphalt, and for certain still exploited, were located to the west of Baghdad in Iraq, on the right bank of the Euphrates: Abu Jir, Hit and Ramadi. One other zone, the ancient country of Madga was situated between the Diyala and Little Zab Rivers in the region of Kirkuk.

More to the east, Luristan and Susiana in the actual Khuzistan to the southwest of Iran counted several sites with bitumen and bituminous calcite. The best known of all is that of Masjid-i-Sulaiman, always exploited. The contorted relief of the region has favored folds and faults carrying an abundance of petroleum sites and favoring the arrival at the surface of hydrocarbons (oil and gas) which accumulate in profundity. These little hydrocarbon sites, known and exploited in antiquity, are most often abandoned today.

Some other sites, less important and more dissiminated geographically, have equally been exploited. In the Near East, outside of Mesopotamia, sites ranged all along the Fertile Crescent as far as Egypt: Baku in the Caucasus, Qalat Shargatt in Assyria to the south of Mosul, Samosate on the west course of the Euphrates. In the Levant, a number are little sites which are abundant along the north-south fault from Lattaquie in Lebanon to the Dead Sea. The bitumen said to be from Judea, collected on the banks of the Dead Sea, has been used from the Neolithic to the third millennium; it was an object of regular commerce across Egypt. One of its characteristics is the sudden appearance in a massive fashion as well as episodically. These appearances are accompanied by odorous emanations and by sulfurous fumes which have hit with astonishment those who have been witnesses. The episode of Sodom and Gomorrah, in Genesis, is probably the echo of one of these manifestations particularly spectacular. Egypt, for its part, appears to have little exploited the resources of the Sinai, of the Jebel Zeit, of Helouan to the south of Cairo, but has made use rather of the bitumen of the Dead Sea to satisfy its proper needs which remained less important.

Dilmun, another site known in antiquity, has not been localized with precision but is found in the Arabian-Persian Gulf, perhaps at Bahrein. At last, at the extremity of the Near Eastern world, the civilizations of the Indus Valley (Mohenjo-Daro, Mergarh) possess local sites at Iskardo (Cachemire), at Hazara, etc.

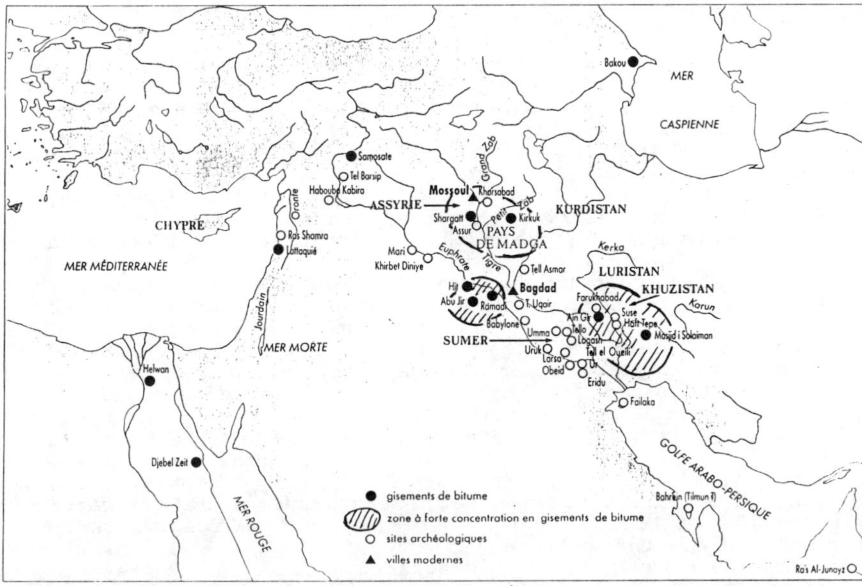

Fig. 2. Map of the Near East showing sites with archaeological bitumen and source areas of bitumen.

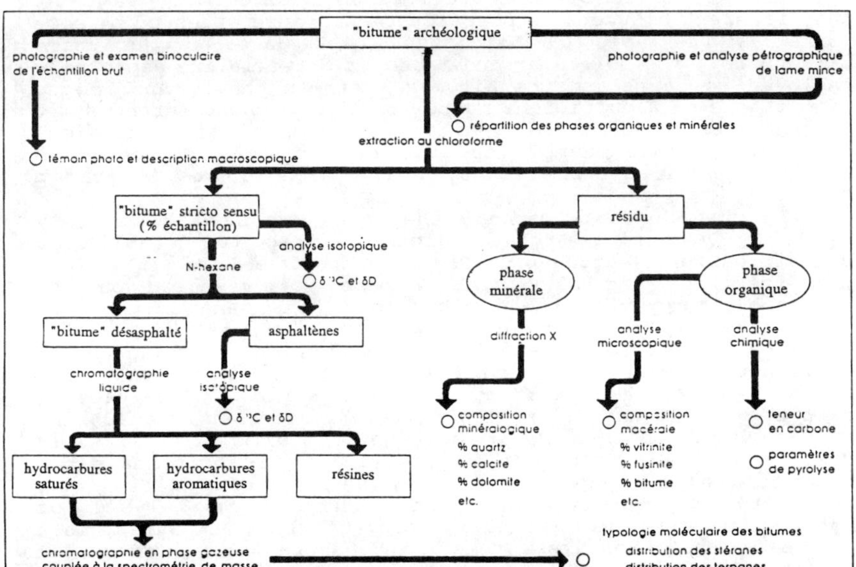

Fig. 3. Protocol used in sample preparation and analysis, involving extraction in chloroform and isotopic and molecular analyses, as well as identification of the organic and mineral phases which permit determination of the nature of the materials and evidence for mixtures.

PHYSICO-CHEMICAL ANALYSES OF BITUMEN

To leave these known sites, can we retrace the geographical origin of archaeological bitumen found on sites in the Near East? It has to us seemed false to place at the beginning a point on the methodology adequate to analyze our samples, which are most often mixtures (asphalt, petroleum or extracts of bituminous rocks) and of other organic elements (vegetal debris, ash) or minerals (clays, calcite, dolomite, ashes and volcanic debris, gypsum) (Fig. 3).

A microscopic and petrographic description is effectuated prior to the chemical analysis of the sample. Having then defined the base constituents, the granulometry and the form of minerals, we investigate the partitioning of bitumen and some elements which serve to reconstitute the processes of fabrication of archaeological bitumen. One proceeds next to the detailed physico-chemical analyses, in particular the molecular analysis of hydrocarbons and the isotopic analysis of the bitumen and asphalt contained in the sample. The chemical examination permits the determination of whether the oils and asphalts utilized to fabricate the archaeological bitumen have or have not biodegraded. If there is diodegradation, one can estimate that the bitumen came from sources of salted water with oil as at Hit and Abu Jir in Iraq or from reservoirs so impregnated as the petroleum sites of Susiana. If to the contrary the bitumen is not biodegraded, one is in the presence of bituminous rock, veritably the mother-rock of the petroleum. The juxtaposition of these two properties, biodegraded and not biodegraded, indicates that it acts as a mixture of at least two different bitumens, a practice furthermore attested by some texts as the epic of Gilgamesh which date to the third millennium. The molecular and isotopic criteria are especially employed to correlate the bitumens between them. And if one can dispose of a sufficient number of geological references, it is possible to find the veritable origin of archaeological bitumen. A case example of this procedure is presented in Appendix I for bitumen found on the site of Ras Shamra in Syria.

By these methods more to the point, we have immediately begun by analyzing archaeological bitumen to establish a classification of archaeological bitumins from Iraq (16). This classification has served as a base for research on the geographic origin of these bitumens. Some tangible results have already been obtained, particuliarly in that which concerns the importance of the production of bitumen in provenance of the region of Hit, Ramadi and Abu Jir. R.J. Forbes, who was the first to realize a series of analyses of natural asphalt and of archaeological bitumen between 1936 and 1969, considered that the majority of the bitumen utilized in Mesopotamia came from this region (17). His hypothesis has been adopted by most contemporary authors but still it is necessary to verify it. More than 100 analyses of archaeological bitumen found in Iraq (Babylon) and Syria, more precisely from the sites of Mari and Habuba Kabira, excavated between 1968 and 1975 by a German team directed by Ernst Heinrich then Eva Stromenger, indicates that only the bitumen of Babylon has for its origin the region of Hit (Fig. 4). These results are not moreover in contradiction with information from known historic texts. In effect, the classical authors making reference to Babylon indicate that the bitumen came from this region.

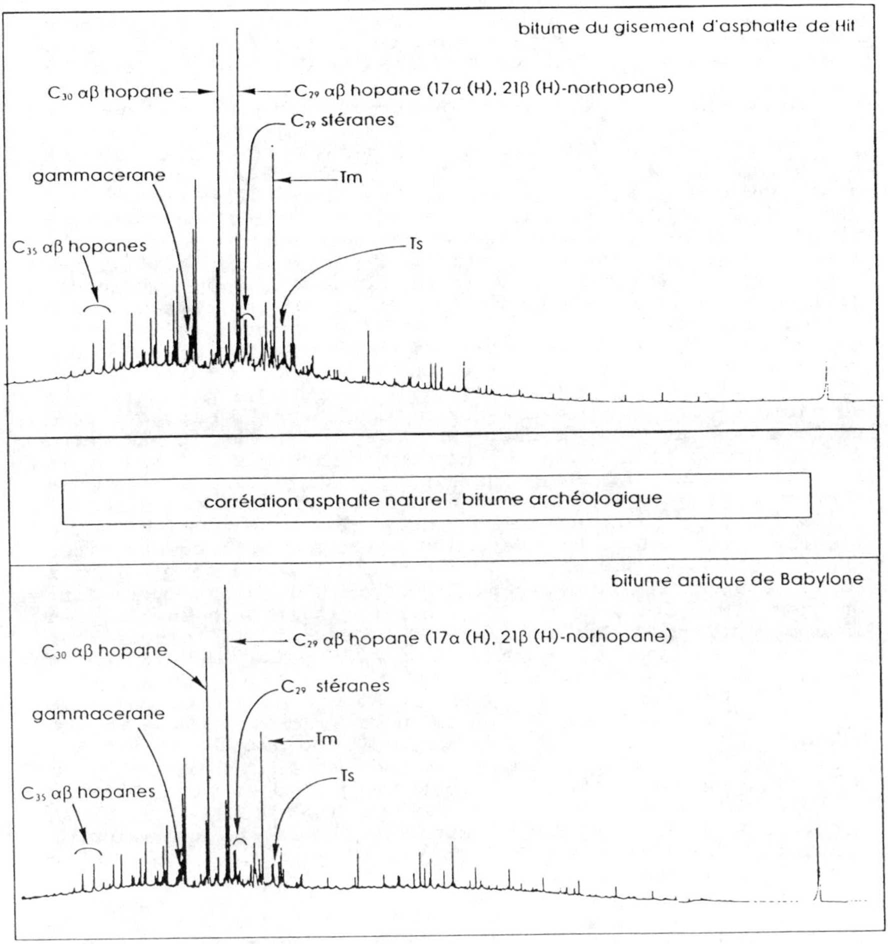

Fig. 4. Sample spectra showing a close relationship between bitumen from Hit and Babylon.

The bitumen from the Sumerian towns, Lagash, Tell el'Oueili, Larsa, Uruk, has a completely different origin. The cuneiform texts dating between 2200 and 1500 B.C., recently commented on by George Petrequin (18), indicate that it is a matter of being from the region of Madga or to be from Dilmun; and the given chemicals, in particular the compatibility of these bitumens with those of Khorsabad or from Failaka, confirm that these two origins are both plausible.

A second important result of our program concerns the site of Ras Shamra on the Syrian littoral. Some bitumin, dating from the Middle and Late Bronze, has been found on some statues or on some of the hafted tools and on the orthostates of the north palace, that is to say on the floor tiles of dressed stone in a field against the base of brick walls, which served at the time as decoration and as protection against the humidity. In comparing the result of analysis of this bitumen with samples of asphalt put at our disposition by our German colleague H. Wehner of the Bundesanstalt fur Geowissenschaften and Rohstoffe (Hanover), we have had some success to determine that this bitumen came from the site of Kfarie (or Jebel Keferie) situated about 38 km to the northeast of Lattaquie and near the town of Babenna (19). Still there, only the comparison with some samples of natural asphalt has permitted us to arrive at this result. Now the difficulty of procuring these samples is a permanent handicap. To this difficulty is added the fact that the properties of natural asphalt vary in the same geographic source as a function of the degree of oxidation and biodegradation. Only one sample is therefore clearly insufficient to characterize the same geological site. It is effectively what we have verified in setting apart a dozen samples from two neighboring localities; Hit and Abu Jir. On these two sites, some of the oil springs forth in globules in some springs of hot water, at the proximity of deposits of most ancient asphalt, altered or solid. The study of these samples shows a certain dispersion of given geochemistry. Meanwhile this dispersion is not of such a nature to put back into question that which has been found, to know that the bitumen of Babylon originates well in the region of Hit-Abu Jir. It is necessary simply to know that the diversity of natural asphalts in the same location, joined to the multitude of possible sources in Iran and in Iraq (20), are given together with which we must compose to obtain an indication of possible origin. To succeed there, the organic geochemistry sets out some trump (21).

In molecular plan, the demarcation consists of analysing in priority the biomarkers not touched by biodegradation. It is that which has been made in Iraq where the steranes and the terpanes, hydrocarbons derived from microalgaes and from bacterias, confirm the origin of the Babylonian bitumens and suggest another origin for those of the site of Tell el'Oueilli for example (22).

The conjugated or interconnected interpretation provided by some organic and petrographic tests has given us some ideas on the fabrication of some archaeological bitumens. The heterogeneous texture of the material is revealing of a melange where the mineral ingredients (quartzite sand, crushed calcite, etc.) and organics (straw, calcined debris and reeds) cohabit with the bituminous binder. The diversity of mineral ingredients suggests that all can be constituted, there comprised of reused material as crushed brick. The bituminous material is the most often an oil or a biodegraded asphalt resembling those from Hit and Abu Jir. But in certain cases, such as at Susa, it is of another nature. The texture of the fashioned objects leaving a bituminous mastic (cement), has allowed one to think of the utilization of a preparation of a base of asphalt or of biodegraded oil which has been intentionally mixed with some ground minerals (23). The geochemical analysis of mastics has produced a different conclusion, for it is established that the objects were made with a powder of bituminous calcite amalgamated with crushed mineral elements. It is the presence of this powder of calcite, having the properties of a veritable mother-rock of petroleum, which has given to the objects their mechanical resistance and which has permitted their sculpture and their polish as a veritable solid rock.

The analysis of objects found at Susa has moreover given other lessons about the mixtures of bitumen of which mention can be found in the texts. These mixtures have been placed as favorable evidence of the discovery, in a sample of archaeological bitumen, of contradictory geochemical characteristics, impossible in natural conditions. In the same sample, we have thus detected the presence of an oil or an extract of nonbiodegraded mother-rock, rich in n-alkanes, hydrocarbons saturated with linear chains, and that of an asphalt extremely biodegraded with very altered steranes. Now, in natural geological conditions, the biodegradation of steranes intervenes always after that of the n-alkanes. Only a mixture of two bitumens has therefore been able to produce this result, confirming thus the given result of texts on the diversity of the commercial products available on the market (Fig. 5). It is to leave these base products, sold in volume or in weights, that the Sumerian asphalters have been able to elaborate the mixtures of which they had need. The archaeological vestiges give moreover some indications of the treatment of bitumen.

BITUMIN AS A LUXURY PRODUCT

The tools, some long stones which were worked by a mixing process, such as kneeding or wedging, the storage amphoras and the basketry vessels abound. At Ur, a cup of bitumen deposited on a mat in an annex of the ziggurat has the appearance of some sort of utilization. At Haft Tepe near Susa or at Sakheri Sughir near Ur, some small stones or lumps of hard bitumen are the ancesters of "ingots" of purified and enriched asphalt that are utilized today by enterprises for the work of waterproofing or making various things water tight. Lastly, the work of Henry Wright at Farukhabad, near the site of natural bitumen at Ain Gir in Khuzistan, has revealed a veritable unity of work in bitumen with at least four intermediate stages of preparation between the raw material and the finished product. Some important quantities of bitumen have been discovered on the site but paradoxically few finished products. The analyses have revealed that the chemical composition of the organic material of the samples from Farukhabad was identical to that of bitumen extract from Ain Gir which actually still produces ten liters of viscous bitumen per day. On the other hand, the bituminous product which composes these samples, after grinding and heating, presently was enriched as a deposit of pure bitumen and soon after rendered very consistant by deposition of the mineral material. Four stages of treatment can be

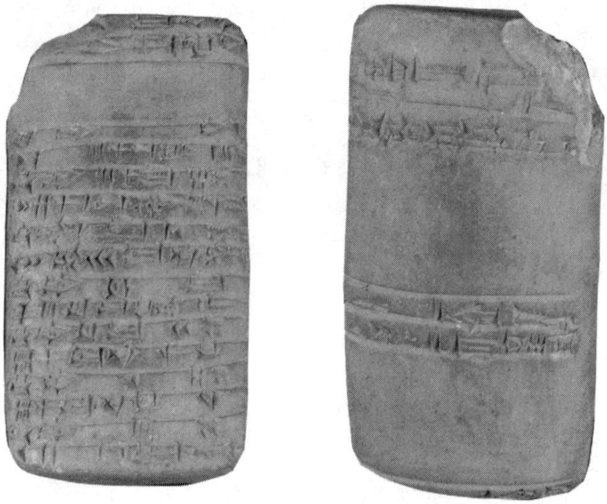

Fig. 5. Tablet from Umma utilized by a merchant named Lukulla in 2039 B.C. which mentions several varieties of bitumen sold at various prices according to quality. It mentions massive and very pure bitumen which were 4 times more expensive than "had", bitumen which was naturally solidified. Mixed bitumen ready to use was more expensive than raw bitumen. Bitumen "of the house", used for construction to join bricks or waterproof terraces, was liquid and could be used in mixtures (J.-Y. Chambault, Coll. Musee du Louvre).

defined and the material obtained presents itself in the form of small, flattened cakes, lumps or stones, perhaps moulded or fashioned by finger and hand. One can therefore suppose that Farukhabad was specialized in the treatment of bitumen and its processing in the form of cakes for transport and subsequent commercialization. The fact that one finds few finished products there signifies that the clearest of the production was exported. And in fact, the bitumen found on the other sites of the region revealed the same organic origin.

This said, can one veritably speak of "civilizations of bitumin"? Not truly in the measure where some centers of very high civilization have been flowering without so much as the employ of some systematic workmanship in bitumen: for example, Ras Shamra, Assyria or the Hittite region in Anatolia. But it is true that in Mesopotamia where the absence of stone conjugates with a natural richness in bitumen, this has played a very important role. It remains meanwhile incontestably a product of luxury in the third millennium where it is utilized however massively. In effect, one can only find it in palaces and temples, only very exceptionally in the private dwellings. To be sure, it is equally bound to domestic usages, but with the exception of the hafting of flint and of the waterproofing of storage vessels, it occurs essentially in the fabrication of objects of luxury. Its usage is rendered commonplace from the middle of the second millennium and especially in the first millennium and at the same time it has strongly diminished until becoming marginal and being only utilized to assure the waterproofing of coarse ceramics. It shall be therefore excessive, in our sense, to believe it as indispensible and common as brick, ceramic or stone tool kits. On the other hand, one can render hommage to the intelligence and the fertile imagination with which the Mesopotamians knew to pull profit from their natural riches. It is false to await the eighteenth century of our era to have an awareness of the potential uses offered by such an ancient material as bitumen (24).

APPENDIX I: A CASE STUDY: THE SOURCE OF ARCHAEOLOGICAL BITUMEN FROM RAS SHAMRA

JACQUES CONNAN, ODILE DESCHESNE AND DANIEL DESSORT

INTRODUCTION

If bitumen forms a part of the materials first readily employed by the artisans and boat-builders of ancient Syrian cities near the Euphrates, it appears to only have been rarely utilized on the site of Ras Shamra, near the Mediterranean (Fig. 6). Compared to their Mesopotamian neighbors, who have known to take the best part of bitumen, ubiquitously found in northern Iraq, the inhabitants of Ras Shamra have apparently not found the same interest for this material, however available in their neighborhood. Is it lack of knowledge of the properties of bitumen or more simply because stone for construction abounds in the region? No one knows this.

Some bitumen was encountered however on the site of Ras Shamra where it has been identified on some objects and monuments dated to the Middle and Late Bronze Age (1600-1200 B.C.). Three samplings of these bitumens have been conferred

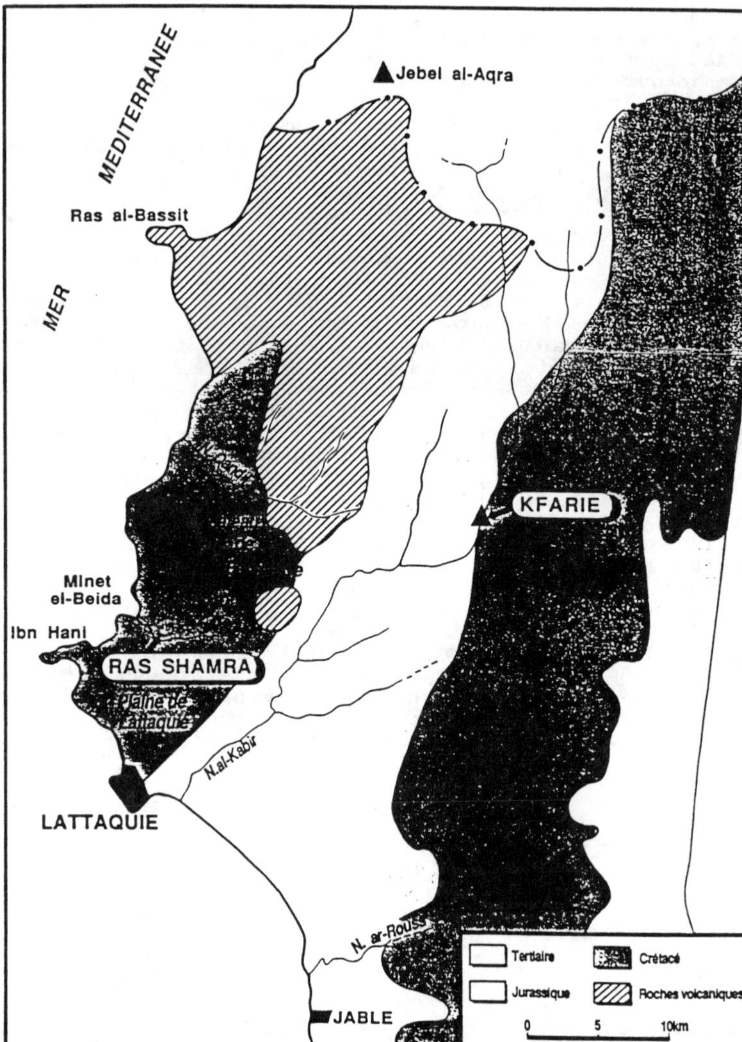

Fig. 6. Geological map of Ras Shamra and environs and of nearby bitumen sources at Kfarie.

on us for analysis in order to try to respond to several questions: how much has one incorporated bitumen in the mixture? What is its composition? From where does it come?

The inventory of occurrences of asphalt-like materials in Syria, published in 1921 by Toll, made note of sites with asphalt to the southeast of Ras Shamra, between Lattaquie and Babenna. Two samples from one of these sites, Kfarie, 38 km to the west of Lattaquie (Fig. 6), have been examined for the purpose of comparison. Their analysis confirms that the archaeological bitumen at Ras Shamra originated effectively from this region (25).

PART I: THE ARCHAEOLOGICAL BITUMENS

The bitumen sampled from blades of flint from the Late Bronze (no. 9) represents a common and strong usage in the antiquity of the Middle East. The blades of flint were mounted in series in the continuous groove of a handle or haft, probably in wood. In order to avoid a very rapid loosening, the groove was previously filled with bitumin, more or less thick, which served as cement, and prevented the blades from moving during use. The sickle from Hassuna (26) of the fifth millennium is a perfect example of this practice which endured widely a very long time, if it is judged by examples attested in the levels of occupation dating to the preceramic neolithic at Qalaat Jarmo (27), or at Nahal Hemar (28). It was lost in all the orient until a recent date, proving that metal has taken a certain time to supplant flint. One has retrieved at Susa in a level of the end of the second millennium, therefore parallel to the occupation of Ras Shamra which occupies us here, a container full of bitumen in which remains the unutilized blades of flint (29). At the same Ras Shamra, the excavations have produced to the day some number of blades carrying traces of bitumen, that the excavators cannot always identify, or that they have sometimes called resin. It remains nevertheless that in regard some quantities of found flint, the middle does not appear as systematically used.

The second sample of bitumen (no. 147) comes from the statuette in stone (Fig. 7) found in 1988 in a pit near the "Temple of Rhytons" (30). It represents a bearded personage and coiffed with a conical tiare, seated on a throne. The personnage has lost his arms and eyes. In the two worked cavities at the level of the elbows, some remains of bitumen allowed one to suppose that the arms, realized in a material probably different, were joined by a tenon, a strong adherence was assured by some bitumen. The setting of the eyes revealed without doubt the same technique.

The identification of this personnage with the god, El, is assured by the resemblance with the other representations of the same god, as the statuette in bronze and gold found in the "South Town" (31), and it was very easy to restore the position of the arms. But there are particularly two statuesses of goddesses in bronze which interest us, found the one and the other at Ras Shamra. The one, which lost the left arm (cast with the rest), is equally amputated from the left forearm, and presents at the level of the elbow an open mortise or slot; the coiffeur has also disappeared but, at the inverse, it ought to join on the visible tenon at the rear of the figure. The other is deprived of two forearms, on which remain the mortises (32). It behaves therefore as a system of fixing perfectly mastered

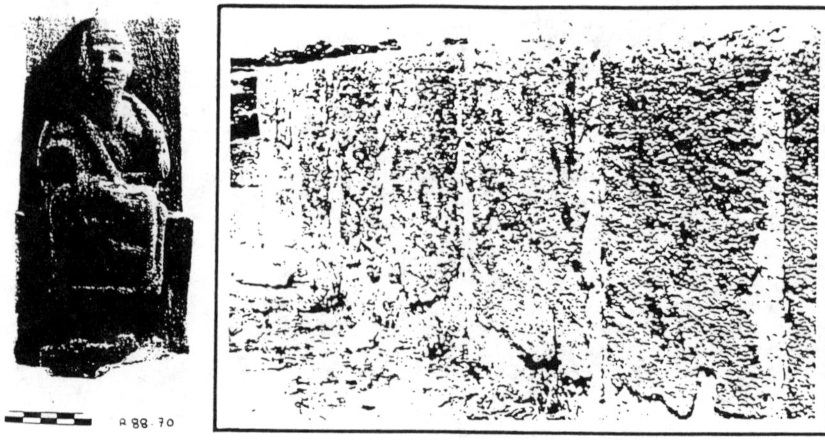

Fig. 7. Statuette of the god El from Ras Shamra-Ougarit (Late Bronze). Eyes and lower arms are missing, but bitumen encrusted cavities where they would have been attached remain.

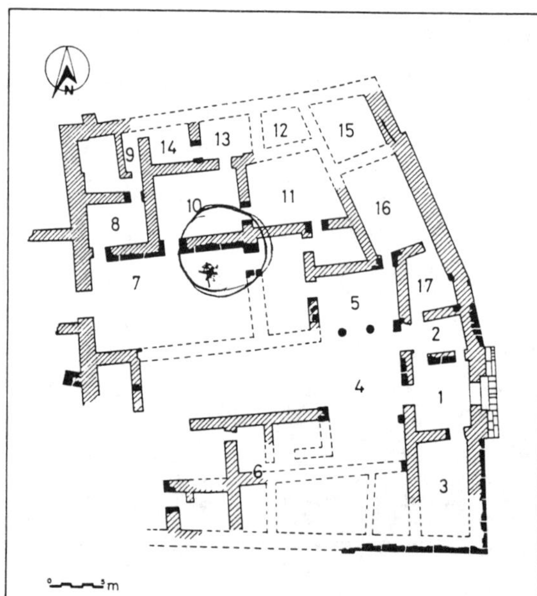

Fig. 8. Orthostates of the North Palace (Middle Bronze) and plan after Saade, 1979, p. 112 with ones coated in bitumen localized.

in the Late Bronze; it offers the double advantage of presenting an attractive aspect by the mixture of materials and therefore colors, and, in the case of the sculpture in wood, of better disengaging or freeing the arms, notably that which makes the gesture of benediction. This was consequently for the sculptors of Ras Shamra, who have never displayed a grand ability in stonework, an indirect way to animate their statues.

This technique of assembly by tenon and mortise was already utilized in the Uruk period. As regards the inlay of the eyes with the aid of bitumen, the statues of Tell es-Sawwan appear to have inaugurated the principle (33). They are followed by a long continuation of statues and statuettes all along the artistic production of the Middle Eastern world. Ras Shamra has not escaped there.

In the "Third Palace" of Ras Shamra, called the "North Palace", discovered among the excavations of 1966-1969, the wall of principal entry and the walls south and west of the principal court were built of orthostates in regular stone (Fig. 8), sort of massive parallelipipeds of about a meter in height, cut and assembled with care, clad with a thick bituminous coating (2 to 4 cm) (34). It was therefore particularly interesting to make the analysis (no. 146) for two reasons. This is, in our knowledge, the only example in the Levant of bitumen utilized in an area which touches the construction. From another source, the palace was well dated to the Middle Bronze; we have there a sample which enlarges our study in time.

The function of this bituminous coating on the walls is an enigma. The bitumen is frequently utilized for the bases of courses, applied in thick layers, applied to seal a pavement in squares of terracotta. It assures also a good discharge of rain water. Sometimes the bitumen paths were designed near the base of courses, assuring to the visitors a certain comfort (35).

The bituminous coatings on the walls are frequent, but one finds them on some interior walls mounted in bricks, coated with bitumen to a certain height--never more than 60 cm--, for the protection from humidity by capillarity. The coating there makes some sort function as paneling, therefore sometimes producing a decorative effect, conjugating thus both usefulness and aesthetic (36). Some pieces, in general most small, sometimes buried, can be able entirely coated with bitumen. They play in this case the role of cisterns or of storage pits, and the bitumen is then an agent of waterproofing. This is not able to be evidently the case of the grand court of Ras Shamra, as so many of the massive slabs of stone withstand the humidity. Only, strictly, some bituminous joints can explain it.

One can also imagine that the bitumen has been applied as a painting on all or part of the wall height, a claim for its use as uniquely decorative, without worry of waterproofing (which has no justification here). The utilization of bitumen for painting is little frequent. It is nevertheless attested in some cases, at Tell Uqair (37) or Susa for example (38). The architects of the north palace at Ougarit deviated to have had equally this objective.

Remaining at last is the possibility of an applied decoration on the coated orthostates: decoration of inlaid mosaics, of glued decorative panels, etc. The reports of excavations never mention this subject, and one can think that over such a length, some fragments of decoration could resist the destruction and attract the attention of excavators. But it is possible also that, to construct the great neighboring palace ("Palace Royal") in the Late Bronze, the "Palace North" was methodically demolished and all the walls were systematically dismantled (39), and that all its decoration was unhappily laid aside only to be reutilized.

The samples from Ras Shamra arise from three very different types of supports the ones from the others. Nevertheless the bitumen has had, for two at least, and perhaps for the third, the same function of adhering agent. For one time the local origin of bitumen utilized at Ras Shamra was determined with certitude and left to the analyses to precise if the compositions (nature of constituents and also proportions) have indicated for the three samples, or similar for the flint and the statuette but different for the revetement of the orthostates: this can confirm the hypothesis of a different utilization, to know purely decorative, and by the same not require the same criteria of composition. It remains at last to know if these compositions are identical on the three or four centuries which cover these objects and if their variations are related to techniques or to different recipes.

PART II: THE SAMPLES, THE ANALYTICAL PROTOCOL AND THE RESULTS

Archaeological Bitumen: The samples from Ras Shamra have been taken from three types of objects: a flint sickle blade (no. 9), orthostates in stone (no. 146), a statuette in wood (no. 147). The quantity of bitumen available was homeopathic, that is to say some grains mixed with mineral debris (Table I).

Natural Asphalt: Two samples from the geological site of Kfarie (or Jevel Keferie or Kfarie, Fig. 6), situated near the village of Babenna about 38 km to the east of Lattaquie (40), we have been given by H. Wehner, of the Bundesanstalt fur Geowissenshaften and Rohstoffe at Hanover.

The two samples were of aspect extremely different. The first (no. 159), ground, resembling a pure bitumen, not associated with some mineral elements. It has been defined as a albertite-wurtzilite (H. Jacob, personal communication). The second (no. 160) is an impregnation of asphalt in a cenomanian calcite (Fig. 6).

Analytical Protocol: The analytical protocol is indicated as that which has been applied previously for the study of archaeological bitumen from Mesopotamia (41). For lack of sufficient material, the study has been focused on the bitumen in the strict sense, as that soluble in chloroform. Only sample no. 146 has been able to be analyzed by x-ray diffraction. It contained 75% calcite and 1% quartz.

The Results: The given analyses have been regrouped in Tables 2, 3 and 4, and are presented below.

1. The Amount of Bitumen and Composition of Bitumen

The three archaeological samples appeared well in the category of bitumens since they contained between 13 and 28% of asphalt extractable in chloroform (Table 2). In this respect,

N° BIT. ARCHEO.	NOM DU SITE	N° LABO	AUTRES REFERENCES	LOCALISATION GEOGRAPHIQUE	DESCRIPTION DES ECHANTILLONS
159	KFARIE	B20595	CO 19369 ALBERTITE S11	PRES DU VILLAGE DE DAMATE (35°35'N - 35°54'E)	ALBERTITE (WURTZILITE) D'APRES JACOBS ASPECT DE BITUME MASSIF BROYE
160		B20596	CO 18370 ASPHALTE S13	PRES DU VILLAGE DE DAMATE (35°35'N - 35°54'E)	IMPREGNATIONS DANS DES CALCAIRES CENOMANIENS
9	RAS SHAMRA		A14LSO 16D BRONZE RECENT	MASTIC D'EMMANCHEMENT DE LAME DE FAUCILLE	TRES FAIBLE QUANTITE D'ECHANTILLON ASPECT DE ROCHE IMPREGNEE. ASSEZ DUR. CONTIENT DE LA CALCITE ET DE LA DOLOMITE
146		B11077	ORTHOSTATE PALAIS NORD BRONZE MOYEN	PRELEVEMENT SUR DE GRANDES DALLES DE PIERRE D'UNE COUR DU PALAIS NORD	BLOC DE «BITUME» - LA SURFACE ETAIT PARSEMEE DE GRAINS DE CALCITE
147		B11955	RS 88-70 PALAIS NORD BRONZE RECENT	TENON DE BRAS DE STATUETTE	ECHANTILLON HETEROGENE - GRAINS DE BITUME A CASSURE CONCHOIDALE AVEC GRAINS DE MINERAUX INSERES DANS LE BITUME

Table 1. Description of the archaeological samples from Ras Shamra and natural asphalts from Kfarie (Syria).

N° BITUME	SITE	E.O. % $CHCl_3$	EXTRAIT ORGANIQUE				E.O.		ASPHALTENES		POINT FUSION	CHROMATOGRAMME DES HYDROCARBURES SATURES					
			SAT %	ARO %	RES %	ASP %	$\delta^{13}C$ ‰/PDB	δD ‰/SMOW	$\delta^{13}C_r$ ‰/PDB	δD ‰/SMOW		PRIST / n-C_{17}	PHYT / n-C_{18}	PRIST / PHYT	A/B	PRIST2 / n-C_{17}	PHYT2 / n-C_{18}
159	KFARIE	88,4	0,1	0,2	10,4	89,3	- 28,1	- 94	- 28,0	- 91	> 310°C	1,04	1,24	0,56	0,84	0,48	0,61
160		15,4	6,2	26,7	44,2	22,8	26,1	- 110	- 26,0	- 87	—	—	—	—	—	—	—
9	RAS SHAMRA	13,7	3,7	2,7	48,5	45,1	—	—									
146		17,7	1,2	0,4	28,8	69,6	25,1	- 100	- 25,0	- 87		—	—	—	—	—	—
147		28,3	1,9	1,0	43,1	53,9	26,4	- 101	- 26,0	- 82		1,07	0,56	0,93	1,89	0,51	0,30

Table 2. Geochemical results of relative base from chloroform extraction for bitumens from Ras Shamra and Kfarie.

Signification des abréviations :

E.O. % $CHCl_3$	: Extrait chloroformique en % en poids/échantillon.
SAT %	: Hydrocarbures saturés C_{15+} en % en poids/E.0.
ARO %	: Hydrocarbures aromatiques C_{15}+ en % en poids/E.O.
RES %	: Résines en % en poids/E.O.
ASP %	: Asphaltènes en % en poids/E.O.
E.O. $\delta^{13}C$‰/PDB	: $\delta^{13}C$ sur extrait chloroformique (en ‰/PDB).
E.O. δD‰/SMOW	: δD sur extrait chloroformique (en ‰/SMOW).
ASPHALTENES $\delta^{13}C$‰/PDB	: $\delta^{13}C$ sur asphaltènes (en ‰/PDB).
ASPHALTENES δD‰/SMOW	: δD sur asphaltènes (en ‰/SMOW).
PRIST/n-C_{17}	: pristane/n-alcane en C_{17}.
PHYT/n-C_{18}	: phytane/n-alcane en C_{18}.
PRIST/PHYT	: pristane/phytane.
A/B	: PRIST/n-C_{17}/PHYT/n-C_{18}.
PRIST2/n-C_{17}	: 2e pic dans le doublet pristane/n-C_{17}.
PHYT2/n-C_{18}	: 2e pic dans le doublet phytane/n-C_{18}.

REFERENCE PARAMETRE	15	16	17	18	19	20	21	22	23	24	25	26	27	28	29	30
PARAMETRES / ECHANTILLONS	27ββS / 27ααS	27ααS / 27ααR	21St / 22St	29ββS / 29ααR	29ααS / 29ααR	224Me / 29ααR	224Me / 22St	% C_{27} / St	% C_{28} / St	% C_{29} / St	27St / 29St	TT / ST	27-30H / 29St	TT / TT+ST	27-30H / 27-30H+29St	27SDIA / 29ααR
159 KFARIE LATTAQUIE	0,69	0,74	1,84	1,01	0,68	0,15	0,23	32	36	31	0,95	2,51	4,19	0,71	0,80	0,23
160 KFARIE LATTAQUIE	—	—	1,76	—	—	—	0,19	—	—	—	—	13,53	—	0,93	1	N. DEF.
9 RAS SHAMRA	—	4,43	1,43	1,19	1,12	0	0	25	20	55	0,48	7,84	7,89	0,88	0,88	0,28
146 RAS SHAMRA	—	—	1,19	—	—	—	0,07	—	—	—	—	11,78	—	0,92	1	N. DEF.
147 RAS SHAMRA	—	—	1,24	—	—	—	0,24	—	—	—	—	2,85	—	0,74	1	N. DEF.

N. DEF. = NON DEFINI

Table 3. Characteristic ratios calculated on steranes (m/z 217).

Signification des abréviations :

15-27ββS : 5α(H), 14β(H), 17β(H), 20S-cholestane.
15-27ααR : 5α(H), 14α(H), 17α(H), 20R-cholestane.
16-27ααS : 5α(H), 14α(H), 17α(H), 20S-cholestane.
17-21St : 5α(H), 14β(H), 17β(H), -pregnane.
17-22St : 5α(H), 14β(H), 17β(H), methyl 20-pregnane.
18-29ββS : 5α(H), 14β(H), 17β(H), 20S-24 ethylcholestane.
18-29ααR : 5α(H), 14α(H), 17α(H), 20R-24 ethylcholestane.
19-29ααS : 5α(H), 14α(H), 17α(H), 20S-24 ethylcholestane.
20-22 4Me : C_{22} 4-methylsterane.
22-% C_{27} St : % C_{27} steranes/total steranes.
23-% C_{28} St : % C_{28} steranes/total steranes.
24-% C_{29} St : % C_{29} steranes/total steranes.
25-27St-29St : C_{27} steranes - C_{29} steranes.
26-TT/ST : Triterpane to sterane datio : ratio of m/z 191 integral for C_{27}-C_{35} terpanes to m/z 217 integral for C_{27}-C_{29} steranes.
27-27-30 H : C_{27}-C_{30} αβ hopanes.
28-TT+ST : C_{27}-C_{35} terpanes + C_{27}-C_{29} steranes (Cf. 26).
30-27Sdia : C27S diasterane.

REFERENCE PARAMETRE		1	2	3	4	5	6	7	8	9	10	11	12	13	'14	31
PARAMETRES / ECHANTILLONS		$C_{29}H$ / $C_{30}H$	Tm / Ts	23/3 / 24/4	23/3 / 21St	29/5 / 29H	Gcrn / $C_{30}H$	2 35/6 / 35H	35H / 33H	29+30 / 35H	28/5 / Tm+Ts	28/5 / 29αβH	$MC_{29}\alpha\beta H$ / $MC_{30}\alpha\beta H$	M HOPANES / HOPANES	SECOH / 29+30H	29BNH / 29H
159	KFARIE LATTAQUIE	1,10	4,50	1,30	0,95	0,12	—	—	0,89	8,47	0,25	0,14	1,10	0,05	0,04	—
160	KFARIE LATTAQUIE	3,76	5,39	1,37	1,51	0,06	—	—	—	—	0,23	0,09	0,68	0,16	0,11	0,05
9	RAS SHAMRA	20,95	5,06	1,19	6,58	0,10	—	—	—	N.D. TRES ELEVE	0,27	0,08	0,79	0,18	0,06	—
146	RAS SHAMRA	7,49	4,32	0,97	3,44	0,15	—	—	—	8,97	0,30	0,11	0,98	0,19	0,09	0,04
147	RAS SHAMRA	2,20	4,32	0,40	1,30	0,11	—	—	—	N.D. TRES ELEVE	0,37	0,13	1,26	0,21	0,18	0,04

N.D. = NON DETERMINE

Table 4. Characteristic ratios calculated on terpanes (m/z 191).

Signification des abréviations :

1-C_{29}H	: 17α(H), 21β(H), norhopane (C_{29}).
1-C_{30}H	: 17α(H), 21β(H) - hopane (C_{30}).
2-Tm	: 18α(H) -22,29,30 trisnorneohopane.
2-Ts	: 17α(H) -22,29,30 trisnorhopane.
3-23/3	: C_{23} tricyclic terpane.
3-24/4	: C_{24} tetracyclic terpane.
4-21St	: 5α(H), 14β(H), 17β(H) -pregnane.
5-29/5	: 17-methyl-28,30-dinorhopane. (TRENDEL et al., 1990).
6-Gcrn	: Gammacerane.
7-35/6	: hexahydrobenzohopane (35/6).
7-35H	: C_{35} αβ hopane.
8-33H	: C_{33} αβ hopane.
9-29+30	: C_{29} H + C_{30} H.
10-28/5	: C_{28} αβ hopane.
11-M C_{29} αβ H	: 2α- methyl - αβ norhopane (C_{30}).
13-MHopanes	: sum of methylhopanes (m/z 205 integral).
13-Hopanes	: sum of hopanes (m/z 191 integral).
14-SECO	: sum of secohopanes (m/z 123 integral).
31-28 BNH	: 28,30-dinor-17α-hopane.

the bitumens of Ras Shamra possesed well the dominant chemical physionomy of archaeological bitumens for their global characteristics are definitively near enough those of Mesopotamian archaeological bitumen (Fig. 9). They act in effect as new samples of bitumen rich is polars, indexed as gilsonite and wurtzilite in the classifications of Cornelius (42). In these samples, the amount of saturated and aromatic hydrocarbons is always weak and attains a value here of 6.4% in sample 9 from Ras Shamra.

The two natural asphalts offer a marked chemical contrast. The first (no. 159) is confirmed as much as quasi-pure bitumen for it was soluble at 88.4% in the chloroform. In this extract, the asphaltanes very largely predominate with an amount of 89%. This bitumen, consisted of the essential composed of polars, melted at elevated temperature, above 310°. Defined by H. Jacob as a albertite/wurtzilite, it effectively possessed the characteristics with 89% of asphaltenes for 0.3% of hydrocarbons.

The other sample (no. 160) is very much classic. It behaves well like an impregnation in calcite cenomanian where the asphalt represents 15.4%. This asphalt possesses much less asphaltene (23%) with by contrast much more hydrocarbon (33%) of which 26.7% was aromatic (Table 2, Fig. 9).

The contrast of chemical properties found in the two asphalts of Kfarie incites prudence; in effect, it suggests directly that it has been desirable to have at one's disposition a more complete lot of samples in order better to know this site. This has unhappily not been possible for the two samples have been the only ones to which we have had access.

2. Isotopy of Extracts and Asphalts

The values of delta^{13}C and deltaD of the chloroform extracts and the asphaltenes conferred to quasi-pure bitumen (no. 159) a character clearly to set out (delta^{13}C near -28.0%o/PDB instead of -25-26%o/PDB) which places it adjoining the oils and asphalts from Hit and Abu Jir to the northwest of Babylon (Fig. 10). The other samples in this study (Ras Shamra and Kfarie) regroup themselves against in a population well individualized by association with all the other archaeological bitumens already studied (Susa, Tell el-Oueili, Larsa, Tello, Uruk, Khirbit Diniye, Babylon, Khorsabad, Failaka, Bahrein). At the regional level, these are the asphalts impregnated in the cenomanian calcite of Kfarie and correlate the best with the archaeological bitumens of Ras Shamra.

3. Molecular Givens: Diagnostic Petroleum

(a) Chromatographic Analyses of Alkanes C_{15+} and Aromatic C_{15+}.

The chromatograms of saturated hydrocarbons C_{15+} (or alkanes C_{15+}) of 5 samples (Fig. 11) shows very diverse imprints.

The sample of albertite/wurtzilite (no. 159) demarks itself as novel or new from the others by a chromatogram of alkanes of non-biodegraded oil where the n-, iso-, antiiso-alkanes are abundant. The report pristane/phytane is nearly of 0.6 (Table 2), that which tends to indicate that the source rock of this asphalt is deposited in a confined environment (carbonates with the evaporitic tendancy?). The polycyclical

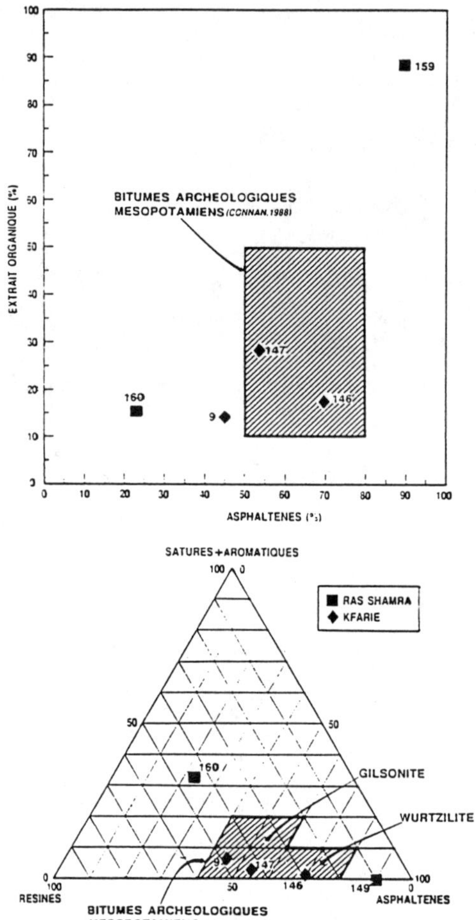

Fig. 9. Quantity of bitumen extractable with chloroform. Composition of this extract. Note that the bitumens from Ras Shamra are compared here to those from Mesopotamia.

hydrocarbons of the family of alpha-beta hopanes are still well represented, that which is able to wish to signify that the maturity of this asphalt is not very elevated.

Two samples, a natural asphalt (no. 160) and an archaeological bitumen (no. 146), display the imprints of total alkanes of extremely biodegraded oils where the n-alkanes have been totally eliminated (Fig. 11). Some analogous situations are "current money" or probable in some number of archaeological bitumens from Syria and Iraq (43).

Between the two extreme situations which come to be equivocal, come to insert themselves some intermediate states (Ras Shamra no. 8 and 147), where the n-alkanes, clearly visible, are all the same attenuated by report of polycyclical alkanes (Tm, C_{29} alpha-beta hopane, etc, Fig. 11). One such situation is in general set at the time the biodegradation of oils not still attained the state of complete elimination of the n-alkanes. It can result in the present case of a mixture of two bitumens of degrees of different biodegradation for it behaves here with archaeological bitumens, fashioned by man, and not with natural asphalts.

The examination of the chromatograms in gaseous phase of the aromatic hydrocarbons C_{15+} (Fig. 12) shows an absence of compounds of low molecular weights (napthalene and homologous methyls, benzothiophene and homologous methyls).

The analysis of these aromatics by coupling CG-MS-ordinator precises the molecular contents in permitting the identification of steroids mono- and triaromatic, some benzohopanes (C_{32} to C_{35}, Fig. 12). The presence of this last molecular category is not surprising in this carbonated context of confined character (44). Certain molecular imprints (Kfarie no. 160, Ras Shamra no. 146, Fig. 12) is all typically made of extremely diodegraded asphalts where the only aromatic structures identifiable are the steroids (mono- and triaromatics) and the benzohopanes, particularly resistant. Near these molecular structures, diagnostic of a confined carbonate environment, the mass spectrometry has delivered some molecular processions (or spectra) attributable to the anthropogenic pollutants. This is just as the wurtzilite is contaminated by the polychlorures of biphenyl (pesticides) and the pthalates, whereas the two archaeological bitumens from Ras Shamra (no. 146-147) are equally so by the pthalates. The sample no. 147 presents another particularity of containing an ensemble of molecules to which we have assigned structures of the type dialkylbenzenes. These structures, not identified with certitude, have not been found in the other samples.

(b) The Detailed Molecular Analyses of Alkenes C_{15+}: Correlation of Natural Asphalt and Archaeological Bitumins.

The mass fragmentograms of steranes (m/z 217) and of terpanes (m/z 191), obtained by coupling gas-phase chromatography and ordinary mass-spectrometry, has been reassembled in the Figures 13, 14 and 15.

With the exception of that which is reported as sample no. 159, not biodegraded, the mass-chromatograms reveal some original distributions, which have permitted us to directly place the bitumens of Ras Shamra in a class with part of the bosom of the population of archaeological bitumins from Syria (Mari, Habuba Kabira), from Iran (Susa) and from Iraq (45).

The originality of imprints was established, remains to uncover the cause by a careful molecular reading studied

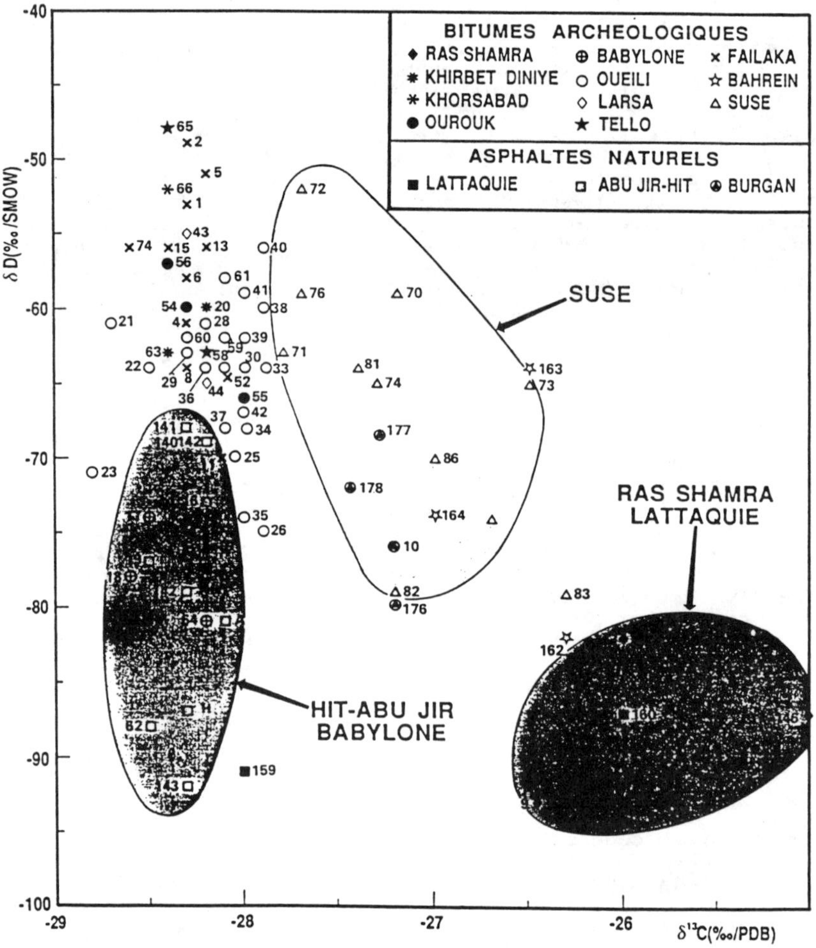

Fig. 10. Isotopic composition (delta[13] and deltaD) of asphaltenes of bitumens from Ras Shamra and Kfarie, compared to those from other sites and sources.

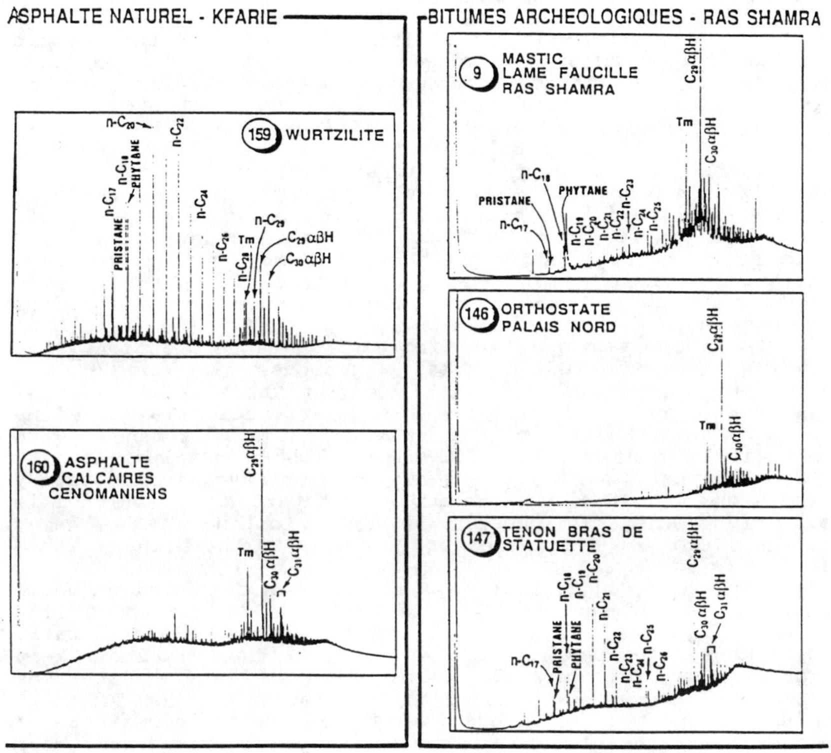

Fig. 11. Chromatograms in gas phase of alkanes C_{15+} of 3 archaeological bitumens from Ras Shamra and of 2 natural asphalts from Kfarie (50 m x 0.21 mm i.d. column, coated with OV1, film thickness 0.11 microns, temperature programmed from 80 to 300°C at 0.6°C per min., H_2 as carrier gas).

deeply. This precise molecular decoding is in effect indispensible to establish that all the bitumens, different in appearance (Fig. 13 and 14) have in fact the links of the same genetic family, diversified by biodegradation more or less intense.

(1) The Distributions of Steranes (m/z 217, Fig. 13)

The spectra of steranes of albertite/wustzilite (no. 159) confirm without ambiguity that the alkanes C_{15+} of this sample are not biodegraded. In effect, one identifies there some steranes of high molecular domaines by some regular steranes (C27, C28, C29, alpha-alpha-alpha R and S, alpha-beta-beta R and S) accompanied by minor disesteranes (C_{27}S and C_{27}R dia). This distribution accords perfectly with that of a carbonated source rock, deposited in an reducing environment, such as has already been proposed. Near regular steranes of high molecular weight, there exist abundant short chain steranes (C_{21} and C_{22}) but few methylsteranes (C_{22}4MeSt, C_{30}MeSt). This distribution made reference figures for the sterane of source rocks for this region. To depart this distribution type, it is possible to engender all the others by biodegradation more and more pronounced.

In this sequence of increasing biodegradation, the distribution of sterane which approaches to most of the distribution type is that of bitumen of the sickle blades of Ras Shamra (no. 9, Fig. 13). This distribution is marked by the preferential elimination of regular steranes with biological configuration (27alpha-alphaR, 28alpha-alphaR, 29alpha-alphaR?) such that it has been observed in laboratory conditions with the stock Nocardia or Arthrobacter (46). This selective alteration goes on an equal footing with a heavy concentration of diesteranes (27 S dia/29alpha-alphaR, Table 3).

The biodegradation of regular steranes can continue until their total elimination. It is the case in the three last bitumens (no. 146, 147, 160) where the ratio triterpanes/steranes attained at the time the values of 12-14 (Table 3). This drastic elimination of regular steranes provoked a concomitant enrichment in diesteranes, reputed to be refractory to the biodegradation (47) and to the appearance of echoes attributable to residual terpanes more and more concentrated in these very altered samples (Fig. 14 and 15).

If these results are in general in accord with the characteristics of altogether the alkenes C_{15+} (Fig. 11), they some problems for sample 147. In effect this archaeological bitumen with extremely degraded steranes (Fig. 13) contains some n-alkanes C_{15+} (Fig. 11). These two bits of information are incompatible in natural conditions for the intense biodegradation of steranes intervenes always downstream, when the n-alkanes have totally disappeared.

To this state, it is necessary to remember that we have concern here with an archaeological bitumen which has thus known to be prepared by mixing an oil and an asphalt, each with a degree of biodegradation very different. This hypothesis has all been made plausible for we have confirmed it elsewhere (at Susa in Iran for example). In all cases, it is not necessarily to be accepted without reserve, for the sample in question was in quantity homeopathic and in addition very heterogeneous. Having had disposition of a small quantity of material, all came to be consummed and mixed. It is not excluded that this is in a state that one has brought about a mixture by

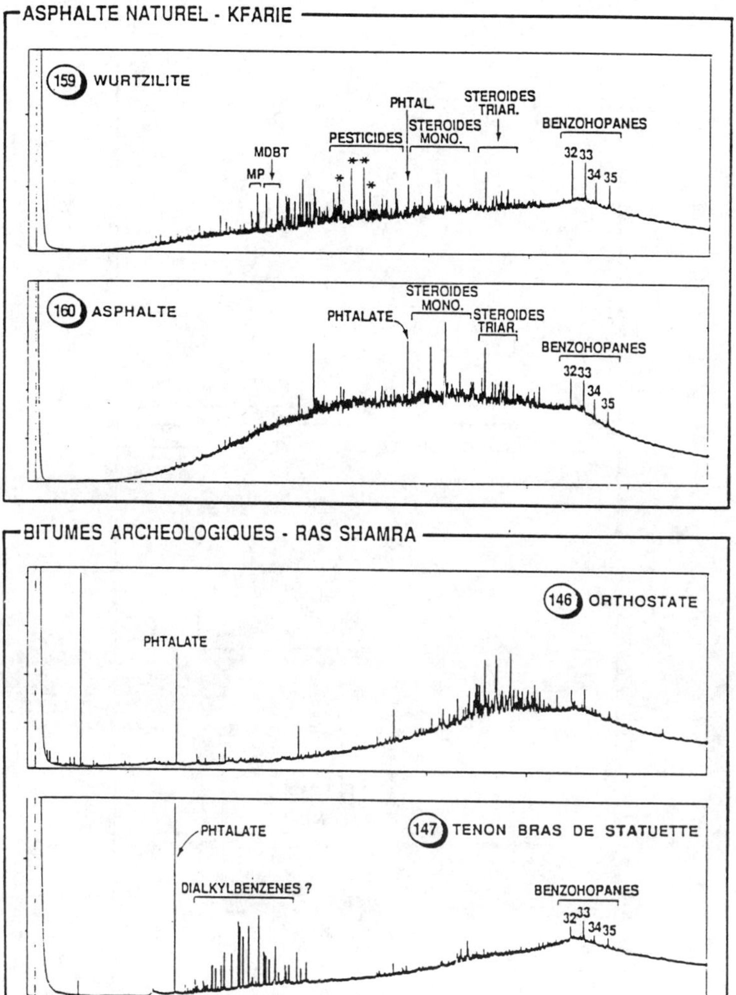

Fig. 12. Chromatograms in gas phase for aromatic hydrocarbons C_{15+} of 2 archaeological bitumens from Ras Shamra and 2 natural asphalts from Kfarie (50 m x 0.25 mm i.d., coated with SE 54, film thickness 0.17 and same program).

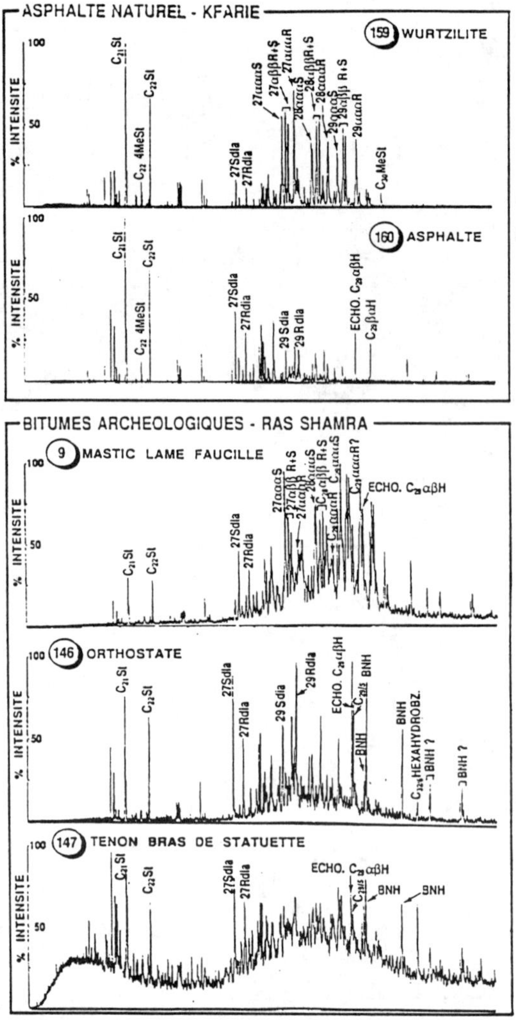

Fig. 13. Distribution of steranes (miz 217) in two natural asphalts from Kfarie and 3 archaeological bitumens from Ras Shamra. This figure shows the progressive transformation of steranes by biodegradation in sequence as 159-9-160, 146, 147. Abbreviations given in Tables 3 and 4.

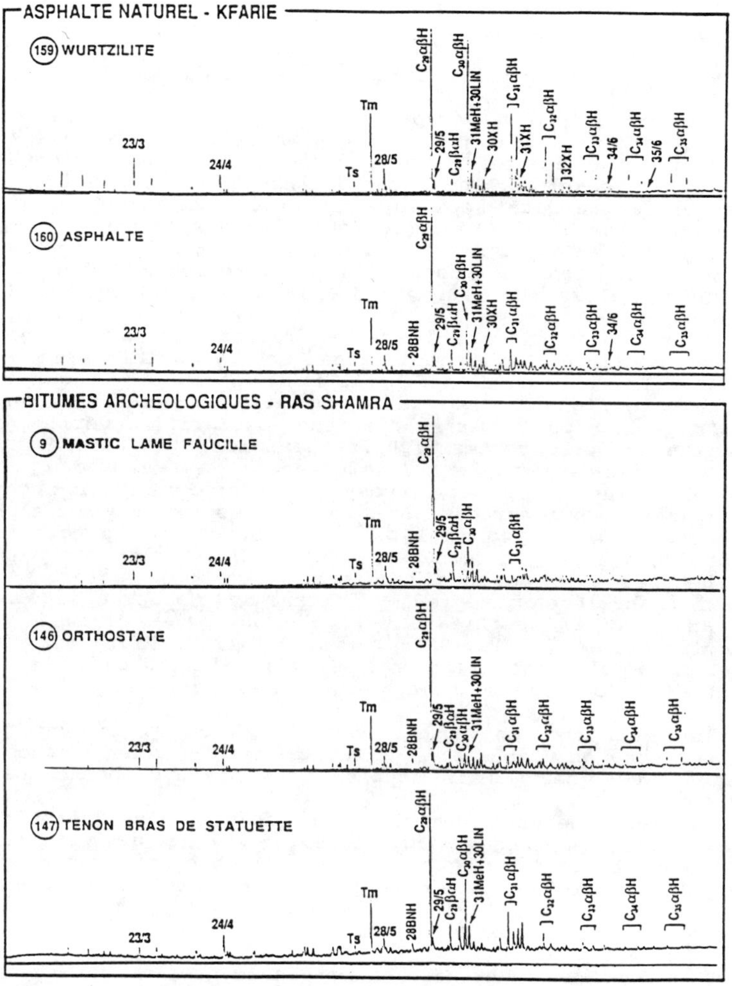

Fig. 14. Distribution of terpanes (miz 191) of 2 natural asphalts from Kfarie and 3 archaeological bitumens from Ras Shamra. This figure shows the progressive transformation of terpanes by preferential biodegradation of C_{30}-C_{35} alphabeta hopanes; the sequence is believed to be 159-147, 160-9, 146 with the same abbreviations as Tables 3 and 4.

juxtaposition of two very different bitumens. In conclusion, it appears reasonable to admit that there has been a mixture, but it is impossible to say if this mixture was intimate and original or if it occurred at the moment of sampling.

(2) The Distributions of Terpanes (m/z, 191 Fig. 14 and 15).

The comparison of terpanes (m\z 191, Fig. 14) made to bring out directly that those of the wurtzilite/albertite (no. 159) can serve as reference because of the nonbiodegraded evidence. These terpanes are dominated by the ubiquitous series of alpha-beta hopanes, common in the archaeological bitumens of the Middle East (48) associated with several minor series (beta-alpha hopanes, methylhopanes, hexahydrobenzohopanes, etc.). In this region of the Middle East nevertheless the terpanes are less abundant by ratio to the steranes - TT/ST = 2.5 instead of 10-25 in Iraq -, that which tends to indicate that the facies of source rocks, although confined, have not attained the evaporitic state.

In effect, in these sediments the contribution eucaryotes (microalgaes rich in steriods) is clearly more marked than in the Iraqi source rocks where the bacterial contribution predominates (procaryotes rich in hopanoids).

To leave the state of references, represented by the terpanes of wurtzilite, it is possible to produce other spectra of terpanes by biodegradation, more or less intense, of alpha-beta hopanes (Fig. 14 and 15). It is necessary to note that the biodegradation of terpanes affects in priority the alpha-beta hopanes of more that 30 atoms of carbon, conforming to that which has been observed by Seifert and Moldowan (cf. ref. 47), and that it modifies little the molecules of more weak molecular weight (Tm/Ts, 28/5/Tm+Ts, 29/5/29H, Table 4). This preferential biodegradation of C_{30}-C_{35} alpha-beta hopanes has the effect of concentrating several molecular families (Fig. 10), minor to depart, to know the methylhopanes (50) - 2alpha-methylhopanes and one other series of methylhopanes, cf. MHopanes/Hopanes, Table 4-, the 28,30-dimer-17alpha-hopanes(C_{28} to C_{33}), the linear chain hopanes, the hexahydrobenzohopanes (51).

All these terpane structures, reputed more resistant that the alpha-beta hopanes of high molecular weight, are not formed by biodegradation but find themselves simply concentrated (52). They are therefore diagnostic to establish the genetic affiliations among the wurtzilite, not altered, and the other bitumens, extremely biodegraded.

4. Molecular Givens: Archaeological Incidences

The originality of distributions of steranes and of terpanes from archaeological bitumins from Ras Shamra has permitted us very rapidly to class them in a population apart, well individualized by relation to all the other antique bitumens already analyzed from Syria (Habuba Kabira, Terqa, Mari) and in Iraq and Iran (Susa). The right resemblance between these distributions and those of cenomanian impregnated calcite (no. 160) of Kfarie orients the research of the origin of these bitumens toward a local source represented by the layer of asphalt at Kfarie. The analysis of molecular contents detailled some three bitumins of Ras Shamra and of two natural asphalts from Kfarie came to conform this hypothesis en carrying some precision.

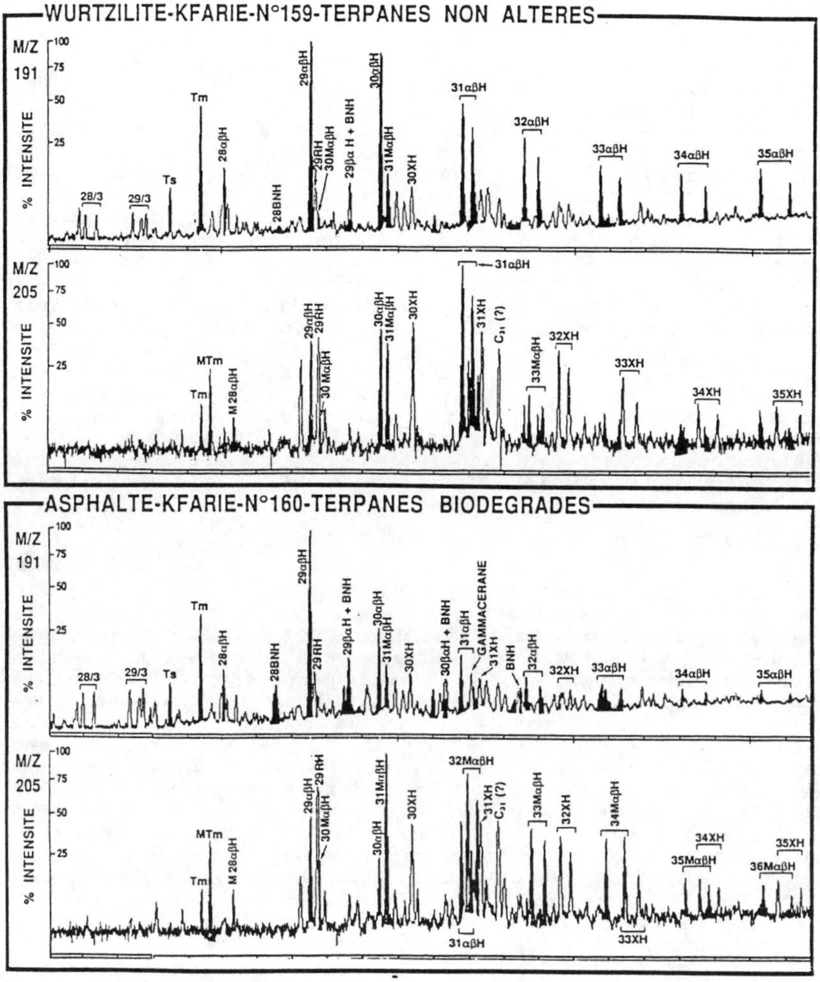

Fig. 15. Identification of the principal families of terpanes in two natural asphalts of the region of Kfarie. Wurtzilite (no. 159) has nonbiodegraded alkanes whereas the impregnation of cenomanien calcites (no. 160) appearn to group extremely biodegraded oils. The figure shows the evolution of terpanes by preferential biodegratation of C_{30}-C_{35} alphabeta hopanes, and the concomitant concentration of very resistant terpanes (see m/z191). Some ones of these families have been identified in the fragtogram of mass 205. It behaves as a 17-methyl-28,30-dimorhopane (noted 29/5), 2alpha-methyl-alphabeta hopanes (C_{28} to C_{36}), of a series of structure "hopane" probably not known but identified here by the letters XH.

The archaeological bitumens were all extremely biodegraded asphalts which have been generated by a source rock of dominant carbonate, deposited in an environment confined apparently non-evaporitic of which the genetic characteristics were near those of wurtzilite/albertite no. 159. This albertite/wurtzilite of these alkenes C_{15+} were not biodegraded had not been encountered in archaeological bitumens.

The three bitumens of Ras Shamra appear or share a same genetic family of which the ends have evolved from the nonbiodegraded oil (type 159) to the extremely biodegraded asphalts (160, 9, 146, 147).

The molecular differences of detail are not attributable to some differences of origin but to more or less intense remodeling of the initial molecular procession by biodegradation. An archaeological bitumen (no. 147, tenon of the arm of the statuette) poses a problem for its molecular properties indicate that there has been necessarily a mixture of two petroleum products of a different degree of biodegradation. Is it a mixture intimately fabricated by man in antiquity or a fortuitous mixture, result of sampling or sample preparation before analysis. One cannot say with certitude, for it was impossible to trace between one or another hypothesis without being at the source of the archaeological sampling (reduced to 6 or 7 grains of bitumen accompanying the mineral debris).

CONCLUSIONS

The three archaeological bitumens from Ras Shamra (sickle blade, orthostate of the North Palace, tenon of the arm of a statuette of a seated man), dated to the Middle to Late Bronze (1600-1200 B.C.), was of extremely biodegraded oil which presented some common global chemical properties, known in this type of samples. Particularly devoid of hydrocarbons (less than 4% saturated and aromatic hydrocarbons), they were essentially constituted of polar products: resins and asphaltenes. In the three samples the amount of extractable bitumen to chloroform varied from 13 to 28 wt%.

From the molecular and isotopic point of view, the characteristics of bitumen from Ras Shamra individualized them by relating to other archaeological bitumen already studied in Syria, Iraq, Iran, etc. They have a local origin which has been found favorable by the establishment of correlations between these bitumens and a natural asphalt, an impregnation in some cenomanian calcites of the region of Kfarie. They have been generated to set out in a source rock, deposited in a confined environment but not evaporitic, which is probably of the same type as those which existed in the Cretaceous superior of Palestine.

REFERENCES

1. R.J. Forbes, "Bitumen and Petroleum in Antiquity", Studies in Ancient Technology, v. 1, E.J. Brill, 1964.

2. G. Hansen, Steinol und Brunnenfeuer, Wintershall Aktiengessellschaft, 1975.

3. R.F. Marschner and H.T. Wright, in G. Carter (ed.), Archaeological Chemistry, v. II, Advances in Chemistry Series,

American Chemical Society, 1978; and R.F. Marschner et al., Paleorient 4 (1978) 97ff.

4. M.I. Venkatesan et al., Nature 295 (1982) 517.

5. S. Lloyd et al., Eridu, Ministry of Culture and Information, Baghdad, 1981; J. de Morgan et al., Memoires de le delegation en Perse, v. 1, 13, 25, E. Leroux, 1900, 1912, 1934; and J. de Morgan et al., Memoires de la Mission Archaeologique en Iran, v. 29, Geuthner 1943.

6. J. Duval, Les Jardins Suspendus de Babylone, Famot-Francois Beauval Editeur, 1980.

7. J.-L. Huot et al., Larsa et Tell el Oueili, Travaux des Campagnes 1978-1981 et 1983, Editions Recherche sur les Civilizations, 1983 and 1987.

8. A. Parrot, Mission Archaeologique de Mari, v. 1, 2, 3, Geuthner, 1956, 1958, 1967; J.C. Margueron et al., Mari, v. 1, 2, 3, 5, Editions Recherche sur les Civilizations, 1982, 1983, 1984 and 1987.

9. H. Frankfort et al., The Gilmilsin Temple and the Palace of the Rulers at Tell Asmar. the University of Chicago Press, 1940.

10. A. Parrot, Tello, Vingt Campagnes de Fouille 1877-1933, Albin Michel, 1948.

11. S.A. Kramer, Iraq 39 (1977) 59ff.

12. E.A. Wallis Budge, "The Babylon Story of the Deluge and the Epic of Gilgamesh", in Publications of the British Museum, Harrison and Sons, London, 1920.

13. S. Cleuziou, personal communication.

14. J. Rulkotter and A. Nissenbaum, Naturwissenschaften 75 (1988) 618.; J. Connan and D. Dessort, C.R. Acad. Sci. Paris 309 (1989) 1665.

15. O. Deschesne, Le Travail du Bitume a Suse, Inventaire Archaeologique, Memoire de l'Ecole du Louvre, 1982.

16. J. Connan, "Quelques Secrets des Bitumes Archeologiques de Mesopotamie Reveles par les Analyses de Geochimie Organique Petroliere", Bull. Centre Rech. Explor. Prod. Elf-Aquitaine, 12 (1988) 759.

17. Cf. note 1.

18. G. Petrequin, Le Bitume en Mesopotamie aux Epoques Neo-Sumerienne et Paleo-Babylonienne d'apres les Sources Cuneiformes, Memoire de Maitrise Universite Paris I, 1989.

19. J. Connan et al., Centre Rech. Explor. Prod. Elf-Aquitaine, 14, 2 (1990) 521.

20. E.H. Pascoe, Memoirs of the Geological Survey of India 48

(1922) 1ff.

21. J. Connan, "La Geochimie Organique au Service de l'Exploration Petroliere", supplement, La Recherche 227 (March 1989) 18-23.

22. Cf. ref. 16.

23. C. Lahanier, Annales du Laboratoire de Recherche des Musees de France, 47 (1977).

24. Other useful references are H. Abraham, "Historical Review", Asphalts and Allied Substances, D. Van Nostrand Co., Inc., 1960, p. 3ff; G. Pile, "Jadis le Petrole", Petrole et Techniques 13 (Aug.-Sept. 1978); J. Connan and O. Deschesne, "Le Bitume dans L'Antiquite", La Recherche 22 (Feb. 1991) 152-159; and J. Connan, O. Deschesne and D. Dessort, "L'Origine des Bitumes Archeologiques de Ras Shamra", Ras Shamra-Ougarit VI: Arts et Industries de la Piere, ADPF, Paris, 1990, pp. 101-125.

25. O. Deschesne, "Exemples d'Utilization dy Bitume a Ras Shamra:, in M. Yon, J. Gauchet, P. Lombard, J. Mallet and O. Deschesne, "Fouilles de la 48e campagne (1988) a Ras-Shamra-Ougarit", Syria 67 (1990) 1-29; and in general, Mission Archeologique de Ras Shamra, Ras Shamra, 1929-1979, CMO, Lyon, 1979.

26. S. Lloyd and F. Safar, "Tell Hassuna: Excavations by the Iraq Government Directorate of Antiquities in 1943 and 1944", Jour. of Near Eastern Studies 4 (1945) 255-289, esp. 268-274 and Fig. 37.

27. J. Mellaart, The Neolithic of the Near East, Thames and Hudson, London, 1975, pp. 81-82.

28. O. Bar-Yosef, A Cave in the Desert: Nahal Hemar, 9000 Year Old Finds, The Israel Museum, Jerusalem, 1985.

29. R. de Mecquenem, "Contribution a l'Etude des Outils en Pierre Trouves dans les Ruines de Suse", L'Anthropologie, 33 (1923) pp. 469-474, figs. 1 and 2.

30. M. Yon and Gauchet, "Une Statuette de Dieu El a Ougarit", Syria 66 (1989) p. 349; Yon et al., 1990, p. 4-5, fig. 2, cf. note 25; M. Yon, "El, Pere des Dieux", Monuments Piot 77 (1990) 1-19.

31. C. Schaeffer, "Nouveaux Temoignages du Culte de El et de Baal a Ras Shamra-Ugarit et Ailleurs en Syrie-Palestine", Syria 43 (1966) 1-19.

32. D. Collon, in Catalogue de l'Exposition Au Pays de Baal et d'Astarte: 10,000 Ans de'Art en Syrie, Petit Palais, Paris, 1983, fig. 2, pl. II, fig. 2, pl. III.

33. Catalogue de l'Exposition Sumer, Assur, Babylone, Petit Palais, Paris, 1981, nos. 3 and 41.

34. C. Schaeffer, "Decouverte d'un Troisieme Palais: Rapport Preliminaire des Fouilles 1968-1969 (XXXe et XXXIe Campagnes)", Syrie 47 (1970) 209-213; J. and E. Lagarce, "Chantier du Paris

aux Orthostates (Palais Nord)", Syrie 49 (1972) 15-25, and the date, pp. 27-33; O. Callot "La Region Nord du Palais Royal d"Ougarit", Comptes Rendus de l'Academie des Inscriptions et Belles Lettres 1986, 735-755, esp. 737 and note 10 on dating.

35. J.-L. Huot, Y. Calvet, D. Charpin, S. Cleuziou and J.-D. Forest, "Larsa: Rapport Preliminaire sur la Sixieme Campagne", Syrie 53 (1976) 1-45, esp. p. 8; A. Parrot, Mission Archeologique de Mari, T. I 1956, p. 13; cf. ref. 36, A. Parrot, 1958, p. 199 and 1967, p. 24.

36. A. Parrot, Mission Archeologique de Mari, T. II, Geuthner, 1958, pp. 15, 66, 73f; A. Parrot, Mission Archeologique de Mari, T. III, Geuthner, Paris, 1967, p. 33; F. Thureau-Dangin and M. Dunand, Til Barsip, Geuthner, Paris, 1936, p. 29f.

37. S. Lloyd and F. Safar, "Tell Uqair: Excavations by the Iraq Government Directorate of Antiquities in 1940-1941", Jour. of Near Eastern Studies 2 (1943) 131-158.

38. R. Ghirshman, "Suse: Campaigne de Fouilles 1965-1966", Arts Asiatiques 15 (1967) Paris, pp. 3-28.

39. Cf. ref. 34, Callot, 1988, p. 749-750.

40. I.M. Toll "The Mineral Resources of Syria", Engineering and Mining Journal 112, no. 22 (1921) 846-852; E. Marchkwald, "Die Asphaltlager von Lattakia", Petroleum Zeitschrift 18 no. 5 (1922) 165-170; H. Abraham, Asphalts and Allied Substances, van Nostrand, Inc., Princeton, 1960, v. 1, pp. 209-211.

41. J. Connan, "Quelques Secrets des Bitumes Archeologiques de Mesopotamie Reveles par les Analyses de Geochimie Organique Petroliere", Bull. Centres Rech. - Explor. Prod. Elf-Aquitaine 12, no. 2 (1988) 759-787.

42. C.D. Cornelius, "Classification of Natural Bitumin: A Physical
and Chemical Approach", Exploration for Heavy Crude Oil and Natural bitumen ed. by R.F. Meyer, AAPG Studies in Geology 25 (1987) 165-174.

43. M.I. Venkatesan, T.W. Linich, H.E. Suess and G. Buccellati, "Asphalt in Carbon 14 Dated Archaeological Samples from Terqa, Syria:, Nature 295, 5849 (1982) 1-3; K.-H. Hellmuth, J. Koch, E. Strommenger-Nagel and H. Wehner, Untersuchungen an Bitumen aus den Ausgrabungen von Habuba Kabira (Syrien), unpublished report, 1989; cf. Connan, ref. 41.

44. G. Hussler, Marqueurs Geochimiques en Series Carbonatees. These de Doctorat es-Sciences, Univ. Louis-Pasteur, Strasbourg, 1985.

45. Cf. ref. 41, Connan 1988.

46. P. Chosson, H. Connan, D. Dessort and C. Lanau, "In Vitro Biodegradation of Steranes and Terpanes: a Clue to Understanding Geological Situations", Biological Markers in Sediments and Petroleum, P. Albrecht, J.M. Moldowan and R.P. Philp, eds., 1990.

47. W.K. Seifert and J.M. Moldowan, "The Effects of Biodegradation on Steranes and Terpanes in Crude Oils", Geochim. Cosmochim. Acta 43 (1979) 11-126.

48. Cf. ref. 41, Connan 1988.

49. Cf. ref. 41, Connan 1988.

50. R.E. Summons and L. Jahnke, "Identification of the Methylhopanes in Sediment and Petroleum", Geochim. Cosmochim. Acta 54 (1990) 247-251; W.K. Seifert, J.M. Moldowan, G.J. Demaison, "Source Correlation of Biodegraded Oils", Org. Geochem. 6 (1984) 633-643; and J.M. Moldowan, P. Sundararaman, T. Salvatori, A. Alajbeg, B. Gjukic, C.Y. Lee and G.J. Demaison, Source Correlation and Maturity Assessment of Select Oils and Rocks from the Central Adriatic Basin (Italy and Yugoslavia), Am. Chem. Soc., Dallas, april 1989.

51. J. Connan and D. Dessort, "Novel Family of Hexacyclic Hopanoid Alkanes (C_{32}-C_{35}) Occurring in Sediments and Oils from Anoxic Palaeoenvironments", Org. Geochem., 11, no. 2 (1987) 103-113; and J.M. Trendel, J. Guilhem, Ph. Crisp, D. Repeta, J. Connan and P. Albrecht, "Identification of Two C-10 Demethylated C_{28} Hopanes in Biodegraded Petroleum", J. Chem. Soc., Chemical Communications, 5 (1990) 424-425.

52. Cf. ref. 46, Chosson et al, 1990.

Editor's Note: P. Vandiver translated this article and errors introduced into the text are my fault. See the last two citations in ref. 24 for the original texts which have been adapted here.

PRELIMINARY ANALYSIS OF THE MASONRY OF THE HAGIA SOPHIA BASILICA, ISTANBUL

RICHARD A. LIVINGSTON*, P.E. STUTZMAN*, R. MARK** AND M. ERDIK***
*Buildings and Fire Research Laboratory, Div. 862, NIST, Gaithersburg, MD 20899
**School of Architecture, Princeton U., Princeton, NJ 08544
***Civil Engineering Dept., Boğaziçi University, 80815 Bebek - Istanbul, Turkey

ABSTRACT

Specimens of brick and mortar taken from the Hagia Sophia Basilica in Istanbul, Turkey, are being analyzed as part of an investigation of the construction history and present state of structural stability of the building. The analysis, covering both chemical composition and physical properties, involves a suite of instrumental methods including quantitative X-ray diffraction, thermal analysis, and automated image analysis of polished sections under SEM. These conventional methods are being supplemented by neutron diffraction and neutron and X-ray small angle scattering. Preliminary results suggest the presence of cementitious materials including plombierite and tetracalcium aluminum carbonate in the mortar as well as calcium carbonate. The mortar contains significant amount of brick dust and also large chunks of brick as coarse aggregate. Given the pozzolanic nature of brick dust, it is possible that the mortar is actually not a conventional carbonated lime material, but rather a proto-concrete. The type of masonry construction involving very wide mortar joints, comparable to the width of the bricks, suggests that the Byzantine architects took advantage of this property of the mortar. This sophisticated materials technology may help to explain how the Basilica was constructed in such a short time period, and also to allow us to understand and model subsequent structural problems, including major deformations of the main piers. Also, the feasibility of dating by radiocarbon and thermoluminescence methods is being investigated to determine which parts of the elaborate buttress system are the original Byzantine structure as opposed to later Ottoman additions.

INTRODUCTION

The Hagia Sophia in Istanbul is one of the world's great architectural monuments(Fig. 1). The enormous domed basilica was constructed by order of the Byzantine Emperor Justinian in the very short time period of 5 years beginning about 532 AD[1]. After the fall of Constantinople in 1453, the Hagia Sophia became an Islamic mosque, and subsequently served as a model for Ottoman architectural design.

The central dome, which is supported on the four main piers, has a span of roughly 32 meters. However, the present dome is considerably taller than the original one, which had to

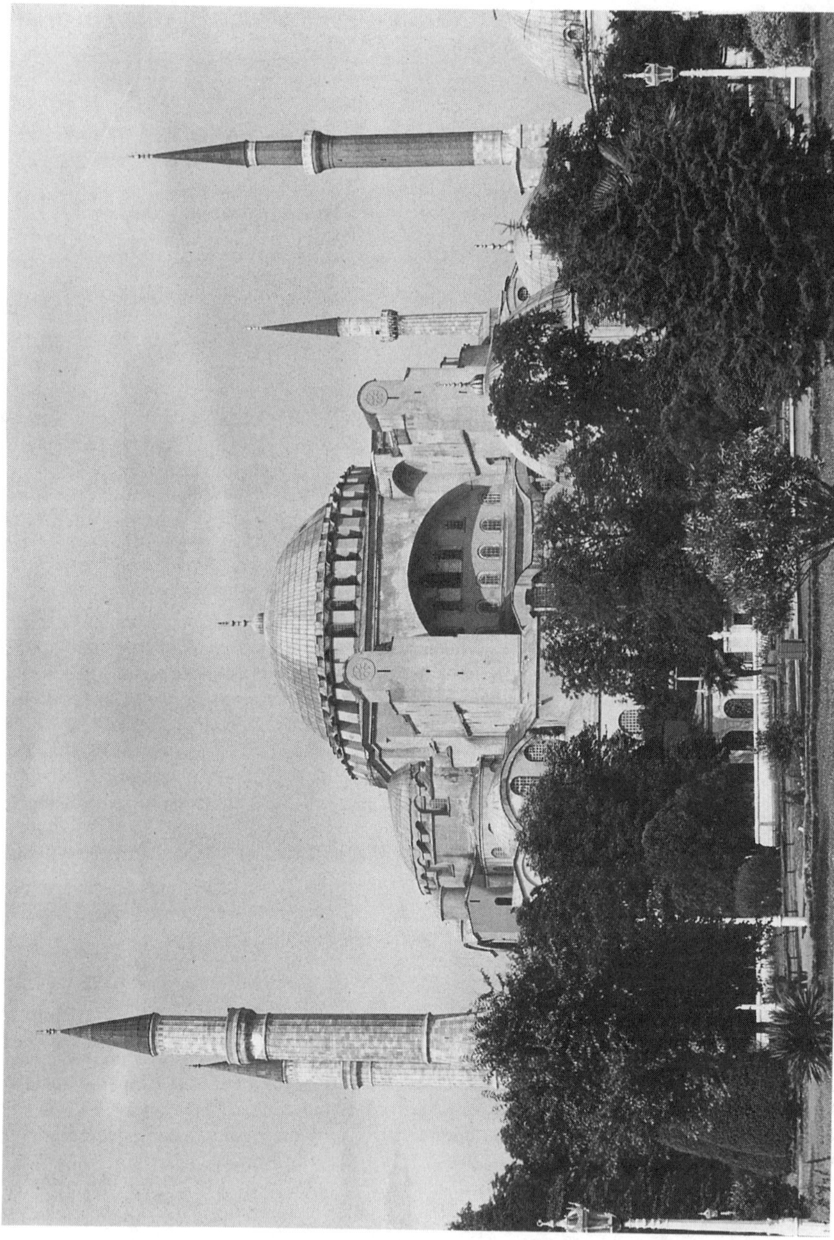

FIGURE 1: The Hagia Sophia, Istanbul, South Elevation (Courtesy R. Mark)

be rebuilt after a partial collapse following a series of earthquakes between 542 and 557 AD. The increase in height is estimated to be 6 meters[2]. Another early structural problem was the deformation of the main piers, which occurred even before construction was complete. This deformation has been measured in this century at 107 cm in the north-south direction and 23 cm in east-west direction. Moreover, the crowns of the main arches have dropped by approximately 70 cm.

The investigation of the structure now underway is being conducted by a team of researchers from Princeton University and from Boğaziçi University in Istanbul[3]. The project has a number of objectives. Among them is a better understanding of the original design principles and construction technology used by the architects and builders. A second objective is an explanation of the present structural state of the building, reflecting its long term structural deformation under the action of gravity, fractures and distortions as a result of short-duration earthquake forces, as well as structural modifications due to several programs of restoration over its long history. The third objective is a structural model that could be used to predict the response of the structure in future earthquakes. A major part of the work consists of the development of a finite element computer model of the structure. This is supported by on-site measurements of the structure using laser surveying techniques. Also an array of sensors has been installed to monitor the Hagia Sophia's dynamic response to local ambient vibrations from such sources as truck and automobile traffic as well as large amplitude vibrations associated with seismic activity.

The materials properties of the mortar are now being investigated because of anomalous results produced by the initial finite element model. For this model it was assumed the mortar was simply a traditional lime-sand mixture. A nominal value for modulus of elasticity of 5×10^9 Pa provided good agreement for the dynamic model in the sense that it predicted natural frequencies of vibration that were very close to measured ones. However, this value was not satisfactory for the static version of the model. Fig. 2a shows a finite element model of a main arch, pier and buttress in an unloaded state. Fig. 2b shows the deformations resulting from adding the deadloads, which produce outward thrusts at the tops of the piers of 400 kN in the east-west direction and 540 kN in the north-south direction. However, the elastic deformations predicted by the model using a modulus of elasticity of 5×10^9 Pa were nearly an order of magnitude lower than those actually measured. Moreover, as can be seen in Fig. 2b, the mortar joints would experience significant bending stresses. This would create tensile stresses in the mortar joints that would at some points exceed the tensile strength assumed for the lime mortar (3.4×10^5 Pa). Consequently, this model leads to the false conclusion that the Hagia Sophia should have collapsed long ago.

Given these anomalies, it was decided to learn more about the actual mortar used in the Hagia Sophia. This has led to collaboration between the Princeton group and researchers in the Building Materials Division of the National Institute of Standards and Technology (NIST). Investigation of the mortar has the main objective of determining the composition of the material, from which the strength properties can be inferred. Other objectives include the direct measurement of strength properties and the dating of samples from various parts of the structure.

Obtaining samples of the mortar for analysis presented some difficulties. Most of the interior surfaces of the Hagia Sophia are covered by a thick layer of plaster, on top of which are frescoes or mosaics that cannot be damaged. The brick and mortar is accessible in interior passages of the buttresses, but it is not always clear whether the material dates to the original construction or to later restorations. The authorities responsible for the Hagia Sophia are also naturally reluctant to permit any large diameter core drilling of the kind that is normally used to obtain cylinders for civil engineering materials testing. Fortunately, there is a significant collection of samples of Hagia Sophia brick and of mortar in the possession of the Dumbarton Oaks Museum in Washington DC. These samples were collected by Robert Van Nice in the course of his decades long study of the building and are thoroughly documented.

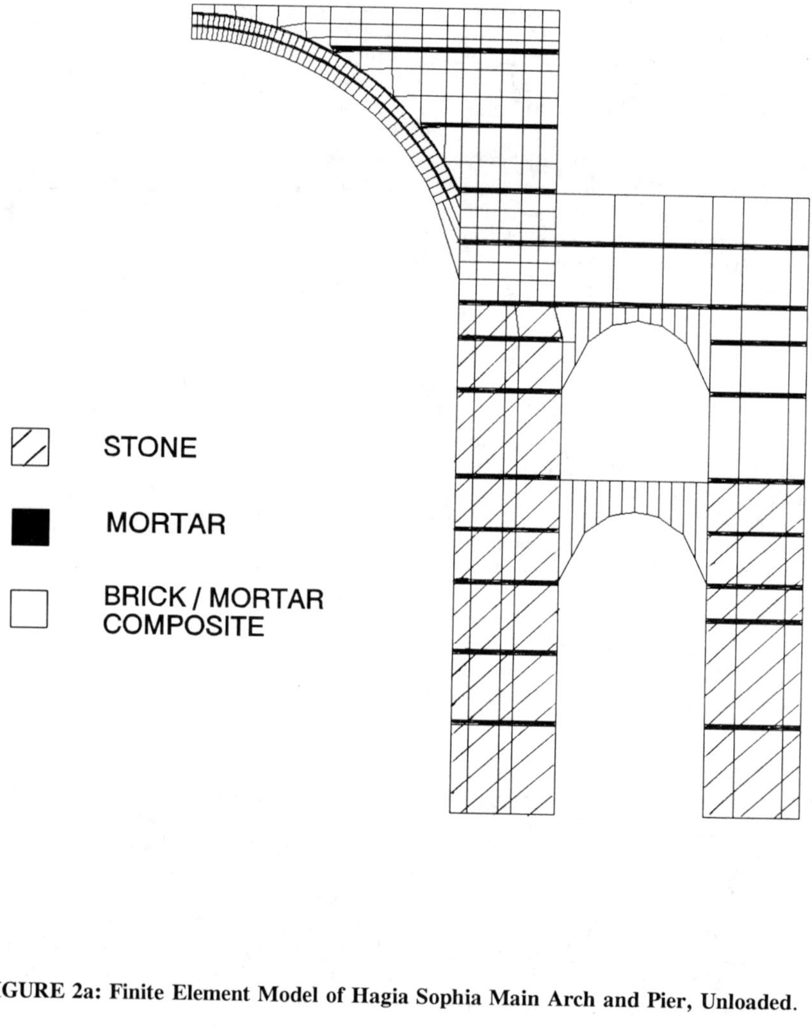

FIGURE 2a: Finite Element Model of Hagia Sophia Main Arch and Pier, Unloaded.

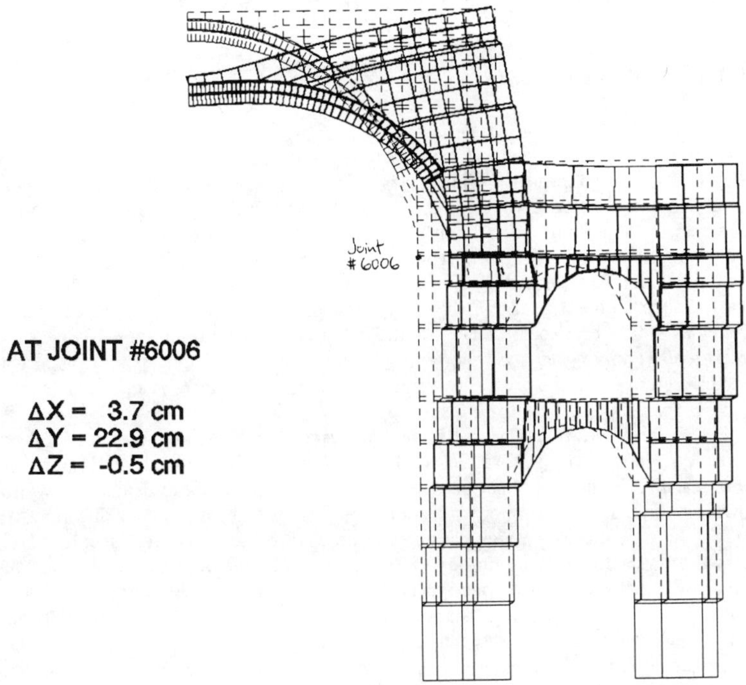

FIGURE 2b: Finite Element Model After Load Applied.

VISUAL ASSESSMENT

Visual inspection of the mortar samples reveals several important features. The mortar joints are typically very thick, on the order of 5-7 cm. The mortar contains fragments of red brick ranging up to a 1 cm in size. The mortar has a characteristic pink color, indicating that crushed brick may have been added. The mortar is generally very porous and friable.

DIFFRACTION ANALYSIS

The conventional method for mineralogical analysis is X-ray diffraction. Figure 3 is an X-ray diffraction pattern of a typical Hagia Sophia mortar. For this analysis, a sample of the mortar from the intrados of the main east arch was used. The sample, including any brick fragments, was handground in an agate mortar. The resulting powder was packed into a standard 16 by 20 mm cavity mount. The diffraction pattern was collected on an automated Philips X-ray powder diffractometer*, using the copper Kα line, operating at 45 kV and 35 mA.

The prominent peaks in the diffraction pattern are mainly associated with quartz and calcite, as would be expected in a traditional carbonated lime mortar. No peaks for $Ca(OH)_2$ were found, implying that the mortar was fully carbonated. Feldspar peaks also were found, perhaps coming from the brick dust admixture. In addition, there is a peak at 11.7° (0.756 nm), which is the principal line for gypsum.

Figure 4 shows another diffraction pattern of the same sample, made by thermal neutron diffraction rather than by X-rays. Neutrons at near room temperatures have wave-lengths on the order of 0.1 nm, and thus are used to investigate interatomic distances by diffraction[4]. However, neutron scattering cross-sections for various elements is very different from those for X-rays. In particular, hydrogen has an extremely large scattering cross section. Hence, neutron diffraction is useful for studying hydrous minerals like those found in cementitious materials. One drawback to neutron diffraction is that there is no extensive library of standard patterns comparable to the JCPDS-International Centre for Diffraction Data Powder Diffraction File for X-rays. Although the X-ray diffraction patterns can give a rough guide for the position of peaks that might be identified by neutron diffraction, the relative intensities of peaks are often very different between the two methods. In fact, prominent peaks in the neutron pattern may not be seen in the X-ray pattern for the same mineral and vice versa. Therefore, it may still be necessary to run standards before neutron diffraction on unknowns can be fully interpreted.

The neutron diffraction patterns were taken at the nuclear reactor at the National Institute of Standards and Technology in Gaithersburg, MD. They were made using BT-1, a single axis diffractometer. The neutron wavelength was selected by a copper monochromator to be 0.154 nm, which is the same as the wavelength of the Cu Kα X-ray used to make that the X-ray diffraction pattern of Figure 3. Thus the two different kinds of diffraction pattern can be directly compared.

The peak at 11.7° also appears in the neutron diffraction pattern for the Hagia Sophia mortar sample. However, comparison with neutron diffraction patterns for calcite and gypsum standards rules out the possibility that gypsum is present. Instead, the peak appears to be due to tetracalcium aluminum carbonate hydrate, $Ca_4Al_2CO_3(OH)_{12}\cdot 5H_2O$, or $C_4A\bar{C}H_{11}$ in cement

* Mention of brand names is made to provide complete details of the experimental procedure. In no case does such identification imply recommendation or endorsement by the National Institute of Standards and Technology, nor does it imply that the products are the best available for the purpose.

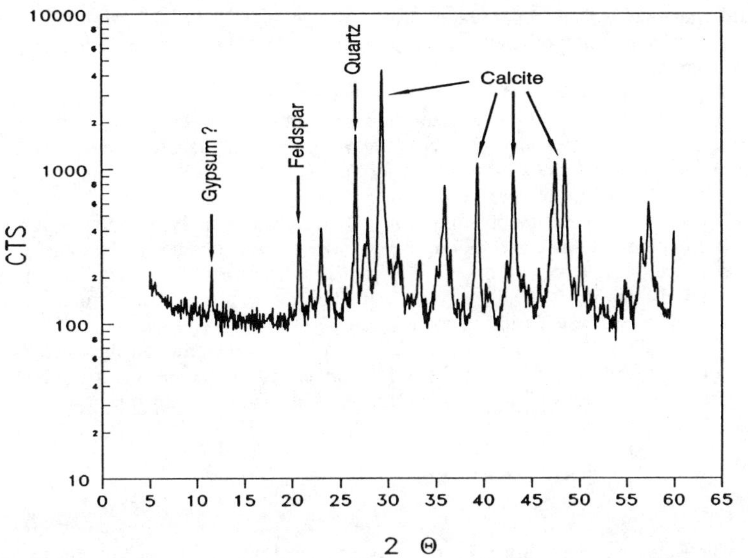

Figure 3: X-ray Diffraction Pattern for Hagia Sophia Mortar Sample.

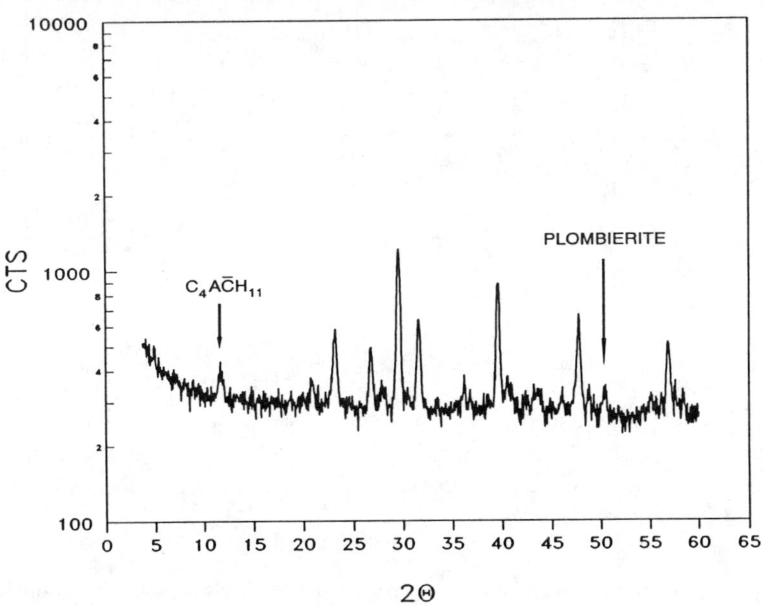

Figure 4: Neutron Diffraction Pattern for Hagia Sophia Mortar Sample.

chemistry notation. The proportion of water of hydration is variable[5]. This mineral is typically found in cementitious materials, especially those that have been in concretes made with limestone aggregate.

Moreover, a peak at 49.8° (0.183 nm) suggests that a calcium silicate hydrate (C-S-H) phase is also present. C-S-H is the main ingredient in concrete made with Portland cement. However, in such cases it tends to be very poorly crystallized. Its structure is often compared with the mineral tobermorite which has the formula $C_5S_6H_5$. In the case of the mortar sample from the Hagia Sophia, however, the prominent tobermorite peak at 1.4 nm is absent. However, this does not rule out the possibility of a C-S-H phase entirely, since there is a mineral, plombierite, which has this composition, but does not show this peak[6]. Plombierite is found in nature where basaltic rock has come into contact with limestone. A similar reaction may occur in this mortar when the lime reacts with the glassy phase in the brick, which has a chemical composition very similar to some basalts[7].

The possible presence of both $C_4A\overline{C}H_{11}$ and plombierite is thus evidence of a cementitious rather than pure lime mortar. However, confirmation awaits the collection of neutron diffraction patterns of standard specimens of both of these minerals.

THERMOGRAVIMETRIC ANALYSIS

The Hagia Sophia mortar sample was also analyzed by thermogravimetric analysis (TGA). The powder sample from the diffraction analysis was heated in a nitrogen atmosphere in a Perkin-Elmer TG-2 instrument at a temperature increase rate of 40° per minute. is given in Fig. 5.

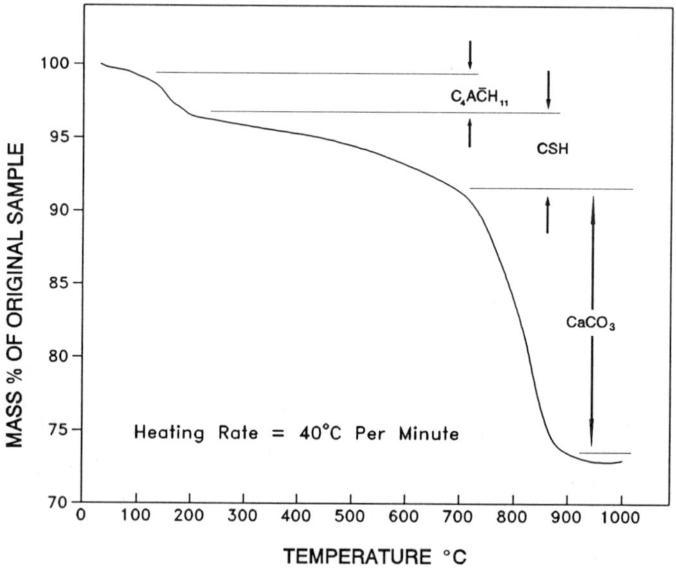

Figure 5: Thermogravimetric Curve for the Hagia Sophia Mortar Sample.

Three major segments of the curve can be identified. The steep initial weight loss around 200°C is associated with dehydration, presumably of the $C_4A\overline{C}H_{11}$. The more gradual weight loss between 300°C and 700°C is typical of the loss of structural water in C-S-H type minerals. Finally, the steeper weight loss above 700°C is characteristic of the loss of CO_2 from the $CaCO_3$ and $C_4A\overline{C}H_{11}$ phases. No break was observed in the curve at 500°C, which would have indicated the presence of $Ca(OH)_2$. This analysis reinforces the results of diffraction that the original lime in the mortar has fully carbonated. Thus the shape of the TGA curve matches those for known mixtures of C-S-H and $CaCO_3$[8].

The amount of CO_2 evolved is approximately 20% of the original sample. This converts to a slaked lime [$Ca(OH)_2$] content of roughly 40% by weight in the original mortar.

The residue of the TGA sample after heating to the maximum temperature of 1000°C was then analyzed by X-ray diffraction. The phases identified are given Table I along with sources materials. The calcium silicate mineral phase larnite (β-C_2S) was identified, which is a diagnostic decomposition product of C-S-H with a low C/S ratio[9]. The other phases in the residue (CA, C_3A and lime) are consistent with the phases detected in the mortar by the other analytical methods.

TABLE I: X-RAY DIFFRACTION ANALYSIS OF MORTAR RESIDUE AFTER THERMOGRAVIMETRIC ANALYSIS

Phase Identified by XRD	**Source**
K-Feldspar	Brick Fragments
Quartz	Brick Fragments + C-S-H
Monocalcium aluminate (CA)	$C_4A\overline{C}H_{11}$
Tricalcium aluminate (C_3A)	$C_4A\overline{C}H_{11}$
Larnite (β-C_2S)	C-S-H(I)
Lime	$CaCO_3$

SCANNING ELECTRON MICROSCOPY

A polished sample of the Hagia Sophia mortar was prepared and coated with carbon for SEM analysis. Figure 6 is a backscattered electron image of the specimen. Visible in this image are brick fragments, in a matrix with calcium carbonate inclusions. X-ray maps of this image, which show that the matrix contains aluminum, calcium and carbon, which is consistent with the picture of a cementitious material composed of $C_4A\overline{C}H_{11}$ and $CaCO_3$. Silicon is found primarily in the brick fragments, along with potassium and aluminum. The eroded edges and reaction rims around the brick fragments indicate a reaction between the brick and the matrix. This would be consistent with the leaching of silica and alumina from the brick to reaction with calcium from the lime to produce the $C_4A\overline{C}H_{11}$, C-S-H and $CaCO_3$ in the matrix.

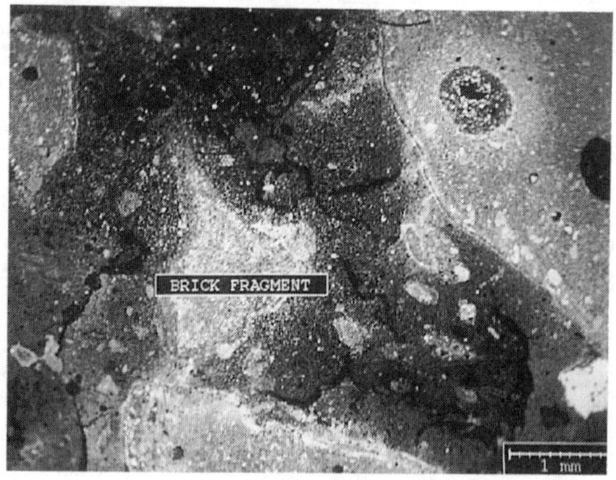

Figure 6: Backscattered Electron Image of Hagia Sophia Mortar Sample (mag. 20 X)

PRELIMINARY FINDINGS

The results thus far are consistent with a pozzolanic mortar rather than a pure lime mortar, and the pozzolana is the crushed brick*. At least some of the aluminum and the soluble silica that comprise the cementitious matrix come from glassy phases of the brick. A pozzolanic reaction with the brick fragments has also been suggested by Mainstone on the basis of chemical analyses of the mortar[10].

There is ample evidence for the use of crushed brick as a pozzolana in ancient Roman masonry. Vitruvius states that

> in using river or sea-sand, the addition of a third part composed of burnt brick, pounded up and sifted, will make your mortar of a better composition to use[11].

However, there has been little published about the use of brick dust in Byzantine mortars. This may simply be due to lack of analyses. Nevertheless, in Ottoman-era masonry, brick dust pozzolana has been used to make what is referred to as Khorosany mortar[12]. It would be reasonable to assume that this practice was continued through the Byzantine period from the Roman era. The proportion of lime in the mortar estimated from the TGA data, 40%, is close

---

* The term "pozzolanic mortar", like many others concerning traditional building materials, has been used over time to mean different things. Its usage here follows that of cement chemistry, which defines a pozzolanic material as one that, unlike Portland cement, does not react by itself with water to yield cementitious phases. Pozzolanas generally contain glassy or cryptocrystalline phases that contribute soluble silica in the presence of lime. Pozzolanas can be natural such as the lava from Pozzuoli, Italy, or artificial such as fly ash, blast furnace slag, or in this case, brick.

to the 1:2 ratio of lime to sand specified by Vitruvius.

Aside from the pozzolanic material, it is also possible that the lime itself may have included some β-C_2S if the limestone raw material contained suitable clay minerals[13]. The hydration of this might have contributed to the C-S-H phase.

The aggregate in the mortar appears to be primarily the brick fragments. The quartz identified in the diffraction patterns does not appear in the SEM images as a separate sand aggregate. Instead, it probably occurs in the brick fragments. Preliminary examination of the samples in the SEM at 1000X magnification revealed that the fine particle sizes of the calcium carbonate had a microstructure typical of carbonated lime rather than limestone[14]. Moreover, no large grains of limestone were seen at lower magnification. Thus limestone apparently was not used as an aggregate.

STRUCTURAL IMPLICATIONS

A major advantage of pozzolanic mortars compared to pure lime ones is higher tensile strength. Table II presents strength data on modern lime-pozzolana mortars. The burnt shale pozzolana entry in this table can be assumed to be roughly equivalent to brick dust. It can be seen that it reaches a tensile strength of 3.5 MPa (500 psi) after a year. There is very little comparable published information on tensile strength of pure lime mortars, but values of 0.21 MPa (30 psi) for medieval mortars have been reported[15]. Using the tensile strength of 3.5 MPa from Table II in the static finite element model of the Hagia Sophia indicates that the structure as modeled would be able to resist the predicted shear and bending stresses. Whether or not this finite element in turn accurately represents the true conditions in the Hagia Sophia itself is an issue beyond the scope of this paper. Nevertheless, it is evident that brick dust mortar would make stronger joints than simple lime mortar.

TABLE II: POZZOLANIC EFFECT OF BURNT SHALE*

MIX PROPORTIONS (BY MASS)			TENSILE STRENGTH, MPa (psi)			
Hydrated Lime	Burnt Shale	Standard Sand	7 days	28 days	90 days	1 year
1	1	6	0.728 (107)	1.43 (207)	2.35 (341)	3.59 (521)
1	2	9	0.905 (133)	2.22 (322)	3.16 (459)	3.86 (560)
1	4	15	1.38 (203)	2.56 (371)	3.54 (514)	3.67 (533)

* Data from Lea[16]

The architects of the Hagia Sophia seem to have been aware of the tensile strength of this mortar, in view of the unusual type of masonry construction they employed. The mortar joints are as wide (5-7 cm), if not wider, than the bricks themselves. In fact, the mortar appears to be the major constituent of the masonry with the bricks acting mostly as reinforcements. This is particularly true in the vertical joints of the main arches. In modern terminology, this material would actually be classified as a concrete instead of a mortar because of the coarse size of the brick aggregate, on the order of 1 cm. Therefore, the Hagia Sophia might be more appropriately considered as an early concrete structure rather than as one of the more familiar brick-and-mortar structures that developed later in medieval Europe.

Given that concrete technology is involved, this raises the question of whether the architects of the Hagia Sophia intentionally used brick fragments rather than limestone, quartz or other stone aggregate to reduce the dead load of the structure. As shown in Table III, brick is the least dense of the materials commonly used for aggregate in concrete.

TABLE III: COMPARISON OF SPECIFIC GRAVITIES OF VARIOUS AGGREGATES

AGGREGATE TYPE	SPECIFIC GRAVITY*
Handmolded brick	1.6 - 2.0
Sandstone	2.0 - 2.6
Limestone	2.2 - 2.6
Quartzite	2.65
Granite	2.6 - 2.7
Basalt	2.8 - 2.9

*Brick data from Baker[17]; stone data from Winkler[18]

Another relevant aspect of pozzolanic mortars is the time required to gain full strength. As shown in Table II, this is on the order of a year for low pozzolana/lime ratios. This may explain the large deformations in the piers during construction that alarmed the architects. Under the very tight construction schedule imposed by Emperor Justinian, the critical phase when the main piers and arches were constructed could not have been much more than 18 months. This means that loads may have been applied to the piers before they had reached their full strength.

Moreover, the development of strength in pozzolanic mortars is more sensitive to ambient temperatures than for Portland cement-type mortars[7]. The tight construction schedule also implies that masonry work may have continued over at least one winter, during which the temperature in Constantinople may have been low enough to affect the strength of the mortar.

MORTAR DATING

Several possibilities exist for dating the mortar samples. Radiocarbon dating of the carbon dioxide in the carbonated lime should be feasible. Since there is apparently little or no limestone aggregate present this should be relatively straightforward. Another radiocarbon method for dating mortar uses the charcoal fragments that are found in the lime as a result of the limestone calcination process.

FURTHER RESEARCH

Mortar Analysis

A final identification of the phases in the mortar by neutron diffraction requires a set of diffraction patterns for the minerals plombierite, tobermorite etc. Efforts are now underway to obtain suitable specimens of these minerals from their type localities in Ireland and Scotland. It may also be possible to produce them artificially[19].

Automated image analysis of SEM images of polished mortar samples is another method of quantifying the phases present. There are a variety of digital techniques for imaging the composition of phases[20]. Once a phase of interest has been highlighted, various parameters characterizing its spatial distribution can be computed.

The use of Small Angle Neutron Scattering (SANS) and Small Angle X-ray Scattering (SAXS) for microstructure characterization is also underway. These techniques together provide information on the fractal dimensions of the mortar, particularly in the length scales of 1 nm to 10 μm, which are critical for determining the physical properties of cementitious materials[21].

After the instrumental analyses are complete, it will be necessary to perform some conventional wet chemistry on the samples. A procedure for analysis of historic mortars has been developed by Middendorf and Knofel[22]. One purpose for this analysis would be to obtain a quantitative estimate of soluble silica, which is an indications of the amount of C-S-H present. This typically done by boiling the sample in a saturated sodium hydroxide solution. The carbonate content can be measured by the amount of CO_2 gas evolved during reaction with hydrochloric acid. Both of these tests are complicated by fact that these constituents can be found in more than one phase. Soluble silica is likely to be found in the brick dust as well as in the C-S-H phase. Carbonate will be found in the $C_4A\overline{C}H_{11}$ as well as in the calcium carbonate. Therefore, prior to the wet chemical analysis, it may prove worthwhile to perform a physical separation of phases, possibly by centrifuge.

Mortar Strength Properties

A primary objective of this study is to determine the tensile strength and elastic modulus of the mortar, which are critical parameters for the finite element structural model. The conventional civil engineering approach, which consists of loading standard size cylinders to failure, is essentially ruled out here because of the lack of large specimens, and the desire to avoid destructive tests as much as possible. Another generic problem with strength testing of brittle materials, regardless of whether they are historic, lies in the problem of a suitable specimen shape and configuration for tensile strength measurement.

An alternative to direct measurement of tensile strength is the modulus of rupture or bending test. This is usually done by applying a known load to the center of a rectangular beam of material that is supported on knife edges at either end, but this requires a specimen with dimensions 10 x 20 x 5.63 cm (4 x 8 x 2¼ in.) [23]. However, a modulus of rupture test that consumes much less material has been developed which consists of loading a circular disk 35 mm in diameter and 4 mm thick by a knife edge across its face[24].

Another indirect method for measuring tensile strength of brittle materials is the split cylinder, or Brazil, test. This involves loading a cylinder of the material in compression on its diameter, which leads to failure by tensile stresses that can be readily calculated[25]. A drawback to this method is that relatively large cylinder sizes (100 mm diameter) are usually specified.

The limiting strength in masonry is often the flexural strength of the interface between the brick and the mortar. In the Van Nice collection at Dumbarton Oaks, there is at least one specimen of brick and mortar together that may be large enough to permit bond strength testing. This can be done by direct shear tests[26] or by applying a torque with the bond wrench method[27].

The dynamic modulus of elasticity can be determined non-destructively by measuring the speed of sound waves through the specimen[25]. An approximate state modulus can be obtained from modulus of rupture measurements on thin disks using a method developed by Wittman and Prim[28].

The measurement of the strength properties could be made considerably simpler if the tests could be made on replica mortar samples rather than on the orginal material. However, producing a suitable replica mortar is not simply a matter of matching the proportions of lime, pozzolana and aggregate in the original. Extensive research on Portland cement materials has shown that the final strength is very sensitive to factors such as water/cement ratio, initial particle size and curing environment, that are very difficult to determine after the fact. However, the SANS/SAXS measurements of mortar microstructure may provide sufficient information to tell if a given replica mortar is a good match to the original.

Brick Analysis

Although this research is focused primarily on the mortar, there are a number of questions about the brick units that could be investigated. One is whether the composition of the units differs from that of brick fragments used as aggregate in the mortar. This might occur if older brick buildings were being recycled to construct the Hagia Sophia. Another question is how the brick units were manufactured. The large dimensions of the units 0.375 x 0.375 x 50 mm and fine texture suggest that methods similar to terracotta work were employed. Finally there is the question of whether the density of the brick units was deliberately varied by the builders in order to lighten the dome.

SUMMARY

At this early stage of the research, we have not yet reached firm conclusions. Our hypothesis is that the mortar used in the construction of the Hagia Sophia was pozzolanic rather than a simple lime mortar. The pozzolana consisted of ground brick dust. Moreover, the use of extremely wide mortar joints and coarse brick aggregate suggest that the builders made use of a concrete technology, instead of conventional brick and mortar masonry design in planning the building.

ACKNOWLEDGEMENTS

The authors are indebted to Dan Neumann of NIST for performing the neutron diffraction measurements and Andrew Allen of NIST for SANS/SAXS measurements.

REFERENCES

1. R.J. Mainstone, Hagia Sophia: Architecture, Structure and Liturgy of Justinian's Great Church, (Thames and Hudson, London, 1988).

2. R. Mark and A. Westagard, in Domes from Antiquity to the Present, edited by I. Mungan (Mirmar Sinan U., Istanbul, 1988), pp. 163-172.

3. R. Mark, A.S. Cakmak and M. Erdik, in The Hagia Sophia from the Age of Justinian to the Present, edited by R. Mark & A.S. Cakmak, (Cambridge U. Press, NY 1992), pp. 120-131.

4. R.B. Von Dreele, in Modern Powder Diffraction, edited by D.L. Bish and J.E. Post, (Mineralogical Society of America, 1989), pp. 333-369.

5. E.T. Carlson and H.A. Berman, J. Res. N.B.S, 64A (4), 333-341 (1960).

6. J.D.C. McConnell, Min. Mag., 30, 293 (1954).

7. M.Regourd, in 8th International Congress on the Chemistry of Cement, (Rio de Janeiro, 1986) 1, pp. 200-229.

8. F.M. Biffen, Anal. Chem, 26 (7) 1133-1136 (1956).

9. H.F.W. Taylor in The Chemistry of Cements, edited by H.F.W. Taylor (Academic Press, London 1964), pp. 221-223.

10. R.Mainstone, op.cit. p. 70.

11. Vitruvius, The Ten Books on Architecture, M.H. Morgan trans. (Dover Publications, NY, 1960), p. 45.

12. A. Gulec in Proceedings of the 4th Experts Meeting of the NATO-CCMS Pilot Study on the Conservation of Historic Brick Structures, edited by S. Fitz (Umweltbundesamt, Berlin, 1991), p. 121.

13. D.L. Rayment and K. Pettifer, Mat. Sci. and Tech., 997-1004 (1987).

14. W.D. Kingery, P.B. Vandiver and M. Prickett, J. Field Arch.,15, 219-243 (1988).

15. R. Mark, Experiments in Gothic Structure, (The MIT Press, Cambridge, MA, 1982), p. 32.

16. F.M. Lea, The Chemistry of Cement and Concrete, (Chemical Publishing Co. NY, 1971) pp. 419-435.

17. I.O. Baker, A Treatise on Masonry Construction, (John Wiley & Sons, NY, 1889) p. 47.

18. E.M. Winkler, Stone: Properties, Durability in Man's Environment, (Springer-Verlag, Vienna, 1975), p. 29.

19. B.L. Raisman, S.A. Scherban and S.K. Nurmagambetova, Kompleksnovye Ispol'zovaniye Mineralnovo Syr'ya, 1, 55 (1986)(in Russian).

20. C.E. Lyman in Microscopy: The Key Research Tool edited by C.E. Lyman, L.D. Peachey and R.M Fisher (The Electron Microscopy Society of America, Woods Hole, MA, 1992), pp. 3-15.

21. A.J. Allen, A.H. Baston and C.R. Wilding, in Pore Structure and Permeability of Cementitious Materials, edited by L.R. Roberts and J.P. Skalny (Mater. Res. Soc. Proc. 137, Pittsburgh, PA, 1988), pp. 119-128.

22. B.Middendorf and D. Knofel, in Proceedings of the 4th Expert Meeting of the NATO-CCMS Pilot Study on the Conservation of Historic Brick Structures, edited by F. Fitz, (Umweltbundesamt, Berlin, 1991), p.75.

23. E. Winkler, op. cit., p. 44.

24. E. Winkler, Bulletin of the Association for Preservation Technology, XVII (2), 35 (1986).

25. F. Berger in Structural Repair and Maintenance of Historical Buildings, edited by C.A. Brebbia, (Computational Mechanics Publications, Southampton, 1989) pp. 83-94.

26. R.H. Atkinson, G.R. Kingsley, R. Saeb, B. Amadei and S. Sture in Proceedings of the 8th International Brick/Block Masonry Conference, edited by J.W. deCourcy (Elsevier, London, 1988), pp. 261-271.

27. J. Matthys in Proceedings of the 8th International Brick/Block Masonry Conference, edited by J.W. deCourcy (Elsevier, London, 1988), pp. 284-291.

28. F.H. Wittman and P.Prim, Materiaux et Constructions, 16, 235 (1983).

PART III C

Technical Analysis, Ethnography and the Written Record

THEOPHILUS AND THE COMPOSITION OF MEDIEVAL GLASS

IAN C. FREESTONE
British Museum, Department of Scientific Research, Great Russell St, London WC1B 3DG, U.K.

ABSTRACT

Current understanding of medieval glass composition is compared with the instructions for the preparation and use of glass outlined by Theophilus in the early twelfth century. The points are illustrated by analyses of glasses from a range of objects and locations. They include the Westminster Retable, the pavement at Cluny Cathedral, Romanesque Mosan enamels on copper, Abbey windows from San Vincenzo in Italy and early enamelled glass vessels which are likely to have been made in Venice.

INTRODUCTION

The writings of Theophilus represent our most important contemporary source on the technology of glass-working in medieval Europe. Theophilus is considered by many authorities to have been Roger, a twelfth century Benedictine monk who worked in metal at the Monastery of Helmarshausen, North West Germany. His treatise, *De Diversis Artibus* [1,2] contains thirty one chapters on glass. Theophilus describes how to build furnaces for making, working and annealling glass; the raw materials for glass; how to make crucibles; how to produce glasses of various colours; how to use glass to produce vessels, windows, enamels and rings.

Many manuscripts list a number of chapters that are now lost. These include accounts of the production of a number of glass colours including blue, green and, most tantalisingly, *gallien* or red.

In the present paper, I compare our current understanding of the composition of medieval glass with the account of Theophilus. In particular, I focus on the intriguing comments on the use of glass for enamelling on metal. Many of the analyses used to illustrate the points are previously unpublished.

ANALYTICAL METHODS

Small fragments, *circa* 1 mm, were removed from the objects, mounted in epoxy resin blocks and polished with diamond pastes down to 1 um. They were coated with a thin layer of carbon and their microstructures examined in the scanning electron microscope (SEM; Cambridge S600 or JEOL JSM840). Their elemental compositions were determined by energy dispersive x-ray analysis in the SEM (LINK Systems x-ray spectrometer) or in the electron microprobe (Cambridge Geoscan or Camebax). As these data have been collected over a decade on a variety of instruments, it is not possible to set generally applicable limits on accuracy and precision. For the purpose of this paper, it is suggested that relative errors of 5% on components with concentrations over 5%, 10% for components in the range 2-5% and 20% for 1-2% are assumed.

WINDOW AND VESSEL GLASS

The instructions on the preparation of glass for windows and vessels are quite specific. The glass was to be made from a mixture of the ash prepared by burning beech wood with washed sand. This type of two-component mixture appears to have been the standard format of glass recipes throughout the pre-industrial world. Glass was produced by mixing a silica-rich component (sand, quartz pebbles) with an alkali-rich flux (mineral natron, saltpetre, plant ash, wood ash). The essential alkaline earth stabilisers required to render the glass insoluble and corrosion resistant were included as incidental components of either the flux (plant ash) or the silica (calcareous sand). Beech ash is lime and potash-rich and when melted with sand produced the typical potash-lime-silica glasses of the medieval period [3,4]. Wood (beech) ash glass appears to have been introduced in the ninth or tenth century, displacing the earlier soda-lime-silica "Roman" type glass. The thirteenth century colourless glass given in Table 1 is typical, containing a substantial range of minor oxides. It is from the Retable of Westminster Abbey, a panel painting which dates to around 1268 A.D.

Theophilus describes how the colour of a batch of glass may be modified by extending the length of time it is held in the furnace, to produce various shades of yellow and purple. This effect is well understood, and is the result of the varying oxidation states of the iron and manganese ions which are incorporated in the glass from the beech ash [5]. These effects have been observed experimentally [6] and were exploited by the medieval glass-makers; Newton has estimated that of 200 analysed fragments of window glass from York Minster, ninety percent are coloured by incidentally included iron and manganese oxides [7].

Table 1. Examples of Medieval Wood Ash Glass

	Colourless Westminster	Green Cluny	Blue Westminster
SiO_2	53.2	45.6	53.3
Al_2O_3	1.0	2.6	1.0
FeO	0.4	0.6	0.9
MnO	1.2	1.3	1.1
MgO	7.1	5.9	6.9
CaO	12.8	16.5	11.9
Na_2O	2.9	1.6	2.6
K_2O	17.1	15.3	17.0
P_2O_5	4.4	6.0	4.4
CuO	<0.4	2.8	0.4
CoO	<0.2	<0.2	0.2
PbO	<0.2	<0.2	0.2

The medieval glass-makers were also able to produce glass of other colours by the deliberate addition of metal oxides. Examples include the green glass, coloured by copper, from a floor in the Cathedral at Cluny, France, dating to around 1120 A.D., and the cobalt blue glass from the Westminster Retable (Table 1).

Theophilus does not provide instructions for the production of opaque glass. The use of antimony- and tin-based opacifiers, common throughout the ancient world and present in Byzantine and Islamic glasses, does not appear to have been within the repertoire of northern European glass-makers as there are few, if any, examples of the presence of these compounds in potash-rich beechwood-type glass. Where such compounds are present the glasses appear to be of recycled Roman or Islamic soda-lime-silica type, or a mixture of northern potash glass with soda-lime-silica glass. Present evidence seems to suggest that the use of tin-based opacifiers did not become established in northern Europe until well after the rise of the Venetian glass-making centre in the fourteenth century.

LEAD-RICH GLASS

Heavily leaded glasses are uncommon in medieval contexts. However, Theophilus describes how small glass finger rings may be made from a lead rich glass. A group of such finger rings have been identified in Britain by Bayley [8]. A group of vessels made of lead-silica glass and containing around 70% PbO have also been recognised [9,10]. These are commonly transparent yellow in colour.

ENAMELS ON METAL

One of the most intriguing chapters refers to the production of enamels on metal:

> *In the ancient buildings of pagans, various kinds of glass are found in the mosaic work - white, black, green, yellow, blue, red and purple. They are not transparent but are opaque like marble, and are like little square stones. From these enamels are made in gold, silver and copper....* [1]

The Mosan group of enamels were made in the metalworking centres of the valley of the River Meuse, such as Maastricht and Liège, in the twelfth century. Some thirty examples of glass from the Mosan enamels in the collections of the British Museum have been sampled and analysed. Typical examples of a range of colours are given in Table 2.

Table 2. Examples of Compositions of Mosan Enamels

	White	Blue	Turquoise	Yellow	Green
SiO2	65.8	63.6	64.9	60.5	66.1
Al2O3	2.3	2.4	2.1	2.4	2.3
FeO	0.6	1.0	0.6	0.6	0.6
MnO	0.4	0.5	<0.2	0.3	0.4
MgO	0.5	0.6	0.6	0.4	0.4
CaO	7.3	7.2	4.8	5.0	5.9
Na2O	15.1	14.9	17.9	15.3	16.8
K2O	0.5	0.5	0.3	0.5	0.5
CuO	0.7	0.7	3.0	<0.3	2.2
PbO	<0.3	1.1	<0.3	9.6	1.9
Sb2O3	6.9	5.9	1.8	0.8	0.6
CoO	<0.1	0.2	<0.1	<0.1	<0.1
SnO2	<0.1	<0.1	0.1	<0.1	<0.1

It is immediately noticed that there is a marked difference between the compositions of the enamel glasses and the wood ash glasses previously described (c.f. Table 1). With the exception of certain red glasses, the enamels are of the soda-lime-silica type, typically with potash and magnesia contents below one percent. They correspond to the low-magnesia glass category of Sayre and Smith [12]. The use of this glass type was particularly prevalent in the Roman period, and it is believed to have been the product of naturally-occuring soda, or natron, from Egypt, and a calcareous sand [13,14]. All of the glasses analysed are opaque. With the exception of the reds, which are coloured and opacified by crystals of cuprous oxide and/or copper, all are opacified by compounds of antimony. Turquoise and blue glasses are opacified by white calcium antimonate crystals in glass matrices coloured by copper and cobalt oxides respectively. Yellows contain yellow crystals of lead antimonate, while greens contain in addition dissolved copper oxide.

Table 3. Examples of Mosan Enamels (this work) compared with Roman Enamels (from ref. 15)

	Roman Turquoise	Mosan Turquoise	Roman Blue	Mosan Blue	Mosan Red
SiO2	66.36	64.9	66.60	66.1	45.9
Al2O3	2.16	2.1	2.42	2.2	2.0
FeO	0.65	0.6	1.17	1.2	2.0
MnO	0.10	<0.2	0.70	0.7	1.1
MgO	0.57	0.6	0.45	0.5	3.2
CaO	4.70	4.8	6.60	6.3	16.5
Na2O	17.20	17.9	15.14	15.5	3.8
K2O	0.60	0.3	0.59	0.5	18.3
CuO	2.83	3.0	0.28	<0.3	2.6
PbO	0.20	<0.3	0.59	1.1	1.8
Sb2O3	3.10	1.8	3.80	3.5	<0.5
CoO	ND	<0.1	0.21	0.2	<0.1
SnO2	0.23	0.1	0.05	<0.1	<0.1

These enamels are typical of the opaque glasses of the Roman period. Indeed, they may be paralleled in virtually every detail by enamels from Roman metalwork (Table 3) [15]. The use of natron as a basis for soda-lime-glass ceased in Egypt itself in the ninth century [16,17]. Throughout the first millennium, tin oxide increasingly replaced antimony oxide as a glass component [18]. Thus it is most unlikely that the production of minor quantities of this glass type continued in northern Europe for the purpose of enamelling. Instead it appears that the Mosan enamellers used recycled Roman glass in their work and that the opaque glasses "like little square stones", described by Theophilus, were the coloured *tesserae* from Roman wall mosaics.

A final point of interest in the enamels is the composition of the some of the red glasses, which are potash-lime-silica glasses typical of the "wood-ash" type (Table 3). They are therefore not recycled Roman glass but are medieval glasses and may represent the "gallien" glass of the missing chapters of Theophilus. It is noted that these are not opacified by either tin or antimony. Opacity and colour are both provided by minute, sub-micrometre, copper-rich particles, probably copper metal. The difficulties in producing opaque copper-red glass have been commented on many times. It seems to have been particularly difficult to obtain in the ancient world and its base composition commonly differs significantly from those of other glasses with which it is associated, suggesting that it was produced in separate, specialised centres which are likely to have been limited in number. It appears that the medieval glass-makers discovered the technique of production of copper-red glass at a relatively early stage and that this was used to supplement the relatively limited availability of old Roman red glass. We must, of course, distinguish between the glass-makers and the enamellers. It is quite possible that the enamellers themselves were unaware of the sources of their glasses. They are likely to have obtained them from merchants or middlemen who would have presented a range of coloured glasses for purchase. The enamellers would have carefully selected the colours they required, a procedure that would have allowed them to exercise close control over their palette.

RECYCLED GLASS IN WINDOWS

In addition to the re-use of Roman mosaic *tesserae* in enamelling, Theophilus also indicates that such opaque glasses were collected and melted with colourless glass to produce (transparent/translucent) coloured glass for use in windows. Abundant evidence for this practice has been found in the current excavations of the Benedictine monastery of San Vincenzo, Molise, Italy under the direction of Dr Richard Hodges of the British School at Rome [19]. Here have been found numerous fragments of crucible, glass drops, discarded rings from the ends of blowing irons and bichrome twisted cane - irrefutable evidence of glass-working dating to the early part of the ninth century. In addition to the finished products of this industrial activity - windows, vessels and lamps - fragments of coloured and opaque Roman vessel glass and abundant mosaic *tesserae* are also present. These appear to have been used as a source of colourant for the window glasses. Analysis (Table 4) reveals that the strongly coloured window glasses contain high concentrations of antimony oxide, the opacifier used in opaque Roman glass. By mixing blue and green opaque glass from old Roman *tesserae* and vessels with colourless glass, the craftsmen produced coloured transparent glass for windows. It should be noted that the ninth century glass-working activity at San Vincenzo precedes the introduction of wood ash glass and that the base glass, as well as the coloured glass, was in the soda-lime-silica tradition.

Table 4. Compositions of Window Glass from San Vincenzo and Early Venetian Glass Enamels from London

	Blue S Vincenzo	Green S Vincenzo	White London	Green London
SiO_2	66.2	65.8	39.6	17.0
Al_2O_3	2.2	2.3	0.5	0.6
FeO	0.8	0.9	0.4	0.4
MnO	0.6	0.3	0.2	<0.2
MgO	0.5	0.5	1.3	<0.2
CaO	7.7	6.2	3.4	0.6
Na_2O	16.4	17.3	4.5	0.7
K_2O	0.9	0.8	0.9	<0.2
CuO	0.6	1.7	<0.3	2.1
PbO	0.9	1.5	29.7	66.5
Sb_2O_3	2.0	0.8	<0.5	<0.5
SnO_2	<0.5	<0.5	17.7	8.6

ENAMELLED GLASS GOBLETS

Theophilus mentions that "the Greeks" produced glass goblets which were gilded and enamelled with the same glasses as those used for enamelling on metals. Unfortunately, surviving Byzantine enamelled glass is very rare, so that it is likely to prove difficult to test this passage analytically. An understanding of the formulae used by the Byzantine enamellers is desirable, as it would have a bearing on our understanding of the development of the enamelling practices of the Venetian glass industry. To what extent were the Venetian enamels developed from the Byzantine tradition, or were they based upon Islamic formulations?

The glass beakers of the "Aldrevandini" group are believed to represent the earliest enamelled glasses from Venice and date to the early fourteenth century [20]. Analysis of a group of these enamels, from excavations at Foster Lane in London, show that they are tin-opacified and lead-rich (Table 4), and differ substantially from the recycled Roman glasses used to enamel on metals. On the other hand, these Venetian enamels are very close in composition to those used on Islamic glass [21]. Thus, if the Theophilus passage is correct, Venetian enamelling technology may owe more to the Islamic than to the Byzantine glass-working tradition. However, this is probably one of the less reliable chapters of Theophilus, for while he is likely to have witnessed many of the other procedures outlined first hand, it is questionable if he witnessed glass-working practices in the Mediterranean area.

CONCLUSIONS

Perhaps the most remarkable aspect of the chapters of Theophilus on glass is just how frequently and in what detail they prove accurate. Far from eroding this impression, modern analytical data reinforce it. One is led to concur with the image of Theophilus as a master craftsman, who was closely familiar with both the materials and the techniques about which he wrote.

ACKNOWLEDGEMENTS

I thank the Dean and Chapter of Westminster Abbey, Mr N Stratford of the department of Medieval and Later Antiquities of the British Museum, Dr R Hodges of the British School at Rome and Mr J Clark of the Museum of London for allowing the sampling and analysis of objects in their care.

REFERENCES

1. Theophilus, De Diversis Artibus, edited and translated by C.R. Dodwell (Thomas Nelson and Sons Ltd 1961).
2. Theophilus, On Divers Arts, edited and translated by J.G. Hawthorne and C.S. Smith (Dover Publications, New York, 1963).
3. J. Barrera, B. Velde, Archeologie Medievale, XIX, 81-130 (1989).
4. A. Cox and K.J.S. Gillies, Glastechn. Ber., 61, 75-84 (1988).
5. W.E.S. Turner quoted in Hawthorne and Smith, ref.2
6. V.C. Sellner, H.J. Oel, B. Camara, Glastechn. Ber. 52, 255-264.
7. R.G. Newton, Glass Tech. 19, 59 (1978)
8. J. Bayley in Object and Economy in Medieval Winchester, edited by M. Biddle (Oxford: Clarendon Oress, 1990), 268.
9. E. Baumgartner, I. Kreuger, Phonix aus Sand und Asche (Klinkhardt and Biermann, Munich).
10. G. Eggert, in Archaeometry 90 edited by E. Pernicka and G.A. Wagner (Birkhauser Verlag, Basel 1991) 247-253.
11. I.C. Freestone in Catalogue of Medieval Enamels II, by Neil Stratford (British Museum Press, 1993).
12. E.V. Sayre and R.W. Smith, Science 133, 1824 (1961).
13. R.H. Brill in Excavations at Jalame, edited by G. Davidson Weinberg (university of Missouri Press, Columbus, 1988) 257-294.
14. E.V. Sayre and R.W. Smith in Archaeological Chemistry edited by M. Levey (Philadelphia, 1967), 279-311.
15. J. Henderson in Materials Issues in Art and Archaeology II, edited by P.B. Vandiver, J. Druzik and G.S. Wheeler (Materials Research Society Proceedings, San Francisco 1990) pp.601-608.

16. E.V. Sayre and R.W. Smith in Recent Advances in the Science and Technology of Materials vol. 3, edited by A. Bishay (New York: Plenum Press, 1974), 47-70.
17. B. Gratuze, J.-N. Barrendon, Archaeometry, 32, 155-162 (1990).
18. I.C. Freestone, M. Bimson, D. Buckton, Annales de l'Association Internationale pour l'Histoire du Verre 11, 271.
19. R. Hodges, C. Coutts, J. Mitchell, Current Archaeology 122, 86.
20. H. Tait, 5000 Years of Glass (British Museum Press, 1991).
21. J. Henderson and J. Allen, Archeomaterials, 4, 167 (1990).

PHYSICAL AND CULTURAL CONSTRAINT OF INNOVATION IN THE LATE PREHISTORIC METALLURGY OF CERRO HUARINGA, PERU

STEPHEN M. EPSTEIN
Department of Anthropology
University of Pennsylvania
Philadelphia, PA 19104

ABSTRACT

The arsenical copper smelting technology reconstructed from the archaeological record at the late prehistoric industrial site of Cerro Huaringa, Peru, remained unchanged for the six centuries preceding the Spanish Conquest. This lack of technological innovation is argued to be a result of the physical constraints imposed by a cultural commitment to human breath as the source of draft.

INTRODUCTION

Cerro Huaringa, on the North Coast of Peru, was the site of an arsenical copper smelting industry belonging to the late prehistoric Sicán Culture (AD 700-1350). The techniques used there reflect a mature technology, not one standing at the dawn of Andean metallurgy. In fact, the metalwork of the Moche Culture, which preceded the Sicán on the North Coast, is unequalled by that of later Andean peoples [1]. Archaeology has not yet uncovered the Moche smelting process, but one can assume that it was ancestral to the Sicán process discovered at Cerro Huaringa, and which continued under the successive occupations of the Chimú (AD 1375-1460 or 1470) and the Inca (AD 1460/70 to 1532) Empires, until it was extinguished by the Spanish conquest.

The metallurgical techniques used at Cerro Huaringa have been reconstructed in detail from the archaeological record and described fully elsewhere [2,3,4,5,6]. This reconstruction is an enumeration of physical behaviors that can be inferred to have taken place if we connect the tangible debris of the archaeological record with the result of copper production. A reconstruction of behavior can provide a firm, objective, and reasonably accurate base from which to venture into the "epistemological murk" of cultural interpretation.

Lechtman [7] calls attention to the fact that a thorough description of technological behavior also contains an implicit description of cultural patterning, which she calls "technological style." Any technological behavior is the enactment of a set of cultural choices. These choices are made from a set of options whose range is constrained by the physical laws of nature and is further constrained by systems of cultural meaning. The laws of nature and systems of meaning both impose patterning in technological knowledge, which patterns technological behavior, which in turn induces patterning in the archaeological record. If one is interested in cultural reconstruction, one must separate the patterns generated by nature from patterns generated by cultural meaning.

Because technological knowledge tends to be cumulative, and technological choices have consequences that affect what choices may be possible in the future, these patterns of cultural meaning extend through time, helping to shape the trajectory of cultural evolution.

At Cerro Huaringa, the trajectory of metallurgical innovation was flat. The process used to produce copper remained the same from the earliest occupation of Cerro Huaringa, about AD 1100, until it was abandoned around the time of the Spanish Conquest in 1532. At the nearby site of Huaca del Pueblo Batán Grande, the same process was in use as early as AD 900. The remarkable fact that copper smelting techniques did not change in this region for six centuries begs for investigation.

This paper will summarize the results of the excavation and reconstruction of the Cerro Huaringa smelting industry and will consider the technological stagnation of that industry by interpreting the data in light of the constraints imposed by the physical world and the Andean cultural perspective.

THE CERRO HUARINGA SMELTING PROCESS

The Cerro Huaringa smelting industry was excavated under the aegis of Izumi Shimada's Sicán Archaeological Project. The following reconstruction is based on three seasons of field investigation, during which more than 20 furnaces were excavated and one entire workshop was cleared.

In the Cerro Huaringa process, copper was smelted from rich malachite ores with an admixture of arsenopyrite, in fairly simple pit or bowl furnaces. Hematite, an iron oxide, was added as a flux, to reduce the melting point of the unwanted country rock in which the ore minerals were embedded. Hard, dense charcoal made from the algarrobo tree fueled the furnaces. Draft was provided by human lung power, delivered through blowpipes of hollowed cane, tipped with ceramic tuyeres.

As the ore was reduced, arsenical copper droplets, or prills, formed in the slag residue. Due to relatively low temperatures inside the furnace, the slag remained so thick, viscous, and unevenly melted throughout the smelting operation that the prills remained trapped in the slag matrix, unable to sink to the bottom of the furnace and coalesce into ingots.

An examination of 18 samples of this slag matrix showed a remarkable heterogeneity on a microstructural scale. Patterns of heat flow and cooling varied over distances of only a few millimeters, indicating that each area had a different thermal history. Cerro Huaringa slag is composed of locally melted material fused together into larger masses. Remelting experiments performed upon samples of the slag indicated that local melting had occurred at temperatures between 1050° and 1200° C. The entire mass, however, had never been liquified and homogenized at a common melting point.

When the furnace had cooled, the slag was dislodged and taken to *batán*-and-*chunga* sets for grinding to extract the trapped prills. A *batán* is a large anvil of tabular basalt, usually with a smooth, deep basin worn into it. A *chunga* is a heavy rocker stone whose movement shattered and crushed the brittle, vitrified slag to the consistency of coarse sand. This treatment left malleable prills, ranging in size from the microscopic to 5 mm in diameter, rolling free in the concavity of the *batán*, ready to be collected.

Shimada [8] noted only minor modifications to the paraphernalia of this procedure in the 600 years separating the earliest excavated Middle Sicán furnaces at about AD 900 from the latest Inca examples at around 1500. The most important of these changes is a general decrease in furnace size. The earliest furnace chambers tended to be larger, with a capacity as great as three or even three-and-a-half liters, while the latest chambers held one-and-a-quarter to one-and-a-half liters. The significance of this decrease in volume will be suggested below.

The physical organization of the copper smelting industry does seem to have responded to the social, political, and economic changes associated with successive conquests by the Chimú and Inca Empires, and with integration into ever wider regional networks. It was, however, the layout of the process that changed, not the techniques of the process.

The Middle Sicán industry of AD 900 to 1100 to 1150 was located in a neighborhood context at the village site of Huaca del Pueblo Batán Grande. By AD 1100, the Late Sicán industry had left its domestic setting and moved to Cerro Huaringa. By AD 1250 or 1300, the industrial precinct there had become a workplace of multiple smelting workshops, built of masonry, and without associated residences. A similar complex of smelting workshops opened at nearby Cerro Sajino, and a metalsmithing complex appeared, situated between the smelting areas on Cerros Huaringa and Sajino. The Chimú (AD 1375 to 1460 or 1470) left a characteristic U-shaped structure on the slope of Cerro Huaringa, overlooking the workshops, and the Inca presence (AD 1460/70 to 1532) is marked by a small administrative building of relatively high quality masonry.

This history of the industry can be summarized as a trend toward greater production capacity achieved by the proliferation of workshop units rather than by any increase in the productivity of each workshop. The industry changed by doing more of the same instead of by developing more efficient techniques.

PHYSICAL CONSTRAINT

This remarkable lack of technological change appears to be an actual suppression of innovation. Even in the absence of a need for change, one would expect some technological drift to occur in the course of the centuries. Some active cultural mechanism seems to have been at work. One might be tempted to ask, "Why did nothing change?" This would be the wrong question.

The mistake would lie in assuming that some cultural pattern had locked everything in stasis, and in overlooking patterns imposed by the practical constraints of nature. A more useful question would be, "Is there one single technical factor that could limit change in the other aspects of the process?" A satisfying answer may be found in the use of lung-powered draft. Given the physical constraints of lung-powered draft, Cerro Huaringa's copper smelting appears to have gone about as far as it could go.

It is well known that the Inca, and presumably their predecessors, never used the bellows. Garcilaso de la Vega, known as El Inca, commented on this fact in 1609, and many other ethnohistoric sources, such as Oviedo (1535), Benzoni (1565), and Raleigh (1596), described the use of the blowpipe as a source of draft [9]. There are no archaeological remains that can be identified as bellows. To the south, prehistoric silver smelting furnaces called "*huayras*," ventilated chimney-like constructions about a meter in height, circumvented the need for bellows by relying on the wind for draft. Although *huayras* are abundantly documented in the ethnohistoric accounts of Cieza de León (1553), Zárate (1555), Capoche (1585), Ramírez (1597), Ovando (1605), Barba (1640), and Cobo (1653) [10], there is no evidence, ethnohistorical or archaeological, that they were used for smelting copper, or used north of the Bolivian altiplano.

The difficulties surrounding lung-powered draft became obvious in an experimental replication of the Cerro Huaringa smelting process. The test was conducted at Cerro Huaringa in 1986, under the direction of Izumi Shimada and the guidance of John Merkel. Locally obtained ores were smelted in two furnaces, one a reconstruction and the other an original, 600 year-old smelter. Human lung power provided draft to the furnaces. The experiment did, in fact, successfully produce copper.

Teams of two to five men blew through canes tipped with ceramic tuyeres for three hours, in shifts ranging from 10 to 30 minutes. The tuyeres each had an 8 mm aperture, as observed in the archaeological samples collected on site. This constriction generates a forceful and economical flow of air. The airblast was directed into the furnace around the edges of a sherd placed over the charge as insulation.

The experiment's most vivid lesson was the extreme difficulty of maintaining the high temperatures required to smelt copper. High combustion temperatures are generated by strong blasts of air. Human lungs can deliver only a limited amount of air. Even allowing for the efficiency and stamina that would come with years of practice, the technology appeared to be operating at its limit.

Temperatures of 1000° C could be achieved and maintained comfortably. Each additional degree was a struggle. An all-out, coordinated effort could occasionally coax a peak reading of 1250° from the thermocouples placed in the furnace chamber. Very little of the material in the furnace was molten at any given time. One can visualize a pea-sized sphere of 1150° heat at the aperture of each tuyere, forming and collapsing with every breath. The smelting process consists of moving these white-hot microenvironments throughout the furnace charge. This phenomenon explains the heterogeneity and localized thermal histories observed in an analysis of slag from Cerro Huaringa.

If, for the sake of argument, the use of lung-powered draft is assumed to be a given, there is little in the rest of the process that can change. The Cerro Huaringa process teeters on the edge of thermal possibility due to its tenuous control over furnace temperatures as they approach the range needed to smelt copper. A stronger airblast is necessary. Adding more tuyeres does not help. The mouth of the typical Cerro Huaringa furnace accommodates three tuyeres most comfortably. More than three only get in each other's way, wasting lung capacity that could be used more profitably if the time were spent in resting.

If the furnace were enlarged to accommodate additional tuyeres, it would require more heat to raise it to smelting temperatures. It is true that this heat would be more efficiently expended because a larger furnace charge would have a more favorable surface-to-volume ratio. As the volume of the mass increases, its surface area increases proportionally less. This means that more mass would be better insulated, radiating heat from less surface area. Unfortunately, the thermal advantage of this economy of scale cannot be put to work in the Cerro Huaringa furnaces. Even small charges were already too large to be thoroughly heated to a uniformly molten state. A larger mass would require still more heat, even though it would be expended more efficiently.

A larger furnace charge would also have been difficult for a lung-powered airblast to penetrate. The constricted apertures of the tuyeres were necessary to increase the force of the relatively weak human wind. Without a strong injection of air, the center of the mass would have remained unheated. In this light, the shrinkage through time of the preferred furnace volume observed by Shimada may have been an effective means of coming to terms with the limits of the human lung.

CULTURAL CONSTRAINT

Innovation is the recombination of old elements into a new configuration, or the introduction of old ideas into a new context [11]. Innovation does not occur in a sterile, random way. It occurs in the human brain, which is trained to a cultural worldview. A cultural worldview can make some potential connections seem obvious or invisible, reasonable or unreasonable, valuable, or meaningless. Necessity may be the mother of invention, but culture is the midwife.

It is difficult to move in this discussion from the general to the specific. To argue that a certain innovation was constrained by some cultural pattern, one would have to demonstrate, ideally, that the innovation should have appeared, and that there is an identifiable cultural pattern that would act to suppress it. This is an almost impossible task. In the example of the bellows, however, one can advance a case that is at least interesting.

Ethnohistoric accounts [12] of Inca religion include descriptions of divination practices that suggest the Inca stood on the very brink of inventing the bag bellows. Soothsayers attempted to divine the future by removing the lungs from a sacrificed animal, blowing into the trachea to inflate them, and reading in the pattern of the veins what was to come. A photograph of an Inca ceramic vessel with such a ritual sculpted on it has been published in Alfred Métraux's *The History of the Incas* [13].

Anyone who has ever inflated a butchered set of animal lungs by blowing down the windpipe would recall the lifelike exhalation that blows back in one's face as soon as pressure on the trachea is relaxed. Inca diviners held single-shot, muzzle-loading bellows in their hands. The intellectual leap needed to apply such a vivid analog to the new context of fire-fanning seems unusually short. It seems even shorter in light of another Inca divining practice [12], that of blowing on a fire through concealed tubes, making the flames dance to accompany oracular pronouncements delivered by ventriloquism. The same group of people were exposed to the potential innovation and its potential application.

For the Inca, breath seems to have had some special significance and effectiveness. These same rituals, viewed from a different perspective, suggest a cultural pattern that might have prevented the innovation of the bellows by blocking the perception that mechanical wind might be of interest or value. The diviner interacts with the cosmological realm through breath, his breath in the animal's lungs, recycled as the animal's breath through its organs of breath.

The Quechua language of the Incas and a recent Quechua ethnographic study provide some clues to the special significance of breath. The Quechua root *cama-* means "life force." The suffix *-yuk* indicates possession, thus a craftsman, *camayuk*, is one who possesses the life force and can impart it to objects [14]. Allen, working in the Peruvian highlands, has noted that "the Quechua cosmos is suffused with creative energy," a life force which "must be shared--with other people, with places and with objects" [15]. One means of sharing that life force is through the breath. In the ritual of *pukuy*, the spiritual essence of an offering, usually coca leaves, is conveyed from the human to the sacred realm by blowing upon it. On the North Coast, Sharon has also observed the use of the breath to share the essence of offerings with the spirit world [16].

The relevance of these clues is based on the assumption that there is and was such a thing as an Andean worldview whose broad patterns transcended local cultural variation within it. This local diversity means that cultural insights should not be transposed exactly and literally from one Andean culture to another. For example, the use of the Quechua lexicon is convenient and suggestive, but it should be noted that the ancient Sicán metallurgists of Cerro Huaringa spoke the extinct language of the North Coast, Muchic or Yunga. Nor were they Incas, so descriptions of Inca ritual may not be perfectly apt. Furthermore, it would be facile to apply modern ethnographic observations literally, to suggest that the ancient metalworkers believed that blowing upon a furnace charge carried its essence into the supernatural dimension of the world. Nevertheless, beneath these caveats of detail, one can clearly see the outline of a broader Andean concept of breath's supernatural importance. One can note that the material system of Andean metal production intersects and interacts with Andean concepts of breath, life force, creation, and the pervasive spiritual aspect of the world. The range of choices fixed by this interaction thus gave a distinctly Andean shape to copper smelting technology at Cerro Huaringa.

TECHNOLOGICAL CONSEQUENCES

If a cultural commitment to human breath was a major factor that prevented Andean metallurgists from being able to maintain high temperatures, one can speculate concerning its important consequences for the trajectory of technological and social evolution.

Without the bellows, they could not melt a furnace charge thoroughly and could not obtain a complete, liquid separation of slag and metal. This meant that they could not tap their furnaces to allow the slag to flow off and a massive metal ingot to be lifted out, but had to continue to rely on prill extraction. This procedure was extremely labor intensive. It required people and equipment to perform the additional task of grinding slag to free the prills, and yet another step to melt the prills into consolidated ingots, consuming still more time, labor, and fuel.

Without the bellows, they could not raise larger furnaces to smelting temperatures and thus could not benefit from an economy of scale. The ability to make copper in larger heats would have saved fuel, would have entailed less start-up, tear-down, and maintenance activity, and would have required fewer workers. These fewer workers would have had to manage the execution of a more precisely balanced process and would thus have needed to develop a more sophisticated level of expertise. One might speculate that the increasing demand for copper which occurred over the Chimú and Inca Periods might have raised the social status of copper smelters as their special knowledge became more valuable. In fact, however, the increased demand for copper was satisfied by building myriad replications of small, prill extraction smelting workshops manned by gangs of semi-skilled laborers who could be easily trained, rotated, and replaced.

CONCLUSIONS

A commitment to the single technological element of human lung-powered draft would provide a parsimonious explanation of the lack of innovation in the Cerro Huaringa smelting process. If the metallurgists believed that their breath made a necessary contribution to the transformation of natural ore to man-made metal, then such a commitment would make perfect sense, and the invention of the bellows would have been rendered unlikely.

Although the importance of this cultural insight is argued here, it is not advanced as the single cause or the only possible explanation for the absence of the bellows in Andean metallurgy. It is presented forcefully because it is a case where the elusive cultural aspect of the complex interactions of the innovation process can be demonstrated with unusual clarity.

ACKNOWLEDGEMENTS

The author gratefully acknowledges the advice of Ursula Franklin of the University of Toronto, where he examined the slag samples described in this paper, and the suggestions of Heather Lechtman, who spent two weeks at Cerro Huaringa during its third field season. This research would not have been possible without the support of Izumi Shimada, the Sicán Archaeological Project, and the University of Pennsylvania's Museum Applied Science Center for Archaeology. Conversations with Eleanor King, Lisa Kealhofer, Kay Candler, and Gil Stein have enriched this paper considerably.

REFERENCES

1. H. Lechtman, in The Coming of the Age of Iron, edited by T. A. Wertime and J. D. Muhly (Yale University Press, New Haven, 1980), pp. 267-334.

2. Shimada, S. M. Epstein, and A. K. Craig, Science, 216, 952-959 (1982).

3. I. Shimada, S. M. Epstein, and A. K. Craig, Archaeology 36(5), 38-45 (1983).

4. S. M. Epstein, MASCA Journal 2, 58-62 (1982).

5. S. M. Epstein and I. Shimada, in Sicán Metallurgy, edited by I. Shimada (Cambridge University Press, Cambridge, in press).

6. S. M. Epstein and I. Shimada, Beiträge zur Allgemeinen und Vergleichenden Archäologie 5, 379-430 (1984).

7. H. Lechtman, in Material Culture: Styles, Organization, and Dynamics of Technology, edited by H. Lechtman and R. S. Merrill (West, New York, 1977), pp. 3-20.

8. I. Shimada, J. F. Merkel, and S. M. Epstein, Willay: Newsletter of the Andean Anthropological Research Group (23/24), 15-17 (1986).

9. W. Bray, Proceedings of the Royal Anthropological Institute of Great Britain and Ireland 1971, 25-43 (1972).

10. E. Boman, Antiquités de la Région Andine de la République Argentine et du Désert d'Atacama (Imprimerie Nationale, Paris, 1908).

11. H. G. Barnett, Innovation: The Basis of Cultural Change (McGraw-Hill, New York, 1953).

12. B. Cobo, Historia del Nuevo Mundo [1653], (Biblioteca de Autores Españoles. Atlas, Madrid, 1956).

13. A. Métraux, The History of the Incas (Schocken Books, New York, 1969), p. 139.

14. G. Taylor, Journal de la Société des Américanistes 63, 231-244 (1974-1976).

15. C. A. Wagner, Ph.D. dissertation, Department of Anthropology, University of Illinois, 1978.

16. D. Sharon, Wizard of the Four Winds: A Shaman's Story (The Free Press, New York, 1978).

FROM THEOPHILUS TO C.S.SMITH: DISCOVERY OF AN ELEVENTH CENTURY AD BELL-CASTING MOLD FROM VENOSA (SOUTHERN ITALY)

M.Vidale *
A.Melucco Vaccaro **
M.R.Salvatore ***
M.Micheli **
C.Balista ****

* Istituto Centrale per il Restauro, Rome, Italy, and Conservation Analytical Laboratory, Smithsonian Institution, Washington, D.C.

** Istituto Centrale per il Restauro, Rome, Italy

*** Istituto Centrale per il Catalogo, Rome, Italy

**** Consultant geoarchaeologist

Abstract

The recent discovery and excavation of the remains of a well preserved mold for bell-casting below the floor of the medieval church of the SS. Trinita' of Venosa (southern Italy) provides a singular opportunity to reconstruct aspects of ancient bell-making technology and to compare the new archaeological data with the textual evidence on the same subject written by the famous monk Theophilus in the XIth chapter of his treatise "De Diversis Artibus". Stratigraphical excavation and video-endoscopic inspection of the mold's interior allowed a preliminary reconstruction of the structure and use of this unique craft installation.

INTRODUCTION

Bell technology has been a subject of considerable scholarly enquiry both in ancient (1-5) and recent times (6-10). Several archaeological excavations in Europe, particularly after 1950 (11), have provided with reports of the discovery of bell-casting pits, usually below the floors of medieval and later churches, within clocktowers, or in nearby open areas. The interior of churches provided an ideal space for casting bells for various reasons: the bell had not to be moved far; the working area was enclosed and easily controlled; the heavy stone foundations of the buildings mantained completely dry the loose sediments usually forming below the floors; the roof sheltered furnaces and molds from weather accidents; and the sanctity of the place, most probably, was seen as a protection against trouble and failures.

A survey of the available bibliography suggests that, in many cases, these pits may be dated to late medieval (XIIIth - XIVth century AD) or Renaissance times (from the XVth century onward). In the XVIth and XVIIth centuries AD similars finds seem to become more common (8, 12-15). Rare instances of bell-casting pits dating to the XIth-XIIth century AD are known, but they are generally very poorly preserved. Often only the outline of the pit and the supporting elements of the mold remained in situ (16-18).

In this paper we present the discovery and partial excavation of an exceptionally well preserved bell-casting mold found below the floor of the church of the SS. Trinita' ("the Holy Trinity") of Venosa in southern Italy (Figs. 1, 2). This find is dated approximately to the XIth century AD.

BACKGROUND

The technology of bells played a major role in the history of European metallurgy. The roots of this technology are still somewhat obscured (11, 19), but uncovering the development of bell-casting between the VIIIth and the XIth century AD may help us bridge the documentary gap existing between the great metallurgical traditions of the classic world and the Renaissance. It is known, for instance, that major artists of this later period, including Donatello, Cellini and Sansovino were assisted by bell-makers in the casting of their bronze statues (20), and that the production of bells was connected to another industry of strategic importance of the Renaissance states, the casting of the first cannons.

Im medieval Europe, bell-casting required substantial investment of capital sponsored by central authorities, as well as, on the artisans' side, high levels of skill. Bell-makers, both in ancient and recent times, may be considered as "classic" examples of attached specialists, closely dependent upon political and religious institutions. Traditionally, this task has been fulfilled by religious specialists, or by families of travelling metalsmiths, temporarily hosted in the yards or in the premises of churches and monasteries (11, 12). With the urban development of the XIIth - XIVth centuries, permanent central workshops are known in the main centers: for example, at York in the XIIth century (18), at Lubeck by 1250 (12), and the same development is reported in Canton Ticino, Switzerland, in the XIVth century (8). These craftsmen had obviously to train their followers or sons through a closely monitored apprenticeship, and one may expect that social behavior, rituals, and technological rules were standardized and in most instaces strictly observed.

The study of ancient bell technology has been based on several different sources of information: epigraphy (on bells or on other media), ancient administrative documents, ancient treatises, archaeology and ethnography (or ethnoarchaeology) (11). We may sistematically compare the archaeological evidence from Venosa with information provided on bell technology by the treatise "De Diversis Artibus" written by the German monk Theophilus, in his XIth chapter entitled "De Campanis Fundendis" ("On the Casting of Bells"). The treatise, most probably, was written between the XIth and the XIIth century AD (2) and may be considered approximately contemporary with the Venosa bell pit. This text contains a very detailed but very vivid description of an operational sequence for bell-making, doubtless the fruit of personal observation.

However, it must be remembered that the mold of the SS. Trinita' and the book probably represent totally indipendent contexts. In this preliminary stage, besides the Latin text, we based our study on the two English translations provided by Hawthorne and Smith (3) and Dodwell (2). Quotations in the text refer to Hawthorne and Smith (3) and appear, for sake of brevity, as DCF (= "De Campanis Fundendis"), followed by the page number.

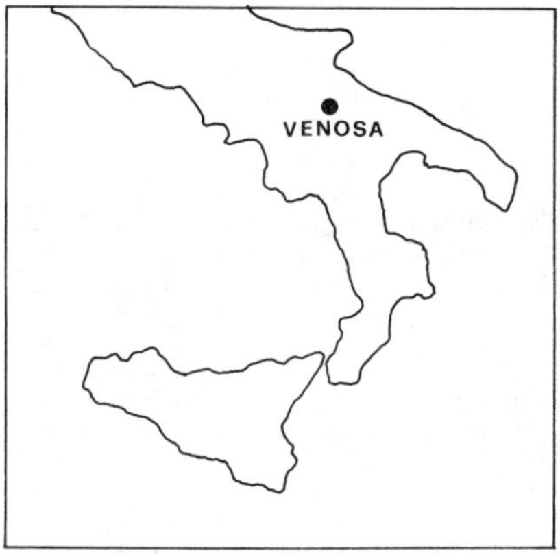

Fig. 1: The location of the ancient city of Venosa in southern Italy.

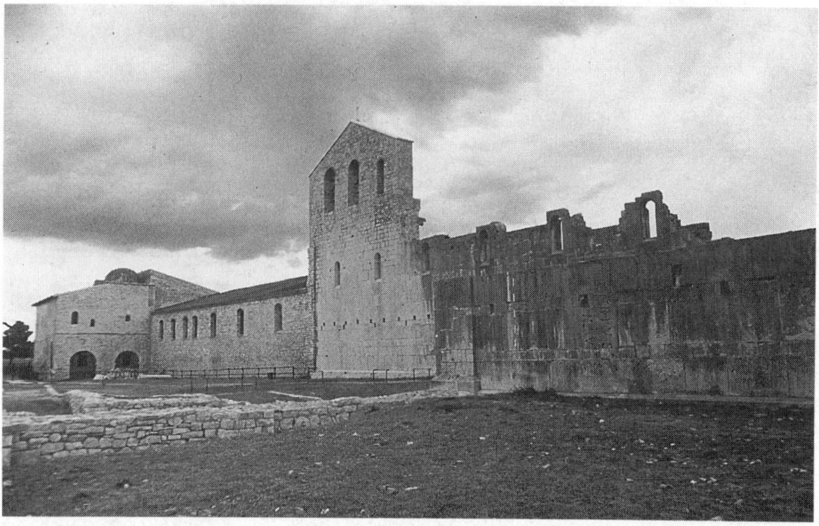

Fig. 2: Venosa. View of the church of the SS.Trinita'.

THE SITE OF VENOSA

In Roman times, Venosa was a city of primary importance, located on the Via Appia, leading from Rome to Brindisi, one ofthe main ports to the eastern Mediterranean (Fig. 1). After the fall of the western empire, Venosa continued to be an important city, its temples transformed into Christian churches. From the beginning of the XIth century onward, the history of Venosa was closely tied to the Normans of the Altavilla house (21). Their power in south Italy soon developed from a loose system of small militaristic chiefdoms connected by kinship relationships to a powerful early state ruled by the two brothers Robert Guiscardus (a surname apparently meaning "The Smart"), Duke of Apulia and Calabria, and Roger, Count of Sicily.

Venosa became Robert's capital, as well as the burial monument of the Altavilla family. This outstanding monument shows a complex sequence of building activities. Probably erected over the foundations of a Roman temple dedicated to the goddess Venus, the church was enriched in 942 with the addition of an important Benedictine monastery. During the XIIth century, at the peak of the Norman power, under Kings Roger and William, the monks of Venosa attempted a very ambitious architectural project: to encorporate their church into a huge, new abbey. But this goal was never achieved. The abbey may still be admired, in its unfinished state, with all its stones removed from ancient Roman monuments.

THE MOLD: CONTEXT AND CHRONOLOGY

The bell-casting mold was discovered below the present floor level, not far from the entrance of the old church. The original surface where the bronze foundry had been installed had been previously erased by later building activities, so that the mold was no longer associated with horizontal living floors. No trace was found of the furnace or the installation used to melt the metal, but some suspect burnt truncated layers were visible on the very edge of the pit, suggesting that the original surface was not much higher. In the filling of the casting pit (Figs. 3, 4) the excavators found a coin which can generically be ascribed to the XIth century AD. Possibly, the bell was cast under the rule of Unfred, Robert Guiscardus or his sons Roger and Boemund: a central aspect of their politics was maitenance of continuous support to the monks of the SS. Trinita'.

THE MANUFACTURE OF THE MOLD'S CORE

A further study will present metallographic and micro-structural analyses of the metal and ceramic residues which are in progress at the Istituto Centrale per il Restauro, Rome, and the Conservation Analytical Laboratory, Smithsonian Institution, Washington, D.C. Here we shall limit ourselves to a description of the macroscopic aspects of the structure of the mold, its casting techniques, and the procedures used for recovering the bell from the casting pit.

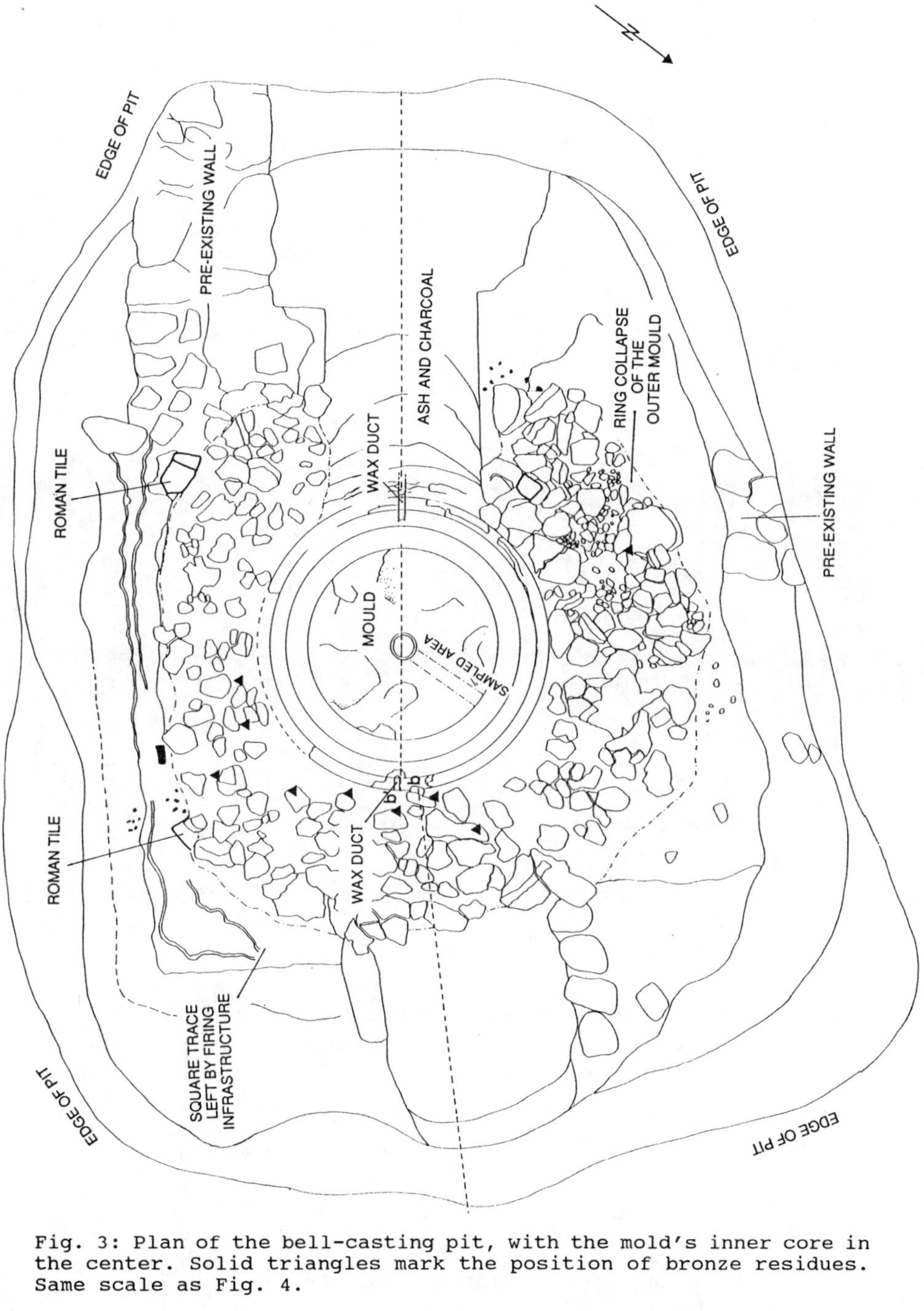

Fig. 3: Plan of the bell-casting pit, with the mold's inner core in the center. Solid triangles mark the position of bronze residues. Same scale as Fig. 4.

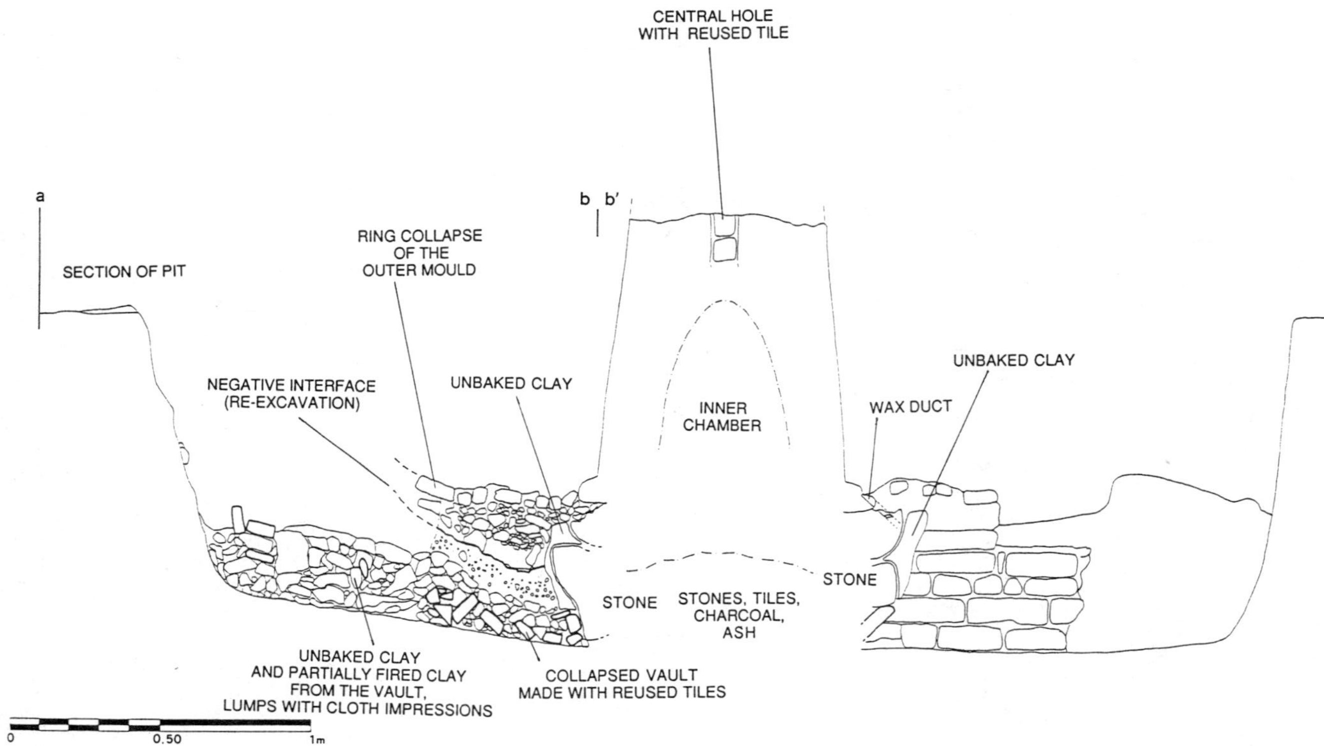

Fig. 4: Section of the bell-casting pit, with the mold's inner core in the center. The probable shape of the inner chamber, graphically reconstructed after video-endoscopy, is represented in dashed lines. The deposit at left of the mold, in the central alley, was excavated by Vidale.

The inner core of the mold (Figs. 4-9) was found still in situ. It stands on top of two short parallel facing walls delimiting a central channel, made with stones and reused Roman bricks and tiles. These ubiquitous Roman tiles provided bell-makers with an excellent building material. The lower channel, only partially excavated by our team, contained layers of ash and charcoal, and had evidently been used for firing the assembled mold from below. The surface of the core is exceedingly smooth (Fig. 10), with regular and fine turning traces, and web-like patterns of thin bronze droppings filling the cracks which formed during the drying and firing phase, and further enhanced by a small amount of shrinking while the cast bell cooled down.

The core has a base diameter of about 1 m, and, although broken at the top, it retains a maximum height of about 95 cm. According to contemporary bell-makers, these proportions suggest a bell weighing not less than 500 kg. This clay element has a truncated conical profile, slightly convex in its lower part, with a regular rim at the base projecting outward for 10 cm. It also has a central cylindrical hole (which turned out to be sealed by a reused Roman tile) visible on the upper fracture surface. How was this inner core built? In order to study the inner structure of the core, we tried to access its interior from above and from the lower firing channel. But the tile piece in the upper cylindrical hole turned out to fit so tightly to be impossible to remove, and the two accesses between the lower parallel facing walls were blocked by two large stones sealed with a thick coat of unbacked clay (Figs. 10, 13, lower right).

As a consequence, it was impossible to observe directly the interior of the core. In order to identify the inner structure of the core without having to excavate and disassemble the installation, we had to use two other methods. First, we cut a radial section of the upper part of the core, on the fracture surface (Figs. 3-5); the section was progressively embedded in layers of resin to avoid breakage, and then removed for microscopic and stratigraphical examination. Second, we exploited a pre-existing large crack to drill a new access. The hole we drilled was 2 cm large, and allowed us to carry out a video-endoscopic inspection of the core's interior. Video-endoscopy is a technique which was frequently used in recent times to explore and document cavities and voids within buildings and artworks (in first place bronze statues). A movable microcamera, placed at the extremity of a flexible cable, is inserted in the cavity to be explored, and the images so obtained are recorded by a video system. The whole station is portable and can be easily installed in less than 1 hour.

A RADIAL SECTION OF THE CORE

The section removed from the top of the core was polished with abrading paper of decreasing size, and finally cleaned with forced air to expose the pore structures (Fig. 5). The graphic interpretation of the section shows a regular sequence of at least 10 parallel layers, varying in thickness from 2 to 4.5 cm. According to our reconstruction, the core was built by applying a sequence of concentric slabs of clay, very rich in organic fibers (including perhaps animal fur). The average thickness of these slabs did not exceed 2-3 cm. The thin organic fibers tend to assume regular, parallel orientations. This iso-orientation of the fine fibers indicates circumferential alignment of construction elements, probably by means of a forming process involving turning.

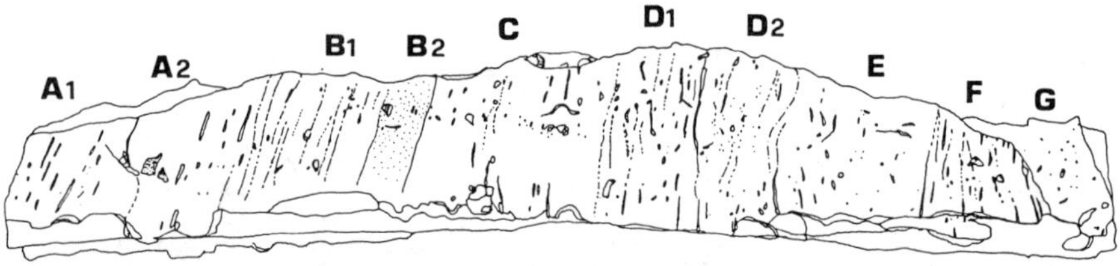

Fig. 5: Radial cross-section cut from the fracture surface of the upper part of the inner core, as shown in Fig. 3. Layer A corresponds to the outer surface. Layers A to F were applied as parallel concentric slabs. The circumferential orientation of the elongated inner pores suggests that slabs were applied by turning. In G, on the contrary, the random structure of the organic fibers and the evidence of forming pressure show that the clay was applied by hand to regularize the central square hole left by the lathe's spindle.

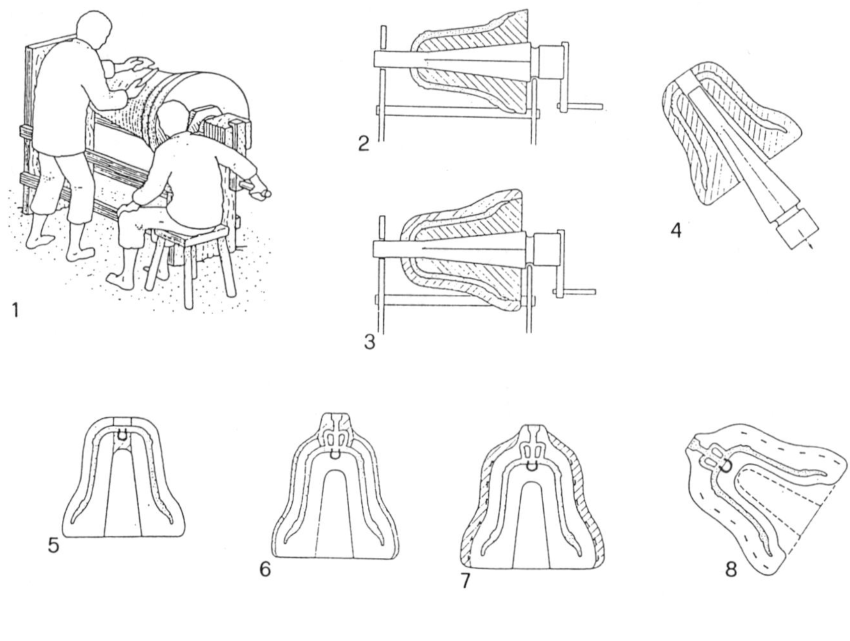

Fig. 6: The operational sequence of bell-making described by Theophilus, according to Donati (8), as follows: 1: building the clay core with a horizontal lathe, 2: the core is covered with a tallow model shaped by turning, 3: the first layers of the outer shell are added onto the tallow model, 4: the lathe's spindle is removed, leaving an inner square hole, 5: the hole is closed with plastic clay to support the U-bent iron staple to be cast intothe bell for hanging the clapper, 6: the upper part of the mold is added to the core, 7: the mold is protected with series of parallel iron hoops and finally covered with an additional layer of clay, 8: the interior of the mold is emptied, to make the mold lighter.

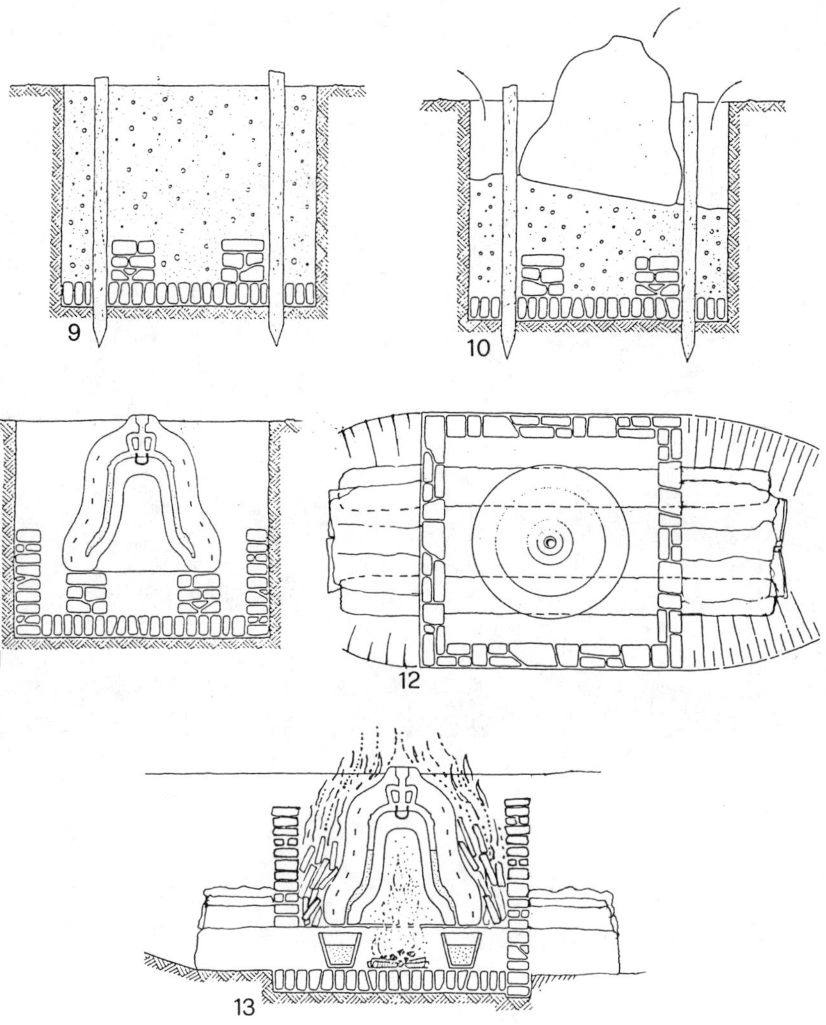

Fig. 7: The operational sequence of bell-making described by Theophilus, according to Donati (8), continuing from Fig. 6, with Figs. 7: 12 and 13 modified: 9: four poles which will serve as guides are sunk aside the two lower facing walls of the base, and the pit is refilled, 10: the mold is placed inside the poles, and the earth is removed, 11: the mold is lowered and set onto the parallel facing walls, 12: plan of the casting pit with the square outer furnace and the mold ready for firing. Note the two underlying vaults made with reused Roman tiles and plastered with clay to provide the two lower accesses, 13: longitudinal section of the casting pit with the square outer furnace and the mold during firing, with fuel providing heat both from below and from above, within the square outer furnace.

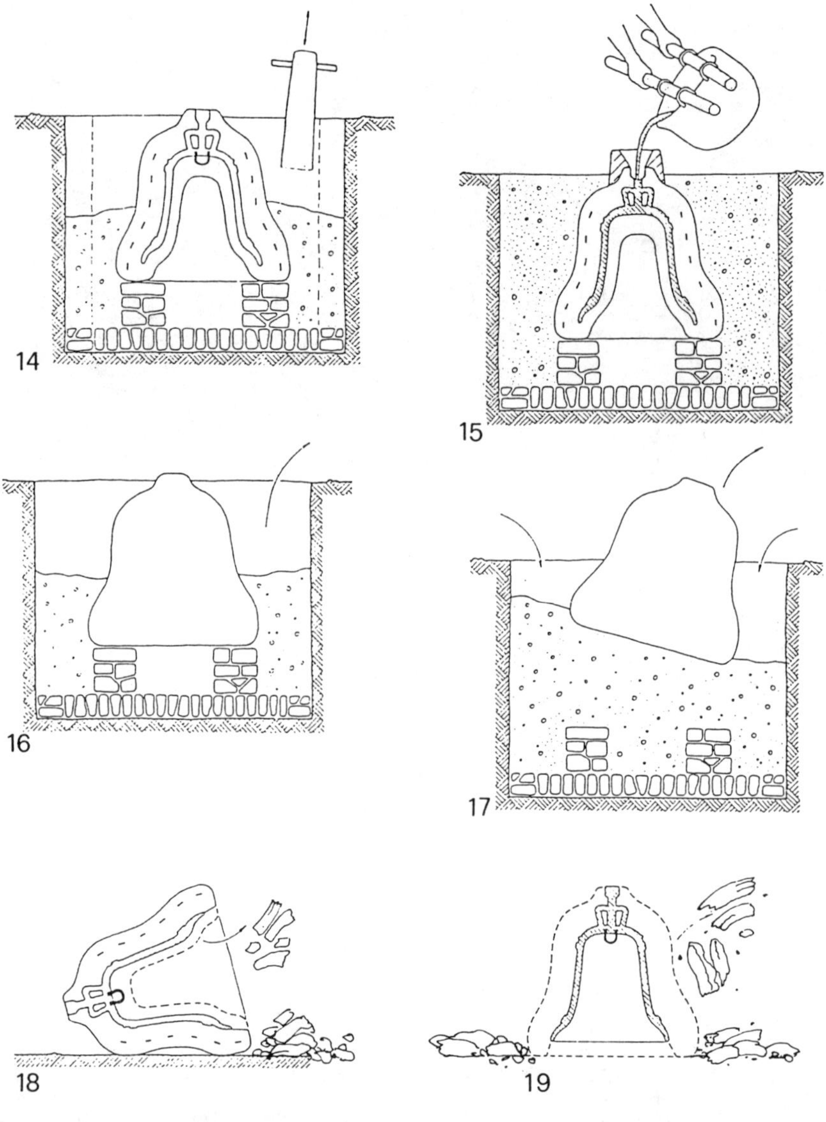

Fig. 8: The operational sequence of bell-making described by Theophilus, according to Donati (8), continuing from Figs, 6 and 7, with Figs 8: 18 and 19 modified: 14: the pit is refilled and the earth packed with blunt poles, 15: casting, 16: the pit is quickly emptied, 17: the pit is refilled and the mold is slowly lifted out of the pit, 18: the inner part of the mold is broken to release the moisture (a detail stressed by Theophilus), 19: the outer mold is broken, and the bell and the iron hoops are recovered.

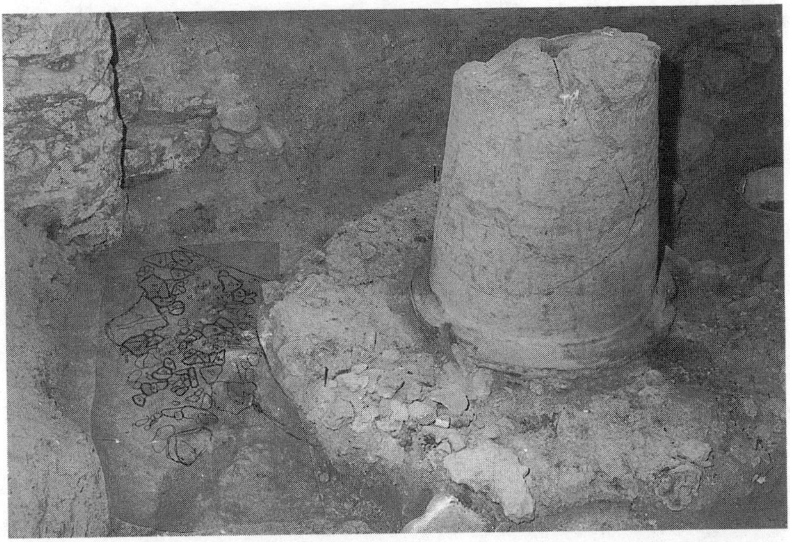

Fig. 9: View of the bell-casting pit surrounded by an annular collapse of broken pieces of the outer mold. Some large pieces were exposed during excavation. At left of the core, a plastic sheet used for mapping and numbering the shell's fragments before removal.

Using the presence and structure of such inclusions and pores as a discriminant marker, it is possible to conclude that the innermost layer, G in Fig. 5, was instead applied by hand, after the outer parallel layers had been completed. In fact, the fibers in this layer are randomly oriented. Video-endoscopic inspection of the interior of the core's wall failed to reveal any structural inclusion, such as bricks, wood or metal supports. This confirmed that the core was a self-supporting structure of concentric slabs of unbaked clay.

These archaeological data are in perfect agreement with Theophilus' and Biringuccio's descriptions. Theophilus (DCF: 168) states that the inner core of the mold had to be constructed by applying superimposed layers of well kneaded clay around the spindle of a revolving horizontal lathe (Fig. 6: 1 and Figs. 6-8 for the entire manufacturing sequence) until the final profile was obtained. Theophilus states that the surface had to be finished by turning it with a wet cloth. In the section of Fig. 5, the superimposed layers are well adhered one to another. Theophilus' rule for obtaining such a compact structure was the following:

> "Take care that you never put on more clay until the layer beneath is completely dry" (DCF: 168).

This rule probably stresses the need of avoiding the formation of even the smallest cracks in the concentric structure of applied slabs because of the drying shrinkage. A collapse of the core under the weight of the molten metal would have been disastrous. If the craftsmen followed strictly this rule, the construction of the core might have required several days; this looks like the type of work that craftsmen, given a careful supervision, would have willingly left to apprentices.

The hole we observe in the core is probably the trace left by the central axis of the lathe, which originally had to be square in section, to grant a better hold in turning (DCF: 167). At present, we cannot exclude a priori that the mold's core was actually formed in situ, in vertical position, with a revolving device such as that later described by Biringuccio and the Encyclopedie. The radial section of Fig. 5, at any rate, would rather support a close match with Theophilus' instructions.

THE MAKING AND ASSEMBLING OF THE OUTER MOLD

After forming, the core was covered by soft pre-formed slabs of tallow (Fig. 6: 2), to be finally shaped by turning, and then sealed by a "shell" formed by multiple layers of clay (Fig. 6: 3-7). Within the walls of this outer shell, series of horizontal parallel iron hoops 2-3 cm wide had to be arranged at regular distances, to strenghten the mold and to contain the weight of the molten bronze (Fig. 6: 7). At Venosa, the fragments of this outer shell were found in a primary context of deposition all around the base of the standing core: they show a typical multi-stratified pattern, with varying colours due to oxidisation and reduction. Many show iron-stained imprints left by the thin metal bands (Figs. 11, 12).

Then, going back to the medieval treatise, the lathe's spindle had to be removed, and, as Theophilus says, the hole was sealed with plastic clay (Fig. 6: 4, 5), to be used as a support to fix the U-bent iron staple for suspending the clapper (DCF: 170). This is confirmed by the core of the Venosa mold, in which, as we have seen, the hole was blocked with a piece of tile, and regularized with hand-pressed fine clay into a round contour. This operation, besides providing a support for the inner component to be cast into the bell's interior, sealed the inner chamber of the core. The firing phase did not require an upper draft hole; gases and smoke should have escaped from the two vaults covering the lower access.

Theophilus cleraly states that the lower part of the mold had to be emptied:

> "Turn the mold on its side, and cut a large, deep cylindrical hole inside the clay (core), so that the remaining clay is more than a foot thick (i.e., little less than 30 cm), since if the mold were solid inside, it could not be lifted because of the excessive weight and it could not be baked throughout because of its thickness" (DCF: 170).

By video-endoscopy we showed that the core is internally empty, and we were able to measure the thickness of its wall. At the point we drilled, this latter was about 20 cm thick (Fig. 6: 7-8). We were also able to visualize the inner surface of the chamber. It appeared carefully smoothed, and at the top the central hole was completely hidden by a new coat of clay. This evidence shows an almost complete match with Theophilus.

SETTING THE MOLD FOR FIRING (Figs. 7: 9-11)

Theophilus gives a detailed description of the excavation of the casting pit and the building of the two parallel walls designed to support the assembled mold. The size of the pit is not specified: obviously, it should correspond to the height of the mold and have an appropriate width. At Venosa, the pit measures 3.80 X 2.60 m. It has an ovoidal contour, with a central axis

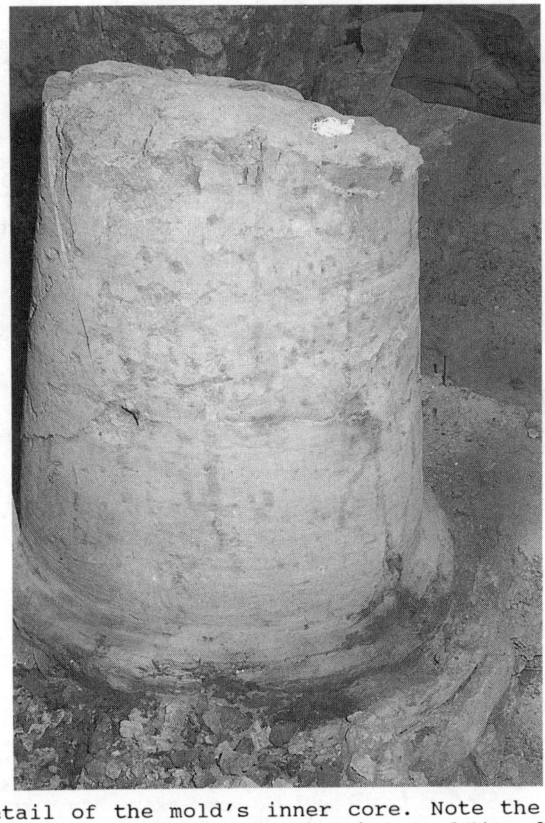

Fig. 10: Detail of the mold's inner core. Note the tall conical shape, typical of the bells of this period, and the slightly convex profile designed for the inetrior of the bell. On the left, the hole drilled for video-endoscopic inspection is also visible.

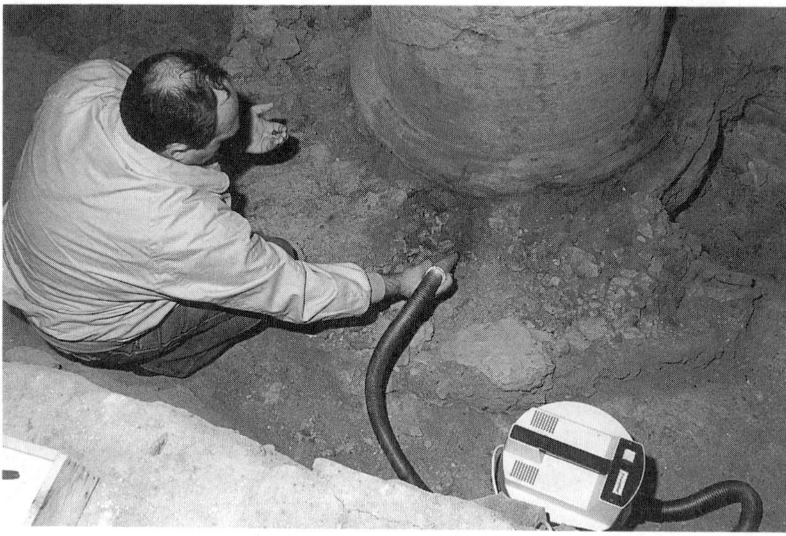

Fig. 11: Excavating the collapse of the outer mold with a vacuum cleaner, the only medium capable of unerathing the soft fragments of outer shell which we believe were intensively trampled in antiquity, avoiding detrimental mechanical contact. Below this layer, the residue of the earth filling piled against the mold for the casting phase, later re-excavated to extract the mold.

oriented towards the north-east (Figs. 3, 4). As correctly remarked by Donati (8) the ovoidal contour of many bell-casting pits found in excavations is explained by the space required by at least two craftsmen working at the sametime in the pit. Blagg (11) reports "8"-shaped bell-casting pits from England which must have been planned for the same purpose.

In the base of the Venosa mold the two parallel facing walls are 30-37 cm high, and the channel in between is 50 cm wide at the mouth of the south-eastern part of the alley: a good match with the measurements prescribed by the treatise (1 X 1.5 feet: DCF, 170) (Figs. 3 and 4). For some reasons, the two parallel walls are not straight, but deviate towards north at an angle.

The procedure illustrated in the text for putting the mold onto the two parallel facing walls at the bottom of the pit is a variant of the old "obelisk" method. 4 vertical poles are sunk into the ground aside the parallel walls, to guide the heavy mold to the desired spot. The pit is refilled, so that only the poles protrude from the surface. The mold is then placed among the four poles, and the pit is re-excavated, first from one side of the poles, then fron another, so that the level is gradually lowered and the heavy mold moves downward (DCF: 170-171) (Fig. 7: 9-11). At Venosa, the ring of collapsed fragments of the outer shell prevented us from searching for post-hole traces. Nonetheless, our observations suggest that the lowermost part of the mold's shell is still partially preserved below its own destruction layers. We are sure that the installation was constructed in two parts, and that the two facing walls are separated from the base of the mold.

Video-endoscopy of the inner chamber of the core unexpectedly showed that below the inner edge of the core the artisans had build some small supporting elements, in form of small contiguous pillars. These elements are not reproduced in Fig. 4, where they should appear at the base of the inner chamber, as we would eventually open the lower entrances and observe directly its interior before confirming their presence. Theophilus does not mention this feature, but similar elements would have supposedly increased support for the mold.

FIRING THE MOLD AND RECOVERING THE TALLOW (Figs. 7: 12, 13)

This step involved the firing of the mold and the recovery of the melting tallow, according to the classic lost-wax system. The core of the mold was fired simultaneously from below (i.e. from the cavity of the inner chamber) and from above. For this purpose, a temporary firing chamber was erected around the mold (Fig. 7: 12, 13). As observed by Hawthorne and Smith (3) the relevant passage of Theophilus is rather obscure; but we think that the find of Venosa may help in clarifying its interpretation.

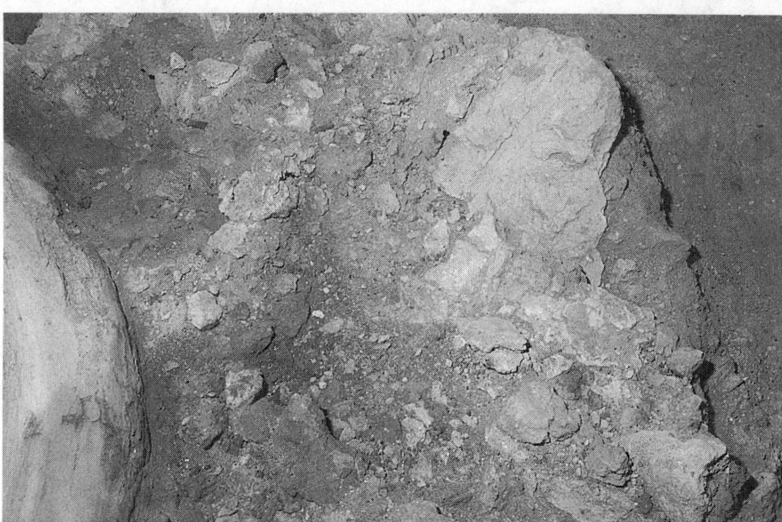

Fig. 12: Detail of the surface of the annular collapse layer, with a large fragment bearing the trace of an iron hoop (top right).

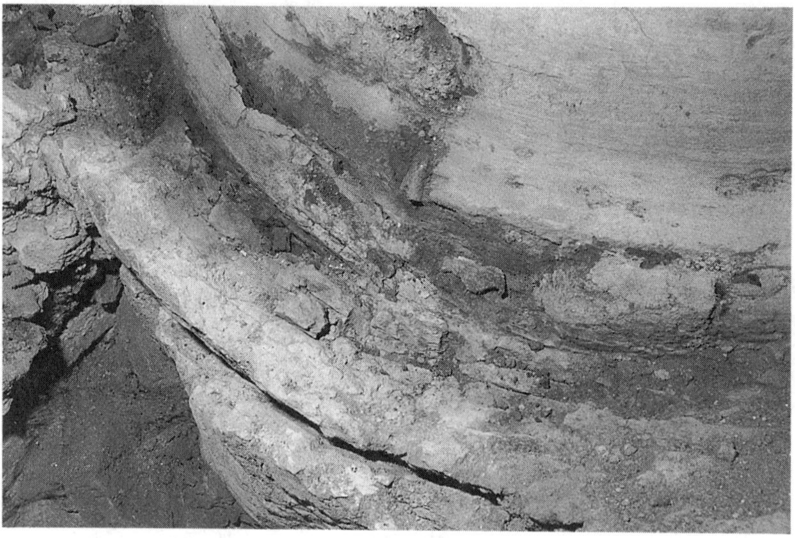

Fig. 13: Detail of the base of the mold in correspondence with the lower channel between the two facing walls. In the center of the picture, the cylindrical hole cutting through the protruding rim of the inner core is one of the escape ducts for the lost-wax method. These ducts were opened with a tool resembling a wimble. The ducts are partially broken, and still closed with stoppers of plastic clay. Note also the broken base of the outer shell, and, below, the slab of unbaked clay applied over one of the two stones blocking the lower accesses to the inner firing chamber.

> "Take some stones that can withstand flame, and some clay and build an arch for the furnace on each side in front of the alley space that you left in the center of the base. Build the furnace all the way around, at a distance of half a foot from the mold. Now, when in building we have reached halfway up the mold, clean out the arches of the furnace and make a hole in the rim of the mold on each side through which the tallow can run out, and put pots beneath. Apply fire and dry wood. While the mold is heating and the tallow begins to escape, gradually finish the surface right to the top of the mold. Place a little cover made of clay or iron over the mouth (of the mold). When the tallow has completely run out, block up both the holes with kneaded clay, using just the right amount so as not to infringe upon the rim of the bell. Put plenty of wood around the mold so that the fire does not fail for a whole day and the following night" (DCF: 170-171).

Archaeological evidence, again, seems to be in literal agreement with Theophilus' instructions. The discussion is presented in four parts, dealing respectively with the "arches" of the furnace, the outer furnace, the escape ducts of the lost-wax system, and the clay or iron "little cover".

1. The "arches" of the furnace. The stratigraphical excavation of the north-eastern part of the channel between the two parallel facing walls (Fig. 4, left) showed an important feature of two "mouths" of the mold (i.e. the two openings below the core). In fact, the filling layer in front of this "mouth" sealed a collapse of large tiles in a primary context of deposition (outlined with a thicker line in the section of Fig. 4, left). These tiles were clearly part of a small vault, perhaps shaped like a roof, originally covered with thick coats of clay, partially turned red by the contact with the flames. We think that these two vaults are the "arches" mentioned by Theophilus. Their setting is conjecturally reconstructed in the plan of Fig. 7: 12 and in the section of Fig. 7: 13. Before infilling again the pit for the casting phase, the two symmetric vaults were destroyed, and the tiles and the clay residues were left in situ. In this deposit we also found lumps of unbaked clay with impressions of a fine cloth, probably from the bags with which the bell-makers brought their clay into the church's yard.

2. The outer furnace. The mold and the pit preserved evidence of the double firing technique previously described. The firing of the inner chamber of the core from the lower channel (Fig. 7: 13) is attested by the fuel and ash found in the channel itself, as well as by the reddened surface of the chamber.

Less obvious was the archaeological evidence of the outer furnace. In the bottom of the pit, around the standing cone of the core and the fragments of the outer shell, we mapped a quadrangular trace formed by various degrees of reddening of the floor sediments and of the local substrata (Fig. 3). This trace was due to the placement around the mold of the wall of the outer furnace described by Theophilus (Fig. 7: 11-13). Some fragment on the floor of the pit suggests that this wall too was made with reused tile pieces. The walls of this square furnace had been completely removed before refilling the pit for the casting phase. The average distance of the reddening limit from the lower edge of the mold's shell is 25-30 cm, a little more than the half foot (about 15 cm) suggested by Theophilus. This evidence shows that the reconstruction provided by Hawthorne and Smith (3) is correct. The only difference is that the outer furnace was square instead of round.

On April 5th, 1992, C.S.Smith, in a letter to P.Vandiver, wrote:

> "Theophilus requires a wall of stones to form the furnace around the mold. Biringuccio is more specific - he recommends "teste di mattoni", brick ends, so placed as to allow air to enter at all heights...Would the stones found in the mold pit at Venosa stack decently when openly packed in this way?"

If we assume that the wall of this furnace was built with broken reused Roman tiles, the pieces would have been well suitable for such an open structure.

3. The escape ducts for the lost-wax method. The tallow, in the Venosa mold, escaped from two symmetric cylindrical ducts, about 12-13 cm long and 1.5 cm thick (Fig. 7: 13; Fig. 13). The tallow ducts, placed exactly over the center of the lower channel, where

pots could fit without problems, run obliquely through the projecting edge of the inner core (Fig. 13). They show another literal correspondence with the medieval treatise. The ducts had been blocked up with two "stoppers" of unbaked clay, occupying about half of the duct "so as not to infringe upon the rim of the bell".

Theophilus gives a very precise statement: the ducts must be cut into the mold's base when the outer furnace is half-built. As the furnace and the duct have no physical or functional interaction, this remark is difficult to understand. Usually, Theophilus provides sound functional explanations for the techniques he observed, but this is left unexplained. We may only think that he reports a traditional rule, perhaps related to the drying times of the base of the outer shell: a very precise state of hardness could have been required for a correct performance of the delicate operation of cutting through the rim of the core. Placing this operation in the middle of the erection of the outer furnace could have been a practical expedient to control its timing.

4. The clay or iron "little cover". This element mentioned by Theophilus is obviously meant for the two lower "mouths" covered by the vaults, as correctly understood by Hawthorne and Smith (3), and not for the top of the outer furnace, as assumed by other authors, for example in (2) and (8).

CASTING AND EXTRACTING THE BELL FROM THE MOLD (Fig. 8: 14-19)

Theophilus writes:

"When the stones and the fire have been completely removed, the earth should be replaced with alacrity, so that the whole pit around the mold is carefully filled up again. Workmen should keep going around ramming fairly lightly with blunt wooden poles and stamping their feet so hard that, as the earth is being put in, it presses against the mold and will prevent it from breaking in any way when the weight of the bell metal is poured in" (DCF: 173).

To remove "the stones and the fire" probably meant to destroy all the firing installations and recover the tiles of the outer furnace. In our mold, in the course of this operation, after the two small vaults had been knocked down, the lower acesses were closed with roughly squared stone blocks, and carefully sealed with clay (Fig. 4; Fig. 13, lower left). In Theophilus there is no mention of these operations. But, considering the hurry with which the craftsmen are supposed to refill the pit after the firing of the mold, this practice would appear as a precaution for retaining, below and within the mold, as much heat as possible. In the bottom of the inner chamber our video-endoscope shoped a pile of cracked rocks, tiles and charcoal (Fig. 4; compare with Fig. 7: 13). This deposit was probably part of "the stones and the fire" of Theophilus, still sealed in the position in which the metalsmiths abandoned it. Embers, ash and red-hot rocks could have been piled in the chamber as a long-standing source of heat.

After this, the pit had to be refilled for the second time in order to support the mold (Fig. 8: 14). The metal was poured into the hot mold (Fig. 8: 15). After casting, the pit, always according to Theophilus, had to be quickly re-excavated for the third time to extract the mold and recover the cooling bell (Fig. 8: 16).

THE TEXT AND THE ARCHAEOLOGICAL EVIDENCE DIVERGE

It is in the extraction phase that we meet the first substantial difference between Theophilus and the Venosa mold. The treatise prescribes the lifting of the mold from the pit with the "obelisk" method (i.e. by refilling the pit for the third time: see Fig. 8: 17), followed by the breakage of the interior of the mold to release the inner residual moisture (Fig. 8: 18), and finally by the breakage of the outer shell (Fig. 8: 19). In this way, the operation was carried out outside the pit. The outer mold fragments were scattered on the surrounding area, tobe eventually thrown back into the final filling of the abandoned pit, in a secondary context of deposition.

On the contrary, at Venosa the standing core of the mold is surrounded, as we have seen, by an annular collapse of fragments of the outer mold abandoned in situ within the pit (Figs. 3, 4, 9-12). We must assume that the craftsmen, after unearthing the mold, broke the outer shell on the spot and only later lifted out the bell. For detaching the bell, they had first to free the rim. This first stage of breakage is attested by a thick basal layer of small trampled fragments and dark powder all around the core. This deposit also contains most of the bronze spills originally captured on the base of the mold (Fig. 3). Our excavation showed that this breakage, trampling and collapse layer formed over the negative interface of the re-excavation of the casting pit.

But the most important point is that the only way of lifting the bell such that the core is left in place is through the use of a reliable crane, an option which contrasts with Theophilus instructions.

CONCLUSIONS

The comparison between Theophilus' "De Campanis Fundendis" and the archaeological evidence from Venosa is limited by the fact that the observed variations might have diachronic, cultural or geographic implications which are very difficult to assess. Nonetheless, both the treatise and the mold provide detailed information which have been integrated in a single synergetic reconstruction.

Originally, we had planned to excavate the bell-casting pit without providing part of the team (in particular geoarchaeologists) with the information of the written text, in order to be able to evaluate indipendently the evidence available in the two contexts. The square trace left by the outer furnace, for example, was recognized without prior knowledge of the existence of an outer furnace; and the two vaults in front of the "mouths" were hypothesized without reference to the text. In the course of excavation, nonetheless, it became clear that this strategy hindered the normal scientific exchange within the group and, as a consequence, made too difficult a proper study of such a complex archaeological object. Some fundamental features of the mold (for example, the remains of the tallow ducts) were found only after the text suggested a careful inspection of the base of the core.

The manufacturing techniques for the mold's core described by Theophilus correspond closely to the archaeological evidence. Equally, the techniques used for assembling the mold and the various parts of the furnace and the firing apparatus appear almost identical.

The differences we have found are the modification of the central hole, the possible presence (to be confirmed) of the supporting elements below the core, and the practice of sealing the inner with stone and clay before casting. Such features would appear to be technological details omitted by Theophilus, and not diverging technological solutions.

We observe that the Venosa mold being laid on its parallel facing walls closely resembles bell-making installations defined as Saxon and Norman by T.F.C.Blagg (12). On the other hand, the Venosa foundry appears quite different from approximately contemporary installations discovered in other Italian churches, such as those of Pieve a Retina (18), S.Andrea di Sarzana (16), or Pavia (17), where the base is formed by an arrangement of 3 to 4 pillars. In such cases, in the words of T.F.C.Blagg

> "It is evident...that the way of building the pit does not conform closely to Theophilus description, even though that is roughly contemporary with the Sarzana and Pavia pits, both datable to the first decades of the twelfth century, or Pavia a little earlier...Already one appears to have divergent local traditions, and this should impose caution against relying too much on Theophilus for precise detail" (11: 427).

Our work demonstrates that, on the contrary, most of the details provided by Theophilus are supported by archaeological evidence: as expected by Blagg, they may simply refer to a technological tradition different from that followed in central-northern Italy. We must consider seriously the hypothesis that the metalsmiths of Venosa were actually Norman like their masters. Given the fact that the treatise and the Venosa mold are indipendent contexts, their close correspondence stresses the existence, between the XIth and the XIIth centuries, of an international body of technological knowledge, shared and reproduced perhaps along kinship ties and religious affiliation among craftsmen, and, most probably, of long-distance relationships among metalsmiths of the same art. The intensive political investments by the early states of that period in gifts and guarantees to churches and monasteries, well documented for the rise of the Norman power (20), would have been an obvious stimulus for the standardization and spreading of this specialized metallurgical know-how.

Based on the text of Theophilus, one is surprised by the contrast between the efficiency and the speed of the manufacturing sequence ("lazy workmen" must be avoided, no one should get "in the way of the other"; after casting, "the earth should be replaced with alacrity", and so on) and the time-wasting, rather primitive method adopted for lowering and lifting the mold within the casting pit. Resort to a good crane, as probably happened at Venosa, could have substantially helped and hastened the whole process. In the church of Pieve a Retina (Tuscany, Italy), bell-making appeared to be a part of more general building activities of the church, and among the post-holes referable to the yard some oblique traces suggested the use of a crane for lifting the bell (18).

A possible explanation for this important divergence is that Theophilus might decsribe the work of a group of travelling craftsmen using a traditional, self-contained tool-kit easily movable from place to place. Bell-makers able to move a mold with such a simple technique would not depend upon machinery and the complexities of building yards and would control, as a consequence, the manufacturing sequence in its completeness - at the price of a relative loss of efficiency, but without external interference in the process.

In the case of Venosa, we might suppose that the manufacture of the bell was a part of a larger yard organization. Following this hypothesis, the most important difference we observed between the treatise and the archaeological data could be explained by contextual forms of organization of work and relationships of production rather than in terms of technological change or technological options.

ACKNOWLEDGEMENTS

The excavations of the SS.Trinita' were carried out by the Archaeological Superintendency of Basilicata, in collaboration with the Department of Historical, Archaeological and Anthropological Sciences of the University "La Sapienza", Rome. The Istituto Centrale per il Restauro, Rome (ICR), was consulted for the paleotechnological analysis of the bell-casting pit and planning of its conservation in situ. The research on the mold is carried out in collaboration with several collegues. Our warmest thanks are due to Prof. A.Bottini, Superintendent of the Basilicata region, M.Tagliente, archaeological curator, M.L.Marchi and G.Sabbatini for their unvaluable support. The project of conservation in situ of the mold is the subject of a thesis to be submitted for graduation at the ICR by M.Morisse and M.Pahor, under the supervision of M.De Luca. Archeometric analysis of ceramic samples is currently carried out by one of the authors (MV) and P.Vandiver at the Conservation Analytical Laboratory, Smithsonian Institution, as well as by C.Meucci, ICR. This research, ultimately, would not have been possible without the enthusiastic help of M.Savarese of the Superintendency of the Venosa castle.

BIBLIOGRAPHY

1. W.Theobald, Technik des KunstHandwerks im zehten Jahrhundert des Theophilus Presbyter Diversarum Artium Schedula. Berlin, 1933.

2. C.R.Dodwell, Theophilus, The Various Arts. Thomas Nelson and Sons Ltd., London, 1961.

3. J.G.Hawthorne and C.S.Smith, On Divers Arts. Dover Publications, Inc., New York, 1963.

4. V.Biringuccio, De La Pirotechnia. Libri X. Venetia, 1540.

5. D.Diderot, L'Enciclopedie, ou Dictionnaire Raisonne' des Sciences, des Artes des Metiers. III. Paris, 1753.

6. V.F.Hanson, J.H.Carlson, K.M.Papouchado, N.A.Nielsen, The Liberty Bell: Composition of the Famous Failure. American Scientist, 64, (6) (1976), p 614-619.

7. A.Delwasse, The New Bells of the Brussels Cathedral. Tin and its Uses, 112 (1977), p 9.

8. P.Donati, Il Campanato. Quaderni d'Informazione, Bellinzona, 1981.

9. C.McCombe, Royal Eijsbouts B.V. - a Dutch Bellfounder Extraordinary. Foundry Trade Journal, 153 (1982), p 661-668.

10. C.McCombe, Whitechapel's Bells Ring Loud and Clear. Foundry Trade Journal, 194 (1983), p 228-242.

11. T.F.C.Blagg, Bell-founding in Italy: Archaeology and History. Papers in Italian Archaeology I, part II, B.A.R. SS 41 (II) (1978), p 423-434.

12. T.F.C.Blagg, H.Blake, A.T.Luttrel, An Umbrian Abbey: San Paolo di Valdiponte: part 2. Papers of the British School at Rome, XLII (1974), p 98-178.

13. T.F.C.Blagg and S.J.Fleming, casting a Bell for the Abbey of San Paolo di Valdiponte. MASCA Journal, 1 (1), p 14-15.

14. B.Johnson, Excavation of the Bell-founding Pit and Restoration of a Bell-mould from Norton Priory, Cheshire. Conservator, 5 (1981), p 20-22.

15. A.Breda, Quingentole (Mantova) Piazza Italia. Edifici lignei d'eta' medioevale e rinascimentale. Notiziario 1987 della Soprintendenza Archeologica della Lombardia (1987), p 86-88.

16. F.Bonora, scavo di una fornace da campana in S.Andrea di Sarzana. Archeologia Medioevale, II (1975), p 123-148.

17. B.Ward-Perkins, Scavi nella Torre Civica di pavia. Archeologia Medioevale, V (1978), p 107-121.

18. E.J.Sheperd, Le Fornaci da Campane. In Various Authors, Architettura in terra d'Arezzo, Vol. 1, Arezzo, 1975, p 208-211.

19: C.Cantu', Schiarimenti e Note alla Storia Universale. Vol. III. Giuseppe Pomba e C., Torino, 1840.

20. B.Boni, Leonardo da Vinci e la tecnica fusoria. La Fonderia Italiana, 12 (1973), p 361-367.

21. G.Crudo, La SS.Trinita' di Venosa. Memorie storiche diplomatiche archeologiche. Trani, 1899.

RESTORATION OF ANTIQUE CHURCH BELLS: A CASE STUDY

G. RAMOS*, M.C. CASTRO*, I. AMADEVILLA*, C. GONZALEZ-TIRADO* and V.M. CASTAÑO#
*E.N.C.R.M., Xicotencatl y Gral. Anaya, México, D.F., MEXICO
#Instituto de Física, U.N.A.M., A.P. 20-364, México, D.F., MEXICO

ABSTRACT

A 17th century bell, from Santuario-Mapethe (Mexico) was restored by using welding techniques. The work included the acoustical characterization of the bell before and after the actual welding as well as the analysis of the composition of the alloy originally employed, so a material as similar as possible to the original would be employed for the restoration. The advantages and disadvantages of the methods are discussed.

INTRODUCTION

Perhaps one of the most interesting memebers of the percussion musical instruments, at least from a physicist's point of view, is the bell. In principle, it can be thought as a simple metal sheet bent around its middle part; but the geometry and the materials used to construct it play a key role on the musical performance of the bell. As for the first aspect, that is, the geometry of the bell, several references explaining the relevance of parameters such as the thickness distribution along the bell, can be found in the literature (1-3). The Materials Science point of view, however, has not been so carefully studied in the recent past by the experts, in spite of some interesting reports on the matter dating from the 16th century (4). Some of the available data, for instance, establish the alloy 78% Cu - 22% Sn as the standard composition for constructing antique bells (3-5), fact to be taken into consideration when reconstructing those instruments. It is also interesting to notice that some of the available literature in this field seem to rely on textual information (4,5) rather than on actual analytical measurements of the composition, structure, etc. Accordingly, the aim here is to add a few more scientific basis to some of the traditional methods (4,5) by employing modern techniques such as X-Ray analysis and acoustical studies (6,7). Thus, in

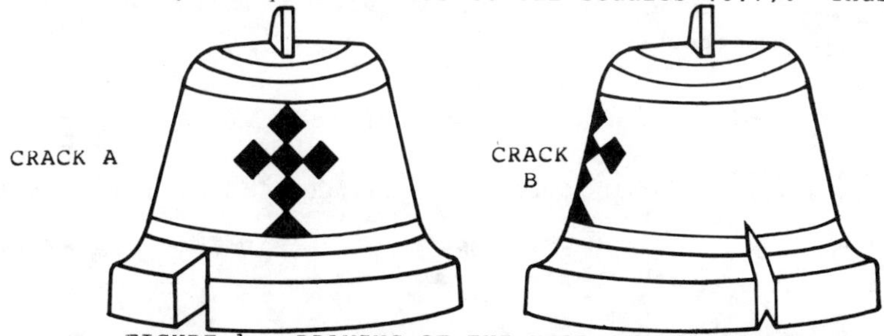

FIGURE 1.- DRAWING OF THE BELL AND THE CRACKS

this present work, a bell from a 17th century church, located in the town of Santuario-Mapethe, municipality of Cardonal, State of Hidalgo, at about one hour to the notheast of Mexico City, was restored by a welding process. The bell presented two big cracks in the lower lip of the body, as shown schematically in Figure 1. The dimensions of the bell are summarized in Table I and Table II contains a description of the two cracks depicted in Figure 1.

TABLE I

total lenght	49.5 cm
lenght of sonorous body	28.5 cm
inner upper diameter	16.5 cm
outer upper diameter	17.5 cm
inner lower diameter	34.5 cm
outer lower diameter	38.5 cm
total weight	35 kg
clappers weight	.6 kg

TABLE II

CRACK	LENGTH	WIDTH	DEPTH
A	9 cm	0.4 cm	2 cm
B	10.9cm	0.8 cm	2.3cm

EXPERIMENTAL

a) Materials characterization. X-Ray diffractometry was used todetermine the composition of the alloy employed in constructing the bell. Small samples from the area of the two cracks were powdered and analyzed in a Siemens D-5000 equipment, which allows quantitative analysis.

b)Acoustics studies. The sound spectrum of the bell, before and after the welding, was carried out with the following equipment:
-Digital recorder Bruel and Kjaer 7400
-High resolution spectra analyzer (2 channels) Bruel and Kjaer 2034
-Microphones Bruel and Kjaer 2807 and 4165
-Microphone gauge
-Audio tapes Denone R-60DT
-Cassette Bruel and Kjaer QR 1009

The recorded frequencies were in the range 0-12 kHz (6-7).

c)Welding. The following procedure was used to heal cracks A and B (Figure 1) on the bell:
1.-A hole was drilled 1 cm above the tip of the cracks to prevent them from growing further.
2.-A chamfer with a "X" shape was made to facilitate the

access to the welding material.
3.-A compact ceramic (brick) furnace was specifically built to obtain a homogeneous heating of the bell (Figure 2).

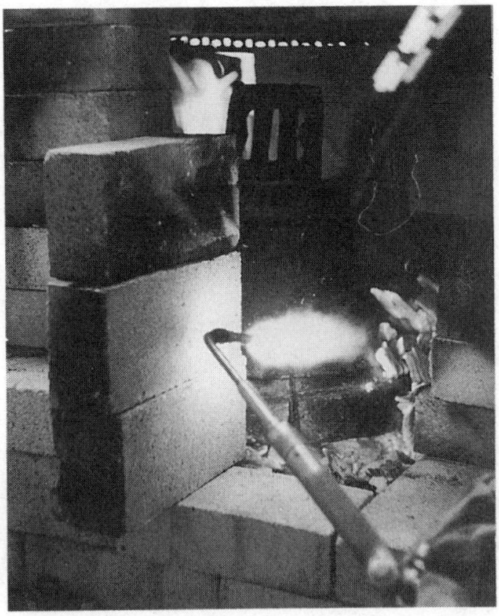

FIGURE 2.- PHOTOGRAPH OF THE FURNACE

4.-The well was heated at 400 °C by using coal and a gas torch.
5.-The actual welding was carried out by using an oxyacetilene equipment. The pressure of the oxygen was of 10 p/cm^2 and the pressure of the acetilene of 5 p/cm^2. the temperature of the flame is of around 3000 °C. A 6 mm burner (number 20) was utilized in each case.
6.-A commercial soldering alloy with the composition 95% Cu, 5% Sn was used in every crack.
7.-The finish was accomplished through a circular silex carbide blade grinder, an industrial moto-tool and metal files, to preserve as well as possible the original shape.

RESULTS AND DISCUSION

Figure 3 shows a typical X-Ray diffractogram from the area of the cracks. From the analysis of the peaks one can observe that, surprinsingly, the bell is made basically of copper and not of the expected Cu-Sn alloy, since no reflections corresponding to that alloy were found. This was also corroborated by atómic absorption analysis. This results is extremely relevant to choose the welding material.

The above finding also explains the poor mechanical strength of the bell and the presence of the cracks, for the recommended alloy ¿78% Cu-22% Sn) provides much better

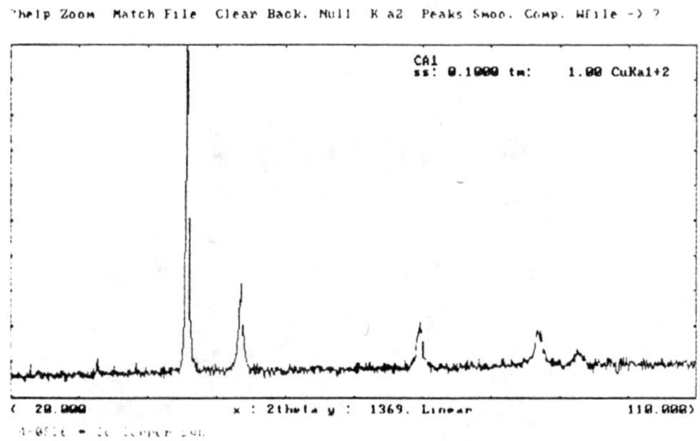

FIGURE 3.- X-RAY DIFFRACTOGRAM OF THE BELL SHOWING ONLY Cu PEAKS

properties, including acoustical performance, as we shall see. We must point out that welding of bells has been reported as early as the 16th century (4). That antique method is in principle similar to what we have described here, the main difference being, besides the obvious lack of analytical resources to identify the alloy and the soldering material, the use of a mould of the inside of the bell, where it is cracked, which is a method suitable for thicker cracks. In any event, it is remarkable the pertinence of that traditional method.

Figure 4 shows the acoustical spectra of the bell recorded in an anechoic chamber to avoid the presence of

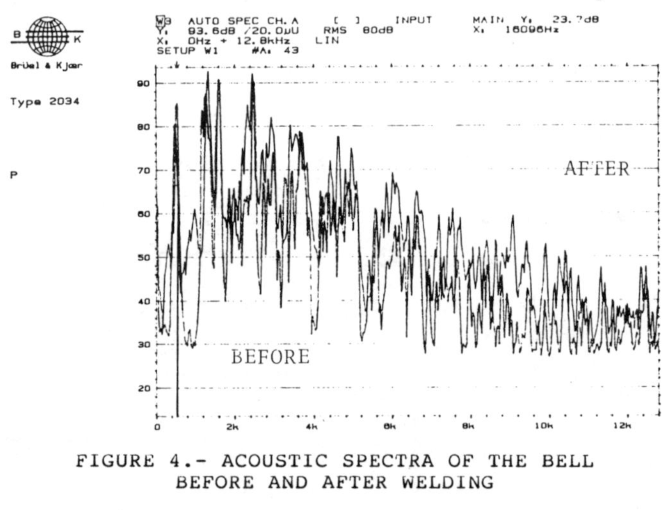

FIGURE 4.- ACOUSTIC SPECTRA OF THE BELL BEFORE AND AFTER WELDING

spurious harmonics. One can notice several features from these plots: first, that the fundamental note, marked by a vertical line, is preserved; second, that the low frequency range is basically the same, whereas at higher frequencies one can observe a recovery in the welded bell as compared to the original. This observations make more sense if one remembers that the modes corresponding to higher frequencies in a bell, are located preciselyin the lip (3), that is, where the cracks where originally located.

In order to test the convenience of welding as a method for restoring the tonal qualities of damaged bell, we artificially produced two cracks in another similar bell, at equivalent positions as in the reconstructed one. The acoustical spectra were obtained in both the untouched bell and in the welded one, obtaining basically the same behavior. The details of the procedure will be reported separately. In our present case, however, one cannot be sure of how close the sound of the welded bell is to the original 17th century bell.

Finally, as one could expect, chlorides and carbonates of various kinds were found forming a film and the patina of the bell, and were eliminated during the analysis and the welding. To homogeneize the color and general aspect of the bell, an artficial patina procedure was performed. This was achieved through the action of ammonium sulfide, which acts as a sort of "colorant" agent for copper and its alloys, by forming a homogeneous protective film of copper sulfate. The ammonium sulfide solution was prepared with destilled water (1:3) and applied to the previously cleaned surface at 300 ^{o}C.

CONCLUSION

Welding seems to be an adequate technique for reconstructing some antique percussion instruments such as bells because its simplicity, low cost and scarcity of research on alternative methods. The esperiments carried out in our laboratories show good evidence of good acoustical recovery by an appropiate choosing of the welding materials and procedures. In any case, more work in this direction has to be performed to bring the technique to sophisticated levels. Finally, one important point that remains to be calrified is taht of the temperature distribution in the bell during the heat treatment. This could probably be done through the analysis of color changes in the metal during the heating and weldings procedures.

ACKNOWLEDGEMENTS

The authors are indebted to Mr. Ricardo Ruíz and to Mrs. Rosa M. Lima for their technical support in various ways. Thanks are due to Dr. Pamela Vandiver and Ms. Martha Goodway for drawing our attention to references 4 and 5.

REFERENCES

1.-H.F. Olson, Musical Engineering, McGraw-Hill, New York (1952).
2.-I. Johnston, Measured Tones, Adam Hilger, New York (1989).
3.-S. Levaire and E. Levy, Tone, Greenwood Press, Westport (1980).
4.-Biringuccio, The Pirotechnia, Book VI, Chapter 15, translated by M. Gnudi and C.S. Smith, Basic Books, New York (1959).
5.-R.W.M. Clouston, Studies in Conservation 2, 170 (1956).
6.-N.H. Fletcher and T.D. Rossing, The Physics of Musical Instruments, Springer Verlag, New York (1991).
7.-A.H. Benade, Fundamentals of Musical Acoustics, Dover, New York (1990).

PART III D

The Use of Artifacts and Their Reuse

PORCELAIN ON DISPLAY
DECORATIVE AND OTHER USES OF ORIENTAL PORCELAIN IN EUROPE

Oliver R. Impey
The Ashmolean Museum, Oxford University, England

Chinese porcelain began to reach Europe in quantity in the late fifteenth century, Japanese in the mid seventeenth. Apart from a brief interlude in the mid seventeenth century, Chinese porcelain was imported until well into the nineteenth century, Japanese only until the 1740s. This porcelain was at first a rarity, to be mounted in precious-metal mounts or kept in the Cabinet of Curiosities. Soon it was to be displayed as part of the decorative effect in a room, while in some cases it became the decorative effect; the porcelain room. Porcelain was imitated in Europe and not much used until it was cheap enough. As the "China-mania" died, porcelain was kept as part of room decoration, often in the new china cupboards.

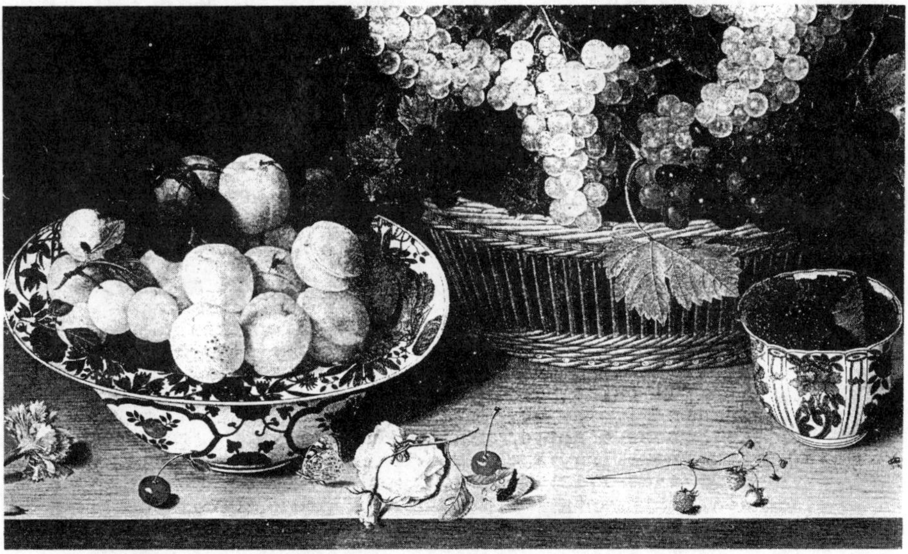

Fig. 1 Detail, oil painting, Still-life with porcelain, by Jan Soreau (active 1615-38). Ashmolean Museum, Oxford.

PORCELAIN ON DISPLAY

In the great painting "The Feast of the Gods" begun by Giovanni Bellini and finished by Tiziano in 1514 (1), the gods eat not off gold or silver plate but off blue and white Chinese porcelain of the late fifteenth century. This marks an implicit recognition of the status of porcelain as a rarity, worthy of use by the gods. Such porcelain was beginning to reach Europe by a process of the filtering through of Chinese goods overland or through Egypt, rather than by direct importation. Porcelain was a great luxury, brought in by diplomatic gifts more than traded (2).

It was therefore to be found in great houses; in the inventory of the Medici in Florence in 1553, there are listed some 373 pieces of porcelain, all of which must have been Chinese or possibly Annamese (3).

Porcelain became much more accessible after the Portuguese had taken Macau and obtained direct trade relations with China, being able to order porcelain to be made in Jingdezhen to conform to European demand, in European shapes and sizes. These wares were markedly different from earlier Chinese porcelain, and the blue and white decorations were in a style now called kraak porselein, for no adequate reason, which usually consisted of a central pictorial device surrounded by a border of radiating lines containing patterns of one sort or another (4). These but rarely reached Northern Europe, which only became aware of them after the capture by the Dutch of Portuguese ships bearing porcelain in 1602 and 1604 (5). The vast profit made on the sale of this porcelain prompted the newly-formed Dutch East India Company, the Vereenigde Oostindische Compagnie (VOC) to compete for the supply of these wares.

Such Chinese porcelains can be seen in the seventeenth century in still-life paintings by innumerable Dutch artists such as Pieter Claesz (6) or Jan Soreau (Fig.1), where they appear as part of displays of richness or of vanity; porcelain was still expensive. It was occasionally actually used; in 1629 in Hatfield House there were, "in the Chamber over the Porter's Lodge... XXVI China dishes to sett out a banquett" (7) . Nevertheless, porcelain was also used as room decoration; plates could be placed in rows over the fireplace, as seen in a painting by Gonsalves Coques (8) (Fig.2), or simply as ornaments on the tops of cupboards, as seen (e.g) in the painting by Pieter de Hooch, "The linen chest" (9).

In the 1640s, the ruling Ming dynasty in China was displaced by the Qing; in the ensuing civil wars, Jingdezhen, the main centre of porcelain manufacture in China was so severely disrupted that iy was virtually out of action for about twenty years. Just before this time, a new style of ceramic decoration had emerged, the so-called "transitional" style, a painterly and very beautiful relaxed style of blue and white painting. An example of a jug in this style can be seen in a painting by Vermeer (10). This disruption at Jingdezhen prevented the Dutch from being able to order the porcelain they needed; they turned to Japan. Thus it was that Japan was

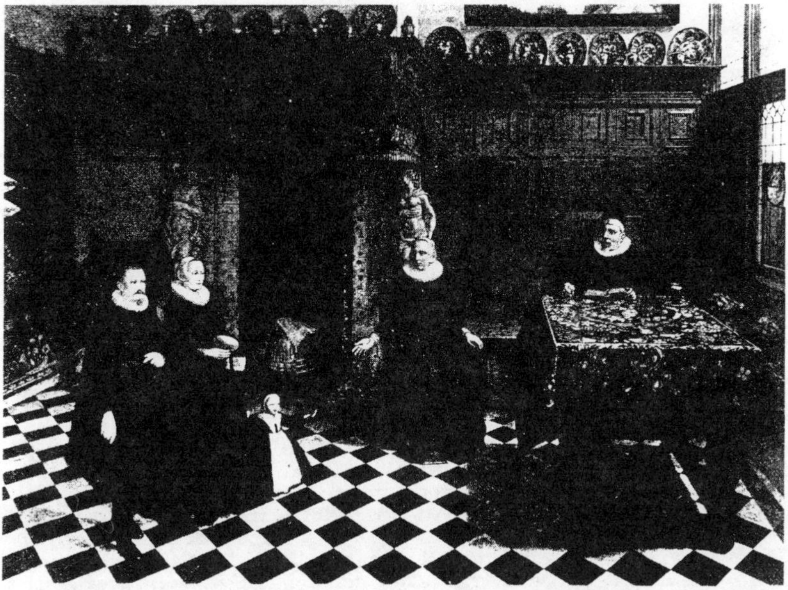

Fig. 2 Oil painting, Interior with porcelain dishes on display, by Gonsalves Coques (1614-84). Musée d'art et d'histoire, Geneva.

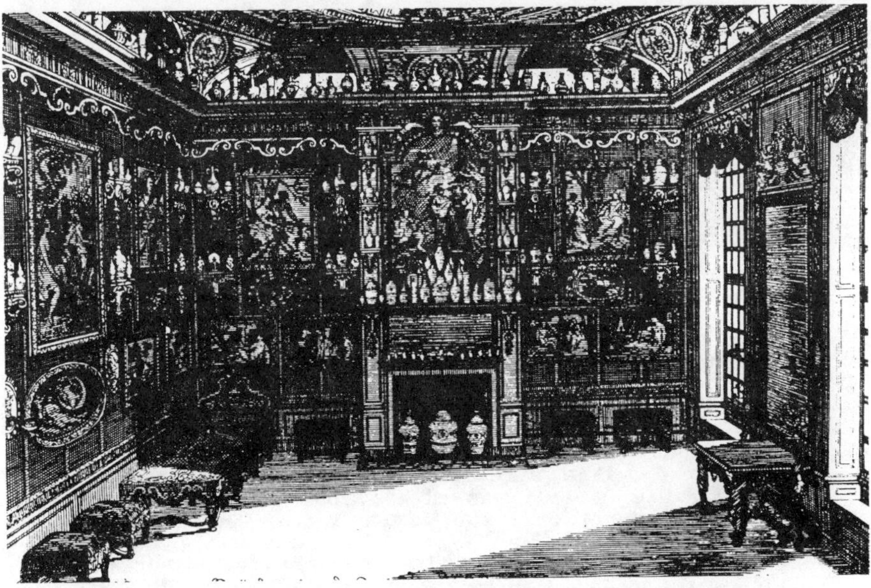

Fig.3 Engraving, A china cabinet, by Daniel Marot, (1663-1752) from <u>Nouveaux livres de partements</u>, before 1702.

first able to compete with China for the western market. By 1659 the Japanese were supplying large quantities of porcelain to Europe, much of it coloured, and to the Near East.

This porcelain is not readily identifiable in Dutch still-life painting, or in interiors, for it was at first made in Chinese styles, the kraak and the "transitional" styles.

These paintings were of the interiors of relatively modest households; for grander houses something more was required. At Somerset House in 1649, the 69 pieces of "purselaine" belonging to King Charles I were almost certainly displayed on "Nyne Woodden hanging Shellves turned & guilt" (11) and such shelves survive at Drayton House (12).

At Burghley House the famous inventory of 1688 (13) lists the porcelain and the places on which each piece stood. Most of these places were "Over the chimney"; in "My Lords Dressing Roome", for instance, there were;

"2 Doggs, 2 Lyons, 2 Staggs, 2 blue & wt Birds
1 heathen Godd with many Armes
2 figures with Juggs at their backs".

Many of these pieces are identifiable in the house today.

Symmetry, as we shall see, was an important part of any arrangement of rooms or objects; doubtless the pairs of pieces of porcelain listed here were placed one each side of the "heathen Godd". Unfortunately it is the chimneypiece that has been changed in this room, not, for once, the porcelain, and we cannot be sure how it looked.

In many such rooms the fireplace was in the corner, so that the chimneypiece would have had a series of shelves, like a quadrant of an etagere, on which to place ornaments; such chimneypieces survive, for example, at Beningborough Hall (14).

Another favourite place for the arrangement of porcelain and other exotic items was over doorcases; at Rosenborg Castle in Denmark, the lacquer room, then King Christian IV's bedroom, had blanc de Chine figures over the doors (15). In 1641, the Earl and Countess of Arundel had in the banqueting room at Tart Hall on shelves over the doors and in the corners, inter alia, "3 white little porcelain figures", almost certainly blanc de Chine (16).

All this is relatively light dressing of a room; the porcelain was part of a scheme of overall decoration, and not necessarily a major part.

The porcelain room is something quite different; here a room is specifically dressed in porcelain as the dominating feature.

The earliest surviving room dressed with oriental porcelain is the mid-17th century ceiling of a room in the Santos Palace in Lisbon (17). Only the ceiling survives, so that we have no knowledge of the function of the room, nor how the walls were decorated nor what the room contained. The ceiling was clearly built especially for the porcelain; not only the wooden frames that hold the individual pieces in position, but the whole pyramidal shape of the roof. There are thus four triangular surfaces tapering towards the apex, which are clad in Chinese late Ming porcelain. No other room remotely like this is known.

Later ceilings of porcelain rooms were either painted, as at the Oranienburg (18), or inlaid with mirror-glass, as at the former porcelain room at Het Loo, later converted to use as a library (19).

Mirror-glass remains a constant feature of porcelain rooms from the time of Daniel Marot right through until the Rococo and later. This, however was upon the walls; Marot got the idea from France and first used it in 1686 at Queen Mary's house in Holland, Honselaardijk, though only a description survives (20). Marot's designs for chimneypieces of about 1695 (21) give some suggestion of how part of it might have looked. In his designs for a China Closet of 1702 (22), the amount of porcelain has increased (Fig. 3); there is, however, a marked lack of case-furniture or of pedestals on which to place porcelain, such as one finds listed in contemporary inventories. This room may represent the Old Water Gallery at Hampton Court, decorated for Queen Mary after 1689, which Daniel Defoe (23) described as "the pleasantest little Thing within Doors that could possibly be made" adding "And here was also her Majesty's fine collection of Delft ware, which indeed was very large and fine; and here was also a vast stock of fine China ware, the like whereof was not to be seen in England".

Queen Mary was a collector; as the owner of a large collection, she had to have somewhere to put it all. Her response and that of others was the origin of the porcelain room.

In the Gallery at Kensington House, in 1696/7, there were 154 pieces of porcelain; the inventory taken then, at the time of Queen Mary's death (24), tells us exactly where each piece stood. Only occasionally are we able, however, to identify the shape described (i.e. a "coloured jar and cover of six squares" was surely the shape described today as a Hampton Court vase (25)) and never are we able to be certain of the individual pieces. The pieces in the gallery were arranged with strict symmetry; there were the same number of corresponding pieces over each of the chimneypieces (there were two) and over each of the two pairs of doors. In some cases we are able to deduce how this might have looked; Shulsky (26) has shown that over the chimney in the Queen's Ante-room, for instance, coloured plates were arranged in a cross-shape against a background of blue and white; some 38 pieces in all.

Queen Mary was a passionate collector of ceramics; her husband William of Orange was not. When Queen Mary died, all the porcelain at Kensington was given to the Earl of Albemarle (27). And yet the love of porcelain ran in the Orange family ; two of William's aunts had porcelain collections. It was for one of these, Luise Henriette, that her husband Frederick William, Elector of Brandenburg built the Oranienburg Palace. Here there were two porcelain rooms; the first was dismantled before 1690 (28) to make the second, the room that is depicted in the famous print by Jean-Baptiste Broebes of 1733 (29). This was a baroque room designed by Christof Pitzler, lined with porcelain plates; on the cornice, and in rows around the door-cases and around panels of mirror-glass in front of which stood massive etageres piled with porcelain jars and bottles. Inset into the ceiling were paintings by Augustin Terwesten, finished before 1695, some of which depicted _putti_ who hold pieces of porcelain. One of these pieces is instantly recognisable as a well-known type of Kakiemon-style Japanese porcelain (30). The vast numbers of pieces of porcelain dressed the room, but the room would have looked satisfactory even if the porcelain had not been there.

Much more committed, as it were, was the porcelain room in the Charlottenburg Palace (Fig. 4), built by Frederick III of Prussia, and designed by Johann Friedrich Eosander von Goethe (31). "Not only are plates and dishes ranked on the cornices, but vases and bottles in sets stand before them and miniature bottles are suspended, like raindrops, below. From the wainscot upwards, specially shaped swirls of giltwood bracketting hold myriads of vases that are clearly selected for size and for shape and colour, all before walls of mirror-glass. Chinoiserie figures wearing upturned basins as hats hold up great bowls, while tiers of bottles and of cups and saucers line the edges of the mirror-plates. Vases stand on pedestals in the corners, in garnitures over the chimney piece and even on the floor. It is all integral; nothing was left to chance and the whole room was specifically designed to show not just porcelain, but this collection of porcelain" (32). In all probability certain pieces of porcelain had been bought with a specific placement in mind, even if no interest was shown in the origin.

This room will be familiar to many Japanese readers through the magnificent reproduction of the room and its contents at the _Oranda-mura_ (33).

The Charlottenburg porcelain room has been rebuilt since the Second World War. The only porcelain room to have survived complete, it that at Schloss Weissenstein, Pommersfelden (34). This is a relatively small room (Fig. 5), built for the Elector-Archbishop of Mainz and Prince-Bishop of Wurzburg, Lothar-Franz von Schönborn, in 1719. This is a mirror-room, a _Spiegel-kabinett_, with superb inlaid wood walls and floors made by Ferdinand Plitzner. The porcelain is arranged in groups of up to five pieces on brackets; originally there were more in each group, as each group was fringed with miniature bottles. The collection is remarkable for being so consistent; most of it is Japanese _ko-Imari_. Those pieces that are not, are either Chinese Imari or Chinese blue and white cold-painted in Europe to resemble Imari. There is little Japanese blue and white and virtually no pieces in Kakiemon style. The intention is quite clear; this was to be a room of Imari (35).

The prints illustrating this room, made by Salomon Kleiner in 1728 (36), give us no chance to identify any

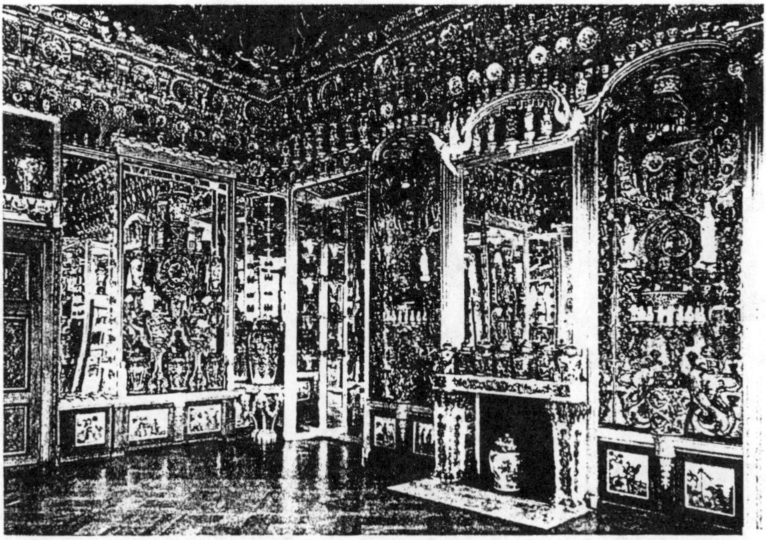

Fig. 4 The porcelain room, the Charlottenburg Palace.

piece of porcelain; Kleiner has simply invented the shapes rather than looking at them. Nor could one identify anything in the print of the Oranienburg; some pieces can be identified in the Charlottenburg prints, especially the great eagle figures. In Dresden, where perhaps the greatest collector of them all, Augustus (the Strong), King of Poland and Elector of Saxony (37), built the Japanisches Palais between 1729 and 1737, more can be identified in the drawings by Zacharius Longelune of about 1735 (38). Among these are the strange bird-cage vases with added lacquer and metal fittings (39), of which there are several still at Dresden.

King Frederik IV of Denmark, who had very little porcelain, wanted a porcelain room at Frederiksberg, so he had one painted, between 1709 and 1711; on trompe l'oeil brackets and cornices are trompe l'oeil pieces of porcelain. Some of these are clearly taken from real pieces (he had some porcelain), others are purely imaginary (40).

At Charlottenburg, Chinese porcelain had been mixed with Japanese solely for the effect; at Pommersfelden the choice had been made to have Imari, though it is extremely doubtful that Schonborn knew the difference between Japanese and Chinese. In the Japanese Palace at Dresden, the Chinese and Japanese porcelain was shown with the Meissen porcelain then being made in Augustus's factory.

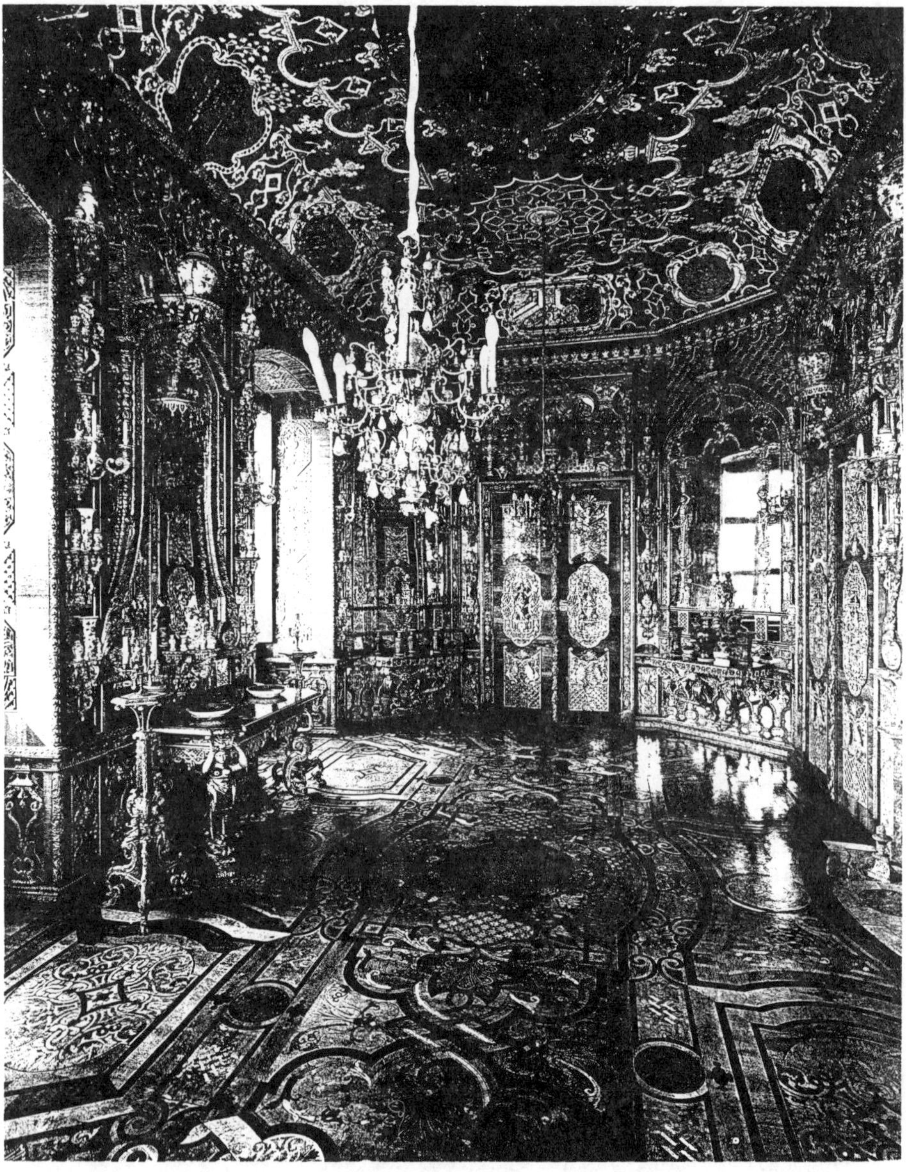

Fig. 5 The Mirror room, Schloss Weissenstein, Pommersfelden. Courtesy Graf von Schönborn.

Augustus did attempt to differentiate between types of porcelain -- usually he got it hopelessly wrong, for instance he called Kakiemon "Old India ware" (41) -- but this was for purposes of the inventory, not for the purposes of display.

Augustus' collection was very large, containing many hundreds of pieces; the collections of most German princely houses were smaller, but most had a porcelain room. At Altenburg, for instance, the porcelain room was small, containing glass-fronted cabinets on each of three sides, with brackets above for the display of individual pieces. In England, at Drayton House in Northamptonshire, there was a tiny porcelain room in one tower, made in the early eighteenth century, where one of the glass fronted cabinets was built into a window embrasure and thus lit from behind, a not entirely satisfactory arrangement. In this room there were also lacquer panels let into the walls, as in a lacquer room, a late seventeenth century craze (42).

The fashion for lacquer, either the fine quality Japanese export lacquer, or its poor relation the Chinese export lacquer or even its derivative the European japanning continued throughout the eighteenth century and later (43). Lacquer, with its rich and glossy appearance was often used in conjunction with porcelain, and it is no surprise to find it in a porcelain room. Lacquer was often cut into pieces for re-setting into currently fashionable shapes of furniture and, of course, much furniture was japanned for the display of porcelain; a fine example of such a cabinet is in the Chinese Pavilion at Drottningholm, in Sweden (44). The black and gold colour-scheme was as fashionable as was the blue and white.

Glass-fronted cabinets, built-in, and thus immovable, also became ancestral to display-cases. The "China Case", made fashionable by such furniture makers as Thomas Chippendale (45) revolutionised the display of porcelain; smaller pieces could be protected from the vandalism of the house-maid with her duster and need not be placed out of reach on the tops of cupboards or on the chimney-piece even in modest households. This, of course, affected what was made, and small figures became the rage in the second half of the eighteenth century (46).

The china cabinet rendered the pyramids of porcelain that have been described at the Oranienburg out of date; yet some do survive, being mostly used for the display of large porcelain vases. At Archangelskoye, near Leningrad, there are half-circle pyramids, and at Pommersfelden, corner pyramids (not in the Speigel-Kabinett); the former are for Chinese and Japanese porcelain, the latter for Japanese porcelain only, in character with the house.

Also at Pommersfelden is a remarkable display of miniatures. Such miniatures were made originally, perhaps, for the dolls' houses that were so fashionable in seventeenth century Holland (47), but were also used for display, as at Charlottenburg. The blue and white pieces are usually Chinese, the coloured, Japanese. At Pommersfelden there were many in the Speigel-Kabinett, but there were more in the Prince-Bishop's bedroom; over the corner stove, of cast iron, was a crown-shaped superstructure with many flat places where these

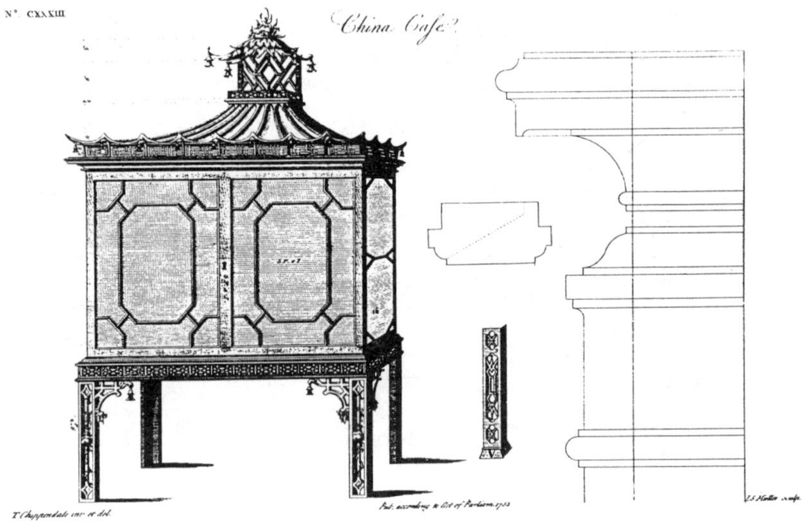

Fig. 6 Engraving, A china case, by Thomas Chippendale, from <u>The Gentleman and Cabinet-maker's Director</u>, third edition, 1762.

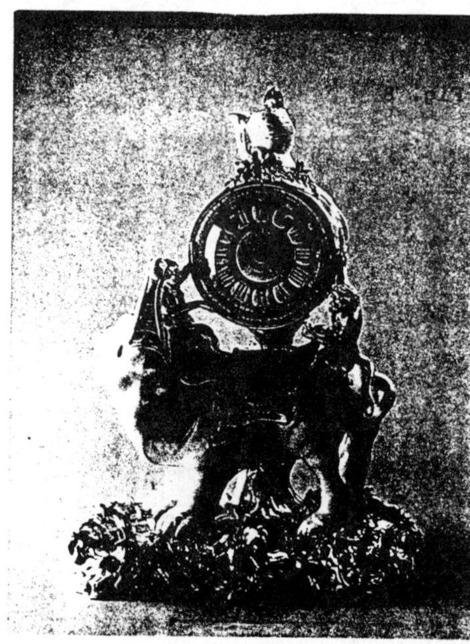

Fig. 7 Japanese porcelain and French ormolu clock, by Etienne Lenoir, Paris. The Residenz Museum, Munich.

miniatures could stand. This in turn was surrounded by brackets. Beside the stove were placed blanc de Chine figures, mostly cold-painted to resemble Imari, while over the stove was a great Imari garniture de cheminée.

Garnitures consisted of three covered vases and two trumpet-shaped vases en suite. These had been much used over doors and on cupboards in the seventeenth century, particularly in Holland, where many pieces of furniture had five brackets as part of the top especially for them, and exist in almost all types of ceramics of that time. They went out of fashion with the coming of the display-cabinet, and were often divided up, each piece to be placed on (or under) a console table or on some other table. Nor, after the beginning of the eighteenth century, were large pieces so often placed in the fire-place during the summer. So there was a problem; what to do with these large pieces. Hence the two solutions found at Pommersfelden, the pyramids and the stove.

Porcelain was well suited to the Rococo styles of the late eighteenth century, finding its place on small brackets on gilt-wood boiseries. Perhaps the most extensive display of this style is at the Amalienborg Palace in Munich (48). Lesser displays were built into English rococo wall-mirrors, especially over-mantels. Chippendale and other cabinet-makers (49) illustrate designs for this; a fine example of about 1765, by Linnel, exists at Osterley in Mrs Child's bedroom (50). This was part of the vogue for chinoiserie, when everthing oriental was mixed with things European supposedly in oriental taste; a visitor to Osterley wrote, in 1772, "the lodging rooms are in the atticks and are furnished with the finest chintzes, painted Taffetys, India paper and decker work, and such a profusion of rich China and Japan, that I could almost fancy myself in Pekin" (51). By this date, the use of the word China to mean all porcelain and the word Japan to mean all lacquer was breaking down, and the word Japan here may well have included porcelain in its meaning; the use of the idea of Pekin was a reference, of course, to the almost totally romantic-ideal world of Cathay, the basis of chinoiserie, not to the harsh reality of China (52).

In the eighteenth century, French taste in porcelain (as in so many other things) differed from that of the English. There it was Chinese monochrome porcelain that was so much in demand (53). Frequently it was mounted in silver; more commonly in ormolu, gilt-bronze. In the process the use was often changed so the vases became jugs and so-on; this meant little as these things were for display and rarely for use. The mounting of porcelain also happened in England and Germany, but usually this would have been the enamelled porcelains of China or Japan, rarely the monochromes (54).

Small cups or jars could be used as ink-wells and could be mounted together with lacquer or with European ceramics, frequently with Vincennes porcelain flowers on gilt-metal trees to form an elaborate and charming mixture (55). Clocks were a favourite; a clock (Fig.7) by Lenoir, Paris, in the Residenz, Munich (56) has the barrel made from a cut-down Kakiemon vase, lying on its side, and is surmounted by an Imari figure of a bird. The clock is

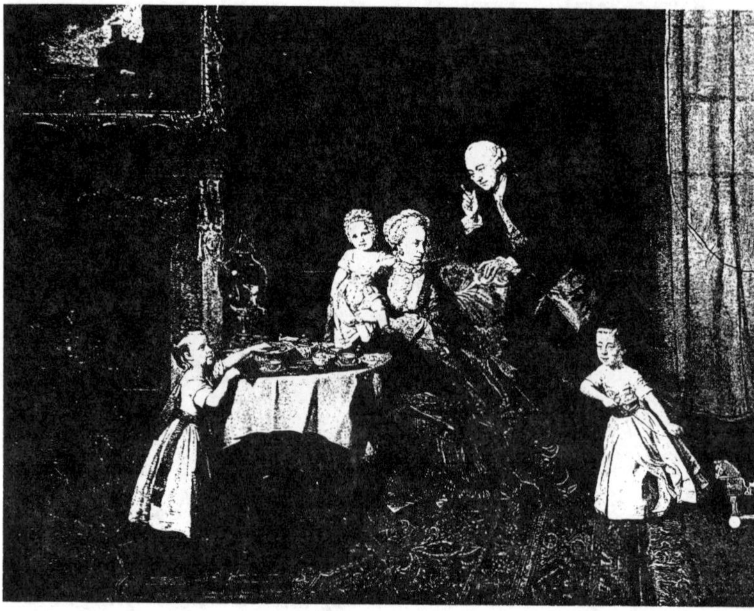

Fig. 8 Oil painting, The family of John, 14th Lord Willoughby de Broke, by John Zoffany (1734-1810). Courtesy Messrs Christies.

supported by a Kakiemon elephant (57) on which sit two gilt-bronze monkeys, while the elephant stands on a gilt-bronze base. This is a typical grand confection for a princely house (58). There is a jewel-case in Vienna (59) made from the sides of Imari and Chinese-Imari square bottles (60) fitted together. Such things were by no means unusual.

Porcelain, though very expensive, could also be used for its proper purposes. Individual items can often be seen portrayed in paintings of domestic interiors, and the presence of quantities of plates of the same pattern, as at Sherborne Castle (61), suggests that they may have been used under special circumstances. Dinner services of large quantities of pieces only appear in the 1730s, and are very rarely Japanese; it was the accoutrements of the tea table that were important. Tea had itself revolutionised the shapes of ceramics (and of other things such as silver) in Europe and the taking of tea became almost a ritual. Sometimes it is possible to distinguish between Chinese and Japanese Imari on the tea-table; in the charming group portrait of the family of John, 14th Lord Willoughby de Broke (Fig. 7) by John Zoffany (62), the bowl that the little girl may well be about to break is probably Japanese Imari, the other pieces Chinese. No-one cared to distinguish between the two.

Porcelain was exotic; it was eastern, and while everyone really knew that it was Chinese or Japanese it still had a vaguely eastern flavour. Thus there is a piece of Imari depicted in the painting by Ingres "Le bain turc" (63) a romantic vision of eastern exoticism (and eroticism) unfamiliar to the painter.

Japanese and Chinese porcelain was first imported into Europe as objects for the Cabinet of Curiosities (64). By the mid seventeenth century it was part of the furnishing of great houses throughout Europe, and by the end of the century was sometimes collected in large quantities. Such collections were displayed, frequently in a room made or adapted specially for the purpose. With the discovery of the secrets of the methods of porcelain manufacture in Europe in the Early eighteenth century, much of the mystery of porcelain was lost and it began to decline in fashion, though it was never entirely abandoned as a room decoration. Smaller pieces were kept in China cabinets made for the purpose. By the mid eighteenth century Japan was no longer exporting porcelain to the west, having been undercut in price-wars with China, and the standard of exported Chinese porcelain varied from very low (65) to the fine quality of the monochrome porcelain for the French market. There was a revival of the taste for mounted enamelled porcelain in Regency Britain at the time of the building of Brighton Pavilion (66). Almost every great house in Europe displays Chinese and Japanese porcelain, now absorbed into "eighteenth century taste", while quantities of lesser pieces can be found in China cabinets in almost every level of European society.

Notes and references

1. Now in the National Gallery, Washington. See Arthur Spriggs, "Oriental porcelain in Western paintings", <u>Transactions of the Oriental Ceramic Society</u>, 36, 1964-6, pp.124-33.
2. For the trade see T. Volker, "Porcelain and the Dutch East India Company", <u>Mededelingen van het Rijksmuseum voor Volkenkunde, Leiden</u>, 11, 1954 and T. Volker, "The Japanese porcelain trade of the Dutch East India Company after 1683", <u>Mededelingen van het Rijksmuseum voor Volkenkunde, Leiden</u>, 13, 1959.
3. R. W. Lightbown, "Oriental Art and the Orient in Late Renaissance and Baroque Italy", <u>Journal of the Warburg and Courtauld Institutes</u>, 32, 1969, pp. 228-79.
4. Maura Rinaldi, <u>Kraak porcelain</u>, London, 1989.
5. Volker, 1954, op. cit. (note 2).
6. Spriggs, 1964-6, op. cit. (note 1), illustrates several examples
7. Manuscript inventory, 1629, Hatfield House.
8. Illustrated in Peter Thornton, <u>Seventeenth-century Interior Decoration in England, France and Holland</u>, New Haven and London, 1978, fig.232. (Musée d'art et d'histoire, Geneva).

9. "The linen chest", the Rijksmuseum, Amsterdam.
10. "Girl interrupted at her music", the Frick Collection, New York. For other examples see Spriggs, 1964-6, op. cit. (note 1), pls 71-6.
11. Quoted in Anna Somers Cocks, "The non-functional use of ceramics in the English Country-house during the eighteenth century", Studies in the History of Art, 25, 1989, pp.195- 216.
12. Illustrated in Mark Hinton and Oliver Impey, Kakiemon porcelain from the English Country House, London, 1989, fig. 2.
13. Manuscript inventory, 1688, Burghley House.
14. For an example see Thornton, op. cit. (note 8), pl.74.
15. See Oliver Impey, "Porcelain for Palaces", in John Ayers, Oliver Impey and J.V.G.Mallet, Porcelain for Palaces, London, 1990, p. 58. Se also Tove Clemmensen and Mogens B. Mackeprang, Kina og Danmark, Copenhagen, 1980, fig. 27.
16. Thornton, op. cit. (note 8), pp.249, 383 and note 22.
17. Daisy Lion-Goldschmidt, "Les porcelaines chinoises du palais de Santos", Arts Asiatiques, 25, 1984, pp. 6-72.
18. Illustrated in L.Reidemeister, "Die Porzellankabinette der Brandenburgisch-Preuzischen Schlösser", Jahrbuch der Preuzischen Kunstsammlungen, Part 1, 54, 1933, pp. 262-72, pl.8.
19. A.M.L.E.Erkelens, "Delfts aardewerk op Het Loo", Nederlands Kunsthistorisch Jaarboek, 31, 1980, pp.263-72.
20. Nicodemus Tessin quoted by Robert J.Charleston "Porcelain as room decoration in eighteenth-century England", Antiques, 96, 1969, p. 894.
21. Daniel Marot, Nouvelles cheminees faittes en plusiers endroits de la Hollande et autres provinces, (Holland c. 1695).
22. Daniel Marot, Nouveaux livres de Partements, London, before 1702.
23. Daniel Defoe, A Tour thro' the Whole Island of Great Britain, ed. G.D.H.Cole, 2 vols., London, 1968, I.p.175.
24. Th. H. Lunsingh Scheurleer, "Documents on the furnishing of Kensington House", Journal of the Walpole Society, 38, 1962, pp.15-58.
25. See Arthur Lane, "Queen Mary II's porcelain collection at Hampton Court", Transactions of the Oriental Ceramic Society, 25, 1949-50, pp. 21-31; but see also Impey 1990, op. cit. (note 15) p.65 and note 41.
26. Linda Shulsky, "Queen Mary's collection of Porcelain and Delft and its display at Kensington Palace", American Ceramic Society Journal, 1990, pp. 51-74.
27. Lunsingh Scheurleer, 1962, op. cit. (note 24).
28. Reidemeister, 1933, op. cit. (note 18), p. 264.
29. Jean-Baptiste Broebes, Vues des palais et maisons de plaisance de Sa Majeste le Roy de Prusse, Augsburg, 1733.

30. For an example see Ayers, Impey and Mallet, 1990, op. cit. (note 15),no. 138.
31. Eosander von Goethe in M. Merian, Theatrum Europeii XVI and XVII, Frankfurt, 1717 and 1718.
32. Impey in Ayers, Impey and Mallet, 1990, op. cit. (note 15), pp. 60-1.
33. The porcelain was made by potters in and around Arita for this replica of the Porzellanzimmer of Charlottenburg.
34. Gerhard Bott, ed., Die Grafen von Schonborn, Nuremburg, 1989.
35. This may be unique; most collections featured those porcelains that were available, though much care was exercised in their symmetrical arrangement.
36. Salomon Kleiner, Representation au naturel des chateaux de Wiessenstein au desus de Pommersfeld, et celui de Geubach.., Augsburg, 1728.
37. Jean-Louis Sponsel, Kabinettstücke der Meissner Porzellan- Manufaktur von Johann Joachim Kändler, Leipzig, 1900.
38. Friedrich Reichel, Early Japanese porcelain, London, 1981, on pp. 120-1, and Ayers, Impey and Mallet, 1990, op. cit. (note 15), fig.16.
39. Ayers, Impey and Mallet, 1990, op. cit. (note 15), fig.17.
40. Mogens Bencard "The glass cabinet at Rosenborg Palace", Journal of the History of Collections, 3, 1991, pp. 1-12. and Clemmensen and Mackeprang, 1980, op. cit. (note 15), fig. 46.
41. Quoted in Reichel, 1981, op. cit. (note 38), p. 62.
42. See Oliver Impey, "Eastern trade and the furnishing of the British Country House" Studies in the History of Art, 25, 1989, pp.177-94.
43. Hans Huth, Lacquer of the West, Chicago, 1971.
44. Åke Setterwall, Kina Slott på Drottningholm, Malmo, 1972, p. 81.
45. Thomas Chippendale, The Gentleman and Cabinet-maker's Director, third edition, London, 1762, Nos. 131-7.
46. Somers Cocks, 1989, op. cit. (note 11).
47. Sometimes dolls houses had miniatures made of glass in imitation of Chinese or Japanese miniature porcelains; see Jet Pijzel-Dommisse, Het Poppenhuis van Petronella de la Court, Utrecht, 1987.
48. See, e.g. Oliver Impey, Chinoiserie, Oxford, 1977, pl. opposite p. 165.
49. For examples see Helena Hayward, Thomas Johnson and English Rococo, London, 1964, Nos. 96-7 and 125.
50. John Hardy and Maurice Tomlin, Osterley Park House, London, 1985, pl. 58.
51. Quoted in Hardy and Tomlin, 1985, op. cit (note 50) pp.105-6 from a letter written by Lady Beauchamp Proctor in 1772.
52. Hugh Honour, Chinoiserie; the vision of Cathay, London, 1961.
53. For examples see F.J.B.Watson and Gillian Wilson, Mounted oriental porcelain in the J. Paul Getty Museum, Malibu, 1982, pls. 12, 15 and 17.
54. For examples see Ayers, Impey and Mallet, 1990, op. cit. (note 15), no.152.

55. See Watson and Wilson, 1982, op. cit. (note 53), pl. 14.
56. Illustrated in Ayers, Impey and Mallet, 1990, op. cit. (note 15), no. 147.
57. This elephant is the same model as those at Burghley house mentioned in the 1688 inventory op. cit. (note 13). See The Burghley Porcelains, catalogue of an exhibition held in the Japan Society Galleries, New York, 1986, no. 93.
58. For other examples see Watson and Wilson, 1982, op. cit. (note 53).
59. Illustrated in Ayers, Impey and Mallet, 1990, op. cit. (note 15), no. 245.
60. The front and the back of the box is of Japanese Imari, the sides of the box and the top, front and back and the sides of the lid are of Chinese Imari.
61. Hinton and Impey, 1989, op. cit. (note 12), p.18.
62. Illustrated in Ayers, Impey and Mallet, 1990, op. cit. (note 15), fig. 4.
63. Jean-Dominique-Auguste Ingres, "Le bain turc", Musée du Louvre, Paris. See, e.g., Ingres, catalogue of an exhibition held at the Petit Palais, Paris, 1967-8, No. 260.
64. John Ayers "The early China Trade" in Oliver Impey and Arthur MacGregor, The Origins of Museums, Oxford, 1985, pp. 259-66.
65. See C.J.A.Jörg, The Geldermalsen; history and porcelain, Groningen, 1986, for a ship's cargo of trade porcelain of the mid eighteenth century.
66. For examples from the Brighton Pavilion see Ayers, Impey and Mallet, 1990, op. cit. (note 15) nos. 71 and 152.

A BASIC STUDY OF USE-WEAR "POLISH" ON EARLY RICE PADDY FARMING TOOLS

Shoh YAMADA
Dept. of Anthropology, Harvard University, Cambridge, MA 02138, and formerly of The Commission of Buried Cultural Properties on Campus, Tohoku University, Katahiracho, Sendai, 980 JAPAN

ABSTRACT

The analysis of use-wear "polish" variation and differences according to rock types contributes to the study of the "polish" formation process. The examination of the polished surface on andesite, rhyolite and a hard variety of slate with the SEM and optical light microscope thows that the polished surfaces are produced by mild wear. This provides a basis for use-wear analysis of stone farming tools made of these rocks, which come from early Japanese rice paddy cultivation contexts.

INTRODUCTION

Rice is the principal food of Japanese people and has a great deal to do with the Japanese life. Over the two thousand years of paddy cultivations, Japanese people have created a special feeling for rice. It has also been connected to Shintoism(their religion). In addition, paddy fields occupy most of the cultivated area on the Japanese Islands and have an important function for controling the flow of water.

Today, however, because of the rapid urbanisation, the area of paddies is decreasing quickly, especially around big cities, this increases the opportunity to excavate ancient paddy fields. The research into the origin of Japanese paddy cultivation can be an interesting topic for archaeologists and the entire Japanese people.

Recent progress of lithic use-wear studies has begun to reveal not only the actual use of morphologically assumed reaping tools in early paddy cultivation but also unknown types of reaping tools. The raw materials of which these reaping tools are made varies greatly by region : They include andesite, rhyolite, sand stone of various grain size, slate, schists, etc. Therefore, the routine method of use-wear analysis developed for "flint"(chert) tools needed some modification. This study provides a basis for the application of use-wear analysis on the various rock types of which these paddy farming tools were made.

First, I will briefly explain about the early neolithic period in the Japan Islands and the use-wear analysis on reaping tools of this period which motivated the present study. Next I will summarize the method of current lithic use-wear studies and my previous studies on use-wear "polish" formation on silicious rocks. Finally I will discuss "plant polish" on andesite, rhyolite and slate.

YAYOI CULTURE IN THE JAPAN ISLANDS

Japanese rice paddy cultivation is assumed to have been introduced from the Asian Continent. Paddy field cultivation reached northern Kyushu in westernmost Japan in the middle of the first millennium B.C.(Fig 1). The diffusion and acceptance

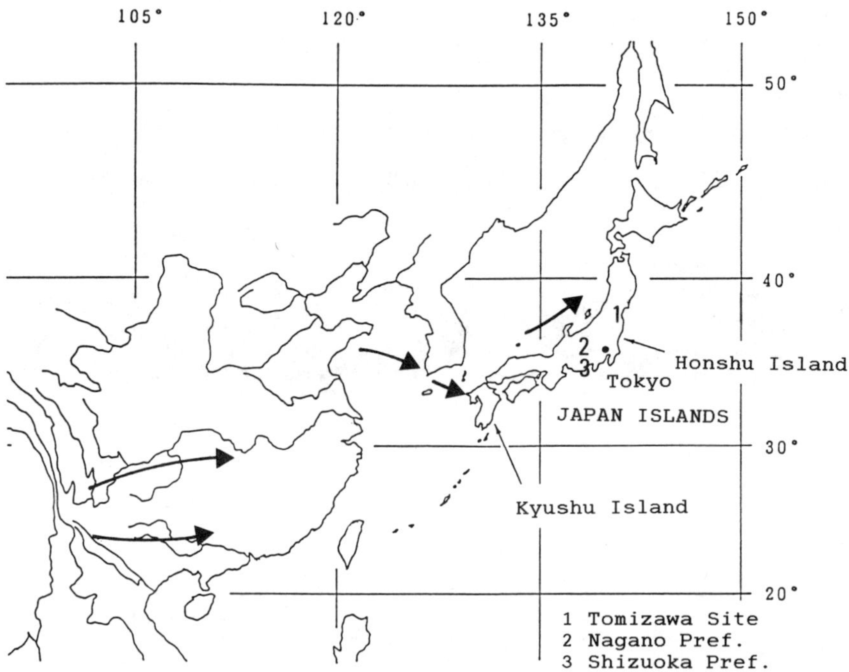

Fig.1. Propagation of Rice Paddy Cultivation to the Japan Islands

of wet rice agriculture marked the change from Jomon culture (hunting and gathering sociey) to Yayoi culture (agricultural society) in the Japan islands. The use of metal implements began in this period, although a number of stone tools continued to be used until the Middle Yayoi period.

Wet rice agriculture was not introduced to northeastern Japan until the middle phase of Early Yayoi. Cold weather in that area is thought to be one of the reasons for this delayed diffusion. Wet rice cultivation eventually reached northernmost Honshu and also the northernmost extension of paddy fields in prehistoric East Asia in 250-150 B.C.. Until 1981 when paddy fields were actually excavated at the northernmost point of the Honshu island, many researchers had doubted the existence of Yayoi period paddy cultivation in this area because of the cold climate. Even today it is not easy to grow rice there.

There are regional variations in the Yayoi culture. There is a distinctive difference between south-west Japan and north-east Japan. In north-east Japan metal implements are not abundant and the style of potteries retain the distinctive tradition of Jomon pottery. The studies on such regional variations in Yayoi Culture, especially those of northernmost area, may bring some insights to the establishment of the early agricultural society in Japan.

USE-WEAR ANALYSIS OF PADDY FARMING TOOLS

The Tomizawa site in Sendai City is representative of the Yayoi paddy field sites in Northern Honshu. Located in a flat backmarsh on the Paciffic Coast, the site yieled 8 superimposed paddy fields. Two types of reaping tools were found in the paddy fields.

One of these, ishi-bocho (Fig.2)is a well-known farming tool typical to East Asia. It is a polished stone tool, half-moon shaped with a convex edge. This tool type is prevalent in southwest Japan, while its number is lower and regionally diverse in northeastern Japan. Based on the analogy from ethnographical examples, the ishi-bocho is believed to have been tied to the fingers and used for snapping off ears of rice, with a string through the two holes. T.Sudo and K.Akoshima[1,2] first made use-wear analysis with the "polish" method developed by L.H.Keeley[3] on ishi-bocho excavated from northern Honshu. The result showed that the previous analogy on function was right.

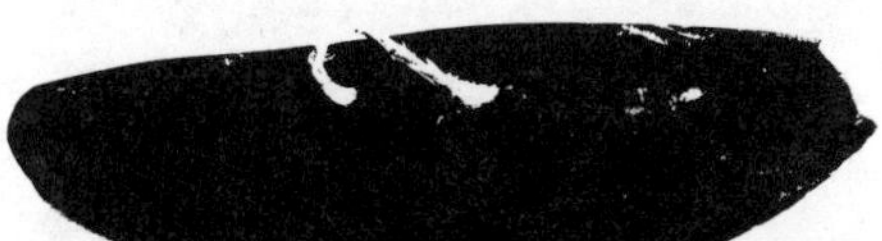

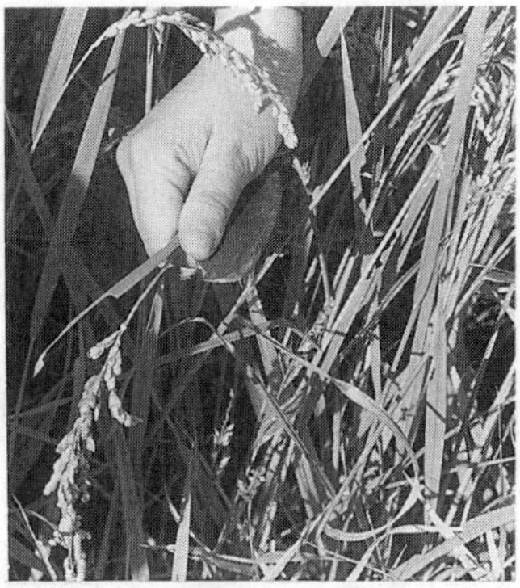

Fig.2. Ishi-Bocho

However, in addition to ishi-bochos, amorphous flake tools which show "corn gloss" on their edges were noticed at the Tomizawa site. The amorphous flake tools can be classified into two types. One is a large type with the maximum length of more than 10cm, mainly detatched along joint-structures of local andesite. The other flake type is a small type, less than 10cm long and mainly made of local rhyolite (Fig.3)[4,5]. My on going use-wear "polish" analysis on 39 tools from Tomizawa proved that they were both used to cut non-woody plant stems. The species of the plant cannot be determined with this method, but the fact that they were recovered from paddy fields where there were few materials dumped, provides strong support that they were somehow involved with paddy farming. Since the ishi-bocho was available and effective for cutting grain heads, the large flake tools might have been used for collectiing the remaining straws for another use, after reaping the ears. Another possibility is that they were used for removing weeds in and around the paddy fields. Also the small types might be supplementary tools of ishi-bocho.

Such combination of snapping tools and cutting tools were also reported by T.Midoshima in Nagano prefecture[6], Middle area of Honshu Island(Fig.1): "Chipped ishi-bocho" is more prevalent than the ordinary polished ishi-bocho in that area. They were both used for snapping ears, while a type of side scraper was used for cutting the stems.

In Shizuoka prefecture, adjoining to Nagano, almost no plant use-wear was detected by me on what is been believed to be "chipped ishi-bocho"[7].

Further use-wear research will reveal the unknown regional variations in paddy technology throughout the entire Japan Islands. Various local rocks are used as raw materials of reaping tools, and the polished surface formed on them is somewhat different from those on flint for which routine analysis of use-wear "polish" has developed. Therefore, we need to modify the routine method.

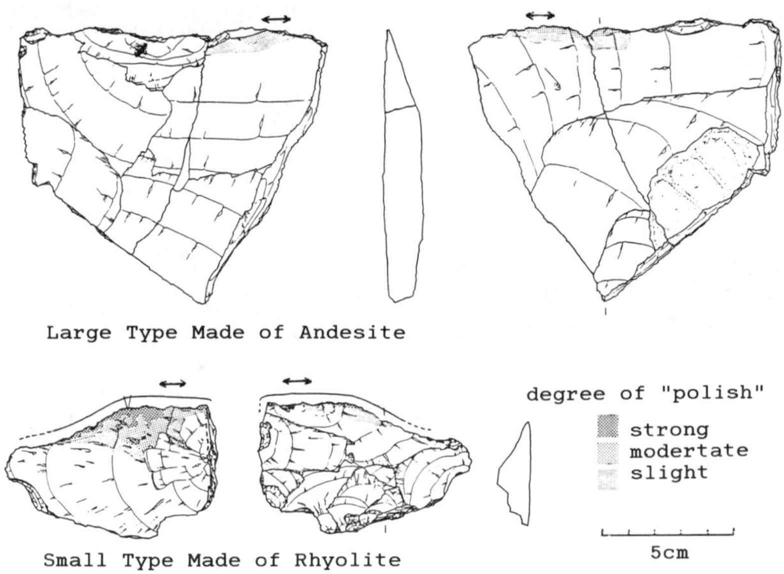

Fig.3. Flake Tools Used for Cutting Plant at the Tomizawa Site [4]

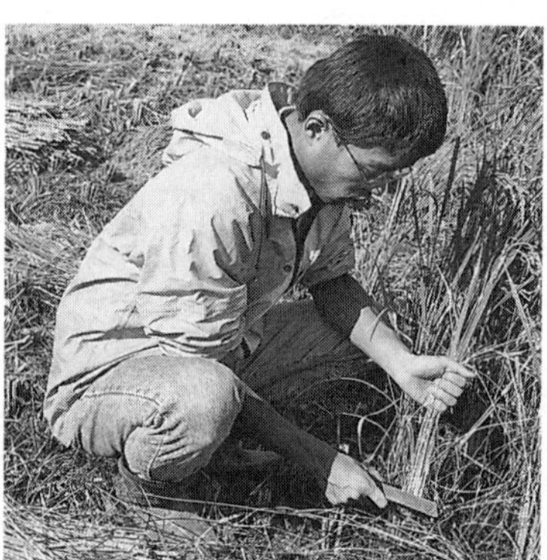

Fig.4. Experiment of Cutting Rice with an Andesite Flake

METHOD OF USE-WEAR "POLISH" ANALYSIS

In the latter half of 1970's, a new method of use-wear studies on stone tools was established by L.H. Keeley[3]. The points of this method are as follows : First, a number of experiments was done with replicative tools made of flint (chert). During the experiments, variables such as the kind of worked materials, the kind of action, and the duration of work was recorded. As a result, several types of "polish" were found formed on the tool edges under a magnifying power of about 200 times of metallurgical microscope. The types of "polish" are believed to correspond to the types of worked materials such as non-woody plants, wood, bone and antler, meat, and hide. These polish types are named after the worked materials and their features are described by optical impressions : e.g., "plant polish" is very bright, smooth, and domed, while "hide polish" is matted and rough. This approach with "polish" has become very popular and many archaeological studies with "polish" analysis have been reported. However, the question of why such "polish" is produced, has not been solved completely.

The problem of understanding the "polish" mechanism is essential for extending this method to various rock types. It will give a theoretical framework to the experimental study, and it will make it possible to minimize the number of experiments on each rock type. Studies of "polish" on various rock types may give some insights to the formation process.

There have been some competing theories on polish formation on flint (chert). The two major ones are wear theory and the silica-gel theory(Fig.5). The former considers that a "polish" is produced by wear, or "progressive loss of the substance"[8] from stone surfaces. While according to the latter, a "polish" is a silica-gel layer produced by hydraiton of a flint surface which can build up to about 10μm thick[9,10]. It is believed, in the case of plant working, transferential silica-gel from worked plants is also merged into the layer. Furthermore, it says, a developed polish becomes fluid during the work, and thus scratches grooved on the surface disappear.

Neither theory has been conclusively proved nor denied.

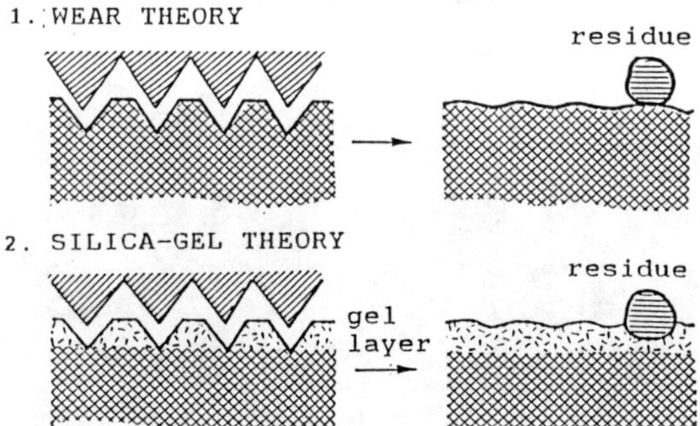

Fig.5. Theories on "Polish" Formation [11]

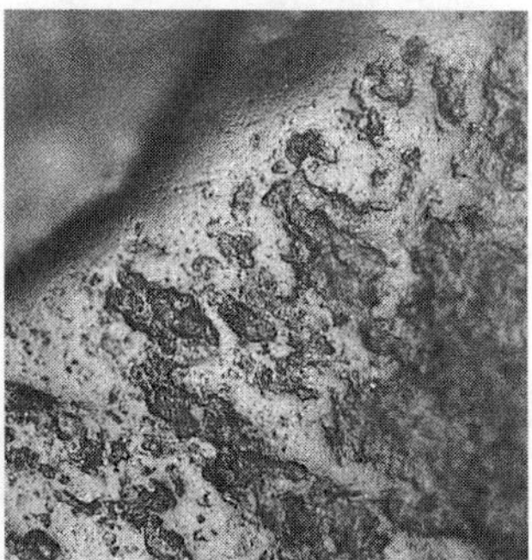

Fig.6. "Plant Polish" [4]

FORMATION PROCESS OF POLISHED SURFACE

In my previous work[11,12], I proved that polish was produced by wear in the meaning defined in tribology. I will summarize the method and result below.

I traced the progressive alteration of the surface at a single point on a specimen using stereophotography of SEM. Silver was used for coating film, which was removed with boiled H_2O_2, 10wt%, 10 minutes after each time of observation and followed by acetone wiping to avoid contamination due to the treatment residues. Silicious shale used in this test is a highly silicified rock which has no lamina structure, so that the results of the present analysis is applicable to chert.

The result is shown in Fig.8. We cannot determine whether the change is additional or attritional from the very minute change found in these photos because the specimen cannot be fixed in precisely the same angle at each time of the observation. However, when we compare no.1 with no.4, and no.2 with no.3, wear is obvious. There is no evidence of thick surface layers being formed nor of surface flow occuring.

In these developmental processes of polished surfaces, two types of wear are confirmed. One is a cutting type which accompanies striation (scratch)(Fig.10). The other is a fracture type which accompanies pitting. In Figs.7 and 8-no.4, new pits are found to be produced. However, we have to note that, with this method, it is impossible to interpret the phenomena occurring on the order of less than a μm. Andersen and Whitlow proved the existence of the amorphous layer of less than 1μm with the ion beam analysis[13].

SEM observation shows that polished surfaces produced by soft materials such as plants tend to be domed and massively invade the depressions on rock surface, while polished surfaces produced by hard, stiff materials such as antler and bone tend to be flat and slightly invade the depressions(Fig.9-11). Dry hard working tends to cause many pits. Therefore, it is concluded that hardness, viscous and elastic properties and surface texture of worked materials are the determining factors for the features of polished surfaces. The feature of optical polish can be understood as features of polished surface produced by wear : invasiveness from edges, invasiveness in depressions, profiles, distribution pattern, invasive pattern, smoothness, and amounts of pits are diagnostic attributes of use-wear polished surface[12].

A polished surface shows a basically similar pattern between 200-2500X. The fractal nature of polished surface is demonstrated by Rees et al.[15,16], although, they are negative about correlation between fractal dimension and worked materials.

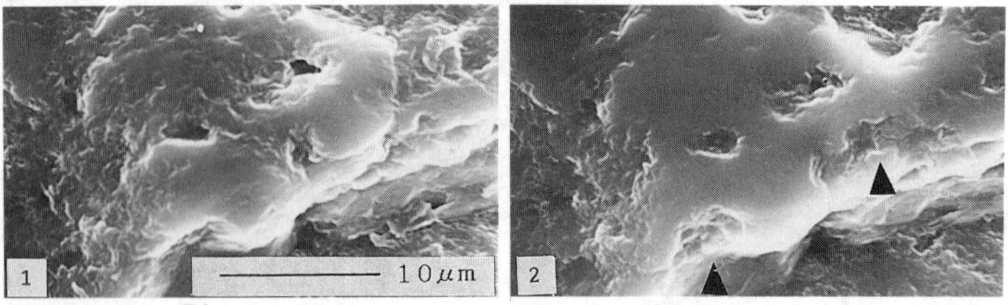

Fig.7. Fracture Type Wear by Cutting Plant [11]

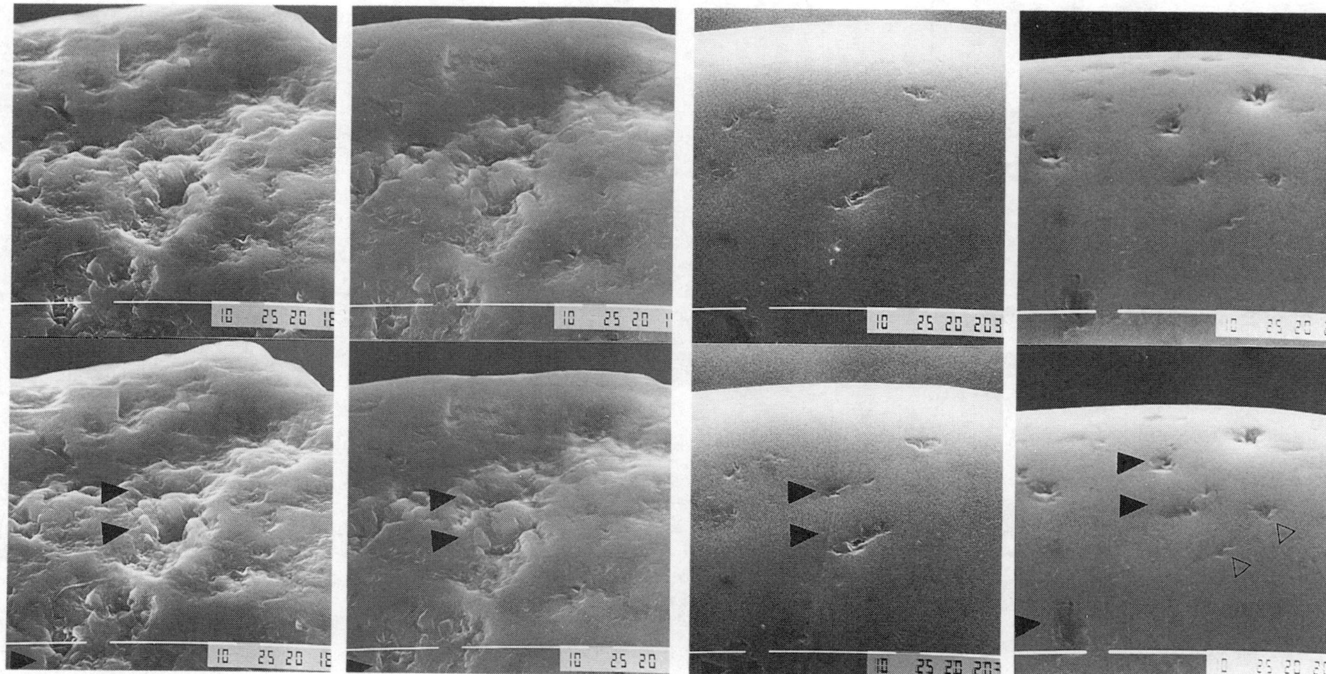

1. 250 strokes 2. 350 strokes 3. 1350 strokes 4. 3350 strokes

⟵ direction of movement

By cutting plant

——— 10μm

Fig.8. Progressive Development of Polished Surface (stereo) [11]

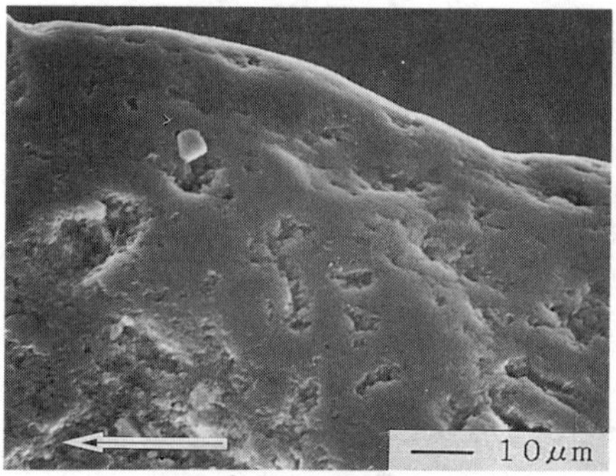

Fig.9. Polished Surface Produced by Cutting Plant [12]

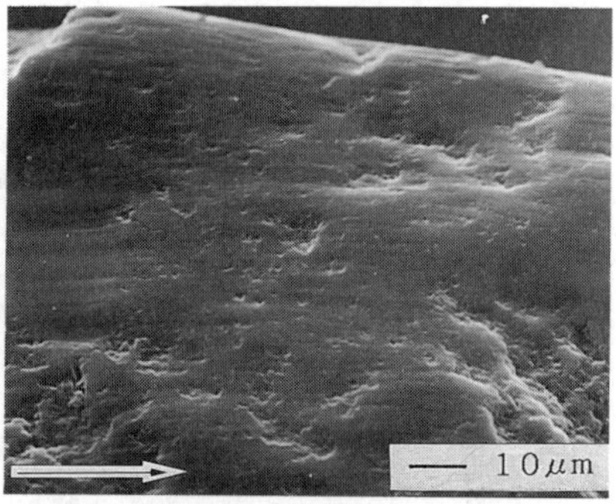

Fig.10. Polished Surface Produced by Cutting Bone [12]

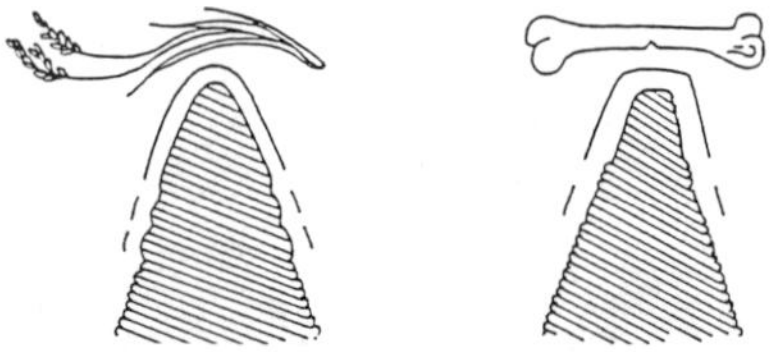

Fig.11. Idealized Profiles of Worn Edges [11]

"PLANT POLISH" FOUND ON REAPING TOOLS

Based on the above results, details of "plant polishes" formed on stone tools from Tomizawa paddy fields were examined.

Andesite

Under an optical microscope, polished surfaces produced on andesite by cutting rice are gapped and do not extend continuously over the edge (Fig.12). However each polished area is smooth and domed and looks similar to "plant polish" on silicious rocks. How are these gapped distribution patterns produced?

The surface texture of andesite is very rough, and observation with a metallurgical microscope whose focus depth is very shallow is not effective for polished surface, especially in undeveloped stages. SEM makes an observation of the details possible and help us understand the nature of polished surface.

The texture of andesite consists of phenocrysts and groundmass, which is very different from the cryptocrystalline structure of flint. Comparing with photographs of thin sections, these gapped polished areas found to correspond to phenochrysts of feldspar and pyroxene (Fig.13). On andesite, optically visible polished surfaces are first produced on big crystals of feldspar and pyroxene(Fig.14). Because the surfaces of these phenocrysts are originally flat, optically distinctive polish is first produced on them, thus forming a gapped pattern. Even if the corner of these square crystals is slightly worn ,it gives us an impression that a large polished area was produced. I named this phenomenon "phenocrysts effect". We often see optical use-wear polish on phenocrysts only when a specimen has been used for a short duration or when it has suffered weathering. In the area where polished surfaces are not detected with an optical microscope, the evidence of wear can be detected with SEM(Fig.15).If the worn surface furhter develops, an optically polished area would cover the groundmass area.

The beveled periphery of phenocryst(Fig.14) proves that the wear is the principal mechanism for the "use-wear polish" here, also.

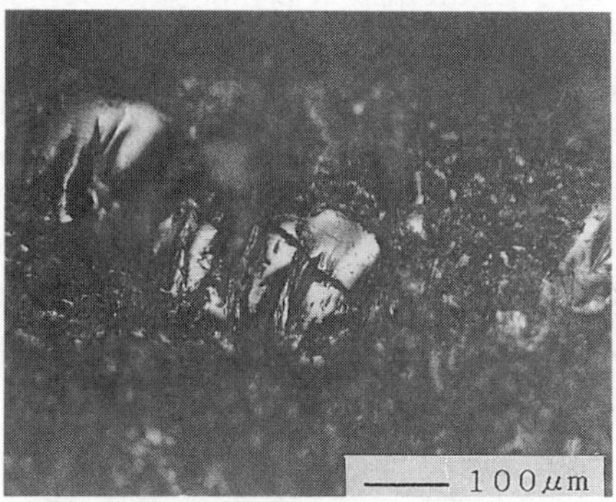

Fig.12. Gapped "Plant Polish" on Andesite

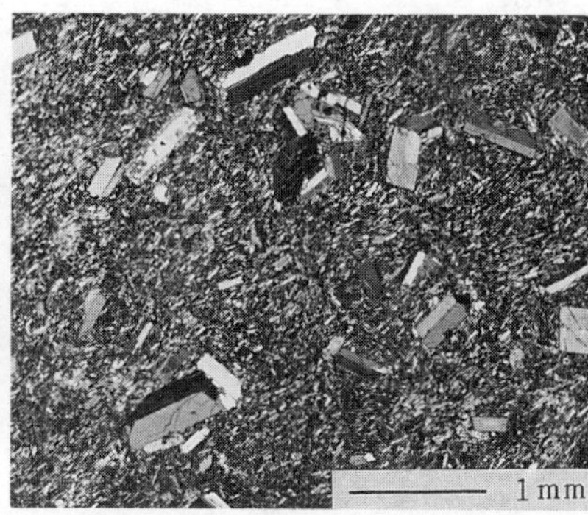

Fig.13 Thin Section of Andesite (cross polar)

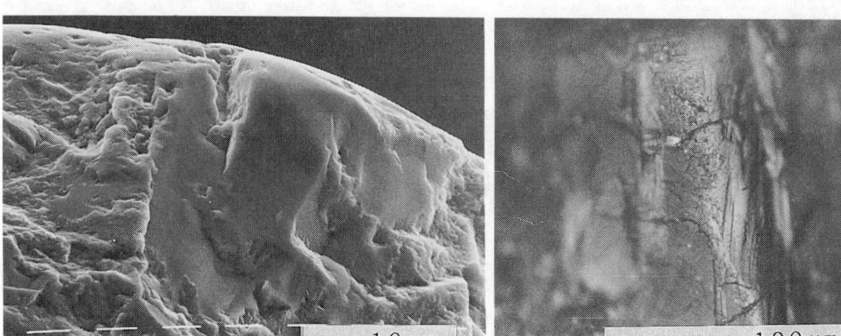

by cutting grass [4] by cutting rice

Fig.14. Worn Phenocryst of Andesite

Rhyolite

The average grain size of rhyolite is between chert and andesite, so polished surfaces produced on it also shows the intermediate characteristics between chert and andesite. Here also we can observe the phenocrysts effect(Fig.16).

Slate

Studies of use-wear "polish" on ishi-bochos made of slate provide more information on the nature of polished surfaces.

Polished surfaces produced on ishi-bochos by rice cutting also show gapped pattern[1,2], which were caused by realtively large inclusions such as chlorite and sericite(Fig.19). However, each polished area is more isolated and rarely extends over the entire surface. This may indicate that the groundmass area is so soft that polished surfaces cannot be formed. The area of polished surface depends on the size and density of polishable grains, so it cannot be a direct index of amount of the work.

Ishi-bochos are usually finished by grindstone. I used artificial grindstones which were standardrized as coarse, medium, fine, in this order, to replicate ishi-bochos for experiments. Polished surfaces produced by grindstone can also show a gapped pattern, but each polished surfaces is flat and distinguishable from domed surface produced by working plants(Fig.18). Only very fine grindstones used for finish can produce polished surfaces[14]. In the case of coarse and intermediate grain-size grindstone, slate surfaces are abraded severely, leaving no polished area, only rough scratches. This indicates that polished surfaces are formed by mild wear.

DISCUSSIONS AND CONCLUSION

The question of use-wear "polish" which archaeologists have been using for the determination of tool function by their optical patterning, can be reduced to the question of the morphology of worn surfaces caused by the nature of the two materials which make contact. Diagnostic features of polished surfaces are determined by hardness, viscous and elastic properties and surface texture of worked materials, and the topography of the rock surface. Especially, the nature of the worked materials which function as a polisher has a significant effect : Polished surfaces produced by plants have a very smooth and domed appearance which is commmon to many rock types.

The process of polishing is not simple abrasive wear. It occurs on the order of less than a µm and only by mild friction. Features of "plant polish" on andesite, rhyolite, slate can be explained by the above wear theory. Conversely, the observation of polished surfaces on these rocks proves the wear theory.

Until now, people had to repeat extensive experiments each time they encountered unfamiliar rock types in archaeological analysis. However, by understanding the mechanism, we can predict the features of polished surfaces caused by a given worked material and a given rock type. Simulation of polished surface formation with computer graphics may be possible. This reduces the number of experiments necessary for each rock type and makes it easier to extend the "polish" method to other materials.

The fact that "plant polish" is distinctive on any rock

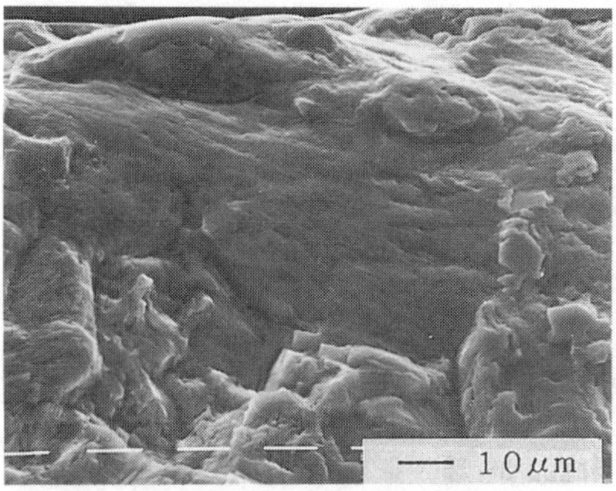

Fig.15. Polished Andesite Surface by Cutting Grass

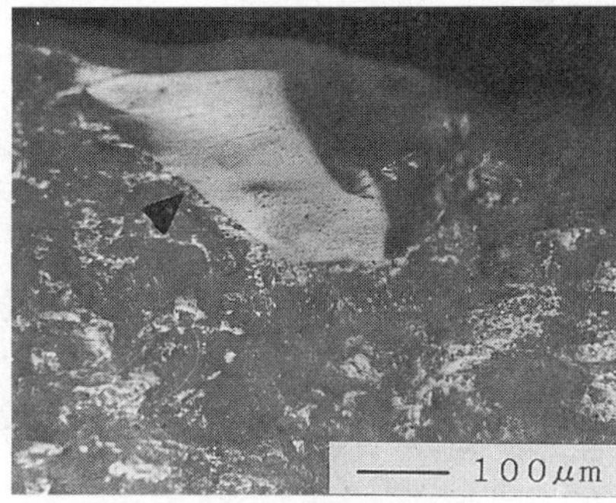

Fig.16. Worn Phenocryst of Rhyolite by Cutting Rice

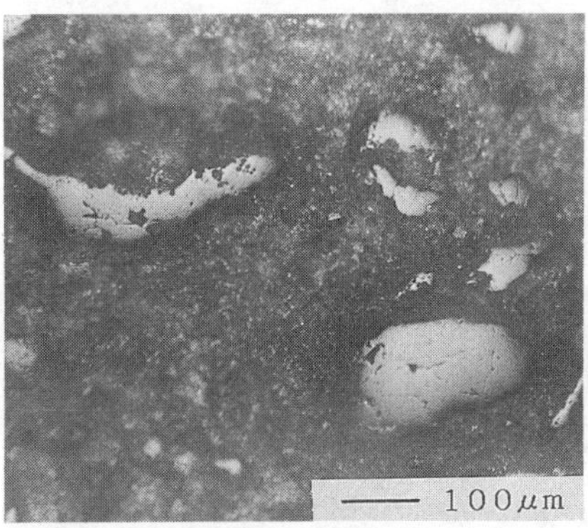

Fig.17. Gapped "Plant Polish" on Ishi-bocho

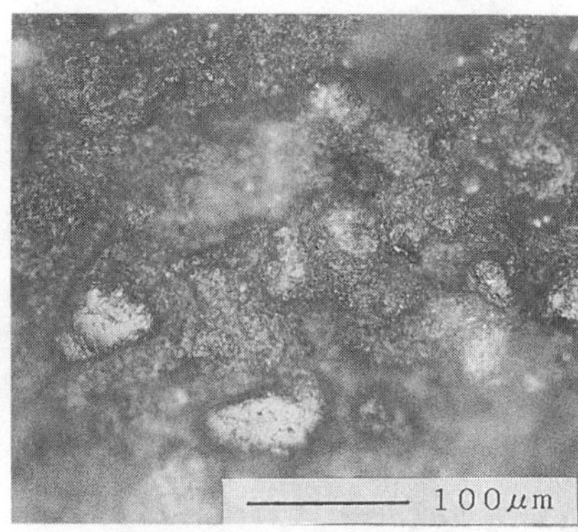

Fig.18. Gapped "Polish" by Grindstone on Ishi-bocho

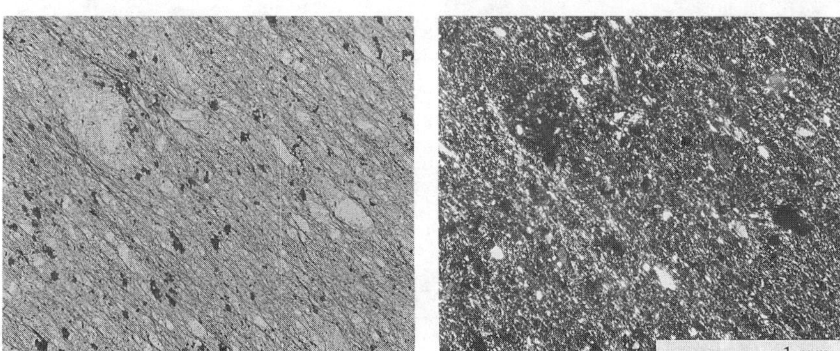

open polar crossed polar

Fig.19. Thin Sections of Slate

type is very advantageous for the study of early paddy cultivation tools. Starting from use-wear analysis on farming tools left by early paddy technology in its northern extension, we may be able to trace its diffusion and acceptance process back to the Asian Continent.

ACKNOWLEDGEMENTS

The lithic use-wear analysis at the Tomizawa site is a cooperative work with Mr.Hirohiko Saino, Sendai Board of Education. I would like to thank Prof.Muneaki Shimura and Dr.Sakae Saito(the Institute of Material Science, Tohoku University) for their help and for permission to do SEM research. Mr.Jun Fujita helped me at the early stage of andesite studies. Students at the Dept. of Archaeology, Tohoku University, helped me with experiments. I wish to express my thanks to Prof. Satoshi Kanisawa for identification of rock minerals, and Mrs.Michelle Kimura for correcting the English of the paper.

REFERENCES

1. K.Akoshima and T.Sudo, in Excavation of the Tomizawa Paddy Field Site (Loc.8) 1, edited by T.Kudo(Sendai Board of Education, Sendai, Japan, 1984), pp.213-216.(in Japanese)
2. T.Sudo and K.Akoshima, Summaries of 51th Meeting, Japan Archaeology Asssociation(1985). (in Japanese)
3. L.H.Keeley, Experimental Determination of Stone Tool Uses : A Microwear Analysis. (Univ.of Chicago Press, Chicago,1980)
4. S.Yamada, in The Tomizawa Site (Loc.15), edited by H.Saino (Sendai Board of Education, Sendai, Japan,1987),pp.461-468. (in Japanese)
5. H.Saino, Study of Local History 39(4),4-18(1989). (in Japanese)
6. T.Midoshima, Cultural Antiqua 41(3),30-43(1989), 41(6),19-27(1989), 41(8),1-15(1989), 43(11),26-35(1991). (in Japanese)
7. S.Yamada, in The Kawai Site, edited by S.Yamada(Inst. of Archaeology,Shizuoka Pref., Shimizu, Japan, in press). (in Japanese)
8. OECD, Glossary of Terms and Definition in the Field of Friction,Wear and Lubrication(1969),p.64.
9. P.Anderson, World Archaeology 12,181-194(1980).
10. M.-E.Mansur-Franchomme, J.Archaeol.Sci. 10,223-230.
11. S.Yamada, Araheology and Natural Science 19,101-123(1986). (in Japanese with English summary)
12. S.Yamada, in Tracéology et Fonction : le Geste Retrouvé, edited by M.Otte(Colloque international de Liège 50, Éditions EURAL, Liège, 1992), in press.
13. H.Andersen and H.Whitlow, Nuclear Instruments and Methods in Phys.Res. 218,468-474.
14. S.Yamada and A.Sawada, in Tracéology et Fonction : le Geste Retrouvé, edited by M.Otte(Colloque international de Liège 50, Éditions EURAL, Liège, 1992), in press.
15. D.Rees, G.Wilkinson, C.Orton, and R.Grace, in Computer and Quantitative Methods in Archaeology, edited by S.P.Q.Rahtz (BAR international series 466(i), Oxford,1988,), pp.177-183
16. D.Rees, G.Wilkinson, R.Grace, and C.Orton, J.Archaeol.Sci. 18,629-640.
17. K.Aoyama, graduation thesis, Tohoku University, 1985.

PRELIMINARY TECHNICAL STUDY OF MEDIEVAL LIMOGES ENAMELS

Pete Dandridge and Mark T. Wypyski
Objects Conservation Department, Metropolitan Museum of Art
1000 Fifth Avenue, New York, NY 10028, U.S.A.

ABSTRACT

Enamels from Northern Spain, Central France, and Limoges, dating from the 10th century to the mid 13th century, were analyzed by energy dispersive x-ray spectrometry and x-ray diffractometry to determine their overall compositions, and the opacifiers and colorants used. The compositions suggest the re-use of Roman tesserae. Calcium and lead antimonates were found as the predominant opacifiers; however, tin oxide and lead-tin yellow were also found in some of the enamels, either singly or in combination with the antimonates. One of the enamels from Limoges varied from the others in that the enamels had overall compositions similar to Islamic or Venetian glasses, and contained only tin oxide as an opacifier.

INTRODUCTION

We have recently had the opportunity to begin a technical and material study of Medieval enamels in the Metropolitan Museum of Art that are attributed to Central and Southern France or Northern Spain, areas in which our collection is very strong and for which there is little technical information available. Eventually our study will incorporate a great amount of material attributed to Limoges, a major center for the development and practice of enameling beginning in the twelfth century and continuing through the Renaissance; however, we thought initially that it would be important to document the practices that preceded and possibly influenced later enamelling in Limoges as well to record the practices in that center. To that end, we selected a group of eight objects which range in date from the tenth century to the end of the thirteenth century and are attributed to South Central France and Northern Spain. This paper discusses the results of our quantitative elemental analyses, the identification of the specific opacifiers and colorants, and the possible relation of these enamels to Roman period and Medieval European vitreous materials.

ANALYTICAL TECHNIQUES

The compositions were analyzed with a modified Amray model 1100 (1600 T) scanning electron microscope using a Kevex model Delta IV energy dispersive X-ray spectrometer. (This model EDS is equipped with a normal beryllium window detector, and does not detect elements below the atomic weight of sodium). Weight percentages of the elements detected were calculated from the counts per second data, compared to well characterized soda-lime-silica reference glasses. Our studies using reference glasses have determined the variation for the major element oxides to be less than two percent for silica, about five percent for sodium, potassium and calcium, and about ten percent for magnesium and aluminum. The minimum detection limits for the elements titanium through zinc were found to be under one tenth of one percent. The minimum detection limits of phosphorus, lead, barium, arsenic, antimony and tin oxides were found to be much higher, about one half of a percent by weight, mainly due to peak overlap problems. Details of this procedure are being prepared for publication [1].

The enamels were sampled, by flaking off very small pieces, on the order of a cubic millimeter in size, from areas of previous loss or damage using a diamond edged scribe. The samples were prepared for analysis by embedding them in epoxy resin, and exposing the interiors by grinding with silicon carbide paper. The cross-sections were then polished with aluminum oxide, and given a high vacuum carbon coating.

One of the objects in this study, the 'St. Matthew', MMA number 17.190.427, was found to be in excellent physical condition, so sampling of the enamels was not considered appropriate; however, the large size of the SEM's sampling chamber allowed for the object's insertion and

Table 1 - The Objects

MMA Accession No.	Description	Attribution	Date
17.190.688	'Luna' copper-gilt cloisonnés set into iron dish	Carolingian, Aquitaine	10th c.
17.190.427	Symbol of 'St. Matthew' cloisonné enamel, copper-gilt	Northern Spain or Southern France	ca. 1100
1983.38	'Seated Prophet' cloisonné and champlevé enamel, copper-gilt	Conques	ca. 1100
17.190.692	'Fantastic Bird' champlevé enamel, copper-gilt	Conques	ca. 1115-1130
17.190.685-.687, .695,.710-.711	'Champagnat Chasse' cloisonné and champlevé enamel, copper-gilt	Northern Spain or Limoges	ca. 1165-1175
17.190.409	Crucifix champlevé enamel, copper-gilt	Limoges, Grandmont	ca. 1185-1190
17.190.833	Crozier champlevé enamel, copper-gilt	Limoges	ca. 1200-1250
32.100.284	Pyxis champlevé enamel, copper-gilt	Limoges	1250-1300

subsequent EDS surface analysis. Because of charging effects, surface irregularities, and possible weathering of the enamel's surface, accurate quantitative analyses were not possible by this technique.

X-ray diffraction analysis was done using a Philips PW 1710 open architecture diffractometer, with Sietronics driving program, on cleaned surfaces of the enamels. (For details of this technique, see discussion in this proceedings [2]). Some enamels were also sampled and analyzed by Debye-Scherrer powder camera x-ray diffraction, using a Philips PW 1729 x-ray generator. XRD patterns were identified against the JCPDS powder diffraction files with the use of the Fein-Marquart Associates micro powder diffraction search/match program. Given the newness of the open architecture system, it seemed advantageous to substantiate our findings with the more traditional powder camera.

RESULTS

Table 2 lists the oxide weight percentages found for the enamels, organized by object. The percentage for copper oxide is reported as CuO, even in the red enamels where most of the copper is presumably present as Cu_2O. Surface analyses of the 'St. Matthew' revealed overall compositions similar to that of all the other enamels analyzed, except for number 17, and 26 to 28, but has not been included in this table given the relative inaccuracy of the technique. Figure 1 is a plot of magnesium oxide weight percent against potassium oxide percent for all 28 enamels compositionally analyzed, except for number 17. Table 3 lists the crystalline colorants/opacifiers identified by XRD, either open architecture (OA) or powder camera (PC).

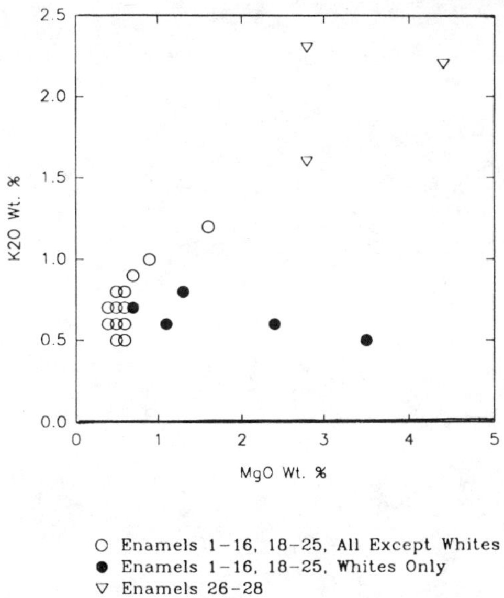

Figure 1 - Wt% MgO vs. Wt. % K_2O for enamels 1 to 27, table 2.

Table 2

Compositions of Selected Enamels - Weight % Oxides

MMA Acc. No.	17.190.688						1983.38				17.190.692		17.190.710	
No.	1 WHI	2 DBL	3 TUR	4 GRN	5 YEL	6 RED	7 WHI	8 DBL	9 TUR	10 GRN	11 DBL	12 TUR	13 TUR	14 GRN
Na_2O	15.1	17.0	17.9	17.4	17.3	15.5	15.0	13.8	16.4	15.1	14.9	18.3	17.7	15.0
MgO	1.1	0.6	0.5	0.7	0.5	0.9	2.4	0.5	0.5	0.6	0.5	0.5	0.7	0.7
Al_2O_3	2.4	2.7	2.5	2.4	2.2	2.9	1.9	2.0	2.0	2.2	2.3	2.0	1.9	2.5
SiO_2	65.7	68.0	68.8	66.8	67.4	58.0	66.1	62.5	65.0	60.8	64.4	62.6	63.0	61.2
K_2O	0.6	0.7	0.6	0.9	0.6	1.0	0.6	0.7	0.6	0.8	0.7	0.5	0.5	0.7
CaO	7.4	7.0	5.2	5.8	5.8	7.1	6.6	7.1	4.6	7.0	6.9	6.0	4.5	5.0
P_2O_5	-	-	-	-	-	-	-	-	-	-	-	-	-	-
SO_3	0.2	0.4	0.4	-	-	-	0.3	0.4	0.4	-	-	-	0.5	-
Cl	0.4	0.9	1.3	1.0	1.2	1.1	0.6	0.6	1.1	0.6	0.7	1.3	1.1	1.1
TiO_2	-	0.1	0.1	0.1	0.1	0.1	0.1	0.1	0.1	0.1	-	0.1	0.1	0.1
MnO	0.2	0.6	-	0.3	0.4	0.3	-	0.6	-	1.8	0.8	-	-	0.4
Fe_2O_3	0.4	1.0	0.5	0.9	0.6	1.6	0.4	1.6	0.6	0.8	1.2	0.8	0.7	0.7
CoO	-	0.1	-	-	-	-	-	0.4	-	-	0.3	-	-	-
CuO	-	0.2	1.9	1.6	-	2.6	-	1.3	4.5	3.9	0.5	2.8	3.8	2.5
ZnO	-	-	-	-	-	0.6	-	-	-	-	-	-	-	-
SnO_2	-	-	-	-	-	0.5	-	-	-	-	-	-	-	-
Sb_2O_3	5.4	0.9	-	<0.6	<0.6	-	5.0	6.5	2.5	2.5	4.1	2.9	3.8	<0.6
PbO	-	-	-	2.6	3.9	7.0	-	-	-	4.0	2.6	0.7	-	10.2

- = Not detected.

Table 2 (cont.)

Compositions of Selected Enamels - Weight % Oxides

MMA Acc. No.	17.190.695			17.190.409			17.190.833					32.100.284		
No.	15 WHI	16 LBL	17 RED	18 WHI	19 LBL	20 DBL	21 WHI	22 DBL	23 GRN	24 YEL	25 RED	26 WHI	27 LBL	28 RED
Na_2O	16.4	15.0	0.9	15.1	14.1	17.2	13.8	15.0	13.6	16.3	14.5	14.0	13.7	16.3
MgO	3.5	0.5	3.5	1.3	0.4	0.4	0.7	0.6	0.6	0.5	1.6	2.8	2.8	4.4
Al_2O_3	1.8	2.1	1.5	2.3	2.5	2.5	2.5	2.8	2.5	2.5	2.9	1.7	1.4	2.0
SiO_2	65.1	65.6	47.3	57.6	63.0	68.3	60.5	64.8	63.3	61.8	58.2	51.9	51.1	56.8
K_2O	0.5	0.8	18.3	0.8	0.6	0.7	0.7	0.7	0.5	0.6	1.2	1.6	2.3	2.2
CaO	6.7	7.3	16.4	7.7	7.2	5.9	8.6	7.2	6.8	5.6	7.3	6.6	7.5	8.3
P_2O_5	-	-	5.4	-	-	-	-	-	-	-	-	-	-	-
SO_3	0.4	-	0.3	-	-	-	-	-	-	-	-	-	-	-
Cl	0.7	0.7	0.3	0.6	0.5	0.9	0.6	0.7	1.0	1.1	0.8	0.9	0.7	0.9
TiO_2	0.1	0.1	0.1	0.1	0.1	-	-	0.1	-	0.1	0.2	-	-	0.1
MnO	-	0.5	0.2	0.5	0.3	0.6	0.3	0.8	0.5	0.3	0.5	0.3	0.3	0.6
Fe_2O_3	0.4	0.9	2.1	0.6	0.8	1.2	0.5	1.4	0.8	0.8	3.3	0.7	1.3	4.6
CoO	-	0.1	-	-	0.1	0.2	-	0.3	-	-	-	-	0.2	-
CuO	-	0.4	1.7	-	0.7	0.4	-	0.5	2.3	0.9	2.1	0.1	0.7	1.4
ZnO	-	-	0.4	-	-	0.2	-	-	0.2	-	0.1	-	0.3	-
SnO_2	-	-	-	2.8	0.6	-	1.5	-	-	-	0.7	5.3	3.9	-
Sb_2O_3	3.9	3.5	-	6.1	7.5	1.7	6.0	3.5	<0.6	<0.6	-	-	-	-
PbO	-	1.8	0.5	4.3	1.8	1.0	3.3	2.5	6.6	9.0	6.0	15.0	13.0	2.6

- = Not detected.

Table 3 - Opacifiers / Colorants Identified BY XRD

Object	Color	Phases Present	Technique	
17.190.688	White	$Ca_2Sb_2O_7$	OA	
Luna	Yellow	$Pb_2Sb_2O_7$	OA	
10th c.				
17.190.427	White	$Ca_2Sb_2O_7$	OA	
St. Matthew	Lt. Blue	$Ca_2Sb_2O_7$	OA	PC
ca. 1100	Dk. Blue	$Ca_2Sb_2O_7$	OA	
	Green	Pb_2SnO_4	OA	
	Flesh	$CaSb_2O_6$ (O,OH), $PbSnO_3$	OA	
1983.38	White	$CaSb_2O_6$(O,OH), $Ca_2Sb_2O_7$	OA	PC
Seated Prophet	Lt. Blue	$Ca_2Sb_2O_7$	OA	
ca. 1100	Dk. Blue	$Ca_2Sb_2O_7$	OA	
	Turquoise	$CaSb_2O_6$(O,OH)	OA	
	Green	$Pb_xSb_{2-z}(O,OH,H_2O)_6$	OA	
	Flesh	$Ca_2Sb_2O_7$	OA	
17.190.692	White	$Ca_2Sb_2O_7$	OA	
Fantastic Bird	Lt. Blue	$Ca_2Sb_2O_7$	OA	
ca. 1115-1130		$CaSb_2O_6$(O,OH)		PC
	Dk. Blue	$CaSb_2O_6$(O,OH)	OA	
	Turquoise	$Ca_2Sb_2O_7$	OA	PC
		$CaSb_2O_6$(O,OH)	OA	
	Green	$Pb_2Sb_2O_6$(O,OH)	OA	
17.190.685-.687,	White	$Ca_2Sb_2O_7$, $CaSb_2O_6$(O,OH)	OA	
.695,.710-.711	Lt. Blue	$Ca_2Sb_2O_7$, $CaSb_2O_6$(O,OH)	OA	
Champagnat Chasse	Turquoise	$Ca_2Sb_2OH_7$	OA	
ca. 1165-1175	Green	$Pb_2Sb_2O_6$(O,OH),	OA	
	Red	Cu, Cu_2	OA	
17.190.409	White	$Ca_2Sb_2O_7$, SnO_2	OA	PC
Grandmont Crucifix	Lt. Blue	$Ca_2Sb_2O_7$, SnO_2	OA	PC
ca. 1185 - 1190	Green	$Pb_2Sb_2O_7$	OA	
	Yellow	$Pb_2Sb_2O_7$	OA	
17.190.833	White	$Ca_2Sb_2O_7$, SnO_2		PC
Crozier	Lt. Blue	$Ca_2Sb_2O_7$		PC
ca. 1200 - 1250	Green	$Pb_2Sb_2O_7$		PC
	Yellow	$Pb_2Sb_2O_7$		PC
32.100.284	White	SnO_2	OA	
Pyxide	Lt. Blue	SnO_2	OA	
ca. 1250 - 1300				

DISCUSSION

We are fortunate in having three Medieval treatises that describe different aspects of the arts of glassmaking and enameling: Theophilus's 'De Diversis Artibus', the 'Mappae Clavicula', and Eraclius's 'De Coloribus et Artibus Romanorum'. The glasses described in these texts were made using wood ash and/or fern ash as sources of alkali, colored by various metal oxides, and opacified using tin oxide. Studies of medieval European glasses have shown that the majority of these glasses can be characterized as potash glass, that is, glass containing large amounts of potassium oxide, usually about ten to twenty percent by weight, with only a small amount of sodium oxide present, and often containing large amounts of calcium and magnesium oxides [3,4,5]. While Henderson and others have documented the existence of 'mixed-alkali' (both sodium and potassium) glasses in Europe from the first and second millenniums B.C. [6], it is thought that the beginning of the use of potassium as the dominant alkali in European glassmaking did not occur until around the tenth century A.D., possibly in response to the increased demand for window glass.

An entirely different technology was recommended in the medieval documents for the fabrication of enamels and imitation gems. In book III, Theophilus recommends re-using the opaquely colored glasses found in the mosaics within ancient buildings [7]. The availability of Roman tesserae apparently constituted a suitable ready source of various colored glasses for early medieval enamelers, who were unable to produce the colors on their own [8].

In contrast to medieval European potash glass, most glass from Roman Europe, the Mediterranean, and the Near East has been found to be soda glass, that is, glass made with sodium as the major alkali. Soda glasses can usually be divided into two categories - low potassium, low magnesium glass, and high potassium, high magnesium glass [9,10]. Roman glass has been found to have remained relatively constant in composition over a broad geographic area and time period, and has been shown, with few exceptions, to contain low levels of potassium and magnesium. Such concentrations are thought to be consistent with the use of natron, a nearly pure sodium carbonate, as the alkali source [11]. Near Eastern and Islamic glass, howvever, was usually made using a plant ash source of soda, and contained significantly higher amounts of magnesium and potassium [12,13].

From ancient times through the Roman period, glass has been opacified with antimony compounds [9,10,12]. The use of these compounds persisted until about the middle of the fourth century A.D., when apparently there was a general changeover to the use of white tin oxide and lead-tin yellow [14]. Bateson and Hedges x-ray fluorescence study of Roman enamels reported substantial quantities of tin in some of the enamels; however, XRD analyses of the enamels detected no tin oxide or lead-tin yellow opacifiers [15]. The presence of lead-tin yellow has been reported by Brill in his analyses of yellow Roman tesserae dating from the second to the fourth centuries A.D. from Kenchrai, Spain [16]; and Henderson has found examples of lead-tin yellow in some British glass dated from the first century B.C. to the first century A.D., as well as several tin oxide opacified glasses from Celtic Europe, dated to the second to first centuries B.C. [17,18].

Several recent compositional studies have helped to clarify the relationship between Roman glass, enamels and tesserae [19,20]. The enamels are similar to the composition generally seen for translucent glasses - soda-lime glass containing only small amounts of magnesium and potassium. The enamels and tesserae were found to contain calcium antimonate and lead antimonate as opacifying agents, as were the opaque decorative elements on vessel glass [21]. Substantial amounts of lead were also seen in many of the enamels and tesserae as opposed to vessel glass which usually contains only trace quantities of lead.

The re-use of Roman tesserae by medieval European glass workers and enamelers has been documented previously. Lundstrom found tesserae in an eighth century Scandinavian bead-making site [22], Englander was able to identify by x-ray fluorescence both antimony and tin compounds as opacifiers and has used the identification of the former as evidence for the re-use of tesserae [23], and Bimson used both x-ray diffraction and x-ray fluorescence to identify as Roman, with the exception of the red enamels, both the opacifiers and the compositions in a group of Mosan enamels [24,25].

The composition of the enamels on the first seven objects in this study were found to be similar to Roman glass. Analyses showed that they are soda glasses with small amounts of magnesium and potassium oxides, generally less than one percent by weight, and varying amounts of lead oxide, up to about ten percent. The white enamels are interesting in that several of them have relatively high amounts of magnesium, but low levels of potassium. Henderson has found in his analyses of Roman enamels that whites often contain elevated magnesium levels, and hypothesizes that they may have been manufactured from a specific recipe using slightly different raw materials which introduced some additional magnesium oxide [19,20].

The white, blue, turquoise, and flesh tone enamels on these seven objects were found to contain calcium antimonate as the main opacifying agent with the exception of the white and blue enamel on the Grandmont Crucifix from the late twelfth century, and the white enamel on the Limoges Crozier from the early thirteenth century, which were found to contain significant amounts of tin oxide intermixed with calcium antimonate. The yellow and green enamels examined were opacified by lead-antimonate, apart from the green enamel on the 'St. Matthew' which contained lead-tin yellow. The flesh colored enamel on this object was also tinted with lead-tin yellow, as well as calcium antimonate. The other colorants are those normally found in Roman glasses and enamels; cobalt in the blues, copper in the turquoise and green, and reduced copper oxide (cuprite) in the reds.

One of the red enamels, table 2, number 18, was found to be a potash glass rather than a soda glass, and appears to be similar in composition to the contemporary glass produced throughout much of medieval Europe for window and vessel glass. Similar results have been found for other red Medieval enamels [25,26,27]. Buckton suggests that this type of glass may be the red glass from Theophilus's treatise known as 'Gallien'[8]. This cannot be ascertained, since the recipe is missing from the treatise, and is only known from the listing in the table of contents.

The composition of the Limoges pyxis attributed to the second half of the thirteenth century is different from the other seven enameled objects. The enamels were found to be soda glass, but unlike the others, had relatively high levels of magnesium and potassium similar to those glasses made with a soda plant ash source of alkali. The white and light blue enamels contained substantial amounts of lead with only tin oxide as the opacifier.

While the majority of medieval European glass has been found to be potash glass, some soda glass has also been found. Studies of the Venetian glass making industry have shown that soda glass was being produced there with a plant ash alkali source imported from the Levant in a tradition that began as early as the eighth century, and continued throughout the medieval period [28]. In their study of window glass from York Minster, Cox and Gillies were able to identify three different compositional groups of soda glass - re-used Roman glass, Roman glass mixed with potash glass, and high magnesium, high potassium soda glass [29]. Further, Foy has reported the presence of soda as a major component in vessel glass from southern France as early as the twelfth century [30]. Our analyses would suggest that the composition of the Limoges pyxis is similar to glass produced in Venice or the Islamic world, rather than the soda glass or mixed-alkali glasses published by either Cox and Gillies or Foy.

CONCLUSION

With this preliminary study, we have begun to elucidate the workshop practices of the medieval enameler in Northern Spain and in Southern and Central France both prior to the development of Limoges as a center of production, as well as during that city's period of activity. Through elemental analysis and x-ray diffraction, we have been able to identify three separate compositional types. The enamels in seven of the objects are, with one exception, very similar to Roman glass both in their overall composition and in their use of antimony compounds as opacifying agents. The occurrence of tin compounds within this group provides further evidence of the Roman use of lead-tin and tin oxide as opacifiers. That several of the objects are decorated with white enamels that contain both calcium antimonate and tin oxide, suggests the presence in the workshop of tesserae opacified by calcium antimonate and by tin oxide that were mixed during grinding either inadvertently or with the intent of altering the color.

The red enamel on the Champagnat Chasse is clearly a different composition, and appears to be have been made using contemporary medieval potash glass, rather than re-used Roman material. It is unclear why a contemporary glass would have been used occasionally in certain medieval enameled objects, but it may have to do with the relative paucity of red Roman tesserae and availability of locally produced red potash glass.

The final variant is found on the Limoges pyxis which appears to be have been made using a soda plant ash alkali. The evidence now available indicates that this type of glass was imported, rather than being made locally. Further analyses may indicate whether the pyxis signals a more broadly based change either in technology or in supply occurring in Limoges in the second half of the thirteenth century.

The small group of enameled objects examined so far has led to some interesting conclusions and speculation. Work is continuing on these objects and others to characterize further enamels from this time, and to gain a better understanding of the relationship between different types of glasses and the development of enamel technology in Limoges.

ACKNOWLEDGEMENTS:

The authors would like to acknowledge Barbara Boehm, Associate Curator of Medieval art at the Metropolitan Museum of Art, for her generous encouragement and her art historical expertise. We would also like to thank Ian Freestone, British Museum Research Laboratory, and Julian Henderson, Department of Archaeology and Prehistory, the University, Sheffield, for sharing with us their research on Medieval enamels.

REFERENCES

1. M. Verita, R. Basso, M.T. Wypyski and R.J. Koestler, 'X-Ray Microanalysis of Ancient Glassy Materials: A Comparative Study of WDS and EDS Techniques'. (Submitted for publication).
2. G.S. Wheeler, J.H. Frantz, and P. Dandridge. 'The Use of Open Architecture X-Ray Diffraction In Qualitative Analysis of Materials in Art and Archaeology.' 1992 MRS Spring Meeting, San Francisco, CA.

3. E.R. Caley, Analyses of Ancient Glasses, 1790-1957, (Corning, New York, 1962).
4. R. Brill, J. Glass Studies, 12, 185-192 (1970).
5. G.A. Cox, O.S. Heavens, R.G. Newton, and A.M. Pollard, J. Glass Studies 21, 54-75 (1981).
6. J. Henderson, Archaeometry 30(1), 77-91. (1988).
7. Theophilus, 'On Divers Arts', translated by C.S. Smith and J. G. Hawthorne (New York, 1979), 59
8. D. Buckton, Trans. 3rd Canadian Conference of Medieval Art Historians, (University of Western Ontario, 1985) pp. 1-6.
9. E.V. Sayre and R.W. Smith, Science 133, 1824-1826 (1961).
10. E.V. Sayre and R.W. Smith, Archaeological Chemistry III, pp. 279-311. (1967).
11. R. Brill, Excavations at Jalame - Site of a Glass Factory in Late Roman Palestine. G. D. Weinberg, editor. (University of Missouri Press, 1988).
12. W.E.S. Turner, J. of the Society of Glass Technology, 40, 277TG-300T (1956).
13. R. Brill, Glass and Glassmaking in Ancient Mesopotamia. A.L. Oppenheim, R. Brill, D. Barag, and A. von Saldern, editors. (Corning Museum of Glass, 1970). pp. 105-128.
14. W. Turner and H. Rooksby, Jahrbuch des Romisch-Germanischen Zentralmuseums, Mainz, 8, 1-6. (1961).
15. J.D. Bateson and R.E.M. Hedges, Archaeometry 17, 177-190 (1975).
16. R. Brill, Proc. 8th ICG Congress, Bruxelles. pp. 47-68.(1968)
17. J. Henderson, Scientific Analysis in Archaeology. Edited by J. Henderson. (Oxford University Committee for Archaeology Monograph no. 19, 1989). pp. 30-62.
18. J. Henderson, Oxford Journal of Archaeolgy 4(3), 267-291 (1985).
19. J. Henderson, Archaeometry '90, 285-294. (1990).
20. J. Henderson, Materials Issues In Art and Archaeology II, edited by P.B. Vandiver, J. Druzik and G.S. Wheeler. (Mater. Res. Soc. Proc., 185, Pittsburgh, PA, 1991) pp. 601-607.
21. I.C. Freestone, J. of Glass Studies, 32, 103-107. (1990).
22. A. Lundstrom, Early Medieval Studies 9, 4-7, (1976).
23. P. Englander, Catalog of Medieval Objects - Enamels and Glass (Museum of Fine Arts, Boston, 1986) pp. xix-xxvi.
24. M. Bimson, Annales du 8e Congres International d'Etude Historique du Verre, Londres-Liverpool, 18-25 Septembre 1979. pp. 161-164. (1981)
25. Buckton, 1985, ref. 10.
26. J. Henderson. Unpublished results presented at the 1991 British Museum Enamel Colloquium.
27. I. Freestone. Unpublished results presented at the 1991 British Museum Enamel Colloquium.
28. M. Verita and T. Toninato, Rivista della Stazione Sperimentale del Vetro, 20(4), 169-175 (1990).
29. G.A. Cox and K.J.S. Gillies, Archaeometry, 28(1), 57-68 (1986).
30. D. Foy, J. Glass Studies, 27, 18-71 (1985).

PART IV

Conservation Science and Preservation

PART IV A

Identification and Physical Chemistry of Post-Depositional Processes

THE DIVERSITY OF NEWLY DISCOVERED DETERIORATION PATTERNS IN ANCIENT EGYPTIAN PIGMENTS: CONSEQUENCES TO ENTIRELY NEW RESTORATION STRATEGIES AND TO THE EGYPTOLOGICAL COLOUR SYMBOLISM

SOLVEIG SCHIEGL*, KARL L. WEINER** AND AHMED EL GORESY*
*Max-Planck-Institut für Kernphysik, P.O. 103980, 6900 Heidelberg 1, Germany
**Ludwig-Maximilians-Universität München, Theresienstr. 41, 8000 München 2, Germany

ABSTRACT

Polished cross sections through more than 1000 pigment samples from ancient Egyptian monuments (5th Dynasty to Roman Times) were investigated in reflected light, with SEM, and with electron microprobe techniques (EMPA). These combined analytical techniques allow many insights into the painting layers and down to the substrate. Textural relations of both the primary paintings and of the deterioration products can then be delineated. They also discern the stages of the various deterioration processes and the degree of damage of the painting layers.

The textural patterns of six deterioration processes not known before are discussed in detail and are documented in BSE-micrographs. Five of these processes commence in synthetic Cu-pigments (Egyptian Blue, glass pigment, and Green Frit) and Fe-bearing glass pigments. The 6th process leads to the formation of complex phosphates and is mainly exogenous. It causes severe damage to natural and synthetic pigment layers. These processes result in destruction of the primary pigments and coloured decorations.

Consequences as to the colour symbolism and future treatment of befallen paintings are discussed.

INTRODUCTION

During the investigation of more than 1000 pigment samples which were mainly collected from wall paintings in several temple sites and necropoles in Egypt (Fig. 1) patterns of six deterioration processes not known before were discerned. This article concentrates especially on the various synthetic pigments since most of the deterioration processes commence in them.

Three of the four synthetic pigments are Cu-bearing melt products. They were produced by melting starting mixtures of SiO_2-rich sand, bronze scrap, Na-, and K-bearing fluxes together with lime. Depending on the amount of lime the ancient Egyptians manufactured two different pigments: Egyptian

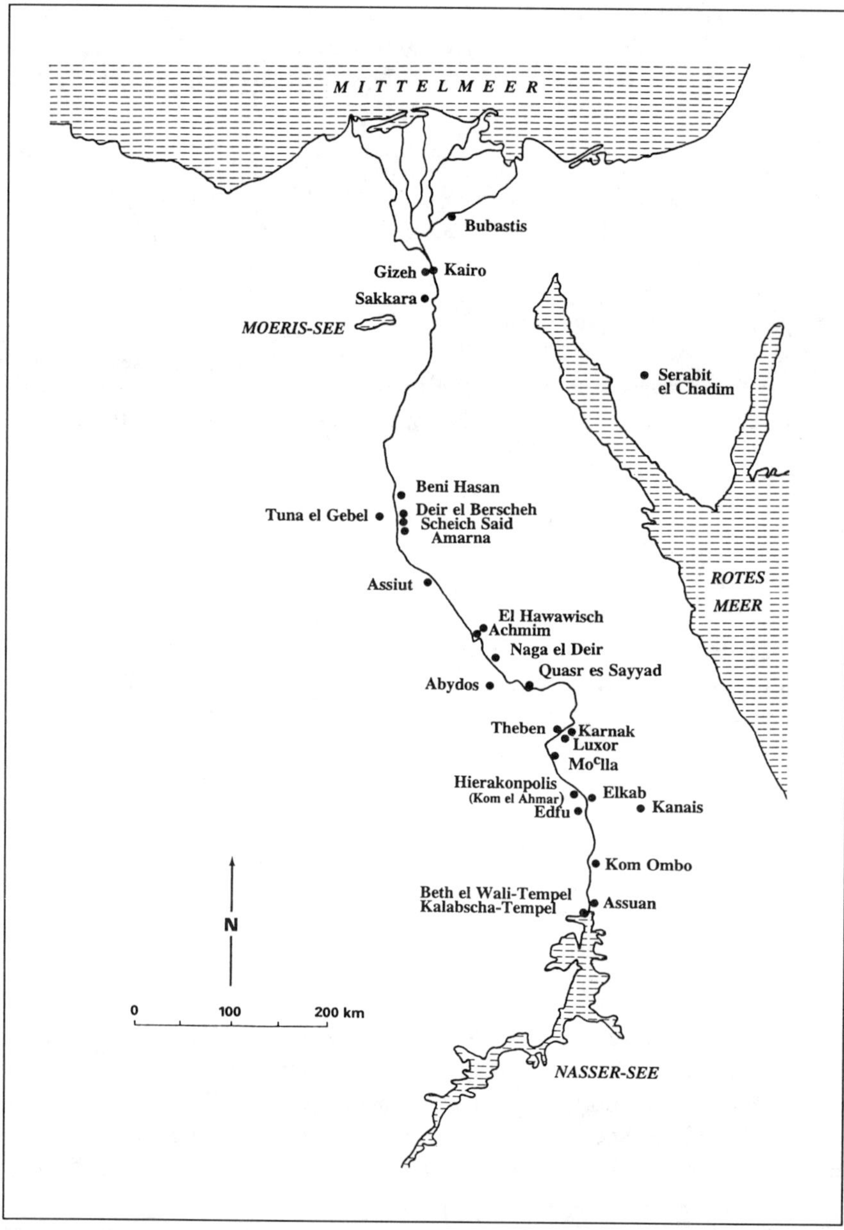

Fig. 1: Map of the Nile valley displaying the provenance of the studied pigment samples.

Blue and Green Frit. Table 1 depicts the compositions of these multiphase materials and their chronological evolution. A major phase in Egyptian Blue is the intensive blue cuprorivaite $CaCu[Si_4O_{10}]$ which dominates the pigment colour. Whereas the Green Frit owes its colour to the green Cu-wollastonite $(Ca,Cu)_3[Si_3O_9]$ (Cu < 1.5 wt.%). Both pigments contain considerable amounts of pale blue silicious glass rich in Cu, Na, K, and Cl. For the production of the Cu-bearing glass pigment - a painting material which was not known before - similar starting mixtures as for Egyptian Blue and Green Frit were used, however, no lime was added. Accordingly, the Cu-glass pigments are free of any Ca-bearing phases and consisted exclusively of pale blue Cu-bearing silicious glass fragments which were rich in alkalies. However, as a result of weathering in most cases the glass is completely depleted in alkalies. In addition the Cu-pigments contain unreacted relic quartz and newly formed SiO_2-phases.

Table 1: *Synthetic pigments used in ancient Egyptian decorations.*

Colour	Pigment Composition	Period of Application
blue	**Egyptian Blue** cuprorivaite + minor Cu-wollastonite + SiO_2 + glass	all epochs 2500 B.C. - 100 A.D.
green	**Green Frit** Cu-wollastonite + minor cuprorivaite + SiO_2 + glass	regularly used since New Kingdom 1500 B.C. - 100 A.D.
pale blue	**Cu-glass pigment** Cu-glass + SiO_2	Old and Middle Kingdoms 2500 - 2000 B.C.
originally green or brown	**Fe-glass pigment** Fe-K-S-bearing glass	found in several decorations of the Old, Middle and New Kingdoms

METHODS OF INVESTIGATION

In order to have deep insights into the pigment layer down to the substrate and identify primary pigment compounds and secondary deterioration products cross sections of the complete sequence of layers were taken in the field. From the cross sections polished surfaces were studied with reflected light microscopy, with scanning electron microscopy (SEM, Cambridge Stereoscan 180S) and the compositions of the individual components - primary or secondary - were then measured quantitatively with the electron microprobe (ARL-SEMQ). In the following the six deterioration processes will be treated sequentially.

DETERIORATION PROCESSES

1. Breakdown of the glass phase in Cu-pigments and of Cu-glazes and secondary formation of basic Cu-chloride

The glass components in all kinds of synthetic Cu-pigments and the Cu-glazes of Egyptian Faiences are affected by devitrification, leaching out of alkalies, dissolution processes and secondary formation of green basic Cu-chloride $Cu_2(OH)_3Cl$ and malachite $Cu_2[(OH)_2/CO_3]$.

Terrace[1] was the first to report findings of basic Cu-chloride in samples of Djehutinacht's sarcophagus (12th Dynasty, Deir el Bersheh). The identity of this phase was ascertained using X-ray diffraction. Using the same method Riederer[2] identified basic Cu-chloride in the paintings of two Theban stone reliefs of the 11th Dynasty. He considered basic Cu-chloride to be a primary pigment although being aware that it also could be a weathering product of another Cu-bearing compound.

Within an extensive and systematic study on the technological and chronological evolution of hundreds of ancient Egyptian pigments Jaksch[3] and El Goresy et al.[4] reported several findings of basic Cu-chloride in decorations of the Old and Middle Kingdoms, and again at the brink of the Ptolemaic Period. On the basis of incomplete textures then encountered in mostly pigment powders these scientists considered basic Cu-chloride to be a synthetic pigment - very probably produced through a complex precipitation process by the ancients.

Based on numerous findings and a detailed survey of the inherent pigment layers we were able to prove these former conclusions with respect to the origin of the basic Cu-chloride to be erroneous[5].

A primary green paste of basic Cu-chloride would be expected to consist of a rather coherent pure basic Cu-chloride layer of approximately 0.5 mm in thickness. This paint was never found in Egyptian decorations and there are several unequivocal textural evidences for the secondary origin of basic Cu-chloride and malachite which will be depicted in a series of BSE-micrographs and discussed below.

Typical Cu-chloride-containing paints in the tombs of the Old and Middle Kingdoms, e.g. Nefer (5th Dynasty, Saqqara), Sefechu (6th Dynasty, El Hawawish), Meru Bebi (6th Dynasty, Schech Said) etc., display a variety of textures which are indicative of secondary formation of basic Cu-chloride and malachite as a result of chemical reactions in the pigment layers and the underlying substrate. These sorts of pigment layers consist of Cu- and Cl-bearing glass fragments of about 100 μm in size (Fig. 2 a). The glass grains display advanced devitrification, microcracks (down to one micron wide and even less) which in turn are partially filled by either basic Cu-chloride and/or malachite and most of the glass fragments are encrusted with thin veneers (< 1 μm) of basic Cu-chloride. These are features that cannot be produced by application of a primary basic Cu-chloride paste. The chemical compositions of the residual glass areas are similar to those obtained for Cu-glazes of Egyptian Faiences - e.g. glass composition of the Cu-glass pigment from Nefer's tomb: 45.4 wt.% SiO_2, 36.8 wt.% CuO, 0.22 wt.% CaO, 0.33 wt.% Na_2O, 0.17 wt.% K_2O, 2.97 wt.% Al_2O_3, 1.07 wt.% MgO, 2.06 wt.% Cl. Characteristic features of the residual glass component in Cu-glass pigments suggestive of devitrification and leaching are: its extreme instability under the electron beam, low totals, and very low alkali contents. These facts altogether are suggestive of the presence of migrating solutions in the pigment layers and underlying substrates (for evidence of substrate alterations see below). As a result the original glass phase had changed to a hydrous gel-like material.

The abundances of basic Cu-chloride and malachite in the Cu-glass pigment samples from Nefer's tomb vary significantly within the grains of the same layer between less than 10 vol.% and more than 95 vol.% (Figs. 2 a - d).

The interstitial glass phase in Egyptian Blue and Green Frit pigment grains (Fig. 3) is also liable to devitrification, dissolution and decomposition to secondary basic Cu-chloride (Fig. 4). An Egyptian Blue particle (Fig. 5) in a sample taken in Petosiris's temple in Tuna el Gebel (about 300 B.C.) displays a far advanced deterioration stage. The interstitial glass phase decomposed completely to fluffy aggregates of minute basic Cu-chloride crystals. The grain size is usually below the resolution of the SEM (< 500 Å).

The precipitation of basic Cu-chloride is by no means confined to individual pigment grains. Secondarily precipitated basic Cu-chloride is found to fill the interstices between the pigment grains as well as fissures in the underlying substrate (Fig. 6). In the substrate beneath an Egyptian Blue sample (Merenptah's tomb, 19th Dynasty, Thebes-Valley of the Kings) basic Cu-chloride occurs as pseudomorphs after rhombohedral crystals of calcite depicting all stages between incipient and complete replacement (Fig. 7). These pseudomorphic textures are clearly indicative for postpainting migration of Cu- and Cl-bearing solutions from the Cu-pigment layer into the carbonate-containing substrate in which secondary precipitation of basic Cu-chloride took place. Even very small amounts of precipitated basic Cu-chloride result in visible green discolourations of the stucco.

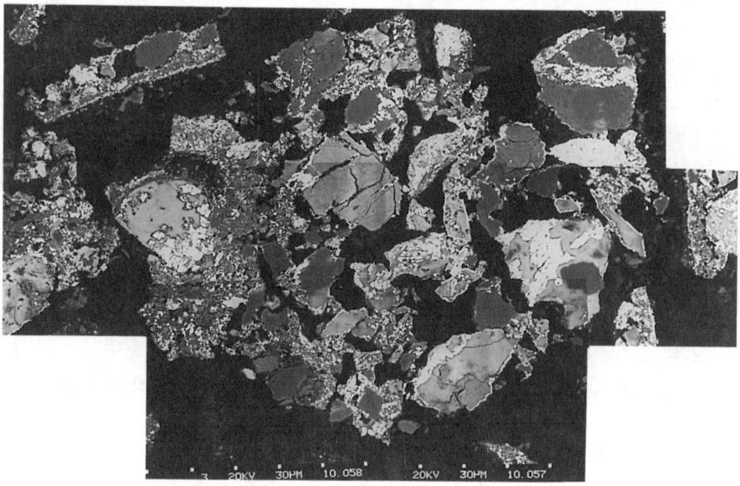

Fig. 2 a: A deteriorated Cu-glass pigment (Nefer's tomb, 5th Dynasty, Saqqara). Grey: heterogenous glass, dark grey: quartz, white: basic Cu-chloride and malachite. For further details see Figs. 2 b - d.

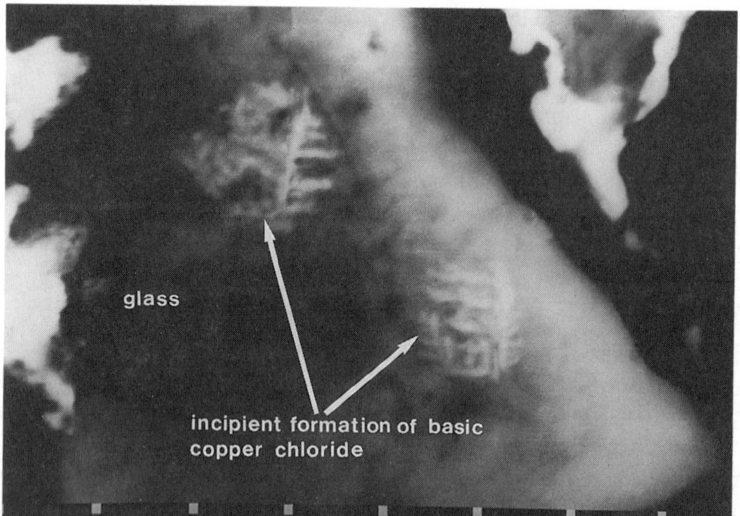

Fig. 2 *b: A detail of the same sample as in Fig. 2 a depicting incipient formation of basic Cu-chloride.*

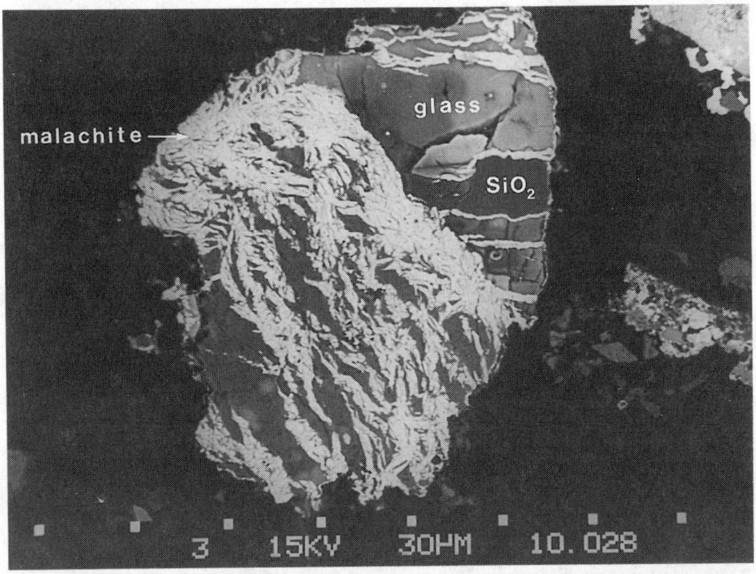

Fig. 2 c: A detail of the same sample as in Fig. 2 a. About half of the glass fragment is replaced by malachite.

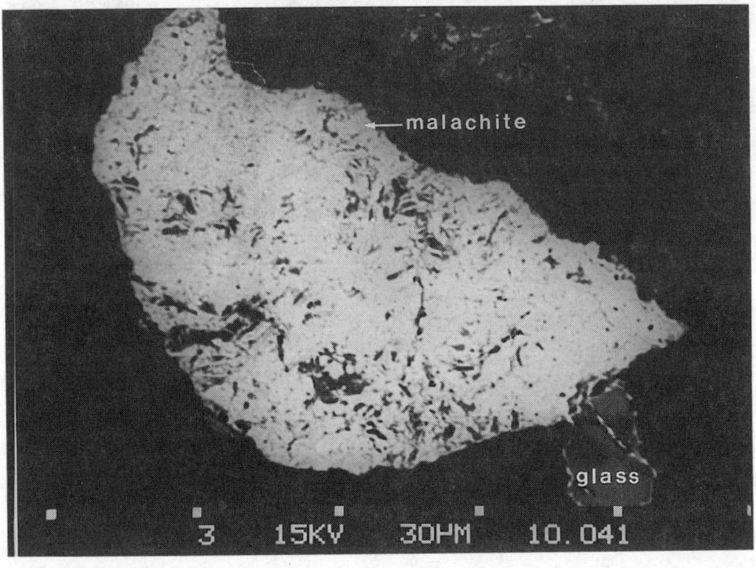

Fig. 2 d: A fragment of the same sample as in Fig. 2 a depicting almost complete decomposition of the glass phase to malachite.

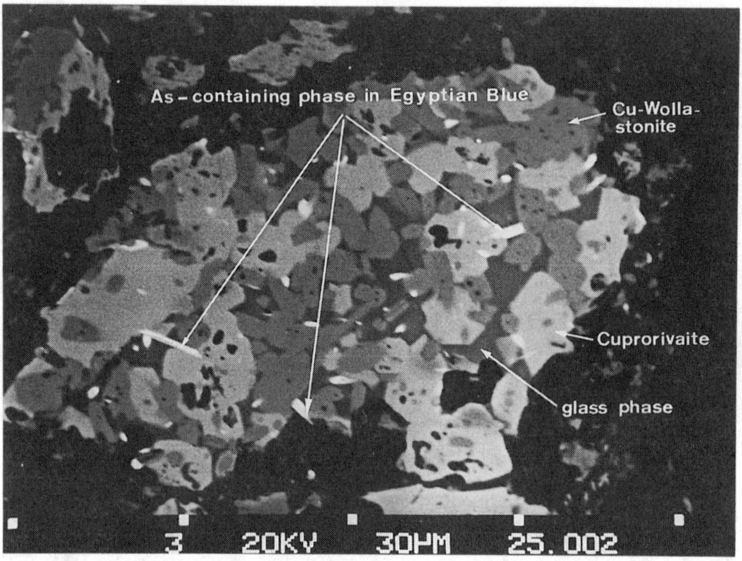

Fig. 3: A detail of an Egyptian Blue sample from Hesi Min's tomb (5th Dynasty, El Hawawish) containing well preserved interstitial glass (dark grey). White: As-bearing phase representing residues of the As-bronze ingredient; bright grey: cuprorivaite, grey: Cu-wollastonite.

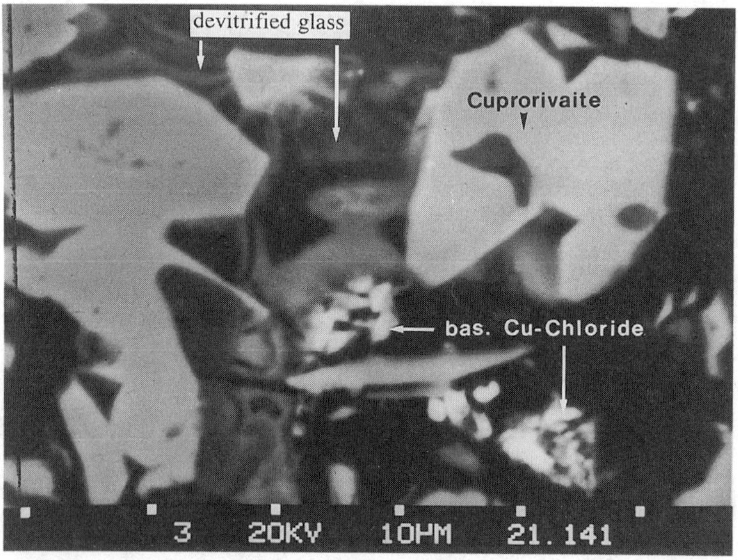

Fig. 4: A detail of Egyptian Blue (Merenptah's tomb, 19th Dynasty, Thebes) displaying devitrified interstitial glass and incipient formation of basic Cu-chloride.

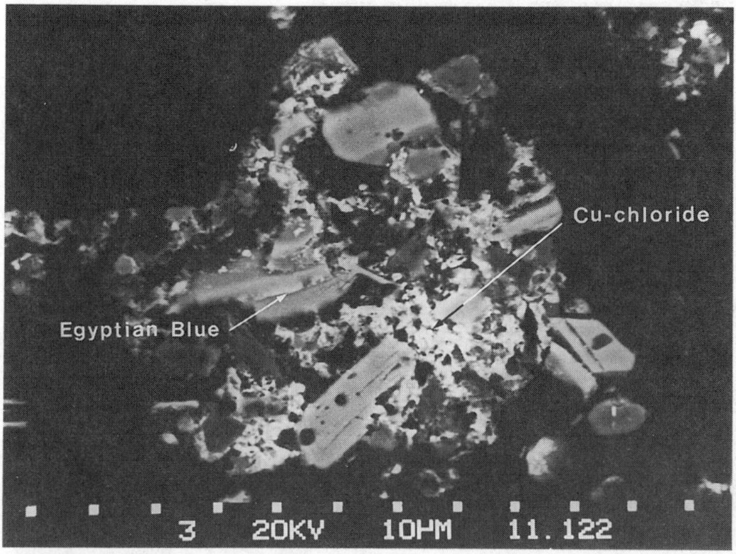

Fig. 5: A decomposed Egyptian Blue fragment (Petosiris's temple, about 300 B.C., Tuna el Gebel). The interstitial glass is completely replaced by secondary basic Cu-chloride.

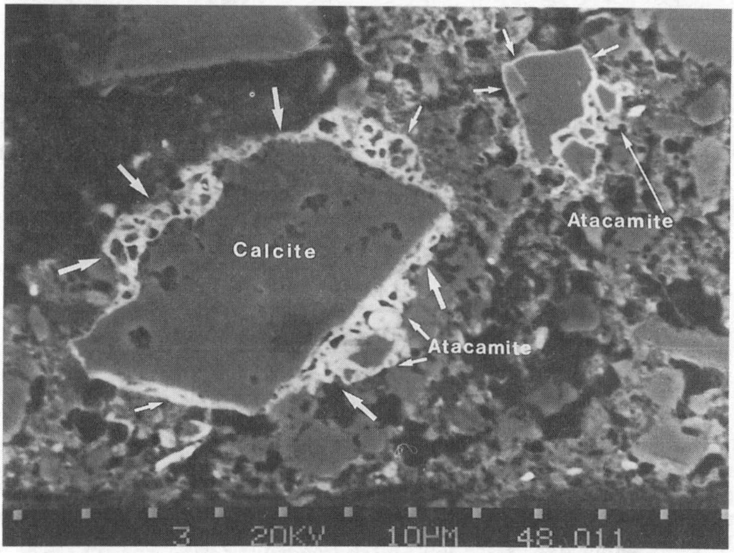

Fig. 6: A detail of stucco (tomb of Tjeti and Kai hep, 6th Dynasty, El Hawawish) in which thin layers of secondary basic Cu-chloride are encrusting the calcite crystals.

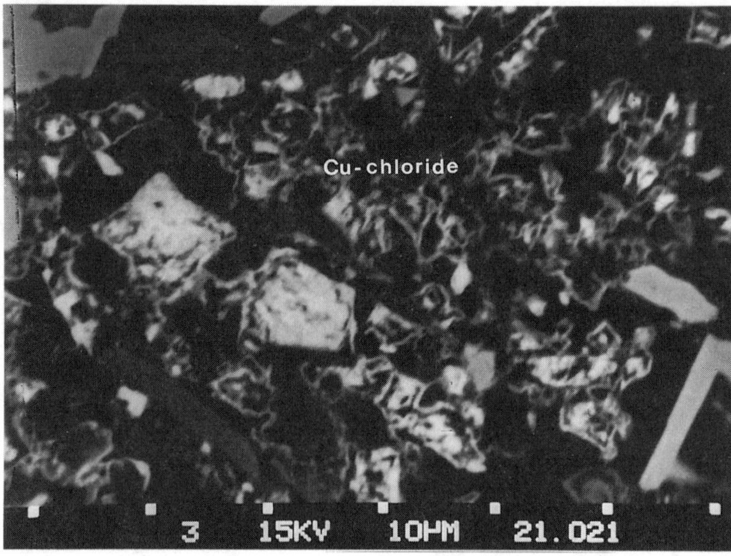

Fig. 7: A detail of substrate (Merenptah's tomb, 19th Dynasty, Thebes) beneath an Egyptian Blue layer. The original rhombohedral carbonate crystals are pseudomorphly replaced by basic Cu-chloride.

A schematic diagram of the successive stages of this deterioration process is depicted in Fig. 8.

Sharkey and Levin[6] experimentally determined the physico-chemical factors governing the formation of the basic Cu-chlorides atacamite and paratacamite. According to their results the precipitation of these phases is favoured if the $CuCl_2$-solutions also contain suspended Ca-carbonate powder.

Thus the substrate layers covering the walls in ancient Egyptian monuments which abundantly consist of calcite or dolomite $CaMg(CO_3)_2$ are excellent media for a runaway reaction resulting in the formation of basic Cu-chloride.

In the course of this deterioration process the originally compact Cu-pigment layer changed to an agglomerate consisting of cuprorivaite and/or Cu-wollastonite, quartz and other SiO_2-phases. Intergranular spaces and interstices primarily filled with glass are now cavities partially filled by tiny clusters of basic Cu-chloride. This kind of alteration results in very spongy and friable textures. Precipitation of basic Cu-chloride can occur within the deteriorating pigment layer, along its surface and in the underlying substrate. In the course of this deterioration process Egyptian Blue and the pale blue Cu-glass pigment change to green, Green Frit becomes paler in shade. As a result of deteriora-

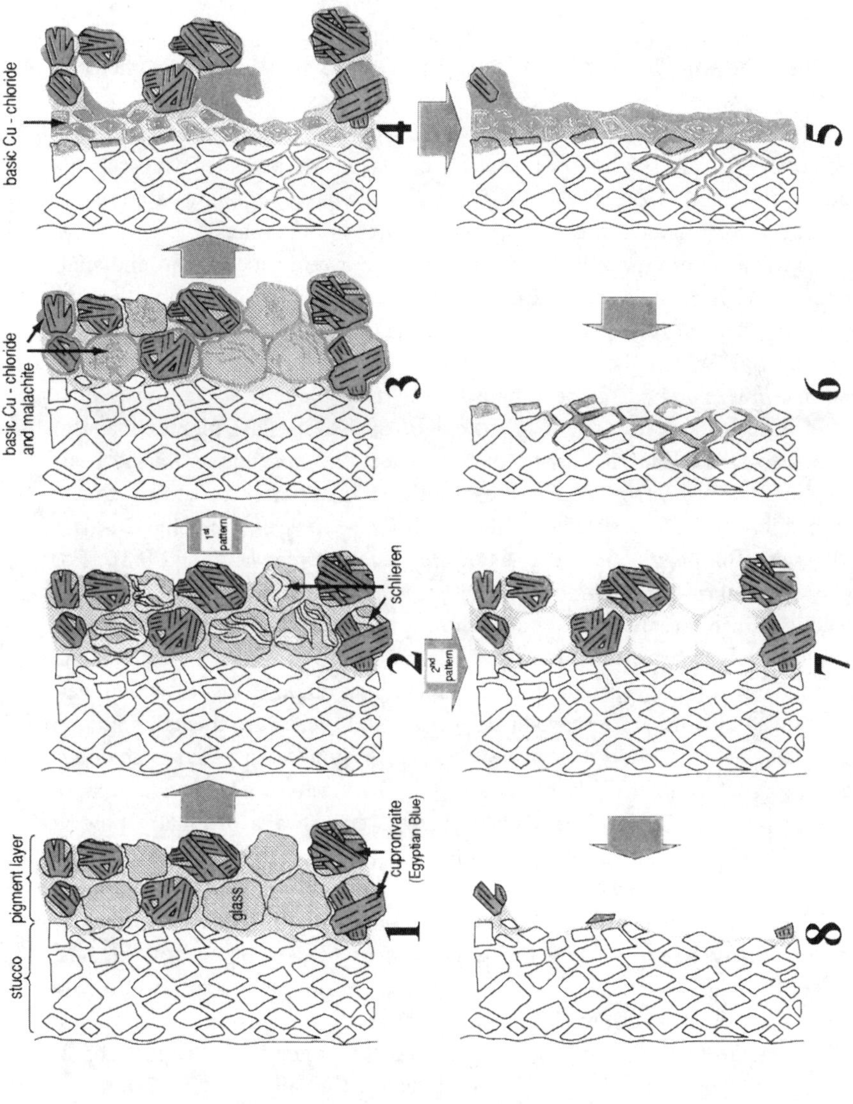

Fig. 8: A schematic diagram depicting the successive stages of formation of secondary basic Cu-chloride (1st pattern, stages 1 - 6) and leaching of Cu-bearing glass (2nd pattern, stages 1,2,7,8).

(See Plate 2 on page vi in the front of this book)

tion and decrease in compactness the pigment layer crumbles. Since green basic Cu-chloride precipitated in the outer part of the stucco this material will be mistakenly believed to be a green primary pigment layer. After spalling off of its altered outer part the stucco is now exposed.

Since the Cu-glass pigments and Cu-glazes are similar in their chemical compositions, we also suspected the Cu-bearing glazes to be very liable to this sort of deterioration process. Green stained patches which are irregularly spread over the dull bluish or turquoise glazed surfaces of several tile fragments collected in Zoser's South Tomb (3rd Dynasty, Saqqara) are considered to be strong indications for secondary formation of basic Cu-chloride. Small fragments from the green areas on the glaze were plucked out and studied with the SEM. Deep cracks and fibrous material which run along the surface and a system of relic glass fibres are clearly indicative of devitrification of the glaze (Fig. 9). On the surface of another fragment (Fig. 10) both devitrification features and partial decomposition to basic Cu-chloride are visible.

In order to ascertain the original colour of the Cu-glass pigments and of Zoser's tiles Dr. E. Hölzl[7] carried out reproduction experiments on the basis of the CuO (wt.%)/SiO_2 (wt.%)-ratios of these materials. According to the electron microprobe measurements the CuO/SiO_2-ratios of the relic Cu-glass vary between 0.43 and 1.63, the ones of the glaze between 0.06 and 0.16. For the glass- and glaze-reproduction varying amounts of Na- and K-bearing fluxes were added since the alkali-oxide concentrations of the original materials were always below 1.0 wt.%, thus indicating severe depletion in these elements. As expected from former reproduction experiments of Cu-bearing Egyptian Faiences[8,9,10,11] also the here reproduced glasses and glazes are intensive blue in colour. Thus we conclude that those tiles which symbolize papyrus mattings were unequivocally intensively blue in origin. Furthermore we conclude that in ancient Egypt pure Cu-bearing silicate glazes were always originally blue in colour and their dull green areas developed in time due to devitrification and precipitation of basic Cu-chloride. A thorough inspection of the tile fragments and of the "blue panel" exhibited in the Egyptian Museum, Cairo, which was reconstructed by Lauer[12] using the same sort of tiles from one of the blue chambers beneath Zoser's Step Pyramid, revealed that numerous had already lost their glaze layer and the white quartz-containing cores are now exposed to the surface. Already in 1938 Lauer expressed the conclusion on the basis of his megascopic inspections that these tiles were originally blue in colour and altered to faint turquoise and green shades as a result of chemical alterations.

According to reproduction experiments the Cu-bearing glass pigments, abundantly used in decorations of the Old and Middle Kingdoms, were very probably pale blue in colour. Obviously there was no green pigment available

Fig. 9

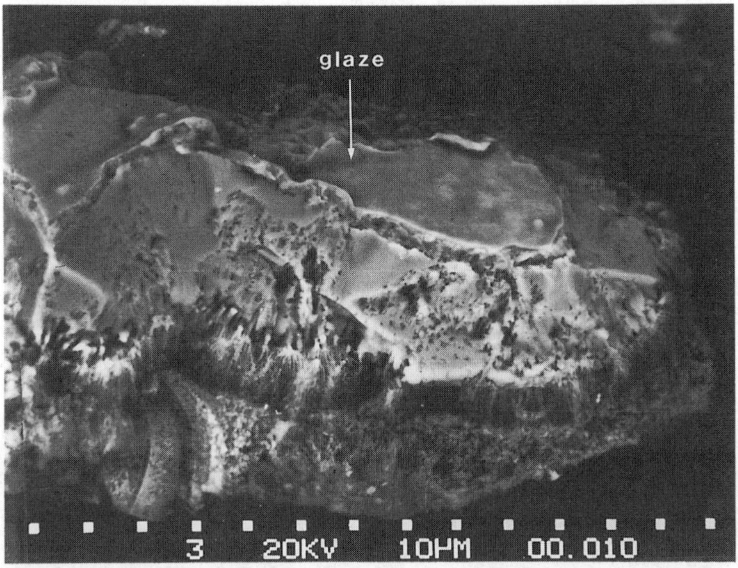

Fig. 10

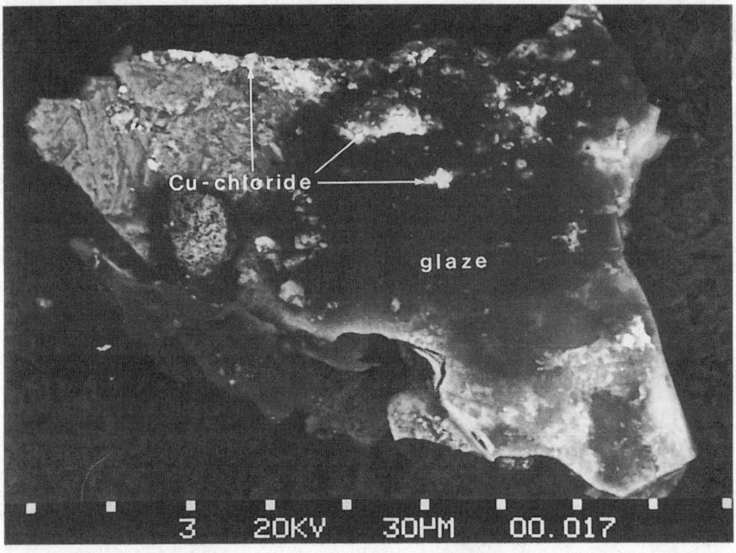

***Figs. 9 and 10:** Particles plucked out of Zoser's tiles from his South Tomb (3rd Dynasty, Saqqara). Fig. 9: Formation of cracks and fibrous material as a result of devitrification of the glaze. Fig. 10: Secondarily formed basic Cu-chloride is spread along the surface of the glaze. In the left upper area also devitrification can be clearly seen.*

to the ancient Egyptians during the epochs preceding the New Kingdom. Instead they used pale blue Cu-glass pigments for painting stems, leaves, and crowns of trees. Since these pigments after deterioration now appear green in colour the logical consequence without further investigation would lead to the wrong conclusion that the green colour is a primary pigment applied by the ancients.

2. Leaching out of the glass in Egyptian Blue and Green Frit

In many samples of Egyptian Blue and less abundantly Green Frit devitrification, leaching and dissolution of the glass phase was observed. However, neither in the pigment layers nor in the underlying substrate of the samples any traces of precipitated basic Cu-chloride were found. Nevertheless, migration of Cu- and Cl-bearing solutions and precipitation of secondary Cu-bearing compounds, e.g. basic Cu-chloride, in deeper zones of the substrate or in neighbouring pigments and painting grounds cannot be excluded.

Fig. 11 displays an individual Egyptian Blue grain (Setau's tomb, 20th Dynasty, El Kab) from which the major part of the interstitial glass had been leached out. Only small amounts of residual glass are still preserved. Fig. 12 depicts a cross section through an Egyptian Blue layer (Gehesa's tomb, 6th Dynasty, El Hawawish) which is full of voids due to dissolution of glass-rich grains. Several pigment grains lost their interstitial glass. As a result of dissolution of the glass component the original compact pigment layer became spongy and friable. Since the colour absorption of the pale blue glass no longer applies the Egyptian Blue layer now consisting mainly of intensive blue cuprorivaite becomes deeper blue in hue. In the course of the advancing deterioration the extremely friable pigment layer starts to crumble and spall off and the substrate is exposed. In this stage the colour changes to pale blue. Fig. 13 is an example for this final stage. A few relic cuprorivaite crystals which are loosely spread on the substrate indicate that the original pigment was Egyptian Blue. Primarily this pigment was certainly as thick as the well preserved Green Frit layer in Fig. 14 - (0.5 - 1 mm). Fig. 8 depicts the successive stages of this deterioration process.

3. Breakdown of Cu-wollastonite

Green Cu-wollastonite is usually a minor component in Egyptian Blue pigments. The amount and size of Cu-wollastonite crystals certainly depend on the $CaCO_3/SiO_2$-ratio of the starting mixture and the cooling rate of the silicate melt. In most Egyptian Blue pigments the wollastonite crystals enclosed in glass phase are filling the interstices between the cuprorivaite crystals (Fig. 3).

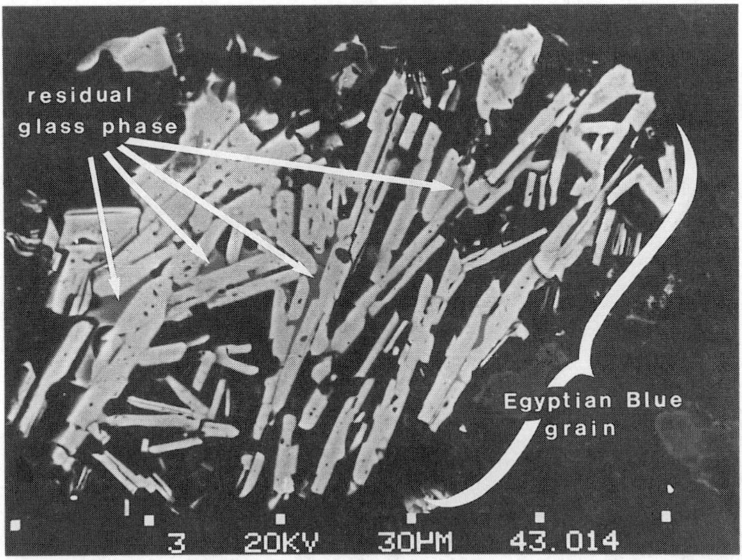

***Fig. 11:** A particle of Egyptian Blue collected in Setau's tomb (20th Dynasty, El Kab) in which the glass phase is leached out to a major part.*

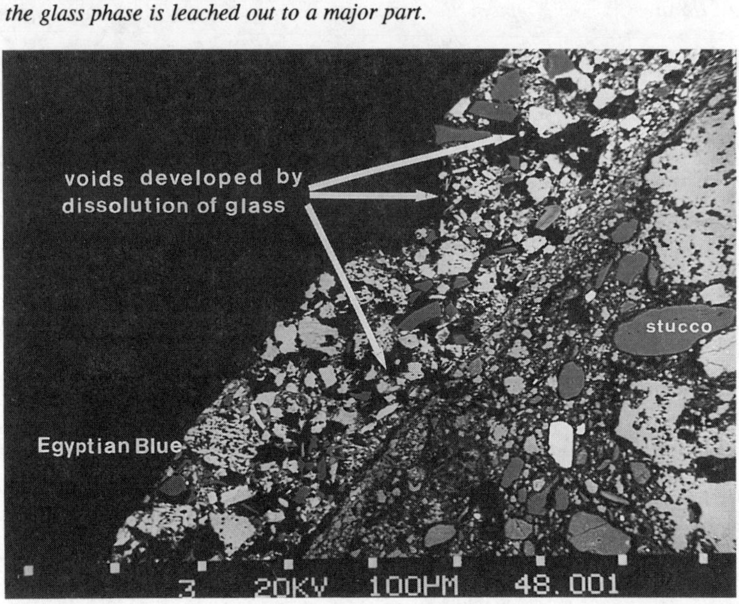

***Fig. 12:** A section through an Egyptian Blue layer (Gehesa's tomb, 6th Dynasty, El Hawawish) which is full of voids as a result of dissolution of glass grains and interstitial glass.*

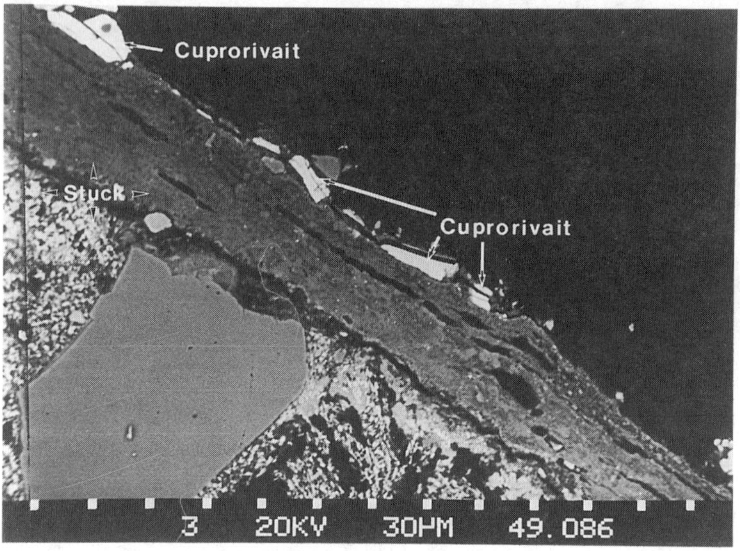

Fig. 13: A section through a sample collected in the tomb of Tjeti and Kai hep (6th Dynasty, El Hawawish). The Egyptian Blue layer had almost spalled off from the stucco.

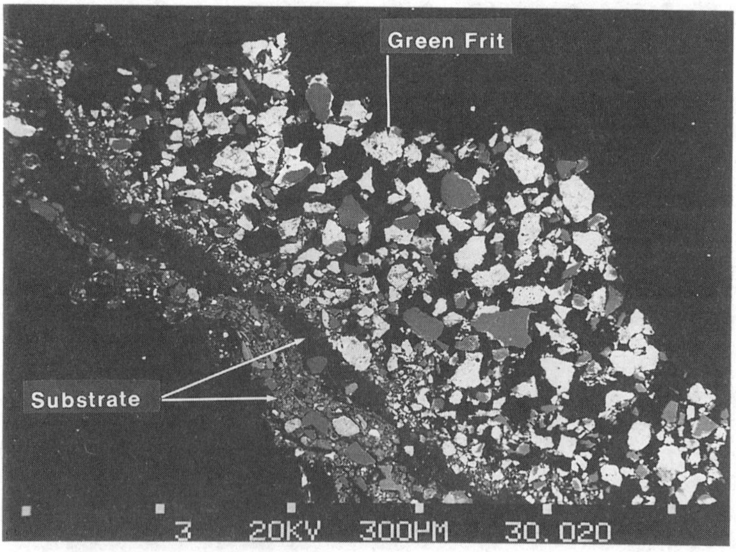

Fig. 14: A section through a well preserved Green Frit layer (Nebamun's tomb, 18th Dynasty, Thebes) depicting a thickness up to 0.6 mm.

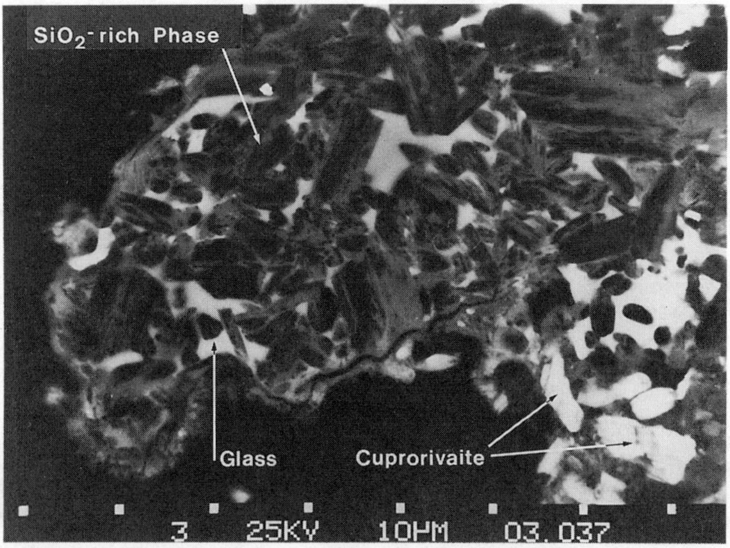

Fig. 15: A fragment originally rich in Cu-wollastonite in Egyptian Blue (Achethotep's tomb, 5th Dynasty, Saqqara). This texture reveals decomposition of wollastonite to SiO_2-rich fribrous material filling the cavities which still depict the outlines of the wollastonite crystals.

Wollastonite is unstable under elevated CO_2-pressures. High CO_2-concentration leads to its breakdown to $CaCO_3$ and silica according to the modified reaction:

$$Ca_3[Si_3O_9] + 3\ CO_2 \rightarrow 3\ SiO_2 + 3\ CaCO_3$$

Elevated CO_2-pressures in tombs can be generated by visitors as well as by CO_2 escaping from open voids and fissures of the carbonate wall rocks.

Fig. 15 depicts a detail of an Egyptian Blue sample which was originally enriched in green Cu-wollastonite. The breakdown of the wollastonite crystals resulted in the formation of cavities now containing residues of SiO_2-rich fibres. The outlines of these cavities still depict the shape of the former wollastonite crystals. Fig. 16 taken from an Egyptian Blue sample from the Amarna Period depicts the complete sequence of stages from still preserved to completely decomposed Cu-wollastonite. Due to the breakdown of green Cu-wollastonite Egyptian Blue pigments become stronger blue in shade.

Especially wollastonite-rich pigments are severely affected by this deterioration process. Their textures get spongy and friable and are liable to spall off from the painting ground.

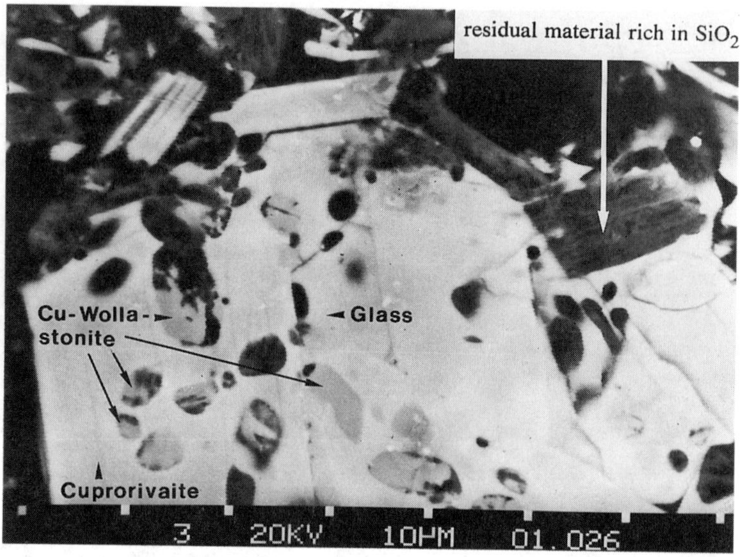

Fig. 16: A detail of Egyptian Blue collected in the tomb of Merire II. (18th Dynasty, Amarna). The whole sequence from preserved wollastonite to completely decomposed wollastonite can be observed.

4. Precipitation of secondary Ca-arsenate in the painting ground

The ancient Egyptians used bronze scrap as Cu-ingredients for the production of their synthetic Cu-pigments. During the epochs preceding the New Kingdom the bronze consisted of Cu-As-bearing alloys. In the 18th Dynasty tin-bearing bronze was introduced and since then the As-Cu-alloys were more and more replaced by Cu-Sn-alloys. The development of the bronze technology in ancient Egypt is exactly reflected by the chronological distribution pattern of As-, Sn and Pb-bearing relict phases in the Egyptian Blue and Green Frit pigments of the various epochs[13]. Fig. 3 displays a particle of Egyptian Blue which was produced in the Old Kingdom and thus contains a considerable amount of tiny Ca-arsenate-phosphate crystals. In the course of the deterioration of As-bearing Egyptian Blue layers the primary relic Ca-arsenate-phosphate crystals are dissolved and As- and P-bearing solutions migrate into the substrate in which then precipitation of secondary Ca-arsenate-phosphate takes place. An example of this process is depicted in Fig. 17. Typically this pigment layer is decomposed to a great extent. Thus it is very poor in glass phase and full of voids. A considerable amount of secondarily precipitated Ca-arsenate-phosphate in the substrate presumably

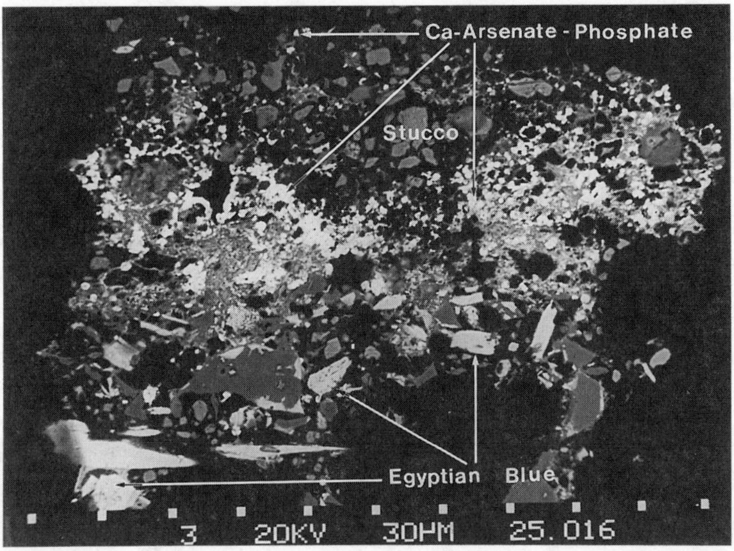

Fig. 17: A section through decomposing Egyptian Blue and underlying substrate (Ahanacht's tomb, 11th Dynasty, Deir el Bersheh). As a result of deterioration of the As-bearing pigment layer secondary Ca-arsenate-phosphate precipitated between pigment and substrate.

accelerates the spalling off of the residual components of the pigment layer as a result of the force of crystallization.

5. Decomposition of Fe-glass pigments to jarosite-rich painting layers

In earlier reports on ancient pigments[14,3,4] rare findings of jarosite-containing layers in wall paintings and on painted ceramics were considered to be intentionally applied primary pigments. However, SEM-investigations of several yellow and red pigmented layers collected from Old, Middle and New Kingdom decorations revealed textures suggestive of advanced decomposition of primary Fe-, K- and probably S-bearing glass pigments to aggregates of yellow or red jarosite crystals $KFe_3[(OH)_6/(SO_4)_2]$. An intact original pigment layer consisting of Fe-glass fragments was not yet found in any sample. However, the primary condition of this pigment can be reconstructed from the textures of decomposed jarosite-containing paints. The grains in Figs. 18 and 19 display a less deteriorated stage. Fig. 18 displays a fragment of a red pigmented layer collected from a wooden figure (Middle Kingdom, Assiut) of the Roemer- und Pelizaeus-Museum, Hildesheim (Germany). Initial formation of

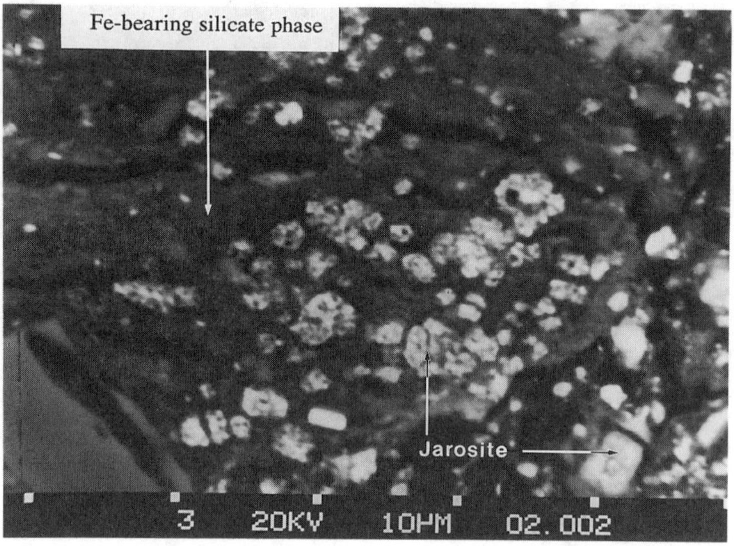

Fig. 18: Incipient formation of jarosite in a fragment consisting of Fe-, K- and S-bearing silicate phase. Detail taken from a red pigmented sample from a wooden figure (Middle Kingdom, Assiut).

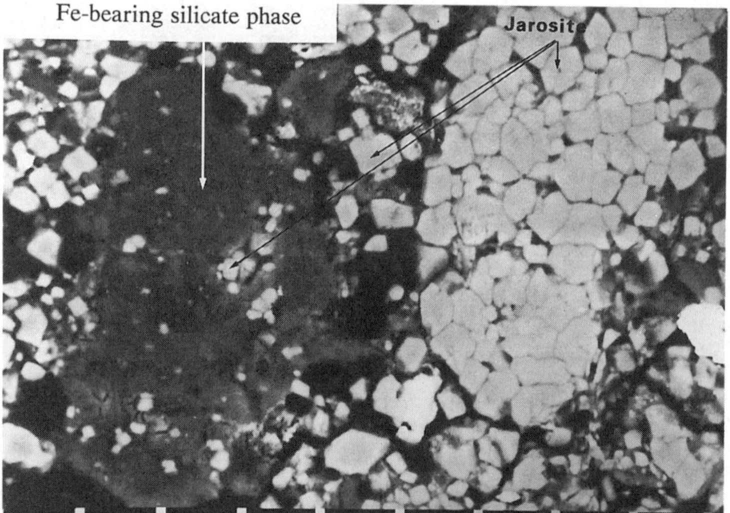

Fig. 19: A detail of a yellow pigmented paint collected in Ahanacht's tomb (11th Dynasty, Deir el Bersheh). The left hand fragment displays incipient formation of jarosite, whereas the right hand fragment is completely decomposed to jarosite. In the course of deterioration jarosite was also formed in the spaces between the former glass fragments.

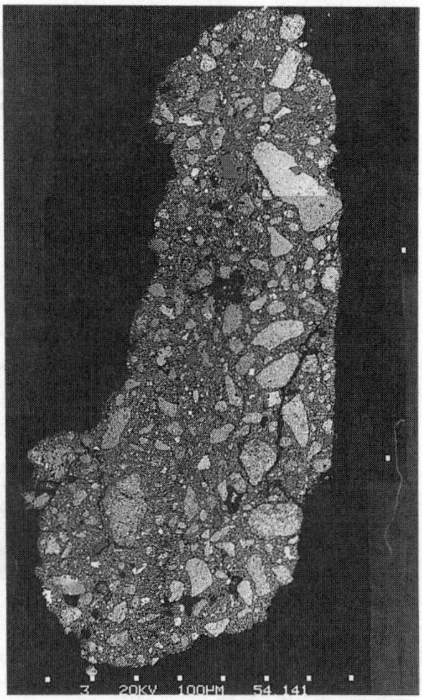

Fig. 20: A mosaic of a series of BSE-micrographs depicting a sample of a yellow inlaid paste of Hemiwnw's statue (4th Dynasty, Gizeh). Aggregates of secondary jarosite still display the outlines of the deteriorated Fe-glass fragments. The intergranular spaces are also filled with jarosite.

idiomorphic jarosite crystals within an iron-, potassium-, and sulphur-bearing silicate phase is visible. A similar stage is found in the left hand grain in Fig. 19, whereas the right hand fragment is completely decomposed to a compact agglomerate of individual jarosite crystals. In the interstices of the original fragments less jarosite crystals were formed and consequentially due to the crystallization freedom they are more idiomorphic than the ones in the fragments. Complete decomposition off all glass fragments (still visible as ghost fragments) to irregularly shaped aggregates consisting of secondary jarosite is shown in Fig. 20 and in detail in higher magnification in Fig. 21. Within the fragments very small areas of Fe-bearing silicious material can be observed. Since the deterioration process is rather advanced in all samples studied and major amounts of jarosite are formed, none of the samples still depict its original colour. We can only assume that these Fe-glass pigments were originally green or brown depending on the oxidation state of iron in the glass. However, we assume they consisted of green glass fragments since the ancients had access to natural brown ochre pigments. The production of green pigments and their availability were obviously problematic in the Old and Middle Kingdoms. Green Frit was not introduced before the New Kingdom.

6. Formation of secondary Ca-phosphate-containing layers

This deterioration process befalls both natural and synthetic pigments. The aggressive phosphate-bearing material penetrates the painted decorations from their surfaces inwards to the substrate (see below). It replaces the pigment particles and carbonate crystals (calcite and dolomite) in the painting

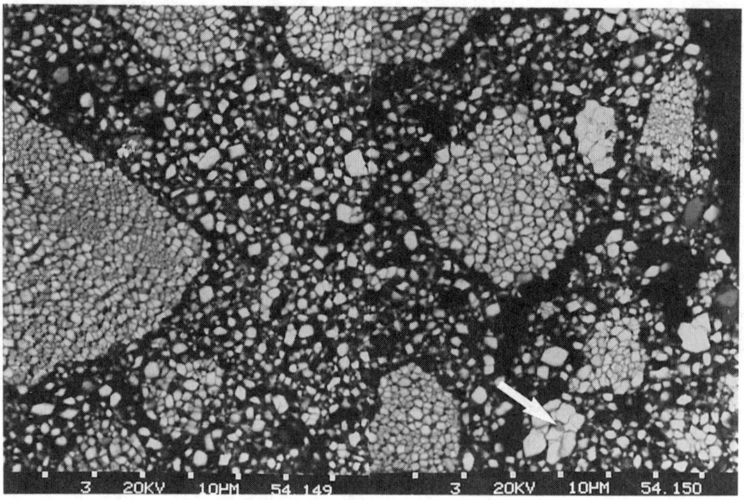

Fig. 21: A detail of Fig. 20 with higher magnification. The individual fragments contain to a major part jarosite. Only a few minute areas of interstitial Fe-bearing silicate phase (arrow) can be observed.

layer and stucco. Repeated addition of such phosphatic material leads to the formation of successive phosphate layers in which pseudomorphs of carbonate rhombohedra are usually well preserved. The secondary material usually has its own colour hue and after complete removal of the primary chromatic layer it is then mistakenly identified as primary material. This process was recognized in tombs in several sites: e.g. El Hawawish, Mo^{c}lla, Deir el Bersheh, Edfu, Thebes West, and Karnak. Replacement of the pigment layers may lead to spongy or fluffy pigment layers. Spongy layers result from leaching out of carbonate crystals thus leaving delicate skeletons of negative carbonate rhombohedra as found in a pigment layer from King Ay's tomb in Thebes (Fig. 22). Due to their fluffy nature such secondary layers crumble in short periods of time. In another case from the same tomb the replacement and repeated precipitation of phosphatic material leads to the formation of coherent and compact layers in which the pseudomorphs of carbonate rhombohedra are well preserved (Fig. 23). Due to their compactness this kind of secondary phosphate layers are resistant to crumbling.

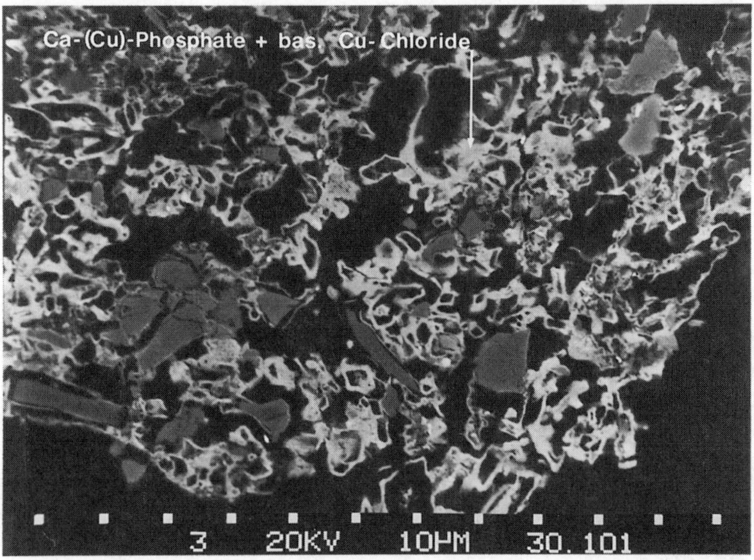

Fig. 22: A fluffy layer consisting of secondary Ca-phosphate and basic Cu-chloride. The Ca-phosphate and basic Cu-chloride enveloped preexisting idiomorphic carbonate rhombohedra which now consist of empty negative forms due to selective leaching of the carbonate crystals (Ay's tomb, 18th Dynasty, Thebes).

Another variety of replacement textures was observed in a red paint layer from Ahanacht's tomb (11th Dynasty, Deir el Bersheh). It consists of alternating compact and undisturbed layers rich in Ca-phosphate (Fig. 24).

From these textures the secondary nature of the pigmented layers is clearly evident. From P-rich solutions Ca-phosphates precipitated on the surface of the stucco. More complex reactions lead to dissolution of primary Cu-bearing pigments and to secondary precipitation of basic Cu-chloride and Ca-phosphate from solutions enriched in Cu, Cl and exogenous phosphorous. Wilson-Young and Burns[15] discussed the possibility that bat droppings, e.g. on the walls in Beni Hasan tombs, are the sources of the phosphorous. Our results agree fully with their conclusion and delineate the various stages of the damage.

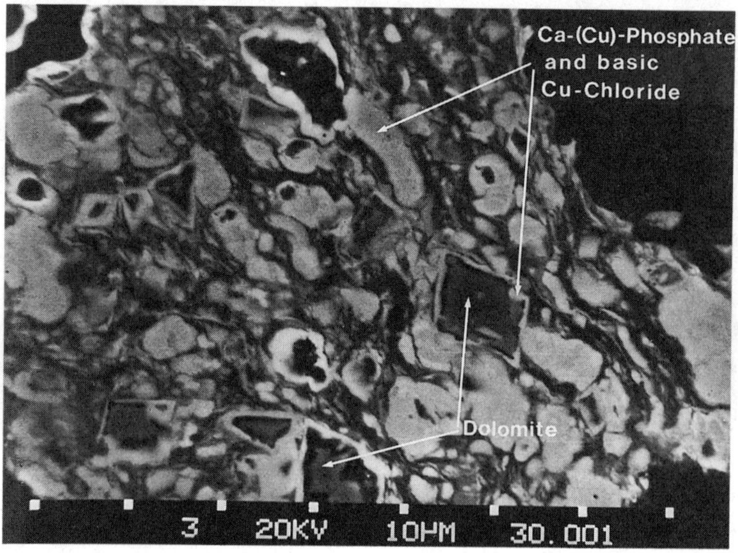

Fig. 23: A detail of a green secondary layer of a sample collected in Ay's tomb (18th Dynasty, Thebes). The whole sequence of partially to complete replacement of carbonate crystals by Ca-phosphate and basic Cu-chloride can be observed. The irregular-shaped phosphate particles very probably replaced former Ca-sulfates which are found in unaltered stucco.

CONSEQUENCES

1. Consequences for conservation and restoration strategies

The reported deterioration processes which have befallen ancient Egyptian painted decorations resulting in chemical and physical alterations of the pigment and substrate layers are multifold and were not known before. Therefore a thorough study of polished sections of the coherent complete sequence of layers using reflected light microscopy, SEM, and electron microprobe techniques to identify primary pigment components and any secondary deterioration products is an indispensable prerequisite before any restoration or conservation plans are developed. The electron microscopic investigation of pigment powders does not lead to reliable results since invaluable information about primary and secondary textural features are lost by the mechanical destruction of the layers. It is mandatory to take a coherent sample comprising of the whole pigment layer sequence and the underlying stucco in order to have

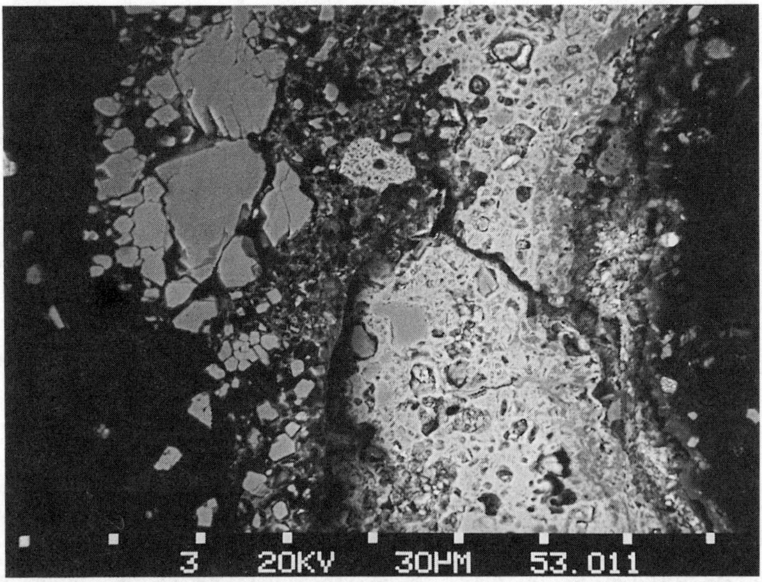

Fig. 24: A pigment layer of previously unknown nature now replaced by secondary phosphate. The texture of this sample (Ahanacht's tomb, 11th Dynasty, Deir el Bersheh) is suggestive of postpainting precipitation of alternating layers of complex Ca-phosphates.

the whole record of the painting technique and the deterioration processes in all their stages.

Commonly X-ray diffractometry is used for identification of pigment samples. This technique alone is obsolete since it can not distinguish between primary and secondary components. Basic Cu-chloride and jarosite were considered erroneously primary pigments as a result of a sole application of this technique[1,2,14].

Deterioration processes also continue after dislocation of decorated fragments from their source monuments to museums and collections. The damage can be seen in many materials on exhibit in the Egyptian Museum in Cairo and in European collections.

Adequate restoration and conservation treatments have yet to be developed through laboratory testing experiments using less valuable original fragments of painted artefacts. No restoration or conservation treatment should be undertaken before the effect of individual treatment procedures on the further development of every deterioration process is experimentally determined.

Conservation and restoration leading to stabilization of the secondary deterioration materials without detailed knowledge of the identity of their primary and secondary components and the stage of the deterioration process will lead to disastrous consequences: in the worst case complete spalling off of the "restored" areas - e.g. Seanchwptah's tomb in Saqqara was discovered with full decoration in 1984, then already in 1989 the decorations were found to be destroyed after "conservation"[16].

No cleansing of the painted decorations regardless whether they consist of primary or secondary components with acids, inorganic or organic solvents or even water should be undertaken till we know through laboratory experiments their effects on the intact and deteriorated pigment layers.

All deterioration processes which befall the pigments and substrates are irreversible. Thus we have to seek for procedures to stop these processes and adhere the chromatic material (primary or secondary) to their substrate with chemically inert adhesives.

2. Consequences to the Egyptological colour symbolism

The here reported chemical reactions lead to formation of secondary materials of different colour hue than in the primary glazes and pigments. Consequently, many decorations display now colours which are entirely different from those intended and originally applied by the ancient artists. Especially there is always the danger of incorrectly considering green chromatic materials as primary pigments.

Green chromatic materials in the decorations of the epochs preceding the New Kingdom consist in all cases of basic Cu-chloride and malachite and were formed by deterioration of Egyptian Blue or pale blue Cu-glass pigments. This does not necessarily mean that the ancients had entirely no access to green pigments during these periods. At that time the artists also used (though less frequently) an iron-bearing glass pigment which was originally probably green in colour and changed to yellow or red as a result of decomposition to jarosite. It seems that this pigment was only applied for specific objects. Stems of plants or crowns of trees were not necessarily painted in green. Instead the ancients frequently used a pale blue Cu-glass pigment for this purpose. Due to secondary formation of green basic Cu-chloride the colour of Green Frit changed to bright green. Also the green discolourations of blue Cu-bearing Faience glazes of the Old and Middle Kingdoms result from decomposition to basic Cu-chloride. Therefore the Egyptological colour symbolism needs a thorough revision.

A list of all the monuments in Egypt whose decorations were found to be befallen by deterioration processes was given in Schiegl et al.[17], and Schiegl[18].

ACKNOWLEDGEMENTS

We are deeply indebted to the Volkswagen-Stiftung for the generous financial support to conduct a detailed research program on ancient Egyptian pigments in time and space. We are grateful to the Egyptian Antiquities Organization (EAO) specifically to its former Presidents Ahmed Kadry† and Abdel Halim Nour El Din and to the following Directors, officials, and inspectors: M. El Sughair, M. Saleh, M. Nasr, Ismail El Masri and other colleagues at various sites in Egypt. We also enjoyed constructive discussions with Ali Radwan Dean of the Faculty of Archaeology, Cairo University. We are also grateful to Shawki Nakhla, Director of the Conservation Centre of the EAO. This work has benefited from discussions with E. Martin-Pardey, I. Blom, and A. Eggebrecht. E. Martin Pardey laid the sampling plan and participated in the field campaign in March 1988. We also would like to express our gratitude to H. Lämmler and F. Künstler for the excellent preparation of hundreds of samples.

REFERENCES

1. E.L.B. Terrace, Egyptian Paintings of the Middle Kingdom, (Braziller, New York, 1968), p. 167-168.
2. J. Riederer, Archaeometry 16, 102 (1974).
3. H. Jaksch, Ph.D. dissertation, Universität Heidelberg, 1985.
4. A. El Goresy, H. Jaksch, M. Abdel Razek and K.L. Weiner, Ancient Pigments in Wall Paintings of Egyptian Tombs and Temples, (Max-Planck-Institut für Kernphysik, Heidelberg, 1986) ; in Interdisziplinäres Gespräch edited by H. Trenkwalder, F. Purtscheller, W. Lukas and H. Seidl (Studia, Innsbruck, 1987), p. 1-57.
5. S. Schiegl, K.L. Weiner and A. El Goresy, Naturwiss. 76, 393 (1989).
6. J.B. Sharkey and S.Z. Lewin, Am. Mineral. 56, 179 (1971).
7. E. Hölzl (private communication).
8. A. Lucas revised by J.R. Harris, Ancient Egyptian Materials and Industries, 4th ed. (Edward Arnold Ltd, London, 1962), p. 172-174.
9. H.E. Wulff, H.S. Wulff and L. Koch, Archaeology 21, 98 (1968).
10. C. Kiefer and A. Allibert, Archaeology 24, 107 (1971).
11. P.B. Vandiver, in Ancient Egyptian Faience, edited by A. Kaczmarzyk and R.E.M. Hedges (Arris & Phillips Ltd., Warminster, England, 1983), Appendix A.
12. J.-P. Lauer, Annales du service des antiquités de l'Égypte 38, 551 (1938).
13. S. Schiegl, K.L. Weiner and A. El Goresy, Erzmetall 43, 265 (1990).
14. W. Noll, Fortschr. Min. 57, 203 (1979).

15. K.M. Wilson-Yang and G. Burns, Can. J. Chem. 66, 2348 (1988).
16. A.G. Shedid (private communication).
17. S. Schiegl, K.L. Weiner and A. El Goresy, Geowiss. 9, 199-209 and 241-246, 1991.
18. S. Schiegl, Ph.D. dissertation, Universität Heidelberg, 1991.

LOSS OF INFRARED LINEAR DICHROISM IN COLLAGEN FIBERS AS A MEASURE OF DETERIORATION IN SKIN AND SEMI-TANNED LEATHER ARTIFACTS

Gregory S. Young
Canadian Conservation Institute, Department of Communications, Ottawa, Canada K1A-0C8

ABSTRACT

Polarized radiation is used in a Fourier transform infrared (FTIR) microspectroscopic technique to detect molecular conformational change in collagen fibers resulting from deterioration. Absorbance spectra of single fibers are produced with polarization oriented first parallel ($A_{\parallel}$) and then perpendicular ($A_{\perp}$) to the fiber axis. The amide III band (1230 - 1300 cm^{-1}) shows strong positive dichroism, with up to six times greater absorbance occurring with parallel polarization. Losses of dichroism result from breakdown in the molecular structure of collagen. Integrations of the amide III bands are used in ratios, $A_{\parallel}/A_{\perp}$, which provide a measure of the degree of molecular order. Ratios for degraded and undegraded material are compared to produce semi-quantitative information about deterioration.

INTRODUCTION

An infrared microspectroscopic method is introduced which probes for changes in molecular conformation to detect deterioration in collagenous materials. The method analyzes single collagen fibers, and, in doing so, provides essentially nondestructive analysis of museum artifacts. It has the potential for providing comparatively rapid analysis in studies of conservation processes, and may also provide detection of environmental effects before changes occur in macroscopic physical properties of artifacts. Although discussed here in the context of ethnographic skins and semi-tanned leathers, the technique should have application to other types of untanned and tanned skin and hide.

Infrared microspectroscopy of molecular conformation

Artifacts composed of untanned and semi-tanned skins derived from mammalian species consist primarily of fibers of type I collagen. The fibers may differ moderately in the quantity of non-collagenous, non-fibrous constituents among species, but the fiber content differs very little in molecular structure [1-3]. In analysis therefore, the structure in fibers of one species is comparable with that in fibers of other species.

The collagen molecule is comprised of three peptides and is almost entirely in the form of a triple helical rod. The rods are laterally assembled in a staggered manner to produce semi-crystalline fibrils of indeterminate length with a common orientation of the molecular axes parallel to the fibril axis. A multitude of fibrils compose a collagen fiber, with the fibrils axially aligned parallel with the fiber axis.

This single, preferred axial orientation of the structural components in fibers produces optical properties which resemblesthose of uniaxial crystals [3-5] and manifests in infrared spectra obtained with linearly polarized radiation.

Infrared absorbance band intensity is dependent, in part, on the degree of coincidence between the direction of the transition dipole moment of the molecular vibration causing the band and the direction of polarization (electric field vector) of the incident radiation [6-8]. By monitoring absorbance intensity, polarized radiation can be used to determine the degree to which a preferred, non-random orientation occurs [6] for specific bonds in the collagen molecule. In this study, polarization was oriented first parallel and then perpendicular to the axis of collagen fibers, and the change in absorbance intensity was compared to produce a semi-quantitative measure of molecular order.

The deterioration of proteinaceous fibers can be grouped into five categories [9], including- physical ageing [10], mechanical stress, and photochemical [11], thermal and chemical [12] degradation. While the mechanisms of degradation are variable among these five, all are expected to alter the native molecular conformation of collagen and, therefore, the dichroism of collagen fibers. Deterioration also causes a loss in the thermal stability of collagen [13], as evidenced by lowered denaturation temperatures in fibers [14] and changes in other thermodynamic properties [15]. This initial study of infrared dichroism was undertaken to investigate the disruption of molecular conformation with losses in thermal stability.

EXPERIMENTAL

Samples

Collagen fibers were excised from— (1) the corium of a deteriorated seal skin sample and an undeteriorated, stretched and dried (but otherwise unaltered) sample of seal skin, (2) the surface of a partly photodegraded segment of seal skin thong, (3) the surfaces of an undeteriorated sample of brain- and smoke-tanned deer skin (manufactured c.1935) and a similarly tanned but also photodegraded deer skin sample [16], and (4) the corium of a sample of brain-tanned deer skin. Microforceps and micro-surgical scissors were used. Care was taken to avoid applying tensile stress to the fibers and to avoid fraying them.

Analytical procedure

Collagen fibers exhibit molecular symmetry in all directions perpendicular to the fiber axis and, thus, exhibit the same infrared absorbance coefficients for unpolarized radiation transmitted in these directions. Consequently, all radial orientations are equally acceptable for analysis using polarized radiation, with the one provision that the fiber axis be oriented perpendicular to the incident radiation. This was automatically achieved by the two preparation methods used in the analyses. Fibers were placed flat in a film of nujol on a single barium fluoride (BaF_2) window or were washed in distilled water and placed between two BaF_2 windows in additional water [17, 18]. A slight momentary pressure was applied to press the two windows against the fibers. The windows supporting the test fibers were then placed flat onto the specimen stage of an infrared microscope.

A Spectra-Tech Inc., IR-Plan Research Microscope accessory was used. The microscope was equipped with a narrow band MCT (mercury-cadmium-telluride) detector and rotatable stage, and was coupled to a Bomem MB120 infrared spectrometer. The infrared was polarized using a zinc selenide wire grid, placed between the globar (silicon carbide) infrared source and the collagen fiber sample. Spectra were collected at a resolution of 4 cm^{-1}, with 100 to 400 coadded scans, using a 15X reflecting Schwarzchild-type Cassegrainian condenser lens and a similar 10X objective lens. The effective numerical aperture of the microscope optics was 0.58.

Fibers were centered in the microscopical field [19] and brought into parallel alignment with the infrared polarizer. Each fiber portion analyzed was isolated using two adjustable remote apertures [20]. The window created by the apertures was slightly smaller than the fiber width to minimize the amount of stray radiation reaching the detector [20,21] and, thus, minimize the effect on the dichroism [22]. The window was left stationary during analysis, and the polarizer was rotated to produce the parallel and perpendicular spectra. Reference spectra were collected under the same instrumental conditions, but in the absence of a fiber.

Effect of minor misalignment on measurements of dichroism

Most collagen fibers exhibit some amount of crimping and misalignment of constituent strands. Control studies (**Figure 1**) in which fibers were repeatedly analyzed at progressively greater misalignment revealed that the dichroic measurements were conveniently insensitive to minor misalignment [23], in the range of approximately +/- 8° from parallel alignment. Fibers to be analyzed were therefore examined with polarized light optics to ensure that crimping and other misalignment were within this limit.

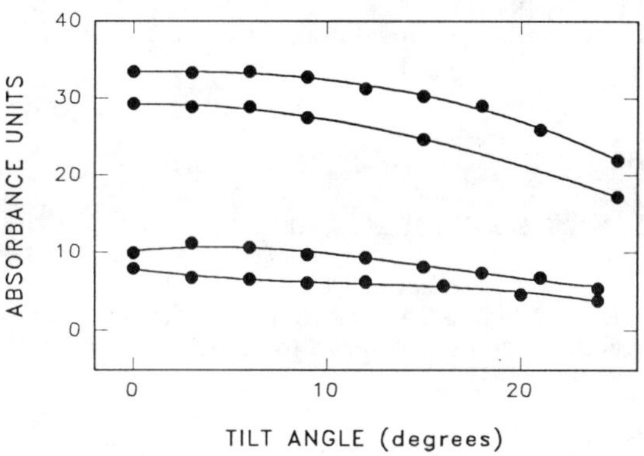

Figure 1 Dichroism versus misalignment (tilt angle)

Effect of instrumental polarization on measurements of dichroism

Even in the absence of an infrared polarizer, the IR radiation may become partially polarized as a result of distortion by the beam splitter in the interferometer and sometimes by aspherical mirrors used to direct the radiation through the spectrometer and microscope [22, 24]. Radiation throughput can therefore be dependent upon the orientation of the polarizer in the system. With the system used in the present experiments, there was a negligible change in intensity of the interferogram of about 2% when the established parallel and perpendicular directions of the wire grid polarizer in the microscope were compared in the absence of a collagen fiber sample [22].

The effect on fiber dichroism was determined for this amount of instrumental polarization. The established method of data collection, whereby the polarizer is rotated to obtain the parallel and perpendicular spectra, was compared with a second method, whereby the fiber sample is rotated and the polarizer is stationary. For the second method, the orientation of the polarizer in the spectrometer was determined at which a maximum of transmittance was produced in the absence of a sample. This orientation was maintained and the fiber was rotated during analysis. When the two methods were compared, no difference in the dichroic ratios was revealed.

Microscopic Shrinkage Temperature Measurements

Collagen fiber samples were prepared and tested in a manner similar to that described by Young [14]. Fibers were soaked in distilled water for approximately ten minutes, then heated in excess water at a rate of 3° C/min in a microscopical hot stage. The temperature range over which shrinkage occurred was recorded.

RESULTS AND DISCUSSION

Dichroic spectra of skins and semi-tanned leathers

Two infrared spectra of an undeteriorated fiber immersed in a nujol film are shown in **Figure 2A**, one produced with the incident radiation polarized parallel and the other perpendicular to the fiber axis. Most of the major absorbance bands in

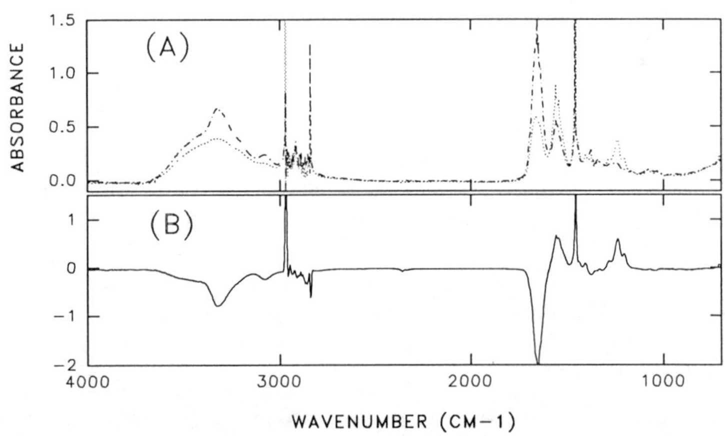

Figure 2 (A) Parallel polarization,dottedline; perpendicular polarization, dotted-and-dashed line. (B) Partial dichroic spectrum (parallel minus perpendicular).

these spectra result from the vibrational modes of amide functional groups [2,3] along the peptide chains of the coiled triple helical molecule of collagen: Amide A (3320 cm^{-1}); Amide B (3080 cm^{-1}); Amide I (1655 cm^{-1}); Amide II (1552 cm^{-1}); Amide III (1245 cm^{-1}). **Figure 2B** is the partial dichroic spectrum of the spectral pair in **2A**, which is obtained by subtracting the perpendicular spectrum ($A_{\perp}$) from the parallel spectrum ($A_{\parallel}$). Most of the amide absorbance bands show sensitivity to polarized radiation. In **2B**, amide A, B, and I show perpendicular dichroism (negative absorbance), whereas amide II and III show parallel dichroism (positive absorbance).

The amide III band, at 1230 - 1300 cm^{-1}, results from C-N stretching and N-H in-plane bending in the planar peptide group. It shows comparatively strong dichroism [2,25,26] in a region of the mid-IR that contains no interference from water or nujol, and it has an absorbance coefficient low enough to allow spectral measurements from unflattened collagen fibers. Dichroism of the amide III band was therefore chosen to detect changes in conformation due to deterioration.

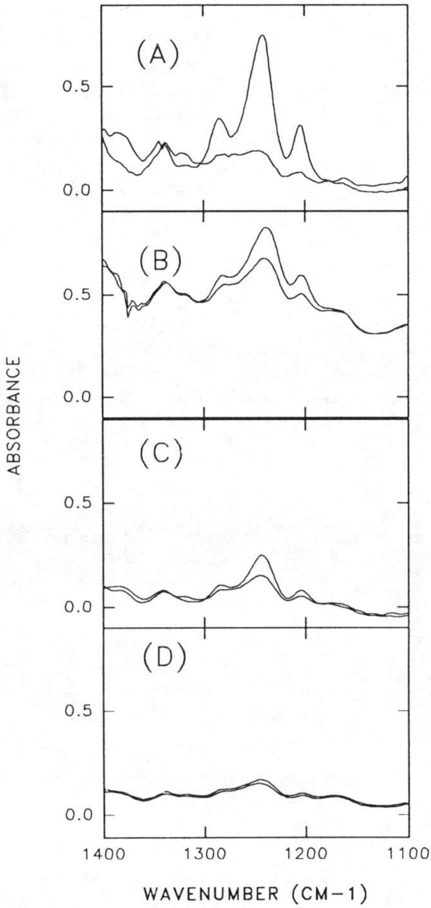

Figure 3 Dichroic amide III band pairs. (A) Seal skin. (B) Deteriorated seal skin. (C) Photodegraded seal skin. (D) Obliteration of dichroism in some photodegraded fibers.

Figure 3 contains parallel and perpendicular spectral pairs of amide III bands for water-soaked fibers taken from the unaltered seal skin, **3A**, and the deteriorated, **3B**, and photodegraded, **3C**, seal skins. The three show a progressive loss of dichroism indicated by the increased similarity of the parallel and perpendicular band intensities. (The parallel band always has the greater intensity.) **Figure 3D** illustrates the near obliteration of dichroism which occurred in some fibers of the photodegraded samples. While the absorbance intensities of the spectra vary with the optical density of the individual fibers (due to differences in thickness among fibers), the relative intensities of the parallel and perpendicular absorbances vary in accordance with changes in molecular conformation.

Dichroic ratios and shrinkage temperatures

Table 1 presents a compilation of 230 spectra in the form of averaged dichroic ratios ($A_{\parallel}/A_{\perp}$) for the samples. The table includes the variance (s^2) for the averages and the limits within which the true mean for the samples should reside at a 95% confidence interval ($\pm ts/N^{1/2}$, where t is the critical value for Student's t distribution with N - 1 degrees of freedom, s is the standard deviation of the mean, and N is the number of replicates). An analysis of variance (F-test, 95% confidence interval) between the average ratios revealed that the dichroic ratios for the samples of the brain-tanned deer skin and the brain- and smoke-tanned deer skin were equivalent. The two ratios were, however, statistically different from the ratios of the three deteriorated samples and from the ratio of the unaltered seal. Finally, while the ratio for the photodegraded brain-and smoke-tanned skin and the degraded seal skin

Table 1:

Dichroic Ratios of Skin and Semi-tanned Leather Samples

SAMPLE TYPE	SAMPLES	AVERAGE D RATIO	$\pm ts/N^{1/2}$ (95%)	s^2	F-TEST $F_{.05}$, df	T_s°C)
SEAL	9	5.50	.32	.274		60-66
					14.93 4.60,1/14	
BRAIN-TANNED DEER	7	4.36	.49	.447		50-58
					01.63* 4.67,1/13	
BRAIN- & SMOKE-TANNED DEER	8	3.97	.34	.263		62-67
					13.90 4.49,1/16	
PHOTO-DEGRADED B- & S- TANNED DEER	10	3.00	.34	.340		35-49
					7.71 4.30,1/22	
SEAL SKIN THONG	14	2.26	.33	.486		33-40
					0.0048* 4.26,1/24	
DEGRADED SEAL	12	2.24	.41	.606		35-39

* Critical value for F-test null hypothesis not exceeded; the two tested dichroic ratios are not significantly different.

were found to be different, the ratios for the degraded seal and photodegraded seal skin thong were not different. The variances are representative of the heterogeneity of dichroism among collagen fibers within each of the six tested samples; they do not describe the precision of the analytical technique. The variance of ratios obtained from spectral replicates was found to be as much as two orders of magnitude smaller. The degree to which sampling influences the variance remains to be determined, but it is expected to be large for material of heterogeneous preparation or deterioration. This will obviously be an advantage under some research circumstances and a disadvantage under others.

Shrinkage temperature (T_S) ranges are also listed in **Table 1**. The deteriorated samples of seal skin and of brain- and smoke-tanned deer skin show substantial losses in thermal stability compared to their undeteriorated counterparts. The reductions in the dichroic ratios are in agreement with lowered thermal stability. Both measurements reveal disruption of molecular conformation with deterioration.

Of particular note in the table is the average shrinkage temperature ranges for the two processed, undeteriorated deer skins. The brain-and-smoke tanned skin is indistinguishable from the sample of unprocessed, undeteriorated seal skin (62° to 67°C and 60° to 66°C respectively), unlike their respective dichroic ratios. The other deer skin which was only brain-tanned had a shrinkage temperature range about 10° below the normal range. These results suggest that the brain-tanning process causes significant disruption to collagen molecular structure. The application of smoke--a tanning method in which aldehyde cross-linking occurs--has apparently raised the shrinkage temperature of the one brain-tanned skin back to a more normal value. However, on the basis of the dichroic measurements, no amount of structural restoration has occurred. More generally, the results show that, in addition to detecting deterioration in skins and semi-tanned leathers, combined measurements of dichroism and shrinkage temperatures also show promise in distinguishing between completely untanned, unprocessed skin and semi-tanned or otherwise processed or altered skin.

ACKNOWLEDGEMENTS

The author would like to give special thanks to Mr. Chuck Gruchy, Director General, CCI, Dr. Kenneth MacLeod, Director of Conservation Research Services, CCI, and Mr. John Taylor, Chief of Analytical Research Services, CCI, for providing the opportunity to conduct this work; to Ms. Elizabeth Pye and Dr. Clifford Price, Institute of Archaeology, University College London; to Mr. Douglas Moffatt, Stacie Institute for Molecular Science, National Research Council of Canada, for initial discussions and for previewing this article, and to colleagues of the Canadian Conservation Institute, including Ms. Elizabeth Moffatt and Mr. R. Scott Williams for instruction and guidance on the instrumentation and Mr. Tom Strang for file conversion software.

REFERENCES

1. R.D.B. Fraser, J. Mol. Biol. 193 115 (1987).

2. R.D.B. Fraser and T.P. Macrae, in Conformation in Fibrous Proteins, Academic Press, London (1973) pp. 344-402.

3. R.D.B. Fraser and T.P. Macrae, in Conformation in Fibrous Proteins, Academic Press, London (1973) pp. 94-125.

4. D.W.L. Hukins, in Connective Tissue Matrix, MacMillan Press Ltd., London (1984) pp. 211-240.

5. A. Elliott, Infra-red Spectra and Structure of Organic Long-chain Polymers, London: Edward Arnold Ltd. (1969) 119pp.

6. N.B. Colthup, L.H. Daly and .E. Wiberley Introduction to Infrared and Raman Spectroscopy, Academic Press, New York (1990) pp. 75-107.

7. I.D. Campbell, and R.A. Dwek, Biological Spectroscopy, Benjamin/Cummings Publishing Company, London (1984) pp. 52-56.

8. H.W. Siesler, and K. Holland-Mortiz, Infrared and Raman Spectroscopy of Polymers, Marcel Dekker Inc., New York (1980) pp. 228-232.

9. R.R. Bresee, J.A.I.C. 25 39 (1986).

10. L.C.E. Struik, Physical Ageing in Amorphous Polymers and Other Materials, Elsevier, New York (1978).

11. D.W. Grattan, J.I.I.C.--C.G. 4 17 (1978).

12. J.H. Bowes, and A.S. Raistrick, J.A.L.C.A. 61 240 (1966).

13. C.V. Horie, Polym. Degradat. Stabil. 29 109 (1990).

14. G.S. Young, in Preprints of 9th Triennial Meeting, ICOM Committee for Conservation, Dresden (1990) p. 626.

15. G.S. Young, Thermal analysis of untanned and semi-tanned hide (forthcoming).

16. C. Miles, Conservation Research Services Project #33, Canadian Conservation Institute, Department of Communicantions, Ottawa (1985).

17. C.I. Marcott, I. Noda and A.E. Dowrey, Anal. Chimica Acta 250 131 (1991).

18. H.H. Mantsch, H.L. Casal and R.N. Jones, in Spectroscopy of Biological Systems, edited by R.J.H. Clark and R.E. Hester, John Wiley and Sons Ltd., New York (1986) pp. 1-46.

19. F. Levy, Anal. Chem. 60 1623 (1988).

20. R.G. Messerschmidt, in Microbeam Analysis--1987, 4B Optical Spectroscopy: Micro-IR, edited by R.H. Geiss, San Francisco Press, San Francisco (1987) pp. 169-172.

21. J.A. Reffner, J.P. Coates and R.G. Messerschmidt, American Laboratory April 105 (1987).

22. B. Chase, in Infrared Microspectroscopy: Theory and Applications, edited by R.G. Messerschmidt and M.A. Harthcock, Marcel Dekker, Inc., New York (1988) pp. 93-102.

23. R.M. Badger and A.D.E. Pullin, J. Chem. Phys. 22 1142 (1954).

24. R.P. Griffiths, and H.A. De Haseth, Fourier Transform Infrared Spectroscopy, New York: Wiley (1986).

25. D. Moffatt, (personal communication).

26. E.M. Bradbury, R.E. Burge, J.T. Randall and G.R. Wilkinson, Discuss. Farad. Soc. 25 173 (1958).

SOME APPLICATIONS OF ION CHROMATOGRAPHY TO THE STUDY OF THE DETERIORATION OF MUSEUM ARTIFACTS.

NORMAN H TENNENT*, BRIAN G COOKSEY**, DAVID LITTLEJOHN** AND BARBARA J OTTAWAY**.

*Department of Chemistry, The University of Glasgow, Glasgow G12 8QQ, Scotland, United Kingdom.
**Department of Pure & Applied Chemistry, University of Strathclyde, Cathedral Street, Glasgow G1 1XL, Scotland, United Kingdom.

ABSTRACT

Ion chromatography has been applied to the investigation of corrosion and efflorescence products on a range of museum artifacts whose deterioration is due to reaction with organic carbonyl pollutants in the storage environment. The potential of the technique and its optimisation for the determination of chloride, acetate, formate, nitrate and sulphate anions and associated cations are reported. Compounds containing acetates and/or formates from ceramic, geological, mollusca and metal collections have been investigated. The utility of ion chromatography for the examination of specimens which show no visible efflorescence, but from which significant acetate and formate levels can be quantified, is also demonstrated. The complementarity of ion chromatography and other techniques, in particular X-ray diffraction and infra red spectroscopy, for the investigation of efflorescence products is emphasised.

INTRODUCTION

The deterioration of museum artifacts due to reaction with organic carbonyl compounds liberated by materials used to fabricate storage and display cabinets was first documented almost a century ago [1] and continues to be a subject of conservation research endeavour [2-6]. Studies have generally focused on (a) the identification of crystalline efflorescence salts produced on the surface of artifacts as a result of chemical reaction with the organic carbonyl pollutants [7], (b) the determination of pollutant concentrations within display and storage cabinets [8], (c) the testing of display materials [9] and (d) the assessment of methods to ameliorate the liberation of pollutants from woods in pre-existing display and storage cabinets [10].

Less attention has been given to the evaluation of factors which predispose specimens to attack by organic carbonyl pollutants. Work in the authors' laboratories has addressed this aspect of the problem and established the composition of the efflorescence products. Various physico-chemical techniques have been applied to characterise efflorescence salts on mollusca collections [5] and bronze antiquities [7]. A range of acetate and/or formate compounds have been identified by combination of X-ray diffraction, infra red spectroscopy, nuclear magnetic resonance spectrometry and elemental micro-chemical analysis. However, difficulties arise when the compounds have not been previously documented. In these instances, the absence of X-ray diffraction patterns or infra red spectra precludes a straightforward identification and necessitates the use of additional means of characterisation.

The potential of ion chromatography for accurate determination of a wide range of cations and anions is ideal for characterisation of novel efflorescence products on museum artifacts. In addition, the low detection limits extend the usefulness of the technique to the quantification of species present on the surface or within the pores of artifacts even when no visible signs of efflorescence are present.

It is the purpose of this paper to describe the application of ion chromatography for determination of a range of cations and anions associated with the formation of efflorescence on museum artifacts. A variety of substrates are susceptible to chemical reaction with organic carbonyl pollutants present in the storage or display environment. Artifacts comprised of calcium carbonate (mollusca collections, carved limestone and terracotta) and bronze antiquities are discussed.

ION CHROMATOGRAPHY

Applications of ion chromatography, a form of liquid chromatography, have not featured in many reports or papers on conservation science. In this technique, ion exchange mechanisms are used to separate ionic species in a solution that is pumped at high pressure through a column containing small spherical beads with ion exchange sites. Various procedures can be used to monitor the elution of the separated ions from the column including, UV absorption spectrometry, fluorescence spectrometry, pulsed amperometry and differential refractive index measurement. However, most analyses by ion chromatography involve conductivity detection.

The first commercial ion chromatograph was produced by Dionex in 1975, following pioneering work by Small, Steven and Bauman [11]. Dionex instruments, like the one used in this study, employ suppressed conductivity detection to reduce the background conductance of the eluent and hence improved the detection limits of the analyte ions. Figure 1 shows a schematic diagram of an ion chromatograph equipped with a suppressor device. The eluent selected for a particular separation is normally degassed with helium and then pumped at a flow rate of eg 2 ml min^{-1} through an injection valve, a separator column, conductivity suppressor and a detection cell. The injection valve contains a small volume loop, typically 50 μl, which is filled with a standard solution or liquid sample in one position and then switched into the eluent flow in a second position. Polystyrene divinylbenzene ion exchange resins are often used for analyte separations. Anion exchange resins have quaternary ammonium functional groups, usually $-CH_2N^+(CH_3)_3A^-$, where A^- is an exchangeable anion (e.g. OH^-). Cation exchangers are normally sulphonated with exchangeable H^+ ions. When a mixture of anions or cations is injected onto a column, separation occurs due to differences in the ion exchange equilibria of the various ions. The separator columns are often used with a guard or precolumn which removes particulate material and prevents poisoning of the ion exchange resin.

If the separated analyte ions pass directly into the conductivity cell, detection would be difficult owing to the high background conductance of the ionic eluent. This problem is overcome by use of a micromembrane suppressor. In this device, the eluent and a regenerant solution flow in opposite directions on either side of a membrane. For the determination of anions, the eluent is often a mixture of $NaHCO_3$ and Na_2CO_3 and the regenerant is H_2SO_4. The anion micromembrane suppressor is surface sulphonated and exchange of Na^+ and H^+ ions occurs across the membrane. This converts the eluent to carbonic acid, H_2CO_3, which is a weak acid of low conductivity. The counter-ion of the analyte anion (eg Na^+) is converted to H^+, which increases the conductivity of the analyte, provided the acid form of the anion is ionized to an appreciable extent. Hence, it is easier to detect the analyte against a background of low conductivity (eg 10-20 µS, rather then 500-700 µS with $NaHCO_3$ - Na_2CO_3).

Eluent:

$$Na^+HCO_3^- + \text{membrane-}SO_3^-H^+ \rightarrow H_2CO_3 + \text{membrane-}SO_3^-\ Na^+$$

Analyte: eg Cl^-

$$Na^+Cl^- + \text{membrane-}SO_3H^+ \rightarrow H^+Cl^- + \text{membrane-}SO_3^-\ Na^+$$

The signal from the conductivity cell will increase when an analyte is eluted, giving a series of peaks which can be recorded on an integrator or a chart recorder. An integrator is preferred as it not only provides a visual display of the chromatogram, but also gives retention times and peak areas for each ion, which are used in quantitative analysis. The detector response is calibrated by injection and separation of standard solutions of the anions or cations to be determined. In this work a linear regression program was used to calculate sample ion concentrations.

EXPERIMENTAL

Small amounts of efflorescence salts (10 - 100 mg) were removed from the artifacts under study and dissolved in a few millilitres of distilled water. A Dionex 4000i ion chromatograph was used to analyse the solutions. Three sets of experimental conditions were required to

TABLE 1. Determination of Acetate, Formate and Chloride Anions by Ion Chromatography with DIONEX 4000i

Column:	HPIC - AS4A with guard column
Eluent:	8 mmol l^{-1} sodium tetraborate
Suppressor:	Anion micromembrane suppressor
Regenerant:	5 mmol l^{-1} H_2SO_4
Conductivity Range:	10 or 30 μS

TABLE 2. Determination of Chloride, Nitrate and Sulphate Anions by Ion Chromatography with DIONEX 4000i.

Column:	HPIC - AS4A with guard column
Eluent:	0.75 mmol l^{-1} $NaHCO_3$ plus 2.2 mmol l^{-1} Na_2CO_3
Suppressor:	Anion micromembrane suppressor
Regenerant:	5 mmol l^{-1} H_2SO_4
Conductivity Range:	10 μS

TABLE 3. Determination of Sodium, Annonium, Potassium, Magnesium and Calcium Cations by Ion Chromatography with DIONEX 4000i.

Column:	HPIC FAST SEP CATION COLUMNS I and II with guard column
Eluent:	20 mmol l^{-1} HCl plus 0.3 mmol l^{-1} *DAP.HCl
Suppressor:	Cation micromembrane suppressor
Regenerant:	0.1 mol l^{-1} tetrabutylammonium hydroxide
Conductivity Range:	10 μS

(*DAP = 2,3-diaminopropionic acid)

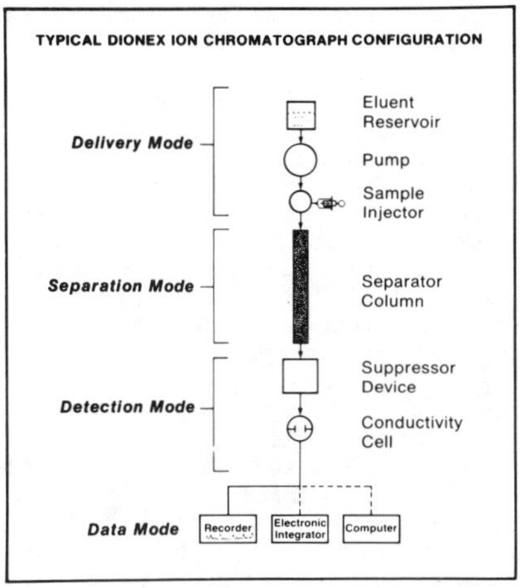

Fig.1 Schematic diagram of ion chromatograph system.

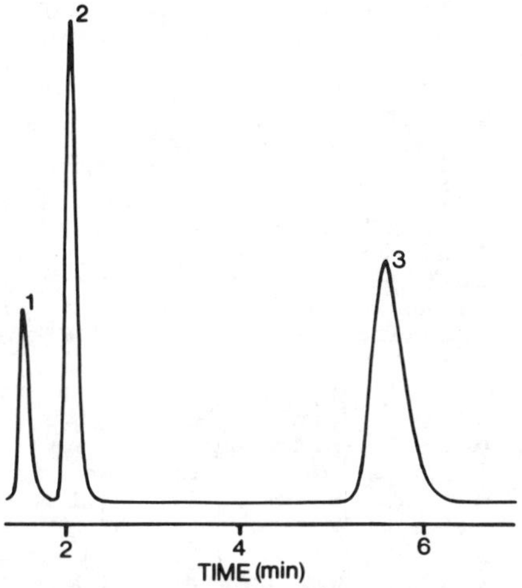

Fig.2 Ion chromatograph showing separation of acetate (10 mg l^{-1}), formate (10 mg l^{-1}) and chloride (10 mg l^{-1}).

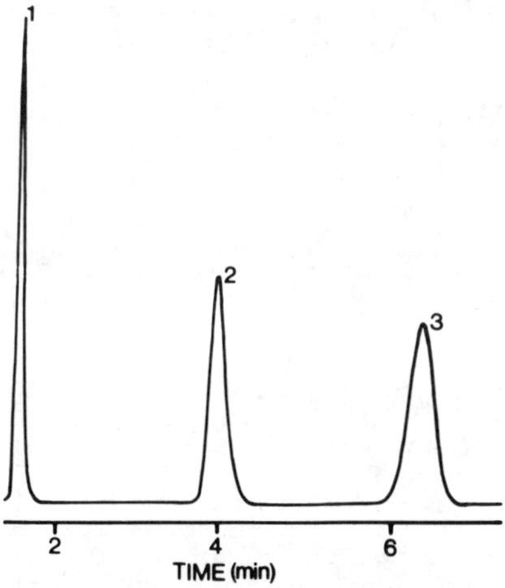

Fig.3 Ion chromatograph showing separation of chloride (1.0 mg l^{-1}), nitrate (2.0 mg l^{-1}) and sulphate (2.0 mg l^{-1}).

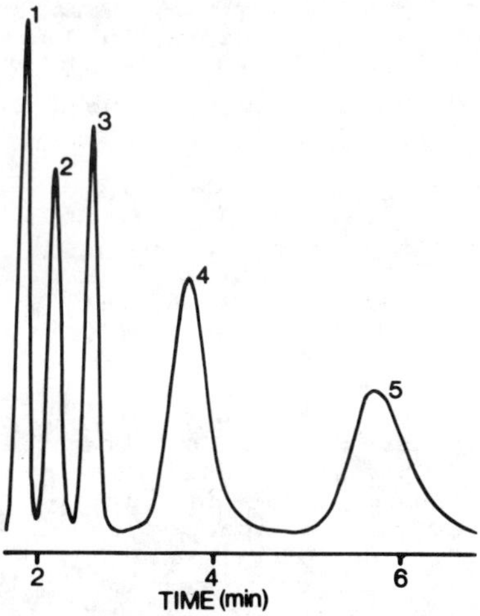

Fig.4 Ion chromatograph showing separation of sodium (2.0 mg l^{-1}), ammonium (3.0 mg l^{-1}), potassium (3.0 mg l^{-1}), magnesium (2.0 mg l^{-1}) and calcium (2.0 mg l^{-1}).

determine (a) acetate, formate and chloride, (b) chloride, nitrate and sulphate and (c) sodium, ammonium, potassium, magnesium and calcium. The details are given in Tables 1 - 3. It was possible to separate the anions with a single column and micromembrane suppressor/regenerant, but different eluents were required. Separation of acetate and formate was only possible with sodium tetraborate, which is a weaker eluent then sodium bicarbonate - sodium carbonate. Chloride could be determined with either eluent. Specimen chromatograms for separation of (a) acetate, formate and chloride, and (b) chloride, nitrate and sulphate are given in Figures 2 and 3, respectively.

To separate the monovalent and divalent cations of interest in a few minutes with a single eluent, a special column switching procedure was used. The FAST SEP CATION columns were arranged in series such that the eluent passed through the CATION II column and then the CATION I column. The monovalent ions were not retained in the CATION II column, which is designed to separate divalent ions, but passed into the CATION I column where separation of K^+, NH_4^+ and Na^+ occurred. After 40 s, and before any of the analyte ions had been eluted from the respective columns, the eluent flow was reversed. As the monovalent ions can pass through the CATION II column faster than the divalent ions, the analytes were eluted from the CATION II column into the conductivity detector cell in the order Na^+, NH_4^+, K^+, Mg^{2+} and Ca^{2+}, as shown in Figure 4.

The solution detection limits achieved for the anions and cations in the specimens analysed were 0.01 mg l^{-1} for Na^+ and Mg^{2+}, 0.02 mg l^{-1} for NH_4^+, K^+, Ca^{2+}, Cl^- $HCOO^-$ and CH_3COO^-, and 0.04 mg l^{-1} for NO_3^- and SO_4^{2-}

RESULTS AND DISCUSSION

Efflorescence on Mollusca Specimens

The formation of white crystalline efflorescence, generally known as Byne's Disease, is a widespread problem affecting tens of thousands of mollusca specimens worldwide. Although the problem was first investigated almost a century ago [1] and a more recent study has characterised, for the first time, the chemical identity of the efflorescence products (one or a combination of calcium acetate hydrate, calcium hemihydrate or a previously unreported calcium acetate formate hydrate [5]), several issues regarding the mode of formation of Byne's Disease remain unclear. As part of an evaluation of the factors which predispose shells to react with organic carbonyl pollutants, the role of residual salts has been investigated by ion chromatography.

It is common to find mollusca collections where specimens from the same species, stored in the same conditions, show a range of deterioration. Shells in pristine condition are often found adjacent to those completely covered with a mass of white crystalline efflorescence (Figure 5). With the knowledge that land snails are generally less badly affected than marine specimens and the evidence that shells affected by Byne's Disease contained significant chloride levels [3], the relationship between the extent of Byne's Disease and the levels of salt residues in the shells has been investigated. These results, to be presented in full elsewhere [12], indicate concordance between the acetate or formate ion concentration and the chloride ion concentration in the shells (Figure 6).

This observation can be ascribed to the fact that salt residues, because of their hygroscopicity, can modify the local moisture content of the shell, thus facilitating attack by acetic acid. (Padfield and Erhardt, in accord with this, found that marble doped with sodium bromide was more susceptible to attack by acetic acid than untreated marble [13].) There is also good circumstantial evidence for the importance of high relative humidity in exacerbating the formation of efflorescence on mollusca collections; specimens stores in conditions where high relative humidity prevails are particularly badly affected by Byne's Disease [12].

Characterisation of Novel Efflorescence Products

The majority of corrosion and efflorescence products which form on museum artifacts can be easily identified. X-ray diffraction and infra red spectroscopy, individually or in combination, are generally successful in characterising the compounds which form on the surface of artifacts during burial or subsequent deterioration in the museum environment. In most instances the compounds formed are the museum counterparts of species which occur elsewhere in nature (see, for example,[14]), although in some cases the incidence on museum artifacts can be very rare [15,16]. Indeed, the corrosion or efflorescence product may be a previously undocumented compound. A well-known example is calclacite, $CaCl(CH_3COO).5H_2O$ first characterised by van Tassel [17,18], and now found ubiquitously as a product of efflorescence derived from the reaction of acetic acid with calcareous specimens.

A related compound was first reported by Gettens and FitzHugh [4] who were unable to characterise it fully, but found it to contain calcium, acetate and chloride ions. They speculated that it may be a pseudomorph of calclacite and referred to it as "Efflorescence X". The authors have found its occurrence is widespread on limestone antiquities and geological material and on terracotta [19]. In one case, a mass of white, "hairy", crystalline material was formed on an Egyptian limestone relief (Figure 7) after storage for approximately 6 years in a cabinet where the levels of acetic acid pollution were estimated to be about 0.14 - 0.32 ppm [20]. By means of ion chromatography it has been ascertained that the principal ionic species are calcium, chloride, acetate and nitrate. Though the stoichiometry can vary somewhat from specimen to specimen, the X-ray diffraction data [4] and infra red spectrum (Figure 8) are

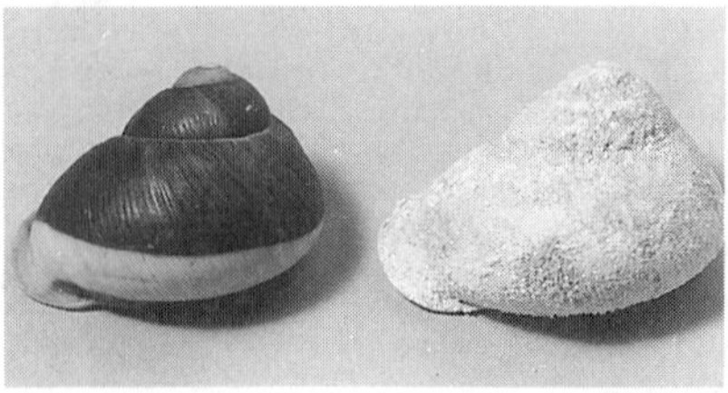

Fig.5 Two shells of *acavus haemastomus* from the same cabinet drawer; one in pristine condition, the other showing complete coverage of efflorescence.

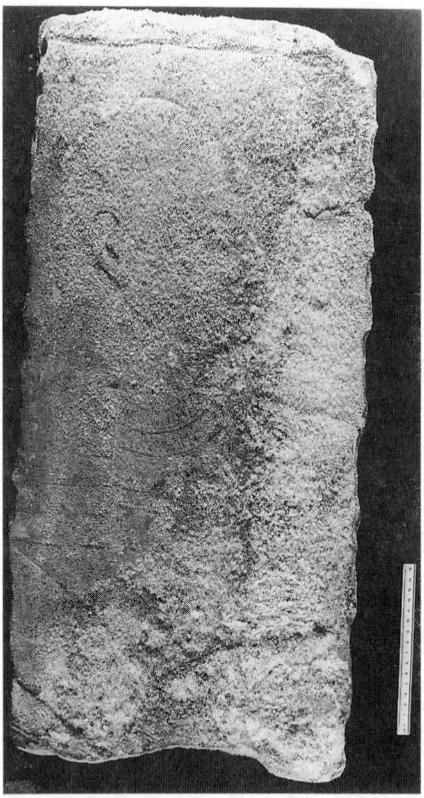

Fig.7 Egyptian limestone relief showing an extensive coverage of "Efflorescence X". (The Burrell Collection, Ref. No. 18/280.)

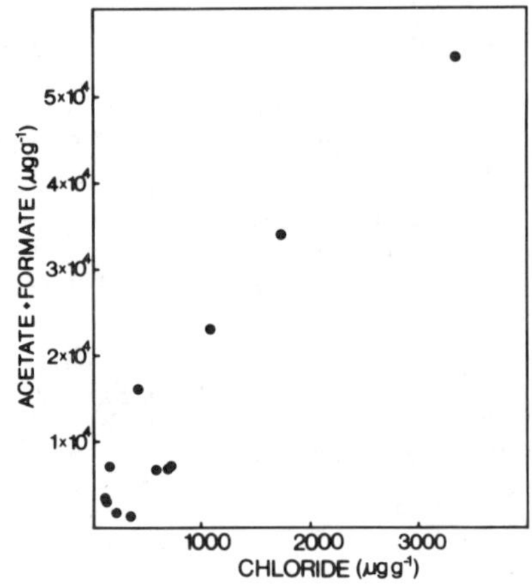

Fig.6 Plot of combined concentrations of acetate and formate against chloride for a series of shells from the same species which display varying degrees of efflorescence.

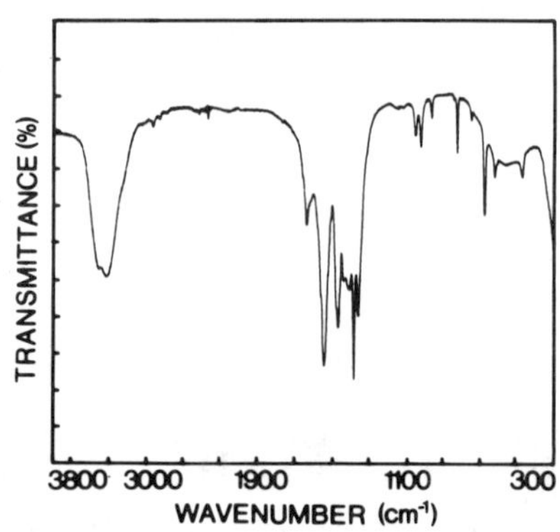

Fig.8 Typical infrared spectrum of "Efflorescence X".

characteristic and permit straightforward identification. The stoichiometry of this compound, for which we propose the name thecotrichite ("a hairy mineral from a storage cabinet"), will be published in full elsewhere [21].

Acetate Efflorescence on Bronze Antiquities

It has recently been shown that bronze antiquities can develop unusual corrosion/efflorescence products as a result of acetic acid pollution in the storage environment [7]. The phenomenon is characterised by the appearance of white crystalline efflorescence, identified as sodium acetate trihydrate, in conjunction with a sky-blue copper corrosion product the identity of which has yet to be established. The factors which predispose bronzes to this mode of deterioration are currently under investigation.

Ion chromatography has been indispensable in providing supportive evidence for the corrosion mechanism. One hypothesis under investigation is that residues of sodium sesquicarbonate from bronze stabilisation treatments [22] may subsequently, in the presence of wood-derived acetic acid, convert to sodium acetate trihydrate with the concomitant formation of a novel blue bronze corrosion product.

TABLE 4. Concentration of Ions Leached from Two Bronze Antiquities After Immersion for 5 h in Distilled Water. (Concentrations are expressed as μg cm^2 i.e. the amount of analyte present per cm^2 of the surface of artifact.)

Bronze	Acetate	Formate	Chloride	Sodium	Ammonium
1 †	75.8±1.4	13.1±0.6	1.9±0.5	28.3±0.9	0.76±0.14
2 †	*	*	1.8±0.4	*	*

Concentrations are quoted as the mean ± one standard deviation of replicate analyses of the solution.

† Bronzes from Ascalon; bronze 1, (B.M. Reg. No. 1913. 5-23, 69) an axehead treated by prolonged immersion in 5 % sodium sesquicarbonate solution; bronze 2, (B.M. Reg. No. 1913. 5-23, 83) a spearhead with no corrosion stabilisation treatment.

* The analyte was present at concentrations less than the detection limit of the ion chromatography method.

In a bid to assess the validity of this hypothesis, two bronze antiquities from the British Museum were selected for study. Both artifacts had the same provenance but only one had been treated by prolonged immersion in a solution of sodium sesquicarbonate [23]. Subsequent storage conditions for both artifacts were the same. Neither showed any visible sign of deterioration, though the stabilised bronze had the pale blue-green appearance of sesquicarbonate-treated material.

These bronzes were immersed in distilled water for five hours and the solutions analysed by ion chromatography. (Aqueous treatment of bronzes is liable to render them susceptible to corrosion. After soaking, the bronzes were stabilised with benzotriazole.) The cation and anion concentrations shown in Table 4 reveal that significant acetate levels were found only on the object which had been treated with sodium sesquicarbonate. The concentration of sodium in this bronze is also higher, indicative of residues from the sodium sesquicarbonate treatment and consistent with the hypothesis that sodium sesquicarbonate retained within the pores of the bronze had reacted with acetic acid to form sodium acetate trihydrate. (Separate experiments have confirmed that this reaction occurs easily [24].) Nonetheless, despite the evidence that this is a viable mechanism, other possibilities exist [7] and are currently under evaluation.

Detection of Acetate and Formate on Artifacts Showing No Signs of Efflorescence

Although acetate was easily detectable by ion chromatography, no visible sign of acetate efflorescence was present in the bronze specimen mentioned above. The potential of ion chromatography to detect trace levels of acetate and formate efflorescence or adsorbed acetic acid and formic acid have therefore also been investigated.

Several museum artifacts, which had been stored for nine years in wooden cabinets where the level of acetic acid is know to be about 0.14 - 0.32 ppm [20], were examined. These artifacts, comprising different materials - bronze, pewter, glass in pristine condition, weathered Roman glass and calcite - displayed no visible sign of efflorescence. Immersion in distilled water for 3 h followed by determination of acetate, formate and chloride by ion chromatography gave the results shown in Table 5.
All the specimens, with the exception of the English wine glass, had measurable levels of acetate, with the bronze arrowheads exhibiting the highest surface concentrations. The arrowheads also had significant levels of formate. It should be noted, however, that the bronze results cannot be compared directly with those in Table 4, as the storage conditions for these artifacts were different. Chloride was present on all the specimens. The very high concentration on the Chinese pottery fragment is of particular interest. It is evident that analysis by ion chromatography could be useful in assessing the removal

of chloride during the conservation of metals, ceramics and stonework.

The result obtained for the English wine glass require further investigation. The absence of detectable acetate and formate might have been due to negligible adsorption of the carbonyl vapours on the glass surface during storage. However, it is also possible that strongly bound acetate and formate was not removed by the immersion procedure used to treat the specimen.

TABLE 5. Acetate, Formate and Chloride Surface Concentrations Determined by Ion Chromatography Following Immersion of a Range of Antiquities in Distilled Water. (Concentrations are expresses as μg cm^{-2} of the surface area of the artifacts.)

Specimen	Acetate	Formate	Chloride
Egyptian bronze arrowhead (Burrell Collection 8/150-7a)	30.0±2.1	36.3±0.3	15.6±0.9
Egyptian bronze arrowhead (Burrell Collection 8/150-7b)	23.8±1.9	36.6±0.3	17.9±0.9
Egyptian bronze arrowhead (Burrell Collection 8/150-7c)	15.4±0.8	8.4±0.1	4.0±0.4
German pewter dish (Burrell Collection 34/35)	7.2±2.1	*	6.8±0.3
Egyptian calcite mortar (Burrell Collection 13/269)	9.8±1.9	0.87±0.27	14.4±0.9
Egyptian calcite bowl (Burrell Collection 38/154)	5.1±1.0	*	15.1±0.5
Chinese pottery fragment (Burrell Collection 38/170)	5.3±1.3	*	453±55
English wine glass (Burrell Collection 15/398)	*	*	0.75±0.25†
Iridescent glass fragments (Burrell Collection 17/63)	2.8±0.4	0.15±0.05	22.1±1.6

Concentrations are quoted as the mean ± one standard deviation of replicate analyses of the solution.

† It was not possible to measure the surface area of this artifact; the results are expressed as μg of analyte per ml of water.

* Not detectable.

CONCLUSIONS

Ion chromatography, although commercially available for more than sixteen years, is not yet a widely utilised method in conservation science research. (A recent survey of conservation science facilities in the U.K. indicated only one laboratory where ion chromatography instrumentation was available [25].) In this paper, the potential of the technique for aiding the characterisation of novel corrosion or efflorescence products has been demonstrated. It has been shown that low levels of pollutant-derived anions (notably acetate and formate) present on the surface of museum artifacts can be determined by ion chromatography, alerting conservators to the possibility of incipient efflorescence even when no visual evidence is available. In addition, the technique has proved useful in assessing the synergistic effect of salt residues in promoting the formation of efflorescence by the reaction of museum artifacts with pollutants in the storage environment.

Future studies will extend the range of artifacts examined, evaluate the case of adsorbed acetic and formic acid and determine the optimal conditions for removal of trace quantities of surface efflorescence for quantification by ion chromatography.

ACKNOWLEDGEMENTS

In the course of this study specimens from several museum collections were investigated and the authors wish to acknowledge their appreciation to the following colleagues who facilitated this work: Mr D Heppel, Royal Museum of Scotland, Edinburgh (mollusc specimens); Miss J Porter, The Burrell Collection, Glasgow (bronze, limestone and other antiquities) and Mr J Spalding, Director, Glasgow Museums & Art Galleries, who encouraged the investigation of these artifacts and the publication of the observations; members of the Department of Conservation, The British Museum, London (bronze antiquities) and Dr J.E. Curtis, Department of Western Asiatic Antiquities, The British Museum, for permission to publish the findings. The provision of ion chromatography results by Angela Nicholson, Helen Marsland and Alan Robertson is also gratefully acknowledged.

REFERENCES

1. L.StG. Byne, `The corrosion of shells in cabinets', J.Conchology 9 (1899) 172-178 & 253-254.

2. M.F. Taboury, `Des modifications chimiques de certaines substances calcaire convervées dans des meubles en bois', Bulletin de la Societé Chimique de France 49 (1931) 1289-91.

3. J.R.Nicholls, `Determination of shells when stored in oak cabinets', J.Soc.Chem.Ind. 53 (1934) 1077-1078.

4. E.W.FitzHugh, and R.J.Gettens, `Calclacite and other efflorescent salts on objects stored in wooden museum cases' in Science and Archaeology (ed. R. Brill), MIT Press, Cambridge MA (1971) 91-102.

5. N.H.Tennent, and T.Baird, `The deterioration of mollusca collections: identification of shell efflorescence', Studies in Conservation 30 (1985) 73-85.

6. P.Hatchfield, and J.Carpenter, `Formaldehyde: How great is the danger to museum collections?', Center for Conservation and Technical Studies, Harvard University Art Museum, Cambridge, MA, USA, (1987).

7. N.H.Tennent and T.Baird, `Identification of acetate efflorescence on bronze antiquities stored in wooden cabinets', The Conservator 16 (1992) 39-47.

8. F.Preusser and J.R.Druzik, `Environmental Research at the Getty Conservation Institute', Restaurator 10 (1989) 160-196.

9. S.M.Blackshaw, and V.D.Daniels, `Selecting safe materials for use in the display and storage of antiquities' in Preprints of ICOM Committee for Conservation, 5th Triennial Meeting, Zagreb (1978) paper 78/23/2.

10. C.Miles, `Wood coatings for display and storage cases', Studies in Conservation 31 (1986) 114-24.

11. H.Small, T.S.Stevens and W.C.Bauman, `Novel ion exchange chromatographic method using conductimetric detection', Anal. Chem. 47 (1975) 1801-1809.

12. N.H.Tennent, B.Cooksey, D.Littlejohn, B.Ottaway and A.Nicolson, `The deterioration of mollusca collections: the role of salt residues in the formation of shell efflorescence', Studies in Conservation, manuscript in preparation.

13. T.Padfield and D.Erhart, `The spontaneous transfer to glass of an image of Joan of Arc', in Preprints of ICOM Committee for Conservation 8th, Triennial Meeting, Sydney (1987) pp 909-913.

14. R.J.Gettens, `Mineral alteration products on ancient metal objects' in Recent Advances in Conservation, IIC, Butterworths, London (1963) pp 89-92.

15. M.Fabrizi, H.Ganiaris, S.Tarling and D.A.Scott, `The occurrence of sampleite, a complex copper phosphate, as a corrosion product on copper alloy

objects from Memphis, Egypt', Studies in Conservation 34 (1989) 45-51.

16. G.S.Wheeler and M.T.Wypysky, `An unusual efflorescence on Greek ceramics', Studies in Conservation, in the press.

17. R.Van Tassel, `Une efflorescence d'acétachlorure calcium sur les roches calcaires dans des collections'. Bulletin du Musee Royal d'Histoire Naturelle de Belgique, 21 (1945) 11pp.

18. R.Van Tassel, `On the crystallography of calclacite, $Ca(CH_3COO)Cl.5H_2O$', Acta Crystallographica, 11 (1958) 745-746

19. N.H.Tennent and D.Littlejohn, unpublished observations.

20. N.H.Tennent, J.Porter and P.Ramsay, unpublished observations.

21. N.H.Tennent, T.Baird, B.Cooksey, D.Littlejohn and L.Gibson, `Thecotrichite: a widespread acetate efflorescence on calcareous specimens stored in wooden cabinets', Studies in Conservation, manuscript in preparation.

22. W.A.Oddy and M.J.Hughes, `The stabilisation of "active" bronze and iron antiquities by the use of sodium sesquicarbonate', Studies in Conservation 15 (1970) 183-9.

23. C.Enderly, personal communication.

24. N.H.Tennent and T.Baird, unpublished observations.

25. J.P.Brown, Survey of current conservation research. University of Wales College, Cardiff, (1992) 10pp.

UNDERSTANDING POST-DEPOSITIONAL PROCESSES THROUGH ELECTRON MICROBEAM ANALYSIS OF ARCHAEOLOGICAL BONE FROM SE MISSOURI

DIANA M. GREENLEE AND ROBERT C. DUNNELL
Department of Anthropology, DH-05, University of Washington, Seattle, Washington 98195

ABSTRACT

Archaic period bone in Southeast Missouri characteristically displays a patchy black and reddish brown staining not found on bone from other-aged sites in the area. The stained areas appear to be post-depositional additions to the bone matrix. Preliminary tests suggest that the stain is inorganic and is rich in Mn and Fe. Here, electron microbeam analysis is used in conjunction with SEM imaging to identify elements specifically linked to the staining. Comparisons are made between the elemental composition of bone and associated sediments from several different-aged occupations. Variability in post-depositional environment between occupations is documented and related to the presence of the staining.

INTRODUCTION

Archaeological bones associated with Archaic period deposits in Southeast Missouri frequently have a distinctive staining (Figure 1). This staining occurs as black and sometimes reddish brown patches that appear to be incorporated into the bone matrix, rather than deposited on the surface. Ashing experiments suggest that the stain is inorganic and bulk neutron activation analyses indicate that manganese and, to a lesser extent, iron may be involved. The limited distribution of the staining suggests the presence of a unique depositional environment which may have entailments for artifact structural and chemical preservation.

To investigate the nature and source of the staining, bone and associated sediment samples were examined from six archaeological loci in Southeast Missouri (Figure 2). The Malden Plain, an early Holocene alluvial terrace, represents the major topographic feature on the landscape. Half of the occupations are situated on the Malden Plain; the other half are located on topographic "islands" in the Eastern Lowlands east of the Malden Plain. Prior to large scale drainage which lowered the water table of this wet, flat environment, slight differences in elevation determined the availability of dry land and amount of seasonal inundation.

The Grey Horse Lake (GHL), Poverty Point Object (PPO), and Hankins (HAN) loci are small middens in the Eastern Lowlands. While these middens appear to have been occupied over a long period of time, the bulk of each probably accumulated during the

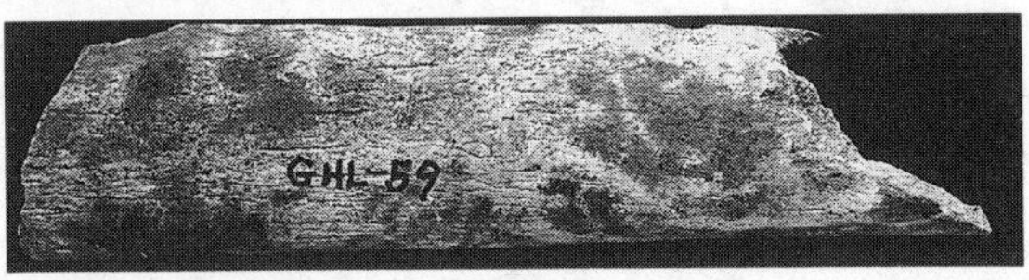

Figure 1. A typical example of a stained bone from an Archaic deposit, Grey Horse Lake.

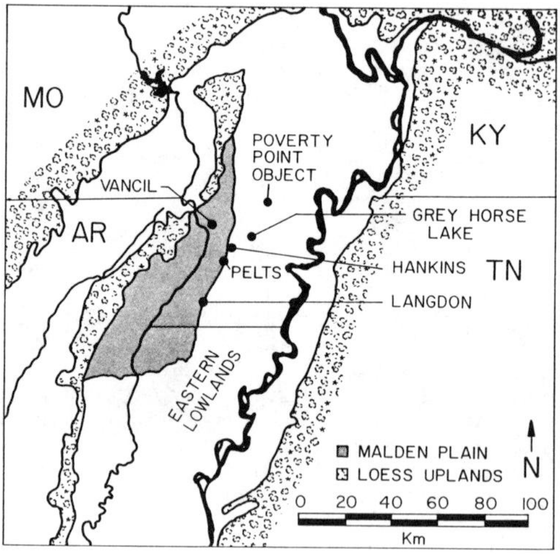

Figure 2. The location of archaeological loci of interest in Southeast Missouri.

Late Archaic/Early Woodland period (5000 B.P. - 2000 B.P.).[1] The middens are areas of neutral to slightly basic pH in an acidic local environment; through time, bone and shell decomposition changed the soil chemistry sufficiently to allow the preservation of later deposits of those materials.

The Langdon (LAN), Vancil (VAN), and Pelts (PEL) localities are situated on the Malden Plain. Artifacts indicate long use of these locations; however, the bulk of the materials are Mississippian in age (1000 B.P. - 500 B.P.). Like the Archaic middens, these deposits are near-neutral pH islands in an acidic environment, but their acidic history prior to the Mississippian occupation precludes the preservation of non-Mississippian bone.

Bones and sediment samples were acquired from surface collections curated at the University of Washington. Human long bone and cranial fragments are analyzed. Reflecting the available material, nineteen bone samples were examined from the three Archaic deposits, while only four samples were available from the three Mississippian deposits. Sediment samples were available for all deposits except Vancil.

Structural and compositional information about the bone and associated soils is necessary to identify the nature and source of the stain. Researchers have shown that backscattered electron (BSE) imaging is useful in studying bone structure,[2] while electron microbeam analysis is invaluable for examining the variability in elemental composition of archaeological materials.[3,4,5] Microprobes, while not routinely used for sediment analysis, are particularly attractive for examining sediment lodged in the medullary canal or in pores of bone. X-ray dot maps clearly show the relationship between elemental composition and structure.

ANALYSIS

Bone and sediment samples were impregnated with epoxy and their surfaces were polished with a series of grits to 0.3 µm. The samples were rinsed in an ultrasonic cleaner and wiped with alcohol between grits and prior to mounting. A JEOL JSM840A scanning electron microscope was used to examine both the structure and composition of bone and soil samples. Backscattered electron images provided structural and gross compositional information. Energy dispersive spectroscopy (EDS) was used to measure the concentration of major and minor (>1-2%) elements in bone and soil.

To facilitate comparisons, the EDS analyses were standardized by accelerating voltage (10 kV), incoming counts per second (2000 cps), acquisition time (100 s), and scan area (0.954 mm^2). Standard analyses were obtained from at least four locations on each sample. Additionally, analyses were performed at higher magnifications and individual spots to document the composition of structural features visible in the BSE images. Multiple analyses were obtained after rotating specimens by 90°.[6]

RESULTS AND DISCUSSION

BSE images of modern and archaeological specimens (Figure 3) provide some measure of the amount of post-depositional structural and compositional alteration that has occurred. Variations in the shades of gray reflect relative differences in the mean atomic number of structural features. Lighter areas have a higher mean atomic number than darker areas. Considerable microbial damage is present in all archaeological specimens examined, with subsequent recrystallization and post-depositional addition of materials.

Figure 4 compares the energy spectra for modern bone and typical examples from Langdon and Hankins. The Langdon sample has two additional peaks not present in the modern bone, aluminum and silica. X-ray dot maps (Figure 5) clearly show that these peaks reflect the post-depositional inclusion of aluminosilicates around osteons rather than contamination during preparation. The Hankins bone has one additional peak, manganese. In contrast to the Mississippian samples, nearly all Archaic bones had significant concentrations of Mn; a few had detectable levels of Fe, as well. The manganese is primarily concentrated at the bone surfaces and around cracks and channels in the interior where soil water has access. Unlike Al and Si in the Langdon example, however, the Mn is often incorporated into the bone matrix, less frequently being adsorbed onto the matrix. In Figure 6, BSE images and x-ray dot maps show that the Mn is concentrated around osteons. The results of spot analyses of grey and white areas in the BSE image are shown in Figure 7. The grey areas do not

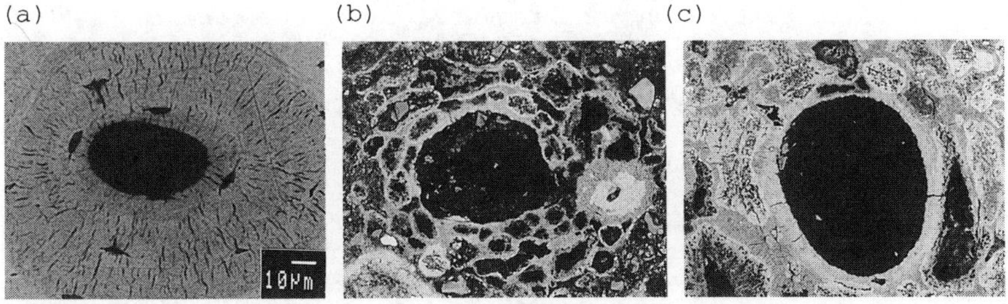

Figure 3. Backscattered electron images of (a) modern, (b) Langdon, and (c) Poverty Point Object bone specimens.

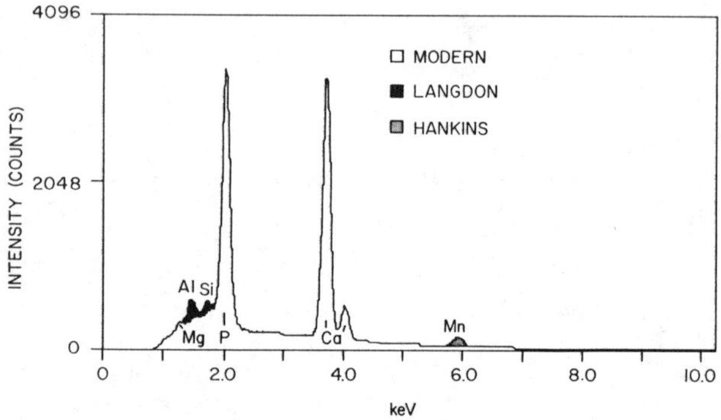

Figure 4. A comparison of the energy spectra of specimens from Langdon and Hankins against modern bone.

(a) (b)

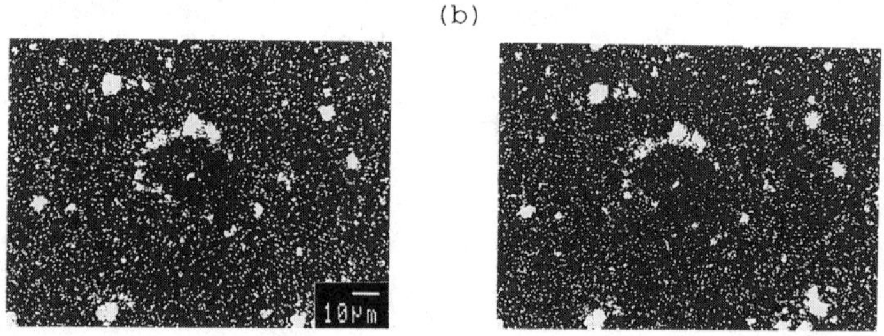

Figure 5. X-ray dot maps showing the distribution of (a) Al and (b) Si around an osteon in the Langdon sample. These images correspond to the Langdon BSE image in Figure 3.

contain detectable amounts of elements other than Ca and P, while the white areas have significant levels of Mn with some Ca and very little P. Clearly, there is a strong microstructural component to the distribution and concentration of diagenetically altered elements.

Figure 8 compares the energy spectra of a PPO bone and its associated sediment sample. The sediment boasts higher counts of Al, Si, K, and Fe; the bone has more P, Ca, and Mn. Consistently, there is a significantly greater level of Mn in bone than in sediments; in fact, iron and manganese concentrations in the sediments are generally not observable as well-defined peaks in the energy spectra and in many cases are below the minimum detection limits of EDS. This suggests that Mn is being concentrated in the bone.

An additional factor which must be considered in explaining the distribution of stained bones is that of soil drainage. Mississippian occupations are located on well- to hyper-drained soils of the Dunklin type which have permeability indices (in/hr) ranging from 0.6 to 20.0[7]; Archaic occupations are located on the poorly drained Wardell loam of the Pemiscot type soils with permeability indices of 0.06 to 0.2.[8] Soil drainage has consequences for oxidation/reduction conditions which are important in determining the solubility of soil minerals.[9]

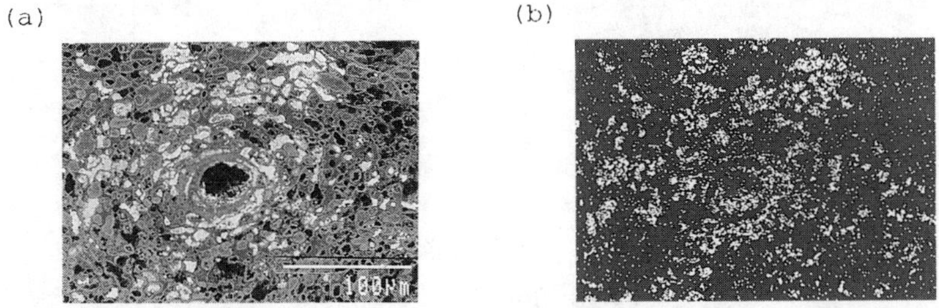

Figure 6. (a) BSE image and (b) associated Mn x-ray dot map of an osteon in the Hankins sample.

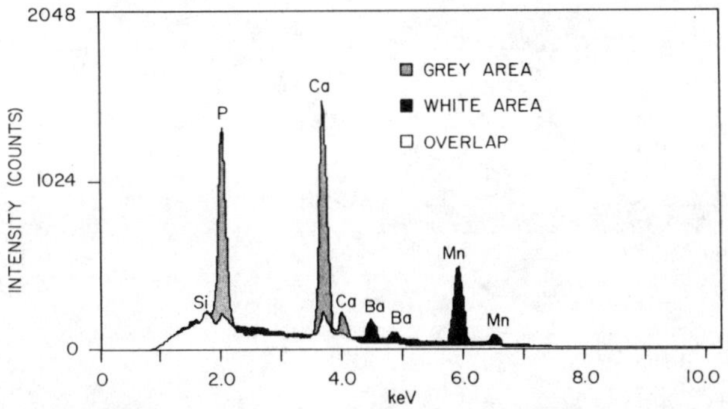

Figure 7. Energy spectra from spot analyses of grey and white areas on the HAN bone (corresponding to BSE image in Figure 6).

Waterlogged soils typically have low redox conditions.

Mn is highly soluble under the acidic pH and low redox conditions characteristic of the environment occupied by the Archaic middens. As a result, Mn is dissolved in the soil solution until it comes into contact with the bone. No longer soluble under basic conditions, the Mn either reacts with recrystallizing calcium phosphate to form manganese phosphate ($MnHPO_4$) or precipitates as minerals such as rhodochrosite ($MnCO_3$) or manganite (MnOOH).[9] Under neutral to slightly basic pH, these minerals are stable. This can account for the apparent concentration of Mn in bone relative to the soil. With better drainage conditions, Mississippian deposits have higher redox levels and the soil Mn remains relatively insoluble. Thus, chemical conditions are not appropriate for the addition of Mn to bones of Mississippian occupations.

CONCLUSIONS

The dark stain characteristic of bones from Archaic deposits is caused by the reaction of manganese with the bone matrix. The Archaic occupations examined here are located in environments with soil conditions that promote Mn solubility and precipitation, namely acidic, waterlogged soils with basic bones. In contrast, Mississippian occupations are associated with acidic

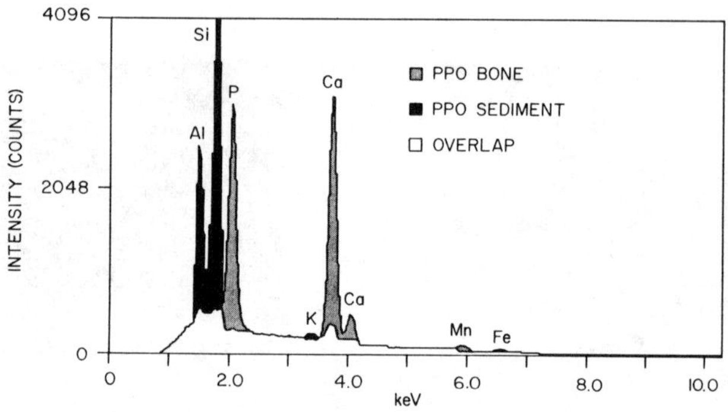

Figure 8. Energy spectra from a PPO bone and its associated sediment.

well-drained soils, conditions that do not support similar chemical reactions.

While this research addresses only one small part of a complex chemical system, it demonstrates the value of a microspatial approach to understanding bone diagenesis. By documenting the relationship between elemental concentrations and bone structure, and linking those data to the soil chemistry, the nature of post-depositional alterations of archaeological bone can be determined. Consequently, we have a better understanding of the archaeological record.

REFERENCES CITED

1. R.C. Dunnell, and F.H. Whittaker, presented at the 50th Southeastern Archaeological Conference, New Orleans, LA, 1988.

2. L.S. Bell, J. Arch. Sci. **17**, 85 (1990).

3. J.B. Lambert, S.V. Simpson, J.E. Buikstra, and D. Hanson, Am. J. Phys. Anth. **61**, 409 (1983).

4. P. Henderson, C.A. Marlow, T.I. Molleson, and C.T. Williams, Nature **306**, 358 (1989).

5. D.M. Greenlee, presented at the 57th Annual Meeting of the Society for American Archaeology, Pittsburgh, PA (1992).

6. J.I. Goldstein, D.E. Newbury, P. Echlin, D.C. Joy, C. Fiori, and E. Lifshin, Scanning Electron Microscopy and X-Ray Microanalysis (Plenum Press, New York, 1981).

7. P.D. Gurley, Soil Survey of Dunklin County, Missouri (USDA, Soil Conservation Service, Washington, D.C., 1979).

8. B.L. Brown, Soil Survey of Pemiscot County, Missouri (USDA, Soil Conservation Service, Washington, D.C., 1971).

9. W.L. Lindsay, Chemical Equilibria in Soils (John Wiley and Sons, New York, 1979).

ANALYSIS OF MEDIEVAL FRESCO PIGMENTS FROM THE GEORGIAN REPUBLIC

MARIE DNEPROVSKAYA
The State Hermitage Museum
The Laboratory of Scientific Examination
Saint Petersburg, Russia

ABSTRACT

The mineral composition of pigments from mural paintings of the 12th and 13th centuries in the grotto monasteries of Bertubani and Udabno (David-Caredga) were investigated. Apart from traditional pigments, many examples of brown-black plattnerite (PbO_2) were found. They were generated by bacteriological oxidation of white lead. This oxidation process has caused great damage to the original painting color. It is necessary to investigate and to remember the potential influence of bacteria on painting pigments while conducting conservation and restoration as well as art historical analysis.

INTRODUCTION

We analyzed representative samples of pigments from mural paintings of the 12th and 13th centuries in the monasteries of Bertubani and Udabno, which are parts of the unique grotto complex known as David-Baredga (1). The mineral composition of the pigments was analyzed by x-ray structure analysis using a Dron-2, and the chemical composition was determined by electron microprobe analysis using a Hitachi-5J.

RESULTS

The colors in the murals are: white, black, red, yellow, brown, green and blue. The white consists of pigments made of hydrocerussite ($2PbCO_3 \cdot Pb(OH)_2$), laurionite ($Pb(OH)Cl$), anglesite ($PbSO_4$), whewellite ($CaC_2O_4 \cdot H_2O$), vedellite ($CaC_2O_4 \cdot 2H_2O$). The brown-black consisted of plattnerite (PbO_2) and carbon (C). The reds consisted of hematite (Fe_2O_3), cinnabar (HgS), and minium (Pb_3O_4). The brown colours consisted of iron ochres on a base of hydrohematite, goethite ($FeO(OH)$) and limonite ($FeO(OH) \cdot nH_2O$). The green was glauconite ($(K,Na)(Fe,Al,Mg)_2(Si,Al)_4O_{10}(OH)_2$). The blue was determined to consist of gauyne-lazurite, structurally closer to gauyne ($Na_6Ca_2(AlSiO_4) \cdot SO_4$).

DISCUSSION

The analysis of the sampled material must be considered in relationship to the tradition of medieval painting technology and to the exploration of the natural, i.e. original as opposed to altered, pigment compositions. These analyses showed that two groups of colorants are present: 1, immutable or unweathered pigments (gypsum, anhydrite, calcite, hydrocerussite, carbon, hematite, cinnabar, minium, iron ochres, glauconite, jarosite, gauyne), and 2, compositions arising as a result of chemical transformation of pigments under the influence of outside or environmental factors. Among the later group, we establish two subgroups: 2a, altered compositions comprising only a small proportion and which do not much influence the color and visual effect of the mural, and 2b, transformation products which change the color. Whewellite and vedellite are generated while burning gypsum in the process of plaster manufacture. We think that laurionite and anglesite are the result of interaction of the lead-based pigments and a ground water containing sulphate, chloride and sodium ions, which is characteristic of this region.

Plattnerite belongs to the 2b subgroup. It is widespread in all colour layers of the murals. Our on-site observation of mural technology permitted us to conclude that $PbO2$ is a transformation product of white lead. It may not completely substitute, and this causes change and great damage to the original colour of the painting. Such great changes of lead-containing pigments were found in mural paintings of the church of St. John the Divine in Rostov, Russia (2) and in Italy (3). Based on the reports of these authors (2,3), we believe that a similar process of oxidation of lead-containing pigments under bacteriological influence may also be the cause of the presence of plattnerite in the murals.

Our investigation shows that various pigments were used to make each of the colors as cited above. However, the visual effect of the medieval murals is greatly changed due to the transformation of the white areas painted using lead white to brown-black. Recognition of this process is a matter of interest both for art history and the conservation and restoration of the monuments.

Acknowledgements

We sincerely thank O.J. Dubik for x-ray structure analyses, E. Sherbovskyi for microprobe analyses and J. Petushkova for discussing the materials.

References

1. G.N. Tchubinachvili, The Grotto Monasteries of David-Garega (Tbilisi, 1948), p. 120.

2. J.P. Petushkova and N.N. Lyalikova, Studies in Conservation 31 (1986), pp. 65-69.

3. M. Matteni and A. Moles, "The Reconversion of Oxidized White Lead in Mural Painting: A Control After a Five Year Period", in ICOM Committee for Conservation, 6th Triennial Meeting, Ottawa (1981), 81/15/1-8.

ROCK VARNISH ON A PRE-COLUMBIAN GREEN JASPER FROM THE TROPICAL RAIN FOREST (THE AHAW PECTORAL)

LEONCIO A. GARZA-VALDES* AND BRIAN STROSS**
*Rosa Verde Tower, 343 W. Houston St., Suite 612, San Antonio, TX 78205
**Anthropology Department, University of Texas at Austin, Austin, TX 78712

INTRODUCTION

With aging, many rocks surfaces not in direct contact with the soil, develop a black coating. This deposit has been given different names: desert varnish [1,2], rock varnish [3], and biogenic rock varnish [4]. The rock varnish is a dark coating, usually 10 to 30 micrometers thick, with a variation from a few micrometers to over 500 micrometers [5]. It is composed of manganese and iron oxides, silica, and clay minerals. A similar type of coating is accreted in pottery found in ancient tombs [6].

The biological origin of the varnish was postulated in 1921 by Francis [7]. Later, Dorn [8], and Dorn and Oberlander [9], described several bacterial groups, which were encapsulated in the rock varnish. Taylor-George et al. [10], and Staley et al. [11,12] described microcolonial fungi (MCF) that are able to deposit the different components of the varnish. One species of this MCF was succesfully isolated and cultured, reported as Lichenothelia convexa [13].

The components of a dark coating on an ancient green jasper carving from Southeast Mesoamerica were analyzed and compared to previous rock varnish analyses.

DESCRIPTION

The Ahaw Pectoral is a carved gem made of an emerald-green jasper. It is composed of five pieces (Plate IA and IB). The pectoral's dimensions are: height 19.5 cm, width 16.5 cm, and weight 323.5 g. It is part of a jade collection in Texas known as the "Ahaw Collection." It was acquired as unprovenienced in 1955. In 1978 the Ahaw Pectoral was studied by two Mesoamerican archaeologists in Texas, they concluded that the pectoral was one of the most important pre-Columbian artifacts known. In 1983 the pectoral was reported to the Society for American Archaeology [14], describing the similarity to a pectoral worn by a ruler (shaman) depicted in La Venta Altar IV, in Tabasco, Mexico [15] and to a design in two earspools found in Tomb C, also in La Venta, Tabasco [16]. In the same year (1983) two pre-Columbian art connoisseurs from New York, indicated that, based on their knowledge of mesoamerican iconography and style, the artifact was a fake, and they offered to buy it for their fake collection.

In 1985 another Mesoamerican archaeologist indicated that the artifact was worn by a king (shaman) at the headdress as a symbol of rulership; in 1990 this piece was reported to the Materials Research Society as the Eight Jaguar Crown [17]. In this paper one of us (Stross) describes the pectoral's symbolism and iconography.

Some part of the pectoral's surface were repolished in modern times. However, some areas were left intact and they contain the dark varnish coating reported in this study.

METHODS OF INVESTIGATION

The dark deposits at the pectoral's surface were studied with the following instruments: (1) optical microscope, (2) Fourier transform infrared spectrography (FT-IR), (3) energy dispersive spectrometry and scanning

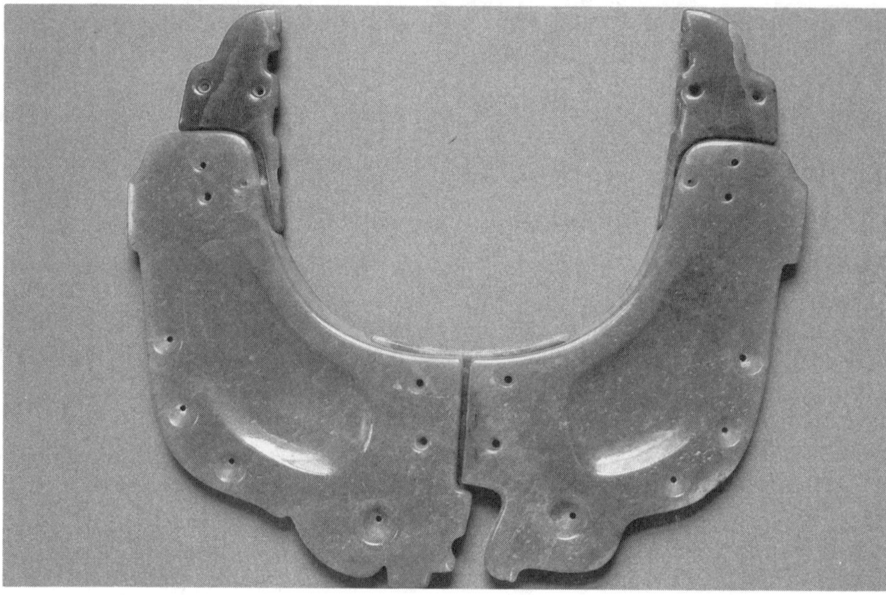

Plate IA. Frontal View of the Ahaw Pectoral

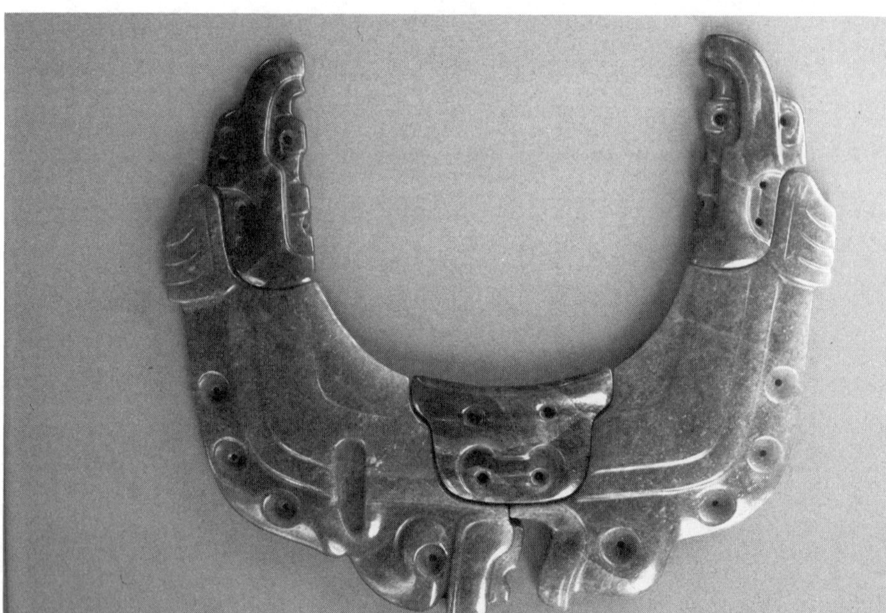

Plate IB. Rear view of the Ahaw Pectoral

(See Plates IA and IB on page vii in the front of this book)

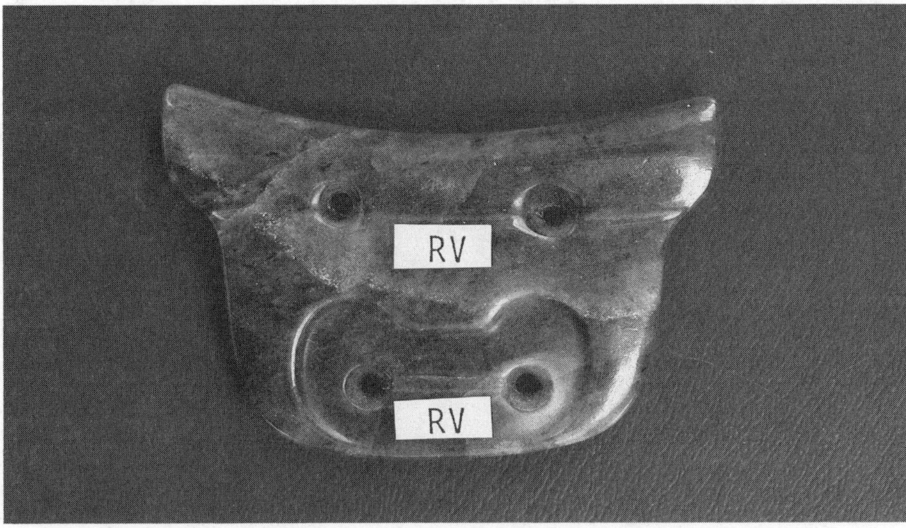

Plate IIA. Central piece of the Ahaw Pectoral (section three).
RV= string grooves with rock varnish remnants.

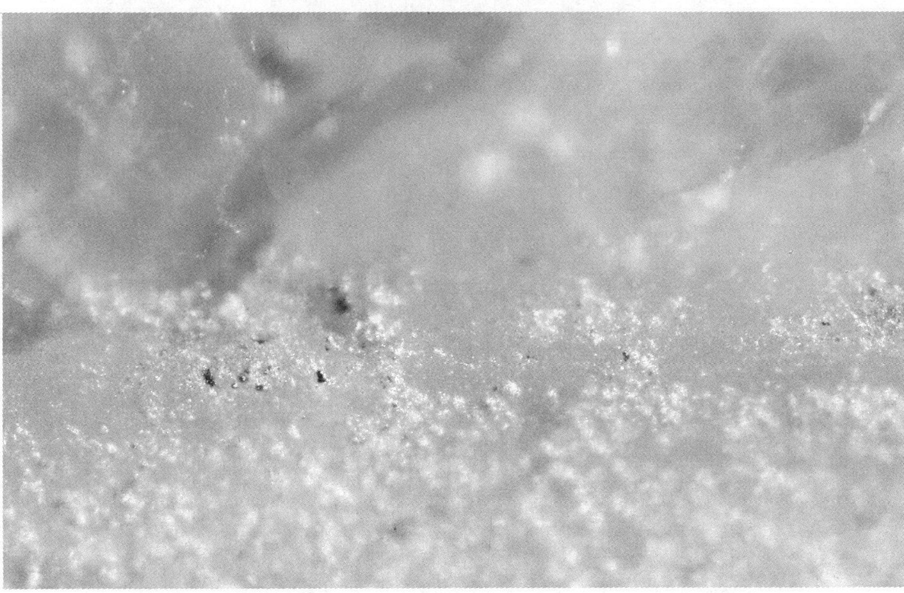

Plate IIB. Photomicrograph of Ahaw Pectoral section three, lower string groove. The upper half shows a repolished area, the lower half conserves the rock varnish. The separation zone between the two areas is well delineated. Visual field 0.81 mm X 0.55 mm.

(See Plates IIA and IIB on page viii in the front of this book)

electron microscope (EDS/SEM), and (4) wavelength dispersive spectrometry (WDS).

Optical microscopy

The varnish on the pectoral's surface was observed in situ using a Nikon Optiphot-Pol microscope, with a Nikon Microflex "FX" Series for photomicrography. Kodachrome 64 and TMAX 100 films were used. The amplification powers used were 20X, 40X, and 80X.

Infrared spectrometry

A few small particles of the dark coating were lifted from the pectoral's surface using a hobby knife. The particles were arranged on the surface of a potassium bromide (KBr) sample disk. This disk was then placed on the sample stage of a SpectroTech, Inc. Spectroscope infrared microscope, which was fitted into the sample compartment of a Bio-Rad Laboratories, Digilab Division FTS-15E / 3240 Fourier transform infrared (FT-IR) spectrophotometer. The FT-IR spectrum of the material was obtained by signal averaging 10,000 scans at a resolution of 4 wavenumbers with a blank area of the KBr disc used as a reference.

Energy dispersive spectrometry/SEM and wavelength dispersive spectrometry.

The dark coating was analyzed after removing scrapings from the pectoral's section one and section three. A hypodermic needle No. 25 was used to remove the samples. Analysis were made on a JEOL 733 superprobe using wavelength dispersive spectrometers. Three synthetic varnish standards [18] were each analyzed several times to check the accuracy of the microprobe calibrations. Analyses were made using a 1 micrometer spot, a 15 keV accelerating voltage, and a beam current of 15 nA. The microprobe was configurated and standardized using the protocol previously described [19]. Counting time for each element ranged from 10 to 20 s. A Bence-Albee matrix correction protocol was used. Chlorine was calibrated on an apatite standard. Analytic totals (oxide weights) have been corrected for the presence of the Cl^- anion.

The "varnish" coating scrapings, after removal from the green jasper, were embedded in epoxy and polished. The resulting slide was carbon-coated and analyzed using energy dispersive spectrometry, scanning electron microscopy, and wavelength dispersive spectrometry.

DATA AND RESULTS

The lapidary technology used in the carving of the Ahaw Pectoral and the bio-cultural remnants found (1) charcoal fragments, (2) cinnabar, (3) copal, (4) cordage traces, (5) blue feather remnants, (6) root deposits, and (7) phytoliths were reported previously [20].

The surfaces where the pieces of the pectoral had been bound together and the string grooves [21] were not repolished (Plate IIA). The division zones between the repolished areas and the rock varnish remnants were analyzed with the optical microscope (Plate IIB) and with the SEM. Many microcolonial fungi (MCF) were found encapsulated by the rock varnish. A sample was removed for SEM analyses. The rock varnish thickness at this area (Fig. 1) is of 20 micrometers.

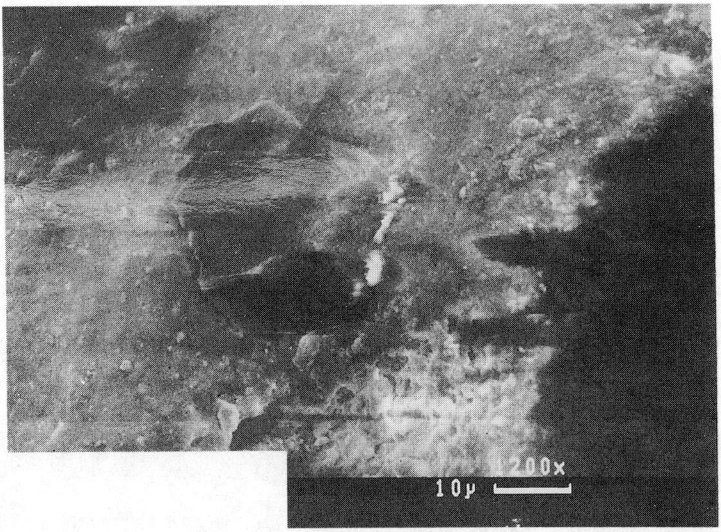

Figure 1. SEM Photomicrograph of the Ahaw Pectoral's rock varnish, the varnish's thickness where sample was obtained (right lower quadrant in photo) is of 20 micrometers.

The FT-IR spectrum (Fig. 2) of the pectoral's section two patina, is a mixture of absorption bands from the different varnish components. The phases present (Table I) are: calcium carbonate, celullose, probably copal, silica, and complex manganese oxides and/or manganese and chlorine silicates (friedelite group).

TABLE I. FT-IR SPECTRUM FROM ROCK VARNISH OBTAINED FROM AHAW PECTORAL'S SECTION TWO

Wavenumbers (1/cm)	Description
3,650 - 3,050	-O-H stretching and or friedelite absorption
3,000 - 2,850	-C-H stretching
1,760 - 1,710	-C=O stretching (esters, carboxylic acids, ketones, aldehydes)
1,710 - 1,550	-C=O stretching
a) 1,710 - 1,680	for acids, esters
b) 1,640	celullose
c) 1,610 - 1,550	-C=O stretching of carboxylic acids
d) 1,590 peak	$-COO^-$ metal salt
e) 1,680 - 1,620	-C=C- stretching (non conjugated)
1,800, 1,430, 880, 720	calcium carbonate
1,160, 1,045, 1,010, 960	consistent with friedelite absorption band
1,080, 800, 780, 520	silica

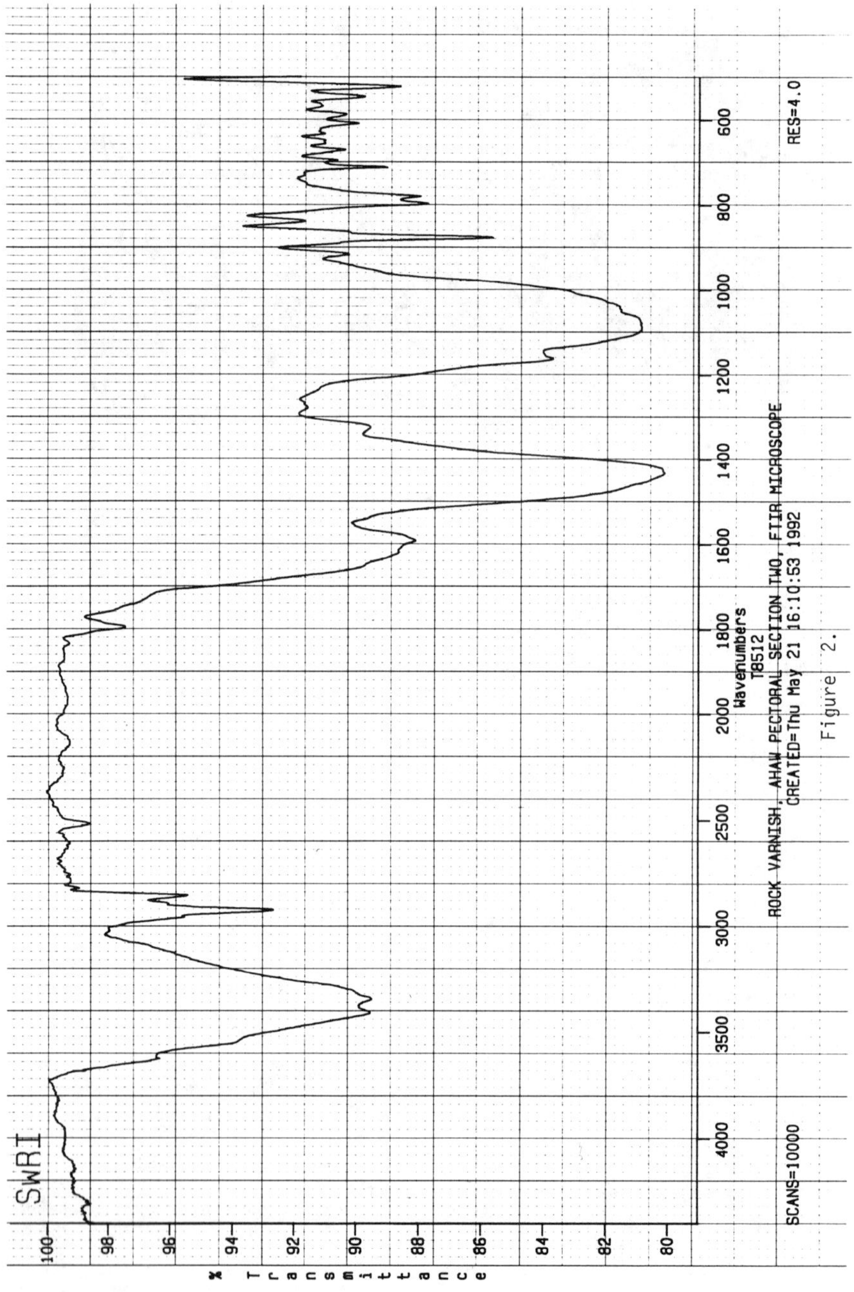
SWRI
% Transmittance
100
98
96
94
92
90
88
86
84
82
80
4000
3500
3000
2500
2000
1800
1600
1400
1200
1000
800
600
Wavenumbers
T8512
ROCK VARNISH, AHAW PECTORAL SECTION TWO, FTIR MICROSCOPE
CREATED=Thu May 21 16:10:53 1992
SCANS=10000
RES=4.0

Figure 2.

TABLE II

WAVE LENGTH DISPERSIVE ANALYSES ON "ROCK VARNISH" MADE ON SCRAPINGS FROM AHAW PECTORAL CENTRAL PIECE
ANALYSES HAVE BEEN NORMALIZED TO 100% OXIDE WEIGHTS AND CORRECTED FOR Cl^- AS AN ANION

Point	SiO_2	Na_2O	Al_2O_3	MgO	P_2O_5	K_2O	CaO	TiO_2	BaO
12	20.30	0.18	0.47	0.17	0.06	0.41	22.30	BDL	BDL
13	25.29	0.45	0.96	0.11	BDL	0.67	18.90	BDL	0.15
14	39.53	0.25	1.99	0.12	BDL	0.34	14.58	BDL	0.20
15	17.91	0.30	1.76	0.14	BDL	0.41	21.28	BDL	BDL
16	21.99	0.25	0.75	0.17	0.11	0.32	21.96	BDL	BDL
AVERAGE	25.00	0.29	1.19	0.14	0.03	0.43	19.80	BDL	0.07

Point	Fe_2O_3	MnO	SO_3	Cl	TOTAL	CORRECTION OF OXIDE TOTAL	TOTAL BEFORE NORMALIZATION
12	0.96	50.99	0.60	4.59	101.04	1.04	68.51
13	1.42	47.57	0.98	4.50	101.02	1.02	49.59
14	2.83	36.53	0.60	3.79	100.86	0.86	45.98
15	2.00	51.44	0.78	5.12	101.16	1.16	45.18
16	1.53	48.29	1.29	4.30	100.97	0.97	53.85
AVERAGE	1.77	46.97	0.85	4.46	101.01	1.01	52.62

Analyses by P. Bierman, University of Washington 9/25/92

The wavelength dispersive analyses (Table II), indicates that the dark color coating is predominantly manganese with lesser amounts of chlorine, calcium, and silica, and trace amounts of potassium, sulfur, iron, magnesium, aluminum, and sodium.

The energy dispersive spectrum in Fig. 3 shows a high concentration of manganese, calcium, chlorine, and silica.

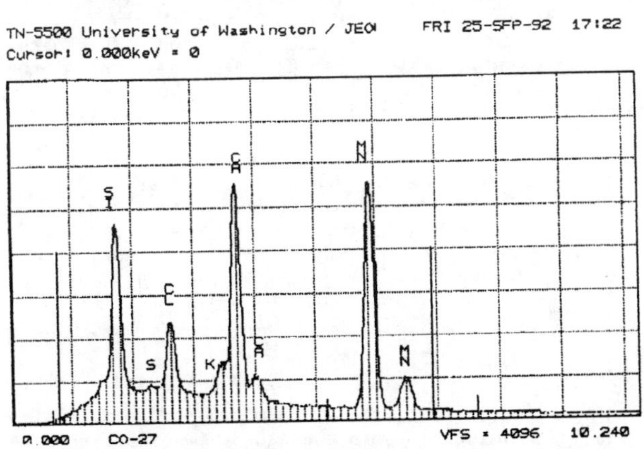

Figure 3. EDS Lower string groove, Ahaw Pectoral central piece, point 15.

PECTORAL'S SYMBOLISM

Iconographic analysis, referring to meanings interpreted from visual elements deliberately constructed by an artifact's makers, can augment artifact examination through physical and chemical tests, and on occasion can provide compelling information concerning the artifact's history and significance. Such is the case with the green Ahaw Pectoral from Guatemala, with its Olmec derived iconography and Maya manufacture [22].

The Ahaw Pectoral's form, as seen specifically in the crocodile heads at either end of the U-shaped artifact (Plate IA), has been shown to represent the reptilian jaws symbolizing a portal to the Otherworld that was part of the Olmec and later the Mayan symbolic traditions [23]. Worn by a Classic Maya ruler, this pectoral representing the cosmic portal, was thus a visible display of the ruler's power deriving from his symbolic control over the portal to the Otherworld.

More than just depicting the jaws of the cosmic portal, the pectoral also has an Olmecoid face at the bottom edge of its U-shape that constitutes also an anthropomorphic image of the sky at times when the Milky Way is gathered along the southern and eastern edges of the sky to form what the Classic Maya recognized as a cosmic portal in the sky [24], a doorway through which gods and shamans could pass from one cosmic plane to another. It is only very recently that the "portal" stage in the Milky Way's transformational cycle has been discovered to have been of importance to the Classic Maya who lived between 200 and 900 AD [25], and so the iconography supports the artifact's authentication as derived from other sources.

Three darker green stones of the Ahaw Pectoral form the apices of an equilateral triangle. The arrangement is surely deliberate and these apical stones are icons of the three stone hearth located at the heart of every Maya house. The stones of the Maya fireplace, with their triangular arrangement, contain the transforming fire of culture and communication. On the cosmic level, in terms of the Maya creation myth, each stone stands for one of the cosmic realms, earth, sky, and sea [26]. The first three stone hearth of cosmic creation mythology is reflected in the sky, as observed in Tedlock's translation of the sacred book of the Quiche Maya [27], and the reflection is made concrete in all likelihood in the triangle formed by Orion's stars Rigel, Saiph, and Alnilam which enclose the fiery Orion Nebula [28].

For the Maya, the human body contained the cosmic hearth in microcosm. Through linguistic and iconographic evidence it is possible to specify where on the human body the three hearthstones and the sacred fire were thought by the Maya to be located, and that is in the pelvic girdle, composed of the sacrum and the two hipbones [29]. These three bones are represented on the Ahaw Pectoral by the central darker stone (Plate IIA), which is shaped like the Maya Ahaw glyph, and by the two lighter larger stones flanking it (Plate IB).

In short, the Ahaw Pectoral illustrates multiple meanings embodied in its different iconographic elements, and these meanings focus on the display of royal power in terms of relating the pectoral and its wearer to creation, to the cosmic doorway, to the sky, and thus ultimately to control over movement between cosmic realms. When a Maya ruler wore the Ahaw Pectoral, he became the center of the universe, placing himself inside the cosmic hearth of creation and transforming himself into "the cosmic flame."

DISCUSSION

The mechanism by which rock varnish forms is still uncertain. The predominant idea is a biological origin by activity of bacteria and or fungi (MCF) capable of oxidizing, concentrating, and depositing manganese. It is a very slow process, to have a natural continuous varnish layer, may require many thousand years [30].

Deserts are the best known ecosystem location of rock varnish, but natural manganese rich varnish may form under different conditions on different stable surfaces. The different phases that form the natural varnish are still in study. The EDS spectra from desert varnish are different from the tomb varnish spectra; in this manganese-rich black coating there is a chlorine bearing phase that is consistent with a mineral in the friedelite group ($[Mn, Fe]_8Si_6O_{15}[OH,Cl]_{10}$).

CONCLUSION

The natural varnish found on the pectoral's surface has high levels of manganese, silica, chlorine, and calcium carbonate. There are different types of natural varnish, two of which are "desert varnish" and "tomb varnish." Research is presently being conducted on the study of the different types of natural manganese rich coatings.

ACKNOWLEDGEMENTS

The authors want to thank Paul Bierman, Department of Geology, University of Washington, Seattle, Washington; Dr. George E. Harlow, Chairman and Curator, Department of Mineral Sciences, American Museum of Natural History, New York; and W. A. McMahon, Senior Research Scientist, Southwest Research Institute, San Antonio, Texas.

REFERENCES

1. C.H. White, Desert varnish, Am. J. Sci. 9, 413-420 (1924).
2. J.D. Laudermilk, On the origin of desert varnish, Am. J. Sci. 21, 51-66 (1931).
3. R.I. Dorn, and T.M. Oberlander, Rock varnish, Prog. Geogr. 6, 317-367 (1982).
4. W.E. Krumbein, and K. Jens, Biogenic rock varnishes of the Negev Desert (Israel): an ecological study of iron and manganese transformation by cyanobacteria and fungi, Oecologia 50, 25-38 (1981).
5. R.I. Dorn, and T.M. Oberlander, op. cit., 317 (1982).
6. M. Aronson, and W.D. Kingery, Patination of ceramic objects, modern and prehistoric: A case from West Mexico, ed. by P.B. Vandiver, J.R. Druzik, and G. Wheeler, Materials Issues in Art and Archaeology II. Mater. Res. Soc. 185, 571-589 (1991).
7. W.D. Francis, The origin of black coatings or iron and manganese oxides on rocks, Proc. R. Soc. Queensl. 32, 110-116 (1921).
8. R.I. Dorn, A biological model of rock varnish formation, Abstracts: Pacific Division, American Association for the Advancement of Science, University of California, Davis, 50 (1980).
9. R.I. Dorn, and T.M. Oberlander, Microbial origin of desert varnish, Science 213, 1245-1247 (1981).
10. S. Taylor-George, F. Palmer, J.T. Staley, D.J. Borns, B. Curtiss, and J.B. Adams, Fungi and bacteria involved in desert varnish formation, Microbiol. Ecol. 9, 227-245 (1983).
11. J.T. Staley, F. Palmer, and J.B. Adams, Microcolonial fungi: common inhabitants of desert rocks, Science 213, 1093-1094 (1982).
12. J.T. Staley, J.B. Adams, and F.E. Palmer, Desert varnish: A Biological perspective, Soil Biochemistry 7, 173-195 (1992).
13. A. Henssen, LICHENOTHELIA, a group of microfungi on rocks. Progress and problems in lichenology in the eighties, Bibl. Lichenol. 25, 257-293 (1987).

14. L.A. Garza-Valdes, Specific iconography of Olmec Rulers, Soc. Amer. Arch. 48, 55 (1983).
15. T. Dickey, V. Muse, and H. Wiencek, The God-Kings of Mexico, Treasures of the World, Stonehenge Press, Inc. Chicago, Ill. 11 (1982).
16. P. Drucker, La Venta, Tabasco, A study of Olmec ceramics and art, Bur. Amer. Ethnol. 153, Fig. 46b (1952).
17. L.A. Garza-Valdes, Technology and weathering of mesoamerican jades as guides to authenticity, ed. by P.B. Vandiver, J.R. Druzik, and G. Wheeler, Materials Issues in Art and Archaeology II. Mater. Res. Soc. 185, 343 (1991).
18. P.R. Bierman, and A.R. Gillespie, Accuracy of rock varnish chemical analyses; implications for cation ratio dating, Geology 19, 196-199 (1991).
19. P.R. Bierman, and K. Harry, Rock varnish cation-ratios may not be a reliable method for dating lithic artifacts, in this volume.
20. L.A. Garza-Valdes, op. cit., 348 (1991).
21. L.A. Garza-Valdes, op. cit., 348 (1991).
22. B. Stross, The Heavenly Portal Carved in Green Stone, unpublished m.s., Dept. of Anthropology, University of Texas, Austin (1992a).
23. B. Stross, op. cit., (1992a).
24. B. Stross, op. cit., (1992a).
25. L. Schele, Workbook for the XVIth Maya Hieroglyphic Workshop at Texas, Dept. of Art and Art History and Institute of Latin American Studies, University of Texas, Austin (1992).
26. L. Schele, op. cit., (1992).
27. D. Tedlock, Popol Vuh, New York, Simon and Schuster, 261 (1985).
28. B. Stross and L.A. Garza-Valdes, The Shamanic Fire of Maya Creation, unpublished m.s., Dept. of Anthropology, University of Texas, Austin (1992).
29. B. Stross, The Sacrum Bone in the Portal, unpublished m.s., Dept. of Anthropology, University of Texas, Austin (1992b).
30. J.T. Staley, et. al., op. cit., 175 (1992).

PART IV B

Deterioration, Treatment and Structure in Architectural and Building Materials

RECENT ECO-ARCHAEOMETRIC STUDIES OF SALINIZATION AND DESALINIZATION OF KARNAK TEMPLES, EGYPT

GEORGE BURNS, K.M. MATSUI*, R.M. POTTER AND T.C. BILLARD†,
Department of Chemistry, University of Toronto, Toronto, Canada M5S 1A1

ABSTRACT

Earlier studies demonstrated quantitatively that recent salinization of the Karnak Temples, Egypt, is due to evaporation of saline groundwaters under the temples. Furthermore, it was determined that the high salinity of groundwaters is due to extensive evaporotranspiration over the adjacent irrigated fields. To abate salinization, in 1986, a desalinization station was installed in the central part of Karnak at Lake Amun. Present studies indicate that by 1987, the desalinization effort was already very successful, with the salinity of the Lake being lowered by a factor of more than 37. Some residual salinity was still evident a year after desalinization stations were installed. It appears that this residual salinity is due primarily to leaching of salts which were deposited around the Lake before 1986. Thus, not only the Lake, but also the area adjacent to it appear to be desalinized as a result of construction of the desalinization station.

INTRODUCTION

Recently, considerable concern has been expressed about the massive deterioration of antiquities in various parts of the Developing World [1-5]. Although much of this deterioration is due to factors which are outside of the domain of science,[3] a considerable amount of the decay can be understood and even predicted using techniques readily available to modern science and technology.[3-5] Since this pathway is often not pursued, and since archaeological explorations are becoming more extensive than ever before, it has been argued that "future generations may look back at the present as that time when the cultural heritage of the human race was most extensively damaged"[3]. This situation is more difficult to avoid than it may seem because fundamental processes responsible for the deterioration, being complex and site-specific, are often not understood.[3] In such cases it is next to impossible to prevent the insidious decay, since it is not even anticipated. Furthermore, when monuments begin to deteriorate, attempts to conserve them may be at least in some cases counter-productive [5].

One way to deal with such problems is to concentrate on studies of fundamental physical and chemical processes which are contributing now, or may possibly contribute in the future, to deterioration of ancient monuments, artifacts and archaeological sites. It was argued that such work should *precede* much conservation effort.[3-5] Furthermore, such studies should be pursued *during and after* conservation and restoration efforts in order to monitor and evaluate its effectiveness and value. To stress the importance of this approach it has been suggested to define it as a separate multidisciplinary field of *eco-archaeometry* [3-5].

One example of the eco-archaeometric approach was the study of the massive destructive salinization processes which occurred in the Karnak Temples, located near the city of Luxor, Egypt.[6] Once the sources of salts were identified and the extent of salinization quantified,[6] it became possible to consider various ways to reduce salinization.

The salinization of the Karnak Temples was particularly extensive in its central part, where the sacred Lake Amun, approximately 2.5 acres in size (Fig. 1), is situated. Lake Amun, excavated in the 19th century, is largely restored; its banks, being vertical, make it look like a giant swimming pool. For the purpose of restoration, high quality limestone blocks were used. However, recently this limestone was severely damaged by salinization, with large pieces of it flaking off at a mere touch (Fig. 2).

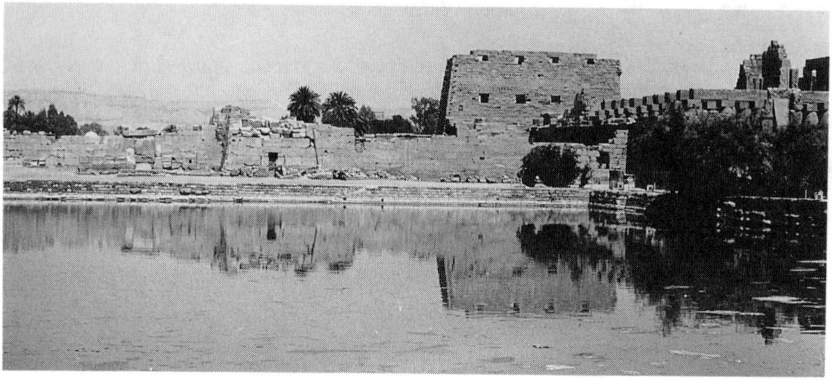

Fig. 1: The Lake Amun of the Karnak Temples.

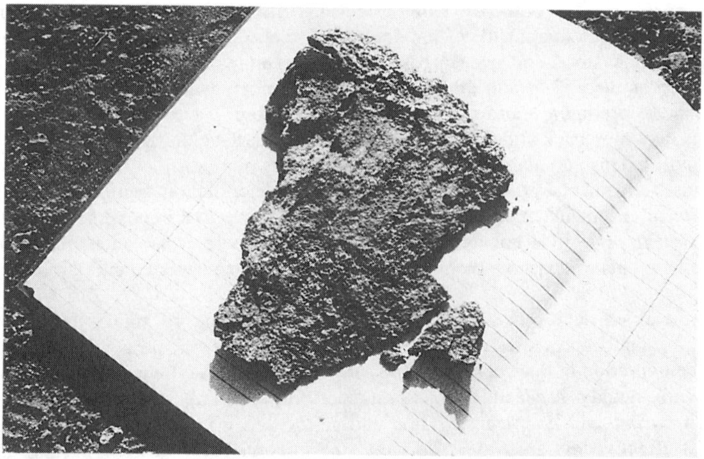

Fig. 2: Flaked sandstone from the banks of the Lake Amun upon a research notebook.

Before the salinization processes were identified in 1980[6], it had been argued that the appearance of large salt deposits at the Lake's shores, and high salinity of the its waters, are beneficial because the Lake attracted large quantities of salt, which otherwise would have spread within the Temples, thus increasing their overall salinization. However, once the sources of salinization were quantified, it became apparent that the Lake should be desalinized because, as it turned out, its high salinity *augments* the salinization, and consequently pollutes, the surrounding antiquities. Subsequently, an extensive project, financed in part by the World Bank, was initiated, and completed by 1986. The project involved pumping the Nile water into Lake Amun and simultaneous removal of contaminated saline water from it, in such a way that the overall water level in the Lake remained time-invariant.

Lake Amun was fed by groundwater which is moving slowly from the irrigated fields located to the east of Karnak. In ancient times, these fields were not irrigated, and therefore, in all probability, the salinity of the groundwater in the Karnak area was low. It was the construction of the Aswan Dams that made possible the cultivation of new fields east of Karnak. Evapotranspiration over these fields concentrated the salts dissolved in the Nile water, so that as it seeped into the ground, the groundwater salinity became about nine times that of the Nile.[7] Before the 1986, salt in the waters of Lake Amun was further concentrated by evaporation of the Lake by yet another factor of approximately ten.[8,9]

The object of the studies described below is the follow-up of our previous work [6,7]: it was necessary to determine to what extent the 1986 desalinization effort was effective. For this purpose in 1987 we assembled samples of (1) the Lake Amun waters near the eastern banks of the Lake, opposite to the clean water inlets, which are located at western banks, (2) the waters from the Canada House well, which is located immediately to the east of the Karnak Temples, and (3) Nile waters. The analyses of these waters and interpretation of data are presented below.

EXPERIMENTAL

K^+ and Na^+ were determined by Atomic Absorption Spectroscopy using a Techtron Model AA-4 spectrophotometer. K^+ samples were run in quintuplicate and Na^+, is triplicate. Ca^{2+} and Mg^{2+} were determined in triplicate by Inductively Coupled Plasma Atomic Emission Spectroscopy, using an ARL 34000 ICP spectrometer. Cl^- and SO_4^{2-} were determined in triplicate by Ion Chromatography, using a Dionex Q/C analyzer. HCO^-_3 and CO_3^{2-} were determined by pH titrations.

RESULTS

The results of the analyses of water samples assembled in 1987 are presented in Tables 1-3. The concentration units in all tables given herein are meq/l and uncertainties in all measurements are given with 90% confidence limits. Two samples from each site were collected. In some cases, the confidence limits for the two samples did not overlap; in such cases, the two samples were assumed to represent real differences in composition. Therefore, the uncertainty limits were calculated on the basis of a sample size of two, using the t-table with degrees of freedom equal to one. Such treatment of the experimental data led to a considerable increase in uncertainties of the confidence limits. In Tables 1-3, L refers to Luxor waters, LA to Lake Amun, CHW, to Canada House well waters, 87 to the year samples were assembled. The digit between letters and 87 refers to the sample number. In Table 4, the ion concentrations in the Nile at Luxor are given; in addition to new 1977 and 1987 data, measurements made in 1969 and 1970, reported earlier,[8,9] are given. The 1977 groundwater salinity data were obtained using samples collected by us from the Canada House well.

Table 1. Ion concentrations in the Nile at Luxor (1987)

Ion	L187	L287	Average
Na^+	0.86 ± 0.01	0.85 ± 0.01	0.86 ± 0.01
K^+	0.14 ± 0.01	0.12 ± 0.01	0.13 ± 0.01
$Na^+ + K^+$	1.00 ± 0.01	0.97 ± 0.01	1.0 ± 0.01
Ca^{2+}	1.32 ± 0.01	1.32 ± 0.01	1.32 ± 0.01
Mg^{2+}	0.65 ± 0.01	0.62 ± 0.01	0.6 ± 0.1
TOTAL CATIONS	2.97 ± 0.02	2.91 ± 0.02	2.9 ± 0.01
Cl^-	0.22 ± 0.01	0.22 ± 0.01	0.22 ± 0.01
SO_4^{2-}	0.35 ± 0.03	0.35 ± 0.03	0.35 ± 0.04
HCO_3^-	1.9 ± 0.4	1.9 ± 0.4	1.9 ± 0.4
CO_3^{2-}	0.19 ± 0.04	0.19 ± 0.04	0.19 ± 0.04
TOTAL ANIONS	2.7 ± 0.4	2.7 ± 0.4	2.7 ± 0.4

Table 2. Ion concentrations in the Lake Amun (1987)

Ion	LA187	LA287	Average
Na^+	1.09 ± 0.03	1.22 ± 0.06	1.2 ± 0.4
K^+	0.30 ± 0.01	0.33 ± 0.01	0.3 ± 0.1
$Na^+ + K^+$	1.39 ± 0.03	1.55 ± 0.06	1.5 ± 0.4
Ca^{2+}	0.81 ± 0.01	0.83 ± 0.01	0.82 ± 0.01
Mg^{2+}	1.03 ± 0.01	1.09 ± 0.01	1.1 ± 0.1
TOTAL CATIONS	3.23 ± 0.05	3.5 ± 0.1	3.4 ± 0.4
Cl^-	0.58 ± 0.04	0.61 ± 0.04	0.60 ± 0.04
SO_4^{2-}	0.75 ± 0.02	0.78 ± 0.04	0.77 ± 0.04
HCO_3^-	1.8 ± 0.4	1.8 ± 0.3	1.8 ± 0.4
CO_3^{2-}	0.18 ± 0.04	0.18 ± 0.04	0.18 ± 0.04
TOTAL ANIONS	3.3 ± 0.4	3.4 ± 0.3	3.4 ± 0.4

Table 3. Ion concentrations in the groundwater (1987)

Ion	CHW 187	CHW 287	Average
Na^+	9.1 ± 0.1	10.4 ± 0.1	10 ± 4
K^+	0.43 ± 0.01	0.41 ± 0.01	0.42 ± 0.01
$Na^+ + K^+$	9.5 ± 0.1	10.8 ± 0.1	10 ± 4
Ca^{2+}	6.62 ± 0.01	4.52 ± 0.01	6 ± 6
Mg^{2+}	3.83 ± 0.01	4.79 ± 0.03	4 ± 3
TOTAL CATIONS	20.0 ± 0.1	20.1 ± 0.1	20 ± 8
Cl^-	4.8 ± 0.7	4.8 ± 0.4	(5 ± 1)
SO_4^{2-}	7.7 ± 0.5	8.1 ± 0.3	8 ± 1
HCO_3^-	5.6 ± 0.7	5.6 ± 0.7	6 ± 1
CO_3^{2-}	0.7 ± 0.1	0.7 ± 0.1	0.7 ± 0.1
TOTAL ANIONS	19 ± 1	19 ± 1	19 ± 2

Table 4. Ion concentrations in the Nile at Luxor 1969-1987

Ion	1969	1970	1977	1987	average
$Na^+ + K^+$	1.0	0.81	1.52	1.0	1.1 ± 0.4
Ca^{2+}	1.0	1.0	1.4	1.32	1.2 ± 0.2
Mg^{2+}	0.79	1.0	0.88	0.6	0.8 ± 0.2
Cl^-	0.80	0.3	0.33	0.22	0.4 ± 0.2
SO_4^{2-}	0.21	0.15	0.35	0.35	0.3 ± 0.1
HCO_3^-	1.6	2.1	2.6	1.9	2.0 ± 0.5
CO_3^{2-}	0.20	0.26	0.21	0.19	0.22 ± 0.04

DISCUSSION

The principal finding of our study is that, as a result of installation of the desalinization station, the salinity of Lake Amun has been reduced from 127 ± 5 meq/L to 3.4 ± 5 meq/L (Table 5), *i.e.* a factor of 37. This finding is valid for the overall salinity in the Lake because the anion to cation ratios for all samples was equal to unity within their 90% confidence limits (Table 6), which implies that all the principal ions had been accounted for. Consequently, the principal conclusion is that desalinization of Lake Amun was very successful.

In order to determine whether there is a difference between the Lake's and Nile, the experimental concentration ratio (ECR) of the various ions in Lake Amun to those of the Nile, *i.e.* $[ion]_{Lake}/[ion]_{Nile}$, were obtained (Table 7).

Table5. Comparison of total cation concentrations in Lake Amun, the groundwater and the Nile

Year	Lake Amun	Groundwater	Nile
1969	1.2×10^2	15	2.8
1970	1.2×10^2	18	2.8
1977	127 ± 5	22 ± 1	
1987	3.4 ± 0.8	20 ± 8	2.9 ± 0.1

Table 6. Total anion to cation gm-equivalent ratios (1987)

Sample	Ratio
L187	0.9 ± 0.1
L287	0.9 ± 0.1
LA187	1.0 ± 0.1
LA287	1.0 ± 0.1
CHW187	0.95 ± 0.05
CHW287	0.95 ± 0.05

Table 7. Ion concentration ratios - Lake Amun/Nile

Ion	Using Ave. Nile Data	Using 1987 Nile Data
$Na^+ + K^+$	1.4 ± 0.7	1.5 ± 0.5
Ca^{2+}	0.7 ± 0.1	0.62 ± 0.01
Mg^{2+}	1.3 ± 0.2	1.7 ± 0.3
Cl^-	1.5 ± 1.2	2.7 ± 0.2
SO_4^{2-}	2.6 ± 0.9	2.2 ± 0.3
HCO_3^-	0.9 ± 0.3	1.0 ± 0.4
CO_3^{2-}	0.8 ± 0.3	1.0 ± 0.4

In the first column of this Table, the ion concentrations of the Nile averaged over time (Table 4) are used to obtain ratios; in the second, 1987 average Nile river data (Table 2) are used. The second column (Table 7) is believed to be more reliable because it provides ratios obtained using samples assembled at the *same* time. In both sets of ratios, for ($Na^+ + K^+$), Mg^{2+}, Cl^- and SO_4^{2-} are higher than unity. The Ca^{2+} ratio is less than unity which, we believe, is due to a higher pH in Lake Amun. Higher pH increases the CO_3^{2-} concentration, precipitating $CaCO_3$; the Lake should be saturated with respect to $CaCO_3$. It was not possible for us to measure the pH at the time the samples were taken; consequently, the Ca^{2+}, HCO_3^- and CO_3^{2-} data cannot be used to assess the extent of desalinization of Lake Amun. Furthermore, ($Na^+ + K^+$) ratios have too high an uncertainty to consider them reliable within 90% confidence limit. However, for Mg^{2+}, Cl^- and SO_4^{2-} ratios are above unity even within this confidence limit.

To explain the latter results it was first surmised that the high Mg^{2+}, Cl^- and SO_4^{2-} ratios might be due to the influx of groundwater only. To check this possibility a dynamical model for the flow of saline water into the Lake was used. This model *assumes* that the influx of salts into the Lake is due to the inflow of saline groundwater only. The first step was to secure the calculated concentration ratios (CCR) based on this model. Calculations were made using the information on (1) the input-output of water and (2) the input-output of salts dissolved in the waters into and out of the Lake. It was furthermore assumed that the water flow in and out of Lake Amun is the steady-state. This assumption is reasonable because the level of the Lake has been maintained; therefore, equations for the flow of saline water is defined by $\Sigma Flow_{in} = \Sigma Flow_{out}$. For the flow of water this equation yields:

$$P_1 + q_1 = P_2 + q_2 + E \qquad (1)$$

where P_1 and P_2 are the mechanical pumping rate of water into and out of Lake Amun, respectively; q_1 and q_2 are the groundwater flow rate into and out of the Lake, and E is the evaporation rate at its surface. In addition, the total amount of Mg^{2+}, SO_4^{2-} and Cl^- that enter or leave the Lake are also conserved within the steady-state approximation. Consequently,

$$P_1 + q_1(C_{gw}) = (P_2 + q_2)\ (CCR) \qquad (2)$$

where C_{gw} is the average concentration ratio of the groundwater. This equation is valid if Mg^{2+}, SO_4^{2-} and Cl^- salts do not precipitate into or redissolve from the Lake and its environment, and if no ions were added or subtracted to water piped into the Lake.

P_1 was calculated from observations of the trajectory of the water which flowed out of the two pipes located at the Lake. It was noted that when the water stream was horizontally approximately 0.6m from the end of the two pipes used it had fallen a distance of 0.25m due to gravity. From this it follows that the linear velocity of the water stream in pipes is 2.7 m/s. The diameter of each pipe was approximately 0.1m; therefore, the total pumping rate into Lake Amun was estimated to be 40 l/s. The evaporation rate from the surface of Lake Amun, E, was taken to be 3 l/s.[12] The groundwater flow rate, q_1 and q_2, are governed by Darcy's Law and therefore they can be expressed in terms of the saturated hydraulic conductivity and the change in hydraulic head (*i.e.* $|\mathbf{q_i}| = |-K_s \nabla P_i|$, where K_s is the hydraulic conductivity of water below the water table). A value of 0.003 cm/min was used for the K_s in the saturated region below the water table.[6] The groundwater flow rates, q_1 and q_2, were obtained by numerically solving the equation using a finite element model [11]. The equation (1) was used to calculate the pumping rate out of Lake Amun (P_2) and was found to be 37 l/s.

In order to obtain CCR for Mg^{2+}, Cl^- and SO_4^{2-} from the equation (2), concentration ratios for the ions in the groundwater at Karnak should be secured. Such data were assembled in 1969 and 1970 from wells drilled in the Karnak area specifically for the purpose of measuring ion concentrations [8], and in 1977 from the Canada House well (Table 2). The CCR calculated in such a way for Mg^{2+}, Cl^- and SO_4^{2-} were found to be 1.1 ± 0.1. Note that the model used requires that the CCR for *all* ions should be the same. However, the measured concentration ratio for the three ions, being 1.7 ± 0.3, 2.7 ± 0.2 and 2.2 ± 0.3 respectively, are ion-specific and are significantly higher than unity (Table 7).

To explain the discrepancy between CCR and ECR data, the initial assumptions used for the CCR calculations have to be reassessed: the discrepancy may be due to ion sources other than those included in the CCR calculations. Indeed, at least three sources are identifiable:

(1) If Nile waters are pre-treated before they are pumped into the Lake, one may expect that [Cl^-] and [SO_4^{2-}] may be marginally increased. This is because, normally, to purify raw water, Cl_2 is used to disinfect it, and aluminum sulphate - to precipitate sediments. It would be reasonable for such procedures to be operative in cases considered herein because the Nile flows fast enough to stir up sediments, and its raw water carries organic materials and parasites, such as dangerous liver fluke (schistosomiasis). Furthermore, it is necessary to use purified water which would be less likely to clog pipes and damage pumps. However, unusually high [Mg^{2+}] in the Lake cannot be explained in this fashion, for Mg^{2+} is not normally used for purification of water.

(2) Between 1965 and 1986 the concentration of ions in the Lake was some ten times higher than in the groundwater under the Temples. However, only $CaCO_3$ is expected to

precipitate under these conditions. The bottom of the Lake is predominately sand with some clay which is a rather porous medium, where soluble salts are not expected to be occluded for a period of about one year (1986-1987). Therefore, diffusion of ions, occluded in the bottom of the Lake, albeit possible, is not likely to be a major source of Mg^{2+}, SO_4^{2-} and Cl^- detected in the 1987 waters of the Lake.

(3) An intercomparison of ECR data suggests that ions usually associated with more soluble salts yield higher ECR. Thus, ECR for Cl^- salts is higher than that for SO_4^{2-} which is most likely to precipitate as slightly soluble calcium sulphate, which in turn is higher than that for Mg^{2+}, which precipitates as a rather insoluble dolomite, $CaMg(CO_3)_2$. This suggests that the source of high ECR are salts in a solid state which dissolve on contact with the purified Lake waters. It is known that between 1965 and 1986 the Lake acted as a source of concentrated saline waters. Such waters diffused out of the Lake first below the water table, and then by capillary action above it depositing copious quantities of salts immediately around the Lake. However, since 1986 the Lake was purified and clean water has been coming in contact by capillary forces with such salt deposits, creating salt concentration gradients. Such processes eventually are leaching salts back into the Lake. Such arguments predict that first the most soluble chlorides would be leached, to be followed to a lesser extent by SO_4^{2-} and so even a lesser extent by the least soluble Mg^{2+} salts. This mechanism is supported by experimental findings and suggests that desalinization of the Lake Amun was more successful than the factor of 37 suggests. This implies that an excess of Cl^-, $SO_4{}^{2-}$ and Mg^{2+} is due largely to desalinization of areas around the Lake.

In conclusion, it can be stated that the approach used in this manuscript suggests new promising avenues to study salinization and desalinization of not only the Karnak Temples, but also other similar archaeological and non-archaeological systems. It appears that meaningful results can be secured with high (90%) confidence limits even if only few samples are available for analysis. Once these facts are appreciated, it should be possible to expand the approach, using a larger number of samples, to secure more extensive and more unequivocal interpretation of experimental data.

This work was supported by the NSERC and UNESCO/CIDA Assistance Progamme.

REFERENCES

* Current address: Ontario Hydro, Box 369, 1082 Lakeshore E., Mississauga, Ont., L5G 4M1
† Current address: Litton Systems Canada Ltd., 25 City View Dr., Etobicoke, Ont., M9W 5A7

1. H.W.M. Hodges (ed.) Proc. Meeting on "*In Situ* archaeological conservation"'; 1986; Mexico City, Instituto Nacional de Antropologic e Historia de Mexico/Getty Concervation Institute.
2. Review of ref. 1., European Cultural Heritage, **3** (1), 24 (1989).
3. George Burns, Nature, **352**, 658 (1991).
4. George Burns, K.M. Matsui, J.E. Smeaton and K.M. Wilson-Yang, Science, Technology and European Cultural Heritage, Proc. of European Symp. 1989. N.S. Baer, C. Sabbioni, A.I. Sors, Eds., Butterworth-Heinemann, Ltd., Oxford, 1991, p. 383.
5. George Burns, Interdisciplinary Science Reviews, **17**, 81, (1992).
6. T.C. Billard and George Burns, Nature, **285**, 654, (1980).
7. George Burns, T.C. Billard and K.M. Matsui, Nature, **344**, 25 (1990).
8. C. Traunecker, Kêmi, **20**, 195 (1970); **20**, 213 (1970).
9. C. Traunecker, Kémi, **21**, 179 (1971).
10. T.C. Billard, Ph.D. dissertation, University of Toronto (1985).
11. K.M.M. Matsui, M.Sc. dissertation, University of Toronto (1990).
12. S. Mahomed, "Hydrology of the Nile Basin", N.Y., Elsevier Science, 1985.

HYDRIC PROPERTIES OF SOME SPANISH BUILDING STONES: A PETROPHYSICAL INTERPRETATION.

LUIS VALDEON , ROSA M. ESBERT and CARLOTA M. GROSSI
Departamento de Geología, Universidad de Oviedo, c/ Jesús Arias de Velasco s/n, 33005 Oviedo, (Spain).

ABSTRACT

The hydric properties of three types of building stone used in certain Spanish monuments of outstanding historical-artistic value have been determined. The stones selected were dolomite, limestone and sandstone used in the cathedrals of Leon (northern Spain), Seville (southern Spain) and Salamanca (central Spain) respectively.

In the work described in this paper, different rock properties related to water were measured (absorption of water and water vapour, evaporation, capillarity and permeability to water vapour). Swelling characteristics of the sandstone were also determined.

The close correlation between the values of the parameters obtained for each property analyzed and the mineralogical-textural features of the rock, notably the porosity and form of the pore system, has been also demonstrated. In these stones, it has been observed that water absorption capacity is linked both to the volume of empty communicating spaces (open porosity) and to the presence and percentage of clay minerals.

INTRODUCTION

The alteration of rocks is caused by the interaction between the rocks themselves and the agents of alteration, among which the presence and circulation of water through the pore system is of particular importance. For this reason, in the study of building stones, analysis of their hydric behaviour provides the key both to understanding the processes that cause their degradation [1] and to the selection and development of conservation treatments [2], [3].

In this paper, the hydric behaviour of three sedimentary rock types used in the construction of some major monuments of the Spanish architectural heritage have been analyzed, including the Cathedrals of Leon, Seville and the Old and New Cathedrals of Salamanca. This behaviour has been interpreted as a function of the petrographic characteristics of the rocks, particularly, the form of the pore system.

The hydric characterisation has been carried out by analyzing a wide range of hydric properties in the three rocks selected: a dolomite, in the case of Leon Cathedral, a limestone, in Seville Cathedral, and a sandstone in Salamanca Cathedrals.

Regarding the petrographic characteristics of the materials studied, the dolomite shows a crystalline texture and a porosity of 9.3% made up of small voids (0.004-60μm). The limestone is coarse grained and poorly coherent; its most notable characteristic is its irregular and poorly developed cement, with an open porosity of 19.4% corresponding mainly to large pores (0.02-1000μm). The sandstone is of uniform fine grain with a phyllosilicate content of 25%, most of which makes up the rock's clayey matrix; 11% of the clays present belong to the smectite group. The sandstone has a porosity of 27% with a wide range of pore sizes (0.004-200μm).

HYDRIC PROPERTIES

The following hydric properties have been determined: free water absorption; absorption under vacuum (saturation water content); desorption of water or evaporation; capillary suction; hygroscopic absorption of water vapour; permeability to water vapour and swelling.

The methods followed in each test, as well as the number and size of the samples have been in accordance with various international standards, specifically, the standards of CNR-ICR [4] for the determination of the water content (W) using free water absorption, evaporation and water vapour absorption properties; RILEM [5] and CNR-ICR [4], for saturation water content; CNR-ICR [6], RILEM [5] and the recommendations of Mamillan [7] for capillary suction; CNR-ICR [8] for permeability to water vapour; and RILEM [5] and ISRM [9] in the case of swelling. The open porosity was determined by means of mercury porosimetry.

In Table I the experimental values of each property plus the open porosity are presented.

TABLE I: Open porosity and hydric properties

Property	Parameter	ROCK TYPE		
		Dolomite	Limestone	Sandstone
Open porosity	n_0 (%)	9.3	19.4	27
Free water Absorption	W (%)	2.5	7.7	14
Saturation water content	W_s (%)	4.2	12.8	17
Evaporation	W (%)	< 0.03	0	0.2
Capillary suction	C kg/m^{2}min$^{1/2}$	2	337	181
Water vapour absorption	W (%)	0.6	0.3	3.5
Permeability to water vapour	Kv x g/m^2 x 24h	58.3	205.2	0.6
Swelling	ϵ_1	---	---	2.3 x 10^{-3}

With the exception of permeability to water vapour, the evolution of each hydric parameter with time has also been represented in order to give an understanding of the kinetics of the hydric processes (Figures 1-5), [10].

With regard to the absorption of water, both the free absorption and absorption under vacuum have been determined. Comparing the three rocks, the dolomite absorbs water the least (W = 2.5%), the limestone absorbs approximately three times as much of the dolomite (W = 7.7%), and the sandstone twice as much as the limestone (W = 14%). Conversely, in terms of the kinetics, the speed of absorption of the limestone is somewhat greater than that of the sandstone and much greater than that of the dolomite (Fig. 1).

The processes of evaporation follow a parallel behaviour to those of absorption: the limestone loses water most rapidly and the dolomite loses water most slowly (Fig. 2). The sandstone still retains a certain amount of water in its pore system at the end of the evaporation test (Table I).

The value of capillary suction (C) varies greatly among these rocks. The dolomite shows a very low coefficient of capillarity compared with the high values for the sandstone and limestone (Table I and Fig.3). The development and strength of the capillary suction in each rock can be likened to the kinetics of absorption and evaporation of water.

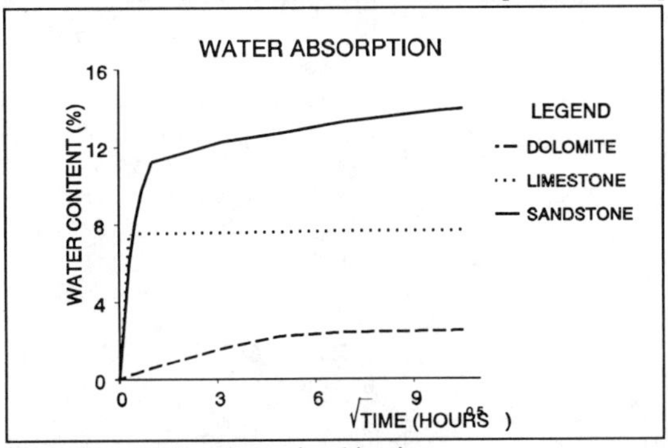

Figure 1.- Water absorption kinetics

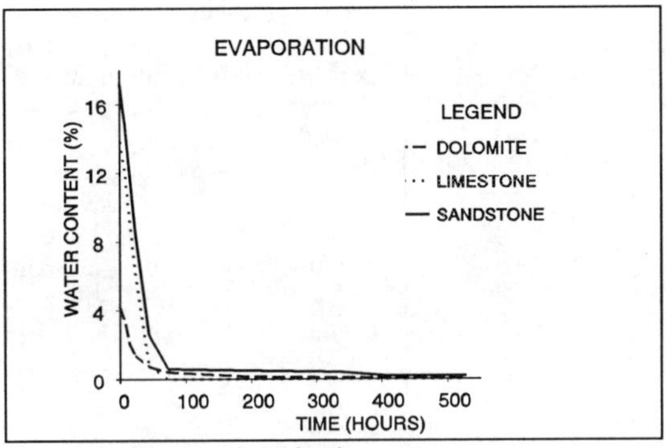

Figure 2.- Evaporation kinetics

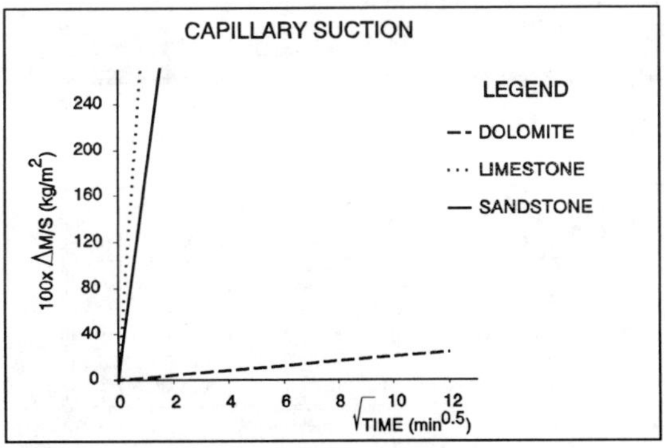

Figure 3.- Capillary suction

Regarding the absorption of water vapour, it can be shown (Table I) that the sandstone has the highest hygroscopic capacity (W = 3.5%) as well as absorbing water vapour most rapidly (Fig. 4). The quantity of water vapour absorbed by this rock represents approximately one fifth of the total capacity of water that can be accommodated in its pore system (W_s = 17%, Table I). The other rocks are noticeably less hygroscopic (Table I). It should be noted that the dolomite, with a lower porosity and lower absorption capacity for liquid water, has a greater hygroscopic capacity than the limestone (Table I).

The results of tests to determine permeability to water vapour are presented in Table I, where great variation can be seen among the different rock types. Most notable is the high permeability to water vapour of the limestone compared with the other rocks tested, and the low value for the sandstone, which is the rock with the highest porosity and greatest absorption capacity.

Finally, the hydric expansion or swelling has been determined. Although swelling tests were started on all three rock types, the results shown relate only to the sandstone, as the dolomite and limestone show no swelling at all. The coefficient of linear expansion (ϵ_l) has been measured in three orthogonal directions. The evolution of this coefficient with time is shown in Fig. 5. It can be seen that the sandstone has a notable linear expansion in all three directions and that it shows a marked directional anisotropy (from highest to lowest: (Y, ϵ_l=2,3 x10^{-3}; X, ϵ_l=1,4 x10^{-3} and Z, ϵ_l=6,7 x10^{-4}). In terms of

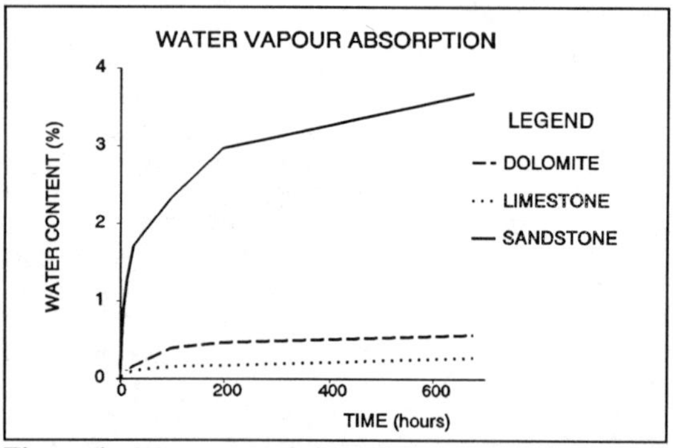

Figure 4.- Water vapour absorption

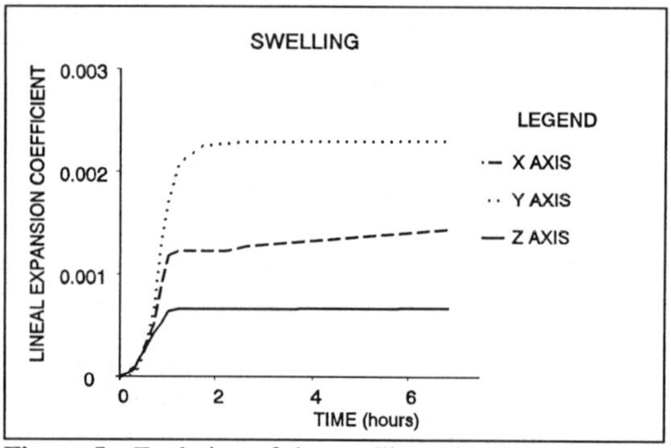

Figure 5.- Evolution of the swelling with time

the kinetics of the process, it can be seen that the swelling develops in a similar way in each of the three directions (Fig. 5): there is first a stage of rapid expansion after which the velocity of swelling decreases markedly until it reaches stability.

PETROPHYSICAL INTERPRETATION

The hydric properties of the rocks studied (values and kinetics) may be interpreted according to their petrographic characteristics, and in particular porosity and the configuration of the pore system. Table II shows, as well as the porosity values for the three types of rock, the size of their voids, the pore throat radius, obtained through mercury porosimmetry, and the pore equivalent radius (pores), obtained by digital image processing.

Table II. Pore throat and pore size.

Rock type	VOLUME		SIZE	
	Open porosity n_0(%)	Trapped porosity (% of n_0)	Pore throat radius (μm)	Pore radius (μm)
Dolomite	9,3	91,4	Range: 0.004-0.1 Mode 0,1	5-60 Mode: 30
Limestone	19,4	61,8	Range: 0.02-70 Mode: 15	10-1000 Mode: 500
Sandstone	27	65,5	Range: 0.004-30 Mode: 10	5-200 Mode: 100

The amount of water absorbed by these rocks, both free and in vacuum, increases with open porosity in the following sequence: dolomite, limestone, sandstone (Table I). However, the kinetics of the absorption, evaporation and capillarity processes is more influenced by pore and pore throat size than by the open porosity value. The larger the rock pores, the more

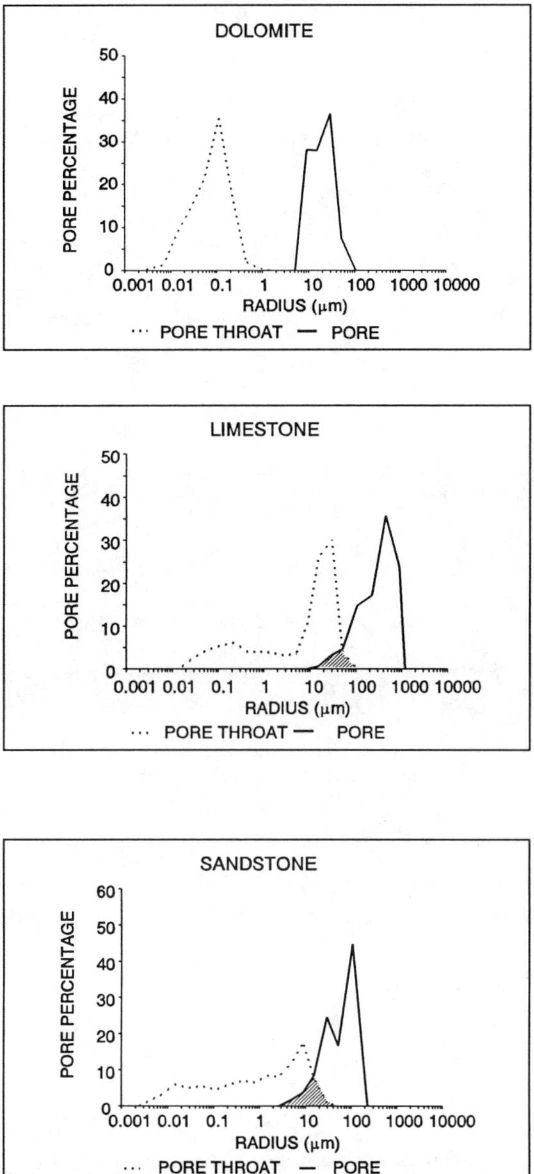

Figs. 6a, 6b, and 6c.- Voids size distribution

rapid the hydric processes. These processes are thus most rapid in limestone, which has the largest pores (Table II, Fig. 6b), and slowest in dolomite, where the pores are smallest (Table II, Fig 6a). The behaviour of sandstone lies between that of dolomite and limestone, although it is closer to the latter. Here too pore size is between that of the other two rocks (Table II, Fig. 6c).

As regards the degree of communication existing within a porous system, the information yielded by the frequency polygons of pore throat and pore size in each rock is analyzed (Figs 6a, 6b, 6c). Comparison of these figures shows that in limestone and sandstone some pore throat and pore sizes coincide, while in dolomite they differ widely, a factor which may contribute to the greater water circulation difficulty found in this rock [11], [12], [1]. This hypothesis is confirmed by the higher "trapped porosity" levels recorded for dolomite (Table II).

As regard absorption of water vapour, sandstone is the most hygroscopic, followed by dolomite and then limestone. The hygroscopicity of sandstone is influenced not only by textural characteristics (high porosity and high percentage of pores of less than 1 μm) but also by the presence of a significant percentage of clay minerals in its crystalline structure, which are capable of retaining water by various mechanisms.

The explanation for the fact that dolomite has a greater hygroscopic capacity than limestone although it has only half the open porosity value (Table II) is that its pores are small in size (Table II), giving it a large "specific surface" to which water vapour molecules may adhere.

Water vapour permeability is four times greater in limestone than in dolomite, while that of limestone is practically nil. Sandstone, although more porous than limestone, is less vapour-permeable due to the presence of expansive clays, which, as they swell with moisture, impede vapour transmission through the rock.

The low permeability of dolomite is due to its low porosity and the configuration of its pore system (Table II). Hydric expansion has only been found in sandstone, due, of course, to the abundance of montmorillonite (11%), an expansive phyllosilicate.

To sum up, it has been demonstrated experimentally that the hydric behaviour of the different rocks depends on their mineralogical-textural characteristics, the importance of each of which varies with the material. Thus the presence of certain clay minerals (montmorillonite), together with the number of pores (porosity) and the three-dimensional configuration of the pore system are the factors which have conditioned the values of each of the hydric properties determined in the three rocks under study (dolomite, limestone and sandstone).

ACKNOWLEDGEMENTS

The authors wish to acknowledge: The "Comisión Interministerial de Ciencia y Tecnología"- Project No. PAT 91-1093-CO3-01.

REFERENCES

1. Valdeón, L. - PhD thesis. Dep. Geología, Universidad de Oviedo, 234 (1989).

2. M. Laurenzi-Tabasso and U. Santamaría, in Vth Int. Cong. on Deterioration and Conservation of Stone, 697-707 (1985).

3. R.M. Esbert, C. Grossi, L. Valdeón, J. Ordaz, F.J. Alonso, and R.M. Marcos, in La Conservacione dei Monumenti nel Bacino del Mediterraneo, Bari, 437-441 (1989).

4. NORMAL: 7/81. CNR - ICR, Roma, 5 (1981).

5. R.I.L.E.M., Matériaux et Constructions, 13 (75), 216-220 (1980).

6. NORMAL: 11/85, CNR-ICR, Roma, 7 (1985).

7. M. Mamillan, in The Conservation of Stone II. Proc. of the Int. Symp., Bologna,. Edit. R.Rossi-Manaresi, 59-72 (1981).

8. NORMAL: 21/85, CNR-ICR, Roma, 5 (1985).

9. ISRM, Int. J. Rock Mech. and Min Sci., 16 (2), 143-156 (1979).

10. F.J. Alonso, R.M. Esbert and J. Ordaz, Boletín Geológico y Minero. T. XCVIII-IV, 555-576 (1987).

11. N.C. Wardlaw and P.P. Taylor, Bull. Canada Pet. Geol. 24 (2), 225-262 (1976).

12. F.M. Pellerin, Bull. Laison Lab. des Ponts des Chaussées, 106, 105-116 (1980).

PORE POTENTIAL AND DURABILITY OF LIMESTONE AT THE SPHINX

K.L. GAURI AND SRINIVAS S. YERRAPRAGADA
Geology Department, University of Louisville, Louisville, KY 40292

ABSTRACT

In mercury porosimetry of limestone two hystereses are generated. A PV-diagram for initial intrusion and extrusion reveals one hysteresis as entrapment of mercury in large cavities that are connected to the exterior through narrow channels. The other hysteresis is a loop (HL) enclosed between the extrusion and reintrusion curves which indicate narrow channels. An HL is unique to a stone.

The thermodynamic work (W) calculated as $\Sigma P \Delta V$, is W_I, W_E, W_{RE}, W_{HY}, and W_T for the initial intrusion, extrusion, reintrusion, the hysteresis loop, and the entrapment, respectively. The pore-potential (U) defined as

$$U = W_I - W_E = W_T + W_{HY} \tag{1}$$

regulates the volume and rate of movement of water through the pores which in turn determines the stone's durability. Microporous stones are highly susceptible to weathering and have a large value for W_{HY}. W_{HY} in association with the W_T can be related to the durability factor (DF) as

$$DF = \frac{W_T}{W_T + W_{HY}} \tag{2}$$

The DF values, thus obtained, accurately reflect the observed condition of the strata of the Sphinx.

INTRODUCTION

Most limestone types that we have studied contain ink-bottle pores, implying that large pores in the interior of the stone are connected to the surface through narrow necks. We have found that the durability of such rocks in outdoor weathering is related to relative proportions of the volumes of the small and large pores because their relationship determines the rate of movement of water into and out of the stone and the degree to which the stone becomes filled with water. These two properties, through the freezing of water and the crystallization of salts determine the quantity of pressure that will be generated in the pores [1,2].

For porosimetric studies, mercury porosimetry provides quantitative accuracy and an extremely efficient means for studying a large number of samples. Conventionally, mercury porosimetry is used to determine the pore volume and pore-size distributions, based upon the classical Washburn equation that uses the approximation that the pores are cylindrical. Further, conventionally, the contact angle of mercury in advance and retreat into and from the stone is taken as constant to calculate the pore size. In the present work, we have been able to reduce any possible effects these approximations may have by calculating the thermodynamic work, PV, from the untreated mercury intrusion and extrusion data. Further, we have formulated an equation, based on the PV-work, that allows calculation of durability factors. These factors correlate well when applied to the Sphinx and certain other limestones whose relative durability in outdoor weathering is known [3].

MERCURY POROSIMETRY OF CERTAIN LIMESTONES

Mercury porosimetry reveals the pore-volume and pore-size distributions through the pressure needed to inject and extrude mercury from the pores and the volume of the mercury intruded or ejected at known pressure. Often this is achieved by injecting mercury in a single procedure of intrusion from low to the high pressure [4,5]. In addition to this intrusion, called initial intrusion, we also extruded the mercury by reducing the pressure, and then reintruded the mercury into the pores, following others, such as, Robertson [5], Wardlaw and Mckellar [6]. A typical diagram obtained by plotting corresponding pressures and volumes for initial intrusion, extrusion, and reintrusion is shown in Fig. 1.

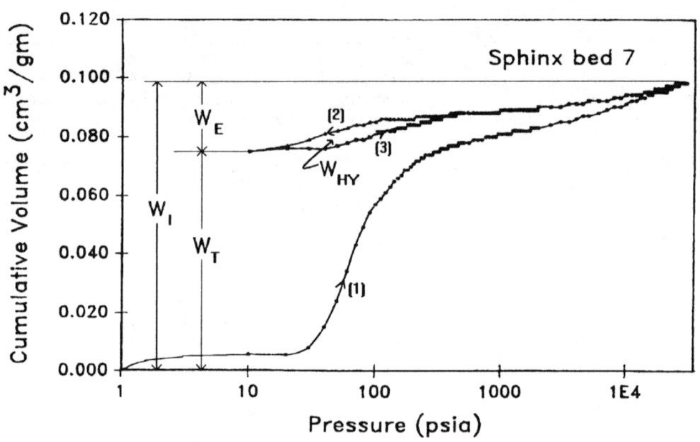

FIG. 1--A typical PV-diagram of a Sphinx limestone showing curves of initial intrusion (1), extrusion (2), and reintrusion (3).

As indicated earlier, mercury porosimetry is based upon the force required to overcome the surface tension to intrude mercury into pores. Assuming that the pores are cylindrical, the Washburn [7] equation expresses the size of pores as follows:

$$r = \frac{2\gamma|\cos\theta|}{P} \tag{3}$$

where r is the radius of the pore intruded at the applied pressure P, and γ and Θ are the surface tension (485 dynes/cm) and the contact angle (130°) of mercury with the pore wall.

In the initial intrusion of an ink-bottle pore system the throats as well as the cavities are filled simultaneously at the threshold pressure of the throats. Therefore, the volume-radii curves for initial intrusion do not reveal the presence of cavities even though the combined volume of the throats and cavities is represented.

In the extrusion process the mercury is ejected from the throats only, as shown by Wardlaw and McKellar [6] following their experimental work on glass models. This is attributed to the snap-off effect, implying that when the pressure is reduced to an extent that a throat begins to extrude, the mercury in the cavity becomes detached from the mercury in the throat. Some mercury thus becomes trapped in the cavities while the throats are emptied at their equivalent pressures. Thus, the trapped mercury represents the volume of large pores.

In the reintrusion process, the mercury enters the small pores only, because the large pores had remained filled after the extrusion process had been completed. However, the extrusion and the reintrusion PV-curves do not coincide but enclose a loop, termed the hysteresis loop (HL). The magnitude of a HL is unique to a given stone (Fig. 2).

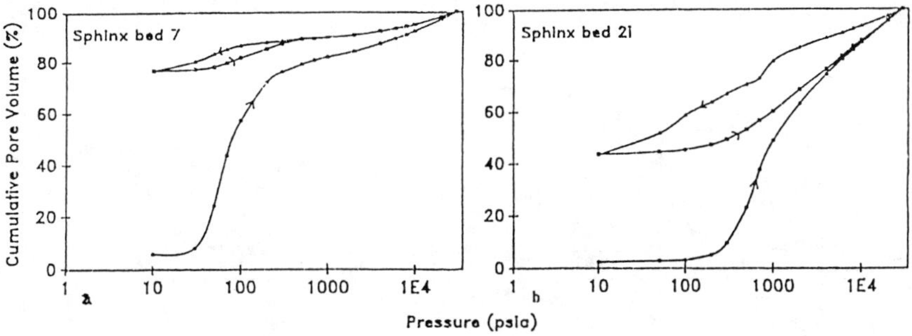

FIG. 2--PV-diagrams showing good (Fig. 2a) and poor (Fig. 2b) Sphinx limestone beds. The DF values for Figs. 2a & b are 0.98 and 0.28 respectively. The bed numbers are same as shown in Fig. 4.

Lowell and Shields [8] have shown from their studies on catalysts and glass-samples that the hysteresis loop is formed due to the changes in the contact angle of mercury from intrusion to the extrusion process. They confirmed this hypothesis by making the extrusion and reintrusion curves coincide when calculated with a reduced contact angle for the extrusion process. We tried to do this for the Sphinx limestone, but we have not been able to find a single value of a withdrawal contact angle that makes the reintrusion and the extrusion curves coincide. Further, stones of different pore-size distributions exhibit systematically different HL; stones with a large volume of micropores have a larger magnitude of HL than highly macroporous stones (Fig. 3). Therefore, we believe that the HL reflects an intrinsic property of stone related to the connectivity of small and large pores.

DURABILITY FACTORS BASED UPON PORE-SIZE DISTRIBUTIONS:

The influence of pore-size distributions upon the durability of stone was recognized long ago. Schaffer [9] and Honeyborne and Harris [10], in an extensive study of a variety of British stones of known durability established an empirical criterion of durability as stone having at least 70% of total volume of pores in the range of diameters greater than 5μm.

Kaneuji et al. [4] developed a correlation equation for frost durability of several Indiana Limestone aggregates in concrete, as determined by freeze-thaw experiments in the laboratory, given as

$$EDF = \frac{0.579}{PV} + 6.12MD + 3.04 \quad (4)$$

where EDF is the expected durability factor; PV and MD are respectively the total volume and the median diameter of pores larger than 0.0045 μm. Correlating these factors with the actual performance of aggregates in outdoor exposure, they concluded that EDF <40, 40 to 50 and >50 characterized stones as non-durable, marginal, and durable.

The durability factors were calculated as per C-666 [11], and were proportionally spread out on a scale from 0 to 100 to obtain EDF. Thus other aggregates, better than those studied by Kaneuji et al. [4], could have EDF greater than 100. Shakoor et al. [12] have used similar schemes for other aggregates with good results. In all these schemes only initial penetration of mercury was carried out to determine pore-size distributions. Thus the volume of large pores could not be determined.

In a study of the limestone beds exposed at the core of the Sphinx, Punuru et al. [3] assigned to three strata, the durability factors (DF) of 100, 98, and 95, based on the degree of disintegration they had experienced during a long time of exposure in the ambient environment (Fig. 4). They also selected three pore ranges, namely <0.5 μm, 0.5-5 μm, and >5 μm, representing respectively, small capillaries, large capillaries, and large pores as maximally influencing the durability of these rocks. Knowing the DF of these

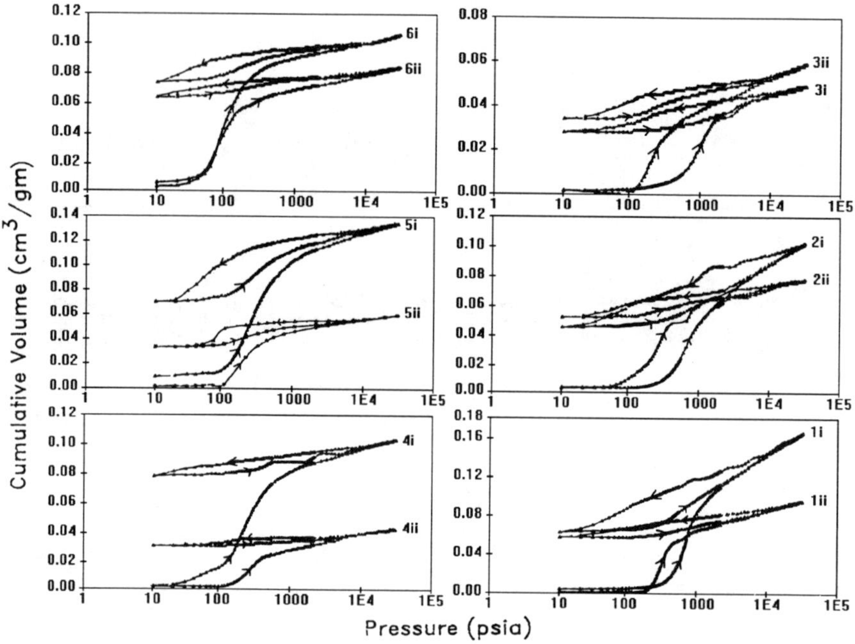

FIG. 3--PV-diagrams of the Sphinx limestones (beds 1-6) showing a systematic change in the magnitude of the hysteresis loop. For actual values of durability factors and hysteresis energies see Table I.

TABLE I--Entrapment (W_T) and hysteresis (W_{HY}) energies, and the calculated durability factors (DF) of the Sphinx limestones.

Limestone Beds (cf. Fig. 4)	W_T (J/gm)	W_{HY} (J/gm)	DF
7	0.4096	0.0076	0.98
6ii	0.3022	0.0848	0.78
6i	0.2794	0.1087	0.71
5ii	0.2103	0.1112	0.65
5i	0.5730	0.2657	0.68
4ii	0.1590	0.0711	0.69
4i	0.2668	0.3373	0.44
3ii	0.2691	0.1328	0.67
3i	0.2099	0.2323	0.48
2ii	0.2846	0.2886	0.50
2i	0.3814	0.9644	0.28
1ii	0.2468	0.3859	0.39
1i	0.3900	0.7222	0.35

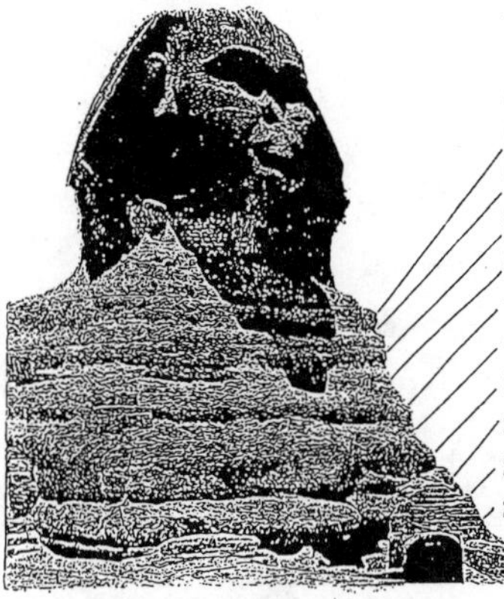

	Na^+	Cl^-	Ca^{2+}	SO_4^{2-}	TOTAL	CLASTICS	DF
6ii	0.07	0.09	0.06	0.99	1.26	4.23	98.9
6i	0.16	0.27	0.27	0.72	1.53	9.46	95.8
5ii	0.09	0.22	0.27	0.73	1.90	3.99	92.5
5i	0.44	0.82	0.10	0.44	1.91	11.45	84.9
4ii	0.13	0.37	0.11	0.38	0.74	4.92	85.8
4i	0.08	0.16	0.59	1.16	2.01	8.14	75.8
3ii	0.30	0.57	0.26	0.55	1.70	2.63	76.3
3i	0.38	0.70	0.08	0.06	1.27	5.23	41.9
2ii	0.84	1.56	0.09	0.13	2.64	3.65	76.9
2i	0.30	0.55	0.48	1.17	2.63	9.31	10.8

FIG. 4--Profile of the Sphinx showing a close correspondence between salt distribution, clastics, and the degree of deterioration of strata seen as projecting and recessed layers.

strata, the volume of those pore ranges were used to obtain the coefficients in the durability equation:

$$DF = 1.23V_1 + 2.62V_2 - 0.98V_3 \tag{5}$$

where V_1, V_2, V_3 are percent volumes in the pore ranges of >5, 0.5 to 5, and <0.5 μm respectively. The DF were then calculated for other strata of the Sphinx, for a stone used during a Pharaonic restoration of the Sphinx nearly 4000 years ago, and for samples of Indiana Limestone used commonly in construction in the United States. The DF for all these rocks reflected accurately the observed condition.

The above studies on pore-size distributions to calculate durability factor depended upon assignment of a certain value of DF to the best sample of the experimental suite, or proportionally adjust the DF of the investigated samples to obtain a value of 100 for the best sample. Also, these studies depended upon the assumption that the pores are cylindrical. The following work on the application of pore energy to calculate durability factor is independent of both these assumptions and should therefore produce accurate predictions for rocks of variable provenances.

MODEL FOR CALCULATION OF PORE POTENTIAL

In a column of volume V, subjected to a pressure, P, work is given as:

$$W = \int PdV \tag{6}$$

where $\int PdV$ is a function to be determined from mercury porosimetry. Since mercury porosimetry requires application of external pressure to force mercury into the pores, the porosimetric data thus may be used in calculating PV-work.

Hysteresis and Pore Potential

According to Lowell and Shields [8], the pressure-volume (PV) work associated with the initial intrusion and extrusion cycle can be expressed as

$$\int dW_1 = \int P_I dV_I + \int P_E dV_E > 0 \tag{7}$$

where P_I and P_E are the initial intrusion and extrusion pressures. V_I, V_E, and V_T are the intrusion, extrusion, and entrapment volumes, respectively. Similarly, the PV-work for the reintrusion and extrusion cycle can be expressed as:

$$\int dW_2 = \int P_{RE} dV_{RE} + \int P_E dV_E > 0 \tag{8}$$

where P_{RE} and V_{RE} are the reintrusion pressures and volumes.

The PV-work, for n radii intervals, can be calculated by numerical summation as:

$$W = \sum P\Delta V \tag{9}$$

The total work of intrusion (W_I) consists of three parts, namely, the work of entrapment (W_T), the work of hysteresis loop (W_{HY}), and the work of extrusion (W_E), such that:

$$W_I = W_T + W_{HY} + W_E \tag{10}$$

The PV work differences between intrusion and extrusion can be explained by means of pore internal energy, U, termed pore potential, and given as:

$$U = \int F_I dL_I - \int F_E dL_E \tag{11}$$

where F_I and F_E are the intrusion and extrusion forces and L_I and L_E are the lengths of the mercury columns. Alternately, the pore potential can be given as:

$$U = \int P_I dV_I - \int P_E dV_E = W_I - W_E \tag{12}$$

Using Eq. (10), Eq. (12) can be written as:

$$U = W_I - W_E = W_T + W_{HY} \tag{13}$$

from which the pore potential may be calculated.

Pore Potential and the Durability Factors

To use pore potential in obtaining durability equation, we know that W_{HY} represents the effect of small pores and W_T the effect of large pores. We also know that in a rock with ink-bottle pores, small pores attract more water and are harmful. The large pores on the contrary are beneficial. Thus, the ratio of W_T to the pore potential, W_T+W_{HY}, should give a good representation for the durability factor. The equation is:

$$DF = \frac{W_T}{W_T+W_{HY}} \tag{14}$$

Table I shows the DF and the appropriate values of W_T and W_{HY} for the Sphinx limestones calculated on the basis of Eq. 14. Fig. 2 shows typical PV-curves for sound and unsound stones along with the DF. All these DF values accurately represent the observed condition of the strata.

The proposed method to calculate DF from Eq. 14 has several advantages over the existing models based upon pore-size distribution, in:

a. it is independent of the pore geometry and the contact angle of mercury which must be used to calculate pore-size,
b. it recognizes all pores, large and small, by obtaining the work in intrusion, extrusion, and reintrusion procedures, and
c. the DF range for all strata is between 0 to 1 compared with other non-normalized DFs where they can have any value beyond 1.

CONCLUSION

A new equation has been developed to obtain durability factors for different limestones. This equation, based upon the thermodynamic PV-work, marks an advance over existing equations by being independent of pore geometry and the angle of contact of mercury with the substrate.

REFERENCES CITED

1. D.W. Everett, Trans. Faraday Soc. 57, 1541 (1961).
2. B. Fitzner and R. Snethlage, G.P. Newsletter 3, 13 (1982).
3. A.R. Punuru, A.N. Chowdhury, and K.L. Gauri, Environ. Geol. Water Sci. 15, 225 (1990).
4. M. Kaneuji, D.N. Winslow, and W.L. Dolch, Cement and Concrete Res. 10, 433 (1980).
5. J. Ordaz and R.M. Esbert in Deterioration and Conservation of Stone, edited by G. Felix (V. International Congress Proc., Lausanne, 1985) pp. 93-102; W.D. Robertson, ibid., pp. 87-92.
6. N.C. Wardlaw and M. McKellar, Powder Technol. 29, 127 (1981).
7. E.W. Washburn, Proc. Nat. Acad. Sci. 7, 115 (1921).
8. S. Lowell and J.E. Shields, J. Colloid and Interface Sci., 83, 273 (1981); Powder Surface Area and Porosity, 2nd ed. (Chapman and Hall Publishers, New York, 1984), p. 234.
9. R.J. Schaffer, The Weathering of Natural Building Stones, Special Report No. 18 (Building Research Establishment, Garston, 1972), p. 149.
10. D.B. Honeyborne and P.B. Harris, Proc. Colston Res. Soc. 10, 343 (1958).
11. Anon, Resistance of Concrete to rapid Freezing and Thawing, American Society for Testing and Materials (ASTM) Standard C 666 - 73.
12. A. Shakoor, T.R. West, and C.F. Scholer, Bull. Assoc. Eng. Geol. 29, 371 (1982).

THE USE OF EPOXY RESINS IN FIELD PROJECTS FOR STONE STABILIZATION

CHARLES M. SELWITZ

The Getty Conservation Institute
4503 Glencoe Avenue, Marina del Rey, CA 90292

ABSTRACT

The direct application of epoxy resins to deteriorating stone often encounters problems of penetration and color formation. Conservation scientists have worked around these problems by using organic solvents as carriers and have successfully consolidated stone with epoxy resins in a large number of projects. Work in Poland, Czechoslovakia, Italy and the United States is described. In Europe the methodology has evolved to treat objects of moderate size and isolated parts of buildings while in the United States procedures have been developed for stabilizing large portions of major buildings.

INTRODUCTION

Many stone conservators who have tried to use epoxy resins to stabilize fragile and decomposing stone, to take advantage of the improvement in strength and durability that these resins can provide, have had problems which have given epoxy resins a bad reputation in the conservation community. These materials do not penetrate well and they are difficult to apply. As a result they tend to form surface crusts and give discolored products. However a small number of conservation scientists has worked hard to overcome these difficulties.[1] The most important improvement was putting the consolidant into an organic solvent. Epoxy resins are not normally used in solution. But, by preparing 10-20% solutions of monomers in light, volatile solvents, the epoxy-hardener system was made available in formulations with both low viscosity and low surface tension. These properties provided deep penetration into many types of stone and offered a way to avoid problems of crust formation and discoloration. In this presentation the discussion will go beyond the research phase and will describe the application of solution technology by preservationists in Europe and the United States to real problems that range from small stone carvings to large sections of major buildings.

THE EUROPEAN APPROACH VS THE AMERICAN APPROACH

The development of the use of epoxy resins in solution followed two distinctly different paths that are outlined in Table 1. The research of Wieslaw Domaslowski in Poland established two principles for the treatment of small objects and minor portions of buildings.[2] The first requirement is that the solvent must be a mixture of an alcohol and an alkylbenzene, such as methanol and toluene. The second is that liquid must be held or contained against the surface of the stone for a period of time in order to get penetration. The application of consolidating solution by brushing was generally not satisfactory. Other stone conservators in Europe, particularly in Italy[3] and Czechoslovakia[4], subsequently confirmed these findings and used them in their own projects. Other aspects of the European approach are that the objects, immediately after treatment, must be wrapped in a plastic vapor barrier to achieve curing and bonding by the epoxy-hardener combination deep inside the stone and that discoloration can be prevented by wiping the surface with solvent. In contrast, in the United States Lal Gauri has developed a protocol that is

TABLE 1 STABILIZATION OF STONE WITH EPOXY RESIN

APPROACH	PROPONENTS	SCALE	FORMULATION	METHOD OF APPLICATION	WRAPPING	ELIMINATION OF COLOR
EUROPEAN	Domaslowski - Poland Kotlik - Czechoslovakia Lazzarini Marinelli Fiori - Italy	Statuary, small objects, minor, isolated portions of buildings	Resin is dissolved in a blend of alcohol and alkylbenzene	Some type of containment where consolidating solution is held against the stone	Yes	Wiping with acetone
AMERICAN	Gauri - USA	Major sections of buildings	Resins dissolved in acetone	Spraying	No	Wiping with acetone Mild abrasion Exposure

designed to treat large sections of major buildings.[5,6] He uses commercial epoxy resins in acetone as solvent instead of alcohol-alkylbenzene blends and makes application by spraying, not by containment. Treated surfaces are not wrapped. Color formation is avoided by wiping with acetone or by using a mild abrasion. Gauri expects surfaces that initially show some yellowing to lose color with exposure to sun and air and the passage of time. Treatment is done by contractors, not stone conservators, who are expected to meet a list of specifications.

CHEMISTRY

The types of epoxy resins used in field projects are shown in Table 2. Most of the projects have been done with the standard commercial bisphenol-A-diglycidyl ether illustrated in the table. Before 1980 triethylenetetramine was commonly used as hardener but this material is moderately toxic and is now generally employed in an amended form. In Italy today a number of stone conservators are avoiding the conventional aromatic epoxy resins and are using cycloaliphatic epoxy compounds and amine curing agents of undisclosed structure. These are products sold by a small chemical company in Milan known as STAC.[7] Aliphatic resins have found favor because of greater stability to oxidative degradation and lower viscosity than corresponding aromatic types. Some idea of the chemical structures of aliphatic compounds, that have been used in other contexts, to provide the ultimate in color stability are shown in the table, specifically the Eponex epoxy resin and a Jeffamine curing agent.

WORK IN POLAND

Details of four projects done in the late 1960's where Domaslowski applied his laboratory results are shown in Table 3. Three of the four involved ornate sandstone portals that frame entrances to historic structures.[8] The deterioration of these doorways had been a matter of widespread public concern. In the two earliest projects, the Esken Palace and the St. Mary Magdalene Church, a cumbersome network of a large number of small cup-like containment dams, attached to the stone surface, was used to maintain solution against the stone. In subsequent projects, like the Old Market House portal and a sandstone statue of St. Francis, Domaslowski switched to a more convenient method of application. The consolidating solution was made to flow continuously against the object through cellulose wadding that was held and wrapped by polyethylene film. Both methods provided deep penetration, from two to four inches, but the flow method was more convenient and wasted less material. A number of independent assessments of these structures through the late eighties found little, if any, sign of continuing deterioration such as powdering or peeling.

WORK IN CZECHOSLOVAKIA

The arrangement used by a conservation scientist in Czechoslovakia, Petr Kotlik, to halt the deterioration of a four foot sandstone statue is shown in Figure 1. The statue was wrapped with polyethylene netting, fitted with rigid inflow and outflow tubes and then tightly sealed inside a sheath of 2 mil polyethylene film. A relatively concentrated formulation consisting of 34% bisphenol-A type resin, 3.4% diethylenetriamine, 45% ethanol and 18% toluene was made to flow in through the bottom inlet tube, and then up and against the statue by pulling a 15 mm Hg vacuum at the top outlet tube. This procedure gave a well stabilized product.

TABLE 2 EPOXY RESIN COMPONENTS

USER	EPOXY COMPOUND	
GAURI KOTLIK DOMAS-LOWSKI	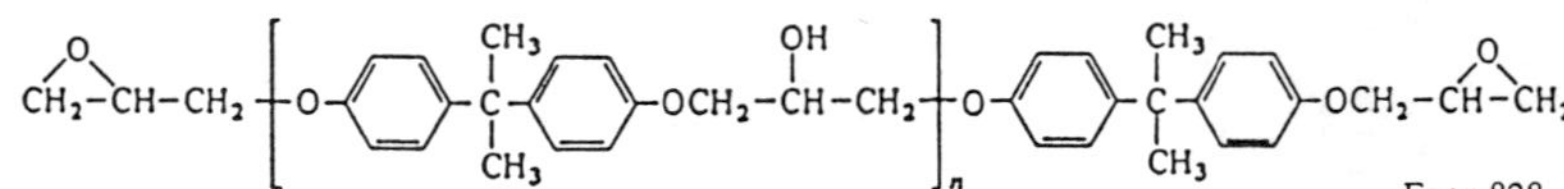	$n = 0$ 87.2% $n = 1$ 11.1% $n = 2$ 1.5% Epon 828, Araldite 6010, Epidian-5
Stable System	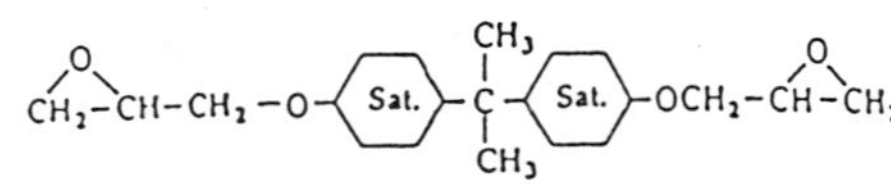	Eponex
LAZZARINI	CYCLOALIPHATIC EPOXY COMPOUND	STAC'S EP2101

USER	AMINE HARDENER	
GAURI KOTLIK DOMAS-LOWSKI	$H_2N\text{-}(CH_2\text{-}CH_2\text{-}NH)_2\text{-}CH_2\text{-}CH_2\text{-}NH_2$	Triethylenetetramine and fatty acid amide derivatives
Stable System	$H_2N\text{-}CH(CH_3)\text{-}CH_2\text{-}[O\text{-}CH_2\text{-}CH(CH_3)]_x\text{-}NH_2$	Jeffamine D-230
LAZZARINI	CYCLOALIPHATIC AMINE COMPOUNDS	STAC'S EP2102

TABLE 3 RESTORATION IN POLAND BY WIESLAW DOMASLOWSKI

DATE	OBJECT	FORMULATION	APPLICATION METHOD
1966	Sandstone portal Esken Palace, Torun	15% (Epidian-5 and TETA) in toluene-methanol	Cups of chalk-paper built onto portal are filled with solution.
1966	Sandstone portal St. Mary Magdalene Church, Wroclaw	10% (Epidian-5 and TETA) in toluene-methanol	Cups of gypsum-sandpaper, built onto portal are filled with solution.
1967	Doorway and portal Old Market House Torun	15% (Epidian-5 and TETA) in toluene-methanol	Solution was wicked against object through cellulose wadding held by polyethylene film.
1969	Sandstone statue of St. Francis, Klodzke	10% Epidian; 1.2% TETA in toluene-methanol	As above

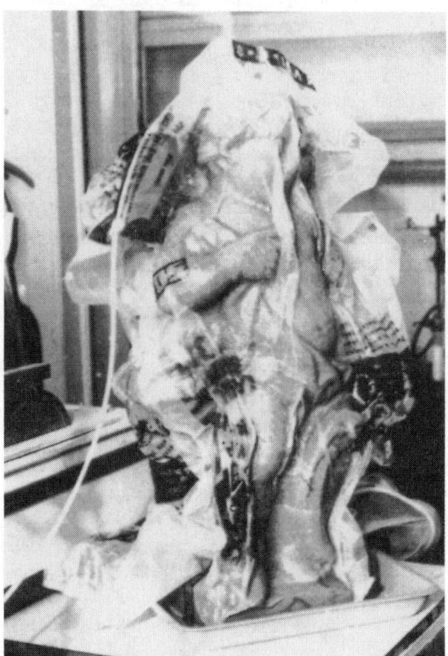

Figure 1. Treatment of limestone, Aug. '91

WORK IN ITALY

More recently, the European approach has been employed by Italian conservators, particularly the Restoration Working Society of Cartura, using cycloaliphatic resins produced in Italy instead of the more conventional phenolic derivatives.[3] A solution of the cycloaliphatic compounds in a toluene-isopropanol co-solvent in the proper ratios cures to a soft, crumbly gel. When an object is impregnated by immersion and kept there until gelation occurs, it can be cleansed from excess chemicals rather easily, by brushing the gel particles away. This is a very convenient procedure. Table 4 outlines eight projects that were carried out using aliphatic resins. It shows that prior to 1985 all treatments were done on site by brushing. These projects included sections of a Roman aqueduct, tower panels and window frames of limestone and a seven hundred year old statue of a lion. Work after 1985 has generally been done by containment procedures. This includes granite and limestone columns, exterior stone panels of historic churches and the base of a statue. In the European approach objects are frequently dismantled and brought to facilities designed to conduct treatments.

The first field project was carried out in 1983 on sections of an ancient Roman aqueduct made of tuff, a porous volcanic rock. The stone was flaking and powdering badly and a significant improvement in mechanical strength was needed. The aliphatic formulation was applied by brushing and the treatment put an end to the process of decay. There was some initial yellowing but with time this faded out while the stone remained strong and stable. Another of the objects treated by brushing is the "Gatta di Santa Andrea", also known as the Venetian Lion. The statue, standing atop a pink column, was presented in 1207 to commemorate a Paduan military victory. It was carved from trachyte, a light colored volcanic rock. Before treatment the stone showed severe scaling and powdering. Nine years after consolidation there was no powdering and fragments were firmly attached to the surface. Discoloration had been avoided by wiping the stone just after treatment with acetone-soaked swabs.

The much deeper penetration that is achieved by holding liquid against substrate for extended periods of time led to the use of containment methods in subsequent projects, even for relatively sizeable structures. Several large columns were consolidated in place by forming a sealed polyethylene sheath around the column and pumping liquid in at the bottom to provide a head of consolidation solution to stand against the stone. In other projects like limestone panels on the Church of Zeno, and columns and bases from the Basilica of Santa Andrea in Montera, the pieces were disassembled and taken to the laboratories of the Restoration Working Society for treatment. Generally the stone piece was placed in a plastic film pouch filled with consolidant which was in turn put in a pouch filled with water to pressure the consolidant into the stone. All of this was held in an appropriately sized wooden box. Ultimately only a simple brushing by hand was required to remove the residual gelled consolidant from the stone surface. An inspection tour of many of these projects in April, 1991 by Lorenzo Lazzarini found the stone surfaces to be in sound shape without appreciable discoloration.

WORK IN THE UNITED STATES

Restoration of stone with epoxy resins in the United States has been done almost exclusively by Lal Gauri. His procedures have evolved from an extensive program of research and development at the University of Louisville on the destructive interaction of acidic pollutants, particularly sulfur dioxide, with marble and limestone and the prevention and repair of damage with conventional epoxy

TABLE 4

EPOXY/STONE CONSOLIDATIONS by the RESTORATION WORKING SOCIETY of CARTURA

	DATE	LOCATION	SUBJECT OF TREATMENT	STONE TYPE	METHOD OF TREATMENT
1.	1983	near Rome	Sections of Roman aqueduct	Vitùrbo Tuff	Brushing
2.	1984	Arzignano	Bell tower panels	Chiampo limestone	Brushing
3.	1984	Trent	Window frames in bell tower	Limestone	Brushing
4.	1984	Venice	Statue: "The Gatta di Santa Andrea"	Eucarean trachyte	Brushing
5.	1985	Venice	Murano column	Granite	Column of consolidant inside a polyethylene containment sheath
6.	1985	Verona	Panels on the Church of Zeno	Oolitic limestone	Plywood containment filled with consolidant
7.	1988	Trent	Base of statue	Limestone	Object consolidated in the facilities of the Restoration Working Society by enclosure in plastic bags filled with consolidant
8.	1989	Santa Andrea in Montera	Twin columns of the Basilica	Limestone	As above

TABLE 5 TREATMENT OF BUILDINGS IN THE UNITED STATES BY LAL GAURI

STRUCTURE, LOCATION	STONE TYPE	HISTORY	PROBLEMS
1. California Building, San Diego	Cast stone	Built 1915; Treated 1975	Cracking, loss of detail of statuary and designs by erosion
2. Crescent Hill Baptist Church, Louisville, Kentucky	Cast stone	Treated 1983	Freeze-thaw deterioration and fracturing
3. Cathedral of the Assumption, Louisville, Kentucky	Brick, limestone	Treated 1991	Water penetration through brick and mortar, low mechanical strength of bricks
4. First Baptist Church Portland, Oregon	Sandstone	Built 1984; Treated 1983	
5. Humana Conference and Fitness Center	Indiana limestone	Treated 1983	

resins. His studies have led to procedures most suitable for the treatment of large surface areas and from the early seventies Gauri has directed the consolidation phase of a number of building restorations. Table 5 provides an overview of his major projects. These include structures as diverse as an anthropology museum in San Diego, the California Building, several churches in Louisville, the First Baptist Church in Portland, Oregon and the Humana Conference and Fitness Center in Louisville. Gauri's approach is characterized by the use of conventional bisphenolic resins in acetone solution and their application by spraying, the removal of discoloration by abrasion, a focus on the restoration of sections of buildings rather than smaller artifacts and the use of specifications to insure performance by contractors. Figure 2 is a photograph of work in 1975 on Gauri's best known effort, the California Building. It shows the scaffolding required to carry out the treatment and provides a picture of the magnitude of the intervention.

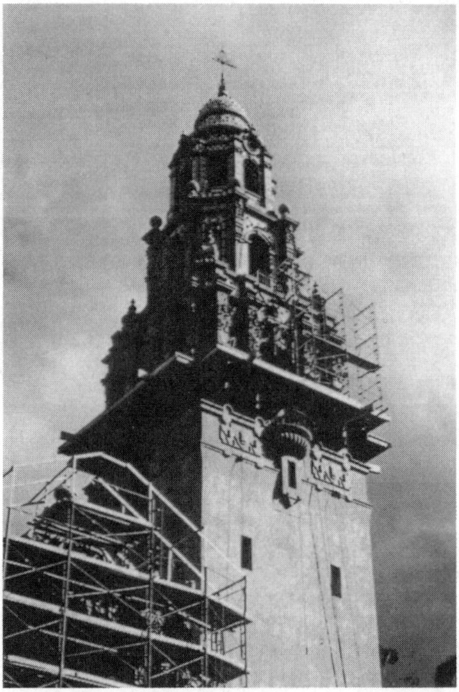

Figure 2. Scaffolding on California Building

More recently, Gauri has developed a program for the preservation of the Cathedral of the Assumption in Louisville. This has included an assessment of the causes of the deterioration of the brick, mortar, cast stone and two types of limestone that make up the building and the design of specifications and treatment protocol. The major problems were excessive water penetration, insufficient mechanical strength both in original bricks and after deterioration, and severe degradation and loss of mortar. The effect of several different consolidants, including commercial alkoxysilanes, on substrates was evaluated. Based on these studies a program of treatment was recommended. To start with, the stone was desalinated by spraying with water, first at 180°F under pressure, and then at room temperature for no longer than 5 minutes in total. A 5:3 ratio of epoxy to aminoamide hardener was dissolved in acetone to make 5%, 10%, and 15% solutions. The selection of concentration, amount and order of solutions used depended on the type of stone and its location. When cracked limestone was encountered, impregnant tended to leak out of the fractures. It was necessary to immediately clean the surface thoroughly with acetone when streaks of liquid polymer appeared. This is current work and it is too early to pass judgement on its effectiveness.

References and Sources

1. C.M. Selwitz in *Materials Issues in Art and Archaeology II.* Edited by P.B. Vandiver, J. Druzik and G.S. Wheeler (Mater. Res. Sc. Proc. 185, Pittsburgh, PA 1990) 181-191
2. W. Domaslowski, Monumentum 4 (1969) 51-64
3. R. Cavalleti, L. Lazzarini, L. Marchesini and G. Marinelli, *Fifth International Congress on the Deterioration and Conservation of Stone,* Lausanne (1985) 769-777
4. P. Kotlik and J. Zelinski, *Studies in Conservation* 28 (1983) 75-79
5. K. L. Gauri, *The International Symposium on the Deterioration and Protection of Stone Monuments,* Paris (1978) 7.5
6. E.F. Szabo and G.T. Szabo, *Fourth International Congress on the Deterioration and Preservation of Stone Objects,* Louisville, KY (1982) 321-333
7. STAC products are available from Bresciani S.r.I., via Padre Semeria 8, 20128 Milan, Italy
8. W. Domaslowski and A. Strzelezyk, *Preprints of the Contributions to the Bologna Congress* (1986) 126-132

SILICONIC AND ACRYLIC RESINS DISPERSED IN WATER AS PROTECTIVES FOR STONE SURFACE

GUIDO BISCONTIN*, P. MARAVELAKI*, E. ZENDRI*, A. GLISENTI°
*Dipartimento di Scienze Ambientali, Universita' di Venezia, Calle Larga S. Marta 2137, 30123 Venezia,(Italy)
° Dipartimento di Chimica Inorganica, Metallorganica ed Analitica, Universita' di Padova, via Loredan 4, 35131 Padova, (Italy)

INTRODUCTION

The aim of the present work is to compare the protective effect of solvent and water dispersed products on marble, Lecce and Istria stones. This choice is justified by the need of products effective also from the ecological and toxicological point of view.

To obtain these informations two tests have been chosen: contact angle and water absorption. The contact angle may, in theory, be considered a measure of the water repellency, since complete wetting implies a contact angle of 0° and absolutely no wetting an angle of 180°. This does not apply to the water absorption by capillarity test because the absorption variation may be caused by either the water repellency action of the treatments or by the pore filling. However this test may give practical information about the general effect of the treatments.

Moreover, the effect of treatments on permeability was investigated by vapour permeability test.

Protective effect was studied after application of the products and at various stages of artificial weathering.

EXPERIMENTAL

Tested Stones

The following types of stone were tested: Carrara marble, Istria stone and Lecce stone.[1,2,3]

The chosen parameter is the porosity: Istria stone is a limestone of compact texture with a porosity of 0.1-1.0 %. This stone is often employed in north-east Italy and usually shows a rather superficial deterioration.

Carrara marble is a crystalline system with a porosity of 2.0-2.5 %. It is employed all over Italian territory. There are lots of different way of deterioration of marble according to the environmental conditions and to the exposure. The deterioration of marble is usually intra-granular and may be spread in depth.

Lecce stone is a very porous limestone (porosity ~ 35%) extensively used in the south of Italy both for building material and for decorative purposes, mainly during the baroque period. This stone suffers strong and deep deterioration with material loss because of the high porosity.

Carrara marble, Istria and Lecce stones are often employed like specimens for testing protective products.

Tested Products

The aim of the present work is to compare the effectiveness of solvent and water dispersed products. The tested products are listed together with the application conditions:

- (**A**): PEMA-PMA copolymer, Paraloid B67 by Rhone-Poulenc, in white spirit 6.5%.
- (**AA**): water dispersed acrylic product ;
- (**SW**): oligomeric polysiloxane product by Wacker in white spirit (7.5 %);
- (**SWA**): water dispersed oligomeric polysiloxane product by Wacker ;

The products were applied by capillarity (short impregnation time according to DIN normative) to cubic sample of 5 cm size (i.e. 5x5x1 and 5x5x2).

I.R. spectrophotometry has been used to characterize the employed products.

Measured Properties

The following tests were chosen to evaluate the protective effects:

1)- water absorption by capillarity according to NORMAL 11/85;

2)- contact angle using special equipment [4],

3)- water vapour permeability to test product-stone interaction and compatibility;

Artificial Weathering

The above measurements were repeated after an artificial weathering aimed to simulate as well as possible some of the actual causes of alteration in the tested stones. The sample were cyclically submitted to salts cycles using NaCl 10%.

RESULTS

Lecce stone

In table I are reported the amounts of absorbed product and the measured values of contact angle. The values of contact angle show that the siliconic products may offer a higher hydrophobic protection than acrylic ones. Moreover water dispersion products exhibit a good behaviour. The values of contact angle obtained after artificial weathering (see table I) show a water repellency loss in samples treated with **AA** (in which the contact angle become 74°) and with **SW** (in which the contact angle become 99°).

Table I

Product	Absorbed amount (g/mq)	Contact angle (°) *not weathered*	Contact angle (°) *weathered*	Traspirability
Lecce stone				
AA	101	98	74	182
A	204	101	100	171
SWA	97	125	114	210
SW	99	133	99	155
NT		0	0	220
Carrara marble				
AA	30	88	-	18
A	32	86	-	18
SWA	26	108	-	76
SW	36	127	-	45
NT		47	-	80
Istria stone				
AA	16	99	75	26
A	20	93	88	23
SWA	9	109	80	33
SW	15	124	79	21
NT		34		37

NT: not treated material

In fig. 1 are reported the results of water absorption by capillarity. It is possible to observe that solvent type products (both siliconic and acrylic) may decrease the amounts of water absorbed. Water dispersed product, on the contrary, show a good behaviour only at the beginning of the test. Artificial weathering always gives rise to an effectiveness loss which is different for the different treatments. Samples treated with water dispersed products absorb like the not-treated ones. Acrylic treated samples keep a good hydrophobic effectiveness.

Water vapour permeability test results are reported in table I. It is easy to see that the not-treated samples, because of high porosity of the stone, are characterized by a high vapour permeability and all the products reduce the vapour permeability. Water dispersed products (both siliconic and acrylic) cause limited permeability loss: the siliconic one, i.e., reduces the transpirability of about 5%. This means that siliconic water dispersed product are not incompatible with the stone. The products in solvent, and particularly **SW**, induce a higher permeability loss.

Carrara Marble

In table I are reported the amounts of absorbed product and the measured values of contact angle. Also for marble, the values of contact angle show that the siliconic products may offer a higher hydrophobic protection than acrylic ones. Moreover water dispersion products (and particularly the silicon one) exhibit a good behaviour; it is important to observe that there is almost no contact angle difference between samples treated with **A** and **AA**.

In fig. 2 are reported the results of water absorption by capillarity. It is possible to observe that all products (both siliconic and acrylic) but **AA** may decrease significatively the the amounts of absorbed water. Four days after there is an increase in absorption of water; this increase is particularly evident for **SWA**. Artificial weathering always gives rise to an effectiveness loss which is particularly evident for samples treated with water dispersed products. Siliconic treatment keeps a good hydrophobic effectiveness.

Water vapour permeability test results are reported in table I. It is easy to see that acrylic treatments cause a high permeability loss whereas siliconic treatments induce a minor loss.

Istria stone

In table I are reported the amounts of absorbed product and the measured values of contact angle. The values of contact angle show that the siliconic products may offer a higher hydrophobic protection than acrylic ones. Moreover, water dispersion products (and particularly the siliconic one) exhibit a good behaviour; it is important to observe that contact angle in the sample treated with **AA** is higher than contact angle observed in the **A** treated ones. Artificial weathering causes a well evident decrease in the values of contact angle in all the samples.

In fig. 3 are reported the results of water absorption by capillarity. It is possible to observe that all the products (both siliconic and acrylic) cause a small decrease of the amounts of water absorbed. However, it is important to observe that Istria stone is characterized by a limited water absorption and that a little decrease may be meaningful. All the treatments seem not to be influenced by the artificial weathering, this is particularly true for water dispersed products. All these results are compatible

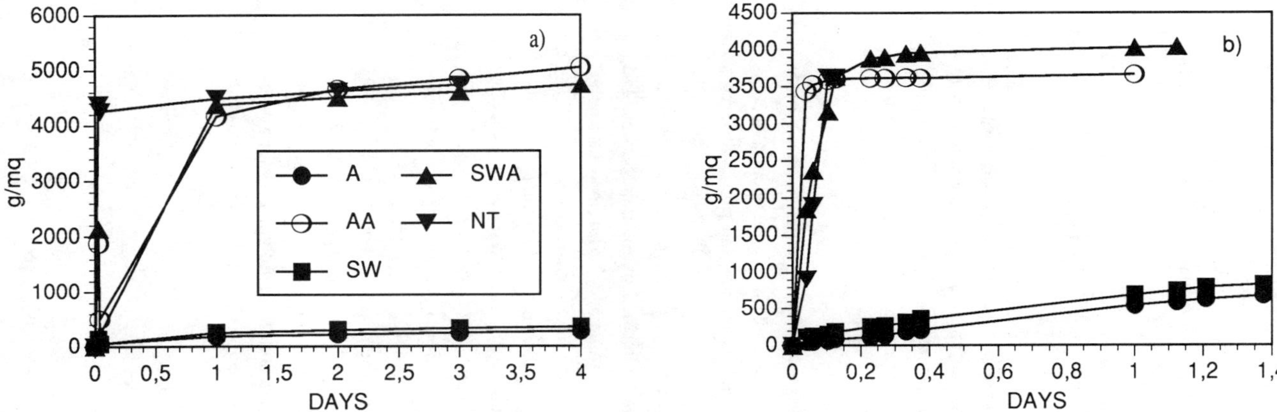

Fig. 1 adsorption by capillarity in the Lecce stone not weathered a) and weathered b).

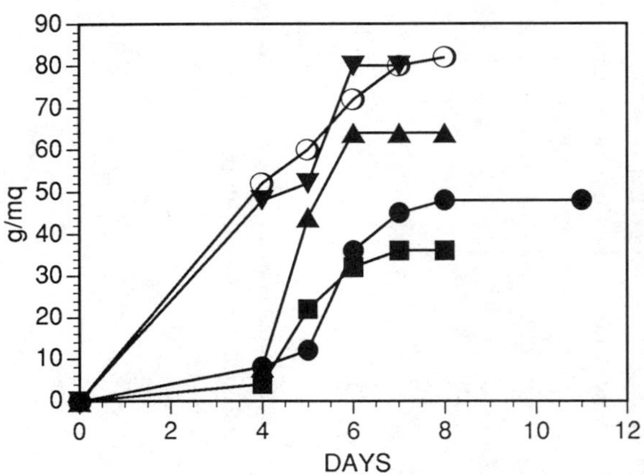

Fig. 2 adsorption by capillarity in the Carrara marble not weathered.

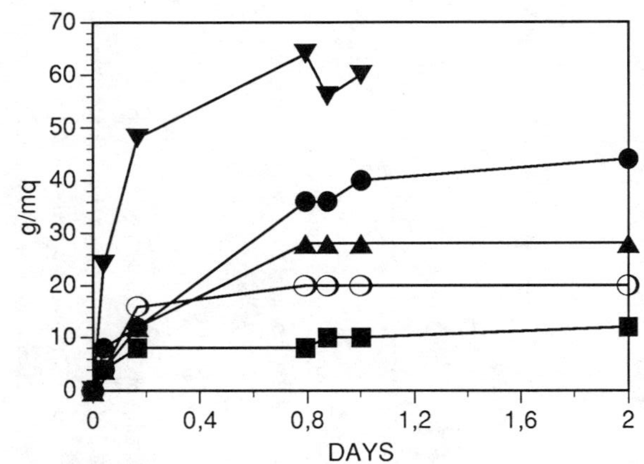

Fig. 3 adsorption by capillarity in the Istria stone not weathered.

with the existence of a surface loss of water repellency but, due to the limited variations in the amounts of absorbed water, there remain a residual hydrofobic effect.

Water vapour permeability test results are reported in table I. It is easy to see that acrylic treatments cause a high permeability loss, whereas siliconic treatments induce a minor loss.

CONCLUSIONS

In this work we have compared the effectiveness of siliconic and acrylic protective products both in solvent and water dispersed. Contact angle and water absorption by capillarity have been chosen to test the products, whereas water vapour permeability has been used to study product-stone compatibility. In spite of the usage of only three tests, the results have given meaningful information about the protective effect of the employed products.

All the products may contribute to the protection of the studied stones even if their effect is not preserving in time. A restrictive parameter for the treatment choice is the induced loss of permeability. In this respect the water dispersed products show a better behavior.

It is important to observe that Lecce stone (which is a very porous stone) which absorbs high amounts of protective products, shows better hydrophobic properties. The lower absorption of protective products by both Istria stone and Carrara marble implies a more difficult rationalization of their water repellency effectiveness.

From the reported results it can also be seen that different products may have different effects on different stones so that it is necessary to study independently the different stones in order to find the adequate treatment. Every treatment must be then improved by choosing the more favourable application conditions (concentration, solvent etc.).

Water dispersed products may be employed when water-stone contact time is limited because they can offer a good water repellency and permeability. When this contact time is long the amount of absorbed water is large. We believe that the effectiveness of water dispersed products could be increased employing water soluble protectives because the dispersed particle size causes a low in depth penetration of the product.

REFERENCES

1 - M. Tabasso Laurenzi, U. Santamaria; Consolidant and protective effects of differents products on Lecce limestone, V Int. Symp. Deteriorat. and Conservation of Stones, Lausanne 1985, 667.

2 - G. Alessandrini, R. Bugini, R. Peruzzi; I materiali lapidei impiegati nei monumenti lombardi e loro problemi di conservazione, Boll. Minist. Beni Culturali, suppl. al N° 41, 1988, 283.

3 - G. Biscontin, G. Driussi, A. Valle; Water vapour transpiration and resin impregnated stone materials, Scientific Methodologies Applied to Works of Art, Firenze (Italy) 1984, 96.

4 - E. De Castro, Determination de l'angle de contact eau-pierre, RILEM 58 V.T.P., 1985

INVESTIGATIONS OF THE SURFACE PROCESSES ON EXPOSED LIMESTONES.

PAGONA MARAVELAKI *, R. BERTONCELLO °, G. BISCONTIN *, G. BATTAGLIN ^, E. ZENDRI * AND E. TONDELLO °.

* Dipartimento di Scienze Ambientali, Università di Venezia, Calle Larga S. Marta 2137, I-30123, ITALY

^ Dipartimento di Chimica-Fisica, Università di Venezia, Calle Larga S. Marta 2137, I-30123, ITALY

° Dipartimento di Chimica Inorganica, Metallorganica ed Analitica, Università di Padova, Via Loredan 4, I-35131, ITALY.

ABSTRACT

Different decay mechanisms, according to the exposure to rain actions, have been identified on a series of Istria stone samples derived from the Venetian buildings. Enquiries have been carried out on these samples by means of porosity measurements and chemical analysis in depth. In order to identify the decay mechanisms and their kinetics, simulation tests have been performed on exposed samples of Istria stone for a period of 19 months. These specimens have been analysed by several techniques in order to verify the chemical dissolution of limestone provoked by the rain action. The initial chemical attack on the exposed samples has been pointed out by XPS and RBS micro-analytical techniques.

Introduction

Present research deals mainly with processes due to environmental action at the surfaces of the exposed limestones. The factors which formerly caused the deterioration on stone works of art have nowadays become more threatening, and new factors have been added. The behaviour of a stone surface exposed to the outside environment depends on factors which refer both to the intrinsic properties of the stone and to the environment [1].

Actual knowledge of mechanisms of decay at present has many unresolved questions and demonstrates that many areas require further sustained study. At first sight, as the stone composition is mainly of calcite, the damage process may appear quite a simple one and most of the research has up to now been limited to disclosing the phenomenon, without going into its mechanism and kinetics.

Even though it is difficult to establish an adequate model for investigating the decay process, two extreme cases can be easily recognized: the material either absorbs the corrosion products or they remain confined at the surface; in the first case, a cumulative effect of the process will be observed, whereas in the second one the effect will be fairly reduced. Moreover, the areas of the monuments located in zones directly exposed to rain or found only where dry particles deposit, represent two situations characterized by quite different morphologies and mechanisms.

The purpose of this paper is the investigation of the behaviour of Istria stone exposed or not to the rain water action.

Mat. Res. Soc. Symp. Proc. Vol. 267.

Investigations have been carried out on a series of Istria stone samples, a microcrystalline limestone widely used in the Venetian buildings, with different macroscopic decay morphologies, as a consequence of the exposure to the environmental conditions.

In order to identify the decay mechanisms and their kinetics, simulation tests have been performed on exposed samples of Istria stone; furthermore, analyses have been carried out on the exposed specimens by means of wet chemical and instrumental techniques and also by micro-analytical and surface characterization techniques.

Experimental

Investigations have been conducted on a series of 40 samples of Istria stone derived from monuments of Venice, while the sampling has been carried out in accordance to the NORMAL standards [2]. The samples were divided in two groups: those which were covered by a white layer derived from parts exposed to rain water, and those with a black crust belonging to not directly exposed locations.

The research was performed by evaluating the most significant physico-chemical parameters of the alteration phenomena, like porosity, quantitative analysis of % C, S and Ca^{++} soluble in the water at different depth of the samples. These investigations contribute to obtain the trend of the process, as well as indications about the thickness of the altered layer.

Porosity measurements have been performed by means of a mercury porosimeter, Ca^{++} determination by means of atomic absorption measurements after dissolving the samples in the water and a carbon-sulphur analyser was employed for the % C and S quantitative analyses. The % lost C (which is shown as Cp on the graphs) was calculated, by subtracting the total % C identified quantity in our samples, from the existent quantity of C in the original Istria stone (12 %); it may be assumed that the actual loss of C carbonate could be higher, since it has been taken into account, that a part of the C content is due to carbonaceous particles, widely present in the black crusts.

In order to relate simulation studies to actual onsite measurements, ten (n. 10) samples of Istria stone were exposed to the Venetian weather from 2/4/90 until 31/10/91. The specimens (with dimensions of 5x5 cm and different thickness from 1-20 mm) were fixed in a glass funnel (angle=30°), which was connected with a conic flask functioning as rain water collector. The upper surface of the specimens was completely exposed to the rain water, whereas the lower one was mainly affected by the dry deposition effects.

Stone deterioration estimates were obtained by direct measurements of changes in rainfall-runoff composition from stone surface dissolution, taking into account measurements of the water soluble salts detected in the runoff water and also after extraction with distilled water from the exposed specimens. Loss of calcium (surface recession) in micrometers is obtained by conversion of the net calcium concentration in the runoff solution to a volume of stone removed. This procedure assumes mass-balance considerations concerning the calcium ion quantities determined both in the rainfall samples and in the soluble salts of the aqueous extractions of the exposed specimens [3].

Chemical analysis of cations and anions were performed by atomic absorption and ion chromatography, respectively, both for the real and naturally aged samples.

XPS spectra were recorded in a VG ESCALAB MK II spectrometer using non-monochromatized MgKa radiation (1253.6 eV-at 15 kV with a current of 20 mA). The working pressure was less than $1x10^{-8}$ Pa. The spectrometer was calibrated by assuming the binding energy (BE) of the Au $4f_{7/2}$ line at 83.9 eV with respect to the Fermi level. Owing to surface charging, samples present an average shift of the signal energies of about 5 eV. As an internal reference for these charging effects the C1s peak of hydrocarbon contamination has been assumed at 285.0 eV [4]. Survey scans (50 eV pass energy; 1 eV step; 0.5 s $step^{-1}$ dwell time) were obtained in the BE range between 0 and 1000 eV. Detailed scans were recorded for the C(1s), O(1s), Ti(2p), N(1s) regions (20 eV pass energy, 0.1 eV step; 1.25 s $step^{-1}$ dwell time). The standard deviation in the BE values of the XPS lines is 0.15 eV. After a Shirley-type background subtraction [5], the raw spectra were fitted using a non-linear least-square fitting program adopting Gaussian-Lorentzian peak shapes for all the peaks [6]. The atomic compositions were evaluated using sensitivity factors as determined from theoretical photoionization cross-sections and asymmetry parameters calculated within the Hartree-Fock-Slater one-electron central potential model [7]. The inelastic mean free path of the analyzed photoelectrons has been assumed proportional to the square root of their kinetic energy.

Depth profiles were carried out by Ar^+ sputtering at 3 kV, 3 mA cm^{-2} beam current density with an argon partial pressure of $3x10^{-4}$ Pa. The beam voltage and current were chosen to minimize preferential sputtering effects [8]. Under these conditions we estimated a sputtering rate of ca. 1 nm min^{-1}.

The near surface composition of the exposed stones was also evaluated by means of "Rutherford backscattering spectrometry" (RBS), using a 2.2 MeV $^4He^+$ beam with an angle $\theta=160^o$. Depth profiles of all the identified elements by the RBS techniques concerned a thickness of ca. 500 nm [9].

Results and Discussion

Porosity measurements, expressed as cumulative volume and pore size distribution, regarding specimens of two ranges of depth (0-5 mm and 5-10 mm) were performed on 10 samples of each group. Different conditions after the environmental attack have been revealed as showed in fig. 1, 2, 3 and 4 concerning porosity and the most important physico-chemical parameters which were taken into account.

On the stone surface which undergoes a continuous rain-washing, like in the case of white crusts, small size capillaries appeared in the outermost layer, even if the larger ones remained practically unchanged in depth. In the external layer the cumulative volume values were larger than those in the inner parts (see fig. 1). The low values of Cp, S and soluble Ca^{++} detected, suggest that white crusts could not be considered as valuable indicators of the decay risk (see fig. 3). Cp variations ascribed to transformed $CaCO_3$ were related to the vacuum volume values, and therefore porosity arises from material loss. The environment affected both physically and chemically the stone surface, giving rise to dissolution products, which partly remain on the stone surface or migrate towards the inner parts occluding consequently the small size pores.

In the case of the dendritic black crusts, as is proved by the lost carbon values, alteration processes had also taken place, which can be dangerous because of their mechanical action. The limited amount of water-humidity led to supersaturated systems of salts of generally high crystallization pressure and large specific volume. The calcium ion amounts mainly found in the first mm of thickness corresponded to those of S; the latter decreased significantly in disagreement with the observed increase of Ca^{++} amounts (see fig. 4). It is very likely that Ca^{++} is present not only as gypsum but also as other salts produced by reactions between $CaCO_3$ and the deposit.

Concerning the dendritic crusts in a depth of 0-5 mm and 5-10 mm, the cumulative volume plot against the pore radius indicated a larger porosity in the external layer, possessing cumulative volume values 5-6 times higher than those observed in the original material. The much lower porosity of the inner parts, even though reaching values higher than those of the original material, provided evidences of a quite compact system (see fig. 2). In the external layer both the large size pores (>10 mm) and the small ones increased; this fact was mainly due to deposition of incoherent material on the surface and to successive dissolution and crystallization of salts.

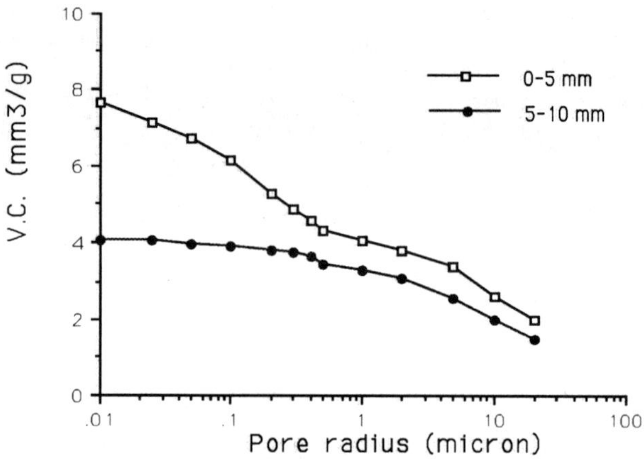

Fig. 1: Porosity (Cum. Vol. mm^3/g) as a function of pore radius (μm) in specimens of two ranges of depth (0-5 and 5-10 mm) in the white crusts.

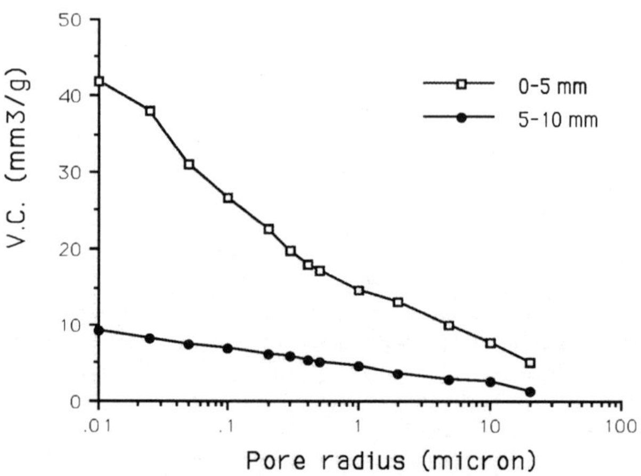

Fig. 2: Porosity (Cum. Vol. mm^3/g) as a function of pore radius (μm) in specimens of two ranges of depth (0-5 and 5-10 mm) in the dendritic black crusts.

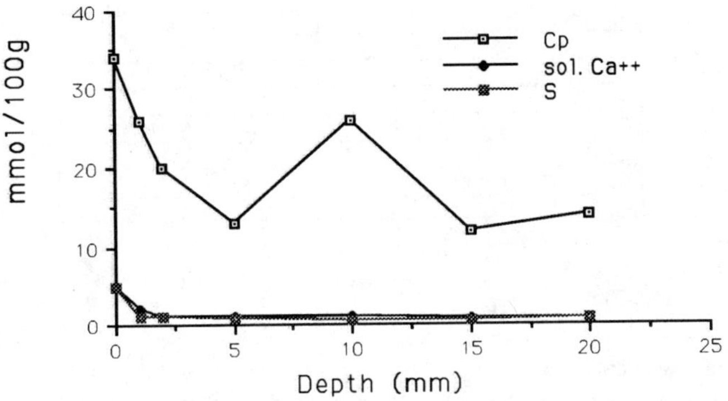

Fig. 3: Depth profiles of Cp, soluble Ca^{++} and S expressed in mmol/100g in the white crusts.

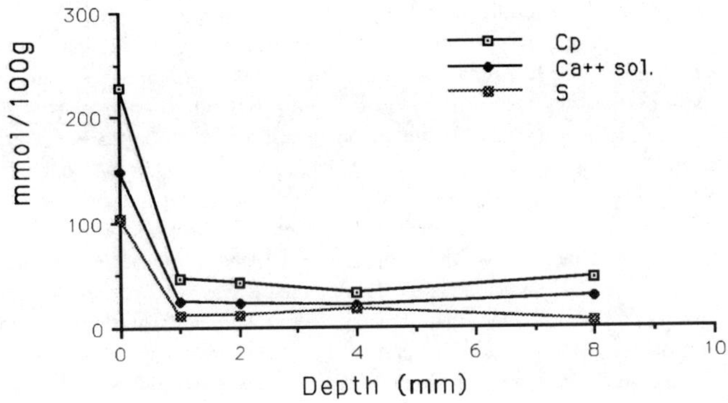

Fig. 4: Depth profiles of Cp, soluble Ca^{++} and S expressed in mmol/100g in the dendritic black crusts.

Table I shows the surface recession in mm of the Istria stone after the two periods of exposure. It is evident that the removed layer does not closely depend on the rainfall amounts, since the different calcium ion quantities detected to the blank runoff samples and the pH rainfall values -not available- have played an important role to the recession estimations. Net sulphate amounts detected in the runoff water provided evidence of gypsum's formation and its successive removal as a consequence of the dissolution effects.

Tab. I: Surface recession average values for 10 samples of Istria stone after exposure.

Rainfall (mm)	1230
Runoff (ml)	5418
Ca^{++} in runoff (mg)	69.64 (±2.36)
Ca^{++} blank (mg)	40.74 (±2.01)
Net Ca^{++} in runoff (mg)	28.89 (±1.23)
Ca^{++} extracted (mg/cm^2)	0.22 (±0.08)
Total Ca^{++} (mg/cm^2)	1.00 (±0.10)
Net $SO_4^{=}$ in runoff (mg)	13.38 (±1.54)
Surface recession (mm)	11.33 (±1.16)

XPS and RBS micro-analytical techniques have been applied to Istria stone specimens after their exposition to the environment.

Fig. 5a, 6a show XPS depth profiles of the identified elements for both faces of the examined samples after their exposure. The elements constituting the carbonate matrix, namely Ca, O, and C were differed going from the inner atomic stratums to the surface. Carbon total values were higher than those provided by the stoichiometry of the material; this increase was due, as is proved by its BE, to the presence of hydrocarbons pollutants at the surface. XPS analyses showed that the values of the carbon (Ch) derived from the hydrocarbons pollutants, which are showed as C1s cont. on the graphs were higher for the first atomic stratums for both faces, folowed by a stable profile (see Fig. 5a, 6a). During the total sputtering time a rather continuously decreasing trend of the Ch profile on the exposed face has been observed; however, on the sheltered face a Ch unchanged profile has been obtained after 10' of sputtering. Interpreting this fact, it is to be assumed that attack mechanisms are slow and superficial for the sheltered face, while they become fast and deep for the exposed one.

Depth profiles of silicon were similar for both faces, while slight differences for aluminum profile have been revealed. On the sheltered areas the aluminum values mainly corresponded to their amounts in the original matrix; however, the high observed values on the exposed face suggest either a contribution of the wet deposition or a selective action on the carbonate matrix which did not involve the aluminum (see Fig. 5b, 6b). Concerning phosphorus depth profiles similar arguments can be assumed. Lead amounts, identified only for the exposed face, together with iron ones detected for both faces represent the detection limits of the XPS technique. The constant present quantity of fluorine for both faces is to be referred to contamination agents deriving from the adjacent to Venice industrial zone.

Depth profiles of sulphur, the most important element of the decay processes presented similar trends for both analysed faces, possessing slightly higher values for the sheltered one (see Fig. 6b). According to the binding energy (BE) values the sulphur is present as sulphates, and also the above mentioned elements as oxides. Sulphur accumulations, observed for the outermost atomic layers of the exposed face, in relation to its constant distribution in the inner stratums of the sheltered face, provided the existence and the extent of the environmental attack.

When dissolution processes had taken place, as is showed from the above reported diagrams, the initial attack was achieved in the first atomic layers (1-2 nm).

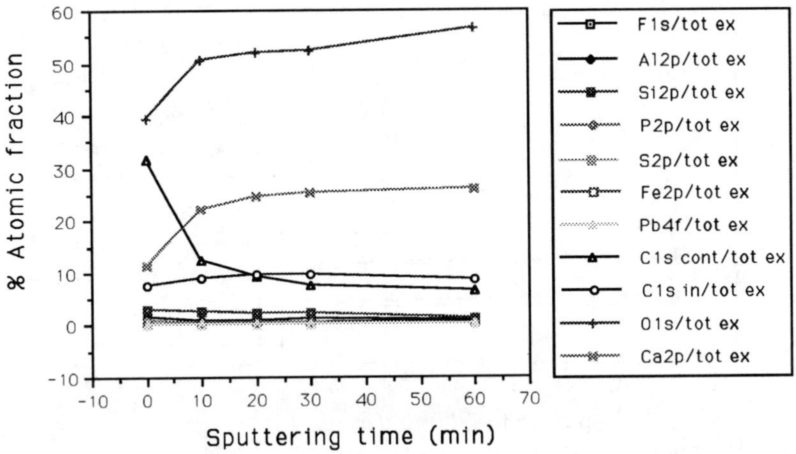

C1s cont/ tot ex: C1s total of contamination (Ch) on the exposed face
C1s in/tot ex: C1s total inorganic on the exposed face

Fig. 5a: XPS depth profiles of all the elements presented on the exposed face of the Istria stone.

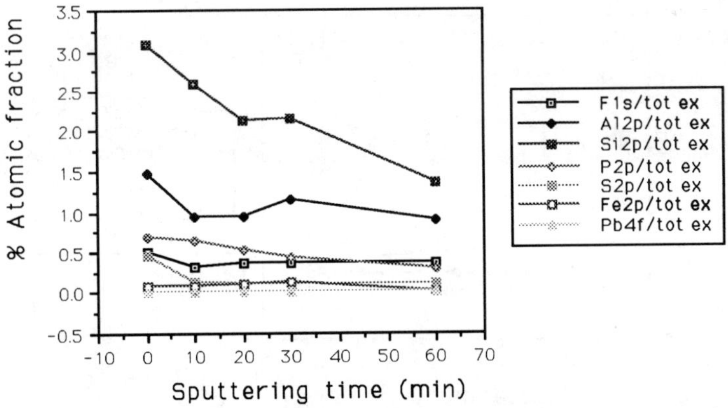

Fig. 5b: XPS depth profiles of all the elements presented in low amounts on the exposed face of the Istria stone.

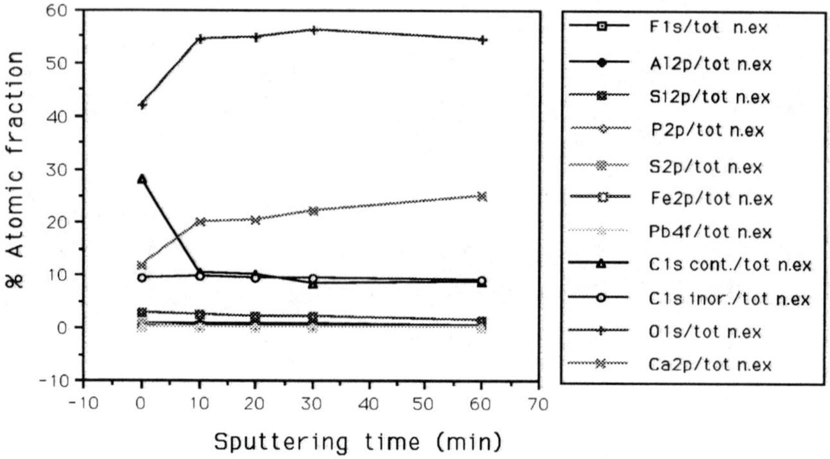

C1s cont/ tot n.ex: C1s total of contamination (Ch) on the sheltered face
C1s inor./tot n.ex: C1s total inorganic on the sheltered face

Fig. 6a: XPS depth profiles of all the elements presented on the sheltered face of the Istria stone.

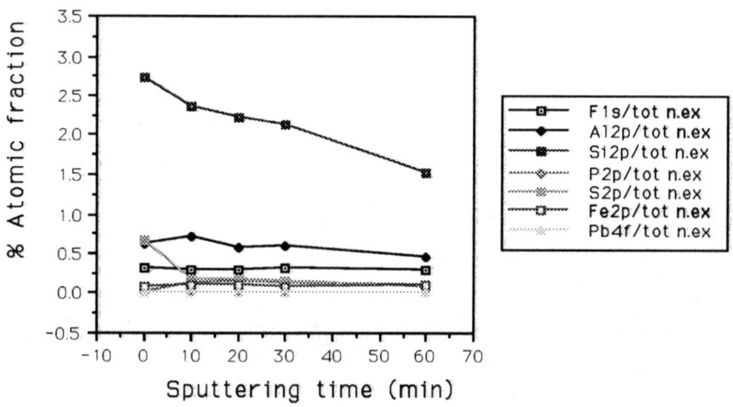

Fig. 6b: XPS depth profiles of all the elements presented in low amounts on the sheltered face of the Istria stone.

Concurrently the same previously analysed sample and also one of a fresh quarry were processed by applying the RBS technique. The latter was taken into account as a reference element for the eventual composition changes observed on the examined specimen.

RBS spectra for the fresh specimen showed that elements which were worthwhile mentioning, besides those which constituted the carbonate matrix, were the zinc and the lead, present in low amounts of 0,08 % and 0,002 % respectively concerning the total atomic concentrations.

Data and spectra reported in tab. II and fig. 7, related to the identified elements after exposure, pointed out that no significant differences between the two analysed faces existed. However, the latter differed from the elements observed on the fresh specimen.

Tab. II: Percent atomic concentrations after the RBS analysis.

% atomic concentrations	Fe	Pb	Zn
Fresh sample	n.q.	0.002	0.080
IS exposed face	0.120	0.002	0.000
IS n.exposed face	0.140	0.004	0.000

n. q. : not quantified, IS: Istria stone

Iron accumulations on both faces have been revealed; this element was presented only in traces on the fresh sample. The almost constant quantity of iron, in a significant amount present in the Venetian aerosol (100 ppt) [10], implies that its origin is in the rain/aerosols or that iron was not involved in dissolution processes.

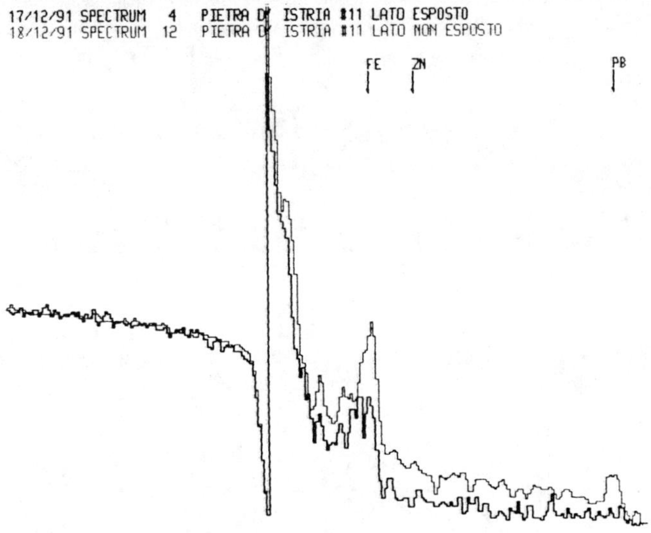

Fig. 7: RBS depth profiles of the identified elements to both faces of the exposed Istria stone.

An absence of zinc has been registrated in the spectrum of the exposed sample. When attempting to interpret this fact two hypothesis could be issued. The first one suggests that the presence of zinc in the fresh sample is by chance; whereas the second one, less easily to be sustained, is related to dissolution processes involving even the zinc. In order to obtain an assessment it is necessary to analyse more samples leading thus to a statistical approach of the total phenomenon.

The low amounts of sulphur did not allow its detection, since the sulphur signal has been superimposed by the one of calcium.

Conclusions

The analyses carried out on samples with the black crusts, where the decay mechanisms were related to little water and especially to high humidity and condensation, indicate that the decay process begins in the outermost layers and extends to significant depths. The first layers corresponding to particulate deposit and to interface, exhibit a remarkable location of the chosen parameters. The latter, in the inner parts, present less homogeneous distributions with accumulations of gypsum.

In the white crusts, as is proved by the low values of gypsum and the other examined parameters, a dissolution of the decay products took place because of the continuous rain-washing. Carbon loss profiles follow an invariable trend until the inner parts implying thus an aggressive action by the loss of the products. Porosity values enough higher in the external layers than the inner ones, in relation to the formation of small size capillaries under of 1000 ', confirm the material decay.

When weathering processes are to be monitored over relatively short time intervals, XPS and RBS techniques proved to be essential for investigating chemical alterations to the top-most layer. Depth profiles of the identified elements by the XPS analysis prove a homogeneous action on the material face exposed to rain. The low porosity of the stone, in this initial face of the process, does not allow the accumulation of the products. A chemical action until layers in the range of 100 ', is assumed by the depth profile of Ch following a decreasing trend. The face of the exposed material sheltered from the rain shows profiles which indicate less dynamic actions related more to particulate deposit.

An evaluation can be issued concerning the attack mechanisms after the evidences provided by the XPS and RBS results and also by the traditional bulk analyses of the samples. At first sight, because of chemical dissolution processes, a homogeneous loss of material was verified leading, as a consequence of an impending physico-chemical action, to formation of a microporosity according to the microcrystalline nature of the material. The presence of small size capillaries can be justified by accumulations of calcium sulphates and carbonates, observed as a white crust on the surface of the monuments constructed of Istria stone.

REFERENCES

1. V. Fassina, A survey on air pollution and deterioration of stonework in Venice, Atmospheric Environment, 12, (1978), pp. 2205-2211.
2. Raccomandazioni NORMAL 3/80, Materiali lapidei: Campionamento, edited by CNR-Centri di Studio Cause di Reperimento e metodi di Conservazione delle Opere d'Arte, ICR-Istituto Centrale del Restauro-Rome.
3. M. M. Reddy, Acid rain damage to carbonate stone: a quantitative assesment based on the aqueous geochemistry of rainfall runoff from stone, Earth surface processes and landforms, 13, (1988), pp. 335-354.
4. M.P. Seah, G.C. Smith in Practical Surface Analysis-Vol 1 Second Edition ed by D. Briggs and M. P. Seah, appendix 1, Wiley, Chichester (UK) (1990), pp. 543-544.
5. D. A. Shirley, Phys. Rev., 55, 4709, (1972).
6. J. Vegh, J. Electron Spectr. Rel. Phen., 46, 411, (1988).
7. J. J. Yeh, I. Lindau, Atomic Data and Nuclear Data Tables, 32, 1, (1985).
8. N. Matsunami et al. in Atomic Data Nuclear Data Tables, 31, 1, (1984).
9. L. C. Feldman, J. W. Mater, in Fundamentals of surface and thin films analysis, edited by L. C. Feldman, J. W. Mater, (Elsevier Science Publishers, New York, 1986).
10. G. Calvelli, D. Ceccato, P. Mittner, E. Schiavuta, A. Zannoni, E. Argese, A. Baldo, A. Rigo, Multielemental pixe analysis of the aerosols and high sensitivity measurements of the SO2 levels in the city of Venice, in Physics in environmental and biomedical research, edited by S. Onori, E. Tabet, (World Scientific, 1986), pp. 83-86.

CLEANING WITH LASER RADIATION ON ISTRIA STONE

PAGONA MARAVELAKI *, G. BISCONTIN*, E. ZENDRI*, R. POLLONI °, W. CECCHETTI °.
* Dipartimento di Scienze Ambientali, Università di Venezia, Calle Larga S. Marta 2137, I-30123, ITALY
° Dipartimento di Chimica-Fisica, Università di Venezia, Calle Larga S. Marta 2137, I-30123, ITALY

ABSTRACT

Cleaning treatments necessary for stone conservation consist of removing compounds that were superimposed on the original material by means of physico-chemical processes. The purpose of this work is to identify the transformation of stone surface by the LASER treatment. Cleaning test on Istria stone with LASER radiation at different energies and in different regimes have been performed. The depletion of the original material has been evaluated by means of SEM microprobe analyses.

Introduction

Exposure of stone surfaces to external conditions can cause significant changes to the original characteristics. Synergistic effects were often observed because of the concurrent action of water and particle deposition. Depending on the conditions of exposure, as well as on the surrounding environmental pollution, different situations, such as previously reported, arise. In the most common case of scarce water weathering, the stone is covered by a layer of variable thickness consisting of airborne particulates and of air-stone interaction products. Both these materials contain gypsum as the main component. The stone surfaces more exposed to water weathering show little presence of deposits, but undergo direct, strong chemical attack leading to the accumulation of erosion products on the surface. Gypsum is found also in this case though to a lesser extent due to its water solubility. The extent of gypsum depends greatly on the stone porosity [1].

Any exposure of the stone to external conditions leads to negative effects and thus remediation is primarily addressed to reestablish physico-chemical stability. Cleaning interventions are undertaken to remove materials other than the original ones. Numerous cleaning techniques were proposed, some of which, based on the application of well established cleaning agents.

In this research the already proposed procedure based on LASER was applied to clean samples of Istria stone taken from Venetian buildings. The effects induced by the LASER radiation on the stone surface were evaluated by scanning electron microscopy (SEM).

Experimental

Cleaning a surface requires caution and adequate know how since some effects can be irreversible. The cleaning process must be carried out gradually and selectively and should avoid formation of harmful by products (i.e. soluble salts) and surface alterations (i.e. microfractures, abrasions) which could accelerate the successive deterioration by the increased specific surface [2].

One of the most promising methods for the removal of black incrustations from stone is based on the application of impulsed LASER (Light Amplification by Stimulated Emission of Radiation). The main advantage of the method arises from its very high selectivity. The highly energetic radiation produced by the LASER can interact effectively with the target materials [3, 4].

Depending on the applied impulse rate, two different effects are to be expected, as shown the fig. 1, 2. The "normal-mode" type radiation (FRM) is absorbed by the black incrustations which reach immediately temperatures of 4,000-5,000 °K, thus burning and vaporizing it. The underlying stone reflects the radiation and does not undergo any damage even after repeatedly application on the same surface (see fig. 1). When radiation fluxes are 10^3 - 10^5 W cm^{-2} and the impulse duration is short (microsecond to millisecond) the heat transfer from the incrustation to the underlying stone is negligible, thus avoiding surface fusion or microfissures.

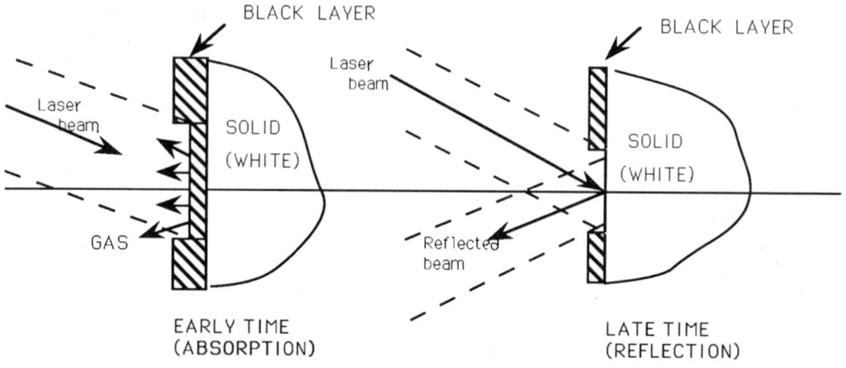

Fig. 1: Effect of long LASER pulses (10^{-3} sec) on material surfaces.

On the other hand, more energetic radiation such as that produced by "Q-switching" can reach fluxes of 10^7 - 10^{10} W cm^{-2}. Under these conditions, the heat tranfer to the stone is very high, the produced plasma propagates thermal wave and pressure and prompts the material to be removed when it expands after the initial compression. "Q-switched" type radiation allows the decomposition of calcium carbonate in carbonatic rocks, the fusion of CaO and the formation of a porous and water permeable crust on the stone surface [5].

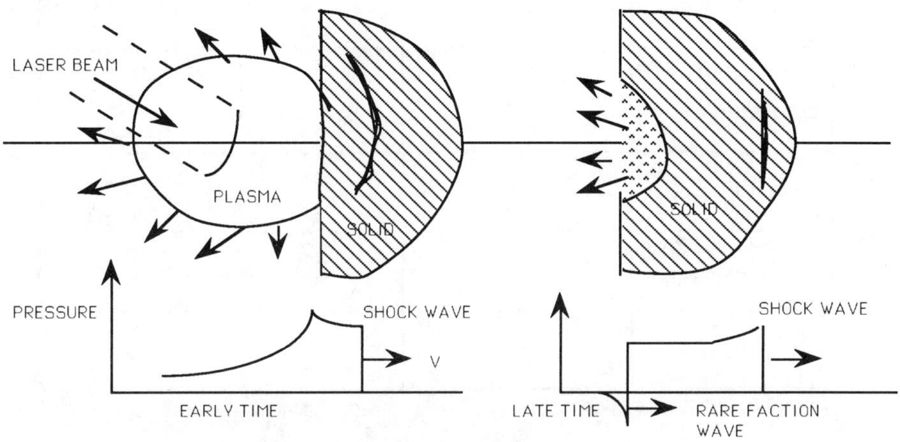

Fig. 2: Effect of very short LASER pulses (10^{-8} sec) on material surfaces.

A big advantage offered only by LASER is its damage-free cleaning of seriously impaired surfaces as well as of surfaces previously treated with protective substances.

Cleaning tests with LASER radiation were performed on Istria stone samples employing a specific laser of Nd: YAG (Neodimium: Yttrium Aluminium Garnet) and operating in the near infrared region (l=1.06 nm) under both the "normal-mode" and "Q-switched" conditions. Energies between 0.5 and 1.0 J per impulse, 1 to 5 pps impulse frequency and 2, 3, 6 mm beam diameters have been employed for the tests. Other operating conditions of the LASER were: 500 ms impulse duration (T) and up to 4 KW power for the "normal-mode" type radiation; T=20 ns and up to 15 KW power for the "Q-switched" type radiation.

All samples cleaned by LASER were later investigated by SEM and optical microscopy. Additional microprobe analysis was performed on some samples.

Results and Discussion

a) Cleaning with LASER of "normal-mode" type radiation.

In Table I the operating conditions of the LASER application are reported.

Tab. I: Operating conditions of the LASER "normal-mode" type radiation.

SAMPLE	ENERGY DENS. (J/cm^2)	BEAM (mm)	IMPULSE
C-1 Dendritic crust	3.5	6	30, 90
	4.8	2	50
	16	2	1, 10
	30	2	10, 50
C-2 Dendritic crust	16	2	1, 5, 10, 20, 40
	8	2	50
	14	3	1, 10, 50
C-3 Thin crust	16	2	30
C-4 Thin crust	11	2.5	20

The SEM analysis showed no differences between the surfaces treated by applying a different number of impulses. This suggests that the first LASER impulse is responsible for the black incrustation removal. The successive radiation is only reflected by the underlying surface material and does not produce any damage.

The treatment of samples with black dendritic incrustations up to 1 mm by the "normal-mode" LASER demonstrates the existence of energy density threshold, equal to 5 J cm^{-2}, below which the formation of gypsum on the treated surface is expected.

The conditions E/A = 14 and 8 J cm^{-2}, respectively, give rise to the diffusion of carbon residues around the black crust application point (see photo 1).

A density of E=16 J cm^{-2} (corresponding to a radiation power of 3.2×10^4 W cm^{-2}) is capable of removing completely the gypsum, as indicated by the microanalysis results. The SEM data, however, show that the treated stone is altered, as expected by such a strong radiation conditions. A "vitrification" appeared on the treated surface which spreads also over the surrounding areas. The presence of fissures and the appearance of small grains document the alteration of the sample C-2 (see photo 2).

In contrast, the sample C-3, with a thin crust, displayed lower alteration after application of the same LASER conditions (E=16 J cm^{-2}). This indicates the relationship between radiation effect and crust morphology. The LASER cleaning does not appear particulary suitable for surfaces covered by thick crusts that can be removed by other available procedures.

Good gypsum removal and no significant surface deterioration as observed for the dendritic samples, were recorded by applying a

LASER energy density of 11 J cm^{-2}. As far as the SEM analysis is incapable of verifying completely the radiation effects, it is desirable to gain additional insight (i.e. through specific surface measurements) to state definitively the extent of the surface alteration.

b) Cleaning with LASER of "Q-switched" type radiation.

Tab. II shows the operating conditions of the LASER "Q-switched" type radiation.

Tab. II: Operating conditions of the LASER "Q-switched" type radiation.

SAMPLE	ENERGY DENS. (J/cm^2)	BEAM (mm)	IMPULSE
C-5 Thin crust	6	2	10
C-6 Thin crust	16	2	10
C-7 Thin crust	22	2	20
C-8 Thin crust	3.3	2	10

The "Q-switched" type LASER radiation was applied only on samples exhibiting black thin incrustations, in order to avoid the above mentioned problems and to verify more precisely the treatment efficiency. The threshold energy density of the "Q-switched" radiation for the black removal was found 0.55 J cm^{-2}, whereas densities of 3.3 and 6.0 J cm^{-2} cause surface fusion though less intense than that observed for the dendritic crusts irradiated by the LASER operating in "normal-mode".

The SEM analysis indicates that energy intensities of 16 and 22 J cm^{-2} induce the diffusion in the treated surface portion of the dirt due to the fusion of organic atmospheric pollutants. A thin vitrified layer covered the treated surface thus preventing the examination of the underlying material. It seems, in this case, that the LASER produces a surface melting of the cleaned Istria stone, but it is not possible to ricognize the main product of the thin vitrified layer, since the SEM microprobe analyses show only the presence of calcium.

The result so far acquired showed no significant differences between the LASER radiations of the "normal-mode" and "Q-switched" types operating with energy densities up to 10 J cm^{-2}, corresponding to powers of $2x10^4$ ("normal-mode") and $5x10^8$ W cm^{-2} ("Q-switched"), respectively.

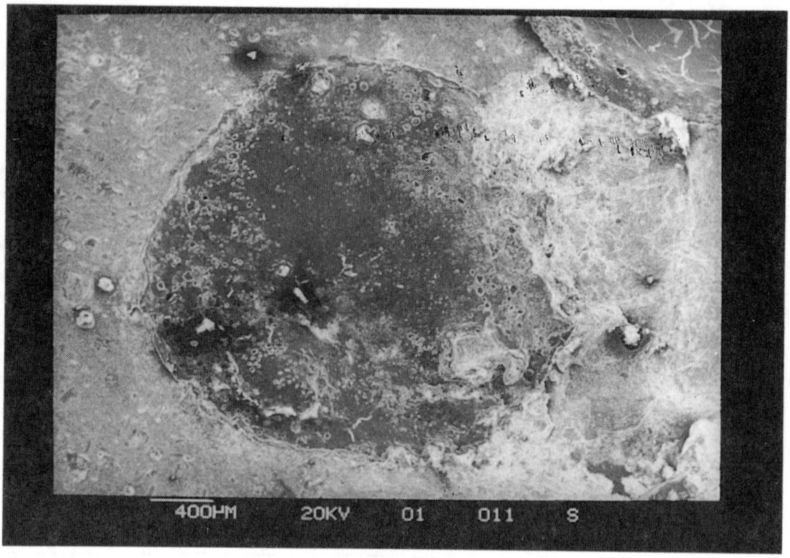

Photo 1: SEM micrograph of a dendritic black crust on Istria stone after LASER cleaning (14 J cm-2, beam 3 mm, 50 impulses).

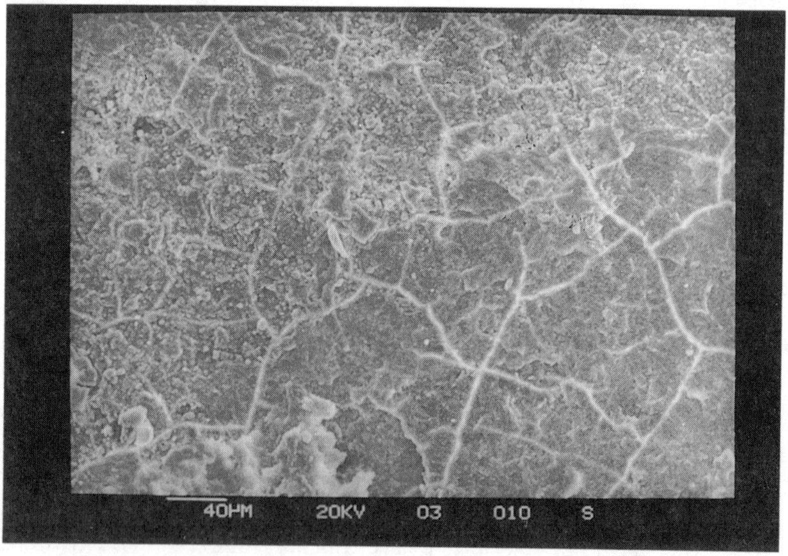

Photo 2: SEM micrograph of a dendritic black crust on Istria stone after LASER cleaning (16 J cm-2, beam 2 mm, 20 impulses).

Conclusions

Although the LASER technique shows undisputable advantages over traditional treatments and succeeds in removing the black incrustations, further research efforts are required to establish the surface characteristics immediately after the treatment and possible alteration of the treated surface over successive time periods. A further application of micro-analytical specific surface techniques such as XPS and RBS may be clarify the products of the thin vitrified layer observed, by employing LASER of several conditions on the cleaned surface. Experimental evaluation of both specific surface and reactivity of the treated surface are needed.

REFERENCES

1. G. Biscontin, C. Félix, P. Maravelaki, E. Zendri, Cheracterisics, Weathering forms and mechanisms of Istria stone in Venice, presented in 2nd Intern. Symp. for the Conservation of Monuments in the Mediterranean Basin, Geneve, 1991, (in press).
2. L. Lazzarini, in La pulitura dei materiali lapidei da costruzione e scultura: metodi industriali e di restauro, Cedam, Padova, 1981.
3. O. Svelto, in Principi del LASER, Boringhieri publishers, Milano, 1972.
4. L. Lazzarini, J. L. Asmus, L. Marchesini, Lasers for the cleaning of statuary: initial results and potentialities, C. R. du Ier Colloque Int. Deterioration des pierres en ouevre, la Rochelle, Chambery, 1972, pp. 89-94.
5. H. Davies, Reflection of electromagnetic waves from rough surfaces, Proc. Inst. Elect. Eng., 101, 209, (1954).
6. J. Asmus, The development of a Laser Cleaner, Proc. Int. Symp. Deterrioration of Buildings Stone, Athens, 1976, pp. 137-141.

MICROSTRUCTURES OF B72 ACRYLIC RESIN/MTMOS COMPOSITES

George Segan Wheeler*, Eugene Wolkow**, and Harry Gafney**

*The Metropolitan Museum of Art, Objects Conservation Department, 1000 Fifth Avenue, NYC 10028

**Queens College, Chemistry Department, 65-30 Kissena Boulevard, Flushing, NY 11367

ABSTRACT

Gels prepared from solutions of B72 acrylic resin and methyltrimethoxysilane (MTMOS) are examined by SEM. Four initial mixtures - 0,2,5 and 10% w/v B72/MTMOS - produced gels whose fractured surfaces exhibit distinct features. Scanning electron microscopy shows that neat MTMOS produces a glass-like gel while the addition of B72 creates MTMOS-derived spheres in a "connective tissue" of acrylic resin. These particle sphere/acrylic resin solids are mechanically weak composites.

INTRODUCTION

Acrylic resin/MTMOS solutions are of interest to the field of art conservation because such solutions have often been used for the preservation of stone sculpture in museums[1][1]. By itself, MTMOS can form a xerogel by reaction with atmospheric moisture; if generated in situ this gel can increase the mechanical strength of stone. Art conservators, in an attempt to improve this strengthening effect and to overcome difficulties in adhering pieces of stone previously consolidated with MTMOS, added acrylic resins such as B72 to MTMOS (B72 dissolves readily in MTMOS). By this means pieces of stone individually consolidated with B72/MTMOS could more easily be glued together at some later time. This paper, which is an extension of the work of Koestler[2][2], reveals the microstructures of fractured surfaces of B72/MTMOS gels through SEM photographs.

MATERIALS AND METHODS

Gels are prepared from MTMOS (Petrarch M9100, Bartram Road, Bristol, PA 19007) which has been double distilled directly before use and as received B72 acrylic resin (B72 is a copolymer of methylacrylate and ethylmethacrylate[3][3] produced by Rohm and Haas, Philadephia, PA 19105). Each gel is derived from an initial volume of 100 ml of MTMOS with the appropriate mass of B72 added to the liquid. In this way solutions of 0,2,5, and 10% w/v B72/MTMOS are prepared and allowed to react in open polypropylene containers (70 mm maximum diameter) under laboratory conditions (23°C and 43% RH on average). These solutions form soft gels after approximately two weeks and continue to react and harden over the ensuing months. After two months the gels cease to lose weight. The operative chemical reactions under conditions of limited water supply are as follows:

Hydrolysis

$$(MeO)_2Si\text{-}OMe \;+\; H_2O \;\text{---}\; (MeO)_2Si\text{-}OH \;+\; MeOH$$

Alcoholic Condensation

$(MeO)_2Si\text{-}OH + MeO\text{-}Si(OMe)_2$ --- $(MeO)_2Si\text{-}O\text{-}Si\text{-}(OMe)_2 + MeOH$

Dehydration Condensation

$(MeO)_2Si\text{-}OH + HO\text{-}Si(OMe)_2$ --- $(MeO)_2Si\text{-}O\text{-}Si\text{-}(OMe)_2 + H_2O$

The hydrolysis and condensation reactions continue up to and beyond gelation to produce stiff albeit brittle gels. The gel derived from neat MTMOS is transparent while those produced from B72/MTMOS solutions are opaque and white.

After twelve months newly fractured surfaces were examined by SEM. A Hitachi S570 was used for all photographs and analyses. A 20 KV energy electron beam was employed for the SEM imaging.

RESULTS

Figure 1 shows the SEM photograph of the fractured surface of a gel derived from neat MTMOS at 200X magnification. The photograph exhibits few features - only the lines of conchoidal fracture of a brittle, glass-like solid.

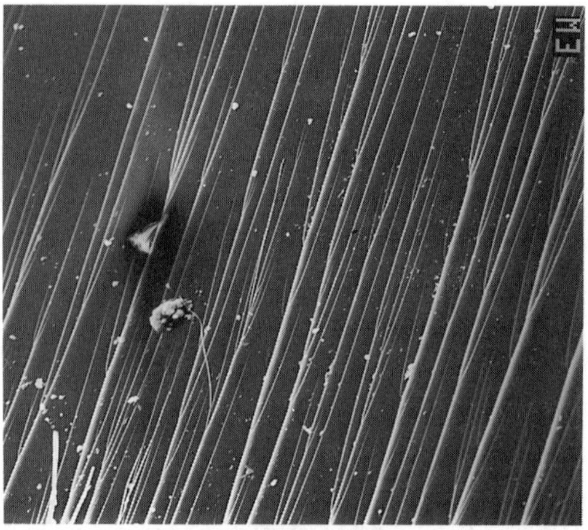

Figure 1. SEM photograph of gel produced by neat MTMOS
Horizontal Width of Photograph = 450 Microns

The sample containing 2% w/v B72 acrylic resin in Figure 2 displays a dramatically different morphology: spheres of one phase are embedded in a second phase. The spheres, which were established by Koestler in a previous paper to be silica-gel derived from MTMOS [2], range in diameter from 15 to 100 microns and average approximately 60 microns. One striking feature of the fractured surface represented in Figure 2 is that the spheres have been pulled wholesale from the acrylic matrix without fracture of the spheres themselves or pulling up of tendrils of acrylic resin in the fracture process. The empty "pockets" where spheres have been removed are clearly evident. These features imply little or no bonding at the interface between the MTMOS-derived silica gel spheres and the B72 acrylic resin.

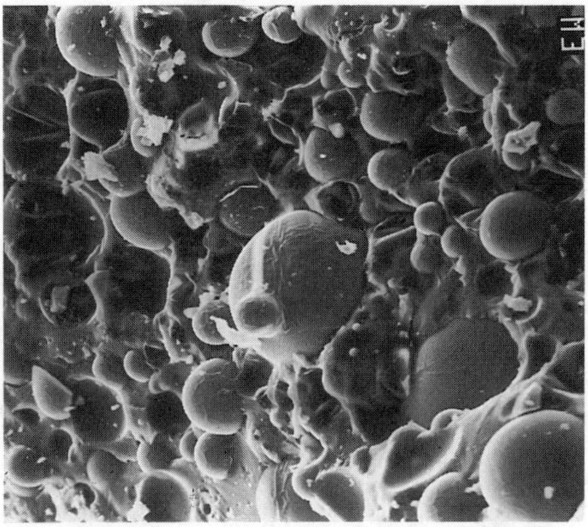

Figure 2. SEM Photograph of Gel Produced by 2% w/v B72/MTMOS
Horizontal Width of Photograph = 450 Microns

Figure 3 shows the fractured surface of a 5% w/v B72/MTMOS gel. It exhibits the same features as does the 2% gel but the particle size is reduced to approximately 45 microns. Once again the intact spheres and pockets are visible.

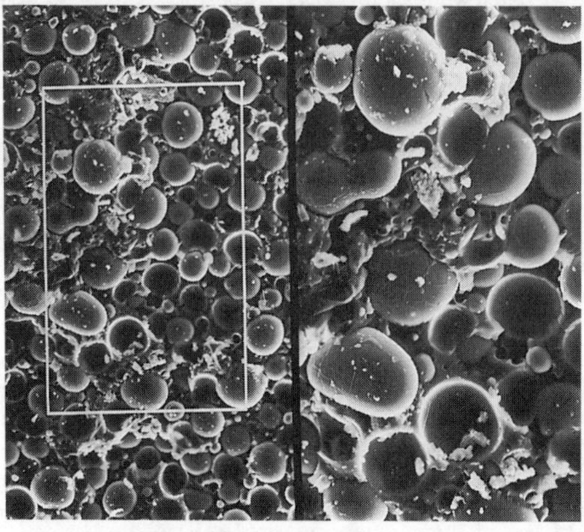

Figure 3. SEM Photograph of gel produced by 5% w/v B72/MTMOS
Horizontal Width of Left = 450 microns, Right 225 microns

Figure 4 is the SEM photograph of the 10% w/v B72/MTMOS gel. The sphere size is yet further reduced to - on average - less than 15 microns and the cavities are once again clear and abundant.

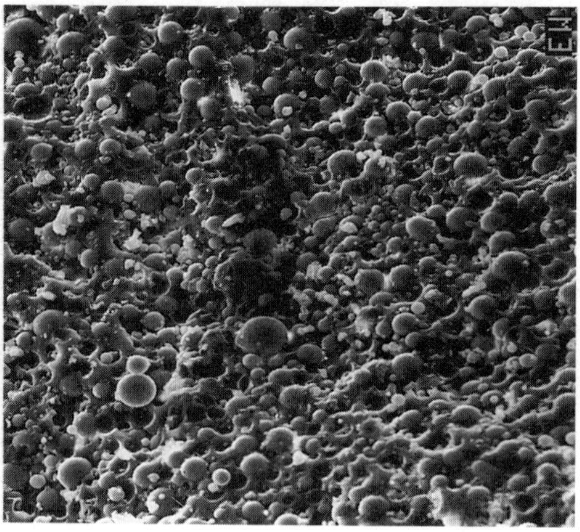

Figure 4. SEM photograph of gel produced by 10% w/v B72/MTMOS
Horizontal Width of Photograph = 450 microns

DISCUSSION AND CONCLUSIONS

The first point which arises from the SEM photographs is why do the samples of B72 exhibit different textures than samples produced by neat MTMOS? As stated previously B72 dissolves readily in MTMOS, however, is does not dissolve in methanol - the abundant by-product of hydrolysis and alcoholic condensation. (Under conditions of limited water supply such as prevails by using atmospheric water to produce hydrolysis, alcoholic condensation is predominant.) It is speculated that B72 phase separates from the initial preparation once sufficient methanol is generated. The phase of methanol/MTMOS-derived oligomers continues to react and later forms silica-gel like solids in the form of spheres. Eventually all of the methanol evaporates.

A second point is that the MTMOS-derived spheres are smaller and more abundant the more B72 is present; this may be due to the fact that the spheres have both less time and less space to develop when surrounded by a more extensive matrix of B72.

Finally, the SEM photographs help to explain the mechanical testing data reported by Fleming [2] in an earlier paper. Fleming demonstrated that the inclusion of B72 to MTMOS reduced the modulus of rupture (MOR) of these gels 60% or more. Neat MTMOS exhibited an MOR of 58 MPa while gels with 1-10% B72 had MORs between 10 and 20 MPa. The photographs in this paper show that the MTMOS phase in the composite gėls is no longer continuous as in a neat MTMOS-derived gel. Not only does the MTMOS form unconnected spheres when B72 is present, but these spheres show little affinity for the acrylic resin, i.e., the spheres are easily pulled from the B72 matrix. Although these two phase gels have not been demonstrated for these consolidants in contact with stone, possibly because in the case of limestone the MTMOS evaporates almost completely or in the case of sandstone it reacts preferentially with the surfaces of the silicate minerals, these composite gels may in fact be less suitable for the preservation of stone than either neat MTMOS-derived gels or B72 acrylic resin deposited by inert, non-reactive solvents such as acetone, diacetone alcohol or toluene.

REFERENCES

1. J.H. Larson, "A museum approach to the techniques of stone conservation," in Fourth International Congress on Deterioration and Preservation of Stone, edited by K.L. Gauri and J.A. Gwinn, 1983, p. 213-238.

2. G.S. Wheeler, G.L. Shearer, S. Fleming, L.W. Kelts, A. Vega, and R.J. Koestler, "Toward a better understanding of B72/acrylic resin/methyltrimethoxysilane stone consolidants," in Materials Issues in Art and Archaeology II, edited by Pamela B. Vandiver, James Druzik, and George Segan Wheeler, 1991, p. 209-226.

3. E. de Witte, M. Goessens-Landrie, E.J. Goethals, and R. Simonds, "The structure of "old" and "new" Paraloid B 72," in ICOM Committee for Conservation 5th Triennial Meeting, Zagreb, 1978, 78/16/3/1-9.

INTERCALATED TWO-DIMENSIONAL CERAMIC NANOCOMPOSITES

E.P. GIANNELIS, V. MEHROTRA, O. TSE, R.A. VAIA, AND T.-C. SUNG
Department of Materials Science and Engineering, Cornell University, Ithaca, NY 14853

ABSTRACT

New generations of polymer-ceramic nanocomposites are being developed by intercalation of layered solids. Synthetic approaches include intercalative polymerization, direct polymer intercalation and molecular dispersion of the host layers in the polymer matrix. The unique physical and mechanical properties of the hybrids are attributed to the molecular dispersion and the synergism between the individual components.

INTRODUCTION

Materials Design

As requirements for high technology applications become more demanding so does the need for developing new materials. Ideally, actual synthesis and characterization must be successfully integrated with the molecular and microstructural design. Thus, new revolutionary design principles and processing approaches that allow control at the molecular level are required. This paper reviews our efforts to develop a new generation of polymer-ceramic nanocomposites using intercalation of layered ceramics [1-3] as illustrated in Figure 1. Our approach is based on the ability of layered solids to ingest (intercalate) a wide variety of chemicals in the gaps (galleries) formed between the layers.

The whole thrust of producing novel composites materials by intercalation of layered ceramics, as opposed to those with the properties of a simple mixture of the components, as shown in Figure 2, is that the molecular features and the apparent synergism in the former could impart unique properties not necessarily limited by the rule of mixtures. Therefore, new composites can be designed that combine the overriding strength, dimensional and thermal stability of ceramics with the fracture properties, processability and dielectric properties of polymers.

This new approach exploits the tendency of the layers in a lamellar solids to organize like a deck of cards, i.e. the layers tend to be organized in the c direction but not necessarily in the a and b axes of individual layers. The self-assembling property of the host is instrumental in organizing literally hundred of thousands of layers in a single processing step with a high degree of ordering. It also distinguishes intercalation from the Langmuir-Blodgett technique where a laborious layer-by-layer deposition is required. Self-assembling components have been used for millions of years by biological organisms to build multifunctional materials with complex structures at many spatial levels [4]. Examples range from strong and tough composites to highly selective sensing membranes.

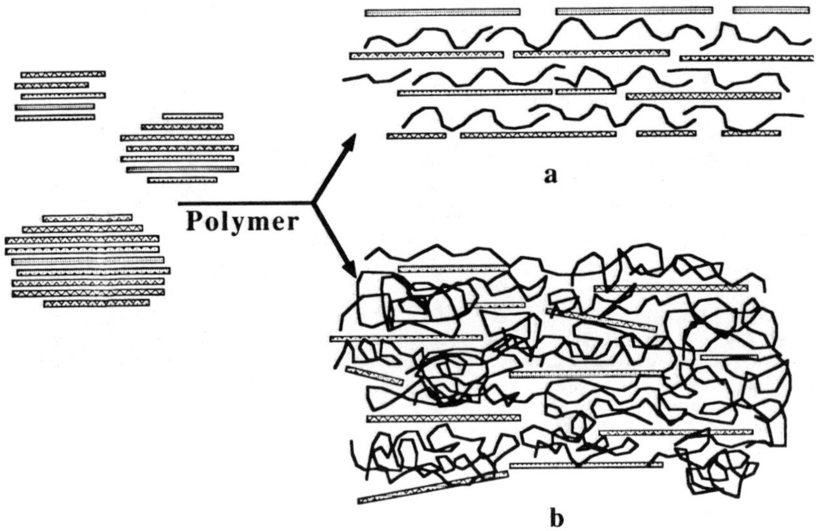

Figure 1. Schematic composite structures obtained using layered ceramics. The rectangular bars represent the ceramic layers. (a) Single polymer layers intercalated in the ceramic galleries; (b) Composites obtained by exfoliation of the ceramic layers.

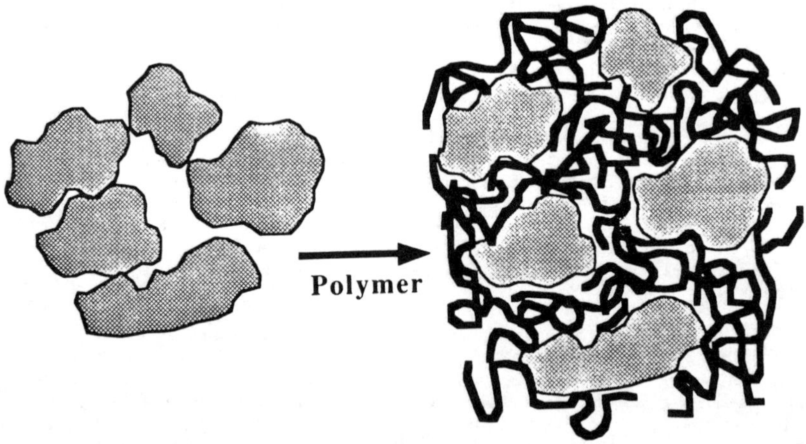

Figure 2. Conventional polymer composite obtained using a particulate ceramic.

Structure and Intercalation Properties of MTSs

Layered solids can ingest various foreign species into the galleries between the layers to form intercalated materials [5]. Graphite and transition metal chalcogenides are some common examples of layered hosts that have been studied quite extensively over the past several decades. In this paper we will introduce a "new" class of host solids, namely, the mica-type silicates (MTSs), that have largely been ignored by the materials and the solid state community. They, as well as the better known talc and micas, belong to the general family of 2:1 layered silicates. With as many as seven atomic planes in individual layers they exhibit a high degree of layer rigidity with respect to transverse layer distortions.

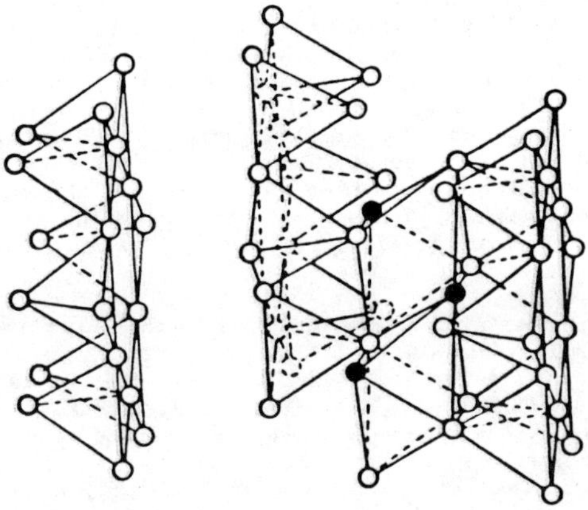

Figure 3. Idealized structure for 2:1 layered silicates showing two tetrahedral sheets sandwiching an octahedral sheet. Open circles represent oxygen atoms; closed circles are hydroxyl groups. In mica the galleries between the layers are usually occupied by potassium cations.

The structure of 2:1 layered silicates consists of two SiO_2 tetrahedral sheets fused to an edge-shared octahedral sheet (Figure 3) [6-8]. The type and population of the cations in the octahedral sheet distinguishes between various layered silicates. For instance, when all octahedral sites are occupied by magnesium, the known filler talc is obtained. Alternatively, in pyrophyllite only two-thirds of the octahedral sites are filled with aluminum. The stacking of the tetrahedral and the octahedral sheets results in a commensurate arrangement eliminating the strain that is often developed in incommensurate structures. In contrast, the strain, which is induced by structural mismatch in crysotile and related asbestos minerals causes the layers to curl and adopt a fiber-like morphology. In crystals of talc and pyrophyllite the layers are held together by weak dipolar and van der Waals forces creating interlayers or galleries between the layers. The sum of a single layer (~1 nm) plus the interlayer represents the repeat unit of the multilayer and is calculated from the 001 harmonics obtained from x-ray diffraction patterns.

In contrast to talc and pyrophyllite where the layers are electrically neutral, the micas exhibit a net charge of negative two per unit cell (equivalent to an Si_8O_{20} unit) due to the isomorphous substitution of silicon by aluminum in the tetrahedral sheet. The charge deficiency is generally balanced by potassium cations that reside in the galleries between the layers of the 2:1 structure. In "ideal" micas the stoichiometry is such that the potassium ions occupy all the ditrigonal cavities formed by the oxygen atoms of the tetrahedral sheet of the layers.

The talc-pyrophyllite group and the micas represent the end members of a hierarchy defined by the degree of layer charge neutrality. Mica-type silicates, otherwise known as swelling silicates, with layer charge densities in the range, $0 < x < 2$, define a separate host family with unique physical and chemical properties not found in the end members. Similarly to micas, gallery ions are incorporated in the structure of swelling silicates to balance the charge deficiency in the layers. But unlike the micas, the swelling silicates can intercalate a large number of polar molecules in their galleries with a substantial increase in the gallery height. When the charge deficiency originates from the octahedral sheet (i.e. Mg^{2+} substituting for Al^{3+}) the negative charge is distributed over all oxygens of the framework resulting in a turbostratic arrangement. In contrast, the layer charge created in tetrahedral sheets tends to be more localized yielding a greater three-dimensional order.

Intercalation of guest species into the host galleries is facilitated by the much weaker interlayer forces compared to the intralayer binding forces [5]. The hydrated gallery cations in MTSs can be readily exchanged with almost any other cation by simple ion exchange reactions, distinguishing them from the end members of the group (talc and micas). In contrast to graphite which is amphoteric, MTSs will accept only positively charged or neutral molecules without any charge exchange mechanism. Since no charge exchange is taking place, intercalation of MTSs is less disruptive and very facile. A diverse class of cations (mono- and polynuclear, organic, inorganic, organometallic and biological) can be introduced in the galleries.

MTSs are also known for their ability to intercalate a large number of polar neutral molecules in their galleries with certain degree of orientation and alignment. The enhancement of orientation and alignment upon intercalation can be exploited to produce, for example, materials with large optical non-linearities. The guest species can enter in stages, i.e. one monolayer at a time and, in some instances, as many as twenty monolayers can be accommodated. Thus, a variety of neutral or positively charged molecules of virtually any size can be introduced in the host galleries by direct intercalation and ion exchange, respectively.

MOLECULAR HYBRIDS PRODUCED BY INTERCALATION OF SINGLE POLYMER CHAINS

Intercalative Polymerization

Intercalative polymerization of aniline in the galleries of layered fluorohectorite (a synthetic silicate) results in highly oriented multilayered films consisting of single chains of polyaniline alternately stacked with the layers of the insulating host [3]. Gallery Cu^{2+} ions, introduced by an intercalative ion exchange process, serve as the oxidation centers for the polymerization of aniline. The reaction can be represented by the following equation, where the horizontal lines identify the layered structure. The polyaniline-silicate hybrid contains 17.5 wt % polymer.

Cu^{2+} + $nC_6H_5NH_2$ → $Cu^{2+}(PANI)$

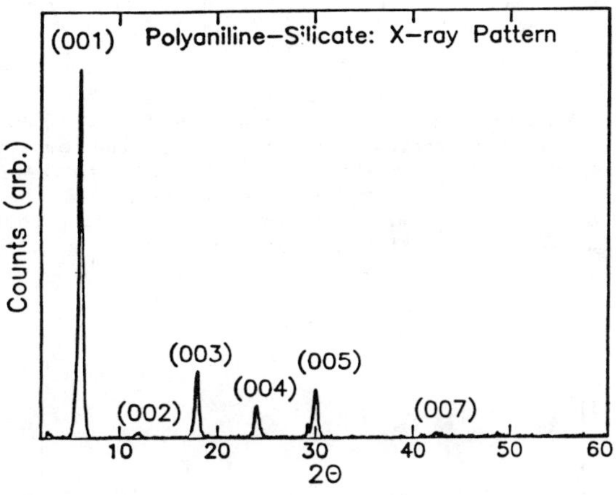

Figure 4. X-ray diffraction pattern of polyaniline-silicate composite.

X-ray diffraction patterns of oriented films confirm that a highly ordered multilayered structure is obtained (Figure 4). Seven (001) harmonics are observed corresponding to a primary repeat unit (d spacing) of 1.49 nm. The difference of 0.53 nm from the corresponding .96 nm for the silicate framework is in agreement with intercalation of single chains of polyaniline in the galleries of the host.

The electronic absorption spectrum of the as prepared material is characteristic of the insulating form of polyaniline. The absorption band at 510 nm (2.4 eV) is attributed to a charge transfer exciton from the benzenoid $[-NH-(C_6H_4)-NH-]$ to quinoid $[-N=(C_6H_4)=N-]$ segments of the emeraldine base similar to bulk polyaniline. Upon protonation, the band at 510 nm gradually disappears and two new bands appear at 374 (3.3 eV) and another centered at 760 nm (1.6 eV) that extends into the near infrared. The changes in the optical spectrum are in accord with the disappearance of the localized quinoid structure and the formation of a polaron lattice.

DC conductivity measurements of the polyaniline-silicate multilayers show a substantial increase in conductivity after protonation. The in-plane electrical conductivity measured at room temperature by the four-probe technique is 0.05 S/cm. The transverse conductivity is about 10^{-7} S/cm which corresponds to an electrical anisotropy $\sigma_{=}/\sigma_{\perp}$ of $5x10^5$. The observed anisotropy is in accord with the film architecture of a conducting phase separated by 1 thick layers of the insulating host. The low, though measurable, transverse conductivity is probably due to the turbostratic ordering of the host lattice as well as to the discontinuities in the insulating layers, which are due to the finite lateral size (~ 10 μ) of the host particles.

The variable temperature in-plane film conductivity is shown in Figure 5. The data is fitted by using the model for limited tunneling between metallic islands (9,10):

$$\sigma = \sigma_0 \exp[-(T_0/T)^{1/2}].$$

Two different temperature regimes are discernible with T_0 7600 and 10,700 K for the low and high temperature ranges respectively. The low temperature T_0 value is in good agreement with that obtained for oriented bulk polyaniline samples (9) implying that the barriers that dominate tunneling are the same in both materials. The different barriers might be the result of a phase transition or structural rearrangement of the conducting phase or the multilayer at ca. 200 K.

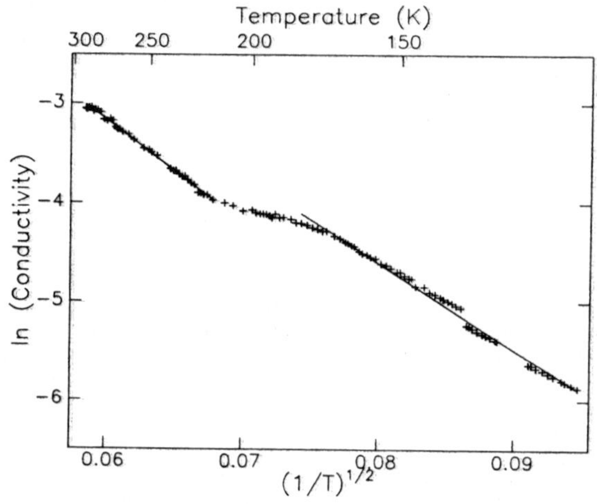

Figure 5. In-plane logσ vs $T^{-1/2}$ for polyaniline-silicate hybrid.

The electronic spectra of the multilayer in both doped (protonated) and undoped forms are blue shifted by about 70 nm with respect to the bulk polyaniline. The energy shifts might be due to differences in conjugation length, different degrees of protonation/oxidation, or, most likely, the effect of molecular confinement in the intercalated state. The two-dimensional inorganic matrix not only can enforce a particular structural arrangement (ring conformation) on the polymer chains but also minimize the extensive electronic interaction that normally occur in the bulk material. The confinement of polyaniline in the two-dimensional host galleries affects the polymer conformation which, in turn, results in increased localization of the charge carriers.

Fracture toughness tests with the crack front oriented perpendicular to the layers yield a fracture toughness of 1.2 MPa·√m for the hybrid while films of pristine fluorohectorite could not be tested because of their brittleness [11]. Thus, at least qualitatively, the fracture toughness of the hybrid is superior to the ceramic matrix. The fracture toughness seems to be insensitive to the strain rate suggesting that plastic (viscoelastic) flow in the polymer does not contribute to any energy dissipative processes. Figures 6 and 7 show the fracture surfaces of the pristine and polyaniline silicate hybrids. The crack tends to follow the plate contours and probably propagates via plate pullout. Since the pristine fluorohectorite films tend to be less oriented than the polymer hybrids, a larger energy dissipation due to an increase in the effective surface area between platelets is expected for the latter consistent with our data. The fracture toughness with the crack front parallel to the layers is about 0.3 MPa·√m in agreement with the anisotropic character of the hybrid and the rather weak bonding at the interface.

The in-plane storage modulus for the hybrid is 13.0 ± 2.7 GPa at room temperature. For comparison, the storage modulus of pristine fluorohectorite is 5.5 ± 1.2 GPa while that for polyaniline films is approximately 3 GPa [12]. The rather low value for the in-plane modulus of fluorohectorite (approximately one order of magnitude lower than mica) is implying that the measured value represents an effective modulus due to the sliding of the silicate platelets and not that due to the in-plane atomic bonding. The sliding of the layers is probably facilitated by the presence of intercalated water molecules that can act as a lubricant. The enhanced in-plane modulus for the hybrid is probably due to the better in plane organization for the composite compared to the pristine silicate host.

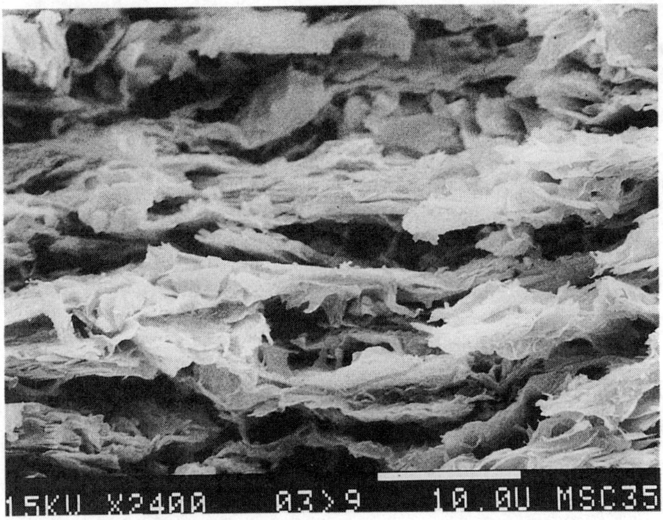

Figure 6. SEM of fracture surface of pristine silicate.

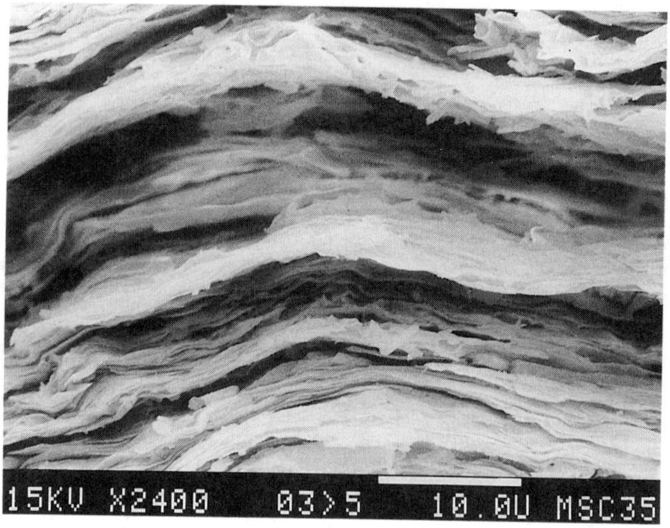

Figure 7. SEM of fracture surface of polyaniline-silicate hybrid.

Polymers in general can be amorphous or semicrystalline depending on the regularity of the polymer chains. Above the glass transition temperature (T_g) the chains in the amorphous phase can undergo segmental motion and the polymer behaves in a rubbery manner while crystallites melt at the melting temperature T_m. Both transitions are accompanied by dramatic changes in the properties of the polymer like specific heat and stiffness.

Differential scanning calorimetry (DSC), dynamic modulus and thermally stimulated current analysis [13] show no glass transition for the polyaniline silicate hybrid. The glass transition temperature of bulk polyaniline is 220 °C [12]. The absence of glass transition temperature for the hybrid is another demonstration of the effect of intercalation and confinement of the polymer and is attributed to the lack of cooperative motion between the polyaniline chains.

Direct Intercalation

Two-dimensional nanocomposites can also be synthesized by direct polymer intercalation. One advantage compared to the intercalative polymerization is that well-characterized polymers of different molecular weights can be used and the effect on the properties of the composites can be studied. Though several neutral or cationically charged polymers can be intercalated, our discussion will focus on the intercalation of polyethylene oxide (PEO).

The synthesis involves mixing an aqueous suspension of montmorillonite (a natural silicate) with a solution of the polymer [14]. X-ray diffraction patterns of oriented films show that a highly ordered multilayered structure is obtained (Figure 8). Six (001) harmonics are observed corresponding to a primary repeat unit (d spacing) of 1.77 nm. The gallery height of 0.81 nm is in agreement with intercalation of single chains of PEO in the host galleries.

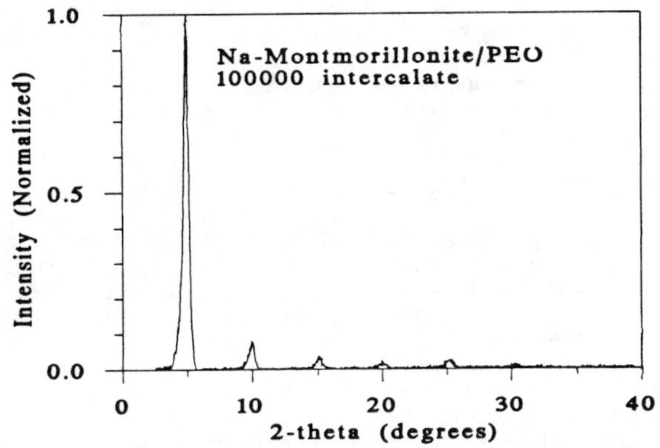

Figure 8. X-ray diffraction pattern of PEO-montmorillonite hybrid.

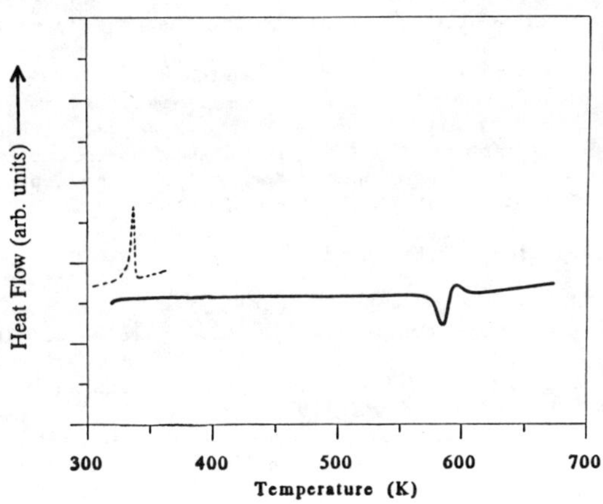

Figure 9. DSC of PEO-montmorillonite hybrid: (———) intercalated form; (- - -) physical mixture.

PEO is about 85 % crystalline at room temperature with a melting temperature of 65 °C, while the glass transition temperature for the amorphous phase is around - 50 °C [15]. Chain complexation with various inorganic salts results in crystalline stoichiometric compounds with melting temperatures depending on the stoichiometry and the nature of the complexing salt. Figure 9 shows the DSC spectrum of PEO intercalate which shows no melting temperature all the way up to the decomposition temperature of 310 °C. In contrast, a PEO-silicate mixture prepared by mechanical mixing of the two powders in the same ratio as in the intercalate clearly shows the characteristic melting of PEO at 64 °C. The lack of melting is attributed to the intercalation and molecular confinement which disrupt the 3D organization of the chains and hinder crystallization. Low temperature DSC experiments are in progress to determine the presence (or absence) of glass transition temperature for the polymer chains in the intercalated state.

MOLECULAR HYBRIDS PRODUCED BY EXFOLIATION OF THE CERAMIC HOST

The last approach, first explored by researchers at Toyota Research Center, is based on an alternative design scheme [16]. It exploits the exfoliation and dispersability of layered ceramics in both aqueous or nonaqueous solvents. The resulting nanocomposites consist of single ceramic layers (1 nm thick) in a continuous polymer matrix (Figure 1b). Because of the poor dispersability of layered silicates in nonaqueous solvents, the host is modified (functionalized) prior to mixing with the polymer. The modification involves replacing the sodium cations in the silicate by an alkyl ammonium cation through a simple ion exchange reaction. This modification improves the dispersability of the ceramic in different organic solvents and it might also affect the structure and chemistry of the interface.

Polyimide composites containing 2 wt% montmorillonite show a 60 and 25 % decrease in the permeability of water and the thermal expansion coefficient, respectively compared to the bulk polymer [16]. In addition, the same composites show an increased modulus while maintaining the dielectric characteristics of polyimide [17]. This combination of properties for the composite makes them very attractive candidates for the dielectric layers in electronic packaging applications. The remarkable improvement in the properties of the composite is probably due to the large aspect ratio and the molecular level dispersion of the ceramic in the polymer. Because of the molecular dispersion a large number of interfaces is created dramatically affecting the properties of the composite.

Surface ordering phenomena in various polymers are already well recognized [18-21]. Some of the systems include aromatic polyimides and block copolymers. In the latter, the morphology near surfaces or interfaces can be dramatically affected by the difference in the surface free energies of the two blocks and their affinity toward the substrate. Using grazing incidence x-ray diffraction the structure of an aromatic polyimide (PMDA-ODA) was shown to exhibit enhanced ordering within 9 nm from the surface [18]. Near the surface, reflections due to electron density correlations parallel to the polymer chain could be observed that were attributed to essentially <u>crystalline</u> order.

If we assume that the polymer becomes crystalline near the ceramic interfaces and that the crystalline order extends 10 nm from the interface we can calculate the amount of silicate required to completely order the polymer for various degrees of layer delamination (Figure 10). n corresponds to the number of layers in the silicate and it reflects the degree of layer delamination. For a completely delaminated system, n = 1, ~8 wt % of the silicate is sufficient to induce complete ordering of the polyimide chains. Also note that if ordering takes place only within 2 nm from the surface the required amount of silicate is 30 wt%. If, on the other hand, delamination is not complete and the ceramic consists of 10 nm thick layers (n = 10) then the weight percent of silicate required is ~50 wt % and increases more as the thickness of the particle increases. Interestingly, it seems that a plateau is reached in the CTE and the water permeability around 8 wt % of silicate [16] which is in very good agreement with the number calculated above for complete ordering of the polymer chains.

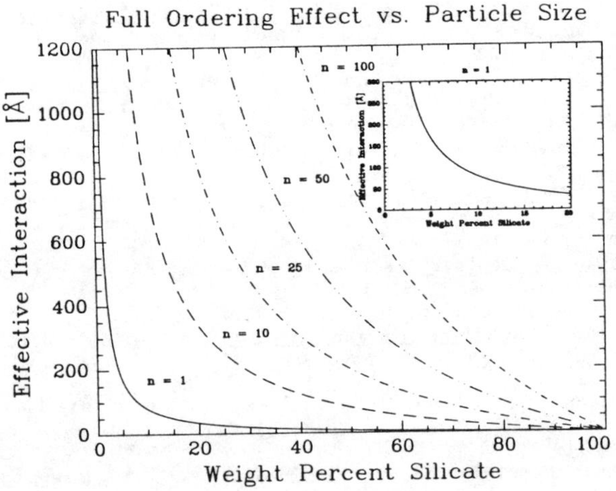

Figure 10. Plot of effective interaction length and the corresponding silicate wt % for various particle thicknesses.

Therefore, the use of layered ceramics as a means to create periodic interfaces in a continuous matrix is an effective approach for developing new polymer composites. Since ordering can only be effected within approximately 10 nm, in a optimal design, interfaces need to be introduced in a periodic fashion throughout the material. This underscores the need of using ceramic compounds with a layered lattice structure that can be dispersed at the molecular level. Furthermore, optimal polymers include those with semi-rigid chains in which the presence of the ceramic layer induces a crystalline or liquid crystalline order over a rather long distance from the interface. For flexible chain polymers the organization is limited to only a few layers from the surface requiring the use of rather large amounts of the inorganic phase in the composite to observe any improvements in properties.

ACKNOWLEDGEMENTS

This work was sponsored by the National Science Foundation (DMR-8818558) through the Materials Science Center at Cornell and by Corning Inc. and IBM. We thank Prof. E.J. Kramer, Prof. W. Sachse, Dr. B.B. Sauer and Dr. J. Smith for experimental assistance and many helpful discussions. O. Tse and R.A. Vaia gratefully acknowledge the support of DoD Fellowships.

REFERENCES

1. V. Mehrotra and E.P. Giannelis, in Polymer Based Molecular Composites, edited by D.W Shaefer and J.E. Mark (Mater. Res. Soc. Proc. 171, Pittsburgh, PA 1990).
2. E.P. Giannelis, V. Mehrotra and M.W. Russell, in Better Ceramics Through Chemistry, edited by C.J. Brinker, D.E. Clark, D.R. Ulrich and B.J.J. Zelinski, (Mater. Res. Soc. Proc. 180 Pittsburgh, PA, 1990).
3. V. Mehrotra and E.P. Giannelis, Solid State Communications, 77, 155 (1991).
4. Materials Synthesis Utilizing Biological Processes, edited by P.C. Rieke, P.D. Calvert and M. Alper (Mater. Res. Soc. Proc. 174, Pittsburgh, PA 1990).
5. Intercalation Chemistry, edited by M.S. Whittingham and A.J. Jacobson (Academic Press, New York, 1982).
6. T.J. Pinnavaia, Science, 220, 365 (1983).
7. S.A. Solin, in Intercalation in Layered Solids, edited by M.S. Dresselhaus (Plenum Press, New York, 1986) pp. 145-154.
8. T.J. Pinnavaia, in Chemical Physics of Intercalation, edited by A.P. Legrand and S. Flandrois (Plenum Press, New York, 1987) pp. 233-252.
9. F. Zuo, M. Angelopoulos, A.G. MacDiarmid and A.J. Epstein, Phys. Rev. B, 36, 3475 (1987).
10. N.F. Mott, Metal-Insulator Transitions, Taylor and Francis, London, 1974.
11. V. Mehrotra, J. Smith, E.J. Kramer and E.P. Giannelis, unpublished results.
12. Y. Wei, G.-W. Jang, K.F. Hsueh, E.M. Scherr, A.G. MacDiarmid and A.J. Epstein, Polym. Mat. Sci. Eng. 61, 916 (1989).
13. V. Mehrotra, B.B. Sauer and E.P. Giannelis, unpublished results.
14. E. Ruiz-Hitzky and P. Aranda, Advanced Materials, 2, 545 (1990).
15. M. Armand, Advanced Materials, 2, 278 (1990).
16. K. Yano, A. Usuki, A. Okada, T. Kurauchi and O. Kamigaito, Polymer Preprint, 65, 1991.
17. R. Vaia, T.-C. Sung and E.P. Giannelis, unpublished results.
18. B.J. Factor, T.P. Russell and M.F. Toney Phys. Rev. Letters 66, 1181 (1991).
19. S.H. Anastadiadis, T.P. Russell, S.K. Satija and C.F. Majkrsak, ibid., 62, 1852 (1989).
20. I. Bitsanis and G. Hadziioannou, J. Chem. Phys. 92, 3827 (1990).
21. B.S. Hsiao and E.J.H. Chen in Controlled Interfaces in Composite Materials, edited by H. Ishida (Elsevier, New York, 1990) pp 613-622.

Editor's Note: This paper has also appeared in MRS Proceedings, volume 249.

EVALUATION OF PROTECTIVE COATINGS FOR THE 'IN SITU' PRESERVATION OF HISTORIC TIMBER BUILDINGS IN A HARSH ANTARCTIC ENVIRONMENT.

JANET HUGHES
Institute for Antarctic and Southern Ocean Studies,University of Tasmania, GPO Box 252C, Hobart, Tasmania 7001, Australia

ABSTRACT

The Australasian Antarctic Expedition base at Commonwealth Bay (67°S 142°E), also known as 'Mawson's Huts' include some of the earliest buildings in Antarctica. The prefabricated timber buildings of Oregon beams and Baltic Pine (Pinus sylvestris) claddings were erected in January 1912 and were occupied for two years: they are thus approximately contemporary with the better known Scott and Shackleton huts near McMurdo Base in the Ross Sea.

The expedition was led by Dr Douglas Mawson, later Professor Sir Douglas Mawson, and its inspiration was scientific rather than a race to the South Pole. The buildings and artefacts remaining on the headland, also called Cape Denison, reflect considerable scientific achievement in geology, upper atmosphere physics, surveying, meteorology and biology and the site constitutes an important monument in the early exploration of this remote and inhospitable continent.

While the huts are of recent date, the rapid rate of deterioration due to the extreme climate means there is some urgency to developing a method of preserving the building. The timbers have been seriously worn, or 'corraded' by snow particles carried in the katabatic winds. These winds can exceed 300km per hour, sometimes for days. The isolation of the site, accessible by sea for only three months of the year adds to the logistical difficulties. There has been considerable public debate in Australia about which method of preservation is appropriate.

The author, a materials conservator and industrial chemist, visited the site in December 1985 and has been studying the problems of the site as a postgraduate student for the past year.

THE PROBLEMS

Methods suggested for treating the huts have included dismantling and re-erection in Australia or alternatively to preserve the huts 'in situ' by methods such as covering with a dome or re-cladding. Each of these approaches has inherent problems, such as changing the microclimate and loss of original material [1].

A unique feature of Mawson's Main Hut is that the exterior of the main hut, while visibly ravaged by the climate, is original and it is an evocative feature of the buildings reflecting the harshness of the place. The use of a protective coating on the timbers is one possible treatment method which is currently being investigated by the author since it offers potential benefits in retaining original material with minimal

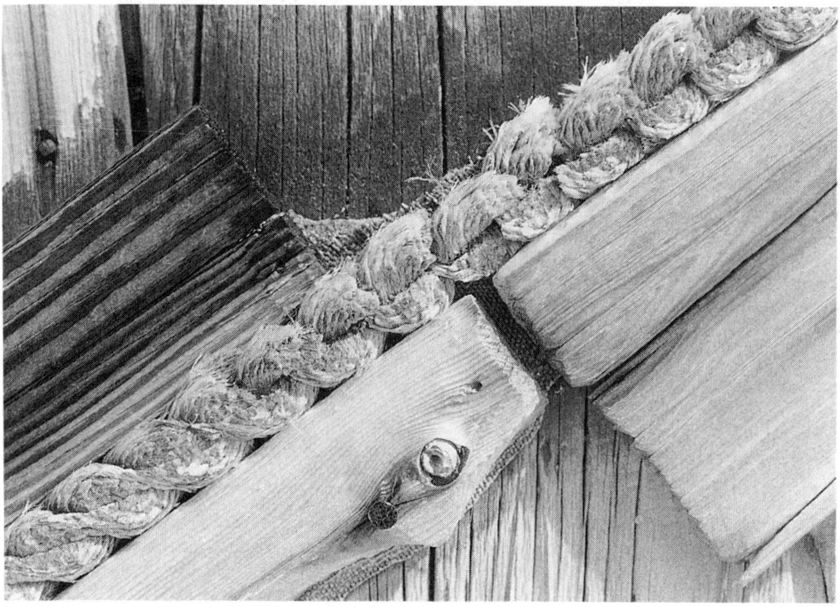

Roof detail of Mawson's Hut: timber battens and rope were used to hold down a sailcloth to prevent wind and snow penetration. All are severely corraded by windborne ice particles.

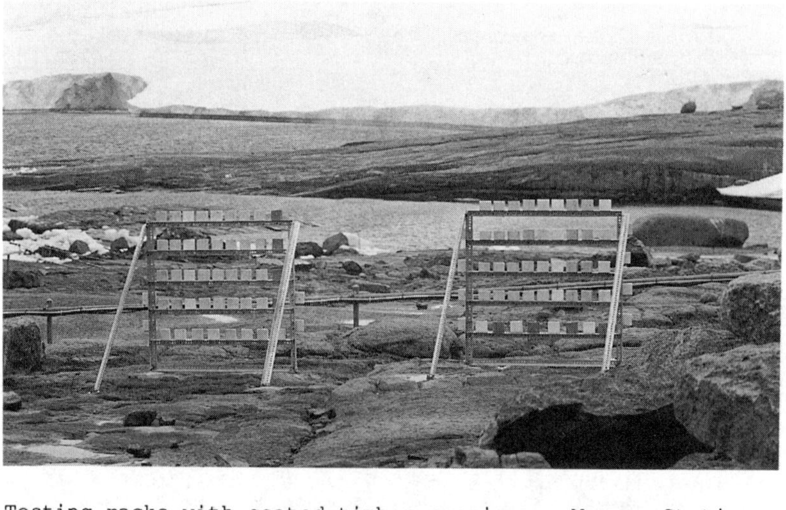

Testing racks with coated timber specimens, Mawson Station, Antarctica. The racks are aligned south to face into the katabatic wind.

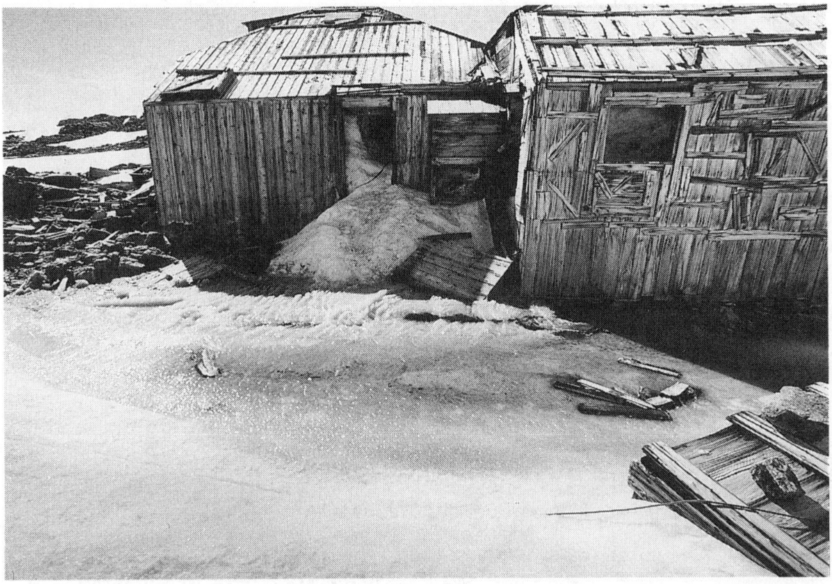

The western wall of Mawson's Huts showing battens (which once held down sailcloth) used to try to keep out blowing snow. Frozen meltwater is seen in the foreground and rising damp is seen at the base of the walls.

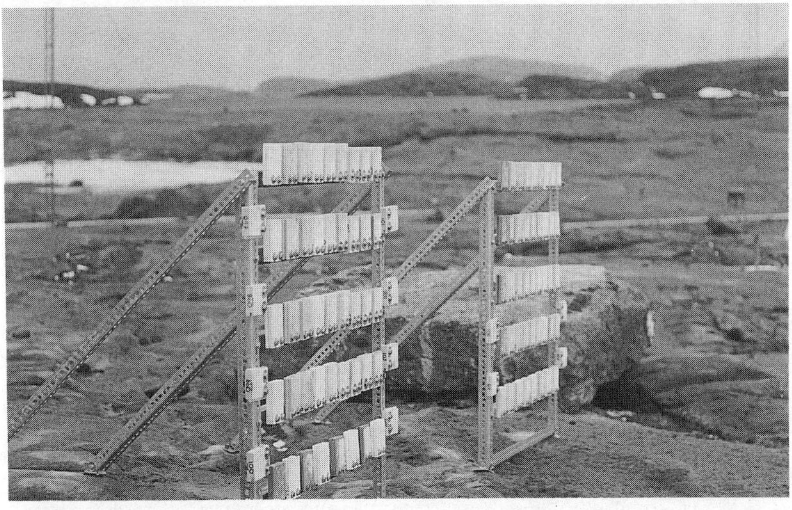

Testing racks at Mawson Station, next to the Stevenson screen. Multiple samples are used to allow for the variation of drift particles with height. Controls are uncoated Baltic pine (8 samples) and primer-coated only (12 sammples). Coatings tested include Commercially abvailable and specially modified silicone rubbers.

interference and easier application. Absorption of the impact of the particles by an elastomeric coating should reduce the rate of corrasion. Elastomeric coatings are prefered to conventional rigid coatings or to ultra-durable coatings such as epoxies, which were found to crack during initial field testing in Antarctica in 1985. 'Reversibility' by peeling is being assessed; thus the treatment could be applied repeatedly or be replaced by a better method if one should be developed in future.

Factors involved in corrasion

The process of corrasion by ice particles is studied by the sciences of geomorphology and meteorology and is important in understanding the materials conservation problem. From a knowledge of the factors involved, principally ice particle size and velocity, it is possible to develop a laboratory simulation of ice corrasion using a recirculating air-abrasive chamber with plastic or glass beads. This is cheaper and easier to use than a wind tunnel where many additional dimensional factors complicate the matter.

A number of theoretical and empirically-derived equations give the rate of snow transport by the wind [2, 3, 4, 5, 6, 7, 8] and hence the rate of corrasion. The rate of corrasion depends on the rate of snow transport. The rate of snow transport varies as the cube of the wind speed and the number and mass of particles carried is given by the following equation [4]:

f= mass flux
F= particle frequency
M= average particle mass
If particle size is non-uniform, then

$$M=m(1+3a^{-1}+2a^{-2}) \qquad (1)$$

where m is the mass of a particle with mean diameter and a is a shape parameter of the gamma function (Budd, 1966)
If density of ice= density of blowing snow(0.9 g cm^{-3}) then

$$m= 0.9\ (\pi/6)D^3 \qquad (2)$$

where D= mean grain diameter (cm) and mass flux,f (gm s^{-1}) is calculated at height z as:

$$f_z = (0.9\ \pi/6)D_z^3\ F_z\ (1+ 3/a_z + 2/ a_z^2) \qquad (3)$$

This gives a formula for total mass of snow particles passing along a 'lane' per unit time.

Thus when considering the impact of the particles on the timber wall of a building, the greatest rate of corrasion will occur near the surface of the snow drift. From photos taken in 1912-13 it is known that snow drifts progressively higher against the walls during the winter, when the katabatic wind speed also increases. This will protect the timbers that are covered but areas that are exposed will experience severe 'blasting' above the level of the snow drift. The shape of the building also influences local wind speed, causing vortices and flow irregularities. These factors cause variations in the rate of corrasion and examination of the building shows that the timber in some areas is more worn than in others. The south wall of the Main Hut, which faces into the full force of the katabatic wind, and the roof of the workshop have suffered the most.

Katabatic winds arise due to the flow of cold, dense air from the polar plateau of the interior of the Antarctic continent, accelerated by gravity as it 'flows' downhill to the coast. The large collection area of cold air and the shape of the catchment area result in Cape Denison having the world's highest annual average wind speed of 80 kilometres (50 miles) per hour [9].

Other factors affecting the corrasion rate include the hardness of the timber and the hardness and angularity of the ice particles. Timber varies throughout the grain and is characterised by bands of thick-walled, denser latewood cells and thinner-walled, less dense earlywood. The hardness of the ice particles varies against temperature and at -20°C (a fairly typical winter temperature in coastal Antarctica) the hardness is equal to Mohs hardness 3, equivalent to calcite [10]. Ice particles in Antarctica are spheroidal as the hexagonal shape characteristic of snow flakes is broken off by tumbling in the wind.

DEVELOPMENT OF THE TESTING METHOD

As the height of the snow drift against the walls varies seasonally, and since the range of factors involved is extensive, it was decided to test the polymers by laboratory simulation of the abrasive environment and to compare these results with field test results.

The following standardisations were used for testing purposes:
i) Baltic Pine (Pinus sylvestris) planking with roughly comparable grain characteristics were selected
ii) moisture content was measured by calculation of weight loss after heating at 105°C for 24 hours
iii) testing hardness of timber was carried out using a modified Janka Hardness Test which measures force required to produce an indentation equal to half the radius of a ball-ended probe; this gives a measure of the elasticity of the 'target' in the collision of the snow particles against the timber
iv) a fixed range of particle sizes were used, equivalent to known ice particle dimensions in Antarctica[4]
v) particle velocity was assumed equal to the air velocity from the orifice of the blasting chamber gun; this was measured using a Pitot tube at a fixed distance equal to five orifice diameters. It is hoped to improve this measurement by adapting instrumentation in the near future.

A major difficulty of the research is to take account of the variability of timber in terms of grain irregularity and hardness. Two further methods of accounting for this have been developed and results of the methods will be compared statistically:

Gravimetric determination of corrasion rate

The timber is weighed, then dried in a laboratory oven at 105°C for 24 hours. After cooling in a desiccator it is re-weighed to determine moisture content and 'bone dry weight'. By weighing accurately before and after blasting and taking into account the moisture content it is possible to determine % weight loss versus time at a range of particle velocities using different sizes and compositions of beads.

Geometric determination of corrasion rate

Measurement of thickness loss of timber after blasting is made by taking a cast of the corraded timber then sectioning the cast across the grain to obtain a thin slice which is viewed under a microscope. The resultant 'hills and valleys' (due to selective loss of the softer earlywood) are measured to determine statistically the rate of dimensional loss of the timber. The effectiveness of different polymers to prevent corrasion can thus be compared objectively.

Field testing

Laboratory testing can only indicate which polymers are effective against individual factors. It is difficult to measure performance against temperature cycling, which is important in Antarctic conditions since diurnal variation is high. Where there are a number of factors which interact it is necessary to conduct field tests. Field testing is being carried out at Mawson Station (which is at the same latitude but is over 2,000km further west of Commonwealth Bay). It is the windiest of the continuously-operated Australian bases, though its annual average wind speed is only half of that Mawson's Huts. It is hoped to conduct further field testing at Commonwealth Bay next time a ship is available- hopefully in December 1992.

The testing racks, made of commercially-available slotted 'Dexion' angular steel were bolted to the rock next to the meteorological station at Mawson Station and the pre-weighed, pre-coated timber specimens were attached. Some specimens were painted as a test of their ease of application and curing ability under Antarctic conditions. 'Controls' were used, one set of four being attached in 'columns' from top centre to bottom centre of each rack. These were either uncoated (bare) timber or timber coated with a primer, of which three were selected (Paraloid B72, Primal AC73, Ketone N). The array of the specimens were carefully arranged so that each top coat was tested both with and without a primer. Specimens coated with the same topcoat were arrayed in columns so the effect of the height above the ground, which determines the number of particles entrained in the wind, can be accounted for.

The protective effect of the various coatings can then be determined by applying the coating to a weighed timber sample and blasting for a measured period to simulate several years of natural 'blasting' at Commonwealth Bay. The coating can be examined under a scanning electron microcope to assess its ability to stand up to the 'blasting' and when the coating is peeled off its reversibility can be assessed. If the coating peels off cleanly without pulling off timber fibres it is judged as successful. Only if the coating satisfies all criteria could it be considered for use on the historic building-

1. It must withstand the Antarctic environment for at least 5 years ('durability').
2. It must be able to be removed by peeling off cleanly from the timber ('reversibility').
3. It must not markedly change the appearance of the timber ('aesthetic').

POLYMER SELECTION

The commercial literature on silicone rubbers was checked and a short list prepared according to the following selection criteria:

<u>Chemical properties</u>

i) non-acid curing -some silicones give off acetic acid as they cure which would corrode metal fittings such as nails on Mawson's huts

ii) must cure at low temperatures (0- 3°C); ie.'low temperature vulcanising'

iii) must have adjustable viscosity so that coating thickness can be controlled

iv) the effect of silane coupling agents must be assessed.

<u>Physical properties</u>

i) must be available as a clear coating- some have pigments which are an essential stabiliser in the formulation; other pigments are not essential to the formulation and may be obtainable as a special unpigmented batch from the manufacturer

ii) colour saturation must be low as this would change the appearance of the timber; darkening increases heat absorption

iii) adhesion must be sufficient to secure the coating firmly in harsh conditions but it must not be so strong as to pull off timber fibres

iv) water vapour transmission must be high or moisture could be trapped inside the building by the coating

The polymers which match these criteria and which were available in Australia include:
Dow Corning Silicones 'All Guard', 1-2577, 730, 734, 3140, 3145
University of South Australia Silicone 2128 Types 1, 2 and 3
Primers selected were Paraloid (Acryloid) B72, Ketone N and Primal AC 73 which are well known to conservators. The meteorologists at Mawson Station monitor progress as they can view the testing racks from their building.

<u>Other research objectives for Mawson's Huts include:</u>

Study of the microclimate inside the hut by remote sensing so the effect of any proposed treatment can be documented and revised accordingly.

Investigation of freeze-thaw damage to timber foundations and the effect of meltwater pools which form at the base of the walls in summer. This is important in assessing the structural preservation of the building.

CONCLUSIONS

Progress toward producing a conservation plan for the site was not possible due to the deficiencies of previously suggested preservation methods. The idea of a reversible elastomeric polymer is a novel approach offering many advantages in reducing interference and loss of original material. It would also result in simpler logistics which is a great advantage in a remote, inhospitable location. While no results are yet available a practical, a low cost testing method is established. The laboratory simulation should yield 'acclerated ageing' data though the importance of field testing is emphasised due to the complexity of the factors interacting in an Antarctic environment.

References

1. J. D. Hughes, Bull. Inst. for the Conservation of Cultural Materials, 14 (3&4): 1-32 (1988); Historic Environment 8 (1&2): 19-24 (1991).

2. R. S. Anderson, Annals of Glacio, 13 : 294 (1989)

3. R. A. Bagnold, The physics of blown sand and desert dunes, (Methuen, London 1941).

4. N.H. Berg, Arctic and Alpine Research 18 (2): 147-16 (1986).

5. R.V. Dietrich, J Glaciology, 18 (78): 148-149 (1977)

6. A. Earl et al, Cold Regions Sci. and Tech. 11 (2) : 199-202 (1985).

7. R. J. Kind, Cold Regions Sci. and Tech. 12 (3): 217-228 (1985).

8. T. Suzuki, K. Takahishi, J. Geology 89 :23-36 (1981)

9. C. T. Madigan, Scientific Reports Series B Vol IV Meteorology- Tabulated and Reduced Records of the Cape Denison Station, Adelie Land, AAE 1911-14, NSW Govt. Printer, 1929.

10. T. R. Butkovich, Hardness of single Ice crystals SIPRE Research Paper 9, May 1954.

Acknowledgements
The assistance of the Australian Antarctic Division in providing transport, clothing and some supplies in Antarctica is gratefully acknowledged. The Australian Antarctic Science Advisory Committee also provided a small grant to assist this project. The assistance of Dr Jim Hasick and Kevin Sheridan at Mawson Station greatly aided the installation of the testing racks. Childcare by Katryna Bilyk, Linda Aldridge and Rosalie Harwood was essential in providing time and peace of mind during my absence in Antactica. Thank you!

TECHNICAL EVIDENCE FOR TEMPORAL PLACEMENT: SCULPTED ADOBE FRIEZES OF CHAN CHAN, PERU

JOANNE PILLSBURY
Sainsbury Research Unit for the Arts of Africa, Oceania and the Americas, University of East Anglia, Norwich, England

ABSTRACT

Recent research on the technical aspects of the architectural reliefs at the site of Chan Chan, Peru, has yielded new evidence with which to assess conflicting relative sequences for the monumental compounds of the imperial city.

ARCHITECTURE AND THE EMPIRE OF CHIMOR

Archeological investigations at the site of Chan Chan, Peru, have provided new data on the nature and development of one of the largest pre-industrial cities in the New World. Chan Chan, capital of the empire of Chimor, flourished on the north coast of Peru between the tenth and fifteenth centuries A.D. At the core of the twenty square kilometer city are ten monumental compounds, or *ciudadelas*, which are believed to have been the palaces, as well as the mortuary structures, of the kings of Chimor. To date, there has been little agreement as to the relative sequence of construction for the *ciudadelas*. During the recent field season, however, research on the technical aspects of the construction of the adobe (mudbrick) architectural reliefs has had important implications for the chronological placement of the *ciudadelas* and the understanding of the development of the city.

In 1990-91, the *Proyecto Reconocimiento Arqueológico de los Frisos de Chan Chan* was organized as a multi-disciplinary project to conserve and document the major known reliefs at the site. Due to the fragility of the materials, the majority of the reliefs are routinely re-buried in order to protect them from climatic and other damaging factors. Investigations were conducted in six of the *ciudadelas*, on eleven friezes. The reliefs range from simple geometric compositions to more complex maritime themes.

The *ciudadelas* at Chan Chan are distinguished by monumental scale, extremely limited access, and complexity and elaboration of architectural forms. The *ciudadelas* range in size from 87,900 square meters to 221,000 square meters, with perimeter walls originally nine - ten meters high.[1] Access to the interior of a *ciudadela* is through a single entryway in the north wall. It has been argued that the *ciudadelas* were occupied by Chimu rulers, and that these enclosures combined the functions of elite residence, centralized storage, redistribution and ultimately royal entombment.[2] According to this model, the *ciudadelas* would have been occupied sequentially; each ruler would build his own new palace.

Although there is evidence for earlier structures at the site, the earliest currently visible structures at Chan Chan date to the 10th - 11th centuries A.D.[3] Construction of monumental architecture is believed to have ceased around the time of the Inca conquest of the north coast, around 1470 A.D.[4] Within these parameters, however, there has been considerable disagreement on the sequence of construction of the *ciudadelas* and the nature of the growth of the city. Numerous chronological schemes have been proposed for the building of the monumental enclosures, each based on a different set of criteria. Attempts to obtain relative and absolute dates from the *ciudadelas* have been frustrated by the Chimu practice of systematic refuse disposal, and renovations. While there are a number of C14 dates for the *ciudadelas*, they are of limited utility for finer chronological ordering. Seriations in order to propose relative chronologies have been placed on individual architectural features,[5] architectural layout,[6] and ceramics in conjunction with other features.[7]

During the course of the field season, it became clear that the variation evident in the construction techniques used to create the reliefs may reflect a temporal difference. Thus, evidence on the materials, tools, and techniques of the construction of the architectural reliefs were studied, both for their value as part of the conservation program of the project, but also, as an independent line of evidence with which to assess conflicting chronological models.

The walls of Chan Chan were made of individual adobe bricks, which were then smoothed over with one of several layers of a clay/sand plaster. The first layer covering the adobes is a thicker, coarser, plaster often referred to as the *torta*. The *torta* contains a higher concentration of larger mineral particles such as gravel, but also contains unidentified organic material and carbonates, elements that are not present in the sculpted layer of the friezes (fig. 1).

Sample #17, S1/R/C (Sample of plaster from sculpted layer, Ciudadela Squier)

Organic material:	negative
Carbonates:	negative
Sulfates:	positive
Irons:	positive
Chlorides:	positive
Silicates:	positive
Phosphates:	positive

Sample #15, S1/E (Sample of plaster from "torta" layer, Ciudadela Squier)

Organic material:	traces
Carbonates:	positive
Sulfates:	positive
Irons:	positive
Chlorides:	positive
Silicates:	positive
Phosphates:	positive

Fig. 1 PRELIMINARY ANALYSIS OF CLAYS

The mixture of sand and clay used for the sculpted layer varies slightly in each sample tested, as would be expected from materials that were hand mixed. The sculpture mix is finer than that of the *torta*, with the sand and clay perhaps being strained through a textile to remove larger particles before mixing. The approximate percentage is 70 - 80 % sand to clay. The reliefs are denser, with particles more compact. A final slip was added to several friezes.

Several of the sixteenth-century Hispanic chroniclers mention that cactus juice was used as a binder for pigments in the Andes.[8] As chemical consolidants have on occasion proved unsatisfactory for the conservation of the reliefs at Chan Chan, conservators of the Institute of Culture, Trujillo, have recently been using the juice of the nopal cactus (*Optunia ficus*) with reported success.[9] In the chemical analysis of a number of samples of the prehistoric reliefs from Chan Chan, evidence for its use was inconclusive. In replication studies, however, it was apparent that the simple clay-sand mixture resulted in a surprisingly strong bond which may have obviated the need for a binder, although it may still have been used due to a perceived efficacy. Beyond the possibility of a chemical binding, the physical effect of pressing together the mixture, as with the *pisé* or rammed earth process and what is evident in the sculpted layer of the reliefs, would have greatly increased the durability of the reliefs, as the particles in close proximity form a stronger bond.

Pigment was used in limited areas in central Chan Chan, and more extensively at the outlying architectural structures known as *huacas*. The most extensive use of pigment in central Chan Chan appears in the Sector Policromado of Ciudadela Gran Chimu. The imagery consists of relief figures with crescent headdresses in yellow, interspersed with geometric motifs, against a black background (fig. 2). The pigment was very thickly applied, and there is evidence of several layers of paint, probably applied on several different occasions. A layer of sediment encountered between layers of paint indicates that the area was re-painted after a flood or heavy rain. Both the yellow and the black pigments are ferric oxide (Fe_2O_3). While limonite (as yellow ferrous oxide is commonly called) is frequently found in ancient Peru in the yellow-ochres, the use of ferric oxide for the black pigment is more unusual.[10] To achieve the black color, the ferrous oxide was burned at 400°or 500° centigrade. It may have been burned with chaff as the combustible, which would thus explain some of the residue of organic material seen in the pigment mixture.

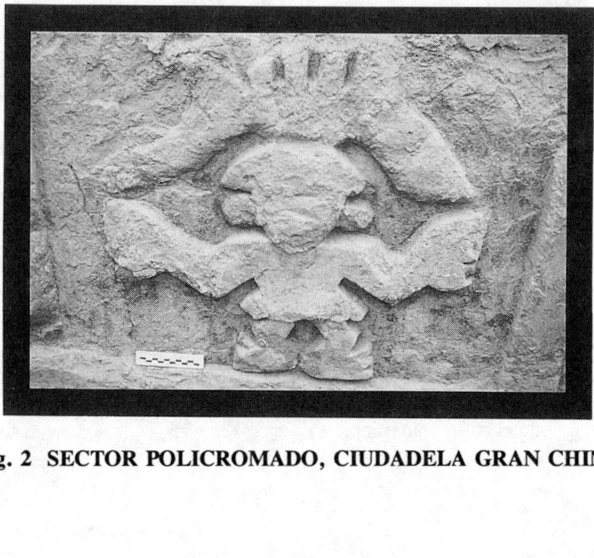

Fig. 2 SECTOR POLICROMADO, CIUDADELA GRAN CHIMU

At the beginning of the project, it was assumed that a single construction technique was used throughout the site. However, during the field season, attempts to determine exactly what technique was used were met with conflicting data. On careful examination, the data fell into patterns that have suggested technical developments over time. In light of disagreements over the chronological placement of the *ciudadelas*, an independent line of evidence, such as that provided by the technical data available with the friezes, will help evaluate conflicting sequences.

Particularly if looked at in conjunction with other factors, the technical developments provide a convincing seriation. As different elements at different rates over time, the most effective chronological schemes examine a set of variables and analyze the overall trends. Thus, several lines of evidence, including ceramic, brick, iconographic, as well as the data from the study of the construction of the reliefs, are examined together to establish the chronological relationships of the reliefs and the monumental compounds.[11]

In general terms, the evolution in relief construction at Chan Chan is from an additive to a subtractive process. The anchor for the early extreme of the sequence lies in evidence available from Moche friezes, a culture which precedes Chimor in roughly the same geographic area. The reliefs from Ciudadela Uhle are closest to the modelling techniques of the Moche period. Over time, there was a switch to a subtractive technique that allows for a greater impact much more quickly. This switch, associated with the friezes from Ciudadela Velarde, allowed for a spectacular flowering of the architectural relief arts. In developing a new technique that essentially compresses several steps into one, artisans were able to cover a much greater expanse of wall with more ornate carvings.

While the additive to subtractive evolution is the overall trend, the situation should not be oversimplified, as the techniques are not mutually exclusive, and details in the earliest friezes may have been cut, as details in the latest friezes were modeled. In all cases, several factors were taken into account, such as types of slips used, construction logic, iconography, as well as supporting archeological data.

Additive

With the additive process, the forms could have been applied to the wall in two ways: modelling the figure *in situ*, or sculpting the form independently and then applying it to the wall. The drawback with the later process is that delicate forms are difficult to transfer in one piece. Slip used as an adhesive, while not absolutely necessary for a good stick, apparently was used in several cases. Evidence for additive processes includes: traces of adhesive slip at the juncture between form and background, greater separation between the form and the wall, more evidence of hand working, few tool marks, few or no cut marks. Generally, but not always, the edges of the forms are more rounded.

The earliest friezes have figures that are relatively more dispersed, with a more carefully worked background. As the subtractive process often leaves tiny fragments of the second layer on the background of the first, the backgrounds of the friezes created by the additive process tend to have a "clearer" appearance. No conclusive evidence was seen to indicate that guidelines were sketched onto the background to indicate where the forms should be placed.

Of the sample of reliefs examined in 1990-91, friezes from Ciudadelas Uhle, Gran Chimu, and Laberinto contain evidence consistent with an additive method of construction. Additional archeological and iconographic evidence supports a placement in the early to middle period of Chan Chan's history.[12]

Subtractive

In the Late Chimu period, there is a change in technique from the additive process to a subtractive technique which enables greater expanses of wall space to be constructed more quickly. Instead of preparing the background and modelling the figures, which is more time-consuming, in the subtractive method, an approximately three centimeter layer of plaster is applied to the wall, and while the plaster is still relatively wet, the forms are excised directly from this layer, allowing for complex imagery more quickly (figs. 3, 4). With the subtractive technique, a greater surface area is taken up by the sculpted forms; there is less free background space. Clearly, as the background was not previously worked, it tends to be rougher. Evidence for the subtractive technique includes: cut marks on the forms and background, overlooked areas where the plaster was not excised, traces of the excised layer on the background, and with certain exceptions, a relatively uniform plane of reliefs.

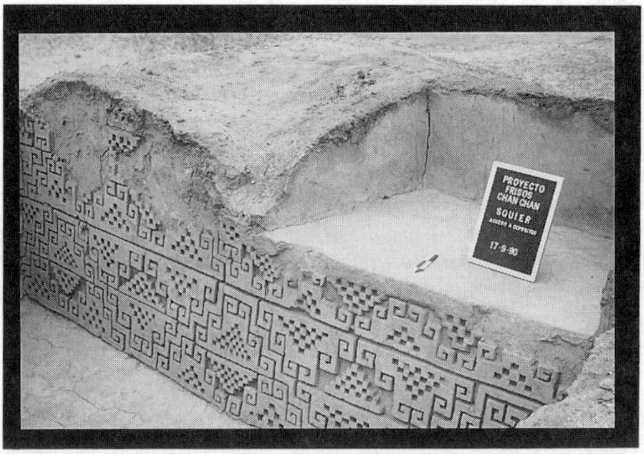

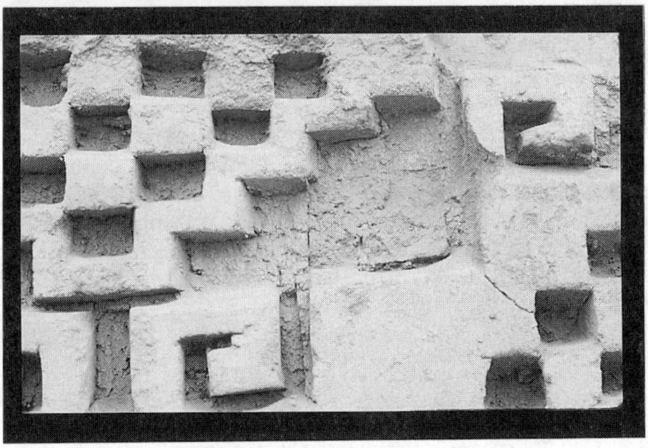

Figs. 2, 3 CIUDADELA SQUIER (view and detail)

The Tule Boat frieze from Ciudadela Velarde may have been one of the first areas where the subtractive technique was used. Large sections of the frieze were sculpted in the subtractive method, based on abundant cut marks apparent on the wall and on the motifs. It may have been a process where the sculptor was forced to cut a difficult form, such as a net, directly on the wall, and extended the process from there. It seems likely that a metal tool was used at Ciudadela Velarde and elsewhere, as the incisions are too thin for standard wooden or even bone tools.

One of the most elaborate friezes known from Chan Chan, the Burr frieze, is located a little over 20 meters north of the Tule Boat frieze in Ciudadela Velarde. A small section of this frieze is all that survives of a composition that once covered at least two sides of a large platform, an extension of close to 80 square meters. It is not surprising that for such an enormous undertaking a technical innovation would be developed. As mentioned above, the construction steps are compressed, with only one layer of plaster added, and the forms excised from this one layer. The Burr frieze was carved when the plaster was still quite damp and malleable. The design was executed panel by panel, with faint vertical lines indicating where the design was separated.

It seems unlikely that molds were ever utilized at Chan Chan to create the friezes, because in even large, repetitive friezes such as the Burr frieze, no two motifs are of exactly the same size. The use of a variety of models or guides, however, seems likely. At Ciudadela Squier, strings may have been used to help lay out the complicated checkerboard wave motif. After it was sketched out, portions would be excised. Several areas that should have been excised according to the design, were overlooked in the final work.

Although the study is still preliminary, it may be possible to observe similar construction evidence in the outlying architecture at Chan Chan, architecture that was not strictly within the parameters of the project. For example, the reliefs from Huaca Esmeralda, to the east of central Chan Chan, show similar traits to the reliefs known from the late *ciudadelas*. Future research on the architectural reliefs will help refine and clarify the chronological relationships at Chan Chan, as well as help in the difficult task of the conservation of the monument.

ACKNOWLEDGEMENTS

The author would like to thank the American Association of University Women, the Fulbright Commission, and the Samuel Kress Foundation for their support of the 1989-91 field season research. Additional study was made possible by a Junior Fellowship in Pre-Columbian Studies at Dumbarton Oaks, Washington D.C. The author would also like to thank the other members of the project, particularly Peter Kvietok, Ricardo Morales and Alberto Barba. Additional assistance was provided by Santiago Uceda and Roberto Sabana of the Universidad Nacional de Trujillo, Peru. A special thanks is due to Ana María Hoyle, of the Instituto Nacional de Cultura, Trujillo, Peru and the members of the department of conservation of the same institution, particularly Carlos del Mar and José Córdova, for their insight and assistance.

REFERENCES

1. Kent C. Day, in Chan Chan: Andean Desert City, edited by M.E. Moseley and K.C. Day (School of American Research/University of New Mexico Press, Albuquerque, 1982) p. 55.

2. M.E. Moseley and K.C. Day, ibid.; G.W. Conrad, American Antiquity 46 (1), 3-26 (1981).

3. A. Kolata in The Northern Dynasties: Kingship and Statecraft in Chimor, edited by M.E. Moseley and A. Cordy-Collins (Dumbarton Oaks, Washington D.C., 1990) pp. 107-144.

4. J. Rowe, Acta Americana VI (1-2), 26-59 (1948).

5. A. Andrews, J. Field Archeology 1 (3/4), 241-264 (1974); G. Conrad in M.E. Moseley and K.C. Day, op cit., pp. 87-117; A. Kolata op cit.

6. K.C. Day, PhD thesis, Harvard University, 1973.

7. J. Topic, PhD thesis, Harvard University, 1977; J. Topic and M.E. Moseley, Ñawpa Pacha 21, 153-172 (1985); R. Cavallaro, PhD thesis, Harvard University, 1988.

8. Juan de Betanzos, Suma y narración de los Incas (Biblioteca Peruana, Primera Serie, Tomo III, Editores Técnicos Asociados, Lima, 1968 [1551]), 271; Bernabe Cobo, Inca Religion and Customs, translated and edited by Roland Hamilton (University of Texas Press, Austin, 1956 [1653]), pp. 203-204.

9. A. M. Hoyle in Sixth International Conference on the Conservation of Earthen Architecture, Adobe '90 Preprints (The Getty Conservation Institute, Los Angeles, 1990), pp. 225-229.

10. Duccio Bonavia, Mural Painting in Ancient Peru, translated by Patricia Lyon (Indiana University Press, Bloomington, 1985), pp. 179-181; Carlos Nuñez Villavicencio in ibid., pp. 203-204.

11. See J. Pillsbury, PhD thesis, Columbia University, 1992 for a complete discussion of the types of data and the chronological issues.

12. See J. Pillsbury ibid.; J. Topic and M.E. Moseley, op cit.; M.E. Moseley, Science 187 (4173), 219-225 (1975); A. Kolata, op cit.

ELECTRON MICROSCOPY STUDIES OF THE CHRONOLOGICAL SEQUENCES OF TEOTIHUACAN PLASTER TECHNIQUE

∞DIANA MAGALONI, £T. FALCON, £ J. CAMA, *R. W. SIEGEL, *R. LEE, ≠R. PANCELLA, °L.BAÑOS, §V CASTAÑO.
∞Instituto de Investigaciones Estéticas, UNAM, § Instituto de Física, UNAM,°Instituto de Materiales UNAM, México city, £ Escuela Nacional de Restauración, INAH, Mexico, City, * Argonne National Laboratory, Ill., USA, ≠Laboratoire de Conservation de la Pierre, EFPL, Lausanne, Switzerland.

ABSTRACT

Over 108 samples of mural paintings from Teotihuacán, México, were studied by scanning electron microscopy, optical microscopy and X-ray diffractometry. The results show a sequence in the techniques employed to produce the mural painting's supporting plasters -made of lime and different sands. The technical stages found can be used as archaeological data to help understand the continuous evolution of 800-year mural paintings. The consequences of our study are discussed in detail.

INTRODUCTION

Mural painting characterizes the art of Teotihuacán. The images represented there are the most important sources of information about the culture that produced them.

The investigation of the significance of the symbols and designs is a key that opens the entrance to a world that flourished 17 centuries ago. Mural painting was not an art buried in tombs, but an art which decorated walls and structures that were inhabited by living persons.

Teotihuacán was a painted city constructed on a grid plan in accordance with the four astronomical directions. The ceremonial heart, where the two great pyramids are located, as well as the peripheral areas, conformed by the so called "room complexes" that probably were public or secular buildings, were all decorated with colour, polychrome figural and abstract designs. During the apogee of the urban centre, about A.D. 600, the entire city covered an area of more than eight square miles [1]. From the beginning, around the first century A.D., to the time of its fall in the eighth century, there is archaeological evidence of continuous dwelling, constant construction and transformation of buildings. As a matter of fact, this is one of the most interesting aspects of the ancient precolumbian city, one can follow the continuous evolution of the urban centre and of its painting tradition over eight hundred years. However, chronological studies of the murals have been a problematic field of research. Since 1960, researchers have pointed out the difficulty to seriate mural paintings based upon the ceramic data associated with the murals, this due to the fact that many of the paintings were removed without recording their archaeological context. In 1962, Clara Millon wrote a doctoral dissertation attempting to give a chronological order to the mural art of Teotihuacán [2]. Her study presented a sereation of murals phases based on the analysis of how the motifs changed throughout time, and incorporated them to the widely accepted chronology established by René Millon [1].

Since then, her sequence has been altered by the discovery of new mural paintings and by other points of view [3], [4]. Arthur Miller wrote in 1973 that "the justification for the Millon mural chronology has not yet been convincingly demonstrated" [5].

In order to help fill this lack of information, we propose a systematic microscopic study of the plasters, which constitute the paintings support. In this manner we can avoid the need for archaeological contexts and the confusing information that evidence from architectonic transformations of the buildings provides. What the present work has attempted is to provide an independent system of seriation for the mural painting based on the supporting plaster itself. To do so, a systematic study of the paintings and their plasters has been undertaken. The work represents an interdisciplinary methodology using scanning electron microscopy and microprobe methods, optical microscopy and stylistic aesthetics analysis [6]. We present here the scanning microscopy analysis of the plasters, which is the basic study for the technical seriation of the mural painting.

EXPERIMENTAL

The samples were collected at places denoting evidence of substructures covered by two later construction levels. The "room complex" of Tetitla, was chosen for its abundance of mural painting. In this place there are two different construction levels and several architectonic transformations, which also indicate different chronological stages. The precise location of the samples taken was recorded as well as the construction level in which the wall painting was placed. All fragments were divided into preliminary chronological groups, according to our stylistic analysis [6] and to the Millon chronology [1].

Over 108 samples were included in polystyrene resin to prepare the microscopic sections. The fragments were placed in both transverse and planar orientations. A complete description of the plaster was made through the observation of, on the one hand, the patterns of distribution of the sands in the calcium carbonate matrix (planar orientation) and on the other, the stratigraphy of the mural painting layers (transverse section), material included in another report [6].

Scanning electron microscopy was used in order to characterize the different plasters in terms of their constituents, observing patterns of distribution of the phases in the matrix, their composition, morphology and sizes. Prior to the observation, the The polystyrene secctions were coated with amorphous carbon to prevent electrostatic charging.

A scanning electron microscope(JXA-S40A) equiped with elemental analysis attachments was used at Argonne National Laboratory. The operation conditions were maintained constant to ensure the reproducibility of the results. All pictures were taken at 15 kv, with a working distance of 39 mm in each case and at 18° incidence angle. .

Backscattered electron images (BEI), secondary electron images (SEI) and energy dispersive spectroscopy (EDS) mapping analysis were done to each sample. All pictures were obtained at 100X magnification and repeated in three different areas of the same sample. The more representative results were chosen to describe a technical stage.

In order to identify the materials, spot analyses were made from each phase found in the matrix and complementary analysis with X-ray diffractometry were also done

RESULTS AND DISCUSSION

Technical period I (A.D.1-200)

Figure 1 is a backscattered electron image (BEI) characteristic of technical period I. In the centre, with a dark gray color, we can observe a silice-aluminum particle, with an irregular geometric form, identifyed by spot analysis. Fragments in lighter gray with rounded profiles were detected as calcium carbonate phases. In black we perceive the porosity of the matrix and several voids. All samples taken from substructures covered by two or more construction levels presented this characteristic pattern: a remarkable lack of silica sands, very small in size, and a larger amount of calcium carbonate chalks included in a porous lime matrix. Identification by X-ray diffraction indicated that the silica particles were potassium feldspars. Technical period I corresponds to the archaeological period Tzacualli-Miccaotli.

Technical period II (A.D. 200-450)

Figure 2 shows a secondary electron image (SEI) where a drastic change in terms of materials and technique is evident, this corresponding to the second technical period. The large rounded particles are volcanic quartz, identified by X-ray diffraction These larger particles, from 200μm to 400μm in diameter, are new materials found in all murals belonging to the Tlamimilolpan period.

Along with the volcanic quartz sands, calcium carbonate round aggregates spread throughout the lime matrix were identified. Technical period II can be distinguished from the first technical period due to the introduction of this new material, even though the tradition of including soft calcium carbonates as charges in the matrix is still in use.

Technical period III (A.D.450-650)

Figure 3 is a BEI picture from a characteristic sample of technical period III. The acicular particles are well distributed among larger irregular forms, and the perceptible density of the calcium carbonate matrix indicates progress

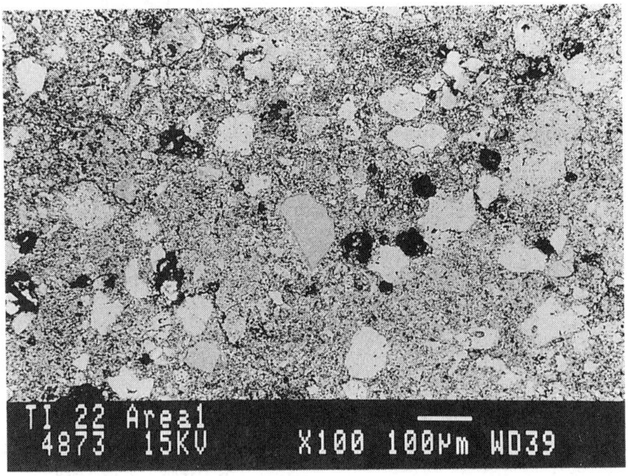

Figure 1. Technical Period I

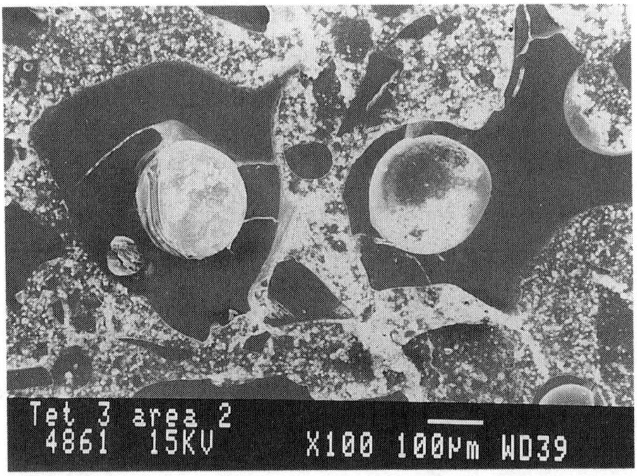

Figure 2. Technical Period II

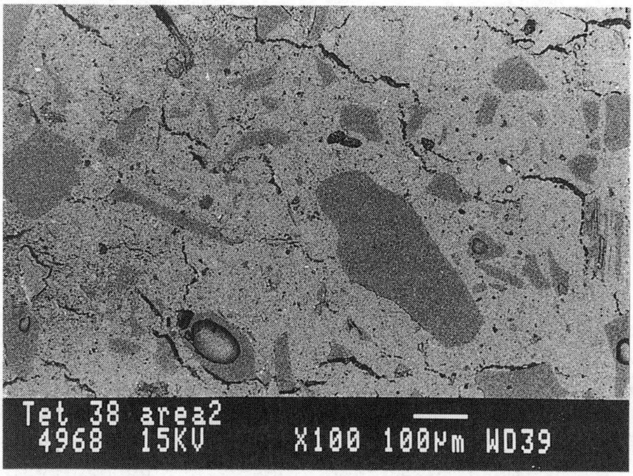

Figure 3. Technical Period III

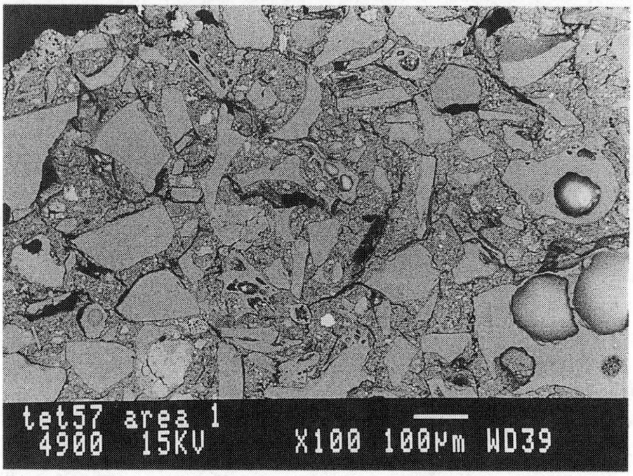

Figure 4. Technical Period IV

in the formulation and work of the support plasters from the Xolalpan period (A.D. 450-650). There are transverse fissures in the matrix that probably are due to the drying process and do not necessarily indicate a lack of compaction of the lime plaster. This stage differentiates it self from period two in virtue of the absence of calcium carbonate round aggregates, and because of the better work upon the feldspathic and quartz sands that results in a more homogeneous distribution. It is worth mentioning the continuity between the inclusion edges and the matrix, which demonstrates the good cohesion of the cementing phase -a sign of quality.

Technical period (AD 650-750)

The last technical stage of mural painting, at Metepec period (A.D.650-750), is found only in the remaining fragments of some of the later structures, they are easily distinguished from the classical Teotihuacán buildings (Angulo, J, 1992, in press). Only the samples coming from the plasters belonging to these structures showed the technical characteristics as now described. It is very interesting to observe how complex the plasters from technical period IV are. As figure 4 shows in the BEI picture, there is a higher proportion of quartz sands in the calcium carbonate matrix. The granulometic principle is now used to construct more durable plasters. The sizes and the distribution of particles goes from only 10μm to *ca.* 400μm. The smaller particles are within the voids left by the larger particles, almost no space is left unfilled with hard quartz pieces of sand, thus improving mechanical resistance.

Energy dispersive spectroscopy analysis (EDS)

EDS was used to obtain an elemental distribution pattern of silicon (Si) and calcium (Ca). In this way the observations of, and distinctions between, the various technical periods was simplifyed. As shown in figures 5 to 9, the distribution of silicon in the left side, permits one to see the morphology, relative sizes and the distribution of the feldspathic and quartz sands The location of calcium, in the right side, allows one to evaluate the density of the lime matrix. Comparing all these figures, one can detect the fundamental differences among the technical periods. Thus enabling one to establish the technical sequences present in the mural painting of Teotihuacán.

Technical period I plasters (figure 5) are composed basically of calcium, thus the charges are calcium carbonate aggregates. The average calcium/silicon ratio in this case was 7.2. Period II, (figure 6) shows a totally different configuration. Large, capricious, rounded-center particles can be seen with the distribution of silicon. Their distribution is not homogeneous, on the contrary, one can detect concentration of sand particles alternated with free spaces of only calcium.The calcium/silicon ratio is 1.2 .

We must point out that a transitional period was found. It corresponds to a specific group of murals, denominated in our study as transitional technical period II/III. This stage is more perceptible in EDS analysis. The characteristic

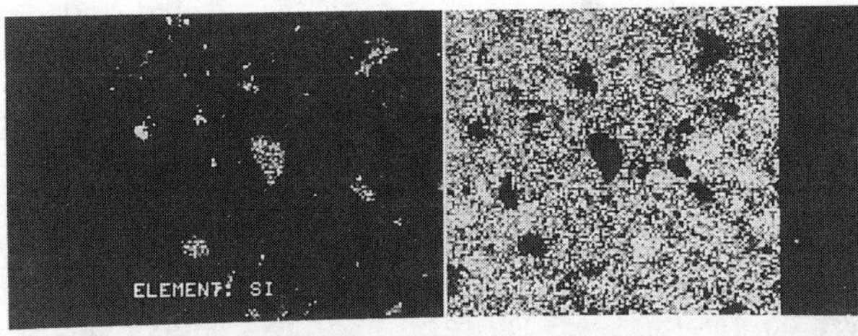

Figure 5. Technical Period I

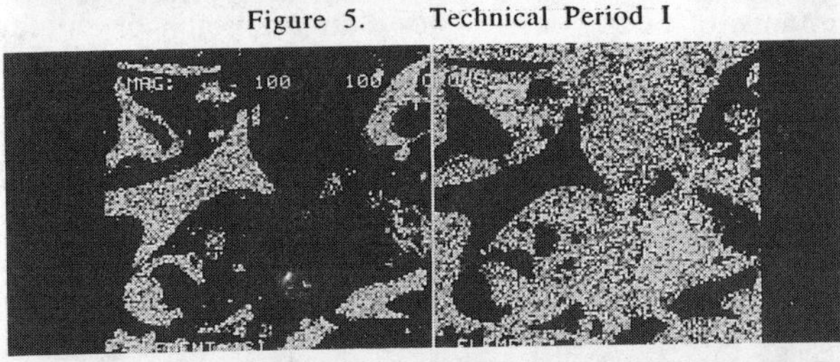

Figure 6. Technical Period II

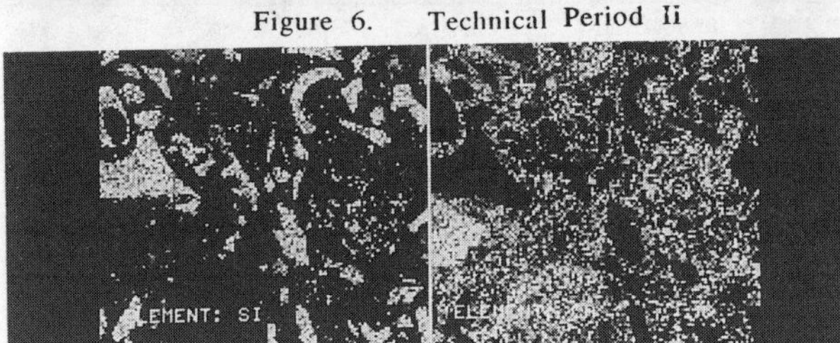

Figure 7. Technical Period II-III

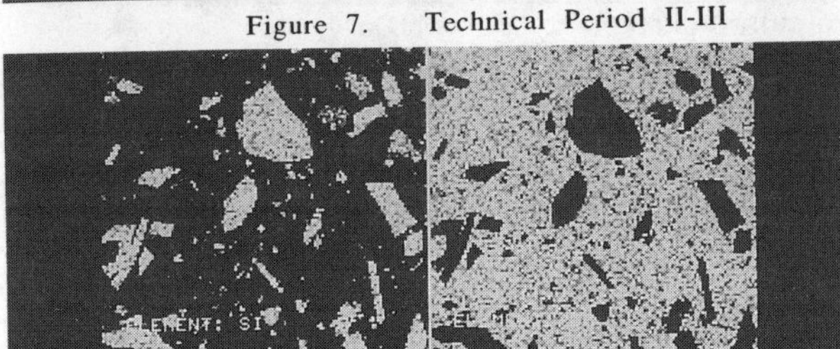

Figure 8. Technical Period III

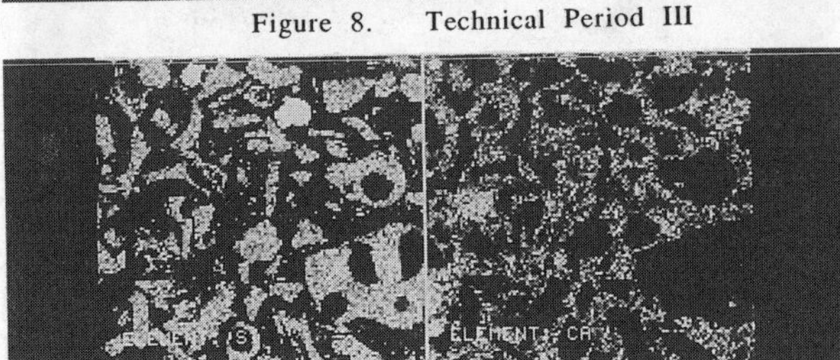

Figure 9. Technical Period IV

pattern of this period shares features of both the second and third technical periods (figure 7). The calcium/silicon ratio is 1.6 .

Period III (figure 8) reveals progress in the technical preparation of plasters. In this case, the distribution of silicon exhibits acicular small particles and irregular larger grains of sand in a quite homogeneous distribution. The density of the calcium carbonate matrix is perceptible in the right side. The relative amount of calcium/silicon is 1.8.

Period IV, (figure 9) is characterized by having a different plaster structure where both composite elements are almost equivalent. They seem to be negative images of one another. The relative amount of calcium/silicon is 0.5.

Technical Periods	Chronology	Archeological Periods
I. Feldspar as sand in small amounts. Calcium Carbonate ($CaCO_3$) aggregates are used as the charge.	A. D. 1 - 200	Tzacualli- Miccaotli
II. Volcanic Quartz as charge is for the first time employed Large capricious forms in an unequal distribution, $CaCO_3$ as aggregates still in use.	A.D. 200 - 300	Early Tlamimilolpan
II-III. Volcanic Quartz and Feldspars in a more equal distribution. The $CaCO_3$ aggregates are out of use.	A.D. 300 - 450	Late Tlamimilolpan
III. A rich calcium matrix. Acicular feldspar particles well distributed among irregular Quartz grains.	A.D. 450- 650	Early and Late Xolalpan
IV. Granulometric principle is now used to build up a more durable plaster. Charges and matrix are equivalent.	A.D. 650-750	Metepec

Table I: technical chronological sequences of the mural painting of Teotihuacán

CONCLUSIONS

1.- The technical periods based on the plaster structure and composition permitted us to propose an independent seriation for the mural painting in Teotihuacán.

2.- The appearance and use of a completely new material (volcanic quartz sands) at the beginning of technical period II (*ca.* A.D. 200), coincides with archaeological evidence which states a drastic social change in ancient

Teotihuacán [7]. This data, together with other material and stylistic changes [6] suggest that a new plaster tradition was introduced during this time.
3.- Technical periods II, II/III, III and IV are basically a unique painting tradition that improves during time. Physical features of the plaster materials show a developing refinement of proportions, shapes and distribution of the sand particles in the calcium carbonate matrix.

ACKNOWLEDGEMENTS

The authors are indebted to the Instituto Nacional de Antropología e Historia (INAH) of México, to Mr. Roberto Gallegos, for the permission to gather samples at Teotihuacán, and to Mr. Alfredo Sánchez for the photographic work.

REFERENCES

1. **Millon**, René, "Cronología y periodificación: datos estratigráficos sobre períodos cerámicos y sus relaciones con la pintura mural, in: XI Mesa Redonda de la SMA, México,1966. p.57
2. **Millon** Clara, A chronological study of the mural art of Teotihuacán, Michigan, Ann Arbor, doctoral dissertation, (University of California, Berkeley, 1962).
3. **Bennyhoff**, James "Chronology and periodization: continuity change in Teotihuacán ceramic tradition", in: XI Mesa Redonda de la SMA, México,1966.
4. **Berrin** , Katherin.in , Feathered serpents and flowering trees: reconstructing the murals of Teotihuacán, edited by Berrin and Millon, (Fine Arts Museum of San Francisco, 1988).p. 24
5. **Miller** Arthur, The mural painting of Teotihuacán, (Washington, Dumbarton Oaks, 1973) .
6. **Magaloni**, Diana, et. al, "El espacio pictórico teotihuacano: tradición y técnica", in: La pintura mural prehispánica en México: Teotihuacán, Enfoque interdisciplinario, edited by B. de la Fuente, (Instituto de Investigaciones Estéticas, UNAM, México, In press, 1992)
7. **Angulo**, Jorge, "Aspectos de la cultura a través de su expresión pictórica", in: de la Fuente, Beatriz, ed. La pintura mural prehispánica en México: Teotihuacán, Enfoque interdisciplinario, In press, 1992.

COMPARATIVE STUDIES OF ANCIENT MORTARS FROM GIZA, EGYPT, AND NEVALI ÇORI, TURKEY

A. Reller and P.-M. Wilde, University of Hamburg, D-2000 Hamburg 13,
H.G. Wiedemann, Mettler Toledo AG, CH-8606 Greifensee,
H. Hauptmann, University of Heidelberg, D-6900 Heidelberg
G. Bonani, ETH, CH-8093 Zürich.

ABSTRACT

Mortars collected from the Cheops pyramid, the Sphinx and the Chephren pyramid, all at Giza, Egypt, have been compared with recently found mortars of a cult building of the aceramic neolithic at Nevali Çori (south east of Urfa, province Hilvan, Turkey), probably one of the earliest sites of civilization. As methods of investigation thermal analysis, X-ray diffraction as well as analytical scanning and transmission electron microscopy have been used. The egyptian mortars have been found to be primarily made up of gypsum, anhydrite and calcite. The comparative results with mortar of the recently detected temple site at Nevali Cori, Turkey, give evidence that the burning or otherwise chemical treatment of limestone as crucial ingredient for the fabrication of mortar has been performed at a very early period (10´000 - 8000 B.C.). In order to further confirm the described findings ^{14}C dating measurements have been carried out. The results of the described studies raise many questions concerning early civilizations and their knowledge of technologies as well as the transfer mechanisms of this knowledge.

INTRODUCTION

The fabrication of mortar is one of the most ancient chemical processes carried out by human civilizations [1]. In this process the calcination of limestone, i.e. the following reversible chemical reaction is of paramount sgnificance:

$$CaCO_3 \rightleftharpoons CaO + CO_2 \qquad (1)$$

If in any culture limestone was calcined, the subsequently occurring back-reaction *via* quicklime and hydroxide, i.e. the re-carbonatization leads to fixation of actual CO_2. Therefore, ^{14}C dating measurements can yield direct informations on the period, at which mortar was fabricated [2]. From literature it is known that for the construction of the pyramids as well as for the sphinx at giza mortar was used [1, 3 - 6]. This type of mortar is made up of processed gypsum, sand and limestone. Although it is known that temperatures around 950°C could be reached in Egypt during the period of the Old Kingdom [4], there seem to be some doubts whether limestone was really calcined and re-carbonatization or even formation of cement by secondary reactions of the CaO could be carried out under well controlled conditions.

Recent excavating on a temple site in Nevali Çori, Turkey, gave evidence for the existence of a highly developped culture long before the development of the cultures in the Euphrates/Tigris region, i.e. 10´000 - 8´000 B.C. [7]. In this most interesting site a well conserved terrazzo floor was found and investigated. It is made up of limestone fragments and limestone mortar. The question arises whether the limestone mortar was fabricated by calcination/re-carbonatization or whether it was the ground mineral mixed with a liquid and applied as a plaster in a mudlike consistency.

RESULTS OF THE EXPERIMENTAL INVESTIGATIONS

The Egyptian mortars have been characterized by X-ray diffraction, analytical scanning calorimetry and thermogravimetry. As the thermogravimetric measurements of the thermal degradation of two samples collected from the Sphinx show (Fig.1), there obviously exist pronounced differences between mortars of one and the same edifice: Depending on the site, where the sampling took place, mortars made up of gypsum/sand/limestone and mortars made up of only sand/limestone could be identified. As a matter of fact the gypsum mortar was collected near the surface, whereas the limestone mortar was collected from a site around 4 m inside the Sphinx. This mortar is nearly pure limestone and the thermal degradation curve accordingly reveals a weight loss of around 40 % (theoretical value: 44%) above 650°C. The relatively low temperature of the degradation can be explained by grain size effects (see also [8]). In the case of mortar containing gypsum an initial loss of water at around 130°C followed by a relatively small weight loss starting at 650°C, i.e. the calcination of limestone, indicates the presence of only small amounts of this material. Thermogravimetric measurements of mortar collected from the Cheops and Chefren pyramids gave evidence for the presence of gypsum only.

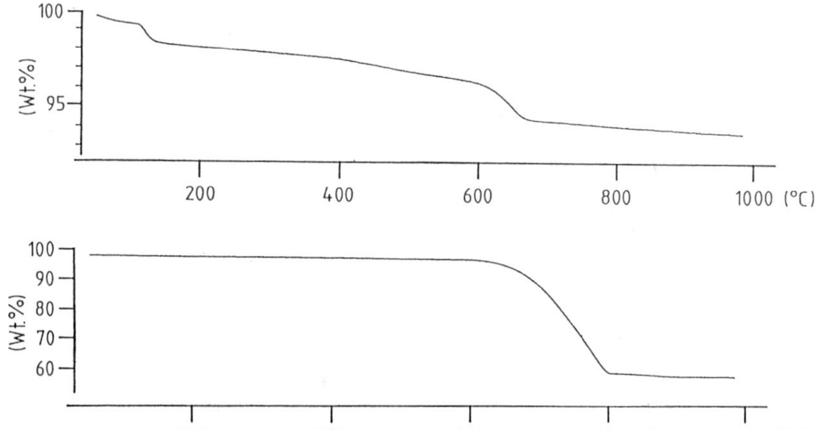

Figure 1

Thermogravimetric curves of the thermal degradation of mortar collected from the Sphinx / Giza. The upper curve characterizes mortar collected near the surface, i.e. gypsum mortar; the lower curve characterizes mortar collected inside the Sphinx, i.e. pure limestone (experimental conditions: heating rate: 4°/min, air).

Mortars collected from the terrazzo floor, Nevali Çori, were first characterized optically and by X-ray diffraction. Optically, the material is made up of large limestone fragments (with sizes up to 5 cm) embedded in a mortar matrix. The thickness of the terrazzo floor is in the range of up to 15 cm and the bottom is partially covered by a layer mortar only. X-ray diffraction patterns were collected from the surface part, from embedded limestone fragments and from the bottom part. As the X-ray patterns in Fig. 2 reveal, the main constituent of all samples is pure limestone, i.e. calcite.

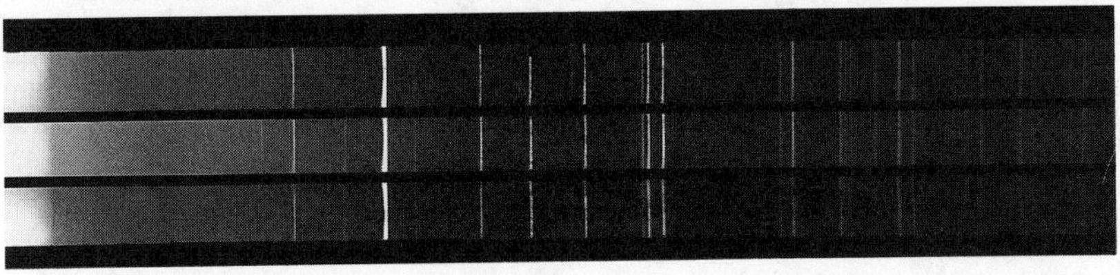

Fig. 2
X-ray powder diffraction patterns of limestone mortar collected from the terrazzo floor, Nevali Çori, Turkey (CuK_{α}-radiation, monochromatized Guinier camera). The samples were isolated from the surface (top pattern), from a large limestone fragment inclusion (central pattern) and the bottom part (bottom pattern). In agreement with conclusions of other studies [4,5], the difference between natural and processed limestone cannot be detected by X-ray diffraction.

Thermogravimetric measurements of embedded limestone, surface and bottom samples were carried out. As the curves in Fig.3 show, the genuine limestone is the most stable compound. The sample collected at the bottom is the least stable and - according to the weight loss - not re-carbonatized up to 100%. This is a first indication that in this early culture limestone was chemically transformed, i.e. calcined or dissolved in acidic liquids under evolution of CO_2.

Fig. 3
Thermogravimetric curves of the degradation of different ground limestone (mortars) samples collected from the terrazzo floor at Nevali Çori.

Curve 1: Embedded limestone fragment (unprocessed limestone);

Curve 2: Mortar collected at the surface (fully carbonat.)

Curve 3: Mortar collected at the bottom (partly carbonat.)

Experimental conditions: heating rate 10°/min, air.

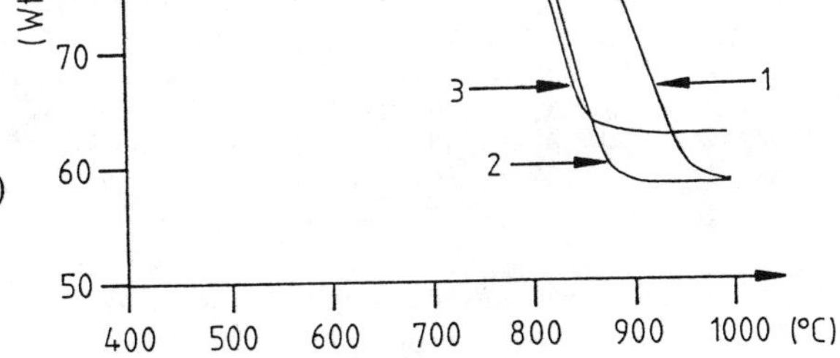

Scanning electron microscopy of the different samples indicates that the natural limestone was formed by organisms such as coccoliths and radiolaria, i.e. by bioinorganic sediments. The problem of differentiation between natural limestone and processed limestone turned out to be rather difficult: As the micrographs in Fig. 4 show there are no clearcut differences between the morphology of natural limestone particles and calcined / re-carbonatized particles (experiment performed in the laboratory using a limestone fragment from Nevali Çori). The re-carbonatized samples obtained from calcined ancient mortar as well as the modern mortar samples reveal common features of sintering effects, but unambiguous characteristics and therefore proofs for the fact that limestone has been calcined for the fabrication of the applied mortar cannot be found at present. Further and more detailed investigations on samples collected at different locii of the site of Nevali Çori - in particular plaster fragments of a wall - are in progress.

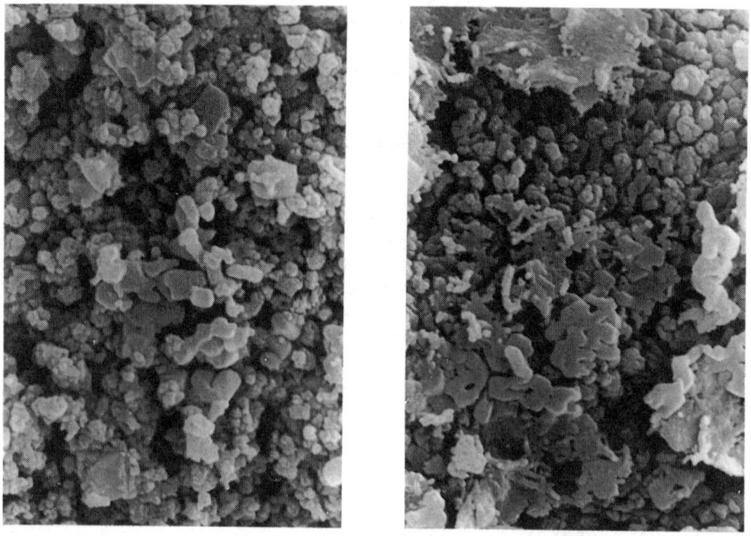

Fig. 4
Scanning electron micrographs of limestone crystallites In the left micrograph crystallites constituting the natural limestone fragments of the terrazzo floor are displayed. In the right micrograph limestone crystallites obtained by calcination and subsequent re-carbonatization of a natural limestone fragment are shown. The white bars at the left side of both micrographs correspond to 10 μm.

^{14}C dating measurements were carried out on the excavated limestone fragments, on mortar samples collected at the surface of the floor and on mortar samples collected at the bottom of the floor. Whereas the measurements of the natural limestone fragment clearly proved an age of >40´000 years, i.e. pure mineralogical calcite sediment, the analogous measurements of the mortar samples yielded an age of around 20´000 years. Considering the fact that slight traces of mineralogical limestone within the mortar, which shifts the dating backwards, these findings clearly prove that the limestone mortar was formed by re-carbonatization, i.e. the natural limestone was chemically altered or calcined by the people of Nevali Çori. Further investigations on different mortars will yield more detailed informations on the kind of fabrication performed as well as a more accurate absolute dating.

DISCUSSION

The presented investigations of mortars collected at edifices of the Old Kindom, Egypt, and at the more ancient site of Nevali Çori reveal, that the calcination of limestone represents one of the crucial chemical processes characterizing the technological know-how of ancient civilizations. Moreover, the fact that calcite undergoes a reversible decarbonatization / re-carbonatization and therefore CO_2 of the period of mortar fabrication is immobilized, in principle allows dating. As it has been shown, however, traces of sedimentary, i.e. not fully calcined or chemically altered limestone leads to shifts in the dating procedure. Applying complementary solid state chemical techniques of characterization such as thermal analysis and analytical electron microscopy, it should be possible to minimize these artefacts. Recent thermogravimetric measurements of the degradation of natural and processed limestone samples allow a semi-quantitative distinction. Since in many cultures the construction of edifices and the concomittant use of mortar is one of the main artifacts of archeological study, detailed and comparative investigations of mortars not only inform the actual technological know-how but also represent a powerful tool for archeometry.

[1] A. Neuburger, "Die Technik des Altertums", R. Voigtländers Verlag, Leipzig (1919).

[2] D.C. Stulik and D.J. Donahue, MRS Bull. 17 (1992) 53.

[3] A. Lucas and J.R. Harris, "Ancient Egyptian Materials and Industries", E. Arnold, London (1962),

[4] W.H. Gourdin and W.D. Kingery, J. Field Archeol. 2 (1975) 133.

[5] W.D. Kingery, P.B. Vandiver and M. Prickett, J. Field Archeol. 15 (1988) 219.

[6] D.D. Klemm and R. Klemm, Archeometry 90, Birkhäuser Verlag, Basel (1990) p. 445 ff.

[7] H. Hauptmann, Sulzer Horizonte 10 (1991) 8; *ibid.* 11/12 (1991) 8.

[8] M. Maciejewski and A. Reller, Thermochim. Acta 142 (1989) 175.

Editor's Note: The site of Nevali Cori is most important, and this article shows that the "plasters" were based on the limestone system, rather than being composed of gypsum. The authors plan further publication of their continued research. Further correspondence with Prof. Reller produced the following additional information: "For the thermogravimetric measurements, the samples were ground to an equivalent size. Based on the experience of former experiments on the decomposition of calcite, microscopic differences of textural features may also influence the temperature range. This fact will be further investigated because it may give us the mentioned semi-quantitative data for the distinction between natural and processed limestone."

PART IV C

Glass Corrosion and Preservation

THEORIES OF CORROSION AS APPLIED TO PRESERVATION OF ARCHAEOLOGICAL CERAMICS

R.L. Schulz, D.E. Clark and D.C. Folz
University of Florida
Department of Materials Science & Engineering
Gainesville, FL 32611

ABSTRACT

An understanding of corrosion processes is beneficial in the preservation of ancient ceramic artifacts. A wealth of information is available on this topic in the materials science and engineering literature. Significant contributions to the corrosion literature have been made over the last 15 years due to the extensive research performed on nuclear wasteform systems. This paper will review the current state of understanding of aqueous corrosion of glasses. We will discuss how this information can be applied to some of the current issues in preservation of archaeological ceramics.

INTRODUCTION

There are major structural differences between glasses and polycrystalline ceramics, even when their compositions are the same. These structural differences which include grain boundaries and porosity (polycrystalline ceramics) and lack of long range order (glasses) can result in differences in corrosion behavior. Although many of the same principles can be applied to polycrystalline ceramics, glass corrosion will be the main focus of this paper.

Glass corrosion has been studied for well over 75 years and today we have a good understanding of the factors (environmental and materials properties) that affect the degradation process. Much of our understanding of the degradation of complex systems has resulted from the nuclear waste program, where glass has been extensively evaluated as a wasteform. Characterization of corrosion layers on artifacts that have been buried for hundreds of years can provide the materials scientist with a better understanding of what to expect from a wasteform after equivalent burial times. Likewise, studies of nuclear wasteforms under controlled laboratory and in burial conditions can be beneficial to the archaeologist in selecting the best preservation techniques.

Factors that affect degradation are shown in Table I. The parallelism between artifacts and modern glasses is clear and the problems of environmental degradation that must be addressed by the archaeologist are similar to those that concern the materials engineer. For the archaeologist, years of burial in a corrosive environment have resulted in alteration of the artifact's surface. His main concerns are to protect the artifact against additional degradation and, in some cases, to restore it to as near an original condition as possible. On the other hand, the materials engineer subjects pristine glass specimens to specified corrosive environments and uses this information to predict long term durability under expected use or storage conditions. Both the archaeologist's and engineer's goal is to minimize degradation during long term storage, display and/or use.

Table I. Degradation Factors

ARTIFACTS*	COMMERCIAL GLASSES
RAW MATERIALS	RAW MATERIALS
Composition	Composition
Homogeneity	Homogeneity
Phases and impurities	Phases and impurities
	Waste products**
MANUFACTURING	MANUFACTURING
Raw materials processing	Raw materials processing
Forming	Melting
Decorating techniques	Casting (SS canister)**
Thermal history	Seal/Interim storage**
	Thermal History
PHYSICAL STATE	PHYSICAL STATE
Amorphous or crystalline	Amorphous or crystalline
Composition	Composition
Homogeneity	Homogeneity
Phases, impurities and their distributions	Phases, impurities and their distributions
USE-WEAR	USE OR STORAGE ENVIRONMENT
Initial uses	Atmosphere
Treatment after recovery	Moisture content
	Temperature
	Environmental cycling
ENVIRONMENT AT DISCOVERY	pH
Land or water	Eh**
pH	Geology**
Moisture content of soil	Radiation**
Temperature	
Environmental cycling	
Energy inputs	
PRESERVATION	
Cleaning	
Restoration	
Storage	
Display	

*Adapted from: reference 23.
**Apply specifically to nuclear waste glass

LABORATORY AND FIELD STUDIES

Laboratory and field studies are carried out for several reasons:

- Determine corrosion mechanisms
- Trace or predict corrosion behaviors
- Determine suitable final environment to preserve specimens.

In order to determine the extent and nature of the surface layers that have formed on glasses as a result of environmental exposure, many analytical techniques are used [3,4]. One of the most informative and quantitative methods is secondary ion mass spectroscopy (SIMS). This technique provides information on the composition of the leached layers in the form of cation percent as a function of leached layer depth. Lodding et al. have been able to discern as many as five distinct layers within the corroded portion of the glass [4-6]. Figure 1 is a SIMS depth profile for a simulated nuclear waste glass after it has been buried in salt in contact with 304L stainless steel for 5 years. In addition to the total depth of corrosion, the composition on a submicron scale can be evaluated. Cations such as Li and Na are leached to considerable depths while cations such as Fe and Al are concentrated in the surface layers. Another interesting feature of this profile is the influx of H, Mg and Cl from the surrounding environment. Analyses such as this can provide the archaeologist with valuable information about the burial environment of the artifact.

The extensive data gathered from the simulated nuclear waste field studies (Chalk River [7], Stripa [8-11], Materials Interface Interaction Test at the Waste Isolation Pilot Plant (MIIT/WIPP) [6,12-14], Mol [15]), numerous laboratory studies of many types of glasses, and studies of natural glass analogs and ancient artifacts have enabled researchers to put together several glass corrosion models.

GLASS CORROSION MODELS

Earlier models concentrated on simple glass systems. The two generally accepted mechanisms of corrosion for simple glasses (two components), ion exchange and matrix dissolution, can be modelled by

$$Q - at^{1/2} + bt \qquad (1)$$

where Q = amount of alkali extracted from the glass
t = exposure time
a,b = experimentally determined constants that depend on glass composition, temperature, other environmental conditions and the element of interest.

Both terms in equation (1) contribute to alkali extraction while only the second term contributes to silica extraction. In the early stage, ion exchange controls the corrosion process and the removal of alkali ions is dependent on the square root of time. This

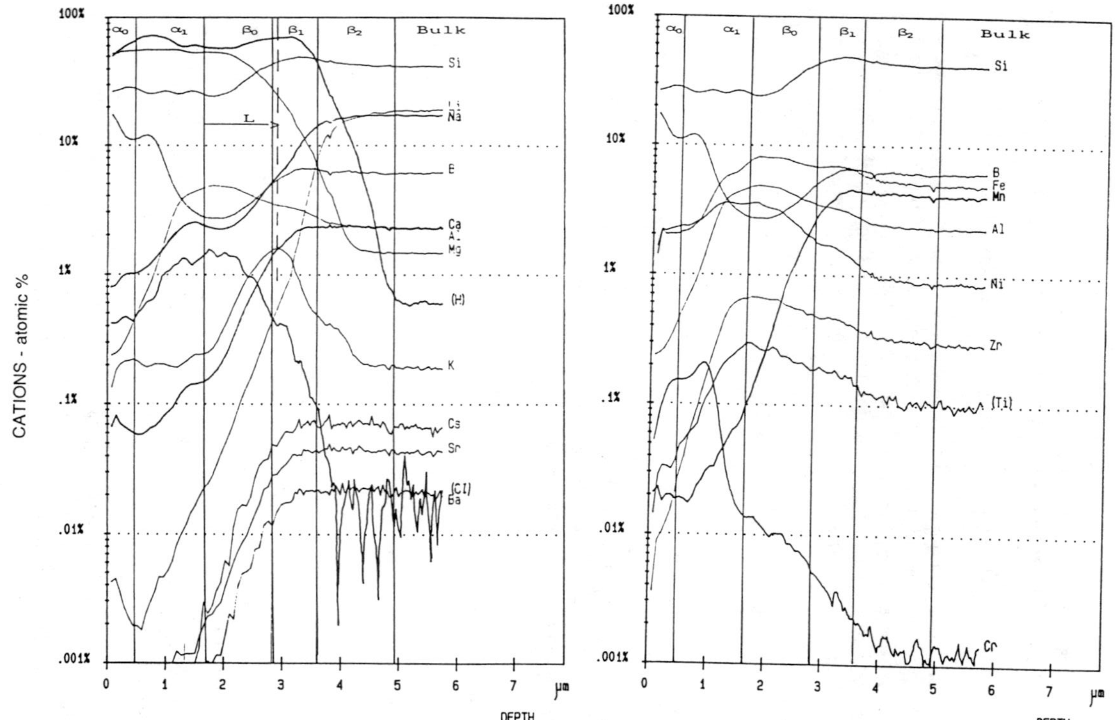

Figure 1. SIMS profile of a glass/metal interface after 5 years burial in salt at the WIPP/MIIT site. H and Cl are not included in the % cation calculations and their profiles are not quantitative. (KEY: α_0 = salt precipitation layer; α_1 = glass precipitation layer; β_0 = principle interaction layer; β_1 = gradient layer; β_2 = interdiffusion layer; L = the effective leached layer depth, begins at the right edge of the precipitated layers and continues to apex of the K peak (thought to mark the reaction front)).

mechanism results in the formation of an alkali-depleted (silica-rich) surface layer with properties that are substantially different from the bulk glass. As the thickness of the surface layer increases, the rate of alkali extracted by ion exchange decreases. Simultaneously, alkali ions are removed from the glass by the slow dissolution of the matrix (silica) which is linearly dependent on time (second term in equation (1)). In the early stage of corrosion the rate of alkali removal due to matrix dissolution is insignificant compared to its rate of removal via ion exchange. However, as the surface layer continues to grow, alkali release due to ion exchange becomes less important and, in the advanced stage, matrix dissolution controls the corrosion process.

When a glass that has undergone ion exchange is removed from the corrosive environment and allowed to dry, dehydration of the surface layer can occur. If the ion exchange is extensive and the surface layer is fairly thick, dehydration can lead to exfoliation of the layer as shown in Figure 2 (22).

The nuclear waste glass composition is complex and new models were developed to explain its corrosion behavior. Several proposed models are discussed briefly below. The interested reader is referred to the literature for more detailed discussions [16-21].

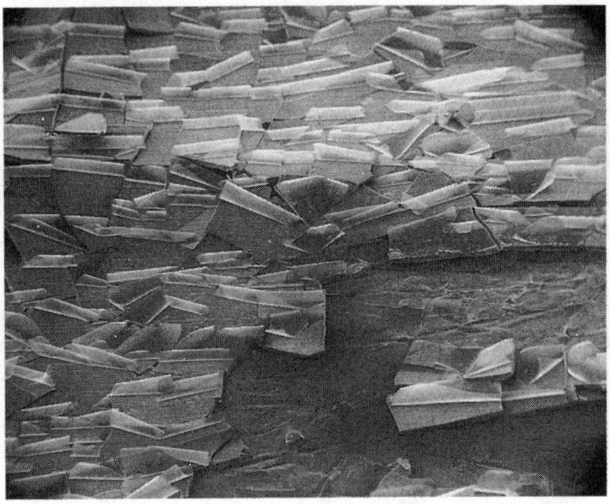

Figure 2. SEM of a corroded 30 mol% Na_2O-10 mol% Ca-60 mol% SiO_2 glass illustrating the exfoliation of the alkali-depleted surface layer (magnification 55x).

SRL Leachability Model[1]. The SRL Leachability Model was developed by R.M. Wallace and G.G. Wicks to describe the corrosion behavior of the complex SRS nuclear waste glass in an aqueous environment [2,16]. It is based on a 3-stage corrosion process as illustrated in Figure 3. In the initial stage of corrosion, water diffuses into the glass and an ion-exchange process takes place between the modifier ions in the glass and protons from the water. Matrix dissolution (Stage II) occurs at intermediate times and is controlled by the pH of the leachate. The novel feature of the model is the third stage. This stage, surface layer formation, can occur at intermediate or long times. Surface layers are formed from the precipitation of nearly insoluble compounds. The surface layers can act as a barrier to further corrosion, or have little or no effect on the continuing corrosion process depending on the chemistry and properties of the layers and the environment in which the sample is placed. All three stages have distinct features that can be readily detected with SIMS. The correlation between Figure 1 and the three stages of corrosion shown in Figure 3 can be explained as follows. Stage I occurs during the initial stages of leaching and is identified by the substitution of H for Li and Na. The β_2 zone of the SIMS profiles most closely resembles this process. Stage II, matrix dissolution, would most closely be identified with the β_0 and β_1 zones. After extended periods of time, Stage III, surface layer formation, has features representative of each the layers, with the α layers representing the two precipitation zones (α_1 = glass precipitation layer, α_0 = salt precipitation layer)

Stages II and III can be expressed mathematically by the following equation [16]:

$$L + (\beta/2)L^2 - KC_s t \qquad (2)$$

where L = amount of silica extracted from the glass
β = constant that is dependent on the diffusion of silica through the layer
K = reaction rate constant
C_s = saturation concentration of silica in the leachate
t = exposure time.

Some important features and assumption of equation (2) should be noted. The equation predicts the extraction of silica (not alkali) from the glass. The equation is valid only for an open system in which the corrosion products are not allowed to accumulate. It is understood that ion exchange does occur but it has negligible effect on silica extraction as long as corrosion products are not allowed to accumulate and cause pH drifts. It should also be emphasized that the surface layers accounted for in the model are due to precipitation reactions and not ion exchange. The equation predicts that the rate of silica extraction will be linearly dependent on time in the early stage and root time dependent in the advanced stage.

[1] Savannah River Laboratory Leachability Model (SRL), R.M. Wallace and G.G. Wicks, Westinghouse Savannah River Co., Aiken, SC 29808

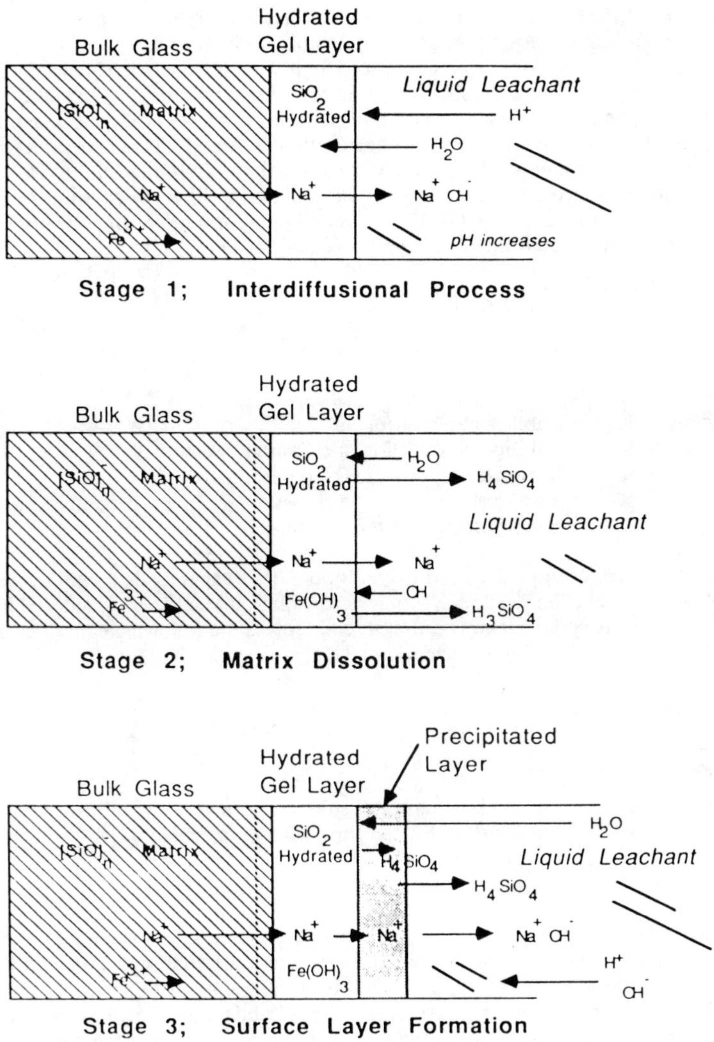

Figure 3. Schematic of surface layer formation as proposed by the SRL model [2].

Geochemical Approach. The geochemical approach for modeling nuclear waste glass was developed by Grambow [17-19]. The modeling for glass/water interactions is accomplished in two steps. The first step is reaction path calculations through the use of computer codes such as PHREEQE, EQ6 or DISSOL. Reaction path modeling describes the transfer of glass constituents to solution or solid reaction products as a function of the amount of glass dissolved per unit volume of solution. The parameters include glass/solution composition, pH, Eh, temperature, glass surface area to solution volume, pCO_2, and flow rate. For the model to be realistic, the formation of metastable phases along the reaction path must be considered along with leachate concentration data and calculated saturation states. The second step, time dependence, is carried out with the assistance of the GLASSOL computer code after an accurate reaction path has been established. For example, the PHREEQE code provides input to the GLASSOL code in the form of activities of ortho-silicic acid and experimentally determined parameters (equation 3) [17].

$$K_{Si} - K^{*} \left(1 - \frac{r_{fin}}{k_{+}}\right) \tag{3}$$

where K_{Si} = stability constant for Si
K^{*} = stability constant for rate limiting reaction
r_{fin} = final rate, experimentally determined
k^{+} = rate constant for the forward reaction, dependent on glass/solution compositions and temperature.

As the activities are calculated from all potential reactions that would consume silica, the effect each reaction would have on glass dissolution kinetics is considered implicitly via the GLASSOL program. The GLASSOL code solves the following equation for time for each step along the established reaction path [17].

$$r_{m} - k_{+}(1 - (a_{HSiO_4}/K^{*})) \tag{4}$$

where r_m = rate of glass matrix dissolution normalized to surface area ($gm^{-2}d^{-1}$)
k_+ = rate constant for the forward rate of reaction ($gm^{-2}d^{-1}$)
a_{HSiO4} = activity of ortho-silicic acid
K^{*} = stability constant for the rate limiting reaction.

Figure 4a shows the results of reaction path calculations for a simulated nuclear waste glass leached in deionized water and Figure 4b shows the time dependence of glass corrosion based on the reaction path model shown in Figure 4a. The codes are flexible enough to allow the user to change parameters to test or validate a variety of experimental conditions and are complementary to the Wallace/Wicks models in that predictions of the most likely constituents in the surface layer formed during Stage III are possible.

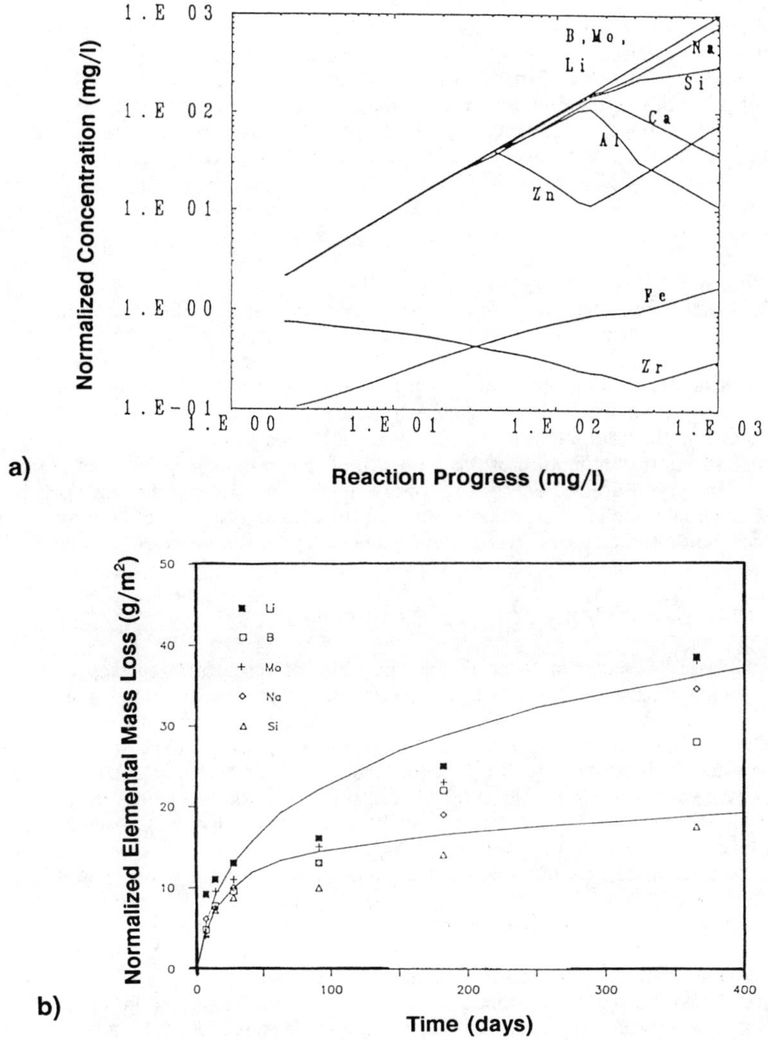

Figure 4. **a)** Plot of reaction path calculations for a simulated nuclear waste glass leached at 90°C in deionized water. **b)** Calculated time dependence of glass corrosion for the R7T7 glass based on reaction path model shown in (4a). Figure 4b shows the release of cations from the glass to solution (experimental and calculated solution data is shown). Adapted from reference 17.

Thermodynamic Approach. The thermodynamic approach to model and predict the long term durability of nuclear waste glasses was developed by C.M. Jantzen and M.J. Plodinec and is based on earlier work done by Newton and Paul on simple glass systems [20,21]. The hydration thermodynamic approach models the relative contributions of ion exchange, matrix dissolution and surface layer formation as a function of glass composition. The unique concept of this model is the empirical correlation between thermodynamics and kinetics of corrosion. In this model, the overall free energy of hydration is calculated as an additive function of the hydration free energies of the silicate and oxide component end members. This is expressed mathematically as [20]

$$\Delta G_{hyd} - \sum x_i \cdot (\Delta G_{hyd})_i \tag{5}$$

where ΔG_{hyd} = total free energy
$\Delta(G_{Hyd})_i$ = free energy change of thermodynamically most stable hydration reaction of component i at mole fraction x.

The assumption here is that the total free energy of hydration is the weighted sum of the hydration energies for the individual species ($SiO_2 + H_2O \leftrightarrow H_2SiO_3$). Figure 5 shows a linear regression plot of the normalized mass loss of Si as a function of the free energy of hydration. An interesting feature of this plot is the linear relationship between the extent of leaching and the hydration energy (The more negative the ΔG_{hyd}, the less durable the glass. For example, the tektites, obsidians and nuclear waste glasses are the most durable at 10^6 years, while the medieval glasses have a durability on the order of 10^3 years).

CURRENT UNDERSTANDING OF GLASS CORROSION

All glasses are subject to corrosion and in most cases form surface layers that influence their appearance. The structure and chemistry of these layers can be complex and there is still much debate about how protective these layers are to the underlying glass. In simple silicate glasses containing a large percentage of alkali ions, a thick porous layer can form that offers little protection. This layer forms due to ion exchange (alkali depletion) and is comprised of a highly porous and hydrated silica network. When removed from the corrosive environment the surface layer usually flakes off during drying. On the other hand, multiple layers can form on the more complex glasses during exposure. These layers form primarily as a result of precipitation reactions that occur in the solution adjacent to the glass. These glasses are chemically durable and the thickness of the surface layers are typically tens of microns versus hundred of microns for the simple glasses. Although these layers may crack when removed from the corrosive environment, they appear to be more mechanically adherent to the underlying glass.

There have been many parametric studies on glass corrosion and important factors that affect it are listed in Table I. We know, for example, how to alter the glass composition to optimize chemical durability. However, from an engineering and manufacturing perspective, composition cannot always be optimized around a single property but must be tailored to provide acceptable values of performance for a range of properties (optical, strength, hardness, density, etc.) as well as processibility.

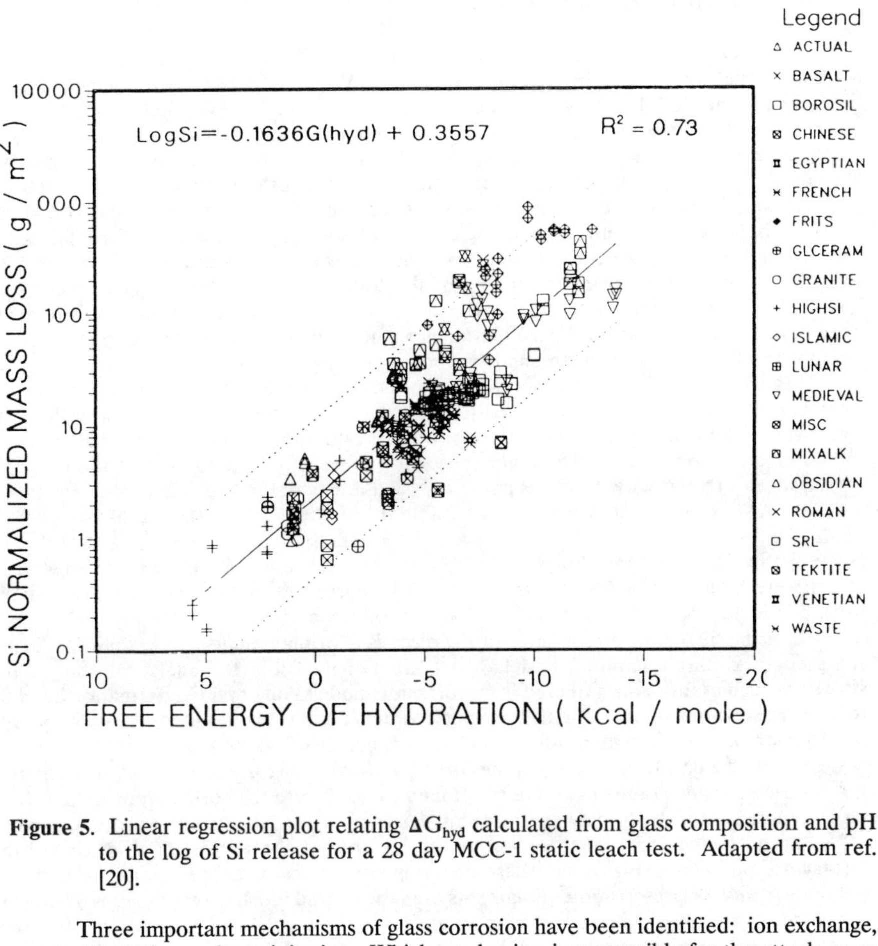

Figure 5. Linear regression plot relating ΔG_{hyd} calculated from glass composition and pH to the log of Si release for a 28 day MCC-1 static leach test. Adapted from ref. [20].

Three important mechanisms of glass corrosion have been identified: ion exchange, matrix dissolution and precipitation. Which mechanism is responsible for the attack, or as is often the case, which mechanism is dominant, depends on the factors listed in Table I. Other mechanisms, either not discussed or not yet identified, may also be important.

Together, the mechanisms, models and analytical tools can provide useful information to both the materials scientist and archaeologist regarding the nature and probable causes of glass corrosion. Furthermore, this information can be useful in designing new materials with improved corrosion resistance or minimize additional corrosion on existing materials (preservation).

ARCHAEOLOGICAL PRESERVATION

How can what we have learned from extensive laboratory and field studies on corrosion behavior be used by the preservationist? Very few artifacts are discovered that have not experienced the ravages of time. Often their true surfaces are obliterated with the by-products of corrosion. After discovery, scientists are faced with the task of deciding the best course of action for a particular piece of history. There are usually several options available: 1) leave the piece in-situ, 2) removal without restoration or special efforts at preservation, 3) removal and partial restoration/preservation, 4) removal and complete restoration/preservation efforts, and 5) removal without any restorative efforts but with plans to preserve the piece. In the first case, the find is studied in-situ, a decision is made (usually due to lack of adequate funding or the fragile nature of the piece) to leave the specimen as it was found. For the second case, a piece is removed from the site, and sent to museums or universities for possible further study, but restoration and display are not attempted. These pieces play an important role in the overall preservation process as they may be sacrificed for characterization which is often destructive. For other pieces of greater significance (cases 3-4), some degree of restoration is attempted. Restoration is a controversial topic and the debate on the merit of restoration projects is one that is likely to continue for some time. The main causes for concern appear to be that the original appearance of the artwork usually is not known and it may not be possible to truly "restore" a piece. Also, restoration may not be advisable in situations where the aesthetic value of the artifact is derived from the surface layers. This discussion also applies to artwork such as the Sistine Chapel and Medieval stained glass cathedral windows where restoration efforts are or were underway. Information gathered from corrosion studies may help restoration work by identifying the line between corrosion products and original material.

For the final case, preservation of the piece is of primary importance. Some artifacts represent a culture's national heritage and are not available for analytical study. In situations such as this, data gathered from corrosion models can provide information helpful to the preservationist in designing proper storage/display environments. With burial environment and artifact composition information, geochemical modeling could be utilized to determine the nature and extent of the corrosion that has already occurred. Furthermore, if a certain element in the environment is found to accelerate the corrosion process, it can be eliminated from the display environment, thus slowing the degradation process.

Some of the steps taken to insure the integrity of the artifact from discovery to display are outlined in Figure 6. These methods are applicable to specimens selected for restoration and/or preservation. Fragments are labeled and ideally, a fragment is available for characterization. Cleaning methods range from simple swabbing with solvents to laser ablation of the corrosion products [23]. Once corrosion products are removed, the piece is stabilized. Traditional methods for stabilization include plaster and plastic fill or mica moisture barriers. Several new methods for stabilization have been proposed, particularly for the preservation of stained glass windows; for example, sol-gel, chemical vapor deposition, ion-exchange techniques and thermal tempering. However, as with any new technique, extensive testing must be carried out to insure these techniques will not have serious detrimental effects [23].

As with modern glass, the factors that cause corrosion should be identified. However, the importance of these factors varies depending on which scenario is selected. Table I lists

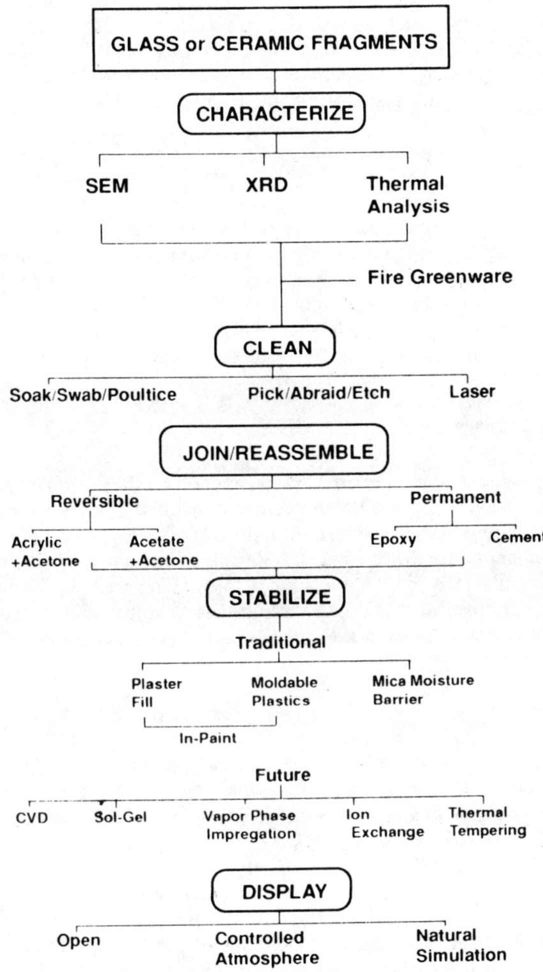

Figure 6. Flow chart of the steps involved in restoration/preservation of an artifact.

a number of causes, but clearly more work needs to be done, particularly with respect to the items under *Preservation*. Large scale testing of artifacts is limited due to the destructive nature of most analytical techniques and inadequate funding. While simulated materials may provide some answers, there is no guarantee that these materials will be entirely accurate predictors of corrosion behavior due to incomplete information on raw materials and processing techniques used at the time the piece was made [24]. If suitable simulated materials could be fabricated, plots similar to those shown in Figures 4 and 5 could be developed for various cleaning and storage environments.

SUMMARY

Experimental procedures and theoretical models have been developed that assist scientists/engineers in predicting the long term corrosion performance of glasses. The study of natural analogs and ancient artifacts has played an important role in the development and verification of these procedures and models. This exchange of information between archaeologists and materials scientists is of critical importance and communication between the two fields of study should be continued and expanded.

ACKNOWLEDGEMENTS

The authors would like to express their sincere gratitude to the following persons without whom the poster and/or written presentation of this work would not have been possible: A.R. Lodding (Chalmers University of Technology), Peter M. Fischer (Scandinavian Archaeometry Centre), Ellis Verink (University of Florida), G.G Wicks (Westinghouse Savannah River Co.), Susan T. Childs (University of Florida), and Pamela Vandiver (Smithsonian Institute). D.E. Clark and R.L. Schulz acknowledge partial financial support from Westinghouse Savannah River Co. under contract #AA-463220.

REFERENCES

1. Wicks, G.G. and Bickford, D., *Technology Review*, pp. 51-58, Nov-Dec. (1989).
2. Wicks, G.G. "Nuclear Waste Glasses: Corrosion Behavior and Field Tests," in **Corrosion of Glass, Ceramics and Ceramic Superconductors**, pp. 218-268 (D.E. Clark and B.K. Zoitos, eds.) Noyes Publications, Park Ridge, NJ (1992).
3. Clark, D.E. and Zoitos, B.K., "Corrosion Testing and Characterization," in **Corrosion of Glass, Ceramics and Ceramic Superconductors**, pp. 51-102 (D.E. Clark and B.K. Zoitos, eds.) Noyes Publications, Park Ridge, NJ (1992).
4. Lodding, A.R., "Characterization of Corroded Ceramics by SIMS," in **Corrosion of Glass, Ceramics and Ceramic Superconductors**, pp. 103-122 (D.E. Clark and B.K. Zoitos, eds.) Noyes Publications, Park Ridge, NJ (1992).
5. Lodding, A.R., E.U. Engström, Zoitos, B.K., Clark, D.E. and G.G. Wicks, "Elemental Depth Profiling of Nuclear Waste Glasses after Two Years Burial in a Salt Geology," to be published in *J. Am. Cer. Soc.*, (1992)

6. Schulz, R.L., Clark, D.E., Wicks, G.G. and Lodding, A.R., "Long Term Field Leaching Studies of Simulated Nuclear Waste Glass in Granite and Salt," **Scientific Basis for Nuclear Waste Management XV**, Vol 257, (1992).
7. Tait, J.C., Hocking, W.H., Betteridge, J.S. and Bart, G., "Field Burial Results and SIMS Analysis of the Chalk River Glass Blocks" **Advances in Ceramics**, Vol. 20; **Nuclear Waste Management II**, pp. 559-566, (D.E. Clark, W.B. White and A.J. Machiels, eds.) American Ceramic Society (1986).
8. Lodding, A.R., E.U. Engström, D.E. Clark, L.O. Werme, and G.G. Wicks, "SIMS Analysis of Leached Layers Formed on SRL Glasses During Burial," **Advances in Ceramics**, Vol. 20; **Nuclear Waste Management II**, pp. 559-566, (D.E. Clark, W.B. White and A.J. Machiels, eds.) American Ceramic Society (1986).
9. Clark, D.E., Zhu, B.F., Robinson, R.S and Wicks, G.G., "Preliminary Report on a Glass Burial Experiment in Granite," **Advances in Ceramics**, Vol. 8; **Nuclear Waste Management** (G.G. Wicks and W.A. Ross, eds.) pp. 324-336, American Ceramic Society (1984).
10. Hench, L.L., Zhu, B.F., A.R. Lodding, and L. Werme, "Stripa Burial: Time Dependence of Leaching," in **Advances in Ceramics**, Vol. 20; **Nuclear Waste Management II**, pp. 583-590, (D.E. Clark, W.B. White and A.J. Machiels, eds.) American Ceramic Society (1986).
11. Zhu, B.F., Clark, D.E., A.R. Lodding and G.G. Wicks, Two-Year Leaching Behavior of Three SRL Nuclear Waste Glasses in Granite," in **Advances in Ceramics**, Vol. 20; **Nuclear Waste Management II**, pp. 591-600, (D.E. Clark, W.B. White and A.J. Machiels, eds.) American Ceramic Society (1986).
12. Wicks, G.G. and Molecke, M.A. "WIPP/SRL In-Situ Testing Program," in **Advances in Ceramics**, Vol. 20; **Nuclear Waste Management II**, pp. 657-667, (D.E. Clark, W.B. White and A.J. Machiels, eds.) American Ceramic Society (1986).
13. A.R. Lodding, E.U. Engström, D.E. Clark and G.G. Wicks, "Quantitative Concentration Profiling and Element Balance in SRL Glass after Two Years in WIPP," in **Ceramic Transactions**, Vol. 9, **Nuclear Waste Management III**, pp. 317-333, (G.B. Mellinger, ed.) American Ceramic Society (1990).
14. Ramsey, W.G. and Wicks, G.G., "WIPP/SRL In-Situ Tests; Compositional Correlations of MIIT Waste Glasses in **Ceramic Transactions**, Vol. 9, **Nuclear Waste Management III**, pp. 257-270, (G.B. Mellinger, ed.) American Ceramic Society (1990).
15. Van Iseghem, P., Timmermans, W., Debruyn, W., Dresselaers, J. and Neerdael, B., "In-Situ Testing of Nuclear Waste Forms in an Underground Laboratory in Clay,' in **Advances in Ceramics**, Vol. 20; **Nuclear Waste Management II**, pp. 649-656, (D.E. Clark, W.B. White and A.J. Machiels, eds.) American Ceramic Society (1986).
16. Wallace, R.M. and Wicks, G.G., "Leaching Chemistry of Defense Borosilicate Glass," in **Mat. Res. Soc. Symp. Proc.**, Vol. 15, pp. 23-28, Elsevier Science Publishing Co., Inc. (1983).
17. Grambow, B., "Geochemical Approach to Glass Dissolution," in **Corrosion of Glass, Ceramics and Ceramic Superconductors**, pp. 124-152 (D.E. Clark and B.K. Zoitos, eds.) Noyes Publications, Park Ridge, NJ (1992).

18. Grambow, B. and Strachan, D.M., "Leach Testing of Waste Glasses Under Near-Saturation Conditions," **Mat. Res. Soc. Symp. Proc.,** Vol. 26, pp. 623-634, Elsevier Science Publishing Co., Inc. (1984).
19. Grambow, B., Hermansson, H.P., Björner, I.K. and Christensen, H., "Reaction of Nuclear Waste Glass with Slowly Flowing Solutions," **Advances in Ceramics**, Vol. 20: **Nuclear Waste Management II**, pp. 465-474, American Ceramic Society, Westerville, OH (1986).
20. Jantzen, C.M., "Thermodynamic Approach to Glass Corrosion," in **Corrosion of Glass, Ceramics and Ceramic Superconductors**, pp. 153-215 (D.E. Clark and B.K. Zoitos, eds.) Noyes Publications, Park Ridge, NJ (1992).
21. Jantzen, C.M. and Plodinec, M.J., "Thermodynamic Model of Natural, Medieval and Nuclear Waste Glass Durability," *J. Non-Cryst. Solids*, 67, pp. 207-233 (1984).
22. Clark, D.E. **A Durability Evaluation of Soda-Lime-Silica Glasses Using Electron Microprobe Analysis, Infrared Reflection Spectroscopy and Other Techniques,** PhD Dissertation, University of Florida, 1976.
23. Vandiver, P.B., "Corrosion and Conservation of Ancient Glass and Ceramics," in **Corrosion of Glass, Ceramics and Ceramic Superconductors**, pp. 393-430 (D.E. Clark and B.K. Zoitos, eds.) Noyes Publications, Park Ridge, NJ (1992).
24. Oppenheim, L.A., **Glass and Glassmaking in Ancient Mesopotamia**, Corning Museum of Glass Press, Corning, NY (1970).

INVESTIGATIONS BY GLASS SENSORS ON THE CORROSIVE ENVIRONMENTAL CONDITIONS AT STAINED GLASS WINDOWS WITH PROTECTIVE GLAZINGS IN EUROPE

JOHANNA LEISSNER AND DIETER R. FUCHS
Fraunhofer-Institut für Silicatforschung, Neunerplatz 2, D-8700 Würzburg, F. R. G.

SUMMARY

At 15 prominent churches and cathedrals in 7 European countries the environmental corrosive stresses at the stained glass windows were assessed by glass sensors. Special emphasis was given to the protective properties of different types of protective glazings. The obtained total stress level data, based on MI-glass sensor measurements, help to conclude on the appropriate choice of the ventilation systems and give information about specific environmental impacts. The results indicate that internally ventilated protective glazings have a higher protective effect than externally ventilated systems. The nonventilated type of protective glazing even raised the corrosive stresses for the stained glass windows.

INTRODUCTION

Medieval stained glass windows belong to the most important cultural heritage of Europe. Like other objects they are acutely menaced by environmental corrosive influences. A widely accepted protection measure are protective glazings. Although many different types are installed throughout Europe, there are nearly no results available about their real effectiveness and few standards for new installations. A survey of temperature and humidity is not sufficient, as glass corrosion is also influenced by noxious gases, dust, microorganisms as well as synergetic interactions [1]. Up to now the knowledge about the precise conclusions between these influences and the resulting impact on the material is still fragmentary. To cope with this restriction, an alternative and integrating concept for total stress level assessments was asked for. Such a solution is offered by glass sensors [2] based on low durable potassium-calcium silicate glasses with high sensitivity against corrosive stresses. The easy-to-handle technique is able to detect the combined impact potential. Due to the sensitivity of the dosimeter glass, judgements about the corrosive stress and thus further damage risks could be made within 12 months. The corrosion progress of the sensors is determined by infrared spectroscopy and checked by microscopy.

FEATURES AND RESULTS OF THE SENSOR STUDY

MI-type glass sensors were placed in a medium height of the windows. Corresponding to CVMA nomenclature the positions are denoted as follows: position 1: front side of historic glazing; position 2: reverse side of historic glazing; position 3: front side of protective glazing; position 4: reverse side of protective glazing.

The glass sensor method is described extensively elsewhere [3]. The difference in intensity (ΔE) of a suitable OH absorption band in the IR spectra of a sample before and after exposure directly corresponds to the degree of corrosion of the sensor and thus to the stress level at the exposure point. In Table I and II the investigated objects are listed, additional data for the interpretations were taken from previous studies [4, 5].

Table I: Features of internally ventilated glazings

object	remarks	window	protective glazing	Delta-E 1	Delta-E 2	Delta-E 3	Delta-E 4	O.P.E.N
St. Jans Gouda (NL)	marine area; city center with heavy traffic; heated church	S	1905 * slots on top/bottom** 2 cm ; 5-6 cm distance***	0,04	0,025		0,21	12,55
Herr. Kerk De Rijp (NL)	rural marine area; low traffic; very humid inside church	S 11	1989 slots on top and bottom 1 cm; 4 cm distance	0,05	0,02		0,17	14,82
St. Dionys Esslingen (D)	industrial area; heavy traffic; heated church	S c8	1976 3,5 x 4,5 cm slots; 5 cm distance	0,025	0,025		0,135	8,14
		N b8		0,02	0,035		0,175	0,5
Klosterkirche Kappel am Albis (CH)	rural area; sometimes heated church	s VII	1986 slots on bottom 1 cm; 3,5 cm distance	0,015	0,02		0,1	4,6
		n VI	1950 only few mm slots on top; 1,5 cm distance	0.025	0,015		0,16*	13,1
St. Bartlmä Friesach (CH)	rural area; heavy traffic; nonheated church	nII 4a	1978/79 slots on top tilted and on bottom 7 cm; 7 cm distance	0,015	0,02	0,02	0,155	5,36
Klosterkirche Königsfelden (CH)		SV		0,03	0,01		0,07	15,16
		Nord LH		0,025	0,025		0,125	8
Brunnenhaus Heiligenkreuz (A)	rural area; extremly humid (through fountain); nonheated church	N IV 1c	1987 slots on top tilted and on bottom 3 cm; 5 cm distance	0,035	0,02	0,02	0,23	13,4
Mariendom Erfurt (D)	city center; heavy traffic; high level of pollution; nonheated church	sII B10	1982/83 50 x 5 cm slots on bottom; 5-7 cm distance	0,09	0,145		>1	2,4
		n VI B10	1989 40 x 5 cm slots on bottom; 5-7 cm distance	0,115	0,12		0,605	7,5
		n II B10	1981/82 50 x 5 cm slots on bottom; 5-7 cm distance	0,1	0,27		0,8	-10,4
Keyenberg (D)	plains; rural area with heavy farming; heated church	nII c	1990 slots on top 1 cm and on bottom 1,5 cm	0,08	0,03		0,325	15,3

Table I: continuation

object	remarks	window	protective glazing	Delta-E 1	Delta-E 2	Delta-E 3	Delta-E 4	O.P.E.N
Stefansdom Vienna (A)	city center; heavy traffic; industrial area; heated church	s II 4b	1988 slots on top tilted and on bottom 7 cm; 8 cm distance	0,035	0,02	0,06	0,15	12,9
Magdalenen-kirche Judenburg (A)	industrial area; heavy traffic; nonheated church	s II 4b	1988 slots on top tilted and on bottom 7 cm; 7 cm distance	0,03	0,01	0,015*	0,085	15,5
Wallfahrts-kirche Straßengel (A)	industrial area; nonheated church; high level of soot	O I 3c	1975 slots on top tilted and on bottom 7 cm; 7 cm distance	0,055	0,04	0,04	0,11	9,0
Köln Dom (D)	plains; industrial area; heavy traffic; nonheated church;	OI	19..? slots on top and on bottom < 0,5 cm 7,5 cm distance	0,025	0,015		0,25	13,6
		S		0,03	0,02		0,23	11,7
Mönchenglad-bach, Hardt (D)	plains; industrial area; medium traffic; heated church	nII	1987 slots on top 1,5 cm and on bottom 0,7 cm; 4,5 cm distance	0,015	0,02		0,28	3,1
Mönchenglad-bach, Hermges (D)	plains; industrial area; heavy traffic; heated church	S	1988 many trees in front of glazing; 4 cm distance	0,025	0,015		0,16	13,5

Table II: Features of nonventilated, externally and mixed ventilated glazings

object	remarks	window	protective glazing	Delta-E 1	Delta-E 2	Delta-E 4	O.P.E.N.
Großmünster Zürich (CH)	industrial area; heavy traffic; heated church	SX	1970 nonventilated; 2 cm distance	0,02	> 1	0,37	-507
		O HId	1970 external ventilation; 6 cm distance	0,005	0,065	0,065	-120
Altenberg Dom (D)	highlands; rural area; humid valley; partly heated church	nIV	1968 mixed ventilation (internal, external); 2,5 cm distance	0,02	0,01	0,13	12,4
		nXV	1972 mixed ventilation; 2,5 cm distance	0,01	0,02	0,185	-1,3
Mönchenglad-bach, Münster (D)	plains; industrial area; heated church	OI	1977 mixed ventilation; 4 cm distance	0,025	0,015	0,255	13,4
Keyenberg (D)	plains; rural area with heavy farming; heated church	nII b	1990 external ventilation; 4 cm distance	0,08	0,24	0,31	-17,4

DISCUSSION

Comparison of indoor and outdoor stress levels

Indoor environments of course are influenced by the local outdoor environment. But the degree of intrusion is modified by various factors, e. g. the specific air conditioning features, the buildings heating system or the frequency and magnitude of visitors coming inside. As shown in Figure 1, there is a general trend of correlation, but also unexpected results are concluded from the glass sensor values.

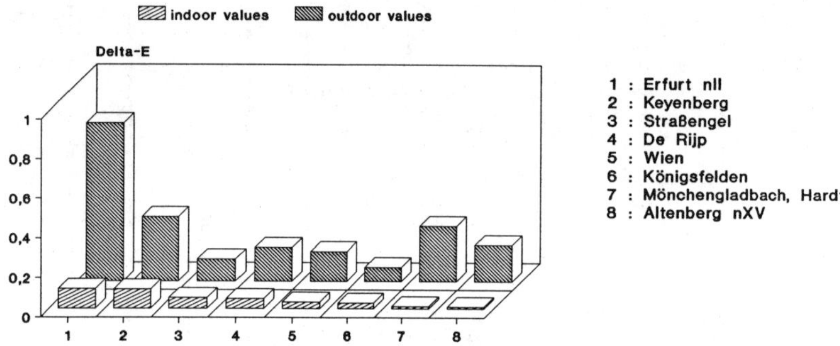

Figure 1: Comparison of indoor/outdoor ΔE values for selected sites

A typical example for a severe influence by the local outdoor environment is Erfurt where very high stress levels are found at all three positions. This result can be well explained by the extremely high air pollution through unfiltered industrial exhaustive gases, lignite firing and thus high sulfur dioxide immission.

The situation at two windows in Kappel am Albis, Switzerland, is quite different, but again the governing effect of the local outdoor stress level is evident. The ΔE values at the outdoor as well as the indoor positions are very low. The church is situated at the Highlands in a rural area without industrial plants and heavy traffic.

On the other side the sensor data obtained at Mönchengladbach, Hardt, prove that indoor environmental stresses are not always directly proportional to the outdoor environmental situation. The ΔE values found at outdoor positions were relatively high (0.3) which makes sense, as the church is situated at an industrial area with medium heavy traffic. Nevertheless, extremely low indoor values at the nII window were registered which is not yet understood at the moment. Anyhow, the results show that quantitative assessments of indoor stress levels give a saver information on the object risks compared to normal outdoor monitoring only.

The preceding results can be explained by the strong effect of air pollution on the total stress. The following example emphasizes that also humidity itself leads to significantly corrosive damages. The pump room of the Zisterzienserstift at Heiligenkreuz, Austria, has a very humid environment and is situated at the rural highlands. The ΔE values (0.230) at position 4 are relatively high compared to a similar pollution situation at the Klosterkirche Königsfelden, Switzerland, where ΔE values of 0.07 were found.

The location of Keyenberg, Germany, is characterized as a rural area in the plains but with very intensive farming. The ΔE values found at the outdoor positions are very high (~ 0.32), even compared to the Großmünster, Zürich (located in the city center with heavy traffic conditions). Another typical example for unusual, but yet very corrosion-effective influences on glass corrosion is given at the pilgrimage church Straßengel, Austria. The extremely high ΔE values at position 1 are due to intensive candle burning and thus increased soot deposition. All these results prove that neither the humidity nor the usually discussed pollutants (SO_2, NO_x) must be studied alone and that assessments by "corrosion dosimeters" like glass sensors are required if the real total stress level is asked for.

Efficiency of protective glazings

Within this study four different types of protective glazings are investigated. During the last years internally ventilated protective glazings were preferred. Only very few samples of nonventilated and externally ventilated systems exist or will be installed. In the Cologne area at three objects a mixed ventilation system was installed with an inside slot on top and bottom and an additional outside slot on bottom.

Overall protective effect numbers (O.P.E.N.)

To perform assessments of the achieved protective effect, so-called "overall protective effect numbers", concluded from the glass sensor data, may be proposed as a criterion:

$$\text{O.P.E.N.} = \left[\frac{\Delta E\,(\text{position 4}) - \Delta E\,(\text{position 2})}{\Delta E\,(\text{position 4})} + \frac{\Delta E\,(\text{position 1}) - \Delta E\,(\text{position 2})}{\Delta E\,(\text{position 1})} \right] \times 10$$

Efficiency of internally ventilated protective glazings

For protective glazings with an internal ventilation the protective effect is contingent upon the indoor environment, a sufficient air circulation and on the distance between original panel and outside glazing. In general, it can be assumed that the protective glazing which has the lowest ΔE values at position 2 (the reverse side of the historic glazing) is the best working one. But the ΔE values of all positions have to be taken into account. There are constructions of protective glazings with high indoor and outdoor values and yet low ΔE values at position 2. These protective glazings have the highest O.P.E.N. and therefore the highest efficiency.

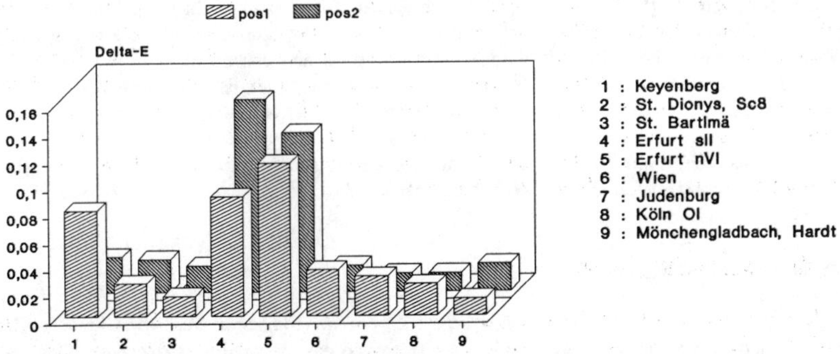

Figure 2: Comparison of ΔE values at position 1 and 2 for selected sites

ΔE (position 2) < ΔE (position 1) << ΔE (position 4)

The installed glazings at Keyenberg (nIIc), Judenburg (SII) and Königsfelden (XV) have the highest O.P.E.N. indicating the good efficiency of those glazings and a lower corrosive stress at position 2 compared to position 1 (see Figure 2). Especially the situation at Keyenberg illustrates best what improvements for historic stained glass windows can be achieved with a proper construction of protective glazings. The protective glazing is capable of reducing the strong corrosive risks given both at the indoor position (ΔE value 0.08) and outdoor position (ΔE value 0.325) for the interspace position 2 (ΔE value 0.03).

Very efficient protective glazings are also installed at De Rijp (O.P.E.N. 14.8), Mönchengladbach, Hermges (O.P.E.N 13.5), window OI in Cologne (O.P.E.N. 13.6), Heiligenkreuz (O.P.E.N. 13.4), window nVI in Kappel am Albis (O.P.E.N. 13.1), and Vienna (O.P.E.N. 12.9). These protective glazings also achieve a significantly lower corrosive stress at position 2 compared to indoor and outdoor values.

ΔE (position 2) ~ ΔE (position 1) << ΔE (position 4)

In Esslingen (Sc8), in Königsfelden (N LH), in St. Bartlmä, in Kappel am Albis (sVII), and in Erfurt (nVI) the protective glazings decrease the outdoor corrosive stresses as well. But at positions 1 and 2 comparable corrosive environments work at the stained glass window. Therefore, their efficiency is in the medium range (O.P.E.N. 5 - 8).

[ΔE (position 2) > ΔE (position 1)] << ΔE (position 4)

The efficiency of the internally ventilated types within this group have very low or even negative O.P.E.N. which stand for a stress level higher (for negative numbers much higher) at position 2 than at position 1.

Examples are the protective glazings from Erfurt. At the investigated windows the strong influence of special circumstances (like a construction site near a window) is evident (see Table I). The nII and sII windows have very low O.P.E.N. (-10.4 and 2.4) although the corrosive stress levels at position 1 are nearly the same for all three windows (ΔE ~ 0.1). The nII window possesses the least efficient protective glazing of all investigated internally ventilated objects due to the special construction of its protective glazing (restricted air circulation through bottom slots, large scale panel with cracks allowing outdoor air exchange [6]. During the exposure time of the sensors, some construction and restoration work inside the church near the nII window was going on and caused lots of dust and dirt raising also the corrosive attack. On the other hand, the considerably higher efficiency observed at the nVI window is a good demonstration that a revised construction of a protective glazing will lead to improved conditions for the satined glass windows. The revision work included improvement of the air circulation through bottom slots and a very tight geometrically patterned glazing with a lead structure.

Nonventilated protective glazings

Only one example of a nonventilated type of protective glazing was investigated. The obtained data, however, show very clearly that this type has the worst protective effect of all investigated objects. At position 2 an extremely high ΔE value (> 1!) was found. The surface of the glass sensor was heavily corroded and this could be seen even macroscopically. This implies that the corrosive stress is significantly risen and is even higher than the outdoor stress. As a consequence, an additional endangering of the stained glass window must be taken into account when only protected by a nonventilated system.

Efficiency of externally ventilated protective glazings

Not that drastically bad, but still unsatisfactory, is the condition for the historic stained glass windows with an externally ventilated protective glazing as installed at the Großmünster in Zürich (O HId) and in Keyenberg (nII b). In the first case the reverse side of the stained glass window is attacked by a stress identical to the corrosive environment at the outdoor position (see Figure 3). Slightly better is the situation at the window nII b in Keyenberg, but still the corrosive impacts on the reverse side of the historic window are very high compared to the outdoor positions. In controversary to these findings, first results of microclimatic measurements like air flows in the interspace, humidity control and daily fluctuations in temperature indicate that the construction details of this type of protective glazing (Keyenberg) seem to be better than those of the internally ventilated type [7]. Again this difference in conclusion shows that results from traditional "factors" monitoring are not sufficient, as the existing total corrosion risk (displayed by the glass sensor response) must be supposed to be closer to the real stained glass endangering.

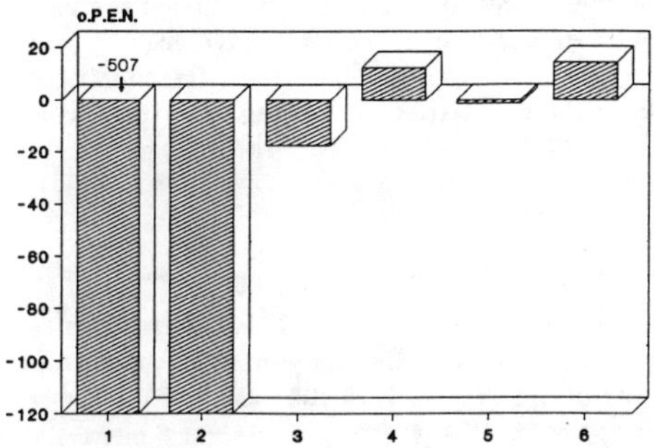

1 : Großmünster Zürich, nonventilated
2 : Großmünster Zürich, external ventilated
3 : Keyenberg, external ventilated
4 : Altenberg Dom, nIV, mixed ventilated
5 : Altenberg Dom, nXV, mixed ventilated
6 : Mönchengladbach Münster, mixed ventilated

Figure 3: Overall protective effect

Efficiency of mixed ventilated protective glazing

The situation for the windows in Altenberg (nIV) and in Mönchengladbach, Münster (OI), which are protected with a mixed ventilation system, is characterized by very high O.P.E.N.s. Also here the first results of microclimatic measurements were interpreted in terms of less effective construction [7], which cannot be agreed now according to the glass sensor data.

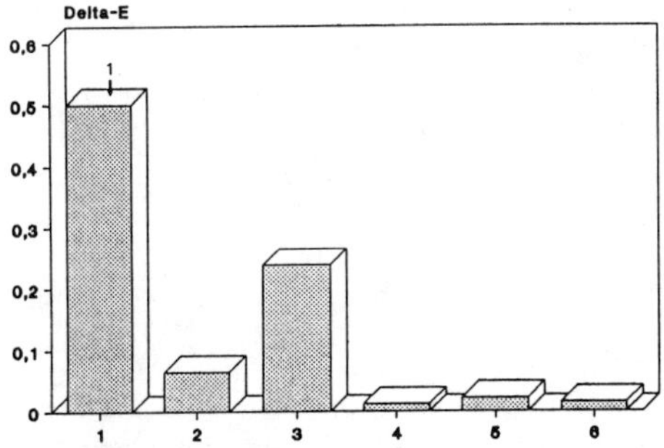

1 : Großmünster Zürich, nonventilated
2 : Großmünster Zürich, external ventilated
3 : Keyenberg, external ventilated
4 : Altenberg Dom, nIV, mixed ventilated
5 : Altenberg Dom, nXV, mixed ventilated
6 : Mönchengladbach Münster, mixed ventilated

Figure 4: ΔE values at position 2

The ΔE values found at the nXV window in Altenberg imply that the protective glazing is less effective since at position 2 the ΔE values are slightly higher than at position 1. But the corrosive impacts which work on the outdoor positions are significantly reduced compared to situations without a protective glazing.

CONCLUSIONS

The information assembled and discussed here proves the potential of glass sensor measurements in the field of estimating the total stress levels at stained glass windows. The results showed clearly that the corrosive risks can be reduced with suitably constructed protective glazings. Internally and mixed ventilated types of protective glazings seem to be first choice. Although large differences in the protective effect of these types occurred, the data are too sparse to draw definite conclusions about their reasons. Here in the future more detailed sensor and joint microclimatic and pollution studies are requested.

Considerations of the very low protective effect of externally ventilated systems indicate that the reverse side of the original is exposed to corrosive environments from which pieces of art should be protected. In the case of nonventilated systems the conditions for the stained glass windows are worse than without a protective glazing. This type leads to an enforced endangering for the originals.

ACKNOWLEDGEMENT

The authors wish to thank C. Cronauer, Th. Böhm and H. Römich for their valuable assistance. This work has been funded by the Bundesministerium für Forschung und Technologie and the Umweltbundesamt (Germany) and various ecclestiastical and conservation sponsors, to whom acknowledgement is made.

REFERENCES

[1] D. R. Fuchs, H. Römich, P. Tur, and J. Leißner, Forschungsbericht UFOPLAN-Nr. 108 07 005/03, März 1991. Ed.: Umweltbundesamt, Bismarckplatz 1, D-1000 Berlin 33.

[2] D. R. Fuchs, H. Patzelt and G. Tünker, Deutsches Patent No. 38 05 495 A1; U. S. Patent No. 07/313,455.

[3] D. R. Fuchs, H. Römich and H. Schmidt, Mat. Res. Soc. Symp. Proc. Vol. 185 (1991) pp. 239 - 251.

[4] J. Leißner and D. R. Fuchs, Congreso Internacional de Rehabilitaction del Patrimonio Arquitectonico y Edificacion, Islas Canarias 1992 (in print).

[5] J. Leißner and D. R. Fuchs, XVI International Congress on Glass, Madrid 1992 (in print) .

[6] M. Jähn (private communication).

[7] S. Oidtmann (private communication).

EVALUATION OF THE EFFECTIVENESS AND THE POTENTIAL DAMAGE OF CLEANING METHODS FOR THE RESTORATION OF STAINED GLASS WINDOWS

HANNELORE RÖMICH AND DIETER R. FUCHS
Fraunhofer-Institut für Silicatforschung, Neunerplatz 2, D-8700 Würzburg, F. R. G.

ABSTRACT

A comparative study on the effects of restorational cleaning methods for stained glass windows was carried out using corrosion sensitive model glasses. Typical deterioration phenomena like crusting, leaching and microcracking could be simulated using accelerated weathering in climatic test cabinets. The sensibilization of such samples caused by cleaning was assessed by infrared spectroscopy and light microscopy after re-weathering.

1. INTRODUCTION

Historic stained glass windows show a variety of corrosion phenomena like crusting or pitting which change the appearance of these works of art enormously. Encrustations of dust and grease with corrosion layers (like gypsum and syngenite) have to be removed from the glass surface for recalling the colourful transparency and as a pretreatment measure for following conservation procedures.

There ist no general recipe for cleaning medieval glass. Traditional cleaning techniques include a variety of "dry" (e. g. brushes, scalpel) and "wet" (e. g. organic solvents, water) methods [1]. Only few scientific investigations are published in this field [2 - 7]. Methods advocated by one author are discarded as unacceptable by another. When chosing the right method for a given problem the conservator has to take into account not only the effectiveness of the technique but also to consider the potential of damage to the sensitive glass, which might occur only years later.

The aim of this project was to investigate the corrosion behaviour of comparable glass samples after cleaning with different techniques in order to obtain basic data for their recommendation in glass conservation.

2. EXPERIMENTAL

Studies of cleaning effects on originals are limited by various reasons. Almost every segment of glass in a window is different from the others concerning its chemical composition and its current surface condition. Evaluation of the risk potential requires observation times of several years. Time-lapse testing in accelerated weathering chambers would shorten the test period but still include an irreversible damage to the glass at the same time. Examples for heavily corroded medieval glasses are shown in Figure 1.

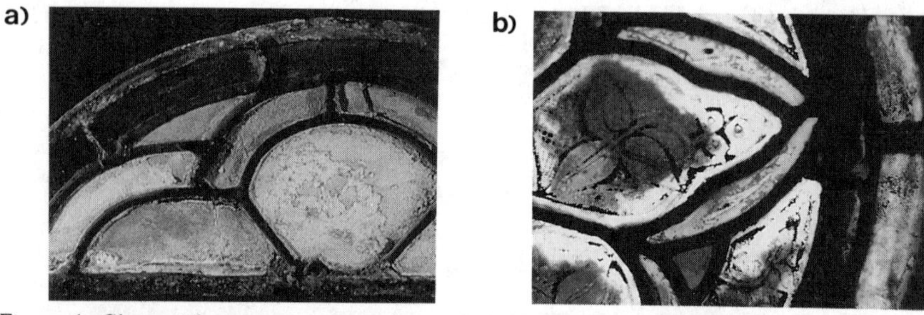

Figure 1: Glass with corrosion crust a) in reflected light (Germany 14th century) and b) in transmitted light (Germany, 13th century)

Corrosion sensitive model glasses present a practicable alternative as substitutes for originals in cleaning experiments [8 - 10]. The chemical composition of the glasses can be chosen in order to simulate corrosion phenomena like encrustations, comparable to those of originals. These simulation materials can be prepared with constant quality and in sufficient quantity and thus guarantee comparable results after extensive testing series of different cleaning methods.

2.1. Model Glasses

In order to clarify the surface effects of various cleaning methods corresponding to the chemical composition of the glasses, three glass types were chosen with different corrosion behaviour (Table I). The preparation of the glass samples (melting, cutting, fire-polishing) is described elsewhere [10].

Table : Composition of model glasses (wt. %)

	SiO_2	Na_2O	K_2O	MgO	CaO	Al_2O_3	P_2O_5
MI	48.0	3.0	25.5	3.0	15.0	1.5	4.0
M2	57.1	-	22.5	-	20.4	-	-
MIII	60.0	-	15.0	-	25.0	-	-

In relation to medieval glass the model glasses are enriched with potassium and thus more sensitive. Various outdoor as well as laboratory weathering tests were carried out to study the comparability of the simulation material with the original [9, 10].

2.2. Corrosion Phenomena

Cracked gel layers, porous crusts and pitting can be simulated an model glasses [9]. For all experiments described below accelerated weathering in a climatic chamber [10] has been used. Examples for corrosion phenomena on originals and on model glasses are given in Figures 2 und 3.

a) b)

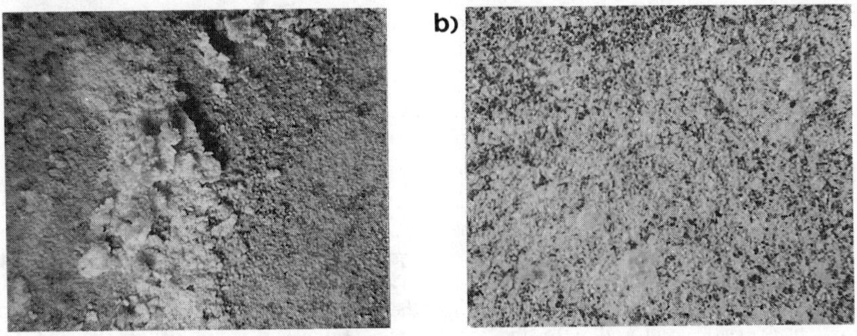

Figure 2: Corrosion crust a) on a medieval glass and b) on a model glass (M2.0 after 1 day weathering in a climatic chamber [10]). Width of each photograph: 5 mm

a) b)

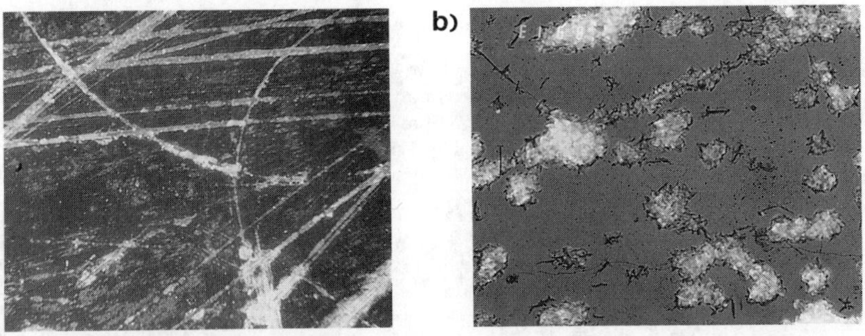

Figure 3: a) Medieval glass with corrosion along by traces of a former restoration, b) model glass with corrosion in scratches of a mechanical treatment (width of micrographs: 500 μm).

2.3. Analytical Methods

The surface condition of the samples (corrosion crust, cracks in the gel layer, scratches from cleaning procedures) was checked with a polarizing microscope (magnification up to 200 x).

The corrosion rate of the glasses was determined by infrared analysis. The difference in the intensity (ΔE) of the OH absorption bands at 3300 cm^{-1} of a sample before and after treatment indicates the degree of corrosion of the glass [10].

2.4. Working program

The successive steps of the experiments described below can be summarized as follows:

- preparation of the model glasses
- simulation of corrosion phenomena (e. g. porous encrustations on the glass M 2.0 are obtained after one day weathering in a climatic chamber [10])
- "Cleaning" of the samples (e. g. with different mechanical procedures)
- accelerated weathering in a climatic chamber (one or two days).

Each step in the working program is accompanied by analytical investigations (light microscopy and IR spectroscopy). The increase in the corrosion rate by weathering (determined by the IR method [10]) indicates the sensibilisation of the glass surface after cleaning.

3. RESULTS AND DISCUSSION

Mechanical cleaning by hand, using scalpels, glass fibre brushes, smooth bristle brushes and emery paper has been used to remove deposits from the surface of three model glasses with different chemical compositions (see Table I). These porous encrustations were produced by one day accelerated weathering: the very sensitive glass MI formed the thickest corrosion crust compared with the thinner but still dense cover of crystals on M2.0, whereas the quite stable glass MIII showed only an incomplete corrosion layer.

3.1. Surface condition after cleaning

The traces of the cleaning procedure have been investigated by light microscopy. Figure 4 shows the results for MI glass. The smooth bristle brush and the scalpel leave some residual crystals on the glass. The glass fibre brush scratches the gel layer in some areas whereas the emery paper attacks the glass by grinding it thoroughly.

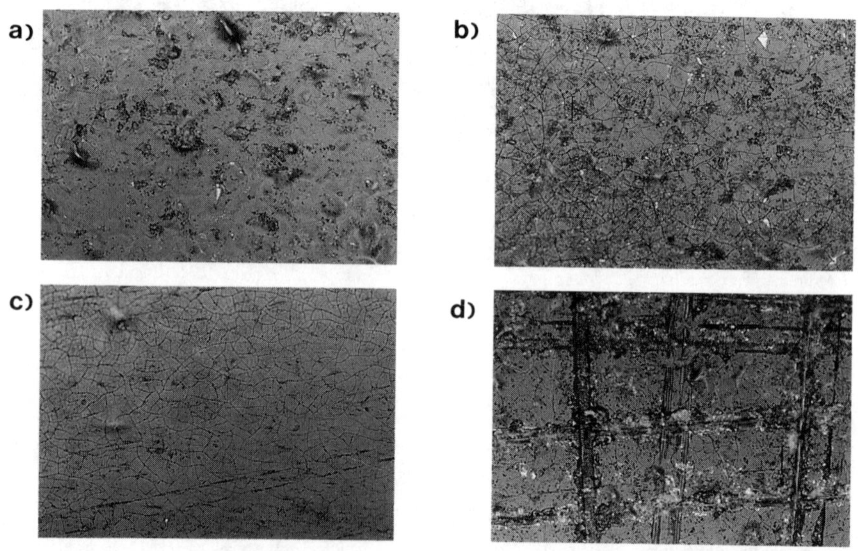

Figure 4: MI glass after mechanical cleaning with a) bristle brush, b) scalpel, c) glass fibre brush, d) emery paper (width of mirographs: 500 μm).

The less corrosion sensitive and more scratch resistant glass M2.0 shows less traces after the same cleaning procedures. On MIII glass there are visible traces only after treatment with emery paper.

These results emphasize the conservators responsibility to make distinctions when cleaning a medieval stained glass window, consisting in general of several types of glasses.

3.2. Risk potential

The sensibilization of the glass surface after cleaning has been examined by assessment of the corrosion rate [10]. IR measurements were carried out on cleaned glass samples before and after weathering in a climatic chamber for one day. The sensitive MI glass was too corroded to be analyzed by this method. The results for M2.0 and MIII glasses are given in Figure 5.

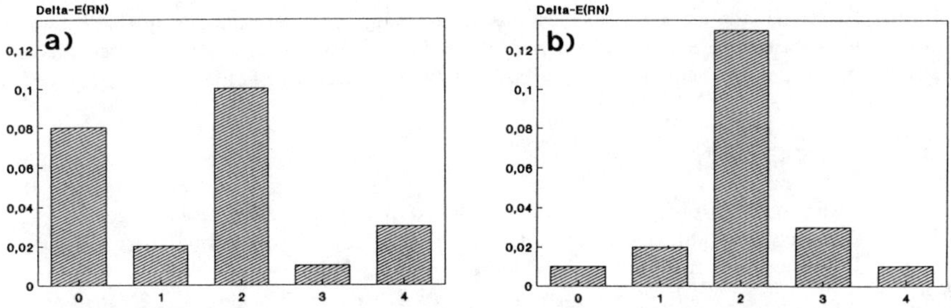

Figure 5: IR data showing the corrosion rates of model glasses after cleaning with different techniques (O: without cleaning, 1: glass fibre brush, 2: emery paper, 3: scalpel, 4: bristle brush) and accelerated weathering;
a) M2.0 glass and b) MIII glass.

The outstanding ΔE-values for both glasses treated with emery paper (no. 2 in Figure 5) underlines the importance of the gel layer for protecting the glass form further corrosion. Deep scratches in the gel layer lead to accelerated corrosion of the glass. The M2.0 glass which did not undergo any cleaning procedure (no. O in Figure 5) shows a very high ΔE value, comparable to the sample with heavy grinding and stands in contrast to the corresponding untreated MIII glass. The thick porous encrustations on M2.0 glass lead to a higher corrosion rate than carefully cleaned samples whereas the incomplete crystal layer on MIII glass is of no significant importance for the further corrosion behaviour of this glass. The chemical composition of the crust should probably be also of importance for the rate of further deterioration.

The ΔE values of the samples cleaned with brushes and scalpels are not significantly different. Like all mechanical methods, the skill and care on the part of the operator are essential for a possible damage. A scalpel, used not in the appropriate way, certainly bears the potential to scratch the surface comparable to emery paper.

4. CONCLUSION AND PERSPECTIVES

Basic investigations concerning the possible damage of different cleaning methods for medieval glass can be carried out on model glasses simulating the corrosion phenomena of the originals. A variety of model glasses with different chemical composition and corrosion behaviour should be involved to correspond to the variety of glass types in a medieval window.

Light microscopy is considered to be an effective tool to visualize the direct effectiveness of a cleaning method for removing surface deposits as well as to demonstrate heavy damage to the gel-layer of the glass, like scratches.

Infrared-spectroscopy was employed to quantify the sensibilization of the glass surface after cleaning for further atmospheric attack. According to these results a difference can be made between glasses with a scratched and an undamaged gel layer. The crust formed by natural weathering has no protective value in reducing the rate of further deterioration. Depending on the thickness and the chemical composition of the crust it may even accelerate the corrosion rate of samples without cleaning, comparable to those with deep scratches.

Further investigations have to be made to improve the understanding of cleaning effects on glass surfaces. The working program will be extended to more sensitive analytical methods to clarify the least possible damage of cleaning procedures, including mechanical, chemical and airbrasive techniques. First results on the potential damage of "wet" techniques (organic solvents, water, complexing agents) are published elsewhere [11].

5. ACKNOWLEDGEMENT

The project was funded by the Bundesministerium für Forschung und Technologie (BMFT), Germany. The authors wish to thank the members of the workshop of Cologne Cathedral, Germany, for their help and encouragement.

6. REFERENCES

[1] R. G. Newton, The deterioration and conservation of painted glass - a critical bibliography. (Oxford University Press 1982).
[2] H. Marschner, Arbeitsheft *32*, Bayer. Landesamt für Denkmalpflege, 44 - 55 (1985)
[3] W. P. Bauer, Verres Réfract. *30*, 62 (1976)
W. P. Bauer, CVMA News Letter *37*/*38*, 21 (1984)
[4] A. Moncrieff, Conservation in Archaeology and the Applied Arts, IIC Congress, Stockholm, 99 - 104 (1975)
[5] H. Marschner, CVMA News Letter *39*/*40*, 24 (1986).
[6] S. Fitz et al., Forschungsbericht UFOPLAN-Nr. 106 08 002, 1984, edited by Umweltbundesamt, Bismarckplatz 1, D-1000 Berlin 33
[7] J. M. Bettembourg, Compte Rendu du 8e Colloque du CVMA, York (U. K.) 1972
J. M. Bettembourg et F. Perrot, Verres Réfract. *30*, 92 - 95 (1976)
[8] D. R. Fuchs, H. Patzelt, H. Schmidt, Kontakt & Studium, Bd. *282*, 174 - 192, Expert Verlag, 1989
[9] D. R. Fuchs, H. Römich, P. Tur, J. Leißner, Forschungsbericht UFOPLAN-Nr. 108 07 005/03, 1991, edited by Umweltbundesamt, Bismarckplatz 1, D-1000 Berlin 33
[10] D. R. Fuchs, H. Römich, H. Schmidt, Mat. Res. Soc. Symp. Proc. Vol. *185*, 239 - 251 (1991)
[11] H. Römich, D. R. Fuchs, Zwischenbericht im Rahmen des Verbundprojektes des Bundesministeriums für Forschung und Technologie BAU C 5026 C/4 (1992)

PART IV D

Corrosion of Metals and Corrosion Protection

SULFATE FORMATION DURING CORROSION OF COPPER ALLOY OBJECTS

Michael B. McNeil[a] and D. W. Mohr[b]
[a]Office of Nuclear Regulatory Research, Nuclear Regulatory Commission, Washington, DC 20555; [b]Naval Coastal Systems Center, Panama City, Florida 32407-5000

ABSTRACT

The copper sulfate minerals chalcanthite ($CuSO_4.5H_2O$), antlerite ($Cu_3SO_4(OH)_4$), and brochantite ($Cu_4SO_4(OH)_6$) are observed as corrosion products on copper alloy objects such as statues which are exposed to weathering. The usual product in aqueous corrosion is brochantite, though posnjakite ($Cu_4SO_4(OH)_6 \cdot H_2O$) is also known. No copper sulfate mineral is commonly found in artifacts recovered from near-surface burial. These observations can be explained in terms of stability diagrams and microbiological activity.

INTRODUCTION

Sulfating corrosion of copper alloys is important to the study of the weathering of objects (particularly bronze statues) exposed to rain in industrial atmospheres[1-4]. It is found that the principal corrosion product formed under these conditions is brochantite ($Cu_4SO_4(OH)_6$), often accompanied by significant quantities of antlerite ($Cu_3SO_4(OH)_4$)[1-4]. Small quantities of chalcanthite ($CuSO_4.5H_2O$) and posnjakite ($Cu_4SO_4(OH)_6.H_2O$[1] are also observed.

Sulfating corrosion of copper alloys also occurs in the pitting of pipe in water distribution systems. Sulfating corrosion of copper can be observed in type 1 [5-8] or type 2[9] pitting corrosion of surfaces (usually the interiors of pipes) exposed to fresh water. The so-called type 3 pitting corrosion of copper[10] appears to be a variety of type 1 corrosion in which, because of peculiarities in the water chemistry, only oxide and sulfate minerals are found. The historic reason for regarding type 3 as a separate type is that the pit crusts first studied in type 1 corrosion were generally of malachite ($Cu_2(OH)_2CO_3$)[5] (Lucey 1967), having formed in high-carbonate waters. However, in waters low in carbonate and relatively high in sulfate, brochantite can form instead[7] in a corrosion process which appears otherwise indistinguishable from carbonating type 1 corrosion.

The authors performed X-ray and optical characterizations of some of the specimen described by Paradies et al.[11] and found the only cupric mineral present in detectable quantities to be posnjakite.

Sulfate minerals very rarely are observed on buried objects[12] though there are isolated observations of brochantite and connellite ($Cu_{19}Cl_4SO_4(OH)_{32} \cdot 3H_2O$)[14]. This is due to the effect of sulfate reducing bacteria, which are almost ubiquitous in deaereated systems

MECHANISMS OF CORROSION

The basic mechanism of type 1 and type 3 pitting corrosion is essentially as put forward by Lucey in 1967. A minor flaw in the copper surface, made anodic by any of a number of possible mechanisms, develops a crust of cuprite (Cu_2O). Under this crust the environment becomes acidic and chloride-rich. Dissolution of copper metal leads to formating a cupric-ion rich electrolyte, and precipitation of nantokite (CuCl) within the pit. Oxidation of cupric ions produces crystals of a copper hydroxychloride, which grow from or sometimes through the cuprite crust, altering to other minerals as they grow. In the classic type 1 pitting corrosion, the resulting mineral is malachite; however, it can be a sulfate if the water has a high sulfate/carbonate ratio. Microscopic examination of the contents of the pits shows[8] that crystals of a copper hydroxychloride (provisionally identified as atacamite because of their acicular structure) grow outward from the cuprite cap on the pit and alter to "hydrated brochantite" (apparently posnjakite) as they grow. The alteration of either brochantite or antlerite in various waters at STP appears to be very slow[2] and posnjakite appears to be stable except at pH below 3.

Type 2 copper pitting corrosion, which also produces sulfate minerals, is characterized by the absence of the sheets of cuprite over the pits. It almost always occurs at temperatures above 60°C and requires trace manganese in the water. This corrosion process is not yet understood, but its dependence on trace manganese and high temperatures suggests that its initiation, at least, depends on manganese-oxidizing bacteria. The observation of type 1 corrosion under conditions which are characteristic of type 2 probably reflects balances between microbe populations in the biofilm[7]:

Biofilms and the associated microbiological activity appear to control type 1 pitting by controlling ionic concentrations at reaction sites. The high copper activities and acidity required to stabilize nantokite can be maintained by the presence of the biofilm, which restricts mixing. The effect of biofilm chemistry on local copper concentrations by biochemical binding has also been explored[17,18]

USE OF STABILITY DIAGRAMS

Sulfating corrosion of copper alloys cannot be understood on the basis of any Pourbaix diagram[19,20] because the Pourbaix diagram is a two-dimensional presentation of phase equilibria controlled by two independent variables, Eh and pH, with all other variables fixed, and is suited to analysis of situations where redox behavior is critical. It is the purpose of this paper to use a different type of stability diagram to present a unified study of weathering and of the corrosion of biofouled surfaces, for which redox equilibria are not critical. It is important to note that in both weathering and pitting high concentrations of ions are possible; in the weathering case because of the limited amount of moisture when the statue is only wet and not actively being rained on (not to mention the entrapment of water in porous corrosion products) and in the fouled-surface case because of the ability of the biofilms to restrict ion motion.

Sulfating corrosion of copper, like bronze disease, is best visualized with the help of a four-dimensional figure representing phase equilibria as functions of the four independent variables pH, Eh, and measures of chemical activity for one metal and one mineral-forming anion. The basic paper by Zen[21] and an analysis of chloride copper corrosion[22] copper give the background for this type of analysis, which will here be applied to corrosion of pure copper in an aqueous medium containing sulfur species.

The stability diagram, which shows regions of stability for various phases and combinations of phases as a function of Eh, pH, activity of the commonest sulfur ion and activity of the commonest copper ion, is four-dimensional. The choice of activity of the commonest ion instead of total activity is important; choice of different measures of sulfur and copper activity would produce diagrams which are equally valid but which are more difficult to understand.

The implications of the phase rule with regard to this four-dimensional diagram are discussed elsewhere[22]. In particular, the use of electron flow as a conjugate flux to electrochemical potential is reviewed. The aqueous phase is everywhere present, so an invariant point is defined by the presence of four nonaqueous phases, a univariant line by three nonaqueous phases, and a bivariant plane by two nonaqueous phases. Invariant and univariant structures can arise in other ways, and these are particularly important in the Cu-H-O-S diagram, which is presented in Figure 1. Consider a univariant line. If a nonredundant chemical equilibrium condition is specified (that is, an equilibrium condition that does not merely reflect the equilibrium between the different phases present), a point on the line is specified. Similarly, two independent nonredundant equilibria will specify a point in a plane, or one nonredundant equilibrium will specify a line in a plane.

In order to have a suitable graphic presentation, it is useful to project structures from the four-dimensional figure onto a plane which is normal to both the pH and Eh axes. In the present analysis, the axes of the plan containing the projections are the activities of the predominant sulfur- and copper-containing ions.

Not all the apparent points and lines in the projection will be truly points and lines. The projection can also contain "degenerate points" (Zen 1966), points in the projection which actually are lines or planes in four dimensions but so oriented as to project as points. In particular, a plane defined by phases and equilibria which are independent of both Eh and pH will project as a point on a plane whose axes are measures of sulfur and copper concentration. Irrespective of this complication, the lines connecting the points in the projection--whether degenerate or not--separate the projection into regions, and points in different regions of the projection correspond to Pourbaix diagrams which differ in structure by the presence of different phase fields or phase boundaries, not just by the shifting of lines. An additional complexity in the projections are "indifferent intersections." These are apparent intersections on the projection of lines that do not intersect at all in four-space.

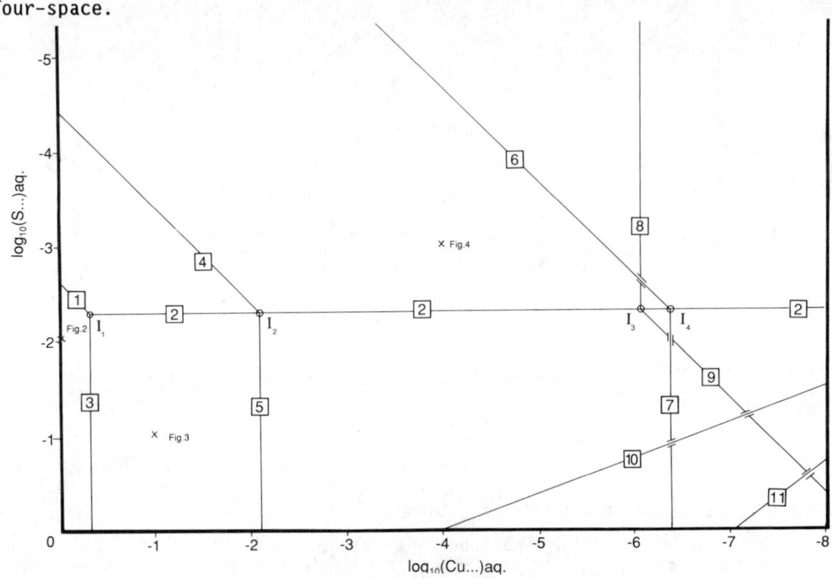

Figure 1: Modified Log-Activity Diagram
Cu-O-H-S System at STP
Crosses Mark Locations of Pourbaix Diagrams, Figures 2, 3, and 4

Figure 1 shows projections from the four-dimensional figure onto a plane whose coordinate axes are the activities of the most common sulfur- and copper-containing ions. Table 1 describes the invariant points in the projection and Table 2 the univariant lines. The indifferent intersections are shown by bridge-like structures. Lines in a projection of this sort separate regions of the projection in which Pourbaix diagrams have different structures. For example, Pourbaix diagrams corresponding to points in the bottom left region of Figure 1 show stability fields for chalcanthite, antlerite, and brochantite. Diagrams corresponding to points in the portion of the diagram containing Figure 3 show stability fields for antlerite and brochantite but not for chalcanthite. Diagrams corresponding to points in other portions of the diagram show either only brochantite or else no solid sulfate fields at all. Figures 2, 3, and 4 show Pourbaix diagrams for points in the different regions. No Pourbaix diagram contains a region which has more than one stable sulfate mineral. Observations of mixed sulfate minerals or of sulfate minerals mixed with malachite may reflect local chemical disequilibrium or chemical equilibria of lower variance due to limited availability of water. The diagrams as drawn do not contain fields for posnjakite (for which no thermodynamic data are available) or langite (which is known from only one observation and which is apparently stable only under water-deficient conditions). The thermodynamic data used are those of the latest National Bureau of Standards publication[23].

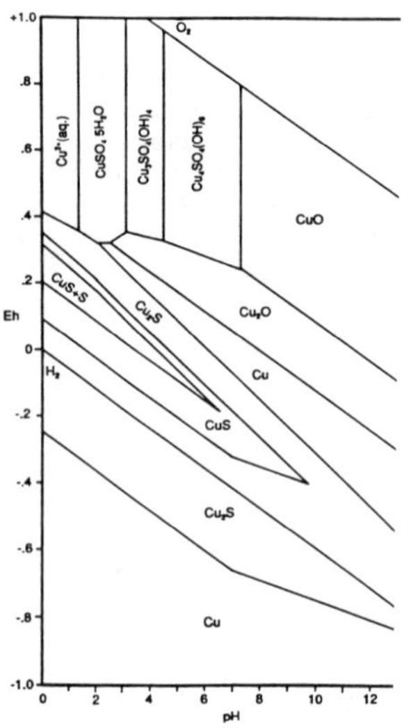

Figure 2: Pourbaix Diagram

Cu-O-H-S System

$log_{10}a(S)=-2$

$log_{10}a(Cu)=0$

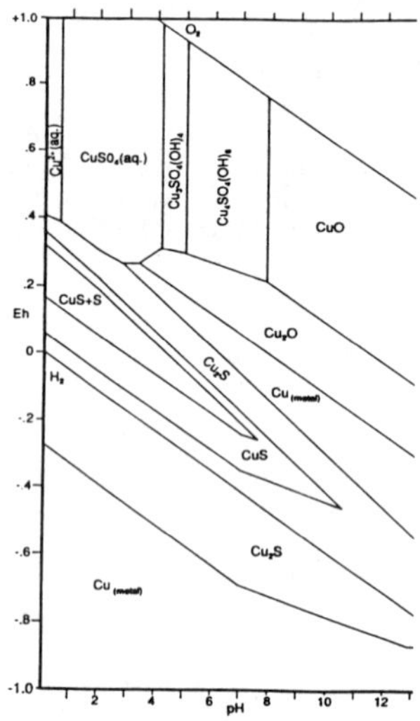

Figure 3: Pourbaix Diagram

Cu-O-H-S System

$log_{10}a(S)=-1$

$log_{10}a(Cu)=-1$

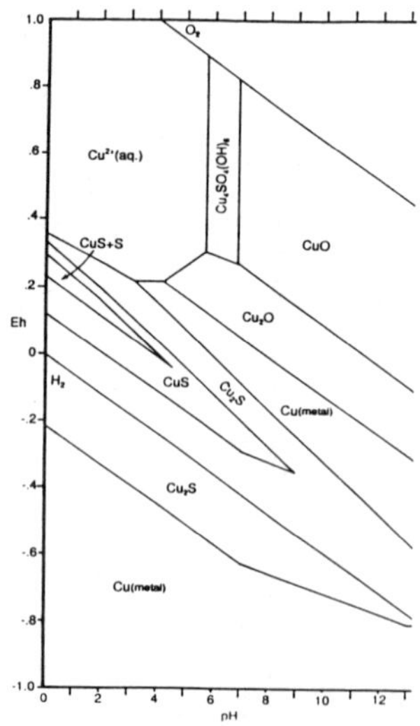

Figure 4: Pourbaix Diagram

Cu-O-H-S System

$log_{10}a(S)=-3$

$log_{10}a(Cu)=-4$

Table 1. Lines in the Cu-H-O-S Figure

L1: Chalcanthite, antlerite, and water are present. This line is a projection of a solid containing the line itself and Eh and pH axes, since the reactions between the phases do not involve protons or electrons.

L2: This line is the projection of a solid defined by the presence of water and the restriction that the activity of Cu^{++} be the same as the activity of $CuSO_4$(aq). The volume contains three axes: copper activity, Eh, and pH.

L3: This line is the projection of a volume whose axes are Eh, pH, and the line itself, which is parallel to the sulfate axis. Two phases are present, chalcanthite and water.

L4: This line is the projection of a plane defined by the presence of water, antlerite, and brochantite.

L5: This is the projection of a plane defined by the presence of water, antlerite, and brochantite.

L6: This is a line. Chalcocite, covellite, and water are present, and the activity of Cu^{++} = activity of $CuSO_4(aq)$.

L7: This is the projection of a plane defined by the presence of water, brochantite, and tenorite. The plane contains the trace of the line and a line perpendicular to the trace and parallel to the Eh axis.

L8: This is the projection of a plane defined by the presence of water, brochantite, and tenorite. The plane contains the trace of the plane and a line which is normal to the trace and parallel to the trace.

L9: This is the projection of a plane defined by the presence of copper and water, and the restriction that activity of Cu^{+} = activity of Cu^{++}.

L10: This is the projection of a plane defined by the presence of copper and water, and by the restriction that the activity of Cu^{+} equals the activity of aqueous $CuSO_4(aq)$. The plane contains the projection and a line normal to the projection and parallel to the pH axis.

Table 2. Points in the Cu-H-O-S Figure

I1: Chalcanthite and water are present. The activity of aqueous Cu^{++} is equal to the activity of aqueous $CuSO_4$. This is a plane in the four-dimensional figure, containing Eh and pH axes, and so projects as a point. Antlerite is also present, and there is a point where antlerite, chalcanthite, cuprite, and water are all present, and the activity of Cu^{++} is equal to the activity of aqueous $CuSO_4$, which is a true invariant point.

I2: Antlerite, brochantite, cuprite, and water are present. The activity of aqueous $CuSO_4$ is equal to the activity of aqueous Cu^{++}. This is a true invariant point.

I3: This is actually a projection of two points connected by a line.
Point I3': Brochantite, copper, cuprite, and water are present. The activity of Cu^{+} and of $CuSO_4$ are equal.
Point I3": Brochantite, copper, chalcocite, and water are present. The activity of aqueous Cu^{+} is equal to the activity of aqueous $CuSO_4$.

I4: Brochantite, cuprite, tenorite, and water are present. The activity of $CuSO_4$ is equal to the activity of aqueous Cu^{++}.

CONCLUSIONS

Considering the projection, one sees that for chalcanthite to form at any Eh/pH combination, one must have very high concentrations of sulfate and copper. This compound could be formed in an open-atmosphere environment only if the surface is wet but has no water circulation. A possible situation would be corrosion by dew or corrosion under a porous, water-saturated layer of another mineral. For antlerite to form, the restrictions on Cu/SO_4 concentrations are much less severe, but they are still indicative of a low-recharge situation. Brochantite can form under Cu/SO_4 conditions which correspond to a wide variety of conditions, including some bulk waters. Consequently, the presence only of brochantite in under-water corrosion is easily understood.

The one observation of a corrosion product composed almost entirely of posnjakite was in a pipe known have a thick, continuous biofilm which contained many gram-negative organisms[11]. In this case the corrosion product mineralogy was presumably controlled by the biofilm in some way[17,18]. The Cu-H-O-S diagram also contains information on the formation of chalcocite (Cu2S) and covellite (CuS) during copper corrosion but is not of great use in studying the sulfiding corrosion of copper and its alloys. At sulfide concentrations as high as 10-6M, the relative stabilities of minerals are primarily controlled by reduced sulfur chemistry, that is, by the variables of the Pourbaix diagram, and consequently, the projection discussed in this paper adds little to corrosion analyses based on ordinary Pourbaix diagrams. When copper alloy objects are buried they are exposed to sulfate reducing bacteria (SRB) which form biofilms with low oxygen activity and high reduced-sulfur ion concentration. This is why sulfate minerals are almost unknown as corrosion products on buried objects.

ACKNOWLEDGEMENTS:

The authors desire to thank Dr. Haim Mathias and Dr. Roy Odom of Florida State University for their experimental assistance, and Professor J. W. Costerton of the University of Calgary for useful discussions. This work was carried out with support from the Office of Naval Research Code 1141MB.

References

[1] Bestek, T., Drys, M., Streakalov, P., Kemkhadze, V., Kozhukarov, V., Cherny, M., Zaydel, M., Protection of Metals (trans. of Zashchita Metallov), vol. 18, pp 528-533 (1982).

[2] Graedel, T. E., Nassau, K., Franey, J.P., vol. 27, pp 639-658 (1987).

[3] Graedel, T. E., Corrosion Science, vol. 27, pp 721-740 (1987).

[4] Graedel, T. E., Corrosion Science, vol. 27, pp 741-770 (1987).

[5] Lucey, V. F., British Corrosion Journal, vol. 2, pp 175-185 (1967).

[6] Campbell, H. S., A Review: Pitting Corrosion of Copper and Its Alloys in Localized Corrosion (eds. R. Staehle, et al.), National Association of Corrosion Engineers, Katy, TX (1974).

[7] Sato, S., Corrosion Engineering in Japan, vol. 31, pp 3-11 (1982).

[8] Shalaby, H. M., Al-Kharafi, F. M., Gouda, V. K., Corrosion, vol. 45, pp 536-547 (1989).

[9] Fujii, T., Kodama, T., Baba, H., Corrosion Science, vol. 24, pp 901-912 (1984).

[10] Lindner, M., Pitting Corrosion of Type 3 in Copper Cold Water Pipes - Causes and Countermeasures, Report number 87:2, Swedish Corrosion Institute, Stockholm (1988).

[11] Paradies, H. H., Hansel, I., Fischer, W., and Wagner, W., Microbiologically Induced Corrosion of Copper Pipes, INCRA Report #404. The x-ray diffraction identification on posnjakite was performed at Florida State University.

[12] Gettens, R. J., Annual Report of the Board of Regents of the Smithsonian Institution, Smithsonian Institution Publication 4530, U.S. Government Printing Office, Washington, DC (1963).

[13] Otto, H., Die Naturwissenschaften, vol. 48, pp 661-664 (1961).

[14] Otto, H., Die Naturwissenschaften, vol. 50, pp 16-17 (1963).

[15] Nassau, K., Miller, A. E., and Graedel, T. E., Corrosion Science, vol 27, pp 703-719 (1987).

[16] Binder, O., Comptes Rendus, vol. 198, pp 653-655 (1935).

[17] Jang, L. K., Harpt, N., Grasmick, D., Vuong, L. N., and Geesey, G., Journal of Physical Chemistry, vol. 94, pp 482-488 (1990).

[18] Costerton, J. W. and Boivin, J., "The Role of Biofilms in Microbial Corrosion," in Microbially Influenced Corrosion and Degradation, N. J. Dowling et al., eds., University of Tennessee, Knoxville, Tennessee (1991).

[19] Garrels, R. M., Christ, C. L., Solutions, Minerals, and Equilibria; Freeman, Cooper and Company, San Francisco, CA (1965).

[20] Pourbaix, M., Atlas of Electrochemical Equilibria in Aqueous Solutions, National Association of Corrosion engineers, Houston, TX (1966).

[21] Zen, E. A., Construction of Pressure-Temperature Diagrams for Multicomponent Systems After the Method of Schreinemakers - A Geometric Approach, U.S. Geological Survey Bulletin #1225 (1966).

[22] Mohr, D. W., McNeil, M. B., Modified Log-Activity Diagrams as Tools for Modelling Corrosion to appear in Journal of Nuclear Materials (1992).

[23] Wagman, D., et al., The NBS Tables of Chemical Thermodynamic Properties, Journal of Chemical and Physical Reference Data, vol. 11, Supplement 2 (1982).

[24] Hamilton, W. A., Annual Reviews of Microbiology, vol. 39, pp 195-219 (1985).

Interpretation of Bronze Disease and Related Copper Corrosion Mechanisms in Terms of Log-Activity Diagrams

Michael B. McNeil
Office of Nuclear Regulatory Research, Nuclear Regulatory Commission, Washington, DC 20555
D. W. Mohr
Naval Coastal Systems Center, Panama City, Florida 32407-5000

ABSTRACT

Thermochemical data are used to construct a modified log-activity diagram for the system Cu-H-O-Cl-(CO_2) at standard temperature and pressure(STP). In these diagrams, coordinate axes represent activities of aqueous copper species and chloride ions other thermodynamic variables are collapsed onto the diagram. Modified log-activity diagrams are used as a supplement to study Pourbaix diagrams for the equilibria relevant to bronze disease and copper pitting and to show the relation between them.

INTRODUCTION

Pourbaix diagrams, or thermodynamic diagrams which display variation of stability of phases and predominance of aqueous species with changes in redox potential (Eh) and acidity (pH) have been widely used in corrosion studies for many years[1,2]. Typically, such diagrams deal with three-element systems that include one metal along with hydrogen and oxygen. Although Pourbaix diagrams may be readily drawn for four-element systems as well, lack of any method to display equilibria for reactions which do not involve either electrons or protons has limited their usefulness to these more complex systems. For example, Pourbaix diagrmas applicable to the four-element system Cu-H-O-Cl cannot display many important equilibria involving the important copper-chloride salts or aqueous species.

This paper develops a modified log-activity diagram in which reaction equilibria are displayed relative to activities of aqueous species for the two elements of choice. The Cu-H-O-Cl system is analyzed and the results related to bronze disease and types 1 + 3 pitting corrosion.

Bronze disease[3] describes the rapid deterioration of a copper alloy object exposed, after long burial, to moisture. This phenomenon is caused by the tendency of chloride ions, during corrosion, to concentrate at anodic sites. This accumulation is a well-understood phenomenon related to the charge balances in corrosion[4] and can result in increases in local chloride concentrations by orders of magnitude[5], leading to the formation, under caps of other copper minerals, of deposits of nantokite (CuCl). When the nantokite is exposed to water and oxygen it reacts to form cupric ions and HCl, leading to an autocatalytic reaction. A similar reaction is known in iron[5,6,7].

Two other copper corrosion phenomena are type 1 and type 3 pitting corrosion of copper[8,9,10,11]. In both of these cases corrosion begins in a pit covered with a fragmented layer of cuprite and containing crystals of nantokite. During pit growth the nantokite deposit remains in the pit, but the surface accumulation of cupric minerals grows. The difference between type 1 and type 3 is that type 1 produces a cupric-mineral cap composed largely of malachite, $Cu_2(OH)_2CO_3$, whereas type 3 produces a cap composed of sulfate minerals.

MODIFIED LOG-ACTIVITY DIAGRAMS

Schreinemakers diagrams[12] are thermodynamic figures which display the possible invariant and univariant equilibria of chemical systems and then project them on a plane defined by two variables, usually temperature and pressure. Such diagrams have been of great use in geology, but are of no direct use in the study of archaeological corrosion phenomena. However, a mathematically parallel construction, termed a log activity diagram, can be used.

To understand the structure of such a diagram, consider the derivation of the phase rule[13]. The phase rule states that the phases plus the number of degrees of variance in a system is equal to the number of independent components plus two. This rule arises from the requirement that the conditions of equilibrium between all the phases be smaller than the number of independent intensive variables by exactly this number.

If no transfer of charge other than on the molecular scale is allowed, the number of phases plus the variance in a Cu-O-H-Cl system is six. In a case where anodic and cathodic reactions can be spatially separated, with the electrons flowing through solids or electrolytes, there is also a flux of electrons, and a "chemical activity" of electrons exactly parallel mathematically to the chemical activity of most molecular species. This adds one effective component to the system, the electron, and one new potential variable, usually measured by Eh. This is in fact the reasoning that makes Pourbaix diagrams useful, and it must be emphasized that the reasoning has nothing to do with electrochemistry or geochemistry. It is a direct consequence of the analysis leading to the phase rule.

Pourbaix diagrams are two-dimensional diagrams, showing regions of species stability as functions of En and pH. To study copper corrosion, a more complex type of diagram is useful. As the first step in construction, consider a four dimensional figure whose axes are Eh, pH, the concentration of the most common copper species, and the concentration of chloride ion.

Because we have five degrees of variance and only a four dimensional figure, there is a degree of indeterminacy. Define a specification as the presence of a phase, or the statement of a nonredundant equilibrium condition. A statement that the activity of copper in one phase is equal to that in another is redundant; the equations corresponding to such equilibria have already been used in deriving the phase rule. An example of a nonredundant equilibrium is the statement that the chemical activity of Cu^{++} and of $CuCl_2^-$ in the aqueous phase are equal. Five specifications define a point. Four specifications define a line. Three specifications define a plane. Two specifications define a 3-space embedded in the 4-dimensional figure.

To produce a log-activity diagram which can be printed on a sheet of paper, imagine a plane whose axes are the activity of the predominant copper species and the activity of chloride ion, and project all these points and lines on it. Planes may not project at all onto the Cu-Cl plane, may project as lines, or may project as points, depending upon their orientation. A line may project as a line, or (if it is normal to the place of projection) as a point. A plane can not project at all (if it is parallel to the place of projection), project as a line, or project as a point.

To each point in this two-dimensional figure corresponds a separate Pourbaix diagram. It is essential to remember that the axes of the diagram are not the total activity of copper and chloride; they are the activity of the predominant species; it is coincidental that the predominant species of chlorine is, for all physically interesting aqueous solutions, chloride. The use of other measures for the intensive variables conjugate to the motion of copper and chlorine or sulfure will produce diagrams equally valid but less easy to use.

DEVELOPMENT OF MODIFIED LOG-ACTIVITY DIAGRAMS FOR CU-H-O-Cl

In the following sections the construction of a modified log-activity diagram will be described for the system Cu-H-O-Cl, and the consequences of introducing CO_2 at equilibrium with the air will be discussed.

Free energy values were taken from the best recent work on malachite[14] and for all other phases from the most recent NBS publication[15]. All free-energy values apply to the standard state of one atmosphere and 298 K. Standard states for solid phases (as well as for water) are the pure compoinds in their most thermodynamically stable forms. Standard states for aqueous ionic species are unit activity in water, and standard state for CO_2 gas is unit fugacity.

Each field bounded by lines corresponds to a unique geometrical relationship of the Pourbaix diagram. This geometrical relationship is altered across each line, either by a change in the pattern of intersection of lines on the Pourbaix diagram or by the appearance or disappearance of a phase or a region of predominance for an aqueous species of copper. Thus the modified log-activity diagram serves as a master key to the set of possible geometries for the Pourbaix diagrams. These geometries are displayed as insets on Figure 1.

In some cases these equilibria result in predetermination of both the redox potential and the pH, resulting in truly univariant equilibria. In other cases, the equilibria are irrelevant to either Eh or pH or both. These cases furnish examples of a condition of degeneracy. Apparently univariant equilibria on the two-dimensional projection are actually of higher variance, but the additional intensive variables are not determined by the equilibria represented by the lines. In geometrical terms, truly univariant equilibria are represented by lines within both the modified log-activity diagram and the four-dimensional figure from which the projection is derived. Degenerate equilibria are linear projections onto the modified log-activity diagram of planes or of solids within the four-dimensional figure. Similarly, intersections of lines may represent either truly invariant degenerate equilibria. In geometrical terms, these intersections may be points in the four-dimensional figure or they may be lines or planes.

Some lines which appear to intersect on the modified log-activity diagram do not, in fact, intersect within the hypothetical four-dimensional diagram. For example, lines 4 and 8 represent equilibria whose redox potentials are of different values. Apparent intersection of these two lines results from collapse of the Eh axis onto the two-dimensional projection. Such apparent intersections do not generate equilibria in which the activities of both chloride anion and aqueous copper are specified. Such apparent intersections are termed "indifferent intersections"[12] and are marked on the log-activity diagram by the engineer's symbol for a bridge.

Figure 1 (from reference 16) shows equilibria among the solid phases copper metal, CuCl, Cu_2O, $Cu_2(OH)_3Cl$, CuO, and water. Equilibria among species of aqueous copper are considered for Cu^{2+}, Cu^{+}, $CuCl^{+}$, $CuCl_2^{-}$, and $CuCl_3^{2-}$. The neutral aqueous species $CuCl_2$ and the basic cuprate and bicuprate anions were not considered. Equilibria involving reaction of the two cuprate anions with $CuCl_2^{-}$ are restricted to activities of aqueous copper whose logarithmic values (base 10) are less than -8: The $CuCl_2$ molecular species becomes the most abundant species of dissolved copper only when the activity of chloride anion is greater than 1. Thus the modified log-activity diagram displayed as Figure 1 does not apply to conditions with pH greater than about 10.

DESCRIPTION OF THE MODIFIED LOG-ACTIVITY DIAGRAM FOR THE SYSTEM Cu-H-O-Cl

Fourteen univariant or apparently univariant lines and eight points arising from intersections of lines result from equilibria pertinent to the Cu-H-O-Cl system. The conditions for all these points and lines are comprehensively reviewed and explained in reference 16, and will not be described in detail here. The construction of one line and three points will be described here for examples.

Line 2 involves equilibria among the phases CuCl, $Cu_2(OH)_3Cl$, Cu_2O, and water, as well as the aqueous ions Cu^{2+} and Cl^{-}. This line separates geometries on the Pourbaix diagram in which the fields for CuCl and $Cu_2(OH)_3Cl$ adjoin, at higher activities of chloride anion and aqueous copper, from those geometries in which the field for Cu_2O is in direct contact with the aqueous field for the predominance of the Cu^{2+} cation. The line is calculated from the set of free-energy data used by simultaneous solution of the following reaction equilibria:

1) $Cu_{2+} + Cl^{-} + e = CuCl$
2) $Eh = .561 + .0592\log_{10}a(Cl^{-}) + .0592\log_{10}a(Cu^{2+})$

3) $2Cu^{2+} + Cl^{-} + 3H_2O = Cu_2(OH)_3Cl + 3H^{+}$
4) $pH = (7.139 - \log_{10}a(Cl^{-}) - 2\log_{10}a(Cu^{2+}))/3$

5) $2Cu^{2+} + H_2O + 2e = Cu_2O + 2H^{+}$
6) $Eh = .207 + .0592pH + .0592\log_{10}a(Cu^{2+})$

7) $Cu_2(OH)_3Cl + H^{+} + 2e = Cu_2O + Cl^{-} + 2H_2O$
8) $Eh = .418 - .0296pH - .0296\log_{10}a(Cl^{-})$

9) $Cu_2(OH)_3Cl + Cl^{-} + 3H^{+} + 2e = 2CuCl + 3H_2O$
10) $Eh = .772 - .0888pH + .0296\log_{10}a(Cl^{-})$

11) $2CuCl + H_2O = Cu_2O + 2H^{+} + 2\,Cl^{-}$
12) $pH = 5.993 + \log_{10}a(Cl^{-})$

In fact, it is necessary to solve only several of the above equations; the others then become identities. Solution by elimination of the terms for Eh and pH yields the following result:

13) $2\log_{10}a(Cl^{-}) + \log_{10}a(Cu^{2+}) = -5.420$

Equation 13 defines line 2 on Figure 1. Substitution of this equation into the set of equations 2/12 fixes values for Eh and pH at any permitted pair of chloride and Cu^{2+} activities.

Results are in accord with the phase rule and presence of five components within the Cu-H-O-Cl system. Since pressure and temperature have been assigned to standard state, presence of the four phases $Cu_2(OH)_3Cl$, CuCl, Cu_2O, and water suffice to establish univariance of the equilibria. It should be noted that if the electron were not regarded as a component, the equilibria would be invariant. Both the chloride and aqueous copper activities would be fixed at values which are in accord with both Equation 13 and the need for electrical neutrality within the aqueous phase. In face, such invariance in the bulk electrolyte would hold for any real system whose elements are restricted to Cu, H, O, and Cl. The lack of requirement for electrical neutrality among the dissolved ions, as well as the implied presence of other ions necessary to stabilize the system, result in univariance on both the Pourbaix diagrams and the modified log-activity diagram which is derived by similar principles.

Addition of other phases and aqueous species of copper results in determination of three invariant points, listed as points $I_{7,8,9}$ at this intersection. All three points share common values for the activities of chloride anion and aqueous species of copper, but their Eh and pH values differ.

ACTION OF CO_2 AT ITS ATMOSPHERIC PARTIAL PRESSURE TO THE MODIFIED LOG-ACTIVITY DIAGRAM

The atmosphere contains carbon dioxide at a partial pressure at $10^{-3.5}$ atmospheres. Addition of CO_2 at this pressure results in replacement of the phase CuO by malachite ($Cu_2CO_3(OH)_2$). The reaction between the two phases consists of addition of CO_2 and H_2O to CuO, and is indifferent to both redox potential and pH. Carbon is considered to exist only in CO_2 gas and in $Cu_2CO_3(OH)_2$).

Inclusion of CO_2 at its atmospheric partial pressure the modified log-activity diagram results in modification of those lines and intersections whose equilibria incorporate the phase CuO. These lines and intersections include points I_4, I_5, and I_6, as well as lines, 8', 9', and 13'. These modifications are shown on Figure 1 by dashed lines.

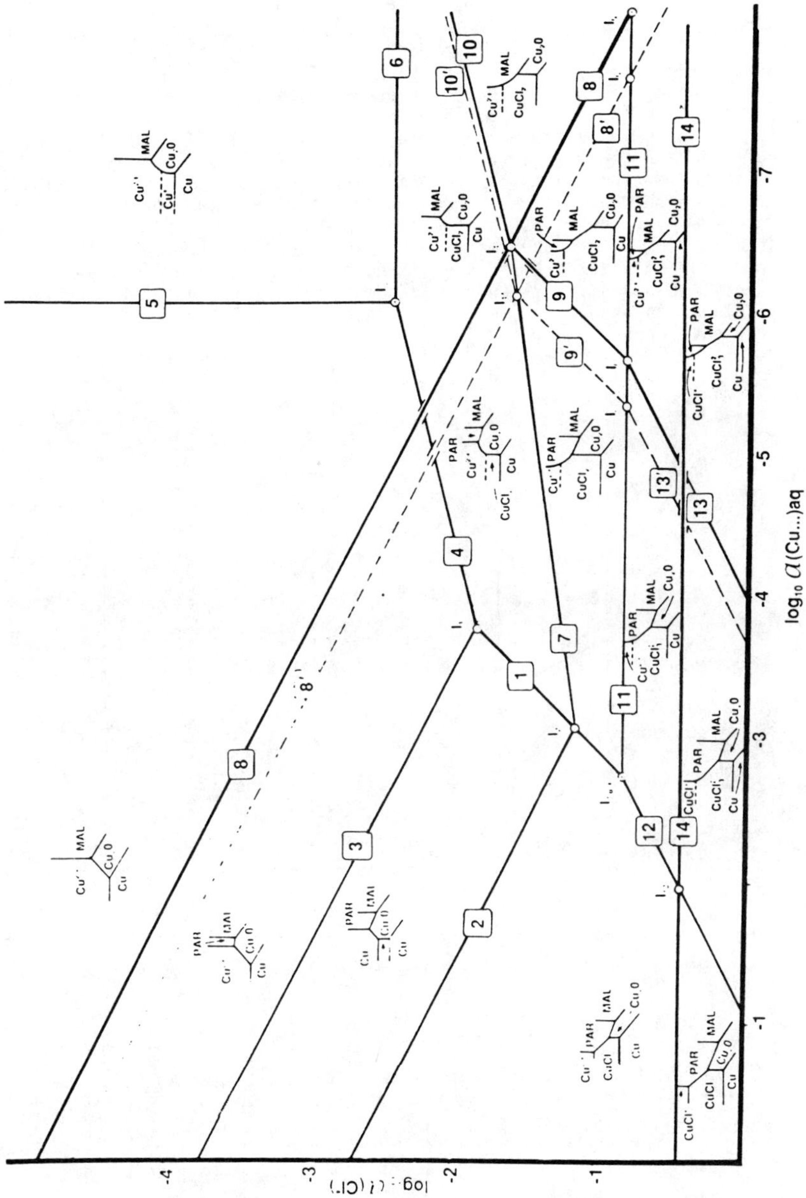

Figure 1. The modified log-activity diagram for the system $Cu-H-O-Cl^-(CO_2)$.

DEVELOPMENT OF A LOG ACTIVITY DIAGRAM FOR THE PITTING CORROSION OF COPPER

Clarification of equilibria pertinent to the development and growth of Type 1 pits and bronze disease[3] may be achieved by display of only those equilibria which involve copper metal. Such equilibria include points I_1, I_3, and I_7, and lines 1, 3, 4, 5, 6, 11, and 12. Figure 2 displays the resulting log-activity diagram. Unlike Figure 1, this figure is not a projection resulting from collapse of the axes for Eh and pH. Instead, all values of redox potential are fixed by the presence of copper metal in equilibrium with chloride and aqueous copper ions. Furthermore, addition of Cu_2O restricts the pH to those values shown in contours on Figure 2. In addition, the ordinate of Figure 2 represents the total activity of aqueous copper, resulting in displacement of points, as well as displacement and distortion of lines.

The highest values for pairs of activity values of chloride and aqueous copper, represented by the field bounded by lines 1, 3, and 12, are not in equilibrium with copper metal. Pairs of activity values which are in equilibrium with both solid CuCl and copper metal are displayed along these lines. Lines 4, 5, 6, and 11, dashed on Figure 2, separate the remaining field of the graph into areas of predominance of the various species of aqueous copper.

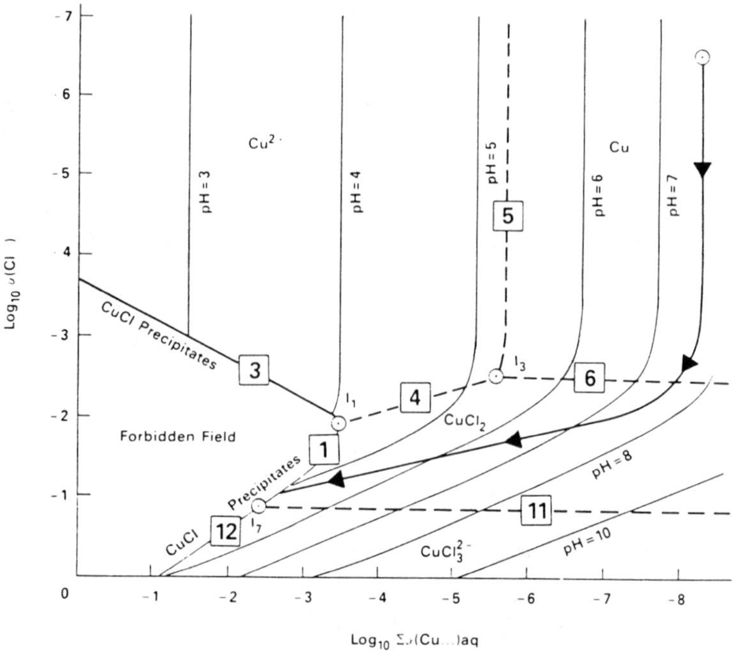

Figure 2. Graphs displaying changes in various thermodynamic parameters as functions of chloride activity for aqueous solutions whose equilibria are governed by presence of copper metal, CuCl, and cuprite.

APPLICATION OF THE MODIFIED LOG-ACTIVITY DIAGRAM TO PITTING CORROSION OF COPPER AND TO BRONZE DISEASE

A common chloride-based corrosion of copper involves formation of pits underneath a Cu_2O membrane. These pits contain both CuCl and Cu_2O in varying amounts. Initiation and growth of such pits, commonly termed Type 1 pitting, occurs within copper exposed to bulk electrolytes whose chloride concentration is insufficient to allow direct formation of either solid CuCl or significant quantities of aqueous copper chloride anions. In the case of water tanks and pipes, premature failure occurs by perforation. This type of corrosion is essentially identical to the "bronze disease" of archeological study, in which copper and copper-based alloy artifacts are pitted and, in some cases, largely replaced by CuCl during long exposure to relatively fresh ground waters.

The model of Lucey[8], which has become widely accepted for Type 1 pitting in copper tanks exposed to certain types of fresh water, is based on observations of the formation of CuCl-Cu_2O pits in the presence of extremely hard well water whose measured chlorinity is 19 ppm, corresponding to a molarity of approximately 2 x 10^{-7}. This model, which analyzes the reactions within a pit covered by a perforated cuprite film which in turn is covered by cupric minerals such as $Cu_2CO_3(OH)_2$, has also been applied in other situations[4,9,11,16,17]. The similar "bronze disease" of archeological study has been discussed by Scott[3].

Use of the modified log-activity diagram (Figure 1) allows some inferences to be made. The measured chlorinity of well water studies by Lucey[8] as well as that of fresh water generally, is insufficient either to support the precipitation of any solid copper chloride salt or to stabilize the important quantities of $CuCl_2^-$ or $CuCl_3^{2-}$ in solution. $Cu_2(OH)_3Cl$ is rarely if ever observed on surfaces exposed to ambient fresh water outside the pits. However, the presence of solid CuCl within the pit demonstrates that significant concentration of chloride anion occurred at that location. Thus, it is generally accepted that initial generation of areas of anodic potential precede development of pits.

Figure 2 shows a possible reaction path for an aqueous solution within a developing pit. Initial conditions, similar to those of the supply water studied by Lucey[8] are marked by a dotted circle. Anodic concentration of chloride anions results in increase in the local activity of Cl^-, shown as a descent of the arrowed line in Figure 2. This increase in chloride activity eventually allows for significant quantities of $CuCl_2^-$ and $CuCl_3^{2-}$. Since both anions would also be attracted to anodic surfaces, its local concentration would increase along with the concentration of Cl^-. This inference is shown on Figure 2 by a diversion of the arrowed line to the left. Eventually, concentration of $CuCl_2^-$ and $CuCl_3^{2-}$ anions would become sufficient for precipitation of solid CuCl. This is shown by termination of the bold arrow along the equilibria which govern precipitation of CuCl from aqueous solutions containing significant quantities of copper chloride anions.

A few general predictions regarding changes in redox potential and pH of the aqueous solution within the pit may be made. The pit potential should have reached its lowest value, approximately 0.03 V (SHE), early in the process, at the time when the concentration of cuprous chloride anions first exceeds that of Cu^{2+}. Concentration of cuprous chloride anions would then result in local increase in Eh. During precipitation of CuCl, conditions within the pit should be characterized by a pH of 5, Eh values in the vicinity of 0.18 V (SHE), $\log_{10}$ chloride activity near -1, and $\log_{10}$ activity of dissolved copper near -3. Some measured thermochemical values are available for Type 1 pitting in copper[4,9,17,18]. These reported values are not in full agreement with either the values which would be predicted by the arguments of this paper or with each other. Although these reported values show an initial drop in redox potential followed by a rise, specific Eh values range from 0.03 V lower to 0.07 higher than those predicted from Figure 2. Inconsistencies may be due to errors within the set of free-energy values or to difficulties in obtaining measurements of conditions within pits.

These surmises concerning the pitting corrosion of copper do not consider kinetically related effects such as overpotentials, reaction times, or diffusion rates. In fact, continued corrosion of the copper within the pit suggests that redox potentials based strictly on thermodynamic equilibria would not correspond exactly to those measured experimentally. Also, the vital question, concerning whatever mechanism allows such extremely strong concentrations of chloride anions to arise within essentially fresh water, is not addressed. Nevertheless, modified log-activity diagrams allow graphical presentation of the relevant equilibria in a manner not otherwise possible, and the thermodynamic inferences resulting from study of such diagrams are in accord with present knowledge of Type 1 pitting and bronze disease.

CONCLUSION

Log-activity diagrams modified by the collapse of the thermodynamic variables represented by Eh and pH help to elucidate reaction equilibria within four-element corrosion systems. They allow display of reaction equilibria which do not involve electron potential or pH, and they simplify search for the possible topologies of the corresponding Pourbaix diagrams. Thus, they are complementary to Pourbaix diagrams for study of metal corrosion. In the system Cu-H-O-Cl-(CO_2), applicable to corrosion of copper metal in chloride-bearing waters, these diagrams allow graphical representation and study of the thermohynamics of formation of CuCl-Cu_2O pits.

ACKNOWLEDGMENTS

Research for this paper was supported by the U.S. Navy Independent Research/Independent Exploratory Development Program, administered by the Naval Coastal Systems Center, Panama City, Florida, and by the Office of Naval Research, Code 1141MB. The authors would like to thank Mrs. Evelyn Queen for her patience with cuts and revisions.

References

[1] Pourbaix, M. J. N. Atlas of Electrochemical Equilibria in Aqueous Solutions. Cebelcor, Brussels, Belgium, second edition. Translated from the French by J. A. Franklin, NACE, Houston (1947).

[2] Garrels, R. M. and Christ, J. L. Solutions, Minerals, and Equilibria. Freeman Cooper, San Francisco (1965).

[3] Scott, D. A. Journal of the American Institute of Conservation, v. 29, p. 193-206 (1990).

[4] Szklarska-Smialowska, Z. Pitting Corrosion of Metals. National Association of corrosion Engineers, Houston, Texas, 431 p. (1986).

[5] Buchwald, V. F. and Clarke, R. S., Jr. American Mineralogist, v. 74, p. 656-667 (1989).

[6] Buddhue, J. D. The Oxidation and Weathering of Meteorites. University of New Mexico Press, Albuquerque (1957).

[7] Zucchi, F., Morigi, G., and Bertolasi, V. Beta Iron Hydroxide Formation in Localized Active Corrosion of Iron Artifacts, in B. F. Brown et al. (eds.) Corrosion of Metal Artifacts, National Bureau of Standards Special Publication 497 (1977).

[8] Lucey, V. F. British Corrosion Journal, v. 2, p. 175-185 (1967).

[9] Campbell, H. S. A Review: Pitting Corrosion of Copper and its Alloys. in Staehle et al., eds., Localized Corrosion, National Association of Corrosion engineers, p. 625-638 (1974).

[10] Sato, S., Corrosion Engineering in Japan, vol. 31, pp 3-11 (1982).

[11] Shalaby, H. M., Al-Kharafi, F. M., Gouda, V. K., Corrosion, vol. 45, pp 536-547 (1989).

[12] Zen, E-An, Construction of Pressure-Temperature Diagrams for Multicomponent Systems after the Method of Schreinemakers - a Geometric Approach. U. S. Geological Survey Bulletin #1225 (1966).

[13] Becker, R., Theorie der Warme, Springer Verlag, Heidelberg, (1966).

[14] Woods, T. L. and Garrels, R. M., Economic Geology, v. 81, p. 1989-2007 (1986).

[15] Wagman, D. D., Evans, W. H., Parker, V. B., Schumm, R. H., Halow, I., Bailey, S. M., Churney, K. L., and Nutall, R. L. The NBS Tables of Chemical Thermodynamic Properties: Selected Values for Inorganic and C_1 and C_2 Organic Substances in SI Units. Journal of Physical and Chemical Reference Data, v. 11, supplement #2, 392 p. (1982).

[16] Mohr, D. W. and McNeil, M. B. , Use of Modified Log Activity Diagrams to Elucidate Type 1 Pitting Corrosion of Copper in Chloride Bearing Waters, Journal of Nuclear Materials, v. 190, p. 329-342 (1992).

[17] Lennox, T. H., Jr. and Peterson, M. H. Potential and pH Relationships in Cathodically Polarized Metal Crevices. in R. Staehle et al., eds., Localized Corrosion, National Association of Corrosion Engineers, p. 173-178 (1974).

[18] Cornwell, F. J., Wildsmith, G., and Gilbert, P. T. Pitting Corrosion of Copper Tubes in Cold Water Service. in Galvanic and Pitting Corrosion - Field and Laboratory Studies, American Society for Testing and Materials, p. 155-179 (1976).

A NEW CORROSION-PROTECTIVE POLYMER FOR ARTIFACT DISPLAY AND STORAGE APPLICATIONS

J. P. Franey, AT&T Bell Laboratories 600 Mountain Ave. Murray Hill, NJ 07974

Keith Donaldson, Baxter Industrial Div. 1919 S. Butterfield Rd. Mundelein, IL 60060

ABSTRACT

Corrosion of artifacts from atmospheric trace gases occur continuously. A new polymeric system which provides atmospheric trace gas corrosion protection has been developed. These polymers are called Reactive Polymers. The polymers consist of a blend of polymer resin (eg. polyethylene, polypropylene, ABS, PVC) with stable solid-state additives. These polymers can be formed into bags, tubing, sheets, bubble-pack, trays, etc. Corrosion protection is provided for artifacts from ancient periods, such as chinese mirrors, and the Freer Chinese Bronzes, to fine silver items from Colonial times in the USA. This protection from corrosion or tarnish by atmospheric H_2S, COS, and HCl is provided for extended periods of time.

I. Introduction

Corrosion of artifacts by trace atmospheric gases is of concern [1,2,3,4]. Although the rate of corrosion may be slow, the effect is cumulative [5]. Artifacts in storage or in shipment are particularly vulnerable to degradation. This problem is most severe for copper and silver which react with hydrogen sulfide (H_2S), carbonyl sulfide (COS), and/or hydrogen chloride (HCl). These trace gases are widespread [6]. Consider hydrogen sulfide: its ambient concentrations (Fig. 1, Museums are considered urban indoors) are generally in the range of 0.1 part per billion (ppb) to 10 ppb , although certain industrial areas, manholes, and areas near oil fields, pulp mills, or sewage treatment plants may have concentrations may exceed 100,000 ppb.

Laboratory measurements [7,8] of H_2S corrosion of copper, silver, and bronze (Fig. 2), show that corrosion product film thickness is proportional to the H_2S concentration times the

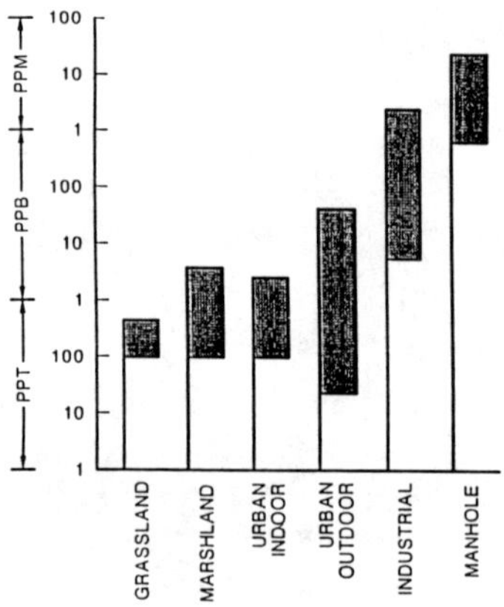

Figure 1. Ambient Concentrations of Hydrogen Sulfide

thickness is proportional to the H_2S concentration times the duration of exposure.

$$\alpha = (C \times T)$$

Where α = Part/Million Hrs., C is gas concentration in ppb/1000, and T is time in hours. One year of exposure of copper to 1 ppb of H_2S can grow a sulfide film on copper about 50 nm thick. Calculations using the above formula indicate that a sulfide film growth of about .0057 nanometers (nm)/hour at 1 ppb of gas concentration. Utilizing this method a determination of sulfide damage can be calculated.

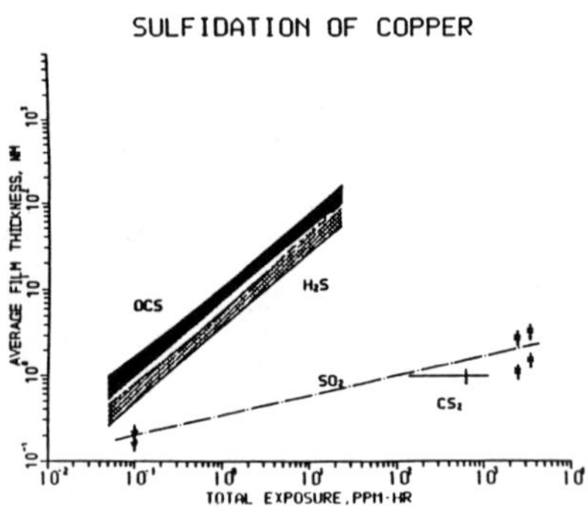

Figure 2. Corrosion of Materials Due to Sulfide Gases.

The traditional method of protecting artifacts from corrosion (other than cleaners and waxes) involves depositing a protective coating called a volatile corrosion inhibitor (VCI) on its surfaces. Typically a volatile organic oil is soaked into a piece of paper and then sealed in a container with the part to be protected. A thin insulating film forms on the material which excludes water and/or gases from the surface. There are two problems with this method. The first is the organic material can evaporate without warning over a period of time, which typically limits its useable life to <3 years from its time of manufacture. The second problem is that VCIs coat the artifact with a film, often requiring an extra cleaning step before viewing.

II. Reactive Polymers

The material described in this paper provides corrosion protection, and has properties which surpass those of the available alternatives. It performs independently of the humidity. It does not contain volatile additives, so it does not contaminate the parts it is used to protect. It consists of a polymer matrix such as polyethylene, polypropylene, ABS, or PVC, containing bonded copper particles. These additives react with corrosive atmospheric gases and thus prevent their penetration through the polymer by neutralizing them. The material is called Reactive Polymer because of the reactivity of bonded copper with corrosive atmospheric gases. Variations of conventional polymer processing techniques have been used to successfully produce Reactive Polymer in various forms.

Corrosion protection from the sulfur and chlorine containing gases H_2S, COS, and HCl results as their diffusion through the Reactive Polymer is stopped by reaction with additives to form stable compounds. The effectiveness of the Reactive Polymer for corrosion protection testing was done by sealing test coupons of copper inside a bag of Reactive Polymer. The bags were placed inside a chamber filled with a measured concentration of H_2S. After completion of exposure, some bags were cross sectioned and examined with a scanning electron microscope with energy dispersive x-ray analysis to determine the location of sulfur. The result (Fig. 3) shows that H_2S has penetrated only part of the way through the polymer. Controls run with polyethylene without the additives show H_2S beginning to penetrate 1000 times sooner.

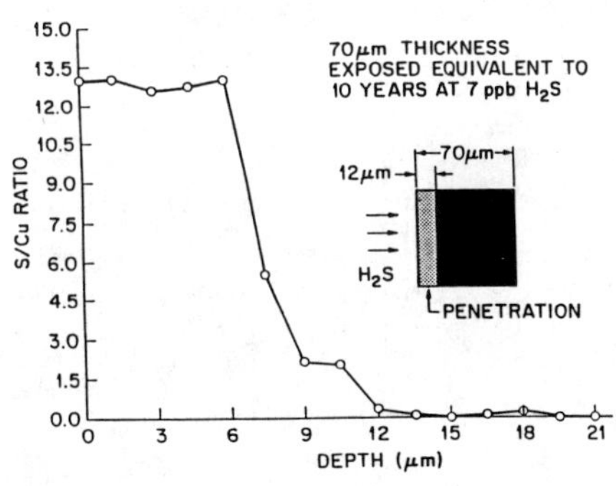

Figure 3. Penetration of Sulfur into Reactive Polymer.

Additional tests were performed using "Clam Shell" film holders, much like the ASTM cup device for measuring water weight loss. With 100% of HCl corrosive gas applied to one side of the film the concentration of permeating gas was measured on the reverse side with a Mast Oxidant Monitor. The time delay of the application of gas to one side to the measurement of the gas on the reverse side was the test criterion. Typical polyethylene time delays of 16 to 18 minutes were measured. For the same resins containing the Reactive Polymer additives time delays were 10,000 to 28,000 minutes.

A similar test for moisture permeation has been carried out by the ASTM cup method. The results (Fig. 4) show that the Reactive Polymer has a water vapor permeability less than half that of a sheet of polyethylene of the same thickness. Samples were run for 5 weeks at 23^0 C.

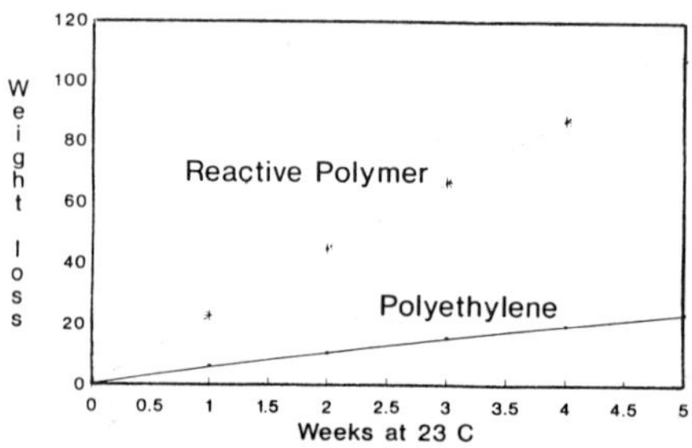

Figure 4. Water Vapor Permeability of Polyethylene and Reactive Polymer Polyethylene.

A very sensitive method for determining the extent of corrosion on a surface is to measure the electrical contact resistance of the surfaces of two mated conductors.

Thirty six copper bars (0.125" x 0.750" x 3") were manufactured, acid dipped, packaged, and stored for 4 weeks in air containing approximately 2 parts per billion of hydrogen sulfide. 12 bars were left bare, and 12 were packaged in a 3 mil bag of Reactive Polymer. The remaining 12 bars were buffed clean with hand held power buffers and No-Ox ® coated (a clear wax/grease). The individual bars were mated, bolted, and torqued to form electrical junctions of each protective scheme. The junctions were then tested for contact resistance using a model #398R Universal Systems 200 Ampere DC power supply to supply current across the junctions. Junction voltage drop was measured with a Fluke #8840A Digital Multimeter. Junction resistance was calculated by R= E/I and dissipated power by I x E.

Figure 5 shows the results. The unprotected bars formed poor junctions. The bars protected by the Reactive Polymer, however, provided junctions as good as the labor intensive buffed and No-Ox ® coated junction.

III. Applications

The corrosion protection characteristics of the Reactive Polymer and the absence of undesirable properties make it

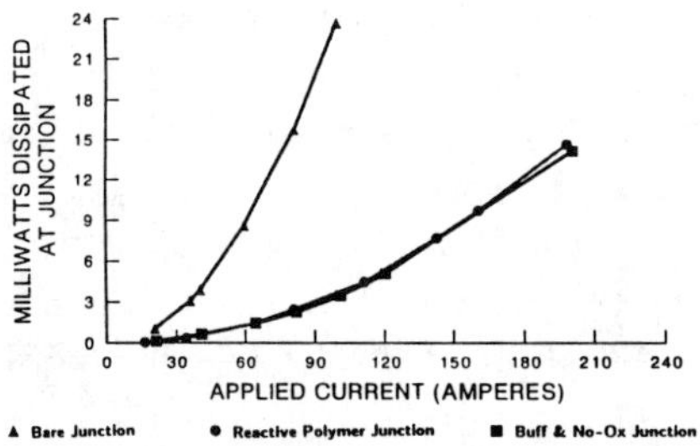

Figure 5. Comparison of Reactive Polymer Protection of Bus Bar Junctions compared With No Protection and Conventional Cleaning and Coating.

Polymer and the absence of undesirable properties make it attractive for use in a wide variety of applications. These fall in the general categories of commercial corrosion prevention, and consumer product corrosion prevention.

Thin polyethylene Reactive Polymer can be used for packaging in the form of bags, sheet, bubble wrap, and pallet wrap. PVC, polycarbonate, and glass filled sheet molding polymers can be used to manufacture many protective products. Flexible polyethylene sheet for one can be used for corrosion protection during and after manufacturing processes. Polypropylene can be used for workbench covers and vacuum formed parts. Other possible products are tote boxes, shipping tubes, storage wrappings, and aluminum foil laminations for dry pack requirements.

Corrosion protection of copper and silver is needed by all materials of a non ferrous nature. One expects that corrosion protection will reduce the number of molecules that materials succumb to as the destructive gaseous chemical interactions occur on a minute to minute basis continually.

ACKNOWLEDGEMENTS

We wish to thank the following for helpful discussions and work on the project: J. Boyle, and T. E. Graedel of AT&T, and H. Huang, H. Gold, and D. Robson of E. I. DuPont Experimental Station, Wilmington, DE.

REFERENCES

1. W. T. Chase, Corrosion and Metal Artifacts, NBS Special Pub. 479, pp. 195-196, U.S. Department of Commerce, Washington, DC (1977)

2. J. Kruger, Corrosion and Metal Artifacts, NBS Special Pub. 479, p. 204, U.S. Department of Commerce, Washington, DC (1977)
3. W. T. Chase and U. M. Franklin, Ars Orientalis 11, 215 (1980)
4. L. Soto, J. P. Franey, T. E. Graedel and G. W. Kammlott, Corrosion Science, 23, pp. 241-250, (1983)
5. J. P. Franey, G. W. Kammlott, and T. E. Graedel, Corrosion Science, 25, 2, 1985
6. R. P. Turco, R. C. Whitten, O. B. Toon, J. P. Pollack, and P. Hamill, Nature 283, 283, 1980
7. T. E. Graedel, G. W. Kammlott, and J. P. Franey, Science 212, 663, 1981
8. T. E. Graedel, and J. P. Franey, JAPCA, Vol 35, No.6, June 1985, 644-648.

THE CONSOLIDATION AND DESALINATION OF CERAMIC IMPREGNATED WITH SODIUM SULPHATE

ALICE B. PATERAKIS
Conservator, Agora Excavations, American School of Classical Studies, 54 Souidias St., Athens 106-76, Greece

ABSTRACT

A calcareous and non-calcareous clay were fired to 900 degrees Celsius and subsequently impregnated with a solution of Na_2SO_4. These test samples were then impregnated with one of three consolidants: Acryloid B72, Tetraethyl Orthosilicate (TEOS), Methyl Triethoxysilane (MTS). Desalination tests using the static immersion and stirred solution methods were made to determine the effect of consolidation on the rate and extent of salt extraction. Desalination was monitored by Ionic Conductivity, Atomic Emission Spectroscopy and Ion Chromatography analysis of the extraction solutions.

INTRODUCTION

Salt-contaminated archaeological pottery was simulated to test conservation materials and treatment methods. The purpose of these tests was to determine: 1) the effect of the consolidant on the amount of salt released in solution, 2) the effect of the consolidant on the rate of salt released in solution, 3) the effectiveness of desalination methods, 4) the effectiveness of consolidants based on degree of desalination achieved, 5) the influence of the physical properties of the ceramic, apparent porosity and liquid absorption, on the desalination. In these tests sound ceramic samples were used since the scope of this project was to determine the effect of the consolidants on the desalination process, not strength and cohesive properties imparted to the host object.

TEST PREPARATION

Two clays similar to those commonly used for pottery manufacture in Greece and the Mediterranean world in antiquity were chosen. A calcareous clay from Crete, Greece and a ferrous clay which was obtained commercially in Toronto, Ontario, Canada. Both clays are combinations of the common clay mineral compounds illite, kaolinite and montmorillonite as determined by X-ray Diffraction analysis. Clay tiles measuring approximately 1.5 x 1.5 x 0.5 cm were prepared and fired in an oxidizing atmosphere to 900 degrees Celsius for a duration of 30 minutes. This temperature was chosen since it falls within the range commonly used for the firing of pottery in ancient Greece. Sodium sulphate was chosen as the salt contaminant based on its occurrence in the burial environment [1]. A 1.5 M solution of sodium sulphate was introduced into the tiles by vacuum impregnation under reduced pressure of approximately 22 mm Hg [2]. The TEOS solution (vol %) consisted on 60% TEOS, 32% ethanol (95%), 8% HCl (1% vol/vol aqueous solution) [3]. The MTS was prepared in the proportion

1 ml MTS to 0.135 ml 0.1N HCl under agitation [12]. The same impregnation method was used to introduce the TEOS and MTS into the salt-laden tiles. The Acryloid B72 as a 10% weight/volume solution in acetone was introduced through the base of the tiles by capillary absorption in a covered chamber to retard evaporation of the solvent.

CONSOLIDANTS

The condition of salt-contaminated ceramics must be determined before salt removal. If the cohesive properties of the object have been weakened by the action of soluble salts or other factors to the extent that the condition of the object will not permit the removal of these salts then a consolidant must be applied which will increase the internal cohesive properties. The volume and distribution of the open pores in the object will influence the quantity of consolidant which may be deposited. The consolidant reduces the diameter of the capillaries and the open pore volume thereby theoretically reducing the rate of water passage and subsequently the rate of salt extraction. Three polymeric materials were tested as consolidants intended for pottery which has been weakened by the action of soluble salts in order to permit the removal of these salts.

Acryloid B72, a methylacrylate-ethyl methacrylate copolymer, was chosen for its long-term stability [4]. The surface tension of the solution influences the degree of penetration, as does the porosity of the object, the volatility of the solvent carrier, the concentration of solids in solution, and the molecular weight of the polymer. The acrylic resin is deposited on the surface of the pore walls and capillaries upon evaporation of the solvent carrier. Acryloid B72 allows the passage of water vapor in porous objects. Passage of liquid water depends on the extent of bulking of the pores by the resin.

Two alkoxysilanes were chosen which have been applied extensively as consolidants for salt-contaminated and salt-damaged stone. Tetraethyl Orthosilicate (TEOS) ($Si(OC_2H_5)_4$) is a tetrafunctional silane derivative and MTS ($CH_3Si(OC_2H_5)_3$) is a trifunctional silane derivative [5,6]. Their low viscosity allows good penetration in the pores. MTS renders the object water-repellent, not waterproof, and allows the passage of water vapour. The increased water-repellency of MTS is afforded by the presence of methyl groups. The polymerization of both is initiated by water and occurs by hydrolysis and condensation [6]. The film of amorphous silica deposited reduces the porosity of the object, retards the penetration of moisture and increases hardness [7]. MTS has been tested as a consolidant for friable ceramics [8] and for the reduction of water absorption in ceramic roof tiles [9]. TEOS is reported to have allowed the migration of salt in limestone [10]. Tests using the alkoxysilanes to consolidate salt-laden brick were found to reduce the porosity, water absorption and permeability of the brick [11].

PHYSICAL PROPERTIES AND WEIGHT GAINS OF THE CERAMIC TILES

The weight gains(%) of the tiles upon impregation with sodium sulphate and the consolidants were determined gravimetrically (Table I). The slightly greater porosity and absorption capacity of the untreated calcareous ceramic were demonstrated by the greater weight gains made upon salt impregnation (Tables I,II). The greater reduction in both properties of the calcareous ceramic upon salt impregnation is reflected by the smaller weight gain made upon consolidation with TEOS and MTS (Tables I,II). The product yield indicates the quantity of silica deposited by TEOS and MTS (Table I). The maximum standard product yield for the fully cured resin is reported to be 37.6% for MTS [12] and 16% for TEOS [3]. The product yield obtained with MTS fell short of the maximum in the low-calcium ceramic but exceeded the maximum in the calcareous ceramic. The product yield with TEOS exceeded the maximum standard yield in both ceramics. The excessive yields with MTS and TEOS may be attributed to incomplete curing of the resin or incomplete evaporation of the solvent carrier. Yields below the maximum standard may be due to loss of the monomer by evaporation from the surface.

TABLE I WEIGHT GAINS (%)

	calcareous ceramic			non-calcareous ceramic			
	w/salt	w/con	product yield	w/salt	w/con	product yield	desal method
control	6.14			4.75			si
	7.04			4.34			ss
w/Acry	9.69	1.15		3.74	0.89		si
	8.18	0.78		3.62	0.49		ss
w/TEOS	4.56	0.74	25.64	3.41	2.92	18.75	si
	4.59	0.88	24.49	3.52	3.05	19.28	ss
w/MTS	4.57	2.44	39.44	3.47	5.22	17.72	si
	4.51	3.37	3.05	3.60	5.25	23.48	ss

con = consolidant,ss = stirred solution,si = static immersion

TABLE II PHYSICAL PROPERTIES OF CERAMIC TEST TILES

tiles	apparent porosity P%	liquid absorption A%	
low-Ca	31.49	17.30	control
high-Ca	33.82	18.44	
low-Ca	23.41	9.22	w/Na_2SO_4
high-Ca	15.91	7.00	

TABLE III ATOMIC EMISSION SPECTROSCOPY RESULTS Na ppm

hrs elapse	control	Na_2SO_4	Acry	TEOS	MTS	desal method	
8	0.9	180	250	170	62.5	si	high-Ca
104	<0.4	8.4	5.9	10.4	8.6	si	
8	1.3	165	275	175	60	ss	
104	<0.4	5.0	3.0	9.8	3.4	ss	
8	0.4	427.5	407.5	385	282.5	si	low-Ca
104	<0.4	0.4	0.4	0.6	2.2	si	
8	0.5	300	450	385	200	ss	
104	<0.4	<0.4	0.4	0.4	1.15	ss	

TABLE IV ION CHROMATOGRAPHY RESULTS SO_4 ppm

	control	Na_2SO_4	Acry	TEOS	MTS	
1st sol	16	1081	640	390	169	high-Ca
last sol	<5	nd	8	22	12	
1st sol	<5	1152	1030	860	580	low-Ca
last sol	nd	nd	<2	<2	<5	

nd = not determined, Acry = Acryloid, sol = solution

METHODS OF DESALINATION

In the static immersion method the tiles were placed on glass stirring rods resting on the bottom of the beaker. In the stirred solution method the tiles were suspended in teflon-coated fiberglass mesh bags in deionized water which was stirred constantly by a magnetic stirrer. Both methods were tested at room temperature. The ionic conductivity in μmhos/cm(μsiemens/cm) of the desalination solution was measured every 2 hours over an 8 hour period daily for the duration of 5 days (104 hours total). The solution was monitored, then replaced with new deionized water, twice daily at the start and finish of the 8 hour day. The conductivity of the solutions were totalled to derive the total ionic conductivity of each desalination test. The first and last extractant solutions of all tests were kept for analysis of sodium (Na ppm) by Atomic Absorption Spectroscopy (Table III) and those of the stirred tests for analysis of sulphate (SO_4 ppm) by Ion Chromatography (Table IV). The stirred solution tests only were monitored by Ion Chromatography.

RESULTS

AES and IC confirm the presence of sodium sulphate in all of the last desalination solutions from the calcareous ceramic and from those of the ferrous ceramic with MTS. AES and IC confirm the removal of all sodium sulphate by the last desalination solution from all the other ferrous ceramic tests. The rate of salt release was determined by calculating each conductivity reading as percent of the total conductivity which was plotted against hours elapsed. Similar curves were shared by like desalination methods and therefore the graphs representing the static immersion method are illustrated (Figures 1,2). More salt was released initially from the non-calcareous ceramic (Figure 1) whereas the release was more gradual from the other ceramic (Figure 2). The percent total conductivities were a bit higher in the stirred method. The slowest release of salt occurred with MTS and the fastest with Acryloid in all graphs. In the non-calcareous ceramic the rate of salt released by those tiles consolidated with Acryloid and TEOS approached that of the unconsolidated tiles 24 hours after the start of desalination whereas the MTS-consolidated tiles never approached the rate of these other tiles (Figure 1). In the calcareous ceramic the rate of salt release from the consolidated tiles closely followed that of the unconsolidated tiles throughout the duration of the tests (Figure 2). The faster rate of desalination of the low-calcium ceramic indicates greater porosity and absorption capacity. Similar results were found in parallel tests using NaCl [13].

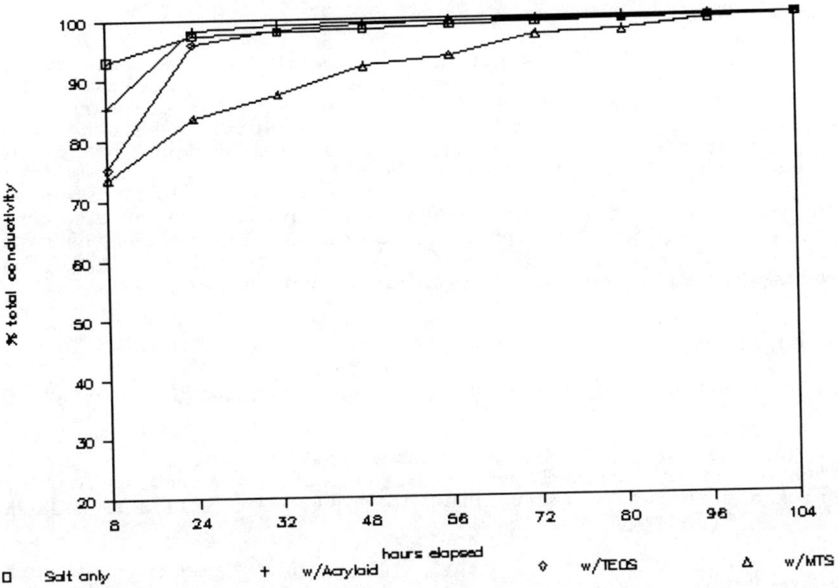

FIGURE 1 RATE OF SALT RELEASE non-calcareous ceramic

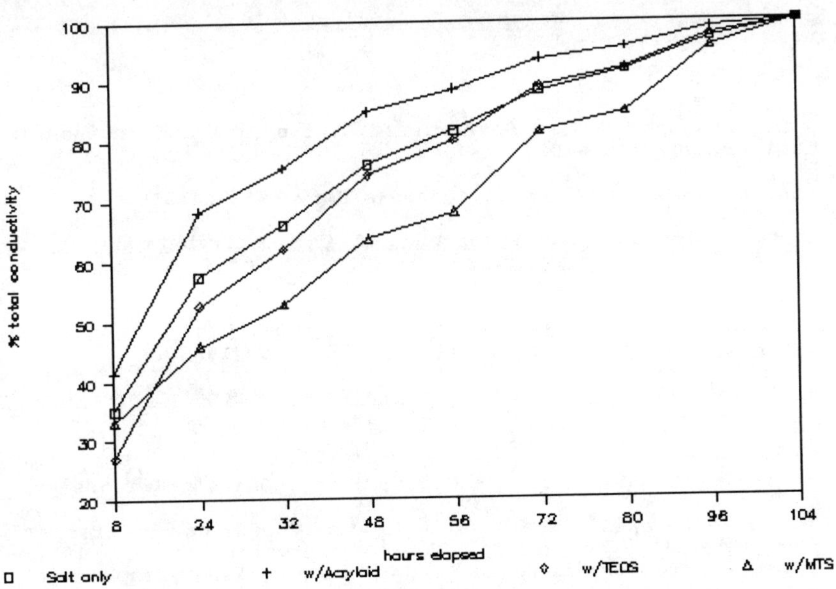

FIGURE 2 RATE OF SALT RELEASE calcareous ceramic

CONCLUSION

There was little hindrance of dissolution and release of salt by Acryloid and TEOS. The increased hydrophobicity of MTS explains the slower rate of salt release with this compound. The stirred solution and static immersion desalination methods were of similar efficiency. Given the slower rate of desalination of those tests which didn't achieve complete removal of sodium sulphate, extending the duration of these tests would be required for complete salt removal. The physical properties of apparent porosity and liquid absorption of the ceramic influenced the rate and extent of salt release, the two ceramics displaying remarkably different patterns of desalination.

BIBLIOGRAPHY

1. E.A.Dowman, Conservation in Field Archaeology, (Methuen, London, 1970), p.28.

2. "Standard Methods for Apparent Porosity, Liquid Absorption, Apparent Specific Gravity, and Bulk Density of Burned Refractory Brick and Shapes by Vacuum Pressure", C830-79, 1983 Annual Book of ASTM Standards, Vol.15.01.

3. Union Carbide Corp., Ethyl Silicates, Chemical Abstracts Registry No. 78-10-4.

4. C.V. Horie, Materials for Conservation, (Butterworths, London, 1987), pp. 106-109.

5. C. Grissom and N. Weiss, AATA Supplement 18(1)(1981), pp. 152-154.

6. C.V. Horie, ibid., pp. 154-156,159.

7. H.D. Cogan and C.A. Setterstrom, Industrial and Engineering Chemistry 39, 1364-1368 (1947).

8. A. Bergeron, MAC report, Queen's University, 1980.

9. J.R. Nishiura, in Preprints of the Contributions to the International Symposium on the Conservation of Stone II, 699-709 (Bologna,1981).

10. S.M. Bradley, in Actes - 5th Congress International Alteration Conservation Pierre (2), 759-768 (1985).

11. T. Perander and T. Raman, in 3rd International Conference on the Durability of Building Material Components (2), 463-477 (1984).

12. J. Hanlan (Dept. of Art Conservation, Queen's University).

13. A.B. Paterakis, "The Desalination of Consolidated Ceramics", International Symposium on The Conservation of Glass and Ceramics, ICOM Committee for Conservation Working Group Ceramics (Amsterdam, 1991), to be published.

PART V

Experiments in Artifact Study and Education

EVALUATION OF RAPID PROTOTYPING TECHNOLOGIES FOR USE IN THE PRODUCTION OF ART AND ARTIFACT STUDY COPIES.

Jerry C. Podany
Department of Antiquities Conservation
The J. Paul Getty Museum, Malibu, California.

ABSTRACT

Many artifacts and works of art are too fragile to be accurately replicated for scholarly study by using traditional methods of plaster or silicone molding techniques. In addition to the potential for loss of surface and fracturing, the use of many modern silicone mold materials may lead to irreversible staining of the original surface.

An evaluation of newly developed rapid prototyping technologies and their potential application to this problem was conducted by the J. Paul Getty Museum Department of Antiquities Conservation in conjunction with Laser Design Inc. of Minnesota and Hughes Aircraft Corp. of California.

Rapid prototyping technology allows accurate copies to be made without surface contact with the original object. This paper discusses the outcome of an initial investigation into one of these processes, stereolithography. A plaster model was submitted for laser scanning using a point laser probe, programmed for a 3.175 mm (0.125") distance between scan lines. The input was filtered and stored in an STL (stereolithography) format which provides the X,Y and Z coordinates of the A,B,C and normal vectors of a predetermined number of surface triangles. This data was then used to produce a three-dimensional copy.

Stereolithography produces hollow or solid three-dimensional forms by feeding data that make up cross sectional slices of the object scanned to a focused laser aimed along the Z axis of the potential copy. The laser polymerizes the outer contour of the 'slice' in a thin layer of photopolymer deposited on a stage located below the laser. Following each slice production the stage lowers approximately 19 mm (0.75") below the surface of the liquid polymer reservoir and upon resurfacing positions approx 0.05mm (0.002") lower than its previous position. A "squeegee" gently crosses over the surface resulting in a thin film of polymer available for the production of the next cross section. The laser is then sent the next cross sectional data and the sequence of section building continues.

The specific challenge presented by this project was to assure that the surface detail of the copy formed presented acceptable fidelity to the original object (model), avoiding any need for follow up bench work or correction. The advantages, drawbacks and newly developed alternatives to this promising technology are discussed.

The Process

Like all new technologies rapid prototyping is known by a myriad of names during this early period of development (Rapid Three-Dimensional Prototyping, Desk Top Manufacturing, Conceptual Modeling, Solid Free-Form Manufacturing, etc.).

Scanned data (including data from CT, computerized tomography) or CAD, CAM or CAE developed data is sorted into planar slice descriptions of the source object or form. The data from these planar slices is used to build a three-dimensional shape by incremental material build up in thin layers. While new approaches to the build up of these layers are being developed daily there now exist four predominant methods: Laser Sintering, Laminated Build Up, Ballistic Particle Manufacture and Stereolithography.

Solid free-form manufacturing is well-suited for the rapid production of complex models obtained through computerized design programs (such as CAD) or from scanned, or tomographic data. The scanned data can be obtained in any number of available ways including three-dimensional coordinate measuring devices and laser scan units.

Once a point model is developed from this data, its form is calculated as a wire frame structure. This structure is then transformed into a surface model by patching each resulting grid with polygons (often triangles). This is then followed by development of a solid model. A set of cross sections is then calculated, using appropriate software employing non-regular set operations [1]. The thickness of each planar cross section, or slice, is determined by the operator's needs or the limitations of the equipment.

The development of this technology has been spurred on by the need for faster, more cost efficient and more reliable ways of producing models or parts in the medical and aerospace fields, as well as in the automotive industry such as a complete Mercedes-Benz exhaust system[2]. Using traditional methods the manufacturing of a prototype involves very long delays between the conceptual design and the finished product, especially when complex shapes are involved. Severe undercutting often causes problems which computerized or robotic technologies, such as computerized mills, cannot overcome due to the limitations on the cutting tool pathways. The creation of a "hand produced" model (clay, assembled, etc.), and subsequent molding, casting and finishing of the part, normally takes weeks or even months to complete. If a change becomes necessary significant loss of time and money result. Rapid prototyping, on the other hand, can verify fit, overall design and proper function of the part prior to mass production and all in a fraction of the time. Additionally new developments allow production using materials that permit immediate application of the part to the system for which it is being designed (such as the production of plastic pressure connectors).

Stereolithography (SLA)

Developed within the aerospace industry, this branch of technology uses a photosensitive polymer for part production. In some units, such as the SLA 250 and SLA 190, produced by 3-D Systems Inc., the polymer is hardened by ultraviolet emissions of a HeCd laser. Other units, such as the 3-D Systems SLA 500 or those manufactured by Quardrax, use full spectrum light from an argon-ion laser for curing a light sensitive polymer. The process has found wide use in industry and recently in the medical field for the production of prosthesis and the study of CT scanned segments. Mankovich, et all [3], reported using SLA for the production of a cranial model from CT data that included full internal detail such as encased sinuses. The normal surface model is one which uses numerous triangular patches of various sizes and configurations. The surface form is then transformed into a laser compatible vector format by slicing the representation at user-defined intervals, normally 0.508-0.1270mm (0.02-0.005"). This slice vector data is then formatted to create a laser path instruction set that guides the laser unit. Figure 1 presents a system schematic.

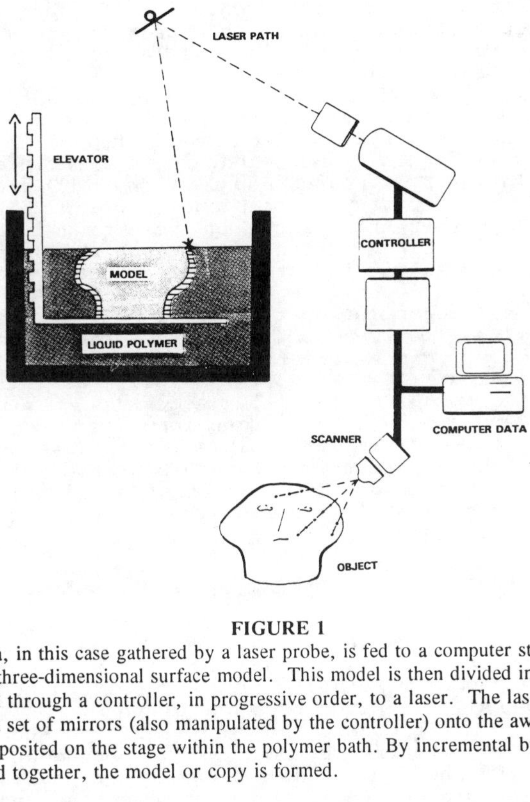

FIGURE 1

The scanned data, in this case gathered by a laser probe, is fed to a computer station where it is used to create a three-dimensional surface model. This model is then divided into "slice" data sets which are passed through a controller, in progressive order, to a laser. The laser beam is guided and focused by a set of mirrors (also manipulated by the controller) onto the awaiting layer of photopolymer deposited on the stage within the polymer bath. By incremental buildup of slices, which are welded together, the model or copy is formed.

The model production begins with the input of the data through a laser device that is pointed to an elevator platform submerged in a vat of photopolymer. A 10 mV HeCd or a 40-50 mV argon laser combined with a shutter and two computer controlled galvanometer mirrors traces the cross sectional path over a thin layer of the photopolymer, polymerizing a small area at a time as it travels. At this point the laser also begins to form the 'footprints' of any supporting structures or 'buttresses' needed to support "floating contours" at a later point in the building of the model. Through proper control of the platform movement, as well as the liquid viscosity and the position of the laser (including the speed of the path), the solid plastic ridge (ie. the first few cross sections) adheres to the platform and forms a foundation for the model.

Following the first slice production the platform lowers into the polymer reserve and rises again. A wiper crosses over the surface of the vat to assure that the new layer of pre-polymer is of the predetermined thickness. The laser then traces the path of the next cross section, polymerizing and 'welding' it to the previous section. The computer instructs the laser to "fill in" near horizontal planes with a cross-hatch pattern to assure a more structurally stable model. Repetition of these actions and production of supports for floating contours (those areas which appear in a given plane but do not 'connect' to the model until later planes are made) results in the construction of the model.

Upon completion the model is removed, cleaned and subjected to a post cure using long-wave UV. This additional cure assures that the unpolymerized resin trapped in the voids of the model walls is fully hardened. Linear shrinkage can be 1.2% maximum in the X, Y planes and 0.2% in the Z plane [3], although these figures vary. Occasional problems have been reported due to differential shrinkage resulting in warpage, especially of flat planar surfaces. However even minimal curvature provides sufficient structural integrity to resist deformation [2].

3-D Systems' second generation model SLA 250 has a drawing speed of 380mm/sec (15"/sec) but a limited object/model size 255 mm (10") cube. The newer SLA 500 increases that model size to 510 X 510 X 610mm (20 X 20 X 24"), as well as increasing production speed to 381 cm/sec (150"/sec) [4].

Quadrax Corp uses a 5 watt argon-ion laser and a low viscosity polymer sensitive to visible light. One advantage of this system is that no uncured polymer is trapped in the honeycomb patterns of the model wall, thus eliminating post production UV cure. Total dimensional tolerances are held at +/- 0.13mm (+/- 0.005") [5].

Selective Sintering

Developed at the University of Texas at Austin,[5] selective sintering technology uses a heat fusible powder that is raised to a temperature just below its melting point. The laser beam, normally supplied by a CO_2 laser of 25-40 watts[6], fuses the powder as it traces the outline of the successive slices and welds it to the previous form. One advantage of the SLS system is that no, or minimal, support structures are necessary during production of the part since the unsintered powder mass itself acts as a natural and overall support. Clean up is faster and there is no need for a post cure, as with SLA. SLS can produce approximately 12.7-25.4mm(0.5-1") per hour using materials suitable for a wide range of demands such as polycarbonate, wax, PVC, ABS plastics, Nylon, low temperature ceramics and metallic powders. Future materials adapted to the technique may minimize or eliminate warpage.

The first commercially available unit was produced by DTM Corp of Austin, Texas. The unit consists of two cylindrical chambers, located side-by-side. Each cylinder holds a vertical stage and a powder leveling device. The "powder cylinder" stage raises, exposing uncured powder to a leveling roller. The roller transfers the powder to a "part cylinder" whose stage had automatically lowered a predetermined increment equal to the thickness of the slice to be polymerized. DTM model 125 SLS system sinters each layer by the 10.6 micrometer wave length beam of a 23 watt CO_2 laser. The model cylinder or chamber is 305 mm in diameter and 380 mm in height (12 X 15"). The SLS system is said to be faster considering that no clean up or post cure are necessary [5].

Applications to Archaeological and Fine Art Research

The Getty Museum has in its collection a rare cruciform Cycladic idol dating to 2500 BC, figure 2. The 39.5cm (15.5") tall idol was carved from soft compact limestone and its surface supports a fragile and porous patina. Requests for copies of this idol for scholarly study and exhibition prompted a search for alternative methods of reproducing the figure.

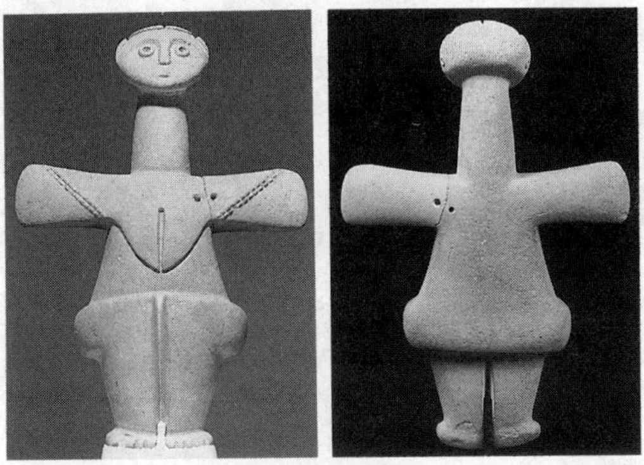

FIGURE 2

Since the surface is extremely fragile traditional methods of mold-making were rejected. Release agents that protect the surface of originals during the mold-making process would have stained or left irreversible residues. Direct silicone molding risked both staining from the unreacted silicone, silicone oil or activator, as well as potential loss of fragile surface detail from physical adherence of the silicone to undercut or porous areas. Rapid prototyping in the form of stereolithography and powder sintering was investigated. The various companies contacted had no previous experience in reproducing such materials. And while all boasted of a high degree of accuracy in overall dimensions, none could predict the accuracy of surface feature reproduction, which is one of the more important aspects. Weathering patterns, surface texture design elements and subtle tool marks all had to be reproduced in acceptable fidelity if the techniques were to be of any use. Obviously the potential of reproducing the idol without coming into physical contact with it was of interest.

It was decided to test the abilities of laser scanning for acquiring the data and stereolithography for generating the model of a similar idol. This second idol is in the collection of the Cyprus Archaeological Museum which kindly provided a plaster copy. Its surface, although more weathered, provided many of the same challenges as the Getty idol. It was decided that for the first trial only the head and neck, to the point of the shoulders, were to be scanned and reproduced.

The head section of the cast was sent to Laser Design Inc. of Minneapolis, Minnesota to be scanned. A grid pattern of approx. 3.175 mm (0.125") between scan lines was used, though some areas were more dense to assure detail fidelity. The data was acquired using a point laser probe, and after consolidation and filtering, was provided as a hollow STL file on 1/4" tape.

The tape was then sent to Allan Lang and Timothy Thelander of Hughes Aircraft for use with their 3-D Systems stereolithography apparatus, SLA 500. The first attempts to manufacture a copy were carried out with a slice division of 0.254mm (0.010"). This model was considered too coarse for the purpose since the 'stepped' contours were too broad, figure 3.

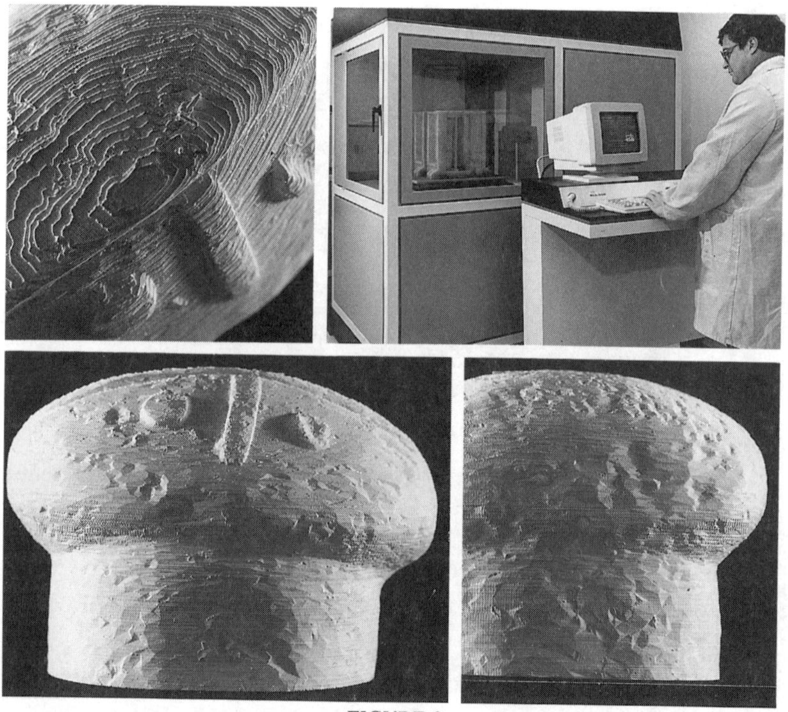

FIGURE 3

3a. A top view of the initial copy showing the stratogrpahic layering of the slice production at 0.254mm definition.

3b. Programming the 3-D Systems Inc. SLA6000 unit.

3c. Frontal and partial back view of the first copy produced using a 0.254mm (0.010") slice thickness. Note the misleading presence of triangular data planes.

The second model was developed using 0.127mm (0.005") slices (approximately the thickness of typing paper) and the accuracy improved dramatically. However, an additional problem began to arise. As explained previously, during the data processing phase the wire model is transformed to a surface model by filling the various point-to-point polygons with triangles. These data triangles began to distinctly appear on the solid model and had the strong potential for being seen as tool marks. The interest of the two scientists from Hughes Aircraft continued as they kindly proceeded to attempt a solution. First they developed methods to reduce the slice size from 0.127mm (0.005") to approximately 0.05mm (0.0025"), which increased the detail of the model and decreased the effect of the stepped contour. The data for the surface generation of the model was then filtered to minimize the effect of the triangular data sets. Working with Laser Design Corp., an increase in the number of triangles from 30,000 to 410,000 was achieved resulting in a increase in file size from 3 megabytes to 69 megabytes. Effectively the triangle size was reduced by a factor of ten, reducing the visually disruptive effect. The final copy, produced with the STL technique, is compared to the "original" plaster cast in figure 4.

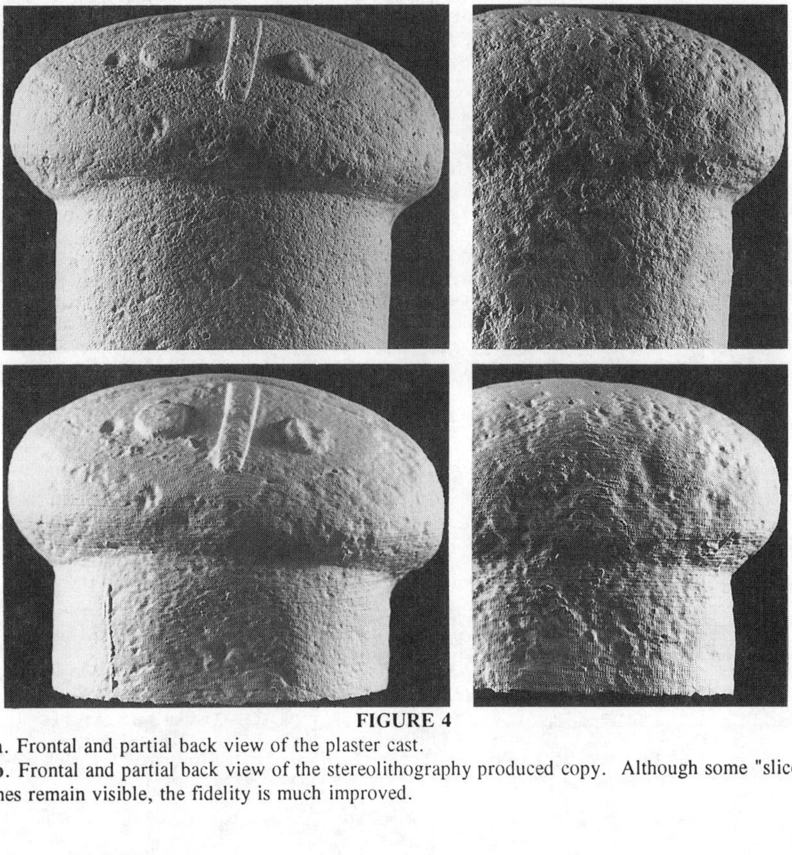

FIGURE 4

4a. Frontal and partial back view of the plaster cast.
4b. Frontal and partial back view of the stereolithography produced copy. Although some "slice" lines remain visible, the fidelity is much improved.

Future Directions

Current models built of straight lines or planar polygons will eventually be replaced by free form curves and surfaces that will reduce if not eliminate the effect of the data point planes discussed earlier in the text. Such improvements will dramatically increase the already promising fidelity of surface reproduction. Additionally new computer algorithms are being developed to automatically create support structures for floating contours [3]. This improvement would minimize operator work, reduce the number of supports and assure that these supports would be placed in the most effective areas.

Conclusions

At the moment unit cost, size limitations and full rendering capabilities place the various techniques of rapid prototyping outside the realm of immediate practicality for the majority of applications within the field of art and archaeology. However, this initial study shows that the technology is promising. Since the difficulties and dangers associated with traditional mold-making techniques are well known, there is little doubt that these newly developed techniques, combined with the rapid pace of development, will soon provide the conservator, scholar and archaeologist with a significantly improved tool for the study and analysis of art and artifacts.

Acknowledgments

The author would like to thank Timothy Thelander and Allan Lang for their assistance, as well as Hughes Aircraft for supporting their involvement in this project. Were it not for the ability of Mr. Thelander and his interest in the application of this technology to the problems of art and archaeology, little exploration would have been possible.
Gratitude is also owed to 3-D Systems Inc. for their interest and generous cooperation, Kent Nutt of DTM Corporation for the production of powder-based rapid prototypes and Laser Design Inc. for completing the contracted scanning work.

Bibliography

1. Y. Choi, L. Weiss, L.E. Gursoz and F.B. Prinz, "Rapid Prototyping from 3D Scanned Data through Automatic Surface and Solid Generation" in Issues in Design/Manufacture Integration, (Proceedings, The Winter Annual Meeting of the American Society of Mechanical Engineers, Texas, Nov. 25-30, 1990) A.S.M.E., New York, N.Y., 1990. pp. 17-23.

2. Hoffmann, Christoph M., Geometric and Solid Modeling; An Introduction, (Morgan Kaufmann, San Mateo, CA., 1989)

3. N.J. Mankovich, A.M. Cheeseman and N.G. Stoker, in Medical Imaging IV: Image Capture and Display, (SPIE International Society for Optical Engineering Proc. Vol.1232, Washington, DC., Feb. 1990) pp. 183-187.

4. Personal communique with Timothy Thelander of Hughes Aircraft (1992).

5. Anon, "Laser Based Methods Speed Up Modelmaking", in Advanced Materials and Processes, V. 138, no. 1 (July 1990), pp. 51-53.

6. K. Nutt, Photonics Spectra, 102-104 (Sept. 1991).

For additional information on rapid prototyping technology:

3-D Systems Inc., Tom Vorgitch, Dir. of Marketing, 26081 Ave. Hall, Valencia, CA. 91355.
3-D Systems first introduced their equipment SLA in 1987 in Detroit.

Quadrax Laser Technologies Inc., Lawrence A. Girouard, VP Marketing and Sales, 300 High Point Ave., Portsmouth, Rhode Island, 02871.

DTM Corp., Denise K. Medler, VP Sales and Marketing, 11002A Metric Blvd., Austin, Texas, 78758.

DeSoto Inc., Des Plaines, Ill. Makes the materials for both processes. The resin is combined with a small amount of photo-initiator with acrylic-functioned monomers and oligomers. When exposed the photo-initiator absorbs the energy and forms a free radical species which in turn initiates polymerization of the acrylic.

SCIENCE OF ART AND ANALYSIS: A MIXTURE FOR UNDERGRADUATE NON-SPECIALISTS

W. STANLEY TAFT* AND JAMES W. MAYER*
*Dept. of Art, Cornell University, Ithaca, NY 14853-7301
*Dept. of Materials Science, Cornell University, Ithaca, NY

ABSTRACT

At Cornell University we are in the third year of teaching an interdisciplinary, undergraduate course on the physical properties and structures of works of Art, and the modern analytical methods used to investigate them: Art, Isotopes, and Analysis. The challenge is to explain concepts familiar to museum scientists and conservators to a group of 150 undergraduate students with a background that ranges from Art History to Computer Science. Painting techniques (Fresco, Tempera, Oil, etc.) are demonstrated to the class. The analytical techniques involve the interactions of electrons, photons, ions and neutrons with pigments and other materials. This instructional approach serves as an introduction to published analyses of works of art.

Introduction

The structure of paintings and scientific methods of analysis are familiar topics to Museum Scientists and Conservators. These topics are also part of the curricula of graduate programs at several Universities. The world behind the doors of the analytical laboratories of the major museums is virtually unknown to most undergraduates. There have been glimpses offered by the Nova TV programs in subjects such as the Restoration of the Sistine Chapel. By and large, however, the university undergraduate is not exposed to the topic of this conference: "Materials Issues in Art and Archaeology."

Over the past five years, there has been increased attention paid to the need for increasing the scientific literacy of the non-scientist/engineer in our society. Cornell University has encouraged the introduction of interdisciplinary courses that reach across the traditional boundaries between Department curricula. The scientific analysis of works of art is an ideal topic for an interdisciplinary course. The art objects themselves are magnificent and hold the attention of the students. The artists were geniuses in their use of paint to describe their view of the world around them. The principles behind the analytical techniques only involve basic concepts and as such can be explained to an interdisciplinary class. The instruments may appear complex but the analysis can be treated at a far simpler, introductory level.

Course Content

The theme of early part of the course is the structure of a painting - a description of the technique - and how light interacts with the paint layers to produce the observed color/ image. The spectrum of visible light is just a small portion of the electromagnetic spectrum. The emphasis is on photons, photon - electron interactions and the photoelectric effect. The perception of color is discussed through photon interactions with the receptors in the eye. The color of objects is due to the subtractive effect.

Interspersed between lectures on photons, demonstrations are given of painting methods and materials employed in fresco, tempera, and oil techniques and lectures given on color dynamics. Slides are shown of a wide range of styles, periods and media including the Sistine Chapel, (before and after restoration) and of painters such as Dieric Bouts and Vincent Van Gogh.

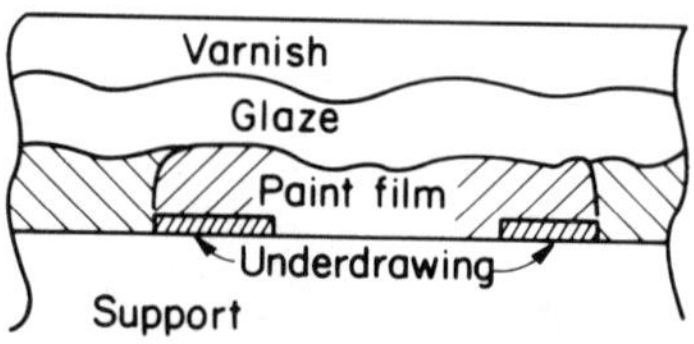

Infra-red reflectography – { charcoal, black underdrawings

X-radiography – { Pentimento; Pb pigments

Polarizing Light Microscopy – visual

X-ray diffraction – { Crystal structure, Pigment identification

X-ray emission
- X-ray fluorescence
- Particle-induced

{ element in pigment identification. $E_{X\text{-}ray} = E_B^1 - E_B^2$

Scanning Electron Microscopy { Pigment morphology

Neutron autoradiography – Pigment ident.

Fig. 1. Science topics in course

The transparency of most paint layers to the infrared is demonstrated with an infrared video camera (short-term loan from Hamamatsu) which reveals charcoal underdrawings that are overlaid with different paint layers. The paint is applied to transparent acetate sheets so that absorption by different pigment combinations is demonstrated. Infrared reflectography introduces the topic of pentimenti (alterations made in the image of the painting by the painter as the painting is developed) and is illustrated with museum studies. Pentimenti discussions lead to x-ray absorption and a return to the photoelectric effect and the electron-volt as the natural unit of energy.

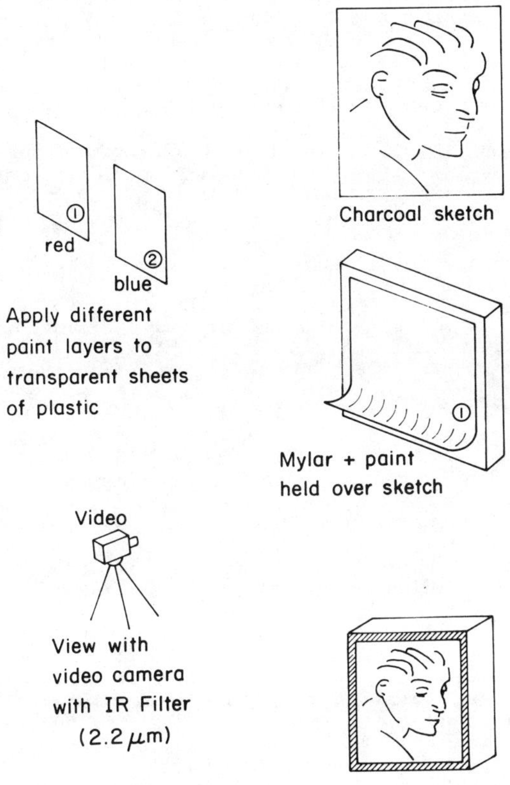

Fig. 2. Infrared Reflectography

At this point, the lectures turn to the microscopic view of a paint film as pigment crystallites suspended in a binding medium. The structure of pigments, the concepts of atomic planes and planar spacing is covered. X-ray diffraction is cited for measurement of interplanar spacings and pigment identification.

Pigment identification leads to the elemental composition, the periodic table, and electron binding energies. X-ray fluorescence takes advantage of the unique spacing between binding energies in order to identify the elements. The use of electron microscopes and electron microprobe follows naturally.

From the keV electron as the incident particle to the incident MeV proton is a natural step. Proton-induced X-ray emission opens up the topic of nuclear reactions and gamma-ray emission. This in turn leads to nuclear structure, isotopes and radioactivity. Neutron-induced autoradiography with examples from museum studies and the Cornell TRIGA reactor provide a link between the nucleus and the painting.

The exposure sequence in autoradiography which is dictated by the nuclear decay lifetime of the elements present, leads to carbon-14 and radiocarbon dating. Dating leads to dendrochronology and examination of oak panels of Rembrandt.

University Framework

The course is cross-listed in five different departments: Materials Science and Engineering, English, Archaeology, Art, and Physics at the 200 level (generally available to sophomores). The inclusion of English arose from the participation of Prof. Donald D. Eddy and the inclusion of paper, inks, and illuminated manuscripts within the course. Then inclusion of Archaeology came from the introduction of dating techniques such as dendrochronology and radiocarbon. The choice of course level arose from the absence of prerequisites for enrollment in the course. The inclusion of Physics and Materials Science and Engineering requires presentation of a substantial amount of basic scientific information that is tied together in a coherent fashion.

In terms of student enrollment and evaluation, the course is successful. The course enrollment for the past two years has averaged 155 students and the student evaluations have been high. The interdisciplinary nature has been met with about one-third of the students from the College of Engineering and the balance primarily from the College of Liberal Arts and Science with majors in Art History, English, Physics, Archaeology and Philosophy, as well as from the College of Architecture, Art and Planning.

The constraint in any introductory level course is the amount of material that can be covered in a fifteen week semester. The world of art represents a vast array of objects-three dimensional (sculptures, vases) as well as two dimensional (paintings, manuscripts). It covers ceramics and metals, as well as pigments, binders and supports. Our choice to look primarily at paintings: how they are made, how light interacts with them, and how their structure is analyzed, albeit restricted, does permit the student to gain an understanding of the methodology applied in the analytical laboratory of a museum.

Course Methodology

The enrollment of 160 students forces a lecture - mode which we supplement with live demonstrations of studio techniques, light refraction (the hand-held laser pointer and a simple prism is an effective demonstration) and radioactivity (a Geiger-counter and a low-level source). Video tapes of laboratory demonstrations are an effective way to illustrate non-portable analytical techniques such as the electron microscope. A 5 to 7 minute tape and a video projector serves to break up the straight lecture format.

To communicate the museum world to the students, we have invited scientists from outside Cornell as lecturers:

Richard Newman, Research Scientist, Museum of Fine Arts, Boston
Dusan Stulik, Head of Analytical Lab., Getty Conservation Inst.
Debora Mayer, Conservator, Winterthur Museum

We have also scheduled broad audience lectures as part of the University colloquium series. Featured speakers have been

Maryan Ainsworth, Metropolitan Museum of Art, New York
Pam Vandiver, Conservation Analytical Lab., Smithsonian

In addition to outside speakers we have also used internal resources and had Cornell Faculty Lecture on their research interests:

Peter Kuniholm - History of Art - Dendrochronology
John Henderson - Archaeology - Obsidian
Ellis Loew - Physiology - Vision & IR
David Clark - Nuclear Studies & Engr. - Nuclear Radiation

Teaching Assistants play a vital role through availability during grading and office hours to give explanations of the text and lectures. At Cornell we use undergraduates as teaching assistants in the course. These undergraduates have taken the course the previous year. They also enroll in a recitation, laboratory project course which serves as a reinforcement of the lecture material. This course, "The Science of Art", has team projects which involve X-radiography (using the livestock x-ray unit of the vet college), optical cross-sections (using the facility of the metallurgical lab), x-ray diffraction in the x-ray lab, electron microscopy and neutron-induced auto-radiography (utilizing the Cornell Triga Reactor). The preparation of written reports and the oral presentations of their research projects prepare the prospective TA's for the onslaught of the next semester's students.

It was necessary to prepare Lecture Notes as the available texts and reference materials tend to address an advanced audience concerned with specialized topics. The heterogeneous nature of the class make it all the more important to establish a base level of understanding. Assignments and problem sets, handed out on a weekly basis, tend to reinforce a standard of performance.

Summary

The Science of Art is an exciting topic that can be taught at the undergraduate level as an interdisciplinary course. In the Cornell approach, class enrollment is large and student response is enthusiastic.

Acknowledgements

We thank the many museum scientists and conservators for their help and patience. We thank Karen Yilmaz for the loan of the Infrared Vidicon Camera.

Author Index

Subject Index